Critical Issues in Alcohol and Drugs of Abuse Testing

Critical Issues in Alcohol and Drugs of Abuse Testing

Second Edition

Edited by

Amitava Dasgupta, PhD
Professor of Pathology and Laboratory Medicine, McGovern Medicine School,
University of Texas Health Science Center at Houston, Houston, TX, United States

Academic Press is an imprint of Elsevier
125 London Wall, London EC2Y 5AS, United Kingdom
525 B Street, Suite 1650, San Diego, CA 92101, United States
50 Hampshire Street, 5th Floor, Cambridge, MA 02139, United States
The Boulevard, Langford Lane, Kidlington, Oxford OX5 1GB, United Kingdom

Copyright © 2019 Elsevier Inc. All rights reserved.

Published In Co operation with AACC.

No part of this publication may be reproduced or transmitted in any form or by any means, electronic or mechanical, including photocopying, recording, or any information storage and retrieval system, without permission in writing from the publisher. Details on how to seek permission, further information about the Publisher's permissions policies and our arrangements with organizations such as the Copyright Clearance Center and the Copyright Licensing Agency, can be found at our website: www.elsevier.com/permissions.

This book and the individual contributions contained in it are protected under copyright by the Publisher (other than as may be noted herein).

Notices
Knowledge and best practice in this field are constantly changing. As new research and experience broaden our understanding, changes in research methods, professional practices, or medical treatment may become necessary.

Practitioners and researchers must always rely on their own experience and knowledge in evaluating and using any information, methods, compounds, or experiments described herein. In using such information or methods they should be mindful of their own safety and the safety of others, including parties for whom they have a professional responsibility.

To the fullest extent of the law, neither the Publisher nor the authors, contributors, or editors, assume any liability for any injury and/or damage to persons or property as a matter of products liability, negligence or otherwise, or from any use or operation of any methods, products, instructions, or ideas contained in the material herein.

British Library Cataloguing-in-Publication Data
A catalogue record for this book is available from the British Library

Library of Congress Cataloging-in-Publication Data
A catalog record for this book is available from the Library of Congress

ISBN: 978-0-12-815607-0

For Information on all Academic Press publications
visit our website at https://www.elsevier.com/books-and-journals

Publisher: Stacy Masucci
Acquisition Editor: Tari K. Broderick
Editorial Project Manager: Megan Ashdown
Production Project Manager: Swapna Srinivasan
Cover Designer: Victoria Pearson

Typeset by MPS Limited, Chennai, India

Contents

List of Contributors — xv
Preface — xvii

1. Alcohol: Pharmacokinetics, Health Benefits With Moderate Consumption and Toxicity
Amitava Dasgupta

Introduction — 1
Alcohol Content of Various Beverages — 2
Pharmacokinetics of Alcohol — 2
  Polymorphism of Aldehyde Dehydrogenase Gene and Alcohol Metabolism — 5
Guidelines for Alcohol Use — 5
Benefits of Consuming Alcohol in Moderation — 6
  Reduced Risk of Cardiovascular Diseases — 6
  Other Physical Health Benefits of Moderate Drinking — 8
  Mental Health Benefits of Moderate Drinking — 9
Health Hazards of Heavy Drinking — 10
  Alcohol Abuse and Liver Damage — 10
  Alcohol Abuse and Brain Damage — 10
  Alcohol Abuse Increases Risk of Cardiovascular Disease and Stroke — 12
  Fetal Alcohol Syndrome — 12
  Other Adverse Effects of Alcohol Abuse — 12
Acute Versus Chronic Alcohol Toxicity — 13
Laboratory Testing for Alcohol — 13
Biomarkers of Alcohol Abuse — 13
Conclusions — 14
References — 14

2. Alcohol Analysis in Various Matrixes: Clinical Versus Forensic Testing
Steven C. Kazmierczak

Introduction — 17
Factors Influencing the Effects of Ethanol — 17
Analysis of Ethanol in Whole Blood Versus Serum or Plasma — 19
Storage of Samples for Ethanol Measurement — 20
Measurement of Ethanol in Body Fluids — 20
Analysis of Ethanol in Expired Breath — 21
Conclusions — 22
References — 22

3. Alcohol Biomarkers: Clinical Issues and Analytical Methods
Joshua A. Bornhorst and Michael M. Mbughuni

Introduction — 25
Alcohol Consumption and Disease Burden — 25
Alcohol Biomarkers: An Overview — 25
Alcohol: Metabolism, Physiology, and Clinical Issues — 27
Toxic Alcohols and Medically Important Alcohol — 28
  Methanol — 29
  Ethylene Glycol — 29
  Isopropanol — 30
  Diethylene Glycol — 30
  Propylene Glycol — 30
Alcohol Biomarkers — 30
  Direct Ethanol Measurements — 31
  Acetaldehyde — 31
  5-Hydroxytryptophol — 31
  Ethyl Glucuronide and Ethyl Sulfate — 32
  Fatty Acid Ethyl Esters — 32
  Indirect Alcohol Biomarkers: Clinical Utility, Analytical Methods, and Analytical Issues — 33
  General Markers of Nutritional Status and Liver Damage — 33
  N-Acetyl-β-Hexosaminidase — 34
  Mean Corpuscular Volume — 35
  Lipid Markers — 35
  Gamma-Glutamyltransferase — 35
Carbohydrate-Deficient Transferrin — 36
  Combination of CDT and GGT or Gamma-CDT — 36
  Sialic Acid and Sialic Index of Apolipoprotein J — 37
Conclusions — 38
References — 38

4. Genetic Markers Related to Alcohol Use and Abuse

Joshua A. Bornhorst and Gwendolyn McMillin

Introduction	43
Role of Laboratory Testing	45
Genetic Markers of Alcohol Dependence	46
Alcohol and Aldehyde Dehydrogenases	46
Neurotransmitter Systems	47
Dopamine	47
GABA	48
Acetylcholine	49
Glutamate	49
Serotonin	49
Neuropeptide Y	49
ACN9	49
Opioids	50
Pharmacogenetics of Alcohol Dependence Treatment	50
Alcohol Detoxification	50
Alcohol Dependence Treatment	51
Conclusions	51
References	52

5. Ethylene Glycol and Other Glycols: Analytical and Interpretation Issues

Uttam Garg, Jennifer Lowry and D. Adam Algren

Introduction	59
Pharmacodynamics and Pharmacokinetics	59
Mechanism of Toxicity	62
Clinical and Laboratory Assessment of Ethylene Glycol Poisoning	62
Laboratory Analysis of Ethylene Glycol and Other Glycols	63
Toxic and Critical Values	64
Challenges and Pitfalls in the Analysis of Ethylene Glycol and Other Related Tests	64
Patient Management	67
Conclusions	67
References	67
Further Reading	69

6. Introduction to Drugs of Abuse

Loralie J. Langman and Christine L.H. Snozek

Introduction	71
Drugs of Abuse Testing in the United States	71
Natural Cannabinoids	73
Synthetic Cannabinoids	73
Cocaine	74
Amphetamine-Type Stimulants	74
Designer Stimulants	75
Opiates and Opioids	75
Benzodiazepines	76
Conclusions	77
References	77

7. Legal Aspects of Drug Testing in US Military and Civil Courts

John F. Jemionek and Marilyn R. Past

Introduction	79
Scientific Evidence	79
Frye Standard for Scientific Evidence or Examination	79
Daubert Standard for Expert Testimony/Opinion	79
Federal Rules of Evidence	80
Laboratory Records and the Court	82
Expert Witness in the Courtroom	82
Role of the Expert Witness Versus Expert Consultant	83
Expert Witness Preparation for Testimony	83
Key Elements of Laboratory Operations and the Legal System	84
Laboratory Certification and Accreditation	84
Personnel Certifications and Licensures	84
Specimen Security, Chain of Custody, and Testing Procedures	85
The Presence of Alcohol or Drug and Impairment	85
Discovery Requests	86
Compliance and Documentation	87
Brady Notices	87
Case Law Pertaining to Forensic Toxicology	88
Constitutional Due Process Clause: Evidentiary Disclosure and Discovery Requests	88
Constitutional Confrontation of Accuser Clause—The Role of Laboratory Personnel and Laboratory Reports	91
Challenges to the Admissibility of Evidence and Inference of Guilt	94
Adherence to Laboratory Procedures and Guidance Documents	97
Marijuana Decriminalization and Medical Use: Federal Versus State Statutes	98
Conclusions	99
Acknowledgment	100
Disclaimer	100
References	100
Appendix	101

8. Pharmacogenomics of Drugs of Abuse

Christine L.H. Snozek and Loralie J. Langman

Introduction	103
Metabolic Enzymes	103

Endogenous Opioid Receptors	107	False-Positive Test Results in Other Drugs-of-Abuse Immunoassays	136
Mu Opioid Receptor Gene (OPRM1)	107	Conclusions	137
Kappa Opioid Receptor Gene (OPRK1)	107	References	138
Delta Opioid Receptor Gene (OPRD1)	108		
Endogenous Opioid Ligands	108		
Monoamine Neurotransmitter System	108		

11. Point of Care Devices for Drugs of Abuse Testing: Limitations and Pitfalls

Veronica Luzzi

Dopamine Receptor Genes	109
Dopamine Transporter Gene (DAT1/SLC6A3)	110
Serotonin Transporter Gene (5-HTT or SLC6A4)	111
Other Monoaminergic-Related Genes	111
Serotonin Receptor Genes (HTR, Multiple Genes)	111
Norepinephrine Transporter Gene (SLC6A2)	112
Dopamine β-Hydroxylase	112
Catechol-O-Methyltransferase	112
Conclusions	112
References	112

Introduction	141
Design of POCT Devices	141
Reliability and Effectiveness of POCT Devices	141
Oral Fluid Testing Using POCT Devices	145
Guidelines for Using POCT Devices	146
Conclusions	147
References	147

9. Immunoassay Design for Screening of Drugs of Abuse

Pradip Datta

Introduction	121
Immunoassays in Drugs of Abuse Testing	121
Formats of Immunoassay Design	122
Polyclonal Versus Monoclonal Antibody	124
Limitation of Immunoassays	124
DAU Assay Components and Protocols	125
Qualitative Versus Quantitative Reporting	126
Specimen Type	126
Sports Doping	127
Conclusions	127
References	127

12. Drugs of Abuse Screening and Confirmation With Lower Cutoff Values

Albert D. Fraser

Introduction	149
Creatinine as a Biomarker of Diluted Urine	149
Criteria for Diluted Urine	150
Specimen Processing in the Correctional Service of Canada Program	150
Correctional Service of Canada Workload and Diluted Urine Specimens	152
Dilution Protocol and Positive Test Results	153
Conclusions	154
References	154

10. Issues of Interferences With Immunoassays Used for Screening of Drugs of Abuse in Urine

Anu S. Maharjan and Kamisha L. Johnson-Davis

Introduction	129
Interferences in Amphetamine Immunoassay	129
Interferences of Sympathomimetic Amines With Amphetamine Immunoassay	130
Vicks Inhaler and False Positive in Amphetamine Immunoassays	131
Dietary Weight Loss Products and Positive Amphetamine	133
Other Drugs That May Cause Positive Amphetamine Screening	134
Interferences With Opiate Immunoassay	135
Interferences With Tetrahydrocannabinol Immunoassay	135
Interferences With Cocaine Immunoassay	136

13. Overview of Analytical Methods in Drugs of Abuse Analysis: Gas Chromatography/Mass Spectrometry, Liquid Chromatography Combined With Tandem Mass Spectrometry and Related Methods

Alec Saitman

Introduction	157
Specimen Preparation	157
Specimen Introduction	159
Gas Chromatography	160
Liquid Chromatography	160
Ionization	161
Electron Ionization	162
Chemical Ionization	162
Electrospray Ionization	162
Atmospheric Pressure Chemical Ionization	163

Quadrupole Mass Spectrometry	163	Benzodiazepines	197
Triple Quadrupole Mass Spectrometry	164	Barbiturates	199
Quantitative Measurement	165	Antidepressants	200
Sensitivity	165	"Z" Drugs	201
Specificity	167	Conclusion	203
Mass Spectrometry Based Methods Versus Immunoassays	168	Acknowledgements	203
Conclusions	169	References	203
References	169		

14. High-Resolution Mass Spectrometry: An Emerging Analytical Method for Drug Testing

Michelle Wood

Introduction	173
Fundamentals of High-Resolution Mass Spectrometry	174
Mass Accuracy	174
Key Benefits of High-Resolution Mass Spectrometry and Exact Mass	175
Mass Resolution and Mass Resolving Power	177
Limitations of Exact Mass and High-Resolution Mass Spectrometry	178
High-Resolution Mass Spectrometer: Technological Aspects	178
Time-of-Flight Mass Spectrometry	179
Orbitrap Analyzer	181
High-Resolution Mass Spectrometry—Acquisition Modes and Types of Data	182
Data-Dependent Acquisition Techniques	183
Data-Independent Acquisition Techniques	183
Targeted, Semitargeted, and Nontargeted Analyses	183
Conclusions	187
References	187

15. Confirmation Methods for SAMHSA Drugs and Other Commonly Abused Drugs

Justin Holler and Barry Levine

Introduction	189
Confirmation Testing: SAMHSA Drugs	190
Gas Chromatography/Mass Spectrometry	190
Liquid Chromatography Combined with Mass Spectrometry or Tandem Mass Spectrometry	193
General Confirmation Requirements	194
Confirmation Tesing: Non-SAMHSA Drugs	195
Narcotic Analgesics	195

16. Critical Issues When Testing for Amphetamine-Type Stimulants: Pitfalls of Immunoassay Screening and Mass Spectrometric Confirmation for Amphetamines, Methamphetamines, and Designer Amphetamines

Larry Broussard

Introduction	207
Issues of Cross-Reactivity	207
Immunoassay Cutoff Concentrations and Cross-Reactivity	208
False-Positive Immunoassay Results	209
Minimizing False-Positive Immunoassay Results	209
True-Positive Results	210
GC/MS and LC-MS/MS Confirmation Procedures	210
Interference and False-Positive Methamphetamine	210
Isomer Resolution	211
Additional Considerations	212
Conclusions	212
References	213

17. Cocaine, Crack Cocaine, and Ethanol: A Deadly Mix

Eric T. Shimomura, George F. Jackson and Buddha Dev Paul

Introduction	215
Routes of Cocaine Administration	215
Cocaine Metabolism	215
Cocaine Toxicity	216
Cocaine Smoking and Toxicity	217
The Combined Toxicity of Cocaine and Alcohol	218
The Role of Cocaethylene in Toxicity	219
Analytical Methods	220
Conclusions	221
Disclaimer	221
References	221

18. Drug-Assisted Sexual Assaults: Toxicology, Fatality, and Analytical Challenge

Matthew D. Krasowski

Introduction	225
γ-Hydroxybutyric Acid and Its Analogs	225
Sources	225
Pharmacology	226
Abuse	226
Analysis	227
Toxicity	227
GHB Associated With Drug-Facilitated Sexual Assault	227
3,4-Methylenedioxymethamphetamine	228
Sources	228
Pharmacology	228
Abuse	228
Analysis	228
Toxicity	229
Association With Drug-Facilitated Sexual Assault	229
Ketamine	229
Sources	229
Abuse	229
Pharmacology	230
Analysis	230
Toxicity	230
Association With Drug-Facilitated Sexual Assault	230
Flunitrazepam	230
Sources	230
Pharmacology	231
Analysis	231
Toxicity	231
Association With Drug-Facilitated Sexual Assault	231
Nonbenzodiazepine Sedative-Hypnotics	231
Sources	231
Pharmacology	231
Analysis	232
Toxicity	232
Association With Drug-Facilitated Sexual Assault	232
Conclusions	232
References	232

19. Overview of Common Designer Drugs

Lilian H.J. Richter, Markus R. Meyer and Hans H. Maurer

Introduction	237
Amphetamine-Derived Designer Drugs	237
PMA, PMMA, and 4-MTA	240
Piperazine-Derived Designer Drugs	240
Alpha-Pyrrolidinophenone-Derived Designer Drugs	241
Beta-Keto-Type Designer Drugs	242
2,5-Dimethoxy Phenethylamine-Type Designer Drugs (2CS)	243
Conclusions	246
References	246

20. New Psychoactive Substances: An Overview

Laura Mercolini

Introduction	247
Synthetic Cannabinoid Receptor Agonists	248
Cathinone Analogs	248
Phenethylamines	250
Tryptamines	250
Miscellaneous New Psychoactive Substances	252
Bioanalytical Strategies for New Psychoactive Substances	254
Conclusions	256
References	256

21. Review of Bath Salts on Illicit Drug Market

Michele Protti

Introduction	259
Pharmacology and Toxicology	260
Bioanalytical Approaches	260
Bioanalysis Using Hematic Matrices	264
Bioanalysis in Urine	266
CA Bioanalysis in Hair	267
Bioanalysis in Oral Fluid	267
Bioanalysis in Miniaturized Dried Samples	268
Conclusions	268
References	269

22. Review of Synthetic Cannabinoids on the Illicit Drug Market

Mahmoud A. ElSohly, Sohail Ahmed, Shahbaz W. Gul and Waseem Gul

Introduction	273
History of Cannabimimetics	273
Indoles	274
Benzoylindoles: Structures and Analysis	277
Naphthoylindoles: Structures and Analysis	289
Phenylacetylindoles: Structures and Analysis	296
Piperazinylindoles: Structures and Analysis	299
Naphthalenyl-Indole-3-Carboxylates: Structures and Analysis	299

Quinolin-8-yl-Indole-3-Carboxylates:
  Structures and Analysis 299
Adamantyl-Indole-3-Carboxamide:
  Structures and Analysis 300
Tetramethylcyclopropylmethanone-
  Pentylindoles: Structures and Analysis 301
MDMB-CHMICA, MDMB-FUBINACA,
  and FUB-AMB: Structures and Analysis 303
ADB-BICA (1-Benzyl-N-(1-Carbamoyl-2,2-
  Dimethylpropan-1-yl)-1H-Indole-3-
  Carboxamide): Structure and Analysis 304
NNEI: Structures and Analysis 304
BzODZ-Epyr: Structures and Analysis 305
JWH-210: Structure and Analysis 305
Indazoles 306
Amino-Methyl-Oxobutan-Indazole-3-
  Carboxamide: Structures and Analysis 306
Adamantyl-Indazole-3-Carboxamide:
  Structures and Analysis 308
Benzo[d]imidazole: Structures and Analysis 309
Naphthalenyl-Indazole: Structures and
  Analysis 310
5F-ADB and 5F-AMB: Structures and
  Analysis 310
SDB-005: Structures and Analysis 310
AMPPPCA: Structures and Analysis 310
NPB-22: Structures and Analysis 311
Pyrroles 311
Naphthoylpyrroles: Structures and Analysis 311
Carbazoles: Structure and Analysis 311
URB-Class: Structures and Analysis 311
Endogenous Cannabinoids: Structures
  and Analysis 312
Cyclohexylphenols: Structures and Analysis 312
Classical Synthetic Cannabinoids:
  Structures and Analysis 313
Miscellaneous: Structures and Analysis 313
  NNL-1: Structure and Analysis 313
  CB-13 313
  WIN-55,212-2 314
Conclusions 314
References 314

23. Application of Liquid Chromatography Combined With High Resolution Mass Spectrometry for Urine Drug Testing

Olof Beck, Alexia Rylski and Niclas Nikolai Stephanson

Introduction 321
History and Fundamentals of HRMS 321
  TOF Analyzer 323
  Orbitrap Analyzer 324

Direct Injection of Urine 324
Chromatography System 325
Features of Immunoassays: What
  are Problems to be Solved? 326
Why HRMS for Drug Testing? 326
  Need for Resolving Power in Urine Drug
    Screening 327
  Application for Urine Drug Screening 329
  Application for Confirmation 331
Conclusions 331
References 331

24. Forensic Toxicology in Death Investigation

Hannah Kastenbaum, Lori Proe and Lauren Dvorscak

Introduction 333
Medicolegal Death Investigation 333
The Utility of Computed Tomography
  in Investigating Drug Poisoning Deaths 334
  Case Report 334
The Role of Autopsy in Drug Poisoning
  Deaths 336
  Case Report 337
Toxicology Specimen Collection at Autopsy 338
Common Postmortem Toxicology Samples 339
  Blood 339
  Vitreous Fluid 339
  Urine 339
  Tissue 339
Ordering Toxicology 339
  Case Report 340
Toxicology Reports 340
Determining Cause of Death 340
  Case Report 341
Determining the Manner of Death 341
  Case Report 341
Conclusions 342
References 342

25. Drug Testing in Pain Management

Gary M. Reisfield and Roger L. Bertholf

Introduction 343
Use of Opioids in Pain Management 344
Analytical Methods for Drug Testing 346
  General Considerations on Immunoassays 346
  Immunoassays Used in Point of Care
    Drug Testing 347
  Homogeneous Immunoassays 349
  Confirmatory Methods 351
  Specimen Tampering 352
  Metabolism of Opioid Analgesics 353
  Challenges in Drug Testing in Pain
    Management 355

Conclusions	356
References	356

26. How Do People Try to Beat Drugs Test? Effects of Synthetic Urine, Substituted Urine, Diluted Urine, and In Vitro Urinary Adulterants on Drugs of Abuse Testing

Shanlin Fu

Introduction	359
Urine Drug Testing Programs	361
How Do People Try to "Beat the Urine Drug Test"?	363
Urine Substitution	363
In Vivo Adulteration and Urine Dilution	363
In Vitro Chemically Adulterated Urine	364
How Do Chemical Adulterants Effect Urine Drug Testing?	365
Hypochlorite-Based Bleach	366
Nitrite	368
Pyridinium Chlorochromate	369
Peroxides	370
Other Oxidants	371
Nonoxidizing Adulterants	372
How Do Laboratories Counteract Urine Adulteration?	375
Urine Integrity Tests	375
Color Tests	376
Dipstick Devices	376
Spectrophotometric Methods	378
Immunoassays	378
Capillary Electrophoresis and Mass Spectrometric Techniques	378
Polyethylene Glycol Urine Marker System	378
Can Targeted Analytes That Have Been Modified With Oxidizing Agents Be Detected?	379
Amphetamine-Type Substances	379
Cannabinoids	379
Opiates/Opioids	381
Conclusions	382
References	385

27. When Hospital Toxicology Report Is Negative in a Suspected Overdosed Patient: Strategy of Comprehensive Drug Screen Using Liquid Chromatography Combined With Mass Spectrometry

Ernest D. Lykissa

Introduction	391
Problem With Limited Drug Testing Protocol and Cutoff Levels	392
Communicating With Physician When Toxicology Report Is Negative	394
Scenario A	394
Scenario B	394
Challenges of Detecting Benzodiazepines	394
Challenges of Detecting Opioids	396
Designer Drugs Are Not Detected in Routine Urine Toxicology	396
Toxicology Screen Using LC–MS/MS	397
Dilute and Shoot Versus Sample Preparation	401
Conclusions	402
References	402
Further Reading	403

28. Testing of Drugs of Abuse in Oral Fluid, Sweat, Hair, and Nail: Analytical, Interpretative, and Specimen Adulteration Issues

Uttam Garg and Carl Cooley

Introduction	405
Drug Testing in Oral Fluid	405
Transfer of Drugs from Plasma to Oral Fluid	407
Sample Collection and Transportation	407
Specimen Analysis	408
Comparison of Positivity of Oral Fluid with Blood and Urine	410
Detection of Amphetamines and Other Sympathomimetic Amines in Oral Fluid	412
Detection of Cannabinoids in Oral Fluid	412
Detection of Cocaine in Oral Fluid	413
Detection of Opiates/Opioids in Oral Fluid	413
Detection of Phencyclidinein Oral Fluid	413
Detection of Barbiturates and Benzodiazepines in Oral Fluid	414
Detection of Other Drugs in Oral Fluid	414
Adulteration Issues in Oral Fluid Specimens	414
Drug Testing in Sweat	415
Sample Collection and Analysis	415
Detection of Amphetamines and Other Sympathomimetic Amines in Sweat	415
Detection of Cannabinoids in Sweat	416
Detection of Cocaine in Sweat	416
Detection of Opiates/Opioids in Sweat	416
Detection of Other Drugs in Sweat	416
Issues of Special Interest in Sweat Drug Testing	417
Hair Drug Testing	417
Specimen Collection and Analysis	417
Analysis of Amphetamines and Other Sympathomimetic Amines in Hair	418
Analysis of Cannabinoids in Hair	418

Analysis of Cocaine in Hair	419
Analysis of Opiates/Opioids in Hair	419
Analysis of Phencyclidine in Hair	420
Analysis of Other Drugs in Hair	420
Environmental or External Contamination of Hair	420
Hair Color	421
Hair Adulteration Issues	421
Drug Testing in Nail	422
Conclusions	422
References	423

29. Advances in Meconium Analysis for Assessment of Neonatal Drug Exposure

Steven W. Cotten

Introduction	429
Meconium as Specimen	429
Analysis of Alcohol and Fatty Acid Ethyl Esters	430
Analysis of Amphetamine	431
Analysis of Antiretrovirals	433
Analysis of Cannabinoids	434
Analysis of Cocaine	434
Analysis of Nicotine	436
Analysis of Opiates	437
Analytical Performance and Outcome Studies	438
Conclusion	439
References	439

30. Analytical True Positive Drug Tests Due to Use of Prescription and Nonprescription Medications

Matthew D. Krasowski and Tai C. Kwong

Introduction	441
Analytical True Positive Versus Clinical False Positive Results	442
True Positive Results Due to the Use of Medications	442
Amphetamines	443
Nonprescription Medication Containing Methamphetamine	443
Prescription Medications Containing Amphetamine or Methamphetamine	444
Substances Known to Metabolize to Methamphetamine and Amphetamine	444
Cocaine	444
Marijuana	445
Opiates	445
Medications Containing Morphine or Codeine	446
Metabolites of Opiates	446
Conclusions	447
References	447

31. Analytical True Positive: Poppy Seed Products and Opiate Analysis

Amitava Dasgupta

Introduction	449
History of Poppy Plant and Opium	449
Opium and Poppy Plants: Legal Issues	450
Variation of Morphine and Other Alkaloids Content in Poppy Seeds	450
Foods Containing Poppy Seeds	452
Reduced Morphine Content in Poppy Seeds During Food Processing	452
Dangers of Consuming Poppy Tea	452
Dangers of Opium-Containing Foods	453
Toxicity From Consuming Poppy Seeds or Poppy Seed–Containing Food	454
Effect of Poppy Seed–Containing Food on Opiate Drug Testing	454
High Morphine Levels After Eating Poppy Seeds	455
Lower Morphine Level in Subjects After Eating Processed Poppy Seed–Containing Foods But High Level After Eating Poppy Seeds	455
Brown Mixture and Opiate Levels in Urine	458
Poppy Seed Defense	458
Acetyl Codeine	458
Morphine/Codeine Ratio	459
Opium Alkaloids as Biomarker of Poppy Seed Ingestion	459
ATM4G: A Biomarker of Heroin Abuse	460
Conclusions	461
References	461

32. Miscellaneous Issues: Paper Money Contaminated With Cocaine and Other Drugs, Cocaine Containing Herbal Teas, Passive Exposure to Marijuana, Ingestion of Hemp Oil, and Occupational Exposure to Controlled Substances

Amitava Dasgupta

Introduction	463
US Paper Money Contaminated With Cocaine and Other Drugs	463
Currency Contaminated With Drugs: Clean Money Versus Drug Money	465
Mechanism of Contamination	467
Can Handling Contaminated Paper Money May Cause Positive Drug Screen?	467

Herbal Teas Contaminated With Cocaine	468
Distinguishing Coca Leaves Chewing From Cocaine Abuse	469
Herbal Tea and Contamination/Adulteration	470
Passive Exposure to Marijuana	**470**
Washing Hair With Cannabio Shampoo	471
Ingestion of Hemp Oil	**472**
Edible Marijuana Products	473
Unwanted Exposure of Marijuana Containing Products to Children	474
Occupational Exposure to Controlled Substances	**474**
Conclusions	**475**
References	**475**

33. Abuse of Magic Mushroom, Peyote Cactus, LSD, Khat, and Volatiles

Amitava Dasgupta

Introduction	**477**
Magic Mushroom	**477**
Prevalence of Magic Mushroom Abuse	477
Active Ingredients of Magic Mushroom	478
Pharmacology and Toxicology of Psilocybin	478
Mechanism of Action of Active Ingredients of Magic Mushroom	480
Therapeutic Potential of Psilocybin	480
Analysis of Active Ingredients of Magic Mushroom	480
Use of Peyote Cactus	**482**
Prevalence of Peyote Cactus Abuse	483
Pharmacology and Toxicology of Mescaline	483
Mechanism of Action of Mescaline	484
Laboratory Determination of Mescaline	484
Abuse of Lysergic Acid Diethylamide	**485**
Analysis of Lysergic Acid Diethylamide and Its Metabolite	486

Abuse of Khat	**487**
Khat: Pharmacology and Toxicology	488
Mechanism of Action	488
Laboratory Analysis of Cathinone	488
Abuse of Volatiles and Glue	**489**
Prevalence of Solvent Abuse	489
Pharmacology and Toxicology of Abused Solvents	490
Laboratory Detection of Solvent Abuse	491
Conclusions	**491**
References	**492**

34. Performance Enhancing Drugs in Sports

Brian D. Ahrens and Anthony W. Butch

Introduction	**495**
The Anti-Doping Movement	**495**
WADA Prohibited List	**496**
Overview of the Testing Procedure	**496**
General Testing Methods	**498**
Anabolic Agents	**500**
Testosterone to Epitestosterone Ratio	503
Longitudinal Steroid Profiling	504
Isotope Ratio Mass Spectrometry	504
Erythropoiesis Stimulating Agents	**504**
Growth Hormone	**506**
Supplements and Prohormones	**506**
Future Challenges	**507**
Designer Steroids	507
Alternative Matrices	508
Results Management	508
Conclusions	**508**
References	**508**
Index	511

List of Contributors

Sohail Ahmed, PhD, Department of Biochemistry, Hazara University, Mansehra, Pakistan; Valor Group of Pharmaceutical Industries, Islamabad, Pakistan

Brian D. Ahrens, UCLA Olympic Analytical Laboratory and Department of Pathology & Laboratory Medicine, David Geffen School of Medicine at UCLA, Los Angeles, CA, United States

D. Adam Algren, MD, Division of Clinical Pharmacology and Medical Toxicology, Children's Mercy Hospital, and University of Missouri School of Medicine, Kansas City, MO, United States

Olof Beck, PhD, Department of Laboratory Medicine, Karolinska Institute, Stockholm, Sweden; Pharmacology Laboratory, Karolinska University Hospital, Stockholm, Sweden

Roger L. Bertholf, PhD, Department of Pathology and Genomic Medicine, Houston Methodist Hospital, Houston, TX, United States

Joshua A. Bornhorst, PhD, Department of Laboratory Medicine and Pathology, Mayo Clinic, Rochester, MN, United States

Larry Broussard, PhD, Department of Clinical Laboratory Sciences, Louisiana State University Health Sciences Center, New Orleans, LA, United States

Anthony W. Butch, PhD, UCLA Olympic Analytical Laboratory and Department of Pathology & Laboratory Medicine, David Geffen School of Medicine at UCLA, Los Angeles, CA, United States, Retired

Carl Cooley, PhD, PRA Health Sciences Lexana, KS, United States

Steven W. Cotten, PhD, Department of Pathology and Laboratory Medicine, University of North Carolina at Chapel Hill, Chapel Hill, NC, United States

Amitava Dasgupta, PhD, Department of Pathology and Laboratory Medicine, University of Texas McGovern Medical School at Houston, Houston, TX, United States

Pradip Datta, PhD, Siemens Healthineers, Newark, DE, United States

Lauren Dvorscak, MD, New Mexico Office of the Medical Investigator, University of New Mexico School of Medicine, Albuquerque, NM, United States

Mahmoud A. ElSohly, PhD, ElSohly Laboratories, Inc., Oxford, MS, United States; National Center for Natural Products Research, Oxford, MS, United States; Department of Pharmaceutics, School of Pharmacy, The University of Mississippi, Oxford, MS, United States

Albert D. Fraser, PhD, Department of Pathology and Pharmacy, Dalhousie University, Halifax, NS, Canada; Horizon Health Network, Saint John, NB, Canada; Government of Canada, Canada

Shanlin Fu, PhD, Centre for Forensic Science, University of Technology Sydney, Ultimo, NSW, Australia

Uttam Garg, PhD, Department of Pathology and Laboratory Medicine, Children's Mercy Hospital, and University of Missouri School of Medicine, Kansas City, MO, United States

Shahbaz W. Gul, ElSohly Laboratories, Inc., Oxford, MS, United States

Waseem Gul, PhD, ElSohly Laboratories, Inc., Oxford, MS, United States; National Center for Natural Products Research, Oxford, MS, United States

Justin Holler, PhD, Chesapeake Toxicology Resources, Frederick, MD, United States

George F. Jackson, PhD, Forensic Toxicologist, Armed Forces Medical Examiner System, Dover, DE, United States

John F. Jemionek, PhD, Division of Forensic Toxicology, Armed Forces Medical Examiner System, Armed Forces Institute of Pathology, Rockville, MD, United States

Kamisha L. Johnson-Davis, PhD, Department of Pathology, University of Utah Health Sciences Center, Salt Lake City, UT, United States; ARUP Institute for Clinical and Experimental Pathology, Salt Lake City, UT, United States

Hannah Kastenbaum, MD, New Mexico Office of the Medical Investigator, University of New Mexico School of Medicine, Albuquerque, NM, United States

Steven C. Kazmierczak, PhD, Department of Pathology, Oregon Health and Science University, Portland, OR, United States

Matthew D. Krasowski, MD, PhD, Department of Pathology, University of Iowa Carver College of Medicine, Iowa City, IA, United States

Tai C. Kwong, PhD, Department of Pathology, University of Iowa Carver College of Medicine, Iowa City, IA, United States; Department of Pathology and Laboratory Medicine, University of Rochester Medical Center, Rochester, NY, United States

Loralie J. Langman, PhD, Department of Laboratory Medicine and Pathology, Mayo Clinic, Rochester, MN, United States

Barry Levine, PhD, Chesapeake Toxicology Resources, Frederick, MD, United States; Chief Medical Examiner, Baltimore, MD, United States

Jennifer Lowry, MD, Division of Clinical Pharmacology and Medical Toxicology, Children's Mercy Hospital, and University of Missouri School of Medicine, Kansas City, MO, United States

Veronica Luzzi, PhD, Providence Saint Joseph's Health and Services, Providence Regional Core Laboratory, Portland, OR, United States

Ernest D. Lykissa, PhD, Expertox Laboratory, Houston, TX, United States

Anu S. Maharjan, PhD, Department of Pathology, University of Utah Health Sciences Center, Salt Lake City, UT, United States

Hans H. Maurer, MD, PhD, Department of Experimental and Clinical Toxicology, Institute of Experimental and Clinical Pharmacology and Toxicology, Center for Molecular Signaling (PZMS), Saarland University, Homburg, Germany

Michael M. Mbughuni, PhD, Department of Pathology, University of Minnesota, Minneapolis, MN, United States; VA Hospital, Minneapolis, MN, United States

Gwendolyn McMillin, PhD, Department of Laboratory Medicine and Pathology, Mayo Clinic, Rochester, MN, United States; ARUP Laboratory and Department of Pathology, University of Utah, Salt Lake City, UT, United States

Laura Mercolini, PhD, Pharmaco-Toxicological Analysis Laboratory, Department of Pharmacy and Biotechnology, University of Bologna, Bologna, Italy

Markus R. Meyer, PhD, Department of Experimental and Clinical Toxicology, Institute of Experimental and Clinical Pharmacology and Toxicology, Center for Molecular Signaling (PZMS), Saarland University, Homburg, Germany

Marilyn R. Past, PhD, Division of Forensic Toxicology, Armed Forces Medical Examiner System, Armed Forces Institute of Pathology, Rockville, MD, United States, Retired

Buddha Dev Paul, PhD, Forensic Toxicologist, Armed Forces Medical Examiner System, Dover, DE, United States, Retired

Lori Proe, DO, New Mexico Office of the Medical Investigator, University of New Mexico School of Medicine, Albuquerque, NM, United States

Michele Protti, PhD, Pharmaco-Toxicological Analysis Laboratory (PTA Lab), Department of Pharmacy and Biotechnology (FaBiT), University of Bologna, Bologna, Italy

Gary M. Reisfield, MD, Department of Psychiatry, University of Florida College of Medicine, Gainesville, FL, United States

Lilian H.J. Richter, Department of Experimental and Clinical Toxicology, Institute of Experimental and Clinical Pharmacology and Toxicology, Center for Molecular Signaling (PZMS), Saarland University, Homburg, Germany

Alexia Rylski, Pharmacology Laboratory, Karolinska University Hospital, Stockholm, Sweden

Alec Saitman, PhD, Providence Regional Laboratories, Portland, OR, United States

Eric T. Shimomura, PhD, Forensic Toxicologist, Armed Forces Medical Examiner System, Dover, DE, United States

Christine L.H. Snozek, PhD, Department of Laboratory Medicine and Pathology, Mayo Clinic Arizona, Scottsdale, AZ, United States

Niclas Nikolai Stephanson, PhD, Pharmacology Laboratory, Karolinska University Hospital, Stockholm, Sweden

Michelle Wood, PhD, Scientific Operations, Waters Corporation, Wilmslow, United Kingdom

Preface

The first edition of "Critical Issues in Alcohol and Drugs of Abuse Testing" was published in 2009 and since then significant changes in the field of alcohol and drugs of abuse testing have occurred. These changes are reflected in the significant expansion of the second edition, adding 13 new chapters covering topics such as bath salts (synthetic cathinone) and synthetic cannabinoids (spices, K2 etc). These were just emerging when the first edition was prepared, but today many cathinone derivatives and over synthetic cannabinoids are described in medical literature. In addition, there also reports of abuse of many other novel psychoactive substances that are covered in new chapters. Chapter 20, Novel Psychoactive Substances: An Overview, provides an overview of novel psychoactive substances while baths salts on illicit drug market are discussed in Chapter 21, Review of Bath Salts on Illicit Drug Market. In Chapter 22, Review of Synthetic Cannabinoids on the Illicit Drug Market, synthetic cannabinoids are reviewed. Drug assisted sexual assaults are a critical problem and are discussed in detail in Chapter 18, Drug-Assisted Sexual Assaults: Toxicology, Fatality, and Analytical Challenge, addressing toxicity, fatality, and analytical challenges associated with date rape drugs.

Gas chromatography/ mass spectrometry (GC/Ms) was originally used for confirmation of drugs of abuse in urine and other biological matrix. This method is still useful today and is widely used (Chapter 15: Confirmation Methods for SAMHSA Drugs and Other Commonly Abused Drugs). Later, liquid chromatography combined with mass spectrometry or tandem mass spectrometry emerged as a useful analytical technique for confirmation of drugs of abuse. This method is superior to GC/Ms because no derivatization is needed (Chapter 13: Overview of Analytical Methods in Drugs of Abuse Analysis: Gas Chromatography/Mass Spectrometry, Liquid Chromatography Combined With Tandem Mass Spectrometry and Related Methods). More recently, liquid chromatography combined with high resolution mass spectrometry emerged as a power method for confirmation of drugs of abuse in urine and body fluids. In Chapter 14, High Resolution Mass Spectometry: An Emerging Analytical Method for Drug Testing, technical aspects of high resolution mass spectrometry are discussed while in Chapter 23, Application of Liquid Chromatography Combined with High Resolution Mass Spectrometry for Urine Drug Testing, application of high resolution mass spectrometry in urine drug analysis is addressed.

Drug testing using point of care devices is becoming more common because such tests can be conducted bedside or in a physician's office. Therefore, a new chapter (Chapter 11: Point of Care Devices for Drugs of Abuse Testing: Limitations and Pitfalls) discusses usefulness and pitfalls of point of care devices in screening for drugs of abuse in urine. Chapter 25, Drug Testing in Pain Management, addresses drug testing in pain management with a chapter on drug testing in death investigation. As expected, all chapters from the 1st edition are included with significant updates by the respective authors. All chapters also include a long list of references so that readers can look into source materials.

I would like to thank my wife, Alice, for her support during long evenings and weekend hours while I devoted time editing this revision. I hope, like the 1st edition, readers will find this 2nd edition valuable as it contains more extensive coverage on this topic.

Respectfully submitted

Amitava Dasgupta
Houston, TX, United States

Chapter 1

Alcohol: Pharmacokinetics, Health Benefits With Moderate Consumption and Toxicity

Amitava Dasgupta

Department of Pathology and Laboratory Medicine, University of Texas McGovern Medical School at Houston, Houston, TX, United States

INTRODUCTION

Ethyl alcohol (also known as ethanol) is commonly referred as "alcohol." Alcohol use by human can be traced back to prehistoric time. Professor Robert Dudley, University of California, Berkeley, proposed an interesting hypothesis known as "Drunken Monkey Hypothesis," which speculates that the human attraction to alcohol may have a genetic basis because primate ancestors of *Homo sapiens* consumed large amounts of fruits. Alcohol produced by yeast from fructose diffused out of the fruit, and alcoholic smell could help a primate to identify fruits as ripe and ready to be consumed. As a result "natural selection" favored monkeys with a keen appreciation for the smell and taste of alcohol. By the time humans evolved from apes approximately 1—2 million years ago, fruits were mostly replaced by roots, tubers, and meat as diet. However, it is still possible that human taste for alcohol arose during our long-shared ancestry with primates. Anecdotally, humans often consume alcohol with food, suggesting that drinking alcohol along with food is a natural instinct [1].

The first historical evidence of alcoholic beverages came from the archeological discovery of Stone Age beer jugs, approximately 10,000 years ago. Egyptians probably consumed wine approximately 6000 years ago. Egyptians used alcoholic beverages (both beer and wine) for pleasure, rituals, medical, and nutritional purposes. Some myths suggest that Egyptian bakers noticed the formation of bubbles when wet grains sat for extended periods before being used to make bread. The earliest evidence of alcohol use in China dated back to 5000 BC, when alcohol was mainly produced from rice, honey, and fruits. In ancient India, alcohol beverages were known as "sura." Use of such drinks was known in 3000—2000 BC and ancient ayurvedic texts concluded that alcohol is a medicine if consumed in moderation, but a poison if consumed in excess. Beer was known to Babylonians as early as 2700 BC. In ancient Greece, wine-making was common in 1700 BC. Hippocrates identified numerous medicinal properties of wine, but was critical of drunkenness. In ancient civilization, alcohol was used primarily to quench thirst because water was unsafe for drinking due to bacterial contamination. Hippocrates specifically cited that water only from springs, deep wells, and rainfall were safe for human consumption. Beer was a drink for common people, while wine was the preferred drink for elites. In ancient Eastern civilization, drinking alcoholic beverages for thirst-quenching was less common than Western civilization because drinking tea was very popular in Asian countries. Tea is a safe drink because during preparation, boiling kills pathogens [2].

Currently, few chemicals such as ethyl alcohol (alcohol) are widely found or generate the degree of debate and controversy. Not only is alcohol readily available in the form of alcoholic beverages, also its chemical properties make it an ideal solvent for flavoring and other compounds used in food industries. Furthermore, it is a product of decomposition by bacteria, giving rise to small, but measurable, amounts in some nonpreserved foods [3,4]. Additional sources of alcohol exposure include mouthwashes, cough and cold preparations, hand cleaners, aftershaves, window cleaners, and many other personal and household products.

ALCOHOL CONTENT OF VARIOUS BEVERAGES

Alcohol content of various alcoholic beverages varies widely, for example, beer contains approximately 4%−7% alcohol while average alcohol content of vodka is 40%−50%. However, due to wide differences between serving size of various alcoholic beverages, one drink (often called one standard drink) contains approximately 0.6 ounces of alcohol, which is equivalent to 14 g of alcohol in each drink. In the United States, a standard drink is defined as a bottle of beer (12 ounces) containing 5% alcohol, 8.5 ounces of malt liquor containing 7% alcohol, a 5-ounce glass of wine containing 12% alcohol, 3.5 ounces of fortified wine containing about 17% alcohol, or one shot of a distilled spirits such as gin, rum, vodka, or whiskey (1.5 ounces) containing 40% alcohol. In general, the average bottle of beer contains an average of 0.56 ounce of alcohol, but a standard wine drink may contain 0.66 ounce of alcohol while distilled spirits may contain up to 0.89 ounce of alcohol [5]. Alcohol content of various popular beverages is given in Table 1.1.

Historically, the alcoholic content of various drinks was expressed as "proof," a term originated in the 18th century when British sailors were paid with money as well as rum. In order to ensure that the rum was not diluted with water, it was "proofed" by dousing gunpowder with it and setting it on fire. If the gunpowder failed to ignite, it indicated that rum was diluted with excess water. In the United States, proof to alcohol by volume is defined as a ratio of 1:2. Therefore a beer which has 4% alcohol by volume is defined as 8 proof. In the United Kingdom, alcohol by volume to proof is a ratio of 4:7. Therefore multiplying alcohol by volume content with a factor of 1.75 would provide the "proof" of the drink.

Currently, in the United States, the alcohol content of a drink is measured as the percentage of alcohol by the volume. The Code of Federal Regulations requires that the label of alcoholic beverages must state the alcohol content by volume. The regulation does not require the "proof" of the drink to be printed. Alcoholic drinks primarily consist of water, alcohol, and variable amounts of sugars and carbohydrates (residual sugar and starch left after fermentation) but negligible amounts of other nutrients such as proteins, vitamins, or minerals. However, distilled liquors such as cognac, vodka, whiskey, and rum contain no sugars. Red wine and dry white wines contain 2−10 g of sugar per liter while sweet wines and port wines may contain up to 120 g of sugar per liter of wine. Beer and dry sherry contain 30 g of sugar per liter [6].

PHARMACOKINETICS OF ALCOHOL

Alcohol is a weakly polar, aliphatic hydrocarbon soluble in both water and lipid, a characteristic that greatly influences the pharmacokinetics once in the body. Alcohol pharmacokinetics in humans are complex and depend on factors such as the amount and type of alcohol ingested, gender, age, body water, and metabolism. After ingestion, alcohol readily diffuses across cellular membranes and is rapidly absorbed within 30−60 min in the duodenum. Coingestion of food, some drugs, and medical conditions that inhibit gastric emptying delay absorption. In one study, 10 healthy men consumed a moderate dosage of alcohol (0.80 g of alcohol per kg of body weight) in the morning after an overnight fast or immediately after breakfast (two cheese sandwiches, one boiled egg, orange juice, and fruit yogurt). The blood alcohol

TABLE 1.1 Alcohol Content of Various Alcoholic Beverages

Alcoholic Beverage	Alcohol Content (v/v)
Fruit juice	<0.1%
Ciders	4%−8%
American beer	4%−7%
Champagne	12%−13%
Table wine	8%−17%
Japanese Sake	14%−16%
Port wine	15%−22%
Whiskey, vodka, rum, and brandy	Usually 40% but much higher alcohol may be present in some brands
Tequila	45%−50%

analysis revealed that the average peak blood alcohol in subjects who consumed alcohol on an empty stomach was 104 mg/dL. In contrast, the average peak blood alcohol in subjects who consumed alcohol after eating breakfast was 67 mg/dL. The time required to metabolize total amount of alcohol was on average 2 h shorter in subjects who consumed alcohol after eating breakfast compared to subjects who consumed alcohol on an empty stomach. The authors concluded that food in the stomach before drinking not only reduces the peak blood alcohol concentration but also increases elimination of alcohol from the body [7]. However, the nature of food such as high fat versus high protein or high carbohydrate has lesser effect of peak alcohol consumption although peak concentration was reached in 30–60 min in volunteers who consumed alcohol in empty stomach versus 30–90 min in volunteers who consumed alcohol after eating a meal rich in fat or rich in carbohydrate [8].

Similar to many other drugs, alcohol absorbed by the stomach and intestines is transported via the mesenteric and portal vein to the liver, where a portion undergoes first-pass metabolism before distribution to the rest of the body. On reaching the circulation, alcohol is widely distributed throughout the body. It penetrates the blood–brain barrier and exerts effects on most organ systems. Because of its chemical nature, alcohol distributes in tissues and fluids according to water content. This is why serum alcohol concentrations are 1.2 times higher than whole blood concentrations. The volume of distribution for alcohol is considered equal to the body water. Total body water content varies with age, gender, and weight, and is greater for males than females (representing ∼50%–60% of the total body weight for males compared to 45%–55% for females). Generally, when a male and female of similar age and weight ingest the same amount of alcohol, the female will achieve a higher peak concentration compared to male simply due to the female's lower body water content.

After ingestion, 85%–95% of alcohol undergoes metabolism within the hepatic and gastric mucosal cells, with only a small amount, 3%–10%, excreted unchanged by kidneys, lungs, and skin. The concentration of alcohol plays a role with the kinetics observed with alcohol metabolism, that is, it follows Michaelis–Menton kinetics in which metabolism changes from first-order to zero-order. At very low (<20 mg/dL) or high (>300 mg/dL) concentrations, alcohol elimination follows first-order kinetics (nonlinear); however, for the concentrations between, metabolism is independent of the dose due to the enzyme saturation and follows zero-order kinetics.

Several enzyme systems are involved in metabolism of alcohol, namely alcohol dehydrogenase (ADH), microsomal alcohol oxidizing system (MEOS), and catalase [9]. The first and most important of these, ADH, is a cytosolic enzyme family found primarily in hepatocytes:

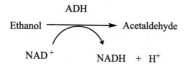

At least five classes of ADH are found in humans. For several classes, multiple isoenzymes can be distinguished by kinetic and structural properties. Lung, stomach, and gastric mucosa also express ADH, but its role in these tissues with respect to alcohol metabolism is unclear. The enzyme is nonspecific and oxidizes several other alcohols and nonalcohols, as well, including methyl alcohol, propyl alcohol, butanediol, and ethylene glycol. The nonspecific nature of ADH has been capitalized based on the clinical treatment of poisonings resulting from the exposure of methyl alcohol, butanediol, and ethylene glycol which are metabolized by ADH to toxic metabolites [10]. When alcohol is administered intravenously it competes with any of these compounds if present in circulation for metabolism. Since the affinity of ADH is greater for alcohol than the other compounds, the production of their toxic products is minimized, thus significantly reducing formations of toxic metabolites after ingestion of methyl alcohol, butanediol, and ethylene glycol. A second therapeutic intervention was the development of the orphan-drug, Fomepizole (4-methylpyrazole, Antizol, United States) [10]. This chemical has long been used in alcohol research because of its ability to inhibit, or block, the action of ADH. In the patient intoxicated with methyl alcohol or ethylene glycol, the effect is similar to the prevention of the production of toxic metabolites.

ADH activity is greatly influenced by the frequency of alcohol consumption. Adults who consume 2–3 alcoholic beverages per week metabolize alcohol at a rate of ∼100–125 mg/kg/h, whereas those considered to be an alcoholic, metabolize the compound at a rate up to 175 mg/k/h. For medium-sized adults, the blood alcohol level declines at an average rate of 15–20 mg/dL/h or a clearance rate of ∼3 ounces of alcohol/h.

The MEOS, located within the smooth endoplasmic reticulum of hepatocytes, includes the cytochrome P-450 isoenzymes CYP2E1, CYP1A2, and CYP3A4. For nonalcoholics, this metabolic pathway is considered a minor, secondary route; but it becomes much more important in alcoholics for whom heavy use of alcohol induces CYP2E1. With induction of the MEOS enzymes, clearance of alcohol from the blood may increase to >30 mg/dL/h. CYP2E1 is also expressed in the brain, where ADH activity is low. Its action leads to the formation of reactive oxygen species such as superoxide, hydroxyl radicals, and hydroxyethyl, which increase the risk of tissue injury.

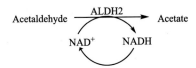

The role of peroxisomal catalase is thought to be quite small under most conditions, contributing to <2% of the oxidation of alcohol:

$$\text{Ethanol} + H_2O \xrightarrow{\text{Catalase}} 2H_2O + \text{Acetaldehyde}$$

The acetaldehyde produced should be subsequently converted to acetate as the result of the action of mitochondrial aldehyde dehydrogenase most commonly by aldehyde dehydrogenase 2 (ALDH2).

$$\text{Acetaldehyde} \xrightarrow[NAD^+ \quad NADH]{ALDH2} \text{Acetate}$$

But under conditions of excessive or chronic ingestion, the production of acetaldehyde may exceed the body's ability to further metabolize it. Because acetaldehyde is a highly reactive molecule, its prolonged presence permits it to bind to numerous enzymes and proteins blocking their normal functions. For example, in the brain, acetaldehyde forms adducts with dopamine giving rise to salsolinol, a compound thought to be involved in alcohol addiction [11]. In other tissues, acetaldehyde readily forms DNA-adducts which may be carcinogenic. Some of the acetaldehyde—protein adducts impair the liver's capacity to synthesize proteins and may contribute to the hepatomegaly encountered in later stages of disease [12]. Pharmacokinetic parameters of alcohol are listed in Table 1.2.

Several nonoxidative pathways also contribute to the metabolism of alcohol and give rise to products that under investigation as potential markers of alcohol abuse. In one such pathway, esterases in liver, pancreas, adipose tissue, brain, and heart catalyze the reaction of alcohol with free fatty acids to form fatty acid ethyl esters (FAEEs) [13]. Esterase also catalyzes the production of cocaethylene when alcohol and cocaine are coingested. Another pathway leads

TABLE 1.2 Pharmacokinetic Parameters of Alcohol

Parameter	Value
Time to peak blood concentration (Cmax)	30–60 min (no food)
	30–90 min (with food)
Volume of distribution	Females: 0.43–0.66 L/kg
	Males: 0.62–0.79 L/kg
Half-life	Half-life varies widely due to zero-order kinetics of alcohol metabolism for most people. Half-life is 2–4 h for blood alcohol level between 60 and 120 mg/dL, but half-life may be substantially higher after consuming higher amount of alcohol.
Elimination	In general 85%–95% of ingested dose is metabolized by ADH and aldehyde dehydrogenase to acetate and water (final products). A small amount of ingested alcohol is excreted unchanged in urine as well as ethyl glucuronide and ethyl sulfate in urine. Moreover, little amount is also excreted in breath and sweat.
Clearance	15–20 mg/dL/h, but clearance can be as high as 23 mg/dL/h in alcoholics.
Serum protein binding	None

to the formation of phosphatidyl alcohol via the enzyme phospholipase D. Poorly metabolized, phosphatidyl alcohol accumulates to detectable levels with chronic alcohol ingestion. The role of this product in alcohol toxicity is unclear, but its formation inhibits the activity of phospholipase D which, in turn, disrupts the formation of phosphatidic acid is inhibited. Without sufficient phosphatidic acid, cell signaling is disrupted [14]. Less than 0.1% of alcohol undergoes conjugation with glucuronic acid or sulfate to form ethyl glucuronide and ethyl sulfate, respectively. The first reaction is mediated by uridine diphosphate-glucuronyl transferases, while the second is mediated by sulfotransferases.

Polymorphism of Aldehyde Dehydrogenase Gene and Alcohol Metabolism

Polymorphisms of *ALDH2* gene that encodes aldehyde dehydrogenase-2 enzyme (responsible for the conversion of acetaldehyde to acetate) play an important role in alcohol metabolism. Currently, there is a strong evidence that variant allele *ALDH2*2* (rs671 G > A) resulting from a single nucleotide exchange causing substitution of lysine for glutamate at position 487 of the mature ALDH2 enzyme protects an individual from alcohol abuse because the enzyme encoded by this defective gene has poor enzymatic activity. As a result, acetaldehyde produced from alcohol by the action of ADH is not effectively removed from blood. This acetaldehyde buildup in blood produces adverse reactions (rapid heartbeat, sweating, etc.) in these individuals thus deterring them from drinking. *ALDH2*2* is a common variant among 45% of East Asians including Han Chinese, Japanese, and Koreans, but rare in other ethnic groups and it has been estimated that 540 million people worldwide (8% of world population) carry this allele. *ALDH2*2* homozygotes (*ALDH2*2/*2*) exhibit essentially no ALDH2 enzymatic activity but heterozygotes (*ALDH2*1/*2*) may show partial ALDH2 activity. Genetic and epidemiological studies have shown that *ALDH2*2* homozygous individuals are almost fully protected from alcohol use disorder, but heterozygous individuals have partial protection (approximately 60%) [15]. Moreover, certain genetic polymorphisms of genes encoding ADH enzyme produce super active enzyme which also results in acetaldehyde buildup in blood thus also deterring these individuals from drinking. Combination of super active ADH and inactive aldehyde dehydrogenase enzyme in an individual provide total protection from drinking.

GUIDELINES FOR ALCOHOL USE

United States Department of Agriculture (USDA) and Department of Health and Human Services jointly publish "Dietary Guidelines for Americans" every 5 years, suggesting Americans what constitutes a balanced diet. These guidelines also include suggestions for drinking in moderation. However, alcohol is not a component in USDA food pattern. If alcohol is consumed, the calories from alcohol must be accounted when other foods are consumed so that daily calorie intake does not exceed the recommended limit (1600–2400 calories per day for women and 2000–3000 calories per day for men). The latest Dietary Guidelines for Americans 2015–2000, eighth edition, suggest that if alcohol is consumed it should be consumed in moderation following these guidelines:

- Up to 1 drink per day for women and up to 2 drinks per day for men and only by adults of legal drinking age (21 years or older).

One drink is defined by the guidelines as containing 14 g (0.6 fluid ounces) of pure alcohol.

Federal Food and Drug Administration has determined that mixing alcohol and caffeine is not a safe practice and recommended four manufacturers of alcoholic beverages containing caffeine to discontinue use of caffeine in these alcoholic drinks. People who mix alcohol and caffeine may drink more alcohol and become more intoxicated than they realize, increasing the risk of alcohol-related adverse events. Energy drinks are gaining popularity among young adults as well as underage drinkers. Energy drinks may contain caffeine which increases craving for alcohol. Studies have indicated that energy drinks may increase the craving for alcohol and binge drinking. Moreover, pleasurable experience of drinking alcohol is also enhanced by consuming energy drinks at the same time [16].

National Institute of Alcohol Abuse and Alcoholism(NIAAA) considers high-risk drinking as consuming 4 or more drinks in any day or 8 or more drinks per week for women and 5 or more drinks in any day or 15 or more drinks per week for men. Binge drinking is defined by NIAAA as the consumption of 5 drinks in 2 h for men and 4 drinks in the same time period for women. However, another government agency Substance Abuse and Mental Health Services Administration (SAMHSA) defines binge drinking as consuming 5 or more alcoholic beverages on the same occasion for the past 30 days. SAMHSA also defines heavy drinking as consuming 5 or more drinks on the same occasion on each of 5 or more days for the past 30 days. Guidelines and various pattern of drinking are listed in Table 1.3.

TABLE 1.3 Definition of Moderate, High-Risk, Heavy, and Binge Drinking

Type of Drinking	Gender	Definition/Guideline	Comments
Moderate drinking (USDA/NIH: 2015–2020) dietary guidelines	Female Male	Up to 1 drink per day. Up to 2 drinks per day.	Many beneficial effects of alcohol can be derived from moderate drinking. Moreover, moderate drinking leads to blood alcohol significantly lower than 0.08%, the legal limit of driving in United States
High-risk drinking (NIAAA)	Female Male	4 or more drinks on any day or 8 or more drinks per week. 5 or more drinks on any day or 15 or more drinks per week for men.	Beneficial effects of alcohol disappear with high-risk drinking.
Heavy drinking (SAMHSA) guideline	Both female and male	Consuming 5 or more drinks at the same occasion on each of 5 or more days in the past 30 days.	Adverse effects of alcohol are observed in heavy drinkers. Moreover, blood alcohol levels often exceed the legal limit.
Binge drinking (NIAAA)	Female Male	4 drinks in 2-h period. 5 drinks in 2-h period.	Even occasional binge drinking is detrimental to health. Moreover, binge drinking often increases blood alcohol levels over legal limit of driving.
Binge drinking (SAMHSA)	Both female and male	Consuming 5 or more drinks at the same occasion at least one day in the past 30 days.	Even occasional binge drinking is detrimental to health. Moreover, binge drinking often increases blood alcohol levels over legal limit of driving.
Underage drinking	Both female and male	Legal age of drinking in all Sates in United States is 21 years. Anyone drinking below that age is considered as underage drinking.	Underage drinking regardless of amount and frequency adversely affects the developing brain and also has many other ill effects.

NIAAA: National Institute on Alcohol Abuse and Alcoholism; SAMHSA: Substance abuse and Mental Health Services Administration.

BENEFITS OF CONSUMING ALCOHOL IN MODERATION

Consuming alcohol in moderation has many health benefits including increased longevity. These health benefits are summarized in Table 1.4. Some of these benefits are attributable to alcohol while many other benefits are due to the combined effect of both alcohol and many beneficial phytochemicals present in beer and wine which are excellent antioxidants. More than 400 different phytochemicals are present in beer; some of these compounds originate from raw materials while others are generated during the fermentation process. Melatonin is generated during the brewing process. Beers with higher alcoholic content usually have higher amounts of melatonin [17]. More than 1600 phytochemicals are present in wine prepared from grapes [18].

Cardiovascular diseases including myocardial infarction are the number one killer in the United States and other industrialized countries, but drinking in moderation reduces the risk of cardiovascular diseases including myocardial infarction.

Reduced Risk of Cardiovascular Diseases

Probably the most notable benefit of drinking in moderation is the reduced risk of cardiovascular diseases. The relationship between alcohol consumption and coronary heart disease was examined in the original Framingham Heart Study which showed a U-shape curve with reduced risk of cardiovascular diseases with moderate drinking but higher risk with heavy drinking. Smoking is a risk factor for coronary heart disease but moderate alcohol consumption may also provide some protection even among smokers [19]. Interestingly, women may get the beneficial effect of alcohol by consuming lower amounts as well as consuming alcohol less frequently than men. [20]. Moderate alcohol consumption can also provide protective effects against heart failure [21]. Gemes et al. based on a study of 58,827 individuals and an

TABLE 1.4 Comparison of Beer, Wine, and Spirits for Various Health Benefits Associated With Moderate Consumption of Alcoholic Beverages

Health Benefit	Beer	Wine	Sprit	Comments
Reduced risk of cardiovascular disease	++	+++	+	Alcohol in moderate concentration increases HDL cholesterol. Resveratrol present in red wine prevents oxidation of LDL. Protection is more significant in men over 50 years of age but women of any age may get benefits of consuming alcohol in moderation.
Reduces mortality in survivors of myocardial infarction	++	++	++	Any type is beverage has protective effect because alcohol is directly responsible for such protection.
Reduced risk of ischemic stroke	++	++	++	Any type of alcoholic beverage (beer, wine, or liquor) can reduce such risk because alcohol is responsible for such effect.
Reduced risk of Type 2 diabetes	++	++	++	Alcohol not only reduces blood glucose after meal by stimulating insulin secretion but also may improve insulin sensitivity. However, for individuals taking sulfonylurea medication alcohol may cause hypoglycemia.
Protection from certain types of cancer	+	++		Oxidative stress increases cancer risk, but both beer and wine contain antioxidants as well as anticarcinogen (for example xanthohumol in beer, resveratrol in red wine) that may reduce risk of certain types of cancer.
Reduced risk of rheumatoid arthritis	++	++	++	Alcohol has antiinflammatory effect. Therefore consuming any type of alcoholic beverage is helpful.
Reduced risk of forming gallstone	++	++	++	Consuming any alcoholic beverage is associated with decreased risk but consuming 5–7 drinks (any type of alcoholic beverage) per week may be needed for such effect as people who drink 1–2 alcoholic beverage per week may not get any benefit.
Protection from common cold		+		Only wine drinking is effective.
Reduced risk of kidney stones	++	+	+	Beer may be more effective than wine.
Better bone health	++	+	+	Silicon which is present in beer but not in wine or liquor has beneficial effect on bone health.
Protection from age-related dementia/Alzheimer's disease	+	++	+	Alcohol has protective effects against dementia and Alzheimer's disease but wine may be superior to beer or liquor in elderly population. Resveratrol found in abundance in grape skin and red wine has protective effects.
Increased longevity	++	+	+	Wine may provide best effect.
Perception of good health		++		Only drinking wine is associated with subjective perception of good health but mechanism is not clear.
Stress relief and better mood	++	++	+	Any alcoholic drink can reduce stress if consumed in moderation but it is preferable to drink beer and or wine due to additional health benefits.

11.6-year follow-up observed that light to moderate consumption of alcohol was associated with lower risk of myocardial infarction [22].

O'Keefe et al. reviewed effect of alcohol on health of the human heart and commented that habitual light to moderate drink lowers the rate from death due to coronary artery disease, diabetes mellitus, congestive heart failure, and stroke. However, excessive alcohol consumption is the third leading cause of premature death in United States. In general, men older than age 50 get more favorable effects of consuming alcohol in moderation than younger men, but women of any age benefit from drinking in moderation. Unfortunately, cardioprotective effect of alcohol has not been documented in epidemiological studies of population from India and China. Interestingly, authors advise individuals who are teetotalers not to initiate light to moderate alcohol consumption for health benefits [23].

There are several hypotheses on how moderate drinking can reduce the risk of heart disease. Many studies have demonstrated increased high-density lipoprotein cholesterol (HDL cholesterol) levels in drinkers compared to nondrinkers and such increased HDL cholesterol is mostly attributable to the alcohol content of the drink. However, level of increased HDL cholesterol in blood may explain approximately 50% of protective effect of alcohol beverages while another 50% may be related to inhibition of platelet aggregation and antioxidant effect of various polyphenolic compounds present in beers and wines. The polyphenolic antioxidant compounds found in abundant in red wine can further reduce platelet activity via other mechanisms [24].

Significant research to understand epidemiological phenomenon known as "French paradox" (low incidence of cardiovascular diseases in French population despite consuming fatty foods) indicates superiority of wine especially red wine in reducing risk of cardiovascular diseases compared to other alcoholic beverages [25]. However, other authors disputed this finding. In an interesting study, Barefoot et al. concluded that the apparent health benefits of wine compared to other alcoholic beverages, as reported by some investigators, may be a result of dietary habits and other lifestyle factors. The authors observed that subjects who preferred wine had healthier diets than those who preferred beer or spirits or had no preference. Wine drinkers also reported eating more servings of fruits and vegetables and fewer servings of red or fried meat. In addition, wine drinkers were less likely to smoke [26]. Light to moderate consumption of alcohol also increases the chance of survival after the first heart attack [27].

Other Physical Health Benefits of Moderate Drinking

Another beneficial effect of consuming alcohol in moderation is reduction in the risk of stroke among both men and women regardless of age or ethnicity. The Copenhagen City Heart Study with 13,329 eligible men and women aged between 45 and 84 years with 16 years of follow-up indicated a U-shaped relation between intake of alcohol and risk of stroke. People who consumed low to moderate alcohol experienced protective effect of alcohol against stroke, but heavy consumers of alcohol were at higher risk of suffering from a stroke compared to moderate drinkers or nondrinkers [28].

Moderate consumption of alcohol reduces the risk of Type 2 diabetes by reducing oxidative stress and increasing insulin sensitivity. Based on the 15 studies conducted in the United States, Finland, Netherland, Germany, UK, and Japan with 369,862 men and women and an average follow-up of 12 years, light drinkers and moderate drinkers had a lower risk of Type 2 diabetes compared to nondrinkers. The risk reduction was approximately 30% in moderate drinkers compared to nondrinkers [29]. Baliunas et al., based on reviewing 20 studies, observed a U-shape relationship between alcohol consumption and risk of Type 2 diabetes, where moderate alcohol consumption decreased the risk of Type 2 diabetes but heavy alcohol consumption increased the risk [30]. Polsky and Akturk commented that light to moderate consumption of alcohol decreases the incidence of diabetes whereas heavy drinkers and binge drinkers are at an increased risk of diabetes. Among people with diabetes, light to moderate alcohol consumption reduces risk of cardiovascular diseases as well as all-cause mortality [31].

Moderate consumption of alcohol may reduce the risk of certain types of cancer because wine and beer contain anticarcinogenic compounds and antioxidants. In the California Men's Health Study using 84,170 men aged between 45 and 69, the authors observed that consumption of one or more glasses of red wine per day was associated with approximately 60% reduced lung cancer risk ever in smokers [32]. Consumption of up to 1 drink per day reduced the risk of head and neck cancer in both men and women, but consuming more than three alcoholic beverages increased the risk of cancer. In an Italian study, the authors observed that moderate consumption of alcohol reduced the risk of renal cell carcinoma in both males and females [33]. However, the relation between moderate alcohol consumption and risk of breast cancer is controversial because there are conflicting reports in the medical literature. Pelucchi et al. concluded that consumption of less than 3 alcoholic drinks per week is not associated with increased risk of breast cancer in women, but consuming 3−6 drinks per week may be associated with a small increase in risk [34]. Beverage type is not associated with increased risk of breast cancer because alcohol acts as weak breast carcinogen. Acetaldehyde, a metabolite of alcohol is also a carcinogen. However, there are many other factors that may increase the risk of breast cancer including family history (*BRCA1* and *BRCA2* gene mutations are most common inherited factors), being older (only 10%−15% breast cancer diagnosis among women 45 years or younger), using hormone replacement therapy, previous or current use of birth control pills, not having a child, or having a dense breast.

Moderate alcohol consumption reduces the risk of rheumatoid arthritis [35]. Nissen et al. reported based on a study using 2908 patients suffering from rheumatoid arthritis that occasional or daily consumption of alcohol reduced the progression of the disease based on radiological studies (X-ray). Best results were observed in male patients [36]. Drinking all types of alcohol (beer wine and liquor) was associated with reduced risk of rheumatoid arthritis [37].

Moderate drinking (5–7 drinks per week) but not light drinking (1–2 drinks per week) may reduce the risk of gallstone disease. All alcoholic beverages types are associated with decreased risk [38]. The relation between moderate alcohol consumption and reduced incidence of common cold has been studied. In a large study using 4272 faculties and staff of five Spanish Universities as subjects, the investigators observed that total alcohol intake from drinking beer and spirits had no protective effect against common cold whereas moderate wine consumption was associated with some protection against common cold [39].

Because moderate consumption of alcohol can prevent many diseases, it is expected that moderate drinkers may live longer than lifetime abstainer. In Physician's Health Study involving 22,071 male physicians in the United States between age 40 and 84 with no history of myocardial infarction, stroke, or cancer, and then 10 years of follow-up, the authors observed that men who consumed 2–6 drinks per week had the most favorable results (20%–28% lower mortality rate) than people who consumed 1 drink per week. In contrast, people who consumed more than 2 drinks per day had approximately 50% chance of higher mortality than people who consumed just 1 drink per week [40]. Interestingly, wine may be superior to beer or spirit in increasing longevity. In a study of 13,064 men and 11,459 women, Gronbaek et al. observed that compared to nondrinkers, light drinkers who avoided wine had a 10% lower chance of mortality but light wine drinkers had 34% lower chance of mortality [41]. In another study, the authors concluded that light drinkers of any kind of wine had lower mortality risk than beer or liquor drinkers [42]. Interestingly only drinking wine in moderation is associated with perception of good health [43].

Mental Health Benefits of Moderate Drinking

Alcohol if consumed in low to moderate amount is effective in reducing stress and increasing affective expression, happiness, euphoria, conviviality, and pleasant as well as carefree feelings. In addition, tension, depression, and self-consciousness have been reported to decrease with low to moderate alcohol consumption [44]. In general, alcohol has a euphoric effect at low to moderate blood alcohol (0.02%–0.05% whole blood alcohol levels; 20–50 mg/dL) concentrations and such effect starts 10–15 min after initiation of drinking. Alcohol is known to cause reduction in inhibition and such effect after drinking is more significant in women compared to men. In one study based on 184 degree-level and postgraduate students (94 females, 90 males), the authors concluded that alcohol at a level of approximately 50 mg/dL (0.05%) facilitated social interaction and communication [45]. Interestingly, moderate alcohol consumption increases happiness even in the unpleasant ambiance [46].

Subjective health may simply be an indicator of actual health status. Studies have shown that moderate consumption of alcohol may provide a rewarding sense of well-being in association with good physical health. Possible link between moderate drinking and success at work may be related to overall better physical health, better psychosocial adjustment as well as greater involvement in employment-related social experiences by moderate drinkers. Moreover, during stressful work environment, nondrinkers are more likely to be absent from work than moderate drinkers. Alcohol may have buffering effect between stress and sickness. In addition, nondrinkers are 27% more likely to be disabled compared to moderate drinkers. Incidentally, studies have shown that moderate drinkers may have higher income than teetotalers [47]. A Spanish study reported that moderate alcohol consumption, especially drinking wine was associated with more active lifestyle and better perception of health in elderly subjects [48]. Studies have shown that moderate alcohol consumption is associated with reduced depression in elderly subjects. However, heavy drinking is associated with higher rate of depression.

Age-related dementia and Alzheimer's disease are neurodegenerative diseases associated with advanced age. Alzheimer's disease is a devastating neurological disorder affecting 1 in 10 Americans over the age of 65 and almost half of all Americans over 85 years old. Moderate alcohol consumption can dramatically reduce the risk of age-related dementia and Alzheimer's disease. The triggering agent for Alzheimer's disease is β-amyloid peptide (Aβ-aggregates), which alters the synaptic activity and disrupts neurotransmission mediated by N-methyl-D-aspartate receptor in the brain. Alcohol in low dosage can prevent formation of Aβ-aggregates thus delaying or preventing onset of Alzheimer's disease. Moreover, low to moderate consumption of alcohol is also associated with reduced risk of other neurodegenerative diseases as evidenced from data of several large clinical trials [49]. Dietary compounds (polyphenols) found in grapes have some protective effect against Alzheimer's disease because these compounds interfere with generation and aggregation of β-amyloid peptide. Resveratrol, a compound found in abundance in red wine and also in grapes provides protection against Alzheimer's disease [50].

HEALTH HAZARDS OF HEAVY DRINKING

All benefits of consuming alcohol in moderation disappear with heavy drinking. In addition, excess consumption of alcohol on a regular basis damages multiple organ systems including brain, heart, bone, immune system, and endocrine system. Major adverse effects of chronic alcohol consumption include decreased life span, increased risk of violent behavior, alcoholic liver diseases including cirrhosis of liver, mood disorders, and significantly increased risk of various cancers. Drinking during pregnancy may be associated with poor outcome in pregnancy including fetal alcohol syndrome. Major adverse effects of heavy drinking are summarized in Table 1.5.

Alcohol Abuse and Liver Damage

In the United States, about 60% of the general population admits alcohol use with 8%–10% reported heavy drinking (more than 2 drinks per day). Alcohol abuse is a leading cause of global morbidity and mortality with majority of alcohol-related disease burden resulting from alcoholic liver diseases because alcohol is hepatotoxic if consumed in excess. The first sign of liver toxicity is fatty changes in the liver which may occur even after few days of heavy drinking, but such changes can be easily reversed by abstinence. However, drinking heavily for a longer period may cause more severe alcohol-related liver injuries such as alcoholic hepatitis. Individuals who continue to abuse alcohol may develop fibrosis and eventually cirrhosis of liver. Women are at greater risk of developing alcoholic liver diseases. Obesity as well as smoking may increase the risk of alcoholic liver diseases [51]. In the United States, alcoholic hepatitis occurs in 35%–40% of all alcohol abusers while approximately 20%–25% of all cases of liver cirrhosis are related to alcohol abuse [52]. Drinking daily increases the risk of alcohol-related liver disease including liver cirrhosis compared to people who drink 2–4 days per week (hazard ratio: 3.65). Moreover, alcohol abuse in recent periods is a better predictor of liver cirrhosis than alcohol abuse in past years. Men consuming 14–28 standard drinks per week also have higher risk than men who consuming less than 14 drinks per week. Interestingly compared to beer and liquor, drinking wine might be associated with lower risk of alcoholic cirrhosis [53]. Welsh and Alexander commented that above a threshold of 7–13 drinks per week for women and 14–27 drinks per week for men there is a risk of developing some alcohol-related liver problem. The greater sensitivity of women towards alcohol toxicity may be related to genetic predisposition of metabolism pattern of alcohol in women where more oxidative byproducts of alcohol are formed compared to men. Consumption of coffee may protect males against alcohol-induced liver damage, but no such data is currently available for females [54].

Risk of alcohol-related liver diseases is significantly higher in individuals infected with hepatitis B or C virus. In one study, the authors observed that even moderate alcohol consumption may adversely affect the progression of liver damage in individuals infected with hepatitis C virus [55]. Mechanism of alcohol-induced liver disease is complex. While in moderate drinkers alcohol is mostly metabolized by ADH in the liver, in alcoholics CYP2E1, a member of the cytochrome P-450 drug metabolizing family of enzymes in the liver becomes activated. In this process, reactive oxygen species are generated. Hydroxyethyl radicals are probably involved in the alkylation of proteins found in hepatocytes causing liver damage.

Alcohol Abuse and Brain Damage

Teenage drinking is a serious public health concern in the United States because alcohol has more damaging effects on the adolescent brain than the adult brain. Onset of drinking at an early age (13 or earlier) has devastating effects on the brain and such adverse effects may last lifetime. Early onset of drinking is also linked to greater risk of alcohol dependance in adult life. Although thiamine deficiency is one of the major factors involved in alcohol-related brain damage, both alcohol and its toxic metabolite acetaldehyde exert toxic effects on neurons. Underage drinkers are also susceptible to immediate ill effects of alcohol use such as blackouts, hangover, and alcohol poisoning and these individuals are also at higher risk of neurodegeneration, impairment of functional brain activity, and neurocognitive deficits. In general, adolescent drinking induces brain structure abnormalities and these changes lead to poor memory, impaired study habits, poor ability to learn, and poor academic performance. Moreover, female adolescent brain is more vulnerable to alcohol exposure than male adolescent brain [56].

Excessive drinking may also damage the adult brain. The two major alcohol-related brain damages are alcoholic Korsakoff's syndrome and alcoholic dementia. Korsakoff's syndrome is a brain disorder caused by deficiency of thiamine and major symptoms are severe memory loss, false memory, lack of insight, poor conversation skills, and apathy. Some heavy drinkers may also have a genetic predisposition to developing this syndrome. In Korsakoff's syndrome,

TABLE 1.5 Adverse Effects of Heavy Consumption of Alcohol

Adverse Outcome	Comments
Fatty liver disease/alcohol-related liver diseases	Consumption of 3 or more drinks per day may increase the risk of fatty liver disease and other alcohol-related liver diseases. Women are higher risk than men. Interestingly wine drinkers are at lower risk than beer/spirit drinkers. Individuals infected with hepatitis have higher risk of alcohol-related liver diseases including liver carcinoma.
Brain damage in adolescents	Early onset of drinking around age 13 has devastating effect on developing brain that may persist lifetime. Teenagers and young adult below age of 21 should not drink at all. Girls are more susceptible than boys.
Brain damage in adults	Smaller brain volume in both men and women with mental impairment and cognitive difficulties. However, women are affected more than men.
Korsakoff syndrome	This disease observed in alcoholics mainly due to thiamine (vitamin B1) deficiency. Severe dementia is the major observation but abstinence from alcohol and proper treatment may be able to reverse some symptoms.
Wernicke-Korsakoff syndrome	In addition to dementia and confusion, visual problems, and muscle weakness (difficult to walk etc.) are also observed. Thiamine deficiency is the major cause of this disease.
Increased risk of cardiovascular diseases including heart attack and heart failure	Consuming more than three drinks each day on a regular basis may cause some damage to the heart. Heavy drinking is also associated with hypertension, heart failure, alcoholic cardiomyopathy, and increased risk of death after heart attack. Women are more susceptible to alcohol induced heart damages than men.
Increased risk of stroke	Consuming 21 or more drinks weekly on a regular basis increases risk of stroke especially hemorrhagic stroke.
Increased risk of cancer	Excessive alcohol consumption is associated cancer of the mouth, pharynx, larynx, and esophagus. Alcoholics may develop cirrhosis of liver which may progress to live cancer.
Damage to the immune system	Alcohol reduces immunity and as result individuals consuming excessive amount of alcohol are more prone to both viral and bacterial infections.
Progression of AIDS	Even consuming 2 or more drinks per day on a regular basis may harm a patient despite receiving treatment.
Damage to endocrine system	High alcohol concentration in blood may interfere with proper secretion of hormones from endocrine glands. Pseudo-Cushing's disease which has all symptoms of Cushing's disease may be observed in alcoholics.
Impaired fertility	Alcohol abuse is associated with fertility problems in both men and women.
Bone damage	Heavy consumption of alcohol may reduce bone mass thus increasing risk of fracture after fall.
Fetal alcohol spectrum disorders and fetal alcohol syndrome	Pregnant women and women plan to be pregnant should not consume any alcohol.
Mood disorder, anxiety, and depression	Alcohol abuse is associated with mood disorders and major depression in both young adults and elderly. Alcohol abuse may also increase the risk of late-life suicide.
Violent behavior	Alcohol abuse may cause aggressiveness and violent behavior. According to US Bureau of Justice approximately 37% state prison inmates and 21% federal prison inmates serving time for violent crimes were under the influence of alcohol when they committed crimes.
Reduced life span	Consuming alcohol exceeding upper limit of moderate drinking and especially binge drinking may reduce life span. People drinking 9 or more drinks in one occasion have high risk of death following injury after such binge drinking episode.

loss of neurons is a common feature including microbleeding in certain regions of gray matter [57]. When Wernicke's encephalopathy accompanies Korsakoff's syndrome in an alcoholic it is called Wernicke–Korsakoff syndrome. Wernicke's encephalopathy and Korsakoff syndrome are the two related diseases, both caused by thiamine deficiency but clinical symptoms may be different. The Royal College of Physicians in London recommends that patients admitted to the hospital who show evidence of chronic misuse of alcohol and poor diet should be treated with B vitamins [58]. Binge drinkers, both males and females, are at higher risk of developing alcohol-related brain damage.

Alcohol Abuse Increases Risk of Cardiovascular Disease and Stroke

Although alcohol if consumed in moderation has cardioprotective effect, drinking more than three drinks a day (any type of beverage) may be harmful to the heart. Chronic alcohol abuse for several years may result in alcoholic cardiomyopathy and heart failure, systematic hypertension, heart rhythm disturbances, and hemorrhagic stroke [59]. Alcoholics who consume 90 g or more of alcohol a day (7–8 drinks) for 5 years are at high risk of alcoholic cardiomyopathy and if they continue drinking alcohol, cardiomyopathy may proceed to heart failure, a potentially fatal medical condition. Without complete abstinence, 50% of these patients will die within the next 4 years of developing heart failure [60]. Heavy drinking also increases the risk of stroke, particularly the risk of hemorrhagic stroke [61].

Fetal Alcohol Syndrome

Fetal alcohol syndrome and fetal alcohol spectrum disorders are due to devastating effect of alcohol on the fetus. A pregnant woman or a woman planning to get pregnant should avoid consuming any amount of alcohol to prevent such birth defects or miscarriage. Approximately 1–4.8 of every 1000 children born in the United States have fetal alcohol syndrome, while as many as 9.1 out of 1000 babies born have fetal alcohol spectrum disorder [62].

Other Adverse Effects of Alcohol Abuse

Alcohol abuse is associated with increased risk of bacterial and viral infection due to impairment of the immune system by excess alcohol in the blood. Exposure to alcohol can result in reduced cytokine production. Alcohol also reduces the viability of mast cells that contributes to the impaired immune system associated with alcohol abuse [63]. Alcohol also accelerates disease progression in patients with HIV infection because of immunosuppression. In one study using 231 patients with HIV infection who were undergoing antiretroviral therapy, the authors observed that even consumption of two or more drinks daily can cause a serious decline in CD4+ cell count (higher CD4+ counts indicates a good response to therapy) [64].

Alcohol abuse can have an adverse effect on the human endocrine system. Alcohol abuse may cause pseudo-Cushing's syndrome which is indistinguishable from Cushing's syndrome characterized by excess production of cortisol causing hypertension, muscle weakness, diabetes, obesity, and a variety of other physical disturbances. Diminished sexual function in alcoholic men has been described for many years. Administration of alcohol in healthy young male volunteers caused a diminished level of testosterone. Alcoholic women often experiences reproductive problems. However, these problems may resolve when a woman practices abstinence from alcohol. To form healthy bone calcium, phosphorus, and active form of Vitamin D is essential. Chronic consumption of alcohol may reduce bone mass through a complex process of inhibition of hormonal balance needed for bone growth including testosterone in men, which is diminished in alcoholics. Alcohol abuse may also interfere with pancreatic secretion of insulin causing diabetes [65].

Epidemiological research has demonstrated a dose-dependent relation between consumption of alcohol and certain types of cancers and the strongest link was found between alcohol abuse and cancer of the mouth, pharynx, larynx, and esophagus. Pancreatitis and gallstone are common among alcohol abusers. In alcoholics endotoxin may be released from gut bacteria by the action of excess alcohol and such process may trigger progression of acute pancreatitis into chronic pancreatitis. Chronic pancreatitis may lead to pancreatic cancer [66].

Although moderate drinking is associated with increased longevity, alcohol abuse is associated with all-cause decreased longevity compared to abstainers. Even occasional heavy drinking may be detrimental to health. Dawson reported an increased risk of mortality among individuals who usually drink more than 5 drinks per occasion but consumed alcohol less than once a month [67]. Binge drinking is a very dangerous practice. In one study, the authors using a population of 1641 men who consumed beer observed that relative risk (RR) of all-cause deaths was 3.10 and RR of fatal myocardial infarction was 6.50 in men who consumed 6 or more bottles of beer per session compared to men who consumed less than 3 bottles of beer per session [68].

Many investigators reported a close link between violent behavior, homicide, and alcohol intoxication. Studies conducted on convicted murders suggested that about half of them were under the heavy influence of alcohol at the time of murder [69]. Alcohol abuse by husband may be related to husband to wife marital violence. Studies have shown the link between alcohol abuse by the husband before marriage and husband to wife aggression in the first year of marriage. The most violence occurs when the husband was a heavy drinker and wife was not [70].

ACUTE VERSUS CHRONIC ALCOHOL TOXICITY

Alcohol is present in a variety of beverage products in which the amount of alcohol varies considerably. Ingestion of one drink generally leads to an approximate blood alcohol concentration of 25 mg/dL. In general, alcohol blood levels between 20 and 50 mg/dL produce pleasurable effects, but even at 80 mg/dL, the legal limit of driving, minor impairments are possible in a person who drinks infrequently. However, significant impairment of motor skill and decreased cognitive abilities are observed at blood alcohol levels between 100 and 150 mg/dL. Symptoms and signs associated with mild acute intoxication, which may occur at a blood alcohol level of 150 mg/dL and higher, include slurred speech, poor coordination, unsteady gait, and uninhibited behavior. However, at blood alcohol level of 200–300 mg/dL, the mental status changes and physical effects become more pronounced and the individual may show signs of dehydration and hypotension which are signs of alcohol intoxication. Severe intoxication (blood alcohol levels over 300 mg/dL) requires aggressive care as respiratory depression, coma, and hypothermia are likely. In general, alcohol poisoning occurs at blood alcohol level of 0.35% but even at blood alcohol concentration over 0.25% (250 mg/dL) some patients may be at a higher risk of coma. However, alcoholics may tolerate higher blood alcohol levels than alcohol-naïve individuals [71]. Celik et al. reported that postmortem blood alcohol level ranged from 136 to 608 mg/dL in 39 individuals who died due to alcohol overdose. Most of these deceased were male [72]. The mechanism of death from alcohol poisoning is usually attributed to paralysis of respiratory and circulatory centers in the brain causing asphyxiation.

LABORATORY TESTING FOR ALCOHOL

Because of its wide distribution into the body water, alcohol is measured in most body fluids: whole blood, serum, breath, oral fluid, urine, sweat, and in vitreous. In the clinical laboratory, gas chromatographic and enzymatic methods are employed with the choice depending upon the laboratory and the setting. Gas chromatography offers the most specific method and there are procedures which allow for the simultaneous measurement of alcohol, methyl alcohol, isopropanol, acetone, and even ethylene glycol. The enzymatic methods are rapid and used by more than 95% of laboratories participating in College of American Pathologists' (CAP) proficiency surveys. The most frequently encountered enzymatic method utilizes a yeast-derived ADH and NAD^+ to convert alcohol to acetaldehyde and reduced NADH. The amount of NADH produced is (absorbs at 340 nm while NAD^+ has no absorption at 340 nm) proportional to the alcohol concentration. Care must be taken to prepare the phlebotomy site correctly when using the enzymatic methods because of the potential for cross-reactivity with any other alcohol which may be present. Please see Chapter 2 (Alcohol Analysis in Various Matrixes: Clinical Versus Forensic Testing) for an in-depth discussion on this topic.

The use of breath as a sample is most commonly found in forensic settings but because blood collection from an intoxicated patient is difficult, the technique has piqued the interest of some emergency departments. The ready partitioning of alcohol from an aqueous solution (blood) into a vapor (inspired air within the lungs) permits the measurement of alcohol in expired air or breath. Techniques most commonly employed today in these devices include infrared spectroscopy, electrochemical sensors, and semiconductors.

BIOMARKERS OF ALCOHOL ABUSE

Given the difficulty of firmly identifying the person who is abusing alcohol, biomarkers of alcoholism have long been sought. The enzyme gamma-glutamyltransferase (GGT) and mean corpuscular volume (MCV) are two of the more popular older tests that have been used as indirect biomarkers of alcohol abuse. GGT activity increases in response to the hepatocellular toxicity of alcohol. However, GGT and MCV may also be elevated for many reasons other than alcohol abuse. Carbohydrate-deficient transferrin (CDT) is an isoform of transferrin deficient in sialic acid residues which can be used as a biomarker of alcohol abuse. However, normal transferrin levels as well as CDT levels are different in men and women. As a result, %CDT is usually used as a biomarker for alcohol abuse because it has the same reference range in both men and women.

Although GGT, MCV, %CDT, etc. are the indirect biomarkers of alcohol abuse, there are also direct alcohol biomarkers. Acetaldehyde−protein adducts, a direct marker of alcohol consumption, are formed due to the interaction of alcohol metabolite acetaldehyde and various proteins present in blood and such adducts persist for days following alcohol ingestion. Ethyl glucuronide and ethyl sulfate are the minor metabolites of alcohol which are detected in biological samples including blood, serum, cerebrospinal fluid, urine, hair, meconium, along with other various organ tissues. Other direct alcohol biomarkers are FAEEs, phosphatidylethonol (measured in whole blood) [73]. Armer et al. commented that although breath alcohol, urine alcohol, urine ethyl glucuronide, and urine sulfate can be measured to monitor patients enrolled in community alcohol treatment programs, urine ethyl glucuronide and ethyl sulfate demonstrated optimum diagnostics performance. The authors recommend routine use of ethyl glucuronide and ethyl sulfate in monitoring patients enrolled in alcohol treatment programs [74]. For an in-depth discussion of alcohol biomarkers please see Chapter 3 (Alcohol Biomarkers: Clinical Issues and Analytical Methods).

CONCLUSIONS

Studies on the impact of alcohol on the body continue to raise as many questions as answers. The compound has such diverse and wide-reaching uses as a solvent that it is virtually impossible to avoid contact. Fortunately, most of these exposures are of low doses and minimal consequences. Clearly, when alcohol is misused or abused as in alcoholism, toxicity is observed in every organ system and there are significant social consequences. However, there is an evidence that when ingested in moderation some alcoholic beverages convey cardioprotective benefits as well as many other benefits. Alcohol testing has expanded from the measurement of alcohol in various body fluids to now include the measurement of various biomarkers of alcohol use.

REFERENCES

[1] Dudley R. Alcohol, fruit ripening and the historical origins of human alcoholism in primate frugivory. Integra Comp Biol 2004;44(4):315−23.
[2] Vallee BL. Alcohol in the Western World. Sci Am 1998;278:80−5.
[3] Pintaric IK. Determination of alcohol in chocolate shell pralines and filled chocolates by capillary gas chromatography. Arh Hig Rada Toksikol 1999;50:23−30.
[4] McLachlan DG, Wheeler PD, Sims GG. Automated gas chromatographic method for the determination of alcohol in canned salmon. J Agric Food Chem 1999;47:217−20.
[5] Kerr WC, Greenfield TK, Tujague J, Brown SE. A drink is a drink? Variation in the amount of alcohol contained in beer, wine and spirits drinks in a US methodological sample. Alcohol Clin Exp Res 2005;29:2015−21.
[6] Liber CS. Relationship between nutrition, alcohol use and liver disease. Alcohol Res Health 2003;2003(27):220−31.
[7] Jones AW, Jonsson KA. Food-induced lowering of blood alcohol profiles and increased rate of elimination immediately after a meal. J Forensic Sci 1994;39:1084−93.
[8] Jones AW, Jonsson KA, Kechagias S. Effect of high fat, high protein and high carbohydrate meals on the pharmacokinetics of a small dose of alcohol. Br J Clin Pharmacol 1997;44:521−6.
[9] Zakhari S. Overview: how is alcohol metabolized by the body? Alcohol Res Health 2006;29:245−54.
[10] Kraut JA, Kurtz I. Toxic alcohol ingestions: clinical features, diagnosis, and management. Clin J Am Soc Nephrol 2008;3:208−25.
[11] McBride WJ, Li TK, Deitrich RA, Zimatkin S, et al. Involvement of acetaldehyde in alcohol addiction. Alcohol Clin Exp Res 2002;26:114−19.
[12] Brooks PJ, Theruvathu JA. DNA adducts from acetaldehyde: implications for alcohol-related carcinogenesis. Alcohol 2005;35:187−93.
[13] Best CA, Laposata M. Fatty acid ethyl esters: toxic non-oxidative metabolites of alcohol and markers of alcohol intake. Front Biosci 2003;8:202−17.
[14] Laposata M. Assessment of alcohol intake current tests and new assays on the horizon. Am J of Clin Pathol 1999;112:443−50.
[15] Lai CL, Yao CT, Chau GY, Yang LF, Kuo TY, Chiang CP, et al. Dominance of the inactive Asian variant over activity and protein contents of mitochondrial aldehyde dehydrogenase 2 in human liver. Alcohol Clin Exp Res 2014;38:44−50.
[16] Marczinski CA. Can energy drinks increase the desire for more alcohol? Ad Nutr 2015;6:96−101.
[17] Garcia-Moreno H, Calvo JR, Maldonado MD. High levels of melatonin generated during the brewing process. J Pineal Res 2013;55:26−30.
[18] Murch SJ, Hall BA, Le CH, Saxena PK. Changes in the levels of indoleamine phytochemicals during veraison and ripening of wine grapes. J Pineal Res 2010;49:95−100.
[19] Friedman LA, Kimball AW. Coronary heart disease mortality and alcohol consumption in Framingham. Am J Epidemiol 1986;124:481−9.
[20] Tolstrup J, Jensen MK, Tjonneland A, Overvad K, Mukamal KJ, Gronbaek M. Prospective study of alcohol drinking patterns and coronary heart disease in women and men. BMJ 2006;332:1244−8.
[21] Bryson CL, Mukamal KJ, Mittleman MA, Fried LP, Hirsch CH, Kitzman DW, et al. The association of alcohol consumption and incident heart failure: the cardiovascular health study. J Am Coll Cardiol 2006;48:305−11.

[22] Gemes K, Janszky I, Laugsand LE, Laszlo KD, et al. Alcohol consumption is associated with a lower incident of acute myocardial infarction: results from a large prospective population based study in Norway. J Intern Med 2016;279:365–75.
[23] O'Keefe JH, Bhatti SK, Bajwa A, DiNicolantonio J, et al. Alcohol and cardiovascular health: the dose makes the poison or the remedy. Mayo Clinic Proc 2014;89:382–93.
[24] Ruf JC. Alcohol, wine and platelet function. Biol Res 2004;37:209–15.
[25] Wu JM, Hiieh TC. Resveratrol: a cardioprotective substance. Ann NY Acad Sci 2011;1215:16–21.
[26] Barefoot JC, Gronbaek M, Feaganes JR, McPherson RS, et al. Alcoholic beverage preference, diet, and health habits in UNC Alumni Heart Study. Am J Clin Nutr 2002;76:466–72.
[27] Rosenbloom JI, Mukamal KJ, Frost LE, Mittleman MA. Alcohol consumption patterns, beverage type, and long term mortality among women survivors of acute myocardial infarction. Am J Cardiol 2012;109:147–52.
[28] Truelsen T, Gronbaek M, Schnohr P, Boyen G. Intake of beer, wine and spirits and risk of stroke: the Copenhagen city heart study. Stroke 1998;29:2467–72.
[29] Koppes LL, Dekker JM, Hendriks HF, Bouter LM, Heine RJ. Moderate alcohol consumption lowers the risk of type 2 diabetes: a meta-analysis of prospective observational studies. Diabetes Care 2005;28(3):719–25.
[30] Baliunas DO, Taylor BJ, Irving H, Roereke M, Patra J, Mohapatra S, et al. Alcohol as a risk factor for type 2 diabetes: a systematic review and meta-analysis. Diabetes Care 2009;2009(32):2123–32.
[31] Polsky S, Akturk HK. Alcohol consumption, diabetes risk and cardiovascular disease within diabetes. Curr Diab Rep 2017;17:136.
[32] Chao C, Slezak JM, Caan BJ, Quinn VP. Alcoholic beverage intake and risk of lung cancer: the California Men's Health Study. Cancer Epidemiol Biomarkers Prev 2008;17:2692–9.
[33] Pelucchi C, Galeone C, Montella M, Polesel J, Crispo A, Talamini R, et al. Alcohol consumption and renal cell cancer risk in two Italian case controlled study. Ann Oncol 2008;19:1003–8.
[34] Pelucchi C, Tramacere I, Boffetta P, Negri E, et al. Alcohol consumption and cancer risk. Nutr Cancer 2011;63 983-980.
[35] Kallberg H, Jacobsen S, Bengtsson C, Pedersen M, et al. Alcohol consumption is associated with decreased risk of rheumatoid arthritis: results from two Scandinavian studies. Ann Rheumatoid Dis 2009;68:222–7.
[36] Nissen MJ, Gabay C, Scherer A, Finchk A. The effect of alcohol on radiographic progression in rheumatoid arthritis. Arthritis Rheum 2010;62:1265–72.
[37] Di Giuseppe D, Alfredsson L, Bottai M, Askling J, et al. Long term alcohol intake and risk of arthritis in women: a population based cohort study. British Med J 2012;345:e4230.
[38] Leitzmann MF, Giovannucci EL, Stampfer MJ, Spiegelman D, et al. Prospective study of alcohol consumption patterns in relation to symptomatic gallstone disease in man. Alcohol Clin Exp Res 1999;23:835–41.
[39] Takkouch B, Regueira-Mendez C, Garcia-Closas R, Figueiras A, Gestal-Oterro JJ, Hernan MA. Intake of wine, beer, and spirits and the risk of clinical common cold. Am J Epidemiol 2002;155:853–8.
[40] Camargo CA, Hennekens CH, Gaziano JM, Glynn RJ, et al. Prospective study of moderate alcohol consumption and mortality in US male physicians. Arch Int Med 1997;157:79–85.
[41] Gronbaek M, Becker U, Johansen D, Gottschau A, et al. Type of alcohol consumed and mortality from all causes, coronary heart disease and cancer. Ann Intern Med 2000;133:411–19.
[42] Klatsky AL, Friedman CD, Armstrong MA, Kipp H. Wine, liquor, beer and mortality. Am J Epidemiol 2003;158:585–95.
[43] Gronbaek M, Mortensen EL, Mygind K, Andersen AT, et al. Beer, wine, spirits and subjective health. J Epidemiol Commun Health 1999;53:721–4.
[44] Baum-Baicker C. The psychological benefits of moderate alcohol consumption: a review of literature. Drug Alcohol Depend 1985;15:305–22.
[45] Clarisse R, Testu F, Reinberg A. Effect of alcohol on psycho-technical test and social communication in a festive situation: a Chrono psychological approach. Chronobiol Int 2004;21:721–38.
[46] Peele S, Brodsky A. Exploring psychological benefits associated with moderate alcohol use: a necessary corrective to assessments of drinking outcomes? Drug Alcohol Depend 2000;60:221–47.
[47] Schrieks IC, Stafleu A, Kallen VL, Grootjen M. The biphasic effects of moderate alcohol consumption with a meal on ambiance-induced mood and autonomic nervous system balance: a randomized crossover trial. PLoS ONE 2014;9:e86199.
[48] Gonzalez-Rubio E, San Mauro I, Lopez-Ruiz C, Diaz-Prieto LE, et al. Relationship of moderate alcohol intake and type of beverage with health behaviors and quality of life in elderly. Qual Life Res 2016;25:1931–42.
[49] Parodi J, Ormeno D, Ochoa-de la Paz LD. Amyloid pore channel hypothesis: effect of alcohol on aggregation state using frog oocytes for an Alzheimer's disease study. BMB Rep 2015;48:13–18.
[50] Pasinetti GM, Wang J, Jo L, Zhao W, et al. Roles of resveratrol and other grape-derived polyphenols in Alzheimer's disease prevention and treatment. Biochem Biophys Acta 2015;1852:1202–8.
[51] Orman ES, Odena G, Bataller R. Alcoholic liver disease: pathogenesis, management and novel targets for therapy. J Gastroenterol Hepatol 2013;28(Suppl. 1):77–84.
[52] Streba L, Vere CC, Streba CT, Ciutra ME. Focus on alcoholic liver disease: from nosography to treatment. World J Gsatroenyerol 2014;20:8040–7.
[53] Askgaard G, Gronbaek M, Kjaer MS, Tjonneland A, et al. Alcohol drinking pattern and risk of alcoholic liver cirrhosis: a prospective cohort study. J Hepatol 2015;62:1061–967.

[54] Walsh K, Alexander G. Alcoholic liver disease. Postgrad Med 2000;281:280—6.
[55] Hezode C, Lonjon I, Roudot-Thorval F, Pawlotsky JM, Zafrani ES, Dhumeaux D. Impact of moderate alcohol consumption on histological activity and fibrosis in patients with chronic hepatitis C, and specific influence of steatosis, a prospective study. Aliment Pharmacol Ther 2003;17:1031—7.
[56] Zeigler DW, Wang CC, Yoast RA, Dickinson BD, McCaffree MA, Robinowitz CB, et al. The neurocognitive effects of alcohol on adolescents and college students. Prev Med 2005;40:23—32.
[57] Kopelman MD, Thomson AD, Guerrini I, Marshall EJ. The Korsakoff syndrome: clinical aspect, psychology and treatment. Alcohol Alcohol 2009;44:148—54.
[58] Harper C. The neurotoxicity of alcohol. Hum Exp Toxicol 2007;26:251—7.
[59] Klatsky AL. Alcohol and cardiovascular health. Physiol Behavior 2010;100:76—81.
[60] Laonigro I, Correale M, Di Biase M, Altomare E. Alcohol abuse and heart failure. Eur J Heart Failure 2009;11:453—62.
[61] Ikehara S, Iso H, Yamagishi K, Yamamoto S. Alcohol consumption, social support and risk of stroke and coronary heart disease among Japanese men: the JPHC study. Alcohol Clin Exp Res 2009;33:1025—32.
[62] Sampson PD, Streissguth AP, Bookstein FL. Incidence of fetal alcohol syndrome and prevalence of alcohol related neurodevelopmental disorder. Teratology 1997;56:317—26.
[63] Numi K, Methuen T, Maki T, Lindstedt KA, Kovanen PT, Kovanen PT, et al. Alcohol induces apoptosis in human mast cells. Life Sci 2009;85:678—84.
[64] Baum MK, Rafie C, Lai C, Sales S, Page JB, Campa A. Alcohol use accelerates HIV disease progression. AIDS Res Hum Retrovir. 2010;26:511—18.
[65] Emanuele N, Emanuele MA. Alcohol alters critical hormonal balance. Alcohol Health Res World 1997;21:53—64.
[66] Apte M, Pirola R, Wilson J. New insights into alcoholic pancreatitis and pancreatic cancer. J Gastroenterol Hepatol 2009;24(3, Suppl. 3): S351—356.
[67] Dawson DA. Alcohol and mortality from external causes. J Stud Alcohol Drug 2001;62:790—7.
[68] Kauhanen J, Kaplan GA, Goldberg DE, Salonen JT. Beer binging and mortality: results from Kuopio ischemic heart disease risk factor study, a prospective population based study. BMJ 1997;315:846—51.
[69] Palijan TZ, Kovacevic D, Radeljak S, Kovac M, Mustapic J. Forensic aspects of alcohol abuse and homicide. Coll Antropol 2009;33:893—7.
[70] Quigley BM, Leonard KE. Alcohol and the continuation of early marital aggression. Alcohol Clin Exp Res 2000;24:1003—10.
[71] Adinoff N, Bone GH, Linnoila M. Acute alcohol poisoning and alcohol withdrawal syndrome. Med Toxicol Adverse Drug Exp 1988;3:172—96.
[72] Celik S, Karapirli M, Kandemir E, Ucar F, Kantarci MN, Gurler M, et al. Fatal ethyl and methyl alcohol related poisoning in Ankara: a retrospective analysis of 10, 720 cases between 2001 and 2011. J Forensic Leg Med 2013;20:151—4.
[73] Nanau RM, Neuman MG. Biomolecules and biomarkers used in diagnosis of alcohol drinking and in monitoring therapeutic interventions. Biomolecules 2015;5:1339—85.
[74] Armer JM, Gunawardana L, Allcock RL. The performance of alcohol markers including ethyl glucuronide and ethyl sulfate to detect alcohol use in clients in a community alcohol treatment program. Alcohol Alcohol 2017;52:29—34.

Chapter 2

Alcohol Analysis in Various Matrixes: Clinical Versus Forensic Testing

Steven C. Kazmierczak
Department of Pathology, Oregon Health and Science University, Portland, OR, United States

INTRODUCTION

Alcohol (ethanol or ethyl alcohol) is one of the most widely used recreational drugs. Consumed in small quantities, alcohol can cause euphoria and increase social interactions. Numerous studies have also demonstrated the cardioprotective effects that drinking small amounts of alcohol can provide in helping prevent heart failure and ischemic stroke. However, when consumed chronically in large amounts, alcohol can lead to a variety of complications including increased risk of pancreatitis, alcoholic liver disease, and certain types of cancer. Consumption of alcohol in large amounts can also cause depression of the central nervous system resulting in impairment of motor skills and cognitive performance. The effects of alcohol on the performance of skilled tasks such as operating a motor vehicle are an important consideration for traffic safety. The enforcement of laws governing drinking and driving are dependent upon the accurate measurement of ethanol. Ethanol measurements may be performed using a wide variety of sample matrixes including breath, urine, and blood. Understanding the differences and limitations that these different matrixes can impose is critical to interpretation of ethanol measurements.

There can be large differences between individuals in the extent of impairment produced by the consumption of alcohol. The most important factor is the frequency and amount of ethanol consumed by an individual. Those who drink more frequently and in greater amounts will develop a greater tolerance and show less impairment when challenged with various tasks to gauge psychomotor performance. The effects of gender on degree of motor and cognitive impairment following alcohol consumption have been inconclusive. Some studies suggest greater degrees of impairment in females following ethanol consumption while others indicate no gender differences in impairment from ethanol.

In addition to ethanol, measurement and quantitation of other volatile compounds including acetaldehyde, methanol, isopropanol, and acetone can be useful as indicators of disease or ingestion. Exposure to acetaldehyde can occur from contact with automobile exhaust and cigarette smoke, and has been identified as a possible carcinogen. Measurements of methanol and isopropanol are important for confirmation of accidental or intentional ingestion. Acetone is useful as a marker of ketoacidosis and can be helpful in the diagnosis of antemortem diabetes mellitus. In addition, blood acetone concentrations have been found to be significantly increased in hypothermia fatalities, with the highest levels found in cases where ethanol was not a contributing factor to the cause of death [1].

FACTORS INFLUENCING THE EFFECTS OF ETHANOL

The effects that an individual experiences following ingestion of alcohol can vary substantially over time. The stimulatory effects of ethanol typically precede the sedative effects of alcohol, although there are significant differences between individuals as to the blood alcohol concentration required to elicit a specific sign or symptom. Table 2.1 shows the typical stages of the clinical symptoms due to ethanol in nontolerant individuals.

An individual's response to the effects of ethanol appears to be more marked when blood ethanol concentrations are rising, than at the same concentration when ethanol concentrations are falling. This phenomenon was noted by Mellanby in 1919 and is known as the "Mellanby effect" [2]. A recent metaanalysis demonstrated that subjective intoxication was greater when individuals were assessed when blood ethanol concentrations were increasing compared with

TABLE 2.1 Typical Symptoms Observed in Individuals As a Function of Blood Ethanol Concentrations

Blood Ethanol Concentration	Clinical Symptoms
10–50 mg/dL	Pleasurable effects, mild euphoria, increased sociability, talkativeness, and self-confidence with decreased inhibition. However, mild impairment may occur around blood alcohol level of 40–50 mg/dL which can only detectable by special testing.
50–80 mg/dL	(30–50 mg/dL). However, at level higher than 60 mg/dL some impairment may be noticed in judgment and attention.
80–120 mg/dL	Legal limit of driving is 0.08% whole blood alcohol (80 mg/dL) where motor may be impaired. Slow information processing and impairment of sensorimotor skills, perception, and memory occurs at blood level exceeding 100 mg/dL.
120–230 mg/dL	Loss of critical judgment and comprehension with decreased sensory response and increased reaction time. With blood alcohol higher than 150 mg/dL, reduced visual acuity and peripheral vision may be observed. In addition, impaired balance and sensorimotor coordination as well as drowsiness may also occur.
230–300 mg/dL	Mental confusion, disorientation, and dizziness as well as vision disturbances (perception of color, motion, and dimensions) may occur at such high blood level. Loss of muscle coordination, staggering, gait, and slurred speech are also noticed as blood alcohol level of 250 mg/dL and higher.
300–400 mg/dL	Almost complete loss of motor function with inability to stand or walk. Marked decrease response to stimuli. Impaired consciousness.
350–500 mg/dL	Depressed or abolished reflexes. Complete unconsciousness as well as possible coma. Impairment of circulation and respiration are also observed. Possible death from respiratory arrest may occur at blood alcohol level of 450 mg/dL.

assessments performed when blood ethanol concentrations were decreasing. However, of greater interest was the finding that objective measures of skills necessary for safe driving, such as response to inhibitory cues and skills measured on driving simulators, were actually worse on the descending part of the ethanol time curve compared to the same blood ethanol concentration on the ascending part of the curve [3].

Ethanol is absorbed in the proximal small intestine and then distributed throughout all body tissues. Distribution in tissues is in proportion to the water content of each tissue. Since adipose tissue contains little water, the volume of distribution is proportional to lean body mass. Individuals with more adipose tissue will have a smaller volume of distribution and achieve greater blood ethanol content when administered the same dose of ethanol as that given to an individual with a greater lean body mass. For example, one study found that individuals with a mean body fat of 18% had a volume of distribution of 0.69, while those with body fat of 26% had a volume of distribution of 0.63 [4].

Following ingestion, ethanol passes readily through the wall of the stomach and then is absorbed into the blood. A variety of factors can influence the rate at which ethanol is absorbed. Ingestion of ethanol results in lower concentrations in blood compared with intravenous administration of an equal amount of ethanol. The difference in blood ethanol observed between these two different routes of administration is due to first-pass metabolism of ethanol.

Initial metabolism of ethanol begins in the stomach by the enzyme gastric alcohol dehydrogenase (ADH). There are a number of different ADH isoenzymes within the gastric mucosa. The rate of gastric emptying can affect the metabolism of ethanol and the rate of emptying can be affected by various factors. Delayed gastric emptying results in increased first-pass metabolism of ethanol by gastric ADH, while enhanced gastric emptying results in decreased first-pass gastric metabolism. Gastric motility can be affected by factors such as age, gender, menopausal status, and cigarette smoking [5]. Fasting and gastrectomy can enhance the rate of gastric emptying resulting in decreased first-pass gastric metabolism of ethanol. While gastric ADH does contribute to first-pass metabolism of ethanol, the magnitude of the contribution of gastric ADH to overall first-pass metabolism of ethanol is relatively small, with the majority of metabolism of ethanol occurring in the liver [5].

Ethanol that is not metabolized within the stomach is metabolized in the liver, with the remaining 2%–10% is excreted unchanged in the breath, sweat, and urine [6]. Within the liver, two different hepatic enzymes catalyze the oxidative metabolism of ethanol. The more important of the two enzymes is ADH Class I, located in the cytosol of hepatocytes. This enzyme is easily saturated with ethanol and metabolism is at a maximum when blood alcohol concentrations reach approximately 20 mg/dL. The second hepatic enzyme involved in the metabolism of ethanol is CYP2E1, a membrane bound enzyme located in the smooth endoplasmic reticulum. In contrast to ADH Class I, CYP2E1 has a higher k_m for ethanol and as a result does not show saturation until the blood alcohol concentration exceeds 80–100 mg/dL. Thus the pharmacokinetic elimination of ethanol is dose dependent and can display zero- or first-order kinetics depending on the blood ethanol concentration. The elimination typically displays zero-order kinetics at concentrations greater than 20 mg/dL. Once blood ethanol concentrations drop below 20 mg/dL the elimination follows first-order kinetics, with the rate of elimination proportional to the blood ethanol concentration [7].

There is interindividual variation in CYP2E1 activity which can be modified by factors such as obesity and previous exposure to alcohol. In individuals that consume ethanol rarely, CYP2E1 metabolizes only a small fraction of the ingested alcohol, the majority of alcohol being metabolized by ADH. However, chronic heavy consumption of ethanol can increase CYP2E1 activity up to tenfold, resulting in a substantial increase in the proportion of alcohol that is metabolized by this enzyme rather than by ADH. Also, because CYP2E1 is involved in the medication of many drugs, upregulation of CYP2E1 in chronic alcoholics can result in increased metabolism for medications that are also metabolized by CYP2E1, requiring increased dosing to achieve the required pharmacologic effect. However, following the ingestion of ethanol, the alcohol competes with the medication for metabolism by CYP2E1 resulting in slowed metabolism of the medication and increased serum drug levels. Thus, achieving relatively stable drug levels in chronic alcoholics can be problematic.

A third enzyme that is involved in the metabolism of ethanol byproducts is aldehyde dehydrogenase (ALDH). This enzyme is responsible for degrading acetaldehyde, a toxic metabolite produced by the action of ADH on ethanol, to nontoxic acetate. The acetate that is produced by ALDH is converted within the liver to acetyl-CoA and eventually resulting in the production of water and carbon dioxide.

ANALYSIS OF ETHANOL IN WHOLE BLOOD VERSUS SERUM OR PLASMA

There are numerous types of specimens that may be submitted to the laboratory for the analysis of ethanol. Blood specimens received for forensic analysis following death are often hemolyzed and may also be clotted, while specimens submitted to clinical laboratories for assessment of ethanol consumption usually consist of unhemolyzed serum or plasma. Also, specimens for postmortem analysis of ethanol often comprise more complex matrices that can make measurement and interpretation even more difficult. Measurement of ethanol in vitreous humor, gastric contents, cerebrospinal fluid, and blood obtained from various vascular sites, bile and liver represent unique challenges. Many of these specimen types are not available from living individuals, making comparison between ethanol in postmortem samples and ethanol measured from a living individual impossible. In addition, there are numerous postmortem factors that can confound ethanol interpretation such as time between death and collection of the sample, temperature extremes that the body may have been exposed to, and postmortem redistribution of ethanol.

Ethanol concentrations measured in plasma or serum and whole blood are not equivalent. Ethanol is distributed in blood components in proportion to their relative water content. The interpretation of ethanol measured in whole blood should take into account the difference in water content between whole blood versus serum and plasma. The water content of serum or plasma is greater than that of whole blood due to the dilutional effect of erythrocytes. The ratio of ethanol in serum or plasma to whole blood has been reported to range from 1.09 to 1.18 [8]. Based on this ratio, the percentage of the whole blood volume occupied by erythrocytes has been proposed as a means for correcting the difference in ethanol measured between whole blood and serum or plasma. However, others have found that the difference between whole blood ethanol and serum or plasma ethanol is independent of hematocrit [9]. Another study that analyzed the ratios of serum to whole blood ethanol over a wide range of ethanol concentrations found that the ratio was concentration dependent and ranged from 1.12 to 1.18 depending on the serum alcohol concentration [10]. Thus, one must be very careful when using the measured plasma or serum ethanol value to estimate the concentration of ethanol in whole blood due to the magnitude of variation in the serum to blood ratio of alcohol. It has previously been recommended that a factor of 1.22:1 be used [9]. This factor is based on the mean plus two standard deviations from factors obtained in a number of different studies. The use of this factor is considered to be more appropriate for forensic work where it will provide a conservative estimate of an individual's blood alcohol concentration. There is no difference in ethanol concentrations measured in plasma versus serum.

STORAGE OF SAMPLES FOR ETHANOL MEASUREMENT

Due to the consequences that a measured ethanol concentration can place on an individual, accurate and reliable methods for measurement are paramount. Accuracy of the technique used to measure ethanol is extremely important and ongoing validation and auditing with a quality assurance program utilizing internal and external quality control materials is preferable. Each laboratory should be aware of the measurement uncertainty associated with the method used to measure ethanol. Knowing the measurement uncertainty allows a laboratory to assess the analytical performance of the method in use. Understanding the source of the measurement uncertainty can be helpful in identifying what aspects of the analytical process is contributing most to any inaccuracy in the measurement of ethanol. For example, one study that investigated measurement uncertainty using a gas chromatographic method for ethanol in blood found that their maximal measurement uncertainty was slightly less than 5% [11]. In breaking down the various sources of measurement uncertainty, the analytical component was found to be the greatest contributor (90%), while traceability to the ethanol standard was the next greatest contributor (<4%). Matrix effects and blood water content were minor contributors to measurement uncertainty (<0.5%). Another study found the total measurement uncertainty for blood ethanol to be 4% [12]. Finally, a more recent study that investigated various assay parameters found that the assay calibrators contributed the most to measurement uncertainty [13]. The second most important parameter contributing to measurement uncertainty was the instability of ethanol in the opened collection tube, highlighting the need to analyze samples as soon as possible once the blood collection tube is opened.

In addition to analytical factors that can influence measurement of ethanol, a variety of preanalytical factors need to be considered. Proper cleaning of the venipuncture site using a disinfectant that does not contain ethanol or other volatile substances is important. The disinfectants that are most commonly used are benzalkonium chloride or povidone-iodine. Proper storage of specimens to prevent false increases or false decreases in ethanol present in the sample is critical. Stability of ethanol over time is a critical consideration if samples are not analyzed immediately or will need to be reanalyzed at a later date. Increases or decreases in ethanol can occur due to loss of ethanol during storage or from metabolism of substrates like glucose by microorganisms present in the sample to produce ethanol. The most important factors impacting long-term storage of samples include the amount of time a sample is stored, the concentration of ethanol in the sample, the types of additives used to preserve the specimen, temperature, the type of cap used to seal the specimen, and air quality above the sample [14]. Also, the type of container can be a factor in loss of ethanol. Another study found that ethanol was lost due to diffusion from polypropylene containers that were used to store blood samples [15]. Storage in glass tubes should prevent loss of ethanol due to diffusion, although a more recent investigation found that loss of ethanol was the same for specimens stored for 1 year in either glass or plastic evacuated tubes [16]. Following 12 months of storage, a decrease of approximately 11 mg/dL occurred in samples with starting ethanol concentrations ranging from 20 to 300 mg/dL.

MEASUREMENT OF ETHANOL IN BODY FLUIDS

The first quantitative method for blood ethanol analysis that proved to be reliable and accurate was known as the Widmark method [17]. Ethanol in the sample was oxidized by a mixture of potassium dichromate and sulfuric acid which were present in excess. The amount of oxidizing agent that remained following the reaction with ethanol was measured by iodometric titration and related to ethanol concentration in the sample. Unfortunately, the Widmark method was not specific for ethanol as other volatile substances such as acetone, methanol, or ether could also be oxidized. This method is now of historical interest and no longer used.

Purification of the enzyme ADH from yeast or horse liver offered a major improvement for measurement of ethanol. One advantage was that acetone, a common interfering substance with the Widmark method, was not oxidized by ADH. Also, this method could be optimized with respect to temperature, pH, and reaction time to minimize oxidation of methanol by ADH, thus allowing this technique to be more specific as compared to the Widmark method. The enzymatic method in use today is based on the oxidation of ethanol to acetaldehyde with the concomitant reduction of NAD to NADH. The formation of NADH is directly proportional to ethanol concentrations and can be measured spectrophotometrically at 340 nm. Serum, plasma, and urine are the most common body fluids measured using this technique.

The enzymatic method is fairly specific for ethanol; however, some studies have found slight cross-reactivity with isopropanol (7%), methanol (3%), and ethylene glycol (4%) [18]. Interference from lactate and lactate dehydrogenase (LD) has also been described for some enzymatic methods. Interference typically requires significantly increased lactate and LD concentrations, so this type of interference has been observed in patients with severe liver disease or those with significant muscle damage and hypoxia. Samples obtained postmortem and analyzed for ethanol using an enzymatic method are also at risk for this type of interference. The mechanism of interference in samples with increased lactate and LD is the oxidation of lactate to pyruvate by the increased LD, with simultaneous reduction of NAD to NADH as shown in Eq. (2.1).

$$NAD + Lactate \xrightarrow{\uparrow LD} NADH + Pyruvate \tag{2.1}$$

However, despite the problems with nonspecificity, the enzymatic method for measurement of ethanol remains the most common method in use for routine analysis. It offers good accuracy with rapid turnaround of test results.

For forensic analysis of ethanol, gas chromatography (GC) is the gold standard. While not as fast as the automated enzymatic methods utilized in most hospital labs, the accuracy and reliability of GC methods make these the method of choice. Laboratories may employ one of two approaches to analysis of ethanol by GC. Volatile substances such as ethanol can escape the liquid medium of blood or other body fluids and be present in sufficient concentrations in the air space above the sample, also referred to as the "head space." Head space analysis of ethanol requires that a sample of the air space above the sample by analyzed. A saturated solution of sodium chloride and internal standard is added to the sample to increase the vapor pressure of the alcohol in the head space. The internal standard typically used is *n*-propanol. The advantage of head space analysis is that the fluid sample itself does not contaminate the injector or column.

Another technique is the direct method whereby the sample is diluted with a solution containing internal standard. After addition of the internal standard, the sample is injected directly into the GC.

The use of GC-mass spectrometry (GC-MS) for the analysis of methanol, acetaldehyde, acetone, and ethanol has been reported [19]. The GC-MS technique for ethanol is more complex than GC-flame ionization detection and requires highly trained personnel.

ANALYSIS OF ETHANOL IN EXPIRED BREATH

Analysis of ethanol in expired breath represents a noninvasive means of estimating blood alcohol concentrations. The physiological basis of the breath alcohol test is based on the equilibrium that is established between alcohol in pulmonary capillary blood and the alveolar air. Alcohols, including ethanol, isopropanol, and methanol are highly water soluble and do not bind to plasma proteins. These attributes make alcohol a reliable indicator of intake.

Breath testing eliminates the need for taking blood samples and can be used in the field for testing drivers suspected to be under the influence of ethanol. More recently, breath alcohol analysis has expanded to use in the work place. When used for evaluating ethanol use in drivers, the concentration of ethanol in breath is used as the means for assessing impairment; there is no attempt to estimate the blood alcohol concentration from breath alcohol measurements. The offence of driving under the influence is based solely on the amount of ethanol in the breath [20].

Handheld screening devices used in the field typically utilize electrochemical sensors that oxidize ethanol to acetaldehyde, producing an electric current in the process. The magnitude of the electric current that is produced is proportional to the amount of ethanol in breath. This method is not affected by the presence of acetone, which can be elevated in patients with diabetic ketoacidosis or alcoholic ketoacidosis. However, other volatiles such as methanol or isopropanol can undergo oxidation and cause a false positive reaction. Alternative methods for measurement of breath alcohol utilize the absorption of infrared energy at wavelengths of 3.4 and 9.5 μm.

In most countries, a waiting time of 15 min must be observed prior to the measurement of breath alcohol. This waiting period is required because immediately following the ingestion of ethanol, the concentration of alcohol is much higher in the saliva and mucous membranes in the mouth and will overestimate the corresponding blood ethanol concentration. Another factor that can affect breath alcohol measurements is due to the contamination of breath samples by ethanol in stomach contents. High concentrations of ethanol can be released into expired breath as a result of belching or regurgitation of stomach contents immediately prior to testing [21].

The ratio of ethanol in measured blood to that measured in the breath, the blood/breath ratio, is generally considered to be 2100:1. This ratio provides a good margin of error for the individual being tested so that false positive results are minimized. Different countries recognize different ratios for legal purposes, with ratios ranging from 2000:1 to 2300:1. Blood/breath ratios that have been reported in the literature range from 1300:1 to 2700:1 [22]. The ratio can vary between individuals, but also within an individual depending on whether the test is performed during the absorption phase or elimination phase of the alcohol curve.

Shortly after ingesting ethanol during the absorption phase, the blood/breath ratio tends to be less than 2100:1. The variation in the blood/breath ratio is due primarily to the difference in ethanol in arterial and venous blood during the early stage of absorption. During the absorption stage, ethanol is present in higher concentrations in arterial blood transporting ethanol from the stomach to other areas of the body. Thus, during the early stages of absorption, the ratio will be low because venous blood is collected for blood alcohol measurement while the breath alcohol is derived from exhaled breath from arterial blood in the lungs.

A number of other factors can also impact the blood/breath ratio. The factors that contribute to this variability include the physiology and dynamics of ethanol uptake and the kinetics of ethanol metabolism [23]. For example, changes in the temperature of expired breath influence the amount of ethanol present in the breath. A change of 1°C in the temperature of expired breath can change the blood/breath ratio by approximately 7% [24]. Hyperventilation can decrease breath alcohol concentrations by up to 25% since rapid breathing does not allow for complete equilibrium to occur between ethanol in the capillary beds of the alveoli and inspired air. The presence and absence of food in the stomach can affect the rate of absorption of ethanol. Consumption of alcohol on an empty stomach can result in ethanol staying in the stomach for a longer period of time and increase the breath alcohol concentration. Consuming ethanol with or following a meal can result in decreased blood alcohol concentrations and decreased bioavailability of ethanol.

CONCLUSIONS

The most common matrix for alcohol analysis is blood. Although alcohol concentrations measured in serum or plasma using enzymatic methods is most commonly used in hospital laboratories for medical treatment purposes, head space GC is the gold standard for forensic alcohol analysis. Alcohol is also measured in urine and other matrices using head space GC analysis. Breath alcohol analysis is also commonly used by police to identify driver suspected of driving under the influence of alcohol.

REFERENCES

[1] Palmiere C, Bardy D, Letovanec I, Mangin P, et al. Biochemical markers of fatal hypothermia. Forensic Sci Int 2013;226:54–61.

[2] Mellanby E. Alcohol. Its absorption into and disappearance from the blood under different conditions. Medical Research Committee, Special Report Series, 1919, number 31.

[3] Holland MG, Ferner RE. A systematic review of the evidence for acute tolerance to alcohol—the "Mellanby effect." J Clin Toxicol 2017;55:545–66.

[4] Cowan JM, Weathermon A, McCutcheon JR, Oliver RD. Determination of volume of distribution for ethanol in male and female subjects. J Anal Toxicol 1996;20:287–9.

[5] Oneta CM, Simanowski UA, Martinez A, Allali-Hassani A, et al. First pass metabolism of ethanol is strikingly influenced by the speed of gastric emptying. Gut 1998;43:612–19.

[6] Kalant H. Pharmacokinetics of ethanol: absorption, distribution, and elimination. In: Beglieter H, Kissin B, editors. The pharmacology of alcohol and alcohol dependence. New York: Oxford University Press; 1996. p. 15–58.

[7] Rangno RE, Kreeft JH, Sitar DS. Ethanol "dose-dependent" elimination: Michaelis–Menten v classical kinetic analysis. Br J Clin Pharmacol 1981;12:667–73.

[8] Penetar DM, McNeil JF, Ryan ET, Lukas SE. Comparisons among plasma, serum, and whole blood ethanol concentrations: Impact of storage conditions and collection tubes. J Anal Toxicol 2008;32:505–10.

[9] Rainey PM. Relation between serum and whole-blood ethanol concentrations. Clin Chem 1993;39:2288–92.

[10] Barnhill MT, Herbert D, Wells DJ. Comparison of hospital laboratory serum alcohol levels obtained by an enzymatic method with whole blood levels forensically determined by gas chromatography. J Anal Toxicol 2007;31:23–30.

[11] Kristiansen J, Petersen HW. Uncertainty budget for the measurement of ethanol in blood by headspace gas chromatography. J Anal Toxicol 2004;28:456–63.

[12] Fung WK, Chan KL, Mok VK, Lee CW. The statistical variability of blood alcohol concentration measurements in drink-driving cases. Forensic Sci Int 2001;110:207–14.

[13] Ince FD, Arslan B, Kaplan YC, Ellidag H. The evaluation of uncertainty for the measurement of blood ethanol. J Clin Anal Med 2016;7:648–51.

[14] Kocak FE, Isiklar OO, Kocak H, Meral A. Comparison of blood ethanol stabilities in different storage periods. Biochem Med 2015;25:57–63.

[15] Brown GA, Neylan D, Reynolds WJ, Smalldon KW. The stability of ethanol in stored blood: Part I. Important variables and interpretation of results. Anal Chim Acta 1973;66:271–83.

[16] Jones AW, Ericsson E. Decrease in blood ethanol concentrations during storage at 4°C for 12 months were the same for specimens kept in glass or plastic tubes. Pract Lab Med 2016;4:76–81.

[17] Widmark EMP. Eine mikromethode zur bestimmung von athyalkohol in blut. Biochem Zt 1922;131:473–84.

[18] Gadsden RH, Taylor EH, Steindel SJ, et al. Ethanol in biological fluids by enzymatic analysis. In: Frings CF, Faulkner WR, editors. Selected methods of emergency toxicology, Vol 11. Washington, DC: AACC Press; 1986. p. 63–5.

[19] Wasfi IA, Al-Awadhi AH, Al-Hatali ZN, Al-Rayami FJ, et al. Rapid and sensitive static headspace gas chromatography–mass spectrometry method for the analysis of ethanol and abused inhalants in blood. J Chromatogr B 2004;799:331–6.

[20] Karch S, Jones R. Simpson's forensic medicine. London, UK: CRC Press; 2003.
[21] Booker JL, Rentroe K. The effects of gastroesophageal reflux disease on forensic breath alcohol testing. J Forensic Sci 2015;60:1516–22.
[22] Jaffe DH, Siman-Tov M, Asher Gopher MA, Peleg K. Variability in the blood/breath alcohol ratio and implications for evidentiary purposes. J Forensic Sci 2013;58:1233–7.
[23] Jones AW. Variability of the blood:breath alcohol ratio in vivo. J Stud Alcohol 1978;39:1931–9.
[24] Dubowski KM. Studies in breath-alcohol analysis: biological factors. Z Rechtsmed 1975;76:93–117.

Chapter 3

Alcohol Biomarkers: Clinical Issues and Analytical Methods

Joshua A. Bornhorst[1] and Michael M. Mbughuni[2,3]

[1]Department of Laboratory Medicine and Pathology, Mayo Clinic, Rochester, MN, United States, [2]Department of Pathology, University of Minnesota, Minneapolis, MN, United States, [3]VA Hospital, Minneapolis, MN, United States

INTRODUCTION

Alcohols are an integral part of modern human life. In fact, human consumption of ethyl alcohol (i.e., ethanol) has been an important aspect of society in the course of recorded history; as a result, ethanol is the most widely abused chemical substance [1–4]. Various types of alcohols are generally available to the public in alcoholic drinks prepared for human consumption or in other products not intended for human consumption but containing alcoholic ingredients [i.e., ethanol, methanol, ethylene glycol, isopropanol, diethylene glycol, and propylene glycol (PG)]. Since ethanol is the main alcohol approved for consumption in recreational beverages, it is considered safe when consumed in moderation [5]. However, it is also one of the most common causes of alcohol intoxication and substance dependence [6]. Acute ethanol consumption initially shows stimulatory effects, but as the blood ethanol concentration increases, central nervous system (CNS) depression and inebriation dominates [7]. The magnitude of these physiological effects usually depends on the blood ethanol concentration. In the United States and other countries, a blood ethanol concentration of ≤ 80 mg/dL (0.08%) is set as a legal limit for operating motorized vehicles, whereas blood concentrations exceeding 300 mg/dL could be lethal [1]. With the physiological effects of ethanol not fully understood, cellular dysfunction is thought to cause depression of the sensorium followed by toxic effects from metabolites leading to organ dysfunction [1,8]. As such, excessive ethanol use can cause significant acute and long-term health risks, including impaired judgement, alcohol dependence, lost productivity, violent crimes, organ failure, cancer, birth defects, and death.

ALCOHOL CONSUMPTION AND DISEASE BURDEN

Even though the deleterious effects of alcohol use are well known, a significant proportion of the world population consumes alcohol, and some individuals abuse it [2,3,9]. In the United States alone, more than 200 billion dollars per year is spent in alcohol-related accidents and illness [5,6,10]. There is evidence that the contribution of alcohol abuse to the global disease burden is larger than all other controlled or illicit substances, with the exception of tobacco [2,3]. It is also believed that up to 3.2% of global deaths and close to 4% of overall global disease can be linked to alcohol use [2]. Reports have also shown about 10% of pregnant women between 15 and 44 years of age admit to alcohol abuse during the first trimester of pregnancy [9]. Alcohol use by pregnant women results in Fetal Alcohol Syndrome (FAS), with other individuals exhibiting some but not all the diagnostic features of FAS categorized as having fetal alcohol spectrum disorders (FASD) [9]. The US Centers for Disease Control and Prevention (CDC) has conducted studies indicating up to 1.5 cases of FAS per 1000 births, with FASD occurring at a rate that is three times higher than FAS [9].

ALCOHOL BIOMARKERS: AN OVERVIEW

Due to the well-established negative outcomes of alcohol abuse or toxic exposures, direct alcohol testing along with alcohol biomarkers are often used in diagnosis, prognosis, and clinical management of alcohol-related incidents or disorders [8,9,11,12]. While genetic markers play an important role, alcohol biomarkers are physiological indicators of alcohol exposure or consumption [13,14]. Their clinical detection often reflects acute, chronic, and/or a high level of alcohol exposure. Current alcohol biomarkers generally fall into two types: direct and indirect biomarkers. Direct biomarkers consist of a given alcohol and its metabolites, all of which directly correlate to alcohol exposure or use and can act as acute markers. Direct markers are specific and can be used to diagnose the type of alcohol involved such as acetaldehyde, acetic acid, ethyl glucuronide (EtG), ethyl sulfate (EtS), fatty acid ethyl esters (FAEEs), and phosphatidylethanol (PEth) for ethanol [9,13,15,16]. On the other hand, indirect alcohol biomarkers reflect the toxic effects of alcohols and their metabolites on organs, tissues, or the chemistry supporting life. They indirectly correlate to alcohol problems and are typically used clinically in issues pertaining to acute or chronic ethanol abuse. These include mean corpuscular volume (MCV), carbohydrate-deficient transferrin (CDT), aspartate aminotransferase (AST), alanine aminotransferase (ALT), lactate dehydrogenase (LDH), gamma-glutamyl transferase (GGT), 5-hydroxytryptophol (5-HTOL), and beta-hexosaminidase (β-Hex) [9,13,16]. Direct and indirect biomarkers discussed in this review are also listed in Table 3.1.

In using biomarkers and other laboratory tests to manage alcohol-related issues, it is important to understand how to properly deploy available tests while managing alcohol disorders related to ethanol abuse and dependence as opposed to exposures to the toxic alcohols [i.e., methanol, isopropanol, ethylene glycol (EG), and diethylene glycol (DEG)] [8,9,12]. Unlike ethanol, toxic alcohols are generally accessible from household products associated with incidental toxic exposure or ingestions. Another medically important alcohol associated with toxic exposures is PG, which is often used in approved pharmaceutical formulations [8,12]. Similar to ethanol-related poisonings, clinical presentation involving the toxic alcohols can include inebriation and toxic effects from one or more metabolites (with the exception of isopropanol, which is directly toxic) [8,12]. Ingestion of toxic alcohols generally leads to cellular dysfunction with nonspecific symptoms, but the course of treatment can sometimes be specific for different types of toxic alcohols, requiring identification of the alcohol involved in order to provide appropriate care [8,12]. Therefore, delay in diagnosis increases the risk of irreversible organ damage and possibly death. For this reason, clinical management of acute alcohol intoxication also involves laboratory assessment for the detection, identification, and quantitation of alcohols and their toxic metabolites, along with monitoring of appropriate alcohol biomarkers, depending on the setting.

In all, alcohol biomarkers introduce a level of objectivity to support other clinical evidence in strategies to guide decision making in clinical, criminal justice, or impaired healthcare provider settings [13]. By no means are alcohol biomarkers a substitute for self-reporting measures, patient history, or physical examination by appropriate health professionals [13]. Alcohol biomarkers simply provide a unique and independent measure which supplements or confirms available clinical information. This allows clinical teams to make informed and accurate decision in screening, diagnosis, prognosis, or monitoring of alcohol-related disorders and incidents. To date, researchers continue to look for improved biomarkers suitable for screening, diagnosis, and prognosis of patients with alcohol-related disorders or monitoring patients in treatment for alcohol abuse [9,16]. The other growing area is alcohol biomarkers for maternal and neonatal screening to manage pregnant women with alcohol use disorders or identify infants at risk for FAS and related disorders [9,15,16]. This chapter will focus on direct and indirect alcohol biomarkers, their clinical issues, and analytical methods.

TABLE 3.1 Direct and Indirect Alcohol Biomarkers

Type of Alcohol Biomarker	Specific Example
Indirect biomarker	• Liver enzymes (mostly commonly used is GGT, but ALT and AST may also be used) • Mean corpuscular volume (MCV) • Carbohydrate-deficient transferrin • Beta-hexosaminidase • Sialic acid index of Apolipoprotein J • 5-Hydroxytryptophol • Alterations in lipids
Direct biomarkers	• Ethanol • Ethyl glucuronide • Ethyl sulfate • Fatty acid ethyl esters • Phosphatidylethanol • Acetaldehyde-hemoglobin adduct

ALCOHOL: METABOLISM, PHYSIOLOGY, AND CLINICAL ISSUES

Ethanol, commonly referred as alcohol is primarily a social drug which can also be used therapeutically for the treatment of toxic alcohol exposure [1,17]. The most common route of exposure is direct ingestion, but ethanol can also be absorbed through the skin or inhaled in industrial exposures [17]. When ingested, ethanol is almost entirely absorbed from the gastrointestinal tract, with peak blood concentrations occurring between 30 and 60 minutes after ingestion and a volume of distribution similar to that of body water (up to 0.65 L/kg body weight) [7,12,17]. A majority of the ethanol ingested is metabolized by the liver and excreted in urine. Only a small amount is excreted unchanged in urine, sweat, breath, and feces [7,17]. Once ethanol is distributed through the body, acute physiological effects include an initial stimulatory effect, but as the blood ethanol concentration rises, CNS depression dominates with onset of inebriation [1,9]. Other physiological disturbances include suppression of mechanisms for temperature regulation, osmotic induction of diuresis, acute pancreatitis, and gastritis [1,7].

Even though there is no established blood ethanol concentration for defining inebriation, signs of intoxication usually occur along with CNS depression after two to three drinks in most individuals [9,17]. The mechanism of CNS depression is not fully understood, but is thought to include activation of the γ-aminobutyric acid dependent pathway to effect CNS inhibition [1,7]. This pathway is also enhanced by other drugs such as barbiturates and benzodiazepines; as a result, combined use of these medications with ethanol can lead to lethal outcomes even at low blood ethanol concentrations [1,7]. Furthermore, chronic ethanol CNS depression is linked to adaptive neurochemical changes associated with alcohol dependence [2,3]. In brief, chronic ethanol use leads to chronic CNS inhibition, but when ethanol is discontinued abruptly, subsequent CNS excitation causes alcohol abstinence syndrome. This syndrome has clinical features including anxiety, increased body temperature, irritability, muscle tremors and cramps, hallucinations, seizures, and possibly death [1]. Another important component of chronic ethanol abuse is organ failure, which is commonly linked to the toxic effects of alcohol metabolism [9,18,19]. The combined effect of alcohol dependence and toxic effects of alcohol metabolism can lead to serious social, professional, criminal justice, and medical complications [13].

The main location for alcohol metabolism is the liver, where nonoxidative and oxidative chemical transformations metabolize ethanol into stable and reactive products respectively [9,17]. Oxidative metabolism accounts for metabolism of 90%–98% of the total ethanol content whereas nonoxidative metabolism accounts for only a small proportion of the overall ethanol content (typically <2% of the total alcohol). Nonoxidative metabolism involves enzyme-catalyzed reactions where ethanol is conjugated to biological molecules such as glucuronide, sulfate, and phospholipids to yield relatively unreactive metabolites [9,13,15,16,20]. The liver enzyme glucuronosyltransferases catalyze conjugation of ethanol and glucuronic acid to yield EtG [21]. Sulfotransferases conjugate ethanol and sulfate to yield EtS [22]. Phospholipase D conjugates ethanol to fatty acids to yield PEth [23], and lastly, fatty acid acyl ester synthase (FAEE synthase) catalyzes ethanol-dependent fatty acid esterification to yield FAEEs [24]. Oxidative metabolism employs a series of sequential steps catalyzed by alcohol dehydrogenase (ADH) and aldehyde dehydrogenase (ALDH), giving rise to toxic metabolites. ADH oxidizes ethanol to acetaldehyde which is highly reactive; then ALDH further oxidizes acetaldehyde to acetic acid, which is correlated with metabolic acidosis (Fig. 3.1). The expression of both ADH and ALDH is induced by alcohol use and the reactions catalyzed require $NAD+$. At high blood alcohol levels, alcohol metabolism can also deplete $NAD+$, resulting in zero-order kinetics, which can have a systemic effect by causing metabolic disturbances in pathways that require $NAD+$ (e.g., tricarboxylic acid cycle) [7,9]. Both products of oxidative ethanol metabolism, acetaldehyde and acetic acid, are associated with toxic effects in ethanol poisoning. Acetaldehyde protein adducts have been shown to preferentially form at lysine residues in proteins including hemoglobin, albumin, tubulin, lipoproteins, and collagen [18,19]. These adducts can impair cellular function, while acetic acid can lead to high anion gap acidosis [1,18]. Acetaldehyde reacts with lipids on cellular membranes leading to lipid peroxidation and eventually generates reactive lipids such as malondialdehyde (MDA) and 4-hydroxy-2-nonenal (HNE) [18,19,25]. The reactive lipids (i.e., MDA and HNE) can subsequently interact with DNA and proteins to form adducts that also impair cellular function.

Another important enzyme which initiates the first step of alcohol oxidation is a microsomal cytochrome P450 2E1, an enzyme which can also be induced by chronic ethanol use [18]. The enzyme 2E1 oxidizes ethanol to acetaldehyde in a reaction that consumes cellular NADPH resources while releasing reactive oxygen species (ROS) such as hydroxyethyl radicals (HER) [18,25]. Furthermore, CYP2E1-dependent NADPH depletion and formation of ROS can cause oxidative stress, with HER indiscriminately reacting with biological molecules to form HER adducts that contribute to cellular dysfunction [18,26,27]. Taken together, oxidative ethanol metabolism by liver ADH and 2E1 creates reactive products that form adducts and induce oxidative stress leading to cell injury or possibly cell death [9,18,25]. It is therefore not surprising that liver failure is prominent in alcoholics and with disturbances in liver structure and function

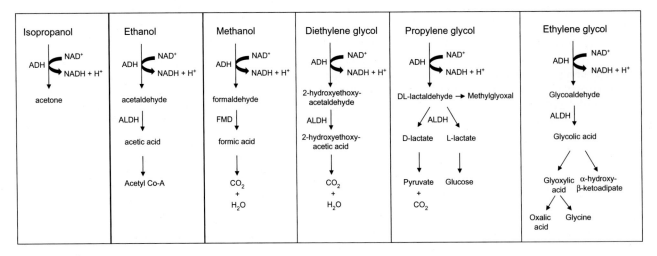

FIGURE 3.1 Metabolic pathways for alcohol metabolism.

directly related to toxic products of oxidative alcohol metabolism [18]. Liver damage associated with alcohol use can therefore result in acute or chronic elevation of biomarkers of liver injury such as AST, ALT, GGT, and LDH, which are also indirect biomarkers of alcohol abuse [9,13,20,28].

TOXIC ALCOHOLS AND MEDICALLY IMPORTANT ALCOHOL

Toxic alcohols are alcohols which are frequently associated with toxic exposures, such as methanol, ethylene glycol, isopropanol, and diethylene glycol. Medically important alcohols include PG, which is formulated in medications and has also been associated with toxic exposures [8,11,12]. Similar to ethanol, clinical issues related to toxic alcohols begin with either acute or chronic exposure. Toxic exposure can in principle occur through skin absorption, inhalation, or ingestion, but common routes of exposure can depend on the type of alcohol involved [1,11,17]. Methanol is commonly used commercially as a solvent in paints and varnishes. It is also present in windshield wiper fluid, antifreeze solutions, and as an alternative form of fuel [11,17]. Ethylene glycol is the principal component in automobile antifreeze solutions. It is also used in deicing products, coolants, detergents, paints, and cosmetics [1,12]. Isopropanol is an industrial alcohol available to the public as a 70% isopropanol solution termed *rubbing alcohol*. Diethylene glycol is an industrial solvent found in brake fluid and some medications, whereas PG is used as a solvent for intravenous, oral, and topical pharmaceutical products [11,17]. PG can also be found in antifreeze and hydraulic fluids [12].

Exposures to methanol, ethylene glycol, isopropanol, and diethylene glycol are typically acute when ingested intentionally, but can also occur from chronic exposure in an occupational setting [8,11,17]. The most common route of exposure for methanol, ethylene glycol, and isopropanol is direct ingestion [18]. This can occur intentionally for individuals looking for ethanol substitutes for the purpose of inebriation, or it can occur unintentionally. Methanol is sometimes ingested inadvertently as a contaminant in moonshine, but both ethylene glycol and methanol are also used by individuals attempting to commit suicide. Diethylene glycol exposures have occurred primarily when it is used as a solvent in medications, and a majority of PG exposures occur from intravenous administration of PG in medications such as lorazepam [12]. Patients with impaired liver or kidney function and receive a continuous infusion of high-dose lorazepam (more than 10 mg/h) for longer than 48 hours are at high risk [8]. Note that these compounds can be ingested inadvertently by children with access to products containing one or more of these compounds [1,8,11,12]. Once absorbed, methanol, ethylene glycol, diethylene glycol, PG, ethanol, and isopropanol are rapidly absorbed from the gastrointestinal tract and distributed through the body [12]. Similar to ethanol metabolism, ADH oxidizes the primary alcohols (i.e., ethanol, methanol, ethylene glycol, diethylene glycol, and PG) to aldehydes, then ALDH converts the aldehydes into corresponding organic acids [8,11]. Secondary alcohols such as isopropanol are oxidized by ADH to give ketones (i.e., acetone). Following metabolism, alcohols are generally excreted through urine, sweat, breath, and feces. The proportion excreted through urine can be important for noninvasive diagnostic clinical testing; this portion amounts to about 50%−70% of diethylene glycol, 25%−45% of PG, 20% of ethylene glycol, up to 20% of isopropanol, 2%−5% of ethanol, and 1%−2% of methanol [12].

Since acute ingestion of toxic alcohols can cause severe cellular dysfunction and increased risk of death, early diagnosis of alcohol toxicity is crucial for initiating therapies that can drastically improve outcomes [8,11,12,29]. Information used to make a diagnosis of alcohol toxicity includes medical history, physical examination, blood chemical profiles, and urine or blood tests to identify direct alcohol biomarkers [8,11,12,29]. Quantitative blood results are also used to aid in clinical management or monitoring of toxic alcohol exposures. Accumulation of the alcohol in blood usually increases the serum osmolar (or osmolal) gap, and organic acids from oxidative metabolism increase the serum anion gap and lower blood pH. In the absence of definitive evidence using direct markers to confirm toxic alcohol exposure, clinicians have to rely on the history of exposure, clinical symptoms, and blood chemistry results that indirectly indicate alcohol exposure (e.g., serum osmolar gap and/or anion gap) [8,11,12,29]. However, the pH value, osmolar and/or anion gaps don't always have the specificity needed to confirm alcohol exposure or guide treatment [8,11,12,29]. Elevation of the serum osmolar gap depends on alcohol concentration and molecular weight of the alcohol, so it is affected differently with each alcohol [8]. Furthermore, if the baseline anion gap is already low, there could be no anion gap increase above the normal range despite toxic accumulation of organic acids [8]. Another consideration is that elevated osmolar or anion gaps along with a low pH don't always indicate toxic alcohol poisoning. Lactic acidosis, ketoacidosis, chronic kidney disease, and the sick cell syndrome may also increase affect these results [8,11]. For these reasons, definitive confirmation of toxic alcohol poisoning often requires quantitation of direct biomarkers to rule out differential diagnosis and hone in on the specific alcohol which is responsible [1,8,11,12]. Direct biomarkers also have the specificity needed to correlate blood concentrations to clinical outcomes. Therefore, monitoring changes in biomarker concentration during treatment is important in guiding therapy, providing clinicians with a quantitative tool for guiding treatment if results are available in a timely manner [1,8,11,12].

Methanol

Methanol has CNS effects which are substantially less severe compared to ethanol [1]. It is oxidized by liver ADH (at approximately 1/10th the rate of ethanol) to yield formaldehyde [1,12]. Formaldehyde is then rapidly metabolized by ALDH to formic acid (or formate) with a half-life of 1−2 minutes in the blood [12]. Formate is converted to carbon dioxide and water by formate dehydrogenase in a reaction that depends on liver tetrahydrofolate concentrations [12,30]. In cases of methanol intoxication, this pathway is easily saturated, contributing to accumulation of formic acid in blood [12]. Formic acid seems to be the major cause of clinical and laboratory abnormalities, including visual abnormalities and metabolic acidosis [12]. Methanol toxicity is associated with decreased vision, blindness, acute kidney injury, confusion, stupor, pulmonary dysfunction, abdominal pain, and coma [8,12]. Clinical findings include optic neuropathy, potentially resulting in blindness; neurologic, pulmonary, and GI effects; or death [1,8,12]. Laboratory tests for diagnosis of methanol intoxication include methanol and formic acid quantitation in blood or urine. Laboratory findings can include elevated osmolal gap and/or elevated anion gap and acidosis with low bicarb. Elevated serum creatinine or low creatinine clearances are also a possibility if there is significant kidney injury. Direct markers of methanol exposure can also play a key role in guiding clinical management of methanol toxicity. The American Academy of Clinical Toxicology suggests initiation of antidotal treatment with fomepizole or ethanol when the serum methanol concentration exceeds 20 mg/dL on a patient with documented history of ingesting toxic alcohols [8,12]. Alternatively, treatment may be considered when the osmolar gap is >10 mOsm/kg or with metabolic acidosis of unknown origin in a patient suspected of ingesting toxic alcohols [8,12].

Ethylene Glycol

Ethylene glycol has CNS effects similar to those of observed from ethanol [1]. Metabolism begins with oxidation by ADH to glycoaldehyde. Glycoaldehyde is metabolized by ALDH to glycolic acid, and then to glyoxylic acid and oxalic acid [1,12]. Glycolic acid can also be metabolized to α-hydroxy-β-keto adipate or glycine [12]. Metabolic acidosis with increased anion gap and organ dysfunction can result from accumulation of glycolic and oxalic acid from ethylene glycol metabolism [8,12]. Oxalic acid readily precipitates with calcium to form insoluble calcium oxalate crystals and hypocalcemia [1,8]. Tissue injury is usually associated with widespread deposition of oxalate crystals in organs along with toxic effects of glycolic and glyoxylic acids, leading to impairment of cellular respiration [1,12]. Oxalate crystals can deposit in the heart, brain, and lungs, causing organ dysfunction [8,12]. Oxalate excretion in the kidneys can result with crystalluria along with severe acute kidney injury or kidney failure [8]. Neurologic dysfunction usually develops first, within 12 hours following exposure, followed by cardiac and pulmonary dysfunction within 1 day and acute kidney injury within 3 days as organic acids are eliminated [8,12]. Laboratory findings for the diagnosis of ethylene glycol intoxication include ethylene glycol and oxalate quantitation in blood and oxalate or crystal analysis in urine. Elevated osmolar gap and/or an elevated anion gap with acidosis and low bicarb are usually indicative. Elevated serum creatinine or low creatinine clearance is possible if there is kidney failure. Direct ethylene glycol biomarkers can also play a key role in guiding clinical management. Similar to methanol, initiation of antidote treatment is recommended at an ethylene glycol concentration >20 mg/dL on patients with a documented history of ingesting [8]. It is also initiated when the osmolar gap is >10 mOsm/kg or with metabolic acidosis of unknown origin in a patient suspected of ingesting toxic alcohols [8].

Isopropanol

Isopropanol has significant CNS depression activity, but it is rapidly oxidized by ADH to yield acetone, a ketone that also suppresses the CNS. Oxidation to a ketone, as opposed to an acid, means isopropanol toxicity usually does not include acidosis due to isopropanol metabolism [1]. Acetone is the stable final product which is eliminated through the kidney with a half-life of approximately 3–6 hours [1]. Due to rapid ADH-dependent oxidation of isopropanol to acetone, serum isopropanol concentrations are usually lower than acetone in cases of isopropanol ingestion. As acetone is eliminated from the body, CNS depression continues, prolonging the apparent CNS effects of isopropanol [1]. Isopropanol intoxication usually presents not only with CNS depression, but also with respiratory dysfunction, cardiovascular collapse, acute pancreatitis, hypotension, and lactic acidosis [8]. Laboratory findings can include an elevated serum osmolar gap from isopropanol and acetone [1]. Other tests used can include isopropanol and acetone quantitation in blood or urine to confirm exposure or to direct treatment. Serum isopropanol concentrations >1500 mg/dL are known to induce coma with a concentration >500 mg/dL suggested as a cutoff for initiating dialysis [8]. However, in most isopropanol intoxication cases, supportive care is the main treatment.

Diethylene Glycol

Diethylene glycol is oxidized by ADH to 2-hydroxyethoxyacetaldehyde, which is further oxidized by ALDH to 2-hydroxyethoxyacetic acid (HEAA) [8,12]. HEAA is subsequently excreted by the kidneys. Toxic effects are thought to result from HEAA accumulation leading to acidosis and organ failure, but the mechanism by which HEAA produces cellular dysfunction and organ failure remains unclear [8,12]. There is also great variability in susceptibility to diethylene glycol poisoning, suggesting pharmacodynamics and individual risk factors are important [12]. Toxic effects can include abdominal pain, vomiting, diarrhea, headache, altered mental status, and acute kidney injury [8,12]. Acute kidney injury is usually the main cause of death, appearing between 8 and 24 hours after exposure to lethal doses of diethylene glycol [8]. Diethylene glycol poisoning can also result in hepatic disease, pancreatitis, and neurologic abnormalities, which appear up to a few days after exposure [8,12]. Laboratory findings in clinical assessment of diethylene glycol can include an elevated osmolar gap, elevated anion gap, elevated creatinine, and acidosis [8,11,12]. Direct quantitation of diethylene glycol and HEAA in serum, urine, and CSF has also been reported, but there is limited clinical guidance on how to use this information beyond confirming exposure [31].

Propylene Glycol

PG is metabolized by ADH to D/L-lactaldehyde, then to D/L-lactic acid by ALDH. Alternatively, D/L-lactaldehyde is metabolized to methylglyoxal which is converted to D-lactate. L-Lactate enters gluconeogenic, whereas D-lactate is metabolized to pyruvate and CO_2 [12]. PG intoxication usually results with acidosis but often with no increased osmolar gap due to rapid conversion of PG to lactic acid [8,12]. Both L- and D-lactic acidosis have been described [12]. PG-induced renal failure due to proximal tubular cell injury has been proposed, but the mechanism of renal toxicity is still unclear [12]. The serum PG concentration associated with toxicity can range from 12 to 520 mg/dL, with toxicity most likely to occur at blood concentrations >100 mg/dL [12].

ALCOHOL BIOMARKERS

The most direct way to confirm alcohol exposure is usually to measure alcohol concentrations in blood, urine, or breath [11]. However, the pharmacokinetics and pharmacodynamics of each alcohol determines the exact time course and concentration of each alcohol in blood, urine, or breath, therefore determining the detection window [12]. The detection window for each alcohol and its metabolite in blood depends on the half-life of elimination and amount consumed for each alcohol. Coingestion of toxic alcohols with ethanol or other competitive substrates for ADH will delay production of the toxic metabolites and prolong elimination times [12]. Each alcohol along with its metabolites or physiological adducts described above are considered direct biomarkers (see Table 3.1) [32]. In general, the detection of alcohol metabolites in urine will extend the time window in which alcohol use can be detected and sensitivity due to concentrated levels of metabolites in urine. Incorporation of direct alcohol biomarkers into hair or meconium will also extend the time window in which alcohol use can be detected.

Indirect biomarkers are usually associated with toxic effects of alcohol metabolism and are commonly measured in blood or urine. Indirect biomarkers of ethanol use include markers of liver damage (e.g., AST, ALT, GGT, and LDH). Other indirect markers of ethanol use include MCV due to the effects of ethanol on red blood cell morphology along with CDT due to ethanol-dependent effect on transferrin glycosylation [33,34]. In principal, the indirect markers of ethanol use which are related to liver damage might also be relevant for acute diethylene glycol poisoning which can induce liver injury [35] Similarly, ethylene glycol and methanol poisoning are associated with kidney injury, which is commonly monitored using serum creatinine or creatinine clearance.

Direct Ethanol Measurements

Direct measurements of circulating ethanol in serum have the advantages of being widely available and are considered relatively specific [14,34]. Ethanol is commonly involved in emergency department visits in the United States, with some emergency departments reporting that one-third of patients admitted exhibited substantially elevated blood alcohol concentrations at certain times of day [36]. National Academy of Clinical Biochemistry laboratory emergency medicine practice guidelines have recommended that ethanol test results be available in under an hour [36,37].

Analytical methods include serum determinations by enzyme assay and gas chromatography, which often exhibit turnaround times of <1 hour. Ethanol is generally cleared in a linear manner (zero-order kinetics) at a rate of approximately 20 mg/dL/h [38]. Analysis using ADH is relatively specific for ethanol, with only slight positive interferences observed in the presence of other alcohols such as isopropanol, methanol, and ethylene glycol [39]. Breath analyzers which measure the vapor pressure of ethanol are still less commonly used at the bedside but suffer a host of potential interferences [40]. Transdermal sensors for the detection of alcohol can also be utilized [41]. Even though the osmolar gap calculation is not a perfect method for screening for the presence of ethanol, it is still at least occasionally utilized as a relatively rapid screen in some emergency situations [42,43].

Rapid blood or serum/plasma ethanol testing can be used to investigate alcohol concentrations in individuals with elevated anion and osmolar gaps. General signs of intoxication may include general CNS depression and extend to symptoms such as coma, respiratory depression, hypotension, and hypothermia. Other potential uses for serum alcohol testing in the emergency room include assessing whether a person should be considered incapacitated in regards to making medical care decisions, or as an indication of potential alcohol dependence [36,37]. It should be noted that ethanol measurement is not always well suited for the detection of alcohol dependence, although high levels of blood alcohol can be associated with alcohol dependence [36].

Acetaldehyde

As discussed previously, acetaldehyde reacts with biological molecules (e.g., DNA, proteins, and lipids) to form adducts. The most potentially useful acetaldehyde adduct is hemoglobin-associated acetaldehyde [28]. Relative concentrations of free and bound acetaldehyde in whole blood have been utilized as a marker of consumption [44]. This circulating marker can be detected for up to 28 days following heavy alcohol ingestion [45]. Autoantibodies against these adducts can also be used to detect heavy alcohol intake [46]. Hemoglobin-bound acetaldehyde continues to increase in the presence of alcohol intake over the course of the 4-month life span of erythrocytes. This adduction process makes whole blood associated acetaldehyde a potential monitor of cumulative long-term alcohol consumption [47].

5-Hydroxytryptophol

Alcohol and acetaldehyde affect the metabolism of serotonin resulting in an increased urinary excretion of the serotonin metabolite 5-HTOL, which can be detected for approximately 5–15 hours longer than ethanol itself [48,49]. This marker is relatively sensitive and specific and may have utility in applications associated with forensic toxicology [50]. The use of the ratio of 5-HTOL to the serotonin metabolite 5-hydroxyindole-3-acetic acid (5-HIAA) has been suggested as a way to monitor recent alcohol intake and intensity [51,52]. Although consumption of low (<10 g) of ethanol may not increase the ratio of 5-HTOL and 5-HIAA in urine, consumption of 50 g of ethanol or more should certainly increase the ratio significantly. Higher ratios are indicative of more ethanol consumption almost in a dose-dependent manner. The specificity of this alcohol biomarker is almost 100% [50]. The sensitivity for consuming 50 g of alcohol or more is 77% at a cutoff value of 15 pmol/nmol [53].

Ethyl Glucuronide and Ethyl Sulfate

EtG/EtS is a marker that can be detected for a period of a few days following alcohol ingestion [54,55]. EtG is a minor metabolite of ethanol resulting from ethanol conjugation with glucuronic acid [56]. Both EtG and EtS are minor products of phase II ethanol metabolism representing <0.1% of total ethanol disposition. EtG is formed by conjugation with glucuronic acid catalyzed by the enzyme UDP-glucuronosyltransferase, while EtS formation is catalyzed by sulfotransferase. Both of these markers can be detected in the blood for ~36 hours and for several days in urine and tissues for several days following cessation of alcohol intake [55,57]. Blood spot analysis has also been shown to be a viable matrix [58]. Consumption of a relatively small quantity of alcohol such as 7 g may result in detectable EtG level in urine up to 6 hours. Detection time is longer after consumption of higher amounts of alcohol.

EtG/EtS species are also present in hair and represent a promising marker for postmortem investigations of alcohol use [35]. In general, the EtG level in hair in 95% of abstainers studied was <1.0 pg/mg of hair, while 30% of abstainers exhibited EtG levels below the detection limit of the highly sensitive liquid chromatography combined with tandem mass spectrometry assay (LC/MS/MS: detection limit: 0.5 pg/mg of hair). Hair color, gender, age, body mass index, smoking, and cosmetic treatment of hair did not appear to influence hair analysis for EtG. Various cutoff concentrations have been proposed for analysis of EtG in hair where value is expressed as pg/mg of hair. Morini et al. stated that 27 pg/mg exhibits a strong sensitivity (92%) and specificity (96%) [59]. A metaanalysis indicated that a cutoff of 30 pg/mg limits the false negatives in differentiating heavy from social drinking and abstinence [60].

EtG in meconium is also measured to investigate possible exposure of a fetus to maternal alcohol use. Bana et al. used a cutoff of 50 ng/gm of meconium for EtG and 1000 ng/gm of meconium for FAEEs for their study and reported that 34.6% women consumed alcohol during pregnancy while 17% women showed positive results with both markers [61]. For hair, EtG sensitivity of 96% and specificity of 99% has been reported at a cutoff concentration of 30 pg/mg of hair to identify individuals who are drinking alcohol chronically at amounts exceeding 60 g/day [62]. Urinary glucuronide at a cutoff of 100 ng/mL, exhibited a sensitivity and specificity was 76% and 93%, respectively. The sensitivity and specificity of urinary EtS at 25 ng/mL cutoff was 82% and 86% respectively when utilized to detect drinking 3–7 days prior to clinic visits [63].

False positive and false negative results have been reported with both EtG and EtS. False positive test results may be due to incidental exposure to alcohol-containing products such as mouthwash and hand sanitizers, especially if a lower cutoff concentration is used. Consuming nonalcoholic beer and wine in larger amounts may also produce false positive results because such products may contain a small amount of alcohol. Eating baker's yeast with sugar, drinking large amounts of apple juice, or even eating ripe bananas may cause detectable amounts of EtG and EtS in urine. Urinary tract infections may also produce false negative test results due to degradation of EtG in urine by the beta-glucuronidase enzyme present in *Escherichia coli*. In contrast, EtS is not affected by this process. In 2006, an advisory was issued due to potentially false positive test results with EtG testing and warned against use of EtG as the sole evidence in determining abstinence in criminal justice, regulatory, or legal settings [64].

Fatty Acid Ethyl Esters

FAEEs represent an alcohol consumption marker with a longer window of detection than direct ethanol concentrations [65,66]. FAEEs are minor metabolites of ethanol which are formed after alcohol consumption in virtually all tissues due to interaction of ethanol with free fatty acids as well as triglycerides, lipoproteins, and phospholipids. This pathway is an enzyme-mediated esterification of fatty acid or fatty acetyl-CoA by ethanol. The FAEE synthase plays a role in formation of FAEEs. In serum, FAEEs appear after alcohol consumption and are bound to albumin and also are also found in the core of lipoproteins along with other neutral lipids. These products are metabolites of ethanol and persist for approximately 24 hours following alcohol ingestion in the serum.

Although serum can be used for determination of concentrations of FAEEs, more recently focus is on determination of hair concentration of FAEEs in both clinical and forensic investigations [67,68]. However, hair from a strictly abstinent person may show a very small amount of FAEE which may be related to trace amount of endogenous ethanol production, nutrition, or use of hair cosmetics. Hair specimens collected from the pubic region, armpit, chest, arm, or thigh show comparable FAEEs to specimens collected from scalp hair. Furthermore, a number of FAEE species can be incorporated into hair where they can be measured after extraction by gas chromatography–mass spectrometry with high specificity and sensitivity [33,69–71]. The sum of these FAEEs in hair specimens varies from <0.2 ng/mg of hair in strict teetotalers to more than 30 ng/mg in samples from alcoholic death. Bertol et al. also proposed a cutoff of 0.5 ng/mg for FAEE concentration in hair (3 cm segment analyzed) to differentiate between social drinking from excessive alcohol consumption (>60 g/day) [72]. Wurst et al. reported that at 0.29 ng/mg cutoff for FAEEs in hair, the sensitivity and specificity was 100% and 90% to identify individuals who abused alcohol chronically [73]. However, in some cases, hair products may alter observed concentrations [74,75]. Gareri et al. studied the effect of hair care products on FAEE concentrations in hair and concluded that regular use of hair products containing as little as 10% alcohol can impact FAEE values in hair. In this case, EtG should be tested because EtG values in hair seems to be unaffected by use of hair products containing alcohol [76].

Additionally, this marker has neonatal hair and meconium sample applications to explore potential prenatal alcohol exposure [74,75,77,78]. The absence of FAEEs in hair does not rule out exposure, and care must be taken to sample hair appropriately [69]. However, small amounts of FAEEs are found in the meconium of neonates without any evidence of maternal use of alcohol. This may arise from endogenous ethanol production or trace amounts of alcohol coming from food. Depending on the combination of FAEEs used for calculation, various cutoff values have been proposed, and a positive cutoff of 2 nmol/g (approximately 600 ng/gm) has been adopted.

Indirect Alcohol Biomarkers: Clinical Utility, Analytical Methods, and Analytical Issues

Indirect biomarkers for the detection of chronic alcohol use can be used to screen for heavy alcohol consumption, identify changes in drinking behavior, and monitoring therapy of alcohol dependence which results in significant cost savings [4,33,34,79,80]. Complexity of testing methods vary from Food and Drug Administration (FDA) approved testing to laboratory originated and validated high complexity "home-brew" testing such as mass spectroscopy. General overviews of the performance characteristics of a number of markers can be found in a number of useful review articles and in Table 3.2 [33,80]. While the direct measurement of alcohol and immediate metabolites in blood and other body fluids can be useful, these types of more direct measurements are typified by relatively small detection windows following alcohol intake. Indirect markers of ethanol are often preferentially utilized for determination of alcohol abuse or dependence [33,75,80–82].

Long-term markers for alcohol use/abuse which remain elevated for long periods of time can be used to investigate the role of alcohol as an etiological factor for other disorders, to initiate detoxification therapy, to treat dependency and also to motivate patients to modify drinking habits, and to monitor abstinence [33,35]. Thus, markers of chronic alcohol intake that can be detected over long periods of time following drinking cessation are often preferentially utilized. Several biochemical markers are available to identify excessive or chronic alcohol use. Some long-term markers of alcohol use include MCV or markers of liver function such as GGT, and CDT. In general, long-term markers are responsive to alcohol intake in the preceding weeks to months.

General Markers of Nutritional Status and Liver Damage

Most alcoholics with liver disease or alcoholic hepatitis suffer from dietary imbalance and protein malnutrition, which may manifest itself in reduced albumin, total protein, or prealbumin [83]. Serum uric acid is often elevated in cases of alcohol dependence [35].

Serum IgG, IgM, and IgA gamma globulins are often elevated in response to antigenic stimulation. Elevations of IgA are observed in most alcohol-dependent individuals while IgG and IgM are elevated, albeit at a lower frequency. These changes are nonspecific, which limits their diagnostic value [28,84].

Liver damage is associated with chronic alcohol use. Total bilirubin is elevated in 60% of patients with alcoholic hepatitis, and prolonged alkaline phosphatase and prolonged prothrombin times are often concurrently observed [80,83]. Hyaluronic acid is a commonly used marker of liver fibrosis and can be elevated in cases of severe alcohol abuse [35]. The serum transaminases aspartate (AST) and alanine (ALT) aminotransferase can also be elevated in liver damage, but appear to have only limited sensitivity for alcohol abuse when looked at it in isolation [85]. ALT has a longer half-life than AST. In acute liver injury, AST levels are higher than ALT; however, after 24–48 hours, ALT levels could be higher than AST. ALT is considered the more specific marker for liver injury, but it is not very specific. Because alcohol has toxic effects to the liver, liver enzymes can be elevated in individuals consuming excessive amounts of alcohol [33,35]. The De Ritis ratio (AST/ALT) of >2 is often considered suggestive of an alcohol etiology of liver disease [35].

TABLE 3.2 Selected Characteristics of Some Markers of Alcohol Use [4,14,34,80]

Alcohol Intake Marker	Specimen	Approximate Period of Detection of Prior Alcohol Use	Comments
Ethanol, blood	Serum or whole blood	<1 day	Direct marker of intake. Elevated anion and osmolar gaps
Fatty acid ethyl esters	Serum, hair, and meconium	<1 day in serum, several months in hair and meconium	High sensitivity for exposure in serum/plasma. Presence in hair and meconium may be clinically useful (~95% sensitivity, ~90% specificity)
5-hydroxytryptophol	Urine	1–3 days	Can be part of a 5-HTOL/5-HIAA ratio (100% specificity)
Ethyl glucuronide	Urine, hair, and meconium	2–3 days in urine but several months in hair	Detected in hair and tissues. Often used in forensic investigations (~75% sensitivity, ~90% specificity)
Aminotransferases, plasma	Serum	3 weeks	AST/ALT >2 suggests alcoholic liver damage
N-acetyl beta-hexosaminidase	Serum, urine	1–2 weeks, serum 2–4 weeks, urine	Has a half-life of <7 days in plasma; elevated in kidney insufficiency, pregnancy, and diabetes (~95% sensitivity, ~90% specificity)
Phosphatidylethanol	Whole blood	2–3 weeks	Is a metabolite of ethanol and with a longer period of detection than ethanol (~95% sensitivity, ~95% specificity)
Gamma-glutamyl transferase, plasma	Serum	2–3 weeks	Nonspecific, but sensitive, widely available
Carbohydrate-deficient transferrin, plasma	Serum	2–3 weeks	Specific indirect marker. Genetic variants of transferrin can cause false positives
GGT and CDT in combination	Serum	2–3 weeks	Increased sensitivity as compared to GGT or CDT alone without loss of specificity (~90% sensitivity, ~95% specificity)
Acetaldehyde adducts (hemoglobin-associated acetaldehyde; HAA)	Whole blood	Variable, 1–4 months for HAA	Titers of IgA autoantibodies can also be utilized as an alcohol exposure marker
Mean corpuscular volume	Whole blood	2–4 months	MCV >100 fl suggestive of alcohol abuse in absence of anemia. B12 and/or folate deficiency, smoking, hypothyroidism, or other liver and hematological disorders can also increase MCV
Sialic acid, Apolipoprotein J	Plasma	Variable, ~8 weeks for Apo J	Ethanol exposure decreases sialylation of Apo J; false positives observed in patients with carcinoma, diabetes, and cardiovascular disease (~90% sensitivity, ~95% specificity)

N-Acetyl-β-Hexosaminidase

N-acetyl-β-Hex is a complex group of glycoprotein lysosomal isoenzymes. Two major isoenzymes of β-Hex have been characterized: Isoenzyme A (one alpha and one beta chain) and B (two beta chains). N-acetyl-β-Hex is an enzyme that can be elevated in heavy drinkers. Alcohol intake reduces biliary excretion of β-Hex, which can be measured in the serum or urine and has a serum half-life of one week. It has been assumed that drinking over 60 g of alcohol (over 4.5 standard drinks) per day for 10 days or more results in an increased level of β-Hex in serum. Kakkaien et al. reported that a mean β-Hex level was 35.0 U/L among drunken men ($n = 25$) but the mean value was 16.8 U/L among healthy males ($n = 16$) who were social drinkers. However, the mean β-Hex level was 19.8 U/L among teetotalers. Levels return to baseline in plasma after about 10 days of cessation, making it a potential intermediate-term marker of alcohol consumption. The authors further reported that sensitivity of β-Hex among heavy drinkers was 85.7% and specificity of β-Hex as alcohol biomarker was 97.6% [86]. While some studies have shown it to be a sensitive marker of heavy drinking, other disorders such as diabetes and hypertension also cause elevations of beta-hex [87,88]. Wehr et al. suggested that urinary β-Hex can be used to monitor sobriety in alcohol-dependent individuals [89]. This marker is more sensitive than GGT, AST, ALT, or MCV as an alcohol biomarker [88]. β-Hex is a large molecule which is not filtered during glomerular filtration. Various conditions may falsely elevate serum and urine β-Hex levels. These can include nonalcoholic liver disease, diabetes, and rheumatoid arthritis.

Mean Corpuscular Volume

Although the exact mechanism remains unknown, red blood cell size (MCV) increases in an apparently dose-dependent relationship with intensity of alcohol intake. MCV is a value routinely produced by automated hematology analyzers during complete blood count analysis. Increased MCV associated macrocytosis in the presence of excess alcohol may occur in the presence of normal folate levels. Because the life span of erythrocyte is 120 days, it may take several months before MCV may return to a normal level after abstinence. An increase in MCV can be noted in patients with <40 g/day consumption [90]. The sensitivity of MCV for alcohol dependence has been estimated at ∼40%, limiting its routine use. Increased MCV can be associated with thyroid disease, folate deficiency, blood loss, pharmaceutical intake, nonalcoholic liver diseases, and various hematological disorders such as megaloblastic anemia. Thus, while about 4% of all adults exhibit elevated MCV, only about 65% of those cases can be attributed to alcohol intake [91,92]. Macrocytosis is usually defined as an MCV value >100 fl and several authors used 100 fl as the cutoff value of MCV to study the relationship between alcohol consumption and MCV [35].

Lipid Markers

Increased concentrations of high-density lipoprotein (HDL) are often associated with prolonged alcohol consumption. These can be elevated even in cases of low daily amounts of chronic intake, although they have limited specificity [93]. Decreased HDL may be useful in the identification of patients in the early phase of alcohol dependence who do not yet have significant liver disease, although it is not commonly utilized in this manner [28]. Chronic alcohol users also can exhibit increased serum triglyceride concentrations.

PEth is a group of phospholipids formed through phospholipase D mediated enzymatic reactions between ethanol and phosphatidylcholine in cell membranes. PEth is measured in whole blood where it is mostly associated with erythrocyte cell membranes. These phospholipids can be detected in whole blood as long as 3 weeks (it has a half-life of ∼4 days) after cessation of drinking [94]. Consumption of one alcoholic drink is not sufficient to produce a detectable level of PEth in blood. In general, consumption of 50 g or more alcohol per day for several weeks is necessary to produce detectable amounts of PEth in blood. While PEth appears to be sensitive to heavy alcohol use and does appear to correlate with amount of ethanol consumed, specimens have poor stability and should be keep frozen prior to analysis [35]. This marker is almost 100% specific because its formation is totally dependent on the presence of ethanol [95].

Gamma-Glutamyltransferase

Enzymatic determination of serum enzyme activities of GGT is perhaps the most commonly known traditional marker for chronic alcohol consumption. In general, a positive correlation exists between GGT serum enzyme activity and alcohol consumption [28]. This biliary enzyme is induced by alcohol consumption and is increased in serum due to increased synthesis as a result of enzyme induction by alcohol or release from hepatocytes due to damage caused by excessive alcohol use. However, elevation of GGT in response to alcohol consumption varies widely between individuals, and levels correlate only moderately with amount of alcohol consumed. Sustained alcohol consumption is needed for elevation in GGT level, but if a person had a history of alcohol abuse, GGT may be elevated after resumption of drinking. In general, consumption of 60 g or more alcohol for 3–6 weeks is needed to observe significant increases in serum GGT levels. Increased activity elevations are seen in cases of severe alcoholic liver disease. Binge drinking seems to elevate GGT activity more than mild chronic drinking. Lower levels of daily alcohol consumption (<40 g/day) and moderate routine alcohol consumption can also result in elevated GGT activities [96,97].

GGT is deemed to be one of the more sensitive markers for alcoholism with sensitivity of 64% and specificity of 72% at a 30 U/L cutoff. Estimates of sensitivity vary widely, ranging from 40% to 80% in alcohol-dependent individuals [28]. In cases of heavy alcohol consumption, GGT is typically 10-fold higher than normal and returns to normal approximately 2 weeks after alcohol intake cessation. The sensitivity of GGT for alcohol consumption has been shown to be higher in men than in women [96]. In response, gender-specific cutoffs (33 U/L for female and 56 U/L for male) had also been proposed. Although average GGT levels are significantly higher in current and former drinkers compared to lifetime abstainers, in one study, the authors observed that men daily drinking showed the highest levels of GGT, whereas in women, highest levels of GGT were observed in weekend drinkers. Women who consumed alcohol without food exhibited higher GGT levels compared to women who consumed alcohol with food. Obesity and hepatitis C infection may increase serum GGT level [98]. False positive results may be encountered in patients receiving therapy with barbiturates, phenytoin, phenazone, dextropropoxyphene, monoamine oxidase inhibitors, tricyclic antidepressants, warfarin, thiazide diuretics, or anabolic steroids. In addition, damage of the liver due to viral infection such as hepatitis or ischemic damage to the liver may also significantly increase serum GGT levels [33]. Other factors that tend to result in mild to moderate increases in average GGT activities include older individuals, postmenopausal women, and women taking hormonal contraceptives [99,100].

CARBOHYDRATE-DEFICIENT TRANSFERRIN

CDT is the collective name of a group of minor isoforms of human transferrin with a low degree of glycosylation. The relative amount of CDT in the serum provides a means for detection of heavy alcohol use in preceding weeks [80,101,102]. Transferrin is an iron transport glycoprotein involved in iron transport and is produced in hepatocytes. Transferrin exists as a heterogeneous population of isoforms that differ in the number of attached charged carbohydrate sialic acid chains. In general, transferrin molecules which are deficient in sialic acid (containing zero to two sialic acid molecules) are minor isoforms present in sera of normal individuals. These minor isoforms include asialo (no sialic acid), monosialo (one sialic acid molecule), and disialotransferrin (two sialic acid molecules).CDT is the collective name of a group of minor isoforms of human transferrin with a low degree of glycosylation. Chronic alcohol (ethanol) use markedly increases the concentrations of the asiolo and disialo CDT isoforms and has little effect on the concentrations of the trisialo, tetrasialo, and pentasialo transferrin isoforms. The half-life of these marker CDT isoforms is approximately 2 weeks after sustained alcohol intake, extending the detection window past many other markers.

There are several analytical methods cleared by the US FDA for including immunoassay, HPLC methods, and capillary gel electrophoresis methods [103]. Consumption of >50 g of ethanol per day (roughly four to five drinks/day) for a period of 1–2 weeks is required to cause a significant measurable increase in the serum CDT fraction by these methods. Thus, the detection of specific CDT isoforms relative to total transferrin concentrations can be utilized to monitor sustained alcohol intake. Furthermore an association between elevated CDT levels and societal problems such as traffic accidents and has been observed [104].

In isolation as a stand-alone marker, CDT is not a considered to be an extremely sensitive marker of alcohol intake. Arndt et al. observed that at a cutoff of 2.4% for CDT, the sensitivity and specificity was 84% and 92%, respectively [101]. However, one should keep in mind that the sensitivity and specificity of CDT percentage was superior to other alcohol biomarkers such as GGT and MCV (6). Although elevated levels of CDT are deemed to be fairly specific for sustained alcohol intake, a number of factors can elevate CDT. Patients suspected of having autosomal congenital disorders of glycosylation may exhibit elevation of CDT isoforms in the absence of alcohol intake. In addition, the presence of rare genetic variants of transferrin (including D, B1, and B2) may interfere with CDT analysis. Advanced liver damage (including biliary cirrhosis, hepatocellular carcinoma, and severe chronic viral hepatitis) can increase observed relative CDT levels, leading to an estimated reduction in specificity to 70–80% for the exclusion of heavy alcohol use. Other potential confounding factors include the presence of monoclonal antibodies, pancreatic or kidney transplantation, and immunosuppressive or antiepileptic drug therapy [101,105,106].

Combination of CDT and GGT or Gamma-CDT

Although no one single marker for alcohol use yields complete or ideal sensitivity and specificity, increased performance can sometimes be achieved by the use of multiple markers. Combining GGT and CDT values to derive a new parameter mathematically may improve the sensitivity and specificity of this calculated parameter as an alcohol biomarker. This is often referred to as gamma-CDT. A gamma-CDT equation has been generated that includes contributions by both GGT and CDT markers and has been shown to correlate to alcohol consumption [102]. It is consistently elevated in patients who consume >40 g/day of ethanol. The mathematical formula to derive this parameter is:

$$\text{GGT-CDT} = 0.8 \times \ln(\text{GGT}) + 1.3 \ln(\text{CDT}).$$

In this case, GGT level is expressed as U/L and CDT level is also expressed as U/L. Sillanukee and Olsson proposed a cutoff value of 6.5 using 95th percentile of the data from the control group for this combined marker. The authors further reported that sensitivity and specificity of GGT-CDT marker were 79% and 93%, respectively which were superior to sensitivity of 65% and specificity of 94% observed with CDT. The sensitivity and specificity of GGT were 59% and 91%, respectively. The authors concluded that combined GGT-CDT is more sensitive and specific than CDT alone as an alcohol biomarker [102].

Sialic Acid and Sialic Index of Apolipoprotein J

Apolipoprotein J (Apo J), or clusterin, is a highly sialylated glycoprotein that is a normal component of plasma HDL particles. Approximately 30% of Apo J molecule is carbohydrate. Synthesis of the mature Apo J molecule requires the addition of sugars to the molecule in a sequential manner and with terminal sialic acid molecules [107]. The activities of enzymes such as sialyltransferase may be reduced in alcoholics, thus inhibiting incorporation of sialic acid. As a result, the sialic acid index of plasma Apo J has been proposed as an alcohol biomarker, where the total amount of sialic acid per mole of Apo J is reduced in alcoholics as compared to healthy individuals [108,109]. In one study, the authors reported that in human subjects, intake of alcohol for 30 days resulted in almost a 50% decrease in sialic acid index. While total serum sialic acid may have sensitivities and specificities approaching CDT, it appears to take longer to return to baseline following cessation, potentially limiting its utility as a marker (2–5 weeks) [110]. The level of sialylation of the glycoprotein Apo J has also been proposed as a marker of heavy alcohol intake [111]. The specificity of sialic acid index of Apo J approached 100% in one study, while sensitivity was ~90–92%. Clinical applications of alcohol biomarkers are listed in Table 3.3.

TABLE 3.3 Selected Potential Clinical Applications of Alcohol Biomarkers [4]

Setting	Indications	Some Common Biomarkers Utilized
Primary care	Screening for alcohol abuse/potential relapse	GGT and CDT percentage are most commonly used
Criminal justice	Behavioral assessment Criminal assessment	Direct blood and urine alcohol level along with ethyl glucuronide, ethyl sulfate, and hair FAEE are commonly used. Point of care devices may also be used including transdermal alcohol sensor or breath direct alcohol sensing devices
Drug/alcohol rehabilitation program	Abstinence	Ethyl glucuronide, ethyl sulfate, along with blood and urine alcohol. Transdermal alcohol sensor devices may also be used
Health and safety screening program	Screening for clinically significant alcohol abuse; long-term assessment	Ethyl glucuronide, ethyl sulfate, GGT, and CDT
Workplace drug/alcohol testing	Short-term abstinence or clinically significant long-term alcohol abuse	Blood and or urine alcohol along with ethyl glucuronide, ethyl sulfate, CDT
Emergency medicine/trauma/transplant service	Short-term clinical assessment; assessment of dependence	GGT, CDT, ethyl glucuronide, ethyl sulfate and possibly hair fatty acid ethyl ester
Reissue of driving license	Abstinence	Ethyl glucuronide in hair and urine, more common in Europe
Pregnant women	Short-/long-term abstinence and clinical assessment	Ethyl glucuronide, ethyl sulfate in urine, or meconium is useful. In addition, testing of fatty acid ethyl ester in maternal hair may provide additional information

CONCLUSIONS

Alcohol testing, both direct and indirect, is useful in a variety of potential clinical settings (see Table 3.3) [80]. Use of alcohol is associated with many negative consequences, contributing to increased accidents, medical costs, lost productivity, and violent crime [6,10,112,113]. While there are other clinical tools, structured interviews using DSM-IV or DSM-V criteria, self-assessment of alcohol intake such as Alcohol Use Disorder Identification Test (AUDIT) score, and alcohol biomarkers are available to help clinicians identify patients who are consuming harmful or hazardous amount of alcohol or are dependent on it [114]. AUDIT, which was developed in collaboration with WHO as a simple screening test for early detection of alcohol use disorders [115,116]. DSM-IV and DSM-V are the Diagnostic and Statistical Manual of Mental Disorders issued by the American Psychiatric Association [117]. According to DSM-IV criteria, 9.7% in the same population would have AUD diagnosis [118,119]. The audit score outperforms CDT and MCV screening for predicting hazardous drinking in some primary settings [120]. Adding CDT screening to the patient's self-report of alcohol consumption in a primary care setting results in significant savings in healthcare cost and may be well received in some patient populations [121,122]. Prenatal alcohol use screening is another societally beneficial application of alcohol marker testing [123,124]. In general, only the minority of patients with alcohol dependence seek treatment. As a patient may not disclose recent alcohol consumption or may underreport alcohol consumption, it is important to have the ability to conduct effective alcohol biomarker testing (direct or indirect) in a patient suspected of alcohol abuse or clinical danger due to acute toxicity.

REFERENCES

[1] Burtis CA, Bruns DE, Sawyer BG, Tietz NW. 3rd ed. Tietz fundamentals of clinical chemistry and molecular diagnostics, xxii. St. Louis, Missouri: Elsevier/Saunders; 2015. p. 1075.

[2] Rehm J, Room R, Monteiro M, Gmel G, et al. Alcohol as a risk factor for global burden of disease. Eur Addict Res 2003;9(4):157–64.

[3] Room R, Graham K, Rehm J, Jernigan D, et al. Drinking and its burden in a global perspective: policy considerations and options. Eur Addict Res 2003;9(4):165–75.

[4] Dasgupta, A. Alcohol and its biomarkers. 2015, Elsevier. p. 91–120.

[5] U.S. Department of Health and Human Services., U.S. Department of Agriculture., and United States. 2015–2020 dietary guidelines for Americans, 8th ed. Washington, DC; 2015,p. 122.

[6] Garbutt JC, West SL, Carey TS, Lohr KN, et al. Pharmacological treatment of alcohol dependence: a review of the evidence. Jama 1999;281(14):1318–25.

[7] Brunton LL, Knollmann BC, Hilal-Dandan R. 13th ed. Goodman & Gilman's the pharmacological basis of therapeutics, xiii. New York: McGraw Hill Medical; 2018. 1419 pages.

[8] Kraut JA, Mullins ME. Toxic alcohols. N Engl J Med 2018;378(3):270–80.

[9] Ingall GB. Alcohol biomarkers. Clin Lab Med 2012;32(3):391–406.

[10] Wiese JG, Shlipak MG, Browner WS. The alcohol hangover. Ann Intern Med 2000;132(11):897–902.

[11] Kraut JA. Diagnosis of toxic alcohols: limitations of present methods. Clin Toxicol (Phila) 2015;53(7):589–95.

[12] Kraut JA, Kurtz I. Toxic alcohol ingestions: clinical features, diagnosis, and management. Clin J Am Soc Nephrol 2008;3(1):208–25.

[13] Substance Abuse and Mental Health Services Administration, The role of biomarkers in the treatment of alcohol use disorders, 2012 revision. Advisory, 2012. 11(2).

[14] McMillin GA, Mellis R, Bornhorst J. Alcohol abuse and dependency genetics of susceptibility and pharmacogenetics of therapy. In: Dasgupta A, editor. Critical issues in drug abuse testing. Humana Press; 2009.

[15] Cabarcos P, Álvarez I, Tabernero MJ, Bermejo AM. Determination of direct alcohol markers: a review. Anal Bioanal Chem 2015;407(17):4907–25.

[16] Nanau RM, Neuman MG. Biomolecules and biomarkers used in diagnosis of alcohol drinking and in monitoring therapeutic interventions. Biomolecules 2015;5(3):1339–85.

[17] Baselt RC. 10th ed. Disposition of toxic drugs and chemicals in man, xxxiv. Seal Beach, California: Biomedical Publications; 2014. 2211 pages.

[18] Tuma DJ, Casey CA. Dangerous byproducts of alcohol breakdown—focus on adducts. Alcohol Res Health 2003;27(4):285–90.

[19] Niemela O. Aldehyde-protein adducts in the liver as a result of ethanol-induced oxidative stress. Front Biosci 1999;4:D506–13.

[20] Das SK, Dhanya L, Vasudevan DM. Biomarkers of alcoholism: an updated review. Scand J Clin Lab Invest 2008;68(2):81–92.

[21] Foti RS, Fisher MB. Assessment of UDP-glucuronosyltransferase catalyzed formation of ethyl glucuronide in human liver microsomes and recombinant UGTs. Forensic Sci Int 2005;153(2–3):109–16.

[22] Kurogi K, Davidson G, Mohammed YI, Williams FE, et al. Ethanol sulfation by the human cytosolic sulfotransferases: a systematic analysis. Biol Pharm Bull 2012;35(12):2180–5.

[23] Gustavsson L, Alling C. Formation of phosphatidylethanol in rat brain by phospholipase D. Biochem Biophys Res Commun 1987;142(3):958–63.

[24] Aleryani S, Kabakibi A, Cluette-Brown J, Laposata M. Fatty acid ethyl ester synthase, an enzyme for nonoxidative ethanol metabolism, is present in serum after liver and pancreatic injury. Clin Chem 1996;42(1):24–7.
[25] Cederbaum AI. Introduction-serial review: alcohol, oxidative stress and cell injury. Free Radic Biol Med 2001;31(12):1524–6.
[26] Moncada C, Israel Y. Protein binding of alpha-hydroxyethyl free radicals. Alcohol Clin Exp Res 2001;25(12):1723–8.
[27] Worrall S, Thiele GM. Protein modification in ethanol toxicity. Adverse Drug React Toxicol Rev 2001;20(3):133–59.
[28] Niemela O. Biomarkers in alcoholism. Clin Chim Acta 2007;377(1–2):39–49.
[29] Megarbane B, Borron SW, Baud FJ. Current recommendations for treatment of severe toxic alcohol poisonings. Intensive Care Med 2005;31(2):189–95.
[30] Kerns W, Tomaszewski C, McMartin K, Ford M, et al. Formate kinetics in methanol poisoning. J Toxicol Clin Toxicol 2002;40(2):137–43.
[31] Perala AW, Filary MJ, Bartels MJ, McMartin KE. Quantitation of diethylene glycol and its metabolites by gas chromatography mass spectrometry or ion chromatography mass spectrometry in rat and human biological samples. J Anal Toxicol 2014;38(4):184–93.
[32] Niemela O. Biomarker-based approaches for assessing alcohol use disorders. Int J Environ Res Public Health 2016;13(2):166.
[33] Hannuksela ML, Liisanantti MK, Nissinen AE, Savolainen MJ. Biochemical markers of alcoholism. Clin Chem Lab Med 2007;45(8):953–61.
[34] Bornhorst J, Stone A, Brown J, Light K. Slate and trait markers of alcohol abuse in Alcohol Biomarkers an Overview In: Dasgupta A, Langman L, editors. Pharmacogenomics of alcohol abuse. CRC Press; 2012.
[35] Niemela O, Alatalo P. Biomarkers of alcohol consumption and related liver disease. Scand J Clin Lab Invest 2010;70(5):305–12.
[36] Church AS, Witting MD. Laboratory testing in ethanol, methanol, ethylene glycol, and isopropanol toxicities. J Emerg Med 1997;15(5):687–92.
[37] Wu AH, McKay C, Broussard LA, Hoffman RS, et al. National academy of clinical biochemistry laboratory medicine practice guidelines: recommendations for the use of laboratory tests to support poisoned patients who present to the emergency department. Clin Chem 2003;49(3):357–79.
[38] Baselt RC, editor. Disposition of toxic drugs and chemicals in man. 5th (ed.) Foster City, CA: CTI; 2000.
[39] Gadsden RH, Taylor EH, Steindel SJ. Ethanol in biological fluids by enzymatic analysis. In: Frings C, Faulhner WR, editors. Selected methods of emergency toxicology. Washington DC: AACC Press; 1986.
[40] Gibb KA, Yee AS, Johnston CC, Martin SD, et al. Accuracy and usefulness of a breath alcohol analyzer. Ann Emerg Med 1984;13(7):516–20.
[41] Leffingwell TR, Cooney NJ, Murphy JG, Luczak S, et al. Continuous objective monitoring of alcohol use: twenty-first century measurement using transdermal sensors. Alcohol Clin Exp Res 2013;37(1):16–22.
[42] Liamis G, Filippatos TD, Liontos A, Elisaf MS. Serum osmolal gap in clinical practice: usefulness and limitations. Postgrad Med 2017;129(4):456–9.
[43] Whittington JE, La'ulu SL, Hunsaker JJ, Roberts WL. The osmolal gap: what has changed? Clin Chem 2010;56(8):1353–5.
[44] Halvorson MR, Campbell JL, Sprague G, Slater K, et al. Comparative evaluation of the clinical utility of three markers of ethanol intake: the effect of gender. Alcohol Clin Exp Res 1993;17(2):225–9.
[45] Peterson CM, Jovanovic-Peterson L, Schmid-Formby F. Rapid association of acetaldehyde with hemoglobin in human volunteers after low dose ethanol. Alcohol 1988;5(5):371–4.
[46] Hietala J, Koivisto H, Latvala J, Anttila P, et al. Igas against acetaldehyde-modified red cell protein as a marker of ethanol consumption in male alcoholic subjects, moderate drinkers, and abstainers. Alcohol Clin Exp Res 2006;30(10):1693–8.
[47] Wozniak MK, Wiergowski M, Namiesnik J, Biziuk M. Biomarkers of alcohol consumption in body fluids - possibilities and limitations of application in toxicological analysis. Curr Med Chem 2017.
[48] Oppolzer D, Barroso M, Gallardo E. Bioanalytical procedures and developments in the determination of alcohol biomarkers in biological specimens. Bioanalysis 2016;8(3):229–51.
[49] Bisaga A, Laposata M, Xie S, Evans SM. Comparison of serum fatty acid ethyl esters and urinary 5-hydroxytryptophol as biochemical markers of recent ethanol consumption. Alcohol Alcohol 2005;40(3):214–18.
[50] Beck O, Helander A. 5-hydroxytryptophol as a marker for recent alcohol intake. Addiction 2003;98(Suppl 2):63–72.
[51] Johnson RD, Lewis RJ, Canfield DV, Blank CL. Accurate assignment of ethanol origin in postmortem urine: liquid chromatographic-mass spectrometric determination of serotonin metabolites. J Chromatogr B Analyt Technol Biomed Life Sci 2004;805(2):223–34.
[52] Borucki K, Schreiner R, Dierkes J, Jachau K, et al. Detection of recent ethanol intake with new markers: comparison of fatty acid ethyl esters in serum and of ethyl glucuronide and the ratio of 5-hydroxytryptophol to 5-hydroxyindole acetic acid in urine. Alcohol Clin Exp Res 2005;29(5):781–7.
[53] Helander A, von Wachenfeldt J, Hiltunen A, Beck O, et al. Comparison of urinary 5-hydroxytryptophol, breath ethanol, and self-report for detection of recent alcohol use during outpatient treatment: a study on methadone patients. Drug Alcohol Depend 1999;56(1):33–8.
[54] Schmitt G, Droenner P, Skopp G, Aderjan R. Ethyl glucuronide concentration in serum of human volunteers, teetotalers, and suspected drinking drivers. J Forensic Sci 1997;42(6):1099–102.
[55] Wurst FM, Skipper GE, Weinmann W. Ethyl glucuronide–the direct ethanol metabolite on the threshold from science to routine use. Addiction 2003;98(Suppl 2):51–61.
[56] Walsham NE, Sherwood RA. Ethyl glucuronide and ethyl sulfate. Adv Clin Chem 2014;67:47–71.
[57] Helander A, Bottcher M, Fehr C, Dahmen N, et al. Detection times for urinary ethyl glucuronide and ethyl sulfate in heavy drinkers during alcohol detoxification. Alcohol Alcohol 2009;44(1):55–61.
[58] Kummer N, Wille SM, Poll A, Lambert WE, et al. Quantification of EtG in hair, EtG and EtS in urine and PEth species in capillary dried blood spots to assess the alcohol aonsumption in driver's licence regranting cases. Drug Alcohol Depend 2016;165:191–7.

[59] Morini L, Politi L, Polettini A. Ethyl glucuronide in hair. A sensitive and specific marker of chronic heavy drinking. Addiction 2009;104(6):915−20.
[60] Boscolo-Berto R, Viel G, Montisci M, Terranova C, et al. Ethyl glucuronide concentration in hair for detecting heavy drinking and/or abstinence: a *meta*-analysis. Int J Legal Med 2013;127(3):611−19.
[61] Bana A, Tabernero MJ, Perez-Munuzuri A, Lopez-Suarez O, et al. Prenatal alcohol exposure and its repercussion on newborns. J Neonatal Perinatal Med 2014;7(1):47−54.
[62] Boscolo-Berto R, Favretto D, Cecchetto G, Vincenti M, et al. Sensitivity and specificity of EtG in hair as a marker of chronic excessive drinking: pooled analysis of raw data and *meta*-analysis of diagnostic accuracy studies. Ther Drug Monit 2014;36(5):560−75.
[63] Stewart SH, Koch DG, Burgess DM, Willner IR, et al. Sensitivity and specificity of urinary ethyl glucuronide and ethyl sulfate in liver disease patients. Alcohol Clin Exp Res 2013;37(1):150−5.
[64] Albermann ME, Musshoff F, Doberentz E, Heese P, et al. Preliminary investigations on ethyl glucuronide and ethyl sulfate cutoffs for detecting alcohol consumption on the basis of an ingestion experiment and on data from withdrawal treatment. Int J Legal Med 2012;126(5):757−64.
[65] Laposata M. Fatty acid ethyl esters: short-term and long-term serum markers of ethanol intake. Clin Chem 1997;43(8 Pt 2):1527−34.
[66] Best CA, Laposata M. Fatty acid ethyl esters: toxic non-oxidative metabolites of ethanol and markers of ethanol intake. Front Biosci 2003;8: e202−17.
[67] Pragst F, Rothe M, Moench B, Hastedt M, et al. Combined use of fatty acid ethyl esters and ethyl glucuronide in hair for diagnosis of alcohol abuse: interpretation and advantages. Forensic Sci Int 2010;196(1−3):101−10.
[68] Suesse S, Pragst F, Mieczkowski T, Selavka CM, et al. Practical experiences in application of hair fatty acid ethyl esters and ethyl glucuronide for detection of chronic alcohol abuse in forensic cases. Forensic Sci Int 2012;218(1−3):82−91.
[69] Pragst F, Balikova MA. State of the art in hair analysis for detection of drug and alcohol abuse. Clin Chim Acta 2006;370(1−2):17−49.
[70] Wurst FM, Yegles M, Alling C, Aradottir S, et al. Measurement of direct ethanol metabolites in a case of a former driving under the influence (DUI) of alcohol offender, now claiming abstinence. Int J Legal Med 2008;122(3):235−9.
[71] Pragst F, Yegles M. Determination of fatty acid ethyl esters (FAEE) and ethyl glucuronide (EtG) in hair: a promising way for retrospective detection of alcohol abuse during pregnancy? Ther Drug Monit 2008;30(2):255−63.
[72] Bertol E, Del Bravo E, Vaiano F, Mari F, et al. Fatty acid ethyl esters in hair: correlation with self-reported ethanol intake in 160 subjects and influence of estroprogestin therapy. Drug Test Anal 2014;6(9):930−5.
[73] Wurst FM, Alexson S, Wolfersdorf M, Bechtel G, et al. Concentration of fatty acid ethyl esters in hair of alcoholics: comparison to other biological state markers and self reported-ethanol intake. Alcohol Alcohol 2004;39(1):33−8.
[74] Dumitrascu C, Paul R, Kingston R, Williams R. Influence of alcohol containing and alcohol free cosmetics on FAEE concentrations in hair. A performance evaluation of ethyl palmitate as sole marker, vs the sum of four FAEEs. Forensic Sci Int 2018;283:29−34.
[75] Pragst F, Suesse S, Salomone A, Vincenti M, et al. Commentary on current changes of the SoHT 2016 consensus on alcohol markers in hair and further background information. Forensic Sci Int 2017;278:326−33.
[76] Gareri J, Appenzeller B, Walasek P, Koren G. Impact of hair-care products on FAEE hair concentrations in substance abuse monitoring. Anal Bioanal Chem 2011;400(1):183−8.
[77] Joya X, Friguls B, Ortigosa S, Papaseit E, et al. Determination of maternal-fetal biomarkers of prenatal exposure to ethanol: a review. J Pharm Biomed Anal 2012;69:209−22.
[78] Morini L, Marchei E, Tarani L, Trivelli M, et al. Testing ethylglucuronide in maternal hair and nails for the assessment of fetal exposure to alcohol: comparison with meconium testing. Ther Drug Monit 2013;35(3):402−7.
[79] Gentilello LM, Ebel BE, Wickizer TM, Salkever DS, et al. Alcohol interventions for trauma patients treated in emergency departments and hospitals: a cost benefit analysis. Ann Surg 2005;241(4):541.
[80] Andresen-Streichert H, Muller A, Glahn A, Skopp G, et al. Alcohol biomarkers in clinical and forensic contexts. Dtsch Arztebl Int 2018;115(18):309−15.
[81] Kummer N, Lambert WE, Samyn N, Stove CP. Alternative sampling strategies for the assessment of alcohol intake of living persons. Clin Biochem 2016;49(13−14):1078−91.
[82] Armer JM, Gunawardana L, Allcock RL. The performance of alcohol markers including ethyl glucuronide and ethyl sulphate to detect alcohol use in clients in a community alcohol treatment programme. Alcohol Alcohol 2017;52(1):29−34.
[83] Ropero-Miller J, Winecker R. Alcoholism, in clinical chemistry theory. In: Kaplan L, Pesce A, Kazmierczak S, editors. analysis, correlation. St Louis: Mosby; 2003.
[84] Mendenhall CL. Alcoholic hepatitis. Clin Gastroenterol 1981;10(2):417−41.
[85] Whitfield JB, Hensley WJ, Bryden D, Gallagher H. Some laboratory correlates of drinking habits. Ann Clin Biochem 1978;15(6):297−303.
[86] Karkkainen P, Salaspuro M. Beta-hexosaminidase in the detection of alcoholism and heavy drinking. Alcohol Alcohol Suppl 1991;1:459−64.
[87] Hultberg B, Isaksson A, Berglund M, Alling C. Increases and time-course variations in beta-hexosaminidase isoenzyme B and carbohydrate-deficient transferrin in serum from alcoholics are similar. Alcohol Clin Exp Res 1995;19(2):452−6.
[88] Javors MA, Johnson BA. Current status of carbohydrate deficient transferrin, total serum sialic acid, sialic acid index of apolipoprotein J and serum beta-hexosaminidase as markers for alcohol consumption. Addiction 2003;98(Suppl 2):45−50.
[89] Wehr H, Habrat B, Czartoryska B, Gorska D, et al. [Urinary beta-hexosaminidase activity as a marker for the monitoring of sobriety]. Psychiatr Pol 1995;29(5):689−96.

[90] Koivisto H, Hietala J, Anttila P, Parkkila S, et al. Long-term ethanol consumption and macrocytosis: diagnostic and pathogenic implications. J Lab Clin Med 2006;147(4):191–6.

[91] Savage DG, Ogundipe A, Allen RH, Stabler SP, et al. Etiology and diagnostic evaluation of macrocytosis. Am J Med Sci 2000;319(6):343–52.

[92] Wymer A, Becker DM. Recognition and evaluation of red blood cell macrocytosis in the primary care setting. J Gen Intern Med 1990;5(3):192–7.

[93] Skinner HA, Holt S, Sheu WJ, Israel Y. Clinical vs laboratory detection of alcohol abuse: the alcohol clinical index. Br Med J (Clin Res Ed) 1986;292(6537):1703–8.

[94] Aradottir S, Asanovska G, Gjerss S, Hansson P, et al. PHosphatidylethanol (PEth) concentrations in blood are correlated to reported alcohol intake in alcohol-dependent patients. Alcohol Alcohol 2006;41(4):431–7.

[95] Hartmann S, Aradottir S, Graf M, Wiesbeck G, et al. Phosphatidylethanol as a sensitive and specific biomarker: comparison with gamma-glutamyl transpeptidase, mean corpuscular volume and carbohydrate-deficient transferrin. Addict Biol 2007;12(1):81–4.

[96] Anton RF, Moak DH. Carbohydrate-deficient transferrin and gamma-glutamyltransferase as markers of heavy alcohol consumption: gender differences. Alcohol Clin Exp Res 1994;18(3):747–54.

[97] Hietala J, Puukka K, Koivisto H, Anttila P, et al. Serum gamma-glutamyl transferase in alcoholics, moderate drinkers and abstainers: effect on gt reference intervals at population level. Alcohol Alcohol 2005;40(6):511–14.

[98] Conigrave KM, Davies P, Haber P, Whitfield JB. Traditional markers of excessive alcohol use. Addiction 2003;98(Suppl 2):31–43.

[99] Puukka K, Hietala J, Koivisto H, Anttila P, et al. Age-related changes on serum ggt activity and the assessment of ethanol intake. Alcohol Alcohol 2006;41(5):522–7.

[100] Puukka K, Hietala J, Koivisto H, Anttila P, et al. Additive effects of moderate drinking and obesity on serum gamma-glutamyl transferase activity. Am J Clin Nutr 2006;83(6):1351–4 quiz 1448–9.

[101] Arndt T. Carbohydrate-deficient transferrin as a marker of chronic alcohol abuse: a critical review of preanalysis, analysis, and interpretation. Clin Chem 2001;47(1):13–27.

[102] Sillanaukee P, Olsson U. Improved diagnostic classification of alcohol abusers by combining carbohydrate-deficient transferrin and gamma-glutamyltransferase. Clin Chem 2001;47(4):681–5.

[103] Bortolotti F, Tagliaro F, Cittadini F, Gottardo R, et al. Determination of CDT, a marker of chronic alcohol abuse, for driving license issuing: immunoassay vs capillary electrophoresis. Forensic Sci Int 2002;128(1–2):53–8.

[104] Bortolotti F, Micciolo R, Canal L, Tagliaro F. First objective association between elevated carbohydrate-deficient transferrin concentrations and alcohol-related traffic accidents. Alcohol Clin Exp Res 2015;39(11):2108–14.

[105] Legros FJ, Nuyens V, Baudoux M, Zouaoui Boudjeltia K, et al. Use of capillary zone electrophoresis for differentiating excessive from moderate alcohol consumption. Clin Chem 2003;49(3):440–9.

[106] Hock B, Schwarz M, Domke I, Grunert VP, et al. Validity of carbohydrate-deficient transferrin (%CDT), gamma-glutamyltransferase (gamma-GT) and mean corpuscular erythrocyte volume (MCV) as biomarkers for chronic alcohol abuse: a study in patients with alcohol dependence and liver disorders of non-alcoholic and alcoholic origin. Addiction 2005;100(10):1477–86.

[107] Gong M, Castillo L, Redman RS, Garige M, et al. Down-regulation of liver Galbeta1, 4Glcnac alpha2, 6-sialyltransferase gene by ethanol significantly correlates with alcoholic steatosis in humans. Metabolism 2008;57(12):1663–8.

[108] Wurst FM, Thon N, Weinmann W, Tippetts S, et al. Characterization of sialic acid index of plasma apolipoprotein J and phosphatidylethanol during alcohol detoxification—a pilot study. Alcohol Clin Exp Res 2012;36(2):251–7.

[109] Ghosh P, Hale EA, Lakshman MR. Plasma sialic-acid index of apolipoprotein J (SIJ): a new alcohol intake marker. Alcohol 2001;25(3):173–9.

[110] Sillanaukee P, Ponnio M, Seppa K. Sialic acid: new potential marker of alcohol abuse. Alcohol Clin Exp Res 1999;23(6):1039–43.

[111] Ghosh P, Hale EA, Lakshman R. Long-term ethanol exposure alters the sialylation index of plasma apolipoprotein J (Apo J) in rats. Alcohol Clin Exp Res 1999;23(4):720–5.

[112] Pidd K, Roche AM. How effective is drug testing as a workplace safety strategy? A systematic review of the evidence. Accid Anal Prev 2014;71:154–65.

[113] Cashman CM, Ruotsalainen JH, Greiner BA, Beirne PV, et al. Alcohol and drug screening of occupational drivers for preventing injury. Cochrane Database Syst Rev 2009;2:CD006566.

[114] Schulte B, O'Donnell AJ, Kastner S, Schmidt CS, et al. Alcohol screening and brief intervention in workplace settings and social services: a comparison of literature. Front Psychiatry 2014;5:131.

[115] Berner MM, Kriston L, Bentele M, Harter M. The alcohol use disorders identification test for detecting at-risk drinking: a systematic review and meta-analysis. J Stud Alcohol Drugs 2007;68(3):461–73.

[116] Hermansson U, Helander A, Huss A, Brandt L, et al. The alcohol use disorders identification test (AUDIT) and carbohydrate-deficient transferrin (CDT) in a routine workplace health examination. Alcohol Clin Exp Res 2000;24(2):180–7.

[117] Hasin DS, Kerridge BT, Saha TD, Huang B, et al. Prevalence and correlates of DSM-5 cannabis use disorder, 2012–2013: findings from the national epidemiologic survey on alcohol and related conditions—III. Am J Psychiatry 2016;173(6):588–99.

[118] Vitesnikova J, Dinh M, Leonard E, Boufous S, et al. Use of AUDIT-C as a tool to identify hazardous alcohol consumption in admitted trauma patients. Injury 2014;45(9):1440–4.

[119] Wade D, Varker T, Forbes D, O'Donnell M. The alcohol use disorders tdentification test-consumption (AUDIT-C) in the assessment of alcohol use disorders among acute injury patients. Alcohol Clin Exp Res 2014;38(1):294–9.

[120] Fujii H, Nishimoto N, Yamaguchi S, Kurai O, et al. The alcohol use disorders identification test for consumption (AUDIT-C) is more useful than pre-existing laboratory tests for predicting hazardous drinking: a cross-sectional study. BMC Public Health 2016;16:379.

[121] Dillie KS, Mundt M, French MT, Fleming MF. Cost-benefit analysis of a new alcohol biomarker, carbohydrate deficient transferrin, in a chronic illness primary care sample. Alcohol Clin Exp Res 2005;29(11):2008–14.

[122] Miller PM, Thomas SE, Mallin R. Patient attitudes towards self-report and biomarker alcohol screening by primary care physicians. Alcohol Alcohol 2006;41(3):306–10.

[123] Chang G. Screening for alcohol and drug use during pregnancy. Obstet Gynecol Clin North Am 2014;41(2):205–12.

[124] Zizzo N, Di Pietro N, Green C, Reynolds J, et al. Comments and reflections on ethics in screening for biomarkers of prenatal alcohol exposure. Alcohol Clin Exp Res 2013;37(9):1451–5.

Chapter 4

Genetic Markers Related to Alcohol Use and Abuse

Joshua A. Bornhorst[1] and Gwendolyn McMillin[1,2]
[1]*Department of Laboratory Medicine and Pathology, Mayo Clinic, Rochester, MN, United States,* [2]*ARUP Laboratory and Department of Pathology, University of Utah, Salt Lake City, UT, United States*

INTRODUCTION

Ethanol (alcohol) use has been an important aspect of many human societies since the beginning of recorded history. However, the excessive use of alcohol is associated with many negative consequences, contributing to increased medical costs, birth defects, lost productivity, and violent crime. In the United States approximately 150 billion dollars per year is spent in alcohol-related accidents and illness [1,2]. There is an evidence that the contribution of alcohol abuse to the global disease burden is larger than that of any other controlled or illicit substances, with the notable exclusion of tobacco [3,4]. It is believed that 3.2%–5.9% of global deaths and as much as 4%–5% of overall global disease can be attributed to alcohol consumption [3,5,6].

A great deal of work has gone into elucidating genetic markers for susceptibility of an individual for alcohol abuse. Although trait markers are generally in research and developmental stages, there are many biochemical markers of alcohol abuse that have been well established [7–9]. Increasingly, there is increasing interest in clinically useful genetic markers [10,11]. This chapter explores genetic markers that could be employed to identify susceptibility to alcohol dependence and to improve therapeutic efforts (see Table 4.1). Alcohol consumption is not always associated with the negative health and welfare impacts, and it should be noted that often consumption may not result in adverse consequences. However, consumption patterns can be linked to disease especially for chronic disease states [6,14].

The definition of "excessive use" of alcohol varies. Greater than 40 g a day of ethanol for men and 20 g a day for women is considered "hazardous" or "harmful" [15]. An acute intake of more than 5–7 drinks in males and 3–5 drinks (assuming 10 g of ethanol per drink) is also considered harmful [16]. Differences in alcohol abuse prevalence are observed for different ages, racial backgrounds, gender, and socioeconomic classes. Alcohol abuse prevalence is generally highest among young white males and unmarried individuals [17,18]. Per capita consumption steadily increased from 1970 to 2007 [19]. Furthermore, between 2006 and 2014 the number of alcohol-related emergency department visits increased 62% [20].

The risk of other adverse health effects is well known to increase with heavy and regular alcohol consumption [16]. Alcohol abuse is known to be directly associated with a number of different medical conditions (Table 4.1), including hepatic, cardiovascular, psychiatric, and neurological disorders [12,15,21,22]. It also should be noted that some studies have shown that drinkers of <30 g of alcohol per day may have a lower mortality rate than those who abstain from drinking altogether [16,23,24]. The World Health Organization estimates that alcohol abuse affects 76.3 million people globally, and has identified reducing alcohol utilization as a priority area in international public health [14].

Alcohol abuse is not, by definition, alcohol dependency or alcoholism. Many environmental, clinical, and genetic factors contribute to the development of alcoholism (see Fig. 4.1). Clinical alcohol dependency is considered as a chronic disease for which diagnosis is defined by *Diagnostic and Statistical Manual of Mental Disorders* (DSM), Cloninger type I–II, and Feighner I–II criteria. Using the diagnostic criteria, as specified in DSM-IV, 19.3 million Americans required treatment for alcohol abuse or dependence in 2007 [25]. Briefly, the DSM-IV differentiates abuse from dependence on the frequency of events resulting in impairment or distress within specific domains of everyday

TABLE 4.1 Selected Health Risks and the Estimated Percent of Worldwide Disease Burden Related to Alcohol Consumption

Disorder	Percent of Disease Burden Related to Alcohol (%)
Cirrhosis of the liver	32
Cancer of mouth and oropharynx	19
Esophageal cancer	29
Liver carcinoma	25
Breast cancer	7
Depression	2
Epilepsy	12
Ischemic heart disease	2
Hemorrhagic stroke	10

Source: Adapted from Room R, Babor T, Rehm J. Alcohol and public health. Lancet 2005;365 (9458):519–530; Rehm J, Room R, Graham K, Monteiro M, et al. The relationship of average volume of alcohol consumption and patterns of drinking to burden of disease: an overview. Addiction 2003;98 (9):1209–1228 and "WHO World Health Report 2002: reducing risks, promoting healthy life," Geneva, World Health Organization.

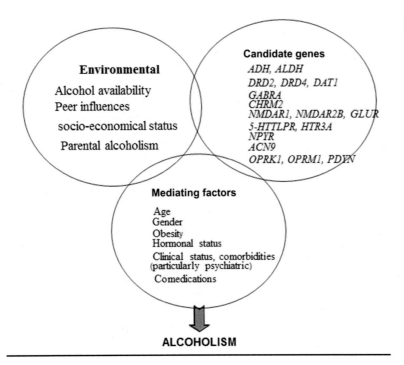

FIGURE 4.1 Factors contributing to alcoholism.

life and the DSM-V presented additional recommendations [26]. Diagnosis of dependence requires three or more events within a 12-month period and evidence of tolerance, withdrawal, or relapse [25,27].

In the United States the prevalence of alcohol dependence is 5.4% in males and 2.3% in females. By way of comparison, only 0.6% of US adults exhibit drug dependence on at a given time, although their lifetime drug dependence prevalence is 10.3% [28,29], yet alcoholism frequently coexists with other addictions, including illicit substance abuse and nicotine dependence more often than be expected by chance [30]. It is estimated that as many as 18% of all adults exhibit alcohol abuse over the course of their lifetime and 12.5% will be dependent on alcohol at some point in their lives. However, alcohol dependency is under-diagnosed, and it is estimated that only about one-tenth of all individuals who exhibit alcohol dependence receive treatment [18,31]. Interestingly, considerable overlap has been recognized between the incidence of alcohol dependency and mental illness. For example, susceptibility to alcoholism may be linked to antisocial personality disorder, conduct disorder, and personality characteristics such as impulsivity, aggressiveness, thrill-seeking, and rebelliousness. Efforts continue to further elucidate the biological changes in specific brain regions associated with the dependence or addiction process [32]. While deemed the most reliable observational criteria for diagnosis of alcohol abuse or dependence, the DSM criteria are often highly subjective.

ROLE OF LABORATORY TESTING

There is a clear need for laboratory testing to identify alcoholism, alcohol abuse, and susceptibility to alcoholism. In addition to biochemical changes or markers associated with alcohol use [33], genetic testing could potentially be used to identify individuals who are predisposed to developing alcoholism or who have clinically relevant abnormalities of ethanol metabolism. A number of the candidate genes and polymorphisms have been identified based on their involvement in alcohol metabolism or the neuronal response to alcohol. Evaluation of variants of these genes and the development of genetic testing may also be useful from a pharmacogenetic perspective, in guiding drug and dose selection for detoxification and abstinence therapy.

Alcohol dependence is a complex psychiatric disorder affected by both genetic and environmental influences. Although alcoholism does not show a clear pattern of direct Mendelian inheritance, studies have shown that the genetic contribution to alcohol use disorder etiology is considerable [34–40]. Controversially, while some studies indicate that the prevalence of alcohol-associated disorders varies among ethnic minorities, other studies indicate no difference in total alcohol overall consumption [41–43].

A number of policy groups have recommended the implementation of alcohol screening and brief intervention strategies in routine health care settings in which it can reduce health care costs, costs associated with emergency room visits and result in a lower overall disease burden [41,43–46]. Thus it appears beneficial to have genetic testing to identify individuals who are predisposed to or otherwise at elevated risk for developing alcoholism. Candidate genes are identified based on their involvement in alcohol metabolism (see Figs. 4.1 and 4.2) such as alcohol dehydrogenase (ADH) and aldehyde dehydrogenase (ALDH), as well as those involved in neuronal response to alcohol such as gamma-aminobutyric acid (GABA) receptor subunits, and serotonin (5-hydroxytryptamine, 5-HT) transporters. These and other genetic tests may also be useful from a pharmacogenetic perspective, in guiding drug and dose selection for detoxification and abstinence therapy.

This overview of the genetic markers that may be employed to identify susceptibility to alcohol dependence, to select therapeutic options to treat abuse, and to monitor therapeutic compliance.

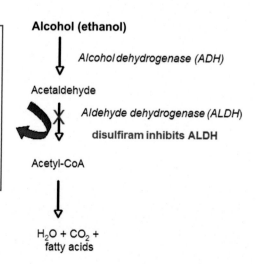

FIGURE 4.2 Schematic of alcohol metabolism.

GENETIC MARKERS OF ALCOHOL DEPENDENCE

As indicated previously, alcohol predisposition and dependence is due to a complex blend and interaction of environmental–behavioral factors and genetic determinants (Fig. 4.1). Social, cultural, and biological factors all contribute to the differences in alcohol use and abuse. An important means for identification of individuals who are at risk for alcoholism may include genetic testing [47]. Based on analysis of large, well-characterized cohorts of twins (nearly 10,000 twin pairs), alcoholism is a moderately to highly heritable psychiatric disease, with heritability of more than 0.5, and about equally across genders [35,48]. Worldwide studies based on diverse technologies, linkage of human polymorphic markers, genetic association studies, and investigation of candidate genes, are accumulating to identify specific genes involved in the development of this disorder. Advances in molecular genetic profiling technology are changing from single-locus methodologies designed to detect single nucleotide polymorphisms (SNPs) to methods designed to detect many SNPs, multigene testing, and even genome-wide analysis. Therefore it is important to identify genes that predispose individuals from diverse populations to alcoholism, genes that alter treatment response, and genes that interact with other environmental factors. In this chapter an overview of the best characterized genetic markers is provided. Original research articles and more comprehensive reviews of specific details describing genetics of alcoholism should be consulted for further information [10,11,49–56]. Potential genetic indicators of alcohol abuse and susceptibility are listed in Table 4.2.

Alcohol and Aldehyde Dehydrogenases

Metabolism of ethanol is a two-stage process of elimination: ADH oxidizes ethanol to acetaldehyde, a toxic intermediate, which is in turn converted to acetate, primarily by the mitochondrial form of ALDH2. A simplified schematic of this pathway is shown in Fig. 4.2. Acetaldehyde is a toxic metabolite which may be involved in the addiction process [57].

ADH exists as a polygene family on chromosome 4, divided into class I and class II genes. Approximately 70% of alcohol metabolism is mediated by isozymes expressed in the liver by class I genes: *ADH1A*, *ADH1B* (previously known as *ADH2*), and *ADH1C*. Clinically significant polymorphisms with altered affinity for ethanol and altered metabolic activity are described, the best studied of which include the *ADH1B*2* (His47Arg) and the *ADH1B*3* (Arg369Cys). These alleles are associated with low affinity for alcohol, but high activity (> 30-fold higher than normal) and rapid conversion of ethanol to acetaldehyde. The *ADH1B*2* is common in Asian populations (allele frequency of approximately 0.9) and is associated with facial flushing due to the accumulation of acetaldehyde. This allele is also associated with resistance to alcoholism; the allele frequency was found to be 0.73 in nonalcoholics and 0.48 in

TABLE 4.2 Potential Genetic Indicators of Alcohol Abuse and Susceptibility

Observational Markers
DSM-IV and V diagnostic criteria
Observation of the use of other substances

Some General Genetic Markers of Alcohol Abuse Susceptibility
Alcohol and aldehyde dehydrogenases
Neurotransmitter systems
Dopamine (DA) associated markers
GABA
ACh
Glutamate
Serotonin
NPY
ACN9
Opioids

Potential Methods for Identification of Markers and evaluation of Alcohol Abuse Susceptibility
Transcriptomics
Genomics
Metabolomics
Proteomics
Epigenomics
Next generation whole-genome sequencing

alcoholics. The *ADH1B*3* allele is most common in Native American and African populations [58–63]. Several of the genes that encode ADH and ALDH enzymes exhibit functional polymorphisms that result in interindividual variation in ethanol metabolism capacity. Genome-wide association studies have linked the number of variants to alcohol phenotypes including relative risk of alcoholism [64–67].

At least 19 putative genes and several pseudogenes are included in the ALDH gene superfamily, yet just *ALDH1* (chromosome 9, cytosolic) and *ALDH2* (chromosome 12, mitochondrial) are associated specifically with acetaldehyde oxidation. The *ALDH2*2* (Glu487Lys), common to several ethnic populations, leads to reduced activity and like the *ADH1B*2*, it is associated with accumulation of acetaldehyde following alcohol intake. Promoter variants in *ALDH1* (*ALDH1A1*2*, a 17 bp deletion, and *ALDH1A1*3*, a 3 bp insertion) are associated with reduced expression of ALDH. All of these ALDH alleles are associated with resistance to alcoholism. People who inherit the combination of a rapid metabolizing ADH allele combined with the impaired metabolizing ALDH alleles had particularly reduced risk for alcoholism. Also of interest is the drug disulfiram which is used in prevention of alcoholism relapse. This drug acts to mimic this phenotype through inhibition of ALDH activity [68,69].

Kuo et al. [59] tested polymorphisms in seven *ADH* genes, along with the *ALDH1A1* and *ALDH2* genes, for association with alcohol dependence. In this study, numerous SNPs in the *ADH* gene were associated with alcohol dependence. Another large case-control study found associations between AD and *ADH5* genotypes; also, diplotypes of *ADH1A*, *ADH1B*, *ADH7*, and *ALDH2* were linked to AD in European- or African-Americans [59]. The *ALDH2*2* (Glu487Lys) polymorphism located in the ADLH2 variant enzyme is prevalent in many Asian ethnic groups and produces acetaldehyde with an associated alcohol-flushing response reaction similar to that observed in the *ADH1B*2* allele (also see http://www.aldh.org) [70–72]. The clinical significance of these and additional *ALDH* genes relative to alcohol use, intolerance, and abuse remains to be defined.

Neurotransmitter Systems

The role of neurotransmitters in the brain is to either stimulate or inhibit the flow of impulse between neurons. Many neurotransmitter systems are involved in the positive, reinforcing, and aversive effects of alcohol. Here the potential involvement of genes involved in dopamine (DA), GABA, cholinergic, glutamate, serotonin, neuropeptide Y (NPY), ACN9, and opioid-mediated neurotransmission is described. Most emphasis is placed on the dopaminergic system, as it has been most extensively studied relative to alcoholism till date. Note that for each system, genes associated with the production, transport, storage, and destruction of neurotransmitters may be relevant, in addition to the genes coding for receptors and signal transduction mechanisms associated with actual neurotransmission. The cumulative and/or compensatory effect(s) of individual genetic variants is not well understood, but here association and potential consequences of affected genes are briefly discussed for selected aspects of each major neurotransmitter system [73].

Dopamine

The DA system is one of the major contributors to the development of anxiety and depression. This neurotransmitter system is widely associated with rewarding, euphoric response to many drugs of abuse, including alcohol, and therefore may contribute to alcoholism predisposition and vulnerability.

DA receptors fall within a class of metabotropic G protein–coupled receptors. There are five subtypes of DA receptors: D_1, D_2, D_3, D_4, and D_5. The D_2 regulates DA synthesis and releases in the presynaptic region, and has been the most studied in relation to alcohol-related phenotypes. This association is based on the thought that the rewarding effects of alcohol are mediated through the mesolimbic DA system [74], called also dopaminergic reward system [75]. In particular, the *TaqI*-A1 polymorphism in the *DRD2* gene has been the subject of numerous association studies to explore the relationship between the A1 allele of *DRD2* with alcohol consumption and dependence, with somewhat conflicting results. Some of the studies report a significantly higher frequency of the A1 allele in alcoholic compared with nonalcoholic populations, suggesting an increased susceptibility to alcohol dependence with this allele [76–79]. A large *meta*-analysis (44 studies with 5273 cases and 3995 controls) considering genotyping data of the *Taq1* polymorphism showed a small but significant association of alcohol dependency [80] in individuals with homozygote or heterozygote for the A1 allele. Other studies have been less supportive of this association [81–83] leaving the role of this polymorphism in susceptibility for alcohol dependence inconclusive. However, these inconsistent results may be explained by the recent discovery that the *TaqI*-A1 polymorphism is in fact located 10 kb downstream from the *DRD2* gene and causes an amino acid substitution in one of the ankyrin repeats in kinase domain 1 of the *ANKK1* gene [84]. This data suggests that the *TaqI*-A1 polymorphism, in combination with polymorphisms of the *ANKK1* and/or *DRD2* could be epistatically associated with psychopathic traits in alcohol-dependent patients [85,86]. Further study will be required to understand this relationship.

The dopamine D_4 receptor (DRD4) is a G protein—coupled receptor with a D_2-like activity, in addition to an inhibitory activity of the adenyl cyclase [87]. A 48-base (*DRD4 E8* 48-bp) of variable number of tandem repeats (VNTR) on exon 3, ranging from 2 to 11 repeats [88], is the most studied polymorphism; 2—5 repeats define the short, and 6—10 repeats define the long allele [89]. The 7-repeat variant of these polymorphisms reduces the ability of the DA to inhibit the cyclic AMP. The "long allele" has been associated by some studies with a "novelty seeking" [90] trait that is often present in the alcoholic phenotype.

The dopamine transporter (SLC6A3) is a membrane-spanning protein that binds to DA terminating the DA signal by removing the neurotransmitter from the synapses into a neuron. A VNTR polymorphism in the 3′ untranslated region of the DA transporter gene (*DAT1*) has been shown to affect the expression of the transporter [91]. Several case-control studies and family-based association studies have reported conflicting associations with the *DAT1* VNTR polymorphism and alcohol-related phenotypes; however, most of these studies had very small sample sizes [92—98]. Elevated frequency of the 7-repeat and a decreased frequency of the 9-allele were found in Japanese alcoholics compared to control subjects [93].

Catechol-*O*-methyltransferase (COMT), dopamine-β-hydroxylase (DβH) and monoamine oxidase are the major enzymes catalyzing metabolism of DA, norepinephrine, and other catecholamines. Genetic polymorphisms in these genes have been described and used as markers for these loci sites. The role of the corresponding genes in alcohol use and dependence is still controversial [99—103]. The COMT gene is found on chromosome 21q11 and contains the functional polymorphism $Val_{158}Met$ in soluble COMT (*S*-COMT) or $Val_{108}Met$ in membrane-bound COMT (*MB*-COMT) [104—106]. The $Val_{158/108}$ enzyme has more activity than the $Met_{158/108}$ form and is linked with decreased amounts of DA [35]. The $Met_{158/108}$ allele is associated with higher consumption of alcohol in men [107] and higher anxiety levels in women [108]. In the COMBINE Alcoholism Treatment Study, relationships between candidate loci and drug metabolism or alcohol addiction were explored [109,110]. In the COMT gene, the $Val_{158}Met$ polymorphism was associated with brain endogenous opioid function, response to stress and anxiety, and differences in cognitive and emotional processes.

Dopamine Metabolizing Enzymes

DβH catalyzes the conversion of DA to norepinephrine. The *DβH* *444 G > A polymorphism, was associated with alcoholism [111]. Two studies did not establish association between a *DβH*-1021 polymorphism and alcoholism [112,113].

GABA

GABA is the major inhibitory neurotransmitter in the central nervous system. $GABA_B$ receptors are a family of G protein—coupled receptors that stimulate the opening of potassium channels. The GABAergic system is thought responsible for producing sedation, anxiolytic, and muscular relaxant effects of ethanol as well as for exhibiting signs of withdrawal. The effects of GABA are mediated through two primary receptor types: $GABA_A$ and $GABA_B$. The $GABA_A$ receptor is a ligand-gated ion channel (modulates chloride conductance into neurons), and is composed of six different subunits, designed as α, β, γ, δ, ρ, and ε; $GABA_B$ receptors are coupled to G proteins. Precise composition of individual receptors affects response to GABA-agonists such as ethanol. Most of the $GABA_A$ genes are organized into clusters located on chromosomes 4, 5, and 15. Linkage and association studies consistently support the association between alcohol dependence and $GABA_A$ $\alpha2$, GABA $\beta1$, and GABA $\beta3$ genes [53,114,115]. Fewer studies have identified a role of $GABA_B$ receptors and alcohol dependence [116,117]. $GABA_C$ receptors, which differ in complexity of structure, abundance, distribution, and function from $GABA_A$ and $GABA_B$ receptors, can be found in retinal, hippocampus, spinal cord, and pituitary tissues [118,119]. The GABA system interacts with the DA system to reinforce the effects of alcohol. Pharmacological treatments that target the inhibitory and excitatory modulators of the CMDS may prevent relapse or reduce heavy drinking have been proposed [118,120]. Clinical trials on the anticonvulsant drug topiramate, which facilitating GABA function through interaction with the $GABA_A$ receptor, have significantly reduced drinking in alcohol-dependent subjects [118,120].

Acetylcholine

Acetylcholine (ACh) is a neurotransmitter found in the brain and autonomic nervous system; it interacts with muscarinic and nicotinic receptors. A region on chromosome 7, including 11 SNPs, within and flanking the gene encoding for the muscarine ACh receptor subtype M2 (*CHRM2*) was significantly associated by linkage with alcohol dependence [121–123]. Several genetic studies have implied that these genes may contribute to alcohol and nicotine coaddiction [124–127]. Chatterjee et al. suggest that FDA-approved ACh receptor ligands, such as varenicline and mecamylamine, may be useful in the treatment of alcohol abuse as well as smoking cessation [128].

Glutamate

Glutamate is the primary excitatory neurotransmitter in the central nervous system and ethanol potently and selectively inhibits glutamate receptors. One receptor, a calcium-conducting ligand-gated glutamate receptor known as the *N*-methyl-D-aspartate (NMDA) receptor, is thought to be particularly involved in development of alcoholism. The inhibitory effect of ethanol is overcome with prolonged ethanol exposure, by way of compensatory "upregulation" of NMDA receptor expression. These changes are believed to underlie the development of ethanol tolerance and dependence as well as acute and delayed signs of withdrawal, particularly agitation and seizures. Allelic variants of the *NMDAR1* and the *NMDAR2B* receptor genes were associated with alcoholism and related traits [129], but not consistently in other studies [130]. Withdrawal from chronic alcohol usage lead to increased activity in the NMDA receptor and increased influx of calcium may be attributed to neurotoxicity and neuronal cell death. Topiramate, which can decrease the release of DA, has shown effectiveness in the management of alcohol dependence [131].

Serotonin

Genes involved in the regulation of the serotonin (5-HT) system provide plausible candidate genes for involvement in alcohol consumption and abuse and dependence [132]. The serotonin transporter is responsible for serotonin reuptake and is a key regulator of serotonin availability in the synaptic cleft. A variant in the gene that codes for the transporter (*5HTT*, on chromosome 17), the *5-HTTLPR* polymorphism, is associated with alcohol consumption. The serotonin transporter (5-HTT) is postulated as a marker for alcohol dependency [133]. Chronic alcohol intake may exert neurotoxic effects upon neurotransmitter systems such as the serotonergic system [134]. The short–long alleles of *5-HTTLPR* result in either 14 (short, S) or 16 (long, L) copies of a 20–23 base pair repeated sequence and affect expression of the gene. The S variant is associated with reduced expression of the transporter protein and was correlated with an increased risk of alcohol dependency [69,70] and may also affect the risk of relapse in abstinent patients, but these associations are somewhat controversial [135,136].

Neuropeptide Y

Several lines of evidence in both human and animal studies suggest that variation in NPY or its receptor genes (*NPY1R*, *NPY2R*, and *NPY5R*) is associated with alcohol dependence as well as alcohol withdrawal symptoms. Reduced activity of the NPY system may be associated with increased alcohol intake [137]. Sequence variations in NPY receptor genes are associated with alcohol dependence, particularly a severe subtype of alcohol dependence characterized by withdrawal symptoms, comorbid alcohol, and cocaine dependence [138].

ACN9

The ACN9 protein is a mitochondrial protein that is involved in gluconeogenesis and metabolism of ethanol or acetate into carbohydrate [77]. Chromosome 7 that encodes ACN9 protein has shown consistent evidence of linkage with a variety of phenotypes related to alcohol dependence in the collaborative study on the genetics of alcoholism (COGA) project. Four of the eight most significant SNPs were located in or very near the *ACN9* gene which encodes for the ACN9 protein, a novel mitochondrial protein involved in gluconeogenesis and the assimilation of ethanol or acetate into carbohydrate [77].

Opioids

The opioid system has been implicated in association with alcoholism, particularly because opioid antagonists are used successfully to manage alcohol dependence. While no significant association has been made between the mu and delta opioid receptor genes and alcohol dependence, the kappa receptor gene (*OPRK1*) and a gene encoding for an opioid ligand, prodynorphin (*PDYN*), were strongly associated with a risk of alcohol dependence [54].

PHARMACOGENETICS OF ALCOHOL DEPENDENCE TREATMENT

Therapeutic approaches to alcohol abuse disorders are based in large part on acute detoxification, followed by long-term measures to prevent relapse. Indeed, addiction, risk of relapse, and remission are commonly compared in parallel to diagnosis and management of chronic disease. Several classes of drugs have been utilized for alleviation of alcohol withdrawal symptoms and dependency treatment [139]. The genetic components associated with alcohol abuse disorders, while not always defined specifically, remain strong, based on familial and twin studies as well as data described in the previous section. It is therefore logical that both the genetic factors responsible for disease and the genetic factors that may predispose therapeutic success or failure should be considered in selection of treatment modalities and management of rehabilitation.

Complimentary to traditional genetics, pharmacogenetics associates genetics with interindividual variability in the two major processes responsible for drug action: pharmacokinetics and pharmacodynamics. Thus, from the perspective of pharmacogenetics, variants in genes associated with either metabolism (pharmacokinetics) or response (pharmacodynamics) of any drug used to treat alcohol dependency could impact efficacy of the drug and optimal dose requirements. Variants in genes associated with pharmacokinetics such as the genes that code for cytochrome P450 enzymes, will impact the time required for drug clearance and time to achieve steady-state concentrations. As a consequence, drug dose and interval of dosing may influence efficacy. Genes associated with pharmacodynamics correlate with the sensitivity or resistance to a particular drug action and hence, drug and drug dose requirements. Pharmacogenetic markers may therefore become important in directing addiction treatment by personalizing both drug and dose selection prior to drug administration [140–143].

There are several classes of drugs applied to the treatment of alcohol abuse and dependence. Here, potential pharmacogenetic markers relative to alcohol detoxification (diazepam) and dependency treatment (disulfiram, naltrexone, acamprosate, ondansetron, topiramate, finasteride, dexmedetomidine, and bromocriptine) are discussed. Mechanism of action of disulfiram is shown in Fig. 4.2. Examples of drugs used for alcohol dependence therapy are listed in Table 4.3.

Alcohol Detoxification

Benzodiazepines are the most commonly used class of drugs to treat withdrawal symptoms associated with ethanol detoxification. While dosing may be titrated to clinical need, it may be important to consider pharmacogenetic testing that could avoid unintentional overdose and fail to respond. Mechanisms of metabolism for individual benzodiazepines should be investigated because some pathways are complex and active metabolites are common. In addition, the kinetics for elimination of each drug and active metabolite vary substantially. Using diazepam as an example, several active metabolites are formed, including nordiazepam, oxazepam, and temazepam with half-lives of approximately 20–40 h for diazepam and nordiazepam, and half-lives of approximately 4–12 h for oxazepam and temazepam. These metabolic reactions are catalyzed by cytochrome P450 (CYP) isozymes, particularly CYP2C19 and CYP3A4. These compounds are further metabolized by formation of glucuronide and other conjugates. Substantial variation in CYP and glucuronyltransferase genes has been associated with altered kinetics and risk of toxicity. Although the safety margin for benzodiazepines is good for acute administration, clinical impact of genetic variation in benzodiazepine metabolism could inadvertently contribute to toxicity if the drugs are administered frequently or for a long period of time [143].

The mechanism of action of benzodiazepines is primarily explained by affinity for the central $GABA_A$ receptor. As such, it is possible that impaired GABAergic function would prevent or impair the clinical sensitivity to benzodiazepine therapy. In the case of benzodiazepine resistance, treatment through a non-GABAergic mechanism may be required. Indeed, dexmedetomidine, a central alpha2-receptor agonist, has also been used successfully to treat alcohol withdrawal [144]. Negative modulators of NMDA glutamate receptors (e.g., acamprosate) may be useful agents for the treating withdrawal signs and symptoms [145,146].

TABLE 4.3 Pharmacogenetics of Treatment for Alcoholism

Generic Name	Trade Name	Indication for Use	Candidate Gene Related to Pharmacokinetics	Candidate Gene Related to Pharmacodynamics
Diazepam	Valium	Prevent withdrawal	CYP2C19	GABRA
			CYP3A4	
Disulfiram	Antabuse	Maintain abstinence	Many CYPs	ADH
Naltrexone	Depade	Maintain abstinence		OPRM1
Acamprosate	Campral	Maintain abstinence		GABRA
				NMDA
Ondansetron	Zofran	Experimental	CYP3A4	HTR3A
Topiramate	Topamax	Experimental		GABRA
				GLUR
Finasteride	Propecia	Experimental	CYP3A4	GABRA
Dexmedetomidine	Precedex	Experimental	CYP2A6	ADRA2A
Bromocriptine	Parlodel	Experimental	CYP3A4	DRD2

Alcohol Dependence Treatment

Many drugs, acting through different or unknown mechanisms have been successfully used to treat alcohol dependence by minimizing cravings, tapering drinking behavior, and preventing recurrence of dependency on alcohol and other substances. Long-term treatment for months to years is often required, so it is important that safety and efficacy of the drugs and dose selected are considered. Pharmacogenetic testing may assist in this selection process, by predicting drug sensitivity or resistance, and optimizing dose and dosing intervals.

Early intervention may benefit from administration of disulfiram, which inhibits ALDH and promotes accumulation of acetaldehyde, leading to an unpleasant reaction (Fig. 4.2). This negative reinforcement approach, while it does not itself prevent craving, may reduce drinking behavior. Genetic variations in the *ADH* and *ALDH* genes can affect response to the alcohol dependence treatment drug disulfiram. Perhaps not surprisingly, patient compliance with disulfiram administration is reported to be poor [147,148]. Pharmacogenetic testing for *ADH* or *ALDH* variants may be considered prior to administration of disulfiram, because persons with impaired ALDH may not respond to the drug [149].

Naltrexone is another well-accepted therapy for long-term prevention of alcoholism [150]. Naltrexone is an opioid receptor antagonist and has been shown most effective for persons who possess the *OPRM1* variants, particularly the Asn40 allele [151]. Craving and anxiety are treated effectively with bromocriptine, a DA antagonist, when the *TaqI*-A1 allele was present with *DRD2*. Acamprosate, which has effects at both opioid and glutamate receptors have also been used successfully to treat alcohol dependence, as might be expected based on speculation of these neurotransmitter system's involvements in the pathogenesis of alcoholism [152].

Finally, antiepileptic drugs, long known to inhibit excessive excitation in the central nervous system, have proven successful applied to dependency treatment, particularly in reducing impulsive behavior [151,153]. Examples of antiepileptic drugs used successfully to treat alcoholism include topiramate, gabapentin, and levetiracetam [154–156].

CONCLUSIONS

Advances in molecular profiling technologies within the disciplines of genomics, transcriptomics, proteomics, and metabolomics will likely play an essential role in the development of prevention strategies and personalized treatments of alcoholism, as well as the identification of new therapeutic targets. In addition to available biochemical markers, an increasing number of gene variant markers have been identified. Continued evolution in array and sequencing technologies, along with the continued establishment of large data and sample repositories are important to achieve progress in a complex disorder like alcoholism.

The identification of genes associated with alcohol dependence can be accomplished by several methods including candidate gene studies, whole-genome association, and linkage studies. In whole-genome studies, broad chromosomal regions are linked to a target phenotypic expression factor by means of a panel of polymorphisms [7,53]. One example of a family-based data set is the COGA that is targeted toward the identification of genes contributing to alcoholism [157,158]. A number of genome-wide association studies have been carried out which have supported many of the genotype–phenotype associations described earlier in this work [55].

Investigations will increasingly need to account for both genomic and epigenomic data [159]. Epigenetic studies often demonstrate substance-specific changes to gene expression that is linked to substance intake patterns [10]. These patterns are based on changes in chemical modifications of DNA and chromatin which alters expression of genetic material. Epigenetic changes can regulate gene expression without changing the DNA code itself and have been shown to be linked to the development of many diseases processes [160–162]. Alcohol intake is now well demonstrated to modify chromatin and appears to be involved in the maintenance of alcohol addiction [163]. This is an emerging and exciting frontier in our understanding and potential treatment of alcohol dependence.

Alcohol abuse and dependency is a tremendous social problem, contributing to disease, crime, and inappropriate behavior. Chemical biomarkers such as mean corpuscular volume, gamma-glutamyltransferase, carbohydrate deficient transferring, ethyl glucuronide, fatty acid ethyl esters, and other chemical biomarkers may provide evidence of alcohol use, abuse, and compliance with abstinence. However, genetic testing may provide additional tools for management of alcohol abuse by detecting individuals at risk for alcoholism. It is well recognized that the risk of developing alcoholism is closely related to both environmental and genetic factors. In fact, it is estimated that 50%–60% of the factors associated with alcoholism are genetic in origin. There are currently 10 tests available for alcohol dependency according to the national genetic test registry [164]. Genetic testing to identify individuals at risk for alcohol abuse may prevent alcohol use disorders, allow for early intervention, or direct therapy. Therapy of alcoholism may include medications that curb craving and anxiety, and may be optimized through the use of pharmacogenetic testing. Indeed, testing genes related to the metabolism of ethanol, the pathogenesis of alcoholism, and the genes linked to successful therapy may be employed to identify susceptibility to alcohol use disorders [13].

REFERENCES

[1] Garbutt JC, West SL, Carey TS, Lohr KN, et al. Pharmacological treatment of alcohol dependence: a review of the evidence. JAMA 1999;281(14):1318–25.
[2] Wiese JG, Shlipak MG, Browner WS. The alcohol hangover. Ann Intern Med 2000;132(11):897–902.
[3] Rehm J, Room R, Monteiro M, Gmel G, et al. Alcohol as a risk factor for global burden of disease. Eur Addict Res 2003;9(4):157–64.
[4] Room R, Graham K, Rehm J, Jernigan D, et al. Drinking and its burden in a global perspective: policy considerations and options. Eur Addict Res 2003;9(4):165–75.
[5] Stickel F, Moreno C, Hampe J, Morgan MY. The genetics of alcohol dependence and alcohol-related liver disease. J Hepatol 2017;66(1):195–211.
[6] World Health Organization. Global status report on alcohol and health-2014 e.G.W.H.O. 2014, Editor.: http://www.who.int/substance_abuse/publications/global_alcohol_report/en/.
[7] Bornhorst J, Stone A, Brown J, Light K. Slate and trait markers of alcohol abuse. In: Dasgupta A, Langman L, editors. Pharmacogenomics of alcohol abuse. CRC Press; 2012.
[8] Wozniak MK, Wiergowski M, Namiesnik J, Biziuk M. Biomarkers of alcohol consumption in body fluids—possibilities and limitations of application in toxicological analysis. Curr Med Chem 2017.
[9] Cabarcos P, Alvarez I, Tabernero MJ, Bermejo AM. Determination of direct alcohol markers: a review. Anal Bioanal Chem 2015;407(17):4907–25.
[10] Prom-Wormley EC, Ebejer J, Dick DM, Bowers MS. The genetic epidemiology of substance use disorder: a review. Drug Alcohol Depend 2017;180:241–59.
[11] Edenberg HJ, Foroud T. Genetics of alcoholism. Handb Clin Neurol 2014;125:561–71.
[12] Room R, Babor T, Rehm J. Alcohol and public health. Lancet 2005;365(9458):519–30.
[13] Rehm J, Room R, Graham K, Monteiro M, et al. The relationship of average volume of alcohol consumption and patterns of drinking to burden of disease: an overview. Addiction 2003;98(9):1209–28.
[14] WHO. Expert committee on problems related to alcohol consumption. Second report. World Health Organ Tech Rep Ser 2007;944:1–53 55–7, back cover.
[15] Das SK, Dhanya L, Vasudevan DM. Biomarkers of alcoholism: an updated review. Scand J Clin Lab Invest 2008;68(2):81–92.
[16] Niemela O. Biomarkers in alcoholism. Clin Chim Acta 2007;377(1–2):39–49.

[17] Delker E, Brown Q, Hasin DS. Alcohol consumption in demographic subpopulations: an epidemiologic overview. Alcohol Res 2016;38(1):7–15.

[18] Hasin DS, Stinson FS, Ogburn E, Grant BF. Prevalence, correlates, disability, and comorbidity of DSM-IV alcohol abuse and dependence in the United States: results from the National Epidemiologic Survey on Alcohol and Related Conditions. Arch Gen Psychiatry 2007;64(7):830–42.

[19] LaVallee, RA, GD Williams, and H Yi, Surveillance report #87: apparent per capita alcohol consumption: national, state, and regional trends: 1970–2007. Bethesda, MD: National Institute on Alcohol Abuse and Alcoholism, Division of Epidemiology and Prevention Research; 2009.

[20] White am, Slater ME, Ng G, Hingson R, et al. Trends in alcohol-related emergency department visits in the United States: results from the nationwide emergency department sample, 2006 to 2014. Alcohol: Clin Exp Res 2018;42(2):352–9.

[21] Lieber CS. Medical disorders of alcoholism. N Engl J Med 1995;333(16):1058–65.

[22] Cargiulo T. Understanding the health impact of alcohol dependence. Am J Health Syst Pharm 2007;64(5 Suppl. 3):S5–11.

[23] Corrao G, Bagnardi V, Zambon A, La Vecchia C. A *meta*-analysis of alcohol consumption and the risk of 15 diseases. Prev Med 2004;38(5):613–19.

[24] Goldberg DM, Hahn SE, Parkes JG. Beyond alcohol: beverage consumption and cardiovascular mortality. Clin Chim Acta 1995;237(1–2):155–87.

[25] The NSDUH report: alcohol treatment: need, utilization and barriers. Rockville, MD: Substance Abuse and Mental Health Services Administration, Office of Applied Studies; 2009.

[26] Hasin Deborah S, Kerridge Bradley T, Saha Tulshi D, Huang Boji, et al. Prevalence and correlates of DSM-5 cannabis use disorder, 2012–2013: findings from the National Epidemiologic Survey on Alcohol and Related Conditions—III. Am J Psychiatr 2016;173(6):588–99.

[27] American Psychiatric Association: diagnostic and statistical manual of mental disorders, fourth edition, text revision. Washington, DC: American Psychiatric Association; 2000.

[28] Compton WM, Thomas YF, Stinson FS, Grant BF. Prevalence, correlates, disability, and comorbidity of DSM-IV drug abuse and dependence in the United States: results from the national epidemiologic survey on alcohol and related conditions. Arch Gen Psychiatry 2007;64(5):566–76.

[29] Schulden JD, Thomas YF, Compton WM. Substance abuse in the United States: findings from recent epidemiologic studies. Curr Psychiatry Rep 2009;11(5):353–9.

[30] Kessler RC, Crum RM, Warner LA, Nelson CB, et al. Lifetime co-occurrence of DSM-III-R alcohol abuse and dependence with other psychiatric disorders in the National Comorbidity Survey. Arch Gen Psychiatry 1997;54(4):313–21.

[31] SAMHSA, Results from the 2009 national survey on drug use and health: volume II. Technical appendices and selected prevalence tables; 2010.

[32] Koob GF. Neurobiological substrates for the dark side of compulsivity in addiction. Neuropharmacology 2009;56(Suppl. 1):18–31.

[33] Wu AH, McKay C, Broussard LA, Hoffman RS, et al. National academy of clinical biochemistry laboratory medicine practice guidelines: recommendations for the use of laboratory tests to support poisoned patients who present to the emergency department. Clin Chem 2003;49(3):357–79.

[34] Nurnberger Jr. JI, Wiegand R, Bucholz K, O'Connor S, et al. A family study of alcohol dependence: coaggregation of multiple disorders in relatives of alcohol-dependent probands. Arch Gen Psychiatry 2004;61(12):1246–56.

[35] Goldman D, Oroszi G, Ducci F. The genetics of addictions: uncovering the genes. Nat Rev Genet 2005;6(7):521–32.

[36] Prescott CA, Kendler KS. Genetic and environmental contributions to alcohol abuse and dependence in a population-based sample of male twins. Am J Psychiatry 1999;156(1):34–40.

[37] Bienvenu OJ, Davydow DS, Kendler KS. Psychiatric 'diseases' vs behavioral disorders and degree of genetic influence. Psychol Med 2011;41(1):33–40.

[38] Strat YL, Ramoz N, Schumann G, Gorwood P. Molecular genetics of alcohol dependence and related endophenotypes. Curr Genomics 2008;9(7):444–51.

[39] Yang BZ, Kranzler HR, Zhao H, Gruen JR, et al. Association of haplotypic variants in DRD2, ANKK1, TTC12 and NCAM1 to alcohol dependence in independent case control and family samples. Hum Mol Genet 2007;16(23):2844–53.

[40] Stacey D, Clarke TK, Schumann G. The genetics of alcoholism. Curr Psychiatry Rep 2009;11(5):364–9.

[41] Schermer CR. Feasibility of alcohol screening and brief intervention. J Trauma 2005;59(3 Suppl.):S119–23 discussion S124–33.

[42] Galvan FH, Caetano R. Alcohol use and related problems among ethnic minorities in the United States. Alcohol Res Health 2003;27(1):87–94.

[43] Glanz J, Grant B, Monteiro M, Tabakoff B. WHO/ISBRA Study on State and Trait Markers of Alcohol Use and Dependence: analysis of demographic, behavioral, physiologic, and drinking variables that contribute to dependence and seeking treatment. International Society on Biomedical Research on Alcoholism. Alcohol Clin Exp Res 2002;26(7):1047–61.

[44] Hannuksela ML, Liisanantti MK, Nissinen AE, Savolainen MJ. Biochemical markers of alcoholism. Clin Chem Lab Med 2007;45(8):953–61.

[45] Kraemer KL. The cost-effectiveness and cost-benefit of screening and brief intervention for unhealthy alcohol use in medical settings. Subst Abus 2007;28(3):67–77.

[46] Gentilello LM, Ebel BE, Wickizer TM, Salkever DS, et al. Alcohol interventions for trauma patients treated in emergency departments and hospitals: a cost benefit analysis. Ann Surg 2005;241(4):541–50.

[47] Mayfield RD, Harris RA, Schuckit MA. Genetic factors influencing alcohol dependence. Br J Pharmacol 2008;154(2):275–87.

[48] Heath AC, Bucholz KK, Madden PA, Dinwiddie SH, et al. Genetic and environmental contributions to alcohol dependence risk in a national twin sample: consistency of findings in women and men. Psychol Med 1997;27(6):1381–96.

[49] Radel M, Goldman D. Pharmacogenetics of alcohol response and alcoholism: the interplay of genes and environmental factors in thresholds for alcoholism. Drug Metab Dispos 2001;29(4 Pt 2):489—94.
[50] Crabbe JC. Alcohol and genetics: new models. Am J Med Genet 2002;114(8):969—74.
[51] Dick DM, Foroud T. Candidate genes for alcohol dependence: a review of genetic evidence from human studies. Alcohol Clin Exp Res 2003;27(5):868—79.
[52] Dick DM, Bierut LJ. The genetics of alcohol dependence. Curr Psychiatry Rep 2006;8(2):151—7.
[53] Ducci F, Goldman D. Genetic approaches to addiction: genes and alcohol. Addiction 2008;103(9):1414—28.
[54] Kohnke MD. Approach to the genetics of alcoholism: a review based on pathophysiology. Biochem Pharmacol 2008;75(1):160—77.
[55] Stickel F, Datz C, Hampe J, Bataller R. Pathophysiology and management of alcoholic liver disease: update 2016. Gut Liver 2017;11(2):173—88.
[56] Available from: https://www.pharmgkb.org/chemical/PA448073/clinicalAnnotation.
[57] Zakhari S. Overview: how is alcohol metabolized by the body? Alcohol Res Health 2006;29(4):245—54.
[58] Li TK. Pharmacogenetics of responses to alcohol and genes that influence alcohol drinking. J Stud Alcohol 2000;61(1):5—12.
[59] Kuo PH, Kalsi G, Prescott CA, Hodgkinson CA, et al. Association of ADH and ALDH genes with alcohol dependence in the Irish Affected Sib Pair Study of alcohol dependence (IASPSAD) sample. Alcohol Clin Exp Res 2008;32(5):785—95.
[60] Williams JT, Begleiter H, Porjesz B, Edenberg HJ, et al. Joint multipoint linkage analysis of multivariate qualitative and quantitative traits. II. Alcoholism and event-related potentials. Am J Hum Genet 1999;65(4):1148—60.
[61] Sherva R, Rice JP, Neuman RJ, Rochberg N, et al. Associations and interactions between SNPs in the alcohol metabolizing genes and alcoholism phenotypes in European Americans. Alcohol Clin Exp Res 2009;33(5):848—57.
[62] Corbett J, Saccone NL, Foroud T, Goate A, et al. A sex-adjusted and age-adjusted genome screen for nested alcohol dependence diagnoses. Psychiatr Genet 2005;15(1):25—30.
[63] Prescott CA, Sullivan PF, Kuo PH, Webb BT, et al. Genomewide linkage study in the Irish affected sib pair study of alcohol dependence: evidence for a susceptibility region for symptoms of alcohol dependence on chromosome 4. Mol Psychiatry 2006;11(6):603—11.
[64] Edenberg HJ. The genetics of alcohol metabolism: role of alcohol dehydrogenase and aldehyde dehydrogenase variants. Alcohol Res Health 2007;30(1):5—13.
[65] Chen CC, Lu RB, Chen YC, Wang MF, et al. Interaction between the functional polymorphisms of the alcohol-metabolism genes in protection against alcoholism. Am J Hum Genet 1999;65(3):795—807.
[66] Crabb DW, Matsumoto M, Chang D, You M. Overview of the role of alcohol dehydrogenase and aldehyde dehydrogenase and their variants in the genesis of alcohol-related pathology. Proc Nutr Soc 2004;63(1):49—63.
[67] Macgregor S, Lind PA, Bucholz KK, Hansell NK, et al. Associations of ADH and ALDH2 gene variation with self report alcohol reactions, consumption and dependence: an integrated analysis. Hum Mol Genet 2009;18(3):580—93.
[68] Neumark YD, Friedlander Y, Thomasson HR, Li TK. Association of the ADH2*2 allele with reduced ethanol consumption in Jewish men in Israel: a pilot study. J Stud Alcohol 1998;59(2):133—9.
[69] Whitfield JB, Nightingale BN, Bucholz KK, Madden PA, et al. ADH genotypes and alcohol use and dependence in Europeans. Alcohol Clin Exp Res 1998;22(7):1463—9.
[70] Luo X, Kranzler HR, Zuo L, Wang S, et al. Diplotype trend regression analysis of the ADH gene cluster and the ALDH2 gene: multiple significant associations with alcohol dependence. Am J Hum Genet 2006;78(6):973—87.
[71] Eng MY, Luczak SE, Wall TL. ALDH2, ADH1B, and ADH1C genotypes in Asians: a literature review. Alcohol Res Health 2007;30(1):22—7.
[72] Wall TL, Johnson ML, Horn SM, Carr LG, et al. Evaluation of the self-rating of the effects of alcohol form in Asian Americans with aldehyde dehydrogenase polymorphisms. J Stud Alcohol 1999;60(6):784—9.
[73] Saccone NL, Kwon JM, Corbett J, Goate A, et al. A genome screen of maximum number of drinks as an alcoholism phenotype. Am J Med Genet 2000;96(5):632—7.
[74] Wise RA, Rompre PP. Brain dopamine and reward. Annu Rev Psychol 1989;40:191—225.
[75] McBride WJ, Bodart B, Lumeng L, Li TK. Association between low contents of dopamine and serotonin in the nucleus accumbens and high alcohol preference. Alcohol Clin Exp Res 1995;19(6):1420—2.
[76] Lawford BR, Young RM, Rowell JA, Gibson JN, et al. Association of the D2 dopamine receptor A1 allele with alcoholism: medical severity of alcoholism and type of controls. Biol Psychiatry 1997;41(4):386—93.
[77] Conner BT, Noble EP, Berman SM, Ozkaragoz T, et al. DRD2 genotypes and substance use in adolescent children of alcoholics. Drug Alcohol Depend 2005;79(3):379—87.
[78] Connor JP, Young RM, Lawford BR, Saunders JB, et al. Heavy nicotine and alcohol use in alcohol dependence is associated with D2 dopamine receptor (DRD2) polymorphism. Addict Behav 2007;32(2):310—19.
[79] Joe KH, Kim DJ, Park BL, Yoon S, et al. Genetic association of DRD2 polymorphisms with anxiety scores among alcohol-dependent patients. Biochem Biophys Res Commun 2008;371(4):591—5.
[80] Smith L, Watson M, Gates S, Ball D, et al. *Meta*-analysis of the association of the Taq1A polymorphism with the risk of alcohol dependency: a HuGE gene-disease association review. Am J Epidemiol 2008;167(2):125—38.
[81] Bolos AM, Dean M, Lucas-Derse S, Ramsburg M, et al. Population and pedigree studies reveal a lack of association between the dopamine D2 receptor gene and alcoholism. JAMA 1990;264(24):3156—60.
[82] Cook BL, Wang ZW, Crowe RR, Hauser R, et al. Alcoholism and the D2 receptor gene. Alcohol Clin Exp Res 1992;16(4):806—9.

[83] Heinz A, Sander T, Harms H, Finckh U, et al. Lack of allelic association of dopamine D1 and D2 (TaqIA) receptor gene polymorphisms with reduced dopaminergic sensitivity to alcoholism. Alcohol Clin Exp Res 1996;20(6):1109–13.

[84] Neville MJ, Johnstone EC, Walton RT. Identification and characterization of ANKK1: a novel kinase gene closely linked to DRD2 on chromosome band 11q23.1. Hum Mutat 2004;23(6):540–5.

[85] Volkow ND, Wang GJ, Telang F, Fowler JS, et al. Profound decreases in dopamine release in striatum in detoxified alcoholics: possible orbitofrontal involvement. J Neurosci 2007;27(46):12700–6.

[86] Ponce G, Hoenicka J, Jimenez-Arriero MA, Rodriguez-Jimenez R, et al. DRD2 and ANKK1 genotype in alcohol-dependent patients with psychopathic traits: association and interaction study. Br J Psychiatry 2008;193(2):121–5.

[87] Van Tol HH, Bunzow JR, Guan HC, Sunahara RK, et al. Cloning of the gene for a human dopamine D4 receptor with high affinity for the antipsychotic clozapine. Nature 1991;350(6319):610–14.

[88] Van Tol HH, Wu CM, Guan HC, Ohara K, et al. Multiple dopamine D4 receptor variants in the human population. Nature 1992;358 (6382):149–52.

[89] Benjamin J, Li L, Patterson C, Greenberg BD, et al. Population and familial association between the D4 dopamine receptor gene and measures of novelty seeking. Nat Genet 1996;12(1):81–4.

[90] Ebstein RP, Zohar AH, Benjamin J, Belmaker RH. An update on molecular genetic studies of human personality traits. Appl Bioinformatics 2002;1(2):57–68.

[91] Vandenbergh DJ, Persico AM, Hawkins AL, Griffin CA, et al. Human dopamine transporter gene (DAT1) maps to chromosome 5p15.3 and displays a VNTR. Genomics 1992;14(4):1104–6.

[92] Samochowiec J, Kucharska-Mazur J, Grzywacz A, Jablonski M, et al. Family-based and case-control study of DRD2, DAT, 5HTT, COMT genes polymorphisms in alcohol dependence. Neurosci Lett 2006;410(1):1–5.

[93] Dobashi I, Inada T, Hadano K. Alcoholism and gene polymorphisms related to central dopaminergic transmission in the Japanese population. Psychiatr Genet 1997;7(2):87–91.

[94] Muramatsu T, Higuchi S. Dopamine transporter gene polymorphism and alcoholism. Biochem Biophys Res Commun 1995;211(1):28–32.

[95] Sander T, Harms H, Podschus J, Finckh U, et al. Allelic association of a dopamine transporter gene polymorphism in alcohol dependence with withdrawal seizures or delirium. Biol Psychiatry 1997;41(3):299–304.

[96] Kohnke MD, Batra A, Kolb W, Kohnke AM, et al. Association of the dopamine transporter gene with alcoholism. Alcohol Alcohol 2005;40 (5):339–42.

[97] Franke P, Schwab SG, Knapp M, Gansicke M, et al. DAT1 gene polymorphism in alcoholism: a family-based association study. Biol Psychiatry 1999;45(5):652–4.

[98] Foley PF, Loh EW, Innes DJ, Williams SM, et al. Association studies of neurotransmitter gene polymorphisms in alcoholic Caucasians. Ann N Y Acad Sci 2004;1025:39–46.

[99] Chen J, Lipska BK, Halim N, Ma QD, et al. Functional analysis of genetic variation in catechol-O-methyltransferase (COMT): effects on mRNA, protein, and enzyme activity in postmortem human brain. Am J Hum Genet 2004;75(5):807–21.

[100] Goldman D, Oroszi G, O'Malley S, Anton R. COMBINE genetics study: the pharmacogenetics of alcoholism treatment response: genes and mechanisms. J Stud Alcohol Suppl 2005;(15):56–64 discussion 33.

[101] Hallikainen T, Lachman H, Saito T, Volavka J, et al. Lack of association between the functional variant of the catechol-o-methyltransferase (COMT) gene and early-onset alcoholism associated with severe antisocial behavior. Am J Med Genet 2000;96(3):348–52.

[102] Vanyukov MM, Moss HB, Yu LM, Tarter RE, et al. Preliminary evidence for an association of a dinucleotide repeat polymorphism at the MAOA gene with early onset alcoholism/substance abuse. Am J Med Genet 1995;60(2):122–6.

[103] Hsu YP, Loh EW, Chen WJ, Chen CC, et al. Association of monoamine oxidase A alleles with alcoholism among male Chinese in Taiwan. Am J Psychiatry 1996;153(9):1209–11.

[104] Badner JA, Gershon ES. *Meta*-analysis of whole-genome linkage scans of bipolar disorder and schizophrenia. Mol Psychiatry 2002;7 (4):405–11.

[105] Lotta T, Vidgren J, Tilgmann C, Ulmanen I, et al. Kinetics of human soluble and membrane-bound catechol O-methyltransferase: a revised mechanism and description of the thermolabile variant of the enzyme. Biochemistry 1995;34(13):4202–10.

[106] Matsumoto M, Weickert CS, Akil M, Lipska BK, et al. Catechol O-methyltransferase mRNA expression in human and rat brain: evidence for a role in cortical neuronal function. Neuroscience 2003;116(1):127–37.

[107] Kauhanen J, Hallikainen T, Tuomainen TP, Koulu M, et al. Association between the functional polymorphism of catechol-O-methyltransferase gene and alcohol consumption among social drinkers. Alcohol Clin Exp Res 2000;24(2):135–9.

[108] Enoch MA, Xu K, Ferro E, Harris CR, et al. Genetic origins of anxiety in women: a role for a functional catechol-O-methyltransferase polymorphism. Psychiatr Genet 2003;13(1):33–41.

[109] COMBINE, S.R.G. Testing combined pharmacotherapies and behavioral interventions in alcohol dependence: rationale and methods. Alcohol Clin Exp Res 2003;27(7):1107–22.

[110] COMBINE, S.R.G. Testing combined pharmacotherapies and behavioral interventions for alcohol dependence (the COMBINE study): a pilot feasibility study. Alcohol Clin Exp Res 2003;27(7):1123–31.

[111] Kohnke MD, Kolb W, Kohnke AM, Lutz U, et al. DBH*444G/A polymorphism of the dopamine-beta-hydroxylase gene is associated with alcoholism but not with severe alcohol withdrawal symptoms. J Neural Transm 2006;113(7):869–76.

[112] Kohnke MD, Zabetian CP, Anderson GM, Kolb W, et al. A genotype-controlled analysis of plasma dopamine beta-hydroxylase in healthy and alcoholic subjects: evidence for alcohol-related differences in noradrenergic function. Biol Psychiatry 2002;52(12):1151–8.

[113] Freire MT, Hutz MH, Bau CH. The DBH-1021 C/T polymorphism is not associated with alcoholism but possibly with patients' exposure to life events. J Neural Transm 2005;112(9):1269–74.

[114] Long JC, Knowler WC, Hanson RL, Robin RW, et al. Evidence for genetic linkage to alcohol dependence on chromosomes 4 and 11 from an autosome-wide scan in an American Indian population. Am J Med Genet 1998;81(3):216–21.

[115] Radel M, Vallejo RL, Iwata N, Aragon R, et al. Haplotype-based localization of an alcohol dependence gene to the 5q34 {gamma}-aminobutyric acid type A gene cluster. Arch Gen Psychiatry 2005;62(1):47–55.

[116] Sander T, Samochowiec J, Ladehoff M, Smolka M, et al. Association analysis of exonic variants of the gene encoding the GABAB receptor and alcohol dependence. Psychiatr Genet 1999;9(2):69–73.

[117] Kohnke M, Schick S, Lutz U, Kohnke A, et al. The polymorphism GABABR1 T1974C[rs29230] of the GABAB receptor gene is not associated with the diagnosis of alcoholism or alcohol withdrawal seizures. Addict Biol 2006;11(2):152–6.

[118] Johnston GA, Chebib M, Hanrahan JR, Mewett KN. Neurochemicals for the Investigation of GABA(C) receptors. Neurochem Res 2010;35(12):1970–7.

[119] Rae C, Nasrallah FA, Griffin JL, Balcar VJ. Now I know my ABC. A systems neurochemistry and functional metabolomic approach to understanding the GABAergic system. J Neurochem 2009;109(Suppl. 1):109–16.

[120] Ait-Daoud N, Lynch WJ, Penberthy JK, Breland AB, et al. Treating smoking dependence in depressed alcoholics. Alcohol Res Health 2006;29(3):213–20.

[121] Wang JC, Hinrichs AL, Stock H, Budde J, et al. Evidence of common and specific genetic effects: association of the muscarinic acetylcholine receptor M2 (CHRM2) gene with alcohol dependence and major depressive syndrome. Hum Mol Genet 2004;13(17):1903–11.

[122] Dick DM, Aliev F, Kramer J, Wang JC, et al. Association of CHRM2 with IQ: converging evidence for a gene influencing intelligence. Behav Genet 2007;37(2):265–72.

[123] Luo X, Kranzler HR, Zuo L, Wang S, et al. CHRM2 gene predisposes to alcohol dependence, drug dependence and affective disorders: results from an extended case-control structured association study. Hum Mol Genet 2005;14(16):2421–34.

[124] Swan GE, Carmelli D, Cardon LR. The consumption of tobacco, alcohol, and coffee in Caucasian male twins: a multivariate genetic analysis. J Subst Abuse 1996;8(1):19–31.

[125] Swan GE, Carmelli D, Cardon LR. Heavy consumption of cigarettes, alcohol and coffee in male twins. J Stud Alcohol 1997;58(2):182–90.

[126] True WR, Xian H, Scherrer JF, Madden PA, et al. Common genetic vulnerability for nicotine and alcohol dependence in men. Arch Gen Psychiatry 1999;56(7):655–61.

[127] Hettema JM, Corey LA, Kendler KS. A multivariate genetic analysis of the use of tobacco, alcohol, and caffeine in a population based sample of male and female twins. Drug Alcohol Depend 1999;57(1):69–78.

[128] Chatterjee S, Bartlett SE. Neuronal nicotinic acetylcholine receptors as pharmacotherapeutic targets for the treatment of alcohol use disorders. CNS Neurol Disord Drug Targets 2010;9(1):60–76.

[129] Wernicke C, Samochowiec J, Schmidt LG, Winterer G, et al. Polymorphisms in the N-methyl-D-aspartate receptor 1 and 2B subunits are associated with alcoholism-related traits. Biol Psychiatry 2003;54(9):922–8.

[130] Schumann G, Rujescu D, Szegedi A, Singer P, et al. No association of alcohol dependence with a NMDA-receptor 2B gene variant. Mol Psychiatry 2003;8(1):11–12.

[131] Johnson BA, Ait-Daoud N, Bowden CL, DiClemente CC, et al. Oral topiramate for treatment of alcohol dependence: a randomised controlled trial. Lancet 2003;361(9370):1677–85.

[132] Kranzler HR, Anton RF. Implications of recent neuropsychopharmacologic research for understanding the etiology and development of alcoholism. J Consult Clin Psychol 1994;62(6):1116–26.

[133] Feinn R, Nellissery M, Kranzler HR. Meta-analysis of the association of a functional serotonin transporter promoter polymorphism with alcohol dependence. Am J Med Genet B Neuropsychiatr Genet 2005;133B(1):79–84.

[134] Halliday G, Ellis J, Heard R, Caine D, et al. Brainstem serotonergic neurons in chronic alcoholics with and without the memory impairment of Korsakoff's psychosis. J Neuropathol Exp Neurol 1993;52(6):567–79.

[135] Dick DM, Plunkett J, Hamlin D, Nurnberger Jr. J, et al. Association analyses of the serotonin transporter gene with lifetime depression and alcohol dependence in the Collaborative Study on the Genetics of Alcoholism (COGA) sample. Psychiatr Genet 2007;17(1):35–8.

[136] Pinto E, Reggers J, Gorwood P, Boni C, et al. The short allele of the serotonin transporter promoter polymorphism influences relapse in alcohol dependence. Alcohol Alcohol 2008;43(4):398–400.

[137] Vengeliene V, Bilbao A, Molander A, Spanagel R. Neuropharmacology of alcohol addiction. Br J Pharmacol 2008;154(2):299–315.

[138] Wetherill L, Schuckit MA, Hesselbrock V, Xuei X, et al. Neuropeptide Y receptor genes are associated with alcohol dependence, alcohol withdrawal phenotypes, and cocaine dependence. Alcohol Clin Exp Res 2008.

[139] McMillin GA, Mellis R, Bornhorst J. Alcohol abuse and dependency genetics of susceptibility and pharmacogenetics of therapy. In: Dasgupta A, editor. Critical issues in drug abuse testing. Humana Press; 2009.

[140] Haile CN, Kosten TA, Kosten TR. Pharmacogenetic treatments for drug addiction: alcohol and opiates. Am J Drug Alcohol Abuse 2008;34(4):355–81.

[141] Ingelman-Sundberg M, Sim SC, Gomez A, Rodriguez-Antona C. Influence of cytochrome P450 polymorphisms on drug therapies: pharmacogenetic, pharmacoepigenetic and clinical aspects. Pharmacol Ther 2007;116(3):496–526.

[142] Kirchheiner J, Seeringer A. Clinical implications of pharmacogenetics of cytochrome P450 drug metabolizing enzymes. Biochim Biophys Acta 2007;1770(3):489–94.
[143] Fukasawa T, Suzuki A, Otani K. Effects of genetic polymorphism of cytochrome P450 enzymes on the pharmacokinetics of benzodiazepines. J Clin Pharm Ther 2007;32(4):333–41.
[144] Darrouj J, Puri N, Prince E, Lomonaco A, et al. Dexmedetomidine infusion as adjunctive therapy to benzodiazepines for acute alcohol withdrawal (November). Ann Pharmacother 2008.
[145] Kotlinska J, Bochenski M. The influence of various glutamate receptors antagonists on anxiety-like effect of ethanol withdrawal in a plus-maze test in rats. Eur J Pharmacol 2008.
[146] Palachick B, Chen YC, Enoch AJ, Karlsson RM, et al. Role of major NMDA or AMPA receptor subunits in MK-801 potentiation of ethanol intoxication. Alcohol Clin Exp Res 2008;32(8):1479–92.
[147] O'Brien CP. Anticraving medications for relapse prevention: a possible new class of psychoactive medications. Am J Psychiatry 2005;162 (8):1423–31.
[148] Fuller RK, Branchey L, Brightwell DR, Derman RM, et al. Disulfiram treatment of alcoholism. A veterans administration cooperative study. JAMA 1986;256(11):1449–55.
[149] Harada S, Misawa S, Agarwal DP, Goedde HW. Liver alcohol dehydrogenase and aldehyde dehydrogenase in the Japanese: isozyme variation and its possible role in alcohol intoxication. Am J Hum Genet 1980;32(1):8–15.
[150] Anton RF, Oroszi G, O'Malley S, Couper D, et al. An evaluation of mu-opioid receptor (OPRM1) as a predictor of naltrexone response in the treatment of alcohol dependence: results from the Combined Pharmacotherapies and Behavioral Interventions for Alcohol Dependence (COMBINE) study. Arch Gen Psychiatry 2008;65(2):135–44.
[151] Quickfall J, el-Guebaly N. Genetics and alcoholism: how close are we to potential clinical applications? Can J Psychiatry 2006;51(7):461–7.
[152] Mann K, Kiefer F, Spanagel R, Littleton J. Acamprosate: recent findings and future research directions. Alcohol Clin Exp Res 2008;32 (7):1105–10.
[153] Ray LA, Hutchison KE. Effects of naltrexone on alcohol sensitivity and genetic moderators of medication response: a double-blind placebo-controlled study. Arch Gen Psychiatry 2007;64(9):1069–77.
[154] Roberto M, Gilpin NW, O'Dell LE, Cruz MT, et al. Cellular and behavioral interactions of gabapentin with alcohol dependence. J Neurosci 2008;28(22):5762–71.
[155] Sarid-Segal O, Piechniczek-Buczek J, Knapp C, Afshar M, et al. The effects of levetiracetam on alcohol consumption in alcohol-dependent subjects: an open label study. Am J Drug Alcohol Abuse 2008;34(4):441–7.
[156] Olmsted CL, Kockler DR. Topiramate for alcohol dependence. Ann Pharmacother 2008;42(10):1475–80.
[157] Bierut LJ, Agrawal A, Bucholz KK, Doheny KF, et al. A genome-wide association study of alcohol dependence. Proc Natl Acad Sci U S A 2010;107(11):5082–7.
[158] Edenberg HJ, Koller DL, Xuei X, Wetherill L, et al. Genome-wide association study of alcohol dependence implicates a region on chromosome 11. Alcohol Clin Exp Res 2010;34(5):840–52.
[159] Haycock PC. Fetal alcohol spectrum disorders: the epigenetic perspective. Biol Reprod 2009;81(4):607–17.
[160] Feinberg AP, Irizarry RA, Fradin D, Aryee MJ, et al. Personalized epigenomic signatures that are stable over time and covary with body mass index. Sci Transl Med 2010;2(49):49ra67.
[161] Satterlee JS, Schubeler D, Ng HH. Tackling the epigenome: challenges and opportunities for collaboration. Nat Biotechnol 2010;28 (10):1039–44.
[162] Portela A, Esteller M. Epigenetic modifications and human disease. Nat Biotechnol 2010;28(10):1057–68.
[163] Palmisano M, Pandey SC. Epigenetic mechanisms of alcoholism and stress-related disorders. Alcohol 2017;60:7–18.
[164] Rubinstein WS, Maglott DR, Lee JM, Kattman BL, et al. The NIH genetic testing registry: a new, centralized database of genetic tests to enable access to comprehensive information and improve transparency. Nucleic Acids Res 2013;41(Database issue):D925–35.

Chapter 5

Ethylene Glycol and Other Glycols: Analytical and Interpretation Issues

Uttam Garg[1], Jennifer Lowry[2] and D. Adam Algren[2]

[1]*Department of Pathology and Laboratory Medicine, Children's Mercy Hospital, and University of Missouri School of Medicine, Kansas City, MO, United States,* [2]*Division of Clinical Pharmacology and Medical Toxicology, Children's Mercy Hospital, and University of Missouri School of Medicine, Kansas City, MO, United States*

INTRODUCTION

Glycol is a chemical compound that is diol containing two hydroxyl groups. In a clinical setting, most commonly encountered glycol is ethylene glycol. Other less frequently encountered glycols include diethylene glycol, propylene glycol, ethylene glycol monobutyl ether (EGBE), and ethylene glycol monomethyl ether (EGME). Glycols are associated with severe morbidity and an increased risk of death. Ethylene glycol is a colorless, odorless, and relatively nonvolatile liquid that is commonly used as an industrial chemical. It is a major ingredient of antifreeze and deicing solutions due to its high boiling point (197°C) and low freezing point. It may be accidentally ingested, particularly by children as it tastes sweet, or intentionally ingested by adults as an ethanol substitute or to inflict self-harm. Although ethylene glycol toxicity is uncommon, when it does occur, it can have a high morbidity or mortality. According to the 2016 Annual Report of the American Association of Poison Control Centers, there were 6325 single exposures to ethylene glycol with 32 fatalities [1]. Similar to ethylene glycol, diethylene glycol is used in antifreeze, brake fluids, and deicing solutions. Propylene glycol is used as a solvent in commercial products as well as a diluent for oral, topical, and intravenous pharmaceutical preparations. Since it is significantly less toxic as compared to ethylene glycol, propylene glycol is the preferred antifreeze used in motor homes and recreational vehicles. Diethylene glycol is used in numerous products including solvents and brake fluids. Other glycol ethers, such as EGME/EGBE, are a group of compounds that are used as organic solvents in various products. Comparison of various glycols is given in Table 5.1 [2–4].

PHARMACODYNAMICS AND PHARMACOKINETICS

Ethylene glycol and other low-molecular-weight glycols are rapidly and completely absorbed through the intestinal tract after oral ingestion. The high-molecular-weight glycols, such as polypropylene glycol and polyethylene glycols, are poorly absorbed through the intestinal tract and are virtually nontoxic. Parent glycols are relatively nontoxic; however, their metabolites are toxic and result in morbidity and mortality. Like ethanol, ethylene glycol reaches a peak concentration within 1–2 h after ingestion. Analogous to ethanol, metabolism of ethylene glycol involves the enzymes alcohol dehydrogenase and aldehyde dehydrogenase (Fig. 5.1). Metabolites of ethylene glycol include glycolic, glyoxylic, and oxalic acids. The primary and probably the most toxic metabolite of ethylene glycol is glycolic acid [5]. It constitutes >90% of the metabolite concentrations. Glycolic acid, along with other metabolites, leads to severe metabolic acidosis. Oxalic acid binds to calcium to form insoluble calcium oxalate crystals that can cause widespread tissue injury, particularly in the kidney.

In healthy individuals, the volume of distribution (Vd) of ethylene glycol is 0.6–0.8 L/kg. The half-life of ethylene glycol is 3–6 h with approximately 20% of the parent compound excreted unchanged by the kidneys. In a patient with renal failure, the half-life is longer due to decreased renal clearance. Ethanol and fomepizole are competitive inhibitors of alcohol dehydrogenase and prevent the metabolism of ethylene glycol to its toxic metabolites [6,7]. When these

TABLE 5.1 Comparison of Different Glycols for Toxicity and Treatment

Glycol	Toxicity and Comments	Treatment
Diethylene glycol	Used as antifreeze, lubricant, brake fluid, and industry solvent. Sweet taste similar to ethylene glycol. More toxic as compared to ethylene glycol. Estimated human lethal dose is 0.014–0.170 mg/kg. Renal failure, coma, metabolic acidosis, and death have been reported after ingestion. Significant toxicity has been reported through repeated dermal application in patients with extensive burn injuries. Like ethylene glycol, it is metabolized by alcohol dehydrogenase and aldehyde dehydrogenase with a half-life of ~3 h.	Supportive, ethanol, and fomepizole may be effective, hemodialysis for severe metabolic acidosis.
Dipropylene glycol	Relatively low toxicity. Human toxicity not well understood.	Supportive care. There is no role for ethanol or fomepizole therapy.
Ethylene glycol	Most commonly used in antifreeze. Central nervous system (CNS) depression similar to ethanol. Toxic metabolites include oxalic, glycolic, and glyoxylic acids.	Ethanol and fomepizole are effective; fomepizole is preferred.
Ethylene glycol monobutyl ether	Toxic effects include lethargy, coma, anion gap metabolic acidosis, hyperchloremia, hypotension, respiratory depression, hemolysis, renal and hepatic dysfunction. Serum levels in poisoning cases have ranged from 0.005 to 432 mg/L.	Ethanol and fomepizole may be effective.
Ethylene glycol monoethyl ether	Studies in animals show that it is metabolized to ethylene glycol. Calcium oxalate crystals have been reported in animals.	Ethanol and fomepizole may be effective.
Ethylene glycol monomethyl ether	Cerebral edema, hemorrhagic gastritis, and degeneration of the liver and kidneys were reported in one autopsy. Toxicity happens after 8–18 h.	Ethanol and fomepizole may be effective, but uncertain.
Polyethylene glycols	Relatively nontoxic. Used as liquid vehicle in cosmetics and topical medications. A group of compounds with molecular weights ranging from 200 to more than 4000. High-molecular-weight compounds (>500) are poorly absorbed and rapidly excreted by the kidneys. Low-molecular-weight compounds (200–400) may result in metabolic acidosis, renal failure, hypercalcemia after massive oral ingestions, or repeated dermal applications in patients with extensive burn injuries.	Supportive care.
Propylene glycol	Used as antifreeze and vehicle solvent for medications. Relatively low toxicity. Occurs in various forms such as 1,2-propanediol and 1,3-propanediol. 1,2-Propanediol form is the most studied one. Metabolizes to lactic acid, pyruvic acid, and acetic acid. Toxicity includes lactic acidosis, CNS depression, coma, hypoglycemia, seizures, and hemolysis.	Supportive care.
Triethylene glycol	Coma, metabolic acidosis with elevated anion gap.	Ethanol and fomepizole may be effective.

treatment modalities are utilized in a person with normal renal function, the elimination half-life of ethylene glycol increases to 17–20 h, and the elimination is entirely renal. Some pharmacokinetic and toxicokinetics properties of ethylene glycol are given in Table 5.2 [8,9].

The toxic dose of ethylene glycol is approximately 0.11 g/kg of body weight. Based on this dose, the toxic volume in mL/kg of an ethylene glycol solution can be estimated by the formula 10.7/product concentration % [10]. In the absence of any treatment, the approximate lethal dose of ethylene glycol is 1.0–1.5 g/kg body weight; however, survivals have been reported with very large amounts of ethylene glycol intake. For example, in one report a 36-year-old man with a history of depression consumed approximately 3 L of ethylene glycol antifreeze in a suicide attempt. The patient's blood ethylene glycol level was 1889 mg/dL [11]. Although the patient developed nausea, emesis, lethargy, metabolic acidosis, and renal failure, the patient survived (after treatment with an ethanol infusion and hemodialysis) without persistent renal failure or other chronic problems. This case highlights the importance of early intervention and

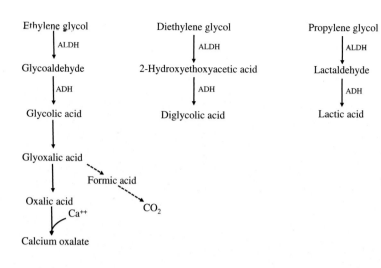

FIGURE 5.1 Metabolism of ethylene glycol, diethylene glycol, and propylene glycol. *ALDH*, Alcohol dehydrogenase; *ADH*, aldehyde dehydrogenase.

TABLE 5.2 Pharmacokinetic Parameters of Ethylene Glycol [8]

Parameter	Range
Volume of distribution (L/kg)	0.83
Toxic level (mg/dL)	>20
Lethal dose (g/kg)[a]	1.0–1.5
Lethal serum concentration (mg/dL)	30–775
Volume of distribution (L/kg)	0.5–0.8
Half-life (h)	3.0–6.0
Half-life (h) during ethanol infusion or fomepizole treatment	17–20
Half-life (h) during hemodialysis	2.5–3.5
Real clearance (mL/kg/min)	3.2
Hemodialysis clearance (mL/min)	156–210

[a]Survivals have been reported at much higher ingestions and serum concentrations.

treatment. In another report, a 34-year-old patient had a very high level of ethylene glycol of 1234 mg/dL [12]. The subject was found unresponsive with two empty 1-L bottles of antifreeze and a suicide note. The patient had a Glasgow Coma Score of 4 with Kussmaul respirations and normal hemodynamics. The patient was endotracheally intubated. Initial arterial blood gas results were pH of 6.79, pCO$_2$ of 37 mmHg, pO$_2$ of 115 mmHg, and bicarbonate of 5.5 mEq/L. Further testing showed an anion gap of 43, plasma osmolality of 828 mOsm/kg, osmolality gap of 483 mOsm/kg, and a serum lactate of >30 mEq/L. Urine showed calcium oxalate crystals. Ethylene glycol poisoning was suspected. The patient received fomepizole (4-methylpyrazole), thiamine, and folate, and was started on dialysis. The patient developed progressive hypotension, apnea, and autonomic dysfunction. The patient expired 22 h after initial presentation.

Diethylene glycol has a half-life of 3–4 h and is primarily metabolized to 2-hydroxyethoxyacetic acid and diglycolic acid. 2-Hydroxyethoxyacetic acid is the major metabolite. Diethylene glycol toxicity is due to metabolic acidosis caused the accumulation of 2-hydroxyethoxyacetic acid. Diglycolic acid is responsible for acute renal injury and neurological sequelae [2].

Propylene glycol is less toxic; therefore, larger doses are needed to cause toxicity. It has a Vd of 0.5–0.8 L/kg and a half-life of 2–5 h. More than half of it is excreted unchanged by the kidneys. It is metabolized by alcohol dehydrogenase and acetaldehyde dehydrogenase to lactic acid. Toxicity of propylene glycol is associated with its acidic metabolites. Toxicokinetic data is rather limited, but it is thought that the glycol ethers are rapidly absorbed and metabolized by alcohol and aldehyde dehydrogenase to acidic metabolites (methoxyacetic acid and butoxyacetic acid).

Mechanism of Toxicity

Ethylene glycol and other glycols as parent compounds are relatively nontoxic; most of their toxic effects are due to their metabolites. After acute ingestion of ethylene glycol, the patient will show signs of central nervous system (CNS) intoxication similar to ethanol; however, its metabolites including oxalic, glycolic, and glyoxylic acids are toxic. These metabolites cause severe metabolic acidosis and increased anion gap. Oxalic acid quickly binds to calcium to form calcium oxalate crystals which can crystallize in the kidneys and cause renal failure [13]. Likewise, calcium oxalate crystals can precipitate in other tissues and cause generalized tissue injury [14]. Glycolic acid can also cause renal tubular damage and renal failure [15]. Other clinical manifestations may include CNS depression, cardiopulmonary failure, convulsions, and coma. These effects are due to metabolic acidosis and hypocalcemia. Metabolic acidosis increases the transport of acidic metabolites into the CNS leading to a deterioration of the CNS function. This leads to a downward spiral of hypoxia and acidemia. Since the parent compound and toxic metabolites have differing onsets of toxicity, the clinical effects generally happen in distinct phases.

- Phase 1 (the first few hours): During this phase the victim may appear intoxicated (similar to ethanol intoxication) and may have ataxia, nausea, vomiting, drowsiness, and slurred speech. Laboratory findings may include an increased osmolal gap.
- Phase 2 (4–24 h): During this phase, the toxic metabolites start accumulating and the clinical effects are due to these metabolites. Symptoms may include hyperventilation, convulsions, coma, and arrhythmias. Renal failure is common but can be reversed with treatment. Pulmonary and cerebral edema may also occur. Laboratory findings include metabolic acidosis, hypocalcemia, calcium-oxaluria, hematuria, and proteinuria.
- Phase 3 (2–4 days): Renal insufficiency with acute tubular necrosis, hematuria, proteinuria, oliguria, or anuria can occur. Generally, no parent or toxic metabolites are detectable during this phase.

Diethylene glycol toxicity is due to its metabolites, 2-hydroxyethoxyacetic and diglycolic acids. Metabolic acidosis is due to an accumulation of 2-hydroxyethoxyacetic acid, whereas renal injury is due to diglycolic acid. The toxicity of propylene glycol is associated with its metabolites lactic, pyruvic, and acetic acids. The acidic metabolites of EGME and EGBE (methoxyacetic acid and butoxyacetic acid) contribute to the development of metabolic acidosis and renal failure.

Clinical and Laboratory Assessment of Ethylene Glycol Poisoning

Most often the diagnosis of ethylene glycol poisoning is based on the patient history of antifreeze ingestion and/or the clinical symptoms of CNS depression and/or metabolic acidosis. Laboratory testing of ethylene glycol testing is not readily available in all clinical laboratories. When samples are sent to a referral laboratory, the results may not be available in the time frame needed to assist in making early clinical decisions. As ethylene glycol and its metabolites affect many other laboratory parameters, assessment of these parameters may be very helpful in the diagnosis and management of a patient [16].

In the early stages of ethylene glycol intoxication (stage 1, first several hours), patient may not present with any significant laboratory abnormalities. However, since ethylene glycol is an osmotically active substance, the osmolal gap may be elevated. The osmolal gap is the difference between the measured and the calculated osmolality. The following formula is commonly used to calculate osmolality; however, many variations have been proposed [17].

$$\text{Calculated osmolality} = [2 \times \text{plasma Na}] + [\text{glucose}/18] + [\text{BUN}/2.8] + [\text{ethanol}/4.6]$$

Ethanol is included in the equation to rule in/out ethanol intoxication. The unit of measurement for Na is mmoles/L, and the units for glucose, blood urea nitrogen (BUN), and ethanol are mg/dL. Glucose, BUN, and ethanol are divided by 18, 2.8, and 4.6, respectively, to convert their concentrations from mg/dL to mmoles/L. Most often, osmolality is measured by freezing point depression. Significantly elevated osmolar gaps (>25) are fairly specific for toxic alcohol ingestion [18]. In addition, the osmolal gap can be used to estimate the serum ethylene glycol concentration:

$$\text{Ethylene glycol concentration (mg/dL)} = \text{osmolal gap multiplied by } 6.2.$$

There are limitations, however, to using the osmolal gap as a screening tool. The "normal" osmol values vary within the population [19]. When measured in both healthy adults and children, the normal osmolal gap is -2 ± 6 [18]. Most clinical laboratories consider an osmolal gap <10 to be normal. A mildly elevated osmolal gap is not specific to toxic alcohol ingestions as other drugs and disease processes may result in an osmolal gap. These conditions include mannitol administration, contrast dye infusions, renal failure, ketoacidosis, and lactic acidosis. The time from ingestion must be taken into account when interpreting the osmolal gap. In late presentations, the osmolal gap may not be elevated. Also, osmolal gap is not sufficiently sensitive to exclude a small ingestion that can be clinically significant.

Another laboratory parameter which may be useful in assessing ethylene glycol toxicity is the anion gap. The anion gap is commonly calculated as follows:

$$\text{Anion gap} = Na^+ - Cl^- - HCO_3^-$$

The metabolites of ethylene glycol (glycolic acid, oxalic acid) are acids and their accumulation will result in an elevated anion gap. As stated previously, the elimination half-life of ethylene glycol is 3–6 h; thus, it will take several hours for acids to be generated after the ingestion and for an anion gap metabolic acidosis to become evident [19].

Ethylene glycol metabolites cause metabolic acidosis and hypocalcemia; therefore, electrolytes including calcium and arterial blood gases including pH should be monitored. In the absence of availability of ethylene glycol testing, examination of urine for calcium oxalate crystals may be helpful in the diagnosis of ethylene glycol exposure; however, calcium oxalate monohydrate crystals may be confused with hippurate crystals. The presence of hypocalcemia and calcium oxalate crystals in the urine is suggestive of ethylene glycol poisoning, although their absence should not be used to exclude toxicity [20]. In addition, some ethylene glycol and other glycol preparations used as antifreeze have fluorescein dye added to it to identify a radiator leak. These fluorescein dyes fluoresce under ultraviolet light; therefore, examining urine or cloths from a victim under ultraviolet light may provide hints about ethylene glycol exposure. The examination of urine for calcium oxalate crystals and fluorescein are neither sensitive nor specific.

Other routine laboratory tests which are helpful in the management of a patient with ethylene glycol and other glycols toxicity include BUN, creatinine, hepatic transaminases, lactate, and glucose. Glucose is helpful to rule out hypoglycemia as the cause of any alteration in mental status. To avoid misinterpretation, it is important to evaluate lactate methods for interference from ethylene glycol metabolites. Given their chemical structural similarities, glycolate can be misinterpreted by some lactate analyzers as lactate [21]. In a study, interference from ethylene glycol metabolites was investigated. Oxalic acid did not interfere in any of the methods studied, whereas glycolic and glyoxylic acids interfered in some assays [22]. Urinary ketones and serum β-hydroxybutyrate levels may help in distinguishing ethylene glycol poisoning from alcoholic or diabetic ketoacidosis.

Although not routinely performed, the measurement of glycolic acid is desirable and may be very helpful. The blood concentration of glycolic acid correlates better with clinical symptoms and prognosis than ethylene glycol levels [5,6,23]. Glycolic acid levels may remain elevated even when ethylene glycol is undetectable. In one study on 41 patients, admission ethylene glycol levels ranged from 6 to 430 mg/dL and did not correlate with mortality. Contrarily, a mortality rate of 33% occurred when the glycolic acid level was >10 mmol/L [23]. Gas chromatography (GC) methods with flame ionization (GC/FID) and mass spectrometry detectors (GC/MS) are available for simultaneous measurement of ethylene glycol and glycolic acid [24–26]. An enzymatic method for the measurement of glycolic acid has also been described [27]. The method can be automated on a chemistry analyzer and can provide fast turnaround time. Toxicity of diethylene glycol and other glycol ethers is associated with the development of metabolic acidosis. Other clinical manifestations include renal failure, altered mental status, seizures, and cranial nerve palsies.

Laboratory Analysis of Ethylene Glycol and Other Glycols

GC/FID or GC/MS are the commonly used methods for the analysis of ethylene glycol and other glycols [24,28–31]. Most GC methods involve protein precipitation by organic solvents such as acetonitrile followed by derivatization of glycols. A commonly used derivatizing agent is phenylboronic acid which reacts with glycols to form cyclic phenylboronate esters which have higher volatility and are suitable for GC. Although methods for underivatized ethylene glycol have been described, they may pose problems due to low volatility and poor chromatographic properties of underivatized glycols. Quantitation is achieved by standard chromatographic calculations by comparing the ratios of peak height or area of an unknown to that of calibrators. Internal standards are generally used and should be selected carefully. A commonly used internal standard is 1,3-propanediol. A capillary column GC method for simultaneous detection of diethylene glycol, ethylene glycol, methanol, isopropanol, acetone, and ethanol has been described [29]. In the past GC/FID had been the most commonly used method for the assay of ethylene glycol and other glycols; however, in

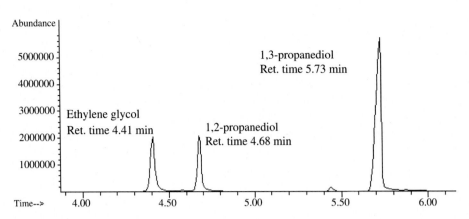

FIGURE 5.2 GC/MS total ion chromatogram of ethylene glycol, 1,2-propanediol, and 1,3-propnaediol. The sample was derivatized with phenylboronic acid. *Author's own data.*

recent years, GC/MS has increasingly been used. GC/MS methods can provide unequivocal identification of different glycols [25,26,28,30]. GC/MS methods for the simultaneous estimation of ethylene glycol and glycolic acid have been described [24,25]. GC/MS total ion chromatogram of phenylboronic acid derivatized ethylene glycol, 1,2-propanediol, and 1,3-pronaediol is shown in Fig. 5.2. Mass spectra of these derivatized glycols are shown in Fig. 5.3.

In addition to GC methods, enzymatic methods have been developed for the screening of ethylene glycol [32−35]. One screening method was developed when it was discovered that ethylene glycol positively interferes with one of the triglyceride methods. In this method, triglycerides are measured by two enzymatic methods and the "triglyceride gap" is used to estimate ethylene glycol. Another method uses lipase and glycerol dehydrogenase. In this method, ethylene glycol causes positive interference. Yet, another method uses lipase, glycerol kinase, glycerol phosphate oxidase, and peroxidase. Ethylene glycol does not interfere in the second method. "Triglycerides" are measured by both methods and the difference in the two methods is used to estimate ethylene glycol concentrations. In this method, methanol, ethanol, *n*-propanol, isopropanol, acetone, 1,3-propanediol, glycolic acid, oxalic acid, glyoxylic acid, and lactic acid did not interfere in the quantitation of ethylene glycol. However, β-hydroxybutyrate and glycerol show a slight positive interference and glycolaldehyde causes a significant positive interference [33].

Another enzymatic method for ethylene glycol screening uses glycerol dehydrogenase purified from *Enterobacter aerogenes*. This enzyme catalyzes the oxidation of ethylene glycol and produces NADH (nicotinamide adenine dinucleotide phosphate) in the presence of NAD (nicotinamide adenine dinucleotide) [35]. This reaction causes an increase in absorbance at 340 nm and is used to quantify ethylene glycol. In this method, methanol, ethanol, *n*-propanol, isopropanol, acetaldehyde, glycolic acid, oxalic acid, glyoxylic acid, and lactic acid do not interfere; however, glycolaldehyde, one of the ethylene glycol metabolites, and glycerol interfere with the assay. Generally, concentration of these compounds is very low and does not cause clinically significant interference.

Due to the limitations of the enzymatic assays discussed earlier, Juenke et al. [34] developed a modified method for the determination of ethylene glycol on automated analyzers. In the original Catachem method, the ethylene glycol concentration was determined by the difference between the absorbance readings at two time points. In this study, the slope of the line was determined by measuring absorbance differences at several points, starting at a later time point than the original two-point design. With this strategy, interferences from propylene glycol, various butanediols, and other related compounds were almost entirely eliminated.

TOXIC AND CRITICAL VALUES

Ethylene glycol levels of > 20 mg/dL are considered toxic and may need therapeutic intervention with ethanol or fomepizole (4-methylpyrazole). The levels > 50 mg/dL are considered critical and indicate a potential need for hemodialysis.

CHALLENGES AND PITFALLS IN THE ANALYSIS OF ETHYLENE GLYCOL AND OTHER RELATED TESTS

Several methods including enzymatic, GC/FID, and GC/MS have been described for the assay of ethylene glycol and other glycols. Enzymatic methods are easy, rapid, and amenable to automation on routine chemistry analyzers; however, enzymatic methods are prone to interferences from a number of compounds including glycerol, propylene glycol,

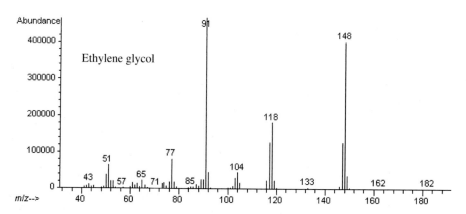
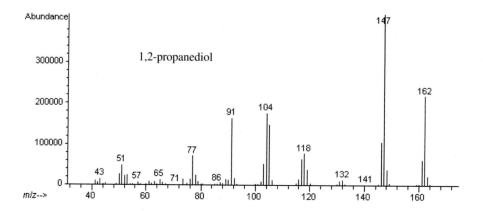
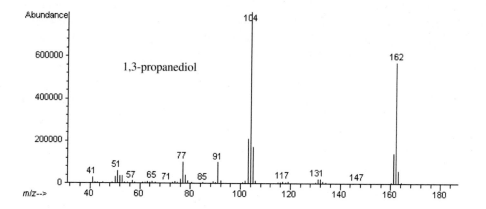

FIGURE 5.3 Mass spectra of ethylene glycol (A), 1,2-propanediol (B), and 1,3-propnaediol (C). *Author's own data.*

2,3-butanediol, lactate, and β-hydroxybutyrate [32,36–39]. Glycerol is generally present in low concentrations and does not cause clinically significant interference. However, glycerol-containing medications can cause erroneous results due to significant accumulation of glycerol in blood. Similarly, propylene glycol is a commonly used vehicle in many medications and may cause a false elevation in ethylene glycol in many enzymatic methods. 2,3-Butanediol is present in the blood of some alcoholics. The presence of high lactate and lactate dehydrogenase produces NADH and can cause false positive values of ethylene glycol [40]. On the other hand, glycolic acid is structurally similar to lactate and several cases have been reported in which chemistry analyzers misinterpreted the presence of glycolic acid as lactic acid [18]. Despite these limitations, enzymatic methods can provide useful information in patient management.

Some laboratories perform what is called "volatile or solvent screen." This test is a spot test and detects volatile alcohols. Since ethylene glycol and other glycols are not volatile and thus are not detected by this method. It is important to communicate with the clinician what is detected and what is not detected by this method. Miscommunication of a negative volatile test or the lack of understanding may result in a false assurance that ethylene glycol is not detected in the patient's sample. Similarly, if GC is used for the assay of volatiles, it is important to clarify which alcohols are detected. In most laboratories, the methods for alcohols or volatiles do not detect glycols.

Since ethylene glycol testing is not readily available, other tests are frequently used in the diagnosis of ethylene glycol exposure; the examination of urine for oxalate crystals is one of the frequently used tests. It is important to keep in mind that oxaluria is a late phenomenon and oxalate crystals may be absent in the early stages of ethylene glycol exposure. In fact, the absence of oxalate crystals in the presence of high concentrations of ethylene glycol has been reported [41]. Garg et al. [42] reported a fatal case in which a very high concentration of ethylene glycol was found in blood and urine; however, no oxalate crystals were found in the urine from the decedent. Furthermore, urine organic acid testing did not show elevated glycolic or oxalic acids; therefore, oxaluria is not very sensitive in the detection of ethylene glycol exposure. In addition, it is not specific for ethylene glycol poisoning. Calcium oxalate crystals are seen in an inherited disorder called primary hyperoxaluria. The excess consumption of foods, such as tomatoes, garlic, and spinach that are rich in oxalate, may show calcium oxalate crystalluria in healthy individuals; therefore, the presence or absence of oxaluria should be interpreted with caution.

Most antifreeze solutions contain fluorescein dye to detect radiator leaks. Some laboratories examine urine under the ultraviolet light to detect fluorescein dye. This test lacks sensitivity, as fluorescein dye may not be present in the solvent consumed by the patient. Also, fluorescein when present appears transiently in the urine following ingestion. Urine fluorescence also lacks specificity as many substances present in urine have fluorescence. This problem particularly has been seen in children. In a study, urines from 150 healthy children were examined under Wood's lamp and 81% of them had fluorescence [43]. The authors concluded that the determination of urine fluorescence is a poor screening tool for suspected antifreeze ingestion in children. In one study, six male volunteers were orally given a 0.6 mg of sodium fluorescein. This amount of sodium fluorescein corresponds to approximately 30 mL of ethylene glycol in typical antifreeze. Urine was collected at 2-h intervals. Using a Wood's lamp, visually detectable fluorescence was seen with 100% reliability for 2 h and 60% reliability for 4 h. In a second group of male volunteers, sodium fluorescein was detected for 6 h using a fluorometer. The authors concluded that florescence detection may be useful in the early hours of ethylene glycol poisoning [44]. Given its constraints, urine fluorescein testing should not be used to confirm or exclude an ethylene glycol exposure.

The osmolal gap has also been used to estimate the amount of ethylene glycol ingestion. Although it may be helpful in early stages of ethylene glycol poisoning, it should be used with great caution [45]. The osmolal gap is not sufficiently sensitive to pick up small ingestions. Also, it may only be useful in the early stages of ingestion when ethylene glycol has not metabolized. Once ethylene glycol has been metabolized, the osmolal gap may not be present because the metabolites that are negatively charged are generally accompanied by positively charged sodium cations. Since sodium is used in the calculation of the osmolal gap, the difference between the measured and calculated osmolarity diminishes. Also, the osmolal gap is nonspecific and may be due to any other cause. In a recent case report, a 60-year-old patient was admitted to the hospital with an altered mental status [46]. Initial testing showed high anion gap metabolic acidosis, elevated osmolal gap, and calcium oxalate crystals in his urine. With these findings, ethylene glycol poisoning was high in the differential. The patient was started on hemodialysis due to suspected ethylene glycol poisoning, but findings did not confirm the presence of ethylene glycol. The patient had increased creatine kinase, aspartate aminotransferase, and alanine aminotransferase. The clinical findings were attributed to the patient's low valproate level leading to seizures and rhabdomyolysis. This case highlights the poor specificity of adjunct laboratory tests.

GC methods are preferred for toxic alcohol detection and quantification but are not readily available in most health care facilities. Despite good specificity of GC, misidentification of other compounds such as ethylene glycol has been reported. In one case, propionic acid was identified as ethylene glycol, leading to the mother being accused of poisoning the child with ethylene glycol. Further testing and investigations showed that the child had methylmalonic acidemia and the GC peak identified as ethylene glycol was, in fact, from propionic acid [47]. In another report, 2,3-butanediol was mistakenly identified as ethylene glycol in plasma specimens from two alcoholic patients [48]. If the presence of ethylene glycol in a sample is questionable, GC/MS should be used for confirmation.

When using GC, the selection of an appropriate internal standard is important to avoid misidentification or an inaccurate quantification of ethylene glycol and other glycols. The use of 1,2-propanediol form of propylene glycol should not be used as it is a commonly used preservative, emollient, and vehicle for oral and intravenous medications. The other form of propylene glycol, 1,3-propanediol, is a commonly used internal standard. It is thought that 1,3-propanediol is not widely available and is, thus, a suitable internal standard; however, this may not be true. Garg et al. [49] reported a fatal case involving 1,3-propanediol. In this report the decedent was found with two antifreeze containers and a suicide note. Ethylene glycol toxicity was suspected and the sample was analyzed for ethylene glycol using GC;

however, no ethylene glycol was detected and the peak area of the internal standard, 1,3-propanediol was very large. It was over 20 times greater than the internal standard peak areas in the calibrators and controls. As the literature failed to show, any report on 1,3-propanediol toxicity, co-elution of 1,2-propanediol was suspected but the testing for 1,2-propanediol was negative. Further studies using GC/MS showed that the large peak was, in fact, 1,3-propanediol. The 1,3-propanediol was quantified using ethylene glycol as an internal standard and quantitated at 445 mg/dL. This was the first fatal case involving 1,3-propanediol. It is possible that cases of 1,3-propanediol toxicity have been missed, as it is a commonly used internal standard. For this reason, it is important to closely scrutinize the internal standard peak height or area to avoid underestimation of ethylene glycol. This report also highlights the effectiveness of GC/MS in the identification of various glycols.

PATIENT MANAGEMENT

Early identification, intervention, and treatment is the key to minimizing glycol's morbidity and mortality. General treatments of ethylene glycol toxicity through supportive measures include maintaining oxygenation/ventilation and treatment of convulsions and cardiac arrhythmias. Metabolic acidosis is treated with sodium bicarbonate and hypocalcemia by calcium gluconate or chloride. Specific antidotes are fomepizole (4-methylpyrazole) and ethanol. Both prevent ethylene glycol and other glycol's metabolism to their toxic metabolites by inhibiting alcohol dehydrogenase. Ethanol inhibits the glycol metabolism by inhibiting alcohol dehydrogenase. Ethanol levels should be maintained >100 mg/dL. At this ethanol level, alcohol dehydrogenase is almost completely saturated and results in clinical symptoms common to ethanol poisoning. However, due to the unpredictable ethanol kinetics and the potential for autoinduction, the dose may need to be increased over time. Since significant CNS depression and hypoglycemia may result, ethanol is no longer routinely recommended in adults and is not advisable in children with toxic glycol ingestions. To this regard, fomepizole is preferable over ethanol treatment because of its relative lack of adverse effects, despite the higher cost. Similar to ethanol, fomepizole dosing should be increased due to autoinduction of alcohol dehydrogenase. Hemodialysis is preferred in severe poisoning indicated by severe metabolic acidosis and plasma ethylene glycol levels of >50 mg/dL; however, treatment with fomepizole alone in patients with significantly elevated levels can be considered in those without acidosis or renal failure [50]. Adjunctive therapies include the administration of pyridoxine and thiamine which serve as cofactors by enzymes that produce nontoxic metabolites.

Treatment of toxicity associated with diethylene glycol and other glycol ethers includes supportive care along with alcohol dehydrogenase inhibitors, such as fomepizole. Hemodialysis can help correct acidosis in addition to enhancing removal of the parent compounds.

CONCLUSIONS

Ethylene glycol and other glycols poisoning are rare but not uncommon. The laboratory plays an essential role in the diagnosis and management of patients with glycols toxicity. As specific methods for glycols analysis are not readily available, physicians rely on many common laboratory tests. These laboratory tests should be interpreted with great caution and in conjunction with the patient's clinical picture. Enzymatic methods for ethylene glycol are fast but prone to interferences. GC methods are better but may pose a number of analytical and interpretive challenges. It is important to understand the limitations of these methods for best patient management.

REFERENCES

[1] Mowry JB, Spyker DA, Brooks DE, Zimmerman A, Schauben JL. Annual Report of the American Association of Poison Control Centers' National Poison Data System (NPDS): 33rd Annual Report. Clin Toxicol (Phila) 2015;2016(54):924–1109.

[2] Anderson IB. In: Olson KR, editor. Ethylene glycols and other glycols. New York: McGraw-Hill; 2018. p. 234–8.

[3] Baselt RC. Diethylene glycol. In: Baselt RC, editor. Disposition of toxic drugs and chemicals in man. Seal Beach, CA: Biomedical Publications; 2017. p. 670–1.

[4] Baselt RC. Ethylene glycol. In: Baselt RC, editor. Disposition of toxic drugs and chemicals in man. Seal Beach, CA: Biomedical Publications; 2017. p. 840–2.

[5] Hewlett TP, McMartin KE, Lauro AJ, Ragan Jr FA. Ethylene glycol poisoning. The value of glycolic acid determinations for diagnosis and treatment. J Toxicol Clin Toxicol 1986;24:389–402.

[6] Brent J, McMartin K, Phillips S, Burkhart KK, Donovan JW, Wells M, et al. Fomepizole for the treatment of ethylene glycol poisoning. Methylpyrazole for Toxic Alcohols Study Group. N Engl J Med 1999;340:832–8.

[7] Kowalczyk M, Halvorsen S, Ovrebo S, Bredesen JE, Jacobsen D. Ethanol treatment in ethylene glycol poisoned patients. Vet Hum Toxicol 1998;40:225–8.

[8] Eder AF, McGrath CM, Dowdy YG, Tomaszewski JE, Rosenberg FM, Wilson RB, et al. Ethylene glycol poisoning: toxicokinetic and analytical factors affecting laboratory diagnosis. Clin Chem 1998;44:168–77.

[9] Barceloux DG, Krenzelok EP, Olson K, Watson W. American Academy of Clinical Toxicology Practice Guidelines on the Treatment of Ethylene Glycol Poisoning. Ad Hoc Committee. J Toxicol Clin Toxicol 1999;37:537–60.

[10] Caravati EM, Erdman AR, Christianson G, Manoguerra AS, Booze LL, Woolf AD, et al. Ethylene glycol exposure: an evidence-based consensus guideline for out-of-hospital management. Clin Toxicol (Phila) 2005;43:327–45.

[11] Johnson B, Meggs WJ, Bentzel CJ. Emergency department hemodialysis in a case of severe ethylene glycol poisoning. Ann Emerg Med 1999;33:108–10.

[12] Erickson HL. Case report of a fatal antifreeze ingestion with a record high level and impressive renal crystal deposition. Case Rep Crit Care 2016;1–3.

[13] Guo C, Cenac TA, Li Y, McMartin KE. Calcium oxalate, and not other metabolites, is responsible for the renal toxicity of ethylene glycol. Toxicol Lett 2007;173:8–16.

[14] Froberg K, Dorion RP, McMartin KE. The role of calcium oxalate crystal deposition in cerebral vessels during ethylene glycol poisoning. Clin Toxicol (Phila) 2006;44:315–18.

[15] Poldelski V, Johnson A, Wright S, Rosa VD, Zager RA. Ethylene glycol-mediated tubular injury: identification of critical metabolites and injury pathways. Am J Kidney Dis 2001;38:339–48.

[16] Church AS, Witting MD. Laboratory testing in ethanol, methanol, ethylene glycol, and isopropanol toxicities. J Emerg Med 1997;15:687–92.

[17] Khajuria A, Krahn J. Osmolality revisited—deriving and validating the best formula for calculated osmolality. Clin Biochem 2005;38:514–19.

[18] Schelling JR, Howard RL, Winter SD, Linas SL. Increased osmolal gap in alcoholic ketoacidosis and lactic acidosis. Ann Intern Med 1990;113:580–2.

[19] Mycyk MB, Aks SE. A visual schematic for clarifying the temporal relationship between the anion and osmol gaps in toxic alcohol poisoning. Am J Emerg Med 2003;21:333–5.

[20] Introna Jr F, Smialek JE. Antifreeze (ethylene glycol) intoxications in Baltimore. Report of six cases. Acta Morphol Hung 1989;37:245–63.

[21] Brindley PG, Butler MS, Cembrowski G, Brindley DN. Falsely elevated point-of-care lactate measurement after ingestion of ethylene glycol. CMAJ 2007;176:1097–9.

[22] Graine H, Toumi K, Roullier V, Capeau J, Lefevre G. Interference of ethylene glycol on lactate assays. Ann Biol Clin (Paris) 2007;65:421–4.

[23] Porter WH, Rutter PW, Bush BA, Pappas AA, Dunnington JE. Ethylene glycol toxicity: the role of serum glycolic acid in hemodialysis. J Toxicol Clin Toxicol 2001;39:607–15.

[24] Porter WH, Rutter PW, Yao HH. Simultaneous determination of ethylene glycol and glycolic acid in serum by gas chromatography-mass spectrometry. J Anal Toxicol 1999;23:591–7.

[25] Hlozek T, Bursova M, Cabala R. Simultaneous and cost-effective determination of ethylene glycol and glycolic acid in human serum and urine for emergency toxicology by GC–MS. Clin Biochem 2015;48:189–91.

[26] Hlozek T, Bursova M, Cabalaa R. Fast determination of ethylene glycol, 1,2-propylene glycol and glycolic acid in blood serum and urine for emergency and clinical toxicology by GC-FID. Talanta 2014;130:470–4.

[27] Hanton SL, Watson ID. An enzymatic assay for the detection of glycolic acid in serum as a marker of ethylene glycol poisoning. Ther Drug Monit 2013;35:836–43.

[28] Perala AW, Filary MJ, Bartels MJ, McMartin KE. Quantitation of diethylene glycol and its metabolites by gas chromatography mass spectrometry or ion chromatography mass spectrometry in rat and human biological samples. J Anal Toxicol 2014;38:184–93.

[29] Williams RH, Shah SM, Maggiore JA, Erickson TB. Simultaneous detection and quantitation of diethylene glycol, ethylene glycol, and the toxic alcohols in serum using capillary column gas chromatography. J Anal Toxicol 2000;24:621–6.

[30] Dasgupta A, Blackwell W, Griego J, Malik S. Gas chromatographic-mass spectrometric identification and quantitation of ethylene glycol in serum after derivatization with perfluorooctanoyl chloride: a novel derivative. J Chromatogr B Biomed Appl 1995;666:63–70.

[31] Orton DJ, Boyd JM, Affleck D, Duce D, Walsh W, Seiden-Long I. One-step extraction and quantitation of toxic alcohols and ethylene glycol in plasma by capillary gas chromatography (GC) with flame ionization detection (FID). Clin Biochem 2016;49:132–8.

[32] Robson AF, Lawson AJ, Lewis L, Jones A, George S. Validation of a rapid, automated method for the measurement of ethylene glycol in human plasma. Ann Clin Biochem 2017;54:481–9.

[33] Blandford DE, Desjardins PR. A rapid method for measurement of ethylene glycol. Clin Biochem 1994;27:25–30.

[34] Juenke JM, Hardy L, McMillin GA, Horowitz GL. Rapid and specific quantification of ethylene glycol levels: adaptation of a commercial enzymatic assay to automated chemistry analyzers. Am J Clin Pathol 2011;136:318–24.

[35] Mahly M, Lardet G, Vallon JJ. Automated Cobas Mira kinetic enzymatic assay for ethylene glycol applied to emergency situations. J Anal Toxicol 1994;18:269–71.

[36] Malandain H, Cano Y. Interferences of glycerol, propylene glycol, and other diols in the enzymatic assay of ethylene glycol. Eur J Clin Chem Clin Biochem 1996;34:651–4.

[37] Jones AW, Hard L. How good are clinical chemistry laboratories at analysing ethylene glycol? Scand J Clin Lab Invest 2004;64:629–34.

[38] Tintu A, Rouwet E, Russcher H. Interference of ethylene glycol with (L)-lactate measurement is assay-dependent. Ann Clin Biochem 2013;50:70–2.

[39] Rooney SL, Ehlers A, Morris C, Drees D, Davis SR, Kulhavy J, et al. Use of a rapid ethylene glycol assay: a 4-year retrospective study at an academic medical center. J Med Toxicol 2016;12:172–9.

[40] Eder AF, Dowdy YG, Gardiner JA, Wolf BA, Shaw LM. Serum lactate and lactate dehydrogenase in high concentrations interfere in enzymatic assay of ethylene glycol. Clin Chem 1996;42:1489–91.

[41] Haupt MC, Zull DN, Adams SL. Massive ethylene glycol poisoning without evidence of crystalluria: a case for early intervention. J Emerg Med 1988;6:295–300.

[42] Garg U, Frazee III C, Johnson L, Turner JW. A fatal case involving extremely high levels of ethylene glycol without elevation of its metabolites or crystalluria. Am J Forensic Med Pathol 2009;30:273–5.

[43] Parsa T, Cunningham SJ, Wall SP, Almo SC, Crain EF. The usefulness of urine fluorescence for suspected antifreeze ingestion in children. Am J Emerg Med 2005;23:787–92.

[44] Winter ML, Ellis MD, Snodgrass WR. Urine fluorescence using a Wood's lamp to detect the antifreeze additive sodium fluorescein: a qualitative adjunctive test in suspected ethylene glycol ingestions. Ann Emerg Med 1990;19:663–7.

[45] Glaser DS. Utility of the serum osmol gap in the diagnosis of methanol or ethylene glycol ingestion. Ann Emerg Med 1996;27:343–6.

[46] Gaddam M, Velagapudi RK, Abu Sitta E, Kanzy A. Two gaps too many, three clues too few? Do elevated osmolal and anion gaps with crystalluria always mean ethylene glycol poisoning? BMJ Case Rep 2017;15:2017.

[47] Shoemaker JD, Lynch RE, Hoffmann JW, Sly WS. Misidentification of propionic acid as ethylene glycol in a patient with methylmalonic acidemia. J Pediatr 1992;120:417–21.

[48] Jones AW, Nilsson L, Gladh SA, Karlsson K, Beck-Friis J. 2,3-Butanediol in plasma from an alcoholic mistakenly identified as ethylene glycol by gas-chromatographic analysis. Clin Chem 1991;37:1453–5.

[49] Garg U, Frazee III CC, Kiscoan M, Scott D, Peterson B, Cathcart D. A fatality involving 1,3-propanediol and its implications in measurement of other glycols. J Anal Toxicol 2008;32:324–6.

[50] Velez LI, Shepherd G, Lee YC, Keyes DC. Ethylene glycol ingestion treated only with fomepizole. J Med Toxicol 2007;3:125–8.

FURTHER READING

Garg U, Lowry J, Algren A. Ethylene glycol and other glycols: testing and interpretation issues. In: Dasgupta A, editor. Alcohol and drugs of abuse testing. Washington, DC: AACC Press; 2009. p. 29–39.

Chapter 6

Introduction to Drugs of Abuse

Loralie J. Langman[1] and Christine L.H. Snozek[2]

[1]*Department of Laboratory Medicine and Pathology, Mayo Clinic, Rochester, MN, United States,* [2]*Department of Laboratory Medicine and Pathology, Mayo Clinic Arizona, Scottsdale, AZ, United States*

INTRODUCTION

The history of poisons dates to earliest times. Ancient man most likely observed toxic effects in nature partly by accident, by noting harmful or fatal effects following the ingestion of some plants or animal products [1]. At the same time, man discovered that nature also created plant and animal products that had pleasurable effects, thus the abuse of "drugs" began. Drug of abuse (DOA) testing is a field that encompasses the difficulties and concerns of areas ranging from clinical pharmacology to forensic toxicology and beyond. The number of pharmacologically active compounds that can be abused is immense and is increasing every day, thanks in large part to the dedicated effort of "garage chemists" seeking both to expand the repertoire of chemically induced effects and to evade the investigations of law enforcement and the laboratory. Later chapters will discuss in detail issues pertaining to individual drug classes, testing methodologies, and strategies for escaping detection; the following is intended as an introduction to the historical philosophy guiding DOA testing and to the future challenges awaiting us.

DRUGS OF ABUSE TESTING IN THE UNITED STATES

Due to concern about the effect of illegal drugs on the combat readiness of the armed forces, in 1971 the US Congress directed the Secretary of Defense to establish the means of identifying and treating drug-abusing military personnel [2]. Thus in the United States, the military was the first to mandate employee drug testing, with the objective of creating a "drug-free" federal workplace. At this time, concerns about drug abuse were increasing nationally, and many employers in the oil, chemical, transportation, and nuclear power industries began their own drug-testing programs. In 1986, the executive branch of the US government became actively involved, when President Ronald Reagan issued an Executive Order (Federal Register 1986) requiring drug testing of federal employees in safety- and security-conscious positions [3]. However, these governmental mandates did not include guidelines for the analytical procedures or standards for workplace drug testing; the resulting variability between testing programs caused much controversy and litigation. For this reason, the American Association of Clinical Chemistry established a task force to address the analytical concerns raised [4].

In 1988, the Department of Health and Human Services (DHHS) published the first mandatory guidelines for drug testing federal employees (CFR part 40), commonly known as the National Institute for Drug Abuse (NIDA) Guidelines. These outlined screening and confirmation cutoffs for the so-called NIDA 5, namely amphetamines, cocaine, marijuana, opiates, and phencyclidine [5]. Chemical structure of phencyclidine is shown in Fig. 6.1. NIDA subsequently became the Substance Abuse and Mental Health Services Administration (SAMHSA), an agency under the US DHHS [6,7], which currently holds responsibility for the guidelines. The initial rules have been amended several times: in 1994, to add alcohol testing and to change the assay cutoff for cannabis; in 1998, to alter the opiate cutoff [8], and most recently to include semisynthetic opioids and revise testing for amphetamine-type stimulants (Table 6.1).

Federal mandates have extended to nonworkplace DOA testing, such as for welfare recipients and within the criminal justice system. Drug testing has also expanded to athletic competitors, both professional and amateur, from worldwide settings like the Olympics down to local college events. Extending drug testing to high school athletes remains controversial, though commercial products are available for parents wishing to test their own children at this age.

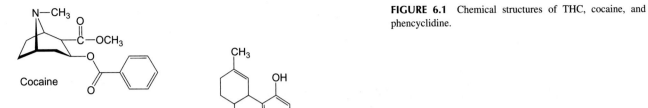

FIGURE 6.1 Chemical structures of THC, cocaine, and phencyclidine.

TABLE 6.1 Suggested Cutoff Concentrations for Workplace Drug Testing

	Urine (ng/mL)
Initial Test	
THC metabolite	50
Cocaine and metabolites	150
Codeine/morphine	2000
Hydrocodone/hydromorphone	300
Oxycodone/oxymorphone	100
6-AM (6-acetyl morphine)	10
PCP (phencyclidine)	25
Amphetamine/methamphetamine	500
MDMA/MDA/MDEA	500
Confirmatory Test	
Δ9-THC-COOH	15
Benzoylecgonine (BE)	100
Morphine	2000
Codeine	2000
Hydrocodone	100
Hydromorphone	100
Oxycodone	50
Oxymorphone	50
6-AM	10
PCP	25
Amphetamine	250
Methamphetamine	250
MDMA (3,4-methylenedioxymethamphetamine)	250
MDA (3,4-methylenedioxyamphetamine)	250
MDEA (3,4-methylenedioxy-N-ethylamphetamine)	250

Finally, testing for illicit drug use has become a commonplace part of diagnostic care in numerous medical settings, such as hospital emergency rooms, prenatal clinics, and delivery rooms; results may be included in a forensic investigation, depending on the situation.

In addition to wider applications outside of the workplace, the number of compounds analyzed in mandatory drug tests has also expanded as well. In recent years, media and government have focused on the ever-growing list of compounds considered illicit drugs, from synthetic cannabinoids to designer opioid, amphetamine, and cathinone derivatives. Although the recognition of a larger scope of compounds available for abuse has not extended to federally regulated DOA testing, clinical and toxicological laboratories have begun the process of expanding their testing repertoires to accommodate the growth.

DOA testing is structured somewhat uniquely among clinically relevant laboratory analyses, because federal regulations for workplace drug testing require two successive assays using different methodologies. This structure has been extended to analyses performed for nonfederal applications, though not uniformly. The initial, screening test is generally based on immunoassays targeted to a particular member of a drug class; e.g., opiate screens make use of morphine-specific antibodies that often do not cross-react with oxycodone. A "presumptive positive" screen therefore only provides a semiquantitative result because of the differences in antibody reactivity to structurally similar compounds and metabolites. Confirmation analysis using a second methodology, most commonly liquid or gas chromatography coupled to mass spectrometry, is required to provide definitive and frequently quantitative evidence of a positive drug test. The advantages of this strategy include cost savings due to elimination of negative samples using a relatively inexpensive screen, and the increased legal defensibility of two independent assays should a positive drug test be challenged. However, the intricacies of which compounds are readily detected and complex situations such as negative confirmation results following a positive screen can be quite confusing to clinicians and others outside the field of DOA testing.

There is an impressive assortment of analytical challenges facing current and future DOA testing laboratories. New "designer" drugs are constantly entering the illicit marketplace, as are readily available adulterants for evading detection by drug testing. At the same time, abuse of prescription and over-the-counter medications is a growing concern. Even within the traditional NIDA 5, there remain several unresolved issues, such as poor reactivity of screening methodologies to compounds within the drug classes. The remaining chapters of this text will discuss in detail various aspects of and challenges in testing for illicit drug use, from the NIDA 5 to designer drugs. As a beginning, we present here brief introductions to some of the most commonly abused compounds.

NATURAL CANNABINOIDS

Psychoactive products obtained from the plant *Cannabis sativa* have been used for euphoric effect for over 4000 years, and are currently the most widely used illicit drugs in the world [9]. Marijuana cigarettes are made from the leaves and flowering tops of the plant, while hashish and hash oil are prepared from a concentrated resin and a lipid-soluble extract, respectively. "Cannabinoids" refer to the >400 chemicals with mental and physical effects found in cannabis, of which delta-9-tetrahydrocannabinol (THC) is the most psychoactive compound [9]. Smoking releases >150 compounds in addition to THC; although most of these components lack psychoactivity, these compounds can have physiological effects [10]. When smoked, THC is quickly absorbed from the lungs into the bloodstream, from which it then rapidly distributes into tissues. THC is extensively metabolized, with the predominant metabolite being the inactive 11-nor-delta-9-tetrahydrocannabinol-9-carboxylic acid (THC-COOH); reporting of results for THC is typically done in terms of THC-COOH equivalents. Chemical structure of THC is shown in Fig. 6.1.

Specific cannabinoid receptors (CB1 and CB2) have been identified in the central and peripheral nervous systems [10]. The most prominent effects of activating CB receptors are euphoria and relaxation; users report feelings of well-being, grandiosity, and altered perception of passage of time. Dose-dependent perceptual changes (e.g., visual distortions), drowsiness, diminished coordination, and memory impairment may occur. Common effects such as increased appetite, attenuation of nausea, decreased intraocular pressure, and relief of chronic pain have led to the use of cannabinoids in medical therapeutics [11].

SYNTHETIC CANNABINOIDS

Originally created as research chemicals and subsequently modified extensively for recreational use, the synthetic cannabinoids comprise a wide array of CB1 agonists, many of which are more potent than THC at this receptor [12]. The compounds are often sprayed onto plant material to be smoked (similar to a marijuana cigarette) or dissolved into liquid for e-cigarette use. The positive effects of synthetic cannabinoids are similar—though often more intense—to those of THC, including relaxation, analgesia, and euphoria. However, in part because of their increased potency at CB1, and in part because of effects on other receptors, many synthetic cannabinoids are also more toxic than natural cannabinoids, with negative effects ranging from agitation and delirium to acute kidney injury, myocardial infarction, and death [13].

Several structural classes of synthetic cannabinoids have been described, and new members of the class appear constantly as suppliers modify existing chemical structures to evade detection and legislation. A variety of methodologies, including high-resolution mass spectrometry, have been utilized for screening and/or confirmation testing of the ever-changing array of new synthetic cannabinoids entering the market. The metabolism of synthetic cannabinoids is complex, but many related compounds have common metabolites which can help with detection. However, the number of synthetic cannabinoids, the lack of reliable pharmacological data, and the rapid evolution of use and abuse patterns make these compounds a particular challenge for clinicians, laboratories, and law enforcement alike. See Chapter 22, Review of Synthetic Cannabinoids on Illicit Drug Market, for more details.

COCAINE

Cocaine is an alkaloid found in the leaves of *Erythroxylon coca*, a shrub indigenous to the Andes. It has been used in clinical medicine for over 100 years, mainly for local anesthesia and vasoconstriction in nasal surgery, and to dilate pupils in ophthalmology. Sigmund Freud famously proposed its use to treat depression and alcohol dependence, but the realities of cocaine addiction quickly brought this idea to an end [11]. Cocaine increases alertness and euphoria through potent stimulation of the central nervous system. It acts by blocking reuptake of the neurotransmitters dopamine and norepinephrine [14], which raises blood pressure, heart rate, and body temperature. Chemical structure of cocaine is shown in Fig. 6.1.

Cocaine metabolism is complex, and occurs via both nonenzymatic hydrolysis and enzymatic transformation in the plasma and liver. The major metabolites, ecgonine methyl ester and benzoylecgonine, are inactive and are formed primarily in the liver. The half-life of cocaine is extremely short (0.5–1.5 h), while that of benzoylecgonine is long enough (4–7 h) to be more amenable to testing [14]. For this reason, screening assays for cocaine actually target benzoylecgonine; the antibodies in these assays generally have good cross-reactivity with cocaine [15–17]. Most confirmation assays offer quantitation of both parent drug and metabolite. Abusing alcohol and cocaine is very dangerous due to formation of cocaethylene, a metabolite of cocaine which is only formed in the presence of alcohol. See Chapter 17, Cocaine, Crack Cocaine, and Ethanol: A Deadly Mix, for more details.

AMPHETAMINE-TYPE STIMULANTS

Several stimulants and hallucinogens chemically related to phenylethylamine are collectively referred as amphetamine-type stimulants. Amphetamines are sympathomimetic amines that are often chiral compounds; generally, D-enantiomers stimulate the central nervous system, while L-enantiomers act peripherally, e.g., appetite suppression. In the past, many individuals using legitimately prescribed amphetamine for appetite control or depression also became unwittingly habituated to its effects, leading to its classification as a Schedule II controlled substance [9]. The use of amphetamine for depression has virtually ceased and the current medical uses include the treatment of narcolepsy and attention deficit disorders. Amphetamines increase synaptic dopamine concentrations, primarily by stimulation of presynaptic release rather than by blockade of reuptake. Increased levels of dopamine in the brain elicit euphoria, contributing to the addictive properties of amphetamines [18].

Most "amphetamine" immunoassays are designed to detect amphetamine or methamphetamine, though some assays are designed to more broadly capture the amphetamine-type stimulant group, which encompasses many of the so-called club drugs, including methylenedioxymethamphetamine (MDMA or ecstasy). Unfortunately, related (e.g., pseudoephedrine) and unrelated (e.g., psychotropic drugs) compounds can cross-react with the antibody producing false positive results [9]. In addition, some older protocols for gas chromatography–based confirmation assays can generate methamphetamine from ephedrine/pseudoephedrine in the specimen. This led to the original SAMHSA requirement for co-detection of amphetamine in order to report a positive methamphetamine result [9]. Most routine screening and confirmation assays do not distinguish between the D- and L-isomers although specially designed chiral derivatization protocols are capable of differentiating L-isomer from D-isomer. See Chapter 16, Critical Issues When Testing for Amphetamine Type Stimulants: Pitfalls of Immunoassay Screening and Mass Spectrometric Confirmation for

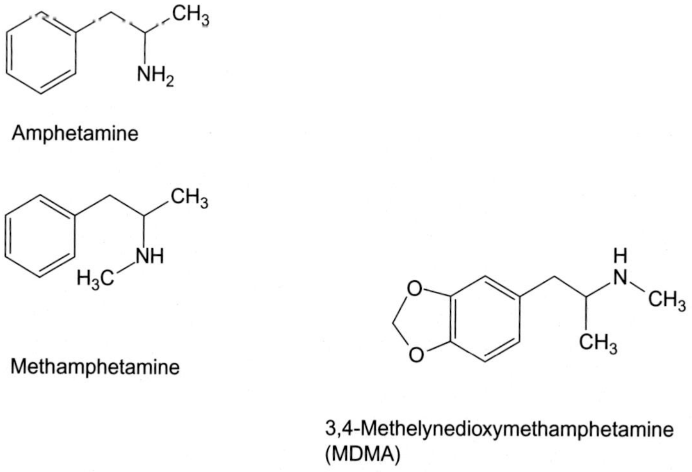

FIGURE 6.2 Chemical structures of amphetamine, methamphetamine, and MDMA.

Amphetamines, Methamphetamines, and Designer Amphetamines, for more details. Chemical structures of amphetamine, methamphetamine, and MDMA are shown in Fig. 6.2.

DESIGNER STIMULANTS

The terms "designer drugs" and "club drugs" originated in the 1980s. These drugs include phenylethylamine, benzylpiperazine, phenylpiperazine, pyrrolidinophenone, and cathinone derivatives. Although they originally were largely associated with all-night dance parties (raves) and nightclubs, use of designer drugs is no longer limited to these settings. Most designer drugs produce euphoria, energy, and a desire to socialize; they also promote social and physical interactions and distort or enhance visual and auditory sensations. Designer drugs previously had the incorrect reputation of being safe; however, experimental and epidemiological studies have revealed risks to humans such as life-threatening serotonin syndrome, hepatotoxicity, neurotoxicity, psychopathology, and abuse potential. See Chapter 19, Overview of Common Designer Drugs, for more details.

OPIATES AND OPIOIDS

Opiates consist of naturally occurring or semisynthetic alkaloids derived from opium, the dried juice from the seeds of the poppy (*Papaver somniferum*) [9]. Morphine, the principal natural opiate, is the structural building block for many of the semisynthetic opioids including heroin, oxycodone, oxymorphone, hydrocodone, hydromorphone, and levorphanol [3]. Opioids interact with the family of opioid receptors (mu, delta, and kappa). Opioid receptor agonists typically produce analgesia, while antagonists block this response. In addition to potent analgesic properties, opiates can also cause sedation, euphoria, and respiratory depression which gives opiates a high abuse potential [9]. Long-term use can lead to tolerance and both physical and psychological dependence [8].

Pain management programs often use urine drug testing to monitor compliance, diversion, or substitution of prescribed opiates, thus it is important to note that not all members of the drug class are reliably detected by opiate immunoassays [9]. Most opiate screening assays target morphine and display varying levels of cross-reactivity toward codeine, hydrocodone, and hydromorphone [19–21]. Oxycodone and oxymorphone demonstrate low cross-reactivity and often cannot be reliably measured by using screening assays for opiates. In addition to traditionally abused opiates such as morphine and heroin, misuse of synthetic opioids such as fentanyl is rapidly gaining in popularity. This presents an analytical challenge, as most synthetic opioids are simply not detectable by opiate assays. Therefore, tests designed specifically for these compounds should be used instead. See Chapter 25, Drug Testing in Pain Management, for in-depth discussion on drug monitoring in pain management.

Methadone, a synthetic opioid, is structurally unrelated to the natural opiates but is capable of binding to opioid receptors. These receptor interactions create many of the same effects as seen with natural opiates, including analgesia and sedation. However, methadone does not produce feelings of euphoria and has substantially fewer withdrawal symptoms than opiates such as heroin [22]. Methadone is used clinically to relieve pain, to treat opioid abstinence syndrome,

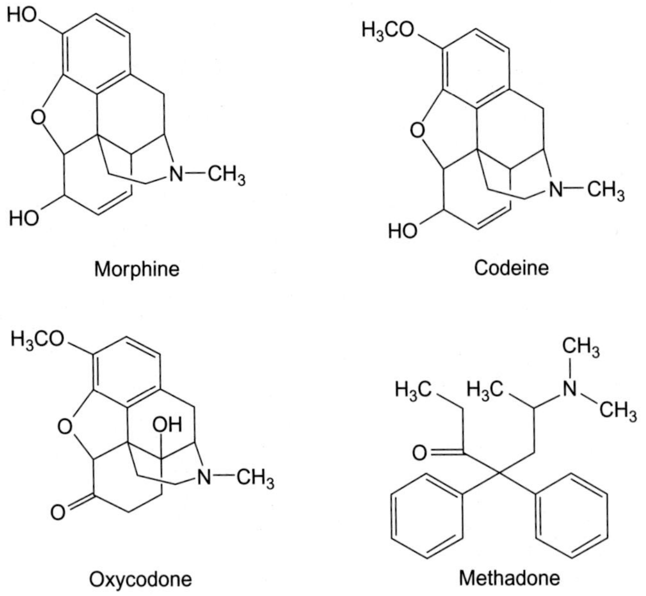

FIGURE 6.3 Chemical structures of morphine, codeine, oxycodone, and methadone.

and to treat heroin addiction in the attempt to wean patients from illicit drug use. Chemical structures of morphine, codeine, oxycodone, and methadone are shown in Fig. 6.3.

Patients taking methadone excrete both the parent drug and the major metabolite 2-ethylidene-1,5-dimethyl-3,3-diphenylpyrrolidine (EDDP) in urine [23–25]. Clinically, it is important to measure both compounds, as methadone excretion varies widely with dose, metabolism, and urine pH [9], while urine EDDP levels are pH-independent and therefore preferable for assessing compliance [9]. Analysis relies on confirmatory methods because in general, opiate immunoassays do not detect methadone or its metabolites while immunoassays for methadone do not detect EDDP and vice versa [22]. Some patients attempt to pass compliance testing by spiking methadone into their urine samples [26]. Under such circumstances a ratio of EDDP to methadone is very low, typically below 0.090, and strongly suggests manipulation of the sample [26].

BENZODIAZEPINES

As a class, benzodiazepines are some of the most widely prescribed drugs with applications such as anxiolytics, muscle relaxants, sleep aids, anesthetic adjuncts, anticonvulsants, and more [10,27] Benzodiazepines are positive modulators of the $GABA_A$ receptor which results in disinhibition of the mesolimbic dopamine system; this may explain the rewarding effects of benzodiazepines. [11]. Benzodiazepines cause sedation, impaired memory and cognition, and disinhibition, and have also been associated with paradoxical effects such as increased agitation and insomnia, especially in pediatric and elderly populations [27,28]. In general, benzodiazepines have a moderate potential for abuse that must be weighed against their beneficial effects [11,29], but the risk is lower than that seen with older sedatives or other recognized drugs of abuse [30]. Some individuals misuse benzodiazepines for the euphoric effects, but abuse more often occurs concomitantly with other drugs, such as anxiolysis during opiate withdrawal. Agents with the shortest half-life, highest potency (e.g., alprazolam, triazolam), and strongest lipophilicity (e.g., diazepam) tend to have the most abuse potential [30]. Chemical structures of some benzodiazepines are listed in Fig. 6.4.

The goal of benzodiazepine immunoassays is to provide sufficient cross-reactivity to detect several parent compounds and metabolites, with minimal reactivity to unrelated drugs. The degree of success in meeting this goal varies considerably between assays and manufacturers. Most immunoassays are calibrated on the common metabolites oxazepam, temazepam, or nordiazepam [31] and because of this assay design compounds with atypical structures (e.g., midazolam, chlordiazepoxide, and flunitrazepam) may not be detected [32,33]. Cross-reactivity with conjugated metabolites, e.g., lorazepam-glucuronide, is generally poor despite these metabolites being the predominant forms in urine. The wide range of potencies adds additional analytical challenge, as detection limits satisfactory for one benzodiazepine may not be enough for another. Detection windows also vary by compound, ranging from a few days up to several weeks after drug exposure.

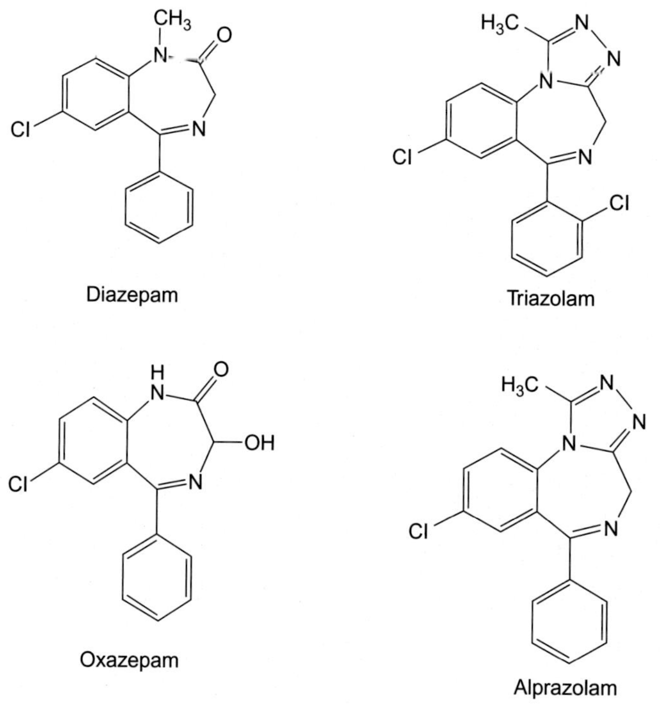

FIGURE 6.4 Chemical structures of some benzodiazepines.

CONCLUSIONS

Although SAMHSA recommends testing of five drugs/drugs class—amphetamines, cocaine, opiates, marijuana, and phencyclidine (PCP)—several other drugs such as benzodiazepines, synthetic cannabinoids, synthetic opioids, and other designer drugs are also abused. Testing of these drugs is important not only in the case of suspected drug overdose but also in preemployment or workplace drug testing in order to ensure a drug-free work environment. Although initial screening tests using immunoassays may cause false positive results, the mass spectrometric confirmation step is very robust and accurate. This confirmation step is mandatory for any legal drug testing protocol including workplace drug testing.

REFERENCES

[1] Fenton JJ. Toxicology: A case-oriented approach. Boca Raton, FL: CRC Press; 2002.
[2] Jenkins AJ, Levine B. Forensic drug testing. Principles of forensic toxicology. Washington, DC: American Association for Clinical Chemistry Press; 1999.
[3] Federal Register. Executive order 12564, Drug-free federal workplace. Federal Register 1986;51:32889−983.
[4] Bernard S, Neville KA, Nguyen AT, Flockhart DA. Interethnic differences in genetic polymorphisms of CYP2D6 in the U.S. population: clinical implications. Oncologist. 2006;11(2):126−35.
[5] Federal Register. Procedures for transportation workplace drug and alcohol programs: final rule, 49 CFR Part 40. Federal Register 2008;65 79462-75579. Subpart F—Drug Testing Laboratories 4087.
[6] Federal Register. Mandatory guidelines for federal workplace drug testing programs. Federal Register 1988;53:11970−89.
[7] Federal Register. Mandatory guidelines for federal workplace drug testing programs. Federal Register 1994;59:29908−31.
[8] Federal Register. Changes to the testing cut-off levels for opiates for federal workplace drug testing programs. Federal Register 1998;60:57587.
[9] Principles of forensic toxicology. In: Levine B, editor. 4th ed Washington, DC: AACC Press; 2017. 533 pp.
[10] Fauci Anthony S, Braunwald Eugene, Kasper Dennis L, Hauser Stephen L, Longo Dan L, Jameson J Larry, editors. Harrison's principles of internal medicine. 17th ed New York City, NY: The McGraw-Hill Companies, Inc; 2008. 2754 p.
[11] http://www.accessmedicine.com/resourceTOC.aspx?resourceID = 16. Basic & Clinical Pharmacology. New York: The McGraw-Hill Companies, Inc. Available from: http://www.accessmedicine.com/resourceTOC.aspx?resourceID = 16.
[12] Kemp AM, Clark MS, Dobbs T, Galli R, Sherman J, Cox R. Top 10 facts you need to know about synthetic cannabinoids: not so nice spice. Am J Med 2016;129(3):240−244 e1.
[13] Riederer AM, Campleman SL, Carlson RG, Boyer EW, Manini AF, Wax PM, et al. Acute poisonings from synthetic cannabinoids—50 U.S. Toxicology Investigators Consortium Registry Sites, 2010−2015. MMWR Morb Mortal Wkly Rep 2016;65(27):692−5.

[14] Langman LJ, Bechtel LK, Meier BMCH. Clinical toxicology. In: Nader Rifai ARH, Wittwer Carl T, editors. Tietz textbook of clinical chemistry and molecular diagnostics. 6th ed St. Louis, MO: Elsevier Saunders; 2018. p. 832–87.
[15] la Porte CJ, Droste JA, Burger DM. False-positive results in urine drug screening in healthy volunteers participating in phase 1 studies with efavirenz and rifampin. Ther Drug Monit 2006;28(2):286.
[16] Rossi S, Yaksh T, Bentley H, van den Brande G, Grant I, Ellis R. Characterization of interference with 6 commercial delta9-tetrahydrocannabinol immunoassays by efavirenz (glucuronide) in urine. Clin Chem 2006;52(5):896–7.
[17] Protonix® (pantoprazole) package insert. Philadelphia, PA: Wyeth-Ayerst Company; 2003.
[18] Paoletti P, Fornai E, Maggiorelli F, Puntoni R, Viegi G, Carrozzi L, et al. Importance of baseline cotinine plasma values in smoking cessation: results from a double-blind study with nicotine patch. Eur Respir J 1996;9(4):643–51.
[19] Reisfield GM, Salazar E, Bertholf RL. Rational use and interpretation of urine drug testing in chronic opioid therapy. Ann Clin Lab Sci. 2007;37(4):301–14.
[20] Liu RH. Evaluation of Common Immunoassay Kits For Effective Workplace Drug Testing. In: Liu R, Goldberger B, editors. Handbook of Workplace Drug Testing. Washington, DC: AACC Press; 1995. p. 67–129.
[21] Smith ML, Hughes RO, Levine B, Dickerson S, Darwin WD, Cone EJ. Forensic drug testing for opiates. VI. Urine testing for hydromorphone, hydrocodone, oxymorphone, and oxycodone with commercial opiate immunoassays and gas chromatography-mass spectrometry. J Anal Toxicol 1995;19(1):18–26.
[22] Yaksh T, Wallace M. Opioids, Analgesia, and Pain Management. In: Brunton LL, Hilal-Dandan R, Knollmann BC, editors. Goodman &; Gilman's: The Pharmacological Basis of Therapeutics. 13th ed New York, NY: McGraw-Hill Education; 2017.
[23] Eap CB, Buclin T, Baumann P. Interindividual variability of the clinical pharmacokinetics of methadone: implications for the treatment of opioid dependence. Clin Pharmacokinet 2002;41(14):1153–93.
[24] Ferrari A, Coccia CP, Bertolini A, Sternieri E. Methadone--metabolism, pharmacokinetics and interactions. Pharmacol Res 2004;50(6):551–9.
[25] Baselt RC. Disposition of Toxic Drugs and Chemicals in Man. 7th ed. Foster City, CA: Chemical Toxicology Institute; 2005.
[26] Galloway FR, Bellet NF. Methadone conversion to EDDP during GC–MS analysis of urine samples. J Anal Toxicol 1999;23(7):615–19.
[27] Brunton LL, Hilal-Dandan R, Knollmann BC, editors. Goodman & Gilman's: The Pharmacological Basis of Therapeutics, 13th ed. New York, NY: McGraw-Hill Education; 2017.
[28] Rothschild AJ, Shindul R, Viguera A, Murray M, Brewster S. Comparison of the frequency of behavioral disinhibition on alprazolam, clonazepam, or no benzodiazepine in hospitalized psychiatric patients. J Clin Psychopharmacol 2000;20(1):7–11.
[29] Woods JH, Katz JL, Winger G. Benzodiazepines: use, abuse, and consequences. Pharmacol Rev 1992;44(2):151–347.
[30] Uhlenhuth EH, Balter MB, Ban TA, Yang K. International study of expert judgment on therapeutic use of benzodiazepines and other psychotherapeutic medications: IV. Therapeutic dose dependence and abuse liability of benzodiazepines in the long-term treatment of anxiety disorders. J Clin Psychopharmacol 1999;19(6 Suppl. 2):23S–29SS.
[31] Green KB, Isenschmid DS. Medical Review Officer Interpretation of Urine Drug Testing Results. Forensic Sci Rev 1995;7:41–60.
[32] Colbert DL. Drug abuse screening with immunoassays: unexpected cross-reactivities and other pitfalls. Br J Biomed Sci 1994;51(2):136–46.
[33] Garretty DJ, Wolff K, Hay AW, Raistrick D. Benzodiazepine misuse by drug addicts. Ann Clin Biochem 1997;34(Pt 1):68–73.

Chapter 7

Legal Aspects of Drug Testing in US Military and Civil Courts

John F. Jemionek and Marilyn R. Past
Division of Forensic Toxicology, Armed Forces Medical Examiner System, Armed Forces Institute of Pathology, Rockville, MD, United States

INTRODUCTION

US military courts are federal criminal courts and as such follow the relevant judicial processes encountered in US civilian criminal courts. The judges and attorneys in US military courts meet the same requirements of board examination, appointment, and qualifications as their civilian court counterparts. US military courts have some unique procedural differences including that witnesses may be questioned by jury members and the accused (defendant) may choose to have jury members, not the judge, render the sentencing decision. But the rules of evidence in US military courts vary little, if at all, from civilian criminal courts. In terms of evidentiary proceedings, the two prominent standards applicable to expert testimony derive from court decisions in *Frye v. United States* [1] and *Daubert v. Merrill Dow Pharmaceuticals* [2].

SCIENTIFIC EVIDENCE

Scientific evidence produced in the court, as well as experts explaining scientific findings to the judge and jury, must meet rigorous standards in order for the court to accept such evidence. These important issues are discussed in this section.

Frye Standard for Scientific Evidence or Examination

The *Frye* standard arose from *Frye v. United States* [1]. In the *Frye* decision, the US Circuit Court for the District of Columbia ruled that expert testimony based upon then novel polygraph (lie detector) results was not permitted. Under the *Frye* standard, scientific evidence or examinations presented to the court must be interpreted as, "a well-recognized scientific principle or discovery, [and] the thing from which the deduction is made must be sufficiently established to have gained general acceptance in the particular field in which it belongs" [3]. In the judicial application of the *Frye* standard, trial or defense counsel had to provide an expert to speak to the validity of the science behind the evidence or examination conducted. The use of novel techniques, procedures, instrumentation, or principles required the courts to evaluate them based upon published papers, books, and judicial precedence on the evidence at issue to determine the reliability and "general acceptance in the particular field in which it belongs." The *Frye* standard has been generally adopted into Rule 702 of the Federal Rules of Evidence as the primary standard for acceptance of scientific evidence or examinations. For most cases, the *Frye* standard of scientific evidence has been superseded by the *Daubert* standard, which addresses not only the scientific analysis or examination procedures but also the admissibility of expert testimony [4] under the Federal Rules of Evidence [5].

Daubert Standard for Expert Testimony/Opinion

The *Daubert* standard was established in 1993 by the US Supreme Court decision in the case of *Daubert v. Merrill Dow Pharmaceuticals*, in which the admissibility of expert witness testimony was addressed [2]. Under the *Daubert* standard, a trial judge must evaluate whether the expected testimony and opinions of expert witnesses are both relevant and reliable [4,6]. First, under relevancy, the judge must determine whether the expert's testimony is applicable to the facts of the case. For example, the moral testimony of an individual with a PhD in theology may not be considered relevant in the determination of the technical expertise of a technician running a mass spectrometer. Second, the testimony of the expert must be based upon scientific methods. The science must be supported by empirical testing that is reliable, reproducible and accurate; it must be subject to peer review and publication in recognized journals in a relevant field of science; it must be based upon theories and analytical techniques that are generally accepted by the relevant scientific community; and, there must be an understanding of any inherent bias, limitations, or error potential in the scientific method employed. In a subsequent ruling, *Kumho Tire Co. v. Carmichael*, the Supreme Court indicated that the judicial oversight of trial proceedings applied to all expert testimony, including testimony both scientific and nonscientific [7]. In response, the United States Congress thereby amended Rules 701 and 702 of the Federal Rules of Evidence. Similar changes were made to the corresponding sections of the Military Rules of Evidence.

Federal Rules of Evidence

It is important for laboratory scientists and expert witnesses to understand the guidelines that govern not only expert testimony but also the basic rules that pertain to the production of laboratory documents for court and the certification of these documents as authentic. Some of the pertinent Federal (and Military) Rules of Evidence are presented, along with a discussion of how they apply to forensic testimony and testing.

Rule 701

Rule 701 applies to opinion testimony by lay witnesses whereby the witness is not testifying as an expert and the testimony consists of opinions or inferences not based on scientific, technical or other specialized knowledge that is covered by Rule 702. This is usually not the rule under which expert testimony regarding drug testing is given, but rather covers normal nonexpert witness opinion testimony.

Rule 702

Rule 702 addresses testimony by experts who have scientific, technical, or other specialized knowledge that will assist in an understanding of evidence or to determine a fact in issue. However, a witness must be qualified before the court as an expert. This determination is based on the individual's knowledge, skill, experience, training, or education and the relevance of this background to address the evidence in question. The expert's testimony and opinions must be (1) based upon sufficient facts or data, (2) the product of reliable principles and methods, and (3) the reliable application of the principles and methods to the facts of the case.

Rule 703

Rule 703 applies to an expert's opinion proffered pertinent to the trial case based on facts or data provided to the expert or personally observed in the course of the expert's professional practice. In the medical field, the expert may base his opinion on facts or data provided in the case that are similar to data or facts the expert would reasonably rely upon in forming an opinion in the expert's professional practice. Under Rule 703, an expert's opinion on facts or data relevant to the case may be attained through (a) first-hand observation of the witness's or defendant's testimony and a professional evaluation of their statements. Expert witness's opinions or evaluations must be supported by published information in peer-reviewed journals or the International Classification of Disease (ICD) or from data or facts pertinent to the case presented to the expert outside the court.

Rule 704a

An expert's opinion is not objectionable based solely on the fact that the expert's opinion encompasses an ultimate issue pertinent to the case.

Rule 704b

This rule limits the extent of an expert's opinion in trial testimony. The expert must not proffer an opinion whether the defendant was or was not able to tell right from wrong or criminal intent that is a pertinent element of the criminal charges. Rule 704b affirms that the "trier of facts," be it a judge or jury, determines questions of fact at trial. Application of Rule 704b was clarified in *United States v. Combs*, 369 F.3d 925, 940 (6th c.2004) [8]. The Sixth

Circuit stated, "Decisions applying Rule 704b to the expert testimony of law enforcement official have found it significant whether the expert actually referred to the intent of the defendant or, instead, simply described in general terms the common practices of those who clearly do possess the requisite intent, leaving unstated the inference that the defendant, having been caught engaging in more or less the same practices, also possessed the requisite intent." In *United States v. Combs*, the officer's testimony did not violate Rule 704b as the officer did not testify as to the intent of the defendant to distribute drugs, but rather the defendant's conduct would be consistent with intent to distribute drugs. This testimony left the "trier of fact," the jury, to determine whether the defendant possessed the required intent [9]. Other case law consistent with Rule 704(b):

United States v. Winbush, 580 F.3d 503, 512 (Seventh Circuit, 2009) [10]
United States v. Freeman, 498 F.3d 893, 900-01, 906-07 (Ninth Circuit, 2007) [11]

Rule 705

Rule 705 applies to the disclosure of facts or data underlying expert opinion. An expert may testify in court, render opinion or interpretation in a case without first testifying to the various underlying scientific principles as found in texts, articles, or publications that form the basis of the testimony, opinion, or interpretation given. However, the court may require the expert to disclose the documentation upon which such testimony or opinion is based.

Rule 706

Rule 706 concerns the court appointment of experts. A court on its own or on the request of either defense or prosecution counsel may appoint an expert witness agreed upon by the counsels or may appoint an expert of the court's own selection. Also, the expert must consent to serve as a witness. After agreeing to serve and being appointed by the court, the witness shall be subject to examination by the counsel requesting expert testimony and cross-examination by opposing counsel.

Rule 801: Hearsay Definitions

Hearsay is a fundamental legal principle that controls what and how some evidences can be entered at trial. In order to understand this area, it is important to review in full the definitions contained in Rule 801. A brief summary of definitions follows:

1. Statement: A statement is either an oral or written assertion or a nonverbal communication of a person (such as a nodding or shaking of the head to indicate affirmation or negation) if intended by the person as an affirmation or negation.
2. Declarant: A declarant is a person who makes a statement, assertion, testimony, or opinion.
3. Hearsay: Hearsay is a statement other than the one made by the declarant while testifying at the trial or hearing that is offered in evidence to prove the truth of the matter asserted.
4. Statements are not hearsay if they are prior statements of witnesses under certain circumstances, statements of identity, statements by a party-opponent or those authorized by that party to make statements, statements of coconspirators, or statements made by employees or agents in the scope of applicable employment or relationship.

Rule 802: Hearsay Rule: Hearsay is not admissible in court except as provided under these rules, or those prescribed by the courts of the United States or by act of Congress.

Rule 803 and 804: Hearsay Exceptions. There are over two dozen exceptions listed in the Rules of Evidence; however, the exceptions most pertinent to the court proceedings that a forensic toxicologist might have to support come from Rule 803 and are as follows:

- Present sense impressions: Statements describing events or conditions perceived during the course of the event or immediately thereafter by the declarant.
- Excited utterances: Statements made while the declarant was under excitement or stress in the course of events taking place.
- Statements for purposes of medical diagnosis or treatment: Statements reflective of the medical history, characterization of symptoms, or the recording of physiological characteristics reasonably related to diagnosis or treatment.
- Recorded recollections: A memorandum of record about a matter made when the matter was fresh in the memory of the declarant and is correctively reflective of the knowledge but about which the declarant now has insufficient recollection.

- Records of regularly conducted activity (also called the "business records exception"): Documents, reports, and records kept in the course of normal business activities and shown by the testimony or certification of the custodian of such records to be complete, accurate and trustworthy.
- Public records and reports.
- Records of vital statistics.
- Statements in ancient documents: Documents in existence 20 years or more whose authenticity is established.
- Market reports and similar commercial publications.
- Learned treatises: Statements contained in published treatises, periodicals, or pamphlets on a subject and established as a reliable authority by the testimony, admission of the declarant or other expert testimony, or by judicial notice.
- Reputation as to character: Reputation of a person's character among associates or in the community.
- Judgment of previous convictions.

Rule 902: Self-Authentication

This rule states that extrinsic evidence of authenticity as a necessary preadmission condition is not required with respect to the 12 examples given under this rule. The example that pertains most to laboratory practices and the court is Rule 902{11}:

> Certified domestic records of regularly conducted activity: The original or a duplicate of a domestic record of regularly conducted activity that would be admissible under Rule 803{6} if accompanied by a written declaration of its custodian or other qualified person, in a manner complying with any Act of Congress or rule prescribed by the Supreme Court pursuant to statutory authority, certifying that the record:
> 1. was made at or near the time of occurrence of the matters set forth by, or from information transmitted by, a person with knowledge of those matters;
> 2. was kept in the course of regularly conducted activity; and
> 3. was made in the regularly conducted activity as a regular practice.
>
> A party intending to offer a record into evidence under this paragraph must provide written notice of that intention to all adverse parties, and must make the record and declaration available for inspection sufficiently in advance of their offer into evidence to provide an adverse party with a fair opportunity to challenge them [5].

LABORATORY RECORDS AND THE COURT

Some of the previously cited Federal Rules of Evidence pertain to laboratory documentation. Record retention and handling procedures are important areas for most forensic laboratories. Generally, lawyers request copies of the chain of custody and testing documentation for a defendant's sample. This compiled information is commonly called a laboratory records package. Documents provided in these packages are generated in the course of regularly conducted business at forensic laboratories and qualify under the Rules of Evidence 803(6) exception of the hearsay rule. As a result, admissibility of the laboratory record package into a legal proceeding does not usually require separate witnesses to testify about processing, testing, and reporting of results. These duplicate records (certified true copies) are admissible under the Rules of Evidence 902{11}, if accompanied by a written declaration from a laboratory records custodian. This written declaration must accompany the release of any package of laboratory records. Use of the specific language in Rule 902{11} will usually prevent any problems with admissibility of the laboratory records package into the legal proceedings.

EXPERT WITNESS IN THE COURTROOM

The acceptance of an individual as an expert before the court is usually authenticated after a copy of the expert witness's curriculum vitae (CV) is placed into evidence. It is important that the CV be concise but document pertinent facts that establish the witness as an expert in the field for which testimony is to be given. This includes advanced education degrees; employment positions to include a brief listing of responsibilities, publications, relevant training, board certification(s), and affiliations with professional organizations; peer-reviewed publications; attendance at professional meetings and so forth. Proficiency in the field is often affirmed orally by listing the various periodicals, journals, and other reference sources routinely reviewed to maintain current knowledge in the applicable field for which testimony is to be given.

Role of the Expert Witness Versus Expert Consultant

As indicated in Rule 702, the introduction of scientific evidence into judicial proceedings may require an individual to be appointed to either the prosecution or defense (Rule 706). Appointment may be either as an expert witness or as an expert consultant. The distinction is important in the roles a witness, as opposed to a consultant, may play in court testimony. As an expert witness, it is anticipated that the individual will be called to testify under oath as an expert for the interpretation of scientific data introduced into evidence. An expert witness may be interviewed by both the prosecution and defense prosecutor prior to court proceedings regarding the expert's anticipated testimony. As an expert witness, the individual may not be permitted to be present in the courtroom to hear the testimony of other witnesses in the case, but such permission to remain should always be asked by the respective counsel. Exceptions may be granted where the expert is permitted to be present in the court during the testimony of other witnesses in order to evaluate the facts of their testimony relative to other evidence introduced at trial. Correspondence, discussions, and documentation between the expert witness and the prosecuting attorneys are discoverable and may be requested by the opposing counsel.

As an expert consultant, it is anticipated that the individual will not be called to testify under oath. Prosecution consultants are usually requested to review pertinent documents, reports, testing data, and so forth to determine whether there is sufficient forensic evidence to charge the defendant with violations of the law. As a prosecution consultant, the written report and opinions will be shared with the defense counsel for possible consideration of plea agreements or citing the specification of law for charges against the defendant. The report may serve as the basis by which the defense counsels will frame a defense strategy to refute the evidence to be presented in trial. The defense counsel, upon receipt of the prosecution's consultant report, may request that an expert consultant be appointed to assist in the defense's separate evaluation of the prosecution's evidence. As a defense consultant, any discussions, reports, documentation between the defense counsel and the defense expert consultant are privileged and not discoverable by the prosecution. Defense consultants may not be interviewed prior to trial by prosecuting attorneys. The defense consultant may sit through the trial proceedings to hear the testimony of witnesses, interpret such testimony for the defense counsel, assist in framing of questions to witnesses, and provide literature references or documentation to challenge the prosecution's witnesses' testimony or substantiate the challenges by the defense counsel. If defense counsel determines that an interpretation rendered by the prosecution's expert must be refuted by another expert, then the role of defense consultant changes to defense expert witness. Upon calling a defense consultant as a defense expert witness, the prosecution counsel will request a recess to interview the expert to determine his or her position and anticipated testimony. All documents, discussions, records relevant to the issues upon which the defense expert witness will be testifying, which were previously provided as a defense consultant to the defense counsel, are no longer privileged and may be requested by the prosecuting attorney as discoverable documents or discussion items.

Expert Witness Preparation for Testimony

A court appearance as an expert witness or consultant requires preparation in multiple areas. It is important to establish communication with the requesting counsel and their paralegal assistants regarding a summary of the case, the expected role as an expert with regard to the evidence in question, and exactly how the expected testimony may be relevant to the case. As an expert, impartiality must be maintained in the interpretation and review of evidence. As part of the preparation, a request should be made for copies of pertinent laboratory reports, investigative reports, and preliminary hearings in the case including preliminary testimony or statements of witnesses, the victim, and the defendant as well as any discovery requests pertinent to expert's appearance and testimony. The review must be complete with attention to the forensic aspects of testing conducted in terms of proper collection and positive identification of the donor, positive labeling, chain of custody, security of the sample, and forensic documentation of testing methodologies and results. Forensic analysis encompasses all documents pertaining to the individuals processing the sample, instrumentation employed in analysis, date and times of analysis, and the scientific review, certification, and reporting of laboratory results. The expert witness must consider the accuracy and the precision of the results reported and any errors or procedures that may call into question the forensic identification or integrity of the sample tested. The expert must understand the underlying science employed and discuss the science in lay terms. Any concerns or limitations in the testing data that preclude the rendering of an opinion on the accuracy or interpretation of evidence being presented must be discussed as soon as possible with the counsel requesting the expert's assistance. This affords counsel time to request any additional testing that may be required, changes to the presentation of evidence in the case, decisions in calling witnesses or introduction of evidence, or even the plea to be entered or accepted.

If called to testify, the expert's testimony must be factual, unbiased, and complete to the questions posed. Unless asked to elaborate, the expert should concisely answer only the questions posed. This is often necessary to avoid giving information that is not allowed under the Rules of Evidence or that has previously been ruled as inadmissible by the judge. The attorneys will usually carefully craft their questions in order to elicit only specific and narrow information. The expert should not attempt to render an opinion or testimony to an ambiguous or misleading question. The expert can state that because of the manner in which the question was asked or phrased, he or she is unable to answer. The judge may instruct the counsel to rephrase the question or may direct the expert to answer the question as a hypothetical situation, or the counsel may decide to withdraw the question. The expert should be prepared to validate his or her opinions or interpretation of the evidence based upon published data and references and should be prepared to produce such documents for presentation as exhibits and evidence in the case. Usually such documents are presented by the expert witness to counsel during pretrial interviews and may be challenged by the opposing counsel during testimony in terms of applicability, interpretation, and any scientific limitations. The expert must be prepared to discuss specifics of the referenced articles, methodologies employed, relevance to the case in question, bias or limitations in the studies, and any exceptions to the science involved. The expert witness's credentials may be challenged as to his or her credibility and experience in the area being discussed. The expert witness must be prepared to address such challenges or limit the scope of his or her testimony and opinions.

KEY ELEMENTS OF LABORATORY OPERATIONS AND THE LEGAL SYSTEM

There are many key elements for proper operation of a forensic or crime laboratory. These issues are discussed in this section.

Laboratory Certification and Accreditation

Forensic evidence goes beyond specimens normally associated with the criminal justice system such as fingerprints, DNA, or hair analysis. Urgent care biological fluids obtained for medical treatment or diagnosis are routinely drawn in emergency room (ER) settings. The results of such samples can be of interest in civil or criminal law to determine whether the individual was under the influence of drugs or medications that affected the person's ability to operate a motor vehicle at the time of the accident. It was also possible that the individual was not taking the prescribed doses of medication as required to prevent seizures. Answers to such questions may be determined in the course of hospital testing and may be of interest in a court trial when personal injury may have occurred.

Clinical laboratories affiliated with hospitals usually conduct testing on biological samples to assist health care providers to ascertain the status of the patient, aid in the diagnosis of underlying medical conditions, and monitor the effectiveness and the progress of patients receiving treatment. Commercial clinical laboratories may provide specialized testing not offered within a hospital or health care facility. Both types of laboratories receive licensure for operation and routine inspection by either state or federal accreditation agencies. Participation in quality assurance (QA) programs to monitor testing accuracy and precision is available through professional organizations such as the College of American Pathologists (CAP), state proficiency testing programs, or commercial QA programs in the testing of blinded samples. All hospitals and commercial clinical laboratories operating within the United States for diagnostic purposes are regulated by the Centers for Medicare and Medicaid Services (CMS) through the Clinical Laboratory Improvement Amendments (CLIA). The exceptions to CLIA are laboratories performing research or testing of nonhuman samples such as veterinary clinics and veterinary testing facilities [12]. Certification and accreditation standards are also important for forensic laboratories that perform workplace drug testing. These laboratories may be certified by the American Board of Forensic Toxicology (ABFT), the Substance Abuse and Mental Health Services Administration (SAMHSA), CAP, Department of Defense (DoD) laboratory certifications, or regulated by state licensure requirements as applicable. Participation in applicable proficiency testing programs, whether by SAMSHA, CAP, or DoD, is also important in maintaining laboratory standards and these records sometimes are the subject of discovery by the court as will be discussed later.

Personnel Certifications and Licensures

Should individuals be called to give testimony regarding testing conducted in their facility, the educational background, training, and technical competency of the staff involved in handling and testing of the sample(s) will be reviewed. Fortunately, in most clinical and commercial laboratories, these aspects of personnel certification and licensure are already addressed by the CLIA legislation and by similar accreditation programs in forensic laboratories. Personnel files training records, certifications, and educational degrees attained must be kept current and on file for inspection. Disciplinary actions, suspensions, and decertification are all discoverable in a discovery request or a court-ordered production of discovery. Clarity of records in the personnel files is important to avoid misinterpretation or misconstruing information regarding personnel actions as to an individual's qualifications, credibility, or competency in performing forensic testing. For example, removal of individual's certification to conduct a testing procedure due to a promotion or the removal of certification due to discontinuation or replacement of a testing procedure within the laboratory should be clearly discerned as an administrative action as opposed to a disciplinary suspension of duties.

Specimen Security, Chain of Custody, and Testing Procedures

Similar to ABFT or SAMHSA laboratory certification, the CLIA regulations govern certification of the clinical testing laboratory, validation of procedures and instrumentation, QA/quality control (QC) programs, and outside peer inspections. There are two areas that separate clinical laboratory operations from forensic testing laboratories: first, in the documentation of chain of custody, secure storage of samples, and controlled access into various laboratory spaces, and second, the two-tiered testing and confirmation analysis for forensic purposes, such as head- space gas chromatography (GC) for blood alcohol concentration (BAC) volatile analysis and GC-mass spectrometry (GC-MS) for positive identification and quantitative analysis of drug(s) present. Many clinical laboratories have procedures for handling samples for medicolegal purposes, such as blood ethanol and drug concentrations from motor vehicle accident victims being treated in ER facilities. In most instances, however, the samples collected from ER patients are treated as routine analysis for drugs and alcohol for diagnostic purposes. In a clinical setting, laboratories analyze blood or urine samples for prescription or illicit drugs by immunoassay or colorimetric assay procedures for generic classes of drugs such as opiates and benzodiazepines. Different cross-reactivity to the drugs within the drug class makes immunoassay drug identification difficult—that is, diazepam vs midazolam—and immunoassays in most instances provide a semi-quantitative result only. The uncertainty of drug identification and concentration may preclude estimates of drug impairment by immunoassay alone. The chain of custody of clinical specimens may also be challenged. Such challenges are covered under Rule 702 and the *Daubert* standard as to admissibility of clinical data in lieu of the forensic standards of evidentiary analysis based upon GC-MS or other forensically accepted methodologies.

In most instances, small hospital-based clinical laboratories cannot afford the capital expense of on-site GC-MS) or liquid chromatography combined with tandem mass spectrometry (LC-MS/MS) equipment and the added expense of maintenance, operation, and training of qualified personnel. An alternative is to establish written procedures defining instances in which duplicate samples are drawn for accident victims arriving in an ER. The first set of samples is forwarded to the clinical laboratory for diagnostic purposes. The second set of samples is obtained and immediately labeled with the patient's name, social security number or other identifier, and the date and time the sample was obtained. A chain of custody document is initiated containing similar identifying information, type and volume of samples obtained, and the individual(s) handling the samples. The samples are placed in a sealable plastic bag with the chain of custody document appended to the bag and stored in a secured, limited-access locker. An example of a secured locker is a wall-mounted locked container that has limited access using a scan-readable identification card and PIN to track access to the securely locked container. In the absence of the earlier example, a log documenting when samples are entered into and retrieved from a secure locker will document the access to and transfer of the sample(s). Also, if an identification card or PIN-activated tracking system is not available, a two-person key access system and log tracking may be considered, with documentation of key transfer between shifts. When the commercial laboratory courier arrives to transfer the samples, the chain of custody document is completed and a copy given to the courier for continuation of the custody chain at the commercial laboratory.

The Presence of Alcohol or Drug and Impairment

As noted earlier, most clinical laboratories use either enzymatic, colorimetric, or immunoassay procedures for the determination of ethanol and drug concentrations for diagnostic purposes. While these assay procedures are highly accurate in most circumstances, it is well documented that an elevated concentration of ethanol or false-positive drug reporting may occur with the immunoassay, enzymatic, and colorimetric testing procedures due to interference or drug-cross-reactivity. Unless positive drug identification and quantitative analysis are confirmed by a forensically recognized procedure, such as GC for analysis of volatile components in blood or GC-MS, in the absence of any other supportive information, clinical drug, and alcohol measurements may be challenged in court proceedings. Exceptions may occur if the laboratory results are corroborated by a field sobriety test, eyewitness accounts as to the degree of incapacitation, forensically acceptable video camera recordings, and so forth. In such situations, the laboratory data may be introduced to substantiate other evidence as to the presence of ethanol or drugs and impairment. The blood alcohol concentration (BAC) or drug is important in ascertaining the level of impairment. Urine alcohol concentrations are valuable for interpreting pre- and post-absorption of BACs. However, the urine drug concentration is influenced by a number of factors, including the degree of hydration, genetic variations in metabolism, drug clearance rates, and the amount and duration of drug consumption. Caution should be exercised in any attempt to relate the degree of incapacitation based solely upon the urine drug concentration.

Discovery Requests

In general, discovery is the exchange of relevant information between parties to the lawsuit. In practice, discovery is the disclosure by the prosecution attorneys to defense counsel of all papers, accompanying charges, convening orders, statements, and related documents under applicable rules and case law. Included are paper copies of charge sheets, interview transcripts, and so forth related to filing charges against the defendant. Under discovery rules, defense counsel may also request discovery of laboratory documentation to include testing documents, personnel records, maintenance records, procedure manuals, incident reports, corrective action reports, administrative or addendum corrections to issued laboratory results, repeat analysis frequency, QA and QC reports, reagent or instrument failures, recalls, complaint reports, inspection reports, and documents related to loss or suspension of licensure, certification or testing. It is not unusual for a discovery request to be broad in scope covering all testing conducted by a laboratory over the course of years. However, just because a broad discovery is requested does not mean that all documents must be mandatorily released to requesting counsel. The laboratory director must work closely with the prosecution counsel, usually through the laboratory's or institution's legal department, to limit the scope of a discovery request(s). The discovery request should be limited to the immediate test(s) pertinent to the case and to a reasonable time frame in which testing of the sample occurred. An elevated blood alcohol in a vehicular manslaughter case should be limited to discovery documentation related to ethanol testing and not all testing conducted by the laboratory. The time duration of the discovery should also be challenged if the discovery is excessive. Under the current common judicial interpretation of existing case law, discovery is often limited to records generated in the course of a particular analyte analysis for a period not to exceed 1 year prior to and 6 months after the reported results in question. Reasonable fiscal reimbursement for generation of such documentation to include copying expenses and related expenses of personnel time required to retrieve, copy, collate, review, and deliver documents should be estimated and the requesting counsel made aware of the anticipated expense.

In the military, the government (prosecuting counsel) must permit the defense counsel to inspect documents, tangible objects (e.g., specimen tubes), and reports. The inspection includes, any books, papers, documents, photographs, tangible objects, building or places, or copies of portions thereof that are material to the preparation of the defense or are intended for use by the prosecution counsel as evidence in the prosecution at trial. This also includes results or reports of scientific tests, experiments, or copies thereof that are also material to the preparation of the defense or are intended for use by the prosecution counsel as evidence at trial. One of the key aspects of the military rule in this area is that such records must be made available to the defense for inspection. If copies of such laboratory documentation requested under discovery are not already in the possession of the government counsel, then the laboratory director may offer to make the applicable original records physically available for inspection without making copies. The documents inspection should be conducted within a specified area of the laboratory with control over the number of records made available for inspection at any given time. Record inspections would normally be attended and supervised by laboratory personnel to ensure that all records are maintained in order and that no records were retained by the defense. Restrictions on the carrying of briefcases, folders, bags, purses, and so forth in the records inspection area may be imposed, but such restrictions should be conveyed as written conditions in the approval of access to original records. Requests for copies of pertinent records may be chargeable to the inspecting counsel, as well as the personnel costs to the laboratory associated with the records inspection process. The laboratory may consider that one laboratory member will be assigned to each inspection member to ensure control over the original records. Requests for documents outside the original scope of the discovery should not be provided to the inspecting counsel during the visit, and such requests should be referred to the prosecuting counsel for review and approval. The laboratory director, in consultation with the legal department, should alert the prosecuting counsel of any additional discovery requests made at the time of inspection. Discovery requests can impact laboratory personnel and operations. Discovery requests cannot be ignored since they may be ordered by the judge and, if so, place the laboratory in possible contempt of court if ignored. It is imperative to immediately alert the laboratory's legal advisors upon receipt of a discovery request.

Compliance and Documentation

Electronic capture and retention of documents have reduced the volume of paper records retained in a laboratory. The ability to capture and retrieve documents has facilitated the evidentiary and discovery processes of records review in legal proceedings. However, data recovery may extend to the raw data whenever replicate testing is conducted and the average of the replicate testing is reported. Such raw data may be important in replicate testing to ensure that the variance in results does not exceed acceptable ranges, that is, results must be within 20% of each other. In such circumstances, replicate blood alcohol measurements of 100 and 115 mg/dL (average 107 mg/dL) would be acceptable, but replicate analysis results of 87 and 127 mg/dL, while yielding the same average of 107 mg/dL, are outside the accuracy and precision of the instrumentation and testing procedure. In the latter situation, the sample would require repeat analysis.

Compliance with laboratory's standard operating procedure protocols normally covers conduct of testing, preparation, and certification of calibrators and controls; instrument calibration and maintenance; reagent preparation and certification; QC and QA procedures; data review, acceptance criterion, and certification processes. The protocols must be current, accurately reflect the laboratory processes, and most importantly, be followed by the laboratory staff. Laboratory processes must be accurately reflected in the procedures manuals. One must avoid overly detailed of statements regarding testing procedures for which some latitude in data interpretation may be required. For example, minor equipment procurement may vary based on price or availability; that is, laboratories may purchase pipettes from a number of manufacturers over the years. Thus it may be inappropriate to cite a specific pipette manufacturer used in testing when the important issue is the volume of sample to be delivered and not necessarily the manufacturer of the pipette. When the required degree of accuracy demands the specific use of a particular pipetting device, a particular manufacturer specification may be warranted. Such minor deviations may seem innocuous, but when the laboratory's accuracy and forensic integrity are being challenged, the cumulative citation of these "errors" by the challenging attorney will call the lay jurors' attention to a possible lack of supervisory oversight and integrity of laboratory operations. This challenge can then extend to calling into question the reliability of the results reported. It is also prudent to allow for discretionary decision processes by supervisory personnel responsible for results reporting that may be at variance with laboratory protocol. For example, in the ethanol measurement example cited at the beginning of this section, the certifying official may choose to report the lower of the two results, namely 87 mg/dL, as there may be insufficient sample to repeat the analysis.

Professional experience and judgment occur throughout the process of forensic testing of a sample. Whenever discretionary variation from written policy or business practice occurs, such variation must be documented. In unusual or nonroutine circumstances, the use of a memorandum for record (MFR) is an easy way to document the basis for such action. The MFR should provide a concise summary of the basis for the variation from written policy or practice. The MFR should cite pertinent information regarding the testing, such as dates, times, sample numbers, case numbers, or batch numbers involved. There must be sufficient clarity in documentation to permit a written reconstruction of the scenario and decision processes, if necessary, at a later time. The MFR must be signed and dated by the individual with authority to certify the final reporting results and variance from policy. The routine use of MFR in laboratory practice facilitates the introduction of such records during expert testimony regarding the testing of a sample under the exemptions to the Hearsay Rule 803(6): records of regularly conducted activity.

Brady Notices

The Supreme Court Ruling of *Brady v. Maryland* [13] required law enforcement agencies and forensic laboratories to notify trial counsel of any potential exculpatory information favorable to the defendant and material to charges preferred, guilt, or punishment. The *Brady* decision requirements for duty to disclose were expanded by another Supreme Court Ruling, *Giglio v. United States* [14]. This case requires prosecutors to provide information to defense counsel that could impeach the credibility of a witness. The duty to document exculpatory and impeachment information requires forensic laboratories to have formal policies in place requiring the routine documentation of laboratory problems/incidents and personnel issues involving procedural irregularities, falsification of documents, and other conduct that could impact credibility. This documentation is called a Brady Notice. Exculpatory information in a Brady Notice contains additional evidence or statements obtained during the course of the investigation, including questionable forensic laboratory practices, QA and QC events that may call into question the accuracy of reported results or credibility of a witness's testimony or laboratory work.

Brady Notices apply whenever an employee is found to have violated forensic practices by altering data, ignoring QC, and QA practices calling into question the forensic and administrative compliance of results already reported. The

Brady Notice is not only limited to technicians' actions, but also instrumentation or procedural limitations in testing protocol resulting in cross-contamination of specimens, misidentification of samples, inaccurate qualitative or quantitative results reported, and personnel issues as already discussed.

The Brady Notice must contain a complete detailed explanation of events, specific time lines, effects upon the forensic accuracy of results if applicable and corrective actions. The military laboratory must submit the Brady Notice to their respective legal higher authority as required for dissemination. Trial counsel must determine whether the Brady Notice information provided is exculpatory to the defendant and must be disclosed to defense counsel.

CASE LAW PERTAINING TO FORENSIC TOXICOLOGY

The number of military trials involving prosecution of service members testing positive for drugs in urinalysis has given rise to case law in a number of areas. All the military services have their own service trial courts, similar to the individual civilian criminal trial courts. There are three service appellate courts for appeal, namely the Army Court of Criminal Appeals, the Navy-Marine Corps Court of Criminal Appeals, and the Air Force Court of Criminal Appeals. In addition, a US Court of Appeals for the Armed Forces (CAAF) is the higher appeal court across all branches of the military. This court is composed of five civilian judges who are appointed by the President of the United States. Only these military courts of appeal and the US Supreme Court render decisions that are binding on lower military courts. Decisions rendered in US district or circuit appeals courts or state courts are considered advisory, but are not binding to decisions rendered in the military courts. This is similar to the decision rendered in one US circuit court not being binding upon the decisions rendered in the US court in a different judicial circuit.

Constitutional Due Process Clause: Evidentiary Disclosure and Discovery Requests

An accused has the constitutional right to a fair trial, and the suppression of exculpatory evidence may violate the defendant's right to due process. Exculpatory evidence is defined as evidence that would be beneficial to the defense at trial in mitigating the defendant's level of guilt or in proving innocence. The failure to disclose exculpatory evidence in the possession or knowledge of the prosecution, or the denial of discovery and admission of evidence of a possible exculpatory nature by the judge, can result in the dismissal or retrial of the case upon appeal. For laboratory scientists and managers, exculpatory evidence may extend to laboratory discovery issues, thus requiring policies that require formal documentation of laboratory incidents and personnel issues that involve procedural irregularities, falsification of documents, and other conduct that could impugn credibility of employees performing the testing.

The following are selected cases issued by civilian and military courts of appeal. The cases cited later should not be construed as all-inclusive. The brief summation of the US Supreme Court decision in *Brady v. Maryland* was seminal in addressing the impact on the defendant's right to a fair trial under the due process clause of the constitution if the government fails to disclose or suppresses exculpatory evidence favorable to the defendant in the determination of guilt or innocence and in the penalty or sentencing phase of the trial [13]. Cases following *Brady* have expanded upon the holding in *Brady* and reinforced that the defense can be entitled to a greater deal of evidence if it is potentially exculpatory.

Brady v. Maryland, US Supreme Court 373 U.S. 83 (1963)

The defendant and a companion (Boblit) were convicted in separate district court trials of first-degree murder and both were sentenced to death [13]. Prior to the trial, Brady's defense attorney had requested permission to examine the pretrial statements of the companion (Boblit) in the murder. Several statements were provided to the defense counsel. However, one statement, in which Boblit admitted to the actual homicide, was withheld. During his trial, Brady admitted under oath to his participation in the crime, but claimed that Boblit actually killed the victim in the commission of a robbery. The defense counsel, in summation to the jury, acknowledged that the defendant was guilty of first-degree murder, but asked that the jury return a guilty verdict without capital punishment. Following the trial, conviction, sentencing, and affirmation of the conviction, the existence and content of the withheld statement came to the attention of the defense counsel. The defense counsel appealed for a new trial. The defense request for a new trial was dismissed by the Maryland Court of Appeals, but the court of appeals returned the case to the lower court for a retrial only on the question of punishment imposed upon the defendant, not on the question of guilt of the defendant. The request for a retrial was appealed to the US Supreme Court. The defense counsel maintained that the withholding of information beneficial to the defendant was a violation of the due process clause of the Fourteenth Amendment of the US Constitution.

The Supreme Court stated, "We agree with the Court of Appeals that suppression of this confession was a violation of the Due Process Clause of the Fourteenth Amendment. .... [T}he suppression by the prosecution of evidence favorable to an accused upon request violates due process where the evidence is material either to guilt or to punishment, irrespective of the good faith or bad faith of the prosecution." However, the Supreme Court also agreed with the Maryland Court of Appeals' decision to order a retrial only in regards to the question of sentencing and not the question of guilt, given the nature of the suppressed evidence.

Giglio v. United States, US Supreme Court 450 U.S. 150 (1972)

The defendant was convicted of passing forged money orders and sentenced to 5 years in prison [14]. While on appeal, defense counsel found new evidence that was not presented in court. The defendant's alleged coconspirator and only witness to the crime (Robert Taliento), was promised immunity from prosecution by a prosecutor if Taliento testified for the grand jury and trial. A different attorney prosecuted the case at trial and was unaware of the affidavit documenting the immunity promise to Taliento. The Supreme Court found that Taliento's credibility was an important issue since he was the only witness to the crime and that the jury should have been made aware of the immunity promise. The Supreme Court reversed and remanded the lower court ruling and required that the defendant be granted a new trial. The *Brady* decision was expanded by *Giglio* to require prosecutors to provide information to defense counsel that could impeach witnesses. This information includes issues that impact the credibility and truthfulness of laboratory employees as already discussed in the Brady Notice section.

United States v. Agurs, US Supreme Court 427 U.S. 97 (1976)

The defendant was convicted in the US District Court for the District of Columbia of second-degree murder for the killing of an individual with a knife during a fight [15]. Evidence showed the victim, just before the killing, had been carrying two knives, including the one that the defendant used to repeatedly stab the victim, but the defendant was uninjured. Following the defendant's conviction, the defense counsel moved for a new trial, asserting that the defense counsel had just discovered that the victim himself had a prior criminal record of assault and carrying a deadly weapon (a knife). The defense's position was that this information would have tended to support the defense argument that the defendant acted in self-defense and that the prosecutor failed to disclose this information to the defense in violation of the right to due process. The district court denied the motion for retrial since the prior conviction gave no new information regarding the victim's character that was not already apparent from the uncontradicted evidence, specifically that the victim carried two knives. The district court also focused on the inconsistency between the claim of self-defense and the fact that the victim had been stabbed repeatedly while the defendant was unscathed. The pathology report indicated that many of the cut wounds on the victim were defensive in nature. The Circuit Court of Appeals for the District of Columbia reversed the decision of the district court, holding that there was no misconduct on the part of the prosecution, but that the evidence of the victim's criminal record was material, and, its nondisclosure required a new trial because the jury might have returned a different verdict if the evidence had been received.

The US Supreme Court reversed the court of appeals' decision and affirmed the district court's initial ruling. The Supreme Court ruled that the prosecutor "…will not have violated his constitutional duty of disclosure unless his omission is of sufficient significance to result in the denial of the defendant's right to a fair trial. …. The mere possibility that an item of undisclosed information might have helped the defense, or might have affected the outcome of the trial does not establish 'materiality' in the constitutional sense. …. The proper standard of materiality must reflect our overriding concern with the justice of the finding of guilt …. Such a finding is permissible only if supported by evidence establishing guilt beyond a reasonable doubt …. This means that the omission must be evaluated in the context of the entire record. If there is no reasonable doubt about guilt whether or not the additional evidence is considered, there is no justification for a new trial. On the other hand, if the verdict is already of questionable validity, additional evidence of relatively minor importance might be sufficient to create a reasonable doubt."

However, the assumption is always in favor of disclosure to the defense, as the Supreme Court stated, "The prudent prosecutor will resolve doubtful questions in favor of disclosure." In review of the totality of evidence presented at the district court trial, the defendant was not denied the guarantee of a fair trial by the due process clause of the Fifth Amendment. Later US Supreme Court cases reinforced the general proposition that "evidence is material only if there is a reasonable probability that, had the evidence been disclosed to the defense, the result of the proceeding would have been different … Reasonable probability" is a probability sufficient to undermine confidence in the outcome" [16].

United States v. Israel, US Court of Appeals for the Armed Forces, 60 MJ 485 (CAAF 2005)

The accused tested positive on a random urinalysis for the presence of cocaine metabolite. During the trial, the accused did not testify on his own-behalf, and his defense counsel's presentation questioned the reliability of the testing program solely on cross-examination of the laboratory expert. The accused was found guilty of illicit drug use and convicted [17]. At trial, the military judge restricted the scope of the defense counsel's cross-examination of the laboratory expert and did not allow the defense to explore several recent testing and procedural errors at the laboratory and irregularities at the base sample collection site. On appeal, the defense argued that the trial judge's ruling on cross-examination unreasonably restricted the defense counsel's ability to cross-examine witnesses and violated the defendant's Sixth Amendment rights. In ruling on the appeal, CAAF stated, "Although the military judge allowed the defense counsel to conduct limited cross-examination regarding the possibility that errors might have occurred in the testing process, he [military judge] excluded most of the evidence that the defense offered in support of that possibility." Based on the Sixth Amendment right to cross-examine witnesses, the CAAF found the military judge abused his discretion in limiting the defense's cross-examination into some areas. The CAAF found that in "... those cases where the Government relies on the general reliability of testing procedures, evidence related to the testing process that is closely related in time and subject matter to the test at issue may be relevant and admissible to attach the general presumption of regularity in the testing process." The CAAF opinion looked at the seriousness of a positive result from an error in testing and found, "It is impossible to say that the members [of the jury] would not have taken evidence of irregularities in the testing process and possible errors in the results into consideration" ...... "We [the CAAF] conclude that the error [by the military judge in limiting cross-examination] was not harmless beyond a reasonable doubt." The CAAF set aside the finding and sentence and returned the record of trial with authorization for a retrial.

United States v. Jackson, US Court of Appeals for the Armed Forces, 59 MJ 330 (CAAF 2004)

The accused tested positive on a random urinalysis for the presence of methamphetamine. During the trial, the accused claimed innocence but was found guilty of the charge of wrongful use of methamphetamine [18]. The defense counsel presentation questioned the reliability of the testing conducted at the laboratory. Prior to trial, the defense counsel submitted an extensive pretrial discovery request to the government for laboratory testing documents and records, memos, reports, and other related documents for the three quarters prior to the defendant's sample testing date and the "available quarters since the accused sample was tested." The discovery submission also stated that the discovery request was a "continuing request" that "includes any information which you may later discover before, during or after trial of his [defendant's] case, or which is not requested in a specific manner." When in turn requesting information from the laboratory, the prosecuting counsel limited the discovery to matters "deemed relevant to litigation," and referred to the time frame of discovery as "months" rather than "quarters" and did not include the continuing nature of the defense request. Approximately 6 weeks prior to the trial date, an error in testing occurred at the testing laboratory involving an internal, blinded QC sample that tested positive for the presence of drug when none should have been found. Under the "continuing request" clause from the first discovery request and the subsequent requests, the defense submitted just prior to trial, the laboratory discrepancy report surrounding this new QA incident should have been provided by the laboratory to the prosecution and defense counsel prior to trial. On appeal, the CAAF ruled that the failure to provide the requested QA discovery information violated the defendant's right to information. The prosecution's case rested primarily on the positive urinalysis. In the CAAF statement, "Although the additional circumstantial evidence introduced by the prosecution regarding Appellant's [the accused] attitude on various occasions might have had some marginal value in rebutting the defense suggestions of innocent ingestion, it did not constitute independent evidence of illegal use." Meanwhile, the defense's focus was on attacking the reliability of the laboratory process. The CAAF found that had the defense possessed the most recent incident of laboratory error prior to trial, the "defense could have argued that the members had been presented with evidence of a specific problem in the testing procedures." The CAAF concluded that the restriction and availability of laboratory testing documentation, "... deprived the defense of information that could have been considered by the members [of the jury] as critical on a pivotal issue in the case—the reliability of the laboratory's report that the Appellant's [defendant's] specimen produced a positive result. Given the significance of this information in the context of the Appellant's trial the error was prejudicial under the "harmless beyond a reasonable doubt standard ....." Since there was error in not producing the information to the defense and there was no other independent evidence of illegal use, the CAAF reversed the lower court ruling, the lower court findings and sentence were set aside, and a retrial was authorized.

United States v. Gonzales, US Court of Appeals for the Armed Forces, 62 MJ 303 (CAAF 2006)

This case is similar to *Jackson* earlier, in that there was error in failing to provide discovery to the defense, but this time there was sufficient independent evidence of illegal drug use to support the conviction to the effect that the error was harmless beyond a reasonable doubt. The accused was found to be in the possession of paraphernalia associated with the use of ecstasy along with fliers for rave parties. The accused also admitted to the "innocent" ingestion of ecstasy at a rave. A urinalysis sample collected from the accused was positive for ecstasy. On a separate incident, on a belief that the accused had stolen a gun, his vehicle was searched and the gun was recovered along with a baggie containing marijuana. At trial, the accused pled guilty to wrongful possession of marijuana but not guilty to wrongful use of ecstasy and carrying a concealed weapon [19]. The defense counsel, aware that the prosecution would rely heavily on the results of his positive urinalysis testing, requested reports, documents, and other records related to the test result; reports of mishandling and misplacement; and any errors in administration, collection, testing, or handling of samples in the several months preceding and following the testing dates of the accused sample. The government produced the requested documents but did not release a discrepancy report for an incident involving a blinded internal negative QC sample that tested positive for the presence of drugs. On review, the CAAF acknowledged the failure of the government in providing all documents under the discovery request. The CAAF stated, "When a defendant makes a specific request for discoverable information, it is error if the Government does not provide the requested information, and the appellant will be entitled to relief unless the Government can show that nondisclosure was harmless beyond a reasonable doubt. …. We [the CAAF] have previously held that the Government's failure to turn over the … discrepancy report should be treated as prejudicial error where the other available evidence [in *United States v. Jackson* 59 MJ 330 CAAF (2004), discussed earlier] does not constitute 'independent evidence of illegal drug use'. … With respect to prejudice, this case differs from our prior cases due to the level and character of the independent evidence of illegal drug use. In addition to the positive drug test, the prosecution introduced independent evidence of drug use including evidence that the accused had drug paraphernalia associated with ecstasy both in his car and at his work station." The accused, "…. also admitted that he attended at least one rave party and had fliers for thirteen rave parties in his car. Although drug use is not a necessary element of a rave, the two are often linked. …. The accused also admitted to prior drug use and to possession. He told investigators that he innocently ingested ecstasy at a rave he attended and testified that he was familiar with its effects. Additionally, the accused pled guilty to possessing marijuana." The CAAF ruled that the accused's "… urine sample was subjected to four different tests, each of which showed positive for ecstasy use. When the missing report is balanced with the evidence arrayed against the accused, the scales tip strongly in favor of conviction. Thus, we [the CAAF] conclude that the Government's error in failing to produce the discrepancy report was harmless beyond a reasonable doubt." The decision of the CAAF affirmed the conviction of the lower courts.

Constitutional Confrontation of Accuser Clause—The Role of Laboratory Personnel and Laboratory Reports

Recent case law has focused on the right of the defendant to call to testimony all laboratory technicians and personnel involved in the testing of a specimen to testify in person in court. The right of the defendant to confront the accuser and witnesses testifying against the defendant is written into law. In forensic and toxicological analyses, what is the role of the technicians performing the testing of the sample, supervisors, and the laboratory director or toxicologists signing out the report of analysis? Are the data from laboratory analysis and the subsequent laboratory report considered testimonial or nontestimonial? Do laboratory personnel performing the test become accusers for testing the defendant's sample?

The selected case laws cited later show that the military and civilian circuit appeals courts across the United States have yielded conflicting interpretations of whether laboratory testing documentation should be considered testimonial or nontestimonial. These questions regarding testimonial and nontestimonial documentation and the role of technicians in the production of laboratory results continue to be addressed in a number of court decisions. The following is a summary of opinions issued by civilian and military courts and should not be construed as all-inclusive. Perhaps some confusion arises from the different types of laboratories for which forensic testing is conducted. The testing conducted at a state or federal crime laboratory is different operationally from testing conducted at a commercial, private, or government forensic testing laboratory. In a crime laboratory, a single forensic technician may take custody of testing samples associated with a given crime scene investigation and conduct testing solely on the sample(s) in question for purposes of criminal investigation. In a forensic testing laboratory, a number of technicians are usually involved in the handling and testing a number of various samples as part of a "batch analysis" with no intended knowledge of association or aspects of the sample to a given investigative incident. A review of the various court decisions is too expansive to cite individually. However, given the number of cases and the disparity in the decisions, intervention by the US Supreme Court may be necessary [20].

These legal decisions involving the right of confrontation have the potential of being extremely disruptive to high-volume forensic laboratories. It is important to work within the legal system to establish possible alternatives to the personal court attendance by technicians, supervisors, and expert toxicologists. The expense of personnel time and travel costs involved in the testimony surrounding the testing of a single sample must be weighed against the impact on the daily laboratory operating processes due to the temporary loss of key personnel. Alternative scenarios may involve telephone or video teleconferencing linkage within the courtroom. Another option is to request the defense counsels to visit the laboratory to conduct interviews or depositions with the staff members involved in testing a particular sample. Any such alternatives should always be coordinated with the prosecuting counsel and/or the laboratory's legal counsel servicing office.

United States of America v. McKinney, US Eighth Circuit Court of Appeals, 631 F2d 569 (8th c.1980)

The defendant was found guilty of assault with intent to murder and conveying a weapon (a hand-constructed "knife" made out of combs and an industrial needle) within a federal penal institution [21]. Blood samples taken from the victim along with blood taken from the needle were subject to analysis at a Federal Bureau of Investigation (FBI) laboratory. The FBI agent testified that he had requested a blood test of the needle. However, there was insufficient blood on the needle to conduct blood type analysis, but the blood sample was shown to be from a human source. The defense argued that the testimony by the FBI agent was hearsay in that the agent did not perform the blood tests on the needle and was merely reporting the results. The United States Eighth Circuit Court of Appeals ruled that the testimony of the FBI Agent was hearsay. However, the court also found that the admission of the hearsay testimony was harmless error since there was overwhelming evidence by witnesses that the defendant did in fact stab the victim with the makeshift knife.

United States of America v. Baker et al., US Eighth Circuit Court of Appeals, 855 F2d 1353 (8th c.1988)

Seven defendants were found guilty of conspiring to distribute a controlled substance (methamphetamine) in addition to charges of transporting firearms in interstate commerce [22]. The defendants contended that the district court committed an error by admitting into evidence several laboratory reports related to the defendants that identified the presence of controlled substances. The defendants also contended that the admission of the laboratory reports violated their rights under the confrontation clause of the Sixth Amendment. The court of appeals denied this argument. "When made on a routine basis, laboratory analyses of controlled substances are admissible as business records under Federal Rule of Evidence 803(6)." The police department laboratory made the reports in the ordinary course of its business and kept the records under its care, custody, and control. The defendants did not challenge the reliability of the reports or claim the reports were made on anything other than a routine basis. The district court acted under a firmly rooted exception to the hearsay rule when it admitted the laboratory reports under the business records exception. The court of appeals concluded that the district court properly admitted the reports.

United States v. Magyari, US Court of Appeals for the Armed Forces, 63 MJ 123 (CAAF 2006)

The specimen of the accused was approximately 1 of 35–40 samples submitted for random urinalysis by a Navy unit [23]. The specimen tested positive for the presence of methamphetamine above the cutoff concentration set by the DoD. The government called several witnesses involved in the collection and handling of the samples for submission to the laboratory. The laboratory's QA officer, who signed out the laboratory result, testified as to the testing and specimen handling within the laboratory. The QA officer was not personally involved in the handling or testing of the specimen from the accused, but was familiar with the testing procedures conducted by the laboratory and interpretation of such laboratory testing results for reporting purposes. No other laboratory witnesses were called, nor were there any challenges at the time by the defense counsel. On appeal, the defense counsel argued that his client's right to confront witnesses was violated and that any statements contained in the laboratory report to indicate the urine sample tested positive for the presence of illicit drug was inadmissible testimonial hearsay and could not be used against the accused at the trial.

In ruling on the appeal, the CAAF cited a prior case of *Crawford v. Washington* (2004) [24], which held that in order for the prosecution to introduce testimonial out-of-court statements at trial, the witness who made the statement must be available and the accused must have had a prior opportunity to cross-examine the witness. While the Supreme Court did not spell out a comprehensive definition of "testimonial," leaving to lower courts the responsibility to determine which statements would qualify as testimonial and falling within the scope of such definition, the Supreme Court did identify three forms of "core" testimonial evidences: (1) ex-parte in-court testimony, (2) extrajudicial statements in formalized trial materials, and (3) statements made under circumstances that would cause a reasonable witness to believe they could be used at trial. The CAAF noted that the specimen of the accused was processed by a technician along with 200 other samples in a test run. The technicians do not correlate a particular sample with a particular individual as each sample is assigned a laboratory identification number. The laboratory technicians handling the specimen of the accused had no reason to suspect the accused of wrongful use and no reason to anticipate that the accused's specimen, among all the other samples, would test positive and become evidence at trial.

The CAAF ruled that because this was a random sample, the laboratory technician had no reason, nor was under any pressure, to reach a particular conclusion about the specimen of the accused. The technician had no reason to test the sample differently from any other sample undergoing analysis. Therefore, the laboratory's reports were inherently nontestimonial because they are business and public records. Rule 803{6} implies that laboratory reports are included in the definition of business records because records from forensic laboratories, hospital laboratories, and DNA laboratories are most likely prepared in the course of routine, regularly conducted business. However, the CAAF also indicated that the government's contention that laboratory reports by definition are not testimonial because they are classified as business and public records may be too broad. In its decision, the CAAF indicated, "The laboratory reports prepared at the behest of law enforcement in anticipation of a prosecution may make the reports testimonial. Thus, laboratory results or other types of routine records may become testimonial where an accused is already under investigation, and where the testing is initiated by the prosecution to discover incriminating evidence. For example, cross-examination may be appropriate where a particular accused is accused of rape and law enforcement conducts and seeks to admit the results from a blood or DNA test. … Cross-examination may also be necessary where a suspect is believed to have operated a vehicle under the influence of drugs or alcohol and a record or affidavit is prepared by hospital personnel for the prosecution's use at trial." In ruling on the facts before it in this case, the CAAF found these factors did not come into play in the testing of the specimen of the accused and the entries in the urine laboratory report were nontestimonial under *Crawford v. Washington* [24]. The record of laboratory testing introduced at trial was simply a record of "regularly conducted" activity of the laboratory and "qualifies as a business record under Military Rules of Evidence 803{6}, a firmly rooted hearsay exception." The CAAF affirmed the conviction of the accused.

United States of America v. Washington, US Fourth Circuit Court of Appeals, 498 F3d 225 (4th c.2007)

In the original trial, the defendant was convicted of driving while under the influence of alcohol or drugs and operating a motor vehicle in an unsafe manner [25]. During the trial, the prosecution placed into evidence the results of laboratory tests conducted on a blood sample from the defendant. The blood tests indicated the presence of alcohol and a Schedule 1 drug, phencyclidine (PCP). The prosecution offered the expert testimony of the laboratory toxicologists who attributed the defendant's observed conduct and unsafe driving to the presence of alcohol and PCP. The defense counsel objected to the testimony by the government's toxicologist, in that the expert relied on the raw data generated by the laboratory's diagnostic machines and therefore the expert's testimony amounted to hearsay of the laboratory technicians who operated the instrumentation producing the test results. The judge in the original trial overruled the defense objections and found the defendant guilty as charged.

On appeal, the defense maintained that the machine-generated data amounted to testimonial hearsay of the testing instrument operators, and the expert was restating the hearsay statements of the instrument operators in violation of the confrontation clause and the hearsay rule found in the Federal Rule of Evidence 802. The Fourth Circuit Court of Appeals found, "Whether the machines properly reported PCP and alcohol is determined by the raw data that the machines generated and its truth is dependent solely on the machine." The court rejected the characterization of the raw data generated by the laboratory instrumentation as statements of the lab technician operating the testing equipment. The raw data generated by the testing instrumentation are the statements of the instrumentation, and not their operators, and statements generated by machines are not out-of-court statements made by declarants that are subject to the confrontation clause. A "statement" is defined by Federal Rule of Evidence 801(a) as an "[1] oral or written assertion or [2] nonverbal conduct of a *person*, if it is intended by the *person* as an assertion." The court of appeals ruled that the raw data generated by the machines do not constitute statements and the machines are not "declarants"; and therefore no out-of-court statements under the confrontation clause were facilitated by the testimony of the laboratory toxicologist. Also, the objection based on hearsay statements made by the laboratory instrumentation were not testimonial in that they did not involve the relation of a past act of history as would be done by a witness. The laboratory instrument's statements related solely to the condition of the blood at the time of analysis without making any links to the past. The expert toxicologist did provide testimony linking the results of the blood sample analysis with the defendant's past behavior. This testimony was presented in the initial trial in conformity with the confrontation clause and not challenged on appeal. The court of appeals ruled, "As the machine's output did not 'establish or prove past events' and did not look forward to 'later criminal prosecution'—the machine could tell no difference between blood analyzed for healthcare purposes and blood analyzed for law enforcement purposes—the output could not be 'testimonial.'... Because raw data printed out by the machines are not testimonial hearsay statements, ...testimony using those data did not violate the confrontation clause, nor the hearsay rule, and the magistrate judge did not err in admitting the testimony...." The court of appeals concluded that the laboratory data on which the laboratory toxicologist relied (1) did not constitute the statements of the laboratory technicians, (2) were not hearsay statements, and (3) were not testimonial. Accordingly, the court of appeals affirmed the conviction of the defendant as guilty.

United States v. Harcrow, US Court of Appeals for the Armed Forces, 66 MJ 154 (CAAF 2008)

The issue as to when a laboratory report becomes testimonial was visited in the CAAF decision [26] regarding testing conducted by a state forensic sciences laboratory. The accused was convicted at the trial for numerous drug offenses including wrongful use of cocaine and wrongful possession of heroin. Articles seized at the accused's home were sent to the Virginia State Division of Forensic Sciences laboratory, and the laboratory's positive findings of drug residue served as the basis for several of the charges. At trial, the defense did not object to admission of the laboratory reports. On appeal, the Navy-Marine Corps Court of Criminal Appeals ruled that the laboratory reports were nontestimonial and admissible under the business records hearsay exception in the Military Rule of Evidence 803{6}. On appeal, the CAAF disagreed, ruling that the state laboratory reports constituted testimonial hearsay under *Crawford v. Washington* [24] and that the admission of the reports was a violation of the Sixth Amendment right to confrontation. However, the CAAF ruled that the error in the admission of the evidence was harmless beyond a reasonable doubt, did not violate a substantial right, and affirmed the findings and sentence. The CAAF basically agreed with the *United States v. Magyari* [23] decision that rejected the premise that laboratory reports are never testimonial and therefore always admissible as business record exemptions under Rule 803. Whereas the laboratory test results in *Magyari* were from a random test with no eye toward prosecution, the laboratory testing conducted in *Harcrow* was conducted specifically at the behest of law enforcement as part of an investigation. In making its decision, the CAAF provided some clarification as to when laboratory reports may be considered testimonial. The court identified three nonexclusive factors to be used when deciding whether evidence is testimonial: (1) whether the statement was elicited by or made in response to law enforcement or prosecutorial inquiry, (2) whether the statement involved more than a routine and objective cataloging of unambiguous factual matters, and (3) whether the primary purpose for making or eliciting the statement was the production of evidence with an eye towards trial. Laboratory reports prepared at the request of law enforcement officers as part of a specific criminal investigation as to the conduct of an identified subject establishes circumstances that make laboratory reports testimonial. The written statements of the laboratory technician in the laboratory report introduced into evidence made for the purposes of producing evidence in anticipation of use at trial constituted testimonial hearsay.

Challenges to the Admissibility of Evidence and Inference of Guilt

A judicial proceeding is the presentation of evidence against a defendant. The inference of guilt is established based upon the weight of evidence presented against the defendant. An example of inference is the observation that upon awakening, one observes that the streets and surrounding area are wet. One may infer that it rained without actually having observed the occurrence of rain. But, are there other possibilities to explain the observation? What is the reliability of our means of assessment? Certitude depends upon the cumulative assessment of information that leads one to the conclusion that an event occurred to the reasonable exclusion of other possibilities. In the drug cases cited later, the admissibility of evidence and inference of guilt were debated as to the inference of drug presence and the illicit use of drugs.

If an individual intended to use cocaine and procured what he believed to be cocaine, but actually was cocaine cut extensively with powdered pseudoephedrine tablets, was he less guilty of cocaine use if a "high" was not attained and the urine cocaine metabolite concentration level was relatively low? If an individual intended to buy heroin but was sold low-grade cocaine and subsequently tests positive for cocaine metabolites, was the intent to use an illicit drug the compelling nature under the law rather than which drug was used or whether the physiological effect of the intended drug was observed by the defendant? Criminal intent is relevant, rather than the amount or type of drug actually used or whether a physiological high was achieved by the defendant. In reaching the current state of military law, several cases were decided in order to establish the level of proof necessary for a conviction of drug use. The following are citations of notable CAAF opinions and decisions.

United States v. Campbell, US Court of Appeals for the Armed Forces, 50 MJ 154 (CAAF 1999)

The accused was found guilty of wrongful use of lysergic acid diethylamide (LSD). The urine sample was tested using immunoassay and GC/tandem mass spectrometry (GC-MS/MS). The defense appealed several points and following oral arguments, the CAAF agreed to take up additional issues [27]. The first issue was that the testing methodology of GC-MS/MS was considered a novel technology for drug testing as opposed to the more commonly used testing methodology of GC-MS, and that the Government failed to show that the testing methodology of GC-MS/MS was sufficiently reliable to ascertain the concentration of the drug in the urine. The second was whether applicable law required a particular cutoff concentration in order to establish beyond a reasonable doubt that the knowing use of LSD by the accused. The third was whether sufficient scientific evidence was presented for the court to draw an inference of wrongful use from the concentration of LSD in the urine specimen of the accused based on the cutoff concentration for reporting a positive GC-MS/MS result. The final issue was whether the expert testimony regarding certification and QC results was sufficient to demonstrate the reliability of the testing procedure of GC-MS/MS under *Daubert* to reasonably exclude the possibility of a false-positive result.

Following presentation of arguments, the CAAF reversed the lower court's decision and dismissed the charges. In discussion, the CAAF attempted to strike a balance between protection of the accused and the harm to the military when service members use drugs knowingly and for illicit purposes. In military case law, the presence of drugs in the body fluids may provide sufficient inference of wrongful drug use. However, in this case, the CAAF expanded on that principle to state that if the prosecution is relying upon a urinalysis result with no other corroboration, the prosecution must also establish the reliability of the testing methodology and explain the significance of the positive test result. In doing so, "…the prosecution's expert testimony must show (a) that the 'metabolite' is 'not naturally produced by the body' or any substance other than the drug in question, (b) that the cutoff level and reported concentration are high enough to reasonably discount the possibility of unknowing ingestion and to indicate a reasonable likelihood that the user at some time would have experienced the physical and any psychological effects of the drug, …and (c) that the testing methodology reliably detected the presence and reliably quantified the concentration of the drug or metabolite in the sample." The CAAF also stated, "… to rely on the permissible inference of knowledge from the presence of the drug in the sample, however, the cutoff level must be such as to rationally permit fact finders to find beyond a reasonable doubt that the accused's use was knowing."

In dissent, one of the judicial members raised the serious question that the CAAF ruling in this case establishes new law by establishing a new requirement for evidentiary sufficiency in urinalysis prosecutions—that the government must show that the cutoff level and reported concentrations are high enough to reasonably discount the possibility of unknowing ingestion and to indicate a reasonable likelihood that the user at some time would have experienced the physical and psychological effects of the drug. Also in dissent, a second judicial member raised serious question that the CAAF ruling is going beyond the original appellant's basis of appeal. The judicial member also stated, "Whether appellant experienced or was observed experiencing the effects of LSD is irrelevant to the issue of whether he knowingly used the drug. A defense witness who had been an emergency medical technician admitted on cross-examination that an individual who takes low dosage of LSD might not exhibit outward signs of drug use. … I would hold that the members logically could infer from the facts in this case that appellant's use of LSD was wrongful beyond a reasonable doubt."

The issues considered by the CAAF in *United States v. Campbell* (1999) ruling were subsequently reviewed by CAAF in 2000, summarized in following sections.

United States v. Campbell, (Campbell II), US Court of Appeals for the Armed Forces, 52 MJ 386 (CAAF 2000)

The CAAF [28] was requested to reconsider its decision in the 1999 verdict. The CAAF granted the government's motion for reconsideration and sought to establish clarification regarding the three-part approach to urinalysis convictions. The CAAF indicated that the 1999 decision was based upon the introduction of a novel procedure for drug analysis, namely GC-MS/MS, and "the deficiency was the absence of evidence establishing the frequency of error and margin of error in the testing process. Lacking such evidence, we [the CAAF] held that the prosecution did not reliably establish that appellant's urine sample tested at or above the Department of Defense cutoff level and did not reasonably exclude the possibility of unknowing ingestion." The CAAF also acknowledged that, "The petition for reconsideration raises the issue of whether the three-part standard is mandatory in all drug testing cases. Given the rapid pace of technological change, we note that the three-part standard does not necessarily constitute the only means of proving knowing use. If the test results, standing alone, do not provide a rational basis for inferring knowing use, then the prosecution must produce other direct or circumstantial evidence of knowing use in order to meet its burden of proof. If the Government relies upon test results, it is not precluded from using evidence other than the three-part standard if such evidence can explain ... a rational basis for inferring knowing, wrongful use."

The CAAF also clarified that evidence that the accused must have felt the effects can be based upon data taken from study in which humans are considered as a class of individuals. The CAAF explained, "The petition for reconsideration also raises the issue of whether, in using the three-part standard, the prosecution must introduce scientific evidence tailored to the specific characteristics of the person whose test results are at issue. We have not established such a requirement in the past, and we do not do so here. These cases involve the type of permissive inference that has been applied for many years to members of the armed forces. It is sufficient if the expert testimony reasonably support the inference with respect to human beings as a class." By issuing this second ruling, CAAF supplemented the previous opinion rendered in 1999. However, this second ruling in *Campbell II* also contained a dissenting opinion that this current CAAF ruling created a new requirement in urinalysis cases for "...evidence that the controlled substance is present in the body of the accused in such a quantity that an expert can opine that the effects of the drug would have been felt." The decision and opinions of the CAAF in the *United States v. Campbell* (1999 and 2000) rulings were subsequently criticized within the legal community as establishing new law as noted in the dissenting opinions [29]. The three-part standard requirements established in *United States v. Campbell* were brought for CAAF review in the following year in *United States. v. Green* in 2001 (see further).

United States v. Green, US Court of Appeals of the Armed Forces, 55 MJ 76 (CAAF 2001)

This case is commonly cited as the basis for the "permissive inference" in urinalysis cases, namely, that the jury "may infer" from the presence of a controlled substance in the body of the accused or from other circumstantial evidence that the accused knew the substance was present. The urine specimen from the accused was collected upon return to his unit following an extended period of leave [30]. The sample tested positive for the presence of cocaine metabolite above the cutoff concentration set by the DoD. In the original trial, the government introduced evidence that included a copy of the laboratory report and the testimony of a witness from the laboratory who testified "as an expert in the field of forensic chemistry." The expert and the laboratory report were admitted with no defense objection. The expert testified that for a person to produce the cocaine metabolite in the urine, the individual would have to have used cocaine and that the cocaine metabolite, benzoylecgonine (BZE), is not a normal product formed within the body. The expert also indicated that cocaine is not used in prescription medications although it has been used in surgical settings. The expert also indicated that it is possible to detect BZE if someone put cocaine directly into the urine sample. The defense contended that the evidence produced in the case was insufficient to prove wrongful use under *United States v. Campbell*. The defense requested that the laboratory results and the testimony of the expert be excluded. The defense motion was denied in the original trial, and the accused was found guilty of drug use. On appeal, the CAAF sustained the conviction.

In discussion, the CAAF stated that "...... opinions in *Campbell* ventured beyond the issue of reliability of the methodology in an effort to provide additional guidance concerning proof in urinalysis cases. In that context, we [the CAAF] described a three-part approach to consideration of urinalysis results. In our opinion, upon reconsideration, we emphasized that the three-part approach did not establish a mandatory standard. We noted that other evidence explaining the test results could be admissible if it met applicable reliability and relevance standards for scientific and specialized knowledge with respect to providing a rational basis for inferring knowing, wrongful use." The CAAF further elaborated that "The military judge has broad discretion as the 'gatekeeper' to determine whether the party offering expert testimony has established an adequate foundation with respect to reliability and relevance. ... In making this determination, the military judge may consider factors such as whether the evidence reasonably discounts the likelihood of unknowing ingestion, or that a human being at some time would have experienced the physical and psychological effects of the drug, but these factors are not mandatory. ... If the military judge determines that the scientific evidence —whether novel or established —is admissible, the prosecution may rely on the permissive inference during its case on the merits. A urinalysis properly admitted under the standards applicable to scientific evidence, when accompanied by expert testimony providing the interpretation ... provides a legally sufficient basis upon which to draw the permissive inference of knowing, wrongful use, without testimony on the merits concerning physiological effects."

In the concurring opinion, one judge went on to state, "In short, contrary to *Campbell I* (*1999*) and II (*2000*) this Court [US CAAF] has never required the Government (for evidentiary sufficiency purposes or for evidentiary admissibility purposes) to introduce evidence that a certain nanogram count discounted innocent ingestion and indicated that the accused, or a user in general, would have experienced the physical or psychological effects of the drug." The concurring judge went on to state, "Put simply, the Government is not required to disprove innocent ingestion or show that a particular accused in a drug case felt the physical and psychological effects of the drug he was accused of taking, or that a user in general would have felt such effects. It is unclear whether the state of the art in drug testing would permit these requirements to be met in many cases."

The various courts, in the earlier cases, displayed considerable circumspection that each trial must be considered individually based upon the merits of evidence. The trial judge as the "gatekeeper" determines the admissibility of evidence. In the conclusion of *United States v. Green*, the CAAF noted that the decisions rendered in both *Campbell* cases caused a great deal of confusion by establishing what appeared to be mandatory requirements to be met by prosecution in drug cases wherein a positive urinalysis is the sole basis for a wrongful use conviction. An example of the limitations imposed upon a court under the *Campbell* decisions would be the defendant's intent. Unless the defendant testifies under oath as to his intent, the intent cannot be ascertained, and urinalysis may be the only evidence of illicit drug use.

Adherence to Laboratory Procedures and Guidance Documents

The federal requirements of clinical laboratories' certification and inspection under CLIA, as well as a forensic laboratory's participation in similar certification or inspection programs, are helpful in assuring compliance with well-written procedural manuals. Inspections and certifications evaluate a laboratory's accuracy and precision in qualitative and quantitative analyses in comparison to peer laboratories. The independent laboratory inspection programs associated with certification provide a critical review of laboratory operations. The following is an example of adherence to laboratory policy and documentation entered into judicial proceedings. This is just an opinion from a county court in the state of Washington; thus it is not a controlling authority over any other courts, but it is an excellent example of how a court can scrutinize the practices of a forensic laboratory and change the entire outcome of a trial.

District Court of King County for the State of Washington, Case No. C00627921 et al., (January 2008)

This court case contested the laboratory adherence to their certification of ethanol standards, compliance with testing procedures, and adherence to testing guidelines. A number of defendants appealed to the King County District Court to suppress the respective defendants' breath alcohol test results based upon fraudulent and scientifically inaccurate testing [31]. The three-judge panel reviewed the breath alcohol testing procedures and in its published motion ruling stated the following:

1. There are multiple checks performed by the breathalyzer instruments located throughout King County and the state of Washington to ascertain the accuracy of the alcohol breath results. One of the checks is the use of an external ethanol simulator standard of known ethanol concentration (field simulator standard) prepared by the Washington State Toxicology Laboratory (WSTL).
2. The instrument is periodically checked, calibrated, and maintained by the Washington State Patrol Breath Test Section using solutions of ethanol and water prepared to known standards (QAP solutions) by the WSTL.
3. The procedures for preparing the QAP and field simulator standard [are] set forth in laboratory protocols developed and implemented by the state toxicologist.
4. Following testing of the solutions, the analyst certifies that [he or she has] performed the test and that the results as published are correct. The certifications are to be used in court in lieu of live testimony by the toxicologist.

The court stated it found many irregularities in the preparation, use, documentation, and certification of the ethanol field simulator solution standard and QAP. First, there was an issue of false certification. The WSTL engaged in a practice of having other toxicologists prepare and test the field simulator standard and QAP for the state toxicologist or WSTL manager, yet the signed documentation certifies that state toxicologist or WSTL manager prepared and tested the simulator solutions.

Second, the court found defective and erroneous certification procedures, including procedural irregularities to the stated protocol. The software used to calculate the concentration of simulator solution standard was "defective" in that the fourth data entry from the fourth toxicologist performing the analysis was omitted in the calculations. The WSTL protocol required the inclusion of all analysts' data in the calculations. No laboratory employee checked the software program to validate the accuracy and compliance of the calculation procedure with the WSTL protocol, and the analysts were not trained or directed to check the calculations performed by the software.

Third, there were problems with software failures, human error, equipment malfunction, and procedural violations. (1) The software used for calculations was not subject to rigorous testing and checking before implementation, resulting in substantial errors and significant data deletions from the calculations. (2) The WSTL had no procedure or protocol requiring the software be validated for accuracy, and no WSTL personnel conducted such testing, nor verified that the data produced using the software were correct. (3) Errors based on software miscalculations existed within almost all field simulator solution certifications issued over a 2-year period. (4) When analysts conducted GC analysis (volatile head-space analysis) of the solutions, the instrumentation printed results automatically. The analysts signed the worksheet results acknowledging the correctness of the data without checking the printout against the original chromatographic data, and incorrect data was inserted into some of the worksheets. These worksheets of field simulator standards were posted on the website and relied upon in determining the accuracy and precision of the breath testing machines employed in the county and state. (5) WSTL utilized several gas chromatograph machines for analysts' use. A machine that malfunctioned was not repaired or maintained adequately, resulting in different operational and measurement characteristics and abnormal variations in readings. The instrument remained online and was used in testing even though individual toxicologists knew that the instrument was not functioning properly.

Fourth, an audit of the WSTL found improper evidentiary procedures, including that simulator solution logbooks were not properly maintained, required self-audits were not performed, and there were possible indications of members of the WSTL not following protocols.

Fifth, there was an issue of inadequate and erroneous protocols and training. The court stated, "Contrary to protocol requirements, toxicologists were trained to discard data generated by the tests if any single data entry lay outside the range for the mean value of the solution as dictated by the protocol. …. Several toxicologists discarded data without identifiable or statistical reasons for doing so. …. Protocols for solution preparation and machine testing were contradictory or inconsistent, resulting in field solutions being used for QAP testing in some cases."

Based upon the above-published finding and other information presented to the three-judge panel, the District Court of King County for the State of Washington East Division ruled in the court order a suppression of the breath tests in each of the defendants' cases. Thus the prosecutors in those cases could not use the breath test evidence in those defendant's trials. In making this ruling, the court stated, "While we agree that trial courts should generally admit scientific evidence if it satisfies the requirements of *Frye*, … we conclude that … the work product of the WSTL is sufficiently compromised by ethical lapses, systemic inaccuracy, negligence and violations of scientific principles that the WSTL simulator solution work product would not be helpful to the trier of fact."

MARIJUANA DECRIMINALIZATION AND MEDICAL USE: FEDERAL VERSUS STATE STATUTES

In 2012 and 2013 Washington State and Colorado passed legislation permitting the personal recreational use of marijuana. Since then, as of March 23, 2017, 26 states and the District of Columbia have passed legislation legalizing the recreational use of marijuana with a number of states licensing the controlled sale of marijuana. The various state legislative initiatives on decriminalization of possession, controlled sale, and personal recreational use of marijuana are conflicted with federal statutes regarding the sale, distribution, and use of marijuana. This is most evident in the application of drug-free workplace standard for state and federal employees or contract personnel. States have imposed limitations and standards for prohibiting state employees' use of drugs in the workplace, especially state employees in critical positions of transportation, health, and safety. However, federal employees whose terms of employment are classified as Testing Designated Positions (TDP), may lose employment following a random urinalysis positive drug test under the Federal Drug-Free Workplace Program. Federal employees in TDP status must maintain security or safety standards as a condition of employment. A drug-positive result warrants the loss of security or safety clearance and therefore violates conditions of employment and results in the individual's employment termination.

In 2008 the Defense Office of Hearings and Appeals (DOHA) in Interagency Security Committee (ISC) Case Number ISCR 07-13294 sustained the government's denial of a security clearance to a DoD contractor for use of marijuana. As stated in the record, based on drug involvement and personal conduct, it is not consistent with the national interest to grant or continue the applicant's security clearance. Clearance was denied and served as basis for removal from federal civil service or contractor employment. The basis of the denial was the Bond Amendment [32]. This

Amendment contains disqualification provisions that apply to those applicants seeking clearances providing access to Special Access Programs (SAP), restricted data (RD) or sensitive compartmented information (SCI). In a memorandum dated June 20, 2008, the Office of the Under Secretary of Defense for Intelligence implemented Section 3002 of Public Law 110-181 (the Bond Amendment) for the adjudication of security clearances.

Specifically, the Bond Amendment repealed 10 U.S.C. Section 996 formerly known as the Smith Amendment, and placed restrictions similar to the Smith Amendment, but which apply to all Federal Government Agencies. The Bond Amendment bars persons from holding a security clearance for access to SAPs, Restricted Data, and SCI if they have been:

- Convicted of a crime and served more than 1 year of incarceration,
- Discharged from the armed forces under dishonorable conditions, and
- Determined to be mentally incompetent by a court or administrative agency.

The Bond amendment also prohibits all Federal Agencies from granting or renewing a security clearance for any covered person who is an unlawful user of a controlled substance or is an addict; this prohibition applies to all clearances. For the purposes of this prohibition:

- An unlawful user of a controlled substance is a person who uses a controlled substance and who has lost the power of self-control with reference to the use of the controlled substance, and any person who is a current user of the controlled substance in a manner other than as prescribed by a licensed physician.
- "Use" includes any use recent enough to indicate that the individual is actively engaged in such conduct and is an addict (lost the power of self-control or habitual user).
    - An "addict" of a controlled substance is any individual who habitually uses any narcotic drug so as to endanger the public moral, health, safety, or welfare;
    - Or is so far addicted to the use of narcotic drugs as to have lost the power of self-control with reference to his or her addiction.

With the expansion of state initiatives on legalization and decriminalization of marijuana, the Office of the Assistant Secretary of Defense for Readiness and Force Management issued on February 4, 2013, a Memorandum entitled *"Prohibition on the Use of Marijuana by Military Service Members and Department of Defense Civilian Employees."* The Memorandum states, "This memorandum reaffirms the federal prohibitions on the use of marijuana by military personnel at all locations in accordance with Article 5, Uniform Code of Military Justice (UCMJ). The provisions of the UCMJ apply regardless of State, District or Territorial Legislation permitting the use of marijuana, to include medical use. Military personnel are subject to prosecution and administrative action for marijuana use, possession or distribution under Article 112a of the Uniform Code of Military Justice (UCMJ). Federal Law supersedes the legislative initiatives of the State, District or Territories of the United States. Legislative initiatives of States, District or Territories are not binding on the military in the administration of military justice under Chapter 47 of Title 10, United States Code."

The Director, United States Office of Personnel Management, expanded the department of defense policy memorandum to all heads of executive departments and agencies dated May 26, 2015. The memorandum expanded the scope to include all federal employees and contractors. In excerpts, "Federal law on marijuana remains unchanged. Marijuana is categorized as a controlled substance under Schedule I of the Controlled Substances Act. Thus knowing or intentional marijuana possession is illegal, even if an individual has no intent to manufacture, distribute or dispense marijuana. In addition, Executive Order 12564, Drug-Free Federal Workplace, mandates that (1) federal employees are required to refrain from the use of illegal drugs; (2) the use of illegal drugs by federal employees, whether on or off duty, is contrary to the efficiency of the service; and (3) persons who use illegal drugs are not suitable for federal employment."

Reinforcing the Federal Government's position on the use of marijuana under the Controlled Substances Act, the United States Department of Justice, Drug Enforcement Administration (DEA) issued a report (July 2016) entitled, "Schedule of Controlled Substances: Maintaining Marijuana in Schedule I of the Controlled Substances Act—Background, Data and Analysis: Eight Factors Determinative of Control and Findings Pursuant to 21 U.S.C. 812(b)." The Department of Justice, Drug Enforcement Administration (DEA) determined, "After consideration of the eight factors discussed earlier and of the HHS's (*Health and Human Services*) recommendation, the DEA finds that marijuana meets the three criteria for placing a substance in Schedule 1 of the Controlled Substances Act under 21 U.S.C. 812(b)(1)."

CONCLUSIONS

As indicated earlier, expert witnesses and laboratory managers need to be knowledgeable about case law. The role of the expert witness is critical to the forensic laboratory as he or she represents the laboratory in court. Preparation for court is essential. There should be no assumptions made as to what may be asked at trial. All aspects of laboratory operations, as well as credibility of laboratory personnel are subject to challenge. Periodic assessment of laboratory operations, adherence to laboratory protocols, ongoing personnel training, and good documentation practices should be established procedures. Timely communications with the counsel requesting expert testimony is imperative in the preparation for court appearance. As recent case law has shown, there have been a number of decisions in military courts and US circuit courts of appeal regarding the confrontation rule. The decisions of these courts and the subsequent opinions rendered will define the role of toxicologists and possibly the role of all associative laboratory personnel involved in sample testing as testimonial witnesses in court.

ACKNOWLEDGMENT

The authors acknowledge the legal assistance of Major Heather N. Larson, USAF, JAG; and Michael L. Smith, PhD and Evelyn M. Jemionek for their excellent review and comments in the preparation of this manuscript.

DISCLAIMER

The opinion or assertions herein are those of the authors and do not necessarily reflect the view of the Department of the Army, Navy, Air Force, or the Department of Defense.

REFERENCES

[1] Frye v. United States, 293 F. 1013 (D.C. Cir 1923). http://caselaw.findlaw.com/dc-court-of-appeals/1199052.html.
[2] Daubert v. Merrill Dow Pharmaceuticals, 509 US 579 (1993). https://supreme.justia.com/cases/federal/us/509/579/case.html.
[3] Frye v. United States, 293 F 1013 (D.C. Cir. 1923): http://law.harvard.edu/publications/evidenceiii/cases/frye.htm.
[4] Jonakait R. Texts, texts or ad hoc determinations: interpretations of the Federal Rules of Evidence. Indiana Law J 1996;71:551–91. Available from: www.lectlaw.com/files/exp22.htm.
[5] Federal Rules of Evidence. https://www.rulesofevidence.org/.
[6] Cornell University Law School, Legal Information Institute, Supreme Court Collection,. Daubert v. Merrell Dow Pharmaceuticals (92–102), 509 US 579 (1993). (http://supct.law.cornell.edu/supct/html/92-102.ZS.html).
[7] Cornell University Law School, Legal Information Institute, Supreme Court Collection. Kumho Tire Co. v. Carmichael (97–1709), 526 US 137 (1999), rev'd, 131 F3d 1433. (http://supct.law.cornell.edu/supct/html/97-1709.ZS.html).
[8] United States v. Combs, 369 F.3d 925, 940 (6th Circuit, 2004). http://openjurist.org/369/f3d/925/united-states-v-combs.
[9] United States v. Combs, 369 F.3d 925, 940 (6th Circuit, 2004). http://federalevidence.com/node/521.
[10] United States v. Winbush, 580 F.3d 503, 512 (Seventh Circuit, 2009). https://www.courtlistener.com/opinion/1204237/united-states-v-winbush/.
[11] United States v. Freeman, 498 F. 3rd 893, 900-01, 906-07 (Ninth Circuit 2007). http://caselaw.findlaw.com/us-9th-circuit/1102564.html.
[12] U.S. Department of Health and Human Services, Centers for Medicare and Medicaid Services. Clinical Laboratory Improvement Amendments (CLIA). http://www.cms.hhs.gov/CLIA/.
[13] Brady v. Maryland, 373 U.S. 83 (1963). http://supreme.justia.com/us/373/83/case.html.
[14] Giglio v. United States, 405 U.S. 150 (1972). http://supreme.justia.com/us/405/150/case.html.
[15] United States v. Agurs, 427 U.S. 97 (1976). http://supreme.justia.com/us/427/97/case.html.
[16] United States v. Bagley, 473 U.S. 667, at 682 (1985). http://supreme.justia.com/us/473/667/case.html.
[17] United States v. Israel, 60 MJ 485 (CAAF2005). www.armfor.uscourts.gov/opinions/2005Term/04-0217.htm.
[18] United States v. Jackson, 59 MJ 330 (CAAF2004). www.armfor.uscourts.gov/opinions/2004Term/03-0336.pdf.
[19] United States v. Gonzales, 62 MJ 303 (CAAF2006). www.armfor.uscourts.gov/opinions/2006Term/03-0394.pdf.
[20] Rogers M.T., Fisher J.L., Karlan P.S., Goldstein T.C. et al. Luis E. Melandez-Diaz v. Massachusetts petition for a writ of certiorari to the Appeals Court of Massachusetts. http://www-personal.umich.edu/~rdfrdman/MDcertreplyfinal1.pdf.
[21] United States v. McKinney, 631 F2d 569 (8th Cir1980). http://bulk.resource.org/courts.gov/c/F2/631/631.F2d.569.80-1263.html.
[22] United States of America v. Dale Baker et al., 855 F2d 1353 (8th Cir1988). http://bulk.resource.org/courts.gov/c/F2/855/855.F2d.1353.86-2199.86-2257.86-2115.86-2112.86-2083.html.
[23] United States v. Magyari, 63 MJ 123 (CAAF 2006). www.armfor.uscourts.gov/opinions/2006Term/05-0300.pdf.
[24] Crawford v. Washington, 541 U.S. 36 (2004). www.supremecourtus.gov/opinions/03pdf/02-9410.pdf.
[25] United States v. Washington, 498 F3d 225. (4th Cir2007). http://search.live.com/results.aspx?FORM = DNSAS&q = pacer.ca4. uscourts.gov%2fopinion.pdf%2f054883.P.pdf.
[26] United States v. Harcrow, 66 MJ 154 (CAAF2008). www.armfor.uscourts.gov/opinions/2008Term/07-0135.pdf.
[27] United States v. Campbell, 50 MJ 154 (CAAF1999). http://www.armfor.uscourts.gov/opinions/1999Term/97-0149.htm.
[28] United States v. Campbell, 52 MJ 386 (CAAF2000). http://www.armfor.uscourts.gov/opinions/2000Term/97-0149R.htm.

[29] Stahlman Michael R. New developments on the urinalysis front: a green light in naked urinalysis prosecutions? Army Lawyer. 2002. DA PAM 27-50-351. www.loc.gov/rr/frd/Military_Law/pdf/04-2002.pdf.
[30] United States v. Green, 55 MJ 76 (CAAF 2001). www.armfor.uscourts.gov/opinions/2001Term/00-0268.pdf.
[31] State of Washington v. Ahmach et. al., (Case No. C00627921 et al.), District Court of King County for the State of Washington, 2008. http://seattletimes.nwsource.com/ABPub/2008/01/30/2004154613.pdf.
[32] Bond Amendment, Public Law 110-181, Section 3002, January 1, 2008 http://www.dss.mil/documents/pressroom/bond_amendment.pdf.

APPENDIX

The following are the more commonly encountered terms that laboratory scientists might expect to hear in the court or in discussions with trial counsel when dealing with legal matters.

Cross-examination: After the initial questioning by the counsel who called the witness to testify, the counsel for the opposing side has the opportunity to question or challenge the witness regarding his or her other testimony, opinions, interpretation of evidence, and answers given under direct examination. The questions used during cross-examination are often very leading and restricted to yes/no answers.

Direct examination: The testimony of the witness for the prosecution or the witness for the defense under initial questioning by the counsel calling the witness. Questions used during direct examination are usually broad, are not restricted to yes/no answers, and allow the witness to answer in a longer narrative fashion.

Discovery: Request by prosecution or defense to opposing counsel or to the judge to have additional documentation made available, such as consultant's reports, oral discussions, papers, and so forth. Discovery may extend to pertinent laboratory QC records, certification reports, adverse action reports, incident reports, personnel disciplinary actions, and other past records believed to have merit in the preparation or presentation of the case. Brady Notices are also part of discovery materials.

Exhibit: Documentation or other physical items offered into court proceeding by the prosecution or defense counsel for purposes of identification and relevance to the court proceedings. Exhibits may also be "demonstrative" in nature, meaning that they are not a piece of substantive evidence (defined later) but are used as an aid to a witness's testimony or a counsel's verbal arguments.

Evidence: Exhibits that have been properly introduced and identified as relevant to the court proceedings that are accepted by the judge are then labeled as evidence for the prosecution or defense. Once evidence is properly admitted, it may be considered by the jury in deliberation.

In camera: A Latin term meaning "in chambers." It refers to a ruling to conduct a hearing or inspection of documents in private, outside of the courtroom, usually in a judge's chambers. The proceeding may be on or off record depending upon the circumstances, but this is usually noted or commented upon by the judge for the record. In camera proceedings are conducted to review or discuss issues of sensitive or proprietary nature and to ensure that a jury is not biased by presentation of certain issues in court.

Merits: Merits of the case involves direct charges, evidence, and testimony that will be presented in trial. It does not include arguments of counsel.

Motions: Motions are requests or challenges by the prosecution and/or defense counsels for consideration by the judge regarding admissibility of evidence, dismissal of charges, limitations in the scope of questioning of witnesses or testimony of witnesses, and a variety of other procedural issues.

Objections: Challenges by the prosecution or defense counsel to a line of questioning, phrasing of a question, or nature of information being solicited during questioning of a witness by opposing counsel. Objections may also be raised in response to information cited, opinion or answers given by a witness.

Publish to the jury: Distribution of evidence for consideration by the jury. This is often accomplished by passing out copies of written documentary evidence to each jury member or passing around a physical item of evidence.

Recross-examination: After redirect examination, the opposing counsel may question, challenge or seek clarification or exception regarding the witness's testimony on issues raised under redirect examination.

Redirect examination: After cross-examination, the counsel who initially called the witness to the stand may ask additional questions or request clarification or elaboration on statements, opinions, etc. that may have arisen as a result of the cross-examination.

Chapter 8

Pharmacogenomics of Drugs of Abuse

Christine L.H. Snozek[1] and Loralie J. Langman[2]
[1]*Department of Laboratory Medicine and Pathology, Mayo Clinic Arizona, Scottsdale, AZ, United States,* [2]*Department of Laboratory Medicine and Pathology, Mayo Clinic, Rochester, MN, United States*

INTRODUCTION

Drug addiction is characterized by both psychological and physiological changes, which appear to be mediated in part by hereditary factors. The use of an addictive substance triggers the onset of the physiological alterations and maintains those neurological changes both during and after chronic abuse, thus, the genetic component of drug abuse and addiction is, by nature, pharmacogenetic [1]. In addition to affecting the onset and progression of the addictive process, genetic differences are known to contribute to interindividual variability in acute drug responses such as toxicity or inefficacy, and might predict likelihood of successful response to rehabilitation. It has been suggested that genetic factors comprise 40%–60% of the variability in the complex nature of addiction [1]. To date, however, much of the promise of addiction-related pharmacogenetics remains speculative; further studies will hopefully highlight variants and future biological targets important for understanding and treating this disease process.

The current goal of pharmacogenetics is working toward prediction of individual differences in the risk for substance abuse, the response to a given compound, and the success of treatment for drug abuse, all based upon knowledge about variation in specific genes. The genes most commonly studied in pursuit of this goal tend to be related to either pharmacokinetic or pharmacodynamic parameters because genes encoding metabolic enzymes are most prominent in these categories, along with genes for drug receptors and transporters. However, many other types of functional proteins have been examined in the contexts of addiction and drug response and are promising future targets for characterization of genetic variation. Included among this latter groups are genes encoding transcription factors, signaling molecules (often downstream of drug receptors), and regulators of bodily processes ranging from circadian rhythm to stress response. This chapter will review specific genes studied in the context of addiction, with some discussion of gene classes that may prove important in future work.

METABOLIC ENZYMES

Most xenobiotics are metabolized at least in part enzymatically; there are many proteins that accomplish this task, but the best known are the enzymes of cytochrome P450 (CYP) family. These proteins comprise a large group of heme-containing monooxygenase proteins that localize to the endoplasmic reticulum and mitochondrial membrane. NADPH (nicotinamide adenine dinucleotide phosphate) is a required cofactor for CYP-mediated biotransformation, and oxygen serves as a substrate [2]. The CYP superfamily is found in many organisms, with over 7700 known members across all species studied. At present, 57 human genes are known to encode CYP isoforms; of these, at least 15 are associated with xenobiotic metabolism (Table 8.1). CYP isoenzymes are named according to sequence homology: amino acid sequence similarity >40% assigns the numeric family (e.g., CYP1, CYP2); >55% similarity determines the subfamily letter (e.g., CYP2C, CYP2D); isoforms with >97% similarities are given an additional number (e.g., CYP2C9, CYP2C19) to distinguish them [3].

The CYP enzymes are expressed primarily in the liver and small intestine, and to a lesser extent in other tissues such as brain, kidney, and lung [3]. Despite the remarkable array of pharmaceutical compounds in current use, only a very small subset of hepatic enzymes perform the vast majority of CYP-mediated biotransformations, with CYP3A4 and CYP2D6 responsible for the largest contribution [2]. For drugs that act on the central nervous system, CYP

TABLE 8.1 Metabolic Enzymes Associated With Common Drugs of Abuse

Abused Drug	Metabolic Enzymes
Amphetamine	CYP2D6
Methamphetamine	CYP2D6, COMT
3,4-Methylenedioxymethamphetamine	CYP2D6
Alprazolam	CYP3A4
Diazepam	CYP3A4 and CYP2C19
Midazolam	CYP3A4
Triazolam	CYP3A4
Cocaine	Serum butylcholinesterase, hepatic carboxylesterase
Fentanyl	CYP3A4 and CYP3A5
Codeine	CYP2D6
Oxycodone	CYP2D6
Hydrocodone	CYP2D6
Heroin	Serum butylcholinesterase, hepatic carboxylesterase
Morphine	UGT2B7, COMT
Methadone	CYP2B6, CYP3A4, and CYP2D6
Tetrahydrocannabinol	CYP3A4, CYP2C9, and CYP2C11
Propoxyphene	CYP2D6
Phencyclidine	CYP3A4

COMT, Catechol-*O*-methyltransferase.

enzymes in the brain are also important to consider. The brain is a heterogeneous organ, so variations in drug metabolism between cell types may create regional microenvironments with different drug concentrations and physiological response [1].

A central aspect of pharmacogenetics involves understanding of how heterogeneity within the CYP system affects drug metabolism and ultimately contributes to treatment response or toxicity. As a class, CYP genes have a great deal of interindividual sequence variability, with differences ranging from single nucleotide polymorphisms (SNPs) to duplication or deletion of entire genes. Although not all of the variant alleles characterized have altered enzymatic function, a large number of them do. *CYP2D6* is perhaps the most notable example of a gene with widely divergent enzymatic activity as a result of sequence variation. The importance of CYP2D6 and other CYP enzymes in drug pharmacokinetics is clear. The role of metabolic enzymes in the addictive process is less well understood, but the evidence that they are involved in substance abuse and addiction is growing.

Despite the fact that CYP2D6 constitutes only 1%–2% of total CYP protein in the liver, this mainly hepatic enzyme metabolizes 20%–25% of drugs, including several opiates and amphetamine-type stimulants [3–8]. *CYP2D6* gene is highly polymorphic, with over 60 variant alleles currently known. Most variants are SNPs or combinations of SNPs, but amplification (*1 N), deletion (*5), truncation (*3, *4), and pseudogene-hybridization (*66) of the *CYP2D6* gene have been described. The absence of a clinical syndrome associated with these variants implies that the enzyme lacks a major endogenous function, although roles in neurotransmitter metabolism [9] and endogenous morphine production have been suggested [10]. *CYP2D6* gene polymorphisms can significantly affect the pharmacokinetics of roughly half the compounds metabolized by that enzyme [9,11–13]. CYP2D6 is unique among the drug-metabolizing CYPs in that it is not inducible, though it can be inhibited by substrate; thus, genetic polymorphisms contribute largely to interindividual variation in CYP2D6 activity.

CYP2D6 is particularly notable for its role in clinical response to opioids, as it converts several of this class of drugs, including codeine, oxycodone, dextromethorphan, and tramadol, into their respective active metabolites. The best-studied example of the influence of CYP2D6 polymorphisms in opioid therapy is its activation of the prodrug, codeine, via demethylation to morphine. The level of CYP2D6 enzymatic activity is strongly dependent upon the alleles expressed; in the case of null alleles, residual activity is essentially zero, while in the case of gene amplification the resultant activity can be many times higher than normal. This variability has a profound effect on codeine therapy because at the same codeine dose, patients with minimal CYP2D6 activity will likely not convert enough of the prodrug to receive adequate analgesia, while those with very strong CYP2D6 function may be at risk for adverse drug reactions due to excessive morphine levels [2,14]. This range of responses to the same compound may therefore affect the risk of addiction by influencing the degree to which individuals react favorably or poorly to drug exposure. The role of CYP2D6 in adverse reactions and poor treatment efficacy for opioids has been reviewed elsewhere [15,16], including guidelines for treatment with codeine [17].

Interestingly, there are some suggestions that CYP2D6 ultrarapid metabolizers may receive less benefit or require higher doses in opiate-addiction therapy with methadone. Studies in Spanish populations indicated lower satisfaction scores [18] and higher methadone doses to achieve response in ultrarapid metabolizers compared to other phenotypes [19]. This is particularly intriguing because CYP2D6 plays only a minor role in methadone metabolism [20,21], although methadone does inhibit CYP2D6 activity [22,23]. However, no differences were seen for reduced-activity phenotypes, and at least one other study showed no effect for CYP2D6 on dose or dropout rate [24], indicating that further study in larger, more diverse populations may be warranted.

The most obvious means by which metabolic enzyme variants might affect the addictive process is by altering the amount of the parent compound relative to active and inactive metabolites, as appears to be the case with CYP2D6 in methamphetamine dependence. CYP2D6 is involved in the formation of several psychoactive metabolites including 4-hydroxymethamphetamine [25,26] and alleles with reduced activity (intermediate metabolizer phenotype) were associated with significantly lower risk of developing addiction to methamphetamine probably due to lesser euphoric or pleasurable effects after drug exposure. There were no CYP2D6 poor metabolizers in this study, but predicted they would show even less likelihood of becoming methamphetamine-dependent [27]. Methamphetamine is a weak inhibitor of CYP2D6, whereas the related drug 3,4-methylenedioxymethamphetamine (MDMA, ecstasy) is a potent inhibitor of the enzyme; both are CYP2D6 substrates. Coadministration of other CYP2D6 inhibitors has been associated with toxic and even life-threatening effects of methamphetamine and MDMA [28]. Several reports have shown that responses to MDMA and other amphetamine-type stimulants vary depending on *CYP2D6* genotype. For example, blood pressure and subjective effects including "drug liking" increased more rapidly in response to MDMA in poor metabolizers [29]. Interestingly, CYP2D6 poor metabolizers were overrepresented compared to nonusers in the nonfatal abuser cohort of a study examining paramethoxymethamphetamine toxicity, which could suggest a protective effect of reduced enzyme activity in response to this drug, and/or a predisposing effect toward drug use [30].

Another member of the CYP family, CYP2B6, has received increasing attention in the context of methadone maintenance therapy after recognition that it is the enzyme primarily responsible for conversion to the major metabolite 2-ethylidene-1,5-dimethyl-3,3-diphenylpyrrolidine (EDDP) [31]. This enzyme acts preferentially on *S*-methadone, which is associated with more cardiac toxicity than *R*-methadone, which has the majority of opioid agonist activity [32]. Individuals with reduced-function *CYP2B6* alleles demonstrate higher methadone concentrations relative to dose in single-dose studies [33] and at steady state [34,35]. *CYP2B6* variants also have been linked to higher postmortem methadone concentrations and methadone/EDDP ratios [36], as well as the need for treatment of neonatal abstinence syndrome [37].

Several other classes of metabolic enzymes have been characterized, though few have been shown to have relevance to drugs of abuse. Human carboxylesterase 1 (hCE1 or CES1) is a serine esterase that hydrolyzes a variety of endogenous and exogenous substrates [38]. It is primarily expressed in hepatocytes, with lesser amounts in several other tissues [39,40]. CES1 is responsible for formation of at least two cocaine metabolites, the major blood metabolite benzoylecgonine [39,41] and the toxic metabolite cocaethylene [39,41,42]. A few CES1 variants causing complete loss of activity have been described [43] and would be predicted to significantly affect cocaine pharmacokinetics and response to the drug. There are currently no reports of these variants in the context of drug use, but CES1 deficiency would likely increase blood concentration and exposure to the drug, and possibly prolongs the effects of a cocaine dose.

Human butyrylcholinesterase (BChE, serum cholinesterase, pseudocholinesterase) hydrolyzes cocaine and its minor metabolite norcocaine to ecgonine methyl ester and norecgonine methyl ester, respectively [41]. BChE is best characterized for its role in metabolizing succinylcholine, a drug given to facilitate muscle relaxation during surgery and tracheal intubation. The drug's duration of effect is usually <10 min; however, patients with BChE deficiency exhibit prolonged

TABLE 8.2 Human Butyrylcholinesterase Variants

	Phenotypic Description	Amino Acid Alteration	Formal Name for Genotype	References
Usual	Normal	None	BCHE	[46] [47]
Atypical	Dibucaine resistant	Asp70Gly	BCHE^{a}70G	[48] [49] [50]
Fluoride 1	Fluoride resistant	Thr243Met	BCHE^{a}243M	[51]
Fluoride 2	Fluoride resistant	Gly390Val	BCHE^{a}390V	[52]
K variant	30% reduction in activity	Ala539Thr	BCHE^{a}539T	[53]
J variant	66% reduction in activity	Glu467Val	BCHE^{a}497V	[53]
Silent	No activity	117 Gly-Frameshift to stop codon at position 129	BCHE^{a}FS117	[54] [55]

^{a}The table omits a number of very rare variants that have been seen in only a few isolated families.
Source: http://www.ncbi.nlm.nih.gov/gene?
Db = omim&DbFrom = gene&Cmd = Link&LinkName = gene_omim&LinkReadableName = OMIM&IdsFromResult = 590.

TABLE 8.3 Receptors and Transporters Involved in Drug Addiction

Receptor/Transporter	Associated Drugs of Abuse
Opioid receptors	Opiates, cocaine, also naltrexone therapy
Opioid ligands (dynorphins, etc.)	Cocaine, opioids
Dopamine receptors	Alcohol, nicotine, cocaine, heroin, also bupropion therapy
Dopamine transporter	Cocaine, amphetamines
Serotonin transporter	Heroin, also high-risk behaviors
Serotonin receptors	Heroin
Norepinephrine transporter	Cocaine

apnea in response to succinylcholine [44]. Various forms of the enzyme were characterized in the 1950s [45], which have subsequently been associated with the genetic mutations (Table 8.2) responsible [48,56].

Although peak cocaine concentrations tend to correlate with the intensity of pharmacological and behavioral effects [57,58], neither cocaine toxicity nor fatality appear to correlate well with dose or concentration [59]. Reduced BChE activity would be expected to prolong the half-life of cocaine, increasing its concentration in the brain and effects on the dopaminergic system, and to divert more cocaine to alternate metabolic pathways to toxic metabolites such as norcocaine [60–62], which could increase the risk of toxicity and even death [63]. Although these predictions are supported to some extent in mouse models [60,64], to date no studies exist confirming these phenomena in humans. One study suggested a link between BChE variants and use of crack cocaine [65]. Interestingly, the existence of BChE variants has led to their characterization and modification in the pursuit of an effective treatment for cocaine toxicity and addiction [66].

Additional enzymes with metabolic activity involved in endogenous regulation of neurotransmitters will be discussed in a later section. In addition to metabolic enzymes, endogenous receptors and transporters regulate various aspects of perception and behavior, and therefore can play significant roles in the mechanism of drug addiction (Table 8.3). We will discuss two such networks with implications in addiction, namely the endogenous opioid and monoaminergic neurotransmission systems.

ENDOGENOUS OPIOID RECEPTORS

Compounds derived from the opium poppy have been used since ancient times as analgesics, antitussives, and soporifics. As a result, opiates have long been recognized as drugs of addiction, and newer synthetic opioids such as fentanyl have joined traditional opiates like morphine as common addictive substances. Endogenous receptors for opiates were reported in 1973 [67–69] and consist of the mu (MOR), delta (DOR), and kappa (KOR) receptors, encoded by *OPRM1*, *OPRD1*, and *OPRK1*, gene respectively. Endogenous ligands have been described, such as endorphins and enkephalins, polypeptides that are involved in innate responses to pain. The opioid receptors mediate both analgesic and rewarding properties of opioid compounds, as well as other physiological effects. Opioid receptors are also important in modulating responses to nonopioid drugs, including cocaine, other psychostimulants, and alcohol [70]. Components of the endogenous opioid system and the genes encoding them have therefore been the focus of research into specific addictions since their discovery.

Mu Opioid Receptor Gene (OPRM1)

The MOR is the site of action of most opioid analgesic medications such as morphine, oxycodone, and fentanyl; the receptor is essential for morphine analgesia, physical dependence, and reward [71–75]. Many variants, particularly SNPs, have been identified throughout *OPRM1* and tested for association with addiction to opiates, cocaine, and other substances. Although the initial molecular targets of cocaine are monoamine transporters, expression and function of the MOR are also affected, particularly in long-term abuse. Chronic experimental administration increases MOR density and mRNA levels in several regions of rat brain [76–79]; similar results have been seen in cocaine-dependent humans.

The best-studied *OPRM1* polymorphism is a base change at position 118 from adenine to guanine (A118G) that alters the protein sequence at residue 40 from asparagine to aspartate [80,81]. The frequency of this polymorphism globally is approximately 20%, but varies widely across populations, occurring in <2% of some groups and in nearly 50% in other groups. The A118G substitution leads to reduced mRNA expression, decreased levels of MOR on the cell surface, and altered binding and downstream signaling in response to endogenous ligands such as ß-endorphin [82,83].

Published literature is inconsistent regarding associations of the A118G polymorphism with pain, analgesia, response to opioids, and addiction [83]. Several studies have shown association of the variant 118G allele with increased susceptibility to opiate (notably heroin) addiction and alcoholism in multiple ethnic groups. However, other reports show no significant link between the A118G polymorphism and addiction. Recent meta-reviews have attempted to address these inconsistencies in much larger populations, with mixed results. A meta-review of 28,689 individuals of European ancestry evaluated dependence on nicotine, alcohol, cannabis, cocaine, and opioids. The overall population demonstrated a small but statistically significant protective effect (odds ratio estimates 0.88–0.92) of the G allele against dependence on those substances in general, although the effect on each substance alone failed to reach statistical significance [84]. Similarly, a meta-review of 9385 individuals found an association of this polymorphism with opioid dependence in Asians [85]. In contrast, a meta-review of 9613 individuals including both European and Asian descent failed to find a significant association of A118G with alcohol dependence [86].

Several other synonymous and nonsynonymous SNPs have been evaluated for their role in drug abuse, with similarly mixed results. Excellent reviews of *OPRM1* genetics in the context of drug use, abuse, and treatment have been published [82,83,87].

Kappa Opioid Receptor Gene (OPRK1)

Similar to the MOR, the KOR is also involved in response to addictive drugs, most notably cocaine and opiates. Several MOR-directed opiates also interact with the KOR, providing a clear mechanism by which polymorphisms in *OPRK1* could influence opioid addiction. The role of the KOR in cocaine addiction is thought to be a consequence of dopamine regulation: signaling through the KOR is associated with reduction of dopamine levels, while its major endogenous ligand, dynorphin, can attenuate cocaine-mediated blockade of dopamine reuptake [88]. Elevation of synaptic dopamine levels provides reinforcement, thus modulation of this neurotransmitter is a key component of addiction to numerous substances [77,89–91]. For this reason, the role of the KOR in modulating dopamine levels and response to substances such as cocaine implicate this gene as a likely target for understanding hereditary predisposition toward addiction.

At least seven SNPs in the human *OPRK1* gene have been reported [92–94], of which only G36T in exon 2 has been extensively studied. Allele frequencies for these variants range from <1% to roughly 15% which have been identified in several populations including Caucasian, African-American, and Hispanic groups [92,94]. Most *OPRK1* polymorphisms to date are silent, i.e., do not affect the amino acid sequence. However, synonymous SNPs can affect mRNA stability and folding, thereby influencing the eventual expression of the gene product. In terms of opiate abuse, *OPRK1* variants have been associated with roles in opiate dependence [95–97], response to methadone treatment [97,98], and severity of naloxone-precipitated withdrawal [99]. Interestingly, in the realm of cocaine abuse and treatment, *OPRK1* polymorphisms appear to influence risk of cocaine relapse after treatment [100] and response to a cocaine vaccine [101].

Delta Opioid Receptor Gene (OPRD1)

The DOR functions directly in pain responses, but also has a role in modulating the effects of MOR-directed compounds [102,103]. There are suggestions that DOR–MOR heterodimerization affects the development of tolerance, which may provide an avenue for DOR to regulate addiction to MOR agonists [104]. One study associated a nonsynonymous variant at position 27 with opioid dependence in European Americans [105]. However, most *OPRD1* polymorphisms are silent; several studies have linked individual variants or haplotypes containing *OPRD1* variants to opioid dependence [106–108], nondependent opioid use [109], and response to opioid dependence treatment [110,111]. There are also preliminary indications that OPRD1 variants might be associated with abstinence from cocaine use [112], although the role of this gene in cocaine dependence is far less extensively studied.

Endogenous Opioid Ligands

The endogenous opioids include endorphins, enkephalins, and dynorphins. These peptides are involved in regulating perceptions of pain and reward, thus it is logical that heritable variations in their respective genes may be important in the addictive process. To date, the most promising endogenous opioid gene in this respect has been *PDYN*. This gene encodes prodynorphin, a precursor polypeptide that encompasses several KOR ligands including dynorphin A, dynorphin B, and α-neoendorphin [88]. Dynorphins appear to counteract the dopaminergic effects of stimulants such as cocaine [113]. Interestingly, dynorphin release and mRNA levels are enhanced by psychostimulants, supporting a role for these peptides in addiction biology [114–120].

The best-characterized *PDYN* polymorphisms are a 68-base variable number of tandem repeats (VNTR, 1–5 copies) in the promoter region where more repeats lead to increased expression [121]. A 3′ untranslated region (3′-UTR) haplotype has also been linked to variable *PDYN* expression [122]. Increased expression appears to provide a protective effect against addiction to cocaine alone or cocaine/alcohol codependence [123,124]. However, other studies have shown no effect of the VNTR polymorphism on dependence on opioids [125,126] or methamphetamine [127]. Interestingly, recent studies have shown sex-specific differences in the importance of *PDYN* variants in drug abuse, including heroin dependence [121,128], and alcohol dependence and related traits [129].

Another endogenous opioid gene has also shown promise as a risk marker, namely the proenkephalin gene *PENK*. Initial studies suggested a link between a $(CA)_n$ repeat polymorphism and opioid dependence, where longer *PENK* alleles (those containing ≥ 81 base pairs) were associated with increased risk of opioid addiction as compared to shorter alleles (≤79 base pairs) [130]. More recent studies have supported an influence of PENK polymorphisms in risk for opioid dependence in an alcohol-dependent cohort [131], as well as risk for cannabis dependence [132].

MONOAMINE NEUROTRANSMITTER SYSTEM

The monoaminergic neurotransmitters act as messengers throughout various regions of the nervous system and function in ways important to addiction including effects on impulse control, behavior modulation, reward response, and positive reinforcement. The major neurotransmitters involved in the addictive process are serotonin and the catecholamines, particularly dopamine and norepinephrine [2,133,134]. There appears to be extensive interaction between the dopaminergic, serotonergic, and opioidergic systems in modulating reward, substance dependence, and drug withdrawal [70,135,136]. Like the endogenous opioid system, specific monoamine receptors play an important role in regulating neurotransmitter functions. In addition, neuronal transporter proteins responsible for controlling release and reuptake are essential in modulating synaptic neurotransmitter levels. Because of this, both monoamine receptors and transporters have been the targets of human genetic analyses in the study of addiction.

The monoamines are of particular interest with respect to abuse and addiction of psychostimulants, because the drugs themselves have substantial effect on these neurotransmitter systems. It has been estimated that the genetic contribution to variability in response to stimulants is approximately 60% [137–140], thus genes related to the monoaminergic systems are a logical place to seek candidate polymorphisms for stimulant dependence. Cocaine produces many of its physiological effects through blockade of dopamine reuptake, resulting in increased synaptic dopamine concentrations. In addition, cocaine can disrupt reuptake of norepinephrine and serotonin [2]. The ability of cocaine to modulate neurotransmitter reuptake is a consequence of its high-affinity binding to the dopamine transporter and lower-affinity binding to the norepinephrine and serotonin transporters [141,142]. In contrast, amphetamine-type stimulants use a different mechanism to increase synaptic monoamine concentrations and create subjective responses similar to those of cocaine. Amphetamines primarily appear to enhance release of neurotransmitters rather than altering the rate or extent of reuptake. Regardless of the means by which it is achieved, elevated synaptic neurotransmitter levels result in sensations of euphoria, which likely contribute to the addictive properties of stimulants [2].

Dopamine Receptor Genes

Drug-induced dopamine elevations appear to mediate reinforcement independently of conscious pleasure perception, increasing the desire to acquire more drug regardless of whether its use is enjoyable [1]. Dopamine receptors therefore play a large role in substance abuse, functioning in both short-term rewarding effects and the long-term development of dependence [2,143]. Signaling through dopamine receptors is implicated in physiological changes following drug exposure, including regulation of transcription and alterations in gene expression [144]. These receptors are encoded by at least five separate genes (*DRD1–DRD5*), the products of which are often categorized as D1-like (DRD1 and DRD5) or D2-like (DRD2, DRD3, and DRD4) based on function. The best studied of these genes in the context of addiction pharmacogenetics is *DRD2*.

A restriction fragment length polymorphism (RFLP), TaqIA, is present some 10,000 base pairs downstream of the *DRD2* gene. The TaqI AI allele has been variably associated with abuse and dependence on numerous substances, including alcohol [145,146], nicotine [147,148], cocaine [149], psychostimulants [150], and heroin [151], as well as outcome following methadone [151] or bupropion therapy [152,153]. Unfortunately, many studies linking the TaqI AI allele to drug addiction had insufficient statistical power to address ethnicity, gender, or other population differences, which may explain contradictory results published by independent groups of investigators. A meta-review conducted by Smith et al. [154] examined the association of the TaqI AI allele with an increased risk for alcoholism, and found that there was indeed a positive, albeit small, association. Another RFLP, TaqIB, affects intron 2 of *DRD2* and some studies have shown a significant association with tobacco use [155], cocaine dependence [149], and polysubstance abuse in Caucasians [156], although as with TaqIA these findings remain somewhat controversial. The TaqIA and TaqIB polymorphisms are in linkage disequilibrium, suggesting that haplotype analysis might provide more definitive results [157]. However, the gene *ANKK1* is also altered by the TaqIA polymorphism and may influence the addictive process independently of *DRD2* [158].

Other *DRD2* polymorphisms have been linked to drug dependence as well. A promoter region variant (-141C Ins/Del) is thought to significantly affect *DRD2* expression. The -141CDel allele is thought to reduce *DRD2* expression and has been associated with heroin use [159], alcoholism [160], and response to bupropion or other nicotine replacement therapies [161]. It has been suggested that low dopamine receptor levels may support use of dopaminergic stimulants as therapy to compensate for the inherent deficiency [162].

Several other *DRD2* variants can also cause decreased gene expression, including the C957T polymorphism. This sequence change, though silent at the amino acid level, alters mRNA folding to cause decreased stability and translational efficiency. In addition, the ability of dopamine to stabilize *DRD2* mRNA was greatly attenuated with the 957T allele [163]. The C957T variant was associated with abstinence following nicotine replacement therapy [161], and increased likelihood of alcoholism in families with a history of alcohol dependence [164]. Several other studies have linked the 957T allele with various neurological and psychological disorders, setting the stage for future work to further examine this variant in addiction and addiction-related behavioral disorders.

The other dopamine receptor genes have been studied as well with variable outcome. Chen et al. [165] examined *DRD2*, *DRD3*, and *DRD4* gene variants in methamphetamine-dependent subjects. The exon 3 VNTR polymorphism in *DRD4* results in individuals carrying 2–11 copies of the repeated sequence and in Caucasians the most common alleles contain either 4 or 7 repeats. Although *DRD2* and *DRD3* showed no association, Chen et al. [165] found that the 7-repeat *DRD4* allele occurred more frequently in methamphetamine abusers than in the controls. This variant has been linked to other addiction-related phenomena such as novelty-seeking [166], heavy drinking [167], and smoking [168], but these results have been challenged by other studies [169]. The existence of numerous *DRD4* alleles and the low prevalence of some variants are undoubtedly confounding factors. Nevertheless, some larger studies [166] have indicated that this gene may exhibit a legitimate association with the addictive process.

There is some evidence of a role for *DRD1* polymorphisms in specific types of addiction [170]: studies have shown associations with a variant in the 5′-UTR of exon 2 (A-48G), which was implicated in severity of opioid addiction [171] as well as alcoholism [172] and sensation seeking in alcoholic males [173]. Several genes, when looked at in combination, have shown associations with opioid addiction, including *DRD2*, *ANKK1*, *DRD4*, *COMT*, and *DAT1* [174–178]. These results suggest that thorough haplotype and gene-combination analyses should be performed before definitive judgment is passed on these genes.

Dopamine Transporter Gene (DAT1/SLC6A3)

The dopamine transporter gene (*SLC6A3* or *DAT1*) encodes a protein involved in two key functions: (1) release of dopamine into the synapse to activate neurotransmitter receptors and (2) reuptake of dopamine into presynaptic neurons to terminate the signal. The former process is stimulated by amphetamines, while the latter is inhibited by cocaine, resulting in similar dopaminergic pharmacological responses and rewarding properties. Several studies indicate that *DAT1* polymorphisms are important in determining susceptibility to addiction and interindividual variations in response to these psychostimulants and other drugs [179,180].

The best-studied *DAT1* variants include a VNTR in the 3′-UTR of the gene; the repeated 40-nucleotide sequence is most commonly present 10 times. It is thought that the 9-repeat variant shows differential transcriptional activity compared to the 10-repeat gene, although studies have yet to conclusively determine the relative in vivo expression of these alleles [181–183]. Another characterized VNTR is a 30-base pair repeat in intron 8 (Int8) [184]. The 5-repeat and 6-repeat alleles are the most common variants at this location; the 6-repeat allele is associated with lower transcriptional activity in vitro [180]. Several other *DAT1* SNPs have also been studied, both alone and in haplotype blocks.

The 3′-UTR VNTR polymorphism has been implicated in individual responses to stimulants, particularly cocaine and amphetamines, although it was not associated with cocaine dependence in a relatively large study of African-Americans [185]. An early study showed increased likelihood of cocaine-induced paranoia in individuals with at least one copy of the 9-repeat allele, although no association between the variant allele and cocaine dependence could be determined [180]. Similar results were reported for increased risk of methamphetamine psychosis with alleles containing 9 or fewer repeats [186]. Interestingly, children receiving methylphenidate for attention-related disorders generally show poorer response in patients with at least one 9-repeat allele [187,188], though other studies report conflicting results [189]. Stein et al. considered homozygous 9-repeat patients independently from heterozygous 9/10 or homozygous 10/10 individuals; this approach showed poorer therapeutic effect of methylphenidate in the former group [190], thus suggesting it may be inappropriate to analyze 9-repeat heterozygotes and homozygotes as a single category. In agreement with this, 9/9 homozygotes showed diminished response to acute amphetamine administration compared to either 9/10 heterozygotes or 10/10 homozygotes [191], which may translate into reduced potential for amphetamine dependence.

There are fewer studies of the Int8 VNTR, but at least one showed an association between cocaine dependence and the homozygous 5/5 genotype [192]. Another study examined the relationship between the 3′-UTR and Int8 variants and response to cocaine in nontreatment-seeking cocaine users [193]. Individuals with at least one copy of the 3′-UTR 9-repeat allele exhibited significantly more pleasurable subjective responses to cocaine compared to those with a 10/10 genotype, whereas individuals with the Int8 6/6 genotype reported more cocaine-induced anxiety compared to those with at least one 5-repeat allele. Subjects who had the 3′-UTR 10/10 genotype and at least one copy of the Int8 5-repeat variant displayed decreased subjective responses to acute cocaine exposure. These findings suggest that the perception of subjective effects may be related to genotype, and that individuals with specific variants might therefore be more vulnerable to relapse.

Relatively few studies have examined non-VNTR *DAT1* polymorphisms in addiction, though several such variants have been described. Two nonsynonymous SNPs, T265C and T1246C, encode valine to alanine substitutions at amino acid residues 55 and 382, respectively [194] and both represent conservative changes but appear to alter the kinetics of dopamine and cocaine interactions with the transporter [195,196]. In addition to stimulant response, *DAT1* polymorphisms appear to be important in risk of dependence, severity of withdrawal, and response to cessation therapy, in both alcoholism and nicotine addiction. As with other abused substances, results vary between studies and populations; however, larger studies and meta-analyses suggest a significant role for this gene in the mechanism of addiction. Stapleton et al. compiled data from several publications regarding *DAT1* alleles in smoking cessation therapy, and showed a small (odds ratio: 1.2), but statistically significant effect of the presence of the 9-repeat allele [197]. Another study considered the VNTR polymorphism and several *DAT1* SNPs in the context of alcohol withdrawal [198]. The strongest link to withdrawal was seen with the 9-repeat allele, but significant associations were also found tying symptom severity with three individual SNPs and with two *DAT1* haplotypes. Haplotype analysis for smoking risk in a group of young women showed a protective effect of certain combinations of SNPs with the 9-repeat allele VNTR [199]. These studies suggest that there is a great deal of potential for a role for *DAT1* variants in addiction, though assessment of multiple polymorphisms is likely to be more informative.

Serotonin Transporter Gene (5-HTT or SLC6A4)

Analogous to the role of the dopamine transporter, the serotonin transporter (encoded by *SLC6A4*) acts in regulation of synaptic serotonin levels, both physiological and in response to exogenous substances such as cocaine [2]. The *SLC6A4* polymorphism most commonly studied in addiction is the presence or absence of a 44-nucleotide sequence in the promoter region, resulting in a long (L) or short (S) allele, respectively. This site is referred to as the 5-HTTLPR, for serotonin (5-HT) transporter linked promoter repeat [200]. The S variant produces lower serotonin transporter levels and activity compared to the L allele, in a dominant fashion [200–202]. Although some reports link the S allele to substance abuse [203], several other studies have shown no correlation between an S genotype and addiction to various substances, including cocaine [204], methamphetamine [205], and heroin [159]. Interestingly, it appears that the 5-HTTLPR polymorphism may be more relevant to behaviors related to the addictive process rather than the actual risk of dependence. For example, Mannelli et al. showed no association between the S allele and prevalence of cocaine dependence in African-Americans, but did find a significant link between the S allele, severity of alcohol use, and lack of response to behavioral therapy [206]. Similarly, the S allele has been tied to smoking behaviors in adolescents [207] and to ever-smoking in adults regardless of dependence level [208]. In agreement with this, one study which did link the S/S genotype to heroin dependence also found a correlation to aggression in the heroin-dependent group [203].

A second *SLC6A4* polymorphism is an intron 2 VNTR, with three known alleles consisting of 9, 10, or 12 copies of a 16–17 nucleotide repeated sequence. In a case–control study, an association of the 10-repeat allele with heroin addiction was found [209]. A trend toward diminished amphetamine response was seen for combined analysis of the 5-HTTLPR and intron 2 VNTR polymorphisms [210], suggesting that variants of this gene may prove more useful in combination with other genetic information. Support for this idea comes from a study showing associations between harm avoidance and reward dependence behaviors with the combined 5-HTTLPR and the *DAT1* VNTR polymorphisms [211]. Likewise, Saiz et al. [212] reported a possible cooperative effect between serotonin transporter and serotonin receptor polymorphisms in heroin dependence. Given the interconnected nature of the monoamine neurotransmitter systems, it is not surprising that combined analysis of multiple genes in the pathway provides greater pharmacogenetic information than single genes in isolation.

Other Monoaminergic-Related Genes

Several other genes involved in neurotransmission by monoamines have been studied in the context of addiction. Most have only received limited attention to date, thus they will be discussed only briefly.

Serotonin Receptor Genes (HTR, Multiple Genes)

There are a number of serotonin receptor subtypes, several of which have been examined for a role in the genetics of addiction because serotonergic responses are implicated in a variety of neurological and psychological phenomena. However, the study of serotonin receptor polymorphisms in addiction lags behind analogous genes like *DRD2*, with occasionally contradictory results in different populations.

The *HTR2A* A-1438G variant was associated with heroin dependence, both alone and in combination with the serotonin transporter gene [212,213]. Certain *HTR1B* haplotypes showed a protective effect against heroin addiction [214], and the *HTR1B* minor allele A1180G was associated with a protective effect from heroin addiction in Europeans [203]. In contrast, results in Han Chinese showed a link between *HTR1B* G861C and A1180G polymorphisms with heroin dependence [215], and a recent meta-analysis of another *HTR1B* SNP concluded that -161A > T is associated with heroin dependence [216].

Norepinephrine Transporter Gene (SLC6A2)

As is the case with other monoamine transporters, the norepinephrine transporter is involved in neurotransmitter reuptake, particularly with norepinephrine and dopamine [217,218]. Several polymorphisms have been identified, including variants affecting the promoter (T-182C), nonsynonymous SNPs, as well as synonymous or intronic variants [219,220]. The 1369C allele encodes for a proline substitution at residue 457, which appears to result in differential response to cocaine, suggesting a possible role in addiction to that drug [221].

Dopamine β-Hydroxylase

The dopamine β-hydroxylase (DβH) enzyme converts dopamine into norepinephrine [2,143], thus altered expression of *DBH* could affect both the dopaminergic and noradrenergic systems. Although several studies have confirmed the ability of various polymorphisms to affect DβH production [222–226], less is known about the functional significance of this differential gene expression. One study did link a 19-nucleotide deletion-c.444A *DBH* haplotype with low DβH levels and cocaine-induced paranoia [224]. Low DβH does not always show correlation with addiction; e.g., the -1021T > C SNP in the 5′-UTR strongly affects DBH expression and was associated with increased propensity toward paranoia during cocaine self-administration [227], but did not show a link to cocaine dependence [192,228].

Catechol-O-Methyltransferase

This enzyme is involved in metabolism of several monoamines; the best-studied variant is a nonsynonymous change resulting in substitution of methionine for valine at residue 158. The 472G > A variant was associated with enhanced response to therapeutic administration of morphine [229] and is associated with opiate addiction in women but not men [230,231]. Other investigators have shown interaction [232] or independently significant effects of the Val158Met allele with *DRD4* polymorphisms in methamphetamine-dependent subjects [194]. Though additional confirmation is needed, these studies indicate that low catechol-O-methyltransferase activity may protect against methamphetamine abuse.

CONCLUSIONS

Pharmacogenetics is still a relatively new science, particularly in application to complex phenomena such as addiction. Studies of promising candidate genes have allowed initial forays into tailoring pharmaceutical therapies for individual patients, but current uses are limited to a small number of gene polymorphisms and an even smaller number of therapeutic agents. Most candidate genes, such as those encoding receptors, transporters, or metabolic enzymes, have been chosen for their involvement in pharmacokinetics or pharmacodynamics. This is certainly a logical starting point, given that individual responses to drugs can clearly affect the addictive process. However, the strong environmental component inherent to drug dependence points to additional biological pathways that contribute to the mechanisms of addiction and may therefore provide additional pharmacogenetic targets. Examples of such pathways include genes involved in transcriptional regulation, response to stress, and even maintenance of circadian rhythm.

Further maturation of genetic analysis is allowing a gradual shift from considering individual polymorphisms in isolation to more sophisticated analyses of multiple variants within a single gene or in combination with other genes. Substance dependence involves numerous factors, ranging from the physiological implications of pharmacokinetic and pharmacodynamic responses to drug exposure to the psychology of impulse control, risk-taking, novelty-seeking, and other behaviors that contribute to the addictive process. Thus it is to be expected that combined consideration of multiple polymorphisms affecting independent pathways in drug addiction would provide more information than analysis of a single variant at a time. As the field continues to advance, studies into the pharmacogenetics of substance addiction will undoubtedly improve our understanding of related areas, such as personalization of therapeutic regimens or treatment of nondrug addictions, and vice versa. However, it is vital to always remember that addiction is a chronic, multifactorial process that goes beyond genetics and treatment or prevention of substance dependence can only be achieved with an approach that considers an individual's genetic background with the physiological, pharmacological, and environmental factors that surround them.

REFERENCES

[1] Rutter JL. Symbiotic relationship of pharmacogenetics and drugs of abuse. Aaps J 2006;8(1):E174–84.

[2] Gutstein HB, Akil H. Opioid analgesics. In: Hardman J, Limbird L, Gilman A, editors. Goodman & Gilman's: The pharmacological basis of therapeutics. 10th ed New York: McGraw-Hill; 2001. p. 569–619.

[3] Evans WE, Relling MV. Pharmacogenomics: translating functional genomics into rational therapeutics. Science. 1999;286(5439):487–91.

[4] Wu D, Otton SV, Inaba T, Kalow W, Sellers EM. Interactions of amphetamine analogs with human liver CYP2D6. Biochem Pharmacol 1997;53(11):1605–12.

[5] Shimada T, Yamazaki H, Mimura M, Inui Y, Guengerich FP. Interindividual variations in human liver cytochrome P-450 enzymes involved in the oxidation of drugs, carcinogens and toxic chemicals: studies with liver microsomes of 30 Japanese and 30 Caucasians. J Pharmacol Exp Ther 1994;270(1):414–23.

[6] Zanger UM, Fischer J, Raimundo S, Stuven T, Evert BO, Schwab M, et al. Comprehensive analysis of the genetic factors determining expression and function of hepatic CYP2D6. Pharmacogenetics 2001;11(7):573–85.

[7] Smith DA, Abel SM, Hyland R, Jones BC. Human cytochrome P450s: selectivity and measurement in vivo. Xenobiotics 1998;28(12):1095–128.

[8] Eichelbaum M, Ingelman-Sundberg M, Evans WE. Pharmacogenomics and individualized drug therapy. Annu Rev Med 2006;57:119–37.

[9] Ingelman-Sundberg M. Genetic polymorphisms of cytochrome P450 2D6 (CYP2D6): clinical consequences, evolutionary aspects and functional diversity. Pharmacogenomics J 2005;5(1):6–13.

[10] Grobe N, Zhang B, Fisinger U, Kutchan TM, Zenk MH, Guengerich FP. Mammalian cytochrome P450 enzymes catalyze the phenol-coupling step in endogenous morphine biosynthesis. J Biol Chem 2009;284(36):24425–31.

[11] Kirchheiner J, Nickchen K, Bauer M, Wong ML, Licinio J, Roots I, et al. Pharmacogenetics of antidepressants and antipsychotics: the contribution of allelic variations to the phenotype of drug response. Mol Psychiatry 2004;9(5):442–73.

[12] Zanger UM, Raimundo S, Eichelbaum M. Cytochrome P450 2D6: overview and update on pharmacology, genetics, biochemistry. Naunyn Schmiedebergs Arch Pharmacol 2004;369(1):23–37.

[13] Bernard S, Neville KA, Nguyen AT, Flockhart DA. Interethnic differences in genetic polymorphisms of CYP2D6 in the U.S. population: clinical implications. Oncologist 2006;11(2):126–35.

[14] Eichelbaum M, Evert B. Influence of pharmacogenetics on drug disposition and response. Clin Exp Pharmacol Physiol 1996;23(10–11):983–5.

[15] Zahari Z, Ismail R. Influence of cytochrome P450, family 2, subfamily D, polypeptide 6 (CYP2D6) polymorphisms on pain sensitivity and clinical response to weak opioid analgesics. Drug Metab Pharmacokinet 2014;29(1):29–43.

[16] Solhaug V, Molden E. Individual variability in clinical effect and tolerability of opioid analgesics—importance of drug interactions and pharmacogenetics. Scand J Pain 2017;17:193–200.

[17] Crews KR, Gaedigk A, Dunnenberger HM, Leeder JS, Klein TE, Caudle KE, et al. Clinical pharmacogenetics implementation consortium guidelines for cytochrome P450 2D6 genotype and codeine therapy: 2014 update. Clin Pharmacol Ther 2014;95(4):376–82.

[18] Perez de los Cobos J, Sinol N, Trujols J, del Rio E, Banuls E, Luquero E, et al. Association of CYP2D6 ultrarapid metabolizer genotype with deficient patient satisfaction regarding methadone maintenance treatment. Drug Alcohol Depend 2007;89(2–3):190–4.

[19] Fonseca F, de la Torre R, Diaz L, Pastor A, Cuyas E, Pizarro N, et al. Contribution of cytochrome P450 and ABCB1 genetic variability on methadone pharmacokinetics, dose requirements, and response. PLoS One 2011;6(5):e19527.

[20] Crettol S, Deglon JJ, Besson J, Croquette-Krokar M, Hammig R, Gothuey I, et al. ABCB1 and cytochrome P450 genotypes and phenotypes: influence on methadone plasma levels and response to treatment. Clin Pharmacol Ther 2006;80(6):668–81.

[21] Shiran MR, Lennard MS, Iqbal MZ, Lagundoye O, Seivewright N, Tucker GT, et al. Contribution of the activities of CYP3A, CYP2D6, CYP1A2 and other potential covariates to the disposition of methadone in patients undergoing methadone maintenance treatment. Br J Clin Pharmacol 2009;67(1):29–37.

[22] Coller JK, Michalakas JR, James HM, Farquharson AL, Colvill J, White JM, et al. Inhibition of CYP2D6-mediated tramadol O-demethylation in methadone but not buprenorphine maintenance patients. Br J Clin Pharmacol 2012;74(5):835–41.

[23] Gelston EA, Coller JK, Lopatko OV, James HM, Schmidt H, White JM, et al. Methadone inhibits CYP2D6 and UGT2B7/2B4 in vivo: a study using codeine in methadone- and buprenorphine-maintained subjects. Br J Clin Pharmacol 2012;73(5):786–94.

[24] Crist RC, Li J, Doyle GA, Gilbert A, Dechairo BM, Berrettini WH. Pharmacogenetic analysis of opioid dependence treatment dose and dropout rate. Am J Drug Alcohol Abuse 2018;1–10 [e-pub ahead of print].

[25] Lin LY, Kumagai Y, Hiratsuka A, Narimatsu S, Suzuki T, Funae Y, et al. Cytochrome P4502D isozymes catalyze the 4-hydroxylation of methamphetamine enantiomers. Drug Metab Dispos 1995;23(6):610–14.

[26] Lin LY, Di Stefano EW, Schmitz DA, Hsu L, Ellis SW, Lennard MS, et al. Oxidation of methamphetamine and methylenedioxymethamphetamine by CYP2D6. Drug Metab Dispos 1997;25(9):1059–64.

[27] Otani K, Ujike H, Sakai A, Okahisa Y, Kotaka T, Inada T, et al. Reduced CYP2D6 activity is a negative risk factor for methamphetamine dependence. Neurosci Lett 2008;434(1):88–92.

[28] de la Torre R, Yubero-Lahoz S, Pardo-Lozano R, Farre M. MDMA, methamphetamine, and CYP2D6 pharmacogenetics: what is clinically relevant? Front Genet 2012;3:235.

[29] Schmid Y, Vizeli P, Hysek CM, Prestin K, Meyer Zu Schwabedissen HE, Liechti ME. CYP2D6 function moderates the pharmacokinetics and pharmacodynamics of 3,4-methylene-dioxymethamphetamine in a controlled study in healthy individuals. Pharmacogenet Genomics 2016;26(8):397–401.

[30] Vevelstad M, Oiestad EL, Bremer S, Bogen IL, Zackrisson AL, Arnestad M. Is toxicity of PMMA (paramethoxymethamphetamine) associated with cytochrome P450 pharmacogenetics? Forensic Sci Int 2016;261:137–47.

[31] Kharasch ED, Stubbert K. Role of cytochrome P4502B6 in methadone metabolism and clearance. J Clin Pharmacol 2013;53(3):305–13.
[32] DePriest AZ, Puet BL, Holt AC, Roberts A, Cone EJ. Metabolism and disposition of prescription opioids: a review. Forensic Sci Rev 2015;27(2):115–45.
[33] Kharasch ED, Regina KJ, Blood J, Friedel C. Methadone pharmacogenetics: CYP2B6 polymorphisms determine plasma concentrations, clearance, and metabolism. Anesthesiology 2015;123(5):1142–53.
[34] Kringen MK, Chalabianloo F, Bernard JP, Bramness JG, Molden E, Hoiseth G. Combined effect of CYP2B6 genotype and other candidate genes on a steady-state serum concentration of methadone in opioid maintenance treatment. Ther Drug Monit 2017;39(5):550–5.
[35] Dennis BB, Bawor M, Thabane L, Sohani Z, Samaan Z. Impact of ABCB1 and CYP2B6 genetic polymorphisms on methadone metabolism, dose and treatment response in patients with opioid addiction: a systematic review and meta-analysis. PLoS One 2014;9(1):e86114.
[36] Ahmad T, Sabet S, Primerano DA, Richards-Waugh LL, Rankin GO. Tell-tale SNPs: the role of CYP2B6 in methadone fatalities. J Anal Toxicol 2017;41(4):325–33.
[37] Mactier H, McLaughlin P, Gillis C, Osselton MD. Variations in infant CYP2B6 genotype associated with the need for pharmacological treatment for neonatal abstinence syndrome in infants of methadone-maintained opioid-dependent mothers. Am J Perinatol 2017;34(9):918–21.
[38] Satoh T. Role of carboxylesterase in xenobiotic metabolism. Rev Biochem Toxicol 1987;8:155–81.
[39] Redinbo MR, Bencharit S, Potter PM. Human carboxylesterase 1: from drug metabolism to drug discovery. Biochem Soc Trans 2003;31(Pt 3):620–4.
[40] Satoh T, Hosokawa M. The mammalian carboxylesterases: from molecules to functions. Ann Rev Pharmacol Toxicol. 1998;38:257–88.
[41] Bencharit S, Morton CL, Xue Y, Potter PM, Redinbo MR. Structural basis of heroin and cocaine metabolism by a promiscuous human drug-processing enzyme. Nat Struct Biol 2003;10(5):349–56.
[42] Pennings EJ, Leccese AP, Wolff FA. Effects of concurrent use of alcohol and cocaine. Addiction (Abingdon, England) 2002;97(7):773–83.
[43] Zhu HJ, Patrick KS, Yuan HJ, Wang JS, Donovan JL, DeVane CL, et al. Two CES1 gene mutations lead to dysfunctional carboxylesterase 1 activity in man: clinical significance and molecular basis. Am J Hum Genet 2008;82(6):1241–8.
[44] Weber WW. Human drug-metabolizing enzyme variants. In: Weber WW, editor. Pharmacogenetics. New York: Oxford University Press; 1997. p. 181–6.
[45] Kalow W, Genest K. A method for the detection of atypical forms of human serum cholinesterase; determination of dibucaine numbers. Can J Biochem Physiol 1957;35(6):339–46.
[46] Whittaker M. Cholinesterase. Monogr Hum Genet 1986;11.
[47] Lockridge O. Genetic variants of human serum cholinesterase influence metabolism of the muscle relaxant succinylcholine. Pharmacol. Therapeut 1990;47(1):35–60.
[48] McGuire MC, Nogueira CP, Bartels CF, Lightstone H, Hajra A, Van der Spek AF, et al. Identification of the structural mutation responsible for the dibucaine-resistant (atypical) variant form of human serum cholinesterase. Proc Natl Aacad Sci USA 1989;86(3):953–7.
[49] Kalow W, Gunn DR. Some statistical data on atypical cholinesterase of human serum. Ann Hum Genet 1959;23:239–50.
[50] Xie W, Altamirano CV, Bartels CF, Speirs RJ, Cashman JR, Lockridge O. An improved cocaine hydrolase: the A328Y mutant of human butyrylcholinesterase is 4-fold more efficient. Mol Pharmacol 1999;55(1):83–91.
[51] Harris H, Whittaker M. Differential inhibition of human serum cholinesterase with fluoride: recognition of two new phenotypes. Nature. 1961;191:496–8.
[52] Nogueira CP, Bartels CF, McGuire MC, Adkins S, Lubrano T, Rubinstein HM, et al. Identification of two different point mutations associated with the fluoride-resistant phenotype for human butyrylcholinesterase. Am J Hum Genet 1992;51(4):821–8.
[53] Bartels CF, James K, La Du BN. DNA mutations associated with the human butyrylcholinesterase J-variant. Am J Hum Genet 1992;50(5):1104–14.
[54] Liddell J, Lehmann H, Silk EA. Silent" pseudo-cholinesterase gene. Nature 1962;193:561–2.
[55] Nogueira CP, McGuire MC, Graeser C, Bartels CF, Arpagaus M, Van der Spek AF, et al. Identification of a frameshift mutation responsible for the silent phenotype of human serum cholinesterase, Gly 117 (GGT----GGAG). Am J Hum Genet 1990;46(5):934–42.
[56] Bartels CF, Jensen FS, Lockridge O, van der Spek AF, Rubinstein HM, Lubrano T, et al. DNA mutation associated with the human butyrylcholinesterase K-variant and its linkage to the atypical variant mutation and other polymorphic sites. Am J Hum Genet 1992;50(5):1086–103.
[57] Cone EJ. Pharmacokinetics and pharmacodynamics of cocaine. J Anal Toxicol J 1995;19(6):459–78.
[58] Isenschmid DS. Cocaine—effects on human performance and behavior. Forensic Sci Rev 2002;14:61–100.
[59] Karch SB, Stephens B, Ho CH. Relating cocaine blood concentrations to toxicity—an autopsy study of 99 cases. J Forensic Sci 1998;43(1):41–5.
[60] Duysen EG, Lockridge O. Prolonged toxic effects after cocaine challenge in butyrylcholinesterase/plasma carboxylesterase double knockout mice: a model for butyrylcholinesterase-deficient humans. Drug Metab Dispos 2011;39(8):1321–3.
[61] Blaho K, Logan B, Winbery S, Park L, Schwilke E. Blood cocaine and metabolite concentrations, clinical findings, and outcome of patients presenting to an ED. Am J Emerg Med 2000;18(5):593–8.
[62] Pellinen P, Kulmala L, Konttila J, Auriola S, Pasanen M, Juvonen R. Kinetic characteristics of norcocaine N-hydroxylation in mouse and human liver microsomes: involvement of CYP enzymes. Arch Toxicol 2000;74(9):511–20.
[63] Hoffman RS, Henry GC, Howland MA, Weisman RS, Weil L, Goldfrank LR. Association between life-threatening cocaine toxicity and plasma cholinesterase activity. Annal Emerg Med 1992;21(3):247–53.

[64] Li B, Duysen EG, Carlson M, Lockridge O. The butyrylcholinesterase knockout mouse as a model for human butyrylcholinesterase deficiency. J Pharmacol Exp Ther 2008;324(3):1146–54.

[65] Negrao AB, Pereira AC, Guindalini C, Santos HC, Messas GP, Laranjeira R, et al. Butyrylcholinesterase genetic variants: association with cocaine dependence and related phenotypes. PLoS One 2013;8(11):e80505.

[66] Lockridge O. Review of human butyrylcholinesterase structure, function, genetic variants, history of use in the clinic, and potential therapeutic uses. Pharmacol Ther 2015;148:34–46.

[67] Pert CB, Snyder SH. Opiate receptor: demonstration in nervous tissue. Science 1973;179(77):1011–14.

[68] Simon EJ, Hiller JM, Edelman I. Stereospecific binding of the potent narcotic analgesic (3H) Etorphine to rat-brain homogenate. Proc Natl Acad Sci USA 1973;70(7):1947–9.

[69] Terenius L. Stereospecific interaction between narcotic analgesics and a synaptic plasma membrane fraction of rat cerebral cortex. Acta Pharmacol Toxicol (Copenh) 1973;32(3):317–20.

[70] Kreek MJ, LaForge KS, Butelman E. Pharmacotherapy of addictions. Nat Rev Drug Discov 2002;1(9):710–26.

[71] Matthes HW, Maldonado R, Simonin F, Valverde O, Slowe S, Kitchen I, et al. Loss of morphine-induced analgesia, reward effect and withdrawal symptoms in mice lacking the mu-opioid-receptor gene. Nature 1996;383(6603):819–23.

[72] Sora I, Takahashi N, Funada M, Ujike H, Revay RS, Donovan DM, et al. Opiate receptor knockout mice define mu receptor roles in endogenous nociceptive responses and morphine-induced analgesia. Proc Natl Acad Sci USA 1997;94(4):1544–9.

[73] Kitanaka N, Sora I, Kinsey S, Zeng Z, Uhl GR. No heroin or morphine 6beta-glucuronide analgesia in mu-opioid receptor knockout mice. Eur J Pharmacol 1998;355(1):R1–3.

[74] Loh HH, Liu HC, Cavalli A, Yang W, Chen YF, Wei LN. mu Opioid receptor knockout in mice: effects on ligand-induced analgesia and morphine lethality. Brain Res Mol Brain Res 1998;54(2):321–6.

[75] Becker A, Grecksch G, Brodemann R, Kraus J, Peters B, Schroeder H, et al. Morphine self-administration in mu-opioid receptor-deficient mice. Naunyn Schmiedebergs Arch Pharmacol 2000;361(6):584–9.

[76] Unterwald EM, Horne-King J, Kreek MJ. Chronic cocaine alters brain mu opioid receptors. Brain Res 1992;584(1–2):314–18.

[77] Unterwald EM, Rubenfeld JM, Kreek MJ. Repeated cocaine administration upregulates kappa and mu, but not delta, opioid receptors. Neuroreport 1994;5(13):1613–16.

[78] Azaryan AV, Coughlin LJ, Buzas B, Clock BJ, Cox BM. Effect of chronic cocaine treatment on mu- and delta-opioid receptor mRNA levels in dopaminergically innervated brain regions. J Neurochem 1996;66(2):443–8.

[79] Yuferov V, Zhou Y, Spangler R, Maggos CE, Ho A, Kreek MJ. Acute "binge" cocaine increases mu-opioid receptor mRNA levels in areas of the rat mesolimbic mesocortical dopamine system. Brain Res Bull 1999;48(1):109–12.

[80] Bergen AW, Kokoszka J, Peterson R, Long JC, Virkkunen M, Linnoila M, et al. Mu opioid receptor gene variants: lack of association with alcohol dependence. Mol Psychiatry 1997;2(6):490–4.

[81] Bond C, LaForge KS, Tian M, Melia D, Zhang S, Borg L, et al. Single-nucleotide polymorphism in the human mu opioid receptor gene alters beta-endorphin binding and activity: possible implications for opiate addiction. Proc Natl Acad Sci USA 1998;95(16):9608–13.

[82] Reed B, Butelman ER, Yuferov V, Randesi M, Kreek MJ. Genetics of opiate addiction. Curr Psychiatry Rep 2014;16(11):504.

[83] Crist RC, Berrettini WH. Pharmacogenetics of OPRM1. Pharmacol Biochem Behav 2014;123:25–33.

[84] Schwantes-An TH, Zhang J, Chen LS, Hartz SM, Culverhouse RC, Chen X, et al. Association of the OPRM1 Variantrs1799971 (A118G) with non-specific liability to substance dependence in a collaborative de novo meta-analysis of European-ancestry cohorts. Behav Genet 2016;46(2):151–69.

[85] Haerian BS, Haerian MS. OPRM1 rs1799971 polymorphism and opioid dependence: evidence from a meta-analysis. Pharmacogenomics 2013;14(7):813–24.

[86] Kong X, Deng H, Gong S, Alston T, Kong Y, Wang J. Lack of associations of the opioid receptor mu 1 (OPRM1) A118G polymorphism (rs1799971) with alcohol dependence: review and meta-analysis of retrospective controlled studies. BMC Med Genet 2017;18(1):120.

[87] Bauer IE, Soares JC, Nielsen DA. The role of opioidergic genes in the treatment outcome of drug addiction pharmacotherapy: a systematic review. Am J Addict 2015;24(1):15–23.

[88] Kreek MJ, Bart G, Lilly C, LaForge KS, Nielsen DA. Pharmacogenetics and human molecular genetics of opiate and cocaine addictions and their treatments. Pharmacol Rev 2005;57(1):1–26.

[89] Maisonneuve IM, Kreek MJ. Acute tolerance to the dopamine response induced by a binge pattern of cocaine administration in male rats: an in vivo microdialysis study. J Pharmacol Exp Ther 1994;268(2):916–21.

[90] Maisonneuve IM, Ho A, Kreek MJ. Chronic administration of a cocaine "binge" alters basal extracellular levels in male rats: an in vivo microdialysis study. J Pharmacol Exp Ther 1995;272(2):652–7.

[91] Unterwald EM, Ho A, Rubenfeld JM, Kreek MJ. Time course of the development of behavioral sensitization and dopamine receptor upregulation during binge cocaine administration. J Pharmacol Exp Ther 1994;270(3):1387–96.

[92] Hollt V. Allelic variation of delta and kappa opioid receptors and its implication for receptor function. In: Harris LS, editor. 1999 Proceedings of the 61st annual scientific meeting of the college on problems of drug dependence national institute of drug abuse. Bethesda, MD: Research Monograph Series. US Department of Health and Human Services, National Institutes of Health. NIH Publication No (ADM)00-4737; 2000. p. 50.

[93] LaForge K, Kreek MJ, Uhl GR, Sora I, Yu L, Befort K, et al. Symposium XIII: allelic polymorphism of human opioid receptors: functional studies: genetic contributions to protection from, or vulnerability to, addictive diseases. In: Harris LS, editor. 1999 Proceedings of the 61st annual scientific meeting of the college on problems of drug dependence national institute of drug abuse. Bethesda, MD: Research Monograph Series. US Department of Health and Human Services, National Institutes of Health. NIH Publication No (ADM)00-4737; 1999, p. 47–50.

[94] Mayer P, Hollt V. Allelic and somatic variations in the endogenous opioid system of humans. Pharmacol Ther 2001;91(3):167–77.

[95] Yuferov V, Fussell D, LaForge KS, Nielsen DA, Gordon D, Ho A, et al. Redefinition of the human kappa opioid receptor gene (OPRK1) structure and association of haplotypes with opiate addiction. Pharmacogenetics 2004;14(12):793–804.

[96] Gerra G, Leonardi C, Cortese E, D'Amore A, Lucchini A, Strepparola G, et al. Human kappa opioid receptor gene (OPRK1) polymorphism is associated with opiate addiction. Am J Med Genet B Neuropsychiatr Genet 2007;144B(6):771–5.

[97] Albonaim A, Fazel H, Sharafshah A, Omarmeli V, Rezaei S, Ajamian F, et al. Association of OPRK1 gene polymorphisms with opioid dependence in addicted men undergoing methadone treatment in an Iranian population. J Addict Dis 2017;36(4):227–35.

[98] Wang SC, Tsou HH, Chung RH, Chang YS, Fang CP, Chen CH, et al. The association of genetic polymorphisms in the kappa-opioid receptor 1 gene with body weight, alcohol use, and withdrawal symptoms in patients with methadone maintenance. J Clin Psychopharmacol 2014;34(2):205–11.

[99] Jones JD, Luba RR, Vogelman JL, Comer SD. Searching for evidence of genetic mediation of opioid withdrawal by opioid receptor gene polymorphisms. Am J Addict 2016;25(1):41–8.

[100] Xu K, Seo D, Hodgkinson C, Hu Y, Goldman D, Sinha R. A variant on the kappa opioid receptor gene (OPRK1) is associated with stress response and related drug craving, limbic brain activation and cocaine relapse risk. Transl Psychiatry 2013;3:e292.

[101] Nielsen DA, Hamon SC, Kosten TR. The kappa-opioid receptor gene as a predictor of response in a cocaine vaccine clinical trial. Psychiatr Genet 2013;23(6):225–32.

[102] Zhu Y, King MA, Schuller AG, Nitsche JF, Reidl M, Elde RP, et al. Retention of supraspinal delta-like analgesia and loss of morphine tolerance in delta opioid receptor knockout mice. Neuron 1999;24(1):243–52.

[103] Nitsche JF, Schuller AG, King MA, Zengh M, Pasternak GW, Pintar JE. Genetic dissociation of opiate tolerance and physical dependence in delta-opioid receptor-1 and preproenkephalin knock-out mice. J Neurosci 2002;22(24):10906–13.

[104] Rozenfeld R, Abul-Husn NS, Gomez I, Devi LA. An emerging role for the delta opioid receptor in the regulation of mu opioid receptor function. Sci World J 2007;7:64–73.

[105] Zhang H, Kranzler HR, Yang BZ, Luo X, Gelernter J. The OPRD1 and OPRK1 loci in alcohol or drug dependence: OPRD1 variation modulates substance dependence risk. Mol Psychiatry 2008;13(5):531–43.

[106] Beer B, Erb R, Pavlic M, Ulmer H, Giacomuzzi S, Riemer Y, et al. Association of polymorphisms in pharmacogenetic candidate genes (OPRD1, GAL, ABCB1, OPRM1) with opioid dependence in European population: a case–control study. PLoS One 2013;8(9):e75359.

[107] Nagaya D, Zahari Z, Saleem M, Yahaya BH, Tan SC, Yusoff NM. An analysis of genetic association in opioid dependence susceptibility. J Clin Pharm Ther 2018;43(1):80–6.

[108] Gao X, Wang Y, Lang M, Yuan L, Reece AS, Wang W. Contribution of genetic polymorphisms and haplotypes in DRD2, BDNF, and opioid receptors to heroin dependence and endophenotypes among the Han Chinese. OMICS 2017;21(7):404–12.

[109] Randesi M, van den Brink W, Levran O, Blanken P, Butelman ER, Yuferov V, et al. Variants of opioid system genes are associated with non-dependent opioid use and heroin dependence. Drug Alcohol Depend 2016;168:164–9.

[110] Clarke TK, Crist RC, Ang A, Ambrose-Lanci LM, Lohoff FW, Saxon AJ, et al. Genetic variation in OPRD1 and the response to treatment for opioid dependence with buprenorphine in European-American females. Pharmacogenomics J 2014;14(3):303–8.

[111] Crist RC, Clarke TK, Ang A, Ambrose-Lanci LM, Lohoff FW, Saxon AJ, et al. An intronic variant in OPRD1 predicts treatment outcome for opioid dependence in African-Americans. Neuropsychopharmacology 2013;38(10):2003–10.

[112] Crist RC, Doyle GA, Kampman KM, Berrettini WH. A delta-opioid receptor genetic variant is associated with abstinence prior to and during cocaine dependence treatment. Drug Alcohol Depend 2016;166:268–71.

[113] Zhang Y, Butelman ER, Schlussman SD, Ho A, Kreek MJ. Effect of the endogenous kappa opioid agonist dynorphin A(1–17) on cocaine-evoked increases in striatal dopamine levels and cocaine-induced place preference in C57BL/6J mice. Psychopharmacology (Berl) 2004;172(4):422–9.

[114] Sivam SP. Cocaine selectively increases striatonigral dynorphin levels by a dopaminergic mechanism. J Pharmacol Exp Ther 1989;250(3):818–24.

[115] Hurd YL, Brown EE, Finlay JM, Fibiger HC, Gerfen CR. Cocaine self-administration differentially alters mRNA expression of striatal peptides. Brain Res Mol Brain Res 1992;13(1–2):165–70.

[116] Hurd YL, Herkenham M. Influence of a single injection of cocaine, amphetamine or GBR 12909 on mRNA expression of striatal neuropeptides. Brain Res Mol Brain Res 1992;16(1–2):97–104.

[117] Herkenham M. Cannabinoid receptor localization in brain: relationship to motor and reward systems. Ann NY Acad Sci 1992;654:19–32.

[118] Daunais JB, Roberts DC, McGinty JF. Cocaine self-administration increases preprodynorphin, but not c-fos, mRNA in rat striatum. Neuroreport 1993;4(5):543–6.

[119] Spangler R, Ho A, Zhou Y, Maggos CE, Yuferov V, Kreek MJ. Regulation of kappa opioid receptor mRNA in the rat brain by "binge" pattern cocaine administration and correlation with preprodynorphin mRNA. Brain Res Mol Brain Res 1996;38(1):71–6.

[120] Spangler R, Unterwald EM, Kreek MJ. Binge" cocaine administration induces a sustained increase of prodynorphin mRNA in rat caudate-putamen. Brain Res Mol Brain Res 1993;19(4):323–7.

[121] Clarke TK, Ambrose-Lanci L, Ferraro TN, Berrettini WH, Kampman KM, Dackis CA, et al. Genetic association analyses of PDYN polymorphisms with heroin and cocaine addiction. Genes Brain Behav 2012;11(4):415–23.

[122] Yuferov V, Ji F, Nielsen DA, Levran O, Ho A, Morgello S, et al. A functional haplotype implicated in vulnerability to develop cocaine dependence is associated with reduced PDYN expression in human brain. Neuropsychopharmacology 2009;34(5):1185–97.
[123] Chen AC, LaForge KS, Ho A, McHugh PF, Kellogg S, Bell K, et al. Potentially functional polymorphism in the promoter region of prodynorphin gene may be associated with protection against cocaine dependence or abuse. Am J Med Genet 2002;114(4):429–35.
[124] Williams TJ, LaForge KS, Gordon D, Bart G, Kellogg S, Ott J, et al. Prodynorphin gene promoter repeat associated with cocaine/alcohol codependence. Addict Biol 2007;12(3–4):496–502.
[125] Zimprich A, Kraus J, Woltje M, Mayer P, Rauch E, Hollt V. An allelic variation in the human prodynorphin gene promoter alters stimulus-induced expression. J Neurochem 2000;74(2):472–7.
[126] Ray R, Doyle GA, Crowley JJ, Buono RJ, Oslin DW, Patkar AA, et al. A functional prodynorphin promoter polymorphism and opioid dependence. Psychiatr Genet 2005;15(4):295–8.
[127] Saify K, Saadat M. Association between VNTR Polymorphism in promoter region of prodynorphin (PDYN) gene and methamphetamine dependence. Open Access Maced J Med Sci 2015;3(3):371–3.
[128] Saify K, Saadat I, Saadat M. Association between VNTR polymorphism in promoter region of prodynorphin (PDYN) gene and heroin dependence. Psychiatry Res 2014;219(3):690–2.
[129] Winham SJ, Preuss UW, Geske JR, Zill P, Heit JA, Bakalkin G, et al. Associations of prodynorphin sequence variation with alcohol dependence and related traits are phenotype-specific and sex-dependent. Sci Rep 2015;5:15670.
[130] Comings DE, Blake H, Dietz G, Gade-Andavolu R, Legro RS, Saucier G, et al. The proenkephalin gene (PENK) and opioid dependence. Neuroreport 1999;10(5):1133–5.
[131] Xuei X, Flury-Wetherill L, Bierut L, Dick D, Nurnberger Jr J, Foroud T, et al. The opioid system in alcohol and drug dependence: family-based association study. Am J Med Genet B Neuropsychiatr Genet 2007;144B(7):877–84.
[132] Jutras-Aswad D, Jacobs MM, Yiannoulos G, Roussos P, Bitsios P, Nomura Y, et al. Cannabis-dependence risk relates to synergism between neuroticism and proenkephalin SNPs associated with amygdala gene expression: case–control study. PLoS One 2012;7(6):e39243.
[133] Molinoff PB, Axelrod J. Biochemistry of catecholamines. Annu Rev Biochem 1971;40:465–500.
[134] Axelrod J. Biochemical pharmacology of catecholamines and its clinical implications. Trans Am Neurol Assoc 1971;96:179–86.
[135] Di Chiara G, Imperato A. Opposite effects of mu and kappa opiate agonists on dopamine release in the nucleus accumbens and in the dorsal caudate of freely moving rats. J Pharmacol Exp Ther 1988;244(3):1067–80.
[136] Herz A. Bidirectional effects of opioids in motivational processes and the involvement of D1 dopamine receptors. NIDA Res Monogr 1988;90:17–26.
[137] Tsuang MT, Lyons MJ, Meyer JM, Doyle T, Eisen SA, Goldberg J, et al. Co-occurrence of abuse of different drugs in men: the role of drug-specific and shared vulnerabilities. Arch Gen Psychiatry 1998;55(11):967–72.
[138] Kendler KS, Karkowski LM, Neale MC, Prescott CA. Illicit psychoactive substance use, heavy use, abuse, and dependence in a US population-based sample of male twins. Arch Gen Psychiatry 2000;57(3):261–9.
[139] Tsuang MT, Lyons MJ, Eisen SA, Goldberg J, True W, Lin N, et al. Genetic influences on DSM-III-R drug abuse and dependence: a study of 3,372 twin pairs. Am J Med Genet 1996;67(5):473–7.
[140] Nishiyama T, Ikeda M, Iwata N, Suzuki T, Kitajima T, Yamanouchi Y, et al. Haplotype association between GABAA receptor gamma2 subunit gene (GABRG2) and methamphetamine use disorder. Pharmacogenomics J 2005;5(2):89–95.
[141] Uhl GR, Hall FS, Sora I. Cocaine, reward, movement and monoamine transporters. Mol Psychiatry 2002;7(1):21–6.
[142] Rothman RB, Baumann MH. Monoamine transporters and psychostimulant drugs. Eur J Pharmacol 2003;479(1-3):23–40.
[143] Moyer TP, Shaw LM. Therapeutic drugs and their management. In: Burtis CA, Ashwood ER, Bruns DE, editors. Tietz textbook of clinical chemistry. 4th ed St. Louis, MO: Elsevier Saunders; 2006. p. 1237–85.
[144] Goodman A. Neurobiology of addiction. An integrative review. Biochem Pharmacol 2008;75(1):266–322.
[145] Noble EP. The D2 dopamine receptor gene: a review of association studies in alcoholism and phenotypes. Alcohol. 1998;16(1):33–45.
[146] Hallikainen T, Hietala J, Kauhanen J, Pohjalainen T, Syvalahti E, Salonen JT, et al. Ethanol consumption and DRD2 gene TaqI a polymorphism among socially drinking males. Am J Med Genet A 2003;119A(2):152–5.
[147] Erblich J, Lerman C, Self DW, Diaz GA, Bovbjerg DH. Stress-induced cigarette craving: effects of the DRD2 TaqI RFLP and SLC6A3 VNTR polymorphisms. Pharmacogenomics J 2004;4(2):102–9.
[148] Bierut LJ, Rice JP, Edenberg HJ, Goate A, Foroud T, Cloninger CR, et al. Family-based study of the association of the dopamine D2 receptor gene (DRD2) with habitual smoking. Am J Med Genet 2000;90(4):299–302.
[149] Noble EP, Blum K, Khalsa ME, Ritchie T, Montgomery A, Wood RC, et al. Allelic association of the D2 dopamine receptor gene with cocaine dependence. Drug Alcohol Depend 1993;33(3):271–85.
[150] Persico AM, Bird G, Gabbay FH, Uhl GR. D2 dopamine receptor gene TaqI A1 and B1 restriction fragment length polymorphisms: enhanced frequencies in psychostimulant-preferring polysubstance abusers. Biol Psychiatry 1996;40(8):776–84.
[151] Lawford BR, Young RM, Noble EP, Sargent J, Rowell J, Shadforth S, et al. The D(2) dopamine receptor A(1) allele and opioid dependence: association with heroin use and response to methadone treatment. Am J Med Genet 2000;96(5):592–8.
[152] David SP, Strong DR, Munafo MR, Brown RA, Lloyd-Richardson EE, Wileyto PE, et al. Bupropion efficacy for smoking cessation is influenced by the DRD2 TaqIA polymorphism: analysis of pooled data from two clinical trials. Nicotine Tob Res 2007;9(12):1251–7.
[153] Swan GE, Valdes AM, Ring HZ, Khroyan TV, Jack LM, Ton CC, et al. Dopamine receptor DRD2 genotype and smoking cessation outcome following treatment with bupropion SR. Pharmacogenomics J 2005;5(1):21–9.

[154] Smith L, Watson M, Gates S, Ball D, Foxcroft D. Meta-analysis of the association of the Taq1A polymorphism with the risk of alcohol dependency: a HuGE gene-disease association review. Am J Epidemiol 2008;167(2):125–38.
[155] Spitz MR, Shi H, Yang F, Hudmon KS, Jiang H, Chamberlain RM, et al. Case–control study of the D2 dopamine receptor gene and smoking status in lung cancer patients. J Natl Cancer Inst 1998;90(5):358–63.
[156] O'Hara BF, Smith SS, Bird G, Persico AM, Suarez BK, Cutting GR, et al. Dopamine D2 receptor RFLPs, haplotypes and their association with substance use in black and Caucasian research volunteers. Hum Hered 1993;43(4):209–18.
[157] Xu K, Lichtermann D, Lipsky RH, Franke P, Liu X, Hu Y, et al. Association of specific haplotypes of D2 dopamine receptor gene with vulnerability to heroin dependence in 2 distinct populations. Arch Gen Psychiatry 2004;61(6):597–606.
[158] Neville MJ, Johnstone EC, Walton RT. Identification and characterization of ANKK1: a novel kinase gene closely linked to DRD2 on chromosome band 11q23.1. Hum Mutat 2004;23(6):540–5.
[159] Li T, Liu X, Zhao J, Hu X, Ball DM, Loh el W, et al. Allelic association analysis of the dopamine D2, D3, 5-HT2A, and GABA(A)gamma2 receptors and serotonin transporter genes with heroin abuse in Chinese subjects. Am J Med Genet 2002;114(3):329–35.
[160] Ishiguro H, Arinami T, Saito T, Akazawa S, Enomoto M, Mitushio H, et al. Association study between the -141C Ins/Del and TaqI A polymorphisms of the dopamine D2 receptor gene and alcoholism. Alcohol Clin Exp Res 1998;22(4):845–8.
[161] Lerman C, Jepson C, Wileyto EP, Epstein LH, Rukstalis M, Patterson F, et al. Role of functional genetic variation in the dopamine D2 receptor (DRD2) in response to bupropion and nicotine replacement therapy for tobacco dependence: results of two randomized clinical trials. Neuropsychopharmacology 2006;31(1):231–42.
[162] Noble EP, Zhang X, Ritchie TL, Sparkes RS. Haplotypes at the DRD2 locus and severe alcoholism. Am J Med Genet 2000;96(5):622–31.
[163] Duan J, Wainwright MS, Comeron JM, Saitou N, Sanders AR, Gelernter J, et al. Synonymous mutations in the human dopamine receptor D2 (DRD2) affect mRNA stability and synthesis of the receptor. Hum Mol Genet 2003;12(3):205–16.
[164] Hill SY, Hoffman EK, Zezza N, Thalamuthu A, Weeks DE, Matthews AG, et al. Dopaminergic mutations: within-family association and linkage in multiplex alcohol dependence families. Am J Med Genet B Neuropsychiatr Genet 2008;147B(4):517–26.
[165] Chen CK, Hu X, Lin SK, Sham PC, Loh el W, Li T, et al. Association analysis of dopamine D2-like receptor genes and methamphetamine abuse. Psychiatr Genet 2004;14(4):223–6.
[166] Ekelund J, Lichtermann D, Jarvelin MR, Peltonen L. Association between novelty seeking and the type 4 dopamine receptor gene in a large Finnish cohort sample. Am J Psychiatry 1999;156(9):1453–5.
[167] Laucht M, Becker K, Blomeyer D, Schmidt MH. Novelty seeking involved in mediating the association between the dopamine D4 receptor gene exon III polymorphism and heavy drinking in male adolescents: results from a high-risk community sample. Biol Psychiatry 2007;61(1):87–92.
[168] Skowronek MH, Laucht M, Hohm E, Becker K, Schmidt MH. Interaction between the dopamine D4 receptor and the serotonin transporter promoter polymorphisms in alcohol and tobacco use among 15-year-olds. Neurogenetics 2006;7(4):239–46.
[169] Luciano M, Zhu G, Kirk KM, Whitfield JB, Butler R, Heath AC, et al. Effects of dopamine receptor D4 variation on alcohol and tobacco use and on novelty seeking: multivariate linkage and association analysis. Am J Med Genet B Neuropsychiatr Genet 2004;124B(1):113–23.
[170] Comings DE, Gade R, Wu S, Chiu C, Dietz G, Muhleman D, et al. Studies of the potential role of the dopamine D1 receptor gene in addictive behaviors. Mol Psychiatry 1997;2(1):44–56.
[171] Levran O, Londono D, O'Hara K, Randesi M, Rotrosen J, Casadonte P, et al. Heroin addiction in African Americans: a hypothesis-driven association study. Genes Brain Behav 2009;8(5):531–40.
[172] Kim DJ, Park BL, Yoon S, Lee HK, Joe KH, Cheon YH, et al. 5′ UTR polymorphism of dopamine receptor D1 (DRD1) associated with severity and temperament of alcoholism. Biochem Biophys Res Commun 2007;357(4):1135–41.
[173] Limosin F, Loze JY, Rouillon F, Ades J, Gorwood P. Association between dopamine receptor D1 gene DdeI polymorphism and sensation seeking in alcohol-dependent men. Alcohol Clin Exp Res 2003;27(8):1226–8.
[174] Vereczkei A, Demetrovics Z, Szekely A, Sarkozy P, Antal P, Szilagyi A, et al. Multivariate analysis of dopaminergic gene variants as risk factors of heroin dependence. PLoS One 2013;8(6):e66592.
[175] Batel P, Houchi H, Daoust M, Ramoz N, Naassila M, Gorwood P. A haplotype of the DRD1 gene is associated with alcohol dependence. Alcohol Clin Exp Res 2008;32(4):567–72.
[176] Comings DE, Gonzalez N, Wu S, Saucier G, Johnson P, Verde R, et al. Homozygosity at the dopamine DRD3 receptor gene in cocaine dependence. Mol Psychiatry 1999;4(5):484–7.
[177] Wiesbeck GA, Weijers HG, Wodarz N, Herrmann MJ, Johann M, Keller HK, et al. Dopamine D2 (DAD2) and dopamine D3 (DAD3) receptor gene polymorphisms and treatment outcome in alcohol dependence. J Neural Transm 2003;110(7):813–20.
[178] Messas G, Meira-Lima I, Turchi M, Franco O, Guindalini C, Castelo A, et al. Association study of dopamine D2 and D3 receptor gene polymorphisms with cocaine dependence. Psychiatr Genet 2005;15(3):171–4.
[179] Vandenbergh DJ, Persico AM, Hawkins AL, Griffin CA, Li X, Jabs EW, et al. Human dopamine transporter gene (DAT1) maps to chromosome 5p15.3 and displays a VNTR. Genomics 1992;14(4):1104–6.
[180] Gelernter J, Kranzler HR, Satel SL, Rao PA. Genetic association between dopamine transporter protein alleles and cocaine-induced paranoia. Neuropsychopharmacology 1994;11(3):195–200.
[181] Fuke S, Suo S, Takahashi N, Koike H, Sasagawa N, Ishiura S. The VNTR polymorphism of the human dopamine transporter (DAT1) gene affects gene expression. Pharmacogenomics J 2001;1(2):152–6.

[182] Heinz A, Goldman D, Jones DW, Palmour R, Hommer D, Gorey JG, et al. Genotype influences in vivo dopamine transporter availability in human striatum. Neuropsychopharmacology 2000;22(2):133–9.

[183] Jacobsen LK, Staley JK, Zoghbi SS, Seibyl JP, Kosten TR, Innis RB, et al. Prediction of dopamine transporter binding availability by genotype: a preliminary report. Am J Psychiatry 2000;157(10):1700–3.

[184] Guindalini C, Howard M, Haddley K, Laranjeira R, Collier D, Ammar N, et al. A dopamine transporter gene functional variant associated with cocaine abuse in a Brazilian sample. Proc Natl Acad Sci USA 2006;103(12):4552–7.

[185] Lohoff FW, Bloch PJ, Hodge R, Nall AH, Ferraro TN, Kampman KM, et al. Association analysis between polymorphisms in the dopamine D2 receptor (DRD2) and dopamine transporter (DAT1) genes with cocaine dependence. Neurosci Lett 2010;473(2):87–91.

[186] Ujike H, Harano M, Inada T, Yamada M, Komiyama T, Sekine Y, et al. Nine- or fewer repeat alleles in VNTR polymorphism of the dopamine transporter gene is a strong risk factor for prolonged methamphetamine psychosis. Pharmacogenomics J 2003;3(4):242–7.

[187] Kirley A, Lowe N, Hawi Z, Mullins C, Daly G, Waldman I, et al. Association of the 480 bp DAT1 allele with methylphenidate response in a sample of Irish children with ADHD. Am J Med Genet B Neuropsychiatr Genet 2003;121B(1):50–4.

[188] Loo SK, Specter E, Smolen A, Hopfer C, Teale PD, Reite ML. Functional effects of the DAT1 polymorphism on EEG measures in ADHD. J Am Acad Child Adolesc Psychiatry 2003;42(8):986–93.

[189] Winsberg BG, Comings DE. Association of the dopamine transporter gene (DAT1) with poor methylphenidate response. J Am Acad Child Adolesc Psychiatry 1999;38(12):1474–7.

[190] Stein MA, Waldman ID, Sarampote CS, Seymour KE, Robb AS, Conlon C, et al. Dopamine transporter genotype and methylphenidate dose response in children with ADHD. Neuropsychopharmacology 2005;30(7):1374–82.

[191] Lott DC, Kim SJ, Cook Jr EH, de Wit H. Dopamine transporter gene associated with diminished subjective response to amphetamine. Neuropsychopharmacology 2005;30(3):602–9.

[192] Fernandez-Castillo N, Ribases M, Roncero C, Casas M, Gonzalvo B, Cormand B. Association study between the DAT1, DBH and DRD2 genes and cocaine dependence in a Spanish sample. Psychiatr Genet 2010;20(6):317–20.

[193] Brewer III. AJ, Nielsen DA, Spellicy CJ, Hamon SC, Gingrich J, Thompson-Lake DG, et al. Genetic variation of the dopamine transporter (DAT1) influences the acute subjective responses to cocaine in volunteers with cocaine use disorders. Pharmacogenet Genomics 2015;25(6):296–304.

[194] Vandenbergh DJ, Rodriguez LA, Hivert E, Schiller JH, Villareal G, Pugh EW, et al. Long forms of the dopamine receptor (DRD4) gene VNTR are more prevalent in substance abusers: no interaction with functional alleles of the catechol-O-methyltransferase (COMT) gene. Am J Med Genet 2000;96(5):678–83.

[195] Lin Z, Uhl GR. Human dopamine transporter gene variation: effects of protein coding variants V55A and V382A on expression and uptake activities. Pharmacogenomics J. 2003;3(3):159–68.

[196] Uhl GR, Lin Z. The top 20 dopamine transporter mutants: structure–function relationships and cocaine actions. Eur J Pharmacol 2003;479(1–3):71–82.

[197] Stapleton JA, Sutherland G, O'Gara C. Association between dopamine transporter genotypes and smoking cessation: a meta-analysis. Addict Biol 2007;12(2):221–6.

[198] Le Strat Y, Ramoz N, Pickering P, Burger V, Boni C, Aubin HJ, et al. The 3' part of the dopamine transporter gene DAT1/SLC6A3 is associated with withdrawal seizures in patients with alcohol dependence. Alcohol Clin Exp Res 2008;32(1):27–35.

[199] Segman RH, Kanyas K, Karni O, Lerer E, Goltser-Dubner T, Pavlov V, et al. Why do young women smoke? IV. Role of genetic variation in the dopamine transporter and lifetime traumatic experience. Am J Med Genet B Neuropsychiatr Genet. 2007;144B(4):533–40.

[200] Lesch KP, Bengel D, Heils A, Sabol SZ, Greenberg BD, Petri S, et al. Association of anxiety-related traits with a polymorphism in the serotonin transporter gene regulatory region. Science. 1996;274(5292):1527–31.

[201] Glatz K, Mossner R, Heils A, Lesch KP. Glucocorticoid-regulated human serotonin transporter (5-HTT) expression is modulated by the 5-HTT gene-promotor-linked polymorphic region. J Neurochem 2003;86(5):1072–8.

[202] Little KY, McLaughlin DP, Zhang L, Livermore CS, Dalack GW, McFinton PR, et al. Cocaine, ethanol, and genotype effects on human midbrain serotonin transporter binding sites and mRNA levels. Am J Psychiatry 1998;155(2):207–13.

[203] Gerra G, Garofano L, Santoro G, Bosari S, Pellegrini C, Zaimovic A, et al. Association between low-activity serotonin transporter genotype and heroin dependence: behavioral and personality correlates. Am J Med Genet B Neuropsychiatr Genet 2004;126B(1):37–42.

[204] Patkar AA, Berrettini WH, Hoehe M, Hill KP, Sterling RC, Gottheil E, et al. Serotonin transporter (5-HTT) gene polymorphisms and susceptibility to cocaine dependence among African-American individuals. Addict Biol 2001;6(4):337–45.

[205] Hong CJ, Cheng CY, Shu LR, Yang CY, Tsai SJ. Association study of the dopamine and serotonin transporter genetic polymorphisms and methamphetamine abuse in Chinese males. J Neural Transm 2003;110(4):345–51.

[206] Mannelli P, Patkar AA, Murray HW, Certa K, Peindl K, Mattila-Evenden M, et al. Polymorphism in the serotonin transporter gene and response to treatment in African American cocaine and alcohol-abusing individuals. Addict Biol 2005;10(3):261–8.

[207] Gerra G, Garofano L, Zaimovic A, Moi G, Branchi B, Bussandri M, et al. Association of the serotonin transporter promoter polymorphism with smoking behavior among adolescents. Am J Med Genet B Neuropsychiatr Genet 2005;135B(1):73–8.

[208] Kremer I, Bachner-Melman R, Reshef A, Broude L, Nemanov L, Gritsenko I, et al. Association of the serotonin transporter gene with smoking behavior. Am J Psychiatry 2005;162(5):924–30.

[209] Tan EC, Yeo BK, Ho BK, Tay AH, Tan CH. Evidence for an association between heroin dependence and a VNTR polymorphism at the serotonin transporter locus. Mol Psychiatry 1999;4(3):215–17.

[210] Lott DC, Kim SJ, Cook EH, de Wit H. Serotonin transporter genotype and acute subjective response to amphetamine. Am J Addict 2006;15(5):327−35.

[211] Kim SJ, Kim YS, Lee HS, Kim SY, Kim CH. An interaction between the serotonin transporter promoter region and dopamine transporter polymorphisms contributes to harm avoidance and reward dependence traits in normal healthy subjects. J Neural Transm 2006;113(7):877−86.

[212] Saiz PA, Garcia-Portilla MP, Arango C, Morales B, Martinez-Barrondo S, Alvarez C, et al. Association between heroin dependence and 5-HT2A receptor gene polymorphisms. Eur Addict Res 2008;14(1):47−52.

[213] Saiz PA, Garcia-Portilla MP, Arango C, Morales B, Bascaran MT, Martinez-Barrondo S, et al. Association study between obsessive-compulsive disorder and serotonergic candidate genes. Prog Neuropsychopharmacol Biol Psychiatry 2008;32(3):765−70.

[214] Proudnikov D, LaForge KS, Hofflich H, Levenstien M, Gordon D, Barral S, et al. Association analysis of polymorphisms in serotonin 1B receptor (HTR1B) gene with heroin addiction: a comparison of molecular and statistically estimated haplotypes. Pharmacogenet Genomics 2006;16(1):25−36.

[215] Gao F, Zhu YS, Wei SG, Li SB, Lai JH. Polymorphism G861C of 5-HT receptor subtype 1B is associated with heroin dependence in Han Chinese. Biochem Biophys Res Commun 2011;412(3):450−3.

[216] Cao J, LaRocque E, Li D. Associations of the 5-hydroxytryptamine (serotonin) receptor 1B gene (HTR1B) with alcohol, cocaine, and heroin abuse. Am J Med Genet B Neuropsychiatr Genet 2013;162B(2):169−76.

[217] Horn AS. Structure−activity relations for the inhibition of catecholamine uptake into synaptosomes from noradrenaline and dopaminergic neurones in rat brain homogenates. Br J Pharmacol. 1973;47(2):332−8.

[218] Raiteri M, Del Carmine R, Bertollini A, Levi G. Effect of sympathomimetic amines on the synaptosomal transport of noradrenaline, dopamine and 5-hydroxytryptamine. Eur J Pharmacol 1977;41(2):133−43.

[219] Stober G, Nothen MM, Porzgen P, Bruss M, Bonisch H, Knapp M, et al. Systematic search for variation in the human norepinephrine transporter gene: identification of five naturally occurring missense mutations and study of association with major psychiatric disorders. Am J Med Genet 1996;67(6):523−32.

[220] Stober G, Hebebrand J, Cichon S, Bruss M, Bonisch H, Lehmkuhl G, et al. Tourette syndrome and the norepinephrine transporter gene: results of a systematic mutation screening. Am J Med Genet 1999;88(2):158−63.

[221] Paczkowski FA, Bonisch H, Bryan-Lluka LJ. Pharmacological properties of the naturally occurring Ala(457)Pro variant of the human norepinephrine transporter. Pharmacogenetics 2002;12(2):165−73.

[222] Wei J, Ramchand CN, Hemmings GP. Possible control of dopamine beta-hydroxylase via a codominant mechanism associated with the polymorphic (GT)n repeat at its gene locus in healthy individuals. Hum Genet 1997;99(1):52−5.

[223] Wei J, Xu HM, Ramchand CN, Hemmings GP. Is the polymorphic microsatellite repeat of the dopamine beta-hydroxylase gene associated with biochemical variability of the catecholamine pathway in schizophrenia? Biol Psychiatry 1997;41(7):762−7.

[224] Cubells JF, Kranzler HR, McCance-Katz E, Anderson GM, Malison RT, Price LH, et al. A haplotype at the DBH locus, associated with low plasma dopamine beta-hydroxylase activity, also associates with cocaine-induced paranoia. Mol Psychiatry 2000;5(1):56−63.

[225] Cubells JF, van Kammen DP, Kelley ME, Anderson GM, O'Connor DT, Price LH, et al. Dopamine beta-hydroxylase: two polymorphisms in linkage disequilibrium at the structural gene DBH associate with biochemical phenotypic variation. Hum Genet 1998;102(5):533−40.

[226] Zabetian CP, Anderson GM, Buxbaum SG, Elston RC, Ichinose H, Nagatsu T, et al. A quantitative-trait analysis of human plasma-dopamine beta-hydroxylase activity: evidence for a major functional polymorphism at the DBH locus. Am J Hum Genet 2001;68(2):515−22.

[227] Kalayasiri R, Sughondhabirom A, Gueorguieva R, Coric V, Lynch WJ, Lappalainen J, et al. Dopamine beta-hydroxylase gene (DbetaH)—1021C-->T influences self-reported paranoia during cocaine self-administration. Biol Psychiatry 2007;61(11):1310−13.

[228] Guindalini C, Laranjeira R, Collier D, Messas G, Vallada H, Breen G. Dopamine-beta hydroxylase polymorphism and cocaine addiction. Behav Brain Funct 2008;4:1.

[229] Oertel B, Lotsch J. Genetic mutations that prevent pain: implications for future pain medication. Pharmacogenomics 2008;9(2):179−94.

[230] Horowitz R, Kotler M, Shufman E, Aharoni S, Kremer I, Cohen H, et al. Confirmation of an excess of the high enzyme activity COMT val allele in heroin addicts in a family-based haplotype relative risk study. Am J Med Genet 2000;96(5):599−603.

[231] Oosterhuis BE, LaForge KS, Proudnikov D, Ho A, Nielsen DA, Gianotti R, et al. Catechol-*O*-methyltransferase (COMT) gene variants: possible association of the Val158Met variant with opiate addiction in Hispanic women. Am J Med Genet B Neuropsychiatr Genet 2008;147B(6):793−8.

[232] Li T, Chen CK, Hu X, Ball D, Lin SK, Chen W, et al. Association analysis of the DRD4 and COMT genes in methamphetamine abuse. Am J Med Genet B Neuropsychiatr Genet 2004;129B(1):120−4.

Chapter 9

Immunoassay Design for Screening of Drugs of Abuse

Pradip Datta
Siemens Healthineers, Newark, DE, United States

INTRODUCTION

Immunoassays are the most common screening method for routine drugs of abuse (DAU) analysis, because of the assays' ease of operation, simplicity, lower cost, and speed. The assays are run on automated systems, which are fast, and often very precise and accurate, providing qualitative or semiquantitative results. The automated immunoassay systems, connected to the laboratory information systems (LIS), are very suitable for total laboratory automation. In addition to the centralized automated systems, DAU assays are also available on various types of point-of-care (POC) or near-patient-testing (NPT) cartridges and systems and such assays are often based on principles of immunoassay. DAU immunoassays are available for the commonly abused drugs allowing more than one cutoff. It is important to note that since many of the assays recognize multiple drugs or their metabolites, the assay kit names may be misleading. Table 9.1 lists some of the commonly available commercial DAU immunoassays and their cutoff analyte concentrations as well as drug/metabolite target of antibody used in assay design.

Commercial immunoassay kits are also available for many other DAU candidates, e.g., caffeine, cotinine (metabolite of nicotine in smokers' urine), lysergic acid diethylamide (LSD), bath salts, etc. Usually, urine is the most commonly used specimen for DAU immunoassays. However, serum, plasma, or other body fluids, including saliva and meconium, have been successfully used as specimens in DAU immunoassays. However, immunoassays can only be used for screening purposes because of the issue of interferences and cross-reactivity. The initial screening results must be confirmed by a more specific analytical technique such as liquid chromatography-tandem mass spectrometry (LC/MS/MS) or gas chromatography-mass spectrometry (GC/MS) for the confirmation of the presence of drug in the specimen. Confirmation of the drug using GC/MS or a comparable analytical technique is mandatory in legal cases.

IMMUNOASSAYS IN DRUGS OF ABUSE TESTING

The main component of the immunoassay reagents is the analyte-specific antibody which is mostly isolated from sera or ascetic fluid of animals like rabbit, mouse, goat, sheep, or chicken after immunization of the animal. Antibodies are generated by conjugating the target analytes (since they are small) to a larger protein, like albumin, and injected into the target animals. The sera from the animals are then repeatedly tested for immunoactivity against the analyte. Once the immunoactivity is deemed suitably specific and sensitive, the antibody now may be used in the immunoassay either as purified immunoglobulins, or even as semipurified plasma protein fractions. Such antibodies are called polyclonal, since they may contain many clones of IgG recognizing different epitopes of the antigen. Conversely, lymphocytes secreting a specified and desired clone can be fused with immortal "cancer" cells to create cell lines which may be grown in vivo or in vitro to produce a very specific single clone of the target antibody which is termed as monoclonal antibody.

The antibody now can be incorporated in myriads of assay designs with various avenues of measurements or tags to generate the most commonly used assay formats used in DAU assays. Most of these immunoassay methods use specimens without any pretreatment. The assays use very small amounts of sample volumes (most <100 μL), reagents are stored in the analyzer, and most have stored calibration curves on the system. Immunoassays offer fast throughput,

TABLE 9.1 Screening Cutoff Concentrations of Commonly Available Immunoassays for DAU Testing in Urine

Drug Tested	Common Cutoff Concentration	Target of Antibody
Amphetamine/methamphetamine	500 ng/mL	Amphetamine or methamphetamine
3,4-Methylenedioxy-methamphetamine (MDMA)		MDMA
Barbiturates	200 ng/mL	Secobarbital
Benzodiazepines	200 ng/mL	Nor-diazepam or nitrazepam or lormetazepam
Cocaine as benzoylecgonine	150 or 300 ng/mL	Benzoylecgonine
Opiates	300 or 2000 ng/mL	Morphine
6-Acetylmorphine (marker metabolite for heroin abuse)	10 ng/mL	6-Acetylmorphine
Oxycodone	100 ng/mL	Oxycodone
Hydrocodone/hydromorphone	300 ng/mL	Hydrocodone
Methadone	300 ng/mL	Methadone or EDDP[a]
Propoxyphene	300 ng/mL	Propoxyphene
Buprenorphine	5 ng/mL	
Phencyclidine	25 ng/mL	Phencyclidine
Marijuana	20 or 50 ng/mL	THC-COOH (11 nor-Δ^9-tetrahydrocannabinol-9-carboxylic acid)
Methaqualone	300 ng/mL	Methaqualone
Lysergic acid diethylamide (LSD)	0.5 ng/mL	LSD
Ketamine	10 ng/mL	Ketamine

[a]EDDP: 2-ethylidine-1,5-dimethyl 3-3-diphenylpyrrolidine, a metabolite of methadone.

specificity, and result reports in a format that the laboratory wants. Immunoassay has the benefits of high sensitivity and relatively good specificity for most DAU except for amphetamines. Commonly used immunoassay formats used for DAU testing are listed in Table 9.2.

FORMATS OF IMMUNOASSAY DESIGN

With respect to assay design, there are two formats of immunoassays: competition and immunometric (commonly referred as "sandwich"). Competition immunoassays work best for small molecules requiring a single analyte-specific antibody. Sandwich immunoassays, on the other hand, are mostly used for large molecules such as proteins or peptides, and use two different specific antibodies. Since most DAU immunoassays involve analytes of small molecular size, they employ the competition format. In this format, the analyte molecules in the specimen compete with analyte (or, its analogues), labeled with a suitable tag and provided in the reagent, for limited number of binding sites provided by, e.g., an analyte-specific antibody (also provided in the reagent). Thus, in these types of assays, higher the analyte concentration in the sample, less of label can bind to the antibody to form the conjugate. If the bound label provides the signal, e.g., fluorescence polarization immunoassay (FPIA), the analyte concentration in the specimen is inversely proportional to the signal produced. However, this assay design has been discontinued. On the other hand, if the signal is generated by the free label then the signal is proportional to the concentration of the abused drug in the specimen. The signals are mostly optical in nature such as absorbance, fluorescence, or chemiluminescence.

TABLE 9.2 Various Types of Commercial DAU Immunoassay Kits Using Homogeneous Assay Format

Immunoassay Types	Example	Assay Signal
Competition	FPIA (TDx from Abbott)	Fluorescence polarization, but this design has been discontinued.
	EMIT (Siemens) on: • ADVIA Chemistry, Dimension, or Viva (Siemens), • Multigent (Abbott), or • SYNCHRON (Beckman-Coulter)	Colorimetry (enzyme modulation) All these assays measure signal at 340 nm due to conversion of NAD to NADH (but SYNCHRON uses particle-enhanced turbidimetric inhibition immunoassay method).
	CEDIA (Microgenics)	Colorimetry (enzyme modulation); at system-dependent visible wavelengths, commonly using beta-galactosidase enzyme (fragments).
	Cobas (Roche), or DRI (Microgenics) KIMS (Online) DAU testing immunoassays	Turbidimetry, latex microparticle assisted; measured at system and assay-dependent wavelengths (e.g., KIMS Cannabinoids at 659 nm).
	EIA kits from various manufacturers (e.g., OraSure)	EIA, batch assay on microtiter plate

There are several variations in this basic format. While the assays can be homogeneous or heterogeneous, most DAU assays are of the former type because of the simplicity of homogeneous format where the bound label has different properties than the free label. For example, in fluorescent polarization immunoassay (FPIA), the free label has different Brownian motion than when the relatively small molecular weight (a few hundreds to thousand Daltons) label is complexed to a large antibody (146,000 Da). This results in difference in the fluorescence polarization properties of the label. This difference is utilized to quantify the bound label [1]. In another type of homogeneous immunoassay, an enzyme is the label, whose activity is modulated differently in the free versus the complexed conditions of the label. This forms the basis of the EMIT (enzyme multiplied immunoassay technique) or CEDIA (cloned enzyme donor immunoassay) technologies [2,3]. In the EMIT method, the label enzyme, glucose 6-phosphodehydrogenase (G6PDH), is active unless in the antigen−antibody complex. The active enzyme reduces NAD (nicotinamide adenine dinucleotide) to NADH, and the absorbance is monitored at 340 nm (NAD has no signal at 340 nm while NADH absorbs at 340 nm). To guard against the interference from a specimen's native G6PDH, the newer improved assays use recombinant bacterial enzymes whose activity conditions are different from the human enzyme. Similarly, in the CEDIA method, two genetically engineered inactive fragments of the enzyme beta-galactosidase are coupled to the antigen and antibody reagents. When they combine, the active enzyme is produced and the substrate, a chromogenic galactoside derivative, produces the assay signal. In a third commonly used format of homogeneous immunoassay (turbidimetric immunoassay or TIA), analytes (antigen) or its analogs are coupled to colloidal particles, e.g., of latex [4]. Since antibodies are bivalent, the latex particles agglutinate in the presence of antibody. However, in the presence of free analytes in the specimen, there is less agglutination. In a spectrophotometer, the resulting turbidity can be monitored as end-point or as rate. In the kinetic interaction of microparticle in solution (KIMS) assay method, in the absence of drug molecules, free antibodies bind to drug microparticle conjugate forming particle aggregates and increase in absorption is observed. When drug molecules are present in urine specimen, these drug molecules bind with free antibody molecules and thus prevent formation of particle aggregates and diminish absorbance in proportion to drug concentration. The online DAU testing immunoassays marketed by Roche Diagnostics (Indianapolis, IN) are based on KIMS format.

In the heterogeneous immunoassay format, the bound label is physically separated from the unbound labels before the signal is measured. The separation is often done magnetically, where the reagent analyte (or its analog) is provided as coupled to paramagnetic particles (PMPs), and the antibody is labeled. Conversely, the antibody may also be provided as conjugated to the PMP, and the reagent analyte may carry the label. After separation and wash, the bound label is reacted with other reagents to generate the signal. This is the mechanism in many chemiluminescent immunoassays (CLIA), where the label may be a small molecule which generates chemiluminescent signal [5]. The label may also be an enzyme [enzyme immunoassay (EIA) or enzyme-linked immunosorbent assay (ELISA)] which generates chemiluminescent, fluorimetric, or colorimetric signal. In older immunoassay formats, the labels used to be radioactive: radioimmunoassay or RIA. But because of safety and waste disposal issues, RIA is rarely used today. Another type of heterogeneous immunoassay uses polystyrene particles. If these particles are microsizes, that type of assay is called microparticle-enhanced immunoassay (MEIA) [6].

Polyclonal Versus Monoclonal Antibody

The main reagent in the immunoassay is the binding molecule which is most commonly an analyte-specific antibody or its fragment. Several types of antibodies or their fragments are now used in immunoassays. There are polyclonal antibodies, which are raised in an animal when the analyte (as antigen) along with an adjuvant is injected to the animal. For a small molecular weight analyte, it is most commonly injected as a conjugate of a large protein. Appearance of analyte-specific antibodies in the animal's sera is monitored, and when sufficient concentration of the antibody is reached, the animal is bled. Although serum can be directly used as the analyte-specific binder in the immunoassay, commonly, the antibodies are purified from serum and then used in the assay. Since there are many clones of the antibodies specific for the analyte, these antibodies are called polyclonal. In newer technologies, however, a mast cell of the animal can be selected as producing the optimum antibody, and then can be fused to an immortal cell. The resulting tumor cell grows uncontrollably producing only the single clone of desired antibody. Such antibodies, called monoclonal antibody, now may be grown in live animals or cell culture. There are several benefits of the monoclonal antibodies over polyclonal antibodies:

1. The characteristics of polyclonal antibodies are dependent on the animal producing the antibodies. Therefore, if the source individual animal must be changed, the resultant antibody may be quite different.
2. Polyclonal antibodies are, in general, less specific for the analyte compared to monoclonal antibodies. Sometimes instead of the entire antibody, a fragment of the antibody such as FAB fragment or dimeric complex of FAB is used as the antibody.

The other major component of the immunoassay is the labeled antigen (or its analog). There are many different kinds of labels, generating different kinds of signals. As described earlier, an enzyme may be used as the label, which in its turn can generate different types of signals depending on the substrate used for the enzyme. Most commercially available reagent kits supply these two reagent components as separate reagents (Reagent 1 and 2); in most cases, they are ready-to-use liquid, though in some cases some reconstitution (of a lyophilized component) or mixing of two or more components to generate the active reagent may be required.

However, for the determination of alcohol, an enzyme (alcohol dehydrogenase) assay is used. Its enzymatic reaction with the alcohol in a sample may either be monitored by photometrically (e.g., absorption increase at 340 nm from the cofactor reaction: NAD $\rightarrow$ NADH), or by coupled reactions that may change absorption at other wavelengths. Obviously, reactions in which absorbance increases than decreases are preferred as signal generators.

LIMITATION OF IMMUNOASSAYS

Even though the immunoassay methods are now widely used, there are limitations and drawbacks of this technique. Antibody specificity is a major limitation of the immunoassays. Many endogenous metabolites of the analyte may have very similar structural recognition motif as the analyte itself. There are also other molecules which are very different from the analyte, but producing comparable recognition motif as the analyte. These molecules are generally called cross-reactants. When present in the sample, these compounds may produce false-positive results [7–9]. Thus, of six THC immunoassays tested with sera from eight subjects on efavirenz (an antiretroviral drug) therapy, the false-positive rates (with 50 ng/mL cutoff) were: 100% (Immunalysis Cannabinoid Direct ELISA kit), 88% (CEDIA), 50% (Triage POC from BioSite), and 0% (EMIT, OraSure, and assays run on the AxSYM platform and also manufactured by the Abbott Laboratories) [10]. Similarly, the presence of fentanyl, an opioid analgesic, in serum caused false-positive LSD results in the CEDIA immunoassay [11].

Amphetamine immunoassays are mostly affected by cross-reactants because a number of structurally similar compounds (see also Chapter 16) such as buflomedil, brompheniramine, chlorpromazine, ephedrine, fenfluramine, isometheptene, mexiletine, N-acetyl procainamide (metabolite of procainamide), perazine, phenmetrazine, phentermine, phenylpropanolamine, promethazine, pseudoephedrine, quinacrine, ranitidine, tolmetin, and tyramine are known to cross-react with various amphetamine assays causing false-positive results [12–15]. Brahm et al. showed in their review that antihistamines/decongestants, antidepressants, antipsychotics—all interfere in amphetamine/methamphetamine immunoassays [14]. Kapur found that in their laboratory CEDIA amphetamine screening immunoassay had only 50.5% results confirmed during a 12-month period in 2005/2006 [15]. In 2011/2012, however, the author found 32.6%

false-positive rate in the same assay; the difference was ascribed to the difference in frequency of occurrence of interfering medications. Dietzen et al. demonstrated that urine specimens containing ranitidine greater than 43 μg/mL interferes with Beckman Synchron amphetamine immunoassay and concluded that due to extreme sensitivity of ranitidine with this assay, Beckman amphetamine assay has little use in author's laboratory because ranitidine concentration in urine specimens routinely exceeds 43 μg/mL in patients taking ranitidine. However, other Beckman assays such as opiate, barbiturates, cocaine metabolite, propoxyphene, and methadone have good specificity while the cannabinoid assay has 100% predictive value based on GC/MS confirmation [16].

Brahm et al. also showed in their review that immunoassays for phencyclidine may be affected by venlafaxine, ibuprofen, and dextromethorphan, while assays for methadone show false-positive results for diphenhydramine, doxylamine, clomipramine, chlorpromazine, quetiapine, thioridazine, and verapamil [14]. Ly et al. reported false-positive urine phencyclidine test result due to the interference of tramadol and its metabolites [17]. Intake of a therapeutic dose of codeine can yield a false-positive CEDIA buprenorphine test result [18].

Tolmetin, a nonsteroidal antiinflammatory drug, can interfere with EMIT assays for urine drug screening if the drug is present in a significant amount (1800 mg/L). Tolmetin has characteristic high molar absorbance at 340 nm, which is the wavelength for detection of signal in EMIT assays. A specimen collected from an arthritic patient receiving 100—to 400 mg tolmetin showed decreased signal when mixed with abused drugs and analyzed by EMIT assays. Similar interference of tolmetin in FPIA assays for DAU was not observed because a different wavelength (525 nm) is used for detecting signal. However, potential false-positive test results using FPIA benzodiazepine assay were observed when urine specimens contained high concentrations of fenoprofen, flurbiprofen, indomethacin, ketoprofen, and also tolmetin [19]. Kunsman et al. reported that vitamin B_2 interferes with TDx DAU assays (Abbott Laboratories) because of an increase in background fluorescence polarization probably caused by interfering fluorophore property of vitamin B_2 or a combination of vitamin B_2 and its metabolites [20].

Interesting immunoassay interference was observed by Cotton et al.; they found several types of commercial baby washes and soaps that could interfere in Vitros cannabinoid immunoassay. When they tested component for the baby washes, several of them showed strong interference in the assay. The false-positive results in newborn and infant specimens had important clinical as well as legal implications in neonatal care [21]. Other components in a specimen, e.g., bilirubin, hemoglobin, or lipid, may interfere in the immunoassay by interfering with the assay signal, and thus produce incorrect assay results. These effects are more commonly found in homogeneous immunoassays which are used for measuring analytes in serum or plasma. Interference of bilirubin or hemoglobin in testing of urine specimens using immunoassays is very rare. Another type of immunoassay interference involves endogenous human antibodies in the specimen, which may interfere with the assay reagent components—the assay antibodies or the antigen labels. Such interference includes that from heterophilic antibodies or various human antianimal antibodies and is more frequently observed in sandwich types of immunoassay. Since the DAU assays use competition format and a single antibody, interferences of heterophilic antibodies in the DAU immunoassays are rare. Moreover, heterophilic antibodies due to large size are usually absent in urine. As a result, interference of heterophilic antibody in DAU testing has never been reported. See Chapter 11 for in-depth discussion on issues of interferences in immunoassays used for DAU testing.

Other than interferences, immunoassays may also suffer from sensitivity issue. Darragh et al. concluded from their study that KIMS, CEDIA, and high-sensitive-CEDIA immunoassays for benzodiazepine are inadequately sensitive for detection of benzodiazepines in urine from patients treated for chronic pain. Out of 299 urine specimens, 141 specimens showed the presence of one or more benzodiazepine/metabolite using LC-MS/MS but CEDIA showed positive results with 78 out of 141 specimens confirmed by LC-MS/MS (55% sensitivity). Similarly, KIMS assay showed 47% sensitivity (66 specimens positive out of 141 positive specimens) but high sensitivity CEDIA assay showed 78% sensitivity [22].

DAU ASSAY COMPONENTS AND PROTOCOLS

Regardless of the format, an immunoassay consists of the reagent kit, calibrators, and controls. In addition to these main components, manual and automated immunoassays may require diluents and supplemental wash or sample pretreatment reagents. The reagents contain the two main components (antibody and label) in buffered protein matrix, which is optimized to generate the best precision and accuracy for the assay. The assay protocol (manual or automated) defines the volumes of sample (or calibrator or controls) to be mixed with volumes of reagents, the incubation time and temperature (37 °C is used most commonly), the signal reading process (read time-points, wavelength, single vs multiple readings, etc.), and the algorithm to convert signal into reportable analyte concentration. If the assay is on, an automated system, both reagent formulations and assay protocols, are optimized to generate best reagent on-board stability and the calibration stability. If a commercial assay kit is used, such performance information, together with typical other assay performances (analytical range, cutoff concentrations, precision, accuracy, cross-reactivity, etc.) are reported on the manufacturer's assay "package insert."

The calibrators define the relationship between analyte concentration and signal generated for a particular lot of reagents (and, if the assay is an automated one, on a particular instrument). The calibrators may be lyophilized or liquid ready-to-use and contain the analyte (or an analog) in a buffered matrix (or, to preserve commutability with the desired specimen type, in serum, plasma, or urine pools). The use of multiconstituent calibrators, containing a group of DAU analytes, is currently preferred over single-component calibrators. Most manufacturers define the "traceability" of the calibrator's assigned analyte concentration; for the DAU assays, such traceability is often defined to USP grade pure analytes. The assay controls confirm the accuracy of the assay in a particular run and also like calibrators, may be lyophilized or liquid ready-to-use, and most commonly contain multiple analytes for convenience.

QUALITATIVE VERSUS QUANTITATIVE REPORTING

Most immunoassays generate signals that do not show excellent linear relationship with the analyte concentrations in the calibrators. Thus, multiple calibrators (four to six levels) are recommended for accurate measurements of the analyte across the entire assay range. Most automated assay systems can store a calibration curve depending on the assay stability on that system. Therefore, when a sample is analyzed during the period denoted by calibration stability, the assay signal is automatically converted into analyte concentration via the stored calibration curve. In manual assays, however, a calibration curve must be run together with the samples for testing.

DAU assays are often used to report "qualitative" results, i.e., positive or negative with respect to a certain analyte concentration (the "cutoff" level). Thus, many of the assays come in qualitative or quantitative formats and in most cases such formats are defined by assay protocol and calibration. In qualitative formats, the calibration can be simplified to only one or two calibrators, centering on the cutoff point thus providing the most accuracy around that point. The algorithm compares the signal observed with a sample with that of the cutoff calibrator and reports the result as positive or negative. Semiquantitative results can be reported with a calibration curve containing a minimum of three calibrator samples; often the combination of the zero-calibrator, together with two or more calibrators at or near cutoff level to generate a calibration curve. Obviously, such assay formats will have increased inaccuracy at analyte concentrations further away from the cutoff points.

SPECIMEN TYPE

While urine is the most commonly used specimen type for DAU assays, serum is used for toxicology screen. Other types of specimens used for DAU assays are saliva, sweat, tears, cerebrospinal, stomach fluids, and bronchial secretions [23–36]. Out of these various specimen types, urine provides the most chances of specimen tampering. Compared to these types of specimens which provide data on current status of drug use, hair and nail specimens may provide longer term history of the drug abuse. While these specimen types may reveal the presence of the abused drug at the sites of action, other types, such as, amniotic fluid, cord blood, meconium, and breast milk are useful for determining fetal and perinatal exposure to drugs. Weak bases, such as cocaine, opiates, benzodiazepines, or nicotine, tend to concentrate in saliva, because its pH is slightly acidic compared with that of plasma. Saliva, which can be collected in commercially available absorbent pads or in plastic tubes, has several advantages as a source of specimen because it requires noninvasive sample collection with minimum discomfort to the patient and it contains more of the parent drug itself with fewer metabolites. Collection of adequate saliva volume can be ensured by stimulation of saliva flow [32,33]. Commercially available saliva collection kits include chewing gums that promote saliva stimulation. Saliva collection is most recommended after wake up, and before the morning ablutions (brushing, flossing, mouthwash, etc.). Disadvantages of saliva are oral contamination from the original drug administration (smoking, snorting, oral ingestion) and the method of sample collection, which may affect salivary drug concentrations as a result of changes in pH and flow rate. Examples of use of such specimens in DAU testing utilizing immunoassays include testing of codeine in sweat and saliva [34], cocaine and amphetamine, ecstasy (3,4-methylenedioxy-methamphetamine: MDMA) in saliva [35–38]. It is important to note that depending on the specimen type and the drug concentration range therein, assay formulations and protocols must be optimized for one specimen type only. More recently, POC devices are also based on immunoassays for rapid detection of drugs in oral fluids for identifying drugged drivers [39]. POC devices is also available for urine drug testing. See Chapter 10 for detail.

Urine specimens, compared to serum or plasma specimens, are rarely less affected by hemolysis or high bilirubin concentration in serum such interferences. Turbidity interference is possible in urine, but the cause is most likely bacterial growth. Preservatives in urine, like acetic acid, boric acid, or alkali, may interfere in some urine assays.

SPORTS DOPING

Even though sports doping is not traditionally considered among the DAU assays, doping abuse has recently been widely spread. Doping includes steroid abuse as well as other pharmaceutical agents to enhance performance. Mostly immunoassays are used to detect these abuses. Like all steroid assays, cross-reactivity of these assays to myriads of natural or synthetic steroids may cause incorrect results. Like the other drugs of use discussed above, the abusers might try to adulterate the specimen to avoid or confuse the assay results.

CONCLUSIONS

Immunoassays are widely used for DAU testing in both hospital setting and also in specialized toxicology laboratories as a screening device for workplace drug testing. Most DAU assays are competitive immunoassays (homogeneous or heterogeneous format) and are subjected to interferences from drug metabolites as well as other drugs with comparable structures to the analyte drug. These assays may be qualitative or quantitative. Alcohol testing employs enzyme assay. In addition, POC devices are also available for NPT or roadside testing. However, it is mandatory to confirm the initial screening result by GC/MS or other specific analytical technique for legal purposes.

REFERENCES

[1] Jolley ME, Stroupe SD, Schwenzer KS, et al. Fluorescence polarization immunoassay III. An automated system for therapeutic drug determination. Clin Chem 1981;27:1575–9.
[2] Urine testing for drugs of abuse. Rockville, MD: National Institute of Drug Abuse (NIDA). In: Hawks RL, Chian CN, editors. Department of Health and Human Services. NIDA Research Monograph; 1986.
[3] Jeon SI, Yang X, Andrade JD. Modeling of homogeneous cloned enzyme donor immunoassay. Anal Biochem 2004;333:136–47.
[4] Datta P, Dasgupta A. A new turbidimetric digoxin immunoassay on the ADVIA 1650 analyzer is free from interference by spironolactone, potassium canrenoate, and their common metabolite canrenone. Ther Drug Monit 2003;25:478–82.
[5] Dai JL, Sokoll LJ, Chan DW. Automated chemiluminescent immunoassay analyzers. J Clin Ligand Assay 1998;21:377–85.
[6] Montagne P, Varcin P, Cuilliere ML, Duheille J. Microparticle-enhanced nephelometric immunoassay with microsphere-antigen conjugate. Bioconjugate Chem. 1992;3:187–93.
[7] Datta P, Larsen F. Specificity of digoxin immunoassays toward digoxin metabolites. Clin Chem 1994;40:1348–9.
[8] Datta P. Oxaprozin and 5-(p-hydroxyphenyl)-5-phenylhydantoin interference in phenytoin immunoassays. Clin Chem 1997;43:1468–9.
[9] Datta P, Dasgupta P. Bidirectional (positive/negative) interference in a digoxin immunoassay: importance of antibody specificity". Ther Drug Monit 1998;20:352–7.
[10] Rossi S, Yaksh T, Bentley H, Brande GVD, Grant I, et al. Characterization of interference with 6 commercial Δ^9-tetrahydrocannabinol immunoassays by efavirenz (glucuronide) in urine. Clin Chem 2006;52:896–7.
[11] Gagajewski A, Davis GK, Kloss J, Poch GK, Anderson CJ, et al. False-positive lysergic acid diethylamide immunoassay screen associated with fentanyl medication. Clin Chem 2002;48:205–6.
[12] Moore KA. Amphetamines/sympathomimetic amines in Principles of Forensic Toxicology (Levine B edited). AACC Press; 2003. p. 341–8.
[13] Grinstead GF. Ranitidine and high concentration phenylpropanolamine cross react in the EMIT monoclonal amphetamine/methamphetamine assay. Clin Chem 1989;35:1998–9.
[14] Brahm NC, Yeager LL, Fox MD, Farmer KC, Palmer TA. Commonly prescribed medications and potential false-positive urine drug screens. Am J Health-Syst Pharm 2010;67:1344–50.
[15] Kapur BM. False positive drugs of abuse immunoassays. Clin Biochem 2012;45:603–4.
[16] Dietzen DJ, Ecos K, Friedman D, Beason S. Positive predictive values of abused drug immunoassay on the Beckman synchron in a veteran population. J Anal Toxicol 2001;25:174–8.
[17] Ly BT, Thornton SL, Buono C, Stone JA, et al. False-positive urine phencyclidine immunoassay screen result caused by interference by tramadol and its metabolites. Ann Emerg Med. 2012;59:545–7.
[18] Berg JA, Schjott J, Fossan KO, Riedel B. Cross-reactivity of the CEDIA buprenorphine assay in drug of abuse screening: influence of dose and metabolites of opioids. Subst Abuse Rehabil 2015;28:131–9.
[19] Joseph R, Dickerson S, Willis R, Frankenfield D, et al. Interference by nonsteroidal anti-inflammatory drugs in EMIT and TDx assays for drugs of abuse. J Anal Toxicol 1995;19:13–17.
[20] Kunsman GW, Levine B, Smith ML. Vitamin B2 interference with TDx drugs of abuse assay. J Forensic Sci 1998;43:1225–7.

[21] Cotton SW, Duncan DL, Burch EA, et al. Unexpected interference of baby wash products with a cannabinoid (THC) immunoassay. Clin Biochem 2012;45:605–9.
[22] Darragh A, Snyder ML, Ptolemy AS, Melanson S. KIMS, CEDIA and HS-CEDIA immunoassays are inadequately sensitive for detection of benzodiazepines in urine from patients treated for chronic pain. Pain Physician 2014;17:359–66.
[23] Kidwell D, Holland J, Athanaselis S. Testing for drugs of abuse in saliva and sweat. J Chromatogr B 1998;713:111–35.
[24] Pichini S, Pacifici R, Altieri I, Pellegrini M, Zuccaro P. Determination of opiates and cocaine in hair as trimethylsilyl derivatives using gas chromatography-tandem mass spectrometry. J Anal Toxicol 1999;5:343–8.
[25] Palmeri A, Pichini S, Pacifici R, Zuccaro P, Lopez A. Drugs in nails: physiology, pharmacokinetics and forensic toxicology. Clin Pharmacokinet 2000;38:95–110.
[26] Van Haeringen NJ. Secretion of drugs in tears. Curr Eye Res 1985;4:485–8.
[27] Bonati M, Kanto J, Tognoni G. Clinical pharmacokinetics of cerebrospinal fluid. Clin Pharmacokinet 1982;7:312–15.
[28] Wong GA, Peierce TH, Goldstein E, Hoeprich PD. Penetration of antimicrobial agents into bronchial secretion. Am J Med 1975;9:219–23.
[29] Pacifici GM, Nottoli R. Placental transfer of drugs administered to the mother. Clin Pharmacokinet 1995;28:235–69.
[30] Pichini S, Basagana X, Pacifici R, Garcia O, Puig C, et al. Cord serum nicotine as a biomarker of fetal exposure to cigarette smoke at the end of pregnancy. Environ Health Perspect 2000;108:1079–83.
[31] Dickson PH, Lind A, Studts P, Nipper HC, Makoid M, et al. The Routine analysis of breast milk for drugs of abuse in a clinical toxicology laboratory. J Forensic Sci 1994;39:2341–5.
[32] Navarro M, Pichini S, Farre M, Ortuno J, Roset PN, et al. Usefulness of saliva for measurement of 3,4-methylenedioxymethamphetamine and its metabolites: correlation with plasma drug concentration and effect of salivary pH. Clin Chem 2001;47:1788–95.
[33] Jenkins AJ, Oyler JM, Cone EJ. Comparison of heroin and cocaine concentrations in saliva with concentrations in blood and plasma. J Anal Toxicol 1995;19:359–74.
[34] Kintz P, Cirimele V, Ludes B. Codeine testing in sweat and saliva with the drugwipe. Int J Legal Med 1998;111:81–4.
[35] Samyn N, Van Haeren C. On-site testing of saliva and sweat with drugwipe and determination of concentration of drugs of abuse in saliva, plasma and urine of suspected users. Int J Legal Med 2000;113:150–4.
[36] Schramm W, Smith RH, Craig PA, Kidwell DA. Drugs of abuse in saliva: A review. J Anal Toxicol 1992;16:1–9.
[37] Moore L, Wicks J, Spiehler V, Holgate R. Gas chromatography-mass spectrometry confirmation of Cozart RapiScan saliva cocaine levels with plasma levels and pharmacologic effects after intravenous cocaine administration in human subjects. J Anal Toxicol 2001;25:520–4.
[38] Kim I, Barnes AJ, Schepers R, Moolchan ET, Wilson L, et al. Sensitivity and specificity of the Cozart microplate EIA cocaine oral fluid at proposed screening and confirmation cutoffs. Clin Chem 2003;49:1498–503.
[39] Edwards LD, Smith KL, Savage T. Drugged driving in Wisconsin: oral fluid versus blood. J Anal Toxicol 2017;41:523–9.

Chapter 10

Issues of Interferences With Immunoassays Used for Screening of Drugs of Abuse in Urine

Anu S. Maharjan[1] and Kamisha L. Johnson-Davis[1,2]
[1]Department of Pathology, University of Utah Health Sciences Center, Salt Lake City, UT, United States, [2]ARUP Institute for Clinical and Experimental Pathology, Salt Lake City, UT, United States

INTRODUCTION

In recent years, immunoassays have evolved to not only detect antigens of interest, but also drugs of abuse. Drugs-of-abuse testing are important both in workplace and in medical settings. Workplace drug testing requires confirmation of a positive specimen by an alternative analytical technique preferably, liquid chromatography-tandem mass spectrometry (LC-MS/MS) or gas chromatography/mass spectrometry (GC/MS), but such confirmations may not be requested by a clinician during medical drug testing. Drug testing is important in medical settings, since most patients do not report their drug use. It has been reported that 66.6% patients who tested positive for amphetamine did not report their amphetamine use to the clinician in the emergency department (ED) [1]. A similar case was seen in 70% of patients who had an opiate positive result, but did not report their opiate abuse. Urine drug screen immunoassay applications are available on various automated analyzers as well as point-of-care devices in health care facilities or in physicians' offices. Some immunoassays have good sensitivity and specificity; however, other assays are subjected to interferences. A study by Johnson-Davis et al. has highlighted the true-positive rate and the false-positive rate of immunoassay performance at a reference laboratory; which supports previous studies which have demonstrated that some immunoassays, such as amphetamine and opiate immunoassays, have poor specificity and a high rate of false positives [2]. As a result, positive screen results for drugs-of-abuse immunoassays with high false-positive rates should be confirmed by GC/MS or LC-MS/MS [3].

Amphetamines immunoassays are subjected to more interference that can lead to a false-positive result compared to other drugs-of-abuse assays. Dietzen et al. reviewed confirmatory results of drugs-of-abuse testing that were positive by immunoassays and reported that there are 175 instances of false-positive screening during 1-year period. Positive predictive values of various immunoassays varied from 0% (amphetamine) to 100% (tetrahydrocannabinol) [4].

INTERFERENCES IN AMPHETAMINE IMMUNOASSAY

3,4-Methylenedioxymethamphetamine (MDMA; Ecstasy) and 3,4-methylenedioxyamphetamine (MDA) are two similar synthetic designer drugs resembling amphetamine, which are widely abused. Immunoassays are the primary screening methods for the detection of amphetamines and in most of these assays, the antibodies appear to be directed toward the amino group of amphetamine, a primary amine, and methamphetamine, a secondary amine. Different commercial immunoassays use either monoclonal or polyclonal antibodies against amphetamine and methamphetamine. Both MDMA and MDA show significant cross-reactivity with commercially available amphetamine immunoassay, but there is a separate immunoassay for MDMA. In one report, the authors evaluated eight commercially available amphetamine immunoassays for their effectiveness for detecting MDMA during immunoassay screening test and observed the cross-reactivity to be lower at high concentrations of these drugs (Table 10.1) [5]. In a similar study, cross-reactivity of

TABLE 10.1 Cross-Reactivities of MDMA and MDA With Various Commercial Amphetamine and Amphetamine/MDMA Immunoassays

Immunoassay	Manufacturer	MDMA Cross-Reactivity	MDA Cross-Reactivity
DRI-amphetamine	Diagnostics Reagents (Sunnyvale, CA)	11%–44%	51%–99%
CEDIA-amphetamine	Microgenics (Fremont, CA)	7%–46%	175%–214%
EMIT d.a.u. monoclonal	Syva (Cupertino, CA)	7%–46%	175%–214%
Synchron-CX-amphetamine	Beckman (Fullerton, CA)	49%–100%	58%–73%
COBAS Integra amphetamine	Roche Diagnostics (Indianapolis, IN)	38%–48%	24%–42%
CEDIA-amphetamine/MDMA	Microgenics (Fremont, CA)	100%	36%–63%
Online amphetamine/MDMA	Roche Diagnostics	111%–142%	75%–100%

MDMA with amphetamine immunoassay showed dependence on concentration and varied from 118% at 150 ng/mL to 18% at 10,000 ng/mL [6].

Amphetamine or methamphetamine immunoassays are created for high-sensitivity screening; therefore, there is a wide range of cross-reactivity to compounds that are structurally or physically similar to phenethylamines. Immunoassays designed for the detection of amphetamines can be classified into three general types based on antibody specificity:

1. Assays highly selective for either amphetamine and MDA or methamphetamine and MDMA.
2. Assays that are able to detect both amphetamine and methamphetamine to varying extent but that also exhibit higher levels of cross-reactivity to various over-the-counter (OTC) cold medications containing sympathomimetic amines. Medications such as bupropion [7], fluoxetine [8], labetalol [8–10], metformin [9–11], methylphenidate [8], promethazine, pseudoephedrine, ranitidine, and trazodone may cause false-positive amphetamine results.
3. Assays for amphetamine and methamphetamine with lower of cross-reactivity to OTC medications [12].

Choosing an optimal amphetamine immunoassay depends on several factors including availability of automated analyzers and intended use. Forensic or workplace laboratories prefer to use amphetamine immunoassays that are specific for amphetamine testing; whereas, hospital laboratories may need a broad spectrum immunoassay that can detect many sympathomimetic amines, including OTC drugs. According to the National Academy of Clinical Biochemistry's (NACB) Laboratory Medicine Practice Guidelines (LMPG), the recommendations for an optimum immunoassays for amphetamine testing in ED patients are those directed toward phenethylamine as a drug class [13]. LMPG also recommends changing the name of the test from "amphetamines" to sympathomimetic amines" or "stimulant amines." The cross-reactivity of various phenethylamine in different immunoassays has been extensively reported [14]. However, it is possible that current assays produced by the manufacturers may have different cross-reactivities from those reported due to changes in antibodies and reagent composition. The analyst should consult immunoassay package inserts or contact the manufacturer for up-to-date cross-reactivity data. It is also possible that potential interfering substances used during studies are lower than those encountered in the clinical setting and that manufacturers often do not test for cross-reactivity of endogenous metabolites [15,16]. Lot-to-lot variability concerning sensitivity to targeted analytes has been reported [17].

Interferences of Sympathomimetic Amines With Amphetamine Immunoassay

Interference of various sympathomimetic amines in amphetamine immunoassay is well recognized. Various OTC cold medications contain these sympathomimetic amines and use of such medications may cause false-positive test results in amphetamine immunoassay screening. In workplace drug testing, taking cold medications are not prohibited, but may cause confusion during medical drug testing, specifically if a confirmation test is not ordered or if the facility does not have the capability to confirm such drug by mass spectrometry. Unfortunately, misidentification of ephedrine/pseudoephedrine as methamphetamine has been reported during GC/MS confirmation [18]. There are several causes of misidentification, which include the following:

1. At an injector port temperature over 185°C in the gas chromatograph of the GC/MS, ephedrine and pseudoephedrine can be thermally dehydrated to methamphetamine.

TABLE 10.2 Electron Ionization and Chemical Ionization Mass Spectral Features of Derivatized Methamphetamine and Ephedrine

Derivative	Electron Impact Base Peak (m/z)	Electron Impact Other Peaks (m/z)	Chemical Ionization Base Peak (m/z)	Chemical Ionization Other Peaks (m/z)
Methamphetamine Propyl carbamate	144	102, 91, 58	236 (M + 1)	176, 144, 119, 58
Ephedrine propyl carbamate	144	102, 77, 58	192	220, 148, 58
Methamphetamine trifluoroacetyl	154	118, 58	246 (M + 1)	154, 119
Ephedrine trifluoroacetyl	154	118, 58	244	276, 158
Methamphetamine 4-carbethoxy	315	118, 91	400(M + 1)	308, 119
Ephedrine 4-carboethoxy	308	118, 91	398	121

2. Electron impact mass spectrum of trifluoroacetyl, pentafluoropropyl, heptafluorobutyl, 4-carbethoxyhexafluorobutyryl, and various carbamate derivatives of methamphetamine are similar to the corresponding derivatized ephedrine/pseudoephedrine. Since ephedrine and pseudoephedrine elutes right after methamphetamine peak, ephedrine/pseudoephedrine can be misidentified as methamphetamine if the mass spectral analysis is not done carefully.

Nevertheless, the misidentification can be completely avoided if the mass spectrometry is performed in the positive chemical ionization mode using methane as the reaction gas [19]. This results in ionization mass spectrum of derivatized methamphetamine that is very different from the derivatized ephedrine/pseudoephedrine. In the electron ionization mode, usually molecular ion of derivatized amphetamine is present as a very weak peak. In contrast, protonated molecular ion is the base peak (100% abundance) in the chemical ionization mass spectra of derivatized amphetamine and methamphetamine. This misidentification problem also exists with propyl carbamate derivatives of methamphetamine. In the electron ionization mode, both methamphetamine propyl carbamate and ephedrine propyl carbamate showed almost identical mass spectral fragmentation patterns. Again using chemical ionization where methamphetamine propyl carbamate showed a base peak at m/z 236 can circumvent this problem and the ephedrine propyl carbamate showed a base peak at m/z 192. The major mass spectral fragmentation pattern differences are given in Table 10.2. Recently, Ojanpera et al. described that using positive chemical ionization mass spectrometry, allowed the authors to observe protonated molecular ion peak of new amphetamine like designer drug 3,4-methylenedioxypyrovalerone in urine at m/z 276 [20].

Ephedrine or pseudoephedrine which is present in many OTC cold medications is responsible for majority of false-positive result in amphetamine immunoassay screening test. These sympathomimetic amines may be present in large amounts in urine specimens which are initially tested positive by amphetamine immunoassays. Stout et al. based on the analysis of 27,400 randomly collected urine specimens reported that 833 out of 1104 urine specimens (92% specimens) that failed to confirm amphetamine, methamphetamine, MDMA, or MDA by GC/MS contained pseudoephedrine [21]. Commonly encountered medications that interfere with amphetamines screening assays are listed in Table 10.3.

Vicks Inhaler and False Positive in Amphetamine Immunoassays

Amphetamine and methamphetamine have optical isomers designated d (or +) for dextrorotatory and l (or −) for levorotatory and the d isomers, the more physiologically active compounds, are the intended targets of immunoassays because d isomers are abused. Ingestion of medications containing the l isomer can cause false-positive results. For example, Vicks inhaler contains the active ingredient l-methamphetamine [22] and extensive use of this product may cause false-positive results for immunoassay screening. Specific isomer resolution procedures must be performed to

TABLE 10.3 Drugs That Interfere With Amphetamine Immunoassays

Drug Class	Individual Drug
Sympathomimetic amines	Brompheniramine, benzphetamine, ephedrine, isometheptene, mephentermine, methylphenidate, pseudoephedrine, phentermine, phenylpropylamine,[a] propylhexedrine, phenylephrine, phenmetrazine, tyramine
Antidepressant/antipsychotic	Chlorpromazine, bupropion, desipramine, fluoxetine, perazine, thioridazine, trimipramine, trazodone
Antihistamine	Promethazine, ranitidine
Antiprotozoal	Quinacrine, chloroquine
Antiemetic	Trimethobenzamide
Beta-blocker	Labetalol
Cardioactive	N-acetyl procainamide (metabolite of procainamide)
	Mexiletine
Nonsteroidal antiinflammatory	Tolmetin
Norepinephrine reuptake inhibitor	Atomoxetine
Vasodilator	Isoxsuprine
Tocolytic agent	Ritodrine

[a]Removed from the U.S. market in 2005.

differentiate the d and l isomers because routine confirmation by gas chromatography mass spectrometry (GCMS) does not determine isomer composition. Poklis et al. reported relatively high concentrations of l-methamphetamine in two subjects (1390 and 740 ng/mL), using chiral GC/MS and after extensively inhaling Vicks inhaler every hour for several hours. However, when such urine specimens were tested by the enzyme multiplied immunoassay technique (EMIT II amphetamine/methamphetamine assay; Dade Behring), results were negative even after such extensive use of Vicks inhaler [23]. In another study, the authors reported that nasal inhalation of Vicks inhaler following recommended dosage should not cause false-positive amphetamine/methamphetamine urine screening results using fluorescence polarization immunoassay (FPIA) for amphetamine/methamphetamine for application on TDxADX/FLx analyzer (Abbott Laboratories, Abbott Park, IL) because antibody used in this immunoassay is stereo-selective and recognized d-methamphetamine or d-amphetamine if present in the urine specimens. However, when two subjects inhaled the recommended dosage twice, positive test result was obtained by the FPIA screening assay. The concentrations of l-methamphetamine in urine specimens of these two subjects were 1560 and 1530 ng/mL, respectively [24]. However, in the case of doubt, l-methamphetamine can be distinguished from d-methamphetamine by using GC/MS and chiral derivatization technique. If greater than 80% l-methamphetamine is found in the specimen, the result is considered as consistent with use of Vicks inhaler. Abbott Laboratories discontinued FPIA assays for amphetamine/methamphetamine assay for application on the TDX analyzer in 2010.

Case Study

An emergency medical team responded to the residence of a 77-year-old man who was experiencing difficulty in breathing and his oxygen saturation was 69%. A breathing treatment of albuterol and ipratropium bromide was administered and his oxygen saturation was improved to 96%. The patient was transported to the ED and was admitted to the intensive care unit. Approximately, 12 h after admission, he was unresponsive and expired. His medical history included heart failure, atrial flutter, and bronchial asthma. His urine drug testing (GC/MS) was positive for acetaminophen, nicotine, cotinine, caffeine,

(Continued)

(Continued)

diltiazem, doxylamine, and methamphetamine. However, chiral analysis of methamphetamine was not initially performed raising possibility of abusing *d*-amphetamine by the man. To address this question further, the authors performed chiral analysis using trifluoroacetyl-*l*-propyl chloride as the chiral derivatizing agent and identified the methamphetamine as the *l*-isomer. Further investigation revealed that the decedent frequently used Vicks inhaler for his bronchial asthma [25].

Dietary Weight Loss Products and Positive Amphetamine

The federal drug administration (FDA) banned ephedra alkaloids containing all products including weight loss products in April 2004. The products that were banned include ma-huang, sida, cordifolia, and Pinellia, but Chinese herbal products, herbal teas and drug containing synthetic ephedrine were not banned. Currently, there is a push for companies' marketing herbal weight loss products to produce ephedra free diet pills. The common ingredients of these ephedra free diet pills may also contain phenylpropanolamine, tyramine, or phentermine, which like ephedrine interferes with the immunoassay screening of amphetamine/methamphetamine in urine specimens. However, other diet loss products which contain hydroxy-citric acid or bitter orange do not interfere with drugs-of-abuse testing.

The FDA warned consumers in January 2006 that the Brazilian dietary supplements Emagrece Sim and Herbathin contain several active drug ingredients including fenproporex which is metabolized to amphetamine [26]. Use of other weight loss or nutritional supplements containing fenproporex has resulted in the detection of *d*-amphetamine in the urines of users [27]. In 2011, the US military removed all dimethylamylamine (DMAA)-containing supplements from all military exchanges worldwide. DMAA, also known as 1, 3-dimethylamylamine, is a synthetic neural stimulant used in dietary supplements that results in false-positive amphetamine [10,28]. DMAA has been linked to cases of increased blood pressure and heart rate, leading to more than two deaths in the US military [10,29]. DMAA is an aliphatic amine that resembles amphetamine and releases norepinephrine; however, it is still unknown if DMAA works similar to amphetamine. When negative drug-free urine was spiked with DMAA, both Roche Amphetamines KIMS (kinetic interaction of microparticle in solution methods) assay and Siemens Syva EMIT II Plus Amphetamines assay gave positive responses at 7500 and 3125 ng/mL [10,28].

Case Study

A 25-year-old woman presented to the ED with persistent abdominal pain of 2 weeks duration and 1-day history of nausea and vomiting. Her medications included high-dose ibuprofen which she discontinued 5 days prior to her presentation and a Brazilian weight loss product, Emagrece purchased from the Internet which she also discontinued 2 weeks ago. The patient discontinued the weight loss product due to feeling mildly tremulous after taking the product and her friends also experienced similar symptoms. On admission to the hospital, results of routine blood test for electrolytes, complete blood count, amylase, lipase, and liver function tests were within normal limits. A urinary toxicology screen was positive for amphetamine. The analysis of weight loss product after extraction using GC/MS confirmed the presence of 5-cyano-amphetamine, an amphetamine derivative used outside the US for weight loss but considered illegal by the FDA. The list of ingredients only mentioned herbal components such as kava kava and ginkgo biloba but there was no mention of the amphetamine-like compound. The patient was treated with metoclopramide, pantoprazole, and acetaminophen and was discharged from the hospital 3 days later when her symptoms resolved [30].

Case Study

A 35-year-old African American female with no prior history of coronary artery disease presented with sudden onset of exertional chest discomfort with radiation to the back. After arrival to the ED, she went into ventricular fibrillation-cardiac arrest (V-fib) for 6 minutes with conversion after electrical cardioversion and subsequent development of PEA arrest for a total of 4 minutes. The patient received tenecteplase and heparin prior to urgent transfer to the catheterization laboratory. A urinary toxicology screen was positive for amphetamine. The patient denied of using Adderall or any other amphetamine, and reported that she increased her physical activity to lose weight and was further supplementing with a natural weight loss dietary supplement [31].

Case Study

A 22-year-old male infantry soldier arrived at the hospital with hyperthermia and dry hot skin [32]. The soldier lost consciousness while running with his battalion for 10 minutes. His laboratory values showed renal insufficiency and increased cardiac and muscle enzymes. The patient received resuscitation along with active core cooling through endovascular device, endotracheal intubation, and fluid restoration. However, the patient went into refractory asystole and passed away 4-hours later. Toxicology analyses on the autopsy showed the presence of DMAA in the antemortem blood at 0.22 mg/L and caffeine at 2.9 mg/L. He had been taking a dietary supplement for about 4 weeks which contained DMAA along with β-alanine, arginine α-ketoglutarate, and creatine. Amphetamine derivatives and creatine together may contribute to dehydration and heatstroke [33].

Other Drugs That May Cause Positive Amphetamine Screening

Ranitidine is a H_2-receptor blocking agent (antihistamine) which reduces acid production by the stomach and is available OTC without any prescription. Dietzen et al. reported that ranitidine if present in urine at a concentration over 43 μg/mL may produce false-positive test results with amphetamine screen using Beckman Synchron immunoassay reagents (Beckman Diagnostics, Brea, CA). This concentration of ranitidine is routinely exceeded in patients taking ranitidine [4]. Poklis also reported that ranitidine interferes with the EMIT d.a.u. amphetamine/methamphetamine assay if ranitidine concentration in urine exceeded 91 μg/mL. The concentrations of ranitidine in urine specimens collected from 23 patients receiving 150–300 mg ranitidine per day varied from 7 to 271 μg/mL [34]. Since September 2014, both Siemens Syva EMIT II Plus Amphetamine assay and Beckman Coulter Synchron AMP assays claim no positive amphetamine result if ranitidine is less than or equal to 1000 μg/mL [35]. However, there were four patient cases from the University of Pittsburgh Medical Center (UPMC) that showed positive amphetamine results due to the interference from ranitidine at a concentration as low as 160 μg/mL on Beckman Coulter platforms [35]. All four samples were negative on the Siemens platform; however, samples with 160 and 320 μg/mL of ranitidine gave positive results for amphetamine in Beckman DxC 800 and DxC600i. Similarly, trazodone interferes with amphetamine immunoassays and also causes false positives in an immunoassay specific for MDMA. A series of patients who tested positive for MDMA (using Ecstasy EMIT II assay) did not show any presence of MDMA in urine when confirmed by a specific liquid chromatography combined with tandem mass spectrometric method. However, all specimens showed the presence of trazodone and its metabolite meta-chlorophenylpiperazine. In addition, another hallucinogen trifluoromethyl phenylpiperazine also cross-reacts with the MDMA assay [36]. Labetalol, a beta-blocker commonly used for control of hypertension in pregnancy, caused a false-positive amphetamine screen by immunoassay [9]. Casey et al. reported that bupropion, a monocyclic antidepressant and an aid for smoking cessation, may cause false-positive screen using amphetamine immunoassay. Out of 234 specimens screened positive by amphetamine immunoassay (EMIT II), 128 specimens did not show any presence of amphetamine during GC/MS confirmation step. Three specimens, where amphetamine could not be confirmed, contained bupropion [7].

After screening, 3571 specimens by EMI II, 389 (10.9%) samples screened positive via EMIT, but confirmed negative by LC-MS/MS, suggesting false-positive immunoassay results [8]. High-resolution mass spectrometry (HRMS) and in silico structure search identified compounds that may cause false-positive results in urine immunoassay. Compounds with same chemical formula with different molecular rearrangements may not be distinguished by immunoassay, therefore, producing false-positive results [8]. For instance, compounds with mass match for erythro-dihyrobupropion and hydroxy bupropion tested positive in immunoassay. Similarly, atomoxetine, an attention deficit hyperactivity disorder (ADHD) medication caused false-positive amphetamine immunoassay due to structure similarities between atomoxetine metabolites and amphetamine.

Case Study

A 27-year-old female with a past medical history of ADHD arrived at the ED following two episodes of tonic–clonic movements [37]. The patient stabilized after arriving in the hospital with no visual or neurological symptoms. The patient was taking atomoxetine for ADHD, and admitted that she takes them occasionally instead of complying to her daily dosage. Prior to her arrival at the hospital, she had taken 120 mg of atomoxetine approximately 12 h before the seizure episode. The dosage was three times higher than her prescribed dose and above the maximum recommended dosage of 100 mg/day. As part of the workup, the patient's urine was screened for illicit drugs and toxins. Her urine screen was negative for drugs such as tetrahydrocannabinol (THC), benzodiazepines, cocaine, and opiates, but screened positive for amphetamines. She denied the use of amphetamines or any other illicit drugs, and a confirmatory GC-MS test was negative for amphetamine, methamphetamine,

(Continued)

> **(Continued)**
> methylene-dioxyamphetamine, 3,4-methylenedioxy-*N*-methylamphetamine (MDMA), 3,4-methylenedioxy-*N*-ethylamphetamine (MDEA). There are many prescription drugs that do result in false-positive amphetamine screen, and the authors believe atomoxetine and its metabolites to be the cause of the false-positive test results in this particular patient.

INTERFERENCES WITH OPIATE IMMUNOASSAY

Certain fluoroquinolone antibiotics may cause false-positive test results with opiate immunoassay screening [38]. Baden et al. evaluated potential interference of 13 quinolones (levofloxacin, ofloxacin, pefloxacin, enoxacin, moxifloxacin, gatifloxacin, trovafloxacin, sparfloxacin, lomefloxacin, ciprofloxacin, clinafloxacin, norfloxacin, and nalidixic acid) with various opiate immunoassays and concluded that levofloxacin, ofloxacin, and pefloxacin administration would most likely cause false-positive result with opiate immunoassays [39]. Rifampicin is used in therapy for tuberculosis. In one report, the authors observed false-positive opiate immunoassay screening result in patients receiving rifampicin. Opiate immunoassay screens were conducted using KIMS assay on the Cobas Integra analyzer (Roche Diagnostics, Indianapolis, IN). The authors observed a 12% cross-reactivity of rifampicin with the opiate assay over a concentration range of 156−5000 ng/mL. A false-positive result may be observed even after 18 hours of administration of a single oral dose of 600 mg of rifampicin [40].

Naloxone, an opioid antagonist drug, is administered to reverse opioid overdose. It binds to opioid receptors and blocks the effects of other opioids. However, recent cases show naloxone cross-reacting with the opiate immunoassay, which could lead to drastic consequences for patients. Physicians order urine tests for opiate immunoassay for compliance and to ensure no illicit drug use, therefore opioid antagonist cross-reactivity may result in a false-positive result. Naloxone cross-reactivity with opiate immunoassay can also be problematic in pediatric patients that receive a dosage as high as 2 mg/dose [10,41]. Siemens' EMIT drugs of abuse urine assays cross-reactivity listnow shows naloxone concentration greater than 11,000 ng/mL can cross-react with the opiate assay. Naloxone and naloxone glucuronide can also cause positive results for oxycodone with homogeneous enzyme immunoassay (HEIA) oxycodone (Immunalysis Corporation) and the DRI oxycodone assay (Microgenics/Thermo Fisher Scientific) [42].

> **Case Study**
> A 3-year-old girl received an intravenous bolus of naloxone after presenting with decreased mental state and probable hypothermia [41]. Her initial urine toxicology report was negative for drugs of abuse; however, the next day after the administration of naloxone, the sample came out positive for opiate, the cutoff for the assay was set at 300 ng/mL. A repeated test on immunoassay screen came out positive. The urine samples were further sent out for confirmation by LC-MS/MS, which showed that naloxone was the compound that corresponded to the positive opiate results.

> **Case Study**
> A 48-year-old man spent 12 days in an inpatient psychiatry facility for treatment of depression which began subsequent to his abuse of cocaine after 90 days of sobriety. He was then transferred to a residential facility. Eleven days after his transfer to the residential facility, the patient was prescribed gatifloxacin for treating urinary tract infection. His urine drug screen before initiation of gatifloxacin therapy was negative. Six days after the start of gatifloxacin therapy, his urine screen was positive for opiate. The patient denied opiate abuse or consuming poppy seed containing food. Moreover, the positive opiate screen did not show the presence of any opiate during GC/MS confirmation. Another opiate drug screen 24 days later was also positive for opiate during drug screen but GC/MS result was negative. However, when another urine sample was analyzed 13 days after cessation of gatifloxacin therapy, the urine immunoassay screen was negative for opiate indicating that the false-positive opiate screen was due to gatifloxacin [43].

INTERFERENCES WITH TETRAHYDROCANNABINOL IMMUNOASSAY

Antiretroviral therapy such as efavirenz (EFV) screens positive for−THC exposure despite the absence of 11-nor-tetahdryocannabinol-9-caobxylic acid (THC-COOH; the metabolite) [44]. Eight individuals who were taking 600 mg EFV/

day were randomized and analyzed for THC using six different instruments that detect THC via immunoassay. Three different immunoassays—Microgenics Corporation (Cedia Dau MultiLevel THC), BioSite Incorporated (Triage TOX Drug Screen), and Immunalysis Corporation (Cannabinoids (THC/CTHC) Direct ELISA Kit) falsely detected THC metabolites at a cutoff value greater than 50 ng/mL even though the patients were not taking THC [44]. A similar study in 2012 identified immunoassays, including Rapid Response (BTNX), CEDIA THC (Microgenics), THCA/CTHC Direct ELISA (Immunoanalysis), Triage TOX (Biosite), and Instant-view THC (Alfa Scientific) that had false-positive THC due to EFV [45]. In some of these immunoassays, cross-reactivity was linked to the position of carrier protein linkage in THC immunoassays. The low molecular weight compounds, cannabinoid haptens, require linkage to carrier proteins to produce immunogens from antibody generation. It was suggested that EFV metabolites interfered with carrier protein linkage with antibodies designed to C1-, C2-, and C5- and caused for a false-positive THC results.

INTERFERENCES WITH COCAINE IMMUNOASSAY

The immunoassay for cocaine detection in drugs-of-abuse screen is designed to detect benzoylecgonine, an inactive metabolite of cocaine. The assay is unable to detect passive inhalation of cocaine; however, false-positive interferences are seen with tea products such as yerba mate or tea coming from Latin Americas. Different types of teas from Latin America are derived from coca plant, and have trace amounts of coca that may be mistaken for cocaine abuse. Five healthy volunteers (4 male and 1 female) consumed coca tealeaf steeped in 8 ounces of water for 15 minutes [46]. Participant 1 consumed 1 cup of 8 ounces tea, the second consumed 2 cups, the third consumed 3 cups, the fourth consumed 4 cups, and the fifth consumed 5 cups of coca tea. Urine from each participant was collected before the consumption and at 2, 12, 24, and 36 h after drinking the tea. Urine collected was analyzed for benzoylecgonine using Abbott AxSym System via immunoassay with a cutoff of 300 ng/mL. Each participant had a positive cocaine assay 2 hours after consuming tea. For three of the participants, the immunoassay for cocaine was positive even after 36 hours. The mean concentration of benzoylecgonine for all the participants after the consumption was 1777 ng/mL. Since yerba mate and tea from Latin American are popular in the United States, patients should abstain from drinking products with coca plants a few days before urine drug testing.

Case Study

A 48-year-old man was hospitalized for a pancreatic carcinoma that was confirmed via biopsy of the pancreatic mass. The patient required intensive pain control, and prior to the hospitalization, he was taking high dosage of long- and short-acting narcotics, OxyContin 40 mg orally twice daily and hydromorphone 6 mg orally every 4 hours as needed for his chronic neck and back pain. Due to his illicit drug abuse in the past (cocaine abuse), urine drug screen was performed. The patient's urine drug test came out positive for both cocaine and oxycodone. The patient was recommended for celiac plexus block for his pain conditions, but the patient refused to take this pain management and the clinician did not think he was suitable for intrathecal pump (ITP) because of his urine drug screen and for the high risk of noncompliance. However, after analyzing the same sample with GC-MS, the results were negative for cocaine and its metabolites. The immunoassay false-positive result would have been detrimental for the patient's pain management [47].

FALSE-POSITIVE TEST RESULTS IN OTHER DRUGS-OF-ABUSE IMMUNOASSAYS

There are other drugs that may cause false-positive test results in various immunoassays used for screening of the presence of abused drugs in urine specimens. Dextromethorphan is an antitussive agent which is found in many OTC cough and cold medications. Dextromethorphan is also abused in high dosages mostly by young adults. Ingesting high amounts of dextromethorphan (over 30 mg) may result in positive false-positive test results with opiate and phencyclidine (PCP) immunoassays. In another report, the authors observed three false-positive phencyclidine tests in pediatric urine specimens using an onsite testing device (Instant-View multitest drugs-of-abuse panel; Alka Scientific, Designs, Poway, CA). The authors concluded that false-positive PCP tests were due to the cross-reactivities from ibuprofen, metamizol, dextromethorphan, and their metabolites with the PCP assay. In addition, pheniramine and methylphenidate also produced a false-positive result with the PCP immunoassay screen [48].

Studies indicate that nonsteroidal antiinflammatory drugs may produce false-positive results in immunoassays screening tests for the presence of drugs of abuse in urine specimens. Rollins et al. studied the effect of such drugs on

TABLE 10.4 Interferences in Various Immunoassays Used for Drugs-of-Abuse Testing

Immunoassay	Interfering Drugs
Benzodiazepines	Oxaprozin, sertraline
Opiates	Diphenhydramine, rifampin, dextromethorphan, verapamil, fluoroquinolones, quinine
Phencyclidine	Dextromethorphan, diphenhydramine, ibuprofen, imipramine, ketamine, meperidine, thioridazine, tramadol, venlafaxine
Tetrahydrocannabinol	Nonsteroidal antiinflammatory drugs, pantoprazole, EFV
Methadone	Diphenhydramine

drugs-of-abuse screening using 60 volunteers and 42 patients taking ibuprofen, naproxen, or fenoprofen. Out of 510 urine specimens collected from 102 individuals, only two specimens showed false-positive test for cannabinoids using enzyme-mediated immunoassay. There was one individual who ingested 1200 mg ibuprofen in three divided dosages and the other patient took naproxen regularly. Two urine specimens were false positive for barbiturates using the FPIA; one patient took ibuprofen and the other patient, naproxen. The authors concluded that there is a small likelihood of false-positive tests results in drugs-of-abuse screening assays in individuals taking nonsteroidal antiinflammatory drugs [48]. Joseph et al. studied 14 nonsteroidal antiinflammatory drugs for potential interference with EMIT and FPIA for various drugs of abuse and observed that tolmetin interferes with EMIT immunoassays at high concentrations (1800 μg/mL and higher) because of high molar absorptivity at 340 nm, the wavelength used for detection in the EMIT technology. Samples containing cannabinoid and benzoylecgonine tested negative in the presence of tolmetin but there was no effect on the FPIA assay because the detection wavelength was 525 nm [49].

Although most reports in the medical literature describe false-positive results in drugs-of-abuse testing due to the presence of OTC medications in urine specimens, Brunk et al. reported a false-negative GC/MS confirmation of marijuana metabolite as methyl derivative due to interference of ibuprofen with the methylation step. The urine specimen tested positive for marijuana metabolite using the EMIT d.a.u. assay but showed negative GC/MS confirmation result. When the specimen was analyzed by thin layer chromatography (TLC), the presence of marijuana metabolite was also observed [50]. Table 10.4 lists the drugs that interfere with various drugs-of-abuse immunoassays.

Case Study

A 9-year-old boy presented to pediatric ED with complaints of altered level of consciousness. He was receiving diphenhydramine for nasal congestion and sleeping difficulty in the previous week before presenting to the ED. The parents denied that he took the medicine on the evening of going to the hospital. His physical examination showed dilated pupils, dry mucous membrane, confusion and occasional hallucination. His urine drug screen showed positive for methadone and serum drug screen was positive for tricyclic antidepressant. The patient denied taking any medication other than diphenhydramine and was admitted to the hospital. His symptoms resolved within 24 h. The authors attributed positive tricyclic antidepressant serum screen due to the interference of diphenhydramine with the assay consistent with the literature reports, but only interference of diphenhydramine with EMIT opiate assay has been previously reported and the assay used (urine drug screen of the boy) was performed by a point-of-care testing device (One-Step, Multi-Drug Multi-Line Screen, ACON Laboratories, San Diego, CA). The authors then investigated the possible interference of diphenhydramine with this assay, and when the authors supplemented drug-free urine specimen with 100 ug/mL of diphenhydramine, the opiate test was positive for opiate. Therefore, the false-positive test result was due to the use of diphenhydramine [51].

CONCLUSIONS

Although false-positive test results are routinely encountered in urine drug testing, such methods are widely used in clinical laboratories for initial screening for the presence of drugs, in urine or serum/plasma in patients admitted to emergency rooms and other clinical settings to support patient care. Melanson et al. commented that directors should be aware of the characteristics of their laboratories' assays and should be able to communicate with ordering physicians regarding proper interpretation of qualitative results. In addition, manufacturer's claim must be interpreted with caution and if possible should also be validated using the patient population of the hospital [52]. However, false-positive results can be avoided in drugs-of-abuse testing by using more specific analytical methods such as chromatography combined with mass spectrometry. Eichhorst et al. described an improved cost-effective high-throughput method for drugs-of-abuse screening by tandem mass spectrometry replacing immunoassay screening [53] or another approach would be to create a large drug testing panel by utilizing mass spectrometry, to replace poor performing drug immunoassays, in combination with drug immunoassays that have a low false-positive rate (<5%) [54].

REFERENCES

[1] Chen WJ, Fang CC, Shyu RS, Lin KC. Underreporting of illicit drug use by patients at emergency departments as revealed by two-tiered urinalysis. Addic Behav 2006;31:2304–8.

[2] Johnson-Davis KL, Sadler AJ, Genzen JR. A retrospective analysis of urine drugs of abuse immunoassay true positive rates at a national reference laboratory. J Anal Toxicol 2016;40:97–107.

[3] Standridge JB, Adams SM, Zotos AP. Urine drug screening: A valuable office procedure. Am Fam Physician 2010;81:635–40.

[4] Dietzen DJ, Ecos K, Friedman D, Beason S. Positive predictive values of abused drug immunoassays on the Beckman synchron in a veteran population. J Anal Toxicol 2001;25:174–8.

[5] Hsu J, Liu C, Liu CP, Tsay WI, Li JH, Lin DL, et al. Performance characteristics of selected immunoassays for preliminary test of 3,4-methylenedioxymethamphetamine, methamphetamine, and related drugs in urine specimens. J Anal Toxicol 2003;27:471–8.

[6] Kunsman GW, Manno JE, Cockerham KR, Manno BR. Application of the syva emit and abbott tdx amphetamine immunoassays to the detection of 3,4-methylene-dioxymethamphetamine (mdma) and 3,4-methylene-dioxyethamphetamine (mdea) in urine. J Anal Toxicol 1990;14:149–53.

[7] Casey ER, Scott MG, Tang S, Mullins ME. Frequency of false positive amphetamine screens due to bupropion using the syva emit II immunoassay. J Med Toxicol 2011;7:105–8.

[8] Marin SJ, Doyle K, Chang A, Concheiro-Guisan M, Huestis MA, Johnson-Davis KL. One hundred false-positive amphetamine specimens characterized by liquid chromatography time-of-flight mass spectrometry. J Anal Toxicol 2016;40:37–42.

[9] Yee LM, Wu D. False-positive amphetamine toxicology screen results in three pregnant women using labetalol. Obstet Gynecol 2011;117:503–6.

[10] Saitman A, Park HD, Fitzgerald RL. False-positive interferences of common urine drug screen immunoassays: A review. J Anal Toxicol 2014;38:387–96.

[11] Fucci N. False positive results for amphetamine in urine of a patient with diabetes mellitus. Forensic Sci Int 2012;223:e60.

[12] Butler D, Guilbault GG. Analytical techniques for ecstasy. Analytical Letters 2004;37:2003–30.

[13] Wu AH, McKay C, Broussard LA, Hoffman RS, Kwong TC, Moyer TP, et al. National academy of clinical biochemistry laboratory medicine practice guidelines: Recommendations for the use of laboratory tests to support poisoned patients who present to the emergency department. Clin Chem 2003;49:357–79.

[14] Ka M. Amphetamines/sympathomimetic amines. In: Levine B, editor. Principles of forensic toxicology. Washington DC: AACC Press; 2003. p. 341–8.

[15] Poklis A. Unavailability of drug metabolite reference material to evaluate false-positive results for monoclonal emit-d.A.U. Assay of amphetamine. Clin Chem 1992;38:2560.

[16] Williams RH, Erickson T, Broussard LA. Evaluating sympathomimetic intoxication in an emergency setting. Lab Medicine 2000;31:497–508.

[17] Singh J. Reagent lot-to-lot variability in sensitivity for amphetamine with the syva emit ii monoclonal amphetamine/methamphetamine assay. J Anal Toxicol 1997;21:174–5.

[18] Hornbeck CL, Carrig JE, Czarny RJ. Detection of a GC/MS artifact peak as methamphetamine. J Anal Toxicol 1993;17:257–63.

[19] Dasgupta A, Gardner C. Distinguishing amphetamine and methamphetamine from other interfering sympathomimetic amines after various fluoro derivatization and analysis by gas chromatography-chemical ionization mass spectrometry. J Forensic Sci 1995;40:1077–81.

[20] Ojanpera IA, Heikman PK, Rasanen IJ. Urine analysis of 3,4-methylenedioxypyrovalerone in opioid-dependent patients by gas chromatography-mass spectrometry. Ther Drug Monit 2011;33:257–63.

[21] Stout PR, Klette KL, Horn CK. Evaluation of ephedrine, pseudoephedrine and phenylpropanolamine concentrations in human urine samples and a comparison of the specificity of dri amphetamines and abuscreen online (kims) amphetamines screening immunoassays. J Forensic Sci 2004;49:160–4.

[22] Solomon MD, Wright JA. False-positive for (+)-methamphetamine. Clin Chem 1977;23:1504.

[23] Poklis A, Jortani SA, Brown CS, Crooks CR. Response of the emit ii amphetamine/methamphetamine assay to specimens collected following use of vicks inhalers. J Anal Toxicol 1993;17:284–6.

[24] Poklis A, Moore KA. Stereoselectivity of the tdxadx/flx amphetamine/methamphetamine ii amphetamine/methamphetamine immunoassay-response of urine specimens following nasal inhaler use. J Toxicol Clin Toxicol 1995;33:35–41.

[25] Wyman JF, Cody JT. Determination of l-methamphetamine: A case history. J Anal Toxicol 2005;29:759–61.

[26] FDA News P06-07, January 13, 2006.

[27] Jemionek JBT, Jacobs A, Holler J, Magluli J, Dunkley C. Five cases of d-amphetamine positive urines resulting from ingestion of "Brazilian nutritional supplements" containing fenproporex have been reported. ToxTalk SOFT newsletter 2006;30:11.
[28] Vorce SP, Holler JM, Cawrse BM, Magluilo Jr. J. Dimethylamylamine: A drug causing positive immunoassay results for amphetamines. J Anal Toxicol 2011;35:183—7.
[29] Lattman Peter, Singer Natasha. Army studies workout supplement after deaths. New York Times 2012;February 2:B1.
[30] Nguyen MH, Ormiston T, Kurani S, Woo DK. Amphetamine lacing of an internet-marketed neutraceutical. Mayo Clin Proc 2006;81:1627—9.
[31] Perez-Downes J, Hritani A, Baldeo C, Antoun P. Amphetamine containing dietary supplements and acute myocardial infarction. Case Rep Cardiol 2016;2016:6404856.
[32] Eliason MJ, Eichner A, Cancio A, Bestervelt L, Adams BD, Deuster PA. Case reports: Death of active duty soldiers following ingestion of dietary supplements containing 1,3-dimethylamylamine (dmaa). Mil Med 2012;177:1455—9.
[33] Bailes JE, Cantu RC, Day AL. The neurosurgeon in sport: Awareness of the risks of heatstroke and dietary supplements. Neurosurgery 2002;51:283—6 discussion6-8.
[34] Poklis A, Hall KV, Still J, Binder SR. Ranitidine interference with the monoclonal emit d.A.U. Amphetamine/methamphetamine immunoassay. J Anal Toxicol 1991;15:101—3.
[35] Liu L, Wheeler SE, Rymer JA, Lower D, Zona J, Peck Palmer OM, et al. Ranitidine interference with standard amphetamine immunoassay. Clin Chim Acta 2015;438:307—8.
[36] Logan BK, Costantino AG, Rieders EF, Sanders D. Trazodone, meta-chlorophenylpiperazine (an hallucinogenic drug and trazodone metabolite), and the hallucinogen trifluoromethylphenylpiperazine cross-react with the emit(r)ii ecstasy immunoassay in urine. J Anal Toxicol 2010;34:587—9.
[37] Fenderson JL, Stratton AN, Domingo JS, Matthews GO, Tan CD. Amphetamine positive urine toxicology screen secondary to atomoxetine. Case Rep Psychiatry 2013;2013:381261.
[38] Zacher JL, Givone DM. False-positive urine opiate screening associated with fluoroquinolone use. Ann Pharmacother 2004;38:1525—8.
[39] Baden LR, Horowitz G, Jacoby H, Eliopoulos GM. Quinolones and false-positive urine screening for opiates by immunoassay technology. JAMA 2001;286:3115—19.
[40] de Paula M, Saiz LC, Gonzalez-Revalderia J, Pascual T, Alberola C, Miravalles E. Rifampicin causes false-positive immunoassay results for urine opiates. Clin Chem Lab Med 1998;36:241—3.
[41] Straseski JA, Stolbach A, Clarke W. Opiate-positive immunoassay screen in a pediatric patient. Clin Chem 2010;56:1220—3.
[42] Jenkins AJ, Poirier III. JG, Juhascik MP. Cross-reactivity of naloxone with oxycodone immunoassays: Implications for individuals taking suboxone. Clin Chem 2009;55:1434—6.
[43] Straley CM, Cecil EJ, Herriman MP. Gatifloxacin interference with opiate urine drug screen. Pharmacotherapy 2006;26:435—9.
[44] Rossi S, Yaksh T, Bentley H, van den Brande G, Grant I, Ellis R. Characterization of interference with 6 commercial delta9-tetrahydrocannabinol immunoassays by efavirenz (glucuronide) in urine. Clin Chem 2006;52:896—7.
[45] Oosthuizen NM, Laurens JB. Efavirenz interference in urine screening immunoassays for tetrahydrocannabinol. Ann Clin Biochem 2012;49:194—6.
[46] Mazor SS, Mycyk MB, Wills BK, Brace LD, Gussow L, Erickson T. Coca tea consumption causes positive urine cocaine assay. Eur J Emerg Med 2006;13:340—1.
[47] Kim JA, Ptolemy AS, Melanson SE, Janfaza DR, Ross EL. The clinical impact of a false-positive urine cocaine screening result on a patient's pain management. Pain medicine 2015;16:1073—6.
[48] Marchei E, Pellegrini M, Pichini S, Martin I, Garcia-Algar O, Vall O. Are false-positive phencyclidine immunoassay instant-view multi-test results caused by overdose concentrations of ibuprofen, metamizol, and dextromethorphan?. Ther Drug Monit 2007;29:671—3.
[49] Joseph R, Dickerson S, Willis R, Frankenfield D, Cone EJ, Smith DR. Interference by nonsteroidal anti-inflammatory drugs in emit and tdx assays for drugs of abuse. J Anal Toxicol 1995;19:13—17.
[50] Brunk SD. False negative GC/MS assay for carboxy thc due to ibuprofen interference. J Anal Toxicol 1988;12:290—1.
[51] Rogers SC, Pruitt CW, Crouch DJ, Caravati EM. Rapid urine drug screens: Diphenhydramine and methadone cross-reactivity. Pediatr Emerg Care 2010;26:665—6.
[52] Melanson SE, Baskin L, Magnani B, Kwong TC, Dizon A, Wu AH. Interpretation and utility of drug of abuse immunoassays: Lessons from laboratory drug testing surveys. Arch Pathol Lab Med 2010;134:735—9.
[53] Eichhorst JC, Etter ML, Rousseaux N, Lehotay DC. Drugs of abuse testing by tandem mass spectrometry: A rapid, simple method to replace immunoassays. Clin Biochem 2009;42:1531—42.
[54] McMillin GA, Marin SJ, Johnson-Davis KL, Lawlor BG, Strathmann FG. A hybrid approach to urine drug testing using high-resolution mass spectrometry and select immunoassays. Am J Clin Pathol 2015;143:234—40.

Chapter 11

Point of Care Devices for Drugs of Abuse Testing: Limitations and Pitfalls

Veronica Luzzi
Providence Saint Joseph's Health and Services, Providence Regional Core Laboratory, Portland, OR, United States

INTRODUCTION

In the last 20 years, point of care testing (POCT), also known as on-site or near-patient testing, has evolved into a new discipline within laboratory medicine. Some of the contributing factors in this evolution have been easy and fast access to tests results, ease-of-use, cost, and advances on instrument connectivity. Drug of abuse testing has been performed using POCT devices for a while. Since the first waived POCT device was approved by the US Food and Drug Administration (FDA), there has been a steady increase in the number of devices being approved [1]. This data is summarized in Fig. 11.1. However there are controversies about testing for drugs of abuse using POCT devices depending on the assay performed, where and who uses the device, and whether or not using POCT contributes to increase cost due to inadequate test utilization. In this chapter, several of the factors contributing to the expansion of POCT for drugs of abuse and the controversies arising from their use will be covered. Whether POCT devices are used in a clinical laboratory, emergency department, or a pain management clinic, understanding their performance and the nuances that can lead to serious errors is paramount for successful patient management.

DESIGN OF POCT DEVICES

Current devices use a solid-phase competitive immunoassay in a lateral flow format. They are often available as strips, cup, or a cassette that can be read manually or using a reader [2,3]. A diagram of a lateral flow immunoassay strip is shown in Fig. 11.2. Close to the sample application line, there are a labeled conjugate and a labeled control antigen pads. Up in the strip, further away from the application line, there are a test and control lines. The test line contains detection antibodies against the target drug or labeled conjugate. The control line contains antibodies against the control antigens. As the sample migrates from the application line to the labeled conjugate and control antigen pads, it carries its contents to the test and control lines. If the concentration of the target drug in the sample is below the cutoff, a color line will appear because the labeled conjugate will bind to the immobilized detection antibodies. On the contrary, if the target drug is above the cutoff, it will compete with the labeled conjugate and inhibit color formation (Fig. 11.3). Result from strips, cups, and cassettes can be evaluated visually by inspecting the presence or absence of color. Testing can be done for a single drug or as a panel. Some strips and cassettes can also be read by semiautomated or automated end point [4] readers. In general, all these devices are easy to use, easy to read, and many times, Clinical Laboratory Improvement Amendments (CLIA) waived. Some users have found difficulty in reading and interpreting competitive immunoassays in the POCT settings and reported unacceptable to use these devices at their institutions [5].

RELIABILITY AND EFFECTIVENESS OF POCT DEVICES

The reliability and effectiveness of drugs of abuse POCT devices has been challenged by several studies [4–11]. The main issues described in the literature are due to the specificity of the antibodies used in manufacturing POCT devices, the inconsistency of cutoffs among reagent lots of the same device, the nomenclature used to name drugs or drug classes, and the difficulty of interpreting the device reporting lines depending on the read time used on the device [12].

142 Critical Issues in Alcohol and Drugs of Abuse Testing

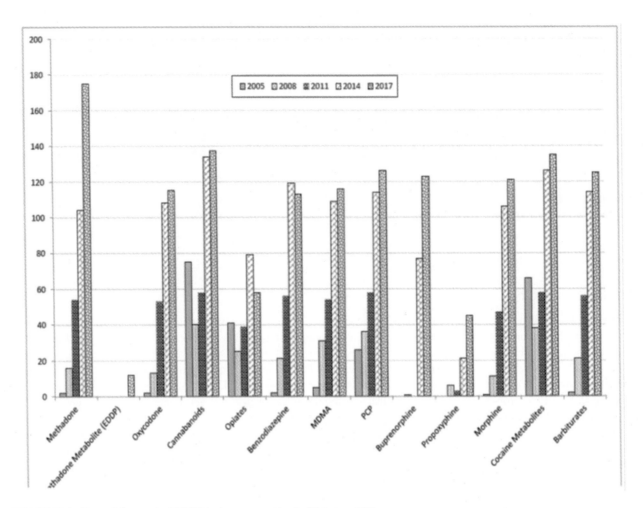

FIGURE 11.1 Drug of abuse waived POCT devices approved by the FDA since 2005.

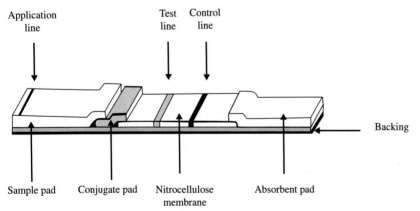

FIGURE 11.2 Diagram of a lateral flow immunoassay strip.

Accuracy issues reported due to antibody specificity is not inherent to POCT devices but rather a characteristic of immunoassays. Results obtained with immunoassays whether they are from a laboratory instrument or a POCT device; often need confirmation with a more specific chemical method. Good laboratory practice recommends that before a patient management decision is made, confirmation of the presence of the drug should be performed [13]. Immunoassays in general use the binding properties of antibodies to capture or detect a molecule, but because antibodies used in clinical assays may be different, drug of abuse POCT devices may cross-react or not with a drug.

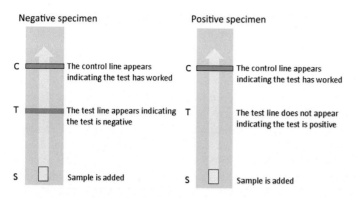

FIGURE 11.3 Diagram of a negative and a positive reaction in a lateral flow immunoassay strip.

In 2002, George and Braithwaite [14] eloquently described some of the potential risks of using POCT devices to measure drugs of abuse. Some of the issues they described then still persist today despite the tremendous growth we have seen in this industry. One of the first methods described in the literature in 1988 used paper chromatography to detect cocaine. It was reported that, three trained individuals read 100 urine tests results, and only correctly identified the presence of cocaine 50% of the time [7]. Several studies performed in the 1990 listed different sensitivity and specificity for POCT drug of abuse assays when compared with other laboratory methods such as Syva EMIT, or gas chromatography/mass spectrometry [8,14]. In all these studies the performance of the assays was highly impacted by factors such as the use of incorrect nomenclature, turbidity of the specimen, reading time, testing settings, and personnel training. For example, a cartridge using latex agglutination inhibition method intended to detect morphine, and also detected other opiates such as codeine and dihydrocodeine [4]. A competitive immunoassay for cannabinoids (accuPINCH) provided negative result in cannabinoid positive turbid specimens. Moreover, if reading time is longer than recommended 5 min, it may increase the rate of false positive test results [7]. Testing settings may also impact accuracy if the personnel reading the device were multitasking, trying to perform patient care at the same time. In one particular report, an essential buffer bead used in a device was prone to loss if personnel not properly trained used the device [8].

Evaluation studies performed on a patient care setting where personnel are not formally trained in laboratory medicine are very valuable. In 2002, Mastrovitch [15] presented evidence that a POCT device used in the emergency department of a tertiary care in the urban medical center (On Trak) performed equivalent to a screening laboratory device used in the central laboratory (Triage). The authors evaluated 170 urine specimens from adult patients (including 34 pediatric patients) analyzed by seven trained nonlaboratory technicians. The drugs evaluated were classified in four groups: cocaine, marijuana, opiates, and amphetamines. The concordance was 97% and 99% for cocaine, and the rest of the drugs respectively. In this study, no confirmation testing was performed. The authors claimed a faster turnaround time and did not elaborate whether or not they encountered specificity issues. A false negative or false positive outcome in a pediatric patient may have implications beyond clinical toxicology, but this fact was not addressed [16,17].

In 2012, a team composed of emergency department and pharmacy personnel in the Netherlands evaluated three POCT devices against a laboratory-based screening method [18]. The POCT devices investigated were TestCard9 (Varina, Middleburg, The Netherlands), Syva RapidTest (SRT) d.a.u. 10 (Dade Behring; Leusden, The Netherlands), and Triage TOX Drug Screen (Biosite; Bunnik, The Netherlands) against the Syva EMIT II immunoassay and chromatographic confirmation. The team tested 80 urine specimens and learned that the sensitivity and specificity for all the devices were at least 93% for amphetamines, cocaine, cannabis, benzodiazepines, and opiates. There were exceptions: for cannabis, the TestCard device sensitivity was 88%, for benzodiazepines, the sensitivities for the SRT and the TestCard were 88% and 80%, respectively. The team concluded favoring the Triage device because it was easy to use and had the option to use an automated reader rather than subjective interpretation used in a manual reading mode.

False negative and false positive findings have been previously described in the literature when drugs of abuse are measured using immunoassays in automated analyzers. Several reviews list the names of different substances interfering with immunoassays, and even the concentrations at which these substances may interfere [6,19,20]. Since POCT devices use immunoassay methodology and the format of the assay in the device may be similar to that of the laboratory immunoassay, interpreting results when using POCT devices for drugs of abuse is just as important as when immunoassays are used in the laboratory. Providing interference information to ordering providers is important for understanding what may be the cause of discrepancies between expected results and findings. The National Academy of Clinical Biochemistry Laboratory Medicine Practice Guidelines: Recommendations for the Use of Laboratory Tests to Support Poisoned Patients Who Present to the Emergency Department, published in 2003, emphasized the importance of providing assay limitation information to ordering providers [21]. Ideally, this information is also available when drugs of abuse are measured at the POCT. The untrained user may contribute to inaccurate results, and in particular this issue becomes very evident when the personnel using the POCT devices are also tasked with other nontechnical duties such as patient care. Training personnel and validating the devices in similar setting to those that are going to be routinely used is very important and have been cited in the literature as contributing to decrease errors [22].

POCT devices report qualitative results, positive or negative, based on a preestablished cutoff (Table 11.1). In comparison with automated chemistry assays, the cutoffs used in POCT devices are very similar. However, during evaluation of a POCT device, the consistency of the cutoff among lot numbers of reagents should be determined. In a recent evaluation, a reference laboratory learned that a commonly used POCT device detects drugs at a different cutoff than those claimed by the manufacturer. For some drugs, the device detected concentrations 20 times smaller than the claimed cutoff, while for other drugs, the amount of drug to be measured needed to be 5%–10% higher than the cutoff to be detected [23]. The study performed as part of the evaluation of this device also showed that the detection of a positive specimen changed to negative if the reading time was extended beyond recommended time. For methadone, extending the reading time to 15 min could possibly lead to false negative results.

Knowing the true positive and false positive rate of the screening tests used in the laboratory or elsewhere is important to understand the limitations of the devices. There are studies describing this rate in POCT devices. One study performed by a reference laboratory in the United States analyzed 8825 urine specimens tested by immunoassay screen (Syva and Microgenics) in a Beckman AU 5810. Positive results were later confirmed using a higher chemical method (gas or liquid chromatography with mass spectrometry or tandem mass spectrometry detection, or flame ionization detection). This group reported that the Beckman AU 5810 immunoassay screening for benzodiazepines, opiates, oxycodone, tetrahydrocannabinol, and amphetamines had a positive rate higher than 5%. 3,4-Methylenedioxymethamphetamine, phencyclidine, and propoxyphene had the lowest number of true positives [24].

TABLE 11.1 Example of Urine Cutoff Used in a POCT Device and an Automated Immunoassay, and Its Corresponding Confirmation Cutoff

Drug or Drug Class	SAMSHA Cutoff	Chemistry Automated Analyzer Example	POCT Detection Devices Cutoff		Confirmation Cutoff (ng/mL)
		AU 5400 EMIT Reagents	NexScreen	RapidCup	GC/MS or LC–MS/MS
D-Amphetamine/D-methamphetamine	500	500	1000	500	200
Barbiturates	200/300	200	300	300	50
Benzodiazepines	200	200	300	300	5–20[a]
Cocaine metabolite (benzoylecgonine)	300	150	300	150	50
Methadone	300	150	300	300	10
Buprenorphine glucuronide	n/a	n/a	n/a	10	2, 5, or 100[a]
Opiates	2000/300	300	300	300	10 or 20[a]
Oxycodone	100	100	100	100	50
Phencyclidine	25	25	25	25	10
Marijuana metabolite (11-nor-Δ^9-THC-COOH)	50	50	50	50	5
MDMA	500	500	500	500	200
Tricyclic antidepressants	500	500	1000	1000	100, or 200[a]

GC, gas chromatography; *LC*, liquid chromatography; *MDMA*, 3,4-methylenedioxymethamphetamine; *MS*, mass spectrometry; *THC*, tetrahydrocannabinol.
[a]It depends on the drug.

Cutoff concentrations are designed to mimic the Substance Abuse and Mental Health Services Administration (SAMHSA). However, in a clinical setting, these cutoff concentrations may be different than cutoff concentrations that are used in forensic drug testing or for preemployment screening. For some patient populations such as pediatric patients, it may be more relevant to know whether or not the substance is present, even if it is below the federally mandated cutoff [25]. The imprecision below the cutoff concentration has been demonstrated to be within 20%. Using a cutoff concentration based on the precision of an assay rather than a cutoff concentration artificially imposed may be more relevant in the pediatric population.

Oral Fluid Testing Using POCT Devices

When performing drugs of abuse testing using POCT devices, body fluids other than urine may be an attractive alternative. For example, oral fluid can be easily available and the presence of a drug may indicate recent drug use [26]. Toennes et al. [27] evaluated the correlation between serum, urine, and oral fluid concentration in drivers suspected of drunk driving. Results from the study demonstrated that impairment symptoms correlated better with alcohol serum and oral fluid concentration than urine. Some countries have incorporated POCT devices that use oral fluid to test for suspected drunk drivers [28–32]. In 2015, SAMHSA, an agency under the Department of Human and Health Services (HHS) has defined oral fluid as an acceptable specimen type to be used in the workplace drug testing program [33]. The HHS has also defined cutoffs for screening and confirmation that are much lower than those of urine specimens (Table 11.2). Alternate methodologies to immunoassays are included as part of the screening in order to reach these lower thresholds. But only HHS-certified laboratories are allowed to test oral fluid specimens for federal agency workplace drug testing programs. Studies have shown that POCT devices are not adequate to meet the HHS cutoff requirements.

There are limitations on using oral fluid in drug of abuse testing when using POCT devices. Oral fluid is usually collected on a device that contains a buffer as preservative. The ratio between the buffer and the oral fluid is not necessarily constant and may preclude from accurate drug concentration measurements [34]. The distribution of a drug in the body will vary and cross-studies comparing different body fluids are needed to interpret oral fluid results accordingly [35]. When evaluating whether or not using oral fluids as an acceptable specimen, rigorous studies need to be performed to demonstrate contaminants, environmental conditions, or collection methods do not alter the drug concentration. Crouch [36] has described the use of oral fluid to replace urine or blood specimens and concluded that collecting a valid and representative specimen is intimately related to the collection device. A more recent systematic meta-analysis of previously published

TABLE 11.2 SAMSHA Oral Fluid Screening and Confirmation Cutoffs

Drug Tested	Screening Cutoff (ng/mL)	Confirmation Cutoff (ng/mL)
Δ^9-Tetrahydrocannabinol	4	2
Cocaine/benzoylecgonine	15	8
Codeine	30	15
Morphine	30	15
Hydrocodone	30	15
Hydromorphone	30	15
Oxycodone	30	15
Oxymorphone	30	15
6-Acetylmorphine	3	2
Phencyclidine	3	2
Amphetamine/methamphetamine	25	15
MDMA/MDA/MDEA	15	15

MDA, 3,4-Methylenedioxyamphetamine; *MDEA*, 3,4-methylenedioxy-*N*-ethylamphetamine; *MDMA*, 3,4-methylenedioxymethamphetamine.

studies evaluating the reliability of detecting five commonly monitored drugs in oral fluid using POCT devices also warns about the variability in specificity and sensitivity of POCT devices. The authors also point out that a standardized study comparing devices, specimen types, and cutoffs may be needed in order to accurately appreciate the differences in devices, when oral fluid is used [37].

GUIDELINES FOR USING POCT DEVICES

The Executive Summary of the National Academy of Clinical Biochemistry: Evidence Based Practice for Point of Care, published in 2004 [13], strongly recommends that POCT device training includes quality monitoring and limitations. Users must be familiar with the devices analytical specifications, and the common interferences or other limitations to prevent misinterpretation errors. If a result from a POCT device is going to trigger a penal or legal decision, confirmation of the results must be mandatory.

The Clinical and Laboratory Standard Institute (CLSI) created several guidelines related to POCT [38–41]. These guidelines describe connectivity for engineers and users (POCT01 and POCT02, respectively), approaches and quality practices to reducing errors at the POCT (POCT07-A, POCT08-A), and a selection criteria for POCT devices (POCT09-A). Guidelines POCT01 and POCT02 were created to address connectivity standardization, and follow the Connectivity Industry Consortium vision to "develop, pilot, and transfer the foundation for a set of seamless 'plug-and-play' POCT communication standard"... These guidelines have helped curved the development of better instrument connectivity, and therefore decreased errors due to manual data entry. However, when POCT devices are used outside a data framework of a university or hospital system, the compliance with these guidelines may not take place. In particular, for pain management monitoring, where the use of POCT devices has been implemented outside the clinical laboratory oversight, data entry errors may still need addressing. Immediate care and primary care clinics where manual entry is often performed may have similar data entry issues. This type of errors will continue to exist unless incorporation of results in the patient medical record is completely automated. Some devices created for monitoring glucose concentration are capable of connecting to a software application via wireless communication. The clinical need of adopting wireless technology for drugs of abuse testing in the POCT setting may not necessarily exist. However, the development of a process free of manual data entry for those locations that are not included in the centralized laboratory network will improve data flow.

Guidelines POCT07-A, POCT08-A, and POCT09-A are all related to error reduction but cover the topic from different perspective. POCT07-A provides information on quality indicators for analytical processes and emphasizes the importance of creating indicators to identify common causes of errors. It also provides guidance for creating error tracking processes. POCT08-A translates good laboratory practices into terms usable by testing personnel that is not trained in the laboratory. Sometimes personnel using POCT devices consist of medical personnel that are required to attend to patients at the same time they perform testing. These are not ideal circumstances and avoiding multitasking may be the only way to prevent common operator errors. Finally, POCT09-A describes important criteria for the selection of POCT devices based on the clinical setting and clinical needs. It also provides information on regulatory and accreditation requirements referring to POCT devices and their use.

In addition to the CLSI guidelines described earlier, the Center for Medicare Services has recently approved a new plan for laboratories aiming at assessing risk of failure and allowing laboratories to customize a quality control plan according to the method and its use, the environment, and personnel competency. This plan is called "individualized quality control plan" or IQCP.

One of the reasons why the use of POCT devices has grown steadily in the last few years is because they appear to provide fast and accurate results without the hassle of sending the specimens to a central laboratory. The perception is that laboratory results correlate really well with POCT device results. As described earlier, there are serious caveats with this perception. Laboratory personnel can be of valuable assistance in implementing, training, and monitoring POCT devices. Laboratory personnel expertise as part of an integrated health care team has value added to the use of POCT devices. POCT devices CLIA defined complexity are waived or non-waived. In general, the nonwaived devices are moderately complex. A waived or nonwaived device may become high complexity if manufacturer's instructions are not followed.

Waived devices are believed to be more reliable and less error prone than high complexity laboratory assays. However, that is hardly the case. Evaluating and monitoring the performance of a waived device is important to understand the limitations of the device and the concomitant implications of those limitations [42].

CONCLUSIONS

Laboratory personnel should be involved in the early adoption of a POCT solution. The activities where the laboratory personnel should be involved are initial evaluation of the analytical specifications of the POCT device, implementing policies, procedures, and training manuals to minimize misinterpretation and operator errors; training and competency assessments to meet regulatory requirements; monitoring device and personnel performance using quality control materials; identifying errors and creating process improvements; and creating an IQCP plan to assess risk and prevent errors. However, it is important to incorporate the routine user of the device in all these activities to be aware of workflows, cognitive errors, or unintended misuse of the device in use [22].

REFERENCES

[1] FDA. CLIA—Tests Waived by FDA from January 2000 to Present 2017 Available from: https://www.accessdata.fda.gov/scripts/cdrh/cfdocs/cfClia/testswaived.cfm.

[2] Weiss A. Concurrent engineering for lateral-flow diagnostics. IVD Technology [Internet]. 1999 January 29, 2018.

[3] Luppa PB, Bietenbeck A, Beaudoin C, Giannetti A. Clinically relevant analytical techniques, organizational concepts for application and future perspectives of point-of-care testing. Biotechnol Adv 2016;34(3):139–60.

[4] George S, Braithwaite RA. A preliminary evaluation of five rapid detection kits for on site drugs of abuse screening. Addiction 1995;90(2):227–32.

[5] Matthews JC, Wassif WS. Potential risk of patient misclassification using a point-of-care testing kit for urine drugs of abuse. Br J Biomed Sci 2010;67(4):218–20.

[6] Reisfield GM, Goldberger BA, Bertholf RL. False-positive" and "false-negative" test results in clinical urine drug testing. Bioanalysis 2009;1(5):937–52.

[7] Jenkins AJ, Darwin WD, Huestis MA, Cone EJ, Mitchell JM. Validity testing of the accuPINCH THC test. J Anal Toxicol 1995;19(1):5–12.

[8] Poklis A, O'Neal CL. Potential for false-positive results by the TRIAGE panel of drugs-of-abuse immunoassay. J Anal Toxicol 1996;20(3):209–10.

[9] Brahm NC, Yeager LL, Fox MD, Farmer KC, Palmer TA. Commonly prescribed medications and potential false-positive urine drug screens. Am J Health Syst Pharm 2010;67(16):1344–50.

[10] Smith MP, Bluth MH. Common interferences in drug testing. Clin Lab Med 2016;36(4):663–71.

[11] O'Kane MJ, McManus P, McGowan N, Lynch PL. Quality error rates in point-of-care testing. Clin Chem 2011;57(9):1267–71.

[12] Melanson SE. Drug-of-abuse testing at the point of care. Clin Lab Med 2009;29(3):503–9.

[13] Nichols JH, Christenson RH, Clarke W, Gronowski A, Hammett-Stabler CA, Jacobs E, et al. Executive summary. The National Academy of Clinical Biochemistry Laboratory Medicine Practice Guideline: evidence-based practice for point-of-care testing. Clin Chim Acta 2007;379(1–2):14–28 [discussion 9-30].

[14] George S, Braithwaite RA. Use of on-site testing for drugs of abuse. Clin Chem 2002;48(10):1639–46.

[15] Mastrovitch TA, Bithoney WG, DeBari VA, Nina AG. Point-of-care testing for drugs of abuse in an urban emergency department. Ann Clin Lab Sci 2002;32(4):383–6.

[16] Cotten SW. Drug testing in the neonate. Clin Lab Med 2012;32(3):449–66.

[17] Cotten SW, Duncan DL, Burch EA, Seashore CJ, Hammett-Stabler CA. Unexpected interference of baby wash products with a cannabinoid (THC) immunoassay. Clin Biochem 2012;45(9):605–9.

[18] Attema-de Jonge ME, Peeters SY, Franssen EJ. Performance of three point-of-care urinalysis test devices for drugs of abuse and therapeutic drugs applied in the emergency department. J Emerg Med 2012;42(6):682–91.

[19] Saitman A, Park HD, Fitzgerald RL. False-positive interferences of common urine drug screen immunoassays: a review. J Anal Toxicol 2014;38(7):387–96.

[20] Moeller KE, Kissack JC, Atayee RS, Lee KC. Clinical interpretation of urine drug tests: what clinicians need to know about urine drug screens. Mayo Clin Proc 2017;92(5):774–96.

[21] Wu AH, McKay C, Broussard LA, Hoffman RS, Kwong TC, Moyer TP, et al. National Academy of Clinical Biochemistry Laboratory Medicine Practice Guidelines: recommendations for the use of laboratory tests to support poisoned patients who present to the emergency department. Clin Chem. 2003;49(3):357–79.

[22] Kranzler HR, Stone J, McLaughlin L. Evaluation of a point-of-care testing product for drugs of abuse; testing site is a key variable. Drug Alcohol Depend 1995;40(1):55–62.

[23] Laboratories A. Summary of NexScreeen cup evaluation; 2012.

[24] Johnson-Davis KL, Sadler AJ, Genzen JR. A retrospective analysis of urine drugs of abuse immunoassay true positive rates at a national reference laboratory. J Anal Toxicol 2016;40(2):97–107.

[25] Luzzi VI, Saunders AN, Koenig JW, Turk J, Lo SF, Garg UC, et al. Analytic performance of immunoassays for drugs of abuse below established cutoff values. Clin Chem 2004;50(4):717–22.

[26] Labay L. Oral fluid: validation of a fit-for-purpose test. Clinical Laboratory News 2017;2017 November 1.

[27] Toennes SW, Kauert GF, Steinmeyer S, Moeller MR. Driving under the influence of drugs—evaluation of analytical data of drugs in oral fluid, serum and urine, and correlation with impairment symptoms. Forensic Sci Int 2005;152(2−3):149−55.
[28] Verstraete AG, Puddu M. Evaluation of different road side drug tests. In: Verstraete A, ed. Rosita. Roadside testing assessment. Gent: Rosita Consortium; 2001. p. 167−232.
[29] Gjerde H, Langel K, Favretto D, Verstraete AG. Detection of illicit drugs in oral fluid from drivers as biomarker for drugs in blood. Forensic Sci Int 2015;256:42−5.
[30] Vandevenne M, Vandenbussche H, Verstraete A. Detection time of drugs of abuse in urine. Acta Clin Belg 2000;55(6):323−33.
[31] Verstraete AG. Oral fluid testing for driving under the influence of drugs: history, recent progress and remaining challenges. Forensic Sci Int 2005;150(2−3):143−50.
[32] Verstraete AG. Detection times of drugs of abuse in blood, urine, and oral fluid. Ther Drug Monit 2004;26(2):200−5.
[33] Services DoHaH. 94 FR 28054—May 15, 2015. In: DHHS, editor; 2015.
[34] Lee D, Vandrey R, Milman G, Bergamaschi M, Mendu DR, Murray JA, et al. Oral fluid/plasma cannabinoid ratios following controlled oral THC and smoked cannabis administration. Anal Bioanal Chem 2013;405(23):7269−79.
[35] Himes SK, Scheidweiler KB, Beck O, Gorelick DA, Desrosiers NA, Huestis MA. Cannabinoids in exhaled breath following controlled administration of smoked cannabis. Clin Chem 2013;59(12):1780−9.
[36] Crouch DJ. Oral fluid collection: the neglected variable in oral fluid testing. Forensic Sci Int 2005;150(2−3):165−73.
[37] Scherer JN, Fiorentin TR, Borille BT, Pasa G, Sousa TRV, von Diemen L, et al. Reliability of point-of-collection testing devices for drugs of abuse in oral fluid: a systematic review and meta-analysis. J Pharm Biomed Anal 2017;143:77−85.
[38] Institute CaLS. Quality management: approaches to reducing errors at the point of care. POCT07-A; 2010.
[39] Institute CaLS. Selection criteria for point of care testing devices. POCT09-A. Wayne, PA: Clinical and Laboratory Standards Institute; 2010.
[40] Institute CaLS. Implementation guide of POCT01 for health care providers. POCT02-A. Wayne, PA: CLSI; 2008.
[41] Institute CaLS. Point-of-care connectivity; approved standard. POCT01-A. Wayne, PA: CLSI; 2006.
[42] Nerenz RD, Song H, Gronowski AM. Screening method to evaluate point-of-care human chorionic gonadotropin (hCG) devices for susceptibility to the hook effect by hCG beta core fragment: evaluation of 11 devices. Clin Chem 2014;60(4):667−74.

Chapter 12

Drugs of Abuse Screening and Confirmation With Lower Cutoff Values

Albert D. Fraser[1,2,3]
[1]Department of Pathology and Pharmacy, Dalhousie University, Halifax, NS, Canada, [2]Horizon Health Network, Saint John, NB, Canada
[3]Government of Canada, Canada

INTRODUCTION

Over the past several decades, clinical and forensic laboratories have used immunoassay-based methods to screen for drugs of abuse in biological fluids such as urine. These assays include cutoff values developed to meet the technical and operational requirements of the reagent systems and the instrumentation used. The cutoff values, however, did not account for the widely varying urine "concentration" from one urine specimen to another. In essence, the inherent assumption of the manufacturers at that time was that the tests would be used for analysis of normally concentrated urine specimens. The analysis of random urine specimens contrasts with other biological matrices such as serum or whole blood, where analyzing a very dilute or concentrated specimen is not a common concern. Drinking excess water may cause diluted specimens. Other rare pathophysiological conditions such as diabetes insipidus also cause diluted urine.

CREATININE AS A BIOMARKER OF DILUTED URINE

In general "creatinine clearance" is widely used as an estimate of renal function. In drugs of abuse testing, creatinine can be used as an indicator of urine dilution as well as intentional adulteration. The normal creatinine concentration of urine varies from ~20 to 400 mg/dL and specific gravity from 1.005 to 1.030. The normal pH of human urine is 4.5–8.0. In 1992 Needleman et al. [1] conducted fluid intake studies in volunteers over 6- and 12-hour time periods. They concluded that creatinine measurements could be used as possible indicators of intentional adulteration by dilution of urine specimens. Needleman stated that any creatinine value in a random urine specimen <10 mg/dL was suggestive of (or consistent with) replacement by water. Other investigators also studied whether creatinine was an acceptable indicator to reduce variation in biological monitoring of random urine specimens. Allesio and Berlin reported large intraindividual and interindividual variations in urinary excretion of xenobiotic substances with urine dilution [2]. They also recommended that one should determine whether correcting a result for creatinine excretion was of value in reducing these variations. It is common in clinical laboratory settings to report analyte results from random or timed urine specimens as analyte concentration/creatinine ratio. Most drug testing laboratories, however, have not chosen to normalize varying drug or drug metabolite excretion by calculating a drug or drug metabolite/creatinine ratio.

Although creatinine concentration below 20 mg/dL is generally considered as a possible indication of intentional dilution of urine, such low creatinine value may also be observed in a small number of healthy normal population. In one study, the authors used cation-pairing, high-pressure liquid chromatography combined with tandem mass spectrometry to determine creatinine concentrations in 4227 urine specimens and observed that 209 specimens (4.94%) had creatinine concentration below 20 mg/dL cutoff. The authors concluded that specimens with creatinine under 20 mg/dL do not always indicate sample adulteration [3]. Chaturvedi et al. using volunteers who drank 800 mL of beverage showed that out of 376 urine specimens analyzed, 36 (10%) fulfilled the criteria of diluted urine (creatinine <20 mg/dL and specific gravity <1.0030 but > 1.0010). For males with at least one diluted sample, body fat was 11% and resting metabolic rate was 29% more than males with no diluted sample. For females with at least one diluted sample, height was

8% less and weight was 20% less than females with no diluted samples. The authors concluded that a diluted urine specimen does not necessarily indicate that it was diluted on purpose [4].

CRITERIA FOR DILUTED URINE

Despite differing views of the merits of incorporating a marker of urine dilution with urine drug screening, it is common practice for clinical and forensic laboratories to include creatinine and/or specific gravity analysis in their drug screening programs. In the Substance Abuse and Mental Health Substance Abuse ((SAMHSA), an agency under United States Department of Health and Human Services) federal workplace drug testing program, urine specimens with a creatinine concentration >2 and <20 mg/dL with a specific gravity >1.0010 and <1.0030 are reported as "Dilute." The SAMHSA Mandatory Guidelines for Federal Workplace Drug Testing Programs in 1994 permitted certified laboratories to conduct tests to ensure specimen validity [5]. The National Laboratory Certification Program (NLCP) issued a guidance document for reporting specimen validity test results in 1998, which defined dilute and substituted urine specimens based on the analysis of creatinine and specific gravity [6]. A specimen was reported as "Dilute" if the creatinine was <20 mg/dL and the specific gravity was <1.003. A specimen was reported as "Substituted" if the creatinine concentration was <5 mg/dL and the specific gravity was <1.001 or >1.020. A substituted specimen is defined as having creatinine and specific gravity values that are so low or so divergent from most urine specimens analyzed that they are not considered to be consistent with normal human urine [7].

An interim final rule published by the United States Department of Transportation in the Federal Register [8] in 2003 lowered the creatinine decision point from 5 to 2 mg/dL. This change was based on a small number of cases in which Federal employees that had legitimate medical explanations for producing urine specimens in the substituted range. This change in reporting was important because prior to this change, specimens reported as substituted were categorized as a refusal to test to the Federal agency. In 2004 revised SAMHSA Mandatory Guidelines were published requiring laboratories to use electronic refractometers that were capable of measuring specific gravity to four decimal places [6]. In addition, the creatinine decision point for identifying a substituted specimen was lowered from 5 to 2 mg/dL. Currently, urine specimens with a creatinine concentration less than 2 mg/dL and a specific gravity of <1.0010 or >1.0200 are reported as "Substituted." Specimens with a creatinine concentration >2 and <5 mg/dL are either reported as "Invalid" or "Dilute" based on specific gravity measurements. If the specific gravity is <1.0010, the specimen is reported as "Invalid"; indicating that it contains an endogenous substance at an abnormal concentration [6]. If the specific gravity is >1.0010 and <1.0030, the urine specimen is reported as "Dilute."

The SAMHSA program considers the impact of much lower creatinine values and specific gravity on urine drug testing operationally, but does not require or permit analysis of dilute urine specimens at lower screening and confirmation cutoff values. The forensic urine drug testing program of the Correctional Service of Canada (CSC) identifies urine specimens that have creatinine concentrations <20 mg/dL and specific gravity measurement ≤ 1.003 as "Dilute" [9–11]. In addition, the CSC program incorporates lower immunoassay screening cutoff values for drugs or drug metabolites that give negative screening findings at the regular protocol cutoff values when urine specimens are deemed to be dilute. The CSC dilution protocol also includes lower confirmation cutoff values as determined by gas chromatography–mass spectrometry (GC-MS) or liquid chromatography–mass spectrometry (LC-MS) for specimens that screen positive in the CSC dilution protocol. The objective of this chapter is to describe the CSC dilution protocol program and experience with this approach in a forensic drug testing program established over 20 years ago. In the early 1990s, offenders in correctional institutions commonly carried water bottles and it was common knowledge that excess fluid consumption prior to providing urine for drug testing would help to "beat the drug test."

SPECIMEN PROCESSING IN THE CORRECTIONAL SERVICE OF CANADA PROGRAM

Fig. 12.1 is a flowchart of CSC urine-specimen testing process. Urine specimens in this program undergo immunoassay screening at CSC defined cutoff values. In Table 12.1, CSC regular protocol for screening cutoff concentrations of drugs in urine are compared with SAMHSA screening cutoff protocol. It is important to note that SAMHSA has no process for testing diluted urine. Specimens that are presumptive positive (above the cutoff concentration of a screening assay) are extracted and subjected to confirmation analysis by GC-MS or LC-MS. In Table 12.2, confirmation cutoff values of drugs according to CSC protocol (undiluted urine specimens) are compared with SAMHSA protocol for drug confirmation. It is important to note that SAMHSA protocol does not include the benzodiazepines that are widely abused worldwide. Moreover, coabuse of benzodiazepines along with opioids is substantial [12].

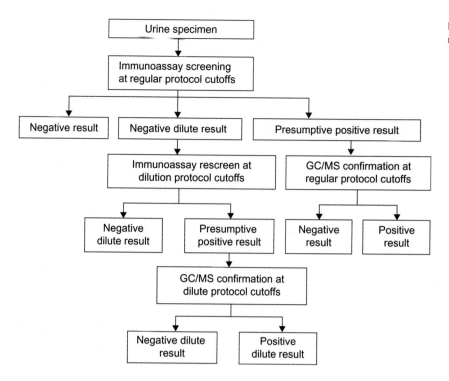

FIGURE 12.1 Flowchart for analysis of specimens in CSC urine drug testing protocol.

TABLE 12.1 CSC Regular Protocol for Screening Cutoff Concentrations Compared With SAMHSA Screening Cutoff Protocol (SAMHSA has no Dilution Protocol)

Drug/Drug Class	CSC Regular Protocol Cutoff (ng/Ml)	SAMHSA Cutoff (ng/mL)[a]
Amphetamine/methamphetamine	500	500
Benzodiazepines	100	Not included in SAMHSA protocol
Cannabinoids	50	50
Cocaine (as benzoylecgonine)	150	150
Opiates	300	2000
6-Monoacetylmorphine	10	10
Methadone metabolite	100	Not included in SAMHSA protocol
Phencyclidine	25	25

[a]SAMHSA recently introduced screening of oxycodone/oxymorphone at 100 ng/mL cutoff. In addition, hydrocodone/hydromorphone is screened at 300 ng/mL cutoff.

Specimens found to be dilute when tested for creatinine along with initial screening tests and presumptive positive for a drug or drug class are also subjected to confirmation analysis using regular protocol GC-MS or LC-MS confirmation cutoff values. Urine specimens, however, found to be dilute (creatinine concentration <20 mg/dL and specific gravity <1.003) and negative for drugs or drug metabolites at regular protocol screening cutoff concentrations, are rescreened at lower immunoassay screening cutoff values (Table 12.3). Specimens that are negative for drugs or drug metabolites after rescreening at lower immunoassay cutoff values are reported as "negative and dilute." Specimens found to be presumptive positive for a drug or drug metabolite(s) are extracted for the specific drug or drug metabolite and the resulting extract is analyzed by GC-MS or LC-MS at lower confirmation cutoff values (Table 12.3). Specimens which are presumptive positive but do not confirm above the dilution protocol confirmation cutoff values are reported as "negative and dilute." Specimens that confirm above the dilution protocol confirmation cutoff values are reported as "positive for that drug or drug metabolite and dilute."

TABLE 12.2 Confirmation Cutoff Concentrations in Regular CSC Protocol and SAMHSA Protocol

Drug/Drug Class	CSC Regular Protocol Confirmation Cutoff (ng/mL)	SAMHSA Confirmation Cutoff (ng/mL)
Amphetamine, methamphetamine, MDMA, MDA	250	250
Benzodiazepines (LC-MS confirmation for) oxazepam, temazepam, diazepam, nordiazepam, alprazolam, lorazepam, triazolam, clonazepam, bromazepam, and N-desalkylflurazepam	50	Not included in SAMHSA protocol
Cannabinoids as Δ^9-THC-11-nor-COOH (THC-COOH)	15	15
Cocaine (as benzoylecgonine)	100	100
Opiates (morphine, codeine)	300	2000
6-Monoacetylmorphine	10	10
Hydrocodone/hydromorphone	300	100
Oxycodone/oxymorphone	300	100
Methadone metabolite (2-ethylidene-1,5-dimethyl-3,3-diphenylpyrrolidine: EDDP)	100	Not included in SAMHSA protocol
Phencyclidine	25	25

TABLE 12.3 CSC Dilution Protocol Cutoff Values

Drug or Drug Metabolite	Screening Cutoff Value (ng/mL)	Confirmation Cutoff Value (ng/mL)
Amphetamines	100	100 (amphetamine, methamphetamine, MDMA, MDA)
Benzodiazepines	50	50 (nordiazepam, oxazepam, temazepam, α-OH alprazolam, lorazepam, α-OH triazolam, bromazepam, N-desalkylflurazepam, clonazepam)
Cannabinoids	20	6 (Δ^9-THC-11-nor-COOH)
Cocaine metabolite	15	15 (benzoylecgonine)
Opiates	120	120 (codeine, morphine, hydromorphone, hydrocodone, oxycodone)
Methadone	50	50 (2-ethylidene-1,5-dimethyl-3,3-diphenylpyrrolidine (EDDP))
Phencyclidine	5	5

CORRECTIONAL SERVICE OF CANADA WORKLOAD AND DILUTED URINE SPECIMENS

The CSC test workload is divided into two categories. Institutional results are for urine specimens collected from offenders living in an institution such as a prison. Community results are from offenders living outside institutions in the community such as at halfway houses, on parole. In the institutional category in December 2017, 30 urine specimens (2.3% of workload) had a low creatinine and low specific gravity on initial testing. Of these 1296 nondilute urine specimens, 287 specimens screened positive for one or more drugs/metabolites using the regular protocol cutoff values. The highest proportions of confirmed positive findings in the institution's category were (1) cannabinoids, (2) methadone metabolite, (3) opiates (codeine and/or morphine), and (4) amphetamines.

The remaining 30 dilute specimens from the institution's group went through the dilution protocol. Nine of these dilute specimens (9/30 or 30.0%) were negative on initial screening at CSC regular protocol cutoff values but screened

TABLE 12.4 Statistics of December CSC 2017 Urine Drug Testing

Urine Specimens	Undiluted	Specimen Confirmed Positive	% Positive	Diluted	Specimen Confirmed Positive	% Positive
Institution	1296	287	22.6	30	9	30.0
Community	1808	735	40.1	69	28	40.5

and confirmed positive for one or more drugs at the lower cutoff values in the CSC dilution protocol. The most prevalent confirmed positive findings in the institutions group of dilute specimens taken through the dilution protocol were (1) cannabinoids and (2) amphetamines.

In the community category in December 2017, 1808 specimens were nondilute but 74 urine specimens had a low creatinine and low specific gravity on initial testing. Of these, five urine specimens screened positive for one or more drugs/metabolites using the CSC regular protocol cutoff values despite being very dilute. The highest number of confirmed positive findings in the community group (undilute and dilute) specimens that confirmed positive with CSC regular protocol cutoff values were (1) cannabinoids, (2) amphetamines, (3) methadone metabolite, and (4) opiates (codeine and/or morphine).

The remaining 69 dilute specimens from the community group in December 2017 went through the CSC dilution protocol. About 28 of these very dilute specimens were negative on initial screening at CSC regular protocol cutoff values but confirmed positive for one or more drugs at the lower cutoff values in the CSC dilution protocol. The most common confirmed positive findings in the community group of dilute specimens taken through the dilution protocol were (1) cannabinoids and (2) amphetamines.

In the institutional group, 44/8339 (0.5%) and in the community group 283/19, 832 (1.4%) of the urine specimens tested in 2007 were reported as "no drugs detected" in the absence of a dilution protocol scheme with lower screening and confirmation cutoff values.

The data from December 2017 indicated 126 specimens out of a total 3325 specimens (all specimens analyzed; institutional plus community) were diluted (3.8%). This data is similar to 2007 workload where 4.2% of all urine specimens analyzed were diluted. Moreover, breakdown of specimens from institutions and community (data from December 2017) also showed similar trends in percentage of urine specimens analyzed which were diluted according to CSC criteria. Among community specimens, 3.81% were dilute specimens where among institutions, 2.3% specimens were dilute specimens. Moreover, presence of illicit drugs was confirmed in 22.1% of all nondilute specimens collected from institutions and 40.6% of nondilute specimens collected from community, indicating that positivity rate was higher among community specimens. This data is presented in Table 12.4.

Various drugs including amphetamine, methamphetamine, MDMA, MDA, benzodiazepines, or metabolites (nordiazepam, oxazepam, temazepam, alprazolam, alpha-OH alprazolam, bromazepam, lorazepam, triazolam, alpha-OH-triazolam, desalkyl-flurazepam, clonazepam, 7-amino clonazepam), cannabinoids (confirmed as THC-COOH), cocaine (as benzoylecgonine), 6-acetyl morphine, opiates (codeine, morphine, hydrocodone, hydromorphone, oxycodone), and EDDP were confirmed in urine specimens. The highest proportions of confirmed positive drugs in the institutions category were (1) cannabinoids, (2) opiates, (3) methadone metabolite, and (4) amphetamines. Although similar to 2017 specimens, cannabinoids were also the highest proportion of all drugs in 2007 institutional specimens but proportions of other drugs in 2017 specimens were different than 2007 findings. The data from 2017 indicates that amphetamine abuse increased significantly compared to 2007 data while cocaine positive findings declined significantly. Percentages of various illicit drugs in CSC drug testing program in December 2017 are given in Table 12.5.

DILUTION PROTOCOL AND POSITIVE TEST RESULTS

Administrative cutoff values for drug screening were based on the immunoassay characteristics when introduced to the marketplace. The most commonly used drug/metabolite cutoff values used today were not based on single or multiple drug doses in human volunteers with urine specimens collected at defined times in a controlled clinical research setting. The major advantage of introducing a dilution protocol into the CSC program was to reduce the frequency of false negative drug tests when analyzing very dilute urine specimens. Disciplinary action was based on the laboratory confirming a drug or metabolite in a urine specimen. There are no disciplinary consequences for providing a very dilute specimen with

TABLE 12.5 Percentage of Specimens Positive for Various Illicit Drugs in CSC Drug Testing in December 2017

Illicit Drug	Positive (Total Specimens) (%)	Positive (Community Specimens) (%)	Positive (Institutional Specimens) (%)
Amphetamines	6.48	7.76	4.42
MDMA/MDA	0.06	0.06	0.08
Benzodiazepines	2.0	3.27	0.39
Cannabinoids	10.3	12.13	7.03
Cocaine metabolite	1.61	2.7	0
Total opiates	6.96	8.34	5.29
6-Monoacetylmorphine	0.10	0.17	0.0
EDDP	6.4	7.52	5.2
Phencyclidine	0	0	0

no drugs or drug metabolites present. There are some limitations to the implementation of dilution protocol. The test menu is more limited than when one is analyzing normally concentrated specimens. Second, almost all published reports of drug metabolite excretion times are from test methods using the SAMHSA or similar cutoff values. The screening and confirmatory cutoff values for cannabinoids were studied extensively in the context of passive inhalation. As one lowers the screening and confirmation cutoff values for cannabinoids and THC-COOH, there is a higher probability of passive inhalation leading to a positive result. The most common positive finding in dilute specimens from institutions was cannabinoids. The most common positive finding in the community specimens was cocaine metabolite in dilute urine specimens. The passive inhalation explanation is not valid in this context since offenders are not to be associated with anyone using an illicit substance such as marijuana or cocaine. In addition, the CSC 2017 program uses a 300 ng/mL screening and confirmation cutoff value for opiates. Whenever one lowers the cutoff values for opiates the probability of nondrug sources of opiates such as consuming poppy seed—containing foods leading to a positive result increases. Only four urine specimens confirmed positive for an opiate in the dilution protocol scheme in 2017. Methadone metabolite (EDDP) was added to the CSC drug testing program in the mid-2007. The positive methadone metabolite results are from offenders who are enrolled in methadone maintenance programs or who obtained methadone from another source. To date, there is limited data on the prevalence of methadone metabolite in very dilute urine specimens.

CONCLUSIONS

In summary, the CSC dilution protocol has been very helpful in reducing the frequency of false negative drug tests when analyzing very dilute urine specimens. Therefore such protocol may be helpful for identifying individuals who are abusing drugs, but their urine specimens may be considered dilute under the standard protocol for analysis. However, such protocol for testing of diluted urine is not currently practiced in the United States. With the recent development of dilute and shoot methods for analysis of drugs and metabolites in urine using high-performance liquid chromatography combined with tandem mass spectrometry, comprehensive drugs screen as well as confirmation is possible even if analytes are present in low quantities [13,14]. Such methodology is very sensitive and can be easily adapted for analysis of diluted urine.

REFERENCES

[1] Needleman SD, Porvaznik M. Creatinine analysis in single collection urine specimens. J Forens Sci 1992;37:1125—33.
[2] Alessio LA, Berlin A. Reliability of urinary creatinine as a parameter used to adjust values of urinary biological indicators. Int Arch Occup Environ Health 1985;55:99—106.
[3] Holden B, Guice EA. An investigation of normal urine with a creatinine concentration under the cutoff of 20 mg/dL for specimen validity testing in a toxicology laboratory. J Forensic Sci 2014;59:806—10.

[4] Chaturvedi AK, Sershon JL, Craft KJ, Cardona PS, et al. Effects of fluid load on human urine characteristics related to workplace drug testing. J Anal Toxicol 2013;37:5−10.
[5] Mandatory Guidelines for Federal Workplace Drug Testing Programs, Federal Register, 129908−129931, June 9, 1994.
[6] Program Document #35 − Guidance for Reporting Specimen Validity Test Results, National Laboratory Certification Program, Research Triangle Park, North Carolina, September 28, 1998.
[7] Mandatory Guidelines for Federal Workplace Drug Testing Programs, Federal Register, 19644−19673, April 13, 2004.
[8] Procedures for Transportation Workplace Drug and Alcohol Testing Programs, Federal Register, 31624−31627, May 28, 2003.
[9] Fraser AD, Zamecnik J, Keravel J, et al. Experience with urine drug testing by the Correctional Service of Canada. Forens Sci Int 2001;121:16−22.
[10] Fraser AD, Zamecnik J. Substance abuse testing by the Correctional Service of Canada. Ther Drug Monit 2002;24:187−91.
[11] Fraser AD, Zamecnik J. Impact of lowering the screening and confirmation cutoff values for urine drug testing based on dilution indicators. Ther Drug Monit 2003;25:723−7.
[12] Jones JD, Mogali S, Comer SD. Polydrug abuse: a review of opioid and benzodiazepine combination use. Drug Alcohol Depend. 2012;125:8−18.
[13] Hegstad S, Hermansson S, Betner I, Spigset O, et al. Screening and quantitative determination of drugs of abuse in diluted urine by UPLC-MS/MS. J Chromatogr B Analyt Technol Biomed Life Sci. 2014;947−948:83−95.
[14] Kong TY, Kim JH, Kim YY, In MK, et al. Rapid analysis of drugs of abuse and their metabolites in human urine using dilute and shoot liquid chromatography-tandem mass spectrometry. Arch Pharm Res 2017;40:180−96.

Chapter 13

Overview of Analytical Methods in Drugs of Abuse Analysis: Gas Chromatography/Mass Spectrometry, Liquid Chromatography Combined With Tandem Mass Spectrometry and Related Methods

Alec Saitman

Providence Regional Laboratories, Portland, OR, United States

INTRODUCTION

Quadrupole mass spectrometry (MS)-based drugs of abuse testing has become an integral component of clinical and forensic laboratories [1]. Parameters such as high sensitivity and high specificity are particularly important when providing positive and negative drug results for clinical and forensic specimens. Quadrupole MS-based testing meet these sensitivity and specificity requirements necessary to reliably detect drugs of abuse in patient and case specimens [2]. MS-based assays also have the potential to distinguish between hundreds of different drugs of abuse from a single data acquisition [3]. Quadrupole MS, however, is a complex methodology. In fact, quadrupole MS is a combination of different separation, ionization, and mass selection methodologies. Individual components can be combined to make a quadrupole MS system that ultimately provides different analytical specifications needed for different drugs of abuse testing applications. This chapter describes these component methodologies most relevant to clinical and forensic drugs of abuse testing using a sequential and comparative approach.

SPECIMEN PREPARATION

Quadrupole MS-based drugs of abuse testing typically begins with sample preparation. Drugs of abuse are tested in a variety of different matrices. These matrices include urine, whole blood, serum/plasma, meconium, and saliva to name a few [4–6]. The drugs contained within these matrices many times cannot be directly introduced into an MS platform for analysis. The sample itself must be purified or conditioned first. This conditioning is designed to maximize sensitivity and specificity while minimizing or eliminating interferences, and contaminants [7]. These contaminants, if introduced into the MS system, can deposit in the MS system and cause unscheduled downtime [7,8]. This process is commonly known as sample preparation or sample cleanup. Many methods exist to perform sample preparation, each having specific benefits and drawbacks.

The simplest form of sample preparation can be accomplished by diluting the sample. This is mainly applicable when testing drugs of abuse that are easily detectable in the mass spectrometer and exist at high concentrations in the specimen. This simple dilution has also been referred to as dilute and shoot [9–11]. The most common matrix for simple dilutions is urine [11]. Diluents added to the matrix normally are aqueous-based mediums. After addition of diluents and any other additives, samples are generally centrifuged and the supernatant used for analysis. This process prevents any undissolved, solid particulate matter from entering the MS system. The concept of dilution is based on reducing the quantity of all dissolved substances that are ultimately introduced into the MS system. Benefits include ease of development and routine use, and more opportunities to automate which, in turn increases speed at which samples are prepared and ultimately lower the cost per sample to prepare [11]. Although dilutions do reduce the overall amount of dissolved substances introduced, none are completely removed. After many introductions of diluted samples, the mass spectrometer system may require more frequent cleanings, exchange of parts, and unscheduled downtime.

An extension of dilution preparations is protein precipitation. This is generally necessary in blood-based matrices such as serum, plasma, and whole blood. Precipitating solvents and reagents are added to the sample to cause a solubility shift of the dissolved proteins to precipitate or crash out of solution [12,13]. The entire mixture is then centrifuged to separate the solid component of the mixture and the liquid supernatant. Precipitating solvents commonly used are methanol and acetonitrile and a common precipitating reagent is aqueous zinc sulfate [12]. With this process, the protein component of the matrix is physically removed from the sample and effectively diluted in a single set of steps. Benefits and drawbacks are similar to simple dilution preparation. It should be noted however, that blood-based matrices even after protein precipitation and dilution typically contain a larger quantity of interfering compounds than urine matrices [14]. This means protein precipitation-based preparations may expedite the need to clean or replace parts on the mass spectrometer system as compared with simple dilutions in urine-based matrices [14].

More robust sample preparation methods involve extraction of desired drugs of abuse out of the original patient matrix while leaving behind a large quantity of undesirable compounds. Extraction procedures are attractive to many laboratories as they provide cleaner samples to be introduced onto the MS system as well as be applicable to a much larger family of drugs of abuse and matrices they are found in [15,16].

The simplest of the extraction methods is called liquid–liquid extraction or LLE. LLE is based on the concept of partitioning soluble compounds of interest between two immiscible liquids. Because all human liquid matrices are aqueous based, an immiscible organic solvent is used that can effectively remove and contain the drug compounds while leaving behind undesirable salts, proteins, and other more polar molecules in the aqueous layer [17]. Typical immiscible liquids for this type of extraction include ethyl acetate, tertiary butyl methyl ether, and dichloromethane to name a few [17]. After extraction into the organic solvent is complete, the two liquid layers are separated from each other. The aqueous layer is then discarded while the organic layer is retained for further preparation. Some MS-based applications can tolerate the original organic extraction solvent for direct introduction into the MS system. However, in other methods, the organic solvent used to extract the compounds of interest is normally too strong of a solvent to be introduced into the MS system. Because of this problem, the organic liquid containing the compounds of interest is generally evaporated to dryness. The dried residue containing the compounds of interest is then reconstituted in an aqueous-based solvent or solvent system suitable for introduction on the mass spectrometer. LLE offers a much better sample cleanup than simple dilutions or protein precipitations [15,17]. However, LLEs require more steps, consumables, and time. They also tend to be more difficult to automate and ultimately generate higher costs to prepare samples [3].

An analogous method to LLE is supported liquid extraction or SLE. Instead of mixing two immiscible solvents together as in LLE, in SLE, first the aqueous portion of the sample containing the compounds of interest is distributed on a solid yet porous matrix [18]. This matrix typically consists of diatomaceous earth [18,19]. After distribution of the aqueous portion, an immiscible organic solvent is added and filtered through the porous matrix. Compounds of interest initially distributed in the porous matrix are extracted out of the aqueous portion into the organic portion. The organic portion containing the compounds of interest has less affinity for the porous matrix and can be removed from the system while leaving behind the aqueous portion and many impurities. This porous matrix is normally contained in small disposable tubes called columns which are open at both ends. This is a different type of column than the ones used in gas and liquid chromatography (LC) (discussed later). Other variations place this porous matrix in individual wells. The wells are connected together to form plates of various sizes, containing various numbers of wells. Plate options may have application in more automated sample preparation platforms. Both of these variations require addition of all liquids from the top and collection of purified liquids from the bottom. Positive pressure (pushing gas on the top of the columns) and negative pressure (vacuuming from the bottom of the columns) is sometimes necessary to facilitate and expedite collection of liquids containing these purified compounds [18]. As with LLE, the organic solvents containing the compounds of interest generally need to be evaporated to dryness and reconstituted with a suitable solvent system before introduction into an MS system. SLE may afford cleaner samples than LLE and tend to be easier to automate for larger volume methods [18]. However, because the columns or plates used in SLE are disposable, this almost always increases the cost per sample which may make it difficult to financially justify [18].

Solid phase extraction or SPE is another type of sample preparation application and uses a similar column or plate-based methodology like SLE. However, instead of a porous matrix, the solid component or solid phase of SPE is normally comprised of many fine particles which have a specific chemical affinity to the compounds of interest [3,20,21]. The liquid portion containing the compounds of interest is added to the solid phase. Depending on the method, organic or aqueous solvents are then added on top of the solid phase and are filtered through. Depending on the chemical affinities of the solid phase, the liquid phase, and the compounds of interest, the liquid portion which passes through the stationary phase is retained for further processing or removed. This depends on whether the eluted liquid portion contains the compounds of interest or undesired interfering compounds. If the liquid portion retained on the solid phase includes the desired analytes, they can then be eluted from the solid phase in a second step. This is accomplished by using a different solvent which the compounds of interest have higher affinity for than the solid phase they are currently interacting with [16,21]. This process ultimately provides chemical separation and purification of the compounds of interest. The columns used in SPE are very similar in methodology to the analytical columns used in LC, the theory of which will be discussed in more detail later. Depending on the solvent used to elute the compounds of interest, the solvent may need to be evaporated to dryness and reconstituted with a more appropriate solvent before introduction into an MS system [15]. SPE tends to provide the purest samples when compared to all other sample preparation methods. This may be beneficial when a high volume of samples are analyzed daily and instrument cleanliness with minimization of unscheduled downtime is especially important for a laboratory [15,22]. Because SPE methods tend to have disposable columns or plates, the cost per sample is generally the highest of the sample preparation methods.

It should be noted that any method where a sample is evaporated to dryness, the subsequent reconstitution liquid used may be a lower volume than the initial sample was. This effectively concentrates the sample allowing more compounds of interest per unit volume to be introduced into the mass spectrometer, potentially increasing the sensitivity.

In certain circumstances, direct analysis of a native compound of interest after sample preparation is not favorable or possible. In this instance, an alteration of the chemical structure of the compound is necessary to afford a new molecule more suitable for analysis. If chemical alteration is necessary, it is generally integrated into the sample preparation process. One method is called chemical derivatization and almost always results in a new molecule which is a heavier mass than the original. Sometimes, chemical derivatization is necessary to convert nonvolatile compounds into volatile ones suitable for gas chromatography (GC) methods [23,24]. In other circumstances, chemical derivatization is necessary to produce larger mass compounds which may be detected in MS methods better than its native counterpart.

Another type of chemical alteration involves cleavage of a portion of the molecule to form a smaller molecular product. This cleavage may produce a molecule which is more volatile and therefore more suitable for GC methods. Other times, the cleavage produces a molecule which has better chromatographic or mass spectrometric characteristics. Typically this cleavage removes a sugar (glucuronide) which was chemically appended to the parent molecule in the liver in order to clear it renally. This cleavage is also known as hydrolysis and can be chemically or enzymatically mediated depending on the chemical makeup of the starting compound [25–27].

SPECIMEN INTRODUCTION

Once sample preparation is completed, purified samples are placed on the mass spectrometer. Depending on the formats available by the MS manufacturer, the sample itself may reside in individual vials or connected plates like a 48 or 96 well plate format. After the samples have been placed on the MS system, a portion of that prepared sample is introduced into the MS system. This introduction of sample occurs in two general stages. In the first stage, a portion of the sample is aspirated from the correct sample position. For GC MS assays, the aspirate may be a portion of the gas above the prepared sample. This technique is commonly used in head space GC MS-based applications [15,28]. For other GC MS and LC MS assays, the aspirate is a liquid portion of the prepared sample. The volume of the sample introduced is often times referred to the injection volume. The volume of aspirate used is dependent on the sensitivity required for the assay and the intrinsic properties of the solution being aspirated. Volumes of aspirate typically fall between 0.1 and 100 μL. The second stage involves injection of the aspirate in either the gas or liquid chromatograph. The exact timing of that injection is recorded in order to identify specific times which compounds of interest are detected by the mass spectrometer after the injection occurred. Detection of compounds of interest (discussed later) is recorded in a two-dimensional digital format known as a chromatogram. The number of specific detected molecular ions is recorded in relation to the time at which they exit either the gas or LC system. A culmination of ion detection counts over a given period of time creates what is known as a chromatographic peak or simply a peak.

GAS CHROMATOGRAPHY

GC is used in some MS applications as an initial physical separation of compounds from one another. Volatility is critically important in GC because compounds of interest can only be separated if they can be sufficiently volatilized at temperatures which do not decompose the compound itself [3,28,29].

GC has three basic components, a solid component (stationary phase), a gas component (mobile phase), and the compounds of interest. Typically, this solid component lines a thin (<0.1 mm) yet very long (>10 m) hollow tube called a column [1]. This is called capillary GC and is the most common type of GC currently employed for drug testing methodologies. The column is wound in a tight coil to minimize space. The column resides in a chamber oven which is heated to temperatures as high as 400°C [2]. The gas component normally consists of inert gasses such as argon or nitrogen.

The interaction of the compounds between the stationary phase and the gas phase is important in determining the time required at a given temperature to exit the column. This is known as the compound's retention time. The less volatile the compound of interest is at a given temperature, the longer it will stay on the column (retain better). The more volatile the compound of interest is, the faster it will leave the column (poorly retained). The temperature of the column can be increased or ramped throughout an analytical run and thus decrease the retention time for compounds to exit the column [2].

LIQUID CHROMATOGRAPHY

LC is widely used in MS to provide an initial separation of compounds from each other. Another term used in place of LC is HPLC (high-pressure or high-performance LC) [12,30]. Both terms tend to be used interchangeably when describing LC drugs of abuse assays. Benefits of LC include: faster separations, separation of nonvolatile compounds, and lower temperatures used compared to GC methods. These properties, in turn, maintain compound integrity better than GC methods [3,12]. Unlike GC, LC methods typically do not depend on derivatization steps during sample preparation [31]. This tends to allow for less labor intensive, more efficient sample preparation methods which increase throughput. LC methods also tend to yield faster sample to sample data acquisition than GC methods because the actual chromatography runtime is shorter [10,11,16].

LC has three basic components: a solid component (stationary phase), a liquid component (mobile phase), and the compounds of interest. The solid component is placed (packed) into a hollow tube called a column. These two components interact with compounds of interest with different affinities.

Fundamentally, if a compound interacts chemically more with the solid phase, the compound will stay on the column longer (retain better). If a compound interacts chemically more with the liquid phase, the compound will leave the column faster (retain poorer) [32]. The stationary phase is considered unchanging so its interaction with a compound can be considered constant throughout the lifetime of the column. However, changing the chemical composition of the liquids in the mobile phase can vary and thus vary the affinity of the compound with the mobile and stationary phases over the analytical run. The temperature of the column also affects how compounds retain along with the flow rate of the mobile phases traveling through the column [33,34].

If three theoretical, chemically different compounds are introduced on the column at the same time, they will all be contained in one tight band of material at the beginning of the column. As mobile phase begins to flow through the column, it interacts with the three different compounds of interest and the solid phase. Since the three compounds are chemically different, the mobile and solid phase interactions are unique to each compound and consequently, each compound travels through the column at a different rate. This difference in travel rate allows for each compound to separate from one another. This separation is the key to LC.

Although many types of LC applications exist, in clinical MS, LC is roughly broken up into two main categories: normal phase chromatography and reverse phase chromatography.

Normal phase chromatography uses polar stationary phases and nonpolar mobile phases. The word "normal" in normal phase chromatography was arbitrarily chosen because it was the first type of separation chromatography developed [33]. Nonpolar compounds tend to exit the normal phase column first followed by more polar compounds. This is true because polar compounds will have a higher affinity to the stationary phase than nonpolar compounds. A popular stationary phase used is silica gel which is a finely divided form of hydrated silicon dioxide [33]. Popular mobile phases for normal phase chromatography are organic solvents such as hexane, ethyl acetate, diethyl ether, and dichloromethane. Increasing polarity of the organic solvents allows for polar compounds to interact more strongly with the mobile phase and move compounds through the column. Hydrophilic interaction chromatography is a variation of normal phase chromatography which has met clinical usefulness for some MS methods [35].

Reverse phase chromatography, as its name suggests is effectively the opposite of normal phase chromatography. The stationary phase is considered nonpolar and mobile phases are considered polar. Polar compounds tend to exit the column first followed by nonpolar compounds [33]. Popular stationary phases comprise silica gel particles similar to normal phase stationary phases which have been chemically linked to hydrocarbons. C-8 stationary phase, for example, indicates that 8 carbon length chains have been chemically linked to the silica gel particles [33]. Similarly C-18 stationary phase indicates that 18 carbon length chains have been chemically linked to the silica gel particles. Primary mobile phases used in reverse phase chromatography are water, methanol, and acetonitrile. Increasing nonpolarity of the solvents allows for nonpolar compounds to interact more strongly with the mobile phase and move compounds through the column [30,33,36]. For drugs of abuse testing in all matrices, reverse phase chromatography is the most common chromatography application used.

Ultimately, LC and GC provide a layer of specificity in identifying particular compounds of interest. Retention time information can be used to confirm the validity of a drug of abuse. Other peak identifiers may be used to confirm the identity of a drug of abuse. Peak shape usually describes how the distribution of the peak on either side of its tallest point looks. Peaks with irregular lumps and bumps in them may indicate an interfering compound with similar mass data, may be complicating interpretation (Fig. 13.1). Peak width at 50% height is another identifier used to confirm the presence of a drug of abuse. A peak that is at the expected retention time but is wider or thinner than expected may indicate the compound eluting may not be the drug of abuse in question. These parameters tend to be very consistent in calibrators, quality control (QC) and true positive patient or case samples. Value limits based on deviations from expected retention time, peak shape, and peak width at 50% are commonly applied to unknown samples to increase specificity and further decrease the chances of a false positive result.

IONIZATION

Before detection is possible in a mass spectrometer, all compounds must transition from neutral molecules to individual ionized molecules. This transition is known as ionization and there are multiple applications available for that process to occur. This transition takes place in an MS instrument component called an ion source or simply source. Although many applications exist to ionize molecules, four are widely used in clinical and forensic MS applications.

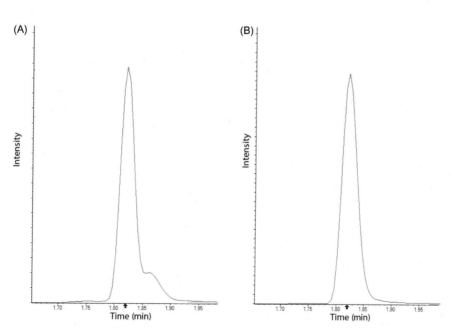

FIGURE 13.1   (A) Peak with irregularities (B) Normal peak shape.

Electron Ionization

Electron ionization or EI, also called "electron impact" in older literature [3,37], is one of the most common ionization method to convert neutral molecules in the gas phase to ionized molecules suitable for detection in GC MS methods [37]. Although other separation methods have been used with EI, GC is typically the separation chromatography for the initial separation technique for this ionization application. This type of ionization is unique because molecules can only be ionized to create positively charged ions. Once gaseous neutral molecules enter the ionization chamber, they are collided with free electrons produced from a resistively heated filament [37]. This filament is usually a very high melting point metal like tungsten. These electrons are accelerated by a potential of around 70 eV when collision with compounds of interest occurs. Ionization is induced via the removal of a single electron from the neutral molecule. Ionization is followed by division of the highly energized molecular ion into smaller, ion fragments [30]. This type of ionization is considered "hard ionization" because the fragments produced are typically small in mass to charge ratio or m/z and rarely are similar in size when compared the original ionized molecular mass. The fragmentation, however, creates a "fingerprint" of smaller mass fragments detected by the mass spectrometer (discussed later) at consistent intensities for a given molecule. This fingerprint pattern combined with the very precise retention times produced by GC methods provides the necessary specificity needed for identification of drugs of abuse.

Chemical Ionization

Chemical ionization or CI is another ionization technique used widely in GC MS applications. CI relies on a reagent gas to assist in the ionization process of the compounds of interest. Typical reagent gasses commonly used in CI are ammonia, methane, and isobutene [30,38]. Unlike EI where the compounds of interest are directly ionized by free electrons, CI uses these reagent gasses as intermediates in the ionization process [38]. The number molecules composing the reagent gas is far greater than the compounds of interest being detected. Initially, the reagent gas becomes ionized and through a cascade of chemical steps, ionizes the compounds of interest. This ionization transfer greatly reduces the energy absorbed by the compounds of interest. Because of this lower energy impact, CI is often considered a "soft ionization" because after ionization, the original molecular mass of the compound of interest is generally left intact [38]. Unlike EI, CI can also produce ions that are either positively or negatively charged. Like EI, CI still requires drugs of abuse compounds to be sufficiently volatile for detection [39].

Electrospray Ionization

Electrospray ionization (ESI) is the most common ionization form used to transition molecules from in-solution to individual ions in LC MS methods. This type of ionization is popular when combined with LC separation methods [40]. Most LC methods add modifiers to the mobile phases which assist compounds of interest to acquire or lose a proton and become negatively or positively charged. These modifiers generally consist of nonmetal containing acids or bases [3,41]. Dissolved metal ions in mobile phases can deposit in the mass spectrometer as insoluble salts which is detrimental to the functionality of the mass spectrometer. Other common chemical adducts which produce positively charged ions other than protons are ammonium ions. The ability for a compound to acquire a charge affects the degree of ionization and the overall sensitivity of that compound in an MS method.

After exiting the LC column, a portion of the liquid containing the compounds of interest is diverted through a small spray needle. The spray needle and the entrance of the mass spectrometer are connected to each other by a high-voltage circuit. In positive mode, positively charged ions are drawn out of the needle contained within fine liquid droplets which are pulled toward the MS entrance. This process occurs in a contained compartment which draws these fine liquid droplets toward the opening of the lower pressure mass spectrometer and the negatively charged mass spectrometer entrance [40]. In the same direction the droplets are traveling, a warm inert gas (typically nitrogen) moves the droplets toward the mass spectrometer and simultaneously aids in evaporating the solvent the droplets are composed of. As the droplets begin to evaporate, the positively charged molecules repel each other with greater force [3]. This occurs because the same amount of positively charged ions must exist in a droplet with ever shrinking surface area. At what is called the Rayleigh limit [42], the charged repulsion forces become greater than the surface tension force holding the droplet together. This causes disintegration of the large droplet into smaller droplets. By dispersing the charged molecules over a larger surface area within multiple smaller droplets, stabilization of the smaller droplets temporarily occurs. As more solvent molecules within the droplets evaporate, the process repeats until only single positively charged ions exist. Compounds of interest may acquire a positive charge at any point while dissolved in the liquid droplets [40,42].

Positive charge acquisition occurs more readily as the volume of the liquid droplets continues to decrease. These positively charged compounds of interest can now be detected by the mass spectrometer. Negatively charged ions can also be produced this way by reversing the voltage circuit described above.

Atmospheric Pressure Chemical Ionization

Atmospheric pressure CI or APCI is a type of ionization used in some LC methods. The application is especially useful when the molecules of interest do not readily acquire or lose charges. APCI methods indirectly ionize compounds of interest through intermediate ions. Compounds in APCI methods begin dissolved in solvent spray droplets very similar to droplets formed in ESI methods [43,44]. However, after desolvation of the liquid droplets to a heated vapor, these compounds of interest generally have not yet acquired any charge and remain neutral. A corona discharge needle is the necessary component needed to convert the neutral molecules of interest into detectable ions in the mass spectrometer. To generate positive ions, for example, the corona discharge needle produces a current strong enough to ionize the drying gas (generally nitrogen) creating positively charged nitrogen adducts. Through a cascade of events, the positively charged adducts bombard and transfer charges to other small neutral molecules present such as H_2O, and eventually the compounds of interest [30]. A vast number of charged adducts are produced in relation to the number of compounds of interest present at any one time. This is coupled with the increased number of collisions occurring because the gasses are at atmospheric pressure. This large excess of charged adducts forces charges onto compounds of interest that would otherwise remain essentially uncharged via ESI methods.

QUADRUPOLE MASS SPECTROMETRY

The power behind MS is the ability for the instrument to not only discriminate two compounds of different masses but also the ability to count each mass it discriminates. In this way, mass spectrometers are simultaneously very specific and also can be quantitative with high sensitivity. Many different types of mass spectrometers exist, each with unique functionality. Quadrupole MS has the largest presence in clinical and forensic drug testing. This type of mass spectrometer consists of one quadrupole or multiple quadrupoles in tandem and a detector. Quadrupoles are four metal rods arranged in a parallel fashion. Each of the two sets of nonadjacent rods is connected together electronically.

A radio frequency (RF) voltage is applied to one pair of opposing rods and a direct current (DC) balance is applied to the other pair of opposing rods. These pairs of opposing rods create an X and a Y-axis with ions traveling through the Z-axis (Fig. 13.2) [2]. Lighter ions (lower mass) in the X-axis, gain energy from the RF field and oscillate at increasingly larger amplitudes. Eventually, these lighter ions will collide with one of the X-axis rods and are removed [45]. The rods in the X-axis effectively act as a high mass filter where high masses can reach the end of the quadrupole without colliding the X-axis rods but low masses will not [45]. For the rods operating in the Y-axis, heavy ions become unstable because of the DC's defocusing capability and will collide with one of the Y-axis rods. Therefore, the rods in the Y-axis serve as a low mass filter where low masses can reach the end of the quadrupole without colliding with the Y-axis rods but high masses will not. The RF voltage and a DC balance are applied in a specific ratio to each of the

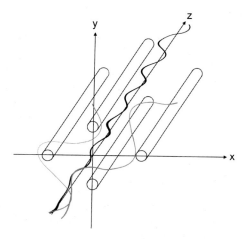

FIGURE 13.2 Design of quadrupole mass analyzer showing stable and unstable ion trajectories.

two sets of opposing rods [2]. This ratio is determined electronically depending on the specific m/z that is selected. This gives the quadrupole the capability of filtering specific compounds with a particular m/z. All ions of all masses enter the quadrupole. However, only the ions of a specific m/z requested will travel through the quadrupole and reach the detector. Any ion with a different m/z other than the m/z selected will have unstable trajectories and collide with one of the four rods. This facilitates the selection of an ion with a specific m/z for a specific time allotment. Another capability of quadrupole mass spectrometers is the ability to scan for a range of m/z values. This is accomplished by constantly changing the voltage applied. Selecting and isolating a single m/z for detection makes the method highly specific.

Applications in GC/MS typically use a single quadrupole by itself to identify molecules and molecular fragments. The fragment ions or product ions generated from the original ion in the ion source called the precursor ion are then scanned for a range of m/z values. A plot of fragments is generated with m/z on the X-axis and the ion counts or intensity on the Y-axis. The specific m/z values detected coupled with the intensities of each m/z provides a "fingerprint" or unique profile of a specific drug of abuse. Combining the very specific retention times achieved by GC and the specific mass spectrometric profile of an individual compound, drugs of abuse can be confidently identified in unknown specimens.

Triple Quadrupole Mass Spectrometry

Specificity can be further increased by addition of more quadrupoles in tandem. The most common setup like this has three quadrupoles and is one of the most common designs for tandem quadrupole mass spectrometers. Many names describe this design such as, triple quadrupole spectrometry, tandem mass spectrometry, triple quadrupole mass spectrometry (MS/MS) among others [12,45,46]. Ions enter the first quadrupole or Q1 and generally represent the m/z of the charged ion from the original mass of the drug of abuse to be identified. This ion is known as the precursor ion with older references calling it the parent ion. The second quadrupole or Q2 is used as a collision cell. A collision cell only has RF voltage applied and provides no selection of masses. Voltage in Q2 is applied along with introduction of an inert gas such as helium, argon, or nitrogen. The precursor ions that are selected in the first quadrupole interact with these inert gasses and the applied RF voltage and experience collision-induced dissociation (CID) causing fragmentation into smaller molecular weight ions. These fragments are called product ions with older references calling them daughter ions. The molecular weight and array of product ions is dependent on the chemical makeup of the precursor ion itself. Generally speaking, at lower voltages, larger product ions predominate whereas, at higher voltages, smaller product ion fragments predominate. This is true because at high voltages, many large product ions that initially fragmented will undergo additional fragmentations, producing smaller product ions. These product ions enter the third quadrupole or Q3 where individual product ions of a specific m/z can be selected or all ions fully scanned. Selection of a precursor ion in Q1, collision of the precursor in Q2, and selection of a product ion in Q3 is called selected ion monitoring or SIM. Each different Q1 to Q3 mass selection is called a mass transition or simply transition. Multiple reaction monitoring or MRM is an extension of SIM in which multiple product ions are filtered from a single or multiple precursor ions in the same timeframe [12,45,47].

Once an ion of a specific m/z has been mass filtered, it reaches the detector and is recorded. In most cases, the detector is an electron multiplier tube. The electron multiplier tube amplifies the electronic signal generated from each ion that reaches it. This amplified electronic signal is converted to a digital signal which is then recorded by the software system that controls the mass spectrometer. For a given timeframe (normally in the millisecond timescale), the quadrupole selects one ion with a specific m/z to be filtered and the detector counts and records each filtered ion and reports the count as a single number in that timeframe.

The specific ion selected for detection can be switched for another during the chromatographic run. In fact, many clinical and forensic assays may select dozens of specific ions representing drugs and/or drug fragments. In a theoretical example, if the mass spectrometer was instructed to select for 10 ions, each of a unique m/z, each for 10 ms with a 5 ms rest between each selection, it takes 150 ms for the mass spectrometer to collect 1 selected ion count for each of the 10 ions. Since most chromatographic compound peaks take 3–10 s to exit the column completely, each chromatographic peak is made up of 20 or more individual data points, with each data point representing a 10 ms ion count for an individual ion. This means that the chromatographic peak is generated from connecting multiple data points in time rather than one continuous acquisition (Fig. 13.3). In this way LCs and GCs are not required to physically separate each compound from one another since multiple compounds can be identified and counted over the timeframe of chromatographic separation.

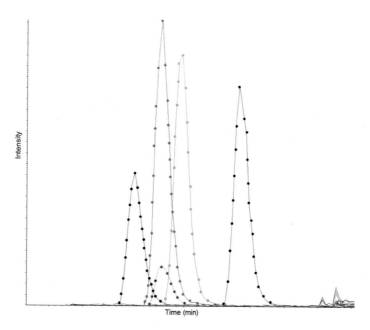

FIGURE 13.3 Chromatographic peaks represented by discrete data points collected.

A separation chromatogram may contain multiple identified compounds overlaid on each other. The X-axis represents time and the Y-axis represents ion count. An area under the curve (AUC) represents the total number of ions detected by the mass spectrometer over the chromatographic run. The AUC is calculated by integrating the ion count over the chromatographic window for a particular compound.

QUANTITATIVE MEASUREMENT

Because the increasing number of ions calculated under the curve is correlated to an increasing concentration, quantitative values of unknown specimens can be achieved. Direct correlation to the AUC to the concentration however, is generally not acceptable [48]. This is due to the fact that small variations in all sample preparation steps, chromatography, and mass acquisition, may alter the final calibrator and/or unknown sample values to unacceptable levels. A functional solution around this problem is the introduction of an internal standard or IS during the initial few steps of sample preparation [49].

ISs can theoretically be any molecule that is structurally different from the compound of interest. However, most ISs used in clinical and forensic assays are isotopically labeled drugs of abuse which retain nearly identical physical and chemical properties as the drug of abuse being tested [49]. Deuterium and carbon-13 labeled ISs are commercially available for most common drugs of abuse being tested in routine laboratories. Because an IS has a different mass than its respective drug of abuse, the IS can be simultaneously analyzed in the same acquisition as the drug of abuse of interest. The same volume with the same concentration of IS is added to each sample early in the sample preparation. An area ratio can then be generated by division of the total amount of ions for the drug of abuse by the total amount of ions for the IS. The area ratio is now considered fixed and any variation in subsequent sample preparation or MS acquisition may alter the overall amount of ions being detected but will not alter the ratio. Generated ratios can then be plotted against the concentration of the calibrators. Unknown samples prepared in the exact same way as the calibrators will have an area ratio generated, which can then be back-calculated to provide a drug concentration. Since the ion ratio cannot be altered, the application is robust in that it buffers variations in the steps preceding ion detection.

Sensitivity

The sensitivity of quadrupole MS methods is important for most drugs of abuse assays. Theoretically, the sensitivity of any MS-based assay is determined by the ability to differentiate a chromatographic peak from the background noise that surrounds it. The peak itself is referred to as the signal and the background noise is generally referred to as just noise. The signal-to-noise (S/N) ratio is the division of the peak height (or area) by height (or area) of a selected section of the noise adjacent to the peak. Fundamentally, sensitivity is the ability for the mass spectrometer to detect enough ions at the retention time of a drug of abuse to produce an S/N ratio of greater than a defined amount. A minimum S/N that has been used to confidently define a peak as a true peak is generally recognized to be 3. Many factors contribute to the number of ions that eventually reach the detector including sample preparation, chromatography, ionization, and mass acquisition.

One of the main goals of initial sample preparation is to purify compounds of interest to maximize the resulting MS signal while decreasing the background noise adjacent to it. Removal of other molecular components potentially accomplishes two actions. First, sample preparation may remove excess ions which may cause ion suppression (discussed later) or lowering of the signal of compounds of interest. Second, sample preparation may remove molecular ions which have similar masses that may increase the baseline noise to unacceptable levels.

Proper chromatography also plays an important role in delivering needed sensitivity to MS assays. Correct peak shape is required to maintain enough peak height in relation to the noise at the low end of sensitivity. Peak symmetry describes how proportioned both sides of the peak are in reference to the tallest point. Ideally, a peak would be look sharp and symmetrical around its tallest point and be Gaussian in shape [50,51]. This is not always achievable. Some chromatographic peaks are either wider on the left side of the tallest point (fronting) or wider on the right side of the tallest point (tailing) (Fig. 13.4). Too much fronting or tailing causes considerable degradation of the peak height, directly influencing the S/N ratio. This may make the peak unacceptable for identification and quantitation in drugs of abuse testing. It may also indicate a design problem with the chromatographic method. Another potential problem with a chromatographic peak is one in which the compound of interest takes too long to elute off of the column. This excess time to elute disperses the compound over a larger timeframe, producing wide and short peaks. The injection volume introduced into the MS system also has a direct effect on the overall S/N.

Another large component in determining the sensitivity of a given drug of abuse is how well it ionizes. When using ESI as an ionization method many contributing factors determine ionization of analytes of interest. Each unique compound also has an intrinsic ability to either gain or lose positive charges. Many drugs of abuse contain nitrogen atoms which are inherently basic and readily acquire an acidic proton in an acidic medium, creating positively charged compounds. Another example are carboxylic acids like the one found on the metabolite of tetrahydrocannabinol (THC), 11-nor-9-carboxy-THC, readily lose a proton in a basic environment creating negatively charged compounds. Many anabolic steroids, however, have few basic or acidic sites on the molecule and therefore do not readily gain or lose positive charges contributing to sensitivity issues. In general, the more likely a compound can gain or lose a positive charge, the more ions will be produced and will eventually be detected by the mass spectrometer.

Because of this, with LC applications, the mobile phase composition may have large influences on the number of potential charged compounds which can be detected. This is why additives such as acids or bases are frequently combined in the mobile phases to promote ionization of compounds of interest [52]. On the other hand, in ESI applications, coeluting compounds can negatively influence the ionization of compounds of interest. This phenomenon is known as ion suppression [53]. Although these coeluting compounds may not necessarily be detected because they are different masses, they acquire or lose positive charges just as the compounds of interest do. All compounds eluting in a given

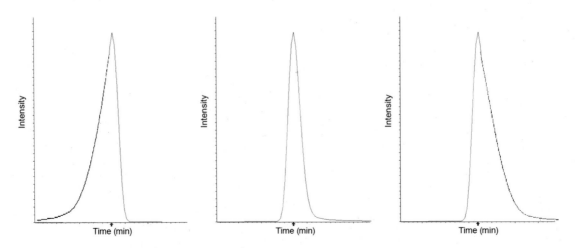

FIGURE 13.4 Chromatographic peaks showing fronting, normal chromatography or tailing.

chromatographic timeframe compete for a fixed number of charges supplied from the mobile phase. If a coeluting compound is in a high quantity and/or is intrinsically better at acquiring a charge, then less molecules of the compound of interest will become charged. This means less ions will ultimately reach the MS detector and the signal of that compound will be suppressed [53].

Fragmentation patterns for compounds of interest also influence the overall signal and ultimate sensitivity of the assay. With MRM acquisitions, the precursor ion must be fragmented into product ions, some of which are then selected for detection. The number of product ion masses generated and the intensity of any one individual ion of a specific mass is dependent on the chemical structure of the original precursor ion. Specific collision voltages along with other electronic modifications can optimize the intensity for a specific product ion [3]. However, even with these optimizations, some precursor ions fail to fragment into product ions with high intensity while others fragment well with a high intensity. In general, precursor ions which fragment well into product ions of high intensity have higher achievable sensitivity than precursor ions which fragment poorly [12,48]. The sensitivity of the assay may define cutoff concentrations at which drugs of abuse can be confidently detected. Other times, clinical or forensic requirements do not require the cutoff concentration to be at the level of the method's sensitivity.

Specificity

Specificity in the context of MS-based methods is the ability for the assay to confidently detect and report a particular drug of abuse. This confident detection must occur in the presence of thousands of other compounds without any influence in the original concentration from any of them. In most LC/MS assays, many precursor ions fragment into multiple different product ions during CID. When electronic conditions for a particular set of MRM experiments for a particular compound are held constant, the ratios of the different product ions produced in relation to each other are also constant. MS/MS-based drug confirmation assays capitalize on this phenomenon to further increase specificity and confidence in a drug result. Most triple quadrupole methods analyze a minimum of two different product ion transitions to identify a single, specific drug of abuse. Because both fragments come from the same precursor ion, their separation chromatograms should have identical retention times and similar if not identical peak shapes (Fig. 13.5). However, each fragment is produced with a unique intensity. If the ion area count from one transition is compared to the ion area count of a second transition, an ion area ratio can be generated. Analyzing calibrators containing an amount of known drug standards, ion ratios for specific drugs of abuse are generated. Using multiple calibrators, an average ion ratio can be generated. If an unknown sample contains a set of peaks at a correct retention time for a particular drug and quantitates above the cutoff, confidence increases that the sample contains the drug in question. Confidence further increases that the sample

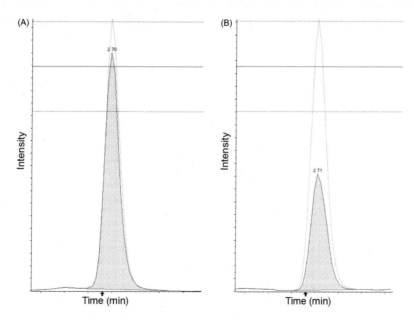

FIGURE 13.5 When both fragments come from the same precursor ion, their separation chromatograms should have identical retention times and similar if not identical peak shapes. Because the ion ratios of compounds A and B are dissimilar, they likely do not originate from the same precursor ion and therefore are not the same compound.

contains the drug in question if an unknown sample then contains a set of peaks, with both mass transitions present in the correct ion ratio as compared to the ion ratio of a calibrator or average of calibrators. The high specificity of LC−MS/MS assays then comes mainly from the combination of peak retention time, and two or more selective MRM transitions at a defined ratio [54].

Because GC/MS-based assays cause compounds to travel such a long distance through a GC column, a large majority of specificity is dependent on the retention time as an identifier of a drug of abuse. However, combinations of different molecular separation methods coupled with different ionization methods and finally MS applications can determine the needed specificity for a particular drug assay.

MASS SPECTROMETRY BASED METHODS VERSUS IMMUNOASSAYS

High sensitivity combined with high specificity of MS methods in general produces more reliable test results than immunoassays. Because of this, MS/MS is normally used as confirmation of presumptively positive immunoassay drug screening results. The antibodies in an immunoassay of a particular drug class are generally raised against one unique drug of abuse. However, many drugs may fit into one family or drug class and may also be detected. For example, many "opiate" immunoassay drug screens are actually designed to detect morphine. However codeine, 6-MAM (6-monoacetyl morphine; heroin metabolite), hydrocodone, and hydromorphone are drugs that have similar structures and are also considered detectable in an opiates immunoassay drug screen [55]. The antibody used is considered to have high sensitivity and lower specificity. This design is intentional and allows for screening of an entire drug class with one unique antibody. The drugs other than morphine all have a particular affinity to the antibody and thus, the concentration of that drug required to yield a signal high enough to screen positive varies [55]. Since many of the drugs (and their metabolites) may be found in the same sample, it is also important to note that each drug is additive to the overall detection signal. That means that if there are multiple different opiate-like drugs in a patient or case sample and are all individually under the cutoff, their additive combination can still screen positive. Many times, a particular drug of interest is too structurally different and has too little affinity to the antibody of interest to be detected [55]. Although it may belong to a specific drug class, it will not screen positive at relevant biological concentrations. In this case, a separate immunoassay must be designed specific to the drug of interest in order for it to be an effective screen. An example of requiring an additional immunoassay for testing the presence of a specific compound is oxycodone which has structural similarities and medical uses to other opiates however requires its own unique screen to capture positive results which are relevant to the analysis.

The ability for a particular antibody to detect multiple compounds can also be troublesome. Non-opiate compounds can also interact with the antibodies in a way to cause a positive result. Because no true opiate drugs exist in the sample, MS will confirm the sample negative and the original positive immunoassay screen result is considered a false positive. This is one of the most important reasons behind MS-based confirmation testing [55]. Immunoassays are designed for screening purpose only. Because they are typically designed with higher sensitivity and lower specificity, many laboratories use immunoassay drug screens to rule out negative samples. This rule out now focuses on the population of presumptive positive results for MS-based confirmation testing [55].

The sensitivity achieved by many MS assays is generally superior to immunoassay drug tests and is sometimes required for certain drug testing applications [32,56]. Some drugs of abuse may exist at therapeutic concentrations that are too low to be detected by current immunoassay methods. Fentanyl may exist at therapeutic concentrations in urine at a concentration less than 5 ng/mL which requires sensitivity that currently only MS methods can provide [57,58].

MS-based confirmation assays are often designed to identify specific drug metabolites as part of the drug confirmation assay [59,60]. These metabolite identifications are especially important when confirming drug results for compliance purposes [61,62]. Compliance testing describes evaluating patient samples from patients supposedly taking prescribed medications. Urine is the most frequent patient matrix collected as it is not invasive, and tends to concentrate drugs for easier immunoassay and MS-based detection methods. This type of testing generally occurs in a clinical setting for patients in pain management programs. Absence of a drug in a patient sample may indicate that a patient is not taking their prescribed medication as indicated. This may be occurring because the patient is selling or giving away their medication to a third party or is hoarding medication to take a higher dose than prescribed [63,64]. To hide this diversion, some patients attempt to mask the absence of drugs in their urine [65]. This is generally attempted by adding an amount of their prescribed drug directly to their urine collection container before or after voiding. Because immunoassays generally do not discriminate metabolites from parent drugs, if enough parent drug was added, the immunoassay screen could be positive. However, once MS analysis completes, only the parent drug will be detectable with undetectable metabolites. This scenario highlights another crucial reason why MS confirmation is an important component in drugs of abuse testing.

Although MS-based assays are generally accepted as a gold standard in drugs of abuse testing, there are drawbacks to the methodology [66]. For clinical laboratories in the United States, Food and Drug Administration (FDA) approved methodologies tend to be far easier to implement than laboratory developed tests (LDTs). In 2017, less than five LC/MS-based clinical assays that are FDA approved were available. This generates a logistical gap between MS methodologies and clinical laboratories that would benefit from the technology. Many laboratories are not willing to, or do not have the resources for, implementing LDTs [67].

Turnaround time for MS-based methods tends to remain longer immunoassay-based methods. This is true for several reasons. For most MS-based methodologies, a unique calibration curve and QC set are prepared during the sample preparation of the patient or case samples. This generally requires that analyses of specific drugs to be batched. Because of this batching, a sample arriving shortly after an MS batch was prepared may require to wait hours or days until analyzed in a subsequent batch. Even within a sample batch, MS-based analysis generally takes longer than other established chemistry and immunoassay methods. With MS-based assays, a sequential analysis of samples occurs. This means the data acquisition for one sample must be completed before the data acquisition from the subsequent sample can be initiated. This is opposed to many traditional chemistry and immunoassay platforms that are random access meaning they can analyze different samples for different drugs of abuse simultaneously. Because of this, fewer samples tend to have completed mass spectrometric analysis per hour than the completion efficiency of random access methodologies.

MS-based methods also require highly trained personnel to complete routine analysis. These technologists are responsible for all components of the MS analysis including: sample preparation, instrument loading, data review, and reporting. Many technologists are also trained in troubleshooting of problem samples as well as troubleshooting instrument failures. This training typically requires more time than technologists trained in other areas of the clinical or forensic laboratories. As such, qualified technologists with prior training in MS-based applications are in short supply and tend to be more difficult to locate and hire.

The lack of FDA approved MS methodologies combined with few highly trained technologists and long sample preparation and analysis time make MS an expensive endeavor to start up and maintain.

CONCLUSIONS

MS can provide confirmatory drug testing that laboratories can be confident in reporting. The methodology accomplishes this by combining necessary sensitivity with necessary selectivity and specificity, which is currently unachievable by other platforms. The power behind the utility of MS-based drugs of abuse testing methods has not gone unnoticed. A PubMed search for "drugs of abuse mass spectrometry" shows over a four fold increase in cited publications from 2001 to 2016. As drugs of abuse usage continues to be a problem in the United States, drug testing methodologies using mass spectrometry will continue to be an integral component of the clinical and forensic laboratory.

REFERENCES

[1] Kwong TC. American Association for Clinical Chemistry. The clinical toxicology laboratory: contemporary practice of poisoning evaluation. 2nd (ed.) Washington, DC: American Association for Clinical Chemistry; 2013. p. xvii. 523 pp.

[2] Jickells S, Negrusz A. Clarke's analytical forensic toxicology. London; Chicago: Pharmaceutical Press; 2008. p. xxiv. 648 pp.

[3] Garg U, Hammett-Stabler CA. Clinical applications of mass spectrometry: methods and protocols. New York: Humana Press; 2010. p. xvi. 538 pp.

[4] Ristimaa J, Gergov M, Pelander A, Halmesmaki E, Ojanpera I. Broad-spectrum drug screening of meconium by liquid chromatography with tandem mass spectrometry and time-of-flight mass spectrometry. Anal Bioanal Chem 2010;398(2):925–35.

[5] Jones J, Tomlinson K, Moore C. The simultaneous determination of codeine, morphine, hydrocodone, hydromorphone, 6-acetylmorphine, and oxycodone in hair and oral fluid. J Anal Toxicol 2002;26(3):171–5.

[6] Weinmann W, Wiedemann A, Eppinger B, Renz M, Svoboda M. Screening for drugs in serum by electrospray ionization/collision-induced dissociation and library searching. J Am Soc Mass Spectrom 1999;10(10):1028–37.

[7] Ortmayr K, Nocon J, Gasser B, Mattanovich D, Hann S, Koellensperger G. Sample preparation workflow for the liquid chromatography tandem mass spectrometry based analysis of nicotinamide adenine dinucleotide phosphate cofactors in yeast. J Sep Sci 2014;37(16):2185–91.

[8] Bourgogne E, Wagner M. Sample preparation and bioanalysis in mass spectrometry. Ann Biol Clin (Paris) 2015;73(1):11–23.

[9] Clark ZD, Cutler JM, Pavlov IY, Strathmann FG, Frank EL. Simple dilute-and-shoot method for urinary vanillylmandelic acid and homovanillic acid by liquid chromatography tandem mass spectrometry. Clin Chim Acta 2017;468:201–8.

[10] Enders JR, McIntire GL. A dilute-and-shoot LC−MS method for quantitating opioids in oral fluid. J Anal Toxicol. 2015;39(8):662−7.
[11] Kong TY, Kim JH, Kim JY, In MK, Choi KH, Kim HS, et al. Rapid analysis of drugs of abuse and their metabolites in human urine using dilute and shoot liquid chromatography−tandem mass spectrometry. Arch Pharm Res 2017;40(2):180−96.
[12] Dass C. Fundamentals of contemporary mass spectrometry. Hoboken, NJ: Wiley-Interscience; 2007. p. xx. 585 pp.
[13] O'Connor D, Clarke DE, Morrison D, Watt AP. Determination of drug concentrations in plasma by a highly automated, generic and flexible protein precipitation and liquid chromatography/tandem mass spectrometry method applicable to the drug discovery environment. Rapid Commun Mass Spectrom 2002;16(11):1065−71.
[14] Polson C, Sarkar P, Incledon B, Raguvaran V, Grant R. Optimization of protein precipitation based upon effectiveness of protein removal and ionization effect in liquid chromatography−tandem mass spectrometry. J Chromatogr B Analyt Technol Biomed Life Sci 2003;785(2):263−75.
[15] Pawliszyn J, Lord HL. Handbook of sample preparation. Hoboken, NJ: Wiley; 2010. p. 491.
[16] Xue YJ, Akinsanya JB, Liu J, Unger SE. A simplified protein precipitation/mixed-mode cation-exchange solid-phase extraction, followed by high-speed liquid chromatography/mass spectrometry, for the determination of a basic drug in human plasma. Rapid Commun Mass Spectrom 2006;20(18):2660−8.
[17] Alders L. Liquid−liquid extraction, theory and laboratory experiments. Amsterdam, Houston: Elsevier Pub; 1955. p. 206.
[18] Zhang Y, Cao H, Jiang H. Supported liquid extraction vs liquid−liquid extraction for sample preparation in LC−MS/MS-based bioanalysis. Bioanalysis 2013;5(3):285−8.
[19] Edel AL, Aliani M, Pierce GN. Supported liquid extraction in the quantitation of plasma enterolignans using isotope dilution GC/MS with application to flaxseed consumption in healthy adults. J Chromatogr B Analyt Technol Biomed Life Sci. 2013;912:24−32.
[20] Niessen WMA. Current practice of gas chromatography−mass spectrometry. New York: M. Dekker; 2001. p. xi. 507 p. p.
[21] Thurman EM, Mills MS. Solid-phase extraction: principles and practice. New York: Wiley; 1998. p. xxvi. pp 344.
[22] Yin P, Zhou L, Zhao X, Xu G. Sample collection and preparation of biofluids and extracts for liquid chromatography−mass spectrometry. Methods Mol Biol 2015;1277:51−9.
[23] Chiang JS, Huang SD. Simultaneous derivatization and extraction of amphetamine and methylenedioxyamphetamine in urine with headspace liquid-phase microextraction followed by gas chromatography−mass spectrometry. J Chromatogr A 2008;1185(1):19−22.
[24] Wu CH, Yang SC, Wang YS, Chen BG, Lin CC, Liu RH. Evaluation of various derivatization approaches for gas chromatography−mass spectrometry analysis of buprenorphine and norbuprenorphine. J Chromatogr A 2008;1182(1):93−112.
[25] Klette KL, Wiegand RF, Horn CK, Stout PR, Magluilo Jr. J. Urine benzodiazepine screening using Roche Online KIMS immunoassay with beta-glucuronidase hydrolysis and confirmation by gas chromatography−mass spectrometry. J Anal Toxicol 2005;29(3):193−200.
[26] Mantovani CC, Silva JPE, Forster G, Almeida RM, Diniz EMA, Yonamine M. Simultaneous accelerated solvent extraction and hydrolysis of 11-nor-Delta(9)-tetrahydrocannabinol-9-carboxylic acid glucuronide in meconium samples for gas chromatography−mass spectrometry analysis. J Chromatogr B Analyt Technol Biomed Life Sci 2018;1074−1075:1−7.
[27] Stidman J, Taylor EH, Simmons HF, Gandy J, Pappas AA. Determination of meprobamate in serum by alkaline hydrolysis, trimethylsilyl derivatization and detection by gas chromatography−mass spectrometry. J Chromatogr 1989;494:318−23.
[28] Kolb B, Ettre LS. Static headspace-gas chromatography: theory and practice. 2nd (ed.) Hoboken, NJ: Wiley; 2006. p. xxv. 349 pp.
[29] Grob RL, Barry EF. Modern practice of gas chromatography. 4th (ed.) Hoboken, NJ: Wiley-Interscience; 2004. p. xi. 1045 pp.
[30] Flanagan RJ. Fundamentals of analytical toxicology. Chichester, England; Hoboken, NJ: John Wiley & Sons; 2007. p. xxxvii. 505 pp.
[31] Mikel C, Almazan P, West R, Crews B, Latyshev S, Pesce A, et al. LC−MS/MS extends the range of drug analysis in pain patients. Ther Drug Monit 2009;31(6):746−8.
[32] Dasgupta A. Advances in chromatographic techniques for therapeutic drug monitoring. Boca Raton, FL: CRC Press/Taylor & Francis; 2010. p. xx. 455 pp.
[33] Dorsey JG, Cooper WT, Wheeler JF, Barth HG, Foley JP. Liquid chromatography: theory and methodology. Anal Chem 1994;66(12):500R−546RR.
[34] Jandera P. Can the theory of gradient liquid chromatography be useful in solving practical problems? J Chromatogr A 2006;1126(1−2):195−218.
[35] Kalogria E, Pistos C, Panderi I. Hydrophilic interaction liquid chromatography/positive ion electrospray ionization mass spectrometry method for the quantification of alprazolam and alpha-hydroxy-alprazolam in human plasma. J Chromatogr B Analyt Technol Biomed Life Sci 2013;942−943:158−64.
[36] Gritti F. General theory of peak compression in liquid chromatography. J Chromatogr A 2016;1433:114−22.
[37] Watson JT, Sparkman OD. Introduction to mass spectrometry: instrumentation, applications and strategies for data interpretation. 4th (ed.) Chichester, England; Hoboken, NJ: John Wiley & Sons; 2007. p. xxiv. 818 pp.
[38] Harrison AG. Chemical ionization mass spectrometry. 2nd (ed.) Boca Raton, FL: CRC Press; 1992. 208 pp.
[39] Maurer HH. Role of gas chromatography−mass spectrometry with negative ion chemical ionization in clinical and forensic toxicology, doping control, and biomonitoring. Ther Drug Monit 2002;24(2):247−54.
[40] Cole RB. Electrospray ionization mass spectrometry: fundamentals, instrumentation, and applications. New York: Wiley; 1997. p. xix. 577 pp.
[41] Saint-Marcoux F, Lachatre G, Marquet P. Evaluation of an improved general unknown screening procedure using liquid chromatography−electrospray−mass spectrometry by comparison with gas chromatography and high-performance liquid-chromatography−diode array detection. J Am Soc Mass Spectrom 2003;14(1):14−22.

[42] Bonhommeau DA. Rayleigh limit and fragmentation of multiply charged Lennard−Jones clusters: can charged clusters provide clues to investigate the stability of electrospray droplets? J Chem Phys 2017;146(12):124314.
[43] Arinobu T, Hattori H, Seno H, Ishii A, Suzuki O. Comparison of SSI with APCI as an interface of HPLC−mass spectrometry for analysis of a drug and its metabolites. J Am Soc Mass Spectrom 2002;13(3):204−8.
[44] Halloum W, Cariou R, Dervilly-Pinel G, Jaber F, Le Bizec B. APCI as an innovative ionization mode compared with EI and CI for the analysis of a large range of organophosphate esters using GC−MS/MS. J Mass Spectrom 2017;52(1):54−61.
[45] Ardrey RE. Liquid chromatography−mass spectrometry: an introduction. New York: J. Wiley; 2003. p. xviii. 276 pp.
[46] Mallet AI. Quantitative biological and clinical mass spectrometry: an introduction. 1st (ed.) Hoboken, NJ: John Wiley & Sons; 2018. 224 pp.
[47] Leaver N. A practical guide to implementing clinical mass spectrometry systems. St Albans, Hertfordshire, UK: ILM; 2011. p. vii. 84 pp.
[48] Duncan MW, Gale PJ, Yergey AL. The principles of quantitative mass spectrometry. 1st (ed.) Denver, CO: Rockpool Productions; 2006. p. ix. 141 pp.
[49] Lin DL, Chang WT, Kuo TL, Liu RH. Chemical derivatization and the selection of deuterated internal standard for quantitative determination−methamphetamine example. J Anal Toxicol 2000;24(4):275−80.
[50] Gray N, Heaton J, Musenga A, Cowan DA, Plumb RS, Smith NW. Comparison of reversed-phase and hydrophilic interaction liquid chromatography for the quantification of ephedrines using medium-resolution accurate mass spectrometry. J Chromatogr A 2013;1289:37−46.
[51] Dai C, Lu P. Peak shape distribution rule of high performance liquid chromatography in the end of column under linear and nonideal conditions. Se Pu 1997;15(5):361−6.
[52] Kato S. Effect of pH of mobile phase on the retention times of preservatives, saccharin and related substances in reversed-phase high performance liquid chromatography. Eisei Shikenjo Hokoku 1982;100:147−9.
[53] Remane D, Meyer MR, Wissenbach DK, Maurer HH. Ion suppression and enhancement effects of co-eluting analytes in multi-analyte approaches: systematic investigation using ultra-high-performance liquid chromatography/mass spectrometry with atmospheric-pressure chemical ionization or electrospray ionization. Rapid Commun Mass Spectrom 2010;24(21):3103−8.
[54] Russo P, Hood BL, Bateman NW, Conrads TP. Quantitative mass spectrometry by isotope dilution and multiple reaction monitoring (MRM). Methods Mol Biol 2017;1606:313−32.
[55] Saitman A, Park HD, Fitzgerald RL. False-positive interferences of common urine drug screen immunoassays: a review. J Anal Toxicol 2014;38(7):387−96.
[56] von Mach MA, Weber C, Meyer MR, Weilemann LS, Maurer HH, Peters FT. Comparison of urinary on-site immunoassay screening and gas chromatography−mass spectrometry results of 111 patients with suspected poisoning presenting at an emergency department. Ther Drug Monit 2007;29(1):27−39.
[57] Verplaetse R, Tytgat J. Development and validation of a sensitive ultra-performance liquid chromatography tandem mass spectrometry method for the analysis of fentanyl and its major metabolite norfentanyl in urine and whole blood in forensic context. J Chromatogr B Analyt Technol Biomed Life Sci 2010;878(22):1987−96.
[58] Lu C, Jia JY, Gui YZ, Liu GY, Shi XJ, Li SJ, et al. Development of a liquid chromatography−isotope dilution mass spectrometry method for quantification of fentanyl in human plasma. Biomed Chromatogr 2010;24(7):711−16.
[59] Allen KR, Azad R, Field HP, Blake DK. Replacement of immunoassay by LC tandem mass spectrometry for the routine measurement of drugs of abuse in oral fluid. Ann Clin Biochem 2005;42(Pt 4):277−84.
[60] Peters FT. Recent advances of liquid chromatography−(tandem) mass spectrometry in clinical and forensic toxicology. Clin Biochem 2011;44(1):54−65.
[61] Krock K, Pesce A, Ritz D, Thomas R, Cua A, Rogers R, et al. Lower cutoffs for LC−MS/MS urine drug testing indicates better patient compliance. Pain Physician 2017;20(7):E1107−13.
[62] Melanson SE, Ptolemy AS, Wasan AD. Optimizing urine drug testing for monitoring medication compliance in pain management. Pain Med 2013;14(12):1813−20.
[63] Forgione DA, Neuenschwander P, Vermeer TE. Diversion of prescription drugs to the black market: what the states are doing to curb the tide. J Health Care Finance 2001;27(4):65−78.
[64] McCabe SE, Teter CJ, Boyd CJ. Medical use, illicit use, and diversion of abusable prescription drugs. J Am Coll Health 2006;54(5):269−78.
[65] Sansone RA, Gaither GA, Righter EL. Prescription diversion among patients in a family practice clinic. Arch Fam Med 2000;9(7):587.
[66] Vogeser M, Seger C. Pitfalls associated with the use of liquid chromatography−tandem mass spectrometry in the clinical laboratory. Clin Chem 2010;56(8):1234−44.
[67] Davis J, Wentz J. How will the FDA impact the laboratory developed test? Clin Lab Sci 2007;20(3):130−1.

Chapter 14

High-Resolution Mass Spectrometry: An Emerging Analytical Method for Drug Testing

Michelle Wood
Scientific Operations, Waters Corporation, Wilmslow, United Kingdom

INTRODUCTION

Liquid chromatography (LC) in combination with mass spectrometry (LC-MS)—particularly tandem mass spectrometry (LC-MS/MS), has become firmly established as a gold standard technique for both qualitative and quantitative analysis in clinical and forensic toxicology laboratories. However, over the last two decades, there has also been steady rise in the popularity of high-resolution mass spectrometry (HRMS).

The adoption of LC combined with HRMS (LC-HRMS) technologies by mainstream toxicology laboratories was initially hampered by the perceived (and often realized) complexity of the systems—not to mention the cost of such analyzers. However, there have been some tremendous leaps in the robustness and ease-of-use, as well as the analytical capabilities and performance of these systems. Furthermore, the cost of acquisition is now within the same range as a standard tandem mass spectrometer.

HRMS has emerged as an extremely powerful technique and has been applied for the analysis of seized materials, doping agents/steroids, veterinary drugs, poisons, medicinal drugs, and drugs of abuse. However it is with toxicological screening in particular, where LC-HRMS is currently demonstrating significant benefits [1−8].

The purpose of general unknown screening (or systematic toxicological analysis) is to screen and identify all drugs or poisons of interest. This has always been a challenging proposition; the panacea—a single, universal technique capable of detecting drugs of abuse, poisons, toxins, prescribed, natural, and over-the-counter medications. In practice however, a combination of analytical techniques are usually applied for the comprehensive examination of biological samples. Even so, in recent years the task has become considerably more challenging owing to the proliferation of novel psychoactive substances (NPS). These synthetic analogs, or derivatives, are designed to mimic the pharmacological effects of traditional recreational drug substances but are intended to avoid routine analytical detection and to circumvent current legislative measures. The rapid emergence of these substances, coupled with the relatively slower availability of corresponding certified reference material (CRM), is certainly contributing to the shift in interest from low-resolution screening techniques to those based on HRMS.

Unlike some of the other traditional screening techniques such as gas chromatography (GC) or LC-MS/MS, which rely on the availability of CRM to establish suitable mass detection characteristics, for example, to select appropriate precursor-to-product ion transitions in a selected reaction monitoring (SRM) experiment, HRMS provides the ability to screen on the basis of exact mass, that is, without reference material. Furthermore, these instruments permit the measurement of the mass-to-charge (*m/z*) of an ion to four or five decimal places instead of just the single-digit integer value, or nominal mass, as provided by their low-resolution counterparts. In many cases, this level of accuracy can allow laboratory scientists to differentiate between isobaric compounds of the same *nominal*, but differing *exact*, masses.

In addition to the enhanced selectivity afforded by accurate mass, these instruments also present the analyst with a number of other analytical advantages including: the ability to collect accurate mass data in a nontargeted (data-independent) manner and, with fewer duty-cycle limitations; high sensitivity in full data acquisition mode and,

Critical Issues in Alcohol and Drugs of Abuse Testing. DOI: https://doi.org/10.1016/B978-0-12-815607-0.00014-9

particularly noteworthy—the possibility to retrospectively interrogate existing data for the presence of new (or other) analytes, of interest, without the need to reextract and reanalyze the sample.

As an increasing number of toxicology laboratories are at a point where they are finally considering acquisition of an HRMS system to complement their other analytical technologies, scientific and procurement teams alike, will be tasked with understanding some of the key differences between low-resolution mass spectrometry and HRMS and particularly keen to understand any additional utility and flexibility of HRMS. In light of this, it is the intention of this chapter to provide readers with an introduction to HRMS—it is particularly aimed at those who are familiar with some form of low-resolution mass spectrometry, for example, GC-MS or LC-MS/MS, but who are new to the concept of HRMS. As such, this chapter provides an overview of the basic principles of HRMS and highlights some of the benefits (and limitations) that HRMS can provide; it illustrates these with specific reference to toxicological analysis.

A number of different types of HRMS instruments (that is to say, instruments capable of delivering resolving power (RP) of $\geq 10,000$ m/Δm) are available these days. Of these, those based on time-of-flight (TOF) and Orbitrap analyzers have emerged as the most useful and commonly used in toxicology. This chapter presents a simple overview of both of these technologies and captures some of the main characteristics of the systems.

HRMS are highly powerful instruments and these days seem to offer an ever-increasing number of data acquisition techniques yielding different types of data. Thus, the final Section of the chapter aims to give a flavor of some of the types of data that can be generated and the opportunities that HRMS might open up to the analytical toxicologist; a small selection of applications are presented which aim to simply represent the diversity of analytical strategies available.

FUNDAMENTALS OF HIGH-RESOLUTION MASS SPECTROMETRY

In 1919, Francis William Aston developed a mass spectrograph that could separate ions that differed in charge-to-mass ratio by as little as 1 part in 130, that is, a mass RP of 130. When Aston introduced Neon into this new mass spectrograph, two distinct lines appeared corresponding to a mass of 20 and 22. Following additional studies, Aston later postulated that Neon comprised two isotopes: one with the mass of 20 with an abundance of around 90%, and one with the mass 22 and an abundance of around 10%. Frederick Soddy and coworkers had previously had shown that the natural decay of radioactive substances gave rise to what they called "isotopes" (other substances, with chemical properties identical to the known element, but with a different atomic mass), therefore Aston's new theory—that stable elements, previously believed to be monoisotopic, might also comprise isotopes—was a major breakthrough.

By 1924, the isotopes associated with 53 of the 80 known stable elements had been detected and their masses measured as whole numbers. This theory remained in place until 1937 when Aston who had developed a much improved mass spectrograph, this time with a mass RP of 2000; demonstrated that, contrary to his earlier suggestion, the masses of elements and their isotopes were *not* exactly whole numbers owing to the energy required to "pack" the protons and neutrons into the atomic nuclei.

Work by John Beynon in the 1960s [9] demonstrated that, by constraining the elements in an organic compound, a molecular chemical formula could be postulated if the mass of a species could be determined with sufficient accuracy. To this day, this remains one of the major advantages of HRMS.

Indeed, if the elemental formula of a compound is known, the exact mass of the molecule can be calculated as the sum of the exact masses of the individual constituent atoms. The usual practice is to use the monoisotopic exact mass of the atom, that is, based on the most prevalent stable isotope. This mass is typically quoted in arbitrary mass units (amu) or Daltons (Da). Monoisotopic masses for a selection of atoms that are present in organic molecules are listed in Table 14.1.

Mass Accuracy

Using the exact mass of the individual elements present in a compound, the exact mass of that compound can be calculated. For example, the exact mass of Amphetamine with the formula of $C_9H_{13}N$, would be $(9 \times 12.0000) + (13 \times 1.007825) + (1 \times 14.00307) = 135.10479$ Da. Therefore, the corresponding protonated molecular ion (assuming electrospray positive ionization mode) should yield a monoisotopic exact mass of *m/z* 136.11208.

However, all mass analyzers incur some small degree of errors during measurement. Consequently, the analysis of amphetamine in a sample may result in a measurement that slightly deviates from this exact mass. Mass accuracy is a measure of the accuracy of the mass information provided by the HRMS and is referred as the difference between the real (or exact) mass and the measured mass.

TABLE 14.1 Monoisotopic Masses for a Selection of Atoms That Are Common in Organic Molecules

Atom	Formula	Integer Mass	Exact Mass (Da)
Carbon	C	12	12.00000
Hydrogen	H	1	1.007825
Nitrogen	N	14	14.00307
Oxygen	O	16	15.99492
Phosphorous	P	31	30.97376
Sulfur	S	32	31.97207
Fluorine	F	19	18.99840
Chlorine	Cl	35	34.96885
Bromine	Br	79	78.91834
Iodine	I	127	126.90448

It can be expressed in two common ways:

$$\text{Mass Difference (Da)} = \text{Mass}_{real} - \text{Mass}_{measured}$$

Or, expressed as an error in parts per million (ppm)

$$\text{Error in ppm} = \text{Difference in mass}/\text{Real mass} \times 10^6$$

For example, if the real mass of benzoylecgonine ($C_{16}H_{19}NO_4$) is m/z 290.1386 but the measured mass is m/z 290.1356—the mass difference is 0.003 Da (or 3 millidaltons, 3 mDa) and the error is 10.34 ppm.

KEY BENEFITS OF HIGH-RESOLUTION MASS SPECTROMETRY AND EXACT MASS

To understand the benefits of exact mass and HRMS, we must consider both the accuracy of the instrument and the resolution—or the RP, of the instrument. Nowadays, most HRMS instruments are capable of routinely achieving mass accuracies within 5 ppm and offer a minimum RP of 10,000 m/Δm. The specific benefits for the analytical toxicologist can be broadly summarized as follows: the ability to accurately and precisely measure the mass of a particular compound increases the certainty and confidence of identification—in other words, the higher the accuracy of the instrument—the lower the ambiguity. It follows that, if we can determine the mass of an unknown substance with high confidence, we can use this information to predict the likely elemental formula based on the mass sufficiency of the constituent atoms. For small molecules up to around 300 Da, it is generally considered that a HRMS instrument with even what might be considered a fairly low RP these days, for example, ~10,000 m/Δm can provide a nonambiguous identification if the atoms are limited to C, H, N, and O owing to the low number of possible formulae. Along the same lines, mass accuracy also can help to rationalize fragmentation pathways which may be performed for structural elucidation studies. For today's analytical toxicologist, exact mass offers the potential to compile libraries without the need for standard reference material.

As the name indicates, HRMS instruments also offer significantly increased RP over their low-resolution counterparts. The combination of high accuracy in mass measurement and high mass resolution provides improved specificity/selectivity which can reduce matrix effects and background interferences—particularly when analyzing complex biological specimens.

As an example, Fig. 14.1 compares the resolution for benzoylecgonine, when observed using a TOF HRMS analyzer and a nominal mass spectrometer, (upper and lower trace, respectively). The figure clearly demonstrates the enhanced accuracy of mass measurement and resolution of HRMS.

Further, the combination of high accuracy and high-resolution also imparts the ability to differentiate between nominally isobaric compounds. As an example, we can consider three toxicologically relevant molecules (amitriptyline,

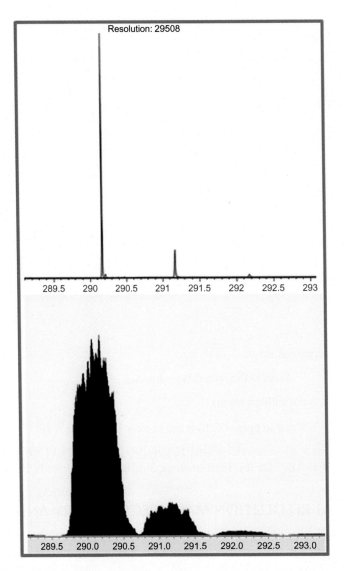

FIGURE 14.1 Comparison of high-resolution time-of-flight (TOF) mass spectrum of benzoylecgonine (upper trace) with standard mass spectrometry (lower trace).

venlafaxine, and entecavir) which share the same nominal mass of m/z 278. The empirically calculated exact mass for each molecule indicates that there are differences between all three, at the second decimal place level. For example, amitriptyline ($C_{20}H_{23}N$) yields m/z of 278.19033 which is 0.0656 Da more than the m/z of 278.12477 produced by entecavir ($C_{12}H_{15}N_5O_3$). Moreover, venlafaxine ($C_{17}H_{27}NO_2$) with a m/z of 278.21146, is 0.0211 Da greater than m/z of 278.19033 produced by amitriptyline. To fully leverage these small differences in mass we require an analyzer that is able to measure masses at this level of accuracy consistently and importantly, is able to differentiate (resolve) one particular mass from other, very close masses.

Figs. 14.2 and 14.3 show the previously discussed analysis of entecavir, amitriptyline, and venlafaxine which are isobaric substances using standard low-resolution electrospray mass spectrometry in positive ionization mode. Analysis of same three isobaric substances, using high-resolution electrospray mass spectrometry in positive ionization mode, is shown in Fig. 14.3. The advantage of high-resolution mass spectrometry (HRMS) over standard low-resolution mass spectrometry is evident from these two figures. Moreover the influence of resolution and RP of HRMS makes this technique very useful in analytical chemistry including analysis of drugs.

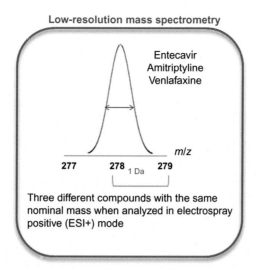

FIGURE 14.2 Analysis of entecavir, amitriptyline, and venlafaxine using standard low-resolution electrospray mass spectrometry in positive ionization mode.

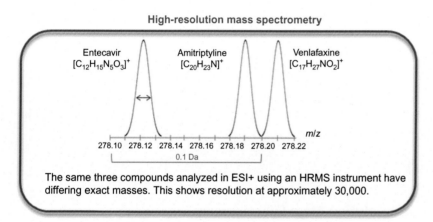

FIGURE 14.3 Analysis of entecavir, amitriptyline, and venlafaxine using high-resolution electrospray mass spectrometry in positive ionization mode.

Mass Resolution and Mass Resolving Power

According to IUPAC, mass resolution is the ability to distinguish between the two neighboring ions of slightly different m/z [10] and represents the "sharpness" of the peak at a particular m/z. In practice, it is a measure of the separation between the two ions and is calculated at a stated m/z; this latter point is an important one, as the resolution can change across the mass range for some types of instruments.

Mass Resolution (RES) can be defined as follows (Fig. 14.4):

$$\text{RES (at a stated } m/z \text{ value)} = M/\Delta M$$

where M is the m/z of $M1$ (or $M2$) and where ΔM is the peak width of $M1$ (or $M2$) when measured at 50% peak of height, that is, full width at half-maximum, FWHM).

Mass RP is the ability of a mass spectrometer to deliver a specified amount of resolution. Although some variants with "enhanced resolution mode" are available—the majority of triple quadrupole and tandem mass spectrometers are considered to be low-resolution instruments and can only distinguish m/z 278 from m/z 277 or m/z 279 but not to any finer discriminating level. These low-resolution instruments will generally achieve peak widths of ~0.7 amu. In comparison, most of the HRMS instruments on the market these days routinely achieve peak widths of ~0.01–0.02 amu and this means that these systems can often differentiate between nominally isobaric compounds.

Using the previous example, we can calculate the minimum resolution requirement to separate (A) entecavir from amitriptyline and the more challenging discrimination of (B) amitriptyline from venlafaxine:

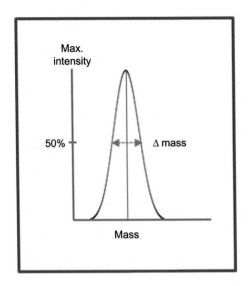

FIGURE 14.4 Calculation of mass resolution—typical peak widths for low- and high-resolution mass spectrometers are ~0.7 and 0.015 amu, respectively
NB: The peak width definition is the more commonly used definition of ΔM these days and applicable to quadrupole, time-of-flight (TOF), Fourier transform and ion trap-based mass spectrometry [10]. A lesser used definition (more commonly used with magnetic sector instruments), is based on two ions and is the 10% valley definition.

(A) RES = 278.12477/(0.0656) = 4,242
(B) RES = 278.19033/(0.0211) = 13,165

Furthermore, typical peak widths for low- and high-resolution mass spectrometers of ~0.7 amu and 0.015 amu respectively, we can see how the RP for each system compares for the ion at m/z 278 and why low-resolution instrumentation cannot discriminate on the basis of mass alone.

RP = 398 (low-resolution instrument)
RP = 18,533 (high-resolution instrument)

Limitations of Exact Mass and High-Resolution Mass Spectrometry

Even with HRMS, in the instance of the chromatographic coelution of two nominally isobaric compounds, if the required resolution to separate them is below the minimum RP of the instrument, then there is a potential for interference and errors in mass measurement leading to incorrect identification and/or elemental composition assignment.

Exact mass can only differentiate isobars, that is, molecules with the same exact mass but different elemental formula. Exact mass alone cannot differentiate between isomers, that is, substances with the same number of constituent atoms but with a different arrangement or stereochemistry. Differentiation of some isomers however may be possible through the use of other analytical parameters such as chromatographic separation or use of other mass spectrometric data, for example, isotopic cluster abundancies and product ion/fragmentation information (Fig. 14.5).

Nowadays, there are several different types of HRMS instruments on the market—each with their own specific advantages and limitations; most of these are *capable* of measuring mass to within 5 ppm, however, all have to be suitably maintained, prepared for use and will have a number of parameters, for example, temperature, acquisition times, etc., that can affect both the accuracy and precision of mass determination and/or resolution.

HIGH-RESOLUTION MASS SPECTROMETER: TECHNOLOGICAL ASPECTS

Although a number of high-resolution analyzers have been applied for toxicological analysis, such as magnetic sector, Fourier transform ion cyclotron resonance (FT-ICR), TOF, and Orbitrap analyzers, it is the latter two technologies that have emerged as the most widely used in toxicology testing today.

Both technologies can be readily hyphenated with additional separation techniques such as GC or LC although this chapter focuses on the latter, for its applicability to toxicological analysis, compatibility with a wider range of

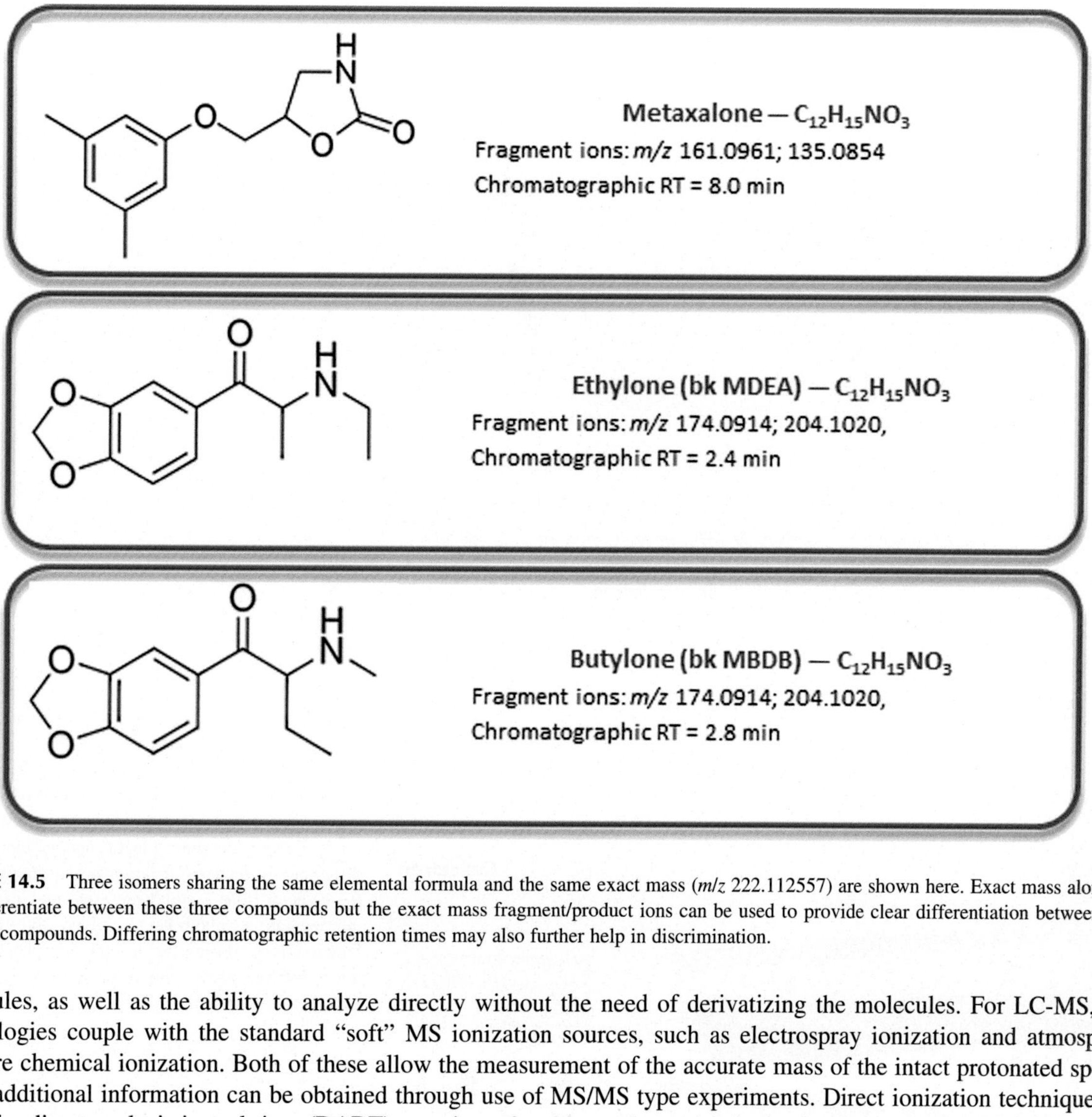

Metaxalone — $C_{12}H_{15}NO_3$
Fragment ions: m/z 161.0961; 135.0854
Chromatographic RT = 8.0 min

Ethylone (bk MDEA) — $C_{12}H_{15}NO_3$
Fragment ions: m/z 174.0914; 204.1020,
Chromatographic RT = 2.4 min

Butylone (bk MBDB) — $C_{12}H_{15}NO_3$
Fragment ions: m/z 174.0914; 204.1020,
Chromatographic RT = 2.8 min

FIGURE 14.5 Three isomers sharing the same elemental formula and the same exact mass (m/z 222.112557) are shown here. Exact mass alone cannot differentiate between these three compounds but the exact mass fragment/product ions can be used to provide clear differentiation between some of these compounds. Differing chromatographic retention times may also further help in discrimination.

molecules, as well as the ability to analyze directly without the need of derivatizing the molecules. For LC-MS, both technologies couple with the standard "soft" MS ionization sources, such as electrospray ionization and atmospheric pressure chemical ionization. Both of these allow the measurement of the accurate mass of the intact protonated species, while additional information can be obtained through use of MS/MS type experiments. Direct ionization techniques, for example, direct analysis in real time (DART), matrix-assisted laser desorption ionization have also been used in combination with HRMS in toxicology.

Time-of-Flight Mass Spectrometry

The concept of TOF was first described in 1946 by Stephens and first commercialized by Bendix Corporation in the late 1950s. Of all MS technologies, this analyzer is perhaps the simplest in terms of design and concept. A fuller description of TOF can be found elsewhere, but a brief overview is provided here.

In this device, mass separation is dependent on ion velocity. As the name suggests the m/z is determined from a measurement of the time it takes an ion to "fly" or traverse a specific path length (the flight tube); the lighter ions, that is, ions of a lower m/z ratio, arrive at the detector faster than ions of a higher m/z. Hence, with a suitably calibrated instrument, the m/z of an unknown substance can be simply calculated from its TOF.

Following ionization, ions are transferred through a variety of lenses before reaching the analyzer region (Fig. 14.6). Once within the analyzer region, the ions are accelerated through an electric field and enter the flight tube. In this "field-free" zone, they drift with the same velocity that they left the acceleration field.

Nowadays, most analyzers utilize an orthogonal acceleration (oa) arrangement where the accelerating field is arranged at a right angle to the direction of the initial ion source. Ideally, TOF analyzers require the ions to be provided in a "pulse" or "packet of ions" in order to allow the flight times of the individual ions to be accurately measured

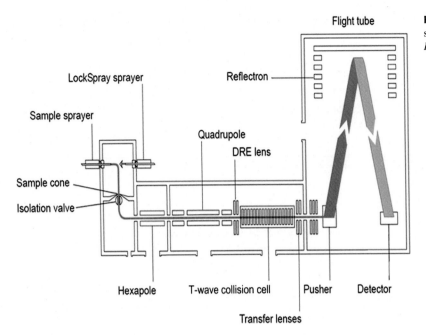

FIGURE 14.6 Schematic design of a QTOF mass spectrometer. *Copyright Waters Corporation, Reprinted with permission.*

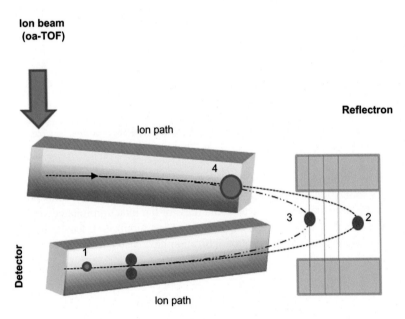

FIGURE 14.7 Schematic illustrating the dual role of the reflectron in a TOF-MS system. The reflectron is a series of electric fields and serves to increase the length of the flight path thereby improving mass resolution. In this schematic, ions 1 and 4 represent molecules of smaller and larger *m/z* respectively; these ions received the same kinetic energy but owing to difference in their *m/z* have differing arrival times at the detector. Ions 2 and 3 represent ions of the same, intermediate *m/z*; both have received slightly different kinetic energy during the acceleration phase however the reflectron corrects for dispersion of kinetic energies thereby improving spectral resolution.

(essentially to provide defined a "start" and "stop" signal). To this end, with oa-TOF analyzers a section of the main beam is selected by a pulsing electrode (the "pusher") and directed to the acceleration region.

All ions, whether they are of the same, or differing, *m/z*, should theoretically receive the same "push" (or kinetic energy) and commence their acceleration at exactly the same time. In practice however, this push is *not* felt by all ions to the same intensity—meaning that even ions of exactly the same *m/z* will not always achieve identical velocity during the initial acceleration. This can effectively lower the resolution of the instrument by creating a small TOF distribution for each *m/z*. To compensate, many of the TOF instruments now include an electrostatic reflector ("reflectron" or "ion mirror"). This device sits at the end of the flight path and lessens the effects of the dispersion of kinetic energies as ions with higher velocity penetrate deeper into the reflectron and spend more time "turning round" than those of the same *m/z* but traveling more slowly (Fig. 14.7). This creates a second focal point for ions of the same *m/z* and positioning the detector at this point minimizes arrival time spread leading to higher spectral resolution.

TABLE 14.2 Key Characteristics of TOF and Orbitrap Analyzers

Analyzer	Mass Range (m/z)	Acquisition Speed	Linear Dynamic Range	Accuracy (ppm)	Resolving Power (RP)[a]
TOF	Up to 100,000	Milliseconds (30 Hz)	10^6	<3	30,000–40,000
QTOF	Up to 10,000	Milliseconds (30 Hz)	10^3–10^4	<3	30,000–40,000
Orbitrap	Up to 4000	Seconds (3 Hz/1.5 Hz)	10^6	<3	100,000/240,000
		Milliseconds (12 Hz)		<3	12,500–30,000

[a]RP varies depending on the actual model.

TABLE 14.3 Comparison of Performance of TOF and Orbitrap

Analyzer	Comments
TOF	• Medium-to-high resolving power, constant across the full mass range • Resolving power is also independent of acquisition rate—providing excellent compatibility with the flow rates associated with ultra-high-performance liquid chromatography; UPLC • High sensitivity (superior to QQQ in full scan mode—no loss of ions associated with "scanning" technologies)
Orbitrap	• Superior resolving power than TOF at slow scan speeds—resolving power is inversely proportional to scan speed, that is, at HPLC compatible acquisition times, resolution is comparable to TOF • Resolution decreases proportionately at mass increases • High sensitivity (superior to QQQ in full scan mode—no loss of ions associated with "scanning" technologies)

The reflectron also serves another key purpose—it can be used to improve mass resolution, by effectively extending the path length of the ions. Once within the reflectron, the ions are retarded and then accelerated back in the opposite direction thus increasing overall distance traveled. Some instruments include multiple reflectrons to further extend path length.

Modern TOF mass spectrometers are available in a number of configurations, for example, TOF-only; two time-of-flight analyzers in succession; or as hybrid instruments in combination with other analyzers, for example, ion mobility separation TOF, or with quadrupoles, for example, quadrupole ion-trap-TOF, or quadrupole TOFs (QTOFs). The latter arrangement is derived from the tandem (triple) quadrupole mass spectrometer however in this design, the TOF analyzer replaces the last quadrupole. This is perhaps the most commonly used TOF configuration in toxicology and permits acquisition of other types of data such as accurate mass MS/MS, MS^E, in addition to the standard full MS data.

TOF instruments have evolved to become one of the most powerful and desirable MS systems for toxicological instruments and offer several advantages including: high m/z accuracy; medium-to-high mass resolution; high speed of analysis (acquisition rate is thousands of spectra per second), high sensitivity over the full mass spectrum (all ions are detected), and an almost unlimited m/z range. Tables 14.2 and 14.3 summarize some of the key characteristics of the analyzer.

Orbitrap Analyzer

As the name suggests, the Orbitrap (invented by Makarov in 1999) is an ion-trapping device. A thorough description of the Orbitrap is detailed elsewhere, but a brief description is provided here.

The actual design was based on the Kingdon Trap of the 1920s which comprised a small filament or wire electrode, pulled along a horizontal axis and enclosed by an outer barrel-shaped casing, which is also an electrode (Fig. 14.8). The space between the two electrodes is linked to the vacuum pumping system to provide high vacuum conditions.

Ions are injected in a pulse into the trap in a perpendicular direction. In the original Kingdon design, a direct current voltage applied between the inner and outer electrodes produced a potential which trapped the ions in the radial

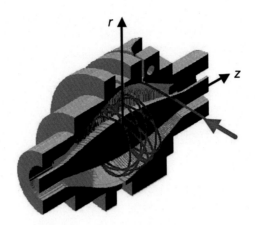

FIGURE 14.8 Cutaway view of the Orbitrap mass analyzer. Ions are injected into the trap at the point indicated by the red arrow. Once trapped, the ions orbit around the central filament and also oscillate in the horizontal (z) direction. *From Hu Q, Noll RJ, Li H, Makarov A, Hardman M, Cooks G. The orbitrap: a new mass spectrometer. J Mass Spectrom 2005;40:430–3. Copyright Wiley, reprinted with permission.*

direction; these days the design is such that the ions are also confined axially. Once trapped, the ions adopt a stable orbit around the central filament (orbital trapping) but also oscillate along the horizontal axis. In this device, the frequency of oscillation is related to the *m/z* ratio of the ion. Detection of the ions is achieved by measuring the image current of the axial motion around the central electrode.

As with TOF analyzers, the Oritraps can exist as MS-only systems or as hybrids for example, with a quadrupole mass filter to perform mass selection prior to analysis of the fragments by the Orbitrap analyzer. This filter makes it possible to perform additional experiments such as MS/MS or MS^n (see below Section). Tables 14.2 and 14.3 summarize some of the key characteristics of the Orbitrap.

HIGH-RESOLUTION MASS SPECTROMETRY—ACQUISITION MODES AND TYPES OF DATA

Regardless of the HRMS system utilized, actual performance is not dependant on instrument specifications, for example, mass accuracy or RP, alone; other factors, such as specimen type and choice of sample preparation, etc., will also play a key role in overall selectivity. Furthermore, in most cases, HRMS is used in combination with chromatographic procedures, utilized for the ability to separate complex mixtures and the components of a sample; however, techniques based on direct analysis methods—such as flow injection analysis [11], DART [12], and paperspray [13], have also been reported in toxicology.

As with nominal mass MS systems, HRMS also offer flexible options for data acquisition. Hybrid HRMS systems can even allow monitoring of precursors and their associated product ions in a manner very similar to product ion scanning on a standard tandem mass spectrometer. For example, the quadrupole of a QTOF can be operated at unit mass resolution to isolate a defined precursor (or precursors), the selected precursor ion can be fragmented in a collision cell and the accurate mass of any product ion(s) subsequently monitored using the high-resolution analyzer. The additional stage of selectivity when operating within this mode, provides significant enhancement in sensitivity over the full MS data acquisition, in the same way that SRM (precursor-to-product transition) or multiple reaction monitoring on a tandem mass spectrometer provides superior sensitivity over full MS or selected ion monitoring (precursor only) modes. Indeed the sensitivity for an HRMS operated in this mode can be comparable to that of a high end tandem MS in SRM mode. The downside is that acquisition rates for HRMS in this mode are usually slower than SRM on a tandem MS, meaning that the option is less conducive for the very large drug panels that have become commonplace with nominal mass tandem MS.

The undeniable appeal of HRMS however, is not with this kind of targeted analysis—but lies with the ability to collect "full" unrestricted MS data, and accordingly, all HRMS systems provide the ability to acquire accurate mass data of all molecular species at high sensitivity. From an analytical toxicologist's point of view, the measured *m/z* can be utilized for the identification and analysis of a known drug substance, however even a mass that is accurate to four or five decimal places may not be sufficient for an unequivocal identification for example, it will not permit differentiation of isomers (same elemental formula, different structure)—so other supporting (or discriminating) information is required

(Fig. 14.5). For unknown drug substances, an accurate mass may allow proposal of a likely elemental formula or formulae; however, this will only take an analyst part way toward an unambiguous identification—other information such as MS/MS data will be required to better elucidate the chemical structure of the molecule. Fortunately, most HRMS systems can also be operated in other acquisition modes that enable collection of MS/MS data during the same analysis; this data may be acquired in a data-dependent [14,15] or data-independent manner [8,16–18].

Data-Dependent Acquisition Techniques

Data-dependent acquisition or, information-dependent acquisition, as it sometimes referred to, usually commences with the acquisition of full MS data ("survey" MS). Within the same analysis, MS/MS can be triggered on the basis of (1) any unspecified precursor ion exceeding a defined abundance/intensity threshold or (2) any predefined ion (m/z of interest are submitted on an "inclusion list") exceeding an abundance/intensity threshold. The former method can be more prone to false positives/false negatives as setting an appropriate threshold is often a delicate balance of applying a threshold low enough to permit detection of low drug concentrations, but not setting this so low that MS/MS is triggered on background/contamination ions. The latter technique—where MS/MS is requested for specific m/z values—usually provides a simpler and more sensitive method.

The disadvantage to data-dependent modes is that while MS/MS data is being acquired, MS is not—thus the data is incomplete and not all available for retrospective data examination.

Data-Independent Acquisition Techniques

For broad screening in toxicology, data-independent MS/MS performed on an HRMS is a more attractive option as a complete dataset for both precursors and product ions are recorded within a single analysis. Vendor examples of data-independent modes include MS^E (Waters Corporation), "All Ions MS/MS" (Agilent Technologies), "All Ion fragmentation" (ThermoFisher), Broadband CID (Bruker), and SWATH (Sciex).

Typically data for precursor and product ions is generated by alternating rapidly between low and high (or ramped) collision energy functions: the low energy data provides information relating the accurate mass of all of the precursor ions, while the high energy provides accurate mass of all product ions (Fig. 14.9). As all product ions for coeluting analytes are present in the same spectrum, sophisticated software is required to rationalize the data. Some approaches also combine a data-independent low energy acquisition with a high collision energy acquisition over sequential quadrupole ranges.

The complete nature of data-independent approaches means that file sizes can be large however, it does permit retrospective analysis and, most importantly, open up the option to process the data using several strategies: targeted, semitargeted, and nontargeted.

Targeted, Semitargeted, and Nontargeted Analyses

In a targeted analysis, the acquired data can be simply compared with libraries prepared through analysis of reference material. This is the most straightforward approach and can be applied to both qualitative screening and quantitative workflows. Starter libraries are available from most HRMS vendors or can be prepared in-house. Where libraries are concerned, attention should be on quality of content rather than quantity, for example, large libraries based on elemental formula only will invariably yield many false positives which will impact on resource downstream (increased requirement to further review tentative identifications; in a forensic setting the need to perform a confirmatory procedure). Libraries comprising multiple parameters for each target and ideally comprising: precursor and product ions (or spectral data), ion ratios, retention time, collision cross section (ion mobility HRMS), etc., will ultimately result in a more confident screening result and fewer false positives.

Semitargeted (or "suspect") screening is a processing technique which leverages the power of HRMS to enable screening on the basis of the exact mass of a compound alone. In recent years, there has been a significant increase in the availability and the abuse of NPS; this situation places a burden on the analytical laboratories—how to ensure that screening methods remain "current" and continue to reflect the trends in drug usage? As availability of reference materials cannot always keep pace with emerging drug substances and that fact that some of the drug trends are short-lived and very transient in nature, an ability to at least screen for these novel drugs without reference material can be beneficial. With HRMS, the analyst can interrogate the acquired data to search for evidence of a response at the appropriate m/z (based on the elemental formula). Some software packages also include in silico fragmentation techniques which

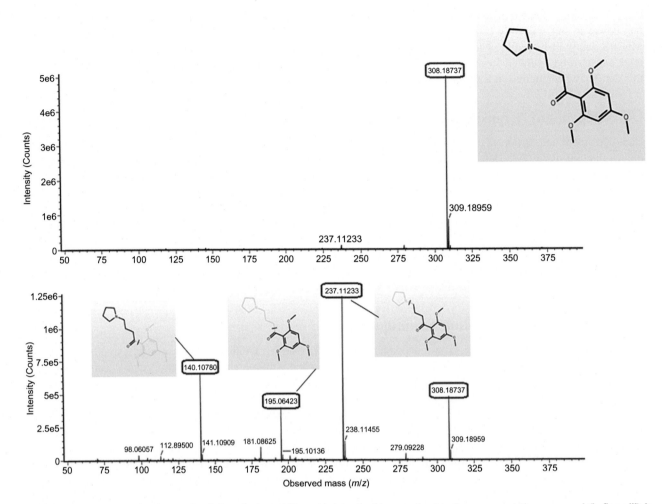

FIGURE 14.9 Example data from a data-independent acquisition technique. In this example, data for a representative compound (buflomedil) is shown; the low energy (upper trace) provides information relating to the intact precursor molecule and comprises measured mass for the molecule and information relating to the isotopic cluster. The high energy data (lower trace) comprises MS/MS information; in this case, three accurate mass fragment ions generated by ramping collision energy are observed.

can facilitate computer generated fragmentation of the precursor molecule to yield proposed theoretical fragment ions, based on simple thermodynamic rules [18,19]. Although evidence of these fragment ions at the same retention time as the precursor molecule, might add weight to a putative identification for the suspected compound, a "positive" response for this type of screen can only be considered a tentative identification at best, and would require further verification through acquisition of suitable reference material to confirm retention characteristics and spectra or actual fragment ions and their ratios.

In a similar manner, a complete dataset can also be interrogated for known common fragment ions or neutral losses. This strategy can aid discovery of new analogs, within the same drug class and which may share a common core structure, as well as metabolites. In other words, although it may not be possible to screen for the precursor mass for an unknown analog (unknown structure/formula means unknown exact mass) it may be possible to find related unknowns by routinely looking for known, characteristic fragment ions [20].

This can be illustrated using some data resulting from recent collaborative project on fentanyl analogs. Consider the current proliferation of fentanyl analogs; many of these analogs have been developed by substituting chemical groups at various positions of the core molecule. However under high energy conditions the precursor molecule of many of these analogs yield the same phenylethyl piperidine moiety ($C_{13}H_{18}N$, m/z 188.1434). Furthermore, this moiety can also subsequently cleave on the N-alkyl chain, between the α − carbon and the piperidine ring, to yield C_8H_9 m/z 105.0699. While analogs methylated on the phenylethyl piperidine moiety tend to yield a m/z 202.1590 fragment. An unidentified molecule within a case sample, giving rise to one or more of these fragments, might be suggestive of a related analog or metabolite and warrant further investigation (Fig. 14.10 and Table 14.4).

True "discovery" or nontargeted screening while certainly possible using HRMS, is more challenging than a targeted or semitargeted analysis; in fact one of the main challenges is the related to actually locating the *appropriate* peaks of interest to elucidate in the midst of all the other data. If an unknown is of sufficient magnitude to be visible in the data, the task is simplified—an accurate mass can be determined and this, along with isotopic cluster information, can be used to propose a possible formula or formulae. From here, chemical databases can be searched to provide a list of compounds with that same formula. It is at this point, that the MS/MS data come into play, especially when used in combination in silico fragmentation tools; measured fragment ions may support (or oppose) the proposed compound. As before, a tentative match from this process would require CRM to confirm identity.

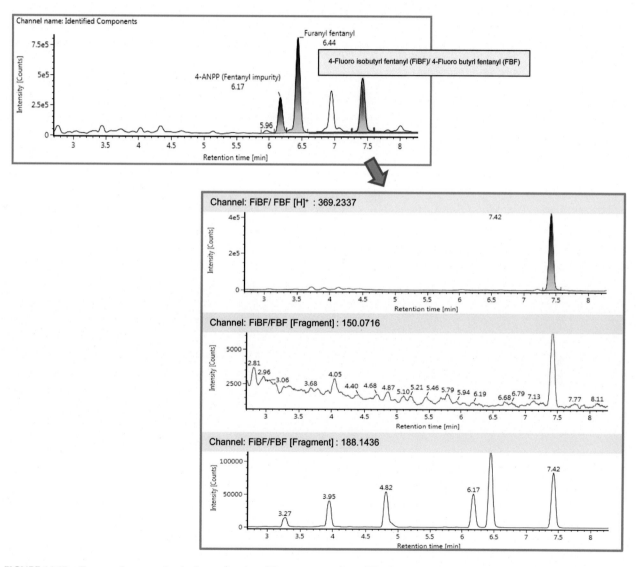

FIGURE 14.10 Common fragments for the fentanyl analogs. The upper trace shows HRMS analysis [8] of a de-intentified case sample that screened positive for furanyl fentanyl (FF), fluoro isobutyryl fentanyl/fluoro butyryl fentanyl (FiBF/FBF), and 4-ANPP, in addition to other drugs. Displayed in the lower panel are the specific details relating to the identification of FiBF/FBF, which eluted at 7.42 min. The data shows the extracted ion chromatograms for the precursor ion and two additional fragment ions as generated using MS^E mode. The lowest trace of this panel displays the extracted ion chromatogram for m/z 188.1434; a fragment ion that is common with other fentanyl analogs and their metabolites. As can be seen from this data, this fragment is also produced during fragmentation of many other analog as well as furanyl fentanyl (FF) and 4-ANPP. The earlier occurrences of the fragment at 3.27, 3.95, and 4.82 min respectively, may be indicative of metabolites of the analogs detected in this sample, or may be other unknown analogs/metabolites and, as such, could be candidates for structural elucidation. Source: *Personal communication - December 2017, Prof Thomas Rosano, National Toxicology Center, Albany, New York.*

TABLE 14.4 A Selection of Data for Fentanyl Analogs Analyzed Using an HRMS Screening Method Demonstrates the Occurrence of Common Fragment Ions

Compound	Elemental Formula	RT (min)	Precursor m/z	Fragment m/z	Fragment m/z	Fragment m/z
3-Methyl fentanyl[a]	$C_{23}H_{30}N_2O$	7.0	351.2431	202.1590	105.0699	
4-ANPP	$C_{19}H_{24}N_2$	6.1	281.2012	188.1434	105.0699	134.0964
4-Fluoro acrylfentanyl	$C_{22}H_{25}FN_2O$	6.3	353.2024	188.1434	105.0699	150.0714
4-Fluoro tetrahydrofuran fentanyl	$C_{24}H_{29}FN_2O_2$	6.0	397.2286	188.1434	105.0699	134.0964
Acetyl fentanyl	$C_{21}H_{26}N_2O$	5.1	323.2118	105.0699	188.1434	202.1226
Acetyl norfentanyl	$C_{13}H_{18}N_2O$	2.0	219.1492	202.1226	136.0757	84.0808
Acrylfentanyl	$C_{22}H_{26}N_2O$	6.0	335.2114	188.1434	105.0699	214.1226
Alpha-methyl acetyl fentanyl[b]	$C_{22}H_{28}N_2O$	5.6	337.2274	202.1590	91.0542	
Alpha-methyl fentanyl[a]	$C_{23}H_{30}N_2O$	6.6	351.2431	202.1590	91.0542	
Butyrfentanyl[a]	$C_{23}H_{30}N_2O$	7.3	351.2431	188.1434	132.08078	105.0699
4-Chlorofentanyl	$C_{22}H_{27}ClN_2O$	7.6	371.1885	188.1434	105.0699	79.0542
Crotonylfentanyl	$C_{23}H_{28}N_2O$	6.7	349.2274	188.1434	105.0699	134.0964
Cyclobutylfentanyl	$C_{24}H_{30}N_2O$	7.6	363.2431	188.1434	242.1539	134.0964
Cyclopentylfentanyl	$C_{25}H_{32}N_2O$	8.5	377.2587	188.1434	105.0699	281.2012
Despropionyl para-fluorofentanyl	$C_{19}H_{23}FN_2$	6.5	299.1912	188.1434	105.0699	134.0964
Fluoro isobutyryl fentanyl/fluoro butyryl fentanyl[c]	$C_{23}H_{29}FN_2O$	7.4/7.5	369.2337	188.1434	105.0699	150.0714
Furanyl fentanyl	$C_{24}H_{26}N_2O_2$	6.5	375.2063	188.1434	105.0699	254.1176
Fentanyl[b]	$C_{22}H_{28}N_2O$	6.3	337.2271	188.1434	105.0699	134.0964
Ocfentanil	$C_{22}H_{27}FN_2O_2$	5.2	371.2126	188.1434	105.0699	134.0964
Phenylacetyl fentanyl	$C_{27}H_{30}N_2O$	8.4	399.2431	188.1434	105.0699	134.0964
Thiofuranyl fentanyl	$C_{24}H_{26}N_2OS$	7.5	391.1839	188.1434	105.0699	134.0964
Valerylfentanyl	$C_{24}H_{32}N_2O$	8.4	365.2587	188.1434	105.0699	281.2012

Where the isomers exist (as indicated by the superscript), other means of differentiation such as retention time (RT) or fragment ions will be crucial. FiBF and FBF are considered too closely eluting to differentiate by this method thus are reported as FiBF/FBF.

Other data mining techniques to locate the peak(s) of interest have been applied such as mass defect filtering [21] or binary comparison, that is, a comparison of all of the m/z values for the components in a matrix blank "reference sample" for example, a single (or pooled) urine, with those observed in the "unknown" urine sample, to reveal unique m/z values [22]. Alternatively an ability to "strip" (subtract) all MS and data associated with known endogenous components that are associated with a particular biological matrix type, and leave possible components for elucidation may assist in discovery workflows.

Discovery of potential metabolites is also an easier task with HRMS than nominal mass instruments; authentic samples that are positive for a known parent drug can be automatically examined for common biotransformations such as phase 1 and phase 2 transformations—in what constitutes a simple mining for expected m/z ratios [23,24]. Again, high-resolution MS/MS data can add support to a tentative identification and even assist in understanding the most likely position of the particular transformation(s).

CONCLUSIONS

In recent years, the application of HRMS to toxicology screening has increased. A key contributing factor in the adoption of this technique has been the proliferation of NPS which cannot easily be addressed using targeted analytical methodologies such as LC-MS/MS. Consequently analysts are looking to integrate more "open," nontargeted screening procedures into their analytical workflow. HRMS provides the ability to perform targeted, semitargeted, and nontargeted analysis and as such, has emerged as a powerful technique in the attempt to keep pace with the highly dynamic field of emerging drugs and analogs.

REFERENCES

[1] Pedersen AJ, Dalsgaard PW, Rode AJ, Rasmussen BS, Muller IB, Johansen SS, et al. Screening for illicit and medicinal drugs in whole blood using fully automated SPE and ultra-high-performance liquid chromatography with TOF-MS with data-independent acquisition. J Sep Sci 2013;36(13):2081−9.

[2] Bidny S, Gago K, Chung P, Albertyn D, Pasin D. Simultaneous screening and quantification of basic, neutral and acidic drugs in blood using UPLC-QTOF-MS. J Anal Toxicol 2017;41(3):181−95.

[3] Stephanson NN, Signell P, Helander A, Beck O. Use of LC-HRMS in full scan-XIC mode for multi-analyte urine drug testing—a step toward a 'black-box' solution? J Mass Spectrom 2017;52(8):497−506.

[4] Ott M, Berbalk K, Plecko T, Wieland E, Shipkova M. Detection of drugs of abuse in urine using the Bruker Toxtyper™: experiences in a routine clinical laboratory setting. Clinical Mass Spectrom 2017;4−5:11−18.

[5] Kinyua J, Negreira N, Miserez B, Causanilles A, Emke E, Gremeaux L, et al. Qualitative screening of new psychoactive substances in pooled urine samples from Belgium and United Kingdom. Sci Total Environ 2016;573:1527−35.

[6] Montesano C, Vannutelli G, Massa M, Simeoni MC, Gregori A, Ripani L, et al. Multi-class analysis of new psychoactive substances and metabolites in hair by pressurized liquid extraction coupled to HPLC-HRMS. Drug Test Anal 2017;9(5):798−807.

[7] Maurer HH, Meyer MR. High-resolution mass spectrometry in toxicology: current status and future perspectives. Arch Toxicol 2016;90(9):2161−72.

[8] Rosano TG, Wood M, Ihenetu K, Swift TA. Drug screening in medical examiner casework by high resolution mass spectrometry (UPLC-MS^E-TOF). J Anal Toxicol 2013;37(8):580−93.

[9] Beynon JH. Qualitative analysis of organic compounds by mass spectrometry. Nature 1954;174(4433):735−7.

[10] Murray KK, Boyd RK, Eberlin MN, Langley GJ, Li L, Naito Y. Definitions of terms relating to mass spectrometry (IUPAC recommendations 2013). Pure Appl Chem 2013;85(7):1515−609.

[11] Alechaga E, Moyano E, Galceran MT. Wide-range screening of psychoactive substances by FIA-HRMS: identification strategies. Anal Bioanal Chem 2015;407(16):4567−80.

[12] Habala L, Valentová J, Pechová I, Fuknová M, Devinsky F. DART-LTQ ORBITRAP as an expedient tool for the identification of synthetic cannabinoids. Leg Med 2016;20:27−31.

[13] Michely JA, Meyer MR, Maurer HH. Paper spray ionization coupled to high resolution tandem mass spectrometry for comprehensive urine drug testing in comparison to liquid chromatography-coupled techniques after urine precipitation or dried urine spot workup. Anal Chem 2017;89(21):11779−86.

[14] Fels H, Dame T, Sachs H, Musshoff F. Liquid chromatography-quadrupole-time-of-flight mass spectrometry screening procedure for urine samples in forensic casework compared to gas chromatography−mass spectrometry. Drug Test Anal 2017;9(5):824−30.

[15] Partridge E, Tobbiani S, Stockham P, Scott T, Kostakis C. A validated method for the screening of 320 forensically significant compounds in blood by LC/QTOF, with simultaneous quantification of selected compounds. J Anal Toxicol 2018;42(4):220−31.

[16] Kinyua J, Negreira N, Ibáñez M, Bijlsma L, Hernández F, Covaci A, et al. A data-independent acquisition workflow for qualitative screening of new psychoactive substances in biological samples. Anal Bioanal Chem 2015;407(2):8773−85.

[17] Arnhard K, Gottschall A, Pitterl F, Oberacher H. Applying 'Sequential Windowed Acquisition of All Theoretical Fragment Ion Mass Spectra' (SWATH) for systematic toxicological analysis with liquid chromatography−high-resolution tandem mass spectrometry. Anal Bioanal Chem 2015;407(2):405−14.

[18] Mollerup CB, Dalsgaard PW, Mardal M, Linnet K. Targeted and non-targeted drug screening in whole blood by UHPLC−TOF−MS with data-independent acquisition. Drug Test Anal 2016;9(7):1052−61.

[19] Colby JM, Thoren KL, Lynch KL. Suspect screening using LC−QqTOF is a useful tool for detecting drugs in biological samples. J Anal Toxicol 2018;42(4):207−13.

[20] Pasin D, Cawley A, Bidney S, Fu S. Characterization of hallucinogenic phenethylamines using high-resolution mass spectrometry for nontargeted screening purposes. Drug Test Anal 2017;9(10):1620−9.

[21] Anstett A, Chu F, Alonso DE, Smith RW. Characterization of 2C-phenethylamines using high-resolution mass spectrometry and Kendrick mass defect filters. Forensic Chem 2018;7:47−55.

[22] Goggin MM, Nguyen A, Janis GC. Identification of unique metabolites of the designer opioid furanyl fentanyl. J Anal Toxicol 2017;4(5):367−75.

[23] Meyer MR, Bergstrand MP, Helander A, Beck O. Identification of main human urinary metabolites of the designer nitrobenzodiazepines clonazolam, meclonazepam, and nifoxipam by nano-liquid chromatography–high-resolution mass spectrometry for drug testing purposes. Anal Bioanal Chem 2016;408(13):3571–91.

[24] Krotulski AJ, Papsun DM, Friscia M, Swartz JL, Holsey BD, Logan BK. Fatality following ingestion of tetrahydrofuranylfentanyl, U-49900 and methoxy-phencyclidine. J Anal Toxicol 2018;42(3):27–32.

Chapter 15

Confirmation Methods for SAMHSA Drugs and Other Commonly Abused Drugs

Justin Holler[1] and Barry Levine[1,2]
[1]Chesapeake Toxicology Resources, Frederick, MD, United States, [2]Chief Medical Examiner, Baltimore, MD, United States

INTRODUCTION

One of the fundamental principles of forensic drug testing is the requirement for confirmation testing. No method based on a single analytical technique is considered sufficient for the forensic identification of a drug. Even two methods, based on the same analytical principle are not viewed as acceptable. A second test based on a different chemical principle is expected before reporting out the presence of a drug. Alternatively, a method that combines two different analytical principles, such as chromatography and mass spectrometry (MS), would satisfy the requirement. Nevertheless, if such a method is used for screening, it is recommended that the test be repeated with a second aliquot of specimen.

The Substance Abuse and Mental Health Services Administration (SAMHSA, an agency under the US Department of Health and Human Services) drug testing program was initiated in 1988 to include the primary drugs of abuse during that time period: amphetamine/methamphetamine, cocaine, marijuana, phencyclidine (PCP), and morphine/codeine [1]. In subsequent revisions, other analytes, including 6-acetylmorphine (6-AM), 3,4-methylenedioxymethamphetamine (MDMA), 3,4-methylenedioxyamphetamine (MDA), and 3,4-methylenedioxyethylamphetamine (MDEA) have been added to the testing panel. Revisions proposed in 2015 add oxycodone, oxymorphone, hydrocodone, and hydromorphone; which will be discussed in further detail under non-SAMHSA drug section [2].

SAMHSA guidelines require all specimens to be tested initially by a screening method. Guidelines also specify the cutoff concentrations for each drug. Any specimen that is at or above the cutoff should be called screened (presumptive) positive (Table 15.1). Currently, the initial test must be an immunoassay test. Generally, immunoassay methods are semiquantitative and the commercial reagents are able to produce consistent positive and negative results at +25% and −25% of cutoff, respectively. Screening reagents with a high throughput instrument are able to process a large number of specimens in a short period of time. Typically, an instrument can test 100 samples for 5 drugs in less than 10 minutes. Sufficient open and blind quality control samples have to be tested with the specimens for an acceptable result [1]. Antigens in the immunoassay reagents are designed to detect a target drug but they are not completely specific. Compounds with similar structures may cross-react with the reagents and may cause presumptive-positive results. The undesired positive rate could vary from <5% for 11-nor-delta-9-tetrahydrocannabinol (THC)-9-carboxylic acid (THC-COOH), the primary metabolite of cannabis use, to as much as 40% for amphetamine. Therefore, a confirmation method is necessary to identify the drug in the presumptive-positive urine. As required by the mandatory guidelines [1], a confirmation method must be a combination of separation by gas chromatography (GC) or liquid chromatography (LC) and MS detection.

Not surprising, misuse and abuse of drugs are not limited to these five classes. As a result, laboratories are often tasked to develop procedures to identify and quantitate new prescription or illicit drugs where patterns of misuse and abuse have developed. There are a number of drugs and drug classes that are not included in the SAMHSA panel, but with significant potential for misuse or abuse. Specifically, the drug classes include opioids and narcotic analgesics, benzodiazepines (BZD), antidepressants, barbiturates and "z" drugs. Although these drugs are not targeted in many

TABLE 15.1 Urine Drug Immunoassay Screening and GC-MS Confirmation Cutoff Concentrations Mandated by US Department of Health and Human Services (DHHS)[a]

Drugs	DHHS Cutoff (ng/mL)	
	Screening	Confirmation
THC-acid[a]	50	15
Benzoylecgonine[b]	150	100
PCP	25	25
Codeine/morphine	2000	2000/2000
6-Acetylmorphine[c]	10	10
Amphetamine/methamphetamine	500	250/250[d]
MDA/MDMA/MDEA[e]	500	250/250/250

[a]THC-acid: 11-nor-delta-9-tetrahydrocannabinol-9-carboxylic acid, a THC metabolite.
[b]A cocaine metabolite.
[c]A heroin metabolite.
[d]DL-amphetamine/DL-methamphetamine; For methamphetamine positive, specimen must have amphetamine ≥ 100 ng/mL.
[e]MDA: 3,4-Methylenedioxyamphetamine; MDMA: N-methyl-MDA; MDEA: N-ethyl-MDA.

regulated programs, their identification in forensic cases must satisfy the same analytical principles. The remainder of this chapter discusses confirmation testing for analytes in the SAMHSA program and for other potentially abused drugs.

CONFIRMATION TESTING: SAMHSA DRUGS

In general GC or LC (also known as high performance LC) is used for separation while appropriate mass spectrometric method is used for detection and confirmation of a particular drug.

Gas Chromatography/Mass Spectrometry

All samples need to be prepared prior to GC-MS confirmation. Sample preparation is a multistep process requiring hydrolysis of the drug conjugates (i.e., THC metabolite and morphine metabolite), extraction from urine (1–5 mL) to a suitable organic solvent, and derivatization (all drugs except for PCP). Drugs are extracted from urine using a liquid–liquid extraction or solid-phase extraction (SPE) technique. Derivatization is important for compounds with polar functionality. The derivatized compounds are volatile and produce better GC-MS chromatograms.

THC is the active compound of marijuana smoke. When ingested, the major metabolites of THC present in urine are THC-COOH and its conjugates. This conjugation is an ester of THC-COOH and hydroxy function of glucuronic acid. Although both base and enzyme hydrolysis can be used, base hydrolysis is preferred because it is simple and efficient. For morphine, most of the compound is present in urine as morphine-3-glucuronide ether. Although acid and enzyme hydrolysis are equally efficient in hydrolyzing the ether linkage, enzymatic hydrolysis is necessary to preserve 6-AM, a metabolite of heroin that has been a required analyte since the 1990s. Hydrolysis allows higher amount of free drug to be present in the sample and leads to better sensitivity.

Except for THC-COOH, all drugs tested under the mandatory guidelines are basic. Base extraction can be used to extract the drugs from the urine into a solvent. Aqueous acid–base purification produces an extract with only basic compounds free from acidic and neutral compounds. Base extractions are simplified by a SPE technique. When the urine samples are passed through the solid-phase column (silica or organic polymer with sulfonic acid), the basic drugs are retained in the column by the ionic bond with sulfonic acid, while other compounds in the urine pass through the column to waste. A linear hydrophobic -C_8H_{17} chain in the silica (-O-Si-O-) also enhances retention of the drug by interacting with the hydrophobic part of the drug molecule. For SPE with styrene-divinylbenzene sulfonic acid (SDVBS) polymer, the polymer itself participates in the enhanced adsorption. Reverse-phase and ion extraction have the major role in SPE methods. For reverse-phase in silica-based extraction, the conditioning of the column with water is important for efficient extraction. Water is added to the column to block the polar sites of the silicic acid by hydrogen bonding (-Si-OH-OH$_2$). When the drug is added, it partitions between the water and C$_8$- on silica. The partition is more in favor of the C$_8$- causing an efficient extraction of the drug from aqueous urine. Otherwise, free silicic acid will act like a normal phase similar to that in silica-based column chromatography. Conditioning is not necessary for SDVBS polymer because it does not contain any functional group that may behave like a normal phase column. SDVBS is more useful in SPE because it is stable from pH 1 to pH 14 whereas silica deteriorates at pH <4 and >10. Opiates in urine after acid hydrolysis at pH <1, can be poured directly onto the SDVBS columns without conditioning the column and adjusting the pH. Also, base elution of drugs produces minimum residue from the SDVBS columns compared to the amount from the silica column.

For extraction of THC-COOH, the urine is adjusted to pH 3.5 after alkaline hydrolysis and the compound is extracted by a solvent or by SPE technique. In SPE, acidic urine is poured onto the SPE column. At this pH, the carboxylic acid is in its nonionized form. Both the C$_8$- in silica and SDVB in SDVBS are efficient in retaining the compound on the column. The THC-COOH is then eluted by a polar solvent (mixture of iso-octane and ethyl acetate). The ease of adsorption, acid−base separation, and solvent elution has made the SPE a method of choice for sample preparation.

All SAMSHA drugs except for PCP (Table 15.2) need to be derivatized for a suitable chromatographic appearance and detection [3−44]. The carboxylic acid in THC-COOH and benzoylecgonine are generally derivatized to the esters by alkyl or perfluoroalkyl alcohols. The hydroxy groups either as phenol (THC-COOH, morphine) or alcohol (morphine, codeine, oxycodone, and oxymorphone) are derivatized to the esters by alkyl acid anhydride or perfluoroalkyl acid anhydrides. Both acid and hydroxy groups can also be derivatized to the trimethylsilyl compounds by trimethylsilyl reagents. The amines with hydrogen (amphetamine, methamphetamine, MDA, MDMA, and MDEA) are derivatized to the amides mostly by a long chain perhaloalkyl acid chloride. Chiral derivatization of amphetamine and methamphetamine using (R)-(-)- or (S)-(+)-alpha-methoxy-alpha-(trifluoromethyl)-phenylacetyl chloride (MTPA) is useful in chiral separation of (S)-(R)-isomers in GC-MS analysis [39,40].

Several mass spectrometric techniques are available for compound identification. The widely used technique in drug testing is the fragmentation of the compound by electron ionization at 70 eV and monitoring the fragment ions. The gaseous compounds from the GC column are introduced into the mass spectrometer and forced through an electron beam. On impact, the molecules gain energy and ionize. The excited molecules immediately undergo fragmentation. The ions are then forced through a set of quadrupoles for separation. When specific DC and AC voltages are applied to the quadrupole rods, only ions of a specific mass/charge ratio will pass through the rods. However, the number of ions passing through the quadrupoles are not enough to get a reasonable signal. To produce more electrons the ions bombard on an electron-rich metal surface (multiplier) to generate a detectable signal.

In mass fragmentation, the fragmentation pattern is characteristic to the molecule. When the energy is distributed in all parts of the molecule, the weakest part (bond) is likely to break. Occasionally, in the excited state, the molecule may rearrange to facilitate fragmentation to stable product ions. Analysis of the fragment ions provides the information on the molecular structure and is the basis of compound identification. Some fragmentation patterns are relatively simple to ascertain, but in a compound like morphine, 40 atoms are connected with 40 bonds. Predicting the fragmentation pattern and assigning structures to all fragment ions is practically impossible. Instead, only a few ions that are characteristic to the compounds are monitored and their structures are analyzed. Most laboratories use selected ion monitoring (SIM) technique to monitor only three ions and compare their abundances with that of a reference compound. The relative abundances are important in compound identification because any unrelated peaks or response at any point on the chromatogram will produce all three ions but their relative abundances are likely to be different. Initially, a standard is injected under a SCAN mode (range 50 and above molecular weight, amu) and the SIM ions that are free from any interference are chosen. Not many laboratories use full SCAN and compare the abundances of only three characteristic ions with that of the reference. In addition, laboratories compare the presence of other ions (fingerprints) and their abundances with that of the reference. A $\geq 85\%$ quality index (QI) match is used to identify the unknown. In compound identification, a comparison of SCAN is more selective than that of SIM. But the sensitivity is less in SCAN because the time (millisecond) spent monitoring each ion in SCAN is less than that when monitoring only three ions in SIM. However, GC retention time along with the relative ion abundances in SIM-MS is unique in compound identification. SAMHSA guidelines [1] require ion abundance matches to be within $\pm 20\%$. Although the guidelines do not mention the specific ions to be monitored, it is the responsibility of the laboratories to monitor ions that provide maximum information on the structure of the compound. With the exception of derivatized amphetamines, the molecular masses of the drugs are more than the masses of the derivatizing agents. But for amphetamines, some of the masses of the derivatizing agents are larger than that of the drugs. So, it is important to select the ions that contain major portions of the amphetamine molecule. Therefore, understanding the structures of the fragment ions is essential in selecting the appropriate ions.

TABLE 15.2 Commonly Used GC-EI/SIM/MS Methods for Selected Drugs Required to be Tested by Laboratories Certified by United States Department of Health and Human Services (DHHS)

Drugs	Derivatives	References
THC-acid[a]	Methyl-	[3–6]
	Trimethylsilyl-	[7–12]
	Perfluoroalkyl-	[13]
Benzoylecgonine[b]	Alkyl-	[14–17]
	Methyl-*tert*-butylsilyl	[18]
	Trimethylsilyl-	[19,20]
	Perfluoroalkyl-	[21]
PCP	None	[22,23]
Codeine/morphine	Alkyl-	[24,25]
	Trimethylsilyl-	[25]
	Perfluoroalkyl-	[24]
6-Acetylmorphine[c]	Propionyl ester-	[26]
	Perfluoroalkyl ester-	[25]
Amphetamine/methamphetamine	4-Carboethoxyhexafluorobutyryl-	[27,29]
	Perfluoroalkyl-	[30,31]
	Trichloroacetyl-	[32,33]
	Methyl-*tert*-butylsilyl-	[34]
	(S)-(-)-N-Trifluoroacetylprolyl-	[35–37]
	(R)-(-)-MTPA[d]	[38–40]
MDA/MDMA/MDEA[e]	4-Carboethoxyhexafluorobutyryl-	[29]
	Perfluoroalkyl-	[30,31]
	Trichloroacetyl-	[33]
	(R)-(-)-MTPA	[39]

[a]THC-acid: 11-nor-delta-9-tetrahydrocannabinol-9-carboxylic acid, a THC metabolite.
[b]A cocaine metabolite.
[c]A heroin metabolite.
[d](R)-(-)-alpha-methoxy-alpha-(trifluoromethyl)phenylacetyl-.
[e]MDA: 3,4-Methylenedioxyamphetamine; MDMA: N-methyl-MDA; MDEA: N-ethyl-MDA.

In most cases, it is difficult to derive the structure of a compound based on the full scan or SIM alone. Some prior information about the compound is essential to compare the results. This could be done either by comparing the full scan with that of a reference that is stored in an electronic library or with that of some structurally related homologs.

There are a number of limitations in the use of GC coupled with electron ionization Ms. For example, for underivatized amphetamines, the mass spectrum does not provide the molecular ion. The major fragment ions of m/z 44 for amphetamine and 58 for methamphetamine are available from the side chain which is common for many other sympathomimetic amines. Other ions are relatively weak. Moreover, the ions are small and subject to background interference. Silylation with *tert*-butyldimethylsilyl is not useful because it does not provide significant response for fragment ions that represent the entire molecule [34]. Derivatizing agents with electronegative perfluoro-groups provide suitable fragment ions. Therefore, appropriate derivatization is important to get a meaningful MS spectrum in drug analysis.

Occasionally, the compound may be thermally labile and rearrange in GC to a new compound (structural artifact). If this occurs, the mass spectral information may not be enough to reveal the structure of the parent compound. As an example, the opiate C-ring may undergo *retro*-Diels-Alder rearrangement in GC and the mass spectrum is identifying only the product. Therefore, in compound identification, it is not always the comparison of the physical property (RT) and mass spectral fragmentation but also the chemical changes that may occur during extraction, derivatization, and GC separation. Analysts must be aware of the possible double bond migration, alkyl migration, rearrangement, and even pyrolytic compounds produced from other compounds (dehydration or group expelled by removed from *beta*-elimination).

MS does not distinguish optical isomers. A suitable chiral derivatizing agent is necessary to separate the compounds in GC and then detect by mass spectrometer because mass spectra of these two derivatized isomers also will be identical. Therefore, one needs to use reference compounds (R or S) to identify the isomers by the GC retention times and not by mass spectrum. In chiral separation of (S)-(R)-isomers of amphetamine and methamphetamine, (R)-(-)-methoxy-trifluoromethylphenylacetyl chloride (MTPA) has been used [39,40]. In the chromatogram, the (R)-drug-(R)-MTPA isomer appears before the (S)-drug-(R)-MPTA isomer. As expected, fragmentation of (R)(R)- and (S)(R) are exactly the same and similar to that of 4-CB or HFB-derivatives of amphetamine and methamphetamine (m/z 91, 119, and M^+ 91) [39]. The strong ion m/z 189 represents only the derivatizing agent and cannot be used. Minor ions m/z 162 for amphetamine and m/z 176 for methamphetamine are significant because they represent the entire drug molecule excluding part of the derivatized group. An ion m/z 200 formed after several rearrangements in the molecule characterizes of methamphetamine. This phenomenon has also been observed with other derivatives of methamphetamine.

Detection of (S) and (R)-isomers after chiral separation with (S)-(-)-trifluoroacetylprolyl chloride (TPC) has limited applicability. The *alpha*-proton on the chiral carbon next to the carbonyl group causes some of the reagent change to the (R)-isomer. In the chromatogram, (S)-drug-(S)-reagent is not separable from the (R)-drug-(R)-reagent (from transformation). For this reason (R)-drug may be erroneously detected as (S)-drug. This transformation of (S)- to (R) of the reagent is generally limited to 10%. Therefore, the (S)- isomer of methamphetamine has to be at least 20% of the total (S)(R)-isomers in order for the specimen to be called positive for (S)-methamphetamine (or D-methamphetamine). Because of these limitations, the TPC method is generally not used as the sole confirmation method. Therefore, the MTPA method is the confirmation method of choice.

LIQUID CHROMATOGRAPHY COMBINED WITH MASS SPECTROMETRY OR TANDEM MASS SPECTROMETRY

Using LC separation prior to MS detection can mitigate many of the drawbacks to GC separation. Since the separation is run at much lower temperatures, there is less chance to alter the structure of thermally labile compounds. Many of the more polar compounds that require derivatization prior to GC separation do not need to be derivatized prior to LC separation. Sample preparation is less extensive; urine specimens often require only dilution prior to LC separation. The mobile phase offers another variable that can be manipulated to affect separation. Optically active compounds can be added to the mobile phase to permit chiral separation of D,L-amphetamine and methamphetamine.

The ionization techniques used after LC separation are more gentle than electron ionization; this reduces the potential for significant rearrangement in the mass spectrometer. For drug analysis, the two most commonly used ionization techniques are electrospray ionization (ESI) and atmospheric pressure chemical ionization (APCI). ESI is the most commonly used ionization technique after LC separation. The interface consists of a nebulizer, a desolvation assembly, and the electrode. As the liquid mobile phase exits the column, it is converted into an aerosol of charged droplets by pressurized nitrogen gas. As the droplets become smaller, they become closer to one another, leading to an electrostatic repulsion. This causes the production of even smaller droplets; this continues until the solvent is evaporated, leaving free ions. ESI is a soft ionization technique, usually producing the M + 1 ion. Both positive and negative ions can be produced [45].

The APCI interface has four components: The nebulizer, the vaporization tube, the corona needle, and the desolvation module. After leaving the column, the mobile phase enters the nebulizer and flows through the needle assembly. High pressure nitrogen gas is blown around the needle, converting the mobile phase and analytes into an aerosol. They are then carried by the nitrogen through the vaporization tube. The mobile phase is evaporated and the analyte is ionized by a discharge from the corona needle. As with ESI, both positive and negative ions may be produced [45].

Today, drug confirmations performed after LC separation typically employ tandem mass spectrometry (MS/MS). After the production of the ions as described above, these ions, called precursor ions, pass through a second quadrupole

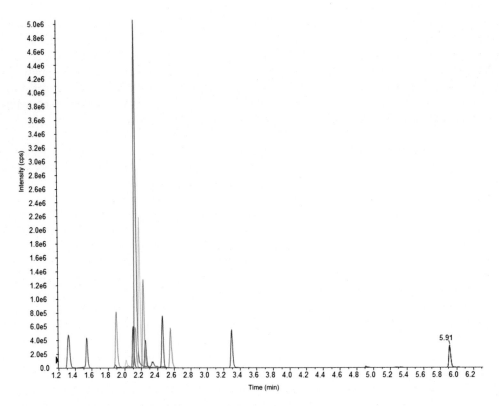

FIGURE 15.1 LC/MS/MS chromatogram of all SAMHSA drugs in a single assay (Retention time in minutes: 1.32 for morphine, 1.90 for codeine, 2.10 for 6-acetylmorphine, 2.13 for amphetamine, 2.18 for MDA, 2.23 for methamphetamine, 2.25 for MDMA, 2.46 for MDEA, 2.55 for benzoylecgonine, 3.30 for phencyclidine, and 5.91 for THC-COOH).

where additional ionization occurs by collision with a collision gas. The ions produced, called product ions, are detected by the third quadrupole. Mass filters can limit the ions passing into the second quadrupole and can be used to detect specific product ions from the third quadrupole. Ion rations of the product ions can be the basis of identification in a manner similar to SIM in GC/MS [45].

Whereas most GC/MS methods are designed to confirm one class of SAMHSA drugs, LC/MS/MS assays can simultaneously detect multiple drug classes [46–55]. In fact, all of the targeted SAMHSA analytes can be confirmed within a single LC/MS/MS chromatographic run following dilution and enzymatic hydrolysis. Fig. 15.1 gives an example of one such method. 6-AM, morphine, codeine, amphetamine, methamphetamine, MDMA, MDA, BE, PCP, and THC-COOH are all detected in the same method. The urine is diluted with an internal standard solution and the conjugated metabolites are hydrolyzed with glucuronidase. After dilution and centrifugation, an aliquot of the supernatant is introduced into the LC/Ms/MS.

General Confirmation Requirements

Regardless of the separation technique used, there are some general requirements for drug confirmations. A suitable deuterated internal standard provides the best quantification of the drug in biological fluids. Fragment ions must not interfere with the selected ions of the drug. Similarly, drug ions must not interfere with the selected ions for the internal standard.

Conditions used in extraction, separation, and MS detection may change compound configuration or may cause transformation to a new compound or artifact. Alkaline extraction of specimens with cocaine may cause hydrolysis of cocaine to benzoylecgonine. If a sample is contaminated by cocaine during collection or sample processing, it may show benzoylecgonine as the artifact. Sometimes, laboratories are requested to test benzoylecgonine-positive samples for *m*-hydroxy benzoylecgonine or *p*-hydroxy benzoylecgonine, two oxidative metabolites. The presence of these metabolites is indicative of cocaine ingestion and not a sample contamination. Concentration profiles of these two compounds have been reported [17]. If a passive exposure from cocaine smoke is an issue, detection of methyl ecgonidine (anhydroecgonidine methyl ester) or ecgonidine (anhydroecgonine) may be indicative of active smoking of cocaine [16,56], but these two compounds may be thermolytic artifacts of cocaine or benzoylecgonine. Controlling the injection port temperature to a minimum may prevent formation of the artifacts. Ephedrine or pseudoephedrine in the GC injection port may produce methamphetamine [57,58]. Periodate oxidation may be necessary to remove the interfering ephedrine and pseudoephedrine from methamphetamine [59]. However, the periodate method must be carefully utilized because it may produce amphetamine as an artifact from methamphetamine [28].

In urine drug testing, the cutoff concentrations required for confirmation are mentioned (Table 15.1). Although currently used instruments can detect a drug far below the cutoff, some of the cutoff concentrations are specifically designed to exclude positive results from unintended exposure of the drugs. THC from skin contact [60] and secondary smoke [61], ingestion of a small amount of THC in hemp products [62–67], cocaine from secondary smoke [68], skin absorption from handling of cocaine-contaminated currency [60,69], ingestion of *Inca Tea* [70,71], and ingestion of small amount of codeine and morphine in poppy seeds [72] are a few of the examples of unintended exposures. Despite the high cutoff concentrations, the urinary concentrations may, sometimes, exceed the cutoff depending on the amount of drugs in *Inca Tea* or poppy seeds. Although the urine cutoff concentrations are far above the instrument sensitivity, the drugs in hair and saliva are tested at concentrations as low as the instrument limits of detection (pg/mg or pg/mL) for special investigations.

CONFIRMATION TESING: NON-SAMHSA DRUGS

In this section, confirmation methods for non-SAMHSA drugs are discussed.

Narcotic Analgesics

Opioid testing has been in place since the beginning of the SAMHSA program in 1988. The two target analytes at that time were morphine and codeine. In 1998, confirmation testing for 6-AM, a specific marker of heroin use, was added. Since positive opioid results could only be reported for these substances, commercially available immunoassays were designed with these analytes in mind [1]. However, use and abuse of opioids and narcotic analgesics is not restricted to heroin, morphine, or codeine. Among other drugs used are the keto-opioids (oxycodone, oxymorphone, hydrocodone, and oxymorphone) and other narcotic analgesics (meperidine, methadone, fentanyl, tramadol, tapentadol, and buprenorphine) [73]. Some of this abuse developed as a result of therapeutic use of these compounds for analgesia. In addition, diversion of these products to the illicit drug market also occurred.

The structure of oxycodone differs from the typical opioid due to the presence of a ketone group. Oxycodone is similar in potency to morphine and is listed as a Schedule II controlled substance. In the 1990s, the release of a sustained release form of oxycodone, OxyContin, with its significantly higher doses contributed to an increase in abuse [74]. Historically, oxycodone was prescribed in 2.5 and 5 mg tablets with doses ranging from 2.5 to 10 mg commonly administered for the treatment of pain in the nonclinical setting [75]. OxyContin tablets are marketed in different dosage forms with each tablet containing oxycodone in doses of 10, 20, 40, 80, or 160 mg [74]. At relatively higher doses, it has become more convenient and less expensive to consume greater doses of oxycodone especially at a dosage of 80 or 160 mg. Colloquially named "Hillbilly Heroin," formulations with higher doses of oxycodone became a viable alternative to heroin.

Oxycodone is well absorbed orally as prescribed and has a quicker onset of action [76]. The duration of action ranges from 4 to 12 hours depending on the formulation [74]. Oxycodone is metabolized via *N*- and *O*-demethylation leading to the formation of an active metabolite oxymorphone (oxymorphone is also available as a drug which is used in pain management and is known as Opana) [73]. The majority of the dose is excreted in the urine as free oxycodone and noroxycodone as well as conjugated oxycodone and oxymorphone.

Hydrocodone is also classified as a keto-opioid, similar in structure to oxycodone except that it does not have a hydroxyl group at carbon 14. At lower doses, it is used as an antitussive and at higher doses; it is prescribed as an analgesic. The duration of action is between 4 and 6 hours and has approximately six times the potency of codeine [77]. Like oxycodone, hydrocodone is *N*- and *O*-demethylated during metabolism. Hydromorphone (Dilaudid) is also an active metabolite that is 8.5 times more potent than morphine [77]. Hydrocodone is also metabolized to dihydrocodeine by reduction of the ketone group.

Meperidine (pethidine, Demerol) is a synthetic narcotic analgesic used for pain management with less potency and shorter duration of action than morphine. It is administered primarily by the oral route, although it is available via parenteral injection [75]. Meperidine is quickly absorbed with high oral bioavailability and is quickly metabolized [78].

The metabolic profile consists of N-demethylation to normeperidine, an active metabolite and hydrolysis to meperidinic and normeperidinic acids as well as conjugation [79]. The metabolic profile is dependent upon pH of the urine.

Methadone (Dolophine, Methadose) is used primarily for maintenance treatment of opiate addicts but can also be used for pain management. Methadone is widely available as a racemic mixture even though the (R)-isomer has ten times the binding affinity to the μ and δ opioid receptors [80]. Methadone works well for withdrawal treatment because it has high oral bioavailability, long elimination half-life, low withdrawal symptoms following discontinuation, and specific antagonists for treatment of overdose [81,82]. Methadone, as mentioned previously, has a relatively long half-life ranging from 15 to 55 hours [75]. It is metabolized by mono and di-demethylation to form unstable compounds that undergo spontaneous cyclization. The resulting metabolites are 2-ethylidene-1,5-dimethyl-3,-3-diphenylpyrrolidine (EDDP) and 2-ethyl-5-methyl-3,3-diphenylpyrrolidine (EMDP), both of which are pharmacologically inactive.

Fentanyl, a synthetic narcotic analgesic, is 80 times more potent than morphine and is used mainly as an anesthetic and an analgesic. Routes of administration available are via intravenous drip, transdermal patch (Duragesic), and an oral lozenge [83]. Fentanyl has a short duration of action due to its short half-life which makes it ideal for surgical anesthesia. Small doses of fentanyl are pharmacologically effective due to its relatively high potency. The parent compound can be detected in the blood and a small percentage is found in the urine. Norfentanyl is the primary metabolite detected in urine along with unknown amounts of the hydroxylated metabolites [75]. Within the past several years, analogs of fentanyl have appeared on the illicit market and have been responsible for fatalities throughout the country. Most of the analogs are found in smaller doses than fentanyl as they are more potent causing detection in blood and urine to be difficult.

Tramadol is a centrally acting analgesic that has been used in the United States since 1995 and is equal in potency to codeine [75,84]. The primary excretion route for tramadol is via the urine and consists of parent compound and the N- and O-demethylated free and conjugated metabolites [75]. Tapentadol is a centrally acting analgesic that is structurally similar to tramadol. It has been available in the United States since 2008. It is metabolized by N-demethylation, alkyl hydroxylation, and conjugation.

Buprenorphine is a partial opioid agonist used both as an analgesic and as a treatment for opioid addiction. It is metabolized by N-dealkylation to norbuprenorphine followed by conjugation and is often prescribed in conjunction with naloxone as suboxone. The use of buprenorphine has increased significantly with the rise of the opioid epidemic in the United States.

Confirmation analysis for narcotic analgesics is straightforward. Liquid/liquid and SPE techniques are effective for non-SAMHSA narcotic analgesics [84–95]. As mentioned previously, conjugated metabolites are present for keto-opiates; therefore, hydrolysis is necessary for quantitation. This can be accomplished using either enzymatic or acid hydrolysis, with acid being the more effective procedure [87]. Analysis of keto-opiates requires a prederivatization step to prevent tautomerization. The ketone can be derivatized with hydroxylamine or methoxylamine to form oximes in conjunction with a second derivatization following extraction (Fig. 15.2). This will inhibit the ketone from converting to an enol in the presence of protons during typical derivatization procedures. Enol formation may cause interference with other opiates [86]. Other issues involved in derivatization are incomplete derivatization, multiple derivatives, and coelution of compounds [88,89]. The multiple functional groups present on the keto-opiates, especially the tertiary hydroxyl group for oxycodone and oxymorphone can lead to the production of multiple derivatives per compound. The use of a deuterated internal standard for each keto-opiate provides accurate quantitation but coelution problems can still persist with formation of multiple derivatives.

The structures of meperidine, tramadol, methadone, hydrocodone, oxycodone, tapentadol, and fentanyl allow for GC/MS analysis without utilizing derivatization. A number of metabolites of these drugs can also be analyzed without derivatization but may result in tailing chromatography, especially normeperidine and norfentanyl. The use of a MS or MS/MS detector has provided better sensitivity leading to accurate quantitation at low concentrations (Fig. 15.3). Sensitivity is critical for analysis of fentanyl because of small doses that are administered. The improved sensitivity has also led to other matrices being analyzed for narcotic analgesics such as oral fluids and hair [93,94]. The GC/MS analysis of methadone can be problematic due to alpha cleavage that occurs producing a base peak of m/z 72 amu (Fig. 15.4). Other fragment ions are available but the response is low compared to the base peak (Fig. 15.5). LC/MS or GC-chemical ionization-MS can be employed to produce a molecular ion as the base peak but little fragmentation will occur [81,95].

LC/MS/MS analysis can be used to overcome many of the challenges associated with the GC/MS analysis of opioids. There is no need for pre or post derivatization of the keto-opiates. Specimen preparation may be a simple dilution with internal standard solution following enzymatic hydrolysis. Since morphine/hydromorphone and codeine/hydrocodone are isobars, the specificity of the tandem mass spectrometer is insufficient for separation; however, chromatographic separation is not difficult. Fragmentation of other narcotic analgesics such as methadone are not an issue with LC/MS/MS either. The use of MS/MS technology also improves sensitivity which allows for identification and quantitation of fentanyl, norfentanyl, and other fentanyl analogs.

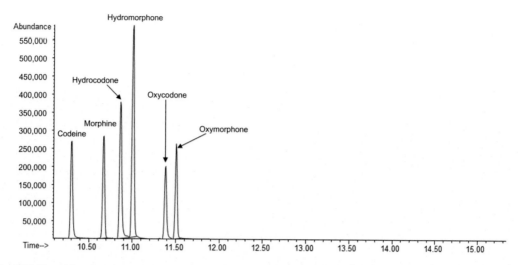

FIGURE 15.2 GC/MS total ion chromatogram for opiates at 0.5 mg/L in blood with dual derivatization using hydroxylamine and BSTFA with 1% TMCS.

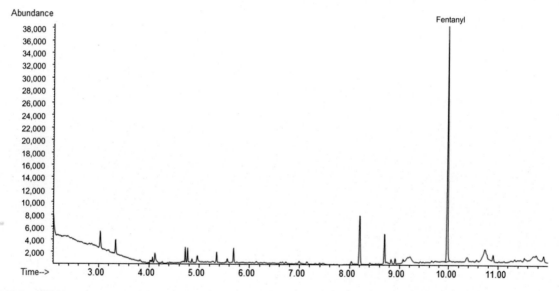

FIGURE 15.3 GC/MS total ion chromatogram for fentanyl at 4 ng/mL in blood.

Benzodiazepines

BZD have become one of the most commonly prescribed drug classes. There are now more than 50 various BZD used for treating anxiety and insomnia. Chlordiazepoxide was the first benzodiazepine introduced in 1961. BZD have a number of therapeutic used, including anxiety reduction, sedation, anesthesia, seizure control, and muscle relaxation. The structural composition of BZD allow for substitutions at five positions leading to the large number of compounds available. They can be divided into three groups: 1,4-benzodiazepines (diazepam), diaxolobenzodiazepines (midazolam), and triazolobenzodiazepines (alprazolam). The structural differences cause changes in polarity, bioavailability, potency, and duration of action. The half-life for BZD range from less than 2 to over 100 hours depending on the drug administered [75]. The metabolic pathways for most BZD are via dealkylation, reduction, and hydroxylation followed by conjugation. Very little, if any parent compound is detected in the urine. Many BZD follow common metabolic routes, leading to the formation of the same metabolites. For example, both diazepam and chlordiazepoxide are metabolized to nordiazepam and oxazepam. As a result, the identification of parent drug from urinary products may be difficult. This is further complicated by the fact that many of these metabolites may themselves be prescribed as drugs. Some common metabolites are oxazepam, temazepam, nordiazepam, and conjugated glucuronides [96,97]. In the urinary profile, 30–80% of a BZD dose is found as free and conjugated metabolites [75]. Stability in biological fluids can be

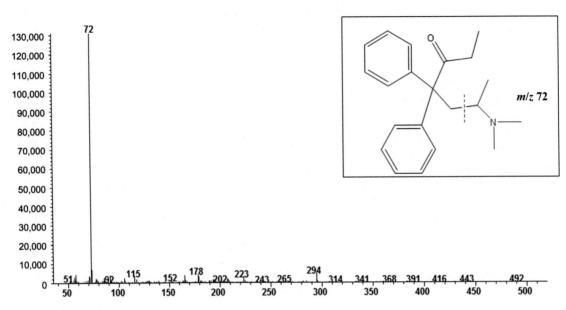

FIGURE 15.4 GC/MS full scan spectrum for methadone with *m/z* range of 50–550 amu.

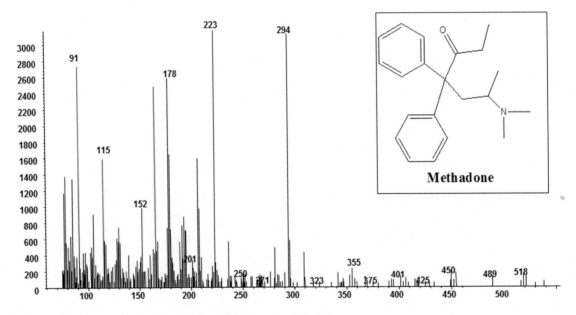

FIGURE 15.5 GC/MS full scan spectrum for methadone with *m/z* range of 75–550 amu.

problematic for some BZD, especially the 7-nitro compounds, such as clonazepam and flunitrazepam. These drugs are both metabolized and break down in vitro to their 7-amino analogs.

Confirmation of BZD is dependent on the specimen being tested. Blood testing requires the targeting of parent compounds while urine testing requires the targeting primarily of metabolites. The large number of BZD that are available is certainly an issue for determining what analytes to report. As the number of BZD continues to increase, this problem becomes very significant. Some laboratories use multiple extraction procedures to quantitate BZD to account for large number of compounds and structural differences. Another reason for multiple extraction procedures is due to the wide range of drug and metabolite concentrations encountered when testing for BZD. For example, alprazolam is prescribed in daily doses ranging from 0.75 to 9 mg while diazepam may be prescribed in doses up to 40 mg [75]. The increasing prevalence of BZD has made testing in both clinical and forensic laboratories almost mandatory.

BZD can be extracted from biological specimens by either liquid/liquid or SPE. Prior to extraction of urine specimens, a hydrolysis step must be completed to cleave the glucuronide conjugation. Most procedures use an enzymatic

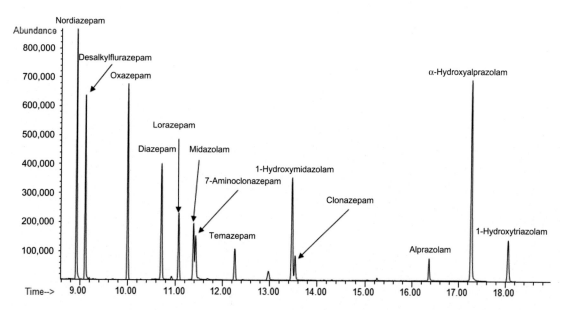

FIGURE 15.6 GC/MS total ion chromatogram for benzodiazepines at 0.1 mg/L in blood after flash derivatization with tert-butyl trimethylsilyl derivatization.

hydrolysis because it is weaker reaction than acid hydrolysis which can convert some BZD to benzophenones. Adjustment of pH following hydrolysis is necessary for both extraction techniques. Derivatization is not required for some BZD like diazepam and midazolam but it is necessary for others, including 7-aminoclonazepam and α-hydroxyalprazolam. Popular derivative choices are alkyl, acyl, and silyl derivatives.

Confirmation testing for BZD is accomplished using either GC or LC with MS or MS/MS detection [96–99]. Deuterated internal standards are commonly used for BZD analysis by MS or MS/MS but are not available for all drugs. Quantitation can be accomplished using a structurally similar deuterated standard. GC can be problematic because of high injection port temperatures affecting thermally labile compounds such as chlordiazepoxide. Derivatization can address the problem and may occur prior to injection or in the injection port. Reconstitution with the derivative (N-methyl-N-tert-butyl-dimethylsilyl-trifluoroacetamide (MTBSTFA)) followed by injection at high temperatures will cause the compounds to flash derivatize, a time-saving process. Separation of BZD is especially important because of the large number of compounds that can be detected. If a laboratory is monitoring 10 or more drugs, separation is a critical part of analysis. A standard phenylmethyl GC column can be used for separation but with compounds that do not derivatize, chromatographic quality may be unacceptable. The use of a more polar stationary phase column containing trifluoropropyl will improve chromatography for underivatized compounds as well as may provide better overall separation (Fig. 15.6). MS is most commonly used for identification and quantitation. Weak fragmentation patterns for BZD are common at typical source parameters of 230°C and 70 eV; therefore it may be necessary to monitor isotopic peaks present because of halogenation. This is a standard practice for BZD analysis [98,99]. LC can alleviate the issues encountered with GC. Derivatization is not required for LC and thermal stability is eliminated by ambient injection temperatures. Standard reverse-phase C8 or C18 columns can be used for separation of BZD. Some methods that detect multiple BZD sometimes do not provide chromatographic resolution and rely on monitoring different ions to distinguish between compounds [98]. LC/MS utilizes a soft ionization technique leading to detection of the molecular ion and poor fragmentation for identification but LC/MS/MS allows for adequate fragmentation. The extraction issues of compound stability can be removed by doing a dilute and shoot followed by LC/MS/MS analysis.

Barbiturates

Although abused widely in the past, barbiturates are infrequently abused in the present time. The presence of barbiturates can be screened using Food and Drug Administration (FDA) approved assays and confirmation can be achieved using GC/MS analysis after alkylation, the most commonly used derivatization method. GC/MS total ion chromatogram of methylated barbiturates at 2 mg/L blood concentration is shown in Fig. 15.7.

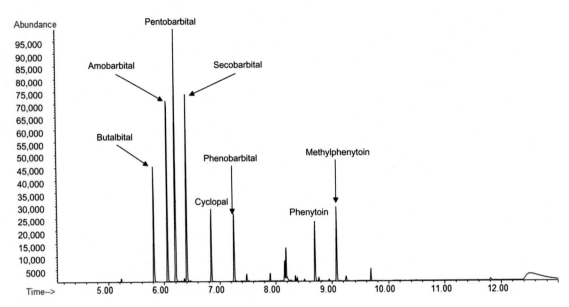

FIGURE 15.7 GC/MS total ion chromatogram of methylated barbiturates at 2 mg/L blood concentration.

Antidepressants

Clinical diagnosis for depression has become a major part of today's society. Studies show that as many as 16% of Americans will suffer from depression at one point in their lifetime. Treatment for depression often involves prescription medication that is classified into three groups: First, second, and third generation antidepressants. The first generation consists of tricyclic antidepressants (TCAs) such as amitriptyline, nortriptyline, imipramine, desipramine, and doxepin. These drugs were first prescribed for depression in the 1950s. All of these compounds contain a backbone structure consisting of three rings, two aromatic and one seven atom ring; they also contain either secondary or tertiary amine groups. TCAs primary mode of action is inhibition of reuptake for catecholamines such as serotonin, dopamine, and norepinephrine. In general, TCAs are not very specific in their actions toward catecholamine reuptake. The lack of specificity is a major disadvantage of TCAs because they can have a high potential for side effects. The major route of metabolism for TCAs is N-demethylation (most form active metabolites), oxidation, and hydroxylation followed by conjugation [75]. Oral administration is the primary route of administration but some TCAs can also be injected intramuscularly. The average daily dose ranges from 75 to 300 mg with a typical half-life of 5−30 hours [75].

The second generation of antidepressants cannot be classified as a group structurally and consists of amoxapine, bupropion, trazodone, and maprotiline. Amoxapine acts similarly to TCAs with oral daily doses ranging slightly higher around 600 mg. The half-life is 8 hours and its major metabolic pathway is hydroxylation followed by conjugation [75]. Bupropion is not related to TCAs or any other antidepressant compound. Bupropion does not inhibit the reuptake of serotonin; it inhibits only norepinephrine and dopamine reuptake. Because of this dopamine reuptake inhibition, bupropion is also used for smoking cessation. Trazodone is also nonstructurally related to most antidepressant drugs. Maprotiline is a tetracyclic antidepressant drug with a long half-life ranging from 26 to 105 hours [75].

The third generation antidepressants have become the most widely used group to treat depression. These drugs are primarily selective serotonin reuptake inhibitors (SSRIs) and as the name suggests are very specific for the inhibition of serotonin reuptake. Some examples of SSRIs are fluoxetine, sertraline, citalopram, and paroxetine. The specificity of SSRIs eliminates the adrenergic, anticholinergic, and antihistaminic side effects that are common with earlier generation antidepressants. SSRIs are much safer due to their specificity but still display some side effects such as anxiety and sleep disturbances. SSRIs are administered orally and have a typical half-life ranging from a half a day up to 3 days [75]. Other drugs included in the third generation antidepressants are duloxetine and venlafaxine, selective norepinephrine, and serotonin reuptake inhibitors. Venlafaxine has a much shorter half-life of 3−7 hours. Venlafaxine is metabolized to N-desmethylvenlafanine and O-desmethylvenlafaxine (ODMV); ODMV is also marketed as an antidepressant.

Confirmation testing for antidepressants also requires a separation step [99−105]. The primary method for extraction is liquid/liquid using a pH 10 buffer and an immiscible organic solvent such a 1-chlorobutane. One step procedures have been used but the addition of a back extraction can help eliminate many interfering compounds in chromatographic analysis [99,100]. However, SPE techniques have increased the efficiency of extractions as well as

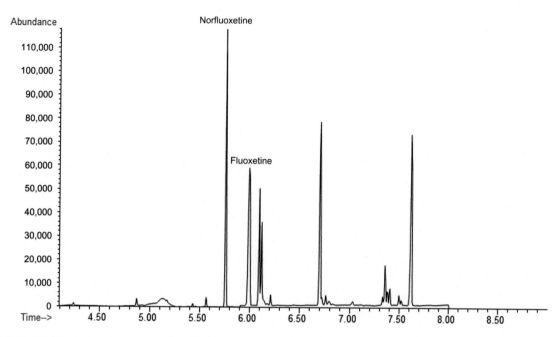

FIGURE 15.8 GC/MS total ion chromatogram for trifluoroacetylated fluoxetine and norfluoxetine at 0.5 mg/L in blood.

decreased the amount or organic waste generated. Antidepressants have been analyzed with unique procedures as well, including solid-phase microextraction, microwave assisted extraction, and direct injection [103–105].

Antidepressants can be confirmed using GC/MS, LC/MS, and LC/MS/MS. GC/MS total ion chromatogram of trifluoroacetylated fluoxetine and norfluoxetine at 0.5 mg/L in blood is shown in Fig. 15.8. All such techniques have proven to be successful in analysis of antidepressants. One disadvantage in the use of electron ionization GC/MS for the underivatized drugs is the poor fragmentation patterns produced for identification. The molecular ion has very low abundance if any at all and the alpha cleavage that occurs in most compounds as well as the thermal instability of antidepressants results in poor identification through fragmentation pattern. Better identification of the antidepressants by GC/MS can be achieved by derivatization, resulting in higher mass ions. Alternatively, the use of chemical ionization allows production of the pseudomolecular ion because it is a softer ionization technique than electron ionization. The use of LC/MS can eliminate the need for derivatization of the metabolites for most compounds. The pseudomolecular ion will be observed due to soft ionization but fragmentation will still be a concern. The relatively high dosage for antidepressants allows a single MS to provide adequate detection limits as well. The use of MS/MS systems can lead to better mass spectrum identification for LC analysis by multiple steps of fragmentation providing parent/daughter ion combinations. LC/MS/MS also alleviates the need for extraction as dilute and shoot is adequate for detection and typically instruments will require detuning to reduce instrument response.

"Z" Drugs

In the 1990s, additional central nervous system depressant drugs were approved in the United States, among them were zaleplon (Sonata), zolpidem (Ambien), and zopiclone (Lunesta). Each of these drugs have short half-lives, less than 7 hours and are used as hypnotic agents. Zaleplon is extensively metabolized by N-desalkylation, oxo-formation, and conjugation. Zolpidem is metabolized by side chain hydroxylation, carboxylation, and conjugation. Zopiclone is metabolized by N-demethylation, N-oxidation, and ester hydrolysis. None of the metabolites of these drugs reportedly are pharmacologically active [75].

Although these drugs are extensively metabolized, confirmation testing of the "Z" drugs targets the parent compounds. Zolpidem and zopiclone can be confirmed after GC separation; the cyano group on zaleplon make GC separation more problematic. All three compounds can be confirmed using LC/MS or LC/MS/MS. One complication in zopiclone analysis is that the compound is unstable in alkaline solutions.

To illustrate the utility of LC/MS/MS in the analysis of non-SAMHSA drugs, Fig. 15.9 shows a total reaction chromatogram of 15 opioid compounds, 7 benzodiazepines, 13 antidepressants, and 2 "z" drugs. Because of the simple specimen preparation and high specificity of the tandem mass spectrometer, more drugs and metabolites can be added as needed.

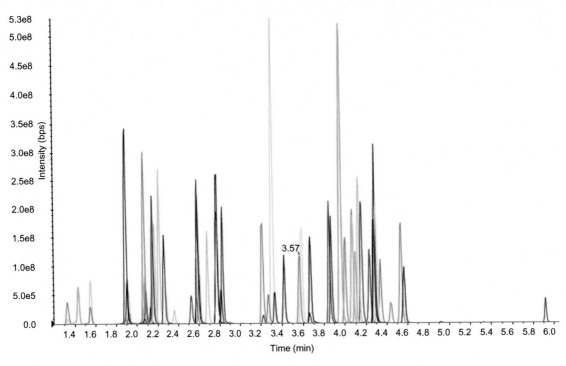

FIGURE 15.9 LC/MS/MS chromatogram of non-SAMHSA drugs in a single assay (Retention time in minutes for individual compounds are listed in Table 15.3).

TABLE 15.3 Retention time (in minutes) of various SAMHSA and non-SAMHSA drugs analyzed in a single LC/MS/MS analysis as described in Fig. 15.9

RT	Compound	RT	Compound
1.36	Morphine	3.26	Norbuprenorphine
1.46	Oxymorphone	3.31	Venlafaxine
1.58	Hydromorphone	3.34	Meprobamate
1.92	Dihydrocodeine	3.37	PCP
1.94	Naloxone	3.45	Fentanyl
1.96	Codeine	3.59	Quetiapine
2.10	Naltrexone	3.62	Citalopram
2.10	Noroxycodone	3.70	Buprenorphine
2.10	Oxycodone	3.71	EDDP
2.10	Pregabalin	3.89	Atomoxetine
2.11	Gabapentin	3.91	Paroxetine
2.16	6-Acetylmorphine	4.00	Cyclobenzaprine
2.18	Norhydrocodone	4.04	Duloxetine
2.19	Hydrocodone	4.11	Amitriptyline
2.20	Amphetamine	4.11	Nortriptyline
2.25	MDA	4.14	Methadone
2.30	Methamphetamine	4.17	Norfluoxetine
2.32	MDMA	4.20	Fluoxetine
2.56	7-Aminoclonazepam	4.27	Hydroxyalprazolam
2.61	Benzoylecgonine	4.32	Norsertraline
2.62	Desmethylvenlafaxine	4.32	Sertraline
2.63	Norfentanyl	4.33	Oxazepam
2.71	Zopiclone	4.34	Carisoprodol
2.80	Tramadol	4.38	Lorazepam
2.84	Methylphenidate	4.47	Alprazolam
2.86	mCPP	4.57	Nordiazepam
2.88	Tapentadol	4.60	Temazepam
3.05	Zolpidem	5.93	Carboxy-THC
3.25	Trazodone		

CONCLUSION

The advancement in instrument technology has allowed the analysis of not only SAMHSA drugs but many non-SAMHSA drugs to be performed within a single assay. LC/MS/MS instruments have become so sensitive that sample preparation can be a simple dilution with internal standard solution and enzymatic hydrolysis to cleave conjugated metabolites. About 50–60 analytes can be completed within 60–80 minutes using LC/Ms/MS.

ACKNOWLEDGEMENTS

The authors wish to acknowledge the contributions of Buddha D. Paul, PhD and Marilyn R. Past, PhD in the section on SAMHSA drug confirmation.

REFERENCES

[1] Department of Health and Human Services (Substance Abuse and Mental Health Services Administration): Mandatory Guidelines for Federal Workplace Drug Testing Programs, Notices; *Fed Reg* 53:11970, 1988; 58:6062, 1993; 59:29908, 1994; 60:57587, 1995; 62:51118, 1997; 69:19644, 2004; 73: 71858 2008 and National Laboratory Certification Program: Manual for Laboratories and Inspectors, Research Triangle Park, NC, November 2006.

[2] Department of Health and Human Services (Substance Abuse and Mental Health Services Administration): Mandatory Guidelines for Federal Workplace Drug Testing Programs, Notices; *Fed Reg80*: 28101 2015.

[3] Whiting JD, Manders WW. Confirmation of 9-carboxy-THC in urine by gas chromatography/mass spectrometry. Aviat Space Environ Med 1983;1031–3.

[4] Paul BD, Mell LD, Mitchell JM, McKinley RM, Irving J. Detection and quantitation of urinary 11-nor-delta-9-tetrahydrocannabinol-9-carboxylic acid, a metabolite of tetrahydrocannabinol, by capillary gas chromatography and electron impact mass fragmentometry. J Anal Toxicol 1987;11:1–5.

[5] Paul BD, Jacobs A. Effects of oxidizing adulterants on detection of 11-nor-delta-9-THC-9-carboxylic acid in urine. J Anal Toxicol 2002;26:460–3.

[6] Jamerson MH, Welton RM, Morris-Kukoski CL, Klette KL. Rapid quantification of urinary 11-nor-delta-9-tetrahydrocannabinol-9-carboxylic acid using fast gas chromatography-mass spectrometry. J Anal Toxicol 2005;29:664–8.

[7] Baker TS, Harry JV, Russell JW, Myers RL. Rapid method for the GC/MS confirmation of 11-nor-9-carboxy-delta-9-tetrahydrocannabinol in urine. J Anal Toxicol 1984;8:255–9.

[8] Dixit V, Dixit V. Solid-phase extraction of 11-nor-delta-9-tetrahydrocannabinol-9-carboxylic acid from human urine with gas chromatography-mass spectrometry. J Chromatogr (B) 1991;567:81–91.

[9] Singh J, Johnson L. Solid-phase extraction of THC metabolite from urine using Empore disk cartridge prior to analysis by GC-MS. J Anal Toxicol 1997;21:384–7.

[10] O'Dell L, Rymut K, Chaney G, Darpino T, Telepchak M. Evaluation of reduced solvent volume solid-phase extraction cartridges with analysis by gas chromatography-mass spectrometry for determination of 11-nor-9-carboxy-delta-9-THC in urine. J Anal Toxicol 1997;21:433–7.

[11] De Cock KJS, Delbeke FT, De Boer D, Van Eenoo P, Roels K. Quantitation of 11-nor-delta-9-tetrahydrocannabinol-9-carboxylic acid with GC-MS in urine collected for doping analysis. J Anal Toxicol 2003;27:106–9.

[12] Abraham TT, Lowe RH, Pirnay SO, Darwin WD, Huestis MA. Simultaneous GC-EI-MS determination of delta-9-tetrahydrocannabinol, 11-hydroxy-delta-9-tetrahydrocannabinol, and 11-nor-9-carboxy-delta-9-tetrahydrocannabinol in human urine following tandem enzyme-alkaline hydrolysis. J Anal Toxicol 2007;31:477–85.

[13] Stout PR, Klette KL. Solid-phase extraction and GC-MS analysis of THC-COOH method optimization for high-throughput forensic drug-testing laboratory. J Anal Toxicol 2001;25:550–4.

[14] Paul BD, McKinley RM, Walsh JW, Jamir TS, Past MR. Effect of freezing on concentration of drugs of abuse in urine. J Anal Toxicol 1993;17:378–80.

[15] Ramacharitar V, Levine B, Smialek JE. Benzoylecgonine and ecgonine methyl ester concentrations in urine specimens. J Forensic Sci 1995;40:99–101.

[16] Paul BD, McWhorter LK, Smith ML. Electron ionization mass fragmentometric detection of urinary ecgonine, a hydrolytic product of methylecgonine, as an indicator of smoking cocaine. J Mass Spectrom 1999;34:651–60.

[17] Paul BD, Lalani S, Bosy T, Jacobs AJ, Huestis MA. Concentration profiles of cocaine, pyrolytic methylecgonine, and thirteen metabolites in human blood and urine: determination by gas chromatography-mass spectrometry. Biomed Chromatogr 2005;19:677–88.

[18] Gerlits J. GC/MS quantitation of benzoylecgonine following liquid-liquid extraction of urine. J Forensic Sci 1993;38:1210–14.

[19] Jenkins AJ, Goldberger B. Identification of unique cocaine metabolites and smoking by-products in postmortem blood and urine specimens. J Forensic Sci 1997;42:824–7.

[20] Williams RH, Maggiore JA, Shah SM, Erickson TB, Negrusz A. Cocaine and its metabolites in plasma and urine samples from patients in an urban emergency medicine setting. J Anal Toxicol 2000;24:478–81.

[21] Aderjan RE, Schmitt G, Wu M, Meyer C. Determination of cocaine and benzoylecgonine by derivatization with iodomethane-D_3 or PFPA/HFIP in human blood and urine using GC/MS (EI or PCI mode). J Anal Toxicol 1993;17:51–5.
[22] ElSohly M, Little TL, Mitchell JM, Paul BD, Mell LD, Irving J. GC/MS analysis of phencyclidine acid metabolite in human urine. J Anal Toxicol 1988;12:180–2.
[23] Stevenson CC, Cibull DL, Platoff GE, Bush DM, Gere JA. Solid phase extraction of phencyclidine from urine followed by capillary gas chromatography/mass spectrometry. J Anal Toxicol 1992;16:337–9.
[24] Paul BD, Mell LD, Mitchell JM, Irving J, Novak AJ. Simultaneous identification and quantitation of codeine and morphine in urine by capillary gas chromatography and mass spectroscopy. J Anal Toxicol 1985;9:222–6.
[25] Paul BD, Shimomura ET, Smith ML. A practical approach to determine cutoff concentrations for opiate testing with simultaneous detection of codeine, morphine, and 6-acetylmorphine in urine. Clin Chem 1999;45:510–19.
[26] Paul BD, Mitchell JM, Mell LD, Irving J. Gas chromatography/electron impact mass fragmentometric determination of urinary 6-acetylmorphine, a metabolite of heroin. J Anal Toxicol 1989;13:2–7.
[27] Czarny R, Hornbeck C. Quantitation of methamphetamine and amphetamine in urine by capillary GC/MS. II. Derivatization with 4-carboethoxyhexafluorobutyryl chloride. J Anal Toxicol 1989;13:257–62.
[28] Paul BD, Past MR, McKinley RM, Foreman JD, McWhorter LK, Snyder JJ. Amphetamine as an artifact of methamphetamine during periodate degradation of interfering ephedrine, pseudoephedrine, and phenylpropanolamine: an improved procedure for accurate quantitation of amphetamine in urine. J Anal Toxicol 1994;18:331–6.
[29] Klette KL, Jamerson MH, Morris-Kukoski CL, Kettle AR, Snyder JJ. Rapid simultaneous determination of amphetamine, methamphetamine, 3,4-methylenedioxyamphetamine, 3,4-methylenedioxymethamphetamine, and 3,4-methylenedioxyethylamphetamine in urine by fast gas chromatography-mass spectrometry. J Anal Toxicol 2005;29:669–74.
[30] Valentine JL, Middleton R. GC-MS identification of sympathomimetic amine in urine: rapid methodology applicable for emergency clinical toxicology. J Anal Toxicol 2000;24:211–22.
[31] Stout PR, Horn CK, Klette KL. Rapid simultaneous determination of amphetamine, methamphetamine, 3,4-methylenedioxyamphetamine, 3,4-methylenedioxymethamphetamine, and 3,4-methylenedioxyethylamphetamine in urine by solid-phase extraction and GC-MS: a method optimized for high-volume laboratories. J Anal Toxicol 2002;26:253–61.
[32] Hornbeck CL, Czarny RJ. Quantitation of methamphetamine and amphetamine in urine by capillary GC/MS. Part I. Advantages of trichloroacetyl derivatization. J Anal Toxicol 1989;13:144–9.
[33] Gan BK, Baugh D, Liu RH, Walia AS. Simultaneous analysis of amphetamine, methamphetamine, and 3,4-methylenedioxymethamphetamine (MDMA) in urine samples by solid-phase extraction, derivatization, and gas chromatography/mass spectrometry. J Forensic Sci 1991;36:1331–41.
[34] Melgar R, Kelly RC. A novel GC/MS derivatization method for amphetamines. J Anal Toxicol 1993;17:399–402.
[35] Westly JW, Halpern B, Karger BL. Effect of solute structure on separation of diasterioisomeric esters and amides by gas-liquid chromatography. Clin Chem 1968;40:2046–9.
[36] Gunne L. The urinary output of *d*- and *l*-amphetamine in man. Biochem Pharmacol 1967;16:863–9.
[37] Fitzgerald RL, Ramos JM, Bogema SC, Poklis A. Resolution of methamphetamine stereoisomers in urine drug testing: urinary excretion of R-(-)-methamphetamine following use of nasal inhalers. J Anal Toxicol 1988;12:255–9.
[38] Gal J. Mass spectra of *N*-[(*S*)-α-methoxy-α-(trifluoromethyl)phenylacetyl] drivatives of chiral amines. Stereochemistry of amphetamine metabolism in the rat. Biomed Mass Spectrom 1978;5:32–7.
[39] Paul BD, Jemionek J, Jacobs A, Searles DA. Enantiomeric separation and quantitation of (+/−)-amphetamine, (+/−)-methamphetamine, (+/−)-MDA, (+/−)-MDMA, and (+/−)-MDEA in urine specimens by GC-EI-MS after derivatization with (R)-(-)- or (S)-(+)-alpha-methoxy-alpha-(trifluoromethyl)phenylacetyl chloride (MTPA). J Anal Toxicol 2004;28:449–55.
[40] Holler JM, Vorce SP, Bosy TZ, Jacobs A. Quantitative and isomeric determination of amphetamine and methamphetamine from urine using a nonprotic elution solvent and R-(-)-alpha-methoxy-alpha-trifluoromethylphenylacetic acid chloride derivatization. J Anal Toxicol 2005;29:652–7.
[41] McKinley S, Snyder JJ, Welsh E, Kazarian CM, Jamerson MH, Klette KL. Rapid quantification of urinary oxycodone and oxymorphone using gas chromatography-mass spectrometry. J Anal Toxicol 2007;31:434–41.
[42] Meatherall R. GC-MS confirmation of codeine, morphine, 6-acetylmorphine, hydrocodone, hydromorphone, oxycodone, and oxymorphone in urine. J Anal Toxicol 1999;23:177–86.
[43] Nowatzke W, Zeng J, Saunders A, Bohrer A, Koenig J, Turk J. Distinction among eight opiate drugs in urine by gas chromatography-mass spectrometry. J Pharm Biomed Anal 1999;20:815–28.
[44] Chen BG, Wang SM, Liu RH. GC-MS analysis of multiply derivatized opioids in urine. J Mass Spectrom 2007;42:1012–23.
[45] Cody J, Vorce S. Mass spectrometry. In: Levine B, editor. Principles of forensic toxicology. 4th ed. Washington, DC: AACC press; 2013. p. 171–92.
[46] Dams R, Murphy CM, Lambert WE, Huestis MA. Urine drug testing for opioids, cocaine and metabolites by direct injection liquid chromatography/tandem mass spectrometry. Rapid Comm Msaa Spec 2003;17:1665–70.
[47] Wood M, Laloup L, del Mar Ramirez Fernandez M, Jenkins KM, Young MS, et al. Quantitative analysis of multiple illicit drugs in preserved oral fluid by solid-phase extraction and liquid-chromatography tandem mass spectrometry. Forensic Sci Int 2005;150:227–38.

[48] Feng J, Wang L, Dai I, Harmon T, Bernert JE. Simultaneous determination of multiple drugs of abuse and relevant metabolites in urine byLC-MS-MS. J Anal Toxicol 2007;31:359–68.
[49] Eichhorst JC, Etter ML, Rousseaux N, Lehotay DC. Drugs of abuse testing by tandem mass spectrometry: a rapid, simple method to replace immunoassays. Clin Biochem 2009;42:1531–42.
[50] Gallardo E, Barroso M, Queiroz JA. LC-MS: a powerful tool in workplace drug testing. Drug Test Anal 2009;1:109–15.
[51] Jagerdeo E, Montgomert M, Karas R, Sibum M. A fast method for screening and/or quantitation of tetrahydrocannabinol and metabolites in urine by automated SPE/LC/MS/MS. Anal Bioanal Chem 2010;398:329–38.
[52] Manchicanti L, Malla Y, Wargo BW, Feoolws B. Comparative evaluation of the accuracy of immunoassay with liquid chromatography tandem mass spectrometry (LC/MS/MS) of urine drug testing (UDT) opioids and illicit drugs in chronic pain patients. Pain Physician 2011;14:175–87.
[53] Wanz J, Yang Z, Lechago J. Rapid and simultaneous determination of multiple classes of drugs and metabolites in human urine by a robust LC-MS/MS method- application to urine drug testing in pain clinics. Biomed Chromatogr 2013;27:1463–80.
[54] Tsai IL, Weng TI, Tseng YJ, Tan HK, Sun HJ, et al. Screening and confirmation of 62 drugs of abuse and metabolites in urine by ultra-high performance liquid chromatography- quadrupole time-of-flight mass spectrometry. J Anal Toxicol 2013;37:642–51.
[55] Shin M, Ji D, Kang S, Yang W, Choi H, et al. Screening of multiple drugs of abuse and metabolites in urine using LC/MS/MS with polarity switching electrospray ionization. Arch Pharm Res 2014;37:760–72.
[56] Shimomura ET, Hodge GD, Paul BD. Examination of postmortem fluids and tissues for the presence of methylecgonidine, ecgonidine, cocaine, and benzoylecgonine using solid-phase extraction and gas chromatography-mass spectrometry. Clin Chem 2001;47:1040–7.
[57] Thurman EM, Pedersen MJ, Stout RL, Martin T. Distinguishing sympathomimetic amines from amphetamine and methamphetamine in urine by gas chromatography-mass spectrometry. J Anal Toxicol 1992;16:19–27.
[58] Hornbeck CL, Carrig JE, Czarny RJ. Detection of a GC/MS artifact peak as methamphetamine. J Anal Toxicol 1993;17:257–63.
[59] ElSohly MA, Stanford DF, Sherman D, Shah H, Bernot D, Turner CE. A procedure for eliminating interferences from ephedrine and related compounds in the GC/MS analysis of amphetamine and methamphetamine. J Anal Toxicol 1992;16:109–11.
[60] ElSohly MA. Urinalysis and casual handling of marijuana and cocaine. J Anal Toxicol 1991;15:46.
[61] Cone EJ, Johnson RE, Darwin WD, Yousefnejad D, Mell LD, Paul BD, et al. Passive inhalation of marijuana smoke: urinalysis and room air levels of delta-9-tetrahydrocannabinol. J Anal Toxicol 1987;11:89–96.
[62] Costantino A, Schwartz RH, Kaplan P. Hemp oil ingestion causes positive urine tests for delta-9-tetrahydrocannabinol carboxylic acid. J Anal Toxicol 1997;21:482–5.
[63] Fortner N, Fogerson R, Lindman D, Iverson T, Armbruster D. Marijuana-positive urine test results from consumption of hemp seeds in food products. J Anal Toxicol 1997;21:476–81.
[64] Lehmann T, Sager F, Brenneisen R. Excretion of cannabinoids in urine after ingestion of cannabis seed oil. J Anal Toxicol 1997;21:373–5.
[65] Callaway JC, Weeks RA, Raymon LP, Walls HC, Hearn WL. A positive urinalysis from hemp (*Cannabis*) seed oil. J Anal Toxicol 1997;21:319–20.
[66] Bosy TZ, Cole KA. Consumption and quantitation of delta-9-tetrahydrocannabinol in commercially available hemp seed oil products. J Anal Toxicol 2000;24:562–6.
[67] Holler JM, Bosy TZ, Dunkley CS, Levine B, Past MR, Jacobs A. Delta-9-tetrahydrocannabinol content of commercially available hemp products. J Anal Toxicol 2008;32:428–32.
[68] Baselt RC, Yoshikawa DM, Chang JY. Passive inhalation of cocaine. Clin Chem 1991;37:2160–1.
[69] Baselt RC, Chang JY, Yoshikawa DM. On dermal absorption of cocaine. J Anal Toxicol 1990;14:383–4.
[70] ElSohly MA, Standford DF, ElSohly HN. Coca tea and urinalysis for cocaine metabolites. J Anal Toxicol 1986;10:256.
[71] Jackson GF, Saady JJ, Poklis A. Urinary excretion of benzoylecgonine following ingestion of Health Inca Tea. Forensic Sci Int 1991;49:57–64.
[72] ElSohly MA, Jones AB. Morphine and codeine in biological fluids: approaches to source differentiation. Forensic Sci Rev 1989;1:13–22.
[73] Physicians Desk Reference 71th Edition. PDR Network LLC. Montvale, NJ 2017.
[74] Moore K, Ramcharitar V, Levine B, Fowler D. Tentative identification of novel oxycodone metabolites in human urine. J Anal Toxicol 2003;27:346–9.
[75] Baselt R. Disposition of toxic drugs and chemicals in man. 11th ed. Seal Beach, CA: Biomedical Publications; 2017.
[76] Oxycontin. Physicians' Desk Reference Electronic Library.
[77] Smith ML, Hughes RO, Levine B, Dickerson S, Darwin WD, Cone EJ. Forensic drug testing for opiates. VI. Urine testing for hydromorphone, hydrocodone, oxymorphone, and oxycodone with commercial opiate immunoassays and gas chromatography-mass spectrometry. J Anal Toxicol 1995;19:18–26.
[78] Ishii A, Tanaka M, Kurihara R, Watanabe-Suzuki K, Kumazawa T, Seno H, et al. Sensitive determination of pethidine in body fluids by gas chromatography-tandem mass spectrometry. J Chromatogr B 2003;792:117–21.
[79] Latta KS, Ginsberg B, Barkin RL. Meperidine: a Critical Review. Am J Therap 2002;9:53–68.
[80] Foster DR, Somogyi AA, Dyer KR, White JM, Bochner F. Steady-state pharmacokinetics of (R)- and (S)-methadone in methadone maintenance patients. J Clin Pharmacol 2000;50:427–40.
[81] Alburges ME, Huang W, Foltz RL, Moody DE. Determination of methadone and its N-demethylation metabolites in biological specimens by GC-PICI-MS. J Anal Toxicol 1996;20:362–8.
[82] Ferrari A, Rosario Coccia CP, Bertolini A, Sternieri E. Methadone-metabolism, pharmacokinetics, and interactions. Pharm Research 2004;50:551–9.

[83] Poklis A. Fentanyl: a review for clinical and analytical toxicologists. J Toxicol: Clin Toxicol 1995;33(5):439–46.

[84] Levine B, Ramcharitar V, Smialek JE. Tramadol distribution in four postmortem cases. Forensic Sci Int 1997;86:43–8.

[85] Fenton J, Mummert J, Childers M. Hydromorphone and hydrocodone interference in GC/MS assays for morphine and codeine. J Anal Toxicol 1994;18:159–64.

[86] Ropero-Miller JD, Lambing MK, Winecker RE. Simultaneous quantitation of opioids in blood by GC-EI-MS analysis following deproteination, detautomerization of keto-analytes, solid-phase extraction, and trimethylsilyl derivatization. J Anal Toxicol 2002;26(7):524–8.

[87] Wang P, Stone JA, Chen KH, Gross SF, Haller CA, Wu AHB. Incomplete recovery of prescription opioids in urine using enzymatic hydrolysis of glucuronide metabolites. J Anal Toxicol 2006;30:570–5.

[88] Broussard LA, Presley LC, Pittman T, Clouette R, Wimbish GH. Simultaneous identification and quantitation of codeine, morphine, hydrocodone, and hydromorphone in urine as trimethylsilyl and oxime derivatives by gas chromatography-mass spectrometry. Clin Chem 1997;43:1029–32.

[89] Cremese M, Wu AHB, Cassella G, O'Connor E, Rymut K, Hill DW. Improved GC/MS analysis of opiates with use of oxime-TMS derivatives. J Forensic Sci 1998;43(6):1220–4.

[90] Goeringer KE, Logan BK, Christian GD. Identification of tramadol and its metabolites in blood from drug-related deaths and drug-impaired drivers. J Anal Toxicol 1997;21:529–37.

[91] Bermejo AM, Seara R, dos Santos Lucas AC, Tabernero MJ, Fernandez P, Marsili R. Use of solid-phase microextraction (SPME) for the determination of methadone and its main metabolite, EDDP, in plasma by gas chromatography-mass spectrometry. J Anal Toxicol 2000;24:66–9.

[92] Coopman V, Cordonnier J, Pien K, Van Varenbergh D. LC-MS/MS analysis of fentanyl and norfentanyl in a fatality due to application of multiple Durogesic® transdermal therapeutic systems. Forensic Sci Int 2007;169:223–7.

[93] Moore C, Rana S, Coulter C. Determination of meperidine, tramadol, and oxycodone in human oral fluid using solid phase extraction and gas chromatography-mass spectrometry. J Chromatogr B 2007;850:370–5.

[94] Lucas ACS, Bermejo AM, Tabernero MJ, Fernandez P, Strano-Rossi S. Use of solid-phase microextraction (SPME) for the determination of methadone and EDDP in human hair by GC-MS. Forensic Sci Int 2000;107:225–32.

[95] Rodriguez-Rosas ME, Medrano JG, Epstein DH, Moolchan ET, Preston KL, Wainer IW. Determination of total and free concentrations of the enantiomers of methadone and its metabolite 2-ethylidene-1,5-dimethyl-3,3-diphenylpyrrolidine in human plasma by enantioselective liquid chromatography with mass spectrometric detection. J Chromatogr A 2005;1073:237–48.

[96] Dickson PH, Markus W, McKernan J, Nipper HC. Urinalysis of α-hydroxyalprazolam, α-hydroxytriazolam, and other benzodiazepine compounds by GC/EIMS. J Anal Toxicol 1992;16:67–71.

[97] Meatherall R. GC-MS confirmation of urinary benzodiazepine metabolites. J Anal Toxicol 1994;18:369–81.

[98] Laloup M, del Mar Ramirez Fernandez M, De Boeck G, Wood M, Maes V, Samyn N. Validation of a liquid chromatography-tandem mass spectrometry method for the simultaneous determination of 26 benzodiazepines and metabolites, zolpidem, and zopiclone, in blood, urine, and hair. J Anal Toxicol 2005;29:616–26.

[99] Martinez MA, Sanchez de la Torre C, Almarza E. Simultaenous determination of viloxazine, venlafaxine, imipramine, desipramine, sertraline, and amoxapine in whole blood: comparison of two extraction/cleanup procedures for capillary gas chromatography with nitrogen-phosphorus detection. J Anal Toxicol 2002;26:296–302.

[100] Martinez MA, Sanchez de la Torre C, Almarza E. A comparative solid-phase extraction study for the simultaneous determination of fluvoxamine, mianserin, doxepin, citalopram, paroxetine, and etoperidone in whole blood by capillary gas-liquid chromatography with nitrogen-phosphorus detection. J Anal Toxicol 2004;28:174–80.

[101] Rana S, Uralets VP, Ross W. A new method for simultaneous determination of cyclic antidepressants and their metabolites in urine using enzymatic hydrolysis and fast GC-MS. J Anal Toxicol 2008;32:355–63.

[102] Titier K, Castaing N, Le-Deodic M, Le-bars D, Moore N, Molimard M. Quantification of tricyclic antidepressants and monoamine oxidase inhibitors by high-performance liquid chromatography-tandem mass spectrometry in whole blood. J Anal Toxicol 2007;31:200–7.

[103] Cantu MD, Toso DR, Lacerda CA, Lancas FM, Carrilho E, Quieroz MEC. Optimization of solid-phase microextraction procedures for the determination of tricyclic antidepressants and anticonvulsants in plasma samples by liquid chromatography. Anal Bioannal Chem 2006;386:256–63.

[104] Wozniakiewicz M, Wietecha-Posluszny R, Garbacik A, Koscielniak P. Microwave-assisted extraction of tricyclic antidepressants from human serum followed by high performance liquid chromatography determination. J Chrom A 2008;1190:52–6.

[105] Santos-Neto AJ, Bergquist J, Lancas FM, Sjoberg PJR. Simultaneous analysis of five antidepressant drugs using direct injection of biofluids in a capillary restricted-access media-liquid chromatography-tandem mass spectrometry system. J Chrom A 2008;1189:514–22.

Chapter 16

Critical Issues When Testing for Amphetamine-Type Stimulants: Pitfalls of Immunoassay Screening and Mass Spectrometric Confirmation for Amphetamines, Methamphetamines, and Designer Amphetamines

Larry Broussard

Department of Clinical Laboratory Sciences, Louisiana State University Health Sciences Center, New Orleans, LA, United States

INTRODUCTION

The original military and workplace drug-screening programs for illegal drugs abuse included the two stimulants amphetamine and methamphetamine, both of which are sympathomimetic amines having a phenethylamine structure. On October 1, 2010 testing for the following designer amphetamines: 3,4-methylenedioxymethamphetamine (MDMA; Ecstasy), 3,4-methylenedioxyamphetamine (MDA), and 3-4-methylenedioxyethylamphetamine (MDEA) was added and on October 1, 2017 MDEA was removed as a drug to be tested [1,2]. Other sympathomimetic amines (see Fig. 16.1) with similar structures include compounds such as ephedrine and pseudoephedrine found in over-the-counter (OTC) medications. Additional designer amphetamines such as methiopropamine, 4-fluoroamphetamine, 4-fluoromethamphetamine, and 4-methylamphetamine are together with cathinones referred to as "bath salts." Immunoassays are the primary screening methods for the detection of amphetamines and in most of these assays, the antibodies appear to be directed toward the amino group of amphetamine, a primary amine, and methamphetamine, a secondary amine. Both monoclonal and polyclonal antibodies have been used in these assays.

ISSUES OF CROSS-REACTIVITY

Because of the structural similarity of the other sympathomimetic amines, cross-reactivity of antibodies is a primary concern when interpreting results of immunoassay screening for amphetamines. Immunoassays designed for the detection of amphetamines have been classified into three general types based on antibody specificity:

Assays highly selective for both amphetamine (and its designer drug counterpart MDA) or methamphetamine (and its designer drug counterpart MDMA) but not for both set of drugs simultaneously.
Assays that are able to detect amphetamine and methamphetamine to varying extent, and also exhibit higher levels of cross-reactivity to the hydroxy amine compounds found in many OTC drugs.
Dual assays for amphetamine and methamphetamine with low levels of cross-reactivity to OTC drugs [3].
An additional class of immunoassays specifically designed to detect MDMA (Ecstasy)/MDA is also available.

The choice of amphetamine immunoassay to use is dependent on several factors, including compatibility with current instrumentation, possible vendor arrangements, and intended use. Laboratories performing workplace drug testing

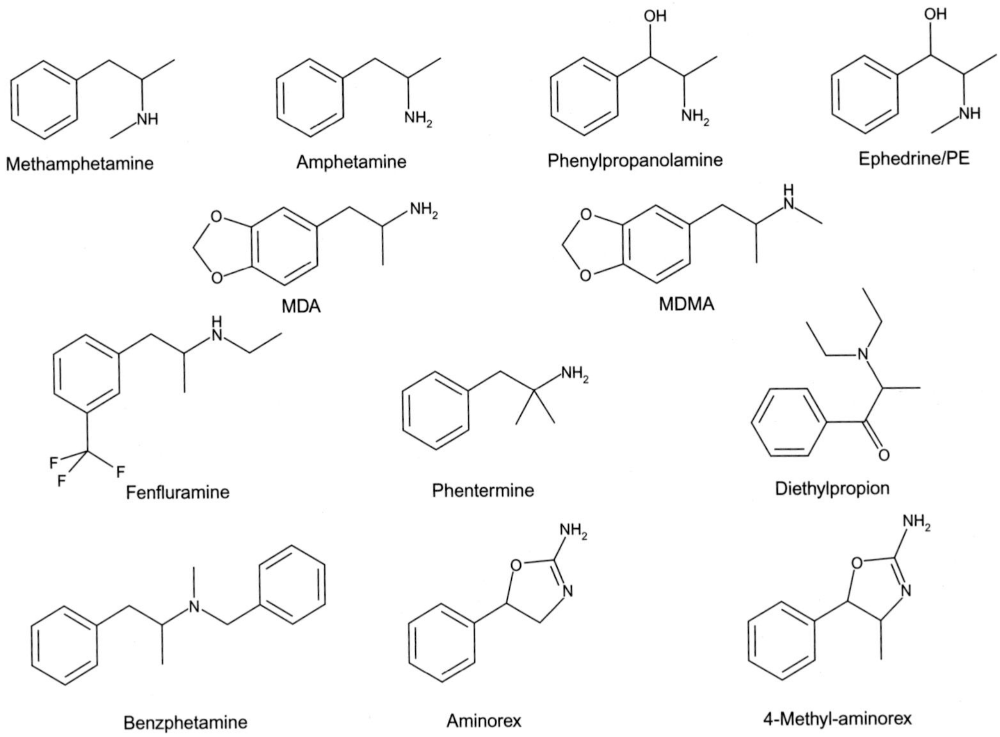

FIGURE 16.1 Chemical structures of amphetamines and sympathomimetic amines.

desire amphetamines immunoassays specific for only amphetamine and methamphetamine and MDMA/MDA, the compounds specified in appropriate legislation or contracts. Laboratories affiliated with emergency departments (ED), pain management monitoring, or forensic agencies desire immunoassays directed toward the broad spectrum of sympathomimetic amines, including OTC ingredients and all types of designer amphetamines. The desire for a broad-spectrum assay is articulated as a recommendation in the National Academy of Clinical Biochemistry's (NACB), Laboratory Medicine Practice Guidelines (LMPG) that the optimum immunoassays for amphetamines testing in ED patients are those directed toward phenethylamines as a class [4]. These guidelines also recommend that the name of the test should be changed from "amphetamines" to "sympathomimetic amines" or "stimulant amines."

Current immunoassays are extremely effective for the detection of the intended compounds such as amphetamine/methamphetamine or MDMA/MDA, but are typically ineffective for use in detecting new amphetamine designer drugs [5]. Immunoassay screening of drug classes including amphetamines is being replaced with definitive screening by liquid chromatography coupled with mass spectrometry in either single (LC-MS) or tandem (LC-MS/MS) mode capable of detecting multiple analytes, including designer amphetamines and other amphetamine-like stimulants. Development of simpler and faster sample preparation procedures, such as "dilute," "shoot," or "dilute and shoot," and the increased capability of LC-MS and LC-MS/MS have facilitated the development of many definitive multianalyte screening methods [6]. Another approach is to combine both immunoassays and mass spectrometry procedures for performing drug screening [7]. For federal workplace drug testing, LC-MS/MS has been recognized as an acceptable alternate technology to immunoassays for initial screening [2,8].

IMMUNOASSAY CUTOFF CONCENTRATIONS AND CROSS-REACTIVITY

The current screening cutoff for amphetamines (amphetamine and methamphetamine) and MDMA/MDA mandated for federal workplace testing is 500 ng/mL [2]. As previously mentioned amphetamines assays are designed to target methamphetamine, amphetamine, both, and/or other sympathomimetic amines and the analyst should be aware of the targeted analyte(s), as well as the concentrations of these analytes in calibrators and controls. Structural similarity to amphetamine or methamphetamine creates potential problems due to cross-reactivity of the antibodies in immunoassays. For federal workplace drug testing the cross-reactivity for analytes within a group (e.g., amphetamines) must be 80% or greater for the analytes within the group (e.g., amphetamine and methamphetamine) [2]. The cross-reactivity of various phenethylamines in different immunoassays has been published [9,10]. This information should only be used as a starting point when comparing immunoassays because many factors influence cross-reactivity. For example, it is possible that current assays produced by the manufacturers may have different cross-reactivities from those reported due to changes in antibodies and reagent composition. The analyst should consult immunoassay package inserts or contact the manufacturer for up-to-date cross-reactivity data. When reviewing cross-reactivity data the reader should realize that studies of potential interfering substances sometimes use concentrations that are lower than those encountered in the clinical setting and that manufacturers often do not test for cross-reactivity of endogenous metabolites [11,12]. Lot-to-lot variability concerning sensitivity to targeted analytes has been reported [13] and it is conceivable that there could be lot-to-lot variability concerning cross-reactivity. Also different instruments may use ratios of reagent to sample different from those used for the manufacturers' cross-reactivity studies. Laboratories, at a minimum, should contact the manufacturer to verify that they have the most recent applicable cross-reactivity information and ideally they should perform cross-reactivity studies of the most common interfering substances on their reagent/instrument system.

FALSE-POSITIVE IMMUNOASSAY RESULTS

Many OTC drugs and prescription medications interfere with amphetamines immunoassays producing false-positive test results (see also Chapter 10: Issues of Interferences With Immunoassays Used for Screening of Drugs of Abuse in Urine). In a study of 100 false-positive immunoassay results, 90 had two or more possible interfering compounds and the authors speculated that the results were produced by the combination of drugs and metabolites with the combined cross-reactivity and not necessarily by a single compound [14]. Investigation of positive amphetamines screening results should also include inquiry about ingestion of diet and herbal preparations since sympathomimetic amines are known appetite suppressants. For example, the Federal Drug Administration (FDA) warned consumers in January 2006 that the Brazilian dietary supplements Emagrece Sim and Herbathin contain several active drug ingredients including fenproporex, which is metabolized to amphetamine [15]. Use of other weight loss or nutritional supplements containing fenproporex has resulted in detection of D-amphetamine in the urines of users [16].

Amphetamine and methamphetamine have optical isomers designated D (or +) for dextrorotatory and L (or −) for levorotatory and the D-isomers, the more physiologically active compounds, are the intended targets of immunoassays. Ingestion of medications containing the L-isomer can cause false-positive results. For example, Vicks inhaler contains the active ingredient L-methamphetamine [17] and extensive use of this product may cause false-positive results for immunoassay screening. Specific isomer resolution procedures must be performed to differentiate the D- and L-isomers because routine confirmation by gas chromatography−mass spectrometry (GC-MS) does not determine isomer composition. Studies have shown that heavy use of the inhaler can cause false-positive results, but when the inhaler was used as directed no false-positive results were obtained [18].

MINIMIZING FALSE-POSITIVE IMMUNOASSAY RESULTS

Several approaches have been taken to minimize the occurrence of false-positive results with amphetamines immunoassays. One approach is to perform a secondary screen of all positive samples following an initial immunoassay test. This secondary screen may consist of repeat analysis utilizing an immunoassay with different cross-reactivity to these compounds produced by a different manufacturer. Another type of secondary screen is reanalysis using the same immunoassay following chemical reaction or modification of the assay in some manner. Pretreatment of samples with sodium periodate in a basic solution eliminates interference from ephedrine, pseudoephedrine, and phenylpropanolamine by oxidative cleavage of the hydroxyl group which is the apparent site of antibody cross-reactivity. This reaction has been utilized in the EMIT amphetamine confirmation kit (Siemens (Dade Behring), San Jose, CA) and is routinely included in most GC-MS confirmation procedures. Another strategy for elimination of false-positive results is neutralization of the signal in a true-positive sample by the addition of antibody to the target analyte. In this situation true positives are distinguished from false positives by the difference in the signal before and after addition of the neutralizing antibody since samples containing high concentrations of cross-reactive substances would show no decrease in the signal after addition of the antibody [19]. Another technique using serial dilution testing and optimal slope cutoffs [determined by Receiver Operating Characteristic (ROC) curve analysis] enabled Woodworth et al. to differentiate samples containing amphetamine/methamphetamine from those containing cross-reacting compounds and to increase the positive predictive value of the immunoassay [20].

Chemical interferences with selected immunoassays have also been reported [18]. These include interference by mefenamic acid with assays measuring absorbance changes, false-negative or invalid results apparently due to lowering of the pH caused by alum, and increased sensitivity for amphetamines with the ONLINE (Roche Diagnostics, Somerville, NJ) immunoassay due to high concentrations of the adulterant nitrite [21].

When analyzing postmortem samples an additional consideration is that false-positive results may be caused by the putrefactive amines, phenethylamine, and tyramine produced by saprogenic bacteria in moderately-to-heavily decomposed bodies. Collecting blood in sodium fluoride and deproteinization using acetone or sulfosalicylic acid are techniques that have been used to reduce or eliminate this interference [22,23].

TRUE-POSITIVE RESULTS

Positive amphetamines results may be also be observed due to ingestion of medications containing amphetamine, methamphetamine, or compounds metabolized to these compounds. Amphetamine [Dexedrine (D-amphetamine), Adderall (D- and L-amphetamine), etc.] and methamphetamine [Desoxyn (D-methamphetamine)] are the active compounds of medications prescribed for appetite suppression, narcolepsy, and attention-deficit disorder and ingestion of these drugs will result in true-positive results due to excretion of these compounds in the urine. Table 16.1 lists compounds whose ingestion can cause positive results due to metabolism to amphetamine and/or methamphetamine.

GC/MS AND LC-MS/MS CONFIRMATION PROCEDURES

The GC/MS and LC-MS/MS are the most frequently used confirmation procedures for the detection of amphetamine, methamphetamine, and other related compounds. The current confirmation cutoffs for federal workplace drug testing for amphetamine and methamphetamine and MDMA/MDA is 250 ng/mL [2]. Current regulations also require that in order to report a sample positive for methamphetamine, in addition to detecting methamphetamine itself, its metabolite amphetamine must be detected using a confirmation cutoff of 100 ng/mL.

All the confirmation procedures for amphetamines and related compounds should include preventative measures to avoid loss of these volatile compounds during the evaporation step of extraction or during analysis. Procedures to reduce/eliminate loss of amphetamines during evaporation include lowering the temperature for evaporation, performing incomplete evaporation, or adding methanolic hydrochloric acid prior to evaporation to produce more stable hydrochloride salts [24–26]. Derivatization using reagents, such as heptafluorobutyric anhydride (HFBA), pentafluoropropionic anhydride (PFPA), trifluoroacetic anhydride (TFAA), 4-carboxyhexafluorobutyryl chloride (4-CB), N-methyl-N-t-butyldimethylsilyl trifluoroacetamide (MTBSTFA), N-trifluoroacetyl-1-prolyl chloride (TPC), $R(-)$-α-methoxy-α-trifluoromethylphenylacetyl chloride (R-MTPAC), chlorodifluoroacetic anhydride (ClF_2AA), 2,2,2-trichloroethyl chloroformate, and propylchloroformate decreases the volatility of amphetamines in addition to improving chromatography and quantitation, and forming higher molecular weight fragments yielding different mass ions and ion ratios than potentially interfering compounds [18]. Use of deuterated internal standard (IS) solution contaminated with nondeuterated analyte can lead to inaccurate concentrations and potential false results. IS solutions should be monitored for the presence of the nondeuterated analyte of interest. Error in addition of IS solution is another potential source of inaccurate results. A recommended quality assurance measure to prevent this type of error is to monitor the relative IS abundance of every sample and establish acceptable limits (such as a requirement to be >50% and <200% of the calibrator IS area) [8].

Although amphetamine and methamphetamine are primarily excreted unchanged in urine, some metabolites of designer amphetamines are excreted as conjugated metabolites [3,24,27,28]. For this reason and because most multianalyte procedures include a hydrolysis step using acid, base, or glucuronidase the efficiency of the hydrolysis should be verified for the initial and periodic evaluations of method parameters.

INTERFERENCE AND FALSE-POSITIVE METHAMPHETAMINE

Contaminants in the heptafluorobutyryl derivatizing reagent have been shown to give methamphetamine peak interferences when ephedrine is present in the sample [29]. In 1993 it was demonstrated that concentrations of methamphetamine <50 ng/mL can be generated from high levels of pseudoephedrine or ephedrine in injection ports at a temperature of 300°C after derivatization with 4-CB, HFBA, and TPC [30]. False-positive results in the federal drug testing program due to generation of methamphetamine resulted in the implementation of the requirement that the metabolite amphetamine must be present at a concentration of 100 ng/mL or higher in order to report a positive methamphetamine [2].

Lowering the injector temperature and periodate pretreatment of samples are two procedures used to prevent generation of methamphetamine. Periodate treatment (0.35 M sodium periodate for 10 min at room temperature) eliminated formation of methamphetamine by selective oxidation of pseudoephedrine, ephedrine, phenylpropanolamine, and

TABLE 16.1 Compounds Causing True-Positive Amphetamine and/or Methamphetamine Confirmation

Drug	Therapeutic Use	Comments
Amphetamine	Narcolepsy, attention-deficit hyperactivity disorder (ADHD) in children	Adderall is one of the prescription medications that contain amphetamine. There are also other products that contain either D-amphetamine or racemic mixture
Lisdexamfetamine	Narcolepsy, ADHD	Inactive water-soluble prodrug (L-lysine-dextroamphetamine) that is readily concerted into D-amphetamine after oral administration
Methamphetamine	ADHD in children	Desoxyn is the brand name of a product that contains methamphetamine
Amphetaminil	Psychotropic drug but mostly withdrawn from the market	Aponeuron is metabolized to amphetamine
Clobenzorex	Anorexic agent used for short-term treatment of obesity	Metabolized to amphetamine
Ethylamphetamine	No known medical use	Metabolized to amphetamine
Fenethylline	ADHD in children	Metabolized to amphetamine
Fenproporex	Anorexic agent used for short-term treatment of obesity	Metabolized to amphetamine
Mefenorex	Anorexic agent used for short-term treatment of obesity	Metabolized to amphetamine
Prenylamine	Treatment of angina	Metabolized to amphetamine
Benzphetamine	Anorexic agent used for short-term treatment of obesity	Metabolized to methamphetamine and then further metabolized to amphetamine
Famprofazone	Antipyretic and analgesic	Metabolized to methamphetamine and then to amphetamine
Selegiline	Early stage of Parkinson's disease	Metabolized to L-methamphetamine
Deprenyl	Parkinson's disease often in combination with levo-dopa	Metabolized to methamphetamine and then to amphetamine
Dimethyl amphetamine	No known medical use	Metabolized to methamphetamine and then to amphetamine
Fencamine	Has been used in the treatment of depression	Metabolized to methamphetamine and then to amphetamine
Furfenorex	Anorexic agent used in the treatment of obesity	Metabolized to methamphetamine and then to amphetamine

norpseudoephedrine at concentrations of 1,000,000 ng/mL in the presence of amphetamine and methamphetamine [31]. Inclusion of a quality control sample containing a high concentration of sympathomimetic amines in each confirmation batch can be used to monitor the effectiveness of the periodate oxidation procedure and is required in the federal drug testing program [8,18].

ISOMER RESOLUTION

Because mass spectrometric procedures using nonchiral derivatives and columns do not differentiate the D (+) or L (−) isomers of amphetamine and methamphetamine, it is necessary to perform isomer resolution to determine that a positive result is due to the presence of the D-isomer. Primary examples are the presence of L-methamphetamine in Vicks inhaler, which cannot be distinguished from the use of illicit methamphetamine (D-isomer or racemic mixture depending on method of production) and the excretion of L-methamphetamine and L-amphetamine by patients taking selegiline for Parkinson's disease. Isomer resolution can be accomplished using a chiral, optically active column, or chiral derivatizing reagents [9]. Use of chiral derivatizing reagents allows analysis on instrument/column systems used for other routine analyses but has the potential disadvantage of possibly obtaining four isomers instead of two if the derivatizing agent is not optically pure. The generally accepted interpretation of isomer resolution results is that >80% of the L-isomer is considered consistent with the use of legitimate medication or conversely >20% of the D-isomer (and total concentration above the cutoff) is considered evidence of illicit use [8,18].

ADDITIONAL CONSIDERATIONS

Performances of initial and periodic evaluations of assay parameters are not only frequently required by accreditation agencies but also demonstrate a commitment to good laboratory practice. Protocols for initial and periodic method evaluations are available from several sources [8,32,33]. Interference studies for both screening and confirmation procedures should include analysis of samples containing compounds with structures similar to amphetamine and methamphetamine. Laboratories that are SAMHSA-certified must perform on an annual basis interference studies for amphetamines confirmation assays by analyzing samples containing interfering substances (phentermine at 50,000 ng/mL, phenylpropanolamine, ephedrine, and pseudoephedrine at 1 mg/mL, and MDMA and MDA at 5000 ng/mL) in the presence of and without amphetamine and methamphetamine at 40% of the cutoff [8]. Interference studies for MDMA/MDA confirmation assays must include analyzing samples containing the same interfering substances listed earlier (phentermine, phenylpropanolamine, ephedrine, and pseudoephedrine) and amphetamine and methamphetamine at 5000 ng/mL. [8]. The Department of Defense Forensic Toxicology Drug Testing Laboratories reported interferences with amphetamine, methamphetamine, MDMA, and MDA testing traced to designer amphetamines/cathinones. They evaluated the potential interferences of methiopropamine, 4-fluoroamphetamine, 4-fluoromethamphetamine, and 4-methylamphetamine on three immunoassay systems (Syva Emit II Plus Amphetamines, Roche KIMS Amphetamines II, and Microgenics DRI MDMA) and GC-MS confirmation analysis using three derivatization procedures (R-MTPAC, HFBA, and ClF_2AA) and reported significant cross-reactivity from all four compounds with the immunoassay systems. They also concluded that laboratories utilizing GC/MS selected ion-monitoring confirmation methods with the three derivatives could experience potential chromatographic and mass spectral interferences from the four compounds in the form of ion ratio and quantitation failures [10]. Hydroxynorephedrine, norephedrine, norpseudoephedrine, phenylephrine, and propylhexedrine are other compounds with structures similar to amphetamine and methamphetamine that may be tested for interference [34].

For MS/MS procedures documentation of the initial validation should include structural information to support the selected ion transitions and systematic evaluation of instrument parameters optimizing transition conditions [8]. For LC-MS and LC-MS/MS procedures the potential effect of components of the particular sample matrix to either suppress or enhance the ionization of the target analyte(s) should be evaluated.

The term matrix effect is often used to refer to ion suppression or enhancement due to particular endogenous urine components and other drugs and/or metabolites that may coelute with the compound of interest. One study demonstrated matrix effects varying from 61% suppression to 139% enhancements [35]. Studies have shown that in general ion suppression is more prevalent for those analytes that elute early and mid-chromatography while late eluting analytes are more likely to cause ion enhancement [35,36]. This effect can be normalized by the inclusion of a stable-isotope IS for each analyte of interest but this can become problematic for multidrug procedures because of the many drugs detected and the absence of readily available internal standards. Rosano et al. developed a novel calibration technique called threshold accurate calibration (TAC) as an alternate way to normalize the matrix effect [35]. This procedure involves analysis of each sample twice, once without the added analyte at a typical cutoff concentration and once with the added analyte and calculation of a TAC ratio for comparison to a calibrator TAC ratio.

When testing for designer amphetamines, such as MDMA and MDA, the analyst must be aware of the metabolism and excretion patterns for these drugs because some of the metabolites (3,4-dihydroxymethamphetamine (HHMA); 3,4-dihydroxyamphetamine (HHA); 4-hydroxy-3-methoxymethamphetamine (HMMA); and 4-hydroxy-3-methoxyamphetamine (HMA)) are excreted as glucuronide and sulfate conjugates. In order to obtain adequate recovery of these metabolites a hydrolysis procedure should be included as part of the confirmation testing [1,24,27].

CONCLUSIONS

Testing for amphetamine, methamphetamine, and other phenethylamine compounds focuses primarily on detection of amphetamine, methamphetamine, and designer drugs often grouped together as Ecstasy and related compounds (MDMA, MDA, and MDEA) and bath salts. One concern for immunoassay screening is false-positive results due to cross-reactivity of the reagent antibodies to other sympathomimetic amines such as ephedrine, pseudoephedrine, phentermine, and phenylpropanolamine. Periodate treatment of samples prior to immunoassay screening or mass spectrometric confirmation analysis can be used to remove interference by sympathomimetic amines. For GC/MS confirmation testing methamphetamine can be generated from the combination of high levels of pseudoephedrine or ephedrine in injection ports at high temperatures after derivatization with certain derivatives. Isomer resolution procedures using chiral columns or derivatization using optically pure chiral derivatizing agents are necessary to distinguish D- and L-stereoisomers, which are not distinguishable by immunoassay screening and most confirmation procedures. Another concern for immunoassay screening is the unavailability of assays that detect many of the newer designer amphetamines such as those classified as bath salts. Screening for multiple drugs using LC-MS/MS is increasing but such methods should be periodically evaluated for interferences due to matrix effects.

REFERENCES

[1] Department of Health and Human Services Substance Abuse and Mental Health Services Administration. Mandatory Guidelines for Federal Workplace Drug Testing Programs; Notice. Fed Regist 2008;73:71858−907. https://www.gpo.gov/fdsys/pkg/FR-2008-11-25/pdf/E8-26726.pdf [Accessed 31.08.18].

[2] Notice from Substance Abuse and Mental Health Services Administration: Revision of Mandatory Guidelines for Federal Workplace Programs notice. Fed Regist 2017; 82: 7920−7950. Document # 2017-00979. https://www.federalregister.gov/documents/2017/01/23/2017-00979/mandatory-guidelines-for-federal-workplace-drug-testing-programs [Accessed 31.08.18].

[3] Butler D, Guilbault GG. Analytical techniques for Ecstasy. Anal Lett. 2004;37:2003−30.

[4] Wu AHB, McKay C, Broussard LA, Hoffman RS, et al. National Academy of clinical biochemistry laboratory medicine practice guidelines: recommendations for the use of laboratory tests to support poisoned patients who present to the emergency department. Clin Chem. 2003;49:357−79.

[5] Nieddu M, Burrai L, Baralla E, Pasciu V, et al. ELISA detection of 30 new amphetamine designer drugs in whole blood, urine and oral fluid using Neogen® "Amphetamine" and "Methamphetamine/MDMA" kits. J Anal Toxicol. 2016;40:492−7.

[6] Rosano TG, Ohouo PY, Wood M. Screening with quantification for 64 drugs and metabolites in human urine using UPLC-MS-MS analysis and a threshold accurate calibration. J Anal Toxicol. 2017;41:536−46.

[7] McMillin GA, Marin SJ, Johnson-Davis KI, Lawlor BG, et al. A hybrid approach to urine drug testing using high-resolution mass spectrometry and select immunoassays. Amer J Clin Path. 2015;143:234−40.

[8] US Department of Health and Human Services. Substance Abuse and Mental Health Services Administration, National Laboratory Certification Program manual for laboratories and inspectors. Rockville, MD: DHHS; 2017.

[9] Moore KA. Amphetamines/Sympathomimetic Amines. In: Levine B, editor. Principles of Forensic Toxicology. Washington, D.C: AACC Press; 2003. p. 341−8.

[10] Holler JM, Vorce SP, Knittel JL, Malik-Wolf B, et al. Evaluation of designer amphetamine interference in GC-MS amine confirmation procedures. J Anal Toxicol. 2014;38:295−303.

[11] Poklis A. Unavailability of drug metabolite reference material to evaluate false-positive results for monoclonal EMIT-d.a.u. assay of amphetamine. Clin Chem. 1992;38:2580.

[12] Williams RH, Erickson T, Broussard L. Evaluating Sympathomimetic Intoxication in an Emergency Setting. Lab Med. 2000;31:497−507.

[13] Singh J. Reagent lot-to-lot variability in sensitivity for amphetamine with the Syva EMIT II monoclonal amphetamine/methamphetamine assay. J Anal Toxicol. 1997;21:174−5.

[14] Marin SJ, Doyle K, Chang A, Concheiro-Guisan M, et al. One hundred false-positive amphetamine specimens characterized by liquid chromatography time-of-flight mass spectrometry. J Anal Toxicol. 2016;40:37−42.

[15] US Food and Drug Administration. FDA Warns Consumers about Brazilian Diet Pills Found to Contain Active Drug Ingredients: Emagrece Sim and Herbathin Dietary Supplements May be Harmful (FDA News P06-07). Rockville, MD: FDA; 2006.

[16] Jemionek J, Bosy TJ, Jacobs A, Holler J, et al. Five cases of D-amphetamine positive urines resulting from ingestion of "Brazilian nutritional supplements" containing fenproporex have been reported. ToxTalkm Society of Forensic Toxicologists Newsletter 2006;30(2):11.

[17] Solomon MD, Wright JA. False-positive for (+)-methamphetamine. Clin Chem. 1977;23:1504.

[18] Broussard L. Interpretation of Amphetamines Screening and Confirmation Testing. In: Dasgupta A, editor. Handbook of Drug Monitoring Methods. Totowa, NJ: Humana Press; 2007. p. 381−95.

[19] Shindelman J, Mahal J, Hemphill G, Pizzo P, et al. Development and evaluation of an improved method for screening of amphetamines. J Anal Toxicol. 1999;23:506−10.

[20] Woodworth A, Saunders AN, Koenig JW, Moyer TP, et al. Differentiation of amphetamine/methamphetamine and other cross-immunoreactive sympathomimetic amines in urine samples by serial dilution testing. Clin Chem. 2006;52:743−6.

[21] Tsai SCJ, ElSohly MA, Dubrovsky T, Twarowska B, et al. Determination of five abused drugs in nitrite-adulterated urine by immunoassays and gas chromatography-mass spectrometry. J Anal Toxicol. 1998;22:474−80.

[22] Hino Y, Ojanpera I, Rasanen I, Vuori E. Performance of immunoassays in screening for opiates, cannabinoids and amphetamines in postmortem blood. Forensic Sci Int. 2003;131:148−55.

[23] Moriya F, Hashimoto Y. Evaluation of Triage screening for drugs of abuse in postmortem blood and urine samples. Jpn J Legal Med. 1997;51:214–19.
[24] Kraemer T, Maurer HH. Determination of amphetamine, methamphetamine and amphetamine-derived designer drugs or medicaments in blood and urine. J Chromatog B Biomed Sci Appl. 1998;713:163–87.
[25] Holler JM, Vorce SP, Bosy TZ, Jacobs A. Quantitative and isomeric determination of amphetamine and methamphetamine from urine using a nonprotic elution solvent and R(-)-α-methoxy-α-trifluoromethylphenylacetic acid chloride derivatization. J Anal Toxicol. 2005;29:652–7.
[26] Blandford DE, Desjardins PRE. Detection and identification of amphetamine and methamphetamine in urine by GC/MS. Clin Chem. 1994;40:145–7.
[27] Pirnay SO, Abraham TT, Lowe RH, Huestis MA. Selection and optimization of hydrolysis conditions for the quantification of urinary metabolites of MDMA. J Anal Toxicol. 2006;30:563–9.
[28] Abraham TT, Barnes AJ, Lowe RH, Spargo EAK, et al. Urinary MDMA, MDA, HMMA, and HMA excretion following controlled MDMA administration to humans. J Anal Toxicol. 2009;33:439–46.
[29] Wu AHB, Wong SS, Johnson KG, Ballatore A, et al. The conversion of ephedrine to methamphetamine and methamphetamine-like compounds during and prior to gas chromatographic/mass spectrometric analysis of CB and HFB derivatives. Biol Mass Spectrom 1992;21:278–84.
[30] Hornbeck CL, Carrig JE, Czarny RJ. Detection of a GC/MS artifact peak as methamphetamine. J Anal Toxicol. 1993;17:257–63.
[31] ElSohly MA, Stanford DF, Sherman D, Shah H, et al. A procedure for eliminating interferences from ephedrine and related compounds in the GC/MS analysis of amphetamine and methamphetamine. J Anal Toxicol. 1992;16:109–11.
[32] Scientific Working Group for Forensic Toxicology (SWGFT). Standard practices for method validation in forensic toxicology. J Anal Toxicol. 2013;37:452–74.
[33] New York State Department of Health. Clinical Laboratory Standards of Practice: Toxicology/Forensic Toxicology. 2017:1–36. https://www.wadsworth.org/sites/default/files/WebDoc/FOTO_September2017.pdf [Accessed 31.08.18].
[34] Goldberger BA, Cone EJ. Confirmatory tests for drugs in the workplace by gas chromatography—mass spectrometry. J Chromatog A 1994;674:73–86.
[35] Rosano TG, Ohouo PY, LeQue JJ, Freeto SM, et al. Definitive drug and metabolite screening in urine by UPLC-MS-MS using a novel calibration technique. J Anal Toxicol. 2016;40:628–38.
[36] Lee HK, Ho CS, Iu YP, Lai PS, et al. Development of a broad toxicological screening technique for urine using ultra-performance liquid chromatography and time-of-flight mass spectrometry. Anal Chim Acta. 2009;649:80–90.

Chapter 17

Cocaine, Crack Cocaine, and Ethanol: A Deadly Mix

Eric T. Shimomura, George F. Jackson and Buddha Dev Paul
Forensic Toxicologist, Armed Forces Medical Examiner System, Dover, DE, United States

INTRODUCTION

The Drug Abuse Warning Network (DAWN) reported over 2.4 million emergency department visits which involved drugs in 2011 [1]. Alcohol was involved in 636,298 visits (26%) either alone or with other drugs. Illicit drugs were involved in 596,475 visits (24%). Cocaine was involved in 505,224 visits (21%), which was 40% of visits and involved illicit drugs. Both alcohol and cocaine were involved in 173,799 visits (7% of total emergency department visits). This represents 34% of visits which involved cocaine. Over the period 1995−2002, emergency department visits increased by 24% for alcohol-related visits and 47% for cocaine-related visits [2]. There has been a change in this trend, as there were no statistically meaningful increases for either alcohol- or cocaine-related emergency department visits for the period 2004−11 [1]. Although cocaine and alcohol use are associated with a large percentage of emergency department visits, the actual number of individuals who abuse cocaine and alcohol is unknown.

ROUTES OF COCAINE ADMINISTRATION

The main routes of administration of cocaine are smoking (crack), insufflation (snorting), and intravenous (IV). In the early 1900s most users reported preferring insufflation as the primary means of cocaine use [3−5]. The cocaine used was the powdered form of the hydrochloride salt. In the early 1980s a free base form of cocaine became available, known as "crack," and quickly gained popularity [3,4]. The free base crack cocaine is prepared by neutralizing the hydrochloride salt. In the free base form vaporization temperature is much lower compared to the hydrochloride salt (80°C vs 180°C, respectively) [6]. This made smoking an easy way of cocaine administration. Smoking is preferred because the effects could be experienced within seconds versus minutes for insufflation. A feeling of "liking" has been observed to be greater for smoking compared to either insufflation or intravenous use [7,8]. Due to vasoconstrictive properties of cocaine, chronic use by insufflation often causes perforated septum and other sinus conditions. These side effects are avoided by smoking. Over the period of 1995− 2005, smoking continued as the predominant route of administration [9]. The observed trends for smoking, insufflation, and all other routes of administration were 79%−73%, 14%−22%, and 7%−5%, respectively.

COCAINE METABOLISM

Benzoylecgonine (BZ) and ecgonine methyl ester (EME) are the most abundant metabolites of over 11 metabolites of cocaine as shown in Fig. 17.1 [10−13]. Cocaine is deactivated to BZ and EME primarily through deesterification (hydrolysis) in the liver. Most other metabolites are pharmacologically inactive or have very low activity. Norcocaine (NC), through an oxidative metabolism, is one of the few metabolites that are pharmacologically active. NC undergoes further metabolism to norcocaine nitroxide, which is hepatotoxic in animals [14−20]. Methylecgonidine (MED or anhydroecgonine methyl ester (AEME)), formed from the smoking (pyrolysis) of cocaine (Fig. 17.2), has also been shown to be active in animals [21−23]. It is metabolized through deesterification to ecgonidine (ED or anhydroecgonine). When cocaine and alcohol are coabused there is a change in the metabolic pathway. Part of the cocaine, instead of

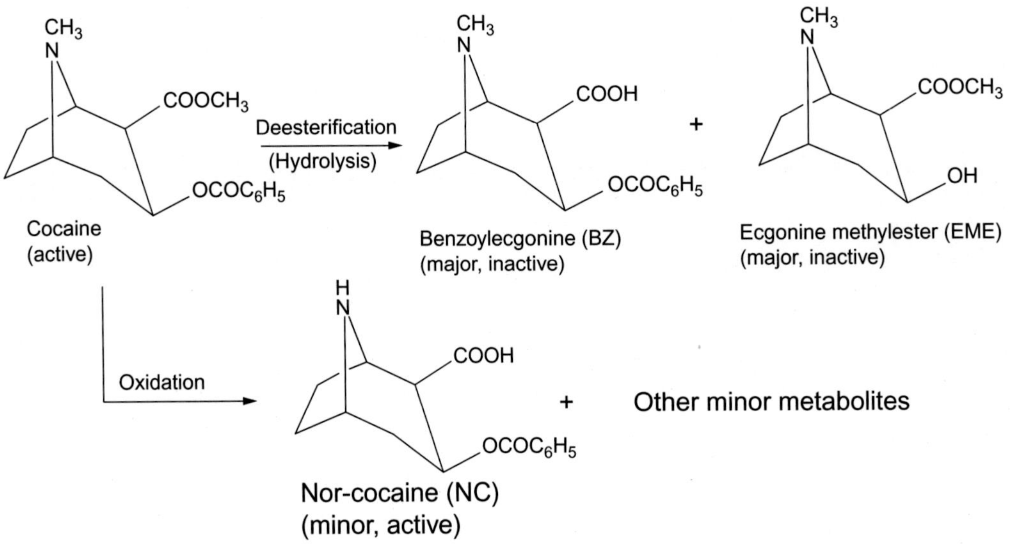

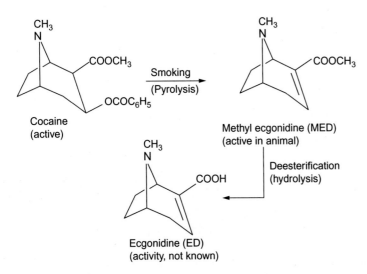

FIGURE 17.1 Metabolism of cocaine to major inactive metabolites, BZ and EME, and to a minor active metabolite NC.

FIGURE 17.2 Smoking (pyrolysis) of cocaine to MED and subsequent metabolic deesterification to ED.

being hydrolyzed to inactive compounds, undergoes transesterification to a pharmacologically active cocaethylene (CE) involving the same nonspecific liver methyl esterase used for hydrolysis (Fig. 17.3) [24]. Alcohol interferes with the detoxification of cocaine to BZ and EME, possibly through inhibition of hepatic esterases [24–28] causing cocaine to remain in the body for a longer period of time. Alcohol is also responsible for increased amount of active NC produced from increased activity of oxidative enzymes in the liver. The presence of alcohol also results in the metabolism of pyrolytic MED to ethylecgonidine (EED or anhydroecgonine ethyl ester) [29,30].

COCAINE TOXICITY

Cocaine abuse is a significant contributor to a variety of physiological conditions, such as coronary heart disease, arrhythmias, myocardial infarctions (MI), aortic dissections, aneurysms, and asthma attacks [31–50]. It is well established that cocaine causes vasoconstriction of blood vessels, which has been linked to health risks, particularly to the heart and brain. This effect combined with its anesthetic effect and observed acceleration of coronary heart disease in chronic cocaine abusers has led to the association of cocaine with cardiotoxicity. The role of cocaine as a significant risk factor in acute MI has been extensively examined [40,43,46,47,51–54]. Vasoconstriction, arrhythmia, acceleration of atherosclerosis, and increased platelet aggregation are all linked to cocaine use and all have been shown to be

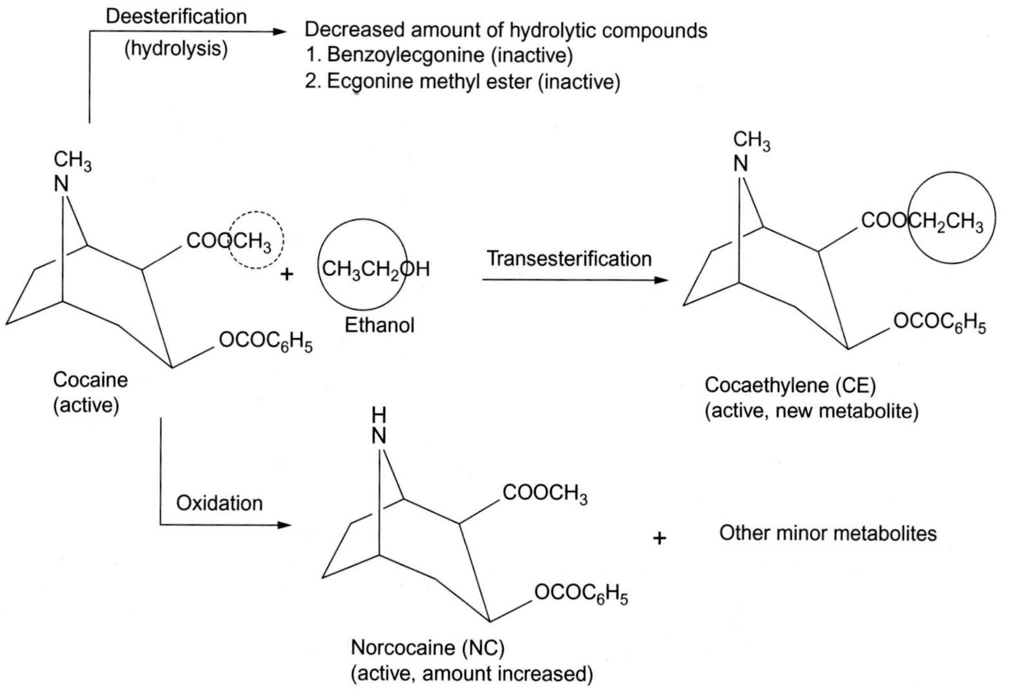

FIGURE 17.3 Alcohol (ethanol) induced metabolic transesterification from cocaine to active CE and metabolic oxidation of cocaine to active NC.

potential contributors to acute MI. In a hospital patient study cocaine was present in 14 out of 38 cases involving aortic dissection, a health risk of cocaine use that is not so widely known or publicized [37]. Of these 14 cases, 11 patients also had hypertension. Thirteen of the fourteen patients indicated that they had been smoking crack cocaine.

A comparative study of deaths due to cocaine, opioid, and nondrug causes examined the correlation between the causes of death and the extent of heart disease [53]. When the data was summarized for the 125 cocaine, 669 opioid, and 360 hanging cases, cardiovascular disease was noticeably higher in the cocaine cases (64 cases/51.2%) compared to either opioid (192 cases/28.7%) or hanging (85 cases/23.6%) cases. Cerebrovascular atherosclerosis was noticeably higher in the cocaine cases (17 cases/13.6%) compared to either the opioid (21 cases/3.1%) or hanging (4 cases/1.1%) cases. Ventricular hypertrophy and ischemic heart disease were associated with a significant number of the cocaine cases. The observed ventricular hypertrophy is consistent with a study of 30 living subjects, where examination with echocardiography revealed higher left ventricular mass in chronic cocaine abusers versus controls ($103 + 24$ vs. $77 + 14$ g/m^2) [33].

The observed cardiotoxicity of cocaine is somewhat mediated due to the rapid elimination of cocaine (half-life, $t_{1/2} \sim 80$ min) [55]. Actual cocaine abusers are likely to take multiple doses to enhance the feeling of euphoria. This can result in higher blood levels of cocaine for an extended period of time. In a multiple dose study a significant heart rate increase was observed (by as much as 10 beats/min higher) and greater euphoric feeling was reported when compared to single-dose cocaine [51]. Physiological changes have been observed for all the routes of administration [6]. For IV, smoked, and insufflation routes, heart rate was observed to be elevated by 46, 32, and 26 beats per minute, respectively. Both systolic/diastolic blood pressure for IV, smoked, and insufflation routes were also elevated by 28/16, 32/22, and 24/11 mm Hg, respectively [6].

However, according to Farre et al., the elevated systolic blood pressure over the placebo was not significant, when cocaine was introduced by insufflation [25]. These differences in findings may be due to different doses used in these studies (2 mg/kg vs 100 mg).

COCAINE SMOKING AND TOXICITY

As a result of smoking cocaine, pyrolysis products may also have an impact on the observed toxicity [10,56,57]. In particular MED may play a role in both cardiac and pulmonary toxicity as evidenced in animal studies, where MED has been observed to relax tracheal rings [23] and, at relatively high inhalation concentrations, induce bronchoconstriction

[21]. Along with cocaine, MED has been observed to decrease arterial blood pressure and heart rate (unlike human responses), while depressing respiratory rate in the hindbrain, cardiac, and respiratory centers of the rabbit [22]. Hypotension and tachycardia have been observed in sheep administered MED [58]. The amount of MED produced from pyrolysis can be as high as 73% of the cocaine dose under certain conditions [57]. However, the typical amount of MED resulting from conditions closer to realistic smoking is about 5% [13,59]. When MED enters the body it is metabolized quickly to ED [12,29,58]. This metabolite is observed at noticeably higher levels than MED in postmortem tissues [60] and in urine in human subjects [11]. Although MED has known toxic effects in animals, there are no human studies investigating toxicity of ED. Another compound, EED, has also been detected in the urine of a crack user coabusing alcohol [30]. Toxicity of EED was observed to be similar to MED in rabbits [22]. Additional studies may be worthwhile to elucidate the toxicity of MED, ED, and EED in humans.

THE COMBINED TOXICITY OF COCAINE AND ALCOHOL

The literature addresses the toxicity of cocaine, alcohol, and both in the combination [34,41−43,47,61−69]. Alcohol abuse is associated with a variety of health complications, including liver toxicity/injury, acute hypertension, and atherosclerosis [61−64,70−84] (see Chapter 1: Alcohol: Pharmacokinetics, Health Benefits With Moderate Consumption and Toxicity). When alcohol is consumed with cocaine there is an immediate impact on the metabolism of cocaine. It is known that when alcohol is consumed concurrent with cocaine, metabolic detoxification of cocaine is altered. In insufflation studies, peak blood levels for cocaine were as much as 30% higher than for cocaine alone [25,44,51,85]. The causes of the elevated levels may result from the increased absorption of cocaine due to the vasodilation effects of alcohol. Since cocaine levels are higher, increased cardiotoxicity would be expected. It should be noted that the studies reporting the elevation of cocaine blood levels have been for cocaine administered through insufflation only. While the blood levels increase for cocaine (insufflation) with alcohol, elimination half-life ($\sim$80 min.) remains the same. Peak heart rate increased for cocaine with alcohol greater than for either cocaine or alcohol alone (Table 17.1) [25]. Peak diastolic blood pressure increased above the placebo group, but was only slightly greater than for cocaine alone. The differences in systolic blood pressure were statistically insignificant when results of placebo, cocaine, alcohol, and cocaine with alcohol are compared. Although statistically insignificant, peak increases of 10 mm Hg compared to placebo were observed for cocaine and cocaine with alcohol. This differs from the reported result of a 24 mm Hg increase in the Jone's study [6]. The increase in systolic blood pressure may have been due to the higher doses (2 mg/kg) in the Jone's study. Cocaine half-life in the presence of alcohol has been reported to increase to as much as 108 min. [27]. In the same study the CE half-life was reported as high as 157 min.

The presence of alcohol increases the blood levels of other cocaine metabolites such as NC [25,26]. NC is active and found in relatively low amounts compared to inactive or low activity metabolites such as BZ and EME [12]. Toxicity of NC has been reported in animals, specifically hepatotoxicity and also immunotoxicity [14−20,86−89]. Hepatotoxicity of alcohol is well known. Combined with cocaine, alcohol is likely to increase liver damage, possibly through alcohol-mediated activity of cytochrome P-450 [79]. The metabolism of NC to the hepatotoxic norcocaine nitroxide is directly related to the activity of cytochrome P-450 in animals [14,16,18,87,90].

Cocaine and alcohol have been examined in relation to learning and performance issues [91−93]. In a study performed by Higgins et al. [92] short-term effects of cocaine coabused with alcohol were found to reduce the impairment of alcohol alone on learning tests involving accuracy and rates of responding. When cocaine was administered with alcohol,

TABLE 17.1 Changes in Peak Heart Rate and Diastolic Blood Pressure of Cocaine (Insufflation) and Alcohol for a 100 mg Dose of Cocaine

Drugs	Changes in Heart Rate[a] (Compared to Placebo) bpm	Changes in Diastolic Blood Pressure[b] (Compared to Placebo) mmHg
Alcohol	+17	−14
Cocaine	+23	+11.5
Cocaine and alcohol	+41	+12

[a]Duration, 2−3 h [25−27].
[b]No statistical difference in systolic blood pressure.

subjects returned to accuracy levels observed in the placebo group. In the rates of responding, cocaine together with alcohol resulted in rates observed in the placebo group for some subjects. In another study subjects were given neurocognitive tests involving intelligence quotient (IQ), verbal skills, psychomotor speed (reaction times), and manual dexterity tests for cocaine with and without alcohol [91]. A decline in performance was observed for either alcohol or cocaine alone on all tests, and in some tests, cocaine combined with alcohol resulted in a greater decline than either alone. Decline in performance was greatest for cocaine with alcohol use in the psychomotor speed tests. These results suggest that cocaine and alcohol together can have an additive negative effect on some neurocognitive functions. Some residual effects persisted for as long as 4 weeks in some subjects after abstinence from cocaine and alcohol. Lewis et al. observed a synergistic effect of cocaine together with alcohol on brain stimulation reward (BSR) responses in a rat study [93]. Increased response rates were observed for cocaine combined with alcohol, compared to either cocaine or alcohol alone.

A survey involving 110 cocaine users was conducted in which the pattern of alcohol use with cocaine is summarized [94]. In this survey it was reported that there were 69 cocaine powder users (insufflation or snorting) and 33 crack cocaine users (smoking). Eight individuals reported using both methods and were excluded from further evaluation. Subjects reported that alcohol use increased when cocaine powder was used and decreased when crack cocaine was used. The average amount of cocaine used also increased for cocaine powder users when alcohol was involved, while there was no significant change for crack cocaine users. The pattern of alcohol use was also different. Powder cocaine users tended to use alcohol continuously, while crack cocaine users tended to use alcohol after smoking. One possible explanation is that since alcohol has a vasodilation effect, it could counter, at least partially, the vasoconstrictive effect of cocaine for intranasal users [94]. This would increase the amount of cocaine powder absorbed and in turn result in greater euphoria. The effects of crack cocaine are felt within seconds versus minutes for powder cocaine, and so the user feels the effects of crack almost immediately. The use of alcohol prior to crack abuse would be expected to reduce the feeling of euphoria [54]. Also the use of alcohol at the end of smoking sessions would be consistent with abusers seeking longer periods of euphoria [7] resulting from sustained levels of cocaine combined with the presence of pharmacologically active CE in the blood.

THE ROLE OF COCAETHYLENE IN TOXICITY

As mentioned earlier CE is a metabolite formed when cocaine and alcohol are consumed together. The primary metabolism of cocaine to inactive BZ and EME is altered to produce predominantly CE along with some BZ [24]. It was observed that CE was also metabolized to BZ. The actual amount of CE produced has been reported to be about one-sixth [27] to one-fifth [85] of the cocaine dose in single-dosing studies in humans. CE appears to have a slower elimination rate [44,85] than cocaine, and the half-life of these compounds have reported values of $148 + 15$ min and $86 + 11$ min, respectively [85]. Pharmacologically active CE parallels cocaine in some aspects of toxicity [28,31,32,52,54,95–100]. The detection of CE has been reported under various conditions [31,37,52,101,102]. In a group of trauma patients, CE was detected in 13 out of 15 patients at the time of admission [31], implying the use of cocaine and alcohol in the recent past. Of these 13 patients both cocaine and alcohol were present in 12 patients. In another group of 451 specimens that tested positive for BZ, 57 specimens contained CE [52].

The exact role of CE in cocaine and alcohol comorbidity is somewhat controversial. Peak plasma levels for CE compared to cocaine in a single-dose study were 28 and 156 ng/mL, respectively, for a 1.25 mg/kg dose of cocaine [44]. The lower level of CE compared to cocaine reported by the authors, cast some uncertainty as to the role CE plays in cocaine and alcohol toxicity. However, in hospital patients [31,32,100,102], CE concentrations were comparable to cocaine. Based on this information CE may contribute toxic effects in combination with those observed for cocaine. In a study involving intranasal administration of cocaine, subjects typically could not tell the difference between ingestion of cocaine and CE [98] for equivalent doses (3 μmol/kg). Cocaine response was somewhat different when cocaine was administered IV (0.5 mg/kg) during CE infusion blood levels of 200 ng/mL. The duration of the cocaine "rush" was shortened by about 66% [54]. One explanation is that acute tolerance resulted from the CE infusion, which reduced the intensity of the cocaine "rush" when administered later. It may be that because of the reduced period of "rush," an abuser would have a tendency to increase subsequent doses of cocaine to achieve the same length of euphoria, and as a result, increase the health risks. Similar increases in heart rate and systolic blood pressure have been observed for equivalent doses of cocaine (0.92 mg/kg) and CE (0.95 mg/kg) [98]. Also the onset and intensity of the "rush" were indistinguishable by the subjects for cocaine and CE. It appears that CE may also have a slower decline in the "rush" than cocaine, although in this study, it was not statistically significant. CE increases the attraction of abusers to cocaine use and also increases the health risks as well.

Although fatality from abuse of cocaine and alcohol has been reported, an accidental death related to cocaine, CE, and caffeine has been reported. The blood alcohol concentration in the subject was only 10 mg/dL, but CE concentration in blood was 0.16 mg/L. The concentration of caffeine in blood was also high (16.40 mg/L). Cocaine was only found in gastric content (0.45 mg/L) along with CE (1.85 mg/L) and caffeine (14.40 mg/L). From the case investigation it was determined that the death was accidental but related to ingestion of cocaine and caffeine tablets [101].

ANALYTICAL METHODS

Cocaine immunoassays utilize antibody targeted to BZ, the major metabolite of cocaine in urine. However, CE also cross-reacts with such antibodies (although cross-reactivities vary widely depending on specific immunoassay) and due to the presence of BZ along with CE in urine of individuals abusing both cocaine and alcohol, immunoassay screening is positive. However, initial positive results must be confirmed using a more specific analytical method, such as gas chromatography/mass spectrometry (GC/MS) of liquid chromatography combined with mass spectrometry (LC-MS), or tandem mass spectrometry (LC-MS/MS).

Chromatographic methods can be used for analysis of cocaine along with its metabolites in various biological matrices including urine, blood, oral fluid, meconium, and hair. Although BZ requires derivatization prior to analysis by GC/MS, CE can be analyzed by GC/MS without derivatization. Electron impact full scan mass spectrum of CE is given in Fig 17.4. Lewis et al. reported that CE accumulates in greater concentration in meconium than urine, and is a useful biomarker for identifying fetal alcohol exposure. The authors also reported a GC/MS protocol for analysis of cocaine, BZ, and CE in meconium and used deuterated cocaine, BZ, and CE as internal standards [103].

Hezinova et al. published a capillary electrophoresis–electrospray mass spectrometric protocol for analysis of cocaine, CE, BZ, NC, and EME in urine [104]. Quantitation of cocaine, BZ, EME, and CE in urine and blood using GC/MS has also been reported. The authors analyzed BZ and EME as pentafluoropropionyl derivatives (derivatization achieved by using pentafluoropropionic acid anhydride and pentafluoropropanol). Cocaine and CE are refractory to derivatization. The authors used cocaine-d_3, BZ-d_3, EME-d_3, and CE-d_3 as internal standards [105].

Chen et al. described an LC-M/MS protocol for simultaneous determination of cocaine and its metabolites including CE in whole blood. However, no derivatization is needed for such analysis. The authors extracted cocaine and its metabolites from whole by solid-phase extraction using methanol/water (95:5 by vol) and used a reverse phase C-18 column

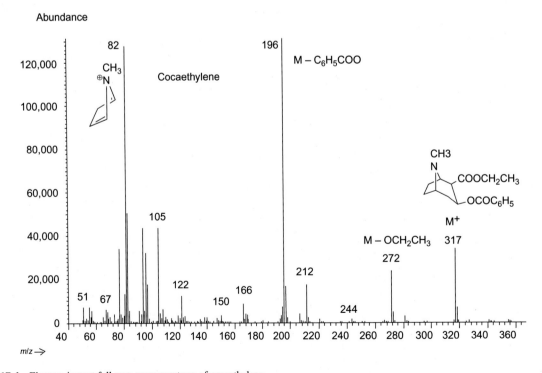

FIGURE 17.4 Electron impact full scan mass spectrum of cocaethylene.

for analysis. Deuterated cocaine, EME, BZ, CE, and ecgonine were used as internal standards. Ion transition m/z 304.1−182.1 using electrospray ionization was monitored for cocaine, ion transition 200.1−182.1 for EME, ion transition 290.1−168.1 for BZ, ion transition 318.2−196.1 for CE, and ion transition 304.2−182.1 was monitored for nor-CE. Other ion transitions were also used for monitoring various metabolites of cocaine [106]. AEME is formed only during the burning (pyrolysis) of cocaine as a result of loss of benzoic acid, in the act of smoking, and hydrolyzed in vivo by means of butyryl cholinesterases and nonenzymatic processes, leading to anhydroecgonine (AEC), both biomarkers of cocaine use in smoked form (crack). These biomarkers can be detected in various biological matrices using GC/MS or LC-MS [107].

CONCLUSIONS

Medical personnel, especially those in emergency departments, continue to witness the dangers of cocaine abuse with and without alcohol. While statistics continue to be reported by DAWN and similar organizations, the actual level of abuse of cocaine and alcohol remains unconfirmed, although most likely, it is significantly higher than those figures. The individual toxicities of cocaine and alcohol have been well established, and the significant health risks are well known. It appears that despite these risks, cocaine and alcohol continue to be abused, both separately, and in combination. Unfortunately the combined use of cocaine and alcohol appears to result in additional health risks, both acute and chronic. Also the increased levels of cocaine and active metabolites when cocaine is abused with alcohol should also be considered in the overall risk and should be studied further.

DISCLAIMER

The opinions or assertions expressed herein are those of the authors and do not necessarily reflect the views of the Department of the Army, the Department of the Navy, the Department of the Air Force, Defense Health Agency, or the Department of Defense.

REFERENCES

[1] Drug Abuse Warning Network. In: SAMHSA, editor. National Estimates of Drug-Related Emergency Department Visits, Vol. Rockville, MD: Department of Health and Human Services; 2011. p. 1−100. 2011.
[2] Emergency Department Trends From the Drug Abuse Warning Network, Final Estimates 1995−2002. In: SAMHSA, ed., Vol. Rockville, MD: Department of Health and Human Services, 2003:1-148.
[3] Hamid A. The developmental cycle of a drug epidemic: the cocaine smoking epidemic of 1981−1991. J Psychoactive Drugs 1992;24:337−48.
[4] Karch SB. Cocaine: history, use, abuse. J R Soc Med 1999;92:393−7.
[5] Petersen RC. Cocaine: an overview. NIDA Res Monogr 1977;13:5−16.
[6] Jones RT. The pharmacology of cocaine smoking in humans. NIDA Res Monogr 1990;99:30−41.
[7] Cone EJ. Pharmacokinetics and pharmacodynamics of cocaine. J Anal Toxicol 1995;19:459−78.
[8] Perez-Reyes M, Di Guiseppi S, Ondrusek G, Jeffcoat AR, Cook CE. Free-base cocaine smoking. Clin Pharmacol Ther 1982;32:459−65.
[9] Cocaine Route of Administration Trends: 1995−2005. The DASIS Report, Office of Applied Studies, Vol. Rockville, MD: Substance Abuse and Mental Health Services Administration; 2007.
[10] Cone EJ, Hillsgrove M, Darwin WD. Simultaneous measurement of cocaine, cocaethylene, their metabolites, and "crack" pyrolysis products by gas chromatography-mass spectrometry. Clin Chem 1994;40:1299−305.
[11] Huestis MA, Darwin WD, Shimomura E, Lalani SA, Trinidad DV, Jenkins AJ, et al. Cocaine and metabolites urinary excretion after controlled smoked administration. J Anal Toxicol 2007;31:462−8.
[12] Paul BD, Lalani S, Bosy T, Jacobs AJ, Huestis MA. Concentration profiles of cocaine, pyrolytic methyl ecgonidine and thirteen metabolites in human blood and urine: determination by gas chromatography-mass spectrometry. Biomed Chromatogr 2005;19:677−88.
[13] Paul BD, McWhorter LK, Smith ML. Electron ionization mass fragmentometric detection of urinary ecgonidine, a hydrolytic product of methylecgonidine, as an indicator of smoking cocaine. J Mass Spectrom 1999;34:651−60.
[14] Rauckman EJ, Rosen GM, Cavagnaro J. Norcocaine nitroxide. A potential hepatotoxic metabolite of cocaine. Mol Pharmacol 1982;21:458−63.
[15] Evans MA. Role of protein binding in cocaine-induced hepatic necrosis. J Pharmacol Exp Ther 1983;224:73−9.
[16] Smolen TN, Smolen A. Developmental expression of cocaine hepatotoxicity in the mouse. Pharmacol Biochem Behav 1990;36:333−8.
[17] Mets B, Virag L. Lethal toxicity from equimolar infusions of cocaine and cocaine metabolites in conscious and anesthetized rats. Anesth Analg 1995;81:1033−8.
[18] Ndikum-Moffor FM, Schoeb TR, Roberts SM. Liver toxicity from norcocaine nitroxide, an N-oxidative metabolite of cocaine. J Pharmacol Exp Ther 1998;284:413−19.

[19] Morishima HO, Whittington RA, Iso A, Cooper TB. The comparative toxicity of cocaine and its metabolites in conscious rats. Anesthesiology 1999;90:1684–90.
[20] Pan WJ, Hedaya MA. Cocaine and alcohol interactions in the rat: effect of cocaine and alcohol pretreatments on cocaine pharmacokinetics and pharmacodynamics. J Pharm Sci 1999;88:1266–74.
[21] Willetts J, Chen LC, Graefe JF, Wood RW. Effects of methylecgonidine on acetylcholine-induced bronchoconstriction and indicators of lung injury in guinea pigs. Life Sci 1995;57:PL225–30.
[22] Erzouki HK, Allen AC, Newman AH, Goldberg SR, Schindler CW. Effects of cocaine, cocaine metabolites and cocaine pyrolysis products on the hindbrain cardiac and respiratory centers of the rabbit. Life Sci 1995;57:1861–8.
[23] Chen LC, Graefe JF, Shojaie J, Willetts J, Wood RW. Pulmonary effects of the cocaine pyrolysis product, methylecgonidine, in guinea pigs. Life Sci 1995;56:PL7–L12.
[24] Dean RA, Christian CD, Sample RH, Bosron WF. Human liver cocaine esterases: ethanol-mediated formation of ethylcocaine. Faseb J 1991;5:2735–9.
[25] Farre M, de la Torre R, Gonzalez ML, Teran MT, Roset PN, Menoyo E, et al. Cocaine and alcohol interactions in humans: neuroendocrine effects and cocaethylene metabolism. J Pharmacol Exp Ther 1997;283:164–76.
[26] Farre M, de la Torre R, Llorente M, Lamas X, Ugena B, Segura J, et al. Alcohol and cocaine interactions in humans. J Pharmacol Exp Ther 1993;266:1364–73.
[27] Harris DS, Everhart ET, Mendelson J, Jones RT. The pharmacology of cocaethylene in humans following cocaine and ethanol administration. Drug Alcohol Depend 2003;72:169–82.
[28] Jatlow P. Cocaethylene: pharmacologic activity and clinical significance. Ther Drug Monit 1993;15:533–6.
[29] Fandino AS, Toennes SW, Kauert GF. Studies on in vitro degradation of anhydroecgonine methyl ester (methylecgonidine) in human plasma. J Anal Toxicol 2002;26:567–70.
[30] Myers AL, Williams HE, Kraner JC, Callery PS. Identification of anhydroecgonine ethyl ester in the urine of a drug overdose victim. J Forensic Sci 2005;50:1481–5.
[31] Bailey DN. Serial plasma concentrations of cocaethylene, cocaine, and ethanol in trauma victims. J Anal Toxicol 1993;17:79–83.
[32] Bailey DN. Plasma cocaethylene concentrations in patients treated in the emergency room or trauma unit. Am J Clin Pathol 1993;99:123–7.
[33] Brickner ME, Willard JE, Eichhorn EJ, Black J, Grayburn PA. Left ventricular hypertrophy associated with chronic cocaine abuse. Circulation 1991;84:1130–5.
[34] Egred M, Davis GK. Cocaine and the heart. Postgrad Med J 2005;81:568–71.
[35] Gorelick DA. Alcohol and cocaine. Clinical and pharmacological interactions. Recent Dev Alcohol. 1992;10:37–56.
[36] Grohe C, Meyer R. The cardiac cocaine connection. Cardiovasc Res 2003;59:805–6.
[37] Hsue PY, Salinas CL, Bolger AF, Benowitz NL, Waters DD. Acute aortic dissection related to crack cocaine. Circulation 2002;105:1592–5.
[38] Karch SB, Green GS, Young S. Myocardial hypertrophy and coronary artery disease in male cocaine users. J Forensic Sci 1995;40:591–5.
[39] Karch SB, Stephens B, Ho CH. Relating cocaine blood concentrations to toxicity—an autopsy study of 99 cases. J Forensic Sci 1998;43:41–5.
[40] Keller KB, Lemberg L. The cocaine-abused heart. Am J Crit Care 2003;12:562–6.
[41] Kloner RA, Hale S, Alker K, Rezkalla S. The effects of acute and chronic cocaine use on the heart. Circulation 1992;85:407–19.
[42] Laposata EA. Cocaine-induced heart disease: mechanisms and pathology. J Thorac Imaging 1991;6:68–75.
[43] Pennings EJ, Leccese AP, Wolff FA. Effects of concurrent use of alcohol and cocaine. Addiction 2002;97:773–83.
[44] Perez-Reyes M, Jeffcoat AR. Ethanol/cocaine interaction: cocaine and cocaethylene plasma concentrations and their relationship to subjective and cardiovascular effects. Life Sci 1992;51:553–63.
[45] Pirwitz MJ, Willard JE, Landau C, Lange RA, Glamann DB, Kessler DJ, et al. Influence of cocaine, ethanol, or their combination on epicardial coronary arterial dimensions in humans. Arch Intern Med 1995;155:1186–91.
[46] Qureshi AI, Suri MF, Guterman LR, Hopkins LN. Cocaine use and the likelihood of nonfatal myocardial infarction and stroke: data from the Third National Health and Nutrition Examination Survey. Circulation 2001;103:502–6.
[47] Rezkalla SH, Kloner RA. Cocaine-induced acute myocardial infarction. Clin Med Res 2007;5:172–6.
[48] Satran A, Bart BA, Henry CR, Murad MB, Talukdar S, Satran D, et al. Increased prevalence of coronary artery aneurysms among cocaine users. Circulation 2005;111:2424–9.
[49] Vasica G, Tennant CC. Cocaine use and cardiovascular complications. Med J Aust 2002;177:260–2.
[50] Hollander JE, Henry TD. Evaluation and management of the patient who has cocaine-associated chest pain. Cardiol Clin 2006;24:103–14.
[51] McCance-Katz EF, Kosten TR, Jatlow P. Concurrent use of cocaine and alcohol is more potent and potentially more toxic than use of either alone—a multiple-dose study. Biol Psychiatry 1998;44:250–9.
[52] Bailey DN. Cocaethylene (ethylcocaine) detection during toxicological screening of a university medical center patient population. J Anal Toxicol 1995;19:247–50.
[53] Darke S, Kaye S, Duflou J. Comparative cardiac pathology among deaths due to cocaine toxicity, opioid toxicity and non-drug-related causes. Addiction 2006;101:1771–7.
[54] Baker J, Jatlow P, Pade P, Ramakrishnan V, McCance-Katz EF. Acute cocaine responses following cocaethylene infusion. Am J Drug Alcohol Abuse 2007;33:619–25.
[55] Jeffcoat AR, Perez-Reyes M, Hill JM, Sadler BM, Cook CE. Cocaine disposition in humans after intravenous injection, nasal insufflation (snorting), or smoking. Drug Metab Dispos 1989;17:153–9.

[56] Martin BR, Lue LP, Boni JP. Pyrolysis and volatilization of cocaine. J Anal Toxicol 1989;13:158−62.
[57] Nakahara Y, Ishigami A. Inhalation efficiency of free-base cocaine by pyrolysis of 'crack' and cocaine hydrochloride. J Anal Toxicol 1991;15:105−9.
[58] Scheidweiler KB, Plessinger MA, Shojaie J, Wood RW, Kwong TC. Pharmacokinetics and pharmacodynamics of methylecgonidine, a crack cocaine pyrolyzate. J Pharmacol Exp Ther 2003;307:1179−87.
[59] Wood RW, Shojaie J, Fang CP, Graefe JF. Methylecgonidine coats the crack particle. Pharmacol Biochem Behav 1996;53:57−66.
[60] Shimomura ET, Hodge GD, Paul BD. Examination of postmortem fluids and tissues for the presence of methylecgonidine, ecgonidine, cocaine, and benzoylecgonine using solid-phase extraction and gas chromatography-mass spectrometry. Clin Chem 2001;47:1040−7.
[61] Ahmed FE. Toxicological effects of ethanol on human health. Crit Rev Toxicol. 1995;25:347−67.
[62] Beilin LJ, Puddey IB. Alcohol and hypertension: an update. Hypertension 2006;47:1035−8.
[63] Marmot MG. Alcohol and coronary heart disease. Int J Epidemiol 2001;30:724−9.
[64] Zahler R, Piselli C. Smoking, Alcohol, and Drugs. New York: Hearst Books; 1992. 12pp.
[65] Cami J, Farre M, Gonzalez M, Segura J, de la TR. Cocaine metabolism in humans after use of alcohol. Clinical and research implications. Recent Dev Alcohol 1998;14:437−55.
[66] Gorelick D. Alcohol and cocaine. Clinical and pharmacological interactions. Recent Dev Alcohol 1992;10:37−56.
[67] Jatlow P, McCance EF, Bradberry CW, Elsworth JD, Taylor JR, Roth RH. Alcohol plus cocaine: the whole is more than the sum of its parts. Ther Drug Monit 1996;18:460−4.
[68] Lange RA, Hillis LD. Cardiovascular complications of cocaine use. N Engl J Med 2001;345:351−8.
[69] Isenschmid DS. Cocaine —Effects on Human Performance and Behavior. Forensic Sci Rev 2002;14:62−100.
[70] Beulens JW, Rimm EB, Ascherio A, Spiegelman D, Hendriks HF, Mukamal KJ. Alcohol consumption and risk for coronary heart disease among men with hypertension. Ann Intern Med 2007;146:10−19.
[71] Bode C, Bode JC. Alcohol's Role in Gastrointestinal Tract Disorders. Alcohol Health & Research World 1977;21:76−83.
[72] Bondi M, Drake A, Grant I. Verbal learning and memory in alcohol abusers and polysubstance abusers with concurrent alcohol abuse. J Int Neuropsychol Soc 1998;4:319−28.
[73] Emanuele NV, Swade TF, Emanuele MA. Consequences of alcohol use in diabetics. Alcohol Health Res World 1998;22:211−19.
[74] Epstein M. Alcohol's Impact on Kidney Function. Alcohol Health Res World 1997;21:84−93.
[75] French S. Mechanisms of alcoholic liver injury. Can J Gastroenterol 2000;14:327−32.
[76] Lieber C. Ethanol metabolism, cirrhosis and alcoholism. Clin Chim Acta. 1997;257:59−84.
[77] Lieber C. Alcohol and the liver: 1994 update. Gastroenterology 1994;106:1085−105.
[78] Lieber C. Hepatic and metabolic effects of ethanol: pathogenesis and prevention. Ann Med 1994;26:325−30.
[79] Lieber C. Hepatic, metabolic and toxic effects of ethanol: 1991 update. Alcohol Clin Exp Res 1991;15:573−92.
[80] Lieber C, De CL. Hepatotoxicity of ethanol. J Hepatol 1991;12:394−401.
[81] Lieber C. Biochemical mechanisms of alcohol-induced hepatic injury. Alcohol Alcohol Suppl 1991;1:283−90.
[82] Mukamal KJ, Chiuve SE, Rimm EB. Alcohol consumption and risk for coronary heart disease in men with healthy lifestyles. Arch Intern Med 2006;166:2145−50.
[83] Rohan TE. Alcohol and ischemic heart disease: a review. Aust N Z J Med 1984;14:75−80.
[84] Saunders JB. Alcohol: an important cause of hypertension. Br Med J (Clin Res Ed.) 1987;294:1045−6.
[85] McCance-Katz EF, Price LH, McDougle CJ, Kosten TR, Black JE, Jatlow PI. Concurrent cocaine-ethanol ingestion in humans: pharmacology, physiology, behavior, and the role of cocaethylene. Psychopharmacology (Berlin) 1993;111:39−46.
[86] Visalli T, Turkall R, Abdel-Rahman MS. Gender differences in cocaine pharmacokinetics in CF-1 mice. Toxicol Lett 2005;155:35−40.
[87] Teaf CM, Freeman RW, Harbison RD. Cocaine-induced hepatotoxicity: lipid peroxidation as a possible mechanism. Drug Chem Toxicol 1984;7:383−96.
[88] Gottfried MR, Kloss MW, Graham D, Rauckman EJ, Rosen GM. Ultrastructure of experimental cocaine hepatotoxicity. Hepatology 1986;6:299−304.
[89] Figliomeni ML, Turkall RM. Developmental immunotoxicity of cocaine and ethanol in postnatal Lewis rats. Immunopharmacology 1997;36:41−8.
[90] Boyer CS, Petersen DR. Potentiation of cocaine-mediated hepatotoxicity by acute and chronic ethanol. Alcohol Clin Exp Res 1990;14:28−31.
[91] Bolla KI, Funderburk FR, Cadet JL. Differential effects of cocaine and cocaine alcohol on neurocognitive performance. Neurology 2000;54:2285−92.
[92] Higgins ST, Rush CR, Hughes JR, Bickel WK, Lynn M, Capeless MA. Effects of cocaine and alcohol, alone and in combination, on human learning and performance. J Exp Anal Behav 1992;58:87−105.
[93] Lewis MJ, June HL. Synergistic effects of ethanol and cocaine on brain stimulation reward. J Exp Anal Behav 1994;61:223−9.
[94] Gossop M, Manning V, Ridge G. Concurrent use and order of use of cocaine and alcohol: behavioural differences between users of crack cocaine and cocaine powder. Addiction 2006;101:1292−8.
[95] Andrews P. Cocaethylene toxicity. J Addict Dis 1997;16:75−84.
[96] Bailey DN. Cocaethylene: a novel cocaine homolog. West J Med 1997;167:38−9.
[97] Jatlow P, Elsworth JD, Bradberry CW, Winger G, Taylor JR, Russell R, et al. Cocaethylene: a neuropharmacologically active metabolite associated with concurrent cocaine-ethanol ingestion. Life Sci. 1991;48:1787−94.
[98] McCance EF, Price LH, Kosten TR, Jatlow PI. Cocaethylene: pharmacology, physiology and behavioral effects in humans. J Pharmacol Exp Ther 1995;274:215−23.

[99] Perez-Reyes M, Jeffcoat AR, Myers M, Sihler K, Cook CE. Comparison in humans of the potency and pharmacokinetics of intravenously injected cocaethylene and cocaine. Psychopharmacology (Berlin) 1994;116:428–32.

[100] Wu AH, Onigbinde TA, Johnson KG, Wimbish GH. Alcohol-specific cocaine metabolites in serum and urine of hospitalized patients. J Anal Toxicol 1992;16:132–6.

[101] Caughlin LJ, O'Halloran RL. An accidental death related to cocaine, cocaethylene, and caffeine. J Forensic Sci 1993;38:1513–15.

[102] Puopolo PR, Chamberlin P, Flood JG. Detection and confirmation of cocaine and cocaethylene in serum emergency toxicology specimens. Clin Chem 1992;38:1838–42.

[103] Lewis DE, Moore CM, Leikin JB. Cocaethylene in meconium specimen. J Anal Toxicol 1994;32:697–703.

[104] Hezinova V, Aturki Z, Kleparnik K, D'Orazio G, Fanali S. Simultaneous analysis of cocaine and its metabolites in urine by capillary electrophoresis-electrospray mass spectrometry using pressurized liquid junction nanoflow interface. Electrophoresis 2012;33:653–60.

[105] Fleming SW, Dasgupta A, Garg U. Quantitation of cocaine, benzoylecgonine, ecgonine methyl ester and cocaethylene in urine and blood using gas chromatography-mass spectrometry (GC-MS). Methods in Mol Biol 2010;603:145–56.

[106] Chen X, Zheng X, Ding K, Zhou Z, Zhan CG, Zheng F. A quantitative LC-MS/MS method for simultaneous determination of cocaine and its metabolites in whole blood. J Pharm Biomed Anal 2017;134:243–51.

[107] D'Avila FB, Limberger RP, Fröehlich PE. Cocaine and crack cocaine abuse by pregnant or lactating mothers and analysis of its biomarkers in meconium and breast milk by LC-MS-A review. Clin Biochem 2016;49:1096–103.

Chapter 18

Drug-Assisted Sexual Assaults: Toxicology, Fatality, and Analytical Challenge

Matthew D. Krasowski
Department of Pathology, University of Iowa Carver College of Medicine, Iowa City, IA, United States

INTRODUCTION

The criminal use of drugs to render a victim susceptible to nonconsensual acts (known as drug-facilitated crime) is not a new phenomenon. Drugs used for this purpose are often rapidly acting central nervous system (CNS) depressants that possess amnestic (impairment of memory) and/or immobilizing properties. The most common approach is intoxication with ethanol, either alone or in combination with other drugs [1]. Drug development in the last 50 years has greatly expanded the pharmaceuticals that have been used for drug-facilitated crimes [2–6]. Common examples include antihistamines, benzodiazepines (especially flunitrazepam and lorazepam), nonbenzodiazepine sedative-hypnotics (zolpidem, zopiclone, and zaleplon), GHB (γ-hydroxybutyrate), ketamine, and methylenedioxymethamphetamine (MDMA, ecstasy). Many of these drugs can be surreptitiously introduced into alcoholic or nonalcoholic beverages without noticeably changing taste or color. This chapter will focus on drugs implicated in drug-facilitated sexual assault (DFSA) and other drug-facilitated crimes that can present analytical challenges: GHB and its analogs, ketamine, MDMA, flunitrazepam, and the nonbenzodiazepine sedative-hypnotics.

γ-HYDROXYBUTYRIC ACID AND ITS ANALOGS

GHB is an endogenous substance found in the mammalian CNS and peripheral tissues with incompletely understood biological functions [7–9]. GHB is a minor metabolite of γ-aminobutyric acid (GABA), the major inhibitory neurotransmitter in the mammalian CNS. GHB binds to the recently characterized excitatory GHB receptor in the brain and is also a weak agonist of the inhibitory $GABA_B$ receptor. GHB was first synthesized in 1960 as a potential anticonvulsant but later abandoned for this clinical purpose [10,11]. GHB now has limited medical use as a treatment for the sleep disorder narcolepsy (trade name Xyrem) [12]. Two compounds structurally related to GHB, γ-butyrolactone (GBL) and 1,4-butanediol (1,4BD), are metabolized to GHB in humans. GHB, GBL, and 1,4BD are used as recreational drugs and for drug-facilitated crimes including DFSA [13–15]. Chemical structures of the drugs covered in this chapter are in Fig. 18.1.

Sources

GHB is sold on the street as a white powder or a colorless liquid with a salty or soapy taste that can be masked in an alcoholic beverage [8]. Street names for GHB include liquid ecstasy, Georgia home boy, grievous bodily harm, and scoop. GHB is also produced as a result of fermentation and is found in small quantities in some alcoholic beverages (especially beers and wines), citrus fruits, and meat.

GBL has widespread commercial and industrial uses and is chemically produced by dehydrogenation of 1,4BD [13,14]. Pure GBL is a hygroscopic liquid miscible in water that has a weak odor and unpleasant taste that can be masked in an alcoholic beverage. Street names for GBL are Blue nitro, Gamma G, and firewater. GBL is an industrial

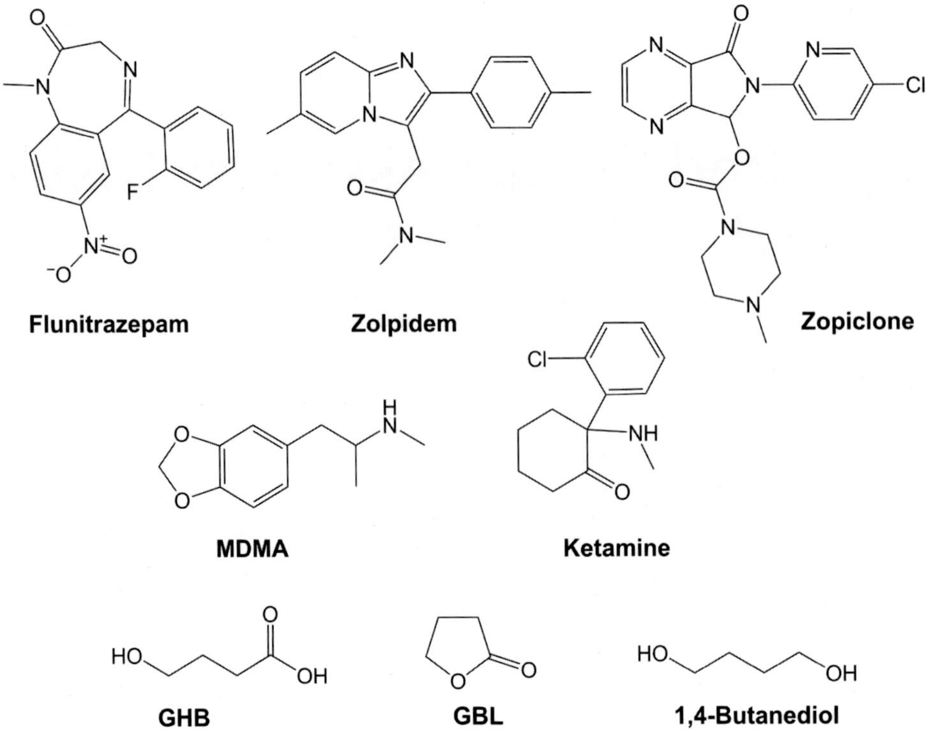

FIGURE 18.1 Chemical structures of flunitrazepam, zolpidem, zopiclone, MDMA, ketamine, GHB, GBL, and 1,4BD.

solvent and can be found in a myriad of products such as glue remover, drilling oils, and paint removers [14,16]. GBL does not have much pharmacologic activity on its own and may thus be thought of as a pro-drug of GHB. GBL has no clinical applications or known endogenous function.

1,4BD is used industrially as a solvent and for the manufacture of plastics and fibers [14,17]. Pure 1,4BD is an odorless and colorless viscous liquid soluble in ethanol and miscible in water. The distinct bitter taste of 1,4BD can be masked by ethanol. Street names for 1,4BD include BD, BlueRaine, pine needle oil, and inner G. Most or all of the pharmacologic actions of 1,4BD are from the conversion to GHB. Thus, similar to GBL, 1,4BD can be considered as a pro-drug of GHB. 1,4BD does not currently have any known endogenous functions or clinical applications.

Pharmacology

GHB is rapidly absorbed with an onset of action within 15 min. Bioavailability is approximately 65%, with a half-life in blood of approximately 25 min. The drug is not protein bound in the plasma. GHB is metabolized by multiple steps ultimately to succinic acid which enters the Krebs cycle for metabolism to carbon dioxide and water. Less than 2% of GHB is eliminated unchanged in the urine, resulting in short detection times for urine screening programs of generally only 8–12 h.

GBL is more lipophilic than GHB and is absorbed faster with higher bioavailability. GBL is rapidly converted into GHB by paraoxonase enzymes found in the blood [13,14]. Levels of paraoxonase have interindividual variation, leading to significant differences in the response to GBL. 1,4BD is converted to GHB by the sequential action of alcohol dehydrogenase and aldehyde dehydrogenase, the same enzymes involved in the metabolism of ethanol and methanol [13,14]. Interindividual differences in levels of these enzymes can influence response and adverse effects of 1,4BD.

Abuse

Beginning in the 1980s, GHB was misleadingly sold in health-food stores as an enhancer for growth hormone, leading to abuse by bodybuilders [11]. During the 1990s, increasing reports of abuse and toxic reactions caused the United States Drug Enforcement Agency (DEA) to restrict GHB as a schedule I drug in March 2000. The specific formulation of GHB used for treatment of narcolepsy (sodium oxybate; marketed as Xyrem) is DEA schedule III. Most of the GHB used in the United States for recreational or criminal purposes is from clandestine production and not from diversion of clinical supplies. GBL and 1,4BD are readily accessible because of their widespread use in industry and thus became alternatives to GHB for recreational or criminal use. GBL and 1,4BD are not officially scheduled by the DEA, but GBL was classified as a list I chemical in 2000 to provide tight regulations for manufacturers of GBL and limit diversion of GBL.

Analysis

The analysis of GHB is technically challenging due to the low concentrations of analyte in biological fluids and its subsequent extensive metabolism [7,18]. Although GHB is an endogenous compound, the amounts present in body fluids during normal health are normally near or below the detection limits of analytical methods, but urine levels as high as 14 μg/mL have been reported and may confound analysis. In addition, concentrations of GHB increase postmortem [7,19–21]. For antemortem specimens, the main problem with analysis for GHB is that the very short elimination half-life of 30–40 min results in many negative samples, with analyte concentrations below detectable limits [18,22,23]. This is particularly true for blood analysis where negative results are expected after 4 h following ingestion. Urine is slightly better, with positive results up to 12 h. Delays in specimen collection may result in concentrations below the lower limit of detection for even the most sensitive methods.

Chromatographic methods, with or without mass spectrometry, have been the typical standard for GHB [18]. Various extraction schemes are in the literature to analyze GHB, GBL, and 1,4BD by gas chromatography/mass spectrometry (GC/MS) [24,25]. Several methods have been reported for the simultaneous analysis of GHB and GBL. One of the most promising has been the use of capillary electrophoresis where GHB can be analyzed without requiring any derivatization or extraction procedures [26]. Hair has become increasingly popular as a specimen for GHB analysis, especially in the workup of potential drug-facilitated crimes [27,28]. Methods for detection of GHB in hair have steadily improved over the last decade [27]. Oral fluid is also emerging as potential specimen type [29,30]. Methods have also been reported for detection of GBL and 1,4BD in urine and whole blood [31].

Interpretation of test results for GHB and its precursors are challenging for several reasons [7,18,21]. First, the blood levels do not correlate well with symptoms. Second, the short half-life results in many patients presenting with concentrations below the limit of detection or in the range of endogenous GHB concentrations [32]. Third, there is no good screening method available for most clinical laboratories. Those clinical laboratories that have GC/MS capability may not have methods for GHB, since it is rarely encountered in emergency department visits for most locations. Fourth, interpretation can be difficult due to endogenous GHB levels and GHB stability in specimens [20,33].

Toxicity

The toxicity of GHB results from its very steep dose–response curve, and overdoses can quickly become life-threatening [8,14,16,19,34]. The user may go in and out of consciousness very rapidly or simply lose control of body muscles and fall to the ground. The toxicity curve is characterized by euphoria, dizziness, visual disturbances, decreased level of consciousness, nausea, vomiting, acute delirium, confusion, agitation, hypothermia, clonic muscle movements, coma, respiratory depression, and death. Loss of consciousness with rapid awaking is the hallmark of a nonfatal GHB overdose. Alcohol use with GHB is particularly dangerous since alcohol potentiates the CNS-depressant effects of GHB [8,14,19]. Coingestion of ethanol and 1,4BD can be especially dangerous, since each drug impacts the metabolism of the other.

Many accidental deaths have been reported from GHB alone and from GHB combined with other drugs such as opioids, ethanol, ketamine, or MDMA [8,14]. All GHB toxicity considerations also apply to abuse of GBL and 1,4BD [8,13,14,16]. For GHB, GBL, and 1,4BD ingestions no specific antidotes are available [35,36]. Flumazenil and naloxone are not effective. Withdrawal reactions can be treated with benzodiazepines. The estimated lethal dose for GHB is 5–10 g, and blood concentrations in fatalities have exceeded 300 mg/L. Lethal doses for GBL and 1,4BD have yet to be determined.

GHB Associated With Drug-Facilitated Sexual Assault

GHB and its analogs have been associated with DFSA. These compounds can be slipped surreptitiously into alcoholic and nonalcoholic beverages and quickly render an unsuspecting victim incapacitated and with little ability to remember events later [4]. Interviews with users of GHB indicate that positive experiences include increased sexual desire, decreased sexual inhibitions, and decreased anxiety; negative experiences include oversedation, loss of consciousness, motor incoordination, and mental confusion [13,14,37]. These effects are exploited when GHB, GBL, or 1,4BD are utilized for DFSA or other crimes [4]. GHB blood levels between 63 and 265 μg/mL often correspond to a state similar to light sleep, with the person coming into and out of consciousness [38]. This state is what is characteristically seen in those under the influence of GHB while driving or in DFSA victims [13,16,39]. The actual numbers of DFSA cases attributable to GHB and its precursors are difficult to estimate due to analytical challenges [40]. The amnesia produced by GHB and its analogs often makes victims unable to serve as valid witnesses [4].

3,4-Methylenedioxymethamphetamine

MDMA is a designer amphetamine that was first synthesized in 1914 and investigated as an appetite suppressant [41]. MDMA became widely available in 1970s when it was used therapeutically as a psychotherapeutic agent and gained popularity as a recreational drug [42,43]. MDMA currently has no approved clinical applications and was classified as a schedule I controlled substance in 1985 in the United States [41].

Sources

MDMA typically comes in pill form, but it has been increasingly been sold in a powder/crystalline form known as "Molly" [44]. One of the challenges with studying the prevalence of MDMA use is that street preparations sold as MDMA vary widely in purity and commonly contain adulterants or may even be another drug altogether. MDMA users commonly abuse other club drugs like GHB and ketamine [45].

Pharmacology

MDMA is absorbed rapidly into the bloodstream and metabolized by two separate pathways in humans [46,47]. The effects of MDMA begin 30–60 min after oral use and last up to 8 h. Although oral use is the most common, inhalation for faster effects is also known. The major metabolism pathway starts with demethylation by cytochrome P450 (CYP) 2D6 to 3,4-dihydroxymethamphetamine (HHMA) and then conversion to 4-hydroxy-3-methoxyamphetamine (HMA) after several additional steps. In the minor metabolism pathway, MDMA is N-demethylated to 3,4-methylenedioxyamphetamine (MDA; itself also a psychoactive drug of abuse) before ending up as HMA. The four MDMA metabolites are excreted in urine as conjugated glucuronide and sulfate metabolites [46]. Parent drug MDMA and metabolite HMMA can be detected as early as 30 min. HMA is not detected until nearly 6 h. Individual urine excretion patterns vary widely and are partially related to the urinary pH. Based on urine test cutoffs of 250 ng/mL for MDMA, urine specimens are not likely to be positive beyond 48 h [46].

Abuse

MDMA is structurally related to mescaline and methamphetamine, other drugs with potent stimulant and hallucinogenic properties [48]. The use of MDMA spread rapidly in the 1990s and was associated with the club scene including rave parties [41–43,49]. The use of MDMA was then leveled off in the United States and European countries before rising again in the 2010s.

Analysis

Urine is the typical specimen for MDMA screening [50]. Specimen types other than urine or blood have been investigated for detection of MDMA use. Hair testing with liquid chromatography/tandem mass spectrometry (LC/MS/MS) and GC/MS has been used to increase the sensitivity and increase the detection time to 8 days following oral ingestion of MDMA [51–53]. Immunoassays have often been used to screen for the presence of MDMA in urine specimens [50]. The last decade has seen an increase in the number of marketed immunoassays specifically targeted for detection of MDMA as opposed to immunoassays marketed as targeting amphetamines, amphetamine/methamphetamine, or methamphetamine. Immunoassays marketed for MDMA detection tend to have good specificity for MDMA, although these assays may cross-react with structurally related drugs such as MDA [54–56]. This cross-reactivity may be helpful since what is sold as MDMA may actually be MDA or related amphetamine derivatives.

Federal guidelines began requiring MDMA testing in 2010 [57]. Screening tests must have a cutoff of 500 ng/mL for MDMA and confirmation cutoff of 250 ng/mL. Confirmatory analysis is typically performed by GC/MS (usually with derivatization) or LC/MS/MS [51,58–60].

Toxicity

MDMA can elevate body temperature, which can produce dehydration and hyperthermia, especially when exacerbated by dancing in a hot environment [61]. Sympathetic stimulation results in tachycardia, mydriasis, diaphoresis, tremor, hyperreflexia, palpitations, and hypertension. Adverse neurological effects include confusion, delirium, paranoia, headache, depression, insomnia, irritability, and nystagmus. Severe reactions can occur, and patients may present with seizures, cerebral edema, Parkinsonism, and the serotonin syndrome [62–64]. Cardiovascular effects of MDMA can progress to atrioventricular block, arrhythmias, and asystole [42]. Musculoskeletal effects such as muscle spasms and bruxism (teeth grinding) have been reported. The combination of the musculoskeletal effects and the elevated body temperature may lead to rhabdomyolysis and renal failure [61]. Cognitive impairment has been demonstrated after long-term exposure to MDMA, which may result in potentially permanent memory impairment [65].

There is no antidote for MDMA intoxication, only supportive care [42,61]. Agitation and anxiety can be controlled with benzodiazepines. If hyperthermia is present, rapid cooling is a priority. For MDMA, the estimated lethal dose is 300 g, with blood concentrations of 3.1–4.2 mg/L blood concentrations in fatalities.

Association With Drug-Facilitated Sexual Assault

Multiple case series have implicated MDMA in DFSA [66,67]. In addition, MDMA has been detected in toxicologic analyses of suspected DFSA cases [67,68]. MDMA pills may be given by the perpetrator to the victim under the guise of being another drug, often in the context of also consuming alcohol. In DFSA legal cases, the ability of MDMA to render a victim vulnerable to sexual assault even while conscious and apparently aware can make it difficult to prove nonconsent [69,70].

KETAMINE

Ketamine was first synthesized in 1962 and gained United States Food and Drug Administration (FDA) approval as a general anesthetic in 1970 (trade name Ketalar). Ketamine is pharmacologically related to phencyclidine (PCP) and eventually displaced PCP as an anesthetic for human clinical use due to a more favorable adverse effect profile [71]. Currently, ketamine is a medication mainly used for the induction and maintenance of general anesthesia [71]. Ketamine is also widely used as a veterinary anesthetic. Ketamine produces dissociative anesthesia resembling a trancelike state while providing sedation, analgesia, and amnesia (memory loss). One of the main clinical advantages of ketamine is minimal effects on heart function, respiration, and airway reflexes, especially relative to other general anesthetics such as propofol or thiopental. This allows ketamine to be a good option for providing anesthesia in low resource or remote settings.

Sources

Ketamine is a chiral aryl-cyclohexylamine derivative. Most pharmaceutical preparations of ketamine are racemic; however, the pharmaceutically more active isomer (S-ketamine) is being investigated for treatment-resistant depression [72]. Ketamine is a core medication for intravenous anesthesia and sedation (along with propofol) in the World Health Organization Model List of Essential Medicines, and thus widely used through the world [73]. The extensive use of ketamine in veterinary anesthesia means that abusers may also divert the drug from veterinary supplies.

Abuse

Reports of ketamine abuse first appeared in 1970s, with rising popularity in the club scene in the 1990s [74–78]. Street names for ketamine include special K, vitamin K, super K, and Kit Kat. The increasing abuse of ketamine, including diversion from medical and veterinary supplies, led to classification as a schedule III controlled substance in the United States in 1999 [74,76]. Ketamine is abused in the club setting, especially to enhance sexual experience. Ketamine abusers may administer the drug by intravenous, intramuscular, intranasal, or smoking. Ketamine may be a preferred drug of abuse or co-abused with other substances such as DXM or GHB [79,80]. Ketamine abusers often aim for what is known as the "k-hole," a state characterized by physical immobility and out-of-body experience [74].

Pharmacology

Ketamine is a noncompetitive antagonist of the *N*-methyl-D-aspartate class of glutamate receptor, a mechanism similar to PCP [81]. In terms of pharmacokinetics, oral ketamine has low bioavailability (15%–20%) due to breakdown by bile acids. Thus, the drug is most commonly administered parenterally. Ketamine is *N*-demethylated by CYP3A4 (major pathway), CYP2B6 (minor), and CYP2C9 to norketamine and dehydronorketamine, the two main metabolites recovered in urine [76,82]. Norketamine contributes significantly to the pharmacologic action of ketamine. Even though this metabolite is only 20%–33% as potent as ketamine as an anesthetic, plasma levels of norketamine may reach three times as that of ketamine. Duration of action of ketamine is 0.5–2 h intramuscularly and 4–6 h orally.

Analysis

Ketamine has been measured in urine, hair, and plasma [76]. Screening immunoassays for ketamine are available but currently not widely employed by clinical laboratories. Ketamine shows little cross-reactivity with immunoassays for commonly used drug abuse screening assays [54]. The most common analytical method for detection of ketamine is GC/MS [53,58,76]. The main analytical challenge with ketamine is the lack of any routinely used rapid screening assays; this may change if ketamine screening immunoassays become more widely adopted. The drug may be reliably detected and quantitated by GC/MS, but this is often not routinely available in many settings. Thus ketamine abuse cases can be missed unless there is a high enough index of suspicion to pursue specialized testing.

Toxicity

Acute adverse effects of ketamine include amnesia, delirium, hallucinations, hyperthermia, altered level of consciousness, tachycardia, hypertension, and impaired muscle tone [74,83]. Ketamine can cause alterations to perception including out-of-body experiences, altered sense of time, and color changes [74]. Death is uncommon if ketamine alone is abused. A retrospective analysis in the United Kingdom from 1993 to 2006 revealed only four deaths attributed to ketamine [74].

There is no specific treatment for ketamine overdose [83]. The minimum lethal dose of ketamine has not been established but is estimated to be approximately 1 g for adults given intravenously or intramuscularly in the absence of other substances. Psychotropic effects of ketamine are observed at concentrations of 0.05–0.3 mg/L [84]. Blood concentrations in two ketamine-associated fatalities were 1.6 and 6.9 mg/L, respectively [85].

Association With Drug-Facilitated Sexual Assault

Ketamine has been implicated in DFSA and other drug-facilitated crimes [2]. In a study in Ontario, Canada, ketamine was detected in 2.3% of suspected DFSA cases [68]. Similar to MDMA, ketamine victims may appear conscious and even aware, complicating the ability to prove nonconsent in legal cases [2]. The potent amnesia properties of ketamine make it difficult or impossible for victims to recall events.

Flunitrazepam

Flunitrazepam (trade name Rohypnol, DEA schedule IV) is a benzodiazepine which was never approved for medical use in the United States, although it is used medically in other parts of the world. Flunitrazepam can cause rapid sedation and is notorious as a vehicle for drug-facilitated crimes [5]. Flunitrazepam is more potent than diazepam (trade name Valium), with a single 1 or 2 mg dose of flunitrazepam producing significant sedative effect.

Sources

Flunitrazepam is not produced for medical purposes in the United States but is marketed in other countries. Flunitrazepam may be diverted from global medical supplies and smuggled into other countries or manufactured illicitly.

Pharmacology

Flunitrazepam enhances the action of the neurotransmitter GABA at brain $GABA_A$ receptors, a mechanism common to other benzodiazepines such as clonazepam, diazepam, and lorazepam [86]. Flunitrazepam is well absorbed after oral ingestion, with peak concentrations occurring in 30—90 min [87]. Flunitrazepam is metabolized by the liver into 7-aminoflunitrazepam (major metabolite) and also to desmethyl-flunitrazepam and 3-hydroxyflunitrazepam (minor metabolites). The effect of flunitrazepam can last up to 12 h. Like other benzodiazepines, prolonged use of flunitrazepam can cause dependence and withdrawal symptoms if the use of the drug is suddenly stopped [37].

Analysis

The presence of flunitrazepam and its metabolites in urine can typically be detected by benzodiazepine immunoassay screening [50]. Confirmation of flunitrazepam and its metabolites in urine can be done by GC/MS or LC/MS/MS method. Concentration of flunitrazepam is often very low in urine, but concentration of 7-aminoflunitrazepam is relatively higher. Urine concentration of 7-aminoflunitrazepam in five flunitrazepam abusers ranged from 455 to 844 ng/mL while the concentrations of the parent drug flunitrazepam ranged from "none detected" to 4.8 ng/mL [88].

Toxicity

Like other benzodiazepines, flunitrazepam possesses a high therapeutic index with relatively minor adverse effects. However, hypotension, tachycardia, palpitations, amnesia, ataxia, and behavioral disturbances can be seen. Respiratory depression may also occur from high doses, especially in children or elderly, or when used in conjunction with other CNS depressants such as ethanol. Flumazenil is a specific antidote for benzodiazepine overdose.

Association With Drug-Facilitated Sexual Assault

The rapid sedative properties of flunitrazepam, especially when combined with other CNS depressants, make it a common drug use for DFSA (hence it is often called *the* "date rape" drug), although other benzodiazepines have been used as well [5]. Flunitrazepam may be added surreptitiously to alcoholic beverages of victims. A significant challenge in DFSA laboratory investigations is whether the analytical methods can detect the low concentrations of flunitrazepam and metabolites often present in body fluids [5]. Thus laboratories involved in analysis of specimens from suspected drug-facilitated crimes should be able to reliably detect these low concentrations [5,89].

Nonbenzodiazepine Sedative-Hypnotics

Three pharmacologically related nonbenzodiazepine drugs (zolpidem, zopiclone, and zaleplon; colloquially referred to as the "Z-drugs") have collectively become the most commonly prescribed hypnotic medications in the United States [90]. The Z-drugs have a generally overall better toxicity profile than the benzodiazepines, including in cases of acute overdose. However, the Z-drugs have also been associated with instances of DFSA and other drug-facilitated crimes.

Sources

Zopiclone is marketed in the United States as the specific isomer eszopiclone (derived from *S*-zopiclone; trade name Lunesta). In other countries, it is predominantly marketed as the racemic mixture generically termed zopiclone. Zolpidem (trade name Ambien) is currently the most widely prescribed of these nonbenzodiazepine hypnotics in the United States [91]. Zaleplon (trade name Sonata) is the third drug in this group. Due to its popularity, the majority of published clinical and toxicologic literature involves zolpidem.

Pharmacology

The nonbenzodiazepine hypnotics have a similar mechanism of action to the benzodiazepines and increase the actions of the neurotransmitter GABA at $GABA_A$ receptors in the brain [90]. Zolpidem is metabolized into inactive metabolites, primarily by the actions of CYP3A4. In adults, the half-life of zolpidem is approximately 2.5 h. This means that even after 7—8 h of sleep (approximately three half-lives), there may still be detectable amounts of zolpidem in the bloodstream. The residual zolpidem that may be present after sleep can potentially cause problems with alertness, cognitive function, and memory [92–94]. The elimination of zolpidem is slower in the elderly and in patients with liver or kidney failure. In comparison to zolpidem, the half-life of zopiclone is longer (approximately 6 h in adults) [95–97]. Like zolpidem, zopiclone is extensively metabolized by the liver. Drug elimination is slowed in the elderly and in patients with kidney or liver failure. Zaleplon has the shortest half-life (1.5 h in adults) of the currently marketed nonbenzodiazepine hypnotics, theoretically resulting in less potential residual cognitive impairment after sleep [95,98,99]. In victims of drug-facilitated crimes, residual Z-drug concentrations may impair memory of the event.

Analysis

The Z-drugs can be detected in hair, urine, or blood by chromatographic methods [100–102]. Unlike benzodiazepines, there are currently no routine laboratory screening tests for detecting zolpidem, zopiclone, or zaleplon in urine or blood. Consequently, the incidence of DFSA using the nonbenzodiazepine hypnotics may be severely underestimated.

Toxicity

Fatalities from Z-drugs are rare. Blood or plasma zolpidem concentrations are usually in a range of 30–300 ng/mL in persons taking the drug therapeutically, 100–700 ng/mL in those arrested for impaired driving, and greater than 1000 ng/mL in acute overdosage [103,104]. A fatality involving zolpidem had blood concentration of 7900 ng/mL and urine concentration of 4100 ng/mL [105]. Zopiclone plasma concentrations are typically less than 100 ng/mL during therapeutic use, but often exceed 100 ng/mL in those arrested for impaired driving ability and greater than 1000 ng/mL in acute overdosage [106,107]. Blood concentrations in zopiclone-associated fatalities have been in the range of 400–3900 ng/mL [106–108]. There is little data on the concentration of zaleplon in body fluids in toxic cases. There is a case report of a death by multidrug overdose that showed zaleplon blood concentration of 2200 and urine concentration of 1400 ng/mL [109].

Association With Drug-Facilitated Sexual Assault

The nonbenzodiazepine hypnotics have been associated with instances of DFSA [110–112]. Within this class of medications, zolpidem is the drug most frequently implicated, although cases involving zopiclone have also been reported [6]. Zolpidem and zopiclone can be slipped into beverages with minimal change in taste. The combination of ethanol or other CNS depressants and nonbenzodiazepine hypnotics increases the risk of severe respiratory depression. The nonbenzodiazepine hypnotics, especially zolpidem, are relatively frequent prescription medications found by GC/MS or LC/MS/MS analysis of urine or blood specimens from female victims of alleged sexual assault, although it is often difficult to ascertain how much the drug contributed to the alleged crime [113,114].

CONCLUSIONS

Diverse assortments of CNS depressants have been associated with drug-facilitated crimes including DFSA. As discussed in this chapter, GHB and its analogs, MDMA, ketamine, flunitrazepam, and the nonbenzodiazepine sedative-hypnotics present analytical challenges and may be missed if specialized analysis is not performed. It is important for health care providers to be aware of such situations so that in the case of a suspected drug overdose or sexual assault where routine toxicology testing is negative, proper steps can be taken to send the specimen to an appropriate reference laboratory for further testing. The increasing use of mass spectrometry–based methods should help in providing better detection of DFSA-associated drugs.

REFERENCES

[1] Kerrigan S. The use of alcohol to facilitate sexual assault. Forensic Sci Rev 2010;22:15–32.
[2] Couper FJ, Saady JJ. The use of miscellaneous prescription medications to facilitate sexual assault. Forensic Sci Rev 2010;22:83–112.
[3] Jenkins AJ, Stillwell ME. The use of over-the-counter medications to facilitate sexual assault. Forensic Sci Rev 2010;22:75–82.
[4] Marinetti L, Montgomery MA. The use of GHB to facilitate sexual assault. Forensic Sci Rev 2010;22:41–59.
[5] Montgomery MA. The use of benzodiazepines to facilitate sexual assault. Forensic Sci Rev 2010;22:33–40.
[6] Stockham TL, Rohrig TP. The use of Z-drugs to facilitate sexual assault. Forensic Sci Rev 2010;22:61–73.

[7] Brailsford AD, Cowan DA, Kicman AT. Pharmacokinetic properties of gamma-hydroxybutyrate (GHB) in whole blood, serum, and urine. J Anal Toxicol 2012;36:88–95.

[8] Brennan R, Van Hout MC. Gamma-hydroxybutyrate (GHB): a scoping review of pharmacology, toxicology, motives for use, and user groups. J Psychoactive Drugs 2014;46:243–51.

[9] Bosch OG, Seifritz E. The behavioural profile of gamma-hydroxybutyrate, gamma-butyrolactone and 1,4-butanediol in humans. Brain Res Bull 2016;126:47–60.

[10] Klunk WE, Covey DF, Ferrendelli JA. Anticonvulsant properties of alpha, gamma, and alpha, gamma-substituted gamma-butyrolactones. Mol Pharmacol 1982;22:438–43.

[11] Wedin GP, Hornfeldt CS, Ylitalo LM. The clinical development of gamma-hydroxybutyrate (GHB). Curr Drug Saf 2006;1:99–106.

[12] Busardo FP, Kyriakou C, Napoletano S, Marinelli E, Zaami S. Clinical applications of sodium oxybate (GHB): from narcolepsy to alcohol withdrawal syndrome. Eur Rev Med Pharmacol Sci 2015;19:4654–63.

[13] Brunt TM, van Amsterdam JG, van den Brink W. GHB, GBL and 1,4-BD addiction. Curr Pharm Des 2014;20:4076–85.

[14] Corkery JM, Loi B, Claridge H, Goodair C, Corazza O, Elliott S, et al. Gamma hydroxybutyrate (GHB), gamma butyrolactone (GBL) and 1,4-butanediol (1,4-BD; BDO): a literature review with a focus on UK fatalities related to non-medical use. Neurosci Biobehav Rev 2015;53:52–78.

[15] van Amsterdam J, Brunt T, Pennings E, van den Brink W. Risk assessment of GBL as a substitute for the illicit drug GHB in the Netherlands. A comparison of the risks of GBL vs GHB. Regul Toxicol Pharmacol 2014;70:507–13.

[16] Wood DM, Brailsford AD, Dargan PI. Acute toxicity and withdrawal syndromes related to gamma-hydroxybutyrate (GHB) and its analogues gamma-butyrolactone (GBL) and 1,4-butanediol (1,4-BD). Drug Test Anal 2011;3:417–25.

[17] Schep LJ, Knudsen K, Slaughter RJ, Vale JA, Megarbane B. The clinical toxicology of gamma-hydroxybutyrate, gamma-butyrolactone and 1,4-butanediol. Clin Toxicol (Phila) 2012;50:458–70.

[18] Ingels AS, Wille SM, Samyn N, Lambert WE, Stove CP. Screening and confirmation methods for GHB determination in biological fluids. Anal Bioanal Chem 2014;406:3553–77.

[19] Busardo FP, Jones AW. GHB pharmacology and toxicology: acute intoxication, concentrations in blood and urine in forensic cases and treatment of the withdrawal syndrome. Curr Neuropharmacol 2015;13:47–70.

[20] LeBeau MA, Montgomery MA, Morris-Kukoski C, Schaff JE, Deakin A. Further evidence of in vitro production of gamma-hydroxybutyrate (GHB) in urine samples. Forensic Sci Int 2007;169:152–6.

[21] Schrock A, Hari Y, Konig S, Auwarter V, Schurch S, Weinmann W. Pharmacokinetics of GHB and detection window in serum and urine after single uptake of a low dose of GBL—an experiment with two volunteers. Drug Test Anal 2014;6:363–6.

[22] Brown SD, Melton TC. Trends in bioanalytical methods for the determination and quantification of club drugs: 2000–2010. Biomed Chromatogr 2011;25:300–21.

[23] Castro AL, Dias M, Reis F, Teixeira HM. Gamma-hydroxybutyric acid endogenous production and post-mortem behaviour—the importance of different biological matrices, cut-off reference values, sample collection and storage conditions. J Forensic Leg Med 2014;27:17–24.

[24] Brown SD, Rhodes DJ, Pritchard BJ. A validated SPME-GC-MS method for simultaneous quantification of club drugs in human urine. Forensic Sci Int 2007;171:142–50.

[25] Lenz D, Kroner L, Rothschild MA. Determination of gamma-hydroxybutyric acid in serum and urine by headspace solid-phase dynamic extraction combined with gas chromatography-positive chemical ionization mass spectrometry. J Chromatogr A 2009;1216:4090–6.

[26] Gong XY, Kuban P, Scholer A, Hauser PC. Determination of gamma-hydroxybutyric acid in clinical samples using capillary electrophoresis with contactless conductivity detection. J Chromatogr A 2008;1213:100–4.

[27] Busardo FP, Pichini S, Zaami S, Pacifici R, Kintz P. Hair testing of GHB: an everlasting issue in forensic toxicology. Clin Chem Lab Med 2018;56:198–208.

[28] Xiang P, Shen M, Drummer OH. Review: drug concentrations in hair and their relevance in drug facilitated crimes. J Forensic Leg Med 2015;36:126–35.

[29] De Paoli G, Bell S. A rapid GC-MS determination of gamma-hydroxybutyrate in saliva. J Anal Toxicol 2008;32:298–302.

[30] De Paoli G, Walker KM, Pounder DJ. Endogenous gamma-hydroxybutyric acid concentrations in saliva determined by gas chromatography-mass spectrometry. J Anal Toxicol 2011;35:148–52.

[31] Johansen SS, Windberg CN. Simultaneous determination of gamma-hydroxybutyrate (GHB) and its analogues (GBL, 1,4-BD, GVL) in whole blood and urine by liquid chromatography coupled to tandem mass spectrometry. J Anal Toxicol 2011;35:8–14.

[32] Mari F, Politi L, Trignano C, Di Milia MG, Di Padua M, Bertol E. What constitutes a normal ante-mortem urine GHB concentration?. J Forensic Leg Med 2009;16:148–51.

[33] Berankova K, Mutnanska K, Balikova M. Gamma-hydroxybutyric acid stability and formation in blood and urine. Forensic Sci Int 2006;161:158–62.

[34] Madah-Amiri D, Myrmel L, Brattebo G. Intoxication with GHB/GBL: characteristics and trends from ambulance-attended overdoses. Scand J Trauma Resusc Emerg Med 2017;25:98.

[35] Dominici P, Kopec K, Manur R, Khalid A, Damiron K, Rowden A. Phencyclidine intoxication case series study. J Med Toxicol 2015;11:321–5.

[36] Dove HW. Phencyclidine: pharmacologic and clinical review. Psychiatr Med 1984;2:189–209.

[37] Gahlinger PM. Club drugs: MDMA, gamma-hydroxybutyrate (GHB), Rohypnol, and ketamine. Am Fam Physician 2004;69:2619–26.

[38] Helrich M, McAslan TC, Skolnik S, Bessman SP. Correlation of blood levels of 4-hydroxybutyrate with state of consciousness. Anesthesiology 1964;25:771–5.

[39] Centola C, Giorgetti A, Zaami S, Giorgetti R. Effects of GHB on psychomotor and driving performance. Curr Drug Metab 2018.

[40] Kapitany-Foveny M, Zacher G, Posta J, Demetrovics Z. GHB-involved crimes among intoxicated patients. Forensic Sci Int 2017;275:23–9.

[41] Rosenbaum M. Ecstasy: America's new "reefer madness.". J Psychoactive Drugs 2002;34:137–42.

[42] Burgess C, O'Donohoe A, Gill M. Agony and ecstasy: a review of MDMA effects and toxicity. Eur Psychiatry 2000;15:287–94.

[43] Edland-Gryt M, Sandberg S, Pedersen W. From ecstasy to MDMA: recreational drug use, symbolic boundaries, and drug trends. Int J Drug Policy 2017;50:1–8.

[44] Parrott AC, Downey LA, Roberts CA, Montgomery C, Bruno R, Fox HC. Recreational 3,4-methylenedioxymethamphetamine or 'ecstasy': current perspective and future research prospects. J Psychopharmacol 2017;31:959–66.

[45] Bruno R, Matthews AJ, Dunn M, Alati R, McIlwraith F, Hickey S, et al. Emerging psychoactive substance use among regular ecstasy users in Australia. Drug Alcohol Depend 2012;124:19–25.

[46] Abraham TT, Barnes AJ, Lowe RH, Kolbrich Spargo EA, Milman G, Pirnay SO, et al. Urinary MDMA, MDA, HMMA, and HMA excretion following controlled MDMA administration to humans. J Anal Toxicol 2009;33:439–46.

[47] Barnes AJ, De Martinis BS, Gorelick DA, Goodwin RS, Kolbrich EA, Huestis MA. Disposition of MDMA and metabolites in human sweat following controlled MDMA administration. Clin Chem 2009;55:454–62.

[48] Simantov R. Multiple molecular and neuropharmacological effects of MDMA (Ecstasy). Life Sci 2004;74:803–14.

[49] Palamar JJ, Mauro PM, Han BH, Martins SS. Shifting characteristics of ecstasy users ages 12–34 in the United States, 2007–2014. Drug Alcohol Depend 2017;181:20–4.

[50] Melanson SE. The utility of immunoassays for urine drug testing. Clin Lab Med 2012;32:429–47.

[51] Cheze M, Deveaux M, Martin C, Lhermitte M, Pepin G. Simultaneous analysis of six amphetamines and analogues in hair, blood and urine by LC-ESI-MS/MS. Application to the determination of MDMA after low ecstasy intake. Forensic Sci Int 2007;170:100–4.

[52] Johansen SS, Jornil J. Determination of amphetamine, methamphetamine, MDA and MDMA in human hair by GC-EI-MS after derivatization with perfluorooctanoyl chloride. Scand J Clin Lab Invest 2009;69:113–20.

[53] Wu YH, Lin KL, Chen SC, Chang YZ. Simultaneous quantitative determination of amphetamines, ketamine, opiates and metabolites in human hair by gas chromatography/mass spectrometry. Rapid Commun Mass Spectrom 2008;22:887–97.

[54] Krasowski MD, Pizon AF, Siam MG, Giannoutsos S, Iyer M, Ekins S. Using molecular similarity to highlight the challenges of routine immunoassay-based drug of abuse/toxicology screening in emergency medicine. BMC Emerg Med 2009;9:5.

[55] Krasowski MD, Siam MG, Iyer M, Pizon AF, Giannoutsos S, Ekins S. Chemoinformatic methods for predicting interference in drug of abuse/toxicology immunoassays. Clin Chem 2009;55:1203–13.

[56] Melanson SE, Baskin L, Magnani B, Kwong TC, Dizon A, Wu AH. Interpretation and utility of drug of abuse immunoassays: lessons from laboratory drug testing surveys. Arch Pathol Lab Med 2010;134:735–9.

[57] Bush DM. The United States Mandatory Guidelines for Federal Workplace Drug Testing Programs: current status and future considerations. Forensic Sci Int 2008;174:111–19.

[58] Lee HH, Lee JF, Lin SY, Chen PH, Chen BH. Simultaneous determination of HFBA-derivatized amphetamines and ketamines in urine by gas chromatography-mass spectrometry. J Anal Toxicol 2011;35:162–9.

[59] Middleberg RA, Homan J. Quantitation of amphetamine-type stimulants by LC-MS/MS. Methods Mol Biol 2012;902:105–14.

[60] Mohamed KM, Bakdash A. Comparison of 3 derivatization methods for the analysis of amphetamine-related drugs in oral fluid by gas chromatography-mass spectrometry. Anal Chem Insights 2017;12 1177390117727533.

[61] Graeme KA. New drugs of abuse. Emerg Med Clin North Am 2000;18:625–36.

[62] Holmes SB, Banerjee AK, Alexander WD. Hyponatraemia and seizures after ecstasy use. Postgrad Med J 1999;75:32–3.

[63] Kolbrich EA, Goodwin RS, Gorelick DA, Hayes RJ, Stein EA, Huestis MA. Physiological and subjective responses to controlled oral 3,4-methylenedioxymethamphetamine administration. J Clin Psychopharmacol 2008;28:432–40.

[64] Sprague JE, Moze P, Caden D, Rusyniak DE, Holmes C, Goldstein DS, et al. Carvedilol reverses hyperthermia and attenuates rhabdomyolysis induced by 3,4-methylenedioxymethamphetamine (MDMA, Ecstasy) in an animal model. Crit Care Med 2005;33:1311–16.

[65] Murphy PN, Wareing M, Fisk JE, Montgomery C. Executive working memory deficits in abstinent ecstasy/MDMA users: a critical review. Neuropsychobiology 2009;60:159–75.

[66] Eiden C, Cathala P, Fabresse N, Galea Y, Mathieu-Daude JC, Baccino E, et al. A case of drug-facilitated sexual assault involving 3,4-methylenedioxy-methylamphetamine. J Psychoactive Drugs 2013;45:94–7.

[67] Scott-Ham M, Burton FC. Toxicological findings in cases of alleged drug-facilitated sexual assault in the United Kingdom over a 3-year period. J Clin Forensic Med 2005;12:175–86.

[68] Du Mont J, Macdonald S, Rotbard N, Bainbridge D, Asllani E, Smith N, et al. Drug-facilitated sexual assault in Ontario, Canada: toxicological and DNA findings. J Forensic Leg Med 2010;17:333–8.

[69] Abondo M, Bouvet R, Baert A, Morel I, Le Gueut M. Sexual assault and MDMA: the distinction between consciousness and awareness when it comes to consent. Int J Legal Med 2009;123:155–6.

[70] Jansen KL, Theron L. Ecstasy (MDMA), methamphetamine, and date rape (drug-facilitated sexual assault): a consideration of the issues. J Psychoactive Drugs 2006;38:1–12.

[71] Lodge D, Mercier MS. Ketamine and phencyclidine: the good, the bad and the unexpected. Br J Pharmacol 2015;172:4254–76.

[72] Williams NR, Schatzberg AF. NMDA antagonist treatment of depression. Curr Opin Neurobiol 2016;36:112–17.
[73] WHO Model List of Essential Medicines; 2015. <http://www.who.int/medicines/publications/essentialmedicines/en/>.
[74] Bokor G, Anderson PD. Ketamine: an update on its abuse. J Pharm Pract 2014;27:582–6.
[75] De Luca MT, Meringolo M, Spagnolo PA, Badiani A. The role of setting for ketamine abuse: clinical and preclinical evidence. Rev Neurosci 2012;23:769–80.
[76] Han E, Kwon NJ, Feng LY, Li JH, Chung H. Illegal use patterns, side effects, and analytical methods of ketamine. Forensic Sci Int 2016;268:25–34.
[77] Smith KM, Larive LL, Romanelli F. Club drugs: methylenedioxymethamphetamine, flunitrazepam, ketamine hydrochloride, and gamma-hydroxybutyrate. Am J Health Syst Pharm 2002;59:1067–76.
[78] Stewart CE. Ketamine as a street drug. Emerg Med Serv 2001;30 30, 32, 34 passim.
[79] Bobo WV, Miller SC. Ketamine as a preferred substance of abuse. Am J Addict 2002;11:332–4.
[80] Britt GC, McCance-Katz EF. A brief overview of the clinical pharmacology of "club drugs.". Subst Use Misuse 2005;40:1189–201.
[81] Peltoniemi MA, Hagelberg NM, Olkkola KT, Saari TI. Ketamine: a review of clinical pharmacokinetics and pharmacodynamics in anesthesia and pain therapy. Clin Pharmacokinet 2016;55:1059–77.
[82] Xu J, Lei H. Ketamine—an update on its clinical uses and abuses. CNS Neurosci Ther 2014;20:1015–20.
[83] Gable RS. Acute toxic effects of club drugs. J Psychoactive Drugs 2004;36:303–13.
[84] Bowdle TA, Radant AD, Cowley DS, Kharasch ED, Strassman RJ, Roy-Byrne PP. Psychedelic effects of ketamine in healthy volunteers: relationship to steady-state plasma concentrations. Anesthesiology 1998;88:82–8.
[85] Peyton SH, Couch AT, Bost RO. Tissue distribution of ketamine: two case reports. J Anal Toxicol 1988;12:268–9.
[86] Mihic SJ, Harris RA. Hypnotics and sedatives. In: Brunton LL, Chabner BA, Knollman BC, editors. Goodman & Gilman's: the pharmacological basic of therapeutics. New York City, NY: McGraw-Hill Co; 2011. p. 457–80.
[87] Moffat AC, Osselton MD, Widdop B. Clarke's analysis of drugs and poisons. 3rd ed. Chicago, IL: Pharmaceutical Press; 2004.
[88] Jourdil N, Bessard J, Vincent F, Eysseric H, Bessard G. Automated solid-phase extraction and liquid chromatography-electrospray ionization-mass spectrometry for the determination of flunitrazepam and its metabolites in human urine and plasma samples. J Chromatogr B Analyt Technol Biomed Life Sci 2003;788:207–19.
[89] LeBeau MA. Laboratory management of drug-facilitated sexual assault cases. Forensic Sci Rev 2010;22:113–19.
[90] Nutt DJ, Stahl SM. Searching for perfect sleep: the continuing evolution of GABAA receptor modulators as hypnotics. J Psychopharmacol 2010;24:1601–12.
[91] Dang A, Garg A, Rataboli PV. Role of zolpidem in the management of insomnia. CNS Neurosci Ther 2011;17:387–97.
[92] Kleykamp BA, Griffiths RR, McCann UD, Smith MT, Mintzer MZ. Acute effects of zolpidem extended-release on cognitive performance and sleep in healthy males after repeated nightly use. Exp Clin Psychopharmacol 2012;20:28–39.
[93] Leger D, Scheuermaier K, Roger M. The relationship between alertness and sleep in a population of 769 elderly insomniacs with and without treatment with zolpidem. Arch Gerontol Geriatr 1999;29:165–73.
[94] Stranks EK, Crowe SF. The acute cognitive effects of zopiclone, zolpidem, zaleplon, and eszopiclone: a systematic review and meta-analysis. J Clin Exp Neuropsychol 2014;36:691–700.
[95] Drover DR. Comparative pharmacokinetics and pharmacodynamics of short-acting hypnosedatives: zaleplon, zolpidem and zopiclone. Clin Pharmacokinet 2004;43:227–38.
[96] Greenblatt DJ, Zammit GK. Pharmacokinetic evaluation of eszopiclone: clinical and therapeutic implications. Expert Opin Drug Metab Toxicol 2012;8:1609–18.
[97] Najib J. Eszopiclone, a nonbenzodiazepine sedative-hypnotic agent for the treatment of transient and chronic insomnia. Clin Ther 2006;28:491–516.
[98] Drover D, Lemmens H, Naidu S, Cevallos W, Darwish M, Stanski D. Pharmacokinetics, pharmacodynamics, and relative pharmacokinetic/pharmacodynamic profiles of zaleplon and zolpidem. Clin Ther 2000;22:1443–61.
[99] Patat A, Paty I, Hindmarch I. Pharmacodynamic profile of Zaleplon, a new non-benzodiazepine hypnotic agent. Hum Psychopharmacol 2001;16:369–92.
[100] Kintz P, Villain M, Dumestre-Toulet V, Ludes B. Drug-facilitated sexual assault and analytical toxicology: the role of LC-MS/MS: a case involving zolpidem. J Clin Forensic Med 2005;12:36–41.
[101] Kintz P, Villain M, Ludes B. Testing for the undetectable in drug-facilitated sexual assault using hair analyzed by tandem mass spectrometry as evidence. Ther Drug Monit 2004;26:211–14.
[102] Salomone A, Gerace E, Di Corcia D, Martra G, Petrarulo M, Vincenti M. Hair analysis of drugs involved in drug-facilitated sexual assault and detection of zolpidem in a suspected case. Int J Legal Med 2012;126:451–9.
[103] Gock SB, Wong SH, Nuwayhid N, Venuti SE, Kelley PD, Teggatz JR, et al. Acute zolpidem overdose—report of two cases. J Anal Toxicol 1999;23:559–62.
[104] Jones AW, Holmgren A, Kugelberg FC. Concentrations of scheduled prescription drugs in blood of impaired drivers: considerations for interpreting the results. Ther Drug Monit 2007;29:248–60.
[105] Lichtenwalner M, Tully R. A fatality involving zolpidem. J Anal Toxicol 1997;21:567–9.
[106] Gustavsen I, Al-Sammurraie M, Morland J, Bramness JG. Impairment related to blood drug concentrations of zopiclone and zolpidem compared to alcohol in apprehended drivers. Accid Anal Prev 2009;41:462–6.

[107] Kratzsch C, Tenberken O, Peters FT, Weber AA, Kraemer T, Maurer HH. Screening, library-assisted identification and validated quantification of 23 benzodiazepines, flumazenil, zaleplone, zolpidem and zopiclone in plasma by liquid chromatography/mass spectrometry with atmospheric pressure chemical ionization. J Mass Spectrom 2004;39:856–72.

[108] Bramness JG, Arnestad M, Karinen R, Hilberg T. Fatal overdose of zopiclone in an elderly woman with bronchogenic carcinoma. J Forensic Sci 2001;46:1247–9.

[109] Moore KA, Zemrus TL, Ramcharitar V, Levine B, Fowler DR. Mixed drug intoxication involving zaleplon ("Sonata"). Forensic Sci Int 2003;134:120–2.

[110] Goulle JP, Anger JP. Drug-facilitated robbery or sexual assault: problems associated with amnesia. Ther Drug Monit 2004;26:206–10.

[111] Maravelias C, Stefanidou M, Dona A, Athanaselis S, Spiliopoulou C. Drug-facilitated sexual assault provoked by the victim's religious beliefs: a case report. Am J Forensic Med Pathol 2009;30:384–5.

[112] Villain M, Cheze M, Tracqui A, Ludes B, Kintz P. Windows of detection of zolpidem in urine and hair: application to two drug facilitated sexual assaults. Forensic Sci Int 2004;143:157–61.

[113] Jones AW, Holmgren A, Ahlner J. Toxicological analysis of blood and urine samples from female victims of alleged sexual assault. Clin Toxicol (Phila) 2012;50:555–61.

[114] Jones AW, Kugelberg FC, Holmgren A, Ahlner J. Occurrence of ethanol and other drugs in blood and urine specimens from female victims of alleged sexual assault. Forensic Sci Int 2008;181:40–6.

Chapter 19

Overview of Common Designer Drugs

Lilian H.J. Richter, Markus R. Meyer and Hans H. Maurer

Department of Experimental and Clinical Toxicology, Institute of Experimental and Clinical Pharmacology and Toxicology, Center for Molecular Signaling (PZMS), Saarland University, Homburg, Germany

INTRODUCTION

Abuse of designer drugs is widespread among young people, especially in the "rave and dance club scene" [1,2]. In the 1990s, consumption of "classic" designer drugs such as ecstasy was peaking but they are still consumed nowadays [3]. For this reason, most of the "classic" designer drugs have been scheduled in many countries [4,5]. As consequence, a wide variety of structural modifications of such drugs have been synthesized and sold in "head shops" or online shops via the Internet [1,6]. These compounds are then called new psychoactive substances (NPS). In this chapter, only classic designer drugs are discussed and their structures are shown in Fig. 19.1. Although designer drugs have the reputation of being safe, several experimental studies in rats and humans and epidemiological studies indicated risks in humans including life-threatening serotonin syndrome, hepatotoxicity, neurotoxicity, psychopathology, and abuse potential [2,6−8]. As metabolites were suspected to contribute to some of the toxic effects and their knowledge is of importance for developing screening approaches, the main metabolic steps of these drugs have to be elucidated [6]. Therefore, this chapter is focused on the chemistry, pharmacology, toxicology, and especially hepatic metabolism of amphetamine-derived, piperazine-derived, pyrrolidinophenone-derived, beta-keto-type, and 2,5-dimethoxy phenethylamine-type designer drugs. Their metabolic pathways are summarized to facilitate the selection of the most suitable target for urine drug testing. Most results were obtained from in vivo studies with rats confirmed by analyses of authentic human urine obtained from case work and/or in vitro by incubation of the drugs with various human liver preparations such as baculovirus-infected insect cell microsomes containing individual human cDNA-expressed cytochrome P450 enzymes (CYP), or human liver microsomes and cytosol [1,2,6,9,10]. Details on the analytical challenge of designer drugs and NPS are published elsewhere [11,12].

AMPHETAMINE-DERIVED DESIGNER DRUGS

The best-known and widespread designer drugs in this class are methylenedioxy-substituted amphetamines, such as MDMA (*R,S*-methylenedioxymethamphetamine, "Adam," "ecstasy"), MDA (*R,S*-methylenedioxyamphetamine, *R,S*-1 (3′4′-methylenedioxyphen-yl)-2-propanamine, "love pills"), MDEA (*R,S*-methylenedioxyethylamphetamine, MDE, "Eve"), BDB (*R,S*-benzo-dioxolylbutanamine, *R,S*-1(3′,4′-methylenedioxyphenyl)-2-butanamine), and MBDB (*R,S*-*N*-methyl-benzo-dioxolylbutanamine), which have a methylenedioxy ($-O-CH_2-O-$) bridge between positions 3 and 4 of the aromatic ring of the amphetamine molecule [1,8].

Their chemistry, pharmacology, and toxicology have been reviewed by Kalant [8] and are summarized in the following section. All these compounds are similar in their chemistry and biological effects. MDMA, as the most frequently consumed drug out of this group, was chosen as an example in the following. Usually, designer drugs are consumed orally in the form of single-dose tablets. The typical dosage ranges of MDMA for recreational use vary between 50 and 150 mg. MDMA does not act by direct serotonin release but by blocking reuptake transporters of serotonin (main effect), noradrenalin, and dopamine. The synthesis products of MDMA and its related compounds are racemic mixtures. The corresponding enantiomers differ, for example, in potency, metabolism, and toxicity. Effects on the users can be divided into single-dose and long-term consumption. The single-dose effects are similar to those of amphetamine with additional entactogenic effects. Physically, they produce a marked increase in wakefulness, endurance and sense of energy, sexual arousal, postponement of fatigue, and sleepiness. According to Kalant [8], the psychological effects are

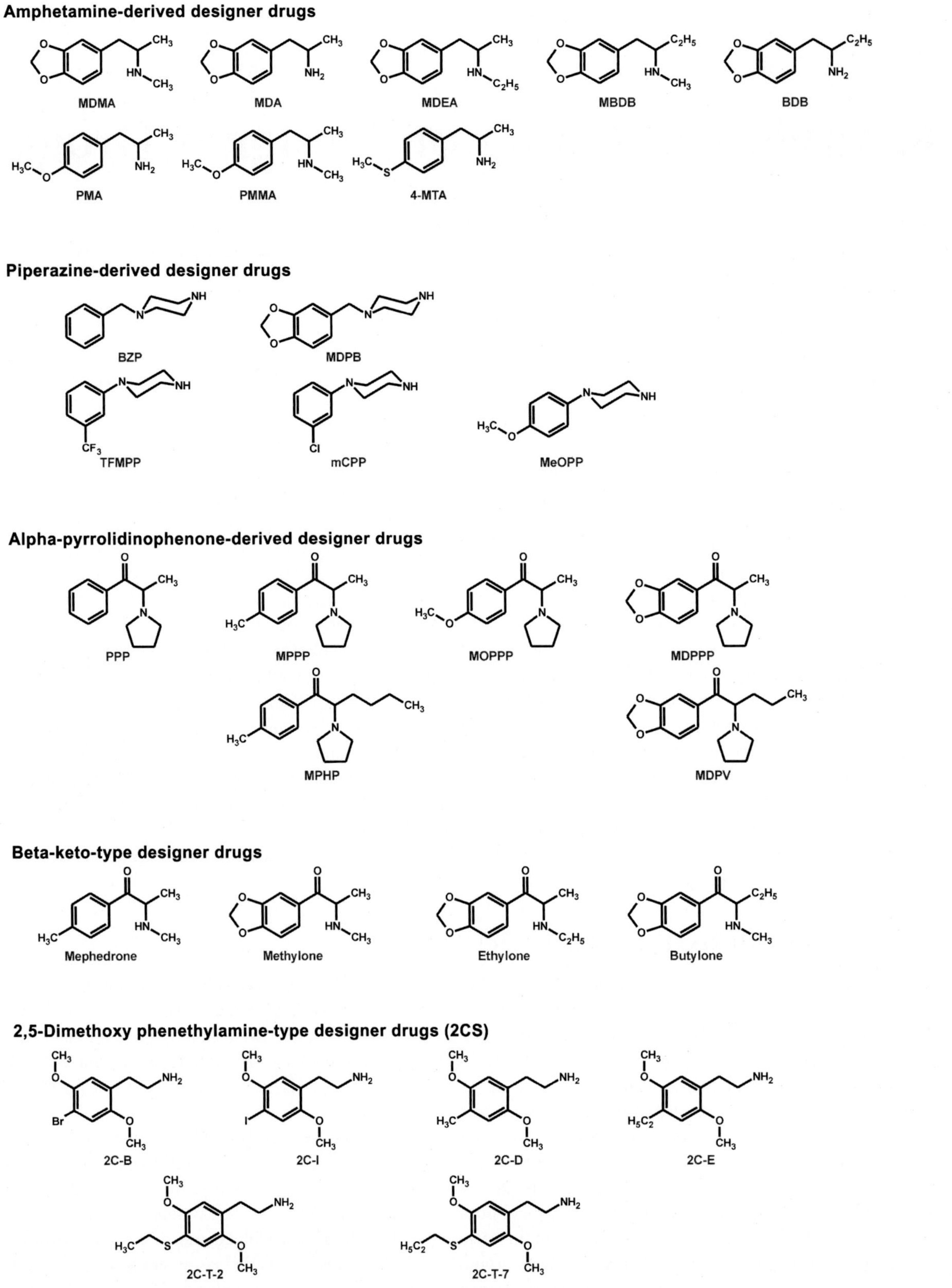

FIGURE 19.1 Chemical structures of amphetamine-derived, piperazine-derived, pyrrolidinophenone-derived, beta-keto-type, and 2,5-dimethoxy phenethylamine-type designer drugs.

euphoria, well-being, sharpened sensory perception, greater sociability, extraversion, heightened sense of closeness to other people, and greater tolerance of their views and feelings. One of the most important long-term effects is the neurotoxicity of these methylenedioxy derivatives. However, the massive release of serotonin at higher doses not only gives rise to acute psychotic symptoms and serotonin syndrome but also causes chemical damage to the cells that released it. It has been suggested [8] that the demonstrated neurotoxic effects of MDMA on the serotonin system may be the possible cause of a variety of mental and behavioral problems that outlast the actual drug experience by months or years. In addition, hepatotoxicity, cardiovascular toxicity, and cerebral toxicity have been described in the context of ecstasy consumption [8].

The metabolism of these compounds was studied in human and rats and was already reviewed by Kraemer and Maurer [9] and is summarized in Fig. 19.2. Two predominant overlapping pathways were postulated for these compounds. First, demethylenation to catechol derivatives, followed by methylation of one hydroxyl group, catalyzed by the catechol-O-methyl-transferase (COMT) and/or glucuronidation via UDP glucuronyltransferases and/or sulfation by sulfotransferases (MDMA, MDA, MDEA, MBDB, BDB). As second pathway, successive degradation of the side chain trough N-dealkylation (MDMA, MDEA, MBDB) and deamination (MDMA, MDA, MDEA, MBDB, BDB) was

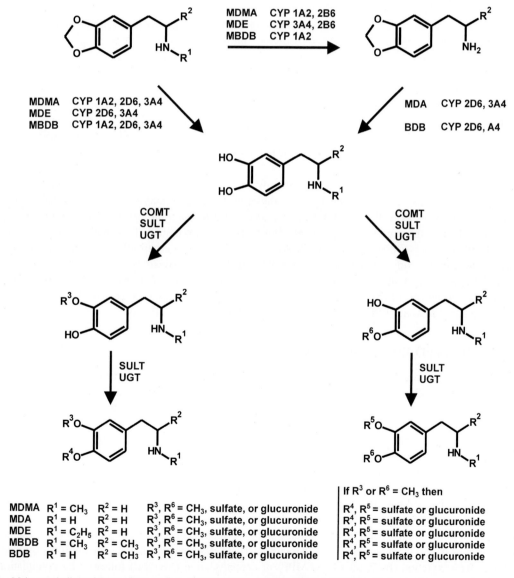

FIGURE 19.2 Main metabolic pathways of the racemic MDMA, MDA, MDE, MBDB, and BDB with involved enzymes. *SULT*, Sulfotransferase; *UGT*, UDP glucuronyltransferase.

described. According to in vitro studies using human liver microsomes and specific inhibitors for the main isoenzymes [7], formation of the catechol derivatives was mainly catalyzed by CYP3A4 and the polymorphically expressed CYP2D6. CYP1A2 was also involved in this metabolic step for MDMA and MBDB, but to a minor extent. In contrast, N-dealkylation of MDMA and MBDB was catalyzed by CYP1A2 and of MDE by CYP3A4. CYP2D6 contributed only to a minor extent, but at this time CYP2B6 was not yet integrated in the CYP testing. Kreth et al. [13] described CYP2B6 as main enzyme catalyzing the N-dealkylation. This was confirmed by Meyer et al. [14] studying the enantioselective metabolism of MDMA. N-demethylation was mainly catalyzed by CYP2B6 (R,S-MDMA), CYP1A2 (R-MDMA), and CYP2B6 (S-MDMA), while for demethylation, the isozyme with the highest contribution to net clearance for R,S-MDMA, R-MDMA, and S-MDMA was CYP2D6.

Schwaninger et al. [15] have investigated the urinary excretion kinetic of MDMA and its phase I and II metabolites in human urine after controlled MDMA administration. According to this study, the main excretion products were MDMA itself and 3,4-dihydroxymethamphetamine and 4-hydroxy-3-methoxymethamphetamine conjugated to sulfates or glucuronides with sulfates present at higher concentrations. This indicated that the O-methylation was preferred at hydroxyl group in position 3 of the aromatic ring [15].

PMA, PMMA, AND 4-MTA

Chemistry, pharmacology, toxicology, and hepatic metabolism of R,S-para-methoxyamphetamine (PMA), R,S-paramethoxy methamphetamine (PMMA), and R,S-4-methylthioamphetamine (4-MTA) were summarized by Staack and Maurer [1] and were reported as follows. During the 1990s, the illicit drug market changed and new compounds such as PMA, PMMA, and 4-MTA appeared and were consumed orally. In the meantime, they were scheduled in most countries. Their hallucinogenic properties were already known since the late 1960s. PMA, PMMA, and 4-MTA showed similar pharmacological and toxicological effects to those of MDMA but with minor stimulating amphetamine-like effects. PMA is a serotonergic compound, evoking serotonin release, inhibiting its uptake with a minor influence of the dopamine system. It was reported that after PMA consumption, a higher rate of lethal complications may occur in comparison to other substituted amphetamine derivatives, what can be probably explained by its pharmacological and toxicological properties.

As depicted in Fig. 19.3, PMA and PMMA have mainly been metabolized by O-demethylation. Aromatic hydroxylation to the corresponding catechol followed by O-methylation by COMT and/or glucuronidation and sulfation were described as minor pathways as well as alteration of the side chain through aliphatic hydroxylation (PMA, PMMA, 4-MAT), N-hydroxylation to the corresponding oxime (PMA, 4-MAT), oxidative deamination followed by oxidation to carboxylic acid (PMA), and N-dealkylation (PMMA). For 4-MTA, no demethylation of the methylthio group but hydroxylation at the aromatic ring was described. Its main metabolic pathway studied in mice and primary human hepatocytes was the degradation of the side chain, comparable to PMA. Incubations using human liver microsomes and baculovirus-infected insect cell microsomes containing individual human cDNA-expressed CYP enzymes indicated that the O-demethylation of PMA and PMMA was exclusively metabolized by CYP2D6.

PIPERAZINE-DERIVED DESIGNER DRUGS

Chemistry, pharmacology, toxicology, and hepatic metabolism of the piperazine-derived designer drugs were already reviewed by Maurer et al. [2] and Staack and Maurer [1] and will be discussed in the following. Since the 1990s, the so-called piperazines emerged on the black market as completely new class of designer drugs. They can be divided into two classes, the benzylpiperazines such as N-benzylpiperazine (BZP) itself, and its methylenedioxy analogue 1-(3,4-methylenedioxybenzyl)piperazine (MDBP, MDBZP), and the phenylpiperazines 1-(3-chlorophenyl)piperazine (mCPP), 1-(3-trifluoromethylphenyl)piperazine (TFMPP), and 1-(4-methoxyphenyl)piperazine (MeOPP). For all compounds except for MeOPP, serotonergic properties and amphetamine-like effects have been described.

Metabolism studies were performed in vivo using rats. Benzylpiperazines were not extensively metabolized and thus excreted mainly as unchanged parent compound in contrast to phenylpiperazines, which were extensively metabolized as shown in Fig. 19.4. It is important to know that mCPP is also a metabolite of therapeutics such as trazodone, nefazodone, etoperidone, and mepiprazol, and that interpretation of the analytical result must consider the presence or absence of unique metabolites of the therapeutics [16]. Piperazine-derived designer drugs were metabolized by alteration of the aromatic ring by hydroxylation (BZP, mCPP, TFMPP) or demethylation (MeOPP) followed by conjugation to sulfates or glucuronides. The metabolic pathways of MDBP were similar to those of methylenedioxy-substituted amphetamines, namely demethylenation followed by O-methylation in position 3 or 4 and/or sulfation or glucuronidation. Further

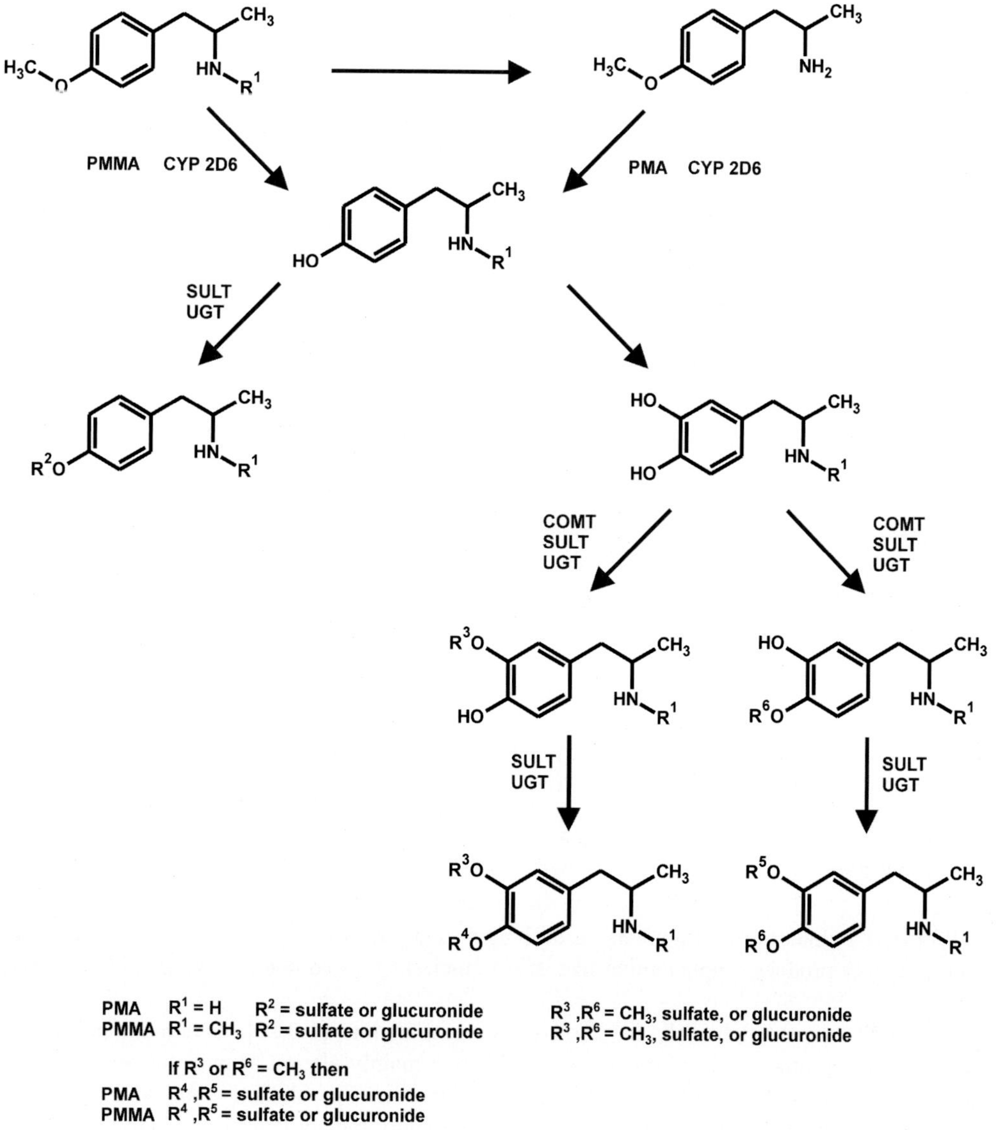

FIGURE 19.3 Main metabolic pathways of the racemic PMA and PMMA with involved enzymes, if they are known. *SULT*, Sulfotransferase; *UGT*, UDP glucuronyltransferase.

metabolic steps were *N*-dealkylation to piperazine and degradation of the piperazine heterocycle to the corresponding ethylenediamine or aniline derivatives followed by *N*-acetylation (BZP, MDBP, mCPP, TFMPP, MeOPP). *O*-Demethylation of MeOPP and hydroxylation of mCPP and TFMPP were mainly catalyzed by CYP2D6. In addition, TFMPP hydroxylation was catalyzed by CYP1A2 and 3A4 to a minor extent.

ALPHA-PYRROLIDINOPHENONE-DERIVED DESIGNER DRUGS

Chemistry, pharmacology, toxicology, and hepatic metabolism of the alpha-pyrrolidinophenone-derived designer drugs was summarized by Maurer et al. [2] and Staack and Maurer [1]. The following pyrrolidinophenone-derived designer drugs have appeared on the illicit drug market: *R,S*-α-pyrrolidinopropiophenone (PPP) as the basic structure of this new class, *R,S*-4′-methoxy-α-pyrrolidinopropiophenone (MOPPP), *R,S*-3′,4′-methylenedioxy-α-pyrrolidinopropiophenone (MDPPP), *R,S*-4′-methyl-α-pyrrolidinopropiophenone (MPPP), *R,S*-4′-methyl-α-pyrrolidinohexanophenone (MPHP), a MPPP derivative with an elongated side chain. In recent years, 3,4-methylene-dioxypyrovalerone (MDPV) became the most common pyrrolidinophenone-derived drugs of abuse and is therefore included in this chapter although being

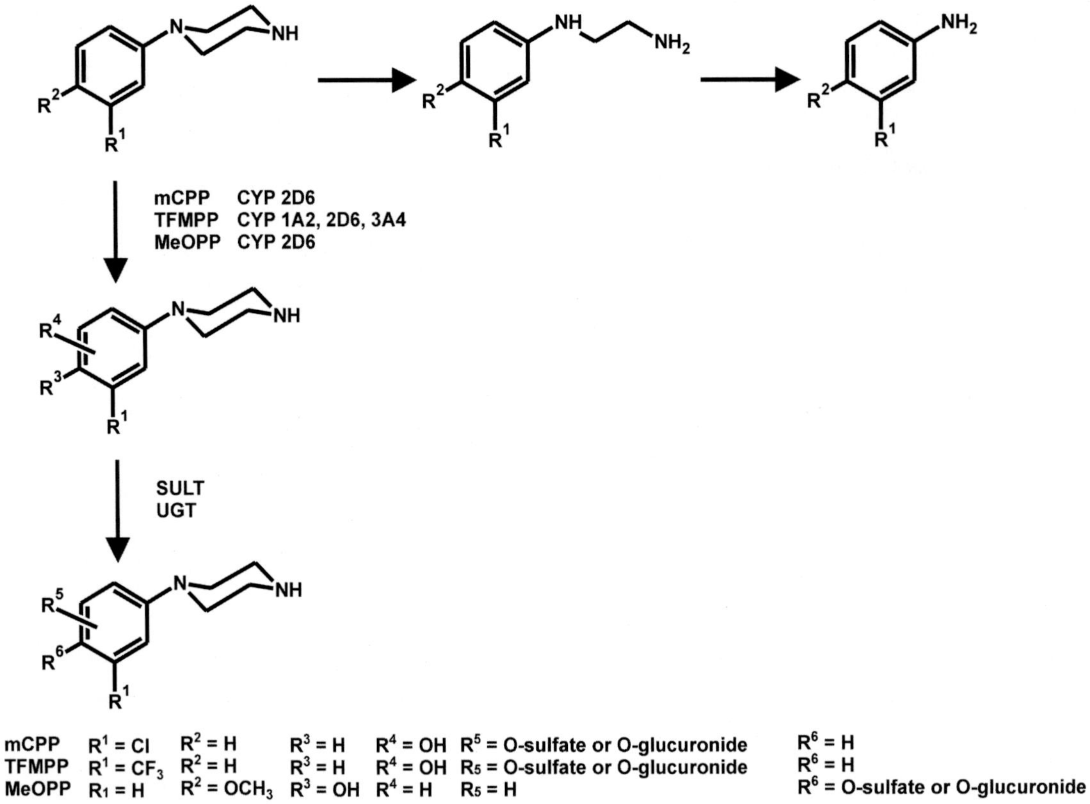

FIGURE 19.4 Main metabolic pathways of the mCPP, TFMPP, and MeOPP with involved enzymes, if they are known. *SULT*, Sulfotransferase; *UGT*, UDP glucuronyltransferase.

classified as NPS [3]. All alpha-pyrrolidinophenone-derived designer drugs have replaced the amino group by a pyrrolidine ring, which probably produce amphetamine-like effects including dopamine release and indirect sympathomimetic properties.

All these compounds were extensively metabolized at least in rats [1,2,17]. The main pathways are shown in Fig. 19.5. In contrast to the other pyrrolidinophenones, PPP was mainly altered at pyrrolidine ring by oxidation to the corresponding lactam or by double dealkylation to cathinone, the main psychoactive alkaloid of Kath (*Catha edulis*), followed by reduction of the keto group. All other drugs were mainly metabolized by oxidation at the tolyl position to the corresponding carboxylic acids (MPPP, MPHP), by *O*-demethylation of the methoxy moiety (MOPPP), or by demethylenation of the methylenedioxy moieties (MDPPP, MDPV). Hydroxylation of the side chain could be observed only for the derivative with elongated side chain (MPHP, MDPV) and hydroxylation of the aromatic ring only for MOPPP. Resulting catechols were methylated and/or glucuronidated or sulfated. In addition, reduction of the keto group (PPP) and oxidation of the pyrrolidine ring to the corresponding lactam (MOPPP, MDPPP, MPPP, MPHP, MDPV) were described as well as oxidative deamination to the corresponding 2-oxo metabolite (MDPPP). Studies on the identification of the CYP isoenzymes involved in the major metabolic steps showed that initial hydroxylation of the 4′-methyl moiety of MPPP and MPHP, *O*-demethylation of MOPPP, and demethylenation of MDPPP and MDPV were catalyzed by CYP2D6 and CYP2C19, with CYP2D6 being the major enzyme according to calculations using the relative activity factor (RAF) approach. The RAF approach can be used for the in vitro-in vivo scaling of pharmacokinetic clearance from in vitro intrinsic clearance measurements in heterologous expression systems. CYP1A2, CYP2B6, and CYP2C9 were additionally involved in MPPP and MPHP hydroxylation of the tolyl methyl group to a minor extent.

BETA-KETO-TYPE DESIGNER DRUGS

Methylone (bk-MDMA), ethylone (bk-MDEA), butylone (bk-MBDB), and mephedrone (4-methylmethcathinone) are beta-keto-substituted analogues of the corresponding amphetamines and appeared as new class on the illicit drug

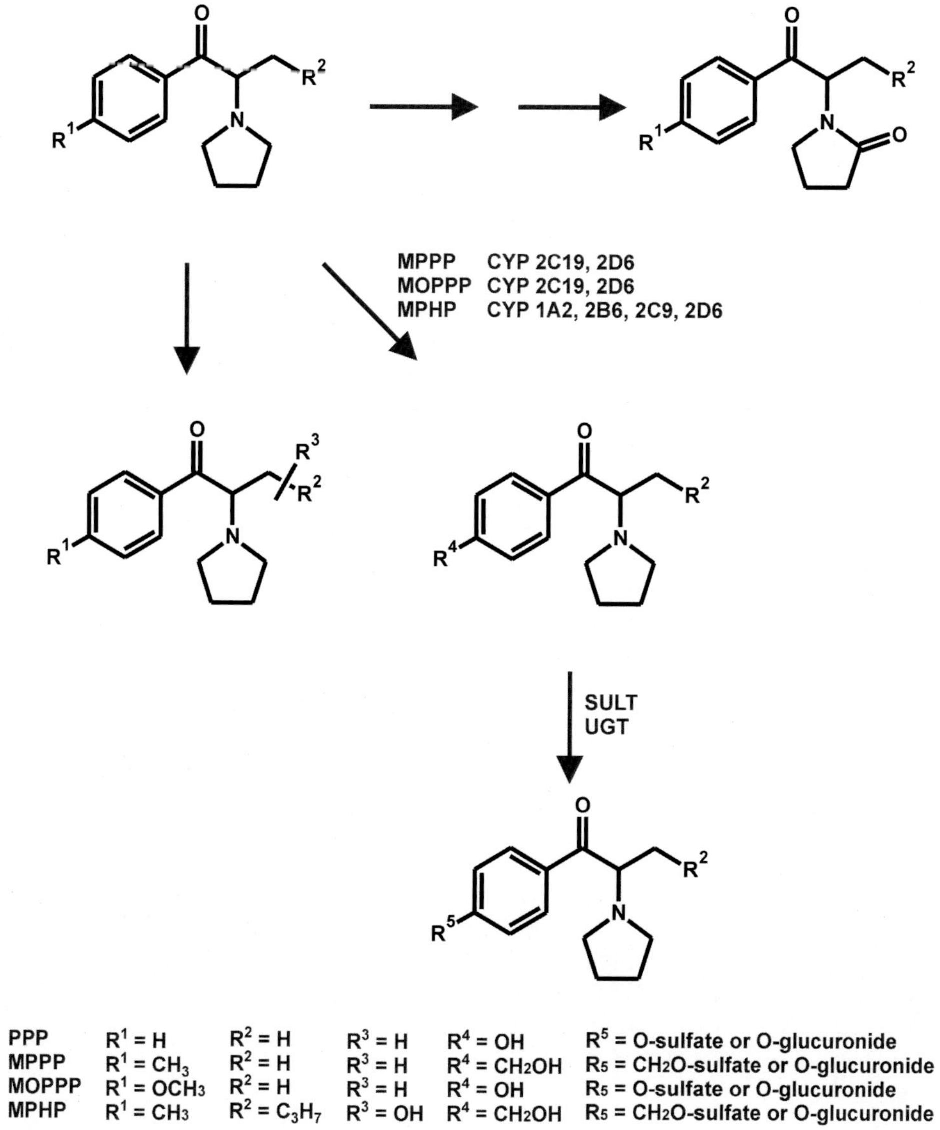

FIGURE 19.5 Main metabolic pathways of the racemic PPP, MPPP, MOPPP, and MPHP with involved enzymes, if they are known. *SULT*, Sulfotransferase; *UGT*, UDP glucuronyltransferase.

MPPP   CYP 2C19, 2D6
MOPPP CYP 2C19, 2D6
MPHP   CYP 1A2, 2B6, 2C9, 2D6

SULT
UGT

PPP	R^1 = H	R^2 = H	R^3 = H	R^4 = OH	R_5 = O-sulfate or O-glucuronide
MPPP	R^1 = CH$_3$	R^2 = H	R^3 = H	R^4 = CH$_2$OH	R_5 = CH$_2$O-sulfate or O-glucuronide
MOPPP	R^1 = OCH$_3$	R^2 = H	R^3 = H	R^4 = OH	R_5 = O-sulfate or O-glucuronide
MPHP	R^1 = CH$_3$	R^2 = C$_3$H$_7$	R^3 = OH	R^4 = CH$_2$OH	R_5 = CH$_2$O-sulfate or O-glucuronide

market. Due to their chemical similarity to amphetamines or methcathinone and the use as alternatives of these drugs, similar pharmacological and toxicological effects as described above could be postulated [6]. Metabolism of methylone and mephedrone was investigated in rat and abuser's urine and the metabolism of ethylone and butylone only in abuser's urine [6]. Methylone, ethylone, and butylone were metabolized similar to their methylenedioxy-substituted amphetamine analogues (Fig. 19.6), namely by *O*-demethylenation followed by methylation of one of the hydroxy groups. As minor metabolic steps, *N*-demethylation and reduction of the keto group were described. Mephedrone was degraded by reduction of the keto group, *N*-demethylation, and oxidation of the tolyl group to the corresponding alcohol and carboxylic acid. Metabolites of all drugs containing hydroxy groups were conjugated to sulfates or glucuronides. CYP enzyme kinetic data of methylone and in vitro incubations of mephedrone with specific CYP inhibitor showed that CYP2D6, CYP1A2, 2B6, and 2C19 were more or less involved in the initial metabolic steps [18,19].

2,5-DIMETHOXY PHENETHYLAMINE-TYPE DESIGNER DRUGS (2CS)

Chemistry, pharmacology, toxicology, and hepatic metabolism of 2,5-dimethoxy phenylamphetamine-type designer drugs (2Cs) were summarized by Meyer and Maurer [6]. The main representatives of this group are 4-bromo-2,5-dimethoxy-β-phenethylamine (2C-B), 4-iodo-2,5-dimethoxy-β-phenethylamine (2C-I), 2,5-dimethoxy-4-methyl-

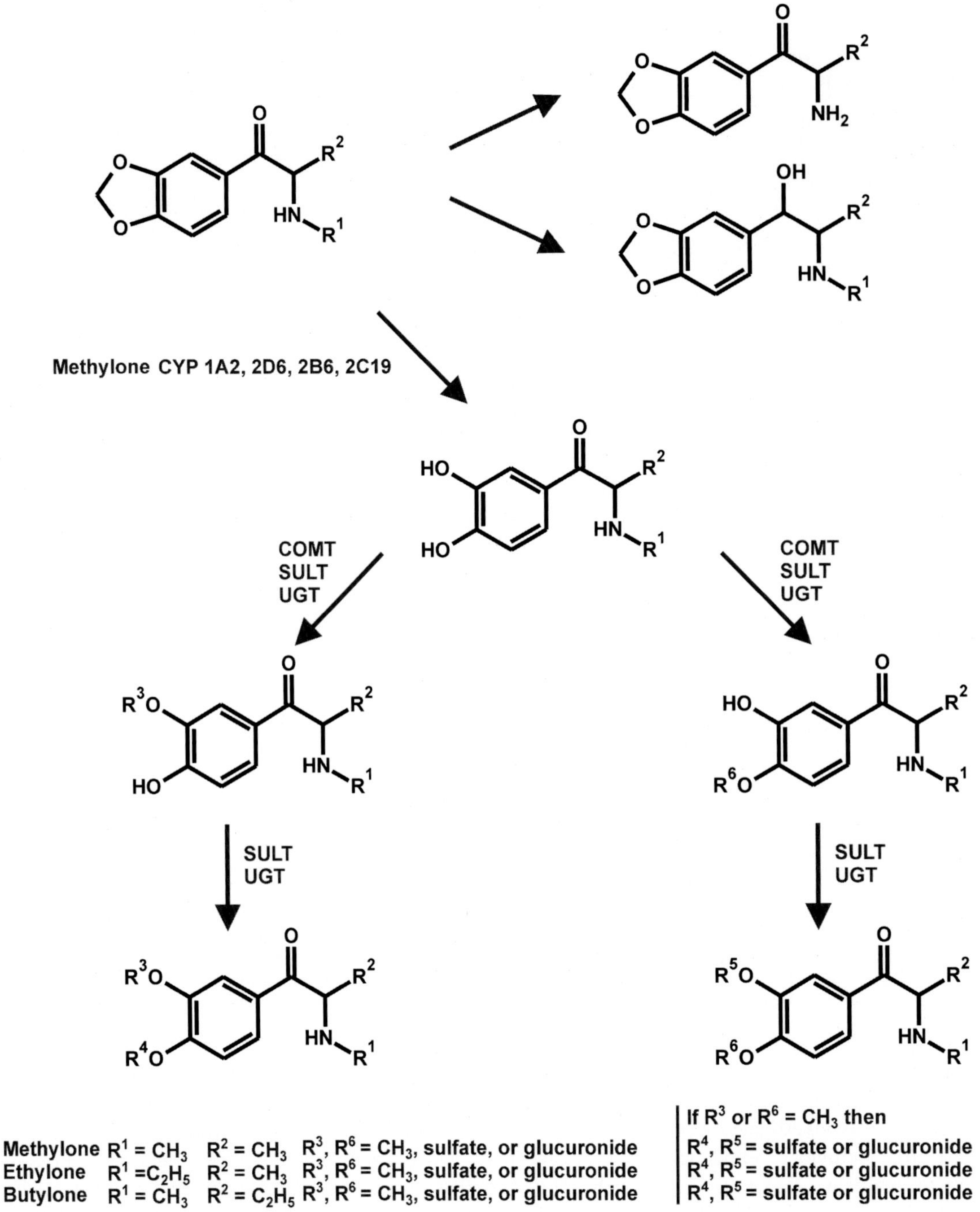

FIGURE 19.6 Main metabolic pathways of the racemic methylone, ethylone, and butylone with involved enzymes, if known. *SULT*, Sulfotransferase; *UGT*, UDP glucuronyltransferase.

β-phenethylamine (2C-D), 4-ethyl-2,5-dimethoxy-β-phenethylamine (2C-E), 4-ethylthio-2,5-dimethoxy-β-phenethylamine (2C-T-2), and 2,5-dimethoxy-4-propylthio-β-phenethylamine (2C-T-7). All these derivatives have a phenethylamine backbone with two methoxy groups in positions 2 and 5 of the aromatic ring and they further contain different lipophilic substituents in position 4. Many 2Cs were first synthesized during the 1970s and 1980s and appeared on the illicit drug market but an enormous increase in consumption was observed during the 1990s after the publication of PiHKAL [20] where their synthesis and pharmacology were described. 2C-B appeared first on the illicit drug market in the mid of 1980s and 2C-T-2, 2C-T-7, and 2C-I a few years later, sold as tablets, powder, or liquids alone or in mixture with other designer drugs. 2Cs shows affinity to $5-HT_2$ receptors and act as agonist or antagonists at different receptor subtypes. The 2Cs have hallucinogen-like effects due to their primary amine functionality separated from the phenyl ring by two carbon atoms (2Cs) and the presence of methoxy groups on positions 2 and 5 of the aromatic ring, and a hydrophobic 4-substituent. 2C-D showed less hallucinogenic properties than the other 2Cs, but eightfold higher potency than mescaline.

Main metabolic pathways are similar for all 2Cs (Fig. 19.7). They were investigated in rat urine and for 2C-B and 2C-E in consumer urine, too with the following results: *O*-demethylation in position 2 or 5 of the aromatic ring, deamination to the corresponding aldehyde, which was not detectable, followed by oxidation to the corresponding carboxylic acid or reduction to the corresponding alcohol, partial glucuronidation or sulfation, and *N*-acetylation. Combinations of

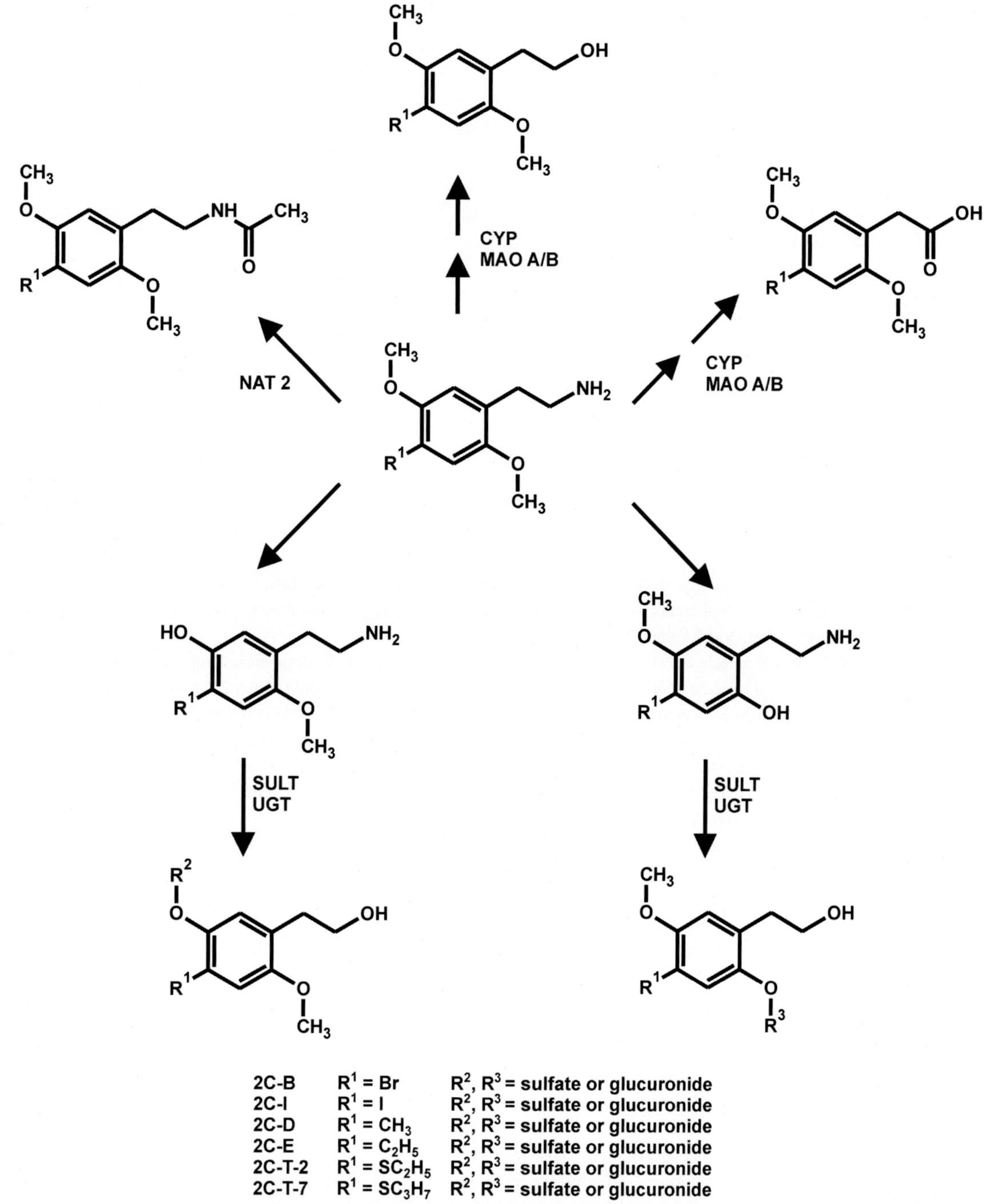
NAT 2
CYP
MAO A/B
CYP
MAO A/B
SULT
UGT
SULT
UGT
2C-B R1 = Br R2, R3 = sulfate or glucuronide
2C-I R1 = I R2, R3 = sulfate or glucuronide
2C-D R1 = CH3 R2, R3 = sulfate or glucuronide
2C-E R1 = C2H5 R2, R3 = sulfate or glucuronide
2C-T-2 R1 = SC2H5 R2, R3 = sulfate or glucuronide
2C-T-7 R1 = SC3H7 R2, R3 = sulfate or glucuronide

FIGURE 19.7 Main metabolic pathways of the 2C-B, 2C-I, 2C-D, 2C-E, 2C-T-2, and 2C-T-7 with involved enzymes. *NAT*, *N*-acetyltransferase; *SULT*, sulfotransferase; *UGT*, UDP glucuronyltransferase.

these steps have also been observed. 2Cs with nonhalogenic substituents in position 4 of the aromatic ring showed hydroxylation at these substituents and, in case of sulfur containing compounds, sulfoxidation. Investigation with baculovirus-infected insect cell microsomes containing individual human cDNA-expressed CYP, monoamine oxidase (MAO) enzymes revealed that the oxidative deamination of all 2Cs was catalyzed by MAO A and B, with higher affinity to MAO A, or to a minor extendt by CYP2D6 for 2C-D, 2C-E, 2C-T-2, and 2C-T-7. The differences of the K_m values between MAO A and B increased with the size of the 4-substituent. Interactions should be considered if 2Cs are taken in combination with therapeutic MAO inhibitors. Studies using recombinant human N-acetyltransferase showed that N-acetylation was mainly catalyzed by the polymorphically expressed N-acetyltransferase 2 [21]. In the meantime, highly potent derivatives occurred on the NPS marked, the so-called NBOMes with an N-2-methoxy-benzyl rest at the primary amine group of the corresponding 2Cs [22].

CONCLUSIONS

Most of the discussed designer drugs have similar pharmacological effects such as stimulating, entactogenic, and partly hallucinogenic effects. The main acute toxicological risk is the serotonin syndrome. They are extensively metabolized with the exception of benzylpiperazines and the polymorphically expressed CYP2D6 is one important enzyme in phase I metabolism. It must be kept in mind that some designer drugs have common metabolites or are metabolites of therapeutic drugs. Thus, differentiation needs detection of unique metabolites.

REFERENCES

[1] Staack RF, Maurer HH. Metabolism of designer drugs of abuse. Curr Drug Metab 2005;6:259–74.
[2] Maurer H, Kraemer T, Springer D, Staack R. Chemistry, pharmacology, toxicology, and hepatic metabolism of designer drugs of the amphetamine (ecstasy), piperazine, and pyrrolidinophenone types—a synopsis. Ther Drug Monitor 2004;26:127–31.
[3] UNODC. World drug report. United Nations Publication; 2017.
[4] UN. United Nations—Single Convention on Narcotic Drugs; 1961.
[5] EMCDDA. Reviewing legal aspects of substitution treatment at international level, 2000.
[6] Meyer MR, Maurer HH. Metabolism of designer drugs of abuse: an updated review. Curr Drug Metab 2010;11:468–82.
[7] Maurer HH, Bickeboeller-Friedrich J, Kraemer T, Peters FT. Toxicokinetics and analytical toxicology of amphetamine-derived designer drugs ("ecstasy"). Toxicol Lett 2000;112:133–42.
[8] Kalant H. The pharmacology and toxicology of "ecstasy" (MDMA) and related drugs. CMAJ 2001;165:917–28.
[9] Kraemer T, Maurer H. Toxicokinetics of amphetamines: metabolism and toxicokinetic data of designer drugs, amphetamine, methamphetamine, and their N-alkyl derivatives. Ther Drug Monitor 2002;24:277–89.
[10] Peters FT, Meyer MR. In vitro approaches to studying the metabolism of new psychoactive compounds. Drug Test Anal 2011;3:483–95.
[11] Meyer MR, Peters FT. Analytical toxicology of emerging drugs of abuse—an update. Ther Drug Monit 2012;34:615–21.
[12] Peters FT, Martinez-Ramirez JA. Analytical toxicology of emerging drugs of abuse. Ther Drug Monit 2010;32:532–9.
[13] Kreth K, Kovar K, Schwab M, Zanger UM. Identification of the human cytochromes P450 involved in the oxidative metabolism of "ecstasy"-related designer drugs. Biochem Pharmacol 2000;59:1563–71.
[14] Meyer MR, Peters FT, Maurer HH. The role of human hepatic cytochrome P450 isozymes in the metabolism of racemic 3,4-methylenedioxymethamphetamine and its enantiomers. Drug Metab Dispos 2008;36:2345–54.
[15] Schwaninger AE, Meyer MR, Barnes AJ, Kolbrich-Spargo EA, et al. Urinary excretion kinetics of 3,4-methylenedioxymethamphetamine (MDMA, ecstasy) and its phase I and phase II metabolites in humans following controlled MDMA administration. Clin Chem 2011;57:1748–56.
[16] Staack RF, Maurer HH. Piperazine-derived designer drug 1-(3-chlorophenyl)piperazine (mCPP): GC-MS studies on its metabolism and its toxicological detection in rat urine including analytical differentiation from its precursor drugs trazodone and nefazodone. J Anal Toxicol 2003;27:560–8.
[17] Meyer MR, Du, Schuster PF, Maurer HH. Studies on the metabolism of the alpha-pyrrolidinophenone designer drug methylenedioxypyrovalerone (MDPV) in rat and human urine and human liver microsomes using GC-MS and LC-high-resolution MS and its detectability in urine by GC-MS. J Mass Spectrom 2010;45:1426–42.
[18] Pedersen AJ, Petersen TH, Linnet K. In vitro metabolism and pharmacokinetic studies on methylone. Drug Metab Dispos 2013;4:1247–55.
[19] Pedersen AJ, Reitzel LA, Johansen SS, Linnet K. In vitro metabolism studies on mephedrone and analysis of forensic cases. Drug Test Anal 2013;5:430–8.
[20] Shulgin A. PiHKAL: a chemical love story. Berkley, CA: Transform Press; 1991.
[21] Meyer MR, Robert A, Maurer HH. Toxicokinetics of novel psychoactive substances: characterization of N-acetyltransferase (NAT) isoenzymes involved in the phase II metabolism of 2C designer drugs. Toxicol Lett 2014;227:124–8.
[22] Kyriakou C, Marinelli E, Frati P, Santurro A, et al. NBOMe: new potent hallucinogens—pharmacology, analytical methods, toxicities, fatalities: a review. Eur Rev Med Pharmacol Sci 2015;19:3270–81.

Chapter 20

New Psychoactive Substances: An Overview

Laura Mercolini

Pharmaco-Toxicological Analysis Laboratory, Department of Pharmacy and Biotechnology, University of Bologna, Bologna, Italy

INTRODUCTION

New psychoactive substances (NPS) represent a challenging issue in the scientific research field, due to the huge and growing number of chemical entities available on the illicit drug market. The term "new" does not necessarily indicate a new compound, but can also refer to those substances synthesized some years ago and recently come back to the limelight in the trafficking of drugs of abuse. International NPS diffusion has become a highly concerning problem, also because they induce frequent conditions of tolerance and dependence although the literature on the subject is still fragmented and under development [1,2]. In Europe, the number of NPS notified to the European Monitoring Centre for Drugs and Drug Addiction (EMCDDA) has increased rapidly since 2008 [3]. More than 700 NPS belonging to several chemical and pharmacological classes are currently monitored and most of them are synthetic cannabinoid receptor agonists (SCRA) or cathinone analogs (CA), thus being the 2 most representative classes of NPS.

Given the huge potential possibilities of chemical structure alterations, the list of NPS considered in this chapter is not exhaustive and is just addressed to provide an overview on the most common seized and identified compounds. Moreover, this chapter does not include those compounds subjected to international controls under the 1961 Convention on Narcotic Drugs or under the 1971 Convention [4,5], like for example, benzodiazepines or opioids.

Preliminary identification of new emerging compounds on the illicit market often derives from emergency room reports or from seized samples; in the latter case, some recommendations on analytical procedures have been published by the Scientific Working Group for the Analysis of Seized Drugs (SWGDRUG) [6]. In the early stages of NPS emergence, new chemical entities circumvented existing legislation and were usually offered as a legal alternative to traditional illicit drugs which are already classified as "controlled substances". NPS consumers could easily obtain them through the World Wide Web, sold as research chemicals, plant food, or bath salts, often labeled as "not for human consumption" [7].

To date in several countries law enforcement agencies adopted faster protocols aimed to classify newly identified NPS among prohibited substances, considering also preventive approaches to include structural analogs, in order to fill any legislative gap [8]. Furthermore, careful controls by governmental and police forces involved in cyber policing aim at identifying and prosecuting websites and online retail stores devoted to illegal product sale. At the same time, however, illicit market has expanded through the so-called deep web or darknet, a submerged platform where customer and payment process anonymity, together with the difficulty of tracing data flows, makes these products widely accessible [9].

As regards NPS bioanalysis, the recent scientific literature is consistent and agrees in recommending the use of mass spectrometry (MS)-based methods for screening and analysis [10–13]. MS methodologies, coupled to a wide range of separative techniques and sample preparation approaches, provide the high levels of sensitivity, selectivity, and also versatility needed for sound and reliable NPS analysis [14]. Several of these compounds are extensively metabolized, therefore analytical strategies must monitor also metabolites, in both plasmatic and urinary analyses [15,16]. According to the main international guidelines, metabolite detection requires a subsequent confirmation using reference standards and the use of mass spectral libraries including also such metabolites is recommended.

Unlike prescription drugs and classic drugs of abuse, pharmacokinetic data on NPS are still largely unknown. Therefore, metabolism studies are crucial for the development of urinary screening methods and key information can be obtained using in vitro model or in vivo ones by investigating samples obtained from users or volunteers under controlled administration [17–19].

MS-based methods for NPS qualitative and quantitative analysis can be classified as targeted or untargeted [14]. The first ones focus on predefined sets of analytes and are performed by means of medium- to low-resolution tandem mass spectrometry (MS/MS) systems, exploiting selected reaction monitoring (SRM) or multiple reaction monitoring (MRM) acquisition modes. These approaches can usually grant high sensitivity and selectivity but, on the other hand, do not detect unexpected or unknown compounds. On the contrary, untargeted strategies cover all the analytes already included in the considered reference library that can be continuously updated by adding new spectral data for newly identified compounds, also allowing a retrospective data mining [20]. High-resolution mass spectrometry (HRMS) data often help in speculating chemical structures, in particular if partial chemical structures are already known. HRMS can be exploited for data-independent acquisition too, since all product ions are scanned without filtering precursor ion, and for data-dependent acquisition mode, where defined precursor ions are fragmented for MS/MS spectral recording.

SYNTHETIC CANNABINOID RECEPTOR AGONISTS

SCRA emerged as recreational drugs of abuse in 2004 and then around 2008 on a larger scale as an alternative to *Cannabis*. Illicit products containing SCRA became popular as "K2" and "Spice" brands, advertised as incenses or smoking mixtures, and typically sold as finely cut inert plant material in order to provide a sort of *Cannabis* likeness [21,22]. While usually promoted not for human consumption, their use is intended by smoking, analogously to *Cannabis*, and producing strong Δ9-tetrahydrocannabinol (THC)-like effects. Compared to natural THC, synthetic cannabinoids are more potent and may have a longer half-life. The effects appear within 10 min after inhalation and may last 2–6 h after use. Several components of these herbal products have been initially investigated with therapeutic intents to act on cannabinoid receptors (CB1 and CB2) [23] and lately extensive investigations gave rise to hundreds of new chemical entities characterized by potent and effective CB receptors activities, but finally abandoned due to unwanted side effects.

SCRA have been classified in five major chemical groups: naphthoylindoles, naphthylmethylindoles, naphthoylpyrroles, napthylmethylindenes and phenylacetylindoles (Fig. 20.1). Despite many SCRA structural classes became controlled under national drug legislations, they are still widely abused, posing very severe human health threats. In fact, in recent years these illicit compounds have often been the subject of several reports as responsible for serious harms, such as intoxications requiring emergency treatment [24].

A comprehensive and complete overview on SCRA pharmacology is not yet available in terms of pharmacokinetics and pharmacodynamics, including psychomotor and physical effects. General pharmacokinetics and toxicology data are reported for the most common SCRA in those cases of intoxication [25–27]; in vitro experiments were carried out by means of human or rat liver microsome incubation, to mimic first phase metabolism [28–30]; in vivo assessments were performed by analyzing urine samples from subjects after self-reported smoking or controlled administration [28,31]. It is reported that SCRA are extensively metabolized in the body into active and inactive compounds.

Chromatographic methods including high-performance liquid chromatography and gas chromatography coupled to MS (LC-MS, GC-MS) are widely listed, with LC-MS seemingly the established technique of choice for SCRA analysis and in general for all NPS identification and quantitation in biological and nonbiological samples [12].

CATHINONE ANALOGS

CA are amphetamine-like cheap alternatives, derived from cathinone, stimulant compound found in khat plant (*Catha edulis*), and possessing pharmacological similarity to amphetamine and methamphetamine. They appeared in the "black" drug market in the mid-2000s and are believed to block the reuptake of norepinephrine, dopamine, and serotonin, while releasing more dopamine and thus suggesting acting similarly to both methamphetamine and cocaine. Their selectivity can vary; mephedrone and methcathinone mainly promote dopamine release, while mephedrone, methylone, ethylone, and butylone can enhance serotonin release. The desired effects of CA use are euphoria, alertness, sexual arousal, and talkativeness, observed within 30–45 min after administration and lasting 1–3 h. They can cause severe cardiovascular and neurological side effects [32].

EMCDDA presented more than 74 new CA between 2005 and 2014, with 30 new ones identified in 2014 alone. These numbers highlight how the development of effective analytical methods for their detection and quantitation is an

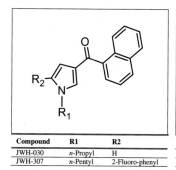

Compound	R1	R2	R3
JWH-015	n-Propyl	Methyl	1-Naphthyl
JWH-018	n-Pentyl	H	1-Naphthyl
JWH-200	2-(4-Morpholinyl)ethyl	H	1-Naphthyl
AM-1220	(1-Methyl-2-piperidinyl)methyl	H	1-Naphthyl
JWH-203	n-Pentyl	H	2-Chloro-benzyl
JWH-250	n-Pentyl	H	2-Methoxy-benzyl
JWH-251	n-Pentyl	H	2-Methoxy-benzyl
RCS-4	n-Pentyl	H	4-Methoxy-phenyl
AB-001	n-Pentyl	H	Adamantyl
AM-1248	(1-Methyl-2-piperidinyl)methyl	H	Adamantyl
AM-679	n-Pentyl	H	2-Iodo-phenyl
AM-694	5-Fluoro-n-pentyl	H	2-Iodo-phenyl
AM-2233	(1-Methyl-2-piperidinyl)methyl	H	2-Iodo-phenyl
AM-1241	(1-Methyl-2-piperidinyl)methyl	H	2-Iodo-5-nitro-phenyl
UR-144	n-Pentyl	H	Tetramethyl-cyclopropyl
XLR-11	5-Fluoro-n-pentyl	H	Tetramethyl-cyclopropyl
XLR-12	4,4,4-Trifluoro-butyl	H	Tetramethyl-cyclopropyl
AB-005	(1-Methyl-2-piperidinyl)methyl	H	Tetramethyl-cyclopropyl

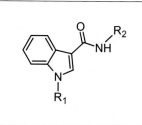

Compound	R1	R2
JWH-030	n-Propyl	H
JWH-307	n-Pentyl	2-Fluoro-phenyl

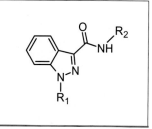

Compound	R1	R2
APICA	n-Pentyl	Adamantyl
5F-APICA	5-Fluoro-n-pentyl	Adamantyl
NNEI	n-Pentyl	1-Naphthyl
5F-NNEI	5-Fluoro-n-pentyl	1-Naphthyl

Compound	R1	R2
APINACA	n-Pentyl	Adamantyl
5F-APINACA	5-Fluoro-n-pentyl	Adamantyl
MN-18	n-Pentyl	1-Naphthyl
5F-MN-18	5-Fluoro-n-pentyl	1-Naphthyl

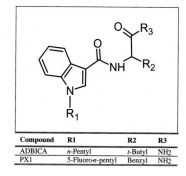

Compound	R1	R2	R3
ADBICA	n-Pentyl	t-Butyl	NH2
PX1	5-Fluoro-n-pentyl	Benzyl	NH2

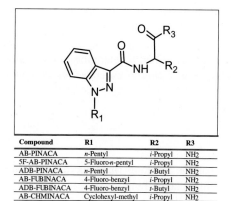

Compound	R1	R2	R3
AB-PINACA	n-Pentyl	i-Propyl	NH2
5F-AB-PINACA	5-Fluoro-n-pentyl	i-Propyl	NH2
ADB-PINACA	n-Pentyl	t-Butyl	NH2
AB-FUBINACA	4-Fluoro-benzyl	i-Propyl	NH2
ADB-FUBINACA	4-Fluoro-benzyl	t-Butyl	NH2
AB-CHMINACA	Cyclohexyl-methyl	i-Propyl	NH2
AMB	n-Pentyl	i-Propyl	Methoxy
5F-AMB	5-Fluoro-n-pentyl	i-Propyl	Methoxy

FIGURE 20.1 Chemical structures of the main synthetic cannabinoid receptor agonists. *Adapted from Figure 1 in Mercolini L, Protti M. Biosampling strategies for emerging drugs of abuse: towards the future of toxicological and forensic analysis. J Pharm Biomed Anal 2016;130:202–19. Copyright Elsevier. Reprinted with permission.*

urgent priority. Mephedrone and methylenedioxypyrovalerone in particular are popular amongst users of "legal high" products and despite their early classification in national prohibited substances lists, they are still being abused. Unlike SCRA, the most common methods for CA consumption are insufflation (snorting) or ingestion; smoking, intramuscular and intravenous injection, and sublingual administration have also been reported but are definitely less common.

A difference between CA and SCRA from the bioanalytical point of view is that the parent molecule cathinone is easily detected in biological samples and often selected as target compounds, since unchanged parent drugs are rapidly excreted in urine. Their blood concentrations are at sufficient high values to be analyzed in hematic fluids, although they can vary because of different administration routes exploited by abusers. A plethora of case reports concerning intoxications and fatalities due to CA consumption is extensive and increasing in the literature, covering a considerable age range within both sexes [33].

From an analytical point of view, several LC-MS and GC-MS methods were implemented for CA analysis [11], also together with phase I and II metabolites. More recently, in light of the often nonenantioselective NPS synthesis performed in clandestine chemical laboratories, chiral separation of racemic mixtures was also performed. For in-depth discussion on CA and chemical structures of most important CA, see Chapter 21, Review of Bath Salts on Illicit Drug Market.

PHENETHYLAMINES

Phenethylamines (PA), together with piperazines, tryptamines (TA), and CA possess stimulant and hallucinogenic effects, constituting the class of the so-called entactogens, psychoactive molecules enhancing empathy and emotional closeness feelings. PA are synthetic compounds commercially known as "party pills". They are active on serotonin receptors, leading to psychedelic effects and in some cases also inhibiting monoamine reuptake [34]. 3,4-Methylenedioxy-methamphetamine (MDMA), known as "Ecstasy," is one of the most popular CA with similar stimulant effects. Amphetamine, methamphetamine, and MDMA are controlled under the 1971 Convention. During the last few years, an increasing use of new PA derivatives on the recreational drug scene has been observed. These relatively new PA include ring substituted molecules such as the 2C series, ring substituted amphetamines such as the D series (e.g., 2,5-dimethoxy-4-iodoamphetamine, DOI and 2,5-dimethoxy-4-chloroamphetamine, and DOC), benzodifurans (1-(4-Bromofuro[2,3-f] [1]benzofuran-8-yl)propan-2-amine, Bromo-DragonFly and 8-bromo-2,3,6,7-benzo-dihydro-difuran-ethylamine, 2C-B-Fly), and MDMA-like drugs (p-methoxy methamphetamine, PMMA) [35]. PA seizures were first reported in the United States and since 2009 NPS belonging to this class started being quite often seized by several countries worldwide; a few examples are 2C-E (2,5-dimethoxyphenethylamine), 2C-I (2-(4-iodo-2,5-dimethoxyphenyl)ethan-1-amine), 4-FA (4-fluoroamphetamine), and PMMA. Other PA increasingly signposted to the United Nations Office on Drugs and Crime (UNODC) since 2011 include 4-FMA (4-fluoromethamphetamine), 5-APB (1-benzofuran-5-ylpropan-2-amin), 6-APB (6-(2-aminopropyl)benzofuran), and 2C-C-NBOMe (2-(4-Iodo-2,5-dimethoxyphenyl)-N-[(2-methoxyphenyl)methyl]ethanamine). Several cases of intoxications described symptoms and side effects, including hypertension, hyperthermia, convulsions, dissociation, hallucinations, respiratory deficits, liver and kidney failure, and death in case of overdose [36]. The lead compound of this wide class, N-benzylpiperazine, acts by enhancing dopamine and norepinephrine release, as well as inhibiting serotonin, dopamine and norepinephrine reuptake [37]. Between 1980s and 1990s, the synthesis of several NPS belonging to this class was described: simple variations on the natural PA mescaline led to the production of powerful psychoactive derivatives [38]. For an overview of the most representative PA, see Fig. 20.2.

TRYPTAMINES

TA are a group of monoamine alkaloids (indolalkylamines), derived from the amino acid tryptophan, that can be found in natural sources including plants, fungi, microbes, and amphibia [39]. Naturally occurring TA derivatives are present in the "magic mushrooms" belonging to the species *Psilocybe cubensis* and containing psilocybin (4-phosphoryloxy-N,N-dimethyltryptamine) and psilocin (4-hydroxy-N,N-dimethyltryptamine). Psilocybin recreational use became widespread in the late 1950s in the United States in particular, whereas synthetic analogs appeared on illicit markets only throughout the 1990s. TA synthesis and their psychoactive effects were thoroughly studied with the identification of 55 compounds and the assessment of their effects on volunteers [40]. In recent years, a group of synthetic TA deriving from N,N-dimethyltryptamine (DMT) and other naturally occurring TA have been listed as NPS, including 5-MeO-DMT (5-methoxy-N,N-dimethyltryptamine), 5-MeO-DPT ("Foxy-Methoxy," 5-methoxy-N,N-dipropyltryptamine), AMT (2-(1H-indol-3-yl)-1-methyl-ethylamine), 4-AcO-DMT (4-acetoxy-N,N-dimethyltryptamine), and 4-AcODiPT (4-Acetoxy-N,N-diisopropyltryptamine). Some of the most important TA are shown in Fig. 20.3. TA use for recreational purposes requires only minimal amounts of these drugs able to produce evident psychoactive effects, thus intoxications, hospitalizations, and fatalities can easily occur. TA are known to be a broad class of drugs able to produce profound changes in sensory perception, mood, and thought. Furthermore, these molecules are not commonly detected by using general drug screening panels and for this reason screening tests performed at the emergency rooms could lead to false negative results, wrong diagnosis, and inappropriate therapies.

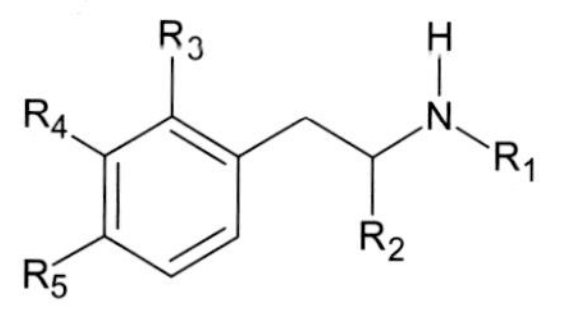

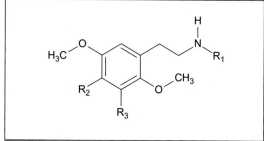

Compound	R₁	R₂	R₃	R₄	R₅
5-APB	H	Methyl	H	–CH=CH–O–	
6-APB	H	Methyl	H	–O–CH=CH–	
2-FA	H	Methyl	F	H	H
3-FA	H	Methyl	H	F	H
3-FMA	Methyl	Methyl	H	F	H
4-FMA	Methyl	Methyl	H	H	F
PMEA	Ethyl	Methyl	H	H	Methoxy
OMMA	Methyl	Methyl	Methoxy	H	H
PMMA	Methyl	Methyl	H	H	Methoxy
MPA	Methyl	Ethyl	Phenyl → Thyophenyl		

Compound	R₁	R₂	R₃
2C-B	H	Br	H
25B-NBOMe	CH₂C₆H₅OCH₃	Br	H
2C-C	H	Cl	H
25C-NBOMe	CH₂C₆H₅OCH₃	Cl	H
2C-D	H	Methyl	H
25D-NBOMe	CH₂C₆H₅OCH₃	Methyl	H
2C-E	H	Ethyl	H
2C-F	H	F	H
2C-G	H	Methyl	Methyl
2C-H	H	H	H
2C-I	H	I	H
2C-IP	H	Isopropyl	H
2C-N	H	NO₂	H
2C-O-4	H	Isopropoxy	H
2C-P	H	Propyl	H
2C-SE	H	SeCH₃	H
2C-T	H	SCH₃	H

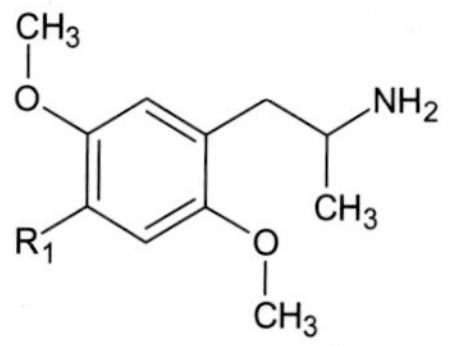

Compound	R₁
DOB	Br
DOC	Cl
DOET	Ethyl
DOF	F
DOI	I
DOM	Methyl
DON	NO₂

Compound	Structure
2C-B-fly	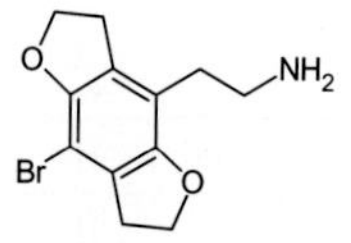
2C-B-butterfly	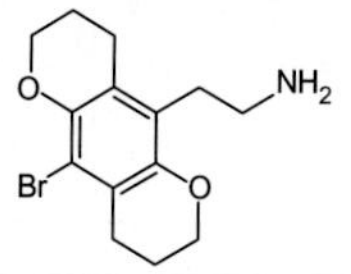
Bromo-dragonfly	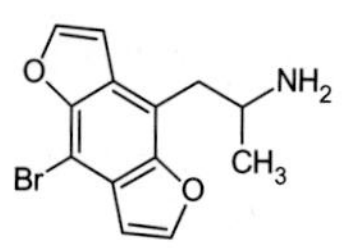
NBOMe-2C-B-fly	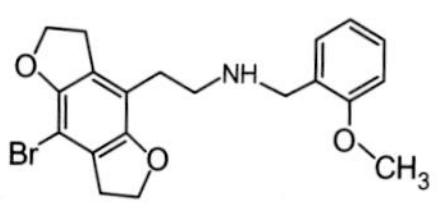
TMF-fly	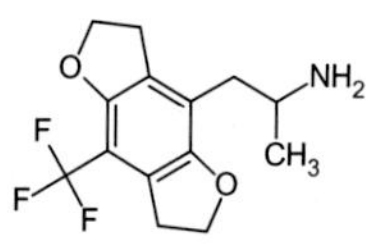

FIGURE 20.2 Chemical structures of the most representative PA. *Adapted from Figure 3 in Mercolini L, Protti M. Biosampling strategies for emerging drugs of abuse: towards the future of toxicological and forensic analysis. J Pharm Biomed Anal 2016;130:202–19. Copyright Elsevier. Reprinted with permission.*

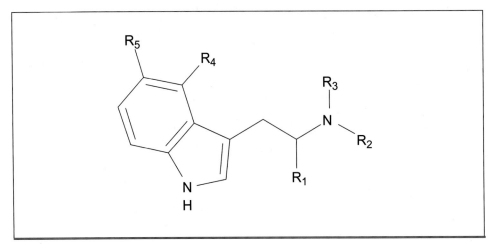

FIGURE 20.3 Chemical structures of some TA. *Adapted from Figure 4 in Mercolini L, Protti M. Biosampling strategies for emerging drugs of abuse: towards the future of toxicological and forensic analysis. J Pharm Biomed Anal 2016;130:202−19. Copyright Elsevier. Reprinted with permission.*

Compound	R_1	R_2	R_3	R_4	R_5
AMT	Methyl	H	H	H	H
DMT	H	Methyl	Methyl	H	H
DET	H	Ethyl	Ethyl	H	H
DPT	H	Propyl	Propyl	H	H
DiPT	H	Isopropyl	Isopropyl	H	H
MiPT	H	Methyl	Isopropyl	H	H
4-OH-DiPT	H	Isopropyl	Isopropyl	OH	H
4-OH-MiPT	H	Methyl	Isopropyl	OH	H
4-OH-MET	H	Methyl	Ethyl	OH	H
5-MeO-AMT	Methyl	H	H	H	Methoxy
5-MeO-DMT	H	Methyl	Methyl	H	Methoxy
5-MeO-DiPT	H	Isopropyl	Isopropyl	H	Methoxy
5-MeO-MiPT	H	Methyl	Isopropyl	H	Methoxy
5-MeO-EiPT	H	Ethyl	Isopropyl	H	Methoxy
5-AcO-DiPT	H	Isopropyl	Isopropyl	H	Acetoxy
5-MeO-DPT	H	Propyl	Propyl	H	Methoxy
5-MeO-DALT	H	Allyl	Allyl	H	Methoxy

MISCELLANEOUS NEW PSYCHOACTIVE SUBSTANCES

More psychoactive compounds that can be still considered NPS and displaying hallucinogenic, anesthetic, and depressant properties are largely used worldwide. Ketamine and γ-hydroxybutyric acid (GHB) consumption, including its precursor γ-butyrolactone, was experienced among subgroups of illicit drug users in the last two decades and there is a growing knowledge on specific health problems related to the abuse of these drugs. Long-term ketamine use can lead to bladder damages [41], while amnesia and persistent hallucinations are linked to GHB continuative use [42,43].

In both North America and Europe, one of the most recent health emergences related to NPS use is represented by new synthetic opioids [44]. In particular, fentanyl derivatives are causing considerable concern: due to their high potency, they are related to serious health risks, not only for abusers but also for those involved in their handling. Consequently, nonfatal intoxications and a large number of deaths caused by synthetic opioids are increasingly occurring since 2012. Due to chemical differences to morphine, synthetic opioid detection by means of standard immunoassays designed for opiates is not possible. Fentanyl-containing prescription drugs are often diverted from the regulated supply chain, thus fentanyl and some of its derivatives (alfentanil, sufentanil, remifentanil, and carfentanil) are subjected to careful international controls, along with some highly potent analogs that are not even commercialized as prescription drugs then illicitly smuggled as drugs of abuse, but they are diffused on purpose exclusively for abuse

(3-methylfentanyl, TMF) [45]. Fentanyl and TMF illicitly produced are sold also as "China white" and in some cases can be marketed as a replacement for heroin, like a "synthetic heroin," the so-called white heroin.

Some of the compounds mentioned above, their analogs and precursors are listed in Fig. 20.4.

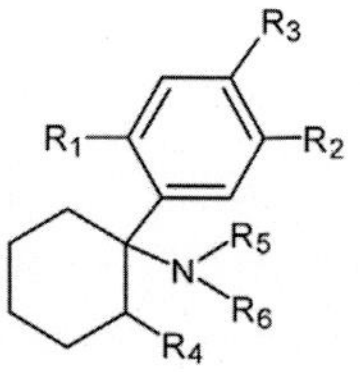

Compound	R_1	R_2	R_3	R_4	R_5	R_6
Ketamine	Cl	H	H	=O	H	Methyl
Methoxetamine	H	Methoxy	H	=O	H	Ethyl
3-MeO-PCE	H	Methoxy	H	H	H	Ethyl
Phencyclidine	H	H	H	H	H	Piperidine
3-MeO-phencyclidine	H	Methoxy	H	H	H	Piperidine
4-MeO-phencyclidine	H	H	Methoxy	H	H	Piperidine

FIGURE 20.4 Ketamine, GHB, and fentanyl, together with their analogs and precursors. *Adapted from Figure 5 in Mercolini L, Protti M. Biosampling strategies for emerging drugs of abuse: towards the future of toxicological and forensic analysis. J Pharm Biomed Anal 2016;130:202–19. Copyright Elsevier. Reprinted with permission.*

GHB	GBL	1,4-BD

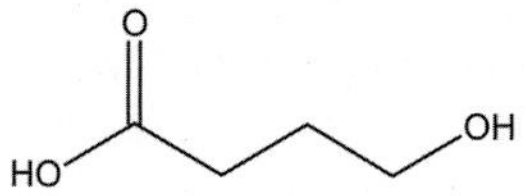

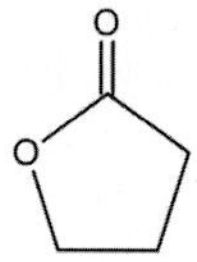

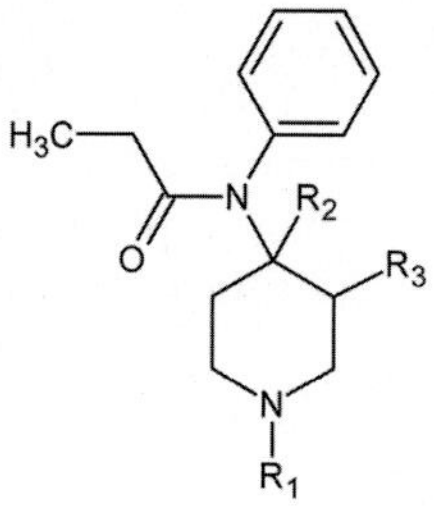

Compound	R_1	R_2	R_3
Fentanyl	(CH₂CH₂-phenyl)	H	H
Sufentanyl	(CH₂CH₂-thienyl)	CH$_2$OCH$_3$	H
Carfentanyl	(CH₂CH₂-phenyl)	COOCH$_3$	H
Norfentanyl	H	H	H
Norsufentanyl	H	CH$_2$OCH$_3$	H
Norcarfentanyl	H	COOCH$_3$	H
Lofentanyl	(CH₂CH₂-phenyl)	COOCH$_3$	CH$_3$
Alfentanyl	(tetrazolinone-ethyl)	CH$_2$OCH$_3$	H
Ramifentanyl	(methoxycarbonyl-ethyl)	COOCH$_3$	H

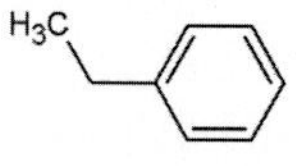

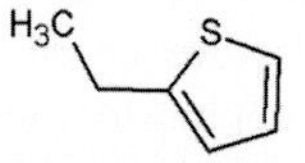

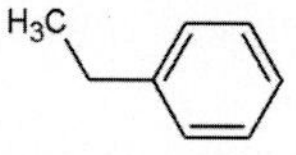

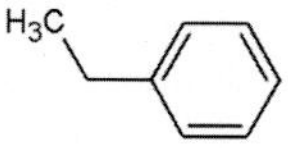

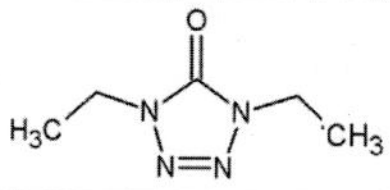

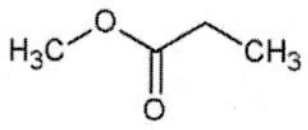

BIOANALYTICAL STRATEGIES FOR NEW PSYCHOACTIVE SUBSTANCES

In the last years, several analytical methods have been developed and published for qualitative–quantitative purposes for a wide range of NPS belonging to different chemical classes. An overview of the most relevant research papers involving multiclass analytical methodologies for NPS in biological samples is reported in Table 20.1. Stimulants, SCRA, TA, and PA were included in the targeted multianalyte approaches discussed below.

A blood screening procedure covering a huge number of analytes, 143 NPS, has been carried out by means of a simple and rapid sample preparation and LC-MS/MS: sensitivity is good, with limits of detection (LOD) in the range 0.01–3.00 ng/mL estimated for 104 compounds. This method was applied to over 1000 real forensic blood samples, collected from drug abusers and drivers over a 3-year period [46]. Another report dealing with a method for the analysis of 64 NPS in whole blood after protein precipitation (PP) was applied to three real cases [47]. Lehmann et al. [48] used automated solid phase extraction (SPE) to analyze 69 compounds in serum samples, by high-throughput sample processing and analysis. This automated approach processed 28 specimens from routine forensic case work, of which 7 ones were positive for NPS consumption. Dispersive liquid–liquid microextraction, using minimal amount of organic solvent as a miniaturized alternative blood sample preparation, was coupled to ultra-high-performance liquid chromatography–tandem mass spectrometry (UHPLC–MS/MS) to monitor NPS in 60 authentic real samples from forensic cases [49]. The same pretreatment procedure was exploited within a study aimed at analyzing 26 NPS including amphetamine-type stimulants, CA, PA, TA, and ketamine and its analogs in both blood and urine, by GC-MS after derivatization [50].

A stability study of 13 molecules among CA and TA in blood specimens was carried out on spiked samples, assessing breakdown products by LC-MS/MS and UHPLC coupled to quadrupole time-of-flight mass spectrometry (QTOF-MS). The presence of degradation products occurring after time was also shown in some real forensic samples [51]. A preliminary study focused on targeted urinary analysis was developed by taking into account only parent compounds without metabolites [52]; 2 years later, a simultaneous urinary quantitation of 40 NPS, including metabolites, by LC-QTOF-MS was presented, using full scan and data dependent mode, and successfully applied to 62 authentic urine specimens [53]. A LC-MS/MS strategy was developed to quantify 56 NPS in blood and urine, including amphetamine derivatives, PA, CA, and TA; SPE extraction was exploited for whole blood analysis, while liquid–liquid extraction (LLE) was applied to urine specimens and comparative data interpretation has been performed from samples obtained from the same subjects [54]. GC coupled to atmospheric pressure chemical ionization and QTOF-MS (GC-APCI-QTOF-MS) is an interesting example of NPS identification by comparing an in-house test library developed ad hoc for 29 compounds (mostly synthetic stimulants) with an LC-QTOF-MS commercial library. This strategy was applied to three postmortem urine samples to evaluate accurate mass measurement potential in those cases when reference standards are not readily available [55].

Regarding multiclass NPS analysis in alternative or complementary biological matrices, some useful data can be found in the recent scientific literature regarding hair, oral fluid, and dried microsample analysis.

Boumba et al. [56] proposed a LC-MS/MS strategy for the analysis of 132 NPS (mainly SCRA, CA, PA, and TA) in hair samples, providing satisfactory LOD values ranging from 0.001 to 0.1 ng/mg. Other methodologies for NPS hair analysis were described, featuring comparable sensitivity [57]. CA together with classic amphetamines, piperazines, and central nervous system prescription drugs were searched, in combination also to some metabolites, in 16 hair samples (30 mg hair/sample), but no positive results were related to NPS consumption [58].

Oral fluid is a very interesting biological matrix where, in recent years, some authors tried to analyze NPS. In one study, 31 new drugs of abuse were determined by UHPLC-MS/MS after PP and microextraction by packed sorbent (MEPS). Unfortunately, no real sample analysis was reported [59]. Another work exploited the same miniaturized sample pretreatment to analyze 11 CA, 6 opiates, scopolamine, cocaine, and metabolites in oral fluid: in this case, 12 real samples from patients on detoxification programs were successfully processed [60]. An advanced comparative approach exploiting miniaturized sampling and pretreatment techniques for biological matrices coupled to LC-MS/MS was presented for the analysis of THC and its main metabolites, together with 10 representative SCRA. In this report, dried blood spot (DBS) and volumetric absorptive microsampling (VAMS) approaches were applied to obtain both capillary and venous whole blood dried microsamples. After development and validation, these innovative methodologies were compared to classic plasma analysis within a monitoring study involving 10 self-reported *Cannabis* and SCRA users and the two datasets where in good agreement, thus proving the high reliability of the new microsampling strategies [12].

The main part of the analyses described above are based on targeted LC-MS/MS, operating in SRM or MRM mode, by means on triple quadrupole and hybrid quadrupole ion trap mass analyzers. Even if equipped with high potential,

TABLE 20.1 Multiclass Bioanalytical Methods for NPS Determination

NPS Category	Biological Sample	Sample Preparation	Analytical Technique	Application to Real Samples	Ref.
SCRA	cDBS, vDBS, cVAMS, and vVAMS	Dried microsampling + solvent extraction	HPLC-MS/MS	Self-reported users	[12]
SCRA, CA	Urine	SPE	UHPLC-QTOF-MS (HRMS)	Subjects from routine toxicological screening	[20]
SCRA, CA, PA, and TA	Blood	PP	HPLC-MS/MS	Subjects from routine forensic screening	[46]
SCRA, CA, PA, TA, and ketamine	Blood	PP	HPLC-MS/MS	Postmortem; intoxication	[47]
SCRA, CA, PA, TA, and ketamine	Serum	SPE	HPLC-MS/MS	Subjects from routine forensic screening	[48]
CA, PA, TA, and ketamine	Blood	DLLME	UHPLC-MS/MS	Postmortem; road accidents; DFSA	[49]
SCRA, CA, PA, TA, and ketamine	Blood, urine	DLLME	GC-MS	Postmortem; road accidents	[50]
CA, TA	Blood, plasma	LLE	HPLC-MS/MS; UHPLC-QTOF-MS (HRMS)	None	[51]
CA, PA, and TA	Urine	Dilution	HPLC-MS/MS	Clinical routine drug testing	[52]
CA	Urine	SPE	HPLC-MS/MS (HRMS)	Self-reported users	[53]
CA, PA, and TA	Blood, urine	SPE, LLE	HPLC-MS/MS	Intoxication; DUI	[54]
CA, TA	Urine	LLE	GC-QTOF-MS HPLC-QTOF-MS (HRMS)	Postmortem	[55]
SCRA, CA, PA, and TA	Hair	Acidic extractive incubation	HPLC-MS/MS	Postmortem	[56]
SCRA, CA	Hair	Accelerated-solvent-extraction + SPE	HPLC-MS/MS (HRMS)	None	[57]
CA	Hair	Acidic extractive incubation + SPE	HPLC-MS/MS	Nonspecified real samples	[58]
SCRA, CA, and PA	Oral fluid	PP + MEPS	UHPLC-MS/MS	None	[59]
CA	Oral fluid	PP + MEPS	HPLC-MS/MS	Patients	[60]
CA, PA, and TA	Urine	Dilution	HPLC-MS/MS (HRMS)	Clinical routine drug testing	[61]
SCRA, CA, and PA	Plasma	PP	HPLC-MS/MS (HRMS)	Nonspecified real samples	[62]

HRMS methods for NPS detection and quantitation still represent a minority [53,57,61]; two published studies exploited both targeted selected-ion monitoring and full scan acquisition and application of full scan acquisition allows an interesting retrospective data mining [20,62].

CONCLUSIONS

The ever-changing plethora of NPS crowding drugs of abuse market is a highly challenging issue in the bioanalytical, clinical, and forensic fields, when considering the number of new substances continuously identified, their availability through online vendors and the still limited toxicological and pharmacological information. Nevertheless, the increasing number of research papers describing bioanalytical methods for NPS detection and quantitation demonstrates how this problem is widely known and tackled by analytical scientists. Among all the strategies adopted to address NPS emergency from an analytical point of view, MS-based multianalyte methods proved to be suitable as identification and determination tools, due to high sensitivity and selectivity. High resolution platforms, in particular, are promising strategies for identification of unknown compounds. Moreover, new approaches to perform qualitative and quantitative analysis of emerging compounds in alternative, innovative and miniaturized biological matrices, in addition to classic fluid blood and urine specimens, could deepen the bioanalytical knowledge about NPS, especially when comparative, multimatrix and multianalyte approaches are exploited.

REFERENCES

[1] Baumeister D, Tojo LM, Tracy DK. Legal highs: staying on top of the flood of novel psychoactive substances. Ther Adv Psychopharmacol 2015;5(2):97–132.
[2] Aarde SM, Taffe MA. Predicting the abuse liability of entactogen-class, new and emerging psychoactive substances via preclinical models of drug self-administration. Curr Top Behav Neurosci 2017;32(1):145–64.
[3] European Monitoring Centre for Drugs and Drug Addiction. European Drug Report 2017: trends and developments. Luxembourg: Publications Office of the European Union; 2017.
[4] United Nations Single convention on narcotic drugs, 1961—As amended by the 1972 Protocol amending the Single Convention on Narcotic Drugs; 1961.
[5] United Nations Convention on Psychotropic substances; 1971.
[6] SWGDRUG, Recommendations for the analysis of seized drugs. Version 7.0; 2016, August 14, 2014.
[7] King LA, Kicman AT. A brief history of 'new psychoactive substances'. Drug Test Anal 2011;3(7–8):401–3.
[8] van Amsterdam J, Nutt D, van den Brink W. Generic legislation of new psychoactive drugs. J Psychopharmacol 2013;27(3):317–24.
[9] Griffiths P, Mounteney J. Disruptive potential of the internet to transform illicit drug markets and impact on future patterns of drug consumption. Clin Pharmacol Ther 2017;101(2):176–8.
[10] Mercolini L, Protti M. Biosampling strategies for emerging drugs of abuse: towards the future of toxicological and forensic analysis. J Pharm Biomed Anal 2016;130(1):202–19.
[11] Mercolini L, Protti M, Catapano MC, Rudge J, et al. LC-MS/MS and volumetric absorptive microsampling for quantitative bioanalysis of cathinone analogues in dried urine, plasma and oral fluid samples. J Pharm Biomed Anal 2016;123(1):186–94.
[12] Protti M, Rudge J, Sberna AE, Gerra G, et al. Dried haematic microsamples and LC-MS/MS for the analysis of natural and synthetic cannabinoids. J Chromatogr B Analyt Technol Biomed Life Sci 2017;1044–1045(1):77–86.
[13] Sundstrom M, Pelander A, Ojanpera I. Comparison between drug screening by immunoassay and ultra-high performance liquid chromatography/high-resolution time-of-flight mass spectrometry in post-mortem urine. Drug Test Anal 2015;7(5):420–7.
[14] Meyer MR, Maurer HH. LC coupled to low- and high-resolution mass spectrometry for new psychoactive substance screening in biological matrices—where do we stand today? Anal Chim Acta 2016;927(1):13–20.
[15] Tsai IL, Weng TI, Tseng YJ, Tan HK, et al. Screening and confirmation of 62 drugs of abuse and metabolites in urine by ultra-high-performance liquid chromatography–quadrupole time-of-flight mass spectrometry. J Anal Toxicol 2013;37(9):642–51.
[16] Maralikova B, Weinmann WJ. Confirmatory analysis for drugs of abuse in plasma and urine by high-performance liquid chromatography–tandem mass spectrometry with respect to criteria for compound identification. Chromatogr B Analyt Technol Biomed Life Sci 2004;811(1):21–30.
[17] Grafinger KE, Hädener M, König S, Weinmann W. Study of the in vitro and in vivo metabolism of the tryptamine 5-MeO-MiPT using human liver microsomes and real case samples. Drug Test Anal 2018;10(3):562–74.
[18] Steuer AE, Williner E, Staeheli SN, Kraemer T. Studies on the metabolism of the fentanyl-derived designer drug butyrfentanyl in human in vitro liver preparations and authentic human samples using liquid chromatography-high resolution mass spectrometry (LC–HRMS). Drug Test Anal 2017;9(7):1085–92.
[19] Wohlfarth A, Roman M, Andersson M, Kugelberg FC, et al. 25C-NBOMe and 25I-NBOMe metabolite studies in human hepatocytes, in vivo mouse and human urine with high-resolution mass spectrometry. Drug Test Anal 2017;9(5):680–98.

[20] Sundstrom M, Pelander A, Ojanpera I. Comparison of post-targeted and pre-targeted urine drug screening by UHPLC-HR-QTOFMS. J Anal Toxicol 2017;41(7):623–30.
[21] Ford BM, Tai S, Fantegrossi WE, Prather PL. Synthetic pot: not your grandfather's marijuana. Trends Pharmacol Sci 2017;38(3):257–76.
[22] Karila L, Benyamina A, Blecha L, Cottencin O, et al. The synthetic cannabinoids phenomenon. Curr Pharm Des 2016;22(42):6420–5.
[23] Huffman JW, Dong D. Design, synthesis and pharmacology of cannabimimetic indoles. Bioorg Med Chem Lett 1994;4(4):563–6.
[24] Abouchedid R, Hudson S, Thurtle N, Yamamoto T, et al. Analytical confirmation of synthetic cannabinoids in a cohort of 179 presentations with acute recreational drug toxicity to an Emergency Department in London, UK in the first half of 2015. Clin Toxicol 2017;55(5):338–45.
[25] Toennes SW, Geraths A, Pogoda W, Paulke A, et al. Pharmacokinetic properties of the synthetic cannabinoid JWH-018 and of its metabolites in serum after inhalation. J Pharm Biomed Anal 2017;140(1):215–22.
[26] Karinen R, Tuv SS, Øiestad EL, Vindenes V. Concentrations of APINACA, 5F-APINACA, UR-144 and its degradant product in blood samples from six impaired drivers compared to previous reported concentrations of other synthetic cannabinoids. Forensic Sci Int 2015;246(1):98–103.
[27] Seely KA, Lapoint J, Moran JH, Fattore L. Spice drugs are more than harmless herbal blends: a review of the pharmacology and toxicology of synthetic cannabinoids. Prog Neuropsychopharmacol Biol Psychiatry 2012;39(2):234–43.
[28] Diao X, Huestis MA. Approaches, challenges, and advances in metabolism of new synthetic cannabinoids and identification of optimal urinary marker metabolites. Clin Pharmacol Ther 2017;101(2):239–53.
[29] Mardal M, Dalsgaard PW, Qi B, Mollerup CB, et al. Metabolism of the synthetic cannabinoids AMB-CHMICA and 5C-AKB48 in pooled human hepatocytes and rat hepatocytes analyzed by UHPLC-(IMS)-HR-MS^E. J Chromatogr B Analyt Technol Biomed Life Sci 2018;1083(1):189–97.
[30] Carlier J, Diao X, Wohlfarth A, Scheidweiler K, et al. In vitro metabolite profiling of ADB-FUBINACA, a new synthetic cannabinoid. Curr Neuropharmacol 2017;15(5):682–91.
[31] Schaefer N, Kettner M, Laschke MW, Schlote J, et al. Simultaneous LC-MS/MS determination of JWH-210, RCS-4, $\Delta(9)$-tetrahydrocannabinol, and their main metabolites in pig and human serum, whole blood, and urine for comparing pharmacokinetic data. Anal Bioanal Chem 2015;407(13):3775–86.
[32] Assi S, Gulyamova N, Kneller P, Osselton D. The effects and toxicity of cathinones from the users' perspectives: a qualitative study. Hum Psychopharmacol 2017;32(3):1–7.
[33] Weinstein AM, Rosca P, Fattore L, London ED. Synthetic cathinone and cannabinoid designer drugs pose a major risk for public health. Front Psychiatry 2017;8(1):156.
[34] Antia U, Lee HS, Kydd RR, Tingle MD, et al. Pharmacokinetics of 'party pill' drug N-benzylpiperazine (BZP) in healthy human participants. Forensic Sci Int 2009;186(1–3):63–7.
[35] King LA. New phenethylamines in Europe. Drug Test Anal 2014;6(7–8):808–18.
[36] Hermanns-Clausen M, Angerer V, Kithinji J, Grumann C, et al. Bad trip due to 25I-NBOMe: a case report from the EU project SPICE II plus. Clin Toxicol 2017;55(8):922–4.
[37] Zwartsen A, Verboven AHA, van Kleef RGDM, Wijnolts FMJ, et al. Measuring inhibition of monoamine reuptake transporters by new psychoactive substances (NPS) in real-time using a high-throughput, fluorescence-based assay. Toxicol In Vitro 2017;45(Pt 1):60–71.
[38] Shulgin AT, Shulgin A. Pihkal: a chemical love story. Berkeley, CA: Transform Press; 1991.
[39] Collins M. Some new psychoactive substances: precursor chemicals and synthesis-driven end-products. Drug Test Anal 2011;3(7–8):404–16.
[40] Shulgin AT, Shulgin A. Tihkal: the continuation. Berkeley, CA: Transform Press; 1997.
[41] Tsai TH, Cha TL, Lin CM, Tsao CW, et al. Ketamine-associated bladder dysfunction. Int J Urol 2009;16(10):826–9.
[42] Schep LJ, Knudsen K, Slaughter RJ, Vale JA, et al. The clinical toxicology of γ-hydroxybutyrate, γ-butyrolactone and 1,4-butanediol. Clin Toxicol 2012;50(6):458–70.
[43] Saracino MA, Catapano MC, Iezzi R, Somaini L, et al. Analysis of γ-hydroxy butyrate by combining capillary electrophoresis-indirect detection and wall dynamic coating: application to dried matrices. Anal Bioanal Chem 2015;407(29):8893–901.
[44] Protti M, Catapano MC, Samolsky Dekel BG, Rudge J, et al. Determination of oxycodone and its major metabolites in haematic in urinary matrices: comparison of traditional and miniaturised sampling approaches. J Pharm Biomed Anal 2018;152(1):204–14.
[45] Zawilska JB. An expanding world of novel psychoactive substances: opioids. Front Psychiatry 2017;8(1):110.
[46] Adamowicz P, Tokarczyk B. Simple and rapid screening procedure for 143 new psychoactive substances by liquid chromatography-tandem mass spectrometry. Drug Test Anal 2016;8(7):652–67.
[47] Vaiano F, Busardò FP, Palumbo D, Kyriakou C, et al. A novel screening method for 64 new psychoactive substances and 5 amphetamines in blood by LC-MS/MS and application to real cases. J Pharm Biomed Anal 2016;129(1):441–9.
[48] Lehmann S, Kieliba T, Beike J, Thevis M, et al. Determination of 74 new psychoactive substances in serum using automated in-line solid-phase extraction-liquid chromatography-tandem mass spectrometry. J Chromatogr B Analyt Technol Biomed Life Sci 2017;1064(1):124–38.
[49] Odoardi S, Fisichella M, Romolo FS, Strano-Rossi S. High-throughput screening for new psychoactive substances (NPS) in whole blood by DLLME extraction and UHPLC-MS/MS analysis. J Chromatogr B Analyt Technol Biomed Life Sci 2015;1000(1):57–68.
[50] Mercieca G, Odoardi S, Cassar M, Strano Rossi S. Rapid and simple procedure for the determination of cathinones, amphetamine-like stimulants and other new psychoactive substances in blood and urine by GC-MS. J Pharm Biomed Anal 2018;149(1):494–501.
[51] Soh YN, Elliott S. An investigation of the stability of emerging new psychoactive substances. Drug Test Anal 2014;6(7–8):696–704.
[52] Al-Saffar Y, Stephanson NN, Beck O. Multicomponent LC-MS/MS screening method for detection of new psychoactive drugs, legal highs, in urine-experience from the Swedish population. J Chromatogr B Analyt Technol Biomed Life Sci 2013;930(1):112–20.

[53] Concheiro M, Castaneto M, Kronstrand R, Huestis MA. Simultaneous determination of 40 novel psychoactive stimulants in urine by liquid chromatography-high resolution mass spectrometry and library matching. J Chromatogr A 2015 Jun 5;1397(1):32–42.

[54] Ambach L, Redondo AH, König S, Angerer V, et al. Detection and quantification of 56 new psychoactive substances in whole blood and urine by LC-MS/MS. Bioanalysis 2015;7(9):1119–36.

[55] Mesihää S, Ketola RA, Pelander A, Rasanen I, et al. Development of a GC-APCI-QTOFMS library for new psychoactive substances and comparison to a commercial ESI library. Anal Bioanal Chem 2017;409(8):2007–13.

[56] Boumba VA, Di Rago M, Peka M, Drummer OH, et al. The analysis of 132 novel psychoactive substances in human hair using a single step extraction by tandem LC/MS. Forensic Sci Int 2017;279(1):192–202.

[57] Montesano C, Vannutelli G, Massa M, Simeoni MC, et al. Multi-class analysis of new psychoactive substances and metabolites in hair by pressurized liquid extraction coupled to HPLC-HRMS. Drug Test Anal 2017 May;9(5):798–807.

[58] Lendoiro E, Jiménez-Morigosa C, Cruz A, Páramo M, et al. An LC-MS/MS methodological approach to the analysis of hair for amphetamine-type-stimulant (ATS) drugs, including selected synthetic cathinones and piperazines. Drug Test Anal 2017;9(1):96–105.

[59] Rocchi R, Simeoni MC, Montesano C, Vannutelli G, et al. Analysis of new psychoactive substances in oral fluids by means of microextraction by packed sorbent followed by ultra-high-performance liquid chromatography–tandem mass spectrometry. Drug Test Anal, 2017;10(5):1–7.

[60] Ares AM, Fernández P, Regenjo M, Fernández AM, et al. A fast bioanalytical method based on microextraction by packed sorbent and UPLC-MS/MS for determining new psychoactive substances in oral fluid. Talanta 2017;174(1):454–61.

[61] Stephanson NN, Signell P, Helander A, Beck O, et al. Use of LC-HRMS in full scan-XIC mode for multi-analyte urine drug testing—a step towards a 'black-box' solution? J Mass Spectrom 2017;52(8):497–506.

[62] Montesano C, Vannutelli G, Gregori A, Ripani L, et al. Broad screening and identification of novel psychoactive substances in plasma by high-performance liquid chromatography–high-resolution mass spectrometry and post-run library matching. J Anal Toxicol 2016;40(7):519–28.

Chapter 21

Review of Bath Salts on Illicit Drug Market

Michele Protti

Pharmaco-Toxicological Analysis Laboratory (PTA Lab), Department of Pharmacy and Biotechnology (FaBiT), University of Bologna, Bologna, Italy

INTRODUCTION

Cathinone analogs (CA) are currently the most frequently reported novel psychoactive substances (NPS) in the literature [1] and the largest NPS category identified by the European Monitoring Centre for Drugs and Drug Addiction (EMCDDA) in 2015 (33%), with a 60-fold increase in seizures since 2008. According to the 2017 European Drug Report, cathinone derivatives were responsible for a half of hospital emergencies involving NPS [2]. CA are sold primarily online as "plant food," "bath salts," or "research chemicals" and labeled as "not for human consumption," in order to avoid potential regulation. The huge number of ways with which these substances can be chemically modified, together with their fast availability, prevent adequate consolidation of analytical methods for their identification and make the comprehension of their effects, toxicology, bioavailability, metabolism, and health effects a particularly tough challenge [3]. Fast availability is mainly due to easiness on modifying cathinone chemical structures to bypass legislations, since routine drug screenings allowing reliable detection are relatively few, both in biological matrices and street samples [4]. Moreover, since NPS, and CA in particular, are often obtained by modifying conventional psychoactive compound molecular structures in illegal laboratories, high concern regarding their safety and the presence of contaminants in distributed products is also raised [5,6].

The development of new strategies for NPS analysis in different biological matrices is a topic of great interest in the bioanalytical field and is currently undergoing a widespread growth. Several methodologies have been proposed for characterization and quantitation of CA. Among the reported strategies, spectroscopic techniques are the most commonly employed. However, most of the already available methods do not allow the prompt identification of novel chemical entities, therefore hampering straightforward field application. In particular, the main problem of most of the approaches described until now is the rapid and continuous emergence of new analogs and derivatives, frequently capable of bypassing existing methods. In order to deal with such a high rate of synthesis of new psychoactive compounds, novel approaches are proposed in order to aim at detecting both scheduled substances together with their derivatives in a preventive way and improving the robustness of existing drug misuse laws [7,8]. For example, targeted quantitative and nontargeted screening methods by high-resolution mass spectrometry (HRMS) could allow retrospective data analysis, when coupled to updated compound spectral reference libraries. The identification and quantitation of CA in biological specimens is essential to assess their intake in clinical and forensic settings. This highlights the need to develop reliable qualitative and quantitative analytical methods and establish CA and metabolite detection windows in biological matrices [9,10]. Moreover, a challenging issue connected to CA detection is their apparent instability in biological matrices, especially in blood and urine [11–15]. Finally, a significant issue is the necessity of available reference standards to incorporate drugs and metabolites into analytical methods and to support forensic identification. Therefore, in vitro and in vivo studies are critical to determine the most important biomarkers of consumption to analyze (parent drug and/or metabolites) [16] and extend detection windows for CA in biological specimens.

PHARMACOLOGY AND TOXICOLOGY

CA are a class of molecules derived from cathinone ((S)-2-amino-1-phenyl-1-propanone), a naturally occurring psychoactive compound present in *Catha edulis*, a plant commonly known as khat. Khat has been used for centuries in East Africa and the Arabian Peninsula for its psychoactive properties [17]. Cathinone is a beta-ketone amphetamine analog that can be structurally modified to give rise to large number of synthetic compounds (Fig. 21.1). In the 1990s, methcathinone was the first reported CA, widely abuse in the United States [18]. While sporadic abuse of methcathinone is still reported, the use of other CA has become epidemic. CA found within "legal high" products mostly include methylone, ethylone, butylone, mephedrone, 4-methylethcathinone (4-MEC), and 3,4-methylenedioxypyrovalerone (MDPV), among hundreds of derivatives available via the internet [19,20]. Among CA derivatives, bupropion is the only cathinone derivative with a reported therapeutic use, with several studies highlighting its beneficial activity in major depressive disorder treatments [21–24], in smoking cessation programs [25], as well as in obesity treatments [26]. In fact, bupropion demonstrated to be useful in modulating psychiatric disorders without the incurrence of some side effects, such as weight gain [27]. On the other hand, all other CA have been strongly associated with abuse for recreational purposes. In particular, mephedrone and MDPV consumption has been massively reported [28] and an alarming number of intoxications, as well as fatalities, were assessed from case studies [29–31]. Nevertheless, it is clear that CA cause numerous severe health effects, from neurological to hepatic damage, among many others [32–34]. CA follow a path common to many drugs of abuse: a rapid onset of psychostimulant effects, including an increased sense of confidence, libido, and energy [35] followed by unpleasant symptoms such as psychosis, anxiety and paranoia [36], and leading, in some cases, to death [37].

Although there are not many studies available, withdrawal symptoms seem to be similar to those observed with cocaine or amphetamines: depression, anxiety, and craving are reported [38–40]. CA were shown to exert their effects by interacting with dopamine, noradrenaline, and serotonin transporters, resulting in an increased brain concentration of such neurotransmitters. However, different affinities toward transporters are observed between cathinone derivatives, which could act by stimulating the release of neurotransmitters, or as uptake blockers, like cocaine, which ultimately leads to different long-term effects. Effects sought by abusers include euphoria, increased empathy and libido, while the most common observed CA side effects result from their cardiovascular and neurological toxicity: tachycardia, arrhythmia, hypertension, hyperthermia, agitation, confusion, psychosis, and coma. Moreover, narrow safety window for CA is reported, thus strongly increasing their harmfulness [41]. Several fatal case reports were attributed to these compounds, often being associated with other substances and resulting in multidrug intoxication [42–44]. It is crucial to clarify CA mechanisms of action in order to deepen the knowledge about their short- and long-term effects, their pharmacokinetic (PK) properties, and the correlation between their biological fluid levels and their activity.

BIOANALYTICAL APPROACHES

Approaches aimed at qualitative as well as quantitative analysis of NPS provide information concerning consumption patterns, metabolism pathways, and correlations between dose and CA and metabolites levels in biological matrices [45]. Given the complexity of samples subjected to CA analysis, separation techniques such as liquid chromatography (LC) can be a valuable tool to identify and accurately quantify these compounds, especially when hyphenated to mass spectrometric (MS) analysis. These strategies can be further optimized and enhanced when coupled to suitable sample preparation techniques developed ad hoc for a wide range of biological samples, including blood, urine, hair, and oral fluid. Despite the growing emergence of new compounds that make the simultaneous CA determination in complex samples a challenging goal for bioanalytical scientists, several studies have been published to date to propose analytical methodologies to address this challenging phenomenon.

Urine is the biological sample of choice for screening analysis, as it can be obtained in a noninvasive way and in large volume. Moreover, drugs of abuse and/or their metabolites can be generally found in higher concentration in urine than in any other biological fluid [46]. However, compound levels in hematic samples (whole blood, plasma, and serum) generally show the best correlation with drug effects [47]. In this section, published methods for CA analysis in biological samples will be summarized, focusing on hematic matrices, urine, hair, oral fluid, and miniaturized alternative matrices. Immunoassays for qualitative screening will not be taken into account, except when expressly compared with quantitative instrumental techniques. The scientific literature offers several toxicological case reports involving the on-purpose development of analytical methodologies for a limited number of compounds and mostly for the application to classical matrices. In this section, only the methodologies involving innovations from the bioanalytical point of view

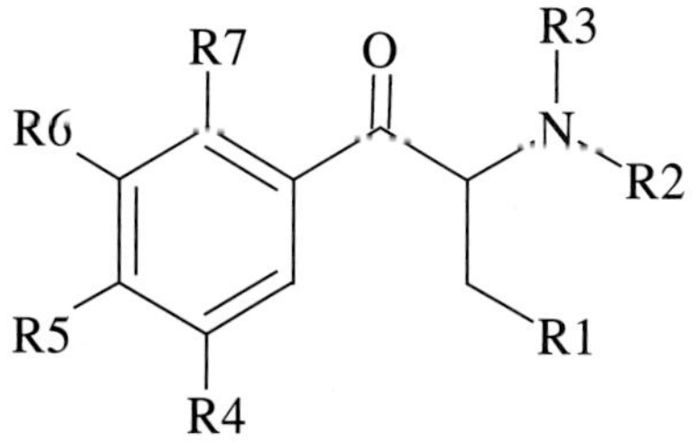

Compound	Other names	R1	R2	R3	R4	R5	R6	R7
Cathinone		H	H	H	H	H	H	H
4-Methylcathinone		H	H	H	H	CH₃	H	H
Dimethylcathinone	Metamfepramone, DMC	H	CH₃	CH₃	H	H	H	H
N-ethyl-N-dimethylcathinone		H	CH₂CH₃	CH₃	H	H	H	H
4-Methylbenzylcathinone	Benzedrone, 4-MBC	H	CH₂Ph	H	H	CH₃	H	H
Methcathinone	Ephedrone	H	CH₃	H	H	H	H	H
2-Methylmethcathinone	2-MMC	H	CH₃	H	H	H	H	CH₃
3-Methylmethcathinone	3-MMC	H	CH₃	H	CH₃	H	H	H
4-Methylmethcathinone	Mephedrone, 4-MMC	H	CH₃	H	H	CH₃	H	H
3,4-Dimethylmethcathinone	3,4-DMMC	H	CH₃	H	CH₃	CH₃	H	H
Ethyl-methcathinone	Pentedrone	CH₂CH₃	CH₃	H	H	H	H	H
2-Ethyl-methcathione	2-EMC	H	CH₃	H	H	H	H	CH₂CH₃
4-Ethylmethcathinone	4-EMC	H	CH₃	H	H	CH₂CH₃	H	H
2-Fluoromethcathinone	2-FMC	H	CH₃	H	H	H	H	F
3-Fluoromethcathinone	3-FMC	H	CH₃	H	F	H	H	H
4-Fluoromethcathinone	Flephedrone, 4-FMC	H	CH₃	H	H	F	H	H
4-Chloromethcathinone	Clephedrone, 4-CMC	H	CH₃	H	H	Cl	H	H
2,3-Dimethyl-methcathinone	2,3-DMMC	H	CH₃	H	H	H	CH₃	CH₃
2,4-Dimethyl-methcathinone	2,4-DMMC	H	CH₃	H	H	CH₃	H	CH₃
4-Methoxymethcathinone	Methedrone	H	CH₃	H	H	OCH₃	H	H
Brephedrone	4-Bromomethcathinone, 4-BMC	H	CH₃	H	H	Br	H	H

FIGURE 21.1 Chemical structures of commonly encountered CA.

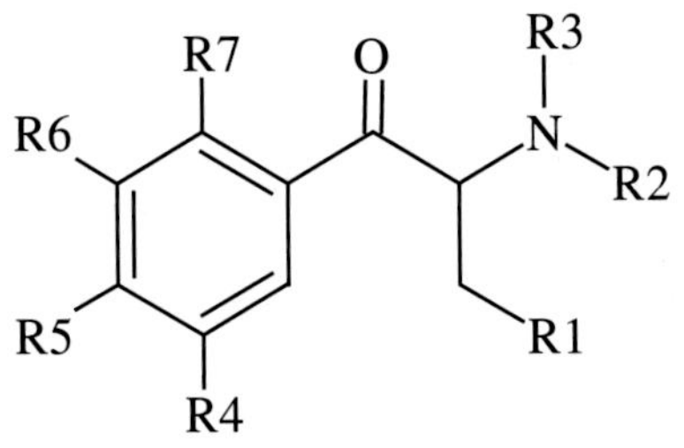

Compound	Other names	R1	R2	R3	R4	R5	R6	R7
Bupropion		H	C(CH$_3$)$_3$	H	Cl	H	H	H
N,N-diethylcathinone	Amfepramone, diethylpropion	H	CH$_2$CH$_3$	CH$_2$CH$_3$	H	H	H	H
Ethylcathinone	Ethcathinone, ethylpropion	H	CH$_2$CH$_3$	H	H	H	H	H
2-Methylethcathinone	2-MEC	H	CH$_2$CH$_3$	H	H	H	H	CH$_3$
3-Methylethcathinone	3-MEC	H	CH$_2$CH$_3$	H	CH$_3$	H	H	H
4-Methylethcathinone	4-MEC	H	CH$_2$CH$_3$	H	H	CH$_3$	H	H
2,3-Dimethylethcathinone	2,3-DMEC	H	CH$_2$CH$_3$	H	H	CH$_3$	H	CH$_3$
3,4-Dimethylethcathinone	3,4-DMEC	H	CH$_2$CH$_3$	H	CH$_3$	CH$_3$	H	H
2-Ethylethcathinone	2-EEC	H	CH$_2$CH$_3$	H	H	H	H	CH$_2$CH$_3$
2-Fluoroethcathinone	2-FEC	H	CH$_2$CH$_3$	H	H	H	H	F
3-Fluoroethcathinone	3-FEC	H	CH$_2$CH$_3$	H	F	H	H	H
4-Fluoroethcathinone	4-FEC	H	CH$_2$CH$_3$	H	H	F	H	H
α-Methylamino-butyrophenone	Buphedrone, MABP	CH$_3$	CH$_3$	H	H	H	H	H
3-Methyl-α-methylamino-butyrophenone	3-Methylbuphedrone	CH$_3$	CH$_3$	H	CH$_3$	H	H	H
4-Methyl-α-methylamino-butyrophenone	4-Methylbuphedrone, 4-MeMABP	CH$_3$	CH$_3$	H	H	CH$_3$	H	H
N-ethylbuphedrone	NEB	CH$_3$	CH$_2$CH$_3$	H	H	H	H	H
4-Methyl-N-methylbuphedrone		CH$_3$	CH$_3$	CH$_3$	H	CH$_3$	H	H
4-Methyl-α-ethylaminobutiophenone		CH$_3$	CH$_2$CH$_3$	H	H	CH$_3$	H	H
Dimethylone	bk-MDDM	H	H	H	H	OCH$_2$O		H
BMDB		CH$_3$	CH$_2$Ph	H	H	OCH$_2$O		H
BMBP	3,4-MDBC	H	CH$_2$Ph	H	H	OCH$_2$O		H

FIGURE 21.1 (Continued)

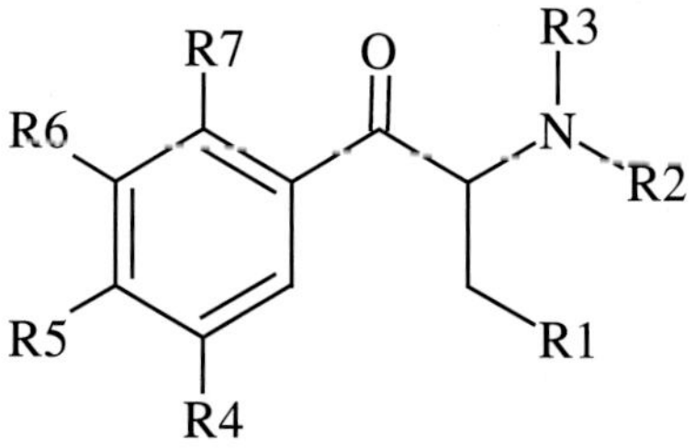

Compound	Other names	R1	R2	R3	R4	R5	R6	R7
Butylone	bk-MBDB	CH₃	CH₃	H	H	OCH₂O		H
Methylbutylone	Dibutylone, bk-DMBDB	CH₃	CH₃	CH₃	H	OCH₂O		H
Ethylone	bk-MDEA	H	CH₂CH₃	H	H	OCH₂O		H
Eutylone		CH₃	CH₂CH₃	H	H	OCH₂O		H
Methylone	MDMC, bk-MDMA	H	CH₃	H	H	OCH₂O		H
Pentylone	bk-MBDP	CH₂CH₃	CH₃	H	H	OCH₂O		H
3-Methylmethcathynone	Metaphedrone, R-MMC	H	CH₂CH₃	H	CH₃	OCH₂O		H
MDPPP		H	Pyrrolidine		H	OCH₂O		H
MDPBP		CH₃	Pyrrolidine		H	OCH₂O		H
MDPV	MDPK	CH₂CH₃	Pyrrolidine		H	OCH₂O		H
3,4-MDPHP		CH₂CH₂CH₃	Pyrrolidine		H	OCH₂O		H
PPP		H	Pyrrolidine		H	H	H	H
2-MePPP		H	Pyrrolidine		H	H	H	CH₃
3-MePPP		H	Pyrrolidine		H	H	CH₃	H
4-MePPP		H	Pyrrolidine		H	CH₃	H	H
3-Fluoro-α-pyrrolidinopropiophenone		H	Pyrrolidine		F	H	H	H
4-Fluoro-α-pyrrolidinopropiophenone		H	Pyrrolidine		H	F	H	H
MPBP		CH₃	Pyrrolidine		H	CH₃	H	H
2-MePBP		CH₃	Pyrrolidine		H	H	H	CH₃
3-MePBP		CH₃	Pyrrolidine		H	H	CH₃	H
MOPP		H	Pyrrolidine		H	OCH₃	H	H

FIGURE 21.1 (Continued)

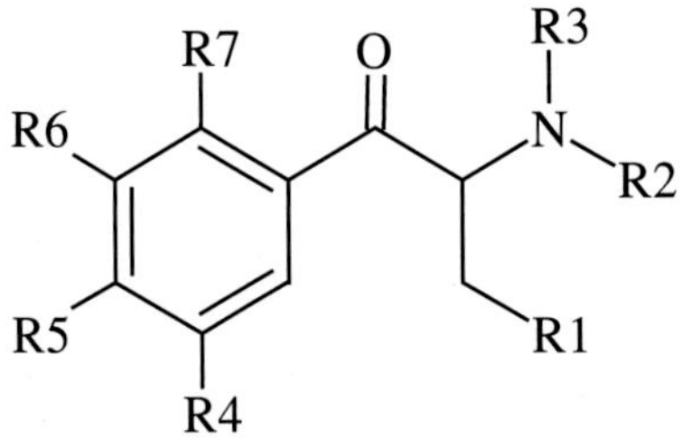

Compound	Other names	R1	R2	R3	R4	R5	R6	R7
4-Methoxy-α-pyrrolidinopentiophenone		CH₂CH₃	Pyrrolidine		H	OCH₃	H	H
3,4-Dimethoxy-α-pyrrolidinopentiophenone		CH₂CH₃	Pyrrolidine		OCH₃	OCH₃	H	H
Pyrovalerone		CH₂CH₃	Pyrrolidine		H	CH₃	H	H
PVP		CH₂CH₃	Pyrrolidine		H	H	H	H
MPHP		CH₃CH₂CH₃	Pyrrolidine		H	H	H	H
α-Pyrrolidinohexiophenone	PV8	C₄H₉	Pyrrolidine		H	H	H	H
α-Pyrrolidinoheptiophenone	PV9	C₅H₁₁	Pyrrolidine		H	H	H	H

FIGURE 21.1 (Continued)

will be named and discussed. Pitfalls and interpretation issues will also be addressed. Key information of selected analytical papers is summarized in Table 21.1.

Bioanalysis Using Hematic Matrices

Traditionally, urine is the sample of choice for the screening and identification of unknown drugs due to high concentrations in urine. However, improvements in sample preparation, separative techniques, and detection have made whole blood and its derivatives (plasma, serum) useful as a biological matrix for both identification and quantitation. As blood physiological parameters can vary within narrow limits, blood is a relatively homogeneous matrix. A great advantage of hematic matrices is that drugs can be detected just after intake prior to metabolism and/or filtration, thus representing actual intoxication states. An overview of the most recent and significant methodologies for CA analysis in hematic matrices is reported in Table 21.1.

Over 30 NPS including CA were analyzed in serum by means of a validated method based on LC coupled to tandem mass spectrometry (LC–MS/MS), after sample processing by solid-phase extraction (SPE). Limits of quantitation (LOQ) were in the range of 1–10 ng/mL for each compound with limits of detection (LOD) near 10 pg/mL. In order to evaluate method applicability in a forensic toxicological setting, the developed method was exploited to investigate postmortem specimens [48].

A gas chromatography–mass spectrometry (GC–MS) method for mephedrone quantitation in human plasma and urine was applied to evaluate its pharmacology and its relative bioavailability and PK. After deproteinization, plasma and urine were subjected to liquid–liquid extraction (LLE) and derivatization before GC–MS analysis. Volunteers were administered oral mephedrone and PK parameters were evaluated in terms of peak plasma concentration, elimination half-life, and intersubject variability [49].

Stability of four NPS including mephedrone and MDPV was assessed in three different biological matrices under different storage conditions: whole blood, serum and urine, each stored at −20, 4, and 22°C for a period of 14 days in the dark in a sealed glass container. Analysis by LC–MS/MS was performed on spiked blank samples on defined time

TABLE 21.1 Bioanalytical Methods for CA Determination

Analyzed CA	Biological Sample	Sample Preparation	Analytical Technique	Application to Real Samples	Ref.
9 CA	Serum	SPE	HPLC–MS/MS	Postmortem	[48]
1 CA	Plasma, urine	PP + LLE + derivatization	GC–MS	Controlled administration study	[49]
2 CA	Whole blood, serum, urine	SPE	HPLC–MS/MS	None	[13]
22 CA	Whole blood, urine	PP + SPE	HPLC–QTOF–MS (HRMS)	Subjects from routine toxicological screening	[50]
2 CA	Whole blood, urine	SPE	UHPLC–MS/MS	Postmortem	[51]
28 CA + metabolites	Urine	SPE	HPLC–MS/MS (HRMS)	Self-reported users	[12]
16 CA + metabolites	Urine	LLE + derivatization	GC–MS	Subjects from routine toxicological screening	[16]
15 CA	Urine	Dilution + filtration	UHPLC–MS, HPSFC–MS	None	[52]
3 CA	Urine	Hydrolysis + dilution	UHPLC–MS/MS	Subjects from routine forensic screening	[53]
14 CA	Urine	Hydrolysis + dilution	MS/MS, LC–MS/MS	Subjects from routine toxicological screening	[54]
3 CA + metabolites	Urine	SPME	DART-MS	None	[55]
12 CA	Urine	SPE	MEKC–MS/MS	None	[56]
132 NPS including CA	Hair	Acidic extractive incubation	HPLC–MS/MS;	Postmortem	[57]
6 CA	Hair	Acidic extractive incubation + SPE	HPLC–MS/MS	Nonspecified real samples	[58]
2 CA + metabolite	Hair	PLE + SPE	CE-DAD	Nonspecified real samples	[59]
13 CA	Oral fluid	PP	HPLC–MS/MS	Subjects from routine toxicological screening	[60]
3 CA	Oral fluid	LLE + derivatization	GC–MS	Nonspecified real samples	[61]
4 CA	Oral fluid	SPE	HPLC–MS/MS, immunoassay	Subjects from routine toxicological screening	[62]
10 CA	Oral fluid	SPE	HPLC–MS/MS	Nonspecified real samples	[63]
10 CA	Oral fluid	SPE	UHPLC–MS/MS	None	[64]
1 CA	DBS	SPE	HPLC–MS/MS	Controlled administration study	[65]
1 CA	DPS	SPE	HPLC–MS/MS	None	[66]
6 CA	VAMS-plasma, VAMS-urine, and VAMS-oral fluid	Dried microsampling + solvent extraction	HPLC–MS/MS	Self-reported users	[67]

points and the results were compared. It became clear that mephedrone was not stable in any of the considered conditions and care must be taken to ensure minimal degradation [13].

A method based on SPE and LC-quadrupole time of flight (LC-QTOF) was developed to identify and quantify 22 CA in urine and blood and leading to LOQ values in the range 0.25–0.50 ng/mL for both matrices. The method was used to analyze CA in 20 authentic samples from routine toxicological analysis, also highlighting data regarding CA stability and the subsequent interpretation [50].

Ultra-high-performance liquid chromatography–tandem mass spectrometry (UHPLC–MS/MS) was used to quantitatively measure 12 illicit drugs including CA in whole blood and urine, using SPE. The research work also reported application to postmortem real samples as a proof of concept for method applicability [51].

Bioanalysis in Urine

Urine can complement blood samples, as it represents a filtrate of blood collecting over some hours, thus containing compounds at concentrations usually higher than those found in hematic matrices. This often means that urine analysis can offer the most sensitive means of detecting drug use. The window of detection for drugs of abuse in urine may be very variable and may be affected by many factors. For example, highly lipophilic drugs are often detectable in urine for weeks, in contrast to other drugs of abuse, some of which are detectable only for a few days. Obviously, urine collection is noninvasive and large sample volumes are available, and its analysis can be relatively simple when compared to blood due to the absence of proteins and cellular material, which can give rise to possible interference if appropriate sample preparation is not applied. Moreover, the presence of stable metabolites in addition to parent drug provides further evidence of drug intake. On the other hand, drug levels in urine do not correlate well with levels in other body fluids, as drug excretion continues after physiologic effect of the drug ceases, resulting in lack of correlation of drug levels with intoxication states. In addition, urine may be unstable if not properly handled and stored and urine specimens could be easily substituted, diluted, or adulterated.

Key information of selected analytical methods published in the literature referred to CA analysis in urine is summarized in Table 21.1.

The simultaneous urinary quantitation by LC-QTOF-MS of 40 NPS, including 28 CA and 4 metabolites was presented, using full scan and data-dependent mode, and successfully applied to 62 authentic urine specimens. By using a sample volume of 100 μL urine, LOQ values of 2.5–5 ng/mL were achieved. This study demonstrated how data-dependent MS acquisition allows including additional compound with minimal method validation steps and how several NPS stimulants, including CA, can be mainly detected as parent compounds in urine [12].

Between 2011 and 2013, a simple GC–MS methodology coupled to LLE and derivatization has been applied to authentic urine specimens coming from more than 30,000 subjects for routine forensic testing. A total of 16 CA found in samples coming from such large population gave the opportunity to investigate the prevalence of CA use, observe metabolic profiles during various stages of urinary excretion, evaluate the extent of CA metabolism, and select parent drugs and/or metabolites as biomarkers of CA abuse [16].

A recent research work was aimed at comparing the performances of ultra-high-performance liquid chromatography (UHPLC) and high-performance supercritical fluid chromatography (HPSFC) both coupled to MS detection, for identification and quantitation of 15 NPS belonging to CA and phenethylamines in urine. A simple dilute-filter-and-shoot protocol was developed, while HPSFC provided better separation of isomers when compared to UHPLC. No method application to real samples was provided in this study [52].

Threshold accurate calibration was exploited within an UHPLC–MS/MS study aimed at the analysis of 64 analytes in urine, including pharmaceutical, classical illicit agents, and NPS [53].

A high-throughput MS/MS system without chromatographic separation was used for the urinary screening of 14 CA and 14 synthetic cannabinoids after minimal sample preparation including urine hydrolysis and dilution. Results from 18,000 urine specimens were compared with a routine methodology based on LLE followed by LC–MS/MS analysis [54].

An interesting methodology based on direct analysis in real-time mass spectrometry (DART-MS) was applied to the characterization and semiquantitative analysis of CA and their metabolites in urine, without extraction steps, taking advantage of the high instrumental mass accuracy. In addition, the implementation within this work of solid-phase microextraction (SPME) led to a further increase in the detectability of both drugs and metabolites [55].

As regards electro-driven separation techniques, a micellar electrokinetic chromatography method coupled to tandem mass spectrometry (MEKC–MS/MS) was proposed for the selective separation, identification, and determination of

12 CA. As proof of concept, the developed MEKC−MS/MS method was applied to spiked urine after SPE clean-up procedure [56].

CA Bioanalysis in Hair

It is well known that hair, and in general keratin matrices, stably incorporate compounds by diffusion from blood capillaries to the growing hair follicle. When compared to blood, keratin matrices offer a wide detection window up to years, allowing a segmental analysis, and thus a retrospective detection of drug intake [68]. Moreover, sample collection is noninvasive and easy to perform, it does not require trained staff and sample adulteration is harder to obtain. Sample storage and transport are feasible and can be performed at room temperature. Compound detection profiles in keratin matrices are similar to blood, with higher concentration of parent compounds compared to metabolites. Hair has become a suitable matrix for both clinical and forensic toxicology, when information over longer periods of time is needed or when sampling cannot be performed immediately after intake (postmortem analysis, drug-facilitated assaults, or when drug habits need to be verified over a wide time frame). Quantitative extraction from hair is a key and complex aspect of the entire bioanalytical process, while one of the main challenges in hair testing is the removal of external contamination in order to effectively assess systemic incorporation of xenobiotics [69].

A promising perspective is offered by the CA determination in hair samples, as parent drugs are the most abundant form and long-term retrospective analysis and consumption patterns could be investigated and detected. Nevertheless, only a few studies have been published to date for CA analysis in keratin matrix and their application to real samples. Boumba et al. [57] proposed an LC−MS/MS strategy for the analysis of 132 NPS (mainly CA together with synthetic cannabinoid receptor agonists, phenethylamines and tryptamines) in hair samples, providing good LOD values ranging from 0.001 to 0.1 ng/mg. CA together with classic amphetamines, piperazines, central nervous system (CNS) prescription drugs, and some metabolites were investigated in 16 hair samples (30 mg hair/sample) [58]. A method based on capillary electrophoresis (CE) has been developed and validated for chiral determination of CA and metabolites in human hair samples: analyte extraction was achieved by using a pressurized liquid extraction (PLE) followed by in-line SPE, while enantioselective separation was achieved, thanks to beta-cyclodextrins (β-CD) used as the chiral selector [59].

Bioanalysis in Oral Fluid

Oral fluid analysis is confirming itself as a valuable tool for toxicological analysis, as it is associated with several advantages over blood [70]. For example, oral fluid concentrations usually reflect the free active compound in blood, theoretically representing the one occurring at the active sites. Nevertheless, oral fluid is collected in a more feasible and less invasive way than blood. Most xenobiotics can be found in oral fluid, based on passive diffusion, in particular nonionizable compounds within the salivary pH range [71]. Disadvantages of oral fluid analysis within toxicological and forensic investigations are mainly represented by low sample volumes, difficult sample handling, and short half-life of some drugs. Moreover, oral fluid collection, either by expectoration, drooling or by using commercial devices such as swabs or pads, induces stimuli affecting salivary pH and production rate, thus also drug disposition and concentration. As a general rule, oral fluid concentrations are lower for acidic drugs than for blood as the equilibrium favors blood, while for basic drugs higher oral fluid concentrations can be observed. Moreover, some compounds showed to be unstable in oral fluid especially without properly controlled storage conditions.

Significant validated analytical methods for CA quantification in oral fluid are presented in Table 21.1.

An UHPLC−MS/MS method for the detection in oral fluid of 32 synthetic stimulant and hallucinogenic drugs, mainly CA commonly sold as bath salts, was developed leading to lower LOQ values of 2.5 ng/mL for all analytes. Sample preparation was carried out by simple protein precipitation with organic solvent and the method was applied to 12 samples from long-term storage which were previously submitted for drug testing [60].

Similar sensitivity was achieved by a simple analytical method based on GC−MS developed for the assessment of 11 amphetamine-like molecules including CA in oral fluid. Samples were collected using a cotton pad collection device, stabilized with citric acid then derivatized in-matrix before extraction and GC−MS analysis. The method has then been applied as proof of concept to two authentic samples received from MDMA and mephedrone users [61].

A study also described the development and validation of an LC combined with tandem mass spectrometry (LC−MS/MS) method for the determination of selected CA and piperazines in oral fluid. Samples were collected by means of commercial oral fluid collection devices and extracted by SPE. The method was applied to 10 real specimens collected from suspected drivers. Moreover, cross reactivity produced by the studied CA with respect to routine immunoassay screening devices was investigated. The authors observe that due to structural similarity, all CA cross-reacted with amphetamine and/or methamphetamine at high concentrations [62].

A method aimed at analyzing 10 CA in oral fluid was developed by implementing a combined approach based on SPE and ultra-high-performance liquid chromatography combined with tandem mass spectrometry (UHPLC−MS/MS). This method was applied for analysis of specimens collected from human subjects. The authors identified five specimens tested positive for CA [63].

The stability of CA in oral fluid was also assessed. In one study, the authors used a SPE and UHPLC−MS/MS method for the determination of 10 CA to evaluate their stability in oral fluid samples collected with several collection devices as well as in neat oral fluid stored under different conditions. Losses up to −100% were observed in neat and preserved samples stored at room temperature. At 4°C, losses up to −88.2% occurred, while all samples were stable if stored at −20°C and after three freeze-thaw cycles [64].

Bioanalysis in Miniaturized Dried Samples

Miniaturized dried sampling approaches are currently on the rise in the bioanalytical field. For example, the use of dried blood spots (DBS) offers several advantages with respect to classical blood specimen, such as minimal invasiveness, streamlined postcollection processing, low biohazard, reduced analysis time and cost, simplified sample storage, and shipment, usually at room temperature [72,73]. These advantages make DBS sampling a convenient tool for blood collection, especially for vulnerable populations and for on-site sampling [74−76]. One limitation of DBS testing is the small sample volume (5−20 μL), implying the use of highly sensitive instrumental techniques. The most relevant and recent methodologies involving the use of miniaturized dried samples for CA analysis are listed in Table 21.1.

DBS sampling has some known drawback, that is, possible volumetric biases due to blood hematocrit (HCT) levels, which can affect quantitative results. Being HCT directly proportional to blood density it can lead to volumetric uncertainty when a subsample is punched out of the spot, therefore correction factors must be determined for populations characterized by different HCT levels [72]. In order to deal with HCT volumetric issues, several approaches have been proposed. One way to potentially address these challenges is represented by the volumetric generation of fixed-volume whole blood samples. In one study, the authors reported development and use of a microfluidic-based sampling device for the creation of volumetric DBS, to be extracted by means of an automated flow-through elution system coupled with online SPE and liquid chromatography-electrospray ionization-tandem mass spectrometry (LC-ESI-MS/MS) for the quantitative determination of mephedrone along with selected stimulants [65]. The research group also proposed a methodological advancement represented by the use of an on-card red blood cell filtration device able to produce plasma in situ from whole blood without the need for centrifugation. The dried plasma spot (DPS) samples obtained were subjected to a similar automated extraction and analysis workflow [66].

Moreover, a feasible dried miniaturized biosampling strategy has been investigated in order to sample accurate and reproducible volumes regardless of the density. Such approach is based on volumetric absorptive microsampling (VAMS) and has the potential to overcome many issues associated with DBS testing, as well as with conventional fluid sampling [77−79]. The only published work to date about the use of VAMS technology for CA analysis deals with the determination of representative CA in dried urine, plasma, and oral fluid microsamples, leading to promising results. In fact, sampling parameters have been thoroughly designed and optimized, with regard to matrix volume, sampling time, drying time, and storage conditions such as temperature, humidity, and light exposure. This approach also demonstrated to produce reliable results also when applied to biological specimens other than whole blood [67].

CONCLUSIONS

In recent years, CA have been the most frequently identified group of NPS. Legislative efforts carried out by many countries aims to eliminate them from legal drug markets by their inclusion to list of prohibited substances. However, illegal laboratories producing NPS can easily circumvent laws by introducing new derivatives which do not appear in such lists and are not easily detected by routine toxicological and forensic analytical strategies. From the structures of the main CA identified in the last decade, it is evident how structural modification of the cathinone molecule is theoretically limitless. Moreover, few data are available on the pharmacology and toxicology of these substances and for each newly identified compound, novel, and updated bioanalytical methodologies are needed to detect CA in biological matrices. Although several analytical methods can be found in the scientific literature on the subject, more targeted and nontargeted methods exploiting HRMS would be advisable, as they allow for retrospective data gathering.

A challenging aspect connected to CA is their low stability and thermal degradation, leading to the need for accurate precautions during the analytical process and to the development of innovative strategies for biological sample collection and pretreatment. Moreover, interpretation of CA levels in biological matrices is hindered by the lack of PK and metabolism studies, as information often is derived only from case reports. In those cases, other substances are typically found in combination with CA, further complicating interpretation. Finally, identification of proper CA abuse biomarkers, through in vitro and in vivo studies and the availability of commercial reference standards must be addressed in order to effectively investigate CA toxicity and extend detection windows. In such a way, bioanalytical laboratories could respond in a rapid and effective way to CA emergence, adapt to changing trends on the drug market, and improve analytical strategies to enhance the detection of these compounds as a valuable tool for medicinal chemists, toxicologists, clinicians, and forensic scientists.

REFERENCES

[1] Smith CD, Williams M, Shaikh M. Novel psychoactive substances: a novel clinical challenge. BMJ Case Rep 2013;2013(1):1–3.
[2] European Monitoring Centre for Drugs and Drug Addiction, European Drug Report 2017: Trends and Developments, Publications Office of the European Union, Luxembourg; 2017.
[3] McNabb CB, Russell BR, Caprioli D, Nutt D, et al. Single chemical entity legal highs: assessing the risk for long term harm. Curr Drug Abuse Rev 2012;5(4):304–19.
[4] Gautam L, Shanmuganathan A, Cole MD. Forensic analysis of cathinones. Forensic Sci Rev 2013;25(1–2):47–64.
[5] Martinotti G, Lupi M, Carlucci L, Cinosi E, et al. Novel psychoactive substances: use and knowledge among adolescents and young adults in urban and rural areas. Hum Psychopharmacol 2015;30(4):295–301.
[6] Guirguis A, Corkery JM, Stair JL, Kirton SB, et al. Intended and unintended use of cathinone mixtures. Hum Psychopharmacol. 2017;32(3): e2598.
[7] Leffler AM, Smith PB, de Armas A, Dorman FL. The analytical investigation of synthetic street drugs containing cathinone analogs. Forensic Sci Int 2014;234(1):50–6.
[8] Smith JP, Sutcliffe OB, Banks CE. An overview of recent developments in the analytical detection of new psychoactive substances (NPSs). Analyst 2015;140(15):4932–48.
[9] Helfer AG, Turcant A, Boels D, Ferec S, et al. Elucidation of the metabolites of the novel psychoactive substance 4-methyl-N-ethyl-cathinone (4-MEC) in human urine and pooled liver microsomes by GC–MS and LC–HR–MS/MS techniques and of its detectability by GC–MS or LC–MS(n) standard screening approaches. Drug Test Anal 2015;7(5):368–75.
[10] Al-Saffar Y, Stephanson NN, Beck O. Multicomponent LC–MS/MS screening method for detection of new psychoactive drugs, legal highs, in urine-experience from the Swedish population. J Chromatogr B Analyt Technol Biomed Life Sci 2013;930(1):112–20.
[11] Busardò FP, Kyriakou C, Tittarelli R, Mannocchi G, et al. Assessment of the stability of mephedrone in ante-mortem and post-mortem blood specimens. Forensic Sci Int 2015;256(1):28–37.
[12] Concheiro M, Anizan S, Ellefsen K, Huestis MA. Simultaneous quantification of 28 synthetic cathinones and metabolites in urine by liquid chromatography-high resolution mass spectrometry. Anal Bioanal Chem 2013;405(29):9437–48.
[13] Johnson RD, Botch-Jones SR. The stability of four designer drugs: MDPV, mephedrone, BZP and TFMPP in three biological matrices under various storage conditions. J Anal Toxicol 2013;37(2):51–5.
[14] Soh YN, Elliott S. An investigation of the stability of emerging new psychoactive substances. Drug Test Anal 2014;6(7–8):696–704.
[15] Kerrigan S, Savage M, Cavazos C, Bella P. Thermal degradation of synthetic cathinones: implications for forensic toxicology. J Anal Toxicol 2016;40(1):1–11.
[16] Uralets V, Rana S, Morgan S, Ross W. Testing for designer stimulants: metabolic profiles of 16 synthetic cathinones excreted free in human urine. J Anal Toxicol 2014;38(5):233–41.
[17] Majchrzak M, Celiński R, Kuś P, Kowalska T, et al. The newest cathinone derivatives as designer drugs: an analytical and toxicological review. Forensic Toxicol 2018;36(1):33–50.
[18] Belhadj-Tahar H, Sadeg N. Methcathinone: a new postindustrial drug. Forensic Sci Int 2005;153(1):99–101.
[19] Brandt SD, Freeman S, Sumnall HR, Measham F, et al. Analysis of NRG 'legal highs' in the UK: identification and formation of novel cathinones. Drug Test Anal 2011;3(9):569–75.
[20] Zawilska J, Wojcieszak J. Pyrrolidinophenones: a new wave of designer cathinones. Forensic Toxicol 2017;35(2):201–16.
[21] Soczynska JK, Ravindran LN, Styra R, McIntyre RS, et al. The effect of bupropion XL and escitalopram on memory and functional outcomes in adults with major depressive disorder: results from a randomized controlled trial. Psychiatry Res 2014;220(1–2):245–50.
[22] Cooper JA, Tucker VL, Papakostas GI. Resolution of sleepiness and fatigue: a comparison of bupropion and selective serotonin reuptake inhibitors in subjects with major depressive disorder achieving remission at doses approved in the European Union. J Psychopharmacol 2014;28(2):118–24.
[23] Seo HJ, Lee BC, Seok JH, Jeon HJ, et al. An open-label, rater-blinded, 8-week trial of bupropion hydrochloride extended-release in patients with major depressive disorder with atypical features. Pharmacopsychiatry 2013;46(6):221–6.

[24] Maneeton N, Maneeton B, Eurviriyanukul K, Srisurapanont M. Efficacy, tolerability, and acceptability of bupropion for major depressive disorder: a meta-analysis of randomized-controlled trials comparison with venlafaxine. Drug Des Dev Ther 2013;7(1):1053–62.

[25] Oliveira P, Ribeiro J, Donato H, Madeira N. Smoking and antidepressants pharmacokinetics: a systematic review. Ann. Gen. Psychiatry 2017;16(1):17.

[26] Gadde KM, Pritham Raj Y. Pharmacotherapy of obesity: clinical trials to clinical practice. Curr Diab Rep 2017;17(5):34.

[27] Carroll FI, Blough BE, Mascarella SW, Navarro HA, et al. Bupropion and bupropion analogs as treatments for CNS disorders. Adv Pharmacol 2014;69(1):177–216.

[28] White CM. Mephedrone and 3,4-methylenedioxypyrovalerone (MDPV): synthetic cathinones with serious health implications. J Clin Pharmacol 2016;56(11):1319–25.

[29] Busardò FP, Kyriakou C, Napoletano S, Marinelli E, et al. Mephedrone related fatalities: a review. Eur Rev Med Pharmacol Sci 2015;19(19):3777–90.

[30] Karila L, Billieux J, Benyamina A, Lançon C, et al. The effects and risks associated to mephedrone and methylone in humans: a review of the preliminary evidences. Brain Res Bull 2016;126(Pt 1):61–7.

[31] Bäckberg M, Lindeman E, Beck O, Helander A. Characteristics of analytically confirmed 3-MMC-related intoxications from the Swedish STRIDA project. Clin. Toxicol 2015;53(1):46–53.

[32] Grecco GG, Kisor DF, Magura JS, Sprague JE. Impact of common clandestine structural modifications on synthetic cathinone "bath salt" pharmacokinetics. Toxicol Appl Pharmacol 2017;328(1):18–24.

[33] Gannon BM, Rice KC, Collins GT. Reinforcing effects of abused 'bath salts' constituents 3,4-methylenedioxypyrovalerone and α-pyrrolidinopentiophenone and their enantiomers. Behav Pharmacol 2017;28(7):578–81.

[34] Abbott R, Smith DE. The new designer drug wave: a clinical, toxicological, and legal analysis. J Psychoactive Drugs 2015;47(5):368–71.

[35] Palamar JJ, Acosta P, Calderón FF, Sherman S, et al. Assessing self-reported use of new psychoactive substances: the impact of gate questions. Am J Drug Alcohol Abuse 2017;43(5):609–17.

[36] Paillet-Loilier M, Cesbron A, Le Boisselier R, Bourgine J, et al. Emerging drugs of abuse: current perspectives on substituted cathinones. Subst Abuse Rehabil 2014;5(1):37–52.

[37] Maskell PD, De Paoli G, Seneviratne C, Pounder DJ. Mephedrone (4-methylmethcathinone)-related deaths. J Anal Toxicol 2011;35(3):188–91.

[38] Scherbaum N, Schifano F, Bonnet U. New psychoactive substances (NPS)—a challenge for the addiction treatment services. Pharmacopsychiatry 2017;50(3):116–22.

[39] Zawilska JB. Mephedrone and other cathinones. Curr Opin Psychiatry 2014;27(4):256–62.

[40] Prosser JM, Nelson LS. The toxicology of bath salts: a review of synthetic cathinones. J Med Toxicol 2012;8(1):33–42.

[41] Dargan PI, Sedefov R, Gallegos A, Wood DM. The pharmacology and toxicology of the synthetic cathinone mephedrone (4-methylmethcathinone). Drug Test Anal 2011;3(7–8):454–63.

[42] Wikström M, Thelander G, Nyström I, Kronstrand R. Two fatal intoxications with the new designer drug methedrone (4-methoxymethcathinone). J Anal Toxicol. 2010;34(9):594–8.

[43] Lusthof KJ, Oosting R, Maes A, Verschraagen M, et al. A case of extreme agitation and death after the use of mephedrone in the Netherlands. Forensic Sci Int 2011;206(1–3):e93–5.

[44] Froberg BA, Levine M, Beuhler MC, Judge BS, et al. Acute methylenedioxypyrovalerone toxicity. J Med Toxicol 2015;11(2):185–94.

[45] Umebachi R, Aoki H, Sugita M, Taira T, et al. Clinical characteristics of α-pyrrolidinovalerophenone (α-PVP) poisoning. Clin Toxicol 2016;54(7):563–7.

[46] Maurer HH. Position of chromatographic techniques in screening for detection of drugs or poisons in clinical and forensic toxicology and/or doping control. Clin Chem. Lab Med 2004;42(11):1310–24.

[47] Kadehjian LJ. Specimens for drugs-of-abuse testing. In: Wong RC, Tse HY, editors. Drugs of abuse. Forensic science and medicine. Totowa, NJ: Humana Press; 2005.

[48] Swortwood MJ, Boland DM, DeCaprio AP. Determination of 32 cathinone derivatives and other designer drugs in serum by comprehensive LC–QQQ–MS/MS analysis. Anal Bioanal Chem 2013;405(4):1383–97.

[49] Olesti E, Pujadas M, Papaseit E, Pérez-Mañá C, et al. GC–MS quantification method for mephedrone in plasma and urine: application to human pharmacokinetics. Anal Toxicol 2017;41(2):100–6.

[50] Glicksberg L, Bryand K, Kerrigan S. Identification and quantification of synthetic cathinones in blood and urine using liquid chromatography-quadrupole/time of flight (LC-Q/TOF) mass spectrometry. J Chromatogr B Analyt Technol Biomed Life Sci 2016;1035(1):91–103.

[51] Zhang L, Wang ZH, Li H, Liu Y, et al. Simultaneous determination of 12 illicit drugs in whole blood and urine by solid phase extraction and UPLC–MS/MS. J Chromatogr B Analyt Technol Biomed Life Sci 2014;955–956(1):10–19.

[52] Borovcová L, Pauk V, Lemr K. Analysis of new psychoactive substances in human urine by ultra-high performance supercritical fluid and liquid chromatography: validation and comparison. J Sep Sci 2018;41(10):2228–95.

[53] Rosano TG, Ohouo PY, Wood M. Screening with quantification for 64 drugs and metabolites in human urine using UPLC–MS–MS analysis and a threshold accurate calibration. J Anal Toxicol 2017;41(6):536–46.

[54] Neifeld JR, Regester LE, Holler JM, Vorce J. Ultrafast screening of synthetic cannabinoids and synthetic cathinones in urine by RapidFire-tandem mass spectrometry. Anal Toxicol 2016;40(5):379–87.

[55] LaPointe J, Musselman B, O'Neill T, Shepard JR. Detection of "bath salt" synthetic cathinones and metabolites in urine via DART-MS and solid phase microextraction. J Am Soc Mass Spectrom 2015;26(1):159–65.

[56] Švidrnoch M, Lněníčková L, Válka I, Ondra PJ, Chromatogr A. Utilization of micellar electrokinetic chromatography–tandem mass spectrometry employed volatile micellar phase in the analysis of cathinone designer drugs. J Chromatogr 2014;1356(1):258–65.
[57] Boumba VA, Di Rago M, Peka M, Drummer OH, et al. The analysis of 132 novel psychoactive substances in human hair using a single step extraction by tandem LC/MS. Forensic Sci Int 2017;279(1):192–202.
[58] Lendoiro E, Jiménez-Morigosa C, Cruz A, Páramo M, et al. An LC–MS/MS methodological approach to the analysis of hair for amphetamine-type-stimulant (ATS) drugs, including selected synthetic cathinones and piperazines. Drug Test Anal 2017;9(1):96–105.
[59] Baciu T, Borrull F, Calull M, Aguilar C. Enantioselective determination of cathinone derivatives in human hair by capillary electrophoresis combined in-line with solid-phase extraction. Electrophoresis 2016;37(17–18):2352–62.
[60] Williams M, Martin J, Galettis P. A validated method for the detection of 32 bath salts in oral fluid. J Anal Toxicol 2017;41(8):659–69.
[61] Mohamed K. One-step derivatization–extraction method for rapid analysis of eleven amphetamines and cathinones in oral fluid by GC–MS. J Anal Toxicol 2017;41(7):639–45.
[62] de Castro A, Lendoiro E, Fernández-Vega H, Steinmeyer S, et al. Liquid chromatography tandem mass spectrometry determination of selected synthetic cathinones and two piperazines in oral fluid. Cross reactivity study with an on-site immunoassay device. J Chromatogr A 2014;1374(1):93–101.
[63] Amaratunga P, Lorenz Lemberg B, Lemberg D. Quantitative measurement of synthetic cathinones in oral fluid. J Anal Toxicol 2013;37(9):622–8.
[64] Miller B, Kim J, Concheiro M. Stability of synthetic cathinones in oral fluid samples. Forensic Sci Int 2017;274(1):13–21.
[65] Verplaetse R, Henion J. Hematocrit-independent quantitation of stimulants in dried blood spots: pipet vs microfluidic-based volumetric sampling coupled with automated flow-through desorption and online solid phase extraction–LC–MS/MS bioanalysis. Anal Chem 2016;88(13):6789–96.
[66] Ryona I, Henion J. A book-type dried plasma spot card for automated flow-through elution coupled with online SPE–LC–MS/MS bioanalysis of opioids and stimulants in blood. J. Anal Chem 2016;88(22):11229–37.
[67] Mercolini L, Protti M, Catapano MC, Rudge J, et al. LC–MS/MS and volumetric absorptive microsampling for quantitative bioanalysis of cathinone analogues in dried urine, plasma and oral fluid samples. J Pharm Biomed Anal 2016;123(1):186–94.
[68] Cuypers E, Flanagan RJ. The interpretation of hair analysis for drugs and drug metabolites. Clin Toxicol 2018;56(2):90–100.
[69] Mercolini L, Mandrioli R, Protti M, Conti M, et al. Monitoring of chronic *Cannabis* abuse: an LC–MS/MS method for hair analysis. J Pharm Biomed Anal 2013;76(1):119–25.
[70] Patsalos PN, Berry DJ. Therapeutic drug monitoring of antiepileptic drugs by use of saliva. Ther Drug Monit 2013;35(1):4–29.
[71] Ruiz ME, Conforti P, Fagiolino P, Volonte MG. The use of saliva as a biological fluid in relative bioavailability studies: comparison and correlation with plasma results. Biopharm Drug Dispos 2010;31(8–9):476–85.
[72] Sadones N, Capiau S, De Kesel PM, Lambert WE, et al. Spot them in the spot: analysis of abused substances using dried blood spots. Bioanalysis 2014;6(17):2211–27.
[73] Mercolini L, Protti M. Biosampling strategies for emerging drugs of abuse: towards the future of toxicological and forensic analysis. J Pharm Biomed Anal 2016;130(1):202–19.
[74] Saracino MA, Catapano MC, Iezzi R, Somaini R, et al. Analysis of γ-hydroxy butyrate by combining capillary electrophoresis-indirect detection and wall dynamic coating: application to dried matrices. Anal Bioanal Chem 2015;407(29):8893–901.
[75] Mercolini L, Mandrioli R, Protti M, Conca A, et al. Dried blood spot testing: a novel approach for the therapeutic drug monitoring of ziprasidone-treated patients. Bioanalysis 2014;6(11):1487–95.
[76] Mercolini L, Mandrioli R, Sorella V, Somaini L, et al. Dried blood spots: liquid chromatography–mass spectrometry analysis of Δ9-tetrahydrocannabinol and its main metabolites. J Chromatogr A 2013;1271(1):33–40.
[77] Protti M, Rudge J, Sberna AE, Gerra G, et al. Dried haematic microsamples and LC–MS/MS for the analysis of natural and synthetic cannabinoids. J Chromatogr B Analyt Technol Biomed Life Sci 2017;1044–1045(1):77–86.
[78] Protti M, Vignali A, Sanchez Blanco T, Rudge J, et al. Enantioseparation and determination of asenapine in biological fluid micromatrices by HPLC with diode array detection. J Sep Sci 2018;41(6):1257–65.
[79] Protti M, Catapano MC, Samolsky Dekel BG, Rudge J, et al. Determination of oxycodone and its major metabolites in haematic and urinary matrices: comparison of traditional and miniaturised sampling approaches. J Pharm Biomed Anal 2018;152(1):204–14.

Chapter 22

Review of Synthetic Cannabinoids on the Illicit Drug Market

Mahmoud A. ElSohly[1,2,3], Sohail Ahmed[4,5], Shahbaz W. Gul[1] and Waseem Gul[1,2]
[1]ElSohly Laboratories, Inc., Oxford, MS, United States, [2]National Center for Natural Products Research, Oxford, MS, United States, [3]Department of Pharmaceutics, School of Pharmacy, The University of Mississippi, Oxford, MS, United States, [4]Department of Biochemistry, Hazara University, Mansehra, Pakistan, [5]Valor Group of Pharmaceutical Industries, Islamabad, Pakistan

INTRODUCTION

Cannabimimetics (synthetic cannabinoids) exhibit agonistic activity on the two cannabinoid receptors, CB1 and CB2. These compounds have been classified into several different major classes: carbazoles, classical cannabinoids, cyclohexylphenols (CPs), endogenous cannabinoids, indoles, indazoles, pyrroles, the URB-class, and others in a miscellaneous class. Every year, new synthetic cannabinoids arise, differing by the addition or removal of a substituent group. Over 130 synthetic cannabinoids and metabolites have been identified, many of which have been detected in multiple herbal incense products. The chemical structures of the synthetic cannabinoids are given in Figs. 22.1–22.30. The information is given according to the classes of the synthetic cannabinoids, and therefore, the only compounds which fall within that particular class are discussed.

HISTORY OF CANNABIMIMETICS

For decades, scientists have developed compounds for stronger agonistic performance on the two cannabinoid receptors, CB1 and CB2. These compounds are designed to have activities similar to those of Δ^9-THC, hence their classification as synthetic cannabinoids (cannabimimetics). John W. Hoffman (JWH) synthesized some of the first cannabimimetics identified in herbal incense products resulting in the "JWH" series. These synthetic cannabinoids were based on the prototypic aminoalkylindole structure. The "HU" series were produced at the Hebrew University, and the "CP" series at Pfizer. The objective was to be able to synthesize compounds that possess high therapeutic value without the adverse side effects. Although the synthesis of these compounds was reported in the literature many years ago, they have only appeared on the illicit market in recent years. Unfortunately, multiple underground laboratories have utilized this research for the production of illicit compounds used as alternatives for marijuana.

Synthetic cannabinoids were first introduced into the market place in 2004 [1] as various herbal incense or potpourri products. These products are sold via multiple venues, including the internet and gas stations, all under different brand names and with labels asserting that the products are "not for human consumption." Due to this disclaimer, there is no regulation on the sale of such products [2,3]. Due to their strength and effectiveness on cannabinoid receptors, these products have become a popular alternative to marijuana use. Cannabimimetics were first identified in such products in 2008.

In March 2011, the US Drug Enforcement Administration (DEA) placed five synthetic cannabinoids (JWH-018, JWH-073, JWH-200, CP-47,497, and CP-47,497-C8 homolog) on the list of Schedule 1 controlled substances. Since then, various European countries have banned multiple herbal products [4,5]. In May 2013, an additional three synthetic cannabinoids [AKB-48 (Fig. 22.15), UR-144 (Fig. 22.18), and XLR-11 (Fig. 22.18)] were classified as Schedule I controlled substances.

Synthetic cannabinoids have a four to five times greater binding affinity to cannabinoid receptors than THC itself, which leads to several harmful effects, including anxiety, high blood pressure, numbness, and seizures [6,7]. In the past

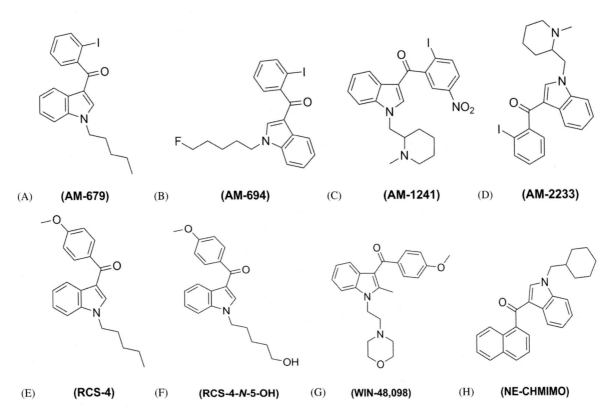

FIGURE 22.1 Chemical structures of benzoylindoles: (A) AM-679, (B) AM-694, (C) AM-1241, (D) AM-2233, (E) RCS-4, (F) RCS-4-*N*-5-OH, (G) WIN-48,098, and (H) NE-CHMIMO.

several years, multiple methodologies have been developed and validated to analyze for the presence of synthetic cannabinoids and/or their metabolites in human biological matrices, including blood, hair, plasma, and serum. These methodologies utilize a wide range of analytical instruments, such as direct analysis in real time mass spectrometry (DART-MS) [8], liquid chromatography coupled with tandem mass spectrometry (LC-MS/MS) [9], high-performance liquid chromatography (HPLC) [10], ultra-performance liquid chromatography-electrospray ionization tandem mass spectrometry (UPLC-MS/MS) [10,11], cheminformatics and immunoassays [11]. In addition to these techniques nuclear magnetic resonance (NMR) [12] was used to identify these compounds in the herbal products.

Since the emergence of synthetic cannabinoids on the illicit market, several reviews have been published, outlining the chemistry, pharmacology, and toxicology of these drugs [13–29], as well as clinical and legal considerations [30]. Multiple reference libraries using different analytical instruments have been compiled in an attempt to identify any drug of abuse present in biological matrices, including oral fluid, urine, plasma, blood, and many more [31–33]. Statistical analyses have been carried out to determine the prevalence of these cannabimimetics in various countries [34], and in an effort to understand how widespread the knowledge of these psychoactive substances is among the general public, questionnaires have been developed for demographics [35]. In addition, the analytical data of existing synthetic cannabinoids has been used to predict new synthetic cannabinoids that may arise on the market [36,37].

Synthetic cannabinoids can be categorized into nine different classes, based on their chemical structures: carbazoles, classical cannabinoids, CP endogenous cannabinoids, indoles, indazoles, pyrroles, the URB-class, and miscellaneous. This book chapter provides basic background information on cannabimimetics and focuses on the categorization of these synthetic cannabinoids and their analysis.

INDOLES

Indoles are the largest class of synthetic cannabinoids. They are classified into eight subgroups depending on the attachment(s) to the indole moeity. These subclasses are: benzoylindoles, naphthoylindoles, phenylacetylindoles, piperazinylindoles, naphthalenyl-indole-3-carboxylates, quinolin-8-yl-indole-3-carboxylates, adamantyl-indole-3-carboxamides, and tetramethylcyclopropylmethanone-pentylindoles. Seven other indoles, MDMB-CHMICA, MDMB-FUBINACA,

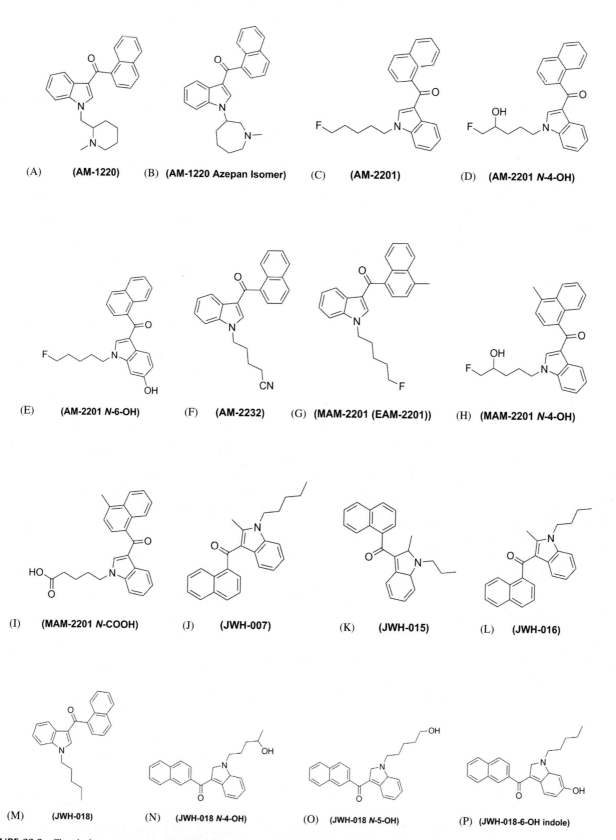

FIGURE 22.2 Chemical structures of naphthoylindoles: (A) AM-1220, (B) AM-1220 Azepan isomer, (C) AM-2201, (D) AM-2201 *N*-4-OH, (E) AM-2201 *N*-6-OH, (F) AM-2232, (G) MAM-2201 (EAM-2201), (H) MAM-2201 *N*-4-OH, (I) MAM-2201 *N*-COOH, (J) JWH-007, (K) JWH-015, (L) JWH-016, (M) JWH-018, (N) JWH-018 *N*-4-OH, (O) JWH-018 *N*-5-OH, (P) JWH-018 6-OH-indole, (Q) JWH-018 *N*-COOH, (R) JWH-019, (S) JWH-019 *N*-5-OH indole, (T) JWH-022, (U) JWH-073, (V) JWH-073 *N*-3-OH-butyl, (W) JWH-073-6-OH-indole, (X) JWH-073 *N*-COOH, (Y) JWH-081, (Z) JWH-122, (AA) JWH-122 *N*-5-OH, (BB) JWH-122 *N*-4-pentenyl, (CC) JWH-200, (DD) JWH-398, (EE) JWH-398 *N*-5-OH, and (FF) JWH-018 and five regioisomeric equivalents.

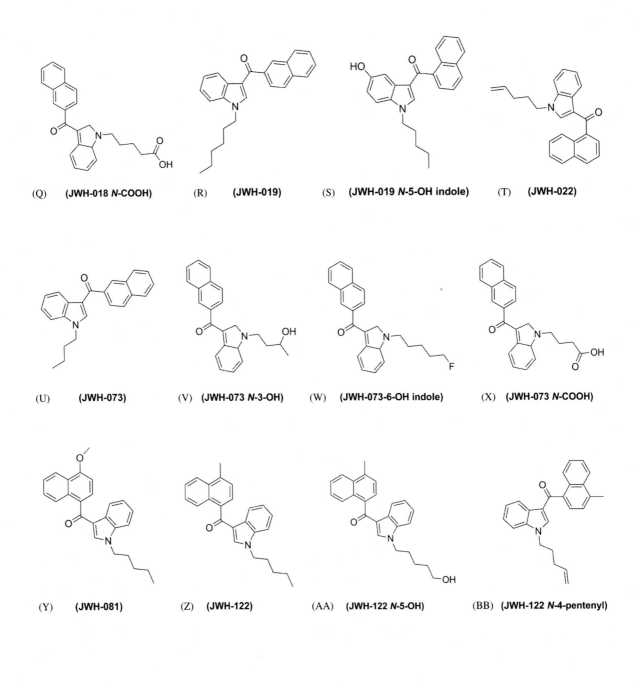
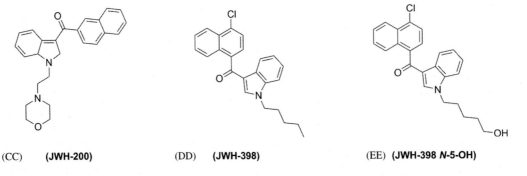

FIGURE 22.2 (*Continued*)

(FF)

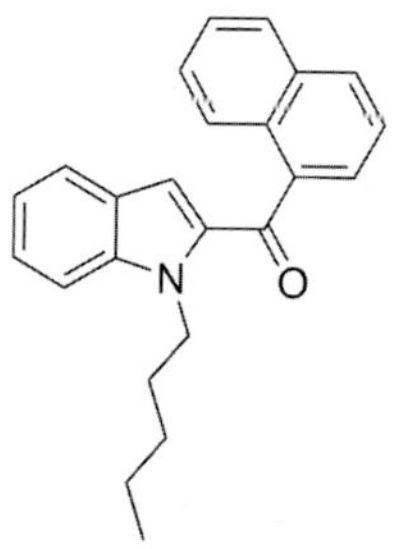

JWH-018

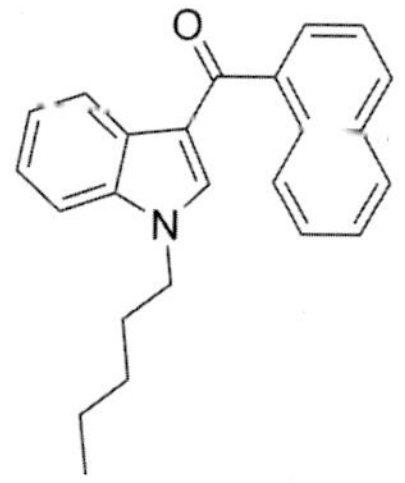

Naphthalen-1-yl(1-pentyl-1H-indol-3-yl)methanone

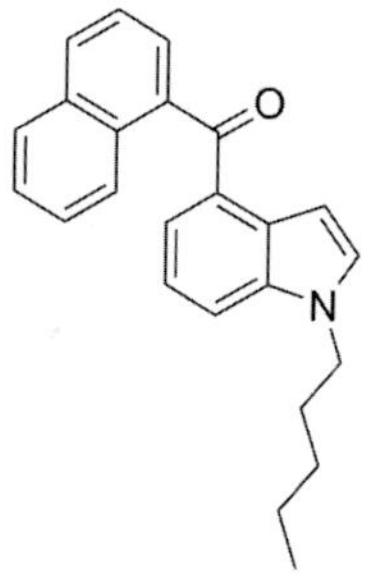

Naphthalen-1-yl(1-pentyl-1H-indol-4-yl)methanone

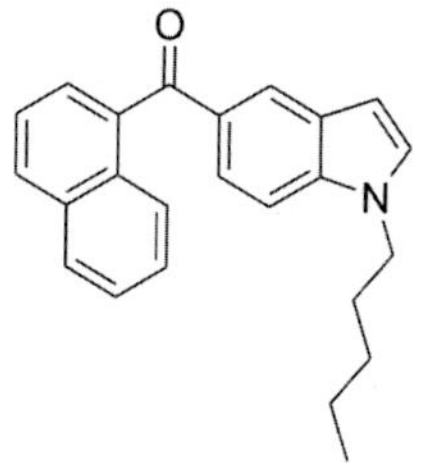

Naphthalen-1-yl(1-pentyl-1H-indol-5-yl)methanone

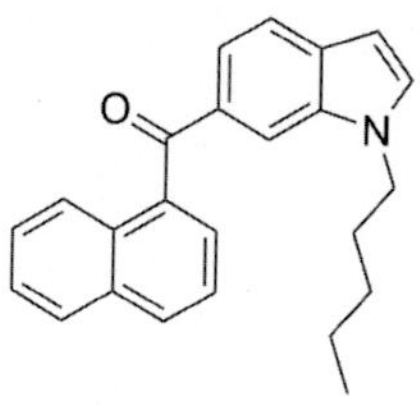

Naphthalen-1-yl(1-pentyl-1H-indol-6-yl)methanone

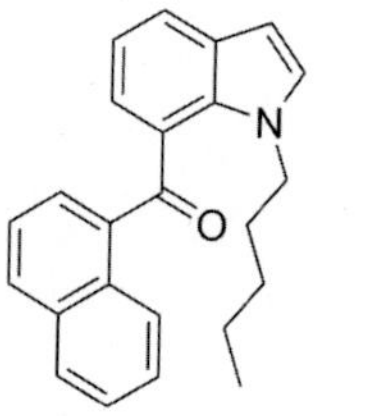

Naphthalen-1-yl(1-pentyl-1H-indol-7-yl)methanone

FIGURE 22.2 *(Continued)*

MDMB-AMB, ADB-BICA, NNEI, BzODZ-Epyr, and JWH-210, were classified separately. These subclasses and cannabimimetics are discussed individually.

Benzoylindoles: Structures and Analysis

The structures of benzoyl indole cannabimimetics are shown in Fig. 22.1.

Various spice products and reference materials, some of which obtained from the European Monitoring Centre for Drugs and Drug Addiction (EMCDDA), personal communications, and customs, were analyzed for various synthetic cannabinoids [17]. The products were tested for AM-694, RCS-4, RCS-4-methylated A, and RCS-4-methylated B using liquid chromatography coupled with mass spectrometry (LC-MS) and gas chromatography coupled with mass spectrometry (GC-MS). These cannabinoids were detected from the samples collected, confirming the efficiency of the method.

Multiple "legal high" products acquired from prefederal ban time frame and postfederal ban time frame were analyzed for multiple synthetic cannabinoids [38], including AM-694, AM-1241, AM-2233, and RCS-4, utilizing ultra-performance liquid chromatography coupled with time of flight mass spectrometry (UPLC-TOF-MS). In samples collected from the prefederal ban time range, 16 out of the 17 analyzed contained only 2 of the now-federally banned

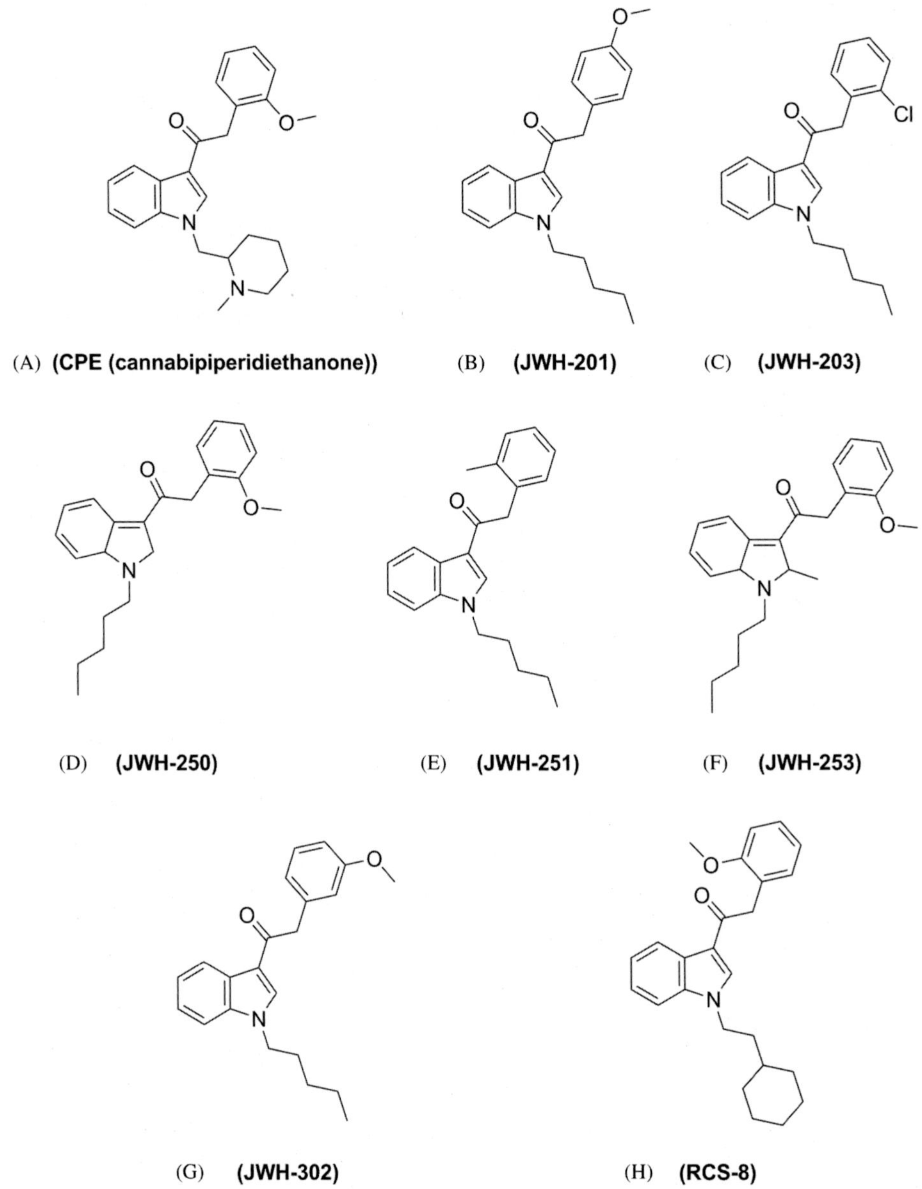

278 Critical Issues in Alcohol and Drugs of Abuse Testing

FIGURE 22.3 Chemical structures of phenylacetylindoles: (A) CPE (cannabipiperidiethanone), (B) JWH-201, (C) JWH-203, (D) JWH-250, (E) JWH-251, (F) JWH-253, (G) JWH-302, and (H) RCS-8.

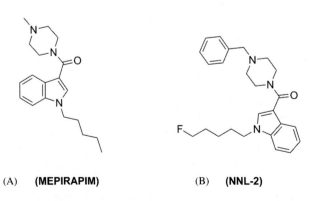

FIGURE 22.4 Chemical structures of (A) MEPIRAPIM and (B) NNL-2.

(A) **(FDU-PB-22)** (B) **(NM-2201)**

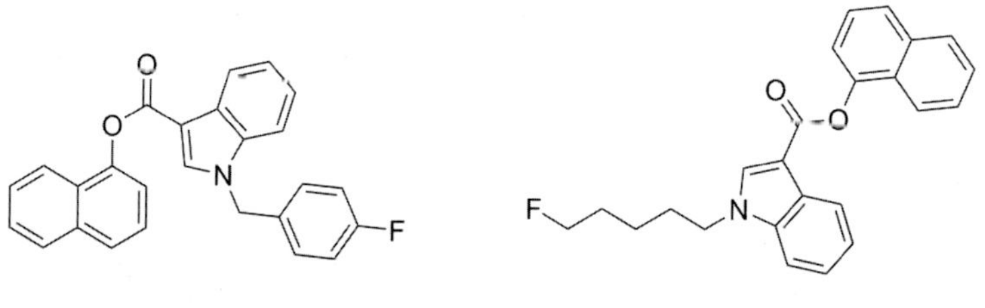

FIGURE 22.5 Chemical structures of (A) FDU-PB-22 and (B) NM-2201.

(A) **(QCBL-018; PB-22; MN-25; LTI-696; QUPIC)** (B) **(QCBL-Bz-F; FUB-PB-22; MN-27)**

(C) **(QCBL-CHM; BB-22; QUCHIC)** (D) **(QCBL-2201; PB-22F; 5F-PB-22; MN-25F)**

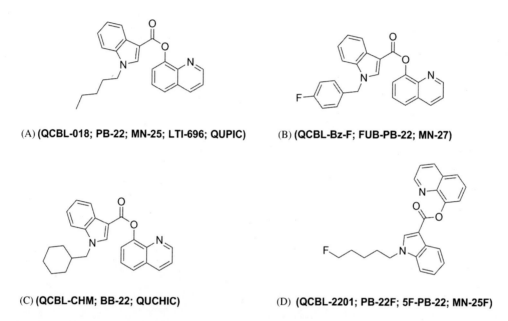

FIGURE 22.6 Chemical structures of (A) PB-22, (B) FUB-PB-22, (C) BB-22, and (D) 5F-PB-22.

(A) **(AB-001)** (B) **(AM-1248)** (C) **(APICA)** (D) **(STS-135)**

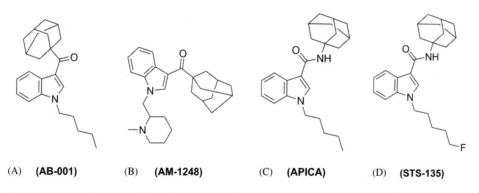

FIGURE 22.7 Chemical structures of (A) AB-001, (B) AM-1248, (C) APICA, and (D) STS-135.

synthetic cannabinoids. Out of the 81 products collected in the postfederal ban time frame, only 4 samples contained a federally controlled substance. Eight nonfederally banned synthetic cannabinoids at the time, including AM-2233, were also detected in the samples collected postfederal ban.

A rapid, reliable headspace gas chromatography coupled with mass spectrometry and using solid-phase microextraction (HS-SPME-GC-MS) method was developed for the analysis of 32 products [39]. In the products received before the DEA ban, AM-694 was identified in only one sample. No samples obtained after the DEA ban contained any benzoylindole cannabimimetics, indicating the shift to other synthetic cannabinoids in the illicit market.

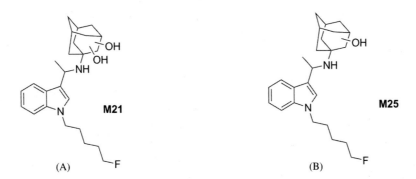

FIGURE 22.8 Chemical structures of (A) monohydroxylated of STS-135 (M25) and (B) dihydroxylated metabolite of STS-135 (M21).

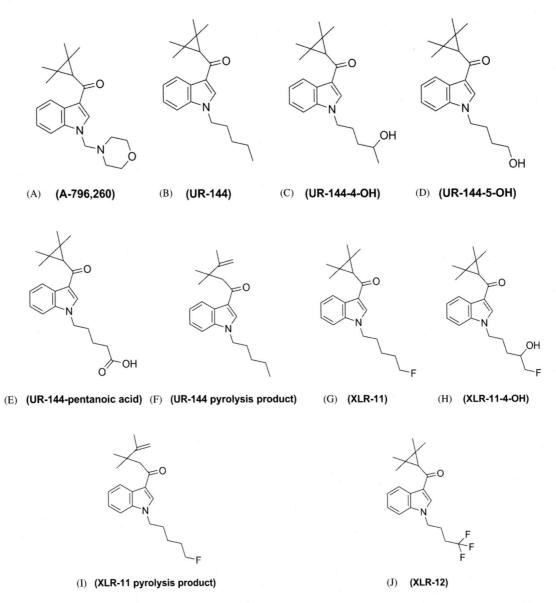

FIGURE 22.9 Chemical structures of (A) A-796,260, (B) UR-144, (C) UR-144-4-OH, (D) UR-144-5-OH, (E) UR-144-pentanoic acid, (F) UR-144 and pyrolysis product, (G) XLR-11, (H) XLR-11-4-OH, (I) XLR-11 pyrolysis product, and (J) XLR-12.

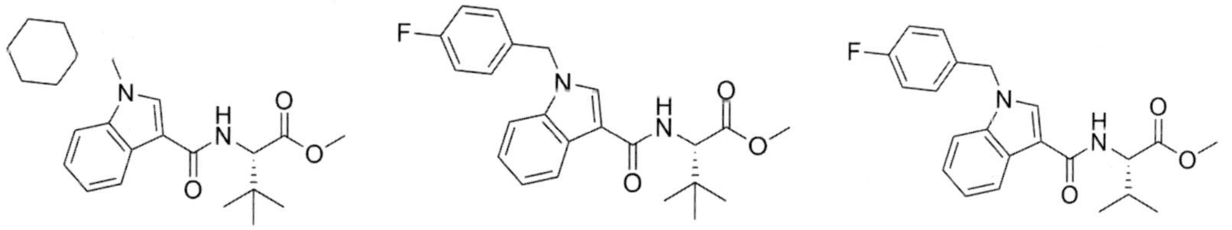

(A) **(MDMB-CHMICA)** (B) **(MDMB-FUBINACA)** (C) **(FUB-AMB)**

FIGURE 22.10 Chemical structures of (A) MDMB-CHMICA, (B) MDMB-FUBINACA, and (C) FUB-AMB.

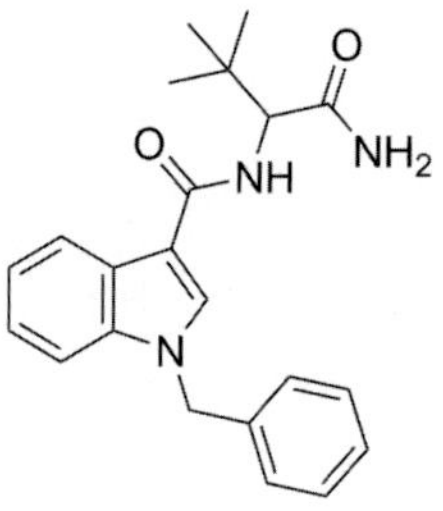

(ADB-BICA)

FIGURE 22.11 Chemical structure of ADB-BICA.

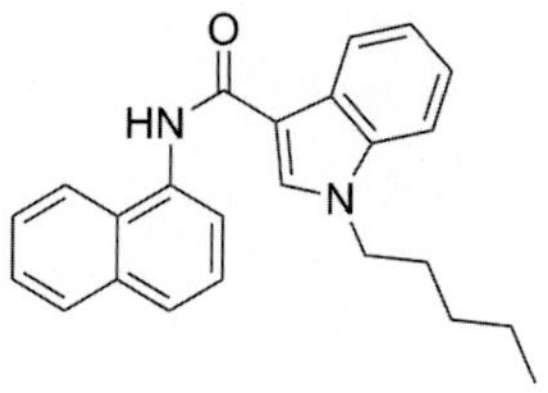

(CBM-018; NNEI; MN24)

FIGURE 22.12 Chemical structures of NNEI.

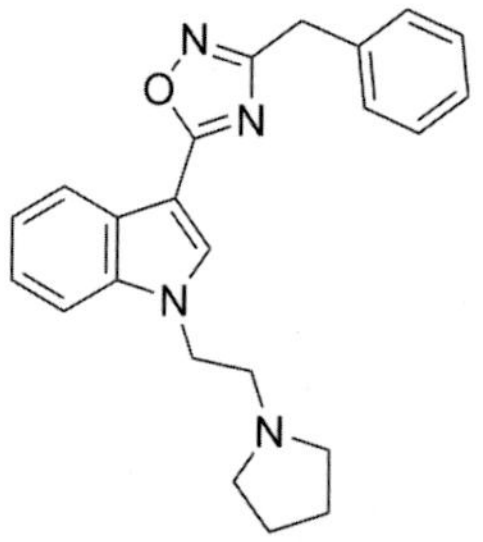

(BzODZ-EPyr)

FIGURE 22.13 Chemical structure of BzODZ-EPyr.

(JWH-210)

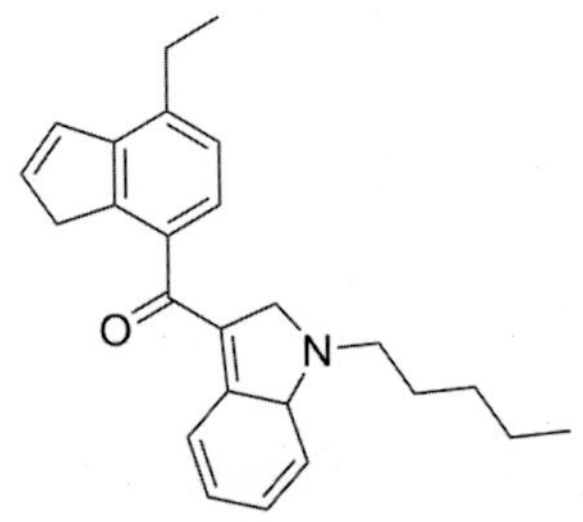

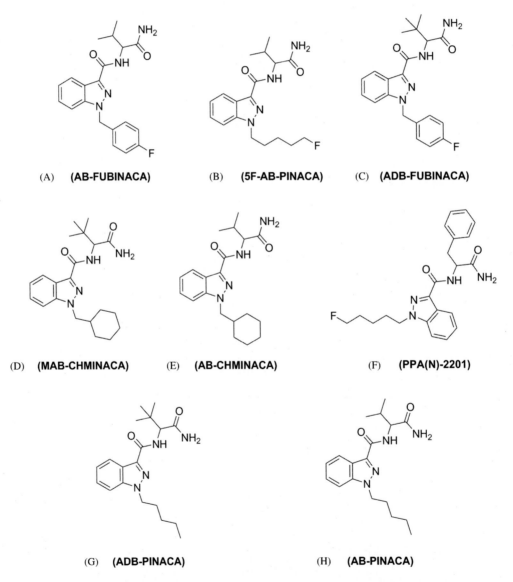

FIGURE 22.14 Chemical structure of JWH-210.

FIGURE 22.15 Chemical structure the amino-methyl-oxobutan-indazole-3-carboxamide class: (A) AB-FUBINACA, (B) 5F-AB-PINACA, (C) ADB-FUBINACA, (D) MAB-CHMINACA, (E) AB-CHMINACA, (F) PPA(N)-2201, (G) ADB-PINACA, and (H) AB-PINACA.

(A) **(APINACA)** (B) **(AKB-48)** (C) **(5F-AKB-48)** (D) **(5Cl-AKB-48)**

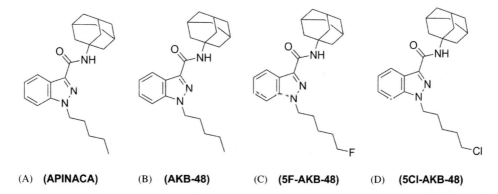

FIGURE 22.16 Chemical structures of adamantyl-indazole-3-carboxamide class: (A) APINACA, (B) AKB-48, (C) 5F-AKB-48, and (D) 5Cl-AKB-48.

(A) **(BIM-018)** (B) **(BIM-2201; FUBIMINA)**

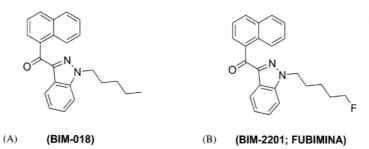

FIGURE 22.17 Chemical structures of benzo[d]imidazole class: (A) BIM-2201 (FUBIMINA) and (B) BIM-018.

(A) **(THJ-018)** (B) **(THJ-2201; AM(N)-2201)**

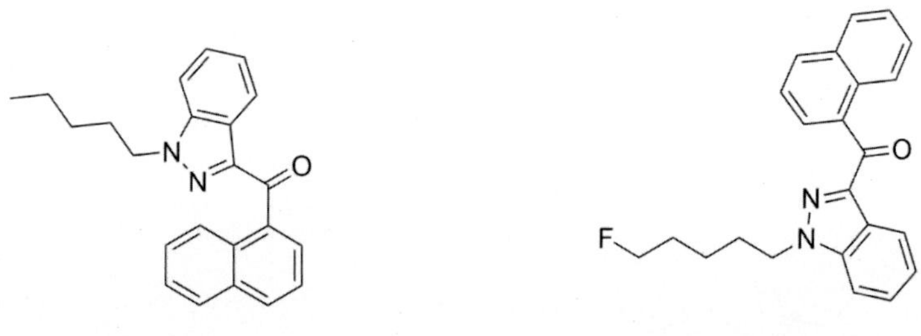

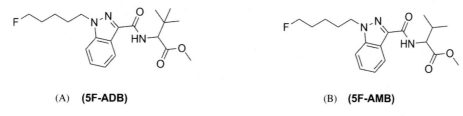

FIGURE 22.18 Chemical structures of the naphthalenyl-indazol class: (A) THJ-018 and (B) THJ-2201.

(A) **(5F-ADB)** (B) **(5F-AMB)**

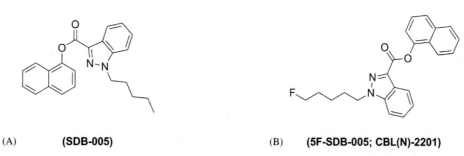

FIGURE 22.19 Chemical structures of (A) 5F-ADB and (B) 5F-AMB.

(A) **(SDB-005)** (B) **(5F-SDB-005; CBL(N)-2201)**

FIGURE 22.20 Chemical structures of (A) SDB-005 and (B) 5F-SDB-005 (CBL(N)-2201).

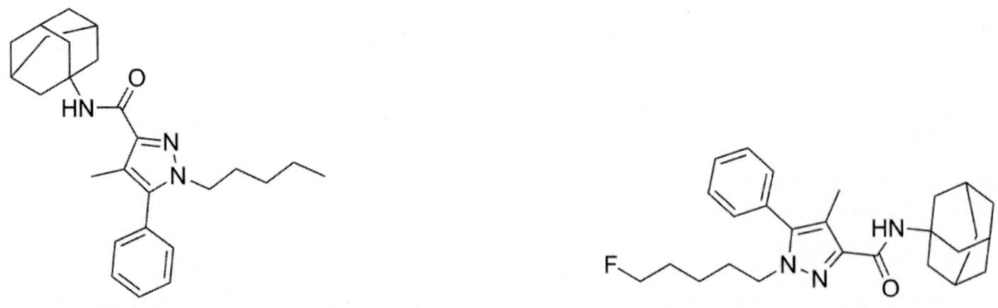

284 Critical Issues in Alcohol and Drugs of Abuse Testing

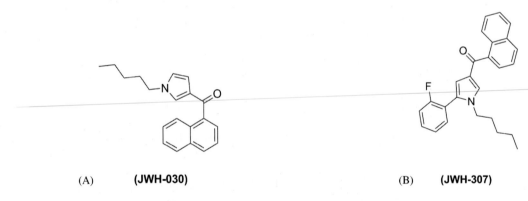

FIGURE 22.21 Chemical structures of (A) AMPPPCA and (B) 5F-AMPPPCA.

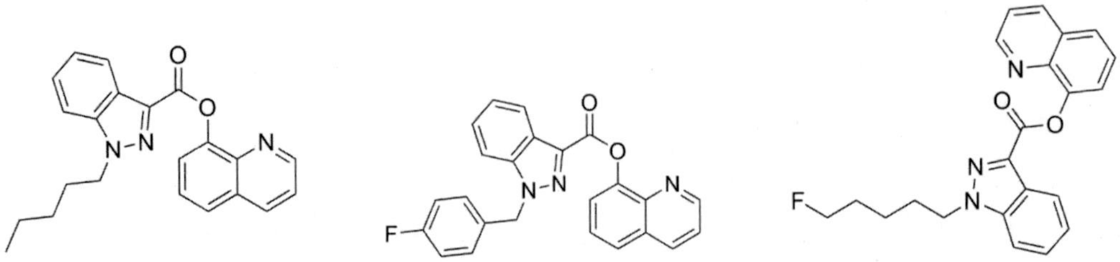

FIGURE 22.22 Chemical structure of (A) NPB-22, (B) FUB-NPB-22, and (C) 5F-NPB-22.

FIGURE 22.23 Chemical structures of (A) JWH-030 and (B) JWH-307.

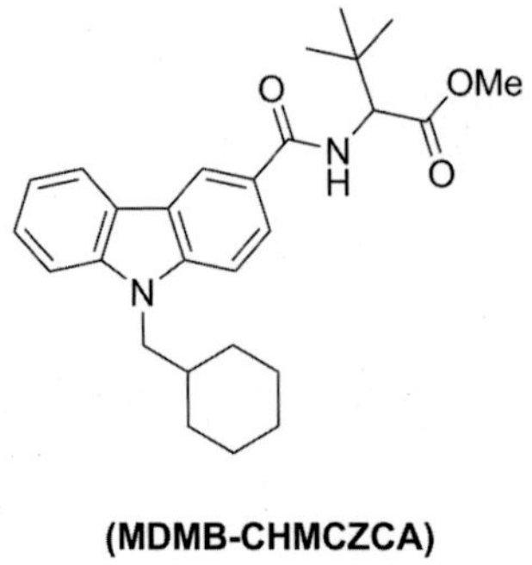

FIGURE 22.24 Chemical structure of MDMB-CHMCZCA.

FIGURE 22.25 Chemical structure of the synthetic cannabinoids in the URB-class: (A) URB-597 and (B) URB-754.

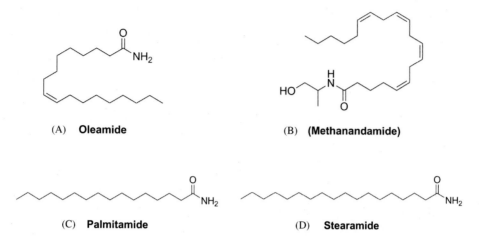

FIGURE 22.26 Chemical structures of (A) oleamide, (B) methanandamide, (C) palmitamide, and (D) stearamide.

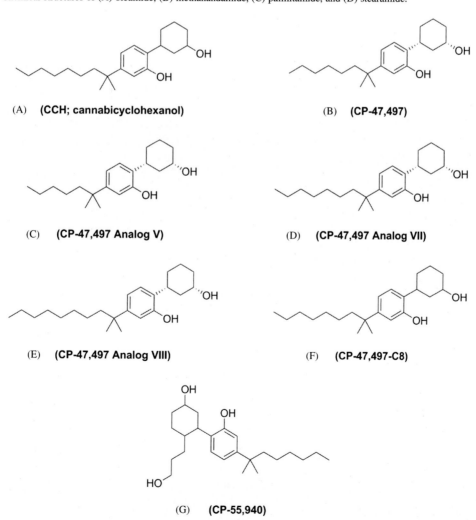

FIGURE 22.27 Chemical structures of (A) CCH (cannabicyclohexanol), (B) CP-47,497, (C) CP-47,497 Analog V, (D) CP-47,497 Analog VII, (E) CP-47,497 Analog VIII, (F) CP-47,497-C8, and (G) CP-55,940.

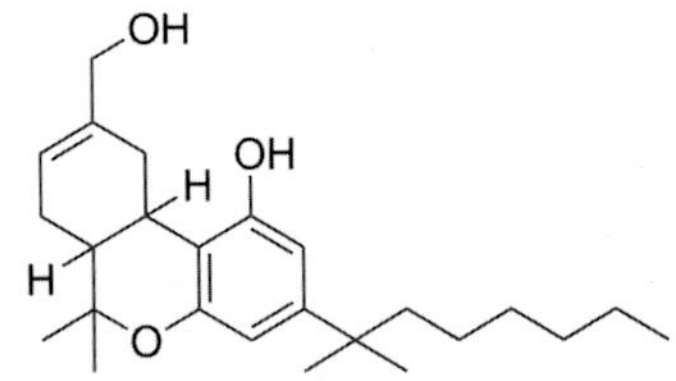

(A) **(HU-210)**

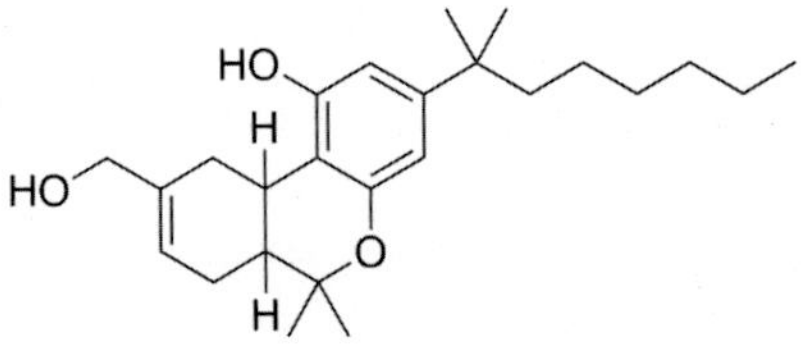

(B) **(HU-211)**

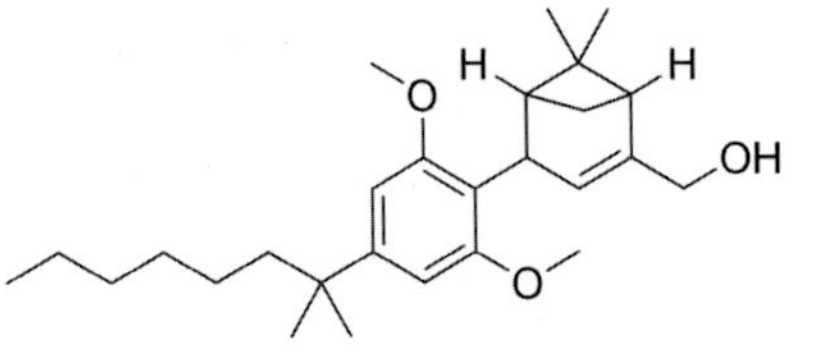

(C) **(HU-308)**

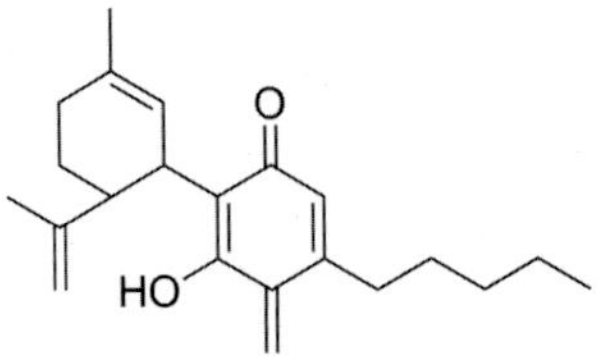

(D) **(HU-311)**

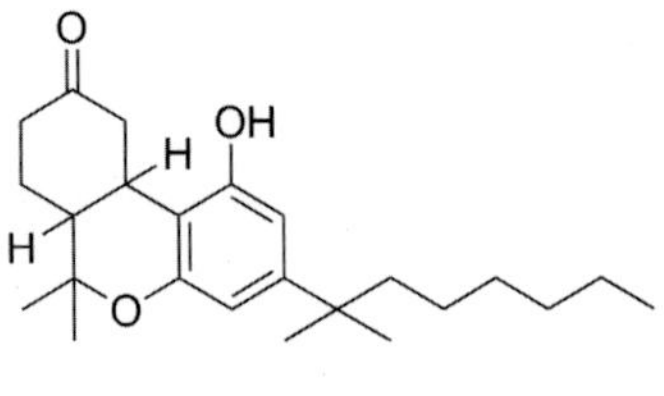

(E) **(Nabilone)**

FIGURE 22.28 Chemical structures of (A) HU-210, (B) HU-211, (C) HU-308, (D) HU-311, and (E) nabilone.

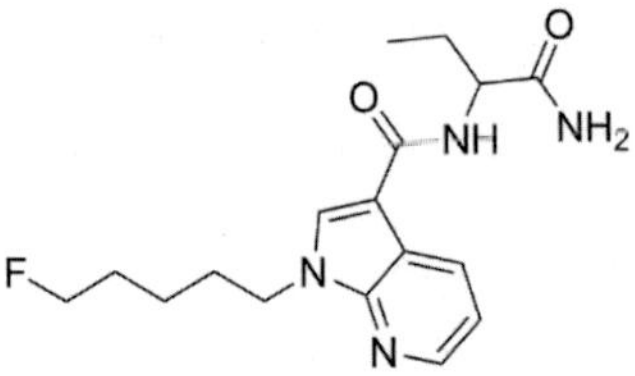

(NNL-1)

FIGURE 22.29 Chemical structures of NNL-1.

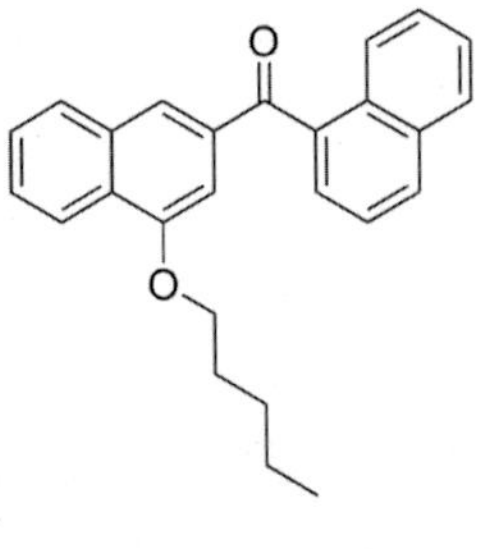

(CB-13; CRA-13)

FIGURE 22.30 Chemical structure of CB-13.

Using capillary separation based on micellar electrokinetic chromatography (MEKC) with diode-array detection (DAD), a method was developed and validated for the analysis of 15 different herbal blend samples to detect synthetic cannabinoids [40]. This method had a limit of detection (LOD) ranging from 1 to 1.5 μg/mL, and a limit of quantitation (LOQ) between 3 and 5 μg/mL. AM-694 was not detected in any of the herbal blend samples. Every positive sample was validated correctly on the LC-MS, proving this method to be an acceptable alternative to GC and LC methods.

A rapid, sensitive, and simple LC-MS method was developed to identify 22 synthetic cannabinoids, including AM-694, RCS-4, and the RCS-4 ortho isomer, in human hair collected from 8 individuals [41]. One individual was bald, necessitating hair collection from the arm and leg of the individual. These eight individuals were all male, in the same age range, and reported last drug use around the same time. None of these samples contained any benzoylindole-type synthetic cannabinoid.

LC-MS, GC-MS, and NMR were used to analyze a chemical product, "Fragrance Powder" and three herbal products [42], each containing approximately 3 g of mixed dried plant material. Many new synthetic cannabinoids, including AM-1241 and AM-2233, were detected.

An LC-MS/MS method was developed to quantitate the concentration of 11 synthetic cannabinoid analytes, including RCS-4-N-(5-OH-pentyl), in human urine [43]. The CV percentage (coefficient of variation) for the 5-OH metabolite of RCS-4 was calculated to be 27.8%. Another LC-MS/MS method was used for the analysis and screening for synthetic cannabinoids in urine samples from patients who had previously consumed these compounds [44]. RCS-4, along with its metabolites, was detected in these samples. An LC-MS/MS method was also developed for the detection of six benzoylindole cannabimimetics (AM-694, AM-1241, RCS-4, RCS-4 C-4 homolog, RCS-4 2-methoxy homolog, and the RCS-4 3-methoxy homolog) in blood samples [45]. The developed method was successful in the identification of these analytes in the blood samples.

A matrix-assisted laser desorption ionization-time-of-flight mass spectrometry (MALDI-TOF-MS) method was developed for the direct and rapid screening of various herbal blends for synthetic cannabinoids [46]. The herbal blends were grounded and loaded onto a MALDI plate for analysis. AM-694 was detected in 21 out of the total 31 herbal blends analyzed.

Thin-layer chromatography (TLC), GC-MS, and HPLC coupled with a time of flight detector (HPLC-TOF) were used in the detection and identification of RCS-4 and AM-694 in herbal products [47]. Some of these products contained multiple cannabimimetics, with concentrations ranging from 5 to 20 mg/g. Using nano-liquid chromatography, many herbal blends were analyzed for the identification of synthetic cannabinoids [48]. A new benzoylindole cannabimimetic, AM-694, was separated and identified.

An LC-MS/MS method was developed for the identification of 29 synthetic cannabinoids, which included AM-695, AM-2233, and RCS-4, in the analysis of severe intoxication cases [49]. The study included 3078 blood samples, as well as case studies for 8 different cases. In the blood samples, the concentrations of AM-694, AM-2233, and RCS-4 were identified to range from 0.05 to 5.07, 0.94, and 0.09 to 3.6 ng/g, respectively. No benzoylindole-type synthetic cannabinoids were identified in the eight different cases studied.

Using multiple reaction monitoring (MRM) mode, a UHPLC-MS/MS method was developed for the analysis of 14 synthetic cannabinoids, including RCS-4, and 15 metabolites, such as RCS-4-N-(5-OH-pentyl) in urine [50]. An Immunalysis immunoassay was purchased for screening, and β-glucuronidase was utilized for hydrolysis. RCS-4-hydroxypentyl was not detected by the immunoassay. However, the developed and validated dilute-and-shoot method was reliable, reproducible, and sensitive for all 29 analytes, with easy identification and determination of the new compounds.

Liquid chromatography-chemiluminescence nitrogen detection (LC-CLND) method was developed for the analysis of 177 samples, seized by the police in 2011–2012 and Finland customs in 2011–2013, for 61 different new psychoactive substances (NPS) [51]. Three synthetic cannabinoids were identified in the samples seized from the police in 2011–2012, and 14 cannabimimetics were detected in the samples collected by Finland customs in 2011–2013. The purity of the three synthetic cannabinoids in the samples seized by the police ranged from 89% to 97%, while the average purity in the samples seized by customs in 2011–2013 for the 14 synthetic cannabinoids ranged from 83% to 105%. The drug content for the Customs 4 sample was identified to be 2.2% of RCS-4.

A DART-MS method was developed [52] for the analysis of five different synthetic cannabinoids in six "Spice products." RCS-4 was detected in "Mr. Sticky Extreme"; however, this cannabimimetic was not detected in any other Spice product.

An LC-MS/MS method was developed for the analysis of human urine for 19 synthetic cannabinoids, including 2 benzoylindoles: RCS-4 and AM-694 [53]. The correlation coefficient (r^2) values ranged from 0.994 to 0.999, and the LODs for the analytes ranged from 0.1 to 0.5 μg/L. There was no evidence of carry over for the synthetic cannabinoids. This method was successfully applied to two anonymous, authentic urine samples, resulting in accurate results.

An MEKC-MS/MS method was developed and validated for the analysis of 12 different synthetic cannabinoids, including AM-2233 and RCS-4 [54]. As part of the micellar pseudostationary phase, the volatile surfactant used was ammonium perfluorooctanoate. The LODs and recoveries for all of the analytes ranged from 6.5 to 76.5 μg/g and 89.5% to 101.7%, respectively. After validation, this method was applied to two herbal products purchased in Japan. AM-2233 and RCS-4 were not identified in either of these samples.

UHPLC-MS/MS analysis was carried out for 344 hair samples for the presence of 23 different synthetic cannabinoids, including 2 benzoylindoles: AM-694 and RCS-4 [55]. All of the hair samples were obtained from the head. Out of the samples tested, 264 samples were previously identified to be positive for at least one, if not more, amphetamines, cocaine, methadone, opiates, and THC. The remaining 80 samples had tested positive for ethyl-glucuronide (EtG), indicating alcohol consumption. All of these samples were retested for the presence of these 23 synthetic cannabinoids, using the developed UHPLC-MS/MS method. No benzoylindoles cannabimimetics were identified in any samples.

NMR spectroscopy was used to screen and quantify synthetic cannabinoids, including RCS-4, in 12 herbal products: Melon Code Black, Cloud Nine, Sweet Leaf, Head Trip, Extremely Legal, Zombie Matter, Moon Spice, Ion Source, Rack City, Nuclear Bomb, Mr. Nice Guy, and K250 [56]. This method has the ability to identify and quantitate these compounds without any chromatographic separation, but retaining high accuracy.

Using a previously published, quantitative LC-MS/MS method [53], over 770 urine specimens from US military personnel were analyzed to identify 53 synthetic cannabinoid analytes, including 20 parent compounds and 33 metabolites [57]. Overall, 10 urine specimens were identified to contain 2 metabolites of RCS-4, namely RCS-4 pentatonic acid and RCS-4 M9 metabolites. In total, 290 urine specimens were identified positive for the presence of at least 1 synthetic cannabinoid.

A method was developed for the identification of five synthetic cannabinoids, including RCS-4, in bulk powder and biofluid samples, using a miniature ion trap mass spectrometry system with two ambient ionization methods: extraction spray ionization and paper spray [58]. The LOD was determined to range from 10 to 20 ng/mL for the analytes. This method proved to be simple, effective, and easily reproducible, with good linearity ($r^2 > 0.99$) for all five synthetic cannabinoids.

A gas chromatography with chemical ionization mass spectrometry (GC-CI-MS) method, operated in MRM mode, was developed and validated for the qualitative analysis of 34 synthetic cannabinoids, including AM-2233, AM-694, and RCS-4, and 7 synthetic cathinones (3-FMC, 4-MMC, 4-MEC, 3,4-DMMC, Methedrone, Methylone, and MDPV) [59]. Seven compounds were chosen to be representative compounds, out of which a benzoylindole-type cannabimimetic, AM-694, was selected. The LOD of AM-694 was determined to be 68.8 ng/mL. The r^2 values for the representative synthetic cannabinoids ranged from 0.994 to 0.998.

A direct analysis in real time coupled with a hybrid ion trap (DART-LTQ ORBITRAP) method was developed for the analysis of synthetic cannabinoids, including a benzoylindole cannabimimetic, AM-2233 [60]. Eight different samples were analyzed, including herbal material (flowers, leaves, and stems) and pure compounds. There is no sample or solvent preparation, making this an alternative to GC and LC analyses.

Two different LC-MS/MS methods were developed and validated for the quantitative analysis of various synthetic cannabinoids in two different matrices [61]. One method dealt with the analysis of fifteen parent synthetic cannabinoids in blood, while the other method analyzed urine samples for seventeen different cannabimimetic metabolites (and two parent compounds qualitatively identified). In blood, the LOD ranged from 0.025 to 0.10 ng/mL, and the LOQ ranged from 0.025 to 0.10 ng/mL. In urine, the LOD ranged from 0.01 to 0.50 ng/mL, and the LOQ ranged from 0.10 to 0.50 ng/mL. RCS-4 was identified to be positive in three blood samples, and RCS N-COOH was detected below LOQ in the urine samples from these three individuals. These methods are useful because when used in combination, they detect the parent compounds in blood and metabolites in urine, indicating the past history of the individual.

Seven herbal and four powdered chemical products were analyzed by NMR and DART-MS for the identification of cannabinoids and synthetic cannabinoids [62]. No benzoylindole cannabimimetic was identified in any of the four powders. Out of the seven herbal products, "Barely Legal" and "Ion Lab" contained AM-2201 and RCS-4, while "Moon Spice" contained JWH-018 and RCS-4. This method is useful, as it provides a rapid and accurate method of analyzing herbal products.

A new benzoylindole-type synthetic cannabinoid (NE-CHMIMO) was identified in an herbal product—"Jamaican Gold Extreme" [63]. This compound was separated using a flash chromatography system, and structural characterization was performed, using GC-MS, gas chromatography coupled with solid-state infrared spectroscopy (GC-sIR), and NMR analysis. This compound was identified to be a cyclohexylmethyl analog of JWH-018. This is the first report on the identification and characterization of NE-CHMIMO.

The pyrolytic products of RCS-4 were identified using GC-MS [64]. 1-(5-Fluoropentyl)-indole, 1-(5-fluoropentyl)-indole-3-carboxaldehyde, 3-benzoylindole, and 1-(5-fluoropentyl)-3-benzoylindole were the pyrolytic products identified for RCS-4. This analysis is useful, as it provides the pyrolytic cannabimimetic metabolites produced from smoking synthetic cannabinoids.

Naphthoylindoles: Structures and Analysis

The structures of the naphthoylindole-type synthetic cannabinoids are shown in Fig. 22.2.

A total of 30 different products were analyzed for multiple synthetic cannabinoids, including numerous napthoylindole cannabimimetics, using LC-MS/MS and GC-MS [17]. These analytes were detected from the samples collected, leading to the conclusion that this method was successful in identifying and detecting these analytes in real samples.

In 2009, 29 herbal mixtures were seized by authorities and analyzed by GC-MS [25]. These herbal products were analyzed for several compounds, including JWH-018 and JWH-073. JWH-018 and JWH-073 were detected in 19 herbal products.

Various "legal high" products were acquired from the prefederal ban time range and postfederal ban time range for over 30 synthetic cannabinoids, including several from the naphthoylindole class, utilizing UPLC-TOF-MS analysis [38]. In samples collected from the prefederal ban time range, 16 out of the 17 products analyzed contained 2 of the now-federally banned synthetic cannabinoids, JWH-018 and JWH-073. None of the samples contained JWH-200. Out of the 81 products in the postfederal ban time range, only four samples contained a now-federally controlled substance. AM-2201, JWH-019, and JWH-122 were also detected in the samples collected postfederal ban.

An HS-SPME-GC-MS method was developed for the analysis of 32 products before and after the DEA ban for the identification of 5 naphthoylindole-type cannabimimetics [39]. In the products received before the DEA ban, 5 of the 27 samples did not contain JWH-018. JWH-019 was only identified in one sample. AM-2201 was not identified in any samples, while JWH-073 and JWH-081 were found in multiple samples each. In the samples obtained after the DEA ban, only one sample contained JWH-018 in conjugation with no other synthetic cannabinoid. Two samples contained AM-2201.

A total of 15 herbal blend samples were analyzed by capillary separation based on MEKC-DAD analysis for the presence of synthetic cannabinoids from different classes, including the naphthoylindole class [40]. JWH-081, AM-2201, JWH-018, and JWH-073 were detected in the herbal blend samples.

A rapid, sensitive, and simple LC-MS/MS method was developed for the analysis of 22 synthetic cannabinoids in human hair collected from 8 male individuals [41]. One of the individuals was bald, making it necessary to take hair from the arm and leg of the individual. The individuals were in the same age range and reported last drug use around the same time. JWH-081 was detected in six individuals, and JWH-073 was determined in seven individuals; JWH-018 was only positive for two individuals. LC-MS, GC-MS, and NMR analyses were conducted for four herbal products for the identification of several synthetic cannabinoids [42]. AM-1220 was identified in Product C.

An LC-MS/MS method was developed for the analysis of 3078 blood samples, as well as case studies for 8 different severe intoxication cases for the detection of 29 synthetic cannabinoids [49]. In the eight different case studies, JWH-018, JWH-081, JWH-201, and JWH-210 were identified as well.

A UHPLC-MS/MS method, in MRM mode, was developed and validated for the detection of 14 synthetic cannabinoids and 15 metabolites in negative urine [50]. β-glucuronidase was utilized for hydrolysis for the Immunalysis immunoassay for screening. JWH-081-5-hydroxypentyl was not able to be detected by the immunoassay. However, the UHPLC-MS/MS method was sensitive to all 29 compounds and metabolites. The LOD for the analytes ranged from 1.0 to 5.0 ng/mL; the LOQ was calculated to range from 3.0 to 5.0 ng/mL.

An LC-CLND method was developed for the analysis of 177 samples seized by the police in 2011–12 and Finland customs in 2011–13 for 61 different NPSs [51]. The percentage of drug content in 12 customs samples and 2 police samples for JWH-018 ranged from 1.7% to 9.2% and 1.6% to 4.7% for JWH-073. The drug content for the Customs 4 sample was identified to be 2.2% of JWH-073, and the drug content in the Customs 5 sample was determined to be 4.3% of JWH-073. There was only one synthetic cannabinoid identified in the Customs 12 sample, JWH-122, at a concentration of 17% of its weight. An LC-MS/MS method was developed for the analysis of human urine for JWH-018, JWH-019, JWH-073, JWH-081, JWH-122, JWH-200, JWH-398, AM-2201, and MAM-2201 [53]. The correlation coefficient (r^2) values ranged from 0.994 to 0.999, and the LODs for the analytes ranged from 0.1 to 0.5 μg/L. There was no evidence of any carry over for the synthetic cannabinoids, proving the accuracy of the method.

An MEKC coupled with mass spectrometry (MEKC-MS/MS) method was developed and validated for the analysis of 12 different synthetic cannabinoids, including AM-1220, JWH-015, JWH-022, and JWH-200 [54]. The LODs and recoveries for all of the analytes ranged from 6.5 to 76.5 μg/g and 89.5% to 101.7%, respectively. After validation, this method was applied to two herbal products purchased in Japan, in which naphthoylindole-type cannabimimetics were detected in these two herbal products.

A UHPLC-MS/MS method was utilized for the analysis of a total of 344 hair samples for the presence of 23 different synthetic cannabinoids [55]. All of these samples were retested for the presence of these 23 synthetic cannabinoids, using the UHPLC-MS/MS method. AM-1220 was only identified in one sample, while JWH-019 was identified in two samples. JWH-018, JWH-073, JWH-081, and JWH-122 were all identified in more than one sample.

NMR spectroscopy was used to screen and quantify synthetic cannabinoids, including AM-2201, JWH-018, and JWH-122, in Melon Code Black, Cloud Nine, Sweet Leaf, Head Trip, Extremely Legal, Zombie Matter, Moon Spice, Ion Source, Rack City, Nuclear Bomb, Mr. Nice Guy, and K250 [56]. This method is clean and simple, without any need for chromatographic isolation, maintaining excellent accuracy.

Using a published, quantitative LC-MS/MS method [53], 777 urine specimens from US military personnel were analyzed to identify 53 synthetic cannabinoid analytes [57]. Overall, 290 urine specimens contained 1 or more of the 22 different synthetic cannabinoid metabolites from 11 parent compounds. JWH-018, JWH-073, JWH-122, and AM-2201 were the most prevalent parent compounds identified. JWH-018-COOH, JWH-018-OH, JWH-073-COOH, JWH-122-OH, and AM-2201-OH were the metabolites identified.

A method was developed for the analysis of bulk powder and/or biofluid samples using a miniature ion trap mass spectrometry system with two ambient ionization methods. The extraction spray ionization and the paper spray ambient ionization methods aided in the identification of five different synthetic cannabinoids (including three naphthoylindoles) [58]. These compounds were detected in the samples, rendering this method effective.

A GC-CI-MS method, run in MRM mode, was developed and validated for the qualitative analysis of 34 synthetic cannabinoids and 7 synthetic cathinones, allowing for the simple identification of these compounds in herbal products [59]. The retention time was locked at 20.765 minutes (proadifen) to provide accurate retention times for the same analytes over time. The method was successful in identifying these NPSs.

A DART-ORBITRAP method was developed for the analysis of eight different samples, all including herbal material and pure compounds for the detection of synthetic cannabinoids, under various experimental conditions [60]. The advantage of this analysis is no sample or solvent preparation. This is a valid alternate to GC and LC analyses.

Two different LC-MS/MS quantitative methods were developed and validated for analyzing blood and urine for various synthetic cannabinoids [61]. One method dealt with the analysis of 15 parent synthetic cannabinoids in blood, while the other method analyzed urine for 17 different cannabimimetic metabolites, with two parent compounds qualitatively identified. In blood, the LOD for the synthetic cannabinoids ranged from 0.025 to 0.10 ng/mL, and the LOQ ranged from 0.025 to 0.10 ng/mL. In urine, the LOD for the synthetic cannabinoids ranged from 0.01 to 0.50 ng/mL, and the LOQ ranged from 0.10 to 0.50 ng/mL. AM-2201, JWH-018, JWH-073, JWH-081, and JWH-122 were identified to be positive in 15 samples, 23 samples, 4 samples, 5 samples, and 16 samples, respectively.

Using NMR and DART-MS analysis, four herbal and/or chemical products were analyzed to identify cannabinoids and synthetic cannabinoids [62]. In the four powders, AM-2201 was identified in "AM-1221," JWH-019 was identified in "AM-2201," and JWH-200 was identified in "JWH-122." Out of the seven herbal products, "Barely Legal" contained AM-2201, "Funky Monkey" contained JWH-122 only, "Ion Lab" contained AM-2201, "Melon Code Black" contained JWH-122, "Moon Spice" contained JWH-018, "Mr. Nice Guy," and "Sweet Leaf" contained JWH-122 and JWH-210. This method is useful, as it provides a rapid and accurate method of analyzing herbal products.

The pyrolytic products of JWH-018 were analyzed, as well as various other synthetic cannabinoids, using GC-MS [64]. The pyrolytic products of JWH-018 were identified to be 1-naphthalenecarbonitrile, 2 H-1-benzopyran-2-one, Methoxynaphthalene or Naphthol, 1 H-indole-3-ethanol, and penten-4-yl-JWH-018; JWH-073: 1-butylindole, 1-butylquinoline, 1-butylindole-3-carboxaldehyde, 1-butyl-3-acetylindole, and buten-3-yl-JWH-073; JWH-081: penten-4-yl-JWH-081; and MAM-2201: 3-(1-methylnaphthoyl)-indole and *N*-propyl-3-(1-methylnaphthoyl)-indole. This analysis is useful, as it provides the pyrolytic cannabimimetic metabolites.

Herbal and chemical products commercially available in Japan were analyzed for the identification and quantification of six synthetic cannabinoids, including four naphthoylindole cannabimimetics: JWH-081, JWH-073, JWH-015, and JWH-200 [65]. LC-MS and GC-MS were used for qualitative and quantitative analysis, AccuTOF-DART was used for the determination of the accurate mass spectrums of the target compounds, HPLC analysis for the isolation of the compounds, and NMR analysis for the accurate determination of the structures of the compounds. JWH-015 and JWH-200 were each found in one product. JWH-081 and JWH-073 were identified in multiple herbal or chemical products.

An herbal product sold in Japan was analyzed by GC-MS, LC-MS, high resolution mass spectrometry (HR-MS), and NMR [66]. The herbal product was found to contain JWH-122 and JWH-081, in addition to a new phenylacetylindole, cannabipiperidiethanone (CPE). Based on NMR analysis, the structure of CPE was characterized and determined to be a mix of two other synthetic cannabinoids: JWH-250 and AM-2233.

An LC-MS/MS method was developed for the analysis of urine samples for JWH-018, JWH-073, and several metabolites of each synthetic cannabinoid [67]. This analytical method does not require solid or liquid phase extraction, only requiring a methanolic dilution step, and hydrolysis with β-glucuronidase, depending on the analyte. This method was successfully able to detect JWH-018, 5 hydroxylated metabolites of JWH-018, one carboxylated derivative of JWH-018, JWH-073, 5 hydroxylated metabolites of JWH-073, and one carboxylated derivative of JWH-073. This method was further applied to urine samples from three subjects. Subjects 1 and 2 had consumed off-white powders, while Subject 3 utilized what was described as "a green-leafy substance" in a cigarette butt. JWH-018 and its metabolites was identified in samples obtained from Subjects 1 and 2, and both JWH-018 and JWH-073, and their respective metabolites, were identified in samples obtained from Subject 3.

An LC-MS/MS method, utilizing solid-phase extraction, was developed and validated for the analysis of oral fluid collected by a Quantisal device for JWH-018 and JWH-073 in samples collected from two naïve subjects [68]. Subject 1 smoked "Blueberry Posh," and Subject 2 smoked "Black Mamba," out of which "Black Mamba" was reported to have a stronger euphoric effect than "Blueberry Posh." Samples were collected at various time points. For the subject who consumed "Blueberry Posh," a peak concentration of 35 µg/L for JWH-018 was quantified; at a sample collected 12 hours after the consumption of "Blueberry Posh," the concentration of JWH-018 was 0.5 µg/L. In the samples collected from Subject 2, 20 minutes after smoking "Black Mamba," the concentration of JWH-018 was identified to be 3 µg/L. Only 5 hours after consumption, the concentration of JWH-018 had fallen below the LOQ, but was still able to be detected. These samples were kept at 4°C for 1 month, and then reanalyzed. The concentrations remained almost identical, confirming stability.

Various Spice aromatic smoking blends available in the Khanty-Mansi Autonomous District (Nizhnevartovsk, Surgut, Khanty-Mansiysk, and Nefteyugansk) were analyzed for JWH-018, JWH-019, and JWH-073, using GC-MS analysis [69]. All of the test samples were found to contain at least one synthetic cannabinoid.

HR-MS and mass defect filtering were used to analyze herbal products for JWH-015, JWH-018, JWH-019, JWH-073, and JWH-200 [70]. At least one synthetic cannabinoid was detected in each sample. This method is highly useful, as it allows for quantification at low concentration levels that would be otherwise unable be quantified; in addition, this method introduces the nontargeted identification of metabolites and prediction of new synthetic cannabinoids.

A UPLC-MS/MS method was developed and validated for the analysis of postmortem whole blood casework for JWH-018 and JWH-073 [71]. The total run time was 2.6 minutes, with an LOD of 0.01 ng/mL for every analyte. The method was applied to postmortem blood samples and proven to be reliable. The concentrations of JWH-018 ranged from 0.1 to 199 ng/mL, while the concentrations for JWH-073 ranged 0.1–68.3 ng/mL.

Using salted-out assisted enzymatic hydrolysis and salted-out assisted liquid–liquid extraction (SALLE), urine samples were analyzed by UPLC-MS/MS for the four urinary metabolites of JWH-018 and JWH-073 (JWH-018-COOH, JWH-018-OH, JWH-073-COOH, and JWH-073-OH) [72]. It was shown that 10 minutes was the best interval to obtain the metabolites of interest from glucuronide conjugates using β-glucuronidase. This method proved to be simple and cost effective. JWH-018, JWH-018-OH, JWH-018-COOH, JWH-073, JWH-073-OH, JWH-200, JWH-250, and JWH-250-OH were extracted from plasma, urine, and whole blood, utilizing supported liquid extraction in cartridge and a 96-well format, by LC-MS/MS analysis [73]. The recoveries for these synthetic cannabinoid parents and metabolites ranged from 70% to 98%, with calibration curves used to quantitate in-house quality control (QC) with accuracies between 90% and 110%.

GC-MS, LC-HR-MS, and TLC analyses were carried out for a research chemical and several herbal mixtures [74]. After analysis, the research chemical was found to contain pure AM-1220 and its azepane isomer. Following this analysis, the herbal mixtures were analyzed; both AM-1220 and its azepane isomer were detected in all of the herbal products. An LC-MS/MS method was developed to quantitate the concentration of eleven synthetic cannabinoid analytes, including JWH-018-*N*-pentanoic acid, JWH-018-*N*-(5-OH-pentyl), JWH-019-*N*-5-OH-indole, JWH-073-*N*-butanoic acid, JWH-073-*N*-(4-OH-butyl), JWH-122-*N*-5-OH-urine, JWH-200-5-OH-indole, and JWH-398-*N*-5-OH pentyl, in human urine [43]. The CV percentage for these analytes were determined to be 11.9, 15.9, 3.8, 8.2, 3.9, 11.2, 12.0, and 14.8, respectively, proving the reproducibility of this method.

Several charred samples containing AM-2201 were analyzed for JWH-018 and JWH-022 as products of combustion, using GC-MS and AccuTOF-DART [75]. JWH-018 and JWH-022 were isolated, and the un-burnt and charred samples were analyzed. The un-burnt sample was found to contain pure AM-1220, with no concentrations of JWH-018 and JWH-022. However, in the burnt samples, both JWH-018 and JWH-022 were identified, confirming the conclusion that JWH-018 and JWH-022 are indeed combustion products of AM-1220.

A liquid chromatography with electrospray ionization mass spectrometry (LC-ESI-MS/MS) method was developed and validated for the analysis of 30 different synthetic cannabinoids in oral fluid [76]. The Dräger DCD 5000 was used for the collection of the oral fluid. This method was applied to 264 real oral fluid samples, out of which 26 oral fluid samples tested positive for parallel serum samples for synthetic cannabinoids. There was a general correlation between positive oral fluid samples and serum samples testing positive; in addition, synthetic cannabinoid concentrations were determined to be greater in oral fluid samples as compared to the serum levels.

The first immunoassay was developed for the detection of JWH-018, JWH-022, JWH-073, JWH-200, AM-1220, and AM-2201 in oral fluids [77]. The oral fluid was collected using the Quantisal device. This assay was then applied to 32 patient samples. The assay cutoff was determined to be 0.25 ng/mL, with the LOD and the upper limit of linearity (ULOL) being 0.1 and 50 ng/mL, respectively. All of the positives were confirmed with LC-MS-MS analysis, with no false-positives were detected. Out of the 32 samples, AM-2201 was detected in 24 samples, and JWH-018 was detected in 5 samples.

The UPLC-TOF-MS was used for the identification of five new synthetic cannabinoids, including MAM-2201, in herbal products [78]. A total of 18 herbal products were analyzed, and no previously controlled synthetic cannabinoids were detected. MAM-2201 was detected in seven herbal products.

Rat and human hair were analyzed for JWH-018, JWH-073, and their metabolites, using an UHPLC-MS-MS method [79]. In the hair of five lean Zucker rats, JWH-073 N-3-OH and JWH-073 N-COOH were both detected in pigmented and nonpigmented hair, while JWH-073 N-4-OH was detected at either less than LOQ concentrations or not detectable at all. In the hair of the human participants, JWH-018, JWH-018 N-5-OH, and JWH-073 were detected. THC-COOH was detected in four samples.

An LC-MS/MS method, using liquid−liquid extraction, was developed for the analysis of various biological matrices (blood, saliva, and urine) for a multitude synthetic cannabinoids [80]. The limits of detections (LODs) for the six synthetic cannabinoids identified in saliva and blood are represented in Table 22.1. In urine, the LODs for the metabolites of JWH-018 and JWH-073 ranged from 0.3 to 0.4 ng/mL.

An ultra-HPLC with photodiode array UV and mass spectrometry (UHPLC-PDA/UV-MS) method was developed for the analysis of 62 NPSs, including various naphthoylindoles [81]. The retention times and LODs were calculated for these synthetic cannabinoids, ranging from 10.086 to 16.64 minutes and 0.3 to 3.0 µg/mL, respectively. This method was also applied to 20 real herbal products, out of which 13 were plant material and 7 powdered material. AM-2201 was only identified in one herbal product, "Barely Legal." The results from the UHPLC-PDA/UV-MS method accurately matched those by GC-MS, proving the reliability of this method and instrument.

A DART-MS method was developed for the analysis of six "Spice products" for five synthetic cannabinoids, including AM-2201, JWH-122, and JWH-210 [52]. "Happy Hour Cherry Bomb," "California Kronic Blueberry," "Relaxinol," "Mr. Nice Guy—Strawberry," and "Dead Man Walking" all contained AM-2201. JWH-122 was identified in "Mr. Nice Guy—Strawberry."

The major metabolites of AM-2201 were identified in human and rat urine, using LC-MS/MS analysis [82]. The metabolites analyzed were JWH-018 N-(4-hydroxypentyl), JWH-018 N-(5-hydroxypentyl), JWH-018 N-pentanoic acid, JWH-018 6-hydroxyindole, JWH-073 N-butanoic acid, AM-2201 N-(4-hydroxypentyl), and AM-2201 6-hydroxyindole. The LODs for the N-OH and 6-OH-indole metabolites of AM-2201 were determined to be 0.1 ng/mL, and the LOQs for these two metabolites were 2.5 ng/mL. These metabolites were quantitated in three samples of rat urine. JWH-018

TABLE 22.1 Synthetic Cannabinoid LODs for LC-MS/MS [80]

Synthetic Cannabinoid	Saliva (ng/mL)	Plasma/Serum (ng/mL)
JWH-007	0.2	0.2
JWH-015	0.3	0.2
JWH-018	0.3	0.2
JWH-073	0.4	0.2
JWH-122	0.5	0.5
JWH-200	0.5	0.5

Source: Adapted from Mazzarino M, de la Torre X, Botrè F. A liquid chromatography-mass spectrometry method based on class characteristic fragmentation pathways to detect the class of indole-derivative synthetic cannabinoids in biological samples. Anal Chim Acta 2014; 87:70−82.

N-4-OH and JWH-073 N-COOH were not detected in rat urine. JWH-018 N-5-OH and AM-2201 N-4-OH were both detected; however, the concentrations were all lower than the LOQ. The concentrations of JWH-018 N-COOH were 35.9 ± 0.1, 132.0 ± 2.7, and 79.5 ± 1.4 ng/mL, and the concentrations of AM-2201 6-OH-indole were determined to be 11.4 ± 0.1, 15.1 ± 0.3, and 17.1 ± 0.4 ng/mL. JWH-073 N-COOH was also detected in all three samples; only one sample was quantifiable at a concentration of 2.8 ± 0.1 ng/mL. This was the first report to identify the different metabolites of JWH-018 and AM-2201.

An LC-QTOF-MS method was developed and validated for the analysis of the metabolites of AM-2201, JWH-018, JWH-019, JWH-073, JWH-081, JWH-122, JWH-398, and MAM-2201, and compared the analysis with results from an Immunalysis Spice K2 homogeneous enzyme immunoassay (HEIA), which was developed to detect metabolites from JWH-018, JWH-073, and AM-2201 [83]. After analysis, it was concluded that there was high cross-reactivity with the MAM-2201 and JWH-122 metabolites in the immunoassay. The LC-QTOF-MS method is more adaptable, reduces the validation, and detects untargeted analytes, while the immunoassay will only detect targeted analytes, proving the LC-QTOF-MS method more efficient, flexible, and reproducible.

GC-MS analysis was used for the comparison of JWH-018 and all five of its regioisomers. The GC-MS used an Rtx-200 stationary phase and separated the six regioisomers [84]. The regioisomers were eluted in following order: 7-(1-naphthoyl)-1-pentylindole, 2-(1-naphthoyl)-1-pentylindole, 6-(1-naphthoyl)-1-pentylindole, 4-(1-naphthoyl)-1-pentylindole, 3-(1-naphthoyl)-1-pentylindole, and 5-(1-naphthoyl)-1-pentylindole. The 7-(1-naphthoyl)-1-pentylindole and 2-(1-naphthoyl)-1-pentylindole elute closely together due to the steric interaction similarities in their structures. The other four regioisomers, 6-(1-naphthoyl)-1-pentylindole, 4-(1-naphthoyl)-1-pentylindole, 3-(1-naphthoyl)-1-pentylindole, and 5-(1-naphthoyl)-1-pentylindole are related structurally, leading to their similar retention times.

HR-MS and LC/QTOF-MS were used for the analysis of postmortem plasma and urine samples for synthetic cannabinoid abuse [85]. Defluorinated metabolites, N-dealkylated metabolites, some hydroxylated metabolites, and further oxidized metabolites of MAM-2201 were detected, based on structure estimation. When quantitated in postmortem plasma samples, the femoral, left ventricle, and right ventricle concentrations were identified to be 140, 438, and 222 ng/mL for AM-1220, 0.86, 1.95, and 1.76 ng/mL for AM-2232, and 16.3, 85.8, and 30.7 ng/mL, respectively, for MAM-2201. In addition, four metabolites (3-naphthoylindole, 3-(4-methyl-1-napthoyl)indole, 5-hydroxypentyl metabolite of MAM-2201, and the carboxy metabolite of MAM-2201) were also quantitated, with concentrations of 9.19, 13.8, and 15.1 ng/mL for the first metabolite in the femoral, left ventricle, and right ventricle plasma samples. For the second, third, and fourth metabolites, the concentrations in the femoral, left ventricle, and right ventricle plasma samples were determined to be 1.79, 9.76, and 4.70 ng/mL, 116, 223, and 178 ng/mL, and 7.30, 14.4, and 25 ng/mL, respectively.

The National Medical Services (NMS) JWH-018 direct enzyme linked immunosorbent assay (ELISA) kit was validated for the detection of synthetic cannabinoids in 2492 urine samples [86]. These results were confirmed by an LC-MS/MS method for the analysis of 29 synthetic cannabinoids. According to the results, the conjugate binds better with the antibodies not in the presence of matrix elements, but rather in the presence of buffer. The results also show how cross-reactivity is increased by the presence of hydroxyl substituents on the alkyl side chain; yet, when the hydroxyl substituent was on the indole moiety, reactivity was decreased significantly. Overall, the 5 µg/L cutoff was the most sensitive and was highly recommended when using the NMS JWH-018 direct ELISA to screen urine samples for synthetic cannabinoids.

An LC-MS/MS method was developed and validated for the analysis of five human urine specimens; one sample was taken from an individual who used a smoking apparatus, one sample from an individual who used dried leaves, and one sample from a cigarette butt, for the identification of 37 synthetic cannabinoids and metabolites [87]. In the fourth sample, JWH-122 N-5-OH, JWH-122 N-4-OH, MAM-2201 N-COOH, and MAM-2201 N-4-OH, as well as THC and two synthetic cannabinoids as parent drugs. No naphthoylindole cannabimimetic was detected in any of the other samples.

An LC-HR-MS/MS method was developed and validated for the analysis of human urine for 47 different synthetic cannabinoids, using nontargeted SWATH acquisition [88]. This nontargeted approach allows for the simple addition of new synthetic cannabinoids, without having to modify the entire method to accommodate for the new synthetic cannabinoids. This method was then applied to eight anonymous urine samples, out of which all were positive for at least one synthetic cannabinoid metabolite, proving the effectiveness and reliability of this method.

A supercritical fluid chromatography coupled with an electron-spray ionization tandem mass spectrometer (SFC-ESI-MS/MS) method was developed for the analysis of three commercially bought dried plant products (labeled A, Y, and G) for eight synthetic cannabinoids [89]. JWH-073 was identified in Product A. JWH-018 was identified in both Product Y and G.

A total of 17 urine specimens were obtained as part of a military law enforcement investigation, out of which 9 specimens were identified to be positive for synthetic cannabinoid usage, using GC-MS [90]. This study included the analysis of two metabolites of JWH-018 and JWH-073 each (JWH-018-OH, JWH-018, COOH, JWH-073-OH, and JWH-073-COOH). A total of 11 out of the 17 samples were hydrolyzed with β-glucuronidase, while 6 samples were not. JWH-073-COOH was identified in seven samples, three of which also contained JWH-073-OH. JWH-018-COOH was identified in seven samples, while JWH-018-OH was identified in eight samples.

A UPLC-MS/MS method was developed and validated for the forensic analysis of JWH-073 and its metabolites JWH-073 N-butanoic acid and JWH-073 N-(4-hydroxybutyl) in blood and urine [91]. The analysis was conducted on 1620 blood samples and 976 urine samples taken from 1856 subjects, leading to a total of 2596 samples. A total of 1428 were identified to be positive for synthetic cannabinoids. A total of 91.1% of the positive samples were from males, while 8.9% were obtained from females. JWH-073-OH, the most prevalent metabolite monitored in the samples, was identified in 58 blood samples and 21 urine samples, and JWH-073-COOH was detected in 27 blood samples and 13 urine samples, while JWH-073 was not identified in any urine or blood sample. JWH-073 metabolites were identified in conjunction with JWH-018 and its metabolites. JHW-073-OH was identified alone in one sample. A total of 5.5% of the samples were identified to be positive for JWH-073 metabolites.

Blood samples from individuals in driving under influence of drug (DUID) and intoxication cases were analyzed [92] using a previously developed and validated LC-MS/MS method [93]. Out of the DUID cases, AM-2201, JWH-018, and JWH-122 were identified with the highest concentrations. In the intoxication cases, the most prevalent synthetic cannabinoids detected were AM-2201, JWH-122, MAM-2201, and JWH-018. JWH-018 and AM-2201 were identified in three autopsy cases each, while AM-2201-4-OH-Pentyl, JWH-018-5-OH-pentyl, JWH-018 pentanoic acid, JWH-073, and JWH-120 were only found in one autopsy case each.

An LC-MS/MS method was developed, validated, and applied to analyze several compounds in wastewater, including JWH-018, JWH-073, and JWH-122 [94]. In the influent wastewater tested, JWH-018 and JWH-073 were not detected. JWH-122 was detected in three samples, all of which were below the LOQ. In the effluent wastewater samples, JWH-018 and JWH-073 were not detected in any sample. JWH-122 was identified in four samples in quantifiable concentrations.

Two synthetic cannabinoids and 11 synthetic cannabinoid metabolites were detected, using liquid–liquid extraction (with an extraction efficiency of greater than 70%) via two instruments: ultra-high-performance supercritical fluid chromatography coupled with mass spectrometry (UHPSFC-MS/MS) and UHPLC-MS/MS [95]. This was the first report of the analysis of synthetic cannabinoids in biological matrices using UHPSFC-MS/MS. After validation, 13 authentic samples were tested using these methods and found positive for 5 different metabolites: AM-2201-N-4-OH-pentyl, JWH-018-N-pentaoic acid, JWH-018 N-5-OH-pentyl, JWH-073 N-butanoic acid, and JWH-122 N-5-OH-pentyl.

Human and rat hair were analyzed for JWH-018 and its acid and hydroxy metabolites, JWH-073 and its acid metabolite, JWH-122 and its hydroxy metabolite, AM-2201 and its hydroxy and hydroxy indole metabolite, MAM-2201, and THC-COOH [96]. This expands the range of synthetic cannabinoids detected in a previously published LC-MS/MS method [80]. JWH-018 was detected in six individuals, and JWH-018-N-COOH was detected in the hair of six individuals, out of which it was quantifiable in only three individuals. JWH-018-N-5-OH was detected in six individuals as well, with concentrations quantifiable for all six individuals. JWH-073 was detected in six individuals, with the hydroxy metabolite detected in only one individual. AM-2201 and MAM-2201 were detected in all of the individuals, while both of AM-2201's metabolites were only detected in two individuals. JWH-122 and its hydroxy metabolite were detected in nine individuals. THC-COOH was tested in six individuals, out of which only three individuals had detectable quantities of THC acid. It is interesting to note that there were four females and five males, all of whose ages ranged from 20 to 30. It is interesting to note that individuals who had blond hair typically had greater concentrations than that of either black or brown colored hair.

An ultra-HPLC with ultra-violet detection (UPLC-UV) and UPLC-MS/MS methods were developed and validated for the analysis of 45 different herbal product samples for THC and six naphthoylindoles-type synthetic cannabinoids [97]. The LOD for the UPLC-UV for the solid compounds ranged from 0.1 to 0.3 μg/mL, while the LOQ ranged from 0.3 to 0.9 μg/mL. The LOD for the UPLC-UV for the liquid samples ranged from 0.1 to 0.4 μg/mL, and the LOQ ranged from 0.3 to 1.2 μg/mL. For the UPLC-MS/MS, the LOD for the solid samples ranged from 0.01 to 0.3 ng/mL, and the LOQ ranged from 0.09 to 1.5 ng/mL. For the liquid samples on the UPLC-MS/MS, the LOD ranged from 0.02 to 0.6 ng/mL; the LOD ranged from 0.06 to 1.8 ng/mL. All 45 samples were tested at 3 different concentrations: Near the LOQ, 10x the LOQ, and 20x the LOQ. This method proved to be successful and can be used to quantify these synthetic cannabinoids in these herbal mixtures.

Two RapidFire tandem mass spectrometry (RF-MS/MS) methods were used for the analysis of 28 different drugs of abuse, including 14 synthetic cannabinoids [98]. This method was validated and proved effective by the analysis of 18,000 urine samples. The LOD and LOQ were determined to be 1.00 ng/mL, and the calibration curve showed good correlation ($r^2 > 0.985$). None of these compounds caused any interference in the negative and low concentration samples. The instrument run time was multiple fold less for the RF-MS/MS than LC-MS/MS analysis time, enabling this method to be quick and efficient.

Using LC-MS/MS, 526 urine samples were analyzed from suspected DUID drivers for the presence of 12 different synthetic cannabinoid metabolites [99]. A total of 19 samples were identified to be positive for any one of these different synthetic cannabinoid metabolites, and 18 of these samples were identified to be positive for alcohol and other illicit drugs, such as PCP, and THC-COOH. AM-2201, JWH-018, and JWH-017 were identified positive in 3 samples. This method was successful for the quantitation of these 12 cannabimimetic metabolites.

An LC-MS/MS method was developed for the identification and determination of three cannabinoids, three cannabinoid metabolites, and four synthetic cannabinoids in urine [100]. The LOD for these 10 compounds ranged from 0.01 to 0.5 ng/mL, and the LOQ ranged from 0.05 to 1 ng/mL. This method was applied to five authentic samples, out of which three samples were identified to be positive. The concentrations of THC-COOH-glucuronide found in these positive samples were 133.51, 201.27, and 54.60, respectively, while the concentrations of THC-COOH were 8.72, 25.96, and 2.81 ng/mL, respectively. Three samples were detected to be positive for JWH-073, while two samples were identified positive for JWH-018.

Electro-oxidation was used for the detection of 11 different synthetic cannabinoids in artificial saliva, as well as in an herbal product [101]. Both analyses resulted in excellent separation and identification. The analysis was carried out on a LC-ESI-QTOF-MS and a gas chromatography by supersonic molecular beam mass spectrometry equipped with triple quadrupole analyzer (GC-SMB-EI-QQQ-MS). The LOD for AM-2201 and JWH-018 for the GC-SMB-QQQ-MS analysis were determined to be 0.633 and 0.486 ng/mL, respectively. For the LC-ESI-QTOF-MS analysis, the LOD was calculated to be 0.009 and 0.005 ng/mL, respectively.

An LC-MS/MS method was developed and validated for the analysis of 32 synthetic cannabinoid metabolites using an Agilent LC-MS-MS. The correlation coefficient (r^2) for all analytes was greater than 0.990, and the calculated concentrations were within ± 20% of the nominal concentrations [102]. This method was then applied to 25 authentic samples from synthetic cannabinoid users, out of which the metabolites of JWH-018 and JWH-073 were detected in the 25 samples.

An LC-MS/MS method was developed and validated for the analysis of blood samples (requiring only 200 μL) for the detection of 143 new synthetic cannabinoids, in a single run [103]. For the validation of this method, 32 compounds were selected; the r^2 values for all 32 cannabimimetics were identified to be greater than 0.97. The recovery percentage for these synthetic cannabinoids ranged from 11.3% to 133%. This method is useful because it allows for the simultaneous detection of 143 synthetic cannabinoids in only 200 μL of a blood sample in a single run.

A method was developed for the simultaneous separation, identification, and quantification of selected synthetic cannabinoids and metabolites by MEKC-MS in serum and urine samples [104]. The r^2 values were greater than 0.99 for all analytes tested; the LOD ranged from 0.9 to 3.0 ng/mL, while the LOQ ranged from 3.0 to 10.0 ng/mL. Only four parent compounds (including JWH-018, JWH-073, and JWH-200) were identified in the serum sample, while the metabolites were identified in the urine samples. This can be an alternate method to GC and LC methods for the detection of cannabimimetics.

A near-infrared spectroscopy (NIR) method was developed for the analysis of six synthetic cannabinoids, including JWH-018, JWH-073, JWH-081, and JWH-122, applied to 13 herbal products [105]. Eight herbal products were identified to be positive for JWH-018, JWH-073, JWH-081, and JWH-122. This method is useful because it provides an efficient and inexpensive technique for the analysis of these illicit compounds.

The performance of the UHPSFC instrument was demonstrated in the separation of various synthetic cannabinoids [106]. Two different mixtures were analyzed: One containing a random collection of 22 various, controlled synthetic cannabinoids, while the other contained JWH-018 and nine of its noncontrolled positional isomers. Using achiral and chiral stationary phases, all of the positional isomers, diastereomer, and 22 synthetic cannabinoids were baseline separated, proving the efficiency of the method. This can be an alternative technique for the detection of these cannabimimetics.

A surface enhanced Raman spectroscopy (SERS) method, using gold nanoparticles, was developed for the identification of four synthetic cannabinoids, including JWH-018, JWH-073, JWH-081, and JWH-122, with each compound producing its own unique Raman spectrum [107]. The analysis was verified in spiked urine samples, proving this method to be successful. In addition to this analysis, the effect of varying MgCl$_2$ concentration on the sensitivity of the Raman spectrum was tested. The efficiency comparison of bench-top Raman analysis versus portable Raman analysis was also tested. This is the first report of SERS analysis of any synthetic cannabinoids.

Eight synthetic cannabinoids were identified in four herbal products, using DART-TOF-MS as a screening method and LC-QTOF-MS as a confirmatory technique [108]. The first herbal product, "Aztec Potpourri," contained both AM-2201 and JWH-122, while the "K2 Pink" herbal product contained JWH-018 and JWH-019. This method proved to be effective and useful in the identification of these eight synthetic cannabinoids in herbal products.

UHPSFC-MS/MS was utilized for the detection of cannabinoid [Δ^9-THC, THC-OH, THC-COOH, THC(OH)$_2$] and synthetic cannabinoid metabolites (JWH-018-OH, JWH-018-COOH, JWH-073-OH, and JWH-122-OH) in three wastewater samples, using liquid–liquid extraction [109]. The coefficient of linearity (r^2) ranged from 0.9957 to 0.9999. THC-COOH was found in levels of 318, 187, and 209 ng/L, for the three wastewater samples, and THC-OH was detected in the first and third wastewater samples at concentrations of 75 and 57 ng/L, respectively. The peaks for Δ^9-THC and THC(OH)$_2$ were superimposed upon one another; the concentrations of these analytes in the three wastewater samples were determined to be 47, 19, and 23 ng/L. The synthetic cannabinoids were not identified in the wastewater samples at quantifiable levels.

An LC-MS/MS method was developed and validated for the analysis of 1142 human whole blood for the detection of two classes of synthetic cannabinoids—aminocarbonyl/carboxamide and arylindole [110]. These synthetic cannabinoids were extracted using two different extraction procedures, each designed to optimize the recovery of the two classes. A third extraction procedure was carried out for the compounds using deuterated internal standards for quantitative analysis. JWH-018, AM-2201, and JWH-122 were identified positive in five samples, two samples, and one sample, respectively. The mean concentration was calculated for some compounds, ranging from 0.21 to 7.0 ng/mL for JWH-018, 0.67 to 1.0 ng/mL for AM-2201, and 0.14 to 0.21 ng/mL for JWH-122.

An HPLC-HR-MS method was developed and validated, using pressurized liquid extraction for various NPSs, including synthetic cannabinoids, in hair samples [111]. The LOD for JWH-200, JWH-018 *N*-(5-hydroxypentyl), MAM-2201 *N*-(5-pentanoic acid), and JWH-081 was determined to be 1, 7, 5, and 3 pg/mg, respectively, while the LOQ for these naphthoylindole-type cannabimimetics was determined to be 5, 30, 40, and 10 pg/mg, respectively. The pressurized liquid extraction is an alternative extraction method to SPE and LLE.

Three GC-MS techniques, GC-EI-MS, positive GC-CI-MS, and negative GC-CI-MS, were used for the structural elucidation of indole-type synthetic cannabinoids [112]. This data was applied to the analysis of seven herbal products. MAM-2201 was identified in three samples (GOU, GOU+, and JIN) but not in any other sample. No other samples contained any naphthoylindole-type synthetic cannabinoids.

An HPLC with charged aerosol detection (HPLC-CAD) method was developed and validated for the quantitation of synthetic cannabinoids in 30 different herbal blends [113]. The LOD and LOQ for JWH-081 were determined to be 0.5 and 1.8 μg/mL, respectively. After validation, the method was applied to four different samples. An LC-QTOF-MS method was utilized for peak identification and purity qualification. In addition, NMR spectroscopy was used for the quantification of the synthetic cannabinoids without a reference material, allowing for structure elucidation and verification. Three different packages of "Mocarz," an herbal mixture, were analyzed. JWH-203, JWH-081, and JWH-019, three structurally similar synthetic cannabinoids, were identified in the samples, with LC-CAD concentrations correlating with those from NMR analysis. This was the first report for utilizing LC-CAD for the quantification of cannabimimetics without any reference standards.

Phenylacetylindoles: Structures and Analysis

The structures of the phenylacetylindole-type synthetic cannabinoids are shown in Fig. 22.3.

Over 25 total samples were collected from different sources and analyzed for 30 synthetic cannabinoids, using LC-MS/MS and GC-MS analysis [17]. JWH-203, JWH-250, and JWH-251 were detected from the samples collected. Several legal high products were acquired from the prefederal ban time range and postfederal ban time range and analyzed for multiple synthetic cannabinoids with UPLC-TOF-MS analysis [38]. In samples collected from the prefederal ban time range, no phenylacetylindole synthetic cannabinoids were detected. Out of the 81 products in the postfederal ban time range, 8 nonfederally banned synthetic cannabinoids, including JWH-203, JWH-250, and RCS-8, were detected.

An HS-SPME-GC-MS method was developed for the analysis of 32 products for 13 synthetic cannabinoids, including JWH-250 [39]. In the products received before the DEA ban, JWH-250 was found in seven samples; however, in the samples obtained after the DEA ban, JWH-250 was not identified in any sample. An MEKC-DAD method was developed and validated for the analysis of 15 herbal blend samples for 10 synthetic cannabinoids, including JWH-250 [40]. The LOD of this method was determined to range from 3 to 5 μg. JWH-250 was not identified in any of the herbal blends samples.

An LC-MS/MS method was developed for the analysis of samples of human hair collected from 8 male individuals for 22 synthetic cannabinoids, including 3 phenylacetylindoles: JWH-203, JWH-250, and JWH-251 [41]. JWH-250 was identified in seven individuals. An LC-MS/MS method was developed for the quantitation of 11 synthetic cannabinoid analytes, including JWH-250-5-OH indole and JWH-250-N-(5-carboxypentyl), in human urine [44]. The CV percentage for these analytes was determined to be 2.9% and 11.4%, respectively, contributing to the excellent reproducibility of this method.

An LC-MS/MS method was developed for the analysis of 3078 blood samples from 8 severely intoxicated individuals for 29 synthetic cannabinoids [49]. The concentrations of JWH-203 and JWH-250 were identified in several samples, ranging from 0.06 to 68 ng/g and 0.14 to 0.73 ng/g, respectively. In the eight different case studies, both naphthoylindole-type synthetic cannabinoids identified. An LC-CLND method was developed for the analysis of 177 samples seized by police in 2011−12 and Finland customs in 2011−13 for 61 different compounds [51]. The drug content in the Customs 5 sample was determined to be 4.3% of JWH-073 + JWH-250, proving the applicability of this method.

A DART-MS method was developed for the analysis of six "Spice products" for five synthetic cannabinoids, including JWH-203 [52]. JWH-203 was identified in "Mr. Sticky Extreme," an herbal product. This method is quick and reproducible, allowing for the quick detection of these cannabimimetics in herbal products. An LC-MS/MS method was developed for the analysis of human urine for 15 synthetic cannabinoids, including JWH-203, JWH-250, RCS-8, and their metabolites [53]. The LODs for the analytes ranged from 0.1 to 0.5 μg/L, and the correlation coefficient (r^2) values for these analytes ranged from 0.994 to 0.999. These parent compounds and metabolites were detected with high accuracies and precision.

An MEKC-MS/MS method was developed and validated for 12 synthetic cannabinoids, including JWH-250 and JWH-251 [54]. This method was applied to two herbal products purchased in Japan. JWH-250 and JWH-251 were not detected in the two products.

A total of 344 hair samples were analyzed by UHPLC-MS/MS for the presence of 23 different synthetic cannabinoids, including JWH-203, JWH-250, and JWH-251 [55]. All 344 samples were retested for the presence of the targeted synthetic cannabinoids using the UHPLC-MS/MS method. JWH-250 was identified in six samples, with concentrations ranging between 4.8 pg/mg and 83.4 pg/mg.

A total of 12 different herbal products, including Melon Code Black, Cloud Nine, Sweet Leaf, Head Trip, Extremely Legal, Zombie Matter, Moon Spice, Ion Source, Rack City, Nuclear Bomb, Mr. Nice Guy, and K250, were analyzed by NMR for the screening and quantification of synthetic cannabinoids [56]. The spectroscopic fingerprints of many cannabimimetics including JWH-250 are an useful method. This analysis was quick and highly accurate, requiring no chromatographic isolation.

Over 700 urine species from US military personnel were analyzed, using a published, quantitative LC-MS/MS method [53] to identify 53 synthetic cannabinoid analytes, including 20 parent compounds and 33 metabolites [57]. Overall, 290 urine specimens contained 1 or more of 22 different synthetic cannabinoid metabolites, including JWH-250 metabolites.

A GC-CI-MS method, run in MRM mode, was developed and validated for the qualitative analysis of 34 synthetic cannabinoids, including 3 phenylacetylindoles: JWH-203, JWH-250, and JWH-251 and 7 synthetic cathinones, allowing for the simple identification of these compounds in herbal products [59]. This method was efficient and successful in the identification of these compounds, identifying this method as an alternative to traditional methods.

Two different LC-MS/MS quantitative methods were developed and validated for analyzing various blood and urine for synthetic cannabinoids [61]. One method analyzed blood for 15 parent synthetic cannabinoids, while the other method analyzed urine for 17 different cannabimimetic metabolites. In blood, the LOD for the synthetic cannabinoids ranged from 0.025 to 0.10 ng/mL, and the LOQ ranged from 0.025 to 0.10 ng/mL. In urine, the LOD for the synthetic cannabinoids ranged from 0.01 to 0.50 ng/mL, and the LOQ ranged from 0.10 to 0.50 ng/mL. JWH-250 was identified in five samples. These methods were successful in detecting the parent cannabimimetics and metabolites.

NMR and DART-MS analyses were used to identify cannabinoids and synthetic cannabinoids in seven herbal products [62]. JWH-203 was identified in "Melon Code Black," and JWH-250 was determined to be present in "Sweet Leaf." This method is useful, as it provides a rapid and accurate method of analyzing herbal products.

Herbal products commercially available in Japan were analyzed for the identification and quantification of six synthetic cannabinoids, including two phenylacetylindoles: JWH-251 and JWH-250 [63]. These products were analyzed through multiple techniques: LC-MS and GC-MS for qualitative and quantitative analysis, AccuTOF-DART analysis for the determination of the accurate mass spectra of the target compounds, HPLC analysis for the isolation of the compounds, and NMR for the accurate determination of the structures of the compounds. JWH-250 was identified in multiple herbal products, while JWH-251 was identified in only two of the products.

An herbal product sold in Japan was analyzed, utilizing GC-MS, LC-MS, HR-MS, and NMR analyses [66]. The herbal product was found to a new cannabimimetic phenylacetylindole, now known as CPE. Based on NMR analysis, the structure of CPE was determined to be a mix of two synthetic cannabinoids, namely JWH-250 and AM-2233.

An LC-MS/MS method was developed and validated for the analysis of oral fluid collected by a Quantisal device for seven synthetic cannabinoids, including JWH-250 [68]. This method was then applied to samples from two naïve subjects; JWH-250 was identified in neither of the samples.

HR-MS and mass defect filtering were used to analyze herbal products for JWH-250 [70]. At least one synthetic cannabinoid was detected in each sample. Using this method, samples with low concentrations can be quantified; in addition, this method predicts new synthetic cannabinoids and provides the nontargeted identification of metabolites. Victor E. Vandell and Frank Kero from Biotage, LLC extracted four parent compounds and their metabolites, including JWH-250 and JWH-250-OH, from plasma, urine, and whole blood utilizing supported liquid extraction in cartridge and a 96-well format, before LC-MS/MS analysis [73]. The recoveries for these parent synthetic cannabinoids and metabolites ranged from 70% to 98%, demonstrating extraction and method efficiencies.

An LC-MS/MS method, using liquid–liquid extraction, was developed for the analysis of blood, saliva, and urine for various indole-derived synthetic cannabinoids, including multiple naphthoylindole-type synthetic cannabinoids [80]. The LODs for JWH-250 and JWH-251 were determined to be 0.1 and 0.4 ng/mL, respectively, in saliva, and 0.5 and 0.1 ng/mL, respectively, in plasma/serum. An SFC-ESI-MS/MS method was developed for the analysis of three commercially bought dried plant products (labeled A, Y, and G) for eight synthetic cannabinoids, including JWH-250 [89]. JWH-250 was not identified in any of the three plant products. A total of 17 urine specimens were analyzed as part of a military law enforcement investigation. Nine specimens were identified to be positive for synthetic cannabinoid usage through GC-MS analysis [90]. JWH-250 and JWH-250 (C4-isomer) were both identified in one sample.

An LC-MS/MS method was developed, validated, and applied for the analysis of wastewater for six different synthetic cannabinoids, including JWH-250 [94]. However, JWH-250 was not identified in the influent and effluent wastewater tested.

UPLC-UV and UPLC-MS/MS methods were developed and validated for the analysis of 45 different herbal product samples for THC and 8 synthetic cannabinoids, including JWH-250 [97]. All 45 samples were tested at 3 different concentrations: Near the LOQ, 10x the LOQ, and 20x the LOQ. The LOD for the UPLC-UV method was calculated to be 0.3 μg/mL for both the solid and liquid samples, while the LOQ for both solid and liquid samples was identified to be 0.9 μg/mL. The LOD for the solid samples using the UPLC-MS/MS method was determined to be 0.3 ng/mL for the solid samples and 0.4 ng/mL for the liquid samples; the LOQ for the solid samples was identified to be 0.9 ng/mL for the solid samples and 1.2 ng/mL for the liquid samples. This method proved to be successful and can be used to quantify these synthetic cannabinoids in these herbal mixtures.

Using LC-MS/MS, 526 urine samples, collected from drivers suspected to be DUID, were analyzed for the presence of multiple metabolites, including JWH-250 N-(5-hydroxypentyl) [99]. A total of 19 samples were identified to be positive for these different synthetic cannabinoid metabolites, and 8 of these samples were identified to be positive for alcohol and other illicit drugs, such as PCP, and THC-COOH. This method was successful for the quantification of these twelve cannabimimetic metabolites.

An LC-MS/MS method was developed and validated for the identification and determination of five samples for three cannabinoids and metabolites, as well as four parent synthetic cannabinoids, including JWH-250, in urine [100]. The LOD for these 10 compounds ranged from 0.01 to 0.5 ng/mL, and the LOQ ranged from 0.05 to 1 ng/mL. JWH-250 was not identified in any of the samples.

A method was developed for the simultaneous separation, identification, and quantification by MEKC-MS of selected synthetic cannabinoids and metabolites in serum and urine samples [104]. The r^2 values were greater than 0.99 for all analytes tested, and the LOD ranged from 0.9 to 3.0 ng/mL, while the LOQ ranged from 3.0 to 10.0 ng/mL. The four parent compounds (including JWH-250) were identified in serum samples, while the metabolites of these cannabimimetics were identified in urine samples, proving the applicability of the method.

An NIR method was developed for the analysis of 13 different herbal products for 6 synthetic cannabinoids, including JWH-250, and 3 phenethylamines [105]. JWH-250 was identified in only one synthetic cannabinoids. This method is useful because it provides an efficient and inexpensive technique for the analysis of these illicit compounds. An HPLC-HR-MS method was developed and validated, using pressurized liquid extraction for various NPSs, including compounds from multiple other classes [111]. The LOD for JWH-250 was determined to be 1 pg/mg, while the LOQ was determined to be 9 pg/mg. This is a quick, easy extraction method for the detection of the NPSs.

A HPLC-CAD method was developed for the quantification of synthetic cannabinoids in 30 herbal blends [113]. The LOD and LOQ for JWH-250 were determined to be 0.5 μg/mL. After validation, the method was applied to four different samples. Three samples of "Mocarz" were analyzed. JWH-203 was identified in all three samples, with the LC-CAD concentrations correlating with those from NMR analysis. This was the first report for LC-CAD analysis of synthetic cannabinoids without reference standards.

Multiple herbal products were analyzed for the synthetic cannabinoid JWH-203 [1-pentyl-3-(2′-chlorophenylacetyl) indole], using multiple spectroscopic analytical procedures, including GC-MS, NMR, and high-performance liquid chromatography coupled with a diode-array detector (HPLC-DAD) [114]. These spectroscopic procedures confirmed the presence of JWH-203 in the herbal products. A UPLC-MS/MS method was developed and validated for the analysis of 80 urine samples and 17 herbal products for JWH-250 [115]. The LOD for JWH-250 was determined to be 0.1 ng/mL, and the LOQ was calculated to be 0.5 ng/mL. JWH-250 was detected in 12 of the dried plant materials. All three powdered herbal products were determined to be negative for the synthetic cannabinoids tested in this method.

Piperazinylindoles: Structures and Analysis

The structures for MEPIRAPIM and NNL-2 are shown in Fig. 22.4.

Four illegal Japanese herbal products were analyzed by LC-MS, GC-MS, HR-MS, and NMR to obtain the chemistry and pharmacological data for MEPIRAPIM [116]. Products A and B consisted of powder-type products, while Products C and D were liquid-type products. MEPIRAPIM was identified in Product B. This is the first report for the chemical and pharmacological data for MEPIRAPIM. The analytical characterizations of four different synthetic cannabinoids, including NNL-2, were identified and determined by LC-MS, Fourier transform infrared spectroscopy (FT-IR), and NMR data [117]. This is the first report on the structure elucidation and analytical characteristics of NNL-2. This data is useful in identifying these compounds in herbal and chemical products, as well as further developing predictions for other various synthetic cannabinoids.

Naphthalenyl-Indole-3-Carboxylates: Structures and Analysis

The structures of FDU-PB-22 and NM-2201 are shown in Fig. 22.5.

A total of 11 different synthetic cannabinoids, including FDU-PB-22, were analyzed by 3 different techniques, including electro-oxidation, in artificial saliva and an herbal product [101]. LC-ESI-QTOF-MS and GC-SMB-EI-QQQ-MS were used to analyze the samples. The LOD for FDU-PB-22 was determined to be 0.025 ng/mL. For the LC-ESI-QTOF-MS analysis, the LOD for FDU-PB-22 was calculated to be 0.124 ng/mL. These analyses resulted in excellent separation and identification.

Three different GC-MS techniques were applied for the detection of several synthetic cannabinoids, including NM-2201 [112]. Using the structure elucidation data, seven herbal products were analyzed. NM-2201 was identified in sample 4—"TRIP." NM-2201 was structurally characterized and identified by GC-MS, GC-HR-MS, UHPLC-HR-MS, NMR, and FT-IR [118]. This was the first report of the characterization of NM-2201 in herbal mixtures via GC-MS and UHPLC-HR-MS analyses.

Quinolin-8-yl-Indole-3-Carboxylates: Structures and Analysis

The structures of the cannabimimetics in this class are shown in Fig. 22.6.

A UHPLC-MS/MS method was developed and validated, in MRM mode, for the detection of 14 synthetic cannabinoids (including PB-22) and 15 metabolites in negative urine [50]. PB-22 was not detected by the immunoassay. The developed and validated dilute-and-shoot method, however, was reliable, reproducible, and sensitive to all 29 compounds and metabolites, easily identifying and analyzing new compounds. Blood samples from individuals in DUID cases and in intoxication cases were analyzed [92], using a previously developed and validated LC-MS/MS method [93]. In the blood samples from individuals in DUID cases, 5F-PB-22 was identified, while in the intoxication cases, 5F-PB-22 was identified in only one autopsy case.

Over 500 urine samples were collected from drivers suspected to be DUID and analyzed for the presence of 12 synthetic cannabinoid metabolites, including 5F-PB-22 3-carboxyindole, BB-22 3-carboxyindole, and PB-22 N-pentanoic acid by LC-MS/MS [99]. A total of 19 samples were identified to be positive for the synthetic cannabinoid metabolites, and 18 of the 19 samples were sampled to be positive for alcohol and other illicit drugs, such as PCP, and THC-COOH. This method was successful for the quantification of these 12 metabolites.

A total of 11 different synthetic cannabinoids, including FUB-PB-22 and 5F-PB-22, were analyzed by LC-ESI-QTOF-MS and GC-SMB-EI-QQQ-MS in artificial saliva, as well as in an herbal product [101]. Both analyses resulted in excellent separation and identification. The LOD for the GC-SMB-QQQ-MS analysis of FUB-PB-22 and 5F-PB-22 were determined to be 1.06 and 0.025 ng/mL, respectively. For the LC-ESI-QTOF-MS analysis of FUB-PB-22 and 5F-PB-22, the LOD was calculated to be 0.002 and 0.001 ng/mL, respectively.

An LC-MS/MS method was developed and validated for the analysis of 32 synthetic cannabinoid metabolites [102]. The correlation coefficient (r^2) for all analytes was greater than 0.990, and the calculated concentrations were within ± 20% of the nominal concentrations. This method was then applied to 25 authentic samples from synthetic cannabinoid users; 5F-PB-22 was identified in five samples.

An LC-MS/MS method was developed and validated for the analysis of synthetic cannabinoids in human whole blood [110]. These synthetic cannabinoids were extracted using two different extraction procedures, each designed to optimize the recovery of the two classes. A third extraction procedure was carried out for the compounds using deuterated internal standards, allowing for quantitative analysis. These methods were applied to the analysis of 1142 forensic blood samples. FUB-PB-22, 5F-PB-22, and BB-22 were identified in five samples, four samples, and one sample, respectively.

An HPLC-CAD method was developed and validated for the quantification of synthetic cannabinoids in 30 different herbal blends [113]. After validation, the method was applied to four different samples. Sample 1 consisted of pure PB-22 and in a powder form. The concentration of PB-22 in this sample was determined to be 945 mg/g, confirmed by NMR analysis. PB-22 was not identified in any other samples. This was the first report for LC-CAD analysis of these synthetic cannabinoids without their reference standards.

Using GC-MS, GC-HR-MS, UHPLC-HR-MS, NMR, and FT-IR, two new synthetic cannabinoids, FUB-PB-22 and MN-27, were identified and structurally characterized [118]. This was the first report of the detection of these synthetic cannabinoids in herbal mixtures via GC-MS and UHPLC-HR-MS analyses.

Four postmortem cases were analyzed for 5F-PB-22 and PB-22, using LC-ESI-MS/MS [119]. In cases 1, 3, and 4, the individuals, all teenagers, died from the intoxication of 5F-PB-22. It is interesting to note that in Case 1, ethanol was also detected, and in Case 2, THC-COOH was detected. The extraction profiles of a range of synthetic cannabinoids were studied to develop an NMR method to quantify 10 different synthetic cannabinoids in a total of 37 herbal smoking blends [120]. Using this method, BB-22 was identified in one product. This is an alternate method that provides specific characterization, allowing for accurate quantification.

DART-MS and LC/QTOF-MS methods were developed and validated for the analysis of 11 different NPSs, including 6 synthetic cannabinoids, such as PB-22 [121]. The LOD for the DART-MS method ranged from 5 to 20 ng/mL, while the LOD for the LC/QTOF-MS was 1.0 ng/mL for all six of the synthetic cannabinoids. PB-22 was not identified in any of the samples. The LC/QTOF-MS method proved to be preferable for more accurate quantitative results. However, the DART-MS method proved to be faster and able to provide a semiquantitative result in 30 seconds, a property useful for scenarios needing quick results.

Using ESI-MS/MS analysis, 13 herbal products were tested for the presence of 8 different synthetic cannabinoids, including 5F-PB-22 [122]. The structures and analytical characteristics for the eight different cannabimimetics were confirmed via NMR analysis. Three samples were identified to contain 5F-PB-22.

An LC-MS/MS method was developed and validated, using liquid−liquid extraction, for the analysis of 93 synthetic cannabinoids in human serum [123]. The LOD ranged from 0.01 to 8.2 ng/mL, while the LOQ ranged from 0.4 to 46.1 ng/mL. This method was then applied to 189 authentic samples, out of which 64 samples were identified to be positive for at least 1 synthetic cannabinoid. The synthetic cannabinoid, 5F-PB-22 was one of the cannabimimetics identified with the most frequency. This method is useful because it includes several new synthetic cannabinoids, keeping up to date with the illicit drug market.

Adamantyl-Indole-3-Carboxamide: Structures and Analysis

The structures of the synthetic cannabinoids in this class are shown in Fig. 22.7.

LC-MS and GC-MS analyses were utilized for the analysis of a chemical product, "Fragrance Powder," as well as three other herbal products, for the presence of seven different synthetic cannabinoids [42].

APICA was detected in the herbal products, making this the first report of the detection of this new synthetic cannabinoid in herbal and chemical products. An LC-MS/MS method was developed for the analysis of 3078 blood samples from 8 severe intoxication cases for 29 synthetic cannabinoids [49]. The concentration of AB-001 was calculated to range from 0.07 to 0.09 ng/g.

A UHPLC-MS/MS method was developed and validated for the detection of 14 synthetic cannabinoids, including AM-1248, and 15 metabolites [50]. The Immunalysis immunoassay was not able to detected AM-1248. An MEKC-MS/MS method was developed for 12 synthetic cannabinoids, including APICA [54]. The LODs of the analytes ranged from 6.7 to 76.5 µg/g. The recoveries were determined to range from 89.5% to 101.7%. After validation, this method was applied to two herbal products purchased in Japan; APICA was detected in neither sample. A validated GC-CI-MS method, in MRM mode was validated for the identification of 34 synthetic cannabinoids (including APICA) and seven synthetic cathinones [59]. This method was successful in the qualitative analysis of these analytes.

Eight different samples [herbal material (flowers, leaves, and stems) and pure compounds] were analyzed via DART-ORBITRAP for the presence of six different synthetic cannabinoids, including AM-1248 and STS-135 [60]. This method is significant due to no sample or solvent preparation required. This is a valid alternate to GC and LC analyses. ESI-MS/MS analysis was carried out for the analysis of 13 herbal products for the identification of 8 different synthetic cannabinoids, including STS-135 [122]. The analytical and structural characteristics were determined and confirmed via NMR analysis. STS-135 was identified in only one herbal product.

Human serum was analyzed for 93 synthetic cannabinoids, using a developed LC-MS/MS method with liquid–liquid extraction [123]. The LOD for these analytes ranged from 0.01 to 8.2 ng/mL, while the LOQ was determined to be in between 0.4 ng/mL and 46.1 ng/mL. After development, this method was applied to 189 forensic samples; STS-135 was detected in only 1 sample. This method is useful because it includes several new synthetic cannabinoids, keeping up to date with the illicit drug market.

Using human hepatocytes, STS-135 was incubated for 3 hours and analyzed for the presence of metabolites via HR-MS analysis [124]. After analysis, 75% of the parent compound had degraded into 29 different metabolites [the monohydroxylated (M25) and dihydroxylated (M21) detected with the highest concentrations, shown in Fig. 22.8)]. This was the first report for the clinical and in vitro studies of the STS-135 metabolites.

Tetramethylcyclopropylmethanone-Pentylindoles: Structures and Analysis

The structures of synthetic cannabinoids in this class are shown in Fig. 22.9.

A UHPLC-MS/MS method, in MRM mode, was developed for the detection of 14 synthetic cannabinoids and 15 metabolites, including UR-144, UR-144-pentanoic acid, XLR-11, XLR-11 6-OH-indole, and XLR-12, in negative urine [50]. These metabolites were not detected by the immunoassay, but they were detected by the UHPLC-MS/MS method. The dilute-and-shoot method was reliable, reproducible, and sensitive to all 29 analytes, with the capability of identifying new compounds.

An LC-MS/MS method was developed and validated for the analysis of human urine for UR-144 and its metabolites [53]. The correlation coefficient (r^2) values ranged from 0.994 to 0.999, and the LODs for all of the analytes ranged from 0.1 to 0.5 µg/L. There was no evidence of any carry over for the synthetic cannabinoids, and the extraction efficiencies for the method ranged from 43.7% to 109.3%. UR-144 N-pentatonic acid and UR-144 N-hydroxypentyl metabolite were detected in two anonymous, authentic urine samples. This method demonstrated accurate and efficient results.

An MEKC-MS/MS method was developed and validated for the analysis of 10 different synthetic cannabinoids, UR-144, and XLR-11 [54]. The LODs for the analytes ranged from 6.5 to 76.5 µg/g. After validation, this method was applied to two herbal products purchased in Japan. XLR-11 was detected in one herbal product; a combination of UR-144 and XLR-11 was both detected in the other herbal product.

NMR spectroscopy was used for the screening and quantification of the synthetic cannabinoids in 12 herbal products [56]. These herbal products included: Melon Code Black, Cloud Nine, Sweet Leaf, Head Trip, Extremely Legal, Zombie Matter, Moon Spice, Ion Source, Rack City, Nuclear Bomb, Mr. Nice Guy, and K250. XLR-11 and UR-144 were two of the synthetic cannabinoids for which a spectroscopic footprint was developed, useful in future analyses. This method is clean and simple, without any need for chromatographic isolation. Using a published LC-MS/MS method [54], over 700 specimens from US military personnel were analyzed for 53 synthetic cannabinoids and metabolites, including UR-144 and UR-144-OH [57]. UR-144-OH was identified to be one of the metabolites with the highest concentrations.

A miniature ion trap mass spectrometry system, with two ambient ionization systems, was used to develop a method of the analysis of bulk powders and/or biofluid samples for five synthetic cannabinoids [58]. This method proved to be simple, effective, and easily reproducible, having an acceptable linearity of $r^2 > 0.99$ and the LOD, ranging from 10 to

20 ng/mL. A GC-CI-MS method, run in MRM mode, was developed and validated for the qualitative analysis of 33 designer drugs, including XLR-11 and 7 synthetic cathinones [59]. This method was used for the qualitative analysis for the compounds.

Two different LC-MS/MS methods were developed and validated for the quantitative analysis of blood and urine for 15 parent synthetic cannabinoids and 17 different cannabimimetic metabolites (and 2 parent compounds) [61]. In blood, the LOD for the synthetic cannabinoids ranged from 0.025 to 0.10 ng/mL, while the LOQ ranged from 0.025 to 0.10 ng/mL. In urine, the LOD for the synthetic cannabinoids ranged from 0.01 to 0.50 ng/mL, with the LOQ ranging from 0.10 to 0.50 ng/mL. UR-144 and XLR-11 were identified to be positive in 5 samples and 17 samples, respectively. These methods are useful because when used in combination, they provide the concentrations of the parent cannabinoids and the metabolites, in two different biological matrices.

The pyrolytic products of UR-144 and XLR-11 were identified using GC-MS analysis [64]. The pyrolysis products for UR-144 and XLR-11 were identified to be UR-144 degradant and 1-(5-Fluoropentyl)-indole-3-carboxaldehyde XLR-11 degradant, respectively. This analysis is useful, as it provides the pyrolytic cannabimimetic metabolites produced from smoking synthetic cannabinoids.

The UPLC-TOF-MS was used for the identification of five new synthetic cannabinoids including A-796, 260, UR-144, and XLR-11 in herbal products [78]. A total of 18 herbal products were analyzed, and no previously controlled synthetic cannabinoids were detected. All five of the cannabimimetics were detected in the products, proving this method effective.

An LC-QTOF-MS method was developed and validated for the detection of 11 synthetic cannabinoid metabolites and UR-144 [83]. In addition, this analysis was compared with results from an Immunalysis Spice K2 HEIA. After analysis, it was concluded that no cross-reactivity with the UR-144 metabolites was present. The LC-QTOF-MS method is more adaptable, reduces validation, and detects untargeted analytes, while the immunoassay only detects targeted analytes. This proves the LC-QTOF-MS method more efficient, flexible, and reproducible.

Blood samples from individuals in DUID cases and intoxication cases were analyzed [92], using a previously developed and validated LC-MS/MS method [93]. Out of the DUID cases, AM-2201, JWH-018, JWH-122, and JWH-210 were identified with the highest concentrations. In the intoxication cases, UR-144 was determined to be one of the most prevalent synthetic cannabinoids detected.

UPLC-UV and UPLC-MS/MS methods were developed and validated for the analysis of 45 different herbal product samples for THC and 8 synthetic cannabinoids, including XLR-11 [97]. For the solid products, on the UPLC-UV, the LOD was calculated to range from 0.1 to 0.3 µg/mL, while the LOQ was ranging from 0.3 to 0.9 µg/mL. In the liquid samples, the LOD for the UPLC-UV was determined to range from 0.1 to 0.4 µg/mL, and the LOQ to range from 0.3 to 1.2 µg/mL. This method proved to be successful and can be used to quantify these synthetic cannabinoids in these herbal mixtures.

Using LC-MS/MS, over 500 urine samples were collected from drivers that were suspected to be DUID and analyzed for the identification of 12 synthetic cannabinoid metabolites, including UR-144 N-pentanoic acid and XLR-11 N-(4-hydroxypentyl) [99]. UR-144 N-pentatonic acid was detected in 16 samples. This method was successful for the quantification of these 12 cannabimimetic metabolites.

LC-ESI-QTOF-MS and GC-SMB-EI-QQQ methods, using electro-oxidation, was developed for the detection of 11 different synthetic cannabinoids, including XLR-11, in artificial saliva and an herbal product [101]. Both analyses resulted in excellent separation and identification. The LOD for XLR-11 for the GC-SMB-QQQ-MS analysis was determined to be 0.324 ng/mL, and for the LC-ESI-QTOF-MS analysis, the LOD was calculated to be 0.001 ng/mL for XLR-11.

An LC-MS/MS method was developed for the analysis of 25 authentic samples from synthetic cannabinoid users for 32 synthetic cannabinoid analytes [102]. The correlation coefficient (r^2) for all of the cannabimimetics was greater than 0.990, and the calculated concentrations were within ± 20% of the nominal concentrations. The metabolites of XLR-11 were detected in 22 out of the 25 samples. This is the first report of a positive drug test for XLR-144 pyrolysis N-pentanoic acid.

Four herbal products were analyzed by DART-TOF-MS as the screening method and by LC-QTOF-MS as confirmation for eight different synthetic cannabinoids, including XLR-11 [108]. UR-144 was the only synthetic cannabinoid identified in Sexy: Exotic Herbal Potpourri, while UR-144 and XLR-11 were both identified in Mad Hatter: Cloud 9. This method proved to be effective and useful in the identification of these eight synthetic cannabinoids in herbal products.

An LC-MS/MS method was developed and validated for the analysis of human whole blood for the detection of multiple cannabimimetics [110]. These synthetic cannabinoids were extracted using two extraction procedures, each designed to optimize the recovery of the two classes. A third extraction procedure was carried out for the compounds using deuterated internal standards, allowing for quantitative analysis. These methods were then applied to the analysis of 1142 forensic blood samples. XLR-11 and UR-144 were detected in 82 samples and 6 samples, respectively. The mean concentration was calculated for only four compounds, ranging from 0.2 to 3.8 ng/mL for XLR-11 and 0.3 to 6.2 ng/mL for UR-144.

An HPLC-HR-MS method, using pressurized liquid extraction, was developed and validated for the detection of various NPSs in hair samples [111]. The LOD for UR-144 was determined to be 6 pg/mg, respectively, while the LOQ was calculated to be 20 pg/mg, respectively. This provides an alternative extraction method other than SPE and LLE. An NMR method was developed from the extraction profiles of a range of cannabimimetics for the quantification of 10 synthetic cannabinoids in 37 herbal smoking blends [120]. UR-144 was identified in eight products. This is an alternate method that provides excellent quantification.

DART-MS and LC-QTOF/MS methods were developed and validated for 3 different samples for 11 new NPSs, including UR-144 [121]. The linearity for the DART-MS for these six cannabimimetics ranged from 0.9906 to 0.9987, while the linearity for the LC/QTOF-MS ranged from 0.9910 to 0.9973. The LOD for the DART-MS method ranged from 5 to 20 ng/mL, while the LOD for the LC/QTOF-MS was 1.0 ng/mL for all six of the synthetic cannabinoids. UR-144 was identified in one sample, at a concentration of 10.10 mg/g by the DART-MS method and 10.24 mg/g by LC/QTOF-MS analysis. The LC/QTOF-MS method proved to be more preferable for more accurate quantitative results. However, the DART-MS method proved to be faster and able to provide a semiquantitative result in 30 seconds, a property useful for scenarios needing quick results.

Human liver hepatocytes and HR-MS analysis have been used to determine the first metabolic profile of XLR-11 [125]. Over 25 different phase I and phase II metabolites were produced, including the carboxylated, hydroxylated, hemiacetal, hemiketal, internal dehydration, and glucuronide metabolites, which were identified by HR-MS analysis. Overall, this was the first data that identified the XLR-11 metabolites, as well as provide mechanisms for XLR-11 metabolism that could be used in future clinical and forensic investigations.

ELISA was developed to analyze human urine for UR-144 and XLR-11 [126]. The assay has a recommended cutoff concentration of 5 ng/mL due to concentrations commonly encountered in real samples. UR-144-COOH and UR-144-5-OH showed a 100% cross-reactivity, while UR-144-4-OH and XLR-11-4-OH showed a 50% reactivity; in contrast, XLR-11 and UR-144 showed a very low cross-reactivity. Overall, 87% positivity for one or more metabolites on the target compounds was detected for the ELISA immunoassay.

Using the US Food and Drug Administration (FDA) and the Scientific Working Group for Forensic Toxicology (SWGFT) guidelines, an LC-MS/MS method was developed and validated for the analysis of 498 oral fluid samples for the presence of XLR-11 and UR-144, as well as their hydroxy metabolites and the pyrolysis products [127]. After analysis, there were 24 samples positive for UR-144 and its metabolites, and 64 samples positive for XLR-11 and its metabolites.

Table 22.2 displays the concentration range, as well as the number of samples for UR-144, XLR-11, and their metabolites, for the samples that were identified to be positive.

An LC-MS/MS method was developed and validated for the analysis of 5 human specimens, to identify 37 synthetic cannabinoids and metabolites [128]. The first three samples were taken from individuals who used a smoking apparatus, dried leaves, and a cigarette butt. Analysis of the first sample showed UR-144 N-COOH, UR-144 N-5-OH, and XLR-11. In the second and third sample, UR-144 N-COOH and UR-144 N-5-OH were identified; however, there were no parent drugs were detected. In the fourth sample, UR-144 N-COOH and UR-144 N-5-OH, and XLR-11 were detected. In the fifth sample, no metabolite of the pentylindole-tetramethylcyclopropylmethanone class was detected.

MDMB-CHMICA, MDMB-FUBINACA, and FUB-AMB: Structures and Analysis

The structures of MDMB-CHMICA, MDMB-FUBINACA, and FUB-AMB are shown in Fig. 22.10.

A total of 13 herbal products were screened for the presence of 8 different synthetic cannabinoids (including MDMB-CHMICA) using ESI-MS/MS [122]. The structural and analytical characteristics for the eight different cannabimimetics were confirmed via NMR analysis. MDMB-CHMICA was identified in two herbal products at concentrations of 12 and 82 mg/g. This is the first report on the identification of MDMB-CHMICA, as well as its structural and physicochemical properties.

An LC-MS/MS method was developed and validated, using liquid–liquid extraction, for the analysis of 93 synthetic cannabinoids in human serum [123]. The LOD ranged from 0.01 to 8.2 ng/mL, while the LOQ ranged from 0.4 to 46.1 ng/mL. After analysis, 64 samples were identified to be positive for at least 1 synthetic cannabinoid. MDMB-

TABLE 22.2 Concentrations of Positive Oral Fluid Samples [127]

Analyte	Number of Positives	Concentration Range (ng/mL)
UR-144	2	<5
	3	5–30
UR-144-PYR	2	<5
	2	5–30
	1	30–60
UR-144-4-OH	5	<5
	7	5–30
	2	30–60
XLR-11	13	<5
	10	5–30
	3	30–60
	2	60–100
XLR-11-PYR	15	<5
	7	5–30
	6	30–60
	1	60–100
	7	>100

Source: Adapted from Amaratunga P, Thomas C, Lemberg BL, Lemberg D. Quantitative measurement of XLR11 and UR-144 in oral fluid by LC-MS-MS. J Anal Toxicol 2014;38:315–321.

CHMICA was one of the cannabimimetics identified with the most frequency. This method is useful because it includes several new synthetic cannabinoids.

Nine herbal products were analyzed, in which eight different synthetic cannabinoids, including MDMB-CHMICA, were identified using a GC-MS method [129]. Products 1, 2A, and 2B contained 31, 25, and 23 mg/g of MDMB-CHMICA, respectively. Product 3 contained five different synthetic cannabinoids, out of which the concentration of MDMB-CHMICA was identified to be 7 mg/gm.

ADB-BICA (1-Benzyl-*N*-(1-Carbamoyl-2,2-Dimethylpropan-1-yl)-1H-Indole-3-Carboxamide): Structure and Analysis

The structure of ADB-BICA is shown in Fig. 22.11.

The analytical characterizations were determined and were used to identify four different synthetic cannabinoids, including ADB-BICA [117]. This is the first report on the structure elucidation and analytical characteristics of ADB-BICA. This data is useful in identifying these compounds in herbal and/or chemical products, as well as further developing predictions for other various synthetic cannabinoids.

NNEI: Structures and Analysis

The structure of NNEI is shown in Fig. 22.12.

Using three different GC-MS for the analysis of various synthetic cannabinoids, the structures for these compounds were elucidated [112]. This structure elucidation was applied to the analysis of seven herbal products. In sample 5 (Family Evolution 01), NNEI was identified and confirmed with its reference standard.

BzODZ-Epyr: Structures and Analysis

The structure of BzODZ-EPyr is shown in Fig. 22.13.

Using GC-MS, GC-HR-MS, UHPLC-HR-MS, FT-IR, proton nuclear magnetic resonance (^{1}H-NMR), and carbon 13 nuclear magnetic resonance (^{13}C-NMR) analyses, 3-benzyl-5-[1-(2-pyrrolidin-1-ylethyl)-1H-indol-3-yl]-1,2,4-oxadiazole (BzODZ-EPyr) was identified an herbal product for the first time [130]. Using these analyses, the analytical characteristics were determined. These characteristics can be used to identify BzODZ-EPyr in future analyses.

JWH-210: Structure and Analysis

The structure of JWH-210 is found in Fig. 22.14.

Over 25 products, including one minor component of JWH-210 reference material, were analyzed for a variety of cannabimimetics, such as JWH-210 and its ethyl derivative, using LC-MS/MS and GC-MS [17]. These analytes were detected from the samples collected, leading to the conclusion that this method was successful in identifying and detecting these analytes.

Multiple legal high products acquired from prefederal ban time range and postfederal ban time range were analyzed for 32 different cannabimimetics, including JWH-210, using UPLC-TOF-MS [38]. In samples collected from the prefederal ban time range, no samples were identified to contain JWH-210. Out of the 81 products in the postfederal ban time range, JWH-210 was detected in 7 samples collected postfederal ban.

An MEKC-DAD method was developed and validated with an LOD ranging from 3 to 5 μg, to analyze 15 different herbal blend samples for synthetic cannabinoids (including JWH-210) [40]. Five synthetic cannabinoids, including JWH-210, were detected in the herbal blend samples. Every positive sample was validated on the LC-MS, proving this method to be a valid alternative to GC and LC methods.

A rapid, sensitive, and simple LC-MS/MS method was developed to identify 22 synthetic cannabinoids, including JWH-210, in human hair collected from 8 individuals [41]. One of the individuals was bald, necessitating taking hair from the arm and leg. The individuals were in the same age range, sex (male), and reported last drug use around the same time. JWH-210 was determined to be positive in three samples. An LC-MS/MS method was developed for the analysis of 3078 blood samples from severe intoxication cases [49]. The study also included case studies for eight different cases. In the blood samples, the concentrations of JWH-210 ranged from 0.05 to 8.1 ng/g. JWH-210 was identified in the first case study, at a concentration of 8.1 ng/g.

An UHPLC-MS/MS method was developed and validated in MRM mode to analyze 14 synthetic cannabinoids and 15 metabolites, including JWH-210-4-OH pentyl, in negative urine [50]. The immunoassay was unable to detect the JWH-210-4-OH metabolite; however, the developed and validated dilute-and-shoot method was reliable, reproducible, and sensitive. A DART-MS method was developed for the analysis of six "Spice products" for five synthetic cannabinoids, including JWH-210 [52]. AM-2201 and JWH-210 were detached in "California Kronic Blueberry" and "Relaxinol"; only JWH-210 was detected in "Dead Man Walking."

An LC-MS/MS method was developed for the analysis of human urine for 15 synthetic cannabinoids (including JWH-210), and their metabolites [53]. The correlation coefficient (r^2) values ranged from 0.994 to 0.999, and the LODs for the analytes ranged from 0.1 to 0.5 μg/L. There was no evidence of any carry over for the synthetic cannabinoids. This method was successfully applied to two anonymous, authentic urine samples.

Using UHPLC-MS/MS, a total of 344 hair samples were analyzed for the presence of 23 different synthetic cannabinoids [55]. Out of the 344 hair samples tested, 264 samples were previously identified to be positive for at least one, if not more, for amphetamines, cocaine, methadone, opiates, and THC. The remaining 80 samples tested positive for ethyl-glucuronide (EtG). All of these samples were retested for the presence of these 23 synthetic cannabinoids, using the UHPLC-MS/MS method. A total of 13 samples were determined to be positive for at least one synthetic cannabinoid (in pg/mg concentrations). JWH-210 was identified in two samples.

Using NMR spectroscopy to screen and quantify synthetic cannabinoids, 12 herbal products (Melon Code Black, Cloud Nine, Sweet Leaf, Head Trip, Extremely Legal, Zombie Matter, Moon Spice, Ion Source, Rack City, Nuclear Bomb, Mr. Nice Guy, and K250) [56]. The spectroscopic identities of multiple synthetic cannabinoids, including JWH-210, were developed. This data can be used in future analyses for the detection of these cannabimimetics.

Over 770 urine specimens from US military personnel were analyzed, using a published, quantitative LC-MS/MS method [53] to identify 53 synthetic cannabinoid analytes, including 20 parent compounds and 33 metabolites [57]. Overall, 290 urine specimens contained 1 or more of 22 different synthetic cannabinoid metabolites, resulting from 11 parent compounds. The 11 parent cannabinoids analyzed included JWH-210, a cannabimimetic that was not detected in very high concentrations.

Two different LC-MS/MS quantitative methods were developed and validated for analyzing various synthetic cannabinoids [61]. One method dealt with the analysis of 15 parent synthetic cannabinoids in blood, while the other method analyzed urine for 17 different cannabimimetic metabolites, with two parent compounds qualitatively identified. In blood, the LOD for the synthetic cannabinoids ranged from 0.025 to 0.10 ng/mL, and the LOQ ranged from 0.025 to 0.10 ng/mL. In urine, the LOD for the synthetic cannabinoids ranged from 0.01 to 0.50 ng/mL, and the LOQ ranged from 0.10 to 0.50 ng/mL. JWH-210 was identified positive in 14 samples. These methods are useful because when used in combination, they provide the parent compounds in blood before being metabolized, as well as the concentration of the metabolites in urine indicating the past history of the individual.

NMR and DART-MS analyses were used to identify cannabinoids and synthetic cannabinoids in herbal and chemical products [62]. JWH-210 was not identified in any of the four powders. Out of the seven herbal products, only one, "Sweet Leaf," contained JWH-210 and JWH-250. This is a rapid and accurate method of analyzing herbal products.

The pyrolytic products were identified for JWH-210, using GC-MS [64]. 1-Pentylquinoline and N-penten-4-yl-JWH-210 were the two pyrolytic products detected for JWH-210. This analysis is useful, as it provides the pyrolytic cannabimimetic metabolites produced from smoking synthetic cannabinoids.

The first immunoassay was developed for the detection of seven different synthetic cannabinoids in oral fluids [77]. The oral fluid was collected using the Quantisal device. This assay was then applied to 32 patient samples, and the assay cutoff was determined to be 0.25 ng/mL, with the LOD and ULOL being 0.1 and 50 ng/mL, respectively. All of the positives were confirmed with LC-MS-MS analysis, with no false-positives. Out of the 32 samples, only 4 oral fluid samples were positive for JWH-210. An LC-MS/MS method was developed, using liquid–liquid extraction, for the analysis of various biological matrices (blood, saliva, and urine) for various indole-derived synthetic cannabinoids, including JWH-210, a naphthoylindole [81]. The LOD and LOQ for JWH-210 were calculated to be 0.5 ng/mL in saliva and 0.4 ng/mL, respectively, in plasma/serum. A total of 17 urine specimens were analyzed as part of a military law enforcement investigation, out of which 9 specimens were identified to be positive for synthetic cannabinoid usage, using GC-MS analysis [91]. JWH-210 was identified in one of the samples. Blood samples were obtained from individuals in DUID and intoxication cases [92] and analyzed using a previously developed and validated LC-MS/MS method [93]. JWH-210 was one of the compounds that was identified with the highest concentrations in the DUID, and in the intoxication cases, JWH-210 was identified in one autopsy case.

An LC-MS/MS method was developed, validated, and applied to analyze several compounds in wastewater, including JWH-210 [94]. In the influent wastewater tested, JWH-210 was detected in two samples, out of which one sample was below the LOQ and the other had a concentration of 3.7 ng/L. In the effluent wastewater samples, JWH-210 was identified in all of the samples, with quantifiable concentrations, except for one sample that was below the LOQ.

A UPLC-MS/MS method was developed and validated for the analysis of 80 urine samples and 17 herbal products [115]. The LOD for JWH-210 was calculated to be 0.1 ng/mL, and the LOQ was determined to be 0.5 ng/mL. A total of 14 herbal products were dried plant material, and 3 were in powder form. JWH-210 was all detected in 12 of the dried plant materials. No synthetic cannabinoids were detected in any of the three powdered herbal products.

INDAZOLES

Indazoles are the second largest class of synthetic cannabinoids. They are classified into seven subgroups depending on the attachment(s) to the indazole moeity. These subclasses are amino-methyl-oxobutan-indazole-3-carboxamide, naphthoylindoles, phenylacetylindoles, piperazinylindoles, adamantyl-indazole-3-carboxamide, benzo[d]imidazole, and naphthalenyl-indazole. Five other indoles: 5F-ADB, 5F-AMB, AMPPPCA, and NPB-22 were classified separately. These subclasses and cannabimimetics are discussed individually.

Amino-Methyl-Oxobutan-Indazole-3-Carboxamide: Structures and Analysis

The structures of the cannabimimetics in this class are shown in Fig. 22.15.

A UHPLC-MS/MS method was developed for the identification of total 29 analytes, 14 parent synthetic cannabinoids and 15 metabolites [50]. In addition, an Immunalysis immunoassay was purchased for screening, using β-glucuronidase for hydrolysis. AB-FUBINACA was not detected by the immunoassay.

The pyrolytic products of various synthetic cannabinoids, including AB-CHMINACA, AB-FUBINACA, and AB-PINACA, were detected by GC-MS [64]. Each cannabimimetic produced its respective metabolites; AB-CHMINACA

TABLE 22.3 Comparison of LODs for LC-ESI-QTOF-MS and GC-SMB-EI-QQQ-MS for the Detection of Synthetic Cannabinoids [101]

Synthetic Cannabinoid	LC-ESI-QTOF-MS	GC-SMB-EI-QQQ-MS
AB-FUBINACA	0.008 ng/mL	0.275 ng/mL
AB-CHMINACA	0.011 ng/mL	0.440 ng/mL
AB-PINACA	0.002 ng/mL	0.045 ng/mL

Source: Adapted from Dronova M, Smolianitski E, Lev O. Electrooxidation of new synthetic cannabinoids: voltammetric determination of drugs in seized street samples and artificial saliva. Anal Chem 2016;88(8):4487–4494.

TABLE 22.4 Number of Positive Samples for the Synthetic Cannabinoids Tested [110]

Synthetic Cannabinoid	Number of Positive Samples
AB-CHMINACA	278
ADB-CHMINACA	153
AB-FUBINACA	66
AB-PINACA	59
ADB-FUBINACA	34
ADB-PINACA	10
5F-ADB-PINACA	8

Source: Adapted from Tynon M, Homan J, Kacinko S, Ervin A, et al. Rapid and sensitive screening and confirmation of thirty-four aminocarbonyl/carboxamide (NACA) and arylindole synthetic cannabinoid drugs in human whole blood. Drug Test Anal 2017;9(6):92.

produced: 3-Hydroxyindazole, N-(1,2-dimethylpropyl)-cinnolinamine, 1-Methylcyclohexanylindazole, 1-Methylcyclohexanyl-N-(valinamidyl)- cinnolinamine, and 1-Methylcyclohexanyl-3-cinnolinamine; AB-FUBINACA produced: 1-Methylbenzylcinnoline; and AB-PINACA produced: 1-Pentylindazole-3-carboxaldehyde and Penten-4-yl-AB-PINACA.

Methods using electro-oxidation were developed in artificial saliva for the analysis of an herbal product for 11 different synthetic cannabinoids, including AB-FUBINACA, AB-CHMINACA, and AB-PINACA [101]. Table 22.3 shows the resulting LODs from the analysis carried out on a LC-ESI-QTOF-MS and GC-SMB-EI-QQQ-MS. Both analyses resulted in excellent separation and identification.

An LC-MS/MS method was developed and validated for the analysis 25 authentic urine samples for the detection of 32 synthetic cannabinoid metabolites [102]. The correlation coefficient (r^2) for all analytes was greater than 0.990, and the calculated concentrations were within ± 20% of the nominal concentrations. In the authentic samples, the metabolites of AB-FUBINACA and AB-PINACA were detected in 22 out of the 25 samples. This is the first report of a positive drug test for AB-FUBINACA, AB-PINACA N-(4-hydroxypentyl), and AB-PINACA N-pentanoic acid.

An LC-MS/MS method was developed and validated for the analysis of 1142 human whole blood samples for the detection of 21 different synthetic cannabinoids [110]. Table 22.4 illustrates the number of samples identified positive for these synthetic cannabinoids. However, the mean concentrations were not calculated for these cannabimimetics.

The structural elucidation of multiple cannabimimetics, including AB-CHMINACA, 5-chloro AB-PINACA, and 5-fluoro AB-PINACA, was conducted, using the GC-EI-MS, positive GC-CI-MS, and negative GC-CI-MS techniques [112]. This data was applied to the analysis of seven herbal products. In sample 1 (Heart Shot Red), 5-fluoro AB-PINACA was detected. In sample 2 (Zonbi Heart), 5-fluoro AB-PINACA and 5-fluro ADB-PINACA were identified, and in sample 6 (Diesel Crown), 5-fluoro AB-PINACA and AB-CHMINACA were detected.

The structural and analytical characterizations of four synthetic cannabinoids, including PPA(N)-2201, were conducted on various instruments [117]. The GC-MS data on PPA(N)-2201 has already been reported; however, this report adds the FT-IR, LC-MS, and NMR data. This data is useful in identifying these four cannabimimetics in herbal and/or chemical products, as well as further developing predictions for other various synthetic cannabinoids.

The extraction profiles of a range of synthetic cannabinoids were studied to develop an NMR data for the quantification of 37 herbal smoking blends to identify 10 synthetic cannabinoids [120]. AB-FUBINACA was identified in 10 products.

DART-MS and LC-QTOF/MS methods were developed and validated for six synthetic cannabinoids, including AB-CHMINACA [121]. The linearity for the DART-MS method for these six cannabimimetics ranged from 0.9906 to 0.9987, while the linearity for the LC/QTOF-MS ranged from 0.9910 to 0.9973. The LOD for the DART-MS method ranged from 5 to 20 ng/mL, while the LOD for the LC/QTOF-MS was 1.0 ng/mL for all six of the synthetic cannabinoids. None of these samples were identified to be positive for AB-CHMINACA.

A total of 13 herbal products were analyzed for the presence of 8 different synthetic cannabinoids, including 5F-AB-PINACA, AB-CHMINACA, and AB-FUBINACA, using ESI-MS/MS [122]. NMR analysis confirmed the analytical and structural characteristics for these eight cannabimimetics. AB-CHMINACA was identified in three products, 5F-AB-PINACA in three products, and AB-FUBINACA in two products.

Human serum was analyzed by a validated LC-MS/MS method, in which over 90 synthetic cannabinoids were targeted [123]. In the method, the LOD was calculated to range from 0.01 to 8.2 ng/mL, while the LOQ was determined to range from 0.4 to 46.1 ng/mL. This method was applied to 189 authentic samples. AB-CHMINACA was the cannabimimetic most frequently identified. This is a very useful method, as it includes various, new cannabimimetics on the illicit drug market. A total of 5 human urine specimens were analyzed by a validated LC-MS/MS method for the identification of 37 synthetic cannabinoids and metabolites [87]. The presence of the AB-PINACA-COOH, 5F-AB-PINACA, and 5F-AB-PINACA-4-OH metabolites were detected in the fifth human urine specimen.

A total of 28 forensic urine samples were analyzed by LC-QTOF/MS for the presence of 92 different metabolites of AB-FUBINACA; however, only 15 metabolites were detected in the samples [131]. Three different metabolites were marked as markers of AB-FUBINACA intake: Hydrolysis of the primary amide, hydroxylation of the amino-oxobutane moiety, and hydroxylation of the indazole ring. In addition to these three metabolites, the parent compound was detected in approximately 54% of the urine samples, indicating that the parent compound is an important testing marker.

Urine samples from three individuals were analyzed by LC-MS/MS [132]. The LOQ for AB-PINACA, AB-FUBINACA, AB-CHMINACA, and MAB-CHMINACA were identified to be 5, 5, 5, and 8 pg/mL, respectively. Each individual's sample contained two ynthetic cannabinoids: AB-PINACA and AB-FUBINACA. In Individual 1, AB-PINACA and AB-FUBINACA had concentrations of 23 and 10 pg/mL, respectively, while in Individual 2, AB-CHMINACA was identified with a concentration of 232 pg/mL. MAB-CHMINACA was identified in Individual 3, with a concentration of 229 pg/mL.

Human hepatocytes are used to study the metabolism of ADB-CHMINACA (MAB-CHMINACA) and for the identification of its major metabolites [133]. For this synthetic cannabinoid, the major metabolites are cyclohexylmethyl metabolite, *tert*-butyl metabolite, and two 4″-hydroxycyclohexyl metabolites. These metabolites were confirmed via LC-MS/MS. Out of the 10 metabolites identified, three metabolites (ADB-CHMINACA hydroxycyclohexylmethyl, ADB-Chminaca 4″-hydroxycyclohexyl, and ADB-CHMINACA hydroxycyclohexylmethyl) were chosen as targets for the identification of ADB-CHMINACA (MAB-CHMINACA) in forensic samples. This was the first report on the metabolism of ADB-CHIMNACA.

Adamantyl-Indazole-3-Carboxamide: Structures and Analysis

The structures of the synthetic cannabinoids of the adamantyl-indazole-3-carboxamide class are shown in Fig. 22.16.

LC-MS, GC-MS, and NMR were used for the analysis of a chemical product, "Fragrance Powder" and three other herbal products containing about 3 g of mixed dried plants for the identification of seven different synthetic cannabinoids [42]. This the first report of APINACA detection in herbal and/or chemical products.

An MEKC-MS/MS method was developed and validated for the analysis of 12 different synthetic cannabinoids, including APINACA [54]. The LODs and recoveries for all of the analytes ranged from 6.5 to 76.5 μg/g and 89.5% to 101.7%, respectively. After validation, this method was applied to two herbal products purchased in Japan; however APINACA was not detected in the two herbal products analyzed.

A GC-CI-MS method, in MRM mode, was developed and validated for the qualitative analysis of 34 synthetic cannabinoids (including APINACA) and synthetic cathinones, allowing for detection of these compounds in herbal products [59].

An NIR method was developed for the analysis of six synthetic cannabinoids, including AKB-48, and three phenethylamines [105]. This method was applied to 13 herbal products. AKB-48 was not identified in any of the herbal products tested. This method is useful because it provides an efficient and inexpensive technique for the analysis of these illicit drugs.

Four herbal products were analyzed, using DART-TOF-MS analysis as screening and LC-QTOF-MS analysis as confirmation [108]. AKB-48 and its metabolite were identified in "Mad Hatter: Cloud 9." This method proved to be effective and useful in the identification of these eight synthetic cannabinoids in herbal products.

The extraction profiles of a range of synthetic cannabinoids were studied to develop NMR data to quantify 10 different synthetic cannabinoids in a total of 37 herbal smoking blends [120]. A total of 11 products contained 5F-AKB-48. This is an alternate method that provides accurate quantification.

DART-MS and LC/QTOF-MS methods were developed and validated for the analysis of 11 different NPSs, including 5F-AKB-48 [121]. The linearity for the DART-MS for these six cannabimimetics ranged from 0.9906 to 0.9987, while the linearity for the LC/QTOF-MS ranged from 0.9910 to 0.9973. The LOD for the DART-MS method ranged from 5 to 20 ng/mL, while the LOD for the LC/QTOF-MS was 1.0 ng/mL for all six of the synthetic cannabinoids. 5F-AKB was not detected in the three authentic samples. Overall, the LC/QTOF-MS method proved to be more preferable for more accurate quantitative results. However, the DART-MS method proved to be faster and able to provide a semiquantitative result in 30 seconds, a property useful for scenarios needing quick results.

Nine herbal products were analyzed for eight different synthetic cannabinoids, using GC-MS analysis [129]. This was the first report of the identification of 5Cl-AKB-48 and 4-pentenyl-AKB-48 in commercially available products. An in-depth characterization of these two compounds was conducted using NMR, EI-MS, ESI-MS/MS, IR, and UV spectroscopy. Product 5 contained 5Cl-AKB-48 and 4-pentenyl-AKB-48 at concentrations of 76 and 1 mg/g, respectively.

A sensitive and reliable LC-MS/MS was developed for the identification of 37 synthetic cannabinoids and metabolites [87]. This method was then applied to five human urine specimens. Three samples were taken from individuals who used a smoking apparatus, the fourth from an individual who used dried leaves, and the fifth sample from a cigarette butt. Analysis of the first sample showed the presence of AKB-48 *N*-5-OH, AKB-48 *N*-COOH metabolites, and 5F-AKB-48. In the second and third sample, the AKB-48 *N*-5-OH and AKB-48 *N*-COOH metabolites were detected, but no parent compounds were identified. There was no detection of AKB-48 metabolites in the fourth and fifth samples.

Human liver microsome (HLM) incubations, as well as 35 urine samples from authentic cases, were analyzed by LC-QTOF-MS for the identification of AKB-48 and 5F-AKB-48 metabolites [134]. The analysis of the HLM identified 41 metabolites of AKB-48 and 37 metabolites of 5F-AKB-48, proving this an efficient method for the identification of these unique metabolites.

The analytical characteristics of three different indazole and pyrazole derivatives, including the APINACA 2*H*-indazole analog, were reported using LC-QTOF/MS, GC-TOF/MS, and NMR spectroscopy [135]. This analytical data is useful for developing methods for the identification of these compounds in herbal or chemical products. This is the first report on these new cannabimimetics.

Benzo[d]imidazole: Structures and Analysis

The structures of the synthetic cannabinoids found in this class are shown in Fig. 22.17.

For the analysis of synthetic cannabinoids in 1142 human whole blood samples, an LC-MS/MS method was developed [110]. These synthetic cannabinoids were extracted using two different extraction procedures, each designed to optimize the recovery of the two classes. A third extraction procedure was carried out for the compounds using deuterated internal standards, allowing for quantitative analysis. FUBIMINA was identified in only one sample.

Four illegal products, purchased in Japan, were analyzed by LC-MS, GC-MS, HR-MS, and NMR to obtain the chemical and pharmacological data for FUBIMINA [118]. Products A and B were powder-type products, while Products C and D were liquid-type products. FUBIMINA was identified in Product A. This is the first report for the chemical and pharmacological data for FUBIMINA.

Smoke mixtures for three new synthetic cannabinoids were analyzed, using GC-MS and UHPLC-HR-MS analysis [136]. Since all three of these cannabimimetics are isomers, NMR analysis was conducted for the structural characterization of these three synthetic cannabinoids. The retention time of for BIM-018 and BIM-2201 (FUBIMINA) was 16.13 minutes and 17.41 minutes, respectively. This is the first report of the detection of 3-naphthoylindazoles and 2-naphthoylbenzoimidazoles in smoke mixtures.

Naphthalenyl-Indazole: Structures and Analysis

The structures of the synthetic cannabinoids of the naphthalenyl-indazole are shown in Fig. 22.18.

Electro-oxidation was utilized for the analysis of 11 different synthetic cannabinoids in artificial saliva, as well as in a herbal product [101]. The analyses were carried out on a LC-ESI-QTOF-MS and GC-SMB-EI-QQQ-MS. The LOD for THJ-2201 was determined to be 1.06 ng/mL for GC-SMB-EI-QQQ-MS, and 0.296 ng/mL for LC-ESI-QTOF-MS analysis. Both analyses resulted in excellent separation and identification. A method was developed and validated for the analysis of synthetic cannabinoid in 1142 human whole blood [110]. THJ-018 and THJ-2201 were identified in 3 samples and 34 samples, respectively.

Nine different herbal products were analyzed for eight synthetic cannabinoids using a GC-MS method [129]. This is also the first report of THJ-018 being identified in herbal products. Product 3 contained THJ-018 and THJ-2201 with concentrations of 1 and 2 mg/g, respectively.

Smoke mixtures for three new synthetic cannabinoids were analyzed with GC-MS and UHPLC-HR-MS [136]. Since all three of the cannabimimetics were isomers, the structural characterization was conducted via NMR analysis. The retention time of THJ-2201 [AM(N)-2201] was determined to be 17.30 minutes.

5F-ADB and 5F-AMB: Structures and Analysis

The structures of 5F-ADB and 5F-AMB are shown in Fig. 22.19.

Electro-oxidation was utilized for the detection of 11 different synthetic cannabinoids in artificial saliva and an herbal product [101]. The analyses were carried out on a LC-ESI-QTOF-MS and GC-SMB-EI-QQQ-MS. The LOD for 5F-AMB on the GC-SMB-QQQ-MS was determined to be 0.356 ng/mL, and for the LC-ESI-QTOF-MS analysis, the LOD was calculated to be 0.007 ng/mL.

An LC-MS/MS method was developed and validated for the analysis of 93 different synthetic cannabinoids in human serum [123]. The LOD for the analytes ranged from 0.01 to 8.2 ng/mL, while the LOQ was determined to range from 0.4 to 46.1 ng/mL. This method was then further applied to 189 authentic samples. A total of 64 samples were identified positive or a synthetic cannabinoid. Only one human serum case contained traces of 5F-AMB. This method is useful because it includes several new synthetic cannabinoids.

Using GC-MS analysis, nine different herbal products were analyzed, for eight different synthetic cannabinoids [129]. Only 5F-ADB was identified in Products 6–8 at concentrations of 71, 120, and 44 mg/g, respectively. An LC-MS/MS method was used for the analysis of urine samples from three individuals, in which two synthetic cannabinoids where identified in each individual [132]. The sample from Individual 2 contained 5F-AMB, while the sample from Individual 3 contained 5F-ADB. The LOQ for 5-AMB and 5F-ADB were identified to be 2 and 3 pg/mL, respectively. In Individual 2, the concentrations of 5F-AMB were quantitated to be 19 pg/mL, and the concentration of 5F-ADB was determined to be 19 pg/mL in Individual 3. This was the first report concerning the successful analysis of parent cannabinoids in such a low concentration in authentic human urine samples.

SDB-005: Structures and Analysis

The structures of SDB-005 and 5F-SDB-005 (CBL(N)-2201) are shown in Fig. 22.20.

The characterization of the structures and identification of five new synthetic cannabinoids, including 5F-SDB-005 [118] was conducted via multiple analyses. This was the first report of the characterization of 5F-SDB-005 in herbal mixtures via GC-MS and UHPLC-HR-MS analyses.

For the analysis of 93 different synthetic cannabinoids in 189 human serum samples, an LC-MS/MS method was developed and validated, for which the LOD was identified to range from 0.01 to 8.2 ng/mL, and the LOQ was determined to be in between 0.4 ng/mL and 46.1 ng/mL [123]. SDB-005 was not identified in any samples. This method is useful because it includes several new synthetic cannabinoids, keeping up to date with the illicit drug market.

AMPPPCA: Structures and Analysis

The structures of AMPPPCA and its fluorinated metabolite are shown in Fig. 22.21.

The analytical characteristics of three synthetic cannabinoids were reported, including AMPPPCA and 5F-AMPPPCA [135]. These samples were all seized from a clandestine laboratory and analyzed using LC-QTOF/MS, GC-

TOF/MS, and NMR spectroscopy. This analytical data is useful for developing methods for the identification of these compounds in herbal and/or chemical products. This is the first report on these new cannabimimetics.

NPB-22: Structures and Analysis

The structures of NPB and its metabolites found in this class are shown in Fig. 22.22.

GC-MS, GC-HR-MS, UHPLC-HR-MS, NMR, and FT-IR analyses were conducted for the characterization of the structures and the identification of two new synthetic cannabinoids: FUB-NPB-22 and 5-fluoro NPB-22 [118]. This was the first report of the detection of these synthetic cannabinoids in herbal mixtures via GC-MS and UHPLC-HR-MS analyses.

PYRROLES

Pyrroles are one of the classes of synthetic cannabinoids. Currently, this class only contains one subclass: Naphthoylpyrroles. The cannabimimetics in this subclass are discussed individually.

Naphthoylpyrroles: Structures and Analysis

The structures of JWH-030 and JWH-307 are shown in Fig. 22.23.

Legal high products acquired from prefederal ban time range and postfederal ban time range were analyzed for 32 different synthetic cannabinoids, including 1 naphthoylpyrrole, JWH-030, utilizing UPLC-TOF-MS analysis [38]. JWH-030 was not detected in any samples acquired from either the prefederal ban time range or the postfederal ban time range.

A total of 344 hair samples were analyzed by UHPLC-MS/MS for the presence of 23 different synthetic cannabinoids, including JWH-307 [55]. Over 260 samples out of the 344 hair samples were previously identified to be positive for at least 1, if not more, for amphetamines, cocaine, methadone, opiates, and THC. The remaining 80 samples had tested positive for ethyl-glucuronide (EtG). JWH-307 was not identified in any samples. An illegal Japanese herbal product was analyzed via UPLC-ESI/MS [136]. Two synthetic cannabinoids, namely, JWH-030 and JWH-307, were identified in the methanolic extract of the product.

CARBAZOLES: STRUCTURE AND ANALYSIS

The structure of MDMB-CHMCZCA is shown in Fig. 22.24.

The analytical characterization data for the new synthetic cannabinoid MDMB-CHMCZCA were determined using multiple spectroscopy techniques, including NMR, vibrational circular dichroism (VCD), and electronic circular dichroism (ECD) spectroscopy [137]. These characterization data can be used in the analysis of herbal/chemical products for the presence of this cannabimimetic.

URB-CLASS: STRUCTURES AND ANALYSIS

The structures of the cannabimimetics in the URB-class are shown in Fig. 22.25.

Using the UPLC-TOF-MS, five new synthetic cannabinoids, including URB-597, were identified in herbal products [78]. A total of 18 herbal products were analyzed, and no previously controlled synthetic cannabinoids were detected. URB-597 was detected in the products, proving this method effective.

A UHPLC-PDA/UV-MS method was developed for the analysis of 62 psychoactive substances [81]. The retention times and LODs were calculated for these synthetic cannabinoids, ranging from 10.086 to 16.64 minutes and 0.3 to 3.0 μg/mL, respectively. The method was applied to 20 real herbal products, out of which 13 were plant material and seven powdered material. Cannabimimetics from the URB-class were not detected in any of the four authentic samples tested. The results from the UHPLC-PDA/UV-MS method accurately matched those given by GC-MS, proving the reliability of this method and instrument.

Several illegal herbal products, sold in Japan, were analyzed [137]. Three synthetic cannabinoids, as well as a new synthetic cannabinoid, 6-methyl-2-{-[(4-methylphenyl)amino]-1-benzoxazin-4-one} (URB-754), were detected.

ENDOGENOUS CANNABINOIDS: STRUCTURES AND ANALYSIS

The structures of oleamide, methanandamide, palmitamide, and stearamide are shown in Fig. 22.26.

A total of 29 herbal mixtures, seized by authorities in 2009, were analyzed by GC-MS [25]. These herbal products were analyzed for three endogenous cannabinoids, oleamide, palmitamide, and stearamide, all of which were of the highest concentrations in the herbal products. Oleamide and palmitamide were detected in relatively similar concentrations; stearamide was detected in concentrations higher than both of the other endogenous cannabinoids.

A rapid, sensitive, and simple LC-MS/MS method was developed and validated to identify 22 synthetic cannabinoids, including methanandamide, in human hair collected from eight individuals [41]. Out of these eight individuals, one of them was bald, necessitating taking hair from the arm and leg of the individual. The individuals were in the same age range and reported last drug use around the same time. Methanadamide was not detected in any of the authentic samples.

An LC-MS/MS method was developed and validated, according to the German Society of Toxicology and Forensic Chemistry (GTFCh), for the detection of one endogenous cannabinoid, methanandamide, in human serum [50]. The correlation coefficients were greater than 0.99 for all of the analytes, and the LLOQ for methanandamide was calculated to be lower than 0.3 ng/mL. This method was applied to more than 100 authentic samples, proving useful in the application of this method. The chemical compositions of various Spice aromatic smoking blends available in the Khanty-Mansi Autonomous District (Nizhnevartovsk, Surgut, Khanty-Mansiysk, and Nefteyugansk) were analyzed for oleamide using a GC-MS method [69]. Oleamide was detected in two samples.

CYCLOHEXYLPHENOLS: STRUCTURES AND ANALYSIS

The structures of the multiple cannabimimetics in the CPs are shown in Fig. 22.27.

Using LC-MS/MS analysis, over 25 samples were analyzed for 29 different synthetic cannabinoids, including 4 CPs—CP-47,497, CP-55,490, cannabicyclohexanol, and the cannabicyclohexanol + C2 variant [17]. These analytes were detected from the samples collected. This method was successful in identifying and detecting these analytes in real samples.

Using GC-MS analysis, 29 herbal products, seized in 2009 by authorities, were analyzed for several compounds, including 2 CPs—cis-CP-47,497-C8 and trans-CP-47,497-C8 [25]. cis-CP-47,497-C8 was identified in 17 herbal products.

Legal high products were acquired from prefederal ban time range and postfederal ban time range and analyzed for 32 different synthetic cannabinoids, including two CPs—CP-47,497 and CP-47,497-C8, using UPLC-TOF-MS [38]. In samples collected from the prefederal ban and postfederal ban time range, none of the samples contained CP-47,497 or the CP-47,497-C8 homolog.

An HS-SPME-GC-MS method was developed for the analysis of CP-47,497, CP-47,497-C8 homolog, and CP-55,940 [39]. A total of 32 products were analyzed for the detection of these synthetic cannabinoids. In the products received before and after the DEA ban, no samples contained any CPs. An LC-MS/MS method was developed for the analysis of human urine for 29 total synthetic cannabinoids, including CP-47,497-C7 and CP-47,497-C8, and their metabolites [53]. The correlation coefficient (r^2) values ranged from 0.994 to 0.999, and the LODs for the analytes ranged from 0.1 to 0.5 µg/L. No carry over was present for the synthetic cannabinoids. This method provided accurate and efficient results.

A GC-CI-MS method, in MRM mode, was developed and validated for the qualitative analysis of 34 synthetic cannabinoids, including 2 CPs (CP-47,497 and CP-47,497-C8), allowing for the simple identification of these compounds in herbal products [59]. The LOD's for CP-47,497 and CP-47,497-C8 were identified to be 19.9 and 43.2 ng/mL, respectively.

An LC-MS/MS method, using solid-phase extraction, was developed and validated for the analysis of oral fluid collected by a Quantisal device for five synthetic cannabinoids, including CP-47,497, and CP-47, 497-C8, in samples collected from two naïve subjects [68]. These two CPs were not detected in any of the samples.

The chemical compositions of various Spice aromatic smoking blends available in the Khanty-Mansi Autonomous District (Nizhnevartovsk, Surgut, Khanty-Mansiysk, and Nefteyugansk) were determined by GC-MS analysis [69]. The method detected multiple synthetic cannabinoids, including cis-CP-47,497-C8, and trans-CP-47,497-C8. These CPs were identified in five herbal products.

An SFC-ESI-MS/MS method was developed for the analysis of three commercially bought dried plant products (labeled A, Y, and G) for eight synthetic cannabinoids, including CCH, CP-47-497, and CP-55,940 [89]. CCH was detected in all three products, with the concentration of *cis*-CCH determined to be much greater than that of *trans*-CCH.

An LC-MS/MS method was developed, validated, and applied to the analysis of wastewater for several compounds, including CP-47,497 [94]. CP-47,497 was not detected in either the influent or effluent wastewater samples.

CLASSICAL SYNTHETIC CANNABINOIDS: STRUCTURES AND ANALYSIS

The structures of HU-210, HU-211, HU-308, HU-311, and nabilone are show in Fig. 22.28.

Over 30 reference materials and herbal products were screened for 30 cannabimimetics, including HU-210 and HU-308 using LC-MS/MS and GC-MS techniques [17]. Legal products bought prefederal ban time range and postfederal ban time range were analyzed for over 30 synthetic cannabinoids, including HU-210, using UPLC-TOF-MS [38]. HU-210 was not detected in any samples from either prefederal ban time range postfederal ban time range. An HS-SPME-GC-MS method was developed for the analysis of 32 herbal products for 13 synthetic cannabinoids, including 2 classical cannabinoids—HU-308 and HU-311 [39]. In the products received before and after the DEA ban, HU-308 and HU-311 were not detected.

An UHPLC-MS/MS method, in MRM mode, was developed for the analysis of negative urine for 14 synthetic cannabinoids (including HU-210) and 15 metabolites [50]. The developed and validated dilute-and-shoot method was reliable, reproducible, and sensitive to all 29 compounds and metabolites, with relative ease of identifying and analyzing new compounds.

An LC-MS/MS method was developed for the analysis of human urine for 32 different analytes, including HU-210 [53]. The correlation coefficient (r^2) values ranged from 0.994 to 0.999, and the LOD's for the analytes ranged from 0.1 to 0.5 µg/L. HU-210 was one of the analytes that was detected. This method was applied to two anonymous, authentic urine samples, which demonstrated accurate and efficient results.

A total of 344 hair samples were analyzed with UHPLC-MS/MS for the presence of 23 different synthetic cannabinoids, including 1 classical cannabinoid: HU-210 [55]. HU-210 was not identified in any of the samples. An LC-MS/MS method was developed and validated for seven different synthetic cannabinoids, including HU-210, in oral fluid [67]. There was no HU-210 detected in the samples collected from the two subjects. An SFC-ESI-MS/MS method was developed for the analysis of three commercially bought dried plant products (labeled A, Y, and G) for eight synthetic cannabinoids, including HU-210 [90]. HU-210 was not identified in any of the three products.

An LC-MS/MS method was developed and validated for the identification and determination of ten analytes, including HU-210 [100]. The LOD for these 10 compounds were determined to range from 0.01 ng/mL to 0.5 ng/mL, and the LOQ was calculated to range from 0.05 to 1 ng/mL. HU-210 was not detected in any of the authentic samples.

A UPLC-MS/MS method was developed and validated for the analysis of 80 urine samples and 17 herbal products for four synthetic cannabinoids, including HU-210 [115]. The LOD for HU-210 was calculated to be 0.5 ng/mL, and the LOQ for HU-210 was determined to be 0.5 ng/mL. None of the samples contained any traces of HU-210.

MISCELLANEOUS: STRUCTURES AND ANALYSIS

In this section, miscellaneous synthetic cannabinoids are discussed.

NNL-1: Structure and Analysis

The structure of NNL-1 is shown in Fig. 22.29.

Four new synthetic cannabinoids were analytically characterized, including NNL-1 [117]. This is the first report on the analytical characteristics and structure elucidation of NNL-1. This data is beneficial in determining the concentrations of these cannabimimetics in herbal and chemical products, including predicting future synthetic cannabinoids.

CB-13

The structure of CB-13 is shown in Fig. 22.30.

LC-MS, GC-MS, and NMR were used to analyze "Fragrance Powder," an herbal product, and three other herbal products containing about 3 g of mixed dried plants [42]. These analytes were used for the identification of 6 synthetic cannabinoids and CB-13. A GC-CI-MS method, in MRM mode, was developed for the qualitative analysis of 34

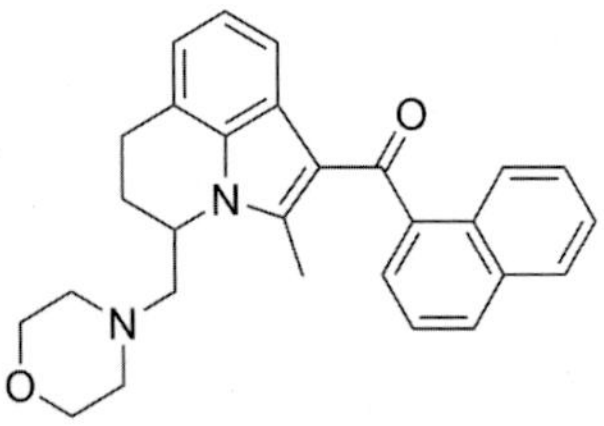

(WIN-55,212-2)

FIGURE 22.31 Chemical structure of WIN-55,212-2.

synthetic cannabinoids, including CB-13, a new synthetic cannabinoid, and 7 synthetic cathinones [59]. The method was successfully validated, with r^2 values all greater than 0.99.

WIN-55,212-2

The structure of WIN-55,212-2 is shown in Fig. 22.31.

Over 25 herbal products and reference materials were analyzed for 30 synthetic cannabinoids, including WIN-55,212-2, using LC-MS/MS and GC-MS analyses [17]. These analytes were successfully detected from the samples. An HS-SPME-GC-MS method was developed for the analysis of 32 different products, both before and after the DEA ban [39]. None of the products analyzed before or after the DEA ban contained WIN-55,212-2.

Over 300 hair samples were analyzed by UHPLC-MS/MS for the presence of 23 different synthetic cannabinoids, including WIN-55,212-2 [55]. After the analysis, 264 samples were identified to be positive for at least 1, noncannabimimetic compound. The other 80 samples tested positive for EtG, indicating alcohol consumption. All of these samples were retested for the presence of these 23 synthetic cannabinoids, using the UHPLC-MS/MS method. None of the hair samples contained any presence of WIN-55,212-2.

An LC-MS/MS method was developed for the analysis of blood, saliva, and urine for various synthetic cannabinoids, including WIN-55,212-2 [80]. The LOD for WIN-55,212-2 was determined to be 0.2 ng/mL in saliva, and 0.1 ng/mL and plasma/serum, respectively.

An HPLC-HR-MS method was developed for the analysis of hair samples for multiple NPSs, including WIN-55,212-2 [111]. The LOD and LOQ for WIN-55,212-2 was calculated to be 8 and 30 pg/mg, respectively. This provides an alternative method other than GC and LC analyses.

CONCLUSIONS

Scientists continue to synthesize and develop new synthetic cannabinoids for medicinal applications; unfortunately, it is inevitable that the majority of these cannabimimetics will be sold illegally, as marijuana substitutes. Without any prior research on the adverse side effects accompanying these drugs, severe complications and fatalities associated with consumption are increasing. Just from 2016 to 2017, over 170 reports and studies were conducted of the analysis on the new synthetic cannabinoids analyzed in various herbal and/or chemical products. After the first DEA ban on five synthetic cannabinoids (JWH-018, JWH-073, JWH-200, CP-47,497, and the CP-47,497-C8 homolog), the frequency of the identification of these cannabimimetics decreased; however, in their place, arose multiple, new synthetic cannabinoids with increasingly stronger agnostic interactions with CB1 and CB2 receptors. Over 130 synthetic cannabinoids and metabolites have been detected and identified in herbal products in the past 10 years. Fortunately, numerous methodologies for the detection and quantitation of these drugs are being developed and put into use. Previously, these products were subjected to GC-MS and LC-MS analyses; as time progresses, an increasing number of products are analyzed by LC-MS/MS and LC-QTOF-MS, providing greater accuracy and precision, which helps monitor and track the synthetic cannabinoids newly added to the illicit market.

REFERENCES

[1] Mustata C, Torrens M, Pardo R, Perez C. Spice drugs: los cannabinoids como nuevas drogas de diseño. Adicciones 2009;21(3).
[2] Carroll F, Lewin A, Mascarella W, Seltzman H, et al. Designer drugs: a medicinal chemistry prespective. Ann N Y Acad Sci 2012;1248:18–38.

[3] Brents LK, Prather PL. The K2/Spice phenomenon: emergence, identification, legislation and metabolic characterization of synthetic cannabinoids in herbal incense products. Drug Metab Rev 2014;46(1):72–85.
[4] ElSohly MA, Gul W, ElSohly K, Murphy TP, Madgula VL, et al. Liquid chromatography-tandem mass spectrometry analysis of urine specimens for K2 (JWH-018) metabolites. J Anal Toxicol 2011;35:487–95.
[5] Wells D, Ott C. The "new" marijuana. Ann Pharmacother 2011;45:414–17.
[6] Seely KA, Brents LK, Radominska-Pandya A, Endres GW, et al. A major glucuronidated metabolite of JWH-018 is a neutral antagonist at CB1 receptors. Chem Res Toxicol 2012;25:825–7.
[7] Hermanns-Clausen M, Kneisel S, Szabo B, Auwärter V. Acute toxicity due to the confirmed consumption of synthetic cannabinoids: clinical and laboratory findings. Addiction 2013;108(3):534–44.
[8] Musah RA, Domin MA, Walling MA, Shepard JRE. Rapid identification of synthetic cannabinoids in herbal samples via direct analysis in real time mass spectrometry. Rapid Commun Mass Spectrom 2012;26:1109–14.
[9] Teske J, Weller JP, Fieguth A, Rothämel T, et al. J Chromatogra B. 2010;878:2959.
[10] Ciolino LA. Quantitation of synthetic cannabinoids in plant materials using high performance liquid chromatography with UV detection (validated method). J Forensic Sci 2015;60(5):1171–81.
[11] Strano-Rossi S, Odoardi S, Fisichella M, Anzillotti L, et al. Screening for new psychoactive substances in hair by ultrahigh performance liquid chromatography-electrospray ionization tandem mass spectrometry. J Chromatogr A 2014;1372:145–56.
[12] Krasowski MD, Ekins S. Using chemiformatics to predict cross reactivity of "designer drugs" to their currently available immunoassays. J Cheminform 2014;6:22.
[13] Rollins C, Spuhler S, Clemens K, Predecki D, et al. Qualitative and quantitative analysis of fluorine containing synthetic cannabinoids using NMR. 245th ACS National Meeting & Exposition, New Orleans, LA, United States, April 7 – 11, 2013; CHED-1398.
[14] Bilici R. Synthetic cannabinoids. Psychiatry 2014;1(2):121–6. Available from: https://doi.org/10.14744/nci.20154.44153.
[15] Casteneto MS, Wohlfarth A, Desrosiers NA, Hartman RL, et al. Synthetic cannabinoids pharmacokinetrics and detection methods in biological matrices. Drug Metab Rev 2015;1–51. Available from: https://doi.org/10.3109/03602532.2015.1029635.
[16] Favretto D, Pascali J, Tagliaro F. New challenges and innovations in forensic toxicology: focus on the "new psychoactive substances". J Chromatogr A 2013;1287:84–95.
[17] Hudson S, Ramsey J. The emergence and analysis of synthetic cannabinoids. Drug Test Anal 2011;3(7–8):466–78.
[18] Namera A, Kawamura M, Nakamoto A, Saito T, et al. Comprehensive review of the detection methods for synthetic cannabinoids and cathinones. Forensic Toxicol 2015;33:175–94.
[19] Nelson ME, Bryant SM, Aks SE. Emerging drugs of abuse. Dis Mon 2014;60:110–32.
[20] Penn HJ, Langman LJ, Unold D, Shields J, et al. Detection of synthetic cannabinoids in herbal incense products. Clin Biochem 2011;44:1163–5.
[21] Presley BC, Jansen-Varnum SA, Logan BK. Analysis of synthetic cannabinoids in botanical material: a review of analytical methods and findings. Forensic Sci Review 2013;350(1):28–62.
[22] Rosenbaum CD, Carreiro SP, Babu KM. Here today, gone tomorrow...and back again? A review of herbal marijuana alternative (K2, spice), synthetic cathinones (bath salts), kratom, *salvia divinorum*, methoxetamine, and piperazines. J Med Toxicol 2012;8:15–32.
[23] Spaderna M, Addy P, D'Souza D. Spicing things up: synthetic cannabinoids. Psychophamracology 2013;228:525–40.
[24] Znaleziona J, Ginterová P, Petr J, Ondra P, et al. Determination and identification of synthetic cannabinoids and their metabolites in different matrices by modern analytical techniques – a review. Anal Chim Acta 2015;874:11–25.
[25] Zuba D, Byrska B. Maciow. Comparison of "herbal highs" composition. Anal Bioanal Chem 2011;400:119–26.
[26] Armenian P, Darracq M, Gevorkyan J, Clark S, et al. Intoxication from the novel synthetic cannabinoids AB-PINACA and ADB-PINACA: a case series and review of the literature.
[27] Aldlgan AA, Torrance HJ. Bioanalytical methods for the determination of synthetic cannabinoids and metabolites in biological specimens. Trends Analyt Chem 2016;80:444–57.
[28] Waters B, Ikematsu N, Hara K, Fujii H, et al. GC-PCI-MS/MS and LC-ESI-MS/MS databases for the detection of 105 psychotropic compounds (synthetic cannabinoids, synthetic cathinones, phenethylamine derivatives). Leg Med 2016;20:1–7.
[29] Meyer MR, Peters FT. Analytical toxicology of emerging drugs of abuse – an update. Ther Drug Monit 2012;34(6):615–21.
[30] Aoun EG, Christopher PP, Ingraham JW. Emerging drugs of abuse: clinical and legal considerations. R I Med J 2014;97(6):41–5.
[31] Wohlfarth A, Weinmann W. Bioanalysis of new designer drugs. Bioanalysis 2010;2(5):965–79.
[32] Wissenbach DK, Meyer MR, Remane D, Philipp AA, et al. Drugs of abuse screening in urine as part of a metabolite-based LC-MS^n screening concept. Anal Bioanal Chem 2011;400:3481–9.
[33] Wohlfarth A, Scheidweiler B, Chen X, Liu H, et al. Qualitative confirmation of 9 synthetic cannabinoids and 20 metabolites in human urine using LC-MS/MS and library search. Anal Chem 2013;85(7):3730–8.
[34] Zuba D, Byrska B. Analysis of the prevalence and coexistence of synthetic cannabinoids in "herbal high" products in Poland. Forensic Toxicol 2013;31:21–30.
[35] Martinotti G, Lupi M, Carlucci L, Cinosi E, et al. Novel psychoactive substances: use and knowledge among adolescents and young adults in urban and rural areas. Hum Psychopharmacol 2015;30:295–301.
[36] Carlsson A, Lindberg S, Wu X, Dunne S, et al. Prediction of designer drugs: synthesis and spectroscopic analysis of synthetic cannabinoid analogs of 1*H*-indol-3-yl(2,2,3,3-tetramethylcyclopropyl)methanone and 1*H*-indol-3-yl)(adamantan-1-yl)methanone. Drug Test Anal 2015;8(10):1015–29.

[37] Paulke A, Proschak E, Sommer K, Achenbach J, et al. Synthetic cannabinoids: in silico prediction of the cannabinoid receptor 1 affinity by a quantitative structure-activity relationship model. Toxicol Lett 2016;245:1−6.
[38] Shanks KG, Dahn T, Behonick T, Terrell A. Analysis of first and second generation legal highs for synthetic cannabinoids and synthetic stimulants by ultra-performance liquid chromatography and time of flight mass spectrometry. J Anal Toxicol 2012;36(6):360−71.
[39] Cox A, Daw RC, Mason MD, Grabenauer M, et al. Use of SPME-HS-GC-MS for the analysis of herbal products containing synthetic cannabinoids. J Anal Toxicol 2012;36:293−302.
[40] Gottardo R, Bertaso A, Pascali J, Sorio D, et al. Micellar electrokinetic chromatography: a new simple tool for the analysis of synthetic cannabinoids in herbal blends and for the rapid estimation of their log P values. J Chromatogr A 2012;1267:198−205.
[41] Hutter M, Kneisel S, Auwärter V, Neukamm MA. Determination of 22 synthetic cannabinoids in human hair by liquid chromatography-tandem mass spectrometry. J Chromato B Analyt Technol Biomed Life Sci 2012;903:95−101.
[42] Uchiyama N, Kawamura M, Kikura-Hanajiri R, Goda Y. Identification of two new-type synthetic cannabinoids, N-(1-adamantyl)-1-pentyl-1H-indole-3-carboxamide (APICA) and N-(1-adamantyl-1-pentyl-1H-indazole-3-carboxamide (APINACA), and detection of five synthetic cannabinoids, AM-1220, AM-2233, AM-1241, CB-13 (CRA-13), and AM-1248, as designer drugs in illegal products. Forensic Toxicol 2012;30:114−25.
[43] de Jager A, Warner JV, Henman M, Ferguson W, et al. LC-MS/MS method for the quantitation of metabolites of eight commonly-used synthetic cannabinoids in human urine − an Australian perspective. J Chromatogr B 2012;897:22−31.
[44] Hutter M, Broecker S, Kneisel S, Auwärter V. Identification of the major urinary metabolites in man of seven synthetic cannabinoids of the aminoalkylindole type present as adulterants in 'herbal mixtures' using LC/MS/MS techniques. J Mass Spectrom 2012;47:54−65.
[45] Ammann J, McLaren JM, Gerostamoulos D, Beyer J. Detection and quantification of new designer drugs in human blood: Part 1- synthetic cannabinoids. J Anal Toxicol 2012;36:372−80.
[46] Gottardo R, Chiarini A, Dal Prá I, Seri C, et al. Direct screening of herbal blends for new synthetic cannabinoids by MALDI-TOF MS. J Mass Spectrom 2012;47:141−6.
[47] Logan BK, Reinhold LE, Xu A, Diamond FX. Identification of synthetic cannabinoids in herbal incense blends in the United States. J Forensic Sci 2012;57(5):1168−80.
[48] Merola G, Aturki Z, D'Orazio G, Gottardo R, et al. Analysis of synthetic cannabinoids in herbal blends by means of nano-liquid chromatography. J Pharm Biomed Anal 2012;71:45−53.
[49] Kronstrand R, Roman M, Andersson M, et al. Toxicological findings of synthetic cannabinoids in recreational users. J Anal Toxicol 2013;37:534−41.
[50] Freijo Jr TD, Harris SE, Kala SV. A rapid quantitative method for the analysis of synthetic cannabinoids by liquid chromatography-Tandem mass spectrometry. J Anal Toxicol 2014;38:466−78.
[51] Rasanen I, Kyber M, Szilvay I, Rintatalo J, et al. Straightforward single-calibrant quantification of seized designer drugs by liquid chromatography-chemiluminescence nitrogen detection. Forensic Sci Int 2014;237:119−25.
[52] Lesiak AD, Musah RA, Domin MA, Shepard JRE. Dart-MS as a preliminary screening method for "herbal incense": chemical analysis of synthetic cannabinoids. J Forensic Sci 2014;59(2):337−43.
[53] Scheidweiler KB, Huestis MA. Simultaneous quantification of 20 synthetic cannabinoids and 21 metabolites, and semi-quantification of 12 alkyl hydroxy metabolites in human urine by liquid chromatography tandem-mass spectrometry. J Chromatogr A 2014;1327:105−17.
[54] Akamatsu S, Mitsuhashi T. MEKC-MS/MS method using a volatile surfactant for the simultaneous determination of 12 synthetic cannabinoids. J Sep Sci 2014;37(3):304−7.
[55] Salamone A, Luciano C, Corcia DD, Gerace E, et al. Hair analysis as a tool to evaluate the prevalence of synthetic cannabinoids in different populations of drug consumers. Drug Test Anal 2014;6:126−34.
[56] Fowler F, Voyer B, Marino M, Finzel J, et al. Rapid screening and quantification of synthetic cannabinoids in herbal products with NMR spectroscopic methods. Anal Methods 2015;7:7907−16.
[57] Casteneto MS, Scheidweiler KB, Gandhi A, Wohlfarth A, et al. Quantitative urine confirmatory testing for synthetic cannabinoids in randomly collected urine specimens. Drug Test Anal 2015;7(6):483−93.
[58] Ma Q, Bai H, Li W, Wang C, et al. Rapid analysis of synthetic cannabinoids using a miniature mass spectrometer with ambient ionization capability. Talanta 2015;142:190−6.
[59] Gwak S, Arroyo-Mora LE, Almirall JR. Qualitative analysis of seized synthetic cannabinoids and synthetic cathinones by gas chromatography triple quadrupole tandem mass spectrometry. Drug Test Anal 2015;7(2):120−30.
[60] Habala L, Valentová J, Pechová I, Fuknová M, et al. Dart-LTQ ORBITRAP as an expedient tool for the identification of synthetic cannabinoids. Leg Med 2016;20:27−31.
[61] Knittel JL, Holler JM, Chmiel JD, Vorce SP, et al. Analysis of parent synthetic cannabinoids in blood and urinary metabolites by liquid chromatography tandem mass spectrometry. J Anal Toxicol 2016;40:173−86.
[62] Marino MA, Voyer B, Cody RB, Dane JA, et al. Rapid identification of synthetic cannabinoids in herbal incenses with Dart-MS and NMR. J Forensic Sci 2016;61(S1):S82−91.
[63] Angerer V, Bisel P, Moosmann B, Westphal F, et al. Separation and structural characterization of the new synthetic cannabinoid JWH-018 cyclohexyl methyl derivative "NE-CHMIMO" using flash chromatography, GC-MS, IR and NMR spectroscopy. Forensic Sci Int 2016;266: e93−8.
[64] Raso S, Bell S. Qualitative analysis and detection of the pyrolytic products of JWH-018 and 11 additional synthetic cannabinoids in the presence of common herbal smoking substrates. J Anal Toxicol 2017;1−8.

[65] Uchiyama N, Kawamura M, Kikura-Hanajiri R, Goda Y. Identification and quantitation of two cannabimimetics phenylacetylindoles JWH-251 and JWH-250, and four cannabimimetics naphthoylindole JWH-081, JWH-015, JWH-200, and JWH-073 as designer drugs in illegal products. Forensic Toxicol 2011;29:25−37

[66] Uchiyama N, Kikura-Hanajiri R, Goda Y. Identification of a novel cannabimimetic phenylacetylindole, cannabipiperidiethanone, as a designer drug in a herbal product and its affinity for cannabinoid CB_1 and CB_2 receptors. Chem Pharm Bull 2011;59(9):1203−5.

[67] Moran CL, Le V, Chimalakonda KC, Smedley AL, et al. Quantitative measurement of JWH-018 and JWH-073 metabolites excreted in human urine. Analy Chem (Washington DC, US) 2011;83(11):4228−36.

[68] Coulter C, Garnier M, Moore C. Synthetic cannabinoids in oral fluid. J Anal Toxicol 2011;35(7):424−30.

[69] Nekhoroshev SV, Nekhoroshev VP, Remizova MN, Nekhoroshev AV. Determination of the chemical compositing of *spice* aromatic smoking blends by chromatography − mass spectrometry. J Anal Chem 2011;66(12):1196−2000.

[70] Grabenauer M, Krol WL, Wiley JL, Thomas BF. Analysis of synthetic cannabinoids using high-resolution mass spectrometry and mass defect filtering: implications for nontargeted screening of designer drugs. Anal Chem 2012;84:5574−81.

[71] Shanks KG, Dahn T, Terrel R. Detection of JWH-018 and JWH-073 by UPLC-MS-MS in postmortem whole blood casework. J Anal Toxicol 2012;36(3):145−52.

[72] Yanes EG, Lovett DP. High-throughput bioanalytical method for analysis of synthetic cannabinoid metabolites in urine using salting-out sample preparation and LC-MS/MS. J Chromatogr B 2012;909:42−50.

[73] Vandell VE, Kero F. Extraction of synthetic cannabinoids parents and metabolites (JWH Sieries) from urine, plasma, and whole blood using supported liquid extraction (ISOLUTE SLE +) in cartridge and 96-well format prior to LC-MS-MS. Lc Gc N AM 2012;S(74).

[74] Kneisel S, Bisel P, Brecht V, Broecker S, et al. Identification of the cannabimimetic AM-1220 and its azepan isomer (N-methylazepan-3-yl)-3-(1-naphthoyl)indole in a research chemical and several herbal mixtures. Forensic Toxicol 2012;30(2):126−34.

[75] Donohue KM, Steiner RR. JWH-018 and JWH-022 as combustion products of AM2201. Microgram Journal 2012;9(2):52−6.

[76] Kneisel S, Auwärter V, Kempf J. Analysis of 30 synthetic cannabinoids in oral fluid using liquid chromatography-electrospray ionization tandem mass spectrometry. Drug Test Anal 2013;5(8):657−69.

[77] Rodrigues WC, Catbagan P, Rana S, Wang G, Moore C. Detection of synthetic cannabinoids in oral fluid using ELISA and LC-MS-MS. J Anal Toxicol 2013;37:526−33.

[78] Shanks KG, Behonick GS, Dahn T, et al. Identification of novel third-generation synthetic cannabinoids in products by ultra-performance liquid chromatography and time-of-flight mass spectrometry. J Anal Toxicol 2013;37:517−25.

[79] Kim J, In S, Park Y, Park M. Deposition of JWH-018, JWH-073 and their metabolites in hair and effect of hair pigmentation. Anal Bioanal Chem 2013;405:9769−78.

[80] Mazzarino M, de la Torre X, Botrè F. A liquid chromatography-mass spectrometry method based on class characteristic fragmentation pathways to detect the class of indole-derivative synthetic cannabinoids in biological samples. Anal Chim Acta 2014;87:70−82.

[81] Li L, Lurie IS. Screening of seized emerging drugs by ultra-high performance liquid chromatography with photodiode array ultraviolet and mass spectrometric detection. Forensic Sci Int 2014;237:100−11.

[82] Jang M, Yang W, Shin I, Choi H, et al. Determination of AM-2201 metabolites in urine and comparison with JWH-018 abuse. Int J Legal Med 2014;128:285−94.

[83] Kronstrand R, Brinkhagen L, Birath-Karlsson C, Roman M, et al. LC-QTOF-MS as a superior strategy to immunoassay for the comprehensive analysis of synthetic cannabinoids in urine. Anal Bioanal Chem 2014;406:3599−609.

[84] Thaxton A, Belal TS, Smith F, DeRuiter J, et al. GC-MS studies on the six naphthoyl-substituted 1-n-pentyl-indoles: JWH-018 and five regioisomeric equivalents. Forensic Sci Int 2015;252:107−13.

[85] Zaitsu K, Nakayama H, Yamanaka M, Hisatsune K, et al. High-resolution mass spectrometric determination of the synthetic cannabinoids MAM-2201, AM-2201, AM-2232, and their metabolites in postmortem plasma and urine by LC/Q-TOFMS. Int J Legal Med 2015;129:1233−45.

[86] Barnes AJ, Spinelli E, Young S, Martin TM, et al. Validation of an ELISA synthetic cannabinoids urine assay. Ther Drug Monit 2015;37(5):661−9. Available from: https://doi.org/10.197/FTD.0.0000000000000201.

[87] Jang M, Shin I, Kim J, Yang W. Simultaneous quantification of 37 synthetic cannabinoid metabolites in human urine by liquid chromatography-tandem mass spectrometry. Forensic Toxicol 2015;33(4):221−34. Available from: https://doi.org/10.1007/s11419-015-0265-x.

[88] Scheidweiler KB, Jarvis MJY, Huestis MA. Nontargeted SWATH acquisition for identifying 47 synthetic cannabinoid metabolites in human urine by liquid chromatographic-high-resolution tandem mass spectrometry. Anal Bioanal Chem 2015;407:883−97.

[89] Toyo'oka T, Kikura-Hanajiri R. A reliable method for the separation and detection of synthetic cannabinoids by supercritical fluid chromatography with mass spectrometry, and its application to plant products. Chemi Pharma Bull 2015;63(10):762−9.

[90] Paul BD, Bosy T. A sensitive GC-EIMS method for simultaneous detection and quantification of JWH-018 and JWH-073 carboxylic acid and hydroxyl metabolites in urine. J Anal Toxicol 2015;39(3):172−82.

[91] Ozturk S, Ozturk YE, Yeter O, Alpertunga B. Application of a validated LC-MS/MS method for JWH-073 and its metabolites in blood and urine in real forensic cases. Forensic Sci Int 2015;257:165−71.

[92] Karinen R, Tuv SS, Oeiestad EL, Vindenes V. Concentrations of APINACA, 5F-APINACA, UR-144, and its degradant product blood samples from six impaired drivers compared to previous reported concentrations of other synthetic cannabinoids. Forensic Sci Int 2015;246:98−103.

[93] Karin Karinen R, Johnson L, Andresen W, Christophersen AS, et al. Stability study of fifteen synthetic cannabinoids of aminoalkylindole type in whole blood, stored in Vacutainer evacuated gas tubes. J Forensic Toxicol Pharmacol 2013;2(1) pages.

[94] Borova VL, Gago-Ferrero P, Pistos C, Thomaidis NS. Multi-residue determination of 10 selected new psychoactive substances in wastewater samples by liquid chromatography-tandem mass spectrometry. Talanta 2015;144:592–603.

[95] Berg T, Kaur L, Risnes A, Havig SM, et al. Determination of a selection of synthetic cannabinoids and metabolites in urine by UHPSFC-MS/MS and by UHPLC-MS/MS. Drug Test Anal 2015;8(7):708–22.

[96] Kim J, Park Y, Park M, Kim E, et al. Simultaneous determination of five napthoylindole-based synthetic cannabinoids and metabolites and their deposition in human and rat hair. J Pharm Biomed Anal 2015;102:162–75.

[97] Heo S, Yoo GJ, Choi JY, Park HJ, et al. Simultaneous analysis of cannabinoid and synthetic cannabinoids in dietary supplements using UPLC with UV and UPLC-MS-MS. J Anal Toxicol 2016;40:350–9.

[98] Neifeld JR, Regester LE, Holler JM, Vorce SP, et al. Ultrafast screening of synthetic cannabinoids and synthetic cathinones in urine by rapidfire-tandem mass spectrometry. J Anal Toxicol 2016;40:379–87.

[99] Davies BB, Bayard C, Larson SJ, Zarwell LW, et al. Retrospective analysis of synthetic cannabinoid metabolites in urine of individuals suspected of driving impaired. J Anal Toxicol 2016;40:89–96.

[100] Dong X, Li L, Ye Y, Zheng L, et al. Simultaneous determination of major phytocannabinoids, their main metabolites, and common synthetic cannabinoids in urine samples by LC-MS/MS. J Chromatogr B 2016;1033–1034:55–64.

[101] Dronova M, Smolianitski E, Lev O. Electrooxidation of new synthetic cannabinoids: Voltammetric determination of drugs in seized street samples and artificial saliva. Anal Chem 2016;88(8):4487–94.

[102] Borg D, Tverdovsky A, Stripp R. A fast and comprehensive analysis of 32 synthetic cannabinoids using agilent triple quadrupole LC-MS-MS. J Anal Toxicol 2016;1–11.

[103] Adamowicz P, Tokarczyk B. Simple and rapid screening procedure for 143 new psychoactive substances by liquid chromatography-tandem mass spectrometry. Drug Test Anal 2016;71(8):794–802.

[104] Švidrnoch M, Přibylka A, Vítězslav M. Determination of selected synthetic cannabinoids and their micellar electrokinetic chromatography – mass spectrometry employing perfluoroheptanoic acid-based micellar phase. Talanta 2016;150:568–76.

[105] Risoluti R, Materazzi S, Gregori A, Ripani L. Early detection of emerging street drugs by near infrared spectroscopy and chemometrics. Talanta 2016;153:407–13.

[106] Breitenbach S, Rowe WF, McCord B, Lurie IS. Assessment of ultra-high performance supercritical fluid chromatography as a separation technique for the analysis of seized drugs: applicability to synthetic cannabinoids. J Chromatogr A 2016;1440:201–11.

[107] Mostowtt T, McCord B. Surface enhanced Raman spectroscopy (SERS) as a method for the toxicological analysis of synthetic cannabinoids. Talanta 2017;164:396–402.

[108] Moore KN, Garvin D, Thomas BF, Grabenauer M. Identification of eight synthetic cannabinoids, including 5F-AKB48 in seized herbal products using DART-TOF-MS and LC-QTOF-MS as nontargeted screening methods. J Forensic Sci 2017;62(5):1151–8.

[109] González-Mariño I, Thomas KV, Reid MJ. Determination of cannabinoid and synthetic cannabinoid metabolites in wastewater by liquid-liquid extraction and ultrahigh performance supercritical fluid chromatography-tandem mass spectrometry. Drug Test Anal 2017.

[110] Tynon M, Homan J, Kacinko S, Ervin A, et al. Rapid and sensitive screening and confirmation of thirty-four aminocarbonyl/carboxamide (NACA) and arylindole synthetic cannabinoid drugs in human whole blood. Drug Test Anal 2017;9(6):924–34.

[111] Montesano C, Vannutelli G, Massa M, Simeoni MC, et al. Multi-class analysis of new psychoactive substance and metabolites in hair by pressurized liquid extraction coupled to HPLC-HRMS. Drug Test Anal 2017;9(5):798–807.

[112] Umebachi R, Saito T, Aoki H, Namera A, et al. Detection of synthetic cannabinoids using GC-EI-MS, positive GC-CI-MS, and negative GC-CI-MS. Int J Legal Med 2017;131:143–52.

[113] Popławska M, Błażewicz A, Kamiński K, Bednarek E, et al. Application of high-performance liquid chromatography with charged aerosol detection (LC-CAD) for unified quantification for synthetic cannabinoids in herbal blends and comparison with quantitative NMR results. Forensic Toxicol 2017;36:122–40.

[114] Bononi M, Belgi P, Tateo F. Analytical data for identification of the cannabimimetic phenylacetylindole JWH-203. J Anal Toxicol 2011;35(6):360–3.

[115] Simões SS, Silva I, Ajenjo AC, Dias MJ. Validation and application of an UPLC-MS/MS method for the quantification of synthetic cannabinoids in urine samples and analysis of seized materials from the Portuguese market. Forensic Sci Int 2014;243:117–25.

[116] Uchiyama N, Shimokawa Y, Matsuda S, Kawamura M, et al. Two new synthetic cannabinoids, AM-2201 benzimidazole analog (FUBIMINA) and (4-methylpiperazin-1-yl)(1-pentyl-1H-indol-3-yl)methanone (MEPIRAPIM), and three phenethylamine derivatives, 25H-NBOMe 3,4,5-trimethoxybenzyl analog, 25B-NBOMe, and 2C-N-NBOMe, identified in illegal products. Forensic Toxicol 2014;32:105–15.

[117] Qian Z, Jia W, Li T, Hua Z, et al. Identification and analytical characterization of four synthetic cannabinoids ADB-BICA, NNL-1, NNL-2, and PPA(N)-2201. Drug Test Anal 2017;9(1):51–60.

[118] Shevyrin V, Melkozerov V, Nevero A, Eltsov O, et al. Synthetic cannabinoids as designer drugs: new representatives of indole-3-carboxylates series and indazole-3-carboxylates as novel group of cannabinoids. Identification and analytical data. Forensic Sci Int 2014;244:263–75.

[119] Behonick G, Shanks KG, Fircahu DJ, Mathur G, Lynch CF, Nashelsky M, et al. Four postmortem case reports with quantitative detection of the synthetic cannabinoid, 5F-PB-22. J Anal Toxicol 2014;38(8):559–62.

[120] Dunne SJ, Rosengren-Holmberg JP. Quantification of synthetic cannabinoids in herbal smoking blends using NMR. Drug Test Anal 2017;9(5):734–43.

[121] Nie H, Li X, Hua Z, Pan W, et al. Rapid screening and determination of 11 new psychoactive substances by direct analysis in real time mass spectrometry and liquid chromatography/quadrupole time-of-flight mass spectrometry. Rapid Commun Mass Spectrom 2016;30(1):141–6.

[122] Langer N, Lindigkeit R, Schiebel H, Papke U. Identification and quantification of synthetic cannabinoids in "spice-like" herbal mixtures: update of the German situation for the spring of 2015. Forensic Toxicol 2016;34:94−107.

[123] Hess C, Murach J, Krueger L, Scharrenbroch L, et al. Simultaneous detection of 93 synthetic cannabinoids by liquid chromatography-tandem mass spectrometry and retrospective application to real forensic samples. Drug Test Anal 2017;9(5):721−33.

[124] Gandhi AS, Wohlfarth A, Zhu M, Pang S, et al. High-resolution mass spectrometric metabolite profiling of a novel synthetic designer drug N-(adamantan-1-yl)-1-(5-fluoropentyl)-1H-indole-3-carboxamide (STS-135), using cryopreserved human hepatocytes and assessment of metabolic stability with human liver microsomes. Drug Test Anal 2015;7(3):187−98.

[125] Wohlfarth A, Pang S, Zhu M, Gandhi AS, et al. First metabolic profile of XLR-11, a novel synthetic cannabinoid, obtained by using human hepatocytes and high-resolution mass spectrometry. Clin Chem 2013;59(11):1638−48.

[126] Mohr ALA, Ofsa B, Kiel AM, Simon JR, et al. Enzyme-linked immunosorbent assay (ELISA) for the detection of use of the synthetic cannabinoid agonists UR-144 and XLR-11 in human urine. J Anal Toxicol 2014;38:427−31.

[127] Amaratunga P, Thomas C, Lemberg BL, Lemberg D. Quantitative measurement of XLR11 and UR-144 in oral fluid by LC-MS-MS. J Anal Toxicol 2014;38:315−21.

[128] Jang M, Kim I, Park YN, Kim J, et al. Determination of urinary metabolites of XLR-11 by liquid chromatography-quadrupole time-of-flight mass spectrometry. Anal Bioanal Chem 2016;408:503−16.

[129] Langer N, Lindigkeit R, Schiebel H, Papke U, et al. Identification and quantification of synthetic cannabinoids in "spice-like" herbal mixtures: update of the German situation for the spring of 2016. Forensic Sci Int 2016;269:31−41.

[130] Shevyrin V, Melkozerov V, Eltsov O, Shafran Y, et al. Synthetic cannabinoid 3-benzyl-5-[1-(2-pyrrolidin-1-ylethyl)-1H-indol-3-yl]-1,2,4-oxadiazole. The first detection in illicit market of new psychoactive substances. Forensic Sci Int 2016;259:95−100.

[131] Vikingsson S, Gréen H, Brinkhagen L, Mukhtar S, et al. Identification of AB-FUBINACA metabolites in authentic urine samples suitable as urinary markets of drug intake using liquid chromatography quadrupole tandem time of flight mass spectrometry. Drug Test Anal 2016;8(9):850−6.

[132] Minakata K, Yamagishi I, Nozawa H, Hasegawa K, et al. Sensitive identification and quantitation of parent forms of six synthetic cannabinoids in urine samples of human cadavers by liquid chromatography-tandem mass spectrometry. Forensic Toxicol 2017;35:275−83.

[133] Carlier J, Dio X, Sempio C, Huestis MA. Identification of new synthetic cannabinoid ADB-CHMINACA (MAB-CHMINACA) metabolites in human hepatocytes. AAPS J 2017;19(2):568−77.

[134] Vikingsson S, Josefsson M, Gréen H. Identification of AKB-48 and 5F-AKB-48 metabolites in authentic human urine samples using human liver microsomes and time of flight mass spectrometry. J Anal Toxicol 2015;39:426−35.

[135] Jia W, Meng X, Qian Z, Hua Z, et al. Identification of three cannabimimetic indazole and pyrazole derivatives, APINACA 2H-indazole analogue, AMPPPCA, and 5F-AMPPPCA. Drug Test Anal 2017;9(2):248−55.

[136] Shevyrin V, Melkozerov V, Nevero A, Eltsov O, et al. 3-Naphthoylindazoles and 2-naphthoylbenzoimidazoles as novel chemical groups of synthetic cannabinoids: chemical structure elucidation, analytical characteristics and identification of the first representatives in smoke mixtures. Forensic Sci Int 2014;242:72−80.

[137] Uchiyama N, Kawamura M, Kikura-Hanajiri R, Goda Y. URB-754. A new class of designer drug and 12 synthetic cannabinoids detected in illegal products. Forensic Sci Int 2013;227:21−32.

Chapter 23

Application of Liquid Chromatography Combined With High Resolution Mass Spectrometry for Urine Drug Testing

Olof Beck[1,2], Alexia Rylski[2] and Niclas Nikolai Stephanson[2]
[1]Department of Laboratory Medicine, Karolinska Institute, Stockholm, Sweden, [2]Pharmacology Laboratory, Karolinska University Hospital, Stockholm, Sweden

INTRODUCTION

Urine drug testing has evolved into a standard investigation technique in clinical and forensic laboratories ever since the development of commercial immunochemical assays for screening of commonly abused illicit drugs in the 1970s. These assays have the feature of being able to directly use the original urine sample as the analytical specimen, that is, no sample preparation is required. The reagents can be applied on fully automated chemistry analyzers capable of processing hundreds of samples per hour. Because of the antibody binding used as the core analytical principle, each test is directed and limited to a specific drug or drug class, for example, cannabis, cocaine, amphetamines. Therefore, a panel of tests is usually performed on each urine specimen being investigated at the laboratory. It is difficult for other techniques to compete with immunoassay screening in urine drug testing due to the turnaround time and cost-effectiveness for a small test panel. However, the number of drugs to be covered have increased in the recent time [1] and the test panel now needs to include new psychoactive substances (NPS) and a range of new therapeutic drugs, for example, sedatives and pain medication. Immunoassay screening methods lack many of these new drugs that have become relevant for the drug testing.

Initially, the role of mass spectrometry in urine drug testing was to produce final evidential results, that is, confirmation for a specific positive finding in the immunoassay screening. The reason for this need to confirm a screening positive finding was due to the fact that the immunoassays give a rather high error rate (1%–15%) of false positives and therefore have no evidential value [2,3]. For this purpose, gas chromatography—mass spectrometry (GC-MS) method was employed. These methods need sample preparation and often derivatization before the instrumental analysis, which make them time consuming and costly [4]. This changed dramatically with the advent of liquid chromatography—mass spectrometry (LC-MS), with the use of tandem mass spectrometers (MS/MS), and with the development of more effective (i.e., ultra-performance) LC-technologies.

LC-MS/MS in selected reaction monitoring (SRM) mode has also been successfully applied to urine drug screening [5]. In the authors experience this is a very useful technology for urine drug screening; however, one disadvantage is that only a limited number of target analytes can be included. Furthermore, the need for optimization experiments and method validation for each analyte makes it a somewhat time-consuming technology. This is especially true when there is a need for continuous updating of the method parameters as for analysis of NPS. This made it of interest to consider alternative mass spectrometry technology, that is, high resolution mass spectrometry (HRMS) [6,7].

HISTORY AND FUNDAMENTALS OF HRMS

The potential value of HRMS for bioanalytical investigations was recognized even before chromatography was combined with MS [8]. The interest at that time was focused on getting information on elemental composition of unknown components for identification purposes. HRMS in combination with gas chromatography for quantitative bioanalytical investigations also has a long history. One such application that used the unique selectivity of HRMS was the environmental analysis of dioxins using gas chromatography and magnetic sector instruments [9]. However, HRMS in combination with LC preferably utilizes two other MS techniques for achieving high resolution data, namely time-of-flight (TOF) and orbitrap (OT). A summary of HRMS features and terminology is given below:

- There is no clear definition or consensus on the resolving power (RP) that makes an instrument qualified to be considered as "high resolution" MS. One group of MS instruments based on quadrupole and ion trap mass analyzers employ unit resolution and monitoring of nominal masses. In high resolution MS the substances with same nominal mass (isobars) but with different elemental compositions are to be resolved. The determination of the accurate mass of an unknown substance is used to estimate its elemental composition. A practical definition of high resolution MS can be a criterion of $RP > 10,000$. This makes it possible to include both the TOF and OT technologies in the concept.
- Mass spectral resolution is defined as the ability to distinguish between two adjacent ions with 50% overlap (FWHM). RP is calculated as: $RP = m/\Delta m$ (Fig. 23.1). It is important to realize that the RP will be different over the mass scale.
- While working with HRMS it is important to consider the relation between exact monoisotopic mass of the molecules studied and the exact masses of the ions being monitored in the instrument. In electrospray ionization, protonated or deprotonated molecules dominate. In the first instance a proton has been added to the molecule but an electron has been lost. In the second case a proton has been lost but an electron has been kept. Mass of proton = 1.00782 amu, electron = 0.00054 amu, $[M+H]^+$ ion weighs: monoisotopic mass + 1.0073, and $[M-H]^-$ ion weighs: monoisotopic mass − 1.0073.
- Mass accuracy is the ability of the instrument to measure the accurate mass of the ion. This ability is related to the RP and mass calibration stability of the instrument. Mass accuracy can be given as absolute error (Δm = exact mass − measured mass), for example, 10 mDa.
- Mass accuracy can also be given as the relative error (ppm = $(\Delta m/m) \times 10^6$), for example, 10 ppm is 1 mDa at m/z 100, but 10 mDa at m/z 1000. Usually the instrument will have a mass accuracy specification in the order of one or a few ppm's.
- Exact mass = calculated theoretical mass.
- Accurate mass = the measured mass in HRMS.
- Resolution power is needed for selectivity because urine contains thousands of low molecular weight substances that will produce response in the analysis. These substances may have either endogenous or exogenous origin, and many of them may share nominal masses. If two substances coelute and have the same nominal mass, they must be resolved by masses otherwise they will fuse together and result in a unified peak with an incorrect mass assignment.
- Chromatographic separation is needed because substances with same elemental composition (isotopomers) have same exact mass. They must be separated chromatographically in order to achieve required selectivity of the measurements.

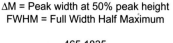
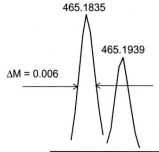

FIGURE 23.1 Resolving power (RP) of HRMS.

TOF Analyzer

TOF is another technique with a long history. It took a long time to achieve the needed high RP. TOF is based on the simple fact that given the same kinetic energy lighter ions will fly faster than heavier ions. The ions are sent to the flight-tube in a pulse-wise manner, and the flight time to the detector is measured and calibrated to mass scale (Fig. 23.2). The optimal function of the technology depends on the advancements in electronics and computer data handling, and can be expected to undergo further developments [10]. Of primary importance for this context is the performance specifications regarding RP and scan time in the mass range below 1000 Da. The RP varies over the mass scale. Examples are given in Table 23.1 where the RPs of two commercial TOF instruments are listed over the mass range 100–1000 Da. The TOF technology has a feature to give higher resolution at higher masses. A TOF instrument can operate with scanning frequencies as high as 50Hz when scanning the mass range of 100–1000 m/z.

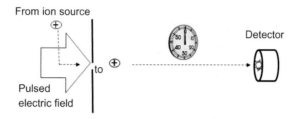

FIGURE 23.2 Schematic presentation of the of the time-of-flight technology. TOF is based on the simple fact that given the same kinetic energy lighter ions will fly faster than heavier ions. The ions are sent pulse-wise into the flight-tube, and the flight time to the detector is measured and calibrated to mass scale.

TABLE 23.1 Performance Specifications of Three High Resolution MS Instruments: Data from Vendors

LC-HRMS Instrument	Waters Xevo G2 XS		Bruker Impact II		Q Exactive	
Technology	TOF		TOF		Orbitrap	
Specified resolving power	40,000		50,000		70,000	
	at m/z		at m/z		at m/z	
	120	21,000			100	99,000
	175	22,000			200	70,000
	333	30,000	322	>28,000	300	57,000
					400	50,000
	480	32,000			500	44,000
	627	36,600	622	>47,500	600	40,000
	684	36,000			700	38,000
	813	39,000			800	35,000
	956	45,000	922	>55,000	900	33,000
	1172	43,900	1222	>60,000	1000	30,000
Dynamic range, orders of magnitude	4		5		6	
Mass accuracy	1 ± ppm		± 0.8 ppm		1 ± ppm	
Scanning frequency Hz	30		50		3.5	
Sensitivity	100 fg reserpine SN > 250:1 RMS		100 fg reserpine with ionbooster SN > 100:1 RMS		50 fg reserpine detected	

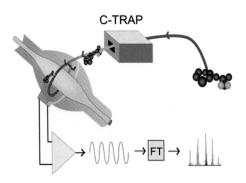

FIGURE 23.3 Schematic presentation of the orbitrap technology. The orbitrap is based on pulsed packages of ions, produced by the C-trap, that are introduced and captured into the analyzer. The ions are kept in an orbitrap movement while the applied electrical field keeps the ions in oscillating movement and produces signals to receiver plates. The signals are Fourier transformed into mass spectral data.

Orbitrap Analyzer

The OT instrument is related to a more recent technology development and the first commercial instrument was not put on the market until 2005. In contrast to TOF, which use the speed of movement of ions, the OT technique is based on orbital trapping of the ions in an electrostatic field (Fig. 23.3), similar to the ion trap technology. The aim of the OT development was to achieve a high-performing research and routine instrument regarding resolution power and sensitivity [11]. The OT is based on pulsed packages of ions, produced by the C-trap, that are introduced and captured into the analyzer. The ions are kept in an OT movement while the applied electrical field keeps the ions in oscillating movement and produces signals to receiver plates [12]. The signals are Fourier transformed into mass spectral data. In order to obtain quality data, the time spent for each ion package in the OT is essential. Table 23.1 presents the performance specifications of one OT instrument, and compares it with two TOF instruments.

The two techniques display somewhat different features. A higher RP is offered with the OT, while the TOF instruments have higher scanning frequency. Instruments of both types can operate in MS and MS/MS mode.

Direct Injection of Urine

One prerequisite for a high-volume application is to automate and simplify the sample preparation, as done in immunoassay screening. A system with direct injection of urine is the best option to design a mass spectrometry system to replace immunoassay screening. Since addition of internal standards are required for optimal analytical performance, a dilution step is included in the procedure, known as "dilute and shoot." The analytical system can include an online solid-phase extraction. This will increase robustness but also discriminate the range of substances covered, and might also limit the capacity.

The possibility of injecting untreated urine into LC-MS system was recognized earlier [13,14], but received skepticism and was criticized at that time [15]. One major reason for this is the matrix effect that occurs in electrospray ionization and can cause severe (99%) ion suppression. An illustrative example is given in Fig. 23.4. In reversed phase chromatography systems, a major ion-suppression effect is seen following elution of the void volume and the most hydrophilic components (e.g., salt ions). Ionization enhancement is often seen later in the gradient system as the fraction of organic solvent increases. The chromatography system must therefore be designed to retain the hydrophilic analytes (e.g., morphine) that might be subjected to signal suppression by choosing the chromatography system appropriately, that is, column material and mobile phase. Another important aspect is to choose appropriate internal standards that can compensate and monitor the matrix effects. Each analyte should preferably have a stable isotope-labeled analog as internal standard. If it is not possible, the internal standards should be spread over the entire range of monitored retention time. Validation for matrix effect should follow the recommendations of Matuziewski and coworkers [16].

The method designed to use direct injection of diluted urine is now well established and accepted [15]. Published methods comprise both screening and confirmation methods for drugs of abuse, including applications for doping control in sports [15]. An alternative approach worth mentioning is to have online solid-phase extraction but this will slow down the capacity [17]. Another emerging technology that could offer increased capacity per instrument is paper spray ionization, which was considered promising despite some reported limitations [18].

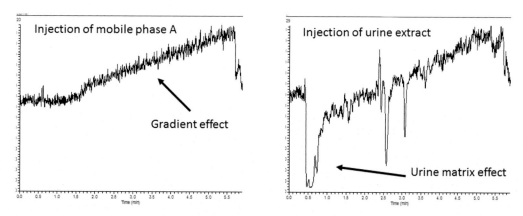

FIGURE 23.4 Example of the influence of matrix on the ionization of amphetamine. Amphetamine was constantly infused together with mobile phase. Mobile phase and a blank urine extract was injected as sample. The drop in response seen after ~0.4 minutes is from the influence from matrix components on the ionization efficiency of the analyte. Increase in response over the gradient is caused by the increasing amount of organic modifier.

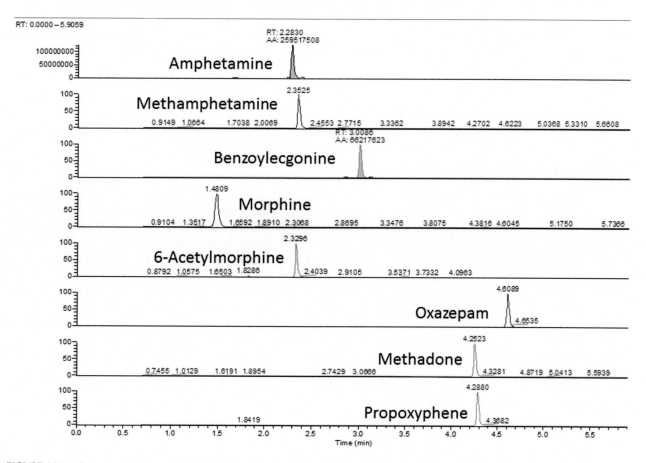

FIGURE 23.5 Chromatogram from the analysis of a urine standard containing amphetamine, methamphetamine, benzoylecgonine, morphine, 6-acetylmorphine, oxazepam, methadone, and propoxyphene. The analysis was performed on a Q Exactive OT system operating at 70,000 resolving power.

Chromatography System

The choice of chromatographic system is always a compromise between the need for chromatographic resolution and a short analysis time. The sub 2 μm particle size column material developed for ultra-high-performance liquid chromatography has offered an increase in chromatographic efficiency. Reversed phase chromatography is most commonly used for drug screening methods. It is important to choose a chromatographic system that retains and separates the hydrophilic analytes from the urine components that cause ion suppression (see Figs. 23.4 and 23.5). Special column materials are available that combine the reversed phase feature with capability to retain the hydrophilic analytes [19] see Fig. 23.5 for an example. Another chromatographic technique that might offer possibilities for urine drug screening with HRMS is capillary electrophoresis [20].

FEATURES OF IMMUNOASSAYS: WHAT ARE PROBLEMS TO BE SOLVED?

If LC-HRMS is to compete with immunoassay screening, it must provide features that make this alternative technology attractive. There are several aspects to consider.

1. A large and flexible test panel: The immunoassay test panel is restricted to commercially available reagents. Developing new tests is a time-consuming process and the development time is often more than reasonable. Even when a new test is developed, it only covers analytes of similar chemical structure. A good example is the synthetic cannabinoids, which comprise hundreds of substances having different structures and no single test can cover them all. Over time this need of new tests will mean that the immunoassay test panel will expand and can eventually become over 10 tests per sample. A mass spectrometry system, however, can easily cover 50 analytes in a method and replace a whole immunoassay test panel. In addition, if new analytes need to be included in the method, it is a matter of having the reference substance and in-house validation, which can be a much more rapid process.
2. Reporting (cutoff) limits: Immunoassay tests are provided with set reporting limits (cutoff levels). These are often not set on basis of clinical demands and needs, or by scientific consideration. Mass spectrometry methods can offer a possibility to tailor the reporting limits to better fulfill the requirement for detection times depending on the purpose of the testing.
3. Information value: There are two important aspects about the value of a positive test reported from a mass spectrometry method. More specific information can be obtained since the exact substance(s) detected and its concentration can be reported. For example, instead of reporting benzodiazepines positive as in an immunoassay the exact substance can be reported, for example, oxazepam. The other and even more important aspect is that the reliability (predictive value) of a positive screening test can be increased so that the incidence of false (unconfirmed) positives can be kept at a minimum (<1%). This will also improve the value of a drug testing service when unconfirmed positives are used (e.g., in clinical testing).
4. Turnaround times: Rapid analyses on high-capacity instruments are a key feature of immunoassay screening. This feature must be conserved when switching to another technology. Depending on the settings and customer requirements, this can become an obstacle for moving to MS screening. An MS system might process 10−20 samples per hour, while an autoanalyzer can report 600 results per hour. If 6 parameters are analyzed per sample, 100 unknowns can be processed per hour, which exceeds the capability of an MS system. If, however, the MS system analyzes 50 parameters, the productivity becomes similar.
5. Cost: Yet another important factor is cost. An MS system is more expensive, has lower capacity, and is less robust than an autoanalyzer for immunoassay screening. This must be fully considered as it represents the most common objection to the idea of using mass spectrometry for drug testing. However, assuming a cost for a MS system to be US$100,000 per year, available instrument time 90%, 250 business days, and a capacity of 150 unknowns per day give an estimated cost of US$3 per sample. If the cost for immunoassay screening is US$1 per parameter, an MS system can actually become a viable alternative also from a cost perspective and allow for multiple instruments.
6. Standardization: One important feature of immunoassay screening is that laboratories use the same methods (reagents) and calibrators. It is important that measurements made at different laboratories produce comparable results. It has been acknowledged that mass spectrometry methods, despite being a sensitive and accurate technology, can produce scattered results in inter-laboratory comparisons. It is therefore of fundamental importance that method validation is done properly and that calibrators are made properly [21]. Proper method validation and participation in proficiency programs is one way to monitor this aspect.
7. Discovery of new substances: HRMS has the feature of providing information on elemental composition of the detected analyte. This feature can be used to tentatively identify new NPS in precious cases [22].

WHY HRMS FOR DRUG TESTING?

LC-HRMS was very early identified as an attractive and promising alternative for multicomponent toxicological urine screening [23]. The authors were attracted by the possibility of using HRMS in full scan mode to detect components based on accurate mass information without having reference material available. However, the use of retention time data was also acknowledged at that time. In subsequent work, the concept was developed by using accurate mass, isotope pattern, and retention time as identification criteria [24]. A mass accuracy error within $\Delta 10$ ppm was reported to be suitable as identification criteria. However, despite using solid-phase extraction for preparation of extracts, both false positive and negative results were obtained. Similar method designs were subsequently used for doping control analysis in sports [25–27]. A later report from Kolmonen and coworkers, still using the same, but improved, method design, stated that the method was suitable for "high-throughput" screening [28].

The first use of the OT technology for screening of doping agents was reported in 2008 by Virus and coworkers [29]. They evaluated a procedure for 29 compounds and used APCI ionization with in-source fragmentation together with a RP setting of 60,000, which was much higher than used before. Doping control methods is another application of the OT technology used during the 2012 London Olympic Games that again involves solid-phase extraction as sample preparation of urine [30]. The total analysis time was 10 minutes, the authors screened 5000 unknowns in one month, and considered this being "fast screening in high throughput doping control" since it was done online during the games. Authors decided to use HRMS with the OT technique since the higher RP was expected to provide more selectivity. Prior to this, another sports doping laboratory reported that sample preparation could be performed with 10 times dilution with internal standards [31]. In these later methods, detection of components was based on accurate mass and retention time data. The value of high RP and excellent accurate mass performance of the instrument was pointed out.

In summary, the LC-HRMS technology, in full scan mode, can offer a generic and very selective acquisition of components, in principle, without a need for setting targets beforehand.

Need for Resolving Power in Urine Drug Screening

Striving for higher RP is a clear tendency in the development of new HRMS instruments. However, what is the need in terms of RP for the application of urine drug screening?

Human urine contains thousands of low molecular components. In addition to endogenous metabolites, a urine sample may contain components of exogenous origin making it a rather undefined specimen with possible background interfering compounds. A database is available on human urine components; http://www.urinemetabolome.ca/. The urine metabolome database contains over 4000 entries and includes components of exogenous origin. Fig. 23.6 plots the distribution of masses of the components in the range of 100–1000 Da extracted from the database. It is evident that most components are in the mass range below 500 Da. Table 23.2 presents a list of 13 common analytes in drug testing along with masses for the monoisotopic species and protonated ions. The closest interfering substance for each analyte is listed in the database and the needed RP for resolving these were calculated. Most of these potentially interfering components are drugs or natural products and are likely to occur more seldom. It should also be noted that all drugs have isotopomers and they must be separated by chromatography or by MS/MS. One example of this is the isotopomers 3,4-methylenedioxymethamphetamine (MDMA) and methedrone, which had to be separated by product ion pattern in MS/MS as they coeluted in the chromatography system [6].

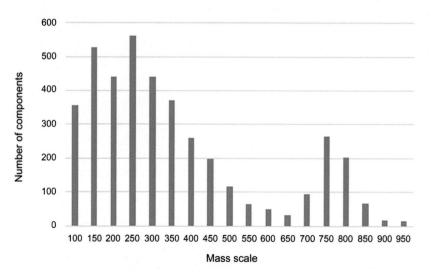

FIGURE 23.6 Distribution over the low molecular mass range of human urine components as taken from the human metabolome database (http://www.urinemetabolome.ca).

TABLE 23.2 Evaluation of Possible Interferences from The Human Urine Metabolome Database

	Exact Mass	[M + H]	Possible Interference	Component	Needed Resolving Power	Comment
Amphetamine	135.1048	136.1121	136.0618	Adenine	2700	Endogenous
Methamphetamine	149.1205	150.1277	150.0774	Methyladenine	3000	Disease metabolite
Benzoylecgonine	289.1314	290.1387	290.1306	Chlophedianol	36,000	Drug
			290.1418	Chloropyramine	92,000	Drug
Morphine	285.1365	286.1438	286.1438	Opiates	Isotopomer	Natural products
			286.1372	Isothipendyl	44,000	Drug
6-Acetylmorphine	327.1471	328.1543	328.1543	Naloxone	Isotopomer	Drug
			328.1212	Desethylamiodaquine	10,000	Drug
Codeine	299.1521	300.1594	300.1594	Hydrocodone	Isotopomer	
			300.1472	N-Depropylpropafenone	25,000	Drug metabolite
M-3-glucuronide	461.1686	462,1759	462.1759	Hydromorphone-3-glucuronide	Isotopomer	Drug metabolite
			462.1806	3-Hydroxyglipizide	99,000	Drug metabolite
THCCOOH	344.1982	345.2055	345.2167	Oxyphencyclimine	32,000	Drug
Diazepam	284.0717	285.0789	285.0789	Mazindol	Isotopomer	Drug
			285.0830	Xanthosine	71,000	Endogenous
			285.0758	Glycetin, Prunetin, Biochanin A	90,000	Natural products
Oxazepam	286.0509	287.0582	287.0582	Norclobazam	Isotopomer	Drug
			287.0762	Diphenolglucuronide	16,000	Drug metabolite
			287.0550	Kaempferol, 3-OH-genistein, Tetrahydroflavone	90,000	Natural products
PCP	243.1987	244.2060	244.1544	Tiglylcarnitine	4,700	Disease metabolite
Methadone	309.1987	310.2165	310.2165	Diphenidol	Isotopomer	Drug
			310.2012	Nadolol, Metipranolol	20,000	Drugs
Propoxyphene	339.2198	340.2271	340.2271	Noracylmetadol, Norlevomethadyl acetat	Isotopomers	Drugs
			340.2384	Disopyramide	30,000	Drug

TABLE 23.3 Evaluation of Interferences for Amphetamine, Morphine, and Oxazepam

	Isotopomers from Kind/Feihn Database	Potential Interferences from Kind/Feihn Database	Comment	In Chromatograms from Authentic Urines	At Retention Time in Chromatograms from Authentic Urines	Comment
Amphetamine	471	4 (RP 10,000)	Most of the potentially interfering compounds are not existing in ChemSpider	Several peaks occur	In one sample of eight	Extracted from full scan data at 70,000 RP with 60 ppm window
Morphine	9971	33 (RP 40,000)		Several peaks occur	No peaks nearby	
		12 (RP 100,000)				
Oxazepam	1538	0 (RP 10,000)	Only chlorinated compounds may interfere	One peak occurs	No peaks nearby	

Another way to estimate the needed RP for drug testing is to do a more theoretical calculation. As a part of the work resulting in the publication of Kind and Fiehn [32], a database was put together with exact masses of all elemental compositions made of the elements C, H, N, O, P, and S using a formula generator. The database contains 1.6 million entities in the mass range 100–500 amu and is available for downloading at http://fiehnlab.ucdavis.edu/projects/identification. An evaluation of possible interfering compounds in this database is presented in Table 23.3 for amphetamine, morphine, and oxazepam. It is clear from Table 23.3 that the largest risk for interference is from isotopomers, which demonstrates the need for and importance of chromatographic resolution. Another conclusion is that for using the full selectivity potential of HRMS in full scan mode, a high RP is needed for some of the analytes.

A third way to evaluate the need for RP is to examine authentic human urine samples for possible interfering components. This was done in eight randomly selected urines from a routine flow of specimens for drug testing for amphetamine, morphine, and oxazepam. As seen from Table 23.3, there are detected peaks in the chromatograms for all three compounds but at other retention times. At the exact retention times of the three analytes, there are nearby compounds in the m/z dimension. When examining the combined spectra over the amphetamine retention time a substance was present in six out of eight samples that appear to be an isotopomer. For morphine, two of the eight samples had interferences within 2.5 ppm. For oxazepam, a component 111 ppm higher in mass appeared in three of eight samples.

In addition to chromatographic separation of possible interfering compounds another approach is to use isotope pattern. For most of the theoretically interfering compounds, the elemental compositions (i.e., number of carbons) are very different from the analytes of interest in drug testing, making this approach attractive.

Application for Urine Drug Screening

The first evaluation of LC-HRMS for urine drug testing in the author's laboratory was done with a TOF instrument [33]. The method was designed to use urine dilution with internal standards and a "rapid" chromatography system (i.e., 4 minutes) and comprised amphetamine, methamphetamine, morphine, codeine, buprenorphine, and some of their metabolites as parameters. A number of 812 authentic urine specimens from routine clinical testing were used to compare its performance with immunochemical screening (CEDIA reagents). The LC-HRMS screening gave much lower rate of false positives, and, in addition, more positives were detected due to lower detection limits. These results were obtained with an instrument with a rather limited RP (>10,000 at m/z 956) and mass accuracy as compared with newer instruments (see Table 23.1). One important observation was that in samples with high concentrations of morphine-3-glucuronide the M + 3 isotope interfered with the peak of the coeluting trideuterated internal standard and the two signals fused into a peak with inaccurate exact mass making both undetected. This example demonstrated the need and importance of a high RP for selectivity and accuracy (Fig. 23.7). The conclusion made from this study was that the

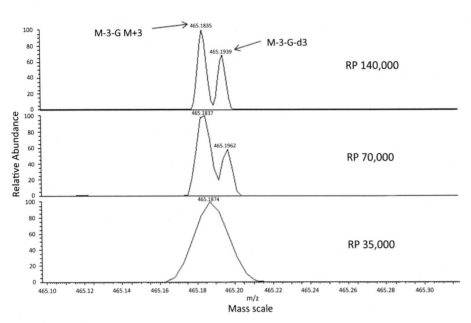

FIGURE 23.7 Importance of mass resolution power for assigning accurate mass. The third morphine-3-glucuronide isotope has a monoisotopic mass close to internal standard morphine-3-glucuronide-d3 internal standard. At high concentrations of morphine-3-glucuronide and to low resolving power these two species will not separate on the mass scale and fuse into one component with a wrong accurate mass.

performance of the screening increased but remaining features required for the future were higher sensitivity and RP, and improved software for automated data evaluation [33].

In our subsequent work, we used the OT technique guided by the conclusion that high RP is of great value [6]. The method was designed to be used as a complement to existing drug screening based on immunoassays and has been used in routine for several years. The method uses a fivefold dilution of urine with internal standards and injection of 2 μL. The applied lower reporting concentration limit was 100 ng/mL. With this method, over 100 unknowns were analyzed per day on one instrument and this was done for several years. The number of analytes was over 100 and the list was updated frequently during the time the STRIDA project was running [34–37]. One part of the method was used for screening of 16 substances being therapeutic drugs, but where no immunoassay tests were available, for example, pregabalin, methylphenidate and zopiclone. The experience from this application strongly supports that the design could be used for a screening method that could replace immunoassays. In order for this to be successful, it was concluded that new software must be developed to allow for automated evaluation and approval of analytical results. If too much manual evaluation is required, it will hamper the development as it is a time- and resource−consuming process.

An LC-HRMS method that can replace immunoassay screening for the common parameters in a high-throughput setting has not yet been demonstrated. Several more publications presenting LC-HRMS methods for urine drug testing are, however, available. McMillin and coworkers reported the use of a TOF LC-HRMS method as a complement to immunoassay screening [38]. The authors called this a hybrid approach, meaning that the best method was used for each parameter considered. The method used liquid/liquid extraction and applied identification criteria were correct retention time (± 0.1 minute) of a peak having correct mass (± 25 ppm). In addition, the isotope pattern was also used to confirm identity. The method was used in routine drug testing for several thousands of samples. The major advantage reported was that a shorter total turnaround time was obtained as compared with the standard screening and confirmation approach. In another paper, the authors using the same method design for serum and plasma, occurrence of false positives was reported [39]. In that report, the performance of three TOF LC-HRMS platforms for urine drug detection were compared using blank urines fortified with 47 compounds [39]. The identification used MS/MS data, and 79%−85% of the compounds were detected. Jagerdeo and Schaff used OT LC-HRMS and solid-phase extraction to screen for 38 compounds in fortified blood and urine samples [40]. The authors reported superior performance regarding false positives and detection limits as compared with immunoassay screening.

Application for Confirmation

Confirmation of screening positive drug tests is well established using LC-MS/MS methods in SRM mode. When applying proper identification criteria on retention time and correct product ion ratio, this application does not have any principle objections and can be considered safe [41]. HRMS in full scan or SIM mode cannot provide the same level of evidential data, and therefore product ion monitoring in MS/MS mode must be used in confirmation methods also for HRMS, which is technically possible. In doing so, the level of confidence will increase even more because of the higher RP. The additional benefit of doing confirmations with HRMS is that the criterion for mass accuracy of the product ions can be applied. However, whether this increased selectivity is needed or not is not known.

CONCLUSIONS

A system based on LC-HRMS that can replace immunoassay screening for urine drug testing has not yet been demonstrated. Available data and technology suggest that this indeed is a possibility. It is important for this to be successful to develop improved software that can reliably and effectively support data evaluation to the point of full automation—a black box solution. When considering the large number of possible isotopomers, we do think that chromatography is required to provide the needed selectivity and that an LC-HRMS system that provides more reliable positive screening results than immunoassays would be a significant contribution to clinical drug testing.

REFERENCES

[1] WHO Drug Report, <https://www.unodc.org/wdr2017/index.html>; 2017 [Accessed 27.02.18].
[2] Schwartz RH. Urine testing in the detection of drugs of abuse. Arch Intern Med 1988;148:2407–12.
[3] Beck O, Villén T. [Drugs of abuse testing safer and safer and more complete]. Lakartidningen 2011;108:2300–3.
[4] Goldberger BA, Cone EJ. Confirmatory tests for drugs in the workplace by gas chromatography-mass spectrometry. J Chromatogr A 1994;674:73–86.
[5] Eichhorst JC, Etter ML, Hall PL, Lehotay DC. LC-MS/MS techniques for high-volume screening of drugs of abuse and target drug quantitation in urine/blood matrices. Methods Mol Biol 2012;902:29–41.
[6] Stephanson NN, Signell P, Helander A, Beck O. Use of LC-HRMS in full scan-XIC mode for multi-analyte urine drug testing - a step towards a 'black-box' solution?. J Mass Spectrom 2017;52:497–506.
[7] Sundström M, Pelander A, Ojanperä I. Comparison between drug screening by immunoassay and ultra-high performance liquid chromatography/high-resolution time-of-flight mass spectrometry in post-mortem urine. Drug Test Anal 2015;7:420–7.
[8] McLafferty FW. High-resolution mass spectrometry. Science 1966;151:641–9.
[9] Špánik I, Machyňáková A. Recent applications of gas chromatography with high-resolution mass spectrometry. J Sep Sci 2017. Available from: https://doi.org/10.1002/jssc.201701016.
[10] Boesl U. Time-of-flight mass spectrometry: introduction to the basics. Mass Spectrom Rev 2017;36:86–109.
[11] Zubarev RA, Makarov A. Orbitrap mass spectrometry. Anal Chem 2013;85:5288–96.
[12] Makarov A, Denisov E, Kholomeev A, Balschun W, Lange O, Strupat K, et al. Performance evaluation of a hybrid linear ion trap/orbitrap mass spectrometer. Anal Chem 2006;78:2113–20.
[13] Henion J, Brewer E, Rule G. Sample preparation for LC/MS/MS: knowing the basic requirements and the big picture of an LC/MS system can ensure success in most instances. Anal Chem 1998;70:650–6.
[14] Dams R, Murphy CM, Lambert WE, Huestis MA. Urine drug testing for opioids, cocaine, and metabolites by direct injection liquid chromatography/tandem mass spectrometry. Rapid Comm Mass Spectrom 2003;17:1665–70.
[15] Beck O, Ericsson M. Methods for urine drug testing using one-step dilution anddirect injection in combination with LC-MS/MS and LC-HRMS. Bioanalysis 2014;6:2229–44.
[16] Matuszewski BK, Constanzer ML, Chavez-Eng CM. Strategies for the assessment of matrix effect in quantitative bioanalytical methods based on HPLC−MS/MS. Anal Chem 2003;75:3019–30.
[17] Helfer AG, Michely JA, Weber AA, Meyer MR, Maurer HH. LC-HR-MS/MS standard urine screening approach: pros and cons of automated on-line extraction by turbulent flow chromatography vs dilute-and-shoot and comparison with established urine precipitation. J Chromatogr B 2017;1043:138–49.
[18] Michely JA, Meyer MR, Maurer HH. Paper spray ionization coupled to high resolution tandem mass spectrometry for comprehensive urine drug testing in comparison to liquid chromatography-coupled techniques after urine precipitation or dried urine spot workup. Anal Chem 2017;89:11779–86.
[19] Denoroy L, Zimmer L, Renaud B, Parrot S. Ultra high performance liquid chromatography as a tool for the discovery and the analysis of biomarkers of diseases: a review. J Chromatogr B 2013;927:37–53.
[20] DiBattista A, Rampersaud D, Lee H, Kim M, Britz-McKibbin P. High throughput screening method for systematic surveillance of drugs of abuse by multisegment injection-capillary electrophoresis-mass spectrometry. Anal Chem 2017;89:11853–61.

[21] Lynch KL. CLSI C62-A: a new standard for clinical mass spectrometry. Clin Chem 2016;62:24—9.
[22] Ojanperä I, Kolmonen M, Pelander A. Current use of high-resolution mass spectrometry in drug screening relevant to clinical and forensic toxicology and doping control. Anal Bioanal Chem 2012;403:1203—20.
[23] Gergov M, Boucher B, Ojanperä I, Vuori E. Toxicological screening of urine for drugs by liquid chromatography/time-of-flight mass spectrometry with automated target library search based on elemental formulas. Rapid Commun Mass Spectrom 2001;15:521—6.
[24] Ojanperä S, Pelander A, Pelzing M, Krebs I, Vuori E, Ojanperä I. Isotopic pattern and accurate mass determination in urine drug screening by liquid chromatography/time-of-flight mass spectrometry. Rapid Commun Mass Spectrom 2006;20:1161—7.
[25] Kolmonen M, Leinonen A, Pelander A, Ojanperä I. A general screening method for doping agents in human urine by solid phase extraction and liquid chromatography/time-of-flight mass spectrometry. Anal Chim Acta 2007;585:94—102.
[26] Georgakopoulos CG, Vonaparti A, Stamou M, Kiousi P, Lyris E, Angelis YS, et al. Preventive doping control analysis: liquid and gas chromatography time-of-flight mass spectrometry for detection of designer steroids. Rapid Commun Mass Spectrom 2007;21:2439—46.
[27] Touber ME, van Engelen MC, Georgakopoulus C, van Rhijn JA, Nielen MW. Multi-detection of corticosteroids in sports doping and veterinary control using high-resolution liquid chromatography/time-of-flight mass spectrometry. Anal Chim Acta 2007;586:137—46.
[28] Kolmonen M, Leinonen A, Kuuranne T, Pelander A, Ojanperä I. Generic sample preparation and dual polarity liquid chromatography-time-of-flight mass spectrometry for high-throughput screening in doping analysis. Drug Test Anal 2009;1:250—66.
[29] Virus ED, Sobolevsky TG, Rodchenkov GM. Introduction of HPLC/orbitrap mass spectrometry as screening method for doping control. J Mass Spectrom 2008;43:949—57.
[30] Musenga A, Cowan DA. Use of ultra-high pressure liquid chromatography coupled to high resolution mass spectrometry for fast screening in high throughput doping control. J Chromatogr A 2013;1288:82—95.
[31] Girón AJ, Deventer K, Roels K, Van Eenoo P. Development and validation of an open screening method for diuretics, stimulants and selected compounds in human urine by UHPLC-HRMS for doping control. Anal Chim Acta 2012;721:137—46.
[32] Kind T, Fiehn O. Metabolomic database annotations via query of elemental compositions: mass accuracy is insufficient even at less than 1 ppm. BMC Bioinformatics 2006;7:234. Available from: https://doi.org/10.1186/1471-2105-7-234.
[33] Saleh A, Stephanson NN, Granelli I, Villén T, Beck O. Evaluation of a direct high-capacity target screening approach for urine drug testing using liquid chromatography-time-of-flight mass spectrometry. J Chromatogr B 2012;909:6—13.
[34] Beck O, Bäckberg M, Signell P, Helander A. Intoxications in the STRIDA project involving a panorama of psychostimulant pyrovalerone derivatives, MDPV copycats. Clin Toxicol 2018;56:256—63.
[35] Helander A, Bäckberg M, Signell P, Beck O. Intoxications involving acrylfentanyl and other novel designer fentanyls — results from the Swedish STRIDA project. Clin Toxicol 2017;5:589—99.
[36] Helander A, Bäckberg M, Beck O. MT-45, a new psychoactive substance associated with hearing loss and unconsciousness. Clin Toxicol 2014;52:901—4.
[37] Helander A, Bäckberg M, Hultén P, Al-Saffar Y, Beck O. Detection of new psychoactive substance use among emergency room patients: results from the Swedish STRIDA project. Forensic Sci Int 2014;243:23—9.
[38] McMillin GA, Marin SJ, Johnson-Davis KL, Lawlor BG, Strathmann FG. A hybrid approach to urine drug testing using high-resolution mass spectrometry and select immunoassays. Am J Clin Pathol 2015;143:234—40.
[39] Marin SJ, Sawyer JC, He X, Johnson-Davis KL. Comparison of drug detection by three quadrupole time-of-flight mass spectrometry platforms. J Anal Toxicol 2015;39:89—95.
[40] Jagerdeo E, Schaff JE. Rapid screening for drugs of abuse in biological fluids by ultra high performance liquid chromatography/Orbitrap mass spectrometry. J Chromatogr B 2016;1027:11—18.
[41] Maralikova B, Weinmann W. Confirmatory analysis for drugs of abuse in plasma and urine by high-performance liquid chromatography-tandem mass spectrometry with respect to criteria for compound identification. J Chromatogr B 2004;811:21—30.

Chapter 24

Forensic Toxicology in Death Investigation

Hannah Kastenbaum, Lori Proe and Lauren Dvorscak
New Mexico Office of the Medical Investigator, University of New Mexico School of Medicine, Albuquerque, NM, United States

INTRODUCTION

The purpose of a medicolegal death investigation is "to investigate deaths deemed to be in the public interest" to serve "justice and public health systems" [1]. The role of county coroner originated in Britain in the late 12th century and this person was responsible for holding inquests into the circumstances of an individual's death and arranging for the disposition of the body and property, the latter primarily back to the king who was considered to own everything [2]. Colonists imported this system of basic death investigation to America and, initially, ordinary citizens were appointed to these posts [3]. In the postrevolutionary period, an attempt to initiate a physician-led, national death investigation system failed to gain widespread support [3]. The movement towards physician-led death investigations gained some traction in the late 19th century and the first medical examiner (ME) system was established in Massachusetts in 1877 [3]. The term ME is commonly used to refer to forensic pathologists, physicians trained in pathology and its subspecialty that deals with examining deceased persons to determine "cause, mechanism, and manner of... death," involved in death investigations [2,4].

Current-day medicolegal death investigations in the United States are overseen and provided for by a mixture of ME's and coroner's offices as dictated by state and local statutes. As of 1995, 22 states had ME systems (including 19 statewide offices), 18 states had mixed ME and coroner systems either with a state ME overseeing county coroners or with individual county MEs or coroners, and 11 states had primarily coroner systems [2]. The qualifications for a coroner vary, but the individual is usually elected, sometimes also a sheriff, and sometimes a licensed physician. The coroner determines which deaths require investigation and what that investigation will entail. If necessary, the coroner arranges for a pathologist to perform an examination of the decedent. The qualifications for an ME typically include a license to practice medicine and training in pathology [2]. The ME's responsibilities for death investigation are the same, but he or she is capable of personally performing the examinations of decedents.

MEDICOLEGAL DEATH INVESTIGATION

A medicolegal death investigation typically begins with the reporting of a death to the ME or coroner office by hospital staff, emergency medical personnel, or law enforcement, depending on the location and circumstances of the death. An investigator gathers information about the decedent's medical and social history and the circumstances under which he or she was found deceased. For out-of-hospital deaths, this typically includes an examination of the scene, or location of death. An individual's history, recent complaints or witnessed symptoms, presence of illicit drug paraphernalia or prescription medications and a count of the contents of the bottles at the scene, and preliminary physical exam findings like frothy fluid exuding from the nose and mouth (foam cone, see Fig. 24.6A) and acute or chronic injection sites (see Fig. 24.5) are often the first facts to suggest a death may be related to a toxic exposure or, in common parlance, an overdose.

The statutes that establish ME and coroner offices may dictate what deaths are required to be reported to the office, but typically do not dictate the extent of investigation into a given death that must occur. A medicolegal death investigation may include investigation of the scene of a death (as described), review of medical history and records, external examination of a decedent, full or partial autopsy (internal examination of the organs), radiology studies, histologic (microscopic) examination of tissues, and other ancillary studies like microbiology and laboratory evaluation including toxicology studies. The extent of investigation is at the discretion of the ME or coroner and there are no federal standards.

The National Association of Medical Examiners (NAME), "a professional organization for medical examiners... and [other] physicians who investigate such deaths," has published a set of Forensic Autopsy Performance Standards that include recommendations regarding the types of deaths and the investigation of which should include a complete autopsy examination [1]. This document states that "the forensic pathologist shall perform a forensic autopsy when... the death is by apparent intoxication by alcohol, drugs, or poison [1]." A forensic autopsy routinely involves collection, labeling, and preservation of blood, urine, and vitreous fluid for possible laboratory testing [1].

THE UTILITY OF COMPUTED TOMOGRAPHY IN INVESTIGATING DRUG POISONING DEATHS

Postmortem computed tomography (PMCT) is gaining popularity as an adjunct to forensic death investigation and may be used as an effective tool for triage in suspected drug intoxication cases [5]. Although different types of drugs result in different physiologic toxicities, common general radiologic findings detected in a broad array of drug fatalities have been described [6]. When radiologic imaging is utilized in combination with toxicological testing, PMCT has been shown to be as reliable as autopsy in subsets of decedents in the detection of drug poisoning deaths [5]. Specifically, in a cohort of 55 cases of known opiate toxicity, the triad of PMCT findings including pulmonary edema, cerebral edema, and urinary bladder distension had a specificity of 100% (Fig. 24.1) [6]. However, such findings individually are nonspecific and must be considered within the entire context of each case.

Additional supportive evidence detectable by PMCT in drug-related deaths includes a gravitational pattern of hyperdensity in the stomach, which may be characteristic of pill debris or sedimentation in the stomach (Fig. 24.2) [7]. However, this must be differentiated from other artifacts or foods that may demonstrate increased density on computed tomography (Fig. 24.3). Other PMCT findings may include incorporated packets of drugs or other foreign bodies, typically within the gastrointestinal tract. Such packets of drugs may burst and cause unintentional toxicity. When compared with other imaging modalities, computed tomography has been shown to have the highest diagnostic accuracy in detecting gastrointestinal foreign bodies [8].

Case Report

A 32-year-old man was in the process of evading law enforcement in a stolen vehicle when he crashed and sustained lethal blunt injuries. A PMCT scan performed in triage protocols detected incidental foreign material in the bowel (Fig. 24.4). Although the material in his gastrointestinal tract did not directly cause or contribute to his death, it provided an example of body stuffing, or hiding of drugs/paraphernalia within a body cavity to evade detection. Usage of

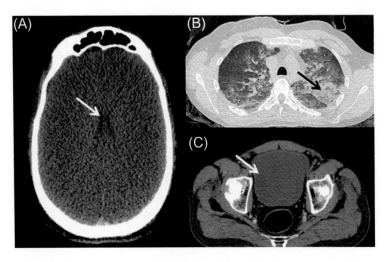

FIGURE 24.1 Triad of PMCT findings that together have high specificity for acute drug poisoning deaths. (A) Cerebral edema—note the hypoattenuation of brain parenchyma overall with loss of gray–white differentiation. The lateral ventricles are decreased in size, indicative of brain swelling (arrow). (B) Pulmonary edema—diffuse ground-glass opacities are within the lung fields with fluid filling the alveolar spaces, greater on the left (arrow). (C) The urinary bladder is significantly distended with an increased axial diameter (arrow).

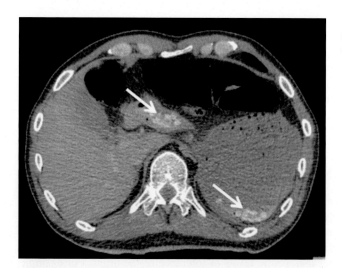

FIGURE 24.2 The arrows point to hyperdense material settling in the stomach and duodenum. Pill fragments were detected in the stomach and small intestine at autopsy.

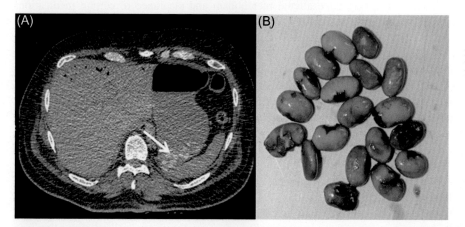

FIGURE 24.3 Hyperdensities in the stomach (A, arrow) must be distinguished from artifacts or other material. In this case, the material turned out to be food particles (B).

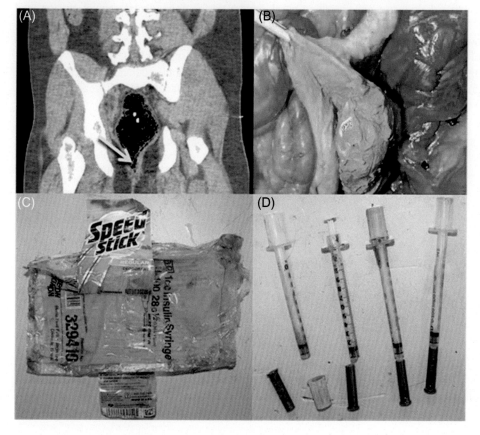

FIGURE 24.4 A case of body stuffing in an attempt to evade law enforcement. (A) Hyperdensities were detected in the rectum on PMCT (arrow). (B) At autopsy, the rectum contained fecal material surrounding a plastic bag. (C) The plastic bag was recovered from the rectum. (D) The plastic bag within the rectum contained syringes caps, and capped needles.

substances such as the methamphetamine and heroin detected in his femoral blood may also help explain his erratic behavior preceding the vehicular collision.

Although PMCT has been demonstrated to be a useful tool in the evaluation of potential drug poisoning deaths, the gold standard of examination in these cases is still autopsy. Current NAME recommendations, as described earlier, reflect this [1].

THE ROLE OF AUTOPSY IN DRUG POISONING DEATHS

Toxicology analysis is an increasingly important component of forensic pathology practice. Performing an autopsy in cases where drug poisoning is thought to play a contributory or causative role in death not only provides a modality for the collection of specimens for testing, but more importantly allows pathologists to rule out other potential causes of death. The autopsy provides context of underlying natural disease processes, allowing for a more accurate interpretation of toxicology findings. Natural disease processes may impact therapeutic levels or tolerance of the drugs. For example, individuals with severe liver and/or renal disease may have altered metabolism and clearance of certain medications or drugs. Furthermore, if the results of postmortem toxicology analysis are unexpectedly negative for significant substances, the autopsy provides valuable information regarding natural diseases and/or injuries that otherwise may explain death.

Cases in which a prolonged hospitalization with documented toxicology testing and evaluation of underlying disease processes precede death may not provide significant additional information by autopsy. The specimens collected at the time of examination in these cases no longer provide accurate information regarding drug presence and concentration at the time the suspected drug toxicity occurred. However, depending on the length of hospitalization, antemortem specimens may be available, requested, and tested to more reliably determine the contribution of drug toxicity to cause and manner of death.

Most often, in cases of suspected drug poisoning, no specific findings to indicate that the death is drug related are uncovered at autopsy [9]. However, several external and internal examination findings may provide important clues to a history of drug use and/or the possibility of intoxication.

External findings such as scarring or linear arrays of puncture marks along superficial vessels often indicate a history of injection drug use (Fig. 24.5). Subcutaneous injections ("skin popping") may cause localized swelling, ulceration, and round scarring of the skin. Inhalational drug use may result in nasal septum perforations, dental enamel erosion, and crystalline material in the nose. Nonspecific external findings in deaths where drug use may be contributory also include foamy fluid at the mouth and nose, evidence of widespread infection (petechial rash), and evidence of prolonged periods of unconsciousness (blistering of the skin or tissue death in areas of compromised blood supply) [10].

Internal examination findings in suspected drug poisoning deaths can include cerebral edema, foamy fluid in the mouth and airways (Fig. 24.6), pulmonary edema, and distension of the urinary bladder. Esophageal and gastric

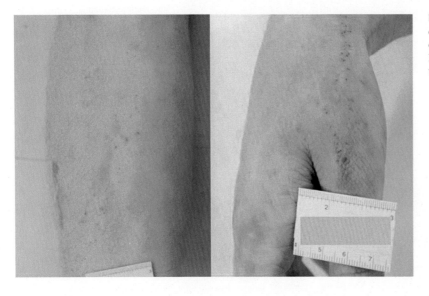

FIGURE 24.5 Evidence of chronic injection drug use. The photographs demonstrate examples of puncture marks and scarring arranged in typical linear arrays overlying superficial veins within the subcutaneous tissues.

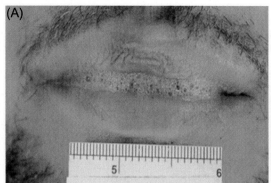 

FIGURE 24.6 Foamy fluid at the mouth (A) and within the airways (B) is associated with, but not specific for, drug-related fatalities. Foamy fluid may also be seen in drowning victims and sometimes in those who die of natural causes.

TABLE 24.1 Selected Drugs With Characteristic Autopsy Findings

Drug	Autopsy Finding Suggestive of Toxicity
Acetaminophen	Centrilobular liver necrosis
Arsenic	Gastrointestinal necrosis, Mees' lines
Carbon monoxide	Cheery red skin coloration, globus pallidus necrosis
Cyanide	Cheery red skin coloration, gastric mucosal hemorrhage, bitter almond odor
Ethylene glycol	Crystals in renal tubules, liver, meninges
Hydrogen sulfide	Green brain discoloration, rotten egg odor
Opiates	Foam cone at nose/mouth
Salicylates	Renal necrosis

contents may provide evidence of pills/capsules, residue, filler material, or ingestion of caustic substances. Incorporated drug packets from body stuffing and/or packing may be detected, most commonly in the gastrointestinal tract. Foreign substances injected, ingested, or inhaled as part of the drug or filler material may be detected on microscopic examination, often associated with an immune reaction.

In cases of prolonged survival or a period of unconsciousness following a suspected drug poisoning, autopsy findings may reflect sequelae of this period of unconsciousness. Findings in such cases may include development of pneumonia, liver necrosis, early myocardial infarction, and rhabdomyolysis, among others. These findings, however, are also nonspecific and typically reflect multiple organ injury due to low oxygen delivery, rather than a direct toxicity of the implicated substance itself.

Certain drugs may result in somewhat more characteristic patterns of findings at autopsy. However, the absence of such findings at autopsy does not necessarily exclude toxicity. Selected potential findings in a sampling of drug intoxications are outlined in Table 24.1.

Case Report

A 46-year-old man had not been seen by family members for several days. A welfare check discovered that he was deceased on his apartment floor, with a suicide note nearby. A number of empty aspirin containers were discovered within the residence (Fig. 24.7A). Pill fragments and vomit were on the floor of the bedroom (Fig. 24.7B). The scene was highly suggestive of an acute toxicity. At autopsy, pill fragments were discovered in the esophagus, stomach, and small intestine, consistent with the scene findings (Fig. 24.7C). Toxicology analysis of the femoral blood revealed high levels of salicylates. The cause of death was salicylate toxicity, and the manner of death was determined to be suicide.

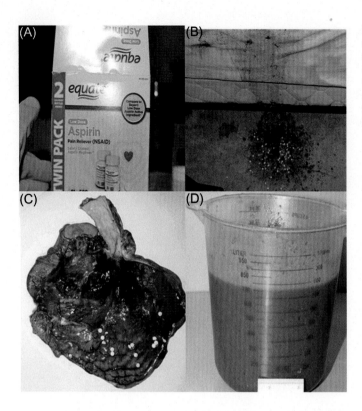

FIGURE 24.7 This case involved acute salicylate toxicity and is an example of how scene investigation may help guide triage decisions as well as toxicology analysis following autopsy. (A) Aspirin containers were found at the scene. (B) Pill fragments and vomit were on the floor of the bedroom. (C) Pill fragments were also detected in the stomach at autopsy. (D) The stomach contents containing gritty pill debris were measured and saved among the collected toxicology samples.

TOXICOLOGY SPECIMEN COLLECTION AT AUTOPSY

To prevent unreliable toxicology results in the postmortem setting, standardized procedures for collection and handling of specimens are essential [9]. Accurate toxicology results are inherently dependent on the quality and quantity of the samples provided to the laboratory for analysis. Routine collection involves utilization of clean, unused needles/syringes/containers, proper labeling of specimens, use of preservative agents when appropriate to reduce degradation, and proper storage and shipping procedures.

Although the availability of specimen types is case-dependent, standard sample collection includes central and peripheral blood samples, urine, and vitreous (eye) fluid. In cases where the state of the remains precludes such sample collection, such as significant decomposition, body cavity fluids, tissue samples, bile, or even maggots may be collected and analyzed. Routine tissues collected at autopsy may include the following: right liver lobe, skeletal muscle, heart, brain, kidney, and spleen. In some instances, gastric contents or cerebrospinal fluid may be useful for analysis. Rarely, for detection of chronic heavy metal poisoning, hair and nails may also be collected.

In routine cases, samples are collected in test tubes both with and without preservative. Unpreserved samples may be used for toxicology screening purposes, for DNA analysis, or for serologic testing [9]. For confirmatory toxicology testing and quantitative analysis, the use of a sodium fluoride preservative is recommended in blood and tissue samples to help prevent enzymatic processes after collection. Without appropriate preservatives, residual esterase activity may cause the postmortem conversion of cocaine to ecgonine methyl ester as well as loss of other esters such as 6-monoacetylmorphine [9]. Additionally, without adequate preservation, substances such as gamma-hydroxybutyric acid may be formed from precursor compounds including endogenous gamma-aminobutyric acid, succinic acid, and putrescine, falsely altering the concentrations of these substances in the postmortem sample [11]. Furthermore, use of sodium fluoride as a preservative prevents micro-organism conversion and alteration of ethanol (alcohol). In cases where volatile substance toxicity is suspected, blood may be collected in an air-tight container with room at the top for headspace analysis.

Sample stability depends not only on collection technique and preservation, but proper storage, packaging, and shipping procedures. The majority of substances in blood samples remain stable if proper storage procedures are followed. For routine, short-term storage, most samples may be refrigerated. Plasma or serum to be analyzed should not be frozen; however long-term storage of other blood or tissue samples requires $-20°C$ or colder [9].

COMMON POSTMORTEM TOXICOLOGY SAMPLES

In this section commonly collected specimens for toxicological analysis are discussed.

Blood

The concentration of drug detected in the blood collected at autopsy is generally considered the best result for determining a drug's potential toxicity and contribution to death. However, after death, the stagnant nature of postmortem blood presents a unique challenge for interpreting quantities of substances detected. Depending on the characteristics of the drug, postmortem diffusion of substances may occur along a concentration gradient, potentially altering the blood level detected upon analysis [12]. Drugs with a large volume of distribution, lipophilic substances, and basic substances are more likely to redistribute after death. Additionally, drugs may diffuse into the blood or accumulate in nearby organs from the gastrointestinal tract, adjacent adipose or soft tissues [9]. To minimize this postmortem redistribution phenomenon, samples are routinely collected from both central and peripheral sources at autopsy, allowing for comparison of drug quantities when necessary. A heart blood to femoral blood ratio for drugs that have a tendency to undergo redistribution is commonly reported in Ref. [13].

Peripheral blood is relatively isolated from internal organs and is, therefore, the preferred specimen for toxicology analysis in the postmortem setting. The femoral veins typically resist movement of blood downward from the abdomen due to the presence of intravascular valves, and are therefore the preferred sites of collection for peripheral blood [9]. If femoral blood is not available, additional acceptable sources of peripheral blood are the subclavian veins.

Vitreous Fluid

The vitreous fluid is relatively protected from microorganisms, and generally lacks high enzyme activity. For these reasons, vitreous fluid is most useful for electrolyte analysis as well as for measuring ethanol concentrations, but also may be helpful in detecting opiates such as 6-monoacetylmorphine [9]. However, because of the relative sequestration of vitreous fluid, distribution of substances into the fluid takes time, and generally lags behind the absorption phase of a drug in the blood. Concentrations of substances such as ethanol detected within vitreous in postmortem samples are typically reflective of the blood concentration in the hours prior to death [9].

Urine

Postmortem urine is also relatively protected from microorganisms, but is useful only in screening for substances, as quantities of drug in the urine have not been shown to accurately reflect blood levels [9]. Due to the temporal delay of a drug's metabolism and circulation before entering the urine, deaths that occur before significant metabolism has taken place may demonstrate falsely negative urine results. Additionally, the urine may detect the presence of drugs that have passed through systemic circulation and are no longer present in the blood.

Tissue

When blood specimens are not available, viable alternative samples for toxicology analysis in the postmortem setting include tissue samples. The most commonly used specimens for such analysis are liver and skeletal muscle. Liver tissue is best collected from the right lobe, where diffusion from other organs such as the stomach is minimized. Skeletal muscle is generally well preserved and physically separated from the internal organs if collected from a site on the extremity (commonly the thigh). Despite the relatively protected nature, concentrations of drug in skeletal muscle can vary and may not accurately reflect the level of drug in the blood at the time of death [9].

ORDERING TOXICOLOGY

No single toxicology test will detect all poisons or drugs and financial resources are limited in many (or most) ME/coroner offices. Consequently, the selection of appropriate toxicology testing requires consideration of the substances suggested by the findings at the scene and autopsy and familiarity with the tests available from the toxicology laboratory used by the certifier of death [14].

Case Report

A 59-year-old woman was found dead in the back of her vehicle with a can of compressed air in one hand and a plastic bag under her head. Multiple other cans of compressed air were also found in the vehicle. The decedent was known to "huff" or inhale volatile substances for their intoxicating effects [15]. Inhalation of volatile substances is associated with death by cardiac arrhythmia [16]. An autopsy was performed and no obvious anatomic cause of death was discovered. Given the decedent's history of inhalant abuse and the scene findings of multiple cans of compressed air containing 1,1-difluoroethane, toxicology panels for both common drugs of abuse and for inhalants were ordered. The sole positive finding was 1,1-difluoroethane. The cause of death was determined to be the toxic effects of 1,1-difluoroethane.

1,1-Difluoroethane was not detected by the basic toxicology panel ordered in this case. Without information from the decedent's medical history and the findings at the scene of death, this substance would not have been suspected or tested. Without scene findings, a medical history or autopsy findings to suggest them, some unusual substances may not be suspected and appropriate testing ordered, leading to missed diagnoses.

TOXICOLOGY REPORTS

The NAME *Forensic Autopsy Performance Standards* document indicates that the toxicology reports used by forensic pathologists must include the source of the sample tested, the type of screen, the test results, and the method of analysis [1].

While the drug levels reported in a toxicology report are useful, they must be interpreted within the context of the decedent's medical history, the autopsy findings, and the other drugs detected [14]. A high level of a drug does not automatically indicate that an individual died from its use. For example, individuals who have used opioid drugs long term will develop tolerance to these drugs, allowing them to go about their daily activities with relatively high levels of these drugs in their systems. Conversely, relatively low levels of opioid drugs could be fatal to individuals who are not accustomed in taking those drugs (the "opioid naïve"). People with heart or lung disease are more vulnerable to injury or death from central nervous system (CNS)—depressant drugs than individuals with healthy cardiopulmonary systems. Stimulant drugs such as cocaine and methamphetamine can be considered contributory to fatal heart arrhythmias at any level and regardless of the degree of cardiovascular disease in the deceased person [17]. Finally, when drugs (including ethanol) with CNS-depressant effects, either of the same class or of different classes, are combined, their effects can be additive, so the detection of low levels of multiple CNS-depressant drugs can be interpreted as combining to lead to fatal CNS depression.

Some drugs detected by toxicology testing could represent metabolites of other drugs or the primary drug ingested. For example, the opioid drug hydrocodone is metabolized to hydromorphone; hydromorphone (commonly sold under the brand name Dilaudid) is an opioid drug in its own right [13]. An understanding of the drugs a decedent has been prescribed, either via documentation of the medications found at the scene of death or by referencing prescription monitoring program records, can clarify which drugs were most likely ingested and which drugs represent metabolites.

6-Monoacetylmorphine is a unique metabolite of heroin and its detection in the blood or urine indicates heroin use. Morphine is also a metabolite of heroin, but in the absence of 6-monoacetylmorphine, determining whether morphine is present as the result of heroin use or morphine ingestion may depend on the decedent's medical history and on the findings at the scene of death. Codeine is commonly detected along with heroin but is not a known metabolite; its presence is thought to be due to impurities in the heroin [13].

DETERMINING CAUSE OF DEATH

The cause of death is "the disease or injury which initiated the train of morbid events leading directly to death, or the circumstances of the accident or violence which produced the fatal injury" [18]. The cause of death is an opinion generated by the certifier of death [19].

In some situations, the cause of death is certain. For example, in an individual with a stab wound of the heart causing blood loss, compression of the heart within the pericardial sac and damage of the heart muscle itself, the certifier of death will have a high level of certainty that the injury is the cause of the death, without which the individual would not have died. The observation of natural disease at the autopsy or detection of ethanol in the blood on toxicology testing is unlikely to alter the determination that "stab wound of chest" is the cause of death.

In other cases, the interplay of natural disease processes, injuries, and the toxic effects of drugs are more complex and the certifier of death must consider the relative contributions of each in order to determine the most likely cause of death. In many cases, a combination of natural disease and injury is considered the most likely cause of death.

The NAME position paper, "recommendations for the investigation, diagnosis and certification of deaths related to opioid drugs" suggests that the generic names of all drugs considered responsible for causing death to be listed in the autopsy report and on the death certificate [14].

Case Report

A 23-year-old man was found collapsed on the floor next to his bed. A syringe and needle were under his body and sticky black material was found on a piece of paper near the body. The decedent's medical history was significant for heroin use. He had recently been released from a drug rehabilitation facility. Autopsy revealed a well-developed young man with an acute needle puncture mark and a "track mark" of one arm. There was no significant natural disease. Toxicology testing of the blood detected alprazolam, codeine, 6-monoacetylmorphine, and morphine. The cause of death was certified as "toxic effects of multiple drugs (alprazolam, codeine, and heroin)."

A 31-year-old woman was found dead on the couch in her home. Drug paraphernalia including needles, syringes, cotton balls, and a piece of aluminum foil with black material on it were at the scene. The decedent's medical history was significant for intravenous drug use and she had reportedly been looking for "a fix" when she was last seen alive. Autopsy revealed an emaciated woman with pneumonia; lung cultures detected *Streptococcus pneumoniae* and *Escherichia coli*. Toxicology testing of the blood detected codeine, ethanol, and morphine. This decedent had significant pneumonia, which would have impaired her breathing and could have been the sole cause of her death. She also had multiple CNS depressants (codeine, ethanol, and morphine) in her system that were considered likely to have further decreased her breathing and contributed to her death. Based on the findings of needles at the scene and the fact that the decedent had not been prescribed morphine, the morphine in the blood was considered to be a metabolite of heroin. The cause of death was certified as "pneumonia due to *Escherichia coli* and *Streptococcus pneumoniae* infection." "The toxic effects of multiple drugs (codeine, ethanol, and heroin)" were listed on the death certificate as an "other significant condition." Other significant conditions are important diseases or conditions present at the time of death that may have contributed to the death but did not lead to the underlying cause of death [19].

DETERMINING THE MANNER OF DEATH

The manner of death is a description of the circumstances surrounding the death [19,20]. This determination is the opinion of the certifier of death and is made by considering the scene investigation, the autopsy findings, and the results of toxicology testing. In most jurisdictions, there are five options for the manner of death classification: natural, accident, suicide, homicide, and undetermined. When a combination of natural disease and external factors contribute to death, preference is given to the nonnatural manner of death [20].

The NAME position paper on opioid deaths recommends classifying opioid deaths in which substances are used or abused without apparent intent for self-harm as "accident" [14]. The NAME Guide to Manner of Death Classification also recommends certifying those "deaths directly due to the acute toxic effects of a drug or poison as Accident (assuming there was no intent to do self harm or cause death)" [20].

Case Report

A 66-year-old man was found dead in bed in a motel room on the day after he checked in. Two empty (recently filled) medication bottles labeled "olanzapine" and one empty (recently filled) medication bottle labeled "lorazepam" were in the room. The decedent's medical history was significant for schizophrenia. He had attempted suicide by ingesting pills several years earlier. Autopsy revealed granular material consistent with partially digested pills in the stomach. Toxicology testing detected lorazepam and olanzapine at very high concentration indicating fatal levels rather than levels expected from prescription use of these drugs. The cause of death was certified as "toxic effects of multiple drugs (lorazepam, olanzapine)." Based on the decedent's history of attempting suicide via ingestion of medications, on the empty pill bottles at the scene and on the excessively high drug levels detected on toxicology testing, the manner of death was certified as suicide.

Suicide is defined by the NAME as a death "resulting from an injury or poisoning as a result of an intentional, self-inflicted act committed to do self harm or cause the death of one's self" [20].

CONCLUSIONS

MEs and coroners are charged with determining the cause and manner of death in cases that fall under their offices' jurisdictions. An increasing number of these deaths are caused by the toxic effects of drugs. MEs and coroners have many tools that they can use to determine the cause and manner of death, including death scene investigation, physical examination of the body, advanced radiological imaging, and toxicology testing. Accurately certifying the cause and manner of death in cases of suspected drug toxicity requires integrating all of the information obtained during the death investigation.

REFERENCES

[1] National Association of Medical Examiners. Forensic Autopsy Performance Standards. The National Association of Medical Examiners website, <https://netforum.avectra.com/eweb/DynamicPage.aspx?Site = NAME&WebCode = StdDocs>; 2016 [accessed 13.01.18].
[2] Spitz DJ. History and development of forensic medicine and pathology. In: Spitz W, editor. Spitz and Fisher's medicolegal investigation of death. 4th ed. Springfield, IL: Charles C Thomas Publisher Ltd; 2006. p. 3–21.
[3] Jentzen JM. Death investigation in America: coroners, medical examiners, and the pursuit of medical certainty. Cambridge, MA: Harvard University Press; 2009. Investigation in America.
[4] Randall BB, Fierro MF, Froede RC, Bennett AT. Forensic Pathology. In: Collins KA, Grover MH, editors. Autopsy performance & reporting. 2nd ed. Northfield, IL: College of American Pathologists; 2003. p. 55–64.
[5] Paul ID, Lathrop SL, Hatch GM, Gerrard CY, Poland V, Zumwalt R, et al. A prospective double-blinded comparison of autopsy and postmortem computerized tomography (PMCT) for the evaluation of potential drug poisoning deaths. Abstract, AAFS annual meeting 2016. Las Vegas, NV, 2016.
[6] Winklhofer S, Surer E, Ampanozi G, Ruder T, Stolzmann P, Elliott M, et al. Postmortem whole body computed tomography of opioid (heroin and methadone) fatalities: frequent findings and comparison to autopsy. Eur Radiol 2014;24:1276–82.
[7] Burke M, O'Donnell C, Bassed R. The use of computed tomography in the diagnosis of intentional medication overdose. For Sci Med Pathol 2012;8:218–36.
[8] Flach P, Ross S, Ampanozi G, Ebert L, Germerott T, Hatch G, et al. Drug Mules as a radiological challenge: sensitivity and specificity in identifying internal cocaine in body packers, body pushers, and body stuffers by computed tomography, plain radiography, and Lodox. Eur J Rad 2012;81(10):2518–26.
[9] Dolinak D. Forensic toxicology: a physiologic perspective. Academic Forensic Pathology; 2013. ISBN: 978–0–9879053–1–4.
[10] Nolte K. Cutaneous and other external manifestations of abuse Unit 29–10 Clinical dermatology, vol. 4. Lippencott-Raven; 1995.
[11] Nishimura H, Moriya F, Hashimoto Y. Mechanisms of gamma-hydroxybutyric acid production during the early postmortem period. Forensic Toxicol 2009;27:55–60.
[12] Pounder DJ, Jones GR. Post-mortem drug redistribution—a toxicological nightmare. Forensic Sci Int. 1990;45(3):253–63.
[13] Baselt Randall. Disposition of toxic drugs and chemicals in man. 9th ed. Seal Beach, CA: Biomedical Publications; 2011.
[14] Davis GG. National Association of Medical Examiners and American College of Medical Toxicology Expert Panel on Evaluating and Reporting Opioid Deaths. Recommendations for the investigation, diagnosis, and certification of deaths related to opioid drugs. The National Association of Medical Examiners website, <https://netforum.avectra.com/eweb/DynamicPage.aspx?Site = NAME&WebCode = PubPositPapr>; 2013 [accessed 13.01.18].
[15] Tormoehlen LM, Tekulve KJ, Nañagas KA. Hydrocarbon toxicity: a review. Clin Toxicol 2014;52(5):479–89. Available from: https://doi.org/10.3109/15563650.2014.923904.
[16] Adgey AA, Johnston PW, McMechan S. Sudden cardiac death and substance abuse. Resuscitation 1995;29(3):219–21.
[17] Dolinak D. Toxicology. In: Dolinak D, Matshes EW, Lew EO, editors. Forensic pathology. Burlington, MA: Elsevier Academic Press; 2005.
[18] International statistical classification of diseases and related health problems, tenth revision. World Health Organization website, <http://www.who.int/topics/mortality/en/>; updated 1992 [accessed 13.01.18].
[19] Centers for Disease Control and Prevention. Medical examiners' and coroners' handbook on death registration and fetal death reporting; 2003 revision.
[20] National Association of Medical Examiners. A guide for manner of death classification. 1st ed.; 2002.

Chapter 25

Drug Testing in Pain Management

Gary M. Reisfield[1] and Roger L. Bertholf[2]

[1]*Department of Psychiatry, University of Florida College of Medicine, Gainesville, FL, United States*
[2]*Department of Pathology and Genomic Medicine, Houston Methodist Hospital, Houston, TX, United States*

INTRODUCTION

The current prescription opioid crisis is not our nation's first [1]. The previous crisis subsided a century ago [2], punctuated by President Woodrow Wilson's signing, in March 1915, of the Harrison Narcotic Act. Nominally a tax act, it required physicians who prescribed or dispensed opioids to register with the federal government and keep detailed records of their transactions. This legislation, in conjunction with changing medical norms and decades of aggressive law enforcement efforts targeting physicians [3], served to suppress the prescribing of long-term opioid therapy for the next three quarters of a century. Although opioids continued to be prescribed for acute pain and, decades later, end-of-life and cancer pain, the prescribing of opioids for chronic noncancer pain occurred largely outside the medical mainstream [4]. Throughout the remainder of the century, opioid crises were attributable chiefly to heroin and were limited, for the most part, to major urban centers [2].

Based on reports in the 1970s of the safe and effective treatment of cancer pain with long-term opioid therapy [5], the practice began to gain medical acceptance [6]. Consequent to the apparent success of the cancer pain experience, the notion of managing chronic noncancer pain with long-term opioid therapy reemerged in the mid-1980s. In a seminal 1986 paper in the eminent journal *Pain*, Portenoy and Foley [7] suggested, based on limited experience and a misinterpretation of extant data, that long-term opioid therapy posed little threat of abuse or addiction in patients without such histories.

This position gained momentum, beginning in the 1990s, with thought leaders offering their support [8–12]. The use of opioids for chronic noncancer pain was endorsed by several professional organizations, including the American Academy of Pain Medicine, the American Pain Society, and the American Geriatric Society [13,14]. The enterprise remained controversial, however, largely because of the absence of long-term outcomes data [15–17].

In 1996, the Purdue Frederick Company introduced OxyContin, an extended-release oxycodone product, for the management of chronic noncancer pain. Purdue promoted this drug aggressively to primary care physicians, who for the most part lacked adequate education and training in pain management and opioid prescribing [18]. In 2001, Purdue spent $200 million marketing the drug. Within 5 years of its release, its sales increased from $48 million to nearly $1.1 billion [19]. Purdue claimed that their product was a safe alternative to traditional immediate-release opioids, and was less prone to abuse and addiction [20]. By 2004, however, OxyContin was the most abused C-II opioid in the nation [21]. Increases in opioid use were not limited to oxycodone. From 1997 to 2006, sales of hydrocodone increased by 244%, hydromorphone by 274%, morphine by 196%, fentanyl by 479%, and methadone by 1177% [22].

Pharmaceutical industry dollars contributed to the opioid crisis in other, less visible ways. For example, they funded patient advocacy groups, such as the American Pain Foundation (APF), which lobbied against state and federal efforts to limit opioid use. The APF received approximately 90% of its $5 million funding in 2010 from the pharmaceutical industry and several of its board members were found to have financial relationships with industry [23]. The industry lobbied the US Congress, which, for example, passed legislation that made it "logistically impossible" for the US Drug Enforcement Administration (DEA), in its efforts to safeguard public health and safety, to impose emergency suspension orders on opioid distributors or manufacturers suspected of suspicious opioid sales [24]. An investigation by the Washington Post and the TV newsmagazine "60 Minutes" revealed that an industry political action committee contributed over $1.5 million to 23 legislators who sponsored versions of the bill [25]. The Federation of State Medical Boards

(FSMB) accepted money from the pharmaceutical industry to produce and distribute opioid prescribing guidelines [26]. The FSMB's 2004 Model Policy for the Use of Controlled Substances for the Treatment of Chronic Pain suggested that state medical boards should consider the undertreatment of chronic pain a departure from acceptable standards of practice [27].

The failure of state legislatures—and the Florida legislature, in particular—was breathtaking. Between 2002 and 2010, Florida was the top oxycodone-dispensing state in the nation [28], earning it the moniker "Oxytourism" capital of the nation. This was enabled by the lack of laws regulating pain clinics and in-office opioid dispensing, as well as the lack of a prescription drug monitoring program [29]. In 2010, 98 of the nation's top 100 dispensing physicians—"pill mill" doctors—were located in Florida [30]. In the first 6 months of 2010, Florida physicians dispensed 41.2 million doses of oxycodone, compared to 4.8 million doses dispensed by physicians in the remaining 49 states combined [31].

Legitimate physicians and patients also played important roles in the opioid crisis. Well-intentioned physicians—under a mandate to assess and treat chronic pain, dealing with complex patients while under intense time pressures, with insufficient education and training in pain management, and often believing they had nothing better to offer—prescribed opioids liberally [32–34]. Patients, often wanting simple, easy solutions to complex problems, pressured physicians to prescribe [35].

The health insurance industry contributed to the opioid crisis by covering the relatively low cost of opioids, but not interdisciplinary pain programs, despite a strong evidence base for their efficacy. Consequently, the number of such programs dwindled to fewer than 100 in 2015 from more than 1000 in 1999 [36].

There were many other contributors to the crisis. The DEA sets annual opioid production quotas, and these quotas have increased dramatically over the past quarter century. For example, between 1993 and 2015, production of oxycodone increased from less than 4 to 150 tons, and production of hydrocodone, hydromorphone, and fentanyl increased 12-, 23-, and 25-fold, respectively [37].

In 1998, the Veterans Health Administration mandated the measurement and documentation of pain as a "fifth vital sign" to be assessed at every clinical encounter with the same vigilance as blood pressure, pulse, respiratory rate, and temperature [38].

In 2001, the Joint Commission on Accreditation of Healthcare Organizations, an accrediting agency for health care facilities, issued its Standards on Pain Management. Although this initiative did not explicitly endorse the prescribing of opioids, it established the assessment and management of pain as a patient right, and mandated that pain management be incorporated into health care organizations' performance measurement and improvement programs [39].

In recent years, there have been a variety of responses to this prescription opioid crisis, including federal investigations of questionable pharmaceutical marketing practices and financial relationships [35]; criminal prosecutions of pharmaceutical companies; legal and regulatory actions against unscrupulous prescribers; state legislative changes (e.g., regulating pain clinics; limiting in-office opioid dispensing; establishing prescription drug monitoring databases); and dissemination of practice guidelines that encourage a more measured approach to the prescribing of opioids [40–43]. These responses have resulted in decreases in opioid prescribing [44], but they have been too little or too late to contain the problem. By the second decade of the 21st century, illicit opioids, including heroin and its fentanyl (and analog) contaminants—potent, inexpensive, and readily available—had supplanted prescription opioids as the source of greatest opioid mortality. As was the case following our previous prescription opioid crisis, the majority of heroin users began their opioid careers with the nonmedical use of prescription opioids [45].

USE OF OPIOIDS IN PAIN MANAGEMENT

Chronic pain is the most common reason for seeking medical care in the United States; it is estimated that one-third of the population is affected [46]. Among the many therapeutic options for managing chronic pain, opioid analgesics remain a frequent choice due to their simplicity, relatively low cost, and patient acceptance. Opiates are isolated from the latex encased in the seed pods of the opium poppy, and include morphine and codeine, along with several nonanalgesic alkaloids. Semisynthetic pharmaceutical derivatives of the opium alkaloids ("opioids") include buprenorphine, hydrocodone, hydromorphone, oxycodone, and oxymorphone. (Heroin is a derivative of morphine [diacetylmorphine], which is used therapeutically in some countries, but is illegal in the United States.) Synthetic opioids include fentanyl, meperidine, methadone, tapentadol, and tramadol. Although opioids are among the most effective analgesics, they also have a high potential for abuse, and recent data indicate that opioids now rival cannabis as the drug of choice for first time illegal drug users [47]. The prevalence of opioid use disorders in chronic pain patients receiving opioid therapy is difficult to determine. In several recent studies, 25%–50% of chronic pain patients exhibited aberrant drug-related behaviors [48–52]. The proportion with opioid use disorders has been estimated to be between 8% and 12% [53,54].

Reflecting the movement toward more aggressive management of chronic noncancer pain, which began in the 1990s, the prescribing of opioid analgesics increased dramatically over the past generation. In 2015, nearly 100 million people—more than one-third of the US population aged 12 or older—were past-year users of prescription opioids [47]. The most commonly reported prescription opioids were hydrocodone (58.3 million people; 21.8% of the population) and oxycodone (27.9 million; 10.4%) [47]. The liberalization of opioid prescribing has been associated with several unintended consequences. For example:

- In 2015, 2.1 million Americans initiated misuse of prescription opioids, exceeded only by the number of those initiating use of cannabis, and exceeding the numbers of those initiating use of any other class of illicit or prescription drug [47].
- In 2015, prescription opioids were the most misused category of prescription psychotherapeutic medications. More than 12 million people misused prescription opioids, comprising nearly 5% of the population aged 12 or older [47]. The major reasons given for past-year opioid misuse included relief of physical pain (62.6%), to feel good or to get high (12.1%), to relax or to relieve tension (10.8%), to help with sleep (4.4%), to help with feelings or emotions (3.3%), to experiment (2.5%), because they were "hooked" (2.3%), and to alter the effects of other drugs (0.9%) [47].
- In 2015, more than 2 million Americans met Diagnostic and Statistical Manual of Mental Disorders (DSM-IV) criteria for prescription opioid use disorder [47], eclipsing the numbers of those addicted to heroin (253,000) and cocaine (959,000) combined [47].
- In 2015, more than 800,000 people aged 12 or older received past-year treatment for prescription opioid use disorder, more than those receiving treatment for prescription sedative, stimulant, and tranquilizer use disorders combined [47].
- Emergency room visits related to the misuse of prescription opioids more than doubled—from 168,379 to 366,181—between 2005 and 2011 [55]. In 56% of these visits, other substances were involved, most commonly benzodiazepines (28%) or alcohol (14%) [55].
- Unintentional overdose deaths involving prescription opioids have increased dramatically in recent years. Prescription opioid deaths increased from 4030 in 1999 to 22,598 in 2015 [56].
- The aggregate societal cost of prescription opioid deaths and use disorders in 2013 was estimated to be $78.5 billion [57].

The sources of prescription opioids destined for nonmedical use are informative. According to Substance Abuse and Mental Health Services Administration's (SAMHSA) 2016 National Survey on Drug Use and Health (NIDA), 54% of those reporting misuse of prescription opioids in the past year indicated that they obtained their most recent opioids from a friend or relative, including by gift (40.5%), by purchase (9.4%), or by theft (3.8%) [47]. In contrast, 36% indicated that they obtained their most recent prescription opioids from a physician, either from a single physician (34.0%) or from multiple physicians (1.7%) [47]. Fewer than 10% indicated that they purchased their most recent opioid from a drug dealer or stranger (4.9%) or procured it in some other way (4.9%) [47]. Thus, according to this government data source, most opioids destined for nonmedical use originate from valid prescriptions from physicians. Moreover, the single largest group of people who got their most recent opioid prescription from a physician were those with opioid use disorders (43.7%) [47]. Physicians thus face the often difficult task of distinguishing between the (nonmutually exclusive) groups of patients who have a legitimate need for opioid analgesia from those who are seeking opioids for nonmedical purposes. Since it has been clearly demonstrated that self-reported drug use and behavioral observations are unreliable indicators of drug abuse, clinicians have increasingly used urine drug testing (UDT) to verify adherence to prescription instructions, detect unauthorized use of other prescription medications, and reveal use of illegal drugs. There is encouraging evidence that UDT in pain management decreases illicit drug use by patients [58], but at the same time, the number of deaths from opioid overdose has nearly quadrupled over the past decade, despite greater surveillance by UDT [59]. Opioid abuse has increased by a similar proportion [60].

Although UDT is now widely used in monitoring patients being treated for chronic pain, as well as in substance abuse recovery programs, questions have been raised over whether the UDT is being ordered appropriately and the results interpreted correctly [46,61,62]. Surveys of family physicians participating in pain management symposia revealed their troubling lack of knowledge about drug testing sensitivity and specificity, drug metabolism, and passive exposure [61]. For example, only 29% of the physicians surveyed were aware that codeine is metabolized to morphine, 22% knew that eating poppy seeds can produce sufficient urinary morphine concentrations to result in a positive opiate screen, and 10% would taper or discontinue opioid therapy in a patient that tested negative for the prescribed drug, without verifying that the assay being used in the laboratory (or at the point of care) was capable of detecting that specific opioid. Disturbingly, 10% of physicians responding to the survey responded that they would consider notifying law enforcement in the latter scenario. Despite physicians' demonstrated deficiencies in UDT interpretation, most feel confident about their interpretive abilities [46].

ANALYTICAL METHODS FOR DRUG TESTING

Laboratory methods for detecting drug use include immunoassays, which are designed to have high sensitivity but may cross-react with structurally similar (and, occasionally, structurally dissimilar) compounds, and confirmatory methods that are highly specific but, in general, more expensive and technically demanding. Among the immunoassays available for UDT are methods adapted for rapid, on-site (point of care) use, and assays designed for use on automated chemistry analyzers. Most confirmatory methods involve mass spectrometry (MS), but there are several permutations of inlet and mass filter options. Historically, gas chromatography (GC) was most often used to separate drugs prior to MS identification, but often required the synthesis of volatile derivatives to produce satisfactory chromatographic results. The development of thermospray and electrospray techniques for interfacing nonvolatile chromatographic methods, such as liquid–liquid chromatography, enhanced the popularity of liquid chromatography (LC)/MS methods, which often could be applied to specimens with minimal pre-analytical manipulation. There are relative merits and limitations of measuring drugs in different body matrices—blood, urine, breath, saliva (or "oral fluid"), nails, and hair [63]. See Chapter 28, Testing of Drugs of Abuse in Oral Fluid, Sweat, Hair, and Nail: Analytical, Interpretative, and Specimen Adulteration Issues, for drugs of abuse testing in oral fluid, sweat, hair, and nail.

General Considerations on Immunoassays

The basic design of the competitive, heterogeneous immunoassay was devised by Rosalyn Yalow and Solomon Bersen. In their landmark 1960 paper [64], they described a method for measuring insulin in blood using radioisotopically labeled bovine insulin and polyclonal anti-insulin antibodies generated in guinea pigs. The vast potential for radioimmunoassay (RIA) was immediately apparent, and the technique was applied to many hormones, vitamins, and enzymes previously too difficult to detect at their low concentrations in blood. Yalow was awarded the 1977 Nobel Prize in Physiology and Medicine for the discovery of RIA, but Bersen died in 1972, and therefore could not share the award, since the Nobel Prize is not given posthumously.

Although immunochemical methods all share a common denominator—use of an antibody to recognize and bind with a specific analyte—there are several analytical strategies that have evolved from this approach. Some immunoassays rely on a ligand modified with a label to measure its concentration in the antibody-bound or -free fraction. Radioactive labels gave rise to RIA, and radioisotopes are easily attached to organic molecules, preserve the structural integrity of proteins, and can be detected at very low concentrations. However, the instability of radioisotopes—shelf-life is limited—and the health risks of exposure to radioactivity have made radioisotopes an unpopular choice for detection labels, and they have been mostly replaced by enzymes and chemiluminescent probes in label-based immunoassays. In label-based methods, either the ligand or the antibody may carry the label. Other immunoassays do not involve a label, but instead depend on the formation of large cross-linked antibody−antigen complexes that alter the light scattering properties of the solution.

Immunoassays are classified as competitive or noncompetitive, depending on whether the antigen or antibody, respectively, is present in excess. In addition, immunoassays are heterogeneous if their design requires separation of the bound and free antigen fractions before measurement of the incorporated label, or homogeneous if separation is not required to distinguish between bound and free antigen fractions. Heterogeneous immunoassays can be either competitive or noncompetitive, whereas virtually all homogeneous immunoassays are competitive. Homogeneous immunoassays are more easily adapted to automated chemistry analyzers because the physical separation of the bound and free fractions is not required. Homogeneous methods typically involve an antigen labeled with probe that has a unique physical or chemical property that changes when the antigen is bound to antibody. A generic scheme for a homogeneous immunoassay is illustrated in Fig. 25.1, and this is the most common configuration of automated methods for detecting drugs. Point of care assays ordinarily are heterogeneous, and may be classified as competitive or noncompetitive, depending on the design; some have features of both. Finally, immunoassays may be one-site, which involves a single antigenic site against which a mixture of polyclonal antibodies, or a monoclonal antibody, is directed, or two-site, in which the analyte has two distinct antigenic sites with antibodies directed at each. Immunoassays for detecting drugs typically are one-site, since drug (and drug metabolite) molecules ordinarily are too small to accommodate two antibodies. Two-site methods are often called "sandwich" methods, since the analyte is sandwiched between two antibodies.

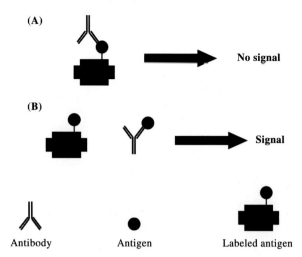

FIGURE 25.1 A general scheme for homogeneous immunoassay. The antigen is covalently attached to a label (or signal molecule), the properties of which are altered when an antibody is bound to the labeled antigen (A). Free antigen displaces the labeled antigen from the antibody (B). Typically, displacement of the labeled antigen results in a detectable signal that is proportional to free antigen concentration, but in some homogeneous immunoassays, the signal is inversely proportional to free antigen concentration. *From Reisfield GM, Salazar E, Bertholf RL. Rational use and interpretation of urine drug testing in chronic opioid therapy. Ann Clin Lab Sci 2007;37(4):301–14. Reprinted with permission.*

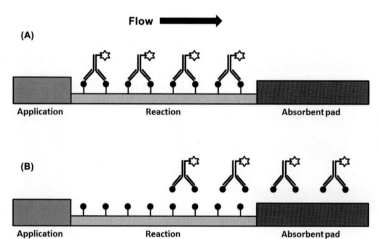

FIGURE 25.2 One design of a lateral-flow immunoassay involves the flow of reagent across a reaction bed that can capture labeled antibodies on immobilized antigen (A). In the presence of free antigen, binding sites on the labeled antibodies are saturated and do not react with the immobilized antigen; the label ends up in the absorbent pad at the end of the device (B). The positive or negative result is determined by whether there is signal in the reaction region; the presence of signal is a negative result, whereas the absence of signal is a positive result.

A vast array of immunoassays have been developed around these basic principles—competitive versus noncompetitive, homogeneous versus heterogeneous, one-site versus two-site—but for practical reasons, only a few have been applied to detecting drugs or drug metabolites. See Chapter 9, Immunoassay Design for Screening of Drugs of Abuse, for more details.

Immunoassays Used in Point of Care Drug Testing

The technology for most point of care drug immunoassays is lateral-flow immunoassay or immuno-chromatographic assay. Although both terms appear in the literature, they refer to the same basic approach, which is the capture of labeled antigens or antibodies moving across a solid support to which antibodies or antigens have been covalently linked. In one scheme, labeled antibodies bind to immobilized antigen as they pass across the test region, unless they have been saturated with free antigens in the specimen applied to the sample pad that promotes capillary flow along the device membrane (Fig. 25.2). Another approach uses immobilized antibodies to capture labeled antigens that have been displaced from soluble antibodies by endogenous unlabeled antigens in the specimen (Fig. 25.3). A third strategy involves antigens covalently bound to gold nanoparticles, mixed with the specimen and free antibodies [65]. In the absence of free antigen, the antigens bound to the gold nanoparticles are obscured by antibodies and cannot be captured

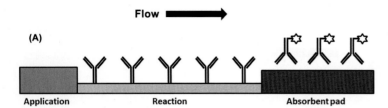

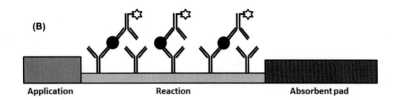

FIGURE 25.3 A two-site, noncompetitive approach to lateral-flow immunoassay involves an immobilized antibody and a soluble labeled antibody. In the absence of antigen to bridge the two antibodies, the label does not appear in the reaction region (A). When antigen is present, the antibodies can form a link with antigen, and signal appears in the reaction region (B). In this scheme, signal is associated with a positive result.

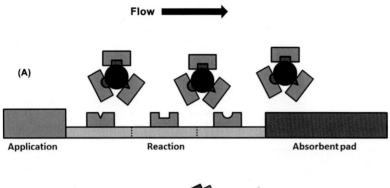

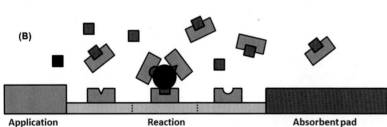

FIGURE 25.4 In the Ascend multianalyte immunoassay technique (Triage, Biosite Diagnostics, Roselle, CA), the specimen, antibodies, and colloidal gold—drug conjugates are applied to the lateral-flow device. In the absence of free drug in the specimen, the antibodies react with the gold-conjugated drugs, preventing their reaction with the immobilized antibodies in the reaction region (A). In the presence of a free drug in the specimen, the antibodies against that drug are saturated, exposing the conjugated drug, which therefore is captured in the reaction region (B). A positive result is indicated by the presence of a red bar in the reaction region for a specific drug.

by antibodies immobilized in the detection regions of the membrane (Fig. 25.4). Also see Chapter 11, Point of Care Devices for Drugs of Abuse Testing: Limitations and Pitfalls.

Lateral-flow immunoassays are difficult to classify precisely. They incorporate elements of competitive immunoassays, in that the antigen usually is present in excess, but only when the free antigen exceeds a threshold concentration. They are mostly heterogeneous, since bound and free fractions are separated, but the separation step is subtle and integral to the lateral-flow design. Both one-site and two-site lateral-flow immunoassay designs have been developed.

Point of care drug testing devices are attractive to clinicians because they offer rapid results and allow contemporaneous discussion with patients with regard to unexpected test results. Many clinicians will recognize that unexpected presumptive positive or negative results, particularly when they conflict with a patient's self-reported drug use, should be confirmed by a more specific laboratory method, but some will not. Several thousand commercially available point of care urine drug testing devices have been granted a waiver from Clinical Laboratory Improvement Amendments requirements by the US Food and Drug Administration (FDA) [66]. One of the conditions under which the FDA grants a waiver to an in vitro diagnostic test is that the manufacturer's instructions for its use must be followed, and product inserts for waived UDT devices usually include a statement variously recommending or requiring confirmation of positive results. In the context of pain management, however, a presumptive negative result may be as potentially important as a presumptive positive result if it is interpreted to mean that a patient has been nonadherent, and may be abusing, hoarding, or diverting prescription medications. As noted in an earlier section of this chapter, clinicians' understanding of the sensitivities and specificities of immunoassays often is limited.

Homogeneous Immunoassays

The most popular methods for automated UDT are homogeneous immunoassays, several of which are adaptable to a variety of commercially available chemistry analyzers. Homogeneous immunoassays are relatively simple to automate, since separation of the bound and free fractions is not required. Depending on the type of label used, the detection system may involve spectrophotometry, turbidimetry, or fluorescence polarization (FPIA). With the exception of FPIA, which was a popular technology through the 1980s and 1990s, but is no longer available, automated UDT screening methods are mostly enzyme immunoassays. Three homogenous immunoassays dominate the market for automated UDT: enzyme-multiplied immunoassay (EMIT; Siemens Healthcare, Deerfield, IL), cloned enzyme donor immunoassay (CEDIA; Thermo Scientific, Waltham, MA), and kinetic interaction of microparticles in solution (KIMS; Roche Diagnostics, Indianapolis, IN).

Enzyme-Multiplied Immunoassay Technique

The patent for the EMIT homogeneous immunoassay was granted to Syva Corporation, a joint venture between Varian Associates and Syntex Corporation (Palo Alto, CA) in 1971. In 1995, Behring Diagnostics, a newly spun-off subsidiary of Dade International, purchased Syva Corporation, and marketed the EMIT products under that corporate name until 1997, when the corporations merged to become Dade Behring. In 2007, Siemens acquired Dade Behring and now it is a part of Siemens Diagnostics. In the design of an EMIT for detecting a drug or metabolite, the antigen is covalently attached to an enzyme, glucose-6-phosphate dehydrogenase (G-6-PD), in a position near the substrate binding site. When antibodies against the drug bind to the enzyme-linked antigen, the active site is sterically hindered by the antibody, and enzyme activity is inhibited. However, if free antigens are present in the specimen, they will compete for the antibody, displacing the enzyme-linked antigen and restoring enzyme activity. Therefore, the enzyme activity is directly proportional to the amount of free antigen (drug or metabolite) in the specimen (Fig. 25.5). The EMIT is an extraordinarily versatile assay because it only requires spectrophotometric measurement of the G-6-PD cofactor, nicotinamide adenine dinucleotide phosphate ($NADP^+$), which can be distinguished from its reduced form (NADPH) by its UV absorption at 340 nm—a characteristic of NADPH, but not $NADP^+$. Hence, an increase in absorption at 340 nm occurs as $NADP^+$ is reduced when glucose-6-phosphate is enzymatically oxidized to 6-phosphogluconate.

Cloned Enzyme Donor Immunoassay

Like the Syva Corporation, Microgenics Corporation (Concord, CA) was founded in 1981 in the biotechnology seedbed that evolved in the San Francisco Bay area in the 1970s and 1980s. Microgenics used recombinant DNA technology to develop a homogeneous immunoassay based on cloned fragments of the β-galactosidase enzyme, which spontaneously

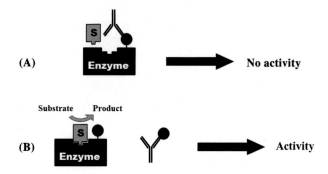

FIGURE 25.5 The EMIT uses an enzyme-labeled antigen to compete with endogenous antigen for binding sites on the antibody. When antibody is bound to the labeled antigen, the substrate binding site on the enzyme is blocked, and enzyme activity is therefore inhibited (A). Free (endogenous) antigens saturate the antibodies and prevent them from binding to the enzyme-labeled antigen, exposing the substrate binding site and restoring enzyme activity (B). *From Reisfield GM, Salazar E, Bertholf RL. Rational use and interpretation of urine drug testing in chronic opioid therapy. Ann Clin Lab Sci 2007;37(4):301–14. Reprinted with permission.*

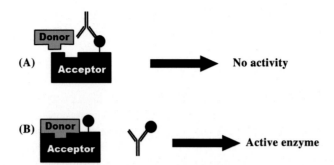

FIGURE 25.6 The CEDIA method uses cloned donor and acceptor fragments of β-galactosidase, which spontaneously associate in solution to form the active enzyme. The analyte molecule is attached to donor (smaller) fragment of the enzyme, and when the labeled analyte is bound to an antibody, association with the acceptor fragment is inhibited (A). Free antigen displaces the labeled antigen from the antibody, allowing association of the donor and acceptor fragments and restoration of enzyme activity (B). *From Reisfield GM, Salazar E, Bertholf RL. Rational use and interpretation of urine drug testing in chronic opioid therapy. Ann Clin Lab Sci 2007;37(4):301−14. Reprinted with permission.*

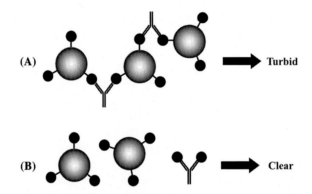

FIGURE 25.7 In the KIMS method, antigens are attached to a microparticle. Antibodies cross-link the antigen-bound microparticles forming large cross-linked complexes that result in turbidity (A). In the presence of free antigen, the microparticles are displaced from the antibody and do not form large complexes, reducing the turbidity (B). In the KIMS method, the signal (turbidity) is inversely proportional to the free antigen concentration. *From Reisfield GM, Salazar E, Bertholf RL. Rational use and interpretation of urine drug testing in chronic opioid therapy. Ann Clin Lab Sci 2007;37(4):301−14. Reprinted with permission.*

associated in solution to reconstitute the active enzyme. The CEDIA method was patented in 1985. CEDIA reagents currently are marketed through a partnership with Thermo Fisher Scientific. In the CEDIA method (Fig. 25.6), the β-galactosidase enzyme is cloned in two sections, a smaller fragment (the "donor") and a larger fragment ("acceptor"). In solution, the donor and acceptor fragments spontaneously associate to form a monomer, and four monomeric units aggregate to form the active tetrameric enzyme. The antigen is covalently attached to the acceptor fragment so that when antibody is attached to the bound antigen, association with the donor fragment is sterically hindered. Free antigens in the specimen compete with the acceptor-labeled antigen for antibodies, releasing the acceptor fragments for association with donor fragments, and resulting in reconstitution of enzyme activity. Therefore, β-galactosidase activity is proportional to free antigen (drug or metabolite) concentration, and is measured spectrophotometrically by enzymatic conversion of dye-labeled substrate. The CEDIA technology has the same attractive feature of the EMIT, which is portability to any clinical analyzer that has the capability of spectrophotometric measurements.

Kinetic Interaction of Microparticles in Solution

A nonenzymatic homogeneous immunoassay for detecting drugs in urine was developed by Roche Diagnostics, one of two principal divisions of F. Hoffman-La Roche Ltd., a health care conglomerate headquartered in Basel, Switzerland. The other division develops and markets pharmaceuticals. The KIMS method (Fig. 25.7) borrowed the approach of earlier latex agglutination assays designed to measure antibodies. For UDT, a drug or metabolite is covalently bound to a latex microparticle and, in the absence of free drug, antibodies cross-link the microparticles to form large complexes that scatter light, measured turbidimetrically. Free antigens occupy binding sites on the antibodies, inhibiting the formation of cross-linked complexes, thereby decreasing the turbidity of the mixture. The scheme is often called "immunoturbidimetric assay." There is no enzyme involved, so the KIMS assays are not susceptible to potential chemical interferences from enzyme inhibitors. On the other hand, turbidimetry is a less specific analytical method because it only detects particles capable of scattering light, as opposed to spectrophotometric measurements, which detect specific compounds based on their absorption of UV or visible light. Turbidimetric methods are susceptible to interference from nonspecific aggregates in solution that scatter light. The analytical performance of immunoturbidimetric methods for UDT is, however, substantially equivalent to enzyme immunoassays.

Confirmatory Methods

MS has been the standard for identifying and measuring organic compounds since a method for detecting methylated derivatives of amino acids was described over a half century ago by Carl-Ove Andersson, an analytical biochemist at the University of Uppsala in Sweden [67]. Mass filters used in MS applications include magnets, quadrupole radio frequency sectors, and time-of-flight instruments. The fundamental design of a mass spectrometer involves an ion source, where a compound is subjected to a high energy flux of electrons (electron ionization) or unstable ionization products (chemical ionization). Exposure to excess energy results in fragmentation of the molecule, with charged fragments separated and quantified when they pass through the mass filter, and detected when they collide with the charged collecting plates in an electron multiplier. The fragmentation pattern of every molecule is unique, because it is related to the stability of individual bonds within the molecule, and even small biomolecules have many intramolecular bonds that contribute to the fragmentation pattern and resulting mass spectrum, which can be considered a chemical fingerprint of the molecule.

A modification of the standard mass spectrometer involves the use of multiple mass filters. In multiple-stage MS, the molecule is ionized and the first mass filter is tuned to select the parent ion for the molecule of interest, which is then exposed to an inert gas such as argon, causing collisional decomposition. The "daughter ions" are used to identify the compound. Multiple-stage mass spectrometers can be highly selective and sensitive, since the first stage provides the ability to eliminate many interfering compounds. A typical configuration of a triple-stage quadrupole mass spectrometer is illustrated in Fig. 25.8.

GC was the first chromatographic method widely adapted to MS detectors. Capillary GC columns used in most GC–MS couplings are much longer than traditional glass columns, improving the chromatographic resolution by a factor of 100–1000, and they do not require an interface to reduce pressure, so the GC effluent could be delivered directly to the ion source. LC is an attractive alternative to GC for coupling with an MS detector because it does not require analytes to be volatile. However, unlike capillary GC, LC involves a large volume of solvent (mobile phase) that must be removed before the column effluent can be introduced into the ion source. Several LC/MS interfaces have been devised. Early attempts to interface an LC to an MS involved a heated conveyer on which the volatile solvent was evaporated before introduction of the column effluent into the ion source. However, the introduction of thermospray and electrospray technologies revolutionized LC/MS applications. These interfaces involve atomizing the LC column effluent into small droplets from which the solvent can be rapidly evaporated (thermospray), abandoning the analyte in the vapor phase, or charging the droplets in an electrical field so their contents can be accelerated by a focusing lens into the ion source. Thermospray and electrospray interfaces are common in LC/MS and LC/MS–MS (LC coupled to a multiple-stage mass spectrometer) instruments available for drug testing.

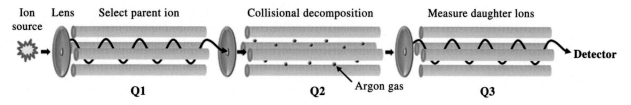

FIGURE 25.8 In a triple-stage quadrupole mass spectrometer, the first mass filter (Q1) selects the parent ion for the molecule to be detected and measured. The second quadrupole (Q2) contains molecules of an inert gas that collide with the parent ion, causing it to decompose into daughter ions. The daughter ions are separated in the third sector (Q3) before being measured at the detector. The fragmentation pattern is specific for a particular daughter ion, and is used to identify and measure the parent compound. *From Reisfield GM, Bertholf RL, Goldberger BA, DuPont RL. The science and clinical uses of drug testing. In Miller SC, Feillin D, Rosenthal RN, and Saitz R, eds. The ASAM Principles of Addiction Medicine, 6e, Wolters Kluwer, 2018. Reprinted with permission from Shannon Miller.*

LC/MS–MS methods are particularly well suited to UDT in pain management, and have become common in laboratories offering those services. One reason for the unique adaptability of this analytical method is that it can detect and quantify multiple drugs and metabolites without concern for volatility or GC chromatographic properties. The method is less susceptible to interferences from co-migrating compounds, due to the greater specificity of parent/daughter ion selectivity of tandem MS. Conjugated metabolites, such as ethyl glucuronide and ethyl sulfate, can be measured without extensive chemical manipulation. In unregulated UDT, confirmatory methods have to be adapted to specific clinical needs, which typically involve confirming adherence to prescribed medications and abstention from commonly abused prescription and illicit drugs.

Specimen Tampering

In forensic UDT, guidelines have been established for detecting tampering with a urine specimen. Acceptable ranges for temperature (immediately post-collection), pH (within certain time limits), creatinine concentration, and specific gravity are intended to reveal specimens that have been substituted (temperature), adulterated (pH, and other specific adulterants such as nitrite), or diluted (creatinine and specific gravity). Substitution with pet urine is also encountered in unobserved collections, and may result in abnormal indices. Chemical warmers are sometimes used to heat a substituted specimen to a temperature within an acceptable range. In general, these specimen integrity checks are appropriate for nonforensic UDT, such as monitoring patients who are prescribed long-term opioid therapy, particularly when the screening is performed for the purpose of ensuring that the patient is not taking nonprescribed drugs.

However, UDT in pain management often serves the additional purpose of verifying adherence to prescribed opioid regimens; a negative screen for a prescribed drug may indicate that the drug has been used up early in the month and hence not administered in the days prior to the test, or that the drug has not been taken, but instead has been hoarded or diverted. Ironically, a nonadherent patient who is tested for a prescribed drug has an incentive to ensure the UDT is positive, rather than negative (in contrast to drug abusers, whose incentive is to produce a negative specimen). Two potential approaches to gaining a positive result are substitution of a urine specimen from someone who is taking the drug, or addition of the drug ("spiking") to the negative urine specimen. The latter approach has two important vulnerabilities:

- Concentrations of drugs in the urine typically fall within predictable limits, although extremes are sometimes encountered. Addition of a few milligrams of drug to a 50-mL urine specimen can produce a urinary drug concentration of more than 100,000 ng/mL, which is unusually high for most opioid dosage ranges.
- Drugs (particularly opioids) are rarely eliminated exclusively as the parent drug. Most opioids undergo some type of oxidative metabolism, as well as conjugation. Therefore, the presence of solely the parent drug, or in some cases, concentrations of parent drug that greatly exceed metabolite(s), in a urine specimen would be suspicious, and metabolites are prohibitively difficult to obtain by patients trying to wangle appropriate UDT results. While critical examination of the UDT results for typical drug/metabolite combination would often reveal a specimen to which drug has been added, there are no widely adopted protocols for this type of assessment, and it is unlikely that most physicians ordering UDT for their patients would scrutinize screening results if they appeared consistent with patient adherence.

There is precedent in forensic drug testing programs for follow-up of positive results with specific tests to look for additional drugs or metabolites. Methamphetamine, for example, is demethylated to produce amphetamine, but only the D-stereoisomer is significantly metabolized by that pathway. Desoxyephedrine, the L-isomer of methamphetamine, which is not a controlled substance and is a constituent of over-the-counter products, produces a mass spectrum nearly identical to methamphetamine, but is not significantly demethylated to amphetamine. Therefore, one protocol for confirmation of methamphetamine-positive specimens requires the presence of amphetamine. In addition, the dextrorotatory (D) stereoisomer of methamphetamine can also be distinguished from the l-isomer by using chiral derivatizing reagents, or on a chromatographic column with a chiral stationary phase. Some nonchiral chromatographic methods with sufficient resolution are capable of separating the stereoisomers.

Heroin is rapidly metabolized to morphine, so the presence of morphine in the urine is consistent with administration of morphine or codeine (of which several legal but controlled pharmaceutical preparations exist) or heroin, which is a Schedule I controlled substance in the United States and is illegal to possess. In some cases, heroin use can be confirmed by the presence of an intermediate metabolite, 6-acetylmorphine (or a contaminant, 6-acetylcodeine), and protocols exist for detecting this metabolite at very low concentrations in urine specimens that have morphine concentrations above the positive threshold.

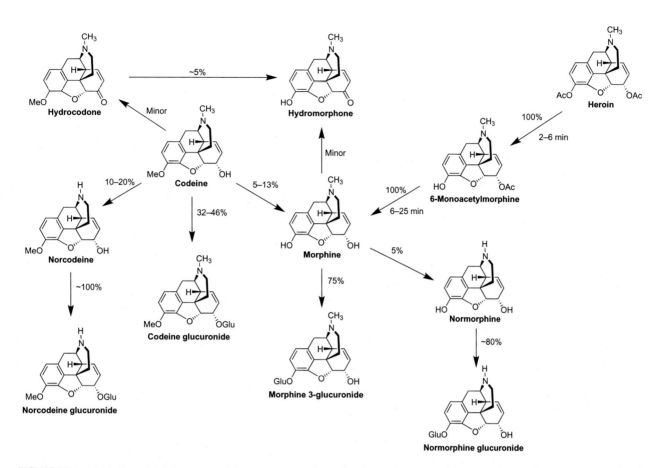

FIGURE 25.9 Metabolism of opioids. *From Reisfield GM, Salazar E, Bertholf RL. Rational use and interpretation of urine drug testing in chronic opioid therapy. Ann Clin Lab Sci 2007;37(4):301−14. Reprinted with permission.*

Metabolism of Opioid Analgesics

Opioid analgesics are widely prescribed for the treatment of acute and chronic pain of moderate to severe intensity. To wit, hydrocodone is one of the most prescribed drugs of any class in the United States. Knowledge of opioid metabolism is critical in clinical urine drug testing because several commercially available opioids are metabolized to other commercially available opioids or are present as process impurities in the manufacture of other commercially available opioids. Failure to recognize these facts can result in false conclusions that patients are using nonprescribed opioids.

Opioid metabolism occurs chiefly in the liver by Phase I and/or Phase II reactions. Phase I reactions are nonsynthetic reactions (e.g., oxidation, reduction), and occur chiefly by means of the cytochrome P450 (CYP) superfamily of enzymes. Fig. 25.9 shows the principal Phase I metabolic pathways for several opioids. Phase II reactions are synthetic reactions in which the drug (or metabolite) is conjugated, resulting in a more hydrophilic and are excreted in urine. In the case of opioids, glucuronic acid is the major conjugate.

The most important oxidative enzymes are CYP2D6 and CYP3A4, which constitute 2%−4% and 30%, respectively, of hepatic CYPs, but are responsible for the oxidative metabolism of 25% and 50%, respectively, of commercially available drugs, including most opioids [68]. CYP enzymes are variously inducible (e.g., CYP3A4) and suppressible (e.g., CYP2D6, CYP3A4) by other drugs, and display a high degree of genetic polymorphism that imbues them with a wide spectrum of activity. CYP2D6 phenotypes comprise poor metabolizers (PM), intermediate metabolizers, extensive metabolizers, and ultra-rapid metabolizers (UM), which account for 5%−10%, 10%−17%, 70%−80%, and 3%−5%, respectively, of the Caucasian population [68]. In contrast, approximately 1% of the Chinese population are PM and 29% of Ethiopians are UM [69].

CYP2D6 metabolizes three pharmaceutically used opioids to other commercially available opioids. The most clinically important example is the conversion of codeine to morphine. Codeine is metabolized primarily by glucuronidation to codeine-6-glucuronide, and secondarily by CYP3A4 to norcodeine. However, approximately 10% of an administered dose is O-demethylated by CYP2D6 to morphine, the metabolite chiefly responsible for codeine's analgesic effects. Two other opioids undergo conversions that, while clinically minor, are toxicologically noteworthy. Oxycodone is a potent opioid analgesic of which approximately 10% is O-demethylated to the more potent oxymorphone [70]. Likewise, hydrocodone is O-demethylated to hydromorphone, with greater concentrations of the metabolite found in UMs [71].

Other toxicologically significant opioid metabolic conversions exist, although the responsible enzymatic pathways have not yet been identified. Several case reports have described the metabolism of morphine to hydromorphone. In each of these cases, the urinary hydromorphone concentration comprised less than 5% of the urinary morphine concentration. A minor pathway exists for the metabolism of codeine to hydrocodone. Hydrocodone can sometimes metabolize to dihydrocodeine, with the latter found in approximately 10% of hydrocodone-positive postmortem cases [72]. It is important to recognize that most commercially available opioids do not metabolize to other opioids. Thus, buprenorphine, the fentanyls, levorphanol, meperidine, methadone, tapentadol, and tramadol are neither metabolized to, nor are metabolites of, other commercially available opioids.

During the manufacture of several opioid products, other opioid(s) may appear as "process impurities," possibly confounding urine drug test interpretation. As can be seen in Table 25.1, allowable limits of these impurities do not exceed 0.5%. This has become an issue over the past two decades because of the increasing prescribing of opioids—particularly high-dose opioids—for the treatment of noncancer pain, together with the increasing use of clinical urine drug testing, an expanded test menu for opioids, and use of lower analyte cutoff concentrations. In 2010, the heroin metabolite 6-acetylmorphine has been reported as a process impurity in the manufacture of morphine, but allowable limits have not yet been established and typically observed percentages have not been reported [73]. 6-Acetylmorphine also has been detected at concentrations above 20 ng/mL in specimens with morphine concentrations below the customary positive threshold of 300 ng/mL [74]. Table 25.1 lists the allowable limits for process impurities in the manufacture of several opioids.

Patients prescribed long-term opioid therapy often use benzodiazepines for various indications [75] including treatment of muscle spasm (diazepam), neuropathic pain (clonazepam), insomnia (temazepam, triazolam), and co-occurring anxiety disorders (e.g., alprazolam, diazepam, lorazepam, oxazepam). Benzodiazepines, like opioids, are also common drugs of abuse, and data indicate that individuals using both opioids and benzodiazepines are more likely to be diagnosed with an opioid use disorder [76]. Furthermore, unintentional drug overdose deaths often involve a combination of opioids and benzodiazepines [77]. Several benzodiazepines, like several opioids, are subject to bioconversion to other commercially available benzodiazepines. For example, temazepam is metabolized to oxazepam (via CYP2C19), diazepam is metabolized to temazepam (via CYP3A4) and thence to oxazepam [78], and chlordiazepoxide is metabolized to nordiazepam and oxazepam [79]. Alprazolam, clonazepam, flunitrazepam, lorazepam, midazolam, and triazolam are neither metabolites of, nor metabolized to, other commercially available benzodiazepines.

TABLE 25.1 Acceptable Impurities Present in Commercial Opioid Drugs

Opioid Drug	Impurity Present	Allowable limit	Comments[a]
Codeine	Morphine	0.15%	Morphine impurity <1%
Morphine	Codeine	0.5%	Codeine impurity <1% for
	6-Monoacetyl morphine	Not established	
Hydrocodone	Codeine	0.15%	Newer preparations may not contain any impurity
Oxycodone	Hydrocodone	1%	Hydrocodone <0.2%
Oxymorphone	Hydromorphone	0.15%	Hydromorphone up to 1%
	Oxycodone		Oxycodone up to 0.5%
Hydromorphone	Hydrocodone	0.1%	Newer preparation may not
	Morphine	0.15%	Contain any impurity

[a]*Manufacture's certificate of analysis of products.*
Source: Medical Review Officer Alert, volume 21, number 3, Triangle Park, NC: Quadrangle Research, LLC, Research; April 2010.
Inclusion of SAMHSA content does not constitute or imply endorsement or recommendation by the Substance Abuse and Mental Health Services Administration, the U.S. Department of Health and Human Services, or the U.S. Government.

Challenges in Drug Testing in Pain Management

The principal focus of the clinical laboratory has always been to provide analytical data that support and promote accurate medical diagnosis and treatment. The benefits of measuring plasma concentrations of drugs were immediately apparent, and clinical laboratories adapted to the emerging need for therapeutic drug monitoring by offering it, typically, within their "toxicology" services. It was a logical designation, since many of the analytical methods for measuring therapeutic drugs were the same as those used for detecting toxins. The advent of homogeneous immunoassays for therapeutic drugs, however, and the application of that same technology for detecting drugs of abuse in urine, blurred the distinction between toxicology and clinical chemistry. Most contemporary UDT assays have been adapted to routine chemistry analyzers.

There was diagnostic value to UDT in the evaluation and treatment of patients presenting to the emergency department, where drug overdose posed urgent diagnostic and therapeutic questions that could often only be resolved by laboratory support. The division between medical and forensic UDT, however, is very clear: screening by UDT immunoassays was usually sufficient for urgent medical evaluation, but forensic UDT must be confirmed to statutory guidelines that specified concentration thresholds for positive results, confirmatory analyses for presumptive positive results, review of UDT results by a medical review officer, and restrictions with regard to collection and documentation of the specimen chain of custody. Laboratories offering forensic UDT services have to conform to state or federal regulations, which were rigorous.

Medical and forensic UDT diverged into two distinct disciplines. Urine toxicology for sports medicine became a third discipline, regulated by its own set of guidelines and regulatory agencies that exceeded the resources of most clinical and forensic drug testing laboratories. For hospital-based clinical laboratories, the division between medical and forensic UDT is clearly apparent; hospitals provide medical UDT, which rarely include confirmatory analyses, and referral laboratories with the resources to comply with NIDA and SAMHSA requirements sought certification to offer forensic UDT under federal contracts.

Drug testing in pain management straddles the divide between medical and forensic UDT. In some respects, it is part of medical evaluation, but in other respects, it is used to determine consequences in a manner similar to forensic UDT. To a patient prescribed opioid therapy for chronic pain, denial of treatment is a consequence tantamount to termination as an employee or removal of privileges to participate in sports activity. Patients face the stigma of being considered a drug abuser, or perhaps worse, a drug dealer, based on the results of UDT. As noted earlier in the discussion of physicians' lack of interpretive knowledge, some physicians may even report the patient to law enforcement authorities based on a UDT result. The severity of these consequences demands that utmost care must be taken to avoid the possibility of misinterpretation of UDT results, but currently there are no safeguards in place to ensure that this does not occur. Laboratories that offer UDT services to pain management practices typically have a toxicologist or medical director qualified to correctly interpret UDT results and perhaps recommend follow-up tests to clarify unexpected or ambiguous results of screening tests. However, effective use of such a resource requires the exchange of information between the clinician and laboratory that is often missing from UDT orders. Other laboratories may request no information about what drug(s) the patient is taking, or what result the clinician expects in order to confirm patient adherence. Without this information, the laboratory does not recognize circumstances that require additional testing. This is a serious deficiency in current practices for using UDT to monitor patients on opioid therapy.

A critical and mandatory feature of forensic UDT is the review of all positive results by a medical review officer certified to interpret UDT. The same approach could be prohibitively expensive and time-consuming for UDT in pain management. In forensic UDT, only positive results have adverse consequences, whereas in pain management, positive or negative UDT results can have adverse consequences, so equal attention must be paid to unexpected negative results. Laboratory tests for drugs and their metabolites in urine have many limitations, and the proper interpretation of positive or negative results requires consideration of several factors, including the specific drug and dose that was administered, frequency and recency of administration, metabolic factors that influence the urinary products characteristic of that drug or metabolite, the presence of other drugs that may influence metabolism, and the sensitivity and specificity of the analytical method used to detect the drug and its metabolites. Clinicians ordinarily do not receive the type of training necessary to resolve these various factors, and therefore the potential for misinterpretation of positive or negative UDT results is disturbingly high.

There is an unaddressed need for standardization, and perhaps regulation, of UDT services applied to patients on monitored opioid therapy, as well as enrollees in substance use disorder recovery programs. Analytical methods exist for sensitive and specific detection of drugs in urine and other biological matrices, and many laboratories offer these services to pain management practices. Used correctly, UDT is an invaluable component of the responsible application of chronic opioid analgesic therapy. However, without precautions against the misinterpretation of results, UDT has the potential to cause harm to individuals who may be responsibly adherent to prescribed and beneficial medication regimens.

CONCLUSIONS

UDT in pain management is widely adopted by clinicians to confirm adherence with controlled substance-based treatment plans. Despite widely available analytical methods, including confirmation of the presence of the prescribed drug(s) in the urine, there are ethical and medical issues regarding the practice of drug monitoring in pain management [80].

REFERENCES

[1] Terry CE. The Harrison anti-narcotic act. Am J Public Health (NY). 1915;5(6):518.
[2] Kolodny A, Courtwright DT, Hwang CS, Kreiner P, Eadie JL, Clark TW, et al. The prescription opioid and heroin crisis: a public health approach to an epidemic of addiction. Annu Rev Public Health 2015;36:559–74.
[3] Musto DF. The American disease. Origins of narcotic control. New Haven, CT: Yale University Press; 1973. 354 pp.
[4] Turk DC, Brody MC, Okifuji EA. Physicians' attitudes and practices regarding the long-term prescribing of opioids for non-cancer pain. Pain 1994;59(2):201–8.
[5] Kanner RM, Foley KM. Patterns of narcotic drug use in a cancer pain clinic. Ann NY Acad Sci 1981;362:161–72.
[6] Seymour J, Clark D, Winslow M. Pain and palliative care: the emergence of new specialties. J Pain Symptom Manage 2005;29(1):2–13.
[7] Portenoy RK, Foley KM. Chronic use of opioid analgesics in non-malignant pain: report of 38 cases. Pain 1986;25(2):171–86.
[8] Melzack R. The tragedy of needless pain. Sci Am 1990;262(2):27–33.
[9] McCarberg BH, Barkin RL. Long-acting opioids for chronic pain: pharmacotherapeutic opportunities to enhance compliance, quality of life, and analgesia. Am J Ther 2001;8(3):181–6.
[10] Passik SD, Weinreb HJ. Managing chronic nonmalignant pain: overcoming obstacles to the use of opioids. Adv Ther 2000;17(2):70–83.
[11] Compton P, Athanasos P. Chronic pain, substance abuse and addiction. Nurs Clin North Am 2003;38(3):525–37.
[12] Fishman SM, Teichera D. Challenges and choices in drug therapy for chronic pain. Cleve Clin J Med 2003;70(2):119–21 [25-7, 31–2 passim].
[13] The use of opioids for the treatment of chronic pain. A consensus statement from the American Academy of Pain Medicine and the American Pain Society. Clin J Pain 1997;13(1):6–8.
[14] The management of chronic pain in older persons. AGS Panel on Chronic Pain in Older Persons. American Geriatrics Society. Geriatrics 1998;53(Suppl. 3):S8–24.
[15] Large RG, Schug SA. Opioids for chronic pain of non-malignant origin—caring or crippling. Health Care Anal 1995;3(1):5–11.
[16] Savage SR. Long-term opioid therapy: assessment of consequences and risks. J Pain Symptom Manage 1996;11(5):274–86.
[17] White PF, Kehlet H. Improving pain management: are we jumping from the frying pan into the fire? Anesth Analg 2007;105(1):10–12.
[18] Meier B. Pain killer: a "wonder" drug's trail of addiction and death. New York: Rodale; 2003.
[19] Van Zee A. The promotion and marketing of OxyContin: commercial triumph, public health tragedy. Am J Public Health 2009;99(2):221–7.
[20] Statement of United States Attorney John Brownlee on the guilty plea of the Purdue Frederick Company and its executives for illegally misbranding OcyContin [press release]; May 10, 2007.
[21] Cicero TJ, Inciardi JA, Munoz A. Trends in abuse of OxyContin and other opioid analgesics in the United States: 2002–2004. J Pain 2005;6(10):662–72.
[22] Manchikanti L, Singh A. Therapeutic opioids: a ten-year perspective on the complexities and complications of the escalating use, abuse, and nonmedical use of opioids. Pain Physician 2008;11(2 Suppl.):S63–88.
[23] Ornstein C, Weber T. The champion of painkillers. ProPublica; 2011 [December 23, 2011].
[24] Mulrooney JJL, Legel KE. Current navigation points in drug diversion law: hidden rock in shallow, murky, drug-infested waters. Marquette Law Review 2017;101:115.
[25] Higham S, Bernstein L. The drug industry's triumph over the DEA. Washington Post; 2017 [October 15, 2017].
[26] Fauber J. Follow the money: pain, policy, and profit. Milwaukee J Sentinel 2012.
[27] Model policy for the use of controlled substances for the treatment of chronic pain. In: Federation of State Medical Boards of the United States I, editor; 2004.
[28] Mack KA, Jones CM, McClure RJ. Physician dispensing of oxycodone and other commonly used opioids, 2000–2015, United States. Pain Med 2017.
[29] Temple J. American pain. Guilford, CT: Rowmann & Littlefield; 2015.
[30] Kuglin FA. Pharmaceutical supply chain: drug quality and security act. Boca Raton, FL: CRC Press; 2016.
[31] Cornwell B. The war rages on, with no ceasefire or surrender in sight. Charlotte Florida Weekly; 2011 [March 24, 2011].
[32] Giannitrapani KF, Ahluwalia SC, McCaa M, Pisciotta M, Dobscha S, Lorenz KA. Barriers to using nonpharmacologic approaches and reducing opioid use in primary care. Pain Med. 2017.
[33] Zgierska A, Miller M, Rabago D. Patient satisfaction, prescription drug abuse, and potential unintended consequences. JAMA 2012;307(13):1377–8.

[34] Katz MH. Opioid prescriptions for chronic nonmalignant pain: driving on a dangerous road. JAMA Intern Med 2013;173(3):178.
[35] Gounder C. Who is responsible for the pain-pill epidemic? New Yorker. 2013; [November 8, 2013].
[36] Schatman ME, Webster LR. The health insurance industry: perpetuating the opioid crisis through policies of cost-containment and profitability. J Pain Res 2015;8:153—8.
[37] Davidson J. Is DEA a bad guy in opioid addiction fight? Washington Post; 2016 [July 5, 2016].
[38] Mularski RA, White-Chu F, Overbay D, Miller L, Asch SM, Ganzini L. Measuring pain as the 5th vital sign does not improve quality of pain management. J Gen Intern Med 2006;21(6):607—12.
[39] Phillips DM. JCAHO pain management standards are unveiled. Joint Commission on Accreditation of Healthcare Organizations. JAMA 2000;284(4):428—9.
[40] Manchikanti L, Kaye AM, Knezevic NN, McAnally H, Slavin K, Trescot AM, et al. Responsible, safe, and effective prescription of opioids for chronic non-cancer pain: American Society of Interventional Pain Physicians (ASIPP) guidelines. Pain Physician 2017;20(2S):S3—92.
[41] Qaseem A, Wilt TJ, McLean RM, Forciea MA. Clinical Guidelines Committee of the American College of Physicians. Noninvasive treatments for acute, subacute, and chronic low back pain: a clinical practice guideline from the American College of Physicians. Ann Intern Med. 2017;166(7):514—30.
[42] Dowell D, Haegerich TM, Chou R. CDC guideline for prescribing opioids for chronic pain—United States. MMWR Recomm Rep 2016;65 (1):1—49.
[43] Franklin GM, American Academy of Neurology. Opioids for chronic noncancer pain: a position paper of the American Academy of Neurology. Neurology 2014;83(14):1277—84.
[44] Guy Jr GP, Zhang K, Bohm MK, Losby J, Lewis B, Young R, et al. Vital signs: changes in opioid prescribing in the United States, 2006—2015. MMWR Morb Mortal Wkly Rep 2017;66(26):697—704.
[45] Compton WM, Jones CM, Baldwin GT. Relationship between nonmedical prescription—opioid use and heroin use. N Engl J Med 2016;374 (2):154—63.
[46] Starrels JL, Becker WC, Alford DP, Kapoor A, Williams AR, Turner BJ. Systematic review: treatment agreements and urine drug testing to reduce opioid misuse in patients with chronic pain. Ann Intern Med 2010;152(11):712—20.
[47] Hughes A., Williams M.R., Lipari, R.N., Bose, J., Copello, E.A.P., Kroutil, L.A. Prescription drug use and misuse in the United States: results from the 2015 National Survey on Drug Use and Health. NSDUH Data Review. In: Substance Abuse and Mental HealthServices Administration CfBHSaQ, editor; 2016.
[48] Reid MC, Engles-Horton LL, Weber MB, Kerns RD, Rogers EL, O'Connor PG. Use of opioid medications for chronic noncancer pain syndromes in primary care. J Gen Intern Med 2002;17(3):173—9.
[49] Katz NP, Sherburne S, Beach M, Rose RJ, Vielguth J, Bradley J, et al. Behavioral monitoring and urine toxicology testing in patients receiving long-term opioid therapy. Anesth Analg 2003;97(4):1097—102 [table of contents].
[50] Passik SD, Kirsh KL. Opioid therapy in patients with a history of substance abuse. CNS Drugs 2004;18(1):13—25.
[51] Ives TJ, Chelminski PR, Hammett-Stabler CA, Malone RM, Perhac JS, Potisek NM, et al. Predictors of opioid misuse in patients with chronic pain: a prospective cohort study. BMC Health Serv Res 2006;6:46.
[52] Wiedemer NL, Harden PS, Arndt IO, Gallagher RM. The opioid renewal clinic: a primary care, managed approach to opioid therapy in chronic pain patients at risk for substance abuse. Pain Med 2007;8(7):573—84.
[53] Fishbain DA, Rosomoff HL, Rosomoff RS. Drug abuse, dependence, and addiction in chronic pain patients. Clin J Pain 1992;8(2):77—85.
[54] Ballantyne JC. Assessing the prevalence of opioid misuse, abuse, and addiction in chronic pain. Pain 2015;156(4):567—8.
[55] Crane E.H. The CBHSQ report: emergency department visits involving narcotic pain relievers. In: Substance Abuse and Mental Health Services Administration CfBHSaQ, editor. Rockville, MD; 2015.
[56] Facts. HJKFFSH. Prescription opioid overdose deaths and death rate per 100,000 population (age adjusted). Available from: https://www.kff.org/other/state-indicator/prescription-opioid-overdose-deaths-and-death-rate-per-100000-population-age-adjusted/?currentTimeframe = 0&sortModel = %7B%22colId%22:%22Location%22,%22sort%22:%22asc%22%7D.
[57] Florence CS, Zhou C, Luo F, Xu L. The economic burden of prescription opioid overdose, abuse, and dependence in the United States, 2013. Med Care 2016;54(10):901—6.
[58] Pesce A, West C, Rosenthal M, Mikel C, West R, Crews B, et al. Illicit drug use in the pain patient population decreases with continued drug testing. Pain Physician 2011;14(2):189—93.
[59] Okie S. A flood of opioids, a rising tide of deaths. N Engl J Med 2010;363(21):1981—5.
[60] Compton WM, Volkow ND. Major increases in opioid analgesic abuse in the United States: concerns and strategies. Drug Alcohol Depend 2006;81(2):103—7.
[61] Reisfield GM, Bertholf R, Barkin RL, Webb F, Wilson G. Urine drug test interpretation: what do physicians know? J Opioid Manag 2007;3 (2):80—6.
[62] Reisfield GM, Webb FJ, Bertholf RL, Sloan PA, Wilson GR. Family physicians' proficiency in urine drug test interpretation. J Opioid Manag 2007;3(6):333—7.
[63] Ropero-Miller JD, Goldberger BA. Handbook of workplace drug testing. 2nd ed Washington, DC: AACC Press; 2008.
[64] Yalow RS, Berson SA. Immunoassay of endogenous plasma insulin in man. J Clin Invest 1960;39:1157—75.
[65] Buechler KF, Moi S, Noar B, McGrath D, Villela J, Clancy M, et al. Simultaneous detection of seven drugs of abuse by the Triage panel for drugs of abuse. Clin Chem 1992;38(9):1678—84.

[66] Watson ID, Bertholf RL, Hammett-Stabler CA. Drugs and ethanol testing at the point of care. Point Care 2007;6:227–30.
[67] Andersson BA. Mass spectrometric and gas chromatographic studies of n-heptafluorobutyryl derivatives of peptide methyl esters. Acta Chem Scand 1967;21(10):2906–8.
[68] Zhou SF, Liu JP, Chowbay B. Polymorphism of human cytochrome P450 enzymes and its clinical impact. Drug Metab Rev 2009;41(2):89–295.
[69] Aklillu E, Persson I, Bertilsson L, Johansson I, Rodrigues F, Ingelman-Sundberg M. Frequent distribution of ultrarapid metabolizers of debrisoquine in an ethiopian population carrying duplicated and multiduplicated functional CYP2D6 alleles. J Pharmacol Exp Ther 1996;278(1):441–6.
[70] Heiskanen T, Olkkola KT, Kalso E. Effects of blocking CYP2D6 on the pharmacokinetics and pharmacodynamics of oxycodone. Clin Pharmacol Ther 1998;64(6):603–11.
[71] Barakat NH, Atayee RS, Best BM, Pesce AJ. Relationship between the concentration of hydrocodone and its conversion to hydromorphone in chronic pain patients using urinary excretion data. J Anal Toxicol 2012;36(4):257–64.
[72] Baker DD, Jenkins AJ. A comparison of methadone, oxycodone, and hydrocodone related deaths in Northeast Ohio. J Anal Toxicol 2008;32(2):165–71.
[73] Available from: http://www.aamro.com/docs/news/27.pdf; 2011.
[74] Crews B, Mikel C, Latyshev S, West R, West C, Pesce A, et al. 6-Acetylmorphine detected in the absence of morphine in pain management patients. Ther Drug Monit 2009;31(6):749–52.
[75] Cone EJ, Caplan YH, Black DL, Robert T, Moser F. Urine drug testing of chronic pain patients: licit and illicit drug patterns. J Anal Toxicol 2008;32(8):530–43.
[76] Hojsted J, Nielsen PR, Guldstrand SK, Frich L, Sjogren P. Classification and identification of opioid addiction in chronic pain patients. Eur J Pain 2010;14(10):1014–20.
[77] Toblin RL, Paulozzi LJ, Logan JE, Hall AJ, Kaplan JA. Mental illness and psychotropic drug use among prescription drug overdose deaths: a medical examiner chart review. J Clin Psychiatry 2010;71(4):491–6.
[78] Mandrioli R, Mercolini L, Raggi MA. Benzodiazepine metabolism: an analytical perspective. Curr Drug Metab 2008;9(8):827–44.
[79] Baselt RC. Disposition of toxic drugs and chemicals in man. 7th ed Foster City, CA: Biomedical Publications; 2004.
[80] Reisfield GM, Maschke KJ. Urine drug testing in long-term opioid therapy: ethical considerations. Clin J Pain 2014;30(8):679–84.

Chapter 26

How Do People Try to Beat Drugs Test? Effects of Synthetic Urine, Substituted Urine, Diluted Urine, and In Vitro Urinary Adulterants on Drugs of Abuse Testing

Shanlin Fu
Centre for Forensic Science, University of Technology Sydney, Ultimo, NSW, Australia

INTRODUCTION

In certain safety-sensitive professions, such as the armed forces, transport industry, mining industry, or any profession that deals with heavy machinery, having a drug-free work environment is of vital importance. In order to achieve this goal, many jurisdictions and workplaces employ drug testing programs [1−3]. Workplaces are not the only environments that implement these strategies, however. Drug testing programs can also be found in use by law enforcement agencies [4,5], rehabilitation and harm minimization programs [6−9] and, of course, by antidoping organizations, such as the World Anti-Doping Agency (WADA), to maintain the integrity of sporting events [10,11].

There are a number of different biological matrices that can be used for drug testing application, including blood, urine, and oral fluid, among others [12−16]. Of the various options available to toxicologists, urine is the most popular matrix [5,17,18]. This matrix is considered the most suitable for a number of reasons, namely its collection is noninvasive and quite simple. Additionally, urine allows for a reasonably lengthy window of detection for most drugs of interest and/or their metabolites [5,17,19]. Given the popularity of this mode of drug testing, many policies and procedures have been developed over the last 30 years or more to control this process and these have been adapted to be suitable in both clinical and medico-legal settings [18−25]. To ensure the reliability of the forensic techniques applied, various government agencies around the world provide guidelines and standards to mandate the execution of these analyses. The Substance and Mental Health Services Administration (SAMHSA) Guidelines [1] and the Military Personnel Drug Abuse Testing Program (MPDATP) [26] control urine drug testing in the United States and the AS/NZS 4308 standard is used in Australia and New Zealand [27]. From an antidoping perspective, WADA has established urine drug testing protocols for doping control [11].

Typically, urine drug testing involves the use of both an initial screening test, followed by a subsequent confirmatory test [27,28]. Generally, the initial screening tests take the form of an immunoassay test [19]. Any specimen that tests positive in the screening tests are then subjected to confirmatory testing via gas chromatography—mass spectrometry (GC−MS) or liquid chromatography—mass spectrometry (LC−MS) [19]. In this chapter, LC−MS refers mainly to liquid chromatography—tandem mass spectrometry (LC−MS/MS) and no distinction between the two terms is made.

One of the major problems facing toxicologists performing urine drug testing is the adulteration of samples. Urine adulteration involves the alteration of collected samples with the intent to create a false negative result. Due to the significant consequences that can follow a positive test result, including legal consequences, such as fines or imprisonment, along with other consequences such as dismissal from employment or suspension from sporting events, individuals who have used prohibited substances are highly motivated to obtain a negative result [29]. Urine adulteration can be very easy as sample collection is often supervised, but not observed, for privacy reasons [19]. The different methods that can be used to adulterate urine samples can be categorized into three separate groups: urine substitution, in vivo alteration and urine dilution, and in vitro adulteration [30].

Urine substitution is, quite simply, the replacement of a urine specimen with a drug-free sample of urine [31], a nonurine liquid (such as water or a saline solution) [32], or synthetic urine which mimics the characteristics of human urine, for example, similar pH, creatinine concentration, and specific gravity [33]. In some cases, individuals who attempt to pass drug screens by urine substitution will use tools such as the Whizzinator, the Urinator, and the Butt Wedge. There have also been reports of catheters being used in some extreme cases [1,18,31,33]. These devices can be used in an attempt to covertly deliver drug-free or synthetic urine in place of an authentic urine sample [34].

Just because a drug screen returns a negative result, it does not mean that there are no drugs of abuse (DOA) present in the urine. It is possible that the amount of drug present was below the cutoff value used for that particular testing protocol [35]. This means that urine dilution, also known as in vivo adulteration, is a simple way to attain a false negative result, especially if the amount of drug present in the urine is just above the cutoff value [35]. This method involves the ingestion of large amounts of fluids, most commonly water, to dilute the urine. Alternatively, certain products, such as diuretic compounds, can be intentionally consumed in an attempt to increase the metabolism and/or excretion of the DOA [31,33]. Urine dilution can also be achieved by the addition of water to an already voided urine sample [18].

On the other hand, in vitro adulteration can be performed by adding exogenous chemicals into a voided sample to obtain a false negative result [18,19]. This form of adulteration can be achieved by the addition of a number of different chemicals that are quite readily available. Common adulterants can include household chemicals such as hypochlorite bleach, table salt, and laundry detergent, as well as commercial products easily sourced from the Internet such as UrinAid, Urine Luck, Stealth, and Klear [31,33]. Many of these adulterants are oxidants that can affect the chemical composition of the analytes of interest.

In an attempt to combat the rising trend of urine adulteration, toxicology laboratories and practitioners have invested a lot of time and effort in creating countermeasures to detect attempted adulteration. Urine substitution can be identified by observing the sample collection procedure, carrying out a series of integrity tests (to monitor variables such as specific gravity, pH, color, smell, and temperature) and even sequencing of the urinary DNA [10,19,36]. In vivo adulteration through the use of diuretic compounds can be detected by screening for the presence of the diuretics themselves in the urine sample [19,37]. This technique is often implemented in various antidoping sports drug testing programs [37]. The concentration of creatinine present in a urine sample can be used as an effective marker of urine dilution, whether as a result of in vivo or in vitro adulteration with water [38]. Finally, in vitro adulteration through the use of exogenous chemicals can be countered through observation of sample collection, urine integrity tests, similar to those used for urine substitution, and/or chemical analysis of any adulterants present within a collected sample [18,19,31,33].

Of the different adulteration techniques possible, the use of chemical adulterants (also known as chemical manipulation) has been the main focus of toxicological research. Many research articles and reviews have been published on this topic (including those written by Jaffee et al. [31] and Dasgupta [33]). Most of the researches in this area focus on the analytical methods used for the detection of the different adulterants [39–43] and the effects of these adulterants on the target drug concentrations in the sample [44–47]. While these different techniques can effectively detect the presence of chemical manipulation, they cannot determine the specific DOA that has been used [18]. Additionally, there can be some uncertainty as to the origin of some adulterants that are detected. For example, an increased nitrite level in a urine specimen (a commonly used adulterant) can be present because of variations in diet, pathology, or even certain medication, rather than exogenous adulterants [48].

The large volume of samples that are received by many toxicology laboratories further limits the capabilities for the detection of urine adulterants. It is not financially viable for laboratories to perform extensive analyses on every urine specimen to detect the presence of adulterants [18]. Therefore, most drug testing laboratories heavily rely on urine integrity test results of suspicious samples, such as odd color or smell, unusual pH, and/or specific gravity parameters, to triage collected specimens and prioritize analysis [1,27]. While this approach is the most practical logistically, it is unable to detect all adulterated samples, especially when the adulterants are used at quite low levels and do not alter the integrity test parameters.

An alternative approach can be taken and has been further developed in recent years in order to address this serious limitation. Degradation and oxidation of the different drug analytes caused by oxidizing chemicals would trigger the formation of altered or oxidized drug analytes [18]. Recent research has identified a series of oxidation products that are unique to a specific combination of DOA and oxidant [49–51]. If the specific products that will be formed by these reactions can be determined, analysts will be able to determine with sufficient accuracy the specific DOA that has been used.

This chapter will review the different methods of adulteration and the effects that these will have on urine drug testing results. The different approaches that can be taken by toxicology laboratories to detect adulteration will be discussed, along with possibilities for the development of new and more effective methods to combat this trend. While this review refers to some new psychoactive substances, the focus will be the urine testing of some of the most popular DOA including amphetamine-type stimulants (ATS), cocaine, cannabis, and opiates.

URINE DRUG TESTING PROGRAMS

In Australia, workplace drug testing is the responsibility of the employer under their duty of care obligations through Workplace Health and Safety (WHS) legislation [52]. All drug testing programs need to follow the Australian standards (AS/NZS 4308) [27] and guidelines for employers to help them set up effective drug testing programs are provided by various government bodies, such as WorkCover NSW [52]. Similarly, in nonfederal workplaces in America, employers may implement workplace drug testing under the guidance of the Division of Workplace Programs from the SAMHSA [53]. SAMHSA is also responsible for overseeing federally regulated drug testing programs in some workplaces such as the Nuclear Regulatory Commission, the Department of Transportation, and the Department of Defense [53].

Typically, urine drug testing carried out for medico-legal purposes follow a similar generic workflow. Initially, a screening test, usually in the form of an immunoassay test, will be conducted to quickly triage samples [18]. The immunoassay screening tests used by toxicology laboratories are based on the application of a specific antibody–antigen binding system [54]. The interaction of this system may indicate the presence of different DOA and/or their metabolites in different biological matrices. The use of immunoassays as a quick and efficient method for the presumptive screening of urine sample has been well established [54]. There are a number of different immunoassay-based methods, which are consistently used including: cloned enzyme donor immunoassay (CEDIA) [55], enzyme immunoassay (EIA) [56], fluorescence polarization immunoassay (FPIA) [57], enzyme-linked immunosorbent assay (ELISA) [56], kinetic interaction of microparticles in solution (KIMS) including Roche Abuscreen ONLINE assays [58], and enzyme-multiplied immunoassay technique (EMIT) [57].

Following a positive screening result, a confirmatory test is required to validate the results that have been obtained. Generally, these tests take the form of MS-based techniques such as GC–MS or LC–MS [19,27,28]. Any DOA and/or metabolites present can be confirmed definitively by the comparison of their chromatographic and mass spectrometric properties to certified reference materials (CRMs) [18]. More recently, GC–MS and LC–MS techniques have also been used for screening tests [59]. The MS program commonly applied is full scan mode for screening tests and selected ion monitoring (GC–MS) or multiple reaction monitoring (MRM) mode (LC–MS) for confirmatory tests [27,59].

There are a number of different drug analytes which are often targeted for different drug classes, and these are listed in Table 26.1. Cocaine is metabolized rapidly in the body [61,62]; therefore, the parent drug itself is not detected and instead it has two major metabolites, ecgonine methyl ester (EME) and benzoylecgonine (BZE) [27,28], are monitored. In the same way, heroin can be detected through the presence of its metabolites 6-monoacetylmorphine (6-MAM) and morphine [62,63]. Cannabis use is determined by the detection of 11-nor-9-carboxy-tetrahydrocannabinol (THC-COOH) [62] which is the major metabolite of tetrahydrocannabinol (THC) which, in turn, is the principal psychoactive component of cannabis [27,28]. On the other hand, ATS use is generally detected through the presence of the parent drug molecules [27,28]. The presence of these parent drugs can be attributed to the fact that most amphetamines are excreted relatively unchanged in urine, although some metabolites are formed [64].

One of the main advantages of screening tests, especially immunoassay-based techniques, is that they require very little to no sample preparation and can often be performed on-site [18]. Conversely, confirmatory tests generally involve at least one sample treatment step, such as hydrolysis, extraction, and/or derivatization [27,28]. When using GC–MS methods, extraction of targeted analytes from the urine matrix, followed by derivatization is used in all confirmatory tests. Hydrolysis is usually not required, however it is used when testing for cannabis and opiates in order to cleave the glucuronide conjugates of THC-COOH and morphine and codeine, respectively [18]. GC–MS techniques are still the most commonly used confirmatory method and are viewed as the "gold standard" in many laboratories [18]. When LC–MS is used for analysis, the sample preparation is very similar to that for GC–MS, with the exception of derivatization, which is not required. Additionally, some CRMs for glucuronide metabolites, including THC-COOH-glucuronide, morphine-3-glucuronide (M3G), morphine-6-glucuronide (M6G), and codeine-6-glucuronide (C6G), are now available [18]. This means that direct analysis of these analytes is now possible with LC–MS, eliminating the need for a hydrolysis step.

More recently, advances in the capabilities of LC–MS techniques have led to the development of "dilute-and-shoot" methods. This approach to analysis involves the dilution of urine samples in an appropriate buffer and injection into the instrument without the need for further sample preparation [65]. Advancements in the techniques used for urine drug

TABLE 26.1 Drug Classes Typically Monitored in Clinical and Medico-Legal Urine Drug Testing Programs [1,18,27,60]

Illicit Drug Class	Commonly Monitored Analytes
Cannabis	11-nor-9-carboxy-tetrahydrocannabinol (THC-COOH)
	Synthetic cannabinoids
Amphetamine-type stimulants (ATS)	Amphetamine
	Methamphetamine
	3,4-methylenedioxyamphetamine (MDA)
	3,4-methylenedioxy-N-ethylamphetmaine (MDEA)
	3,4-methylenedioxymethamphetamine (MDMA)
	Benzylpiperazine
	Ephedrine/pseudoephedrine
	Phentermine
Opiates	6-monoacetylmorphine (6-MAM)
	Morphine
	Codeine
Cocaine	Benzoylecgonine (BZE)
	Ecgonine methyl ester (EME)
Phencyclidine	Phencyclidine (PCP)
Benzodiazepines (Benzos)	Alprazolam/α-hydroxy-alprazolam
	Clonazepam/7-amino-clonazepam
	Diazepam
	Flunitrazepam/7-amino-flunitrazepam
	Nitrazepam/7-amino-nitrazepam
	Oxazepam
	Temazepam
Anesthetic agents	Ketamine/norketamine
	Propofol
Synthetic/Semisynthetic opioids	Fentanyl/norfentanyl
	Hydromorphone
	Methadone
	Oxycodone
	Pethidine/norpethidine
	Tramadol
Hallucinogens	Lysergic acid diethylamide (LSD)/nor-LSD
	NBOMe derivatives

Source: Adapted from Bush DM. The US Mandatory Guidelines for Federal Workplace Drug Testing Programs: current status and future considerations. Forensic Sci Int 2008;174(2−3):111−19; Fu S. Adulterants in Urine Drug Testing. In: Makowski GS, editor. Advances in clinical chemistry, vol. 76. Burlington: Academic Press; 2016. p. 123−63; AS/NZS 4308:2008 Procedures for specimen collection and the detection and quantitation of drugs of abuse in urine. New Zealand: Standards Australia/Standards; 2008; Australian Health Practitioner Regulation Agency, Drug and Alcohol Screening Protocol: Registrant Information. Melbourne: Australian Health Practitioner Regulation Agency; 2017.

testing is not the focus of this chapter. Readers are, therefore, encouraged to refer to relevant research and review articles for more information on that topic.

HOW DO PEOPLE TRY TO "BEAT THE URINE DRUG TEST"?

There are a number of different ways that people can try to beat urine drug tests. These can range from very simple techniques including diluting a sample with water, to more complex methods of substituting a urine specimen with clean urine or synthetic urine, or even chemically altering the urine to mask the presence of any DOA. It is still unknown how much urine adulteration occurs on a regular basis. A survey conducted of 6,800,000 federal and federally regulated specimens tested in the United States by the Department of Health and Human Services (DHHS) from May 2004 to April 2005 found that about 140,000 (2.1%) specimens were drug positive, and about 10,000 (0.15%) were found to be manipulated [1–3]. This accounts for approximately 7% of the positive cases. It is generally accepted that the actual number of successful urine adulteration cases would be higher due to the changing nature of the adulterant market and the existence of drug testing facilities and protocols that do not comply with accepted standards [18,31,33]. This chapter will discuss three different categories of urine adulteration: Urine substitution, in vivo adulteration, and in vitro chemical adulteration.

Urine Substitution

Urine substitution can be carried out through the use of authentic, drug-free urine, or synthetic urine. Samples of synthetic urine are available from many Internet sites [34]. This synthetic urine has similar characteristics as normal urine and is a popular choice for people trying to beat workplace drug testing. This kind of substitution is generally quite easy because of concerns for individual's privacy. In many drug testing programs, especially workplace and clinical drug testing environments, the collection of a urine sample is supervised but not directly observed [18,27]. Many different brands and sources of synthetic urine can be found on the Internet, retailing at anywhere between A$31 and A$100 [34].

The composition of normal human urine varies dramatically from person to person due to factors including diet, fluid intake, some diseases such as diabetes, and a number of other factors. In addition, there are often some trace macromolecules in normal urine specimens that may not be present in synthetic urine [34]. In general, synthetic urine mimics the specific gravity, pH, and creatinine content present within a human urine sample and dyes may even be added so that the color is also similar. Some kits that can be purchased even included devices and instructions to maintain the sample at the temperature that would be expected for a freshly voided urine sample [34]. This makes detecting the presence of synthetic urine with specimen integrity tests difficult, except through careful visual inspection [34].

Sophisticated devices can be used to assist with the substitution of urine specimens even when the collection is observed, such as in sports drug testing programs [18]. Prosthetic penis devices, such as the Whizzinator, can be used to assist with the delivery of synthetic urine, even in a monitored collection process [34]. In extreme cases, drug users have employed a catheter to inject clean synthetic or authentic urine into their bladder prior to taking a drug test [34]. While this method is effective for beating a drug test, use of nonsterilized plastic tubes rather than proper medical catheters can lead to some major health problems, such as urinary tract infections [34]. Corrupted doping officials have also been previously implicated in the manipulation of athletes' urine specimens [18,19,36].

In Vivo Adulteration and Urine Dilution

There are many commercially available products that, when taken with a large quantity of fluids, can help to flush out DOA and their metabolites [34,35]. These products include both fluids, such as Detox XXL Drink, Absolute Carbo Drinks, and Ready Clean Drug Detox Drink, and capsules or tablets, such as Fast Flush Capsules and Ready Clean Gel Capsules [35]. The purpose of these products is to dilute the collected urine specimen and, therefore, push the concentration of any remaining DOA or metabolites below the cutoff value [34,35]. This method of adulteration is easy to accomplish as the actual adulteration itself occurs off-site from the specimen collection [18].

Cone et al. [66] investigated the effects of excess fluid intake on false negative marijuana and cocaine urine drug tests. It was found that metabolite levels of these drugs measured by EMIT were dramatically reduced [66]. In addition, it was found that, while consumption of excess water was effective, consumption of some herbal tea products diluted urine faster than water alone [66]. In sports antidoping, diuretics have been used to help flush out previously ingested restricted substances by forced diuresis [35]. As a result, the presence of diuretics in urine specimens has also been prohibited [35]. Generally, if a screening test returns a negative result for a urine specimen, confirmatory testing is not conducted. This means that this technique of adulteration can be very useful for specimens where the original drug concentration is only slightly higher than the cutoff, however it is ineffective for heavy drug users [34,35].

Diluting urine with hot tap water can also be completed after the sample has been voided [34]. The use of hot water ensures that the specimen remains within the correct temperature limits so as to not arouse suspicion. As such, many collection facilities have removed any sources of hot water from where specimen collection takes place [34]. Another factor that needs to be considered is hyperhydration or water intoxication. This can be a potential health risk for those trying to avoid a positive result by drinking copious amounts of water, as it can lead to potentially fatal disturbances of brain function and electrolyte imbalance disease [34]. On the other hand, Finkel presented a case where a patient who submitted a urine specimen that was considered to be diluted was later found to have chronic water intoxication from a weight loss regime [34,67]. This highlights the importance of considering case-specific details when assessing whether or not a specimen has been adulterated.

In Vitro Chemically Adulterated Urine

One of the most problematic forms of urine adulteration is the in vitro adulteration of a voided urine specimen through the addition of various different exogenous chemicals. This type of adulteration can still be performed even when the specimen collection is monitored [18]. For example, a number of different proteases were found to be used by athletes to provide false negatives when tested for prohibited substances such as erythropoietin or other peptide hormones [18]. In some cases this adulteration was achieved by placing protease powder underneath their fingernail and urinating over their fingers [11], or even inserting protease pellets into the urethra prior to providing a urine specimen [68].

The library of different chemicals that can be used to chemically adulterate urine samples is very broad and encompasses both oxidizing and nonoxidizing chemicals. Some of the different adulterants that have been reported by drug testing laboratories are summarized in Table 26.2. These adulterants can affect the screening immunoassay tests, the confirmatory tests, or both, resulting in the return of a false negative test result [18].

TABLE 26.2 Examples of Various Chemicals That Are Often Used as Urine Adulterants

Commercial Name	Active Ingredient (Nonoxidizing)	References
Drano	Sodium hydroxide	[69]
Visine eye drops	Benzalkonium chloride	[70]
UrinAid	Glutaraldehyde	[71]
Meat tenderizer	Papain	[72]
Table salt	Sodium chloride	[69]
Vinegar	Acetic acid	[69]
Zinc powder	Zinc	[73]
Detergent/soap		[69]
Golden-seal tea		[69]
Commercial Name	**Active Ingredient (Oxidizing)**	**References**
Bleach	Sodium hypochlorite	[74]
Betadine	Iodine	[75]
Klear	Potassium nitrite	[76,77]

(Continued)

TABLE 26.2 (Continued)

Commercial Name	Active Ingredient (Oxidizing)	References
Stealth	Peroxide and peroxidase	[41]
Urine Luck	Potassium chlorochromate (PCC)	[45,78]
Whizzies	Sodium nitrite	[79]
Other common chemicals	Ammonium cerium nitrate	[75,80]
	Potassium chlorate	[46]
	Potassium chromate	[46]
	Potassium dichromate	[80]
	Potassium perchlorate	[46,75,80]
	Potassium permanganate	[46,75,80]
	Sodium iodate	[46,75]
	Sodium metaperiodate	[75,80]

HOW DO CHEMICAL ADULTERANTS EFFECT URINE DRUG TESTING?

The goal of any urine adulterant is the same; they are used to mask the presence of any DOA present in a urine specimen. The mechanisms of how each adulterant achieves this goal may be different, however. Table 26.3 summarizes the various mechanisms that have been proposed for the masking effect of adulterants on both screening immunoassays and confirmatory mass spectrometric-based techniques.

In general, immunoassay tests rely on the recognition and binding of DOA and/or their metabolites in urine specimens with their specific antibodies present within a test kit [54]. Therefore, any chemical that can interfere with these antibody−antigen interactions will have an effect on the accuracy of the immunoassay tests applied and invalidate the results. The most common mechanisms for the interference of the adulterants include the formation of insoluble analyte−adulterant complexes [81], denaturing of the protein antibody structure rendering it ineffective [83] and altering the epitope of an antigen (drug analyte), affecting the antibody−antigen recognition [49,84]. As previously identified, specimens that pass a screening test will not be submitted for further analysis. Therefore, use of an effective chemical adulterant can prevent further investigation of the sample through the provision of a false negative result on the immunoassay test and, as such, the presence of that adulterant may go unnoticed [18,44].

Confirmatory testing methods are also at risk of adverse effects from adulterants present in urine specimens [33]. Exogenous chemicals added to a urine sample may result in a large change in the ionic strength of the specimen, significantly alter the pH of the sample, or otherwise form micelles or insoluble complexes with the analytes of interest [18,31,82]. These kinds of interactions have the potential to reduce the efficacy of sample extraction procedures, possibly leading to a false negative test result. Of particular note, oxidizing adulterants may oxidize and destroy the targeted analytes, making them completely undetectable by current analytical methods due to their altered chemical identity [18,19]. Compounding this problem is that fact that oxidizing chemicals present in a sample may also destroy any internal standards added to samples for quantitative analysis, invalidating the test results [76,85−87]. On the other hand, the significant reduction in concentration of an internal standard might serve to flag the sample as suspicious. This may act as a trigger for the laboratory to perform adulterant testing on the sample, especially if the immunoassay tests have returned positive results [18].

TABLE 26.3 Proposed Mechanisms of Action of Common Urine Adulterants

Mechanism	Effect on Laboratory Analysis	
	Screening Tests	**Confirmatory Tests**
1. Binding of adulterants to the targeted analytes	Adulterants may form micelles or insoluble complexes with the targeted analytes that may not be detected by immunoassays [81]	Formation of adulterant–analyte complex may reduce extraction efficiency [31,82]
2. Increased ionic strength of urine specimen	An increase in the ionic strength of the specimen could alter the protein structure of the antibody–antigen pair, resulting in a decrease of immunoassay sensitivity [82,83]	An increase in the ionic strength of the specimen may affect analyte extraction efficiency [31,82]
3. Modification of the specimen pH	Adulterants may increase or decrease the pH of the specimen, potentially altering the binding and reaction rates of the immunoassay. Changes in the pH could also reduce the solubility of the targeted analytes in the urine matrix [84]	A change in specimen pH could affect analyte extraction efficiency [31,82] and stability [49]
4. Interactions of oxidizing adulterants with antibody proteins or enzymes	Such interactions could cause the denaturation of antibody/protein structures and may also lead to a significant decrease in protein/enzyme binding capacity [83]	N/A
5. Oxidation of targeted analytes by oxidizing adulterants	Oxidizing adulterants may react with the targeted analytes present in the specimen. Major oxidative modifications of the target analyte structure could inhibit the enzyme or antigen–antibody binding [49,84]	Targeted analytes could be altered in structure and no longer detectable by confirmatory tests [19]. Oxidizing adulterants may degrade the internal standards used for quantitative analyses and invalidate test results [46,76]

Due to the long-standing practice of adulterating urine specimen with exogenous chemicals, a large variety of methods and products are commercially available. The effects of these different adulterants on both screening and confirmatory testing techniques will be detailed.

Hypochlorite-Based Bleach

Hypochlorite-based bleach is a chemical commonly found in many households. This chemical has proven to be one of the most effective urine adulterants currently in use [18]. It has been shown previously that hypochlorite adulteration has the ability to produce false negative results for a number of drugs of interest across both immunoassay-based screening tests and MS-based confirmatory tests (Table 26.4).

Hypochlorite is among the most popular exogenous chemicals used to adulterate urine specimens that are positive for cannabis in order to return a false negative result. The effect of this chemical, along with 15 other common adulterating agents, on the analysis of urine by FPIA was studied by Schwarzhoff and Cody [94]. Of the different DOA examined in this work, the cannabinoid test was the most susceptible to this method of adulteration, causing false negative test results. Baiker et al. [88] found that hypochlorite-based adulterants were effective for causing false negative results with both the FPIA screen and the KIMS assay. Additionally, Baiker et al. [88] also reported a significant decrease in the levels of THC-COOH which were detectable using GC–MS analysis techniques.

The use of hypochlorite to produce interference with the response for amphetamine with the CEDIA test has been reported by Wu et al. [69]. More recently, Pham et al. [51] reported that high concentrations of hypochlorite could be used to effectively mask the presence of 3,4-methylenedioxymethamphetamine (MDMA) in CEDIA at 300 ng/mL (a concentration cutoff which is widely accepted by toxicological laboratories [27]). Hypochlorite-based adulterants, as reported by Cody and Schwarzhoff [95], can also render a concentration-dependent inhibition on radioimmunoassay (RIA). At a concentration of 10% hypochlorite-based bleach, false negative results were obtained from specimens positive for amphetamines. At lower concentrations of 1% and 5%, however, the results could not be replicated and the specimens remained positive [95]. Confirming the concentration-dependent nature of hypochlorite as an adulterant, Mikkelsen and Ash [84] stated that signal inhibition for the EIA assay of amphetamine was possible with concentrations

TABLE 26.4 Summary of the Effects of Hypochlorite as a Urine Adulterant on Results of Both Screening and Confirmatory Analysis Techniques

Drug Type[a]	Technique[b]	Influence of Adulterant	References
Cannabis	CEDIA	Strong interference with immunoassay	[69]
	EMIT	False negative returned	[84]
	FPIA	Decreased immunoassay sensitivity	[88,89]
	RIA	False positive returned	[89]
	GC–MS	Complete degradation of target analyte	[46]
ATS	CEDIA	Strong interference with immunoassay	[51,69]
	EMIT	False negative returned	[84]
	FPIA	Decreased immunoassay sensitivity	[90]
	GC–MS	Major decrease in target analyte recovery	[91]
	LC–MS	Strong decrease in target analyte recovery and unstable reaction product detected	[51]
Opiates	CEDIA	Strong interference with immunoassay	[69]
	EMIT	False negative returned	[84]
	FPIA	Decreased immunoassay sensitivity	[92]
	KIMS	Major reduction in immunoassay sensitivity	[93]
Cocaine	CEDIA	Strong interference with immunoassay	[69]
	EMIT	Oxidant concentration-dependent interference	[84]
PCP	CEDIA	Strong interference with immunoassay	[69]
	FPIA	Decreased immunoassay sensitivity	[89]
Benzos	CEDIA	Strong interference with immunoassay	[69]

[a]ATS, Amphetamine-type stimulants; PCP, phencyclidine; Benzos, benzodiazepines.
[b]CEDIA, Cloned enzyme donor immunoassay; EMIT, enzyme-multiplied immunoassay technique; FPIA, fluorescence polarization immunoassay; KIMS, kinetic interaction of microparticles in solution; RIA, radioimmunoassay.

ranging from 520 to 4650 ng/mL. Chou and Giang [90] found that hypochlorite had a moderate to high potential to return false negative readings for amphetamines with FPIA. Along with the various screening methods outline, hypochlorite has also been shown to interfere with the results of confirmatory GC–MS techniques [51,91]. Experiments conducted by Chou et al. [91] found that adulteration with hypochlorite led to a 36%–63% decrease in a sample's initial ATS urinary concentration. Additionally, Pham et al. [51] reported that hypochlorite reduced the MDMA concentration in a sample spiked at 150 ng/mL to a concentration well below the recommended cutoff value for confirmatory testing [27]. The samples analyzed in this study showed loss of MDMA concentration ranging between 53% and 64% [51].

Both screening and confirmatory testing techniques commonly utilized to detect the presence of opiates have been found to be affected by hypochlorite. Hypochlorite has been reported to directly affect different immunoassay reagents, resulting in the possible return of false negative results for FPIA [92], EMIT assay [84], CEDIA [69], and KIMS assay [93]. Additionally, hypochlorite has been found to cause degradation of opiate analytes in urine that was analyzed by GC–MS [32,33].

Various immunoassays for cocaine were reported to suffer from interference in the presence of hypochlorite. Mikkelsen and Ash [84] documented the concentration-dependent signal reduction for the EMIT assay for cocaine in samples that had been adulterated with hypochlorite-based bleach. The major targeted metabolite of cocaine, BZE, was present in different samples at concentrations of up to 1180 ng/mL and was effectively masked by bleach at a concentration of 42 μL/mL. If the concentration of bleach was increased to 125 μL/mL, BZE concentrations of up to 1820 ng/

mL could be masked and return a false negative result [84]. Wu et al. [69] reported that hypochlorite also decreased the cocaine response for CEDIA. It is worth noting, however, that bleach concentrations of up to 10% did not affect the levels of BZE when tested using the RIA [95].

Hypochlorite-based bleach at 10% produced strong interference for phencyclidine (PCP) and benzodiazepines (Benzos) in EMIT assay [69].

Nitrite

Nitrite is a powerful oxidizing agent, and the influence that it has on drug testing results can be found summarized in Table 26.5. Nitrite has been found to be extremely effective in masking the presence of cannabinoids by rendering false negatives for both screening and confirmatory tests. The effects of nitrite adulteration resulting in the degradation of THC-COOH and its internal standard TCH-COOH-d_6 was first studied by ElSohly et al. [76] via GC–MS analysis. Following on from this, Tsai et al. [85] explored the effects of nitrite adulteration on the KIMS immunoassay test. The sensitivity of this particular immunoassay to cannabinoids was found to decrease moderately by nitrite at concentrations of up to 1000 mM. On the other hand, nitrite concentrations as low as 30 mM were found to dramatically decrease the response for THC-COOH through GC–MS analysis [85]. Lewis et al. [96] discovered that an unstable nitroso-derivative of THC-COOH was formed as a by-product of the destruction of THC-COOH present in urine specimens. It is important to consider the fact that the efficacy of nitrite is affect by the pH of the environment. The oxidizing effect of this adulterant is most apparent under acidic conditions, and, therefore, the reaction between nitrite and THC-COOH will occur more rapidly in acidic rather than neutral or basic samples, as confirmed by Paul and Jacobs [46] and Tsai et al. [44]. In a healthy population, the normal urine pH falls within a range of pH 4.5–8.0 [22]. Theoretically, this means that a person providing a neutral or basic urine specimen (pH 7–8) would not see effective results from nitrite adulteration. In practice, however, most sample preparation procedures used for confirmatory testing include an acidification step to allow the acidic THC-COOH to be extracted from the urine [18]. This means that the pH of the sample will be lowered sufficiently to enable the reaction between nitrite and any THC-COOH present within the sample to

TABLE 26.5 Summary of the Effects of Nitrite as a Urine Adulterant on Results of Both Screening and Confirmatory Analysis Techniques

Drug Type[a]	Technique[b]	Influence of Adulterant	References
Cannabis	EMIT, KIMS	Decreased immunoassay sensitivity	[44]
	KIMS	Decreased immunoassay sensitivity	[85]
	GC–MS	Complete degradation of target analyte	[46,54]
	LC–MS	Significant decrease in recovery of target analyte and unstable reaction product detected	[96]
ATS	Monitect PC11	Not affected by adulterant	[97]
	KIMS	Decreased immunoassay sensitivity	[85]
	GC–MS	Slight increase in target analyte recovery	[54]
Opiates	KIMS	No significant effect	[85]
	GC–MS	Slight increase to target analyte recovery	[54]
	LC–MS	Strong decrease in target analyte recovery and stable reaction products detected	[49,50]
Cocaine	KIMS	No significant effect	[85]
	GC–MS	Slight decrease in target analyte recovery	[54]
PCP	KIMS	No significant effect	[85]

[a]ATS, Amphetamine-type stimulants; PCP, phencyclidine.
[b]EMIT, Enzyme-multiplied immunoassay technique; KIMS, kinetic interaction of microparticles in solution.

occur. In order to counteract this process, the addition of a reducing agent, namely a hydrosulfite or bisulfite, into an adulterated specimen prior to acidification was investigated with the purpose of destroying any nitrite present within the sample and therefore preserving the targeted analyte [44,76]. Unfortunately, this countermeasure has only shown limited efficacy in preserving the THC-COOH analyte.

Tsai et al. [85] also found that nitrite can be used to mask the presence of amphetamine within urine samples when analyzed using the KIMS assay, however it does not affect results obtained through GC–MS analysis. Conversely, nitrite was found to not exhibit adverse effects on the Monitect PC11 lateral flow immunoassay when screening for amphetamines in urine [97].

Commercially available adulterants, such as Klear and Whizzies, which contain nitrite, have been reported to be effective at masking the presence of opiates [32,33,44]. It has also been found that the presence of 6-MAM can be obscured from the CEDIA when potassium nitrite is added to a sample in acidic conditions (pH < 7) [49]. Nitrite-based adulterants have also been demonstrated to destroy 6-MAM, morphine and M6G, but not M3G and codeine in samples subjected to LC–MS analysis [49,50].

Finally, Tsai et al. [85] also reported that nitrite could be used to mask the presence of BZE via different immunoassays; however it did not cause interference with GC–MS confirmatory analysis.

Pyridinium Chlorochromate

The effects of pyridinium chlorochromate (PCC) on urine drug testing are summarized in Table 26.6. Wu et al. [78] reported that PCC dramatically decreased the response for all EMIT II assays. This study also found that adulteration with PCC resulted in a decrease in the response rate for cannabinoid in KIMS assays and a significant loss of THC-COOH in GC–MS analysis techniques [78]. The effects of commercially available urine adulterant "Urine Luck" was investigated by Paul et al. [45]. A concentration of 2 mM of PCC was able to mask the presence of THC-COOH quite effectively, resulting in the detection of only 0%–42% of the initial concentration remaining in the urine samples. This loss of the targeted analyte was found to increase with decreasing pH and increasing reaction time (0–3 days) [45].

There have been conflicting reports regarding the effects of PCC as an adulterant for the concealment of ATS in urine. PCC has been reported to produce an increased response with the KIMS assay for amphetamine; however, the

TABLE 26.6 Summary of the Effects of Pyridinium Chlorochromate (PCC) as a Urine Adulterant on Results of Both Screening and Confirmatory Analysis Techniques

Drug Type[a]	Technique[b]	Influence of Adulterant	References
Cannabis	EMIT	Major decrease in immunoassay response	[78]
	GC–MS	Significant decrease in recovery of target analyte	[78]
ATS	Monitect PC11	Decreased immunoassay sensitivity	[97]
	KIMS	Increased immunoassay sensitivity	[78]
	GC–MS	No effect on recovery of target analyte	[78]
Opiates	EMIT	Major decrease in immunoassay response	[78]
	KIMS	Decrease in immunoassay response	[31]
	GC–MS	Significant decrease in target analyte recovery	[78]
	LC–MS	Strong decrease in target analyte recovery and stable reaction products detected	[49,50]
Cocaine	EMIT	Moderate decrease in immunoassay sensitivity	[78]
	GC–MS	No effect on recovery of target analyte	[78]
PCP	EMIT	No effect	[78]

[a]ATS, Amphetamine-type stimulants; PCP, phencyclidine.
[b]EMIT, Enzyme-multiplied immunoassay technique; KIMS, kinetic interaction of microparticles in solution.

response of methamphetamine was unaffected with GC–MS confirmation [35,78]. On the other hand, Wong [97] found that, using the Monitect PC11 lateral flow immunoassay, false negatives were produced by PCC for amphetamine.

The interference of PCC with the detection of opiates, due to its oxidative nature, in urine has been well documented. Firstly, PCC was found to lower the responses of morphine through the KIMS assay [31]. At a concentration of 2 mM, similar to the effective concentration for adulterating cannabinoids, the loss of a large amount of morphine was registered. The loss of the targeted analyte, however, was dependent on the pH of the urine specimen [45,78]. In one particular study, the concentrations of both free morphine and free codeine were drastically lowered in specimens that had a lower pH. On the other hand, opiate analytes in urine were seen to be resistant to oxidation with PCC in urine that had a relatively high pH reading (pH 5–7) [33]. In a study conducted by Luong et al. [86], the responses to a CEDIA test for opiates were found to be reduced by 9%–15% when adulterated with 20 mM PCC and by almost 50% at a concentration for 100 mM. This finding was corroborated by a previous study by Wu et al. [78], where PCC resulted in a significant drop in responses for opiates with the EMIT II and KIMS assays.

In addition, Luong et al. [87] reported that PCC at concentrations of 20 or 100 mM caused a significant decrease in the response of morphine and codeine through GC–MS analysis. The response for the internal standards, deuterated forms of the analytes tested, was also found to decrease, with morphine-d_6 found to be more susceptible to degradation or transformation by PCC. Morphine-d_6 was found to be completely destroyed with the adulterant present at a concentration of 20 mM, while codeine-d_6, on the other hand, only saw complete destruction ($>98\%$) at a higher concentration of 100 mM PCC [87]. Wu et al. [78] concurred with the results and found that the loss of morphine was more prominent than the loss of codeine. Finally, the addition of PCC to a urine sample was found to effectively produce false negatives for cocaine using the Monitect PC11 lateral flow immunoassay [97].

Peroxides

The effects of the peroxide/peroxidase system on urine drug testing are summarized in Table 26.7. Oxidizing adulterants, which contained either peroxide or peroxidase, were found to be quite ineffective in furnishing false negative results for the detection of amphetamines when using the Monitect PC11 lateral flow immunoassay [97] as well as the CEDIA [58], however it was highly effective for the obscuration of cocaine [97] and opiates [58].

There have been numerous attempts in the literature to demonstrate the effects of the commercially available adulterant Stealth, which contains peroxide, on masking the detection of opiates in urine. Cody and Valtier [58] documented false negative results on both the KIMS and CEDIA when Stealth was used to adulterate urine specimens containing morphine at 2500 ng/mL. Conversely, urine specimens spiked with particularly high levels (6000 ng/mL) of M3G and codeine tested positive to both immunoassays used. The results from this study show that the success of Stealth as an adulterant is to some extent inversely related to the concentration of the opiate analyte in a sample. The use of peroxide-based adulterants also managed to result in the nondetection of morphine and codeine, as well as their deuterated counterparts when used as internal standards, through GC–MS confirmatory analysis. Given that the internal

TABLE 26.7 Summary of the Effects of Peroxide/Peroxidase as a Urine Adulterant on Results of Both Screening and Confirmatory Analysis Techniques

Drug Type[a]	Technique[b]	Influence of Adulterant	References
Cannabis	CEDIA, KIMS	False negative returned	[58]
ATS	CEDIA	Not affected by adulterant	[58]
	Monitect PC11	Not affected by adulterant	[97]
Opiates	CEDIA, KIMS	Substantial negative effect	[58,98]
	GC–MS	Complete degradation of target analyte	[98]
Cocaine	CEDIA, KIMS	Not affected by adulterant	[58]
PCP	CEDIA, KIMS	Not affected by adulterant	[58]

[a]ATS, Amphetamine-type stimulants; PCP, phencyclidine.
[b]CEDIA, Cloned enzyme donor immunoassay; KIMS, kinetic interaction of microparticles in solution.

standards were also lost, the theory that the mechanism of action for this type of adulterant might relate to its ability to alter the opiate molecule seems to be supported [58,98]. Cody and Valtier [58] also investigated the possibility of using sodium hydrosulfite or sulfamic acid added to urine specimens that had been adulterated to Stealth, which showed to have a positive result on the recovery of the opiate analytes detected with GC–MS analysis. It has been suggested that these chemicals can assist to remove any excess oxidizing agent and, therefore, prevent the further oxidation of the remaining opiate analytes [32,46].

Other Oxidants

There are a number of other chemicals that possess oxidizing properties have been used to adulterate urine specimens and have been found to interfere with drug testing results. These oxidizing chemicals may include iodine (in products like Betadine), ammonium cerium nitrate, many potassium-containing chemicals, such as potassium chlorate, potassium dichromate, potassium perchlorate, potassium permanganate, and sodium-containing chemicals, such as sodium iodate and sodium metaperiodate [18]. The effects of these chemicals on the results of drug tests are listed in Table 26.8.

A study by Charlton and Fu [75] investigated the effects of six different oxidizing adulterants on the detection of THC-COOH in urine by LC–MS analysis. Potassium permanganate and potassium perchlorate were found to result in a moderate decrease in THC-COOH recovery, while sodium iodate and sodium metaperiodate caused the significant impact on the analyte concentration, while ammonium cerium nitrite and Betadine showed the greatest impact.

Chou and Giang [90] revealed that adulteration with potassium dichromate was effective for masking the presence of ATS in urine. The study showed that a concentration of 10% (w/w) of potassium dichromate resulted in interference for the screening of ATS by FPIA and reduced the sensitivity of that assay by 11%–27%. Additionally, results for

TABLE 26.8 Summary of the Effects of Other Oxidizing Chemicals as Urine Adulterants on Results of Both Screening and Confirmatory Analysis Techniques

Adulterant	Drug type[a]	Technique[b]	Influence of Adulterant	References
Betadine	Cannabis	LC–MS	Major decrease in target analyte recovery	[75]
Ammonium cerium nitrate	Cannabis	LC–MS	Major decrease in target analyte recovery	[75]
Potassium chlorate	Cannabis	GC–MS	Minor decrease in recovery of target analyte	[46]
Potassium chromate	Cannabis	GC–MS	Complete degradation of target analyte	[46]
Potassium dichromate	ATS	FPIA	Decreased immunoassay sensitivity	[90]
		GC–MS	Major decrease in target analyte recovery	[91]
	Opiates	FPIA	Slight decrease in immunoassay sensitivity	[92]
Potassium perchlorate	Cannabis	LC–MS	Moderate decrease in recovery of target analyte	[75]
Potassium permanganate	Cannabis	GC–MS	Complete degradation of target analyte	[46]
		LC–MS	Moderate decrease in recovery of target analyte	[75]
Sodium iodate	Cannabis	LC–MS	Significant decrease in recovery of target analyte	[75]
Sodium metaperiodate	Cannabis	LC–MS	Significant decrease in recovery of target analyte	[75]

[a] ATS, Amphetamine-type stimulants.
[b] FPIA, fluorescence polarization immunoassay.

confirmatory testing using GC–MS when urine specimen was adulterated with potassium dichromate showed a significant reduction in the initial concentration of ATS [91].

Nonoxidizing Adulterants

Apart from the various oxidizing agents detailed previously, various nonoxidizing common household chemicals such as table salt (sodium chloride), Drano (a toilet cleaning product containing sodium hydroxide), hand soap, Visine eye drops, vinegar, as well as commercially available adulterants including "UrinAid" (glutaraldehyde) [69,84]. These adulterants have been used for a long time and the effects of these chemicals are summarized in Tables 26.9 and 26.10. Kim and Cerceo [100] published the first paper detailing the effects of table salt on drug testing with the EMIT assay in

TABLE 26.9 Summary of the Effects of Some Common Nonoxidizing Household Chemicals as Urine Adulterants on Results of Both Screening and Confirmatory Analysis Techniques

Adulterant	Drug Type[a]	Technique[b]	Influence of Adulterant	References
Drano (sodium hydroxide)	Cannabis	EMIT	Decreased immunoassay sensitivity	[84]
		FPIA	Increased immunoassay sensitivity	[89]
		CEDIA	Major decrease in immunoassay sensitivity	[69]
	ATS	EMIT	Decreased immunoassay sensitivity	[84]
		CEDIA	Decreased immunoassay response	[69]
	Opiates	EMIT	Decreased immunoassay sensitivity	[84]
		CEDIA	Decreased immunoassay response	[69]
	Cocaine	EMIT	Decreased immunoassay sensitivity	[84]
		CEDIA	Decreased immunoassay response	[69]
	PCP	CEDIA	Decreased immunoassay response	[69]
		RIA	Increased immunoassay sensitivity	[89]
		FPIA	Decreased immunoassay response	[89]
	Benzos	CEDIA	Decreased immunoassay response	[69]
		EMIT	Decreased immunoassay response	[84]
Visine eye drops	Cannabis	EMIT	Decreased immunoassay sensitivity	[84]
		CEDIA	Major decrease in immunoassay response	[69]
	ATS	CEDIA	No effect	[69]
	Opiates	CEDIA	No effect	[69]
	Cocaine	CEDIA	No effect	[69]
	PCP	CEDIA	No effect	[69]
	Benzos	CEDIA	Decreased immunoassay sensitivity	[69]
		EMIT	Decreased immunoassay response	[84]
Papain/meat tenderizer	Cannabis	EMIT	Decreased immunoassay sensitivity	[72]
		FPIA	Decreased immunoassay sensitivity	[72,99]
		KIMS	Increased immunoassay sensitivity	[72]
		GC–MS	Decreased analyte concentration	[72,99]

(Continued)

TABLE 26.9 (Continued)

Adulterant	Drug Type[a]	Technique[b]	Influence of Adulterant	References
Table salt	Cannabis	CEDIA	No effect	[69]
		EMIT	Decreased immunoassay sensitivity	[84]
	ATS	CEDIA	No effect	[69]
		EMIT	Decreased immunoassay sensitivity	[84]
	Opiates	CEDIA	No effect	[69]
		EMIT	Decreased immunoassay sensitivity	[84]
	Cocaine	CEDIA	No effect	[69]
		EMIT	Decreased immunoassay sensitivity	[84,100]
	PCP	CEDIA	No effect	[69]
	Benzos	CEDIA	No effect	[69]
Vinegar	Cannabis	CEDIA	Minor decrease in immunoassay sensitivity	[84]
		EMIT	Decreased immunoassay sensitivity	[69]
	ATS	CEDIA	No effect	[69]
	Opiates	CEDIA	No effect	[69]
	Cocaine	CEDIA	Minor decrease in immunoassay sensitivity	[69]
	PCP	CEDIA	No effect	[69]
	Benzos	CEDIA	No effect	[69]
		EMIT	Decreased immunoassay response	[84]
Zinc	Cannabis	ELISA	Effective in interfering with test results	[101]
		EMIT	Potential production of false negative results	[73]
	ATS	ELISA	Effective in interfering with test results	[101]
		EMIT	Potential production of false negative results	[73]
	Opiates	EMIT	Potential production of false negative results	[73]
	Cocaine	ELISA	Effective in interfering with test results	[101]
		EMIT	Potential production of false negative results	[73]
Detergent/soap	Cannabis	CEDIA	Major decrease in immunoassay response	[69]
		EMIT	Decreased immunoassay sensitivity	[84]
		RIA	Decreased immunoassay sensitivity	[89]
	ATS	CEDIA	Major decrease in immunoassay response	[69]
	Opiates	CEDIA	Major decrease in immunoassay response	[69]
	Cocaine	CEDIA	Major decrease in immunoassay response	[69]
	PCP	CEDIA	Strong interference with immunoassay	[69]
	Benzos	CEDIA	No effect	[69]

(Continued)

TABLE 26.9 (Continued)

Adulterant	Drug Type[a]	Technique[b]	Influence of Adulterant	References
Golden-seal tea	Cannabis	CEDIA	Major decrease in immunoassay response	[69]
		EMIT	Decreased immunoassay sensitivity	[84]
	ATS	CEDIA	Major decrease in immunoassay response	[69]
	Opiates	CEDIA	No effect	[69]
	Cocaine	CEDIA	No effect	[69]
	PCP	CEDIA	No effect	[69]
	Benzos	CEDIA	No effect	[69]

[a]ATS, Amphetamine-type stimulants; PCP, phencyclidine; Benzos, benzodiazepines.
[b]CEDIA, cloned enzyme donor immunoassay; ELISA, enzyme-linked immunosorbent assay; EMIT, enzyme-multiplied immunoassay technique; FPIA, fluorescence polarization immunoassay; KIMS, kinetic interaction of microparticles in solution; RIA, radioimmunoassay.

TABLE 26.10 Summary of the Effects of Glutaraldehyde as a Urine Adulterant on Results of Both Screening and Confirmatory Analysis Techniques

Drug Type[a]	Technique[b]	Influence of Adulterant	References
Cannabis	EMIT	Decreased immunoassay sensitivity	[71,102]
	FPIA	Interference with the assay	[71]
	KIMS	Increased immunoassay sensitivity	[71]
	CEDIA	Strong interference with test results	[69]
ATS	EMIT	Decreased immunoassay sensitivity	[71,102]
	FPIA	Interference with the assay	[71]
	KIMS	Highly variable results	[71]
	CEDIA	No effect	[69]
Opiates	EMIT	Decreased immunoassay sensitivity	[71,102]
	FPIA	Interference with the assay	[71]
	CEDIA	No effect	[69]
Cocaine	EMIT	Decreased immunoassay sensitivity	[71,102]
	FPIA	Interference with the assay	[71]
	CEDIA	Strong interference with test results	[69]
PCP	CEDIA	Strong interference with test results	[69]
	KIMS	Increased immunoassay sensitivity	[71]
Benzos	CEDIA	No effect	[69]
	EMIT	Decreased immunoassay sensitivity	[102]

[a]ATS, Amphetamine-type stimulants; PCP, phencyclidine; Benzos, benzodiazepines.
[b]CEDIA, Cloned enzyme donor immunoassay; EMIT, enzyme-multiplied immunoassay technique; FPIA, fluorescence polarization immunoassay; KIMS, kinetic interaction of microparticles in solution.

1976. Since that time, many different research papers have been published which investigate the adulteration of urine specimens with these substances.

Mikkelsen and Ash [84] tested the effects of several common household chemicals and discovered that the sensitivity of the EMIT immunoassay was decreased for the detection of cannabis by hand soap, Drano, golden-seal tea, table salt, and Visine. In the case of the testing for ATS, cocaine, and opiates, the decrease in sensitivity was caused by Drano and table salt. Decreased immunoassay response for Benzos was observed for Drano, Visine, and vinegar. PCP assay was affected by Drano only.

Wu et al. [69] reported that the presence of glutaraldehyde, detergent, or high concentrations of Drano (~20%) was able to produce strong interference in the CEDIA for most targeted DOA (cannabis, ATS, opiates, cocaine, and PCP). On the other hand, minimal interference, or interference that was selective for particular analytes, was observed when samples were adulterated with golden-seal tea, Visine or Drano at lower concentrations (0.1%). When table salts or vinegar were used as adulterants, no significant interference was observed [69].

The effects of the commercially available adulterant "UrineAid" (containing the active ingredient glutaraldehyde) on various immunoassay screening tests, including EMIT, FPIA, and KIMS, was explored by Goldberger and Caplan [71]. The analysis of cannabis, ATS, opiates, and cocaine by FPIA and KIMS showed interference in the presence of this adulterant. It was also noted that the sensitivity of the EMIT assay for all drugs tested was adversely affected. This finding was corroborated by George and Braithwaite [102], who documented the ability of glutaraldehyde to furnish false negatives with the EMIT screen at concentrations of 0.75% and 2.00% (v/v) in urine samples.

Papain, a cysteine protease that is commonly used as the main ingredient in meat tenderizers, has also been suggested to be an effective adulterant to obscure the presence of THC-COOH in urine [72,99]. The use of this adulterant has been found to decrease the sensitivity of both the EMIT and FPIA screening assay for the detection of THC-COOH, as well as lowering the detectable concentration by GC−MS confirmatory analysis.

A comparatively recent development in urine adulteration has seen the use of zinc as an adulterant to reduce the detection of cannabis, methamphetamine, opiates, and cocaine by immunoassays such as ELISA and EMIT [73,101].

HOW DO LABORATORIES COUNTERACT URINE ADULTERATION?

Given the huge potential for different method of adulteration to produce false negative results, countermeasures are needed to be able to detect urine manipulation. To address this, toxicology laboratories have developed a series of countermeasures to flag the potential adulteration of collected urine specimens. These countermeasures include urine integrity tests, color tests, dipstick devices, various spectrophotometric methods, immunoassay tests, capillary electrophoresis, mass spectrometric techniques (namely electrospray ionization MS), and polyethylene glycol urine marker systems.

Urine Integrity Tests

Urine integrity tests are a vital step in the screening process of urine samples. Many of the other adulterant detection techniques can be time consuming and costly, meaning it is not realistically viable for laboratories to submit all samples to these tests. Therefore, urine integrity tests can act as an early warning system for urine adulteration. Urine integrity tests generally monitor a series of basic parameters such as temperature, pH, specific gravity, and creatinine levels. These parameters form part of endogenous urine characteristics [18]. The normal physiological ranges for each of these parameters in a freshly collected human urine specimen are presented in Table 26.11. If the measured parameters differ significantly from the values presented in the table, it can indicate that the sample has been manipulated. For example, a very low creatinine level may indicate excessive hydration or that the sample has been subjected to in vitro adulteration [108]. As a result, testing for creatinine levels in a sample is common practice for urine drug testing laboratories. Sittiwong and Unob [109] developed a paper-based colorimetric technique for the detection of creatinine, in which the concentration of creatinine can be determined through the intensity of the color change. This testing method was able to detect creatinine at levels as low as 4.6 mg/L and demonstrated a high degree of accuracy. Wyness et al. [110] investigated the use of a handheld refractometer for the measurement of urine specific gravity. It was found that this technique was a fast, simple, and accurate way to measure the specific gravity and was able to demonstrate analytical validity [110].

One of the main advantages of urine integrity tests is that they can be performed on-site where the specimen collection is being carried out. This makes the personnel who are overseeing the specimen collection the first line of defense against urine adulteration [18]. The initial visual inspection of samples, as well as the completion of on-site integrity

TABLE 26.11 Human Urine Integrity Test Parameters

Parameter	Expected Range	References
Creatinine concentration	800–200 mg/dL (7.0–17.8 mM)	[103,104]
pH	4.7–7.8	[105,106]
Specific Gravity	1.003–1.035 g/mL	[103,105]
Temperature	32.5–37.7°C	[107]

tests, is the responsibility of these supervisors. These days the specimen cups themselves often incorporate testing panels capable of measuring many of these basic parameters; however, the efficacy of these countermeasures is reliant on the type of specimen cup used and the experience of the testing personnel [18]. In the case of some common adulterants, such as bleach, the strong odor can be an obvious sign of adulteration [32]. These integrity tests, while very useful, are only really suitable for the identification of obviously manipulated and, therefore, invalid specimens.

At low concentrations, the use of Stealth can result in a urine sample having a slightly darker yellow color, while PCC adulteration can result in a more intense yellow. The use of routine integrity tests for the detection of oxidizing adulterants, however, can run into problems due to the natural variations in urine color. The color of a urine sample can vary based on a number of legitimate factors, such as diet, medication, physiological, and pathological conditions [32,48,58]. Many oxidizing agents commonly used are also odorless, even at relatively high concentrations [18]. In cases where urine adulteration is suspected, laboratories can conduct additional tests to determine the presence of adulterants. The criteria for the detection of adulterants are listed in Table 26.12.

Color Tests

Spot tests can be advantageous because they provide very quick results and are easy to perform. On the other hand, a major drawback to these kinds of tests is their lack of specificity. While PCC, nitrite, and Stealth can all be detected with various spot tests, false positive results are quite common [18].

The Cr^{6+} ion present within PCC can be detected by taking 1 mL of a urine sample and adding two drops of a 1% (w/v) 1,5-diphenylcarbazide solution in methanol. A positive result will be indicated by a color change to a reddish purple color. Interference have been observed, however, when other ions, such as mercury, vanadium, and molybdenum, are present within a sample as they provide a similar color change [78]. Adulteration of samples with PCC can also be identified through the reaction with an acidified potassium iodide reagent or a hydrogen peroxide reagent, as both of these reagents provide rapid color changes in a positive sample [42]. Additionally, a dark brown precipitate will form after reaction with hydrogen peroxide [33,35].

The presence of nitrite in a sample can be detected by a reaction with acidified potassium permanganate. This reagent is initially pink in color, however samples containing nitrite will immediately discolor and effervescence will be observed. If a sample contains high levels of glucose, a false positive may be returned, however the color change and effervescence will occur at a much slower rate [33,35,113]. Adulteration with nitrite can also be detected with acidified potassium iodide.

The peroxidase enzyme that is present in the commercially available adulterant Stealth is usually the target for color tests to detect this compound. An immediate color change will be observed for samples positive for this substance when subjected to a solution of tetramethylbenzidine with 100 mM phosphate buffer. Additionally, an acidified potassium dichromate reagent will furnish a color change to a deep blue color, which will fade over time, in samples containing Stealth [33,35,113].

Dipstick Devices

There are many different types of dipstick devices commercially available, and their main advantage is the fact that they are very portable and allow for on-site testing for adulterants. Most dipstick devices consist of a plastic strip with a chemically treated pad attached. Each different pad will act as an assay for different specimen variables. One such device, the MASK Ultrascreen (Kacey Inc.) device, contains a large array of different assays, which can test for a

TABLE 26.12 Criteria for Identification of Adulterated and Substituted Urine [18,22,107,111,112]

Criteria	Interpretation
1. Urine pH	
pH < 3 or ≥ 11 (outside of endogenous range)	Adulterated
pH ≥ 3 and < 4 or pH ≥ 10 and < 11	Invalid result
2. Creatinine Level and Specific Gravity	
Creatinine < 5 mg/dL (0.44 mM) Specific gravity < 1.002 g/mL	Substituted
Creatinine < 5 mg/dL (0.44 mM) Specific gravity ≥ 1.002 g/mL	Substituted
Creatinine ≥ 5 mg/dL (0.44 mM) and < 20 mg/dL (1.77 mM) Specific gravity < 1.003 g/mL	Diluted
Creatinine ≥ 5 mg/dL (0.44 mM) Specific gravity = 1.000 g/mL	Invalid result
Creatinine ≥ 5 mg/dL (0.44 mM) and < 20 mg/dL (1.77 mM) Specific gravity ≥ 1.020 g/mL	Invalid result
3. Presence of Chemicals	
Nitrite ≥ 11 mM	Adulterated
Nitrite ≥ 4 mM and < 11 mM	Invalid result
Chromate (IV) > the laboratory's limit of detection	Adulterated
Halogen > the laboratory's limit of detection	Adulterated
Glutaraldehyde > the laboratory's limit of detection	Adulterated
4. Other Indication	
Decreased GC–MS internal standard (≥70%)	Invalid result

Source: Adapted from Fu S. Adulterants in urine drug testing. In: Makowski GS, editor. Advances in clinical chemistry, vol. 76. Burlington: Academic Press; 2016. p. 123–63; Cook JD, Caplan YH, Lodico C, Bush DM. The characterization of human urine for specimen validity determination in workplace drug testing: a review. J Anal Toxicol 2000;24(7):579–88; National Institute on Drug Abuse, Mandatory Guidelines for Federal Workplace Drug Testing Programs, National Institute on Drug Abuse, editor; 1988. p. 11970–89; Edgell K, Caplan YH, Glass LR, Cook JD. The defined HHS/DOT substituted urine criteria validated through a controlled hydration study. J Anal Toxicol 2002;26(7):419–23; Dasgupta A, A health educator's guide to understanding drugs of abuse testing. Sudbury, MA: Jones & Bartlett Learning; 2011.

number of different criteria, including creatinine, pH, specific gravity, nitrite, glutaraldehyde, PCC, and Stealth. Unfortunately, this device can only detect these adulterants at concentrations that are much higher than the recommended levels for urine adulteration, limiting the usefulness of this device [77]. Both Jaffee et al. [31] and Dasgupta [33] have extensively reviewed the efficacy of a number of different dipstick testing devices for urine adulteration.

A selection of urinalysis test strips, namely MultiStix (Bayer), and Combur-Test (Roche Diagnostics), are widely available for the monitoring of nitrite, pH and specific gravity of urine specimens [18]. These devices also have the ability to indirectly monitor the presence of Stealth in a sample. Samples adulterated with Stealth will display strong positive reading for glucose and blood; however, the test is actually detecting the peroxidase activity of the oxidant that is present within Stealth [18]. These particular devices do not have the capacity to differentiate between diluted and substituted urine specimens, as they are not sensitive enough to detect the differences in specific gravity at relevant cut-off levels. Additionally, these devices are not specifically designed to detect the presence of adulterants, meaning that nitrite can be detected at clinically significant ranges (<3 mM), which are significantly lower than the concentrations that are generally observed in cases of adulteration with substances such as Klear, which are normally in the range of 40–300 mM [48]. This means that variations in diet, pathology or medication used by the subject can result in positive detection for nitrite with these devices [48].

There have been a few devices, including the Adultacheck 4 and 6 (Sciteck Diagnostics), which have been designed specifically for forensic toxicological purposes. These dipstick devices can be used to measure the creatinine and pH levels of a sample over a large range, meaning that unusually high or low levels can be detected. The downside to this is that it is difficult to determine precise values for these parameters [77,79,114]. It has been reported that the Intect 7 (Branan Medical Corporation) is the most sensitive and economical device currently available on the market, and it can be used to test the creatinine and pH levels over a wide range. Similar to the Adultacheck devices, however, precise values cannot be determined with this device [79]. The Adultacheck devices can detect the presence of PCC, glutaraldehyde and nitrite, however, it is important to note that these devices detect nitrite levels above the clinically relevant cutoffs. The Intect 7 also tests for these adulterants and includes additional test strips for specific gravity and exogenous hypochlorite. The main advantage of this device is its ability to detect hypochlorite-based bleach at levels of only 10 µL/mL in urine [77,79]. This is particularly significant as only small amounts of bleach are used for urine adulteration specifically to avoid suspicion in urine integrity tests.

Spectrophotometric Methods

Spectrophotometric analysis is a generally accepted method for the analysis of the peroxidase enzyme activity and, therefore, can be used to detect the presence of Stealth in urine samples [41]. As identified previously, the presence of PCC can be identified by a color reaction with 1,5-diphenylcarbazide, and this can be monitored with spectrophotometry [45]. Six other spectrophotometric techniques have been developed in order to detect the presence of other oxidizing chemicals in urine which include nitrite, ferric, oxychloride, permanganate, and hydrogen peroxide [43]. Although these methods for adulterant detection have been developed, they are not generally used as part of routine testing.

Immunoassays

The Microgenics DRI General Oxidant-Detect Test is an immunoassay-based technique developed specifically for the detection of oxidizing urine adulterants [115]. This assay relies on the reaction between tetramethylbenzidine and any oxidant present within a sample and can be performed on an automated clinical chemistry analyzer. The product of this reaction is a colored complex, which can be viewed under 660 nm light.

Capillary Electrophoresis and Mass Spectrometric Techniques

Capillary electrophoresis techniques can be used to detect the Cr^{6+} ions present in PCC and the NO_2^- ions present within nitrite-based oxidants [40,82,116]. The chromium species found in PCC can also be detected by a variety of different mass spectrometric-based techniques such as LC–MS, GC–MS and inductively coupled plasma-MS [33,117]. The detection of active Cr^{6+} and NO_2^- ions present in commercial adulterants has been demonstrated using high-performance LC coupled to MS or conductivity detectors [33].

More recently, Steuer et al. [118] presented a proof-of-concept study for the use of a metabolomics-based untargeted MS technique for the detection of adulteration with potassium nitrite. Principal component analysis was used to monitor a large number of endogenous features in urine samples and found significant concentration changes between blank and adulterated urine samples. This type of untargeted MS analysis allows for the identification of a broad spectrum of markers for urine adulteration. If more potential markers are identified for different adulterants, these can then be incorporated into routine MS screening procedures [118].

Mass spectrometric techniques can also be used to detect the presence of substituted urine. A study by Goggin et al. [29] identified three different markers, namely benzisothiazolinone and two ethylene glycols (triethylene glycol and tetraethylene glycol), which were present within synthetic urine that are not present within authentic samples. Conversely, they also identified four new endogenous markers, namely uric acid, 3-methylhistidine, normetanephrine, and urobilin, present within authentic samples, which can be used as markers for an authentic sample [29].

Polyethylene Glycol Urine Marker System

Schneider et al. [119] and Jones et al. [120] explored the efficacy of using a polyethylene glycol marker system for drug screening in an opiate substitution outpatient clinic. The use of this marker system involves the patient ingesting biologically inert, low molecular weight polyethylene glycols that are then excreted in the urine. This process limits the possibility of urine substitution, as the specimens are readily identifiable by the detection of the specific marker substance administered to the patient. The authors suggested that the use of a marker system such as this would allow for greater capacity to detect urine adulteration [119]. This method is particularly effective for the detection of substituted urine, as there will be a lack of the expected polyethylene glycols. This method is not as effective at detecting the possibility of in vitro adulteration, however, as the exogenous chemicals can be added after the sample has already been

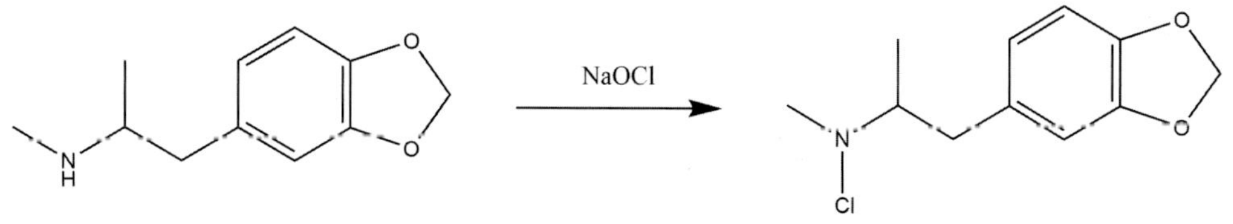

FIGURE 26.1 Oxidation product of MDMA when exposed to sodium hypochlorite (NaOCl).

voided, meaning that the marker compounds may still be present. Therefore, the presence of the marker in a urine specimen does not rule out the possibility that the sample has been chemically manipulated.

CAN TARGETED ANALYTES THAT HAVE BEEN MODIFIED WITH OXIDIZING AGENTS BE DETECTED?

Currently, the procedures that are used by drug testing laboratories are able to determine whether a sample has been adulterated by oxidizing chemicals, provided other sources such as medication and food intake can be accounted for. These current methods do not have the capacity to identify the specific drug that has been taken, due to the modification that has taken place because of the oxidant. In most scenarios, the mere presence of an adulterant in the sample is enough evidence for a sanction or caution to be issued, however, ideally the specific drug species present within a sample should be identified. It has been suggested that, in instances where oxidizing adulterants have been detected, adulterant testing should focus on the oxidation products that would form from reactions with common adulterants [18,19]. It is assumed that the original drug compounds will have been oxidized and, therefore, no longer detectable, so if stable oxidation products could be identified, the detection of these substances would not only indicate that adulteration with oxidizing chemicals has been attempted, but also which drug species were present within the sample originally. Some oxidation products for ATS, cannabinoids, and opiates have already been identified and will be discussed in the following sections.

Amphetamine-Type Substances

Pham et al. [51] explored the reaction products formed from the interaction of MDMA with sodium hypochlorite. The major product of this reaction was proposed to be N-chloroMDMA (Fig. 26.1, structure 1) based on the MRM transitions that were observed from LC−MS analysis (Table 26.13).

When stored at 4°C, the oxidation product N-chloroMDMA was stable for roughly 10 h, however at room temperature (20°C) it decomposes rather quickly, reverting to MDMA. This study has highlighted the possibility of this product, N-chloroMDMA, as a marker for urine adulteration with hypochlorite [51].

Cannabinoids

Lewis et al. [96] made the first attempt at identifying oxidation products in adulterated urine samples. This study investigated the effects of potassium nitrite on THC-COOH. The reaction between the adulterant and drug metabolite produced a nitroso-derivative of THC-COOH. This reaction product was deemed unsuitable as a marker for cannabis use, however, as it was found to be unstable [96]. A more recent study using LC−MS analysis found the same unstable nitroso-THC-COOH (Fig. 26.2, structure 2) was identified. Additionally, an extra, stable nitro-THC-COOH derivate was also identified (Fig. 26.2, structure 3) [122]. It was unable to be determined whether the substitution occurs at the *ortho-* or *para-* positions relative to the phenolic group located on the aromatic ring. Further studies will be required to determine the suitability of this oxidation product as a marker for nitrite adulteration of urine samples containing cannabis.

When urine positive for cannabis is adulterated with hypochlorite, three different chlorinated products were found by LC−MS analysis, due to the chlorination of the aromatic ring of the THC-COOH structure (Fig. 26.3): two different isomers of mono-chloro-THC-COOH and di-chloro-THC-COOH [123]. These results were later supported by a study from González-Mariño et al. [124] who identified the same three analytes when THC-COOH was reacted with

TABLE 26.13 MRM Parameters During LC–MS Analysis for the Detection of Oxidation Products of Some Drugs and/or Drug Metabolites in Urine Following Adulteration of Samples by Oxidizing Chemicals[a]

Oxidation Products	Precursor Ion [M + H]$^+$ (m/z)	Product Ion (m/z)	Collision Energy (eV)	References
N-chloroMDMA (1)	228	163	6	[51]
		135	20	
		105	25	
Di-iodo-THC-COOH (9)	597	470	8	[121]
		455	24	
		344	20	
		329	8	
2-Nitro-MAM (10)	373	327	30	[49]
		209	35	
2-Nitromorphine (11)	331	152	70	[50]
		115	80	
2-Nitro-M6G (12)	507	331	35	[50]
		285	45	

[a]LC–MS analysis was conducted either on an Agilent 1290 LC system coupled to an Agilent 6490 triple quadrupole mass spectrometer [50,121] or on an Afilent 1200 LC system coupled to an Agilent 6460 triple quadrupole mass spectrometer [49,51]. MS was operated in positive electrospray ionization mode. Chromatographic separation was achieved on C18 columns. Mobile phases used included isocratic elucidation of 80% acetonitrile in 20 mM formic acid [121], or gradient elution of acetonitrile and ammonium formate at 2 mM [51] or 20 mM [49,50]. Urine samples were passed through a 0.2 μm membrane syringe filter before LC–MS analysis [49–51,121].

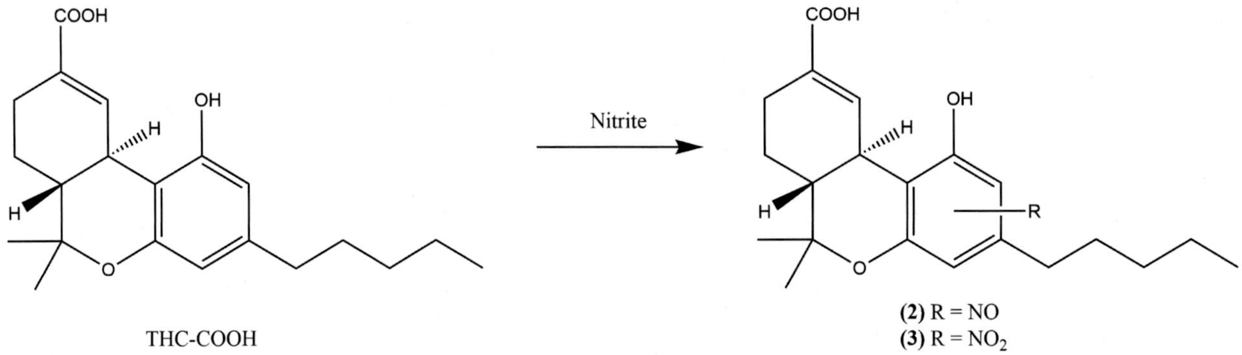

FIGURE 26.2 Oxidation products from the reaction of THC-COOH in urine adulterated with potassium nitrite. (2) Nitroso-THC-COOH and (3) nitro-THC-COOH.

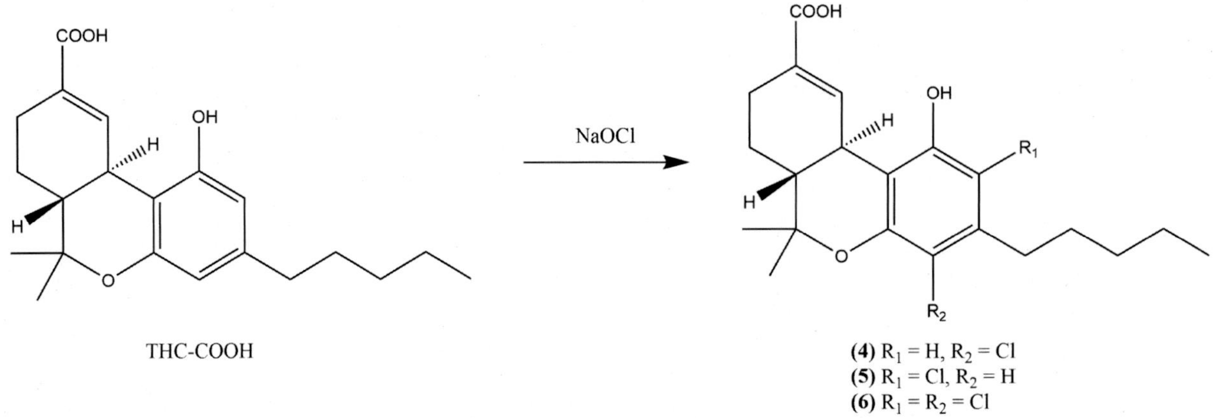

FIGURE 26.3 Oxidation products from the reaction of THC-COOH in urine adulterated with sodium hypochlorite (NaOCl). (4 and 5) Mono-chloro-THC-COOH and (6) di-chloro-THC-COOH.

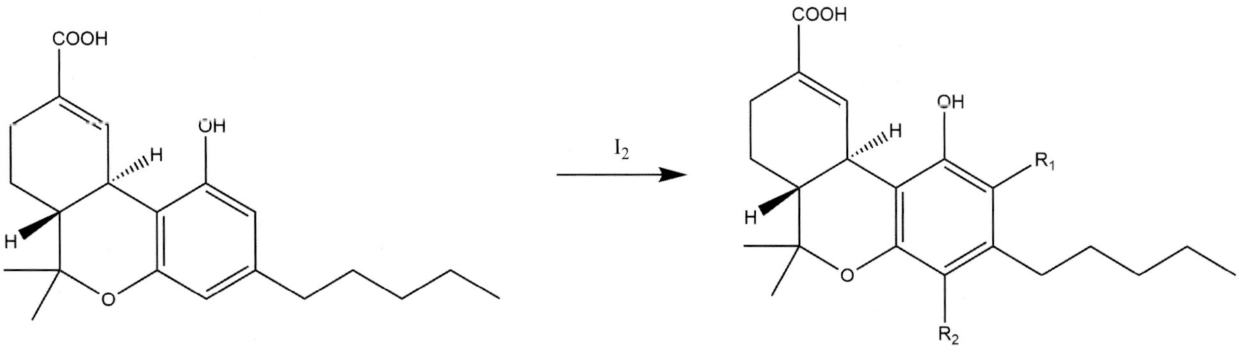

FIGURE 26.4 Oxidation products from the reaction of THC-COOH in urine adulterated with I2 in Betadine. (7 and 8) Mono-iodo-THC-COOH and (9) di-iodo-THC-COOH.

chlorinated water. In addition to the adulteration with hypochlorite-based adulterants, the effects of Betadine on urinary THC-COOH have also been explored [121]. This adulteration follows the same mechanism as for hypochlorite, two isomers of mono-iodo-THC-COOH (Fig. 26.4, structures 7 and 8) and di-iodo-THC-COOH (Fig. 26.4, structure 9) are formed. Although the formation of these products has been documented, the potential of these different halogenated species to be used as markers for the adulteration of THC-COOH needs to be assessed further.

Opiates/Opioids

As detailed previously, nitrite can effectively mask the presence of 6-MAM from CEDIA and LC−MS analysis when in an acidic environment (pH < 7) [49]. Studies into the mechanism of this reaction have revealed a single oxidation product that is formed, namely 2-nitro-6-monoacetlymorphine (2-nitro-MAM, Fig. 26.5, structure 10) [49]. This particular product was found to be relatively stable in the experimental conditions used in the study, and was able to be detected for at least 11 days following adulteration. 2-Nitro-MAM was able to be detected in the urine of a heroin user which had been adulterated with nitrite, which demonstrates the suitability of this product as a marker under these circumstances. Another study conducted by Luong and Fu [50] investigated the effects of potassium nitrite adulteration on morphine, codeine, M3G, and M6G. Through LC−MS analysis, stable oxidation products were identified for both morphine and M6G, 2-nitromorphine (Fig. 26.5, structure 11), and 2-nitromorphine-6-glucuronide (Fig. 26.5, structure 12) respectively (Table 26.13). Codeine and M3G, on the other hand, were found to be unaffected by nitrite adulteration [50]. These findings highlight the necessity for the phenolic OH group at the C_3 position of the chemical structure in order for nitration to occur at C_2. In addition, the trimethylsilyl derivative of 2-nitromorphine was still detectable with GC−MS after enzymatic hydrolysis of the urine specimens tested [50]. Review of the available literature suggests that these nitro-derivatives of the different opiate substances are not endogenous in human urine, as they were not detected in blank pooled urine substances from both male and female donors with varying diets and ethnicities. These factors suggest that the identified oxidation products are promising candidates for nitrite adulteration markers [49,50].

Following adulteration of urine samples with PCC, a number of opiate oxidation products have been identified. Exposing codeine in urine to PCC has revealed the formation of four different, stable oxidation products which have been characterized with both high-resolution MS analysis and, where possible, nuclear magnetic resonance spectrometric analysis [125]. The four identified products are codeinone, 14-hydroxycodeinone, 6-O-methylcodeinone, and 8-hydroxy-7,8-dihydrocodeinone (Fig. 26.6). Similarly, three oxidation products of C6G were identified upon exposure to PCC: Codeinone, codeine, and a lactone of C6G (this is a tentative assignment, however) (Fig. 26.7) [125].

Urine specimens positive for opiates were adulterated with PCC at different concentrations (20 and 100 mM) and analyzed with the CEDIA and GC−MS [87]. A number of urine samples were spiked with 6-MAM, morphine and its glucuronides at a concentration of 10 µg/mL and adulterated with PCC (0.02−100 mM), following which they were analyzed with LC−MS and the products were characterized. It was reported that the concentrations of morphine,

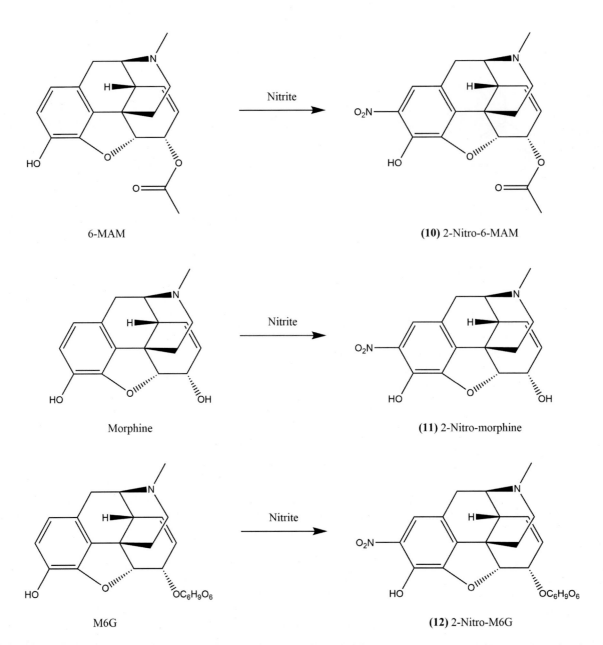

FIGURE 26.5 Oxidation products from the reactions of 6-monoacetlymorphine, morphine, and morphine-6-glucuronide, respectively, in urine adulterated with potassium nitrite.

codeine, and the internal standards were significantly reduced [87]. The samples which contained 6-MAM were found to produce oxidation products of 7,8-diketo-6-MAM and 7,14-dihydroxy-6-MAM (Fig. 26.8). On the other hand, samples containing morphine and M3G showed products of 7,8-diketo-morphine (tentative assignment) and morphinone-3-glucuronide, respectively, upon exposure to PCC (Fig. 26.9) [87].

CONCLUSIONS

Toxicological laboratories around the world continue to face many challenges from urine adulteration, in addition to the difficulties posed by the constant emergence of new designer drugs. Urine integrity tests and oxidant tests can be very helpful in monitoring the adulteration of specimens; however, they are not guaranteed to detect all possible oxidants at the concentrations used to adulterate urine. In most medico-legal settings, subsequent specimens may be acquired where adulteration is suspected, however by the time that this is completed, drugs that may have been present might be

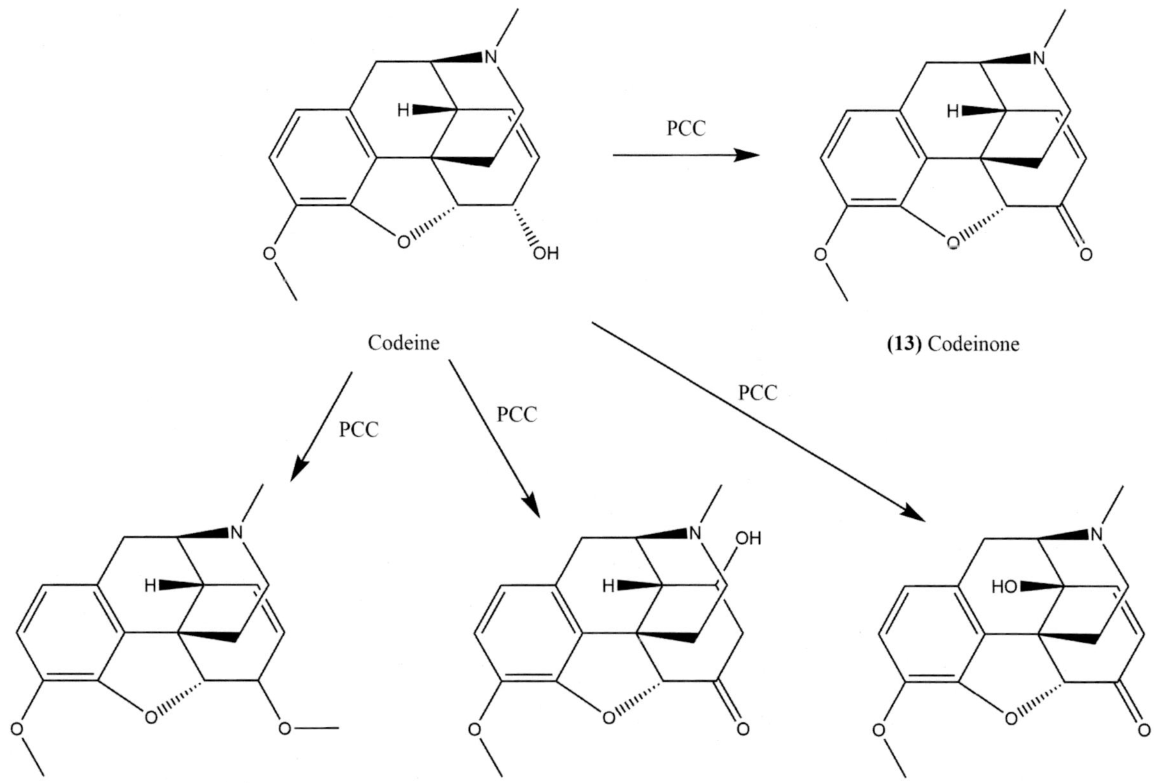

(14) 6-O-Methylcodeine (15) 8-Hydroxy-7,8-dihydrocodeinone (16) 14-Hydroxycodeinone

FIGURE 26.6 Oxidation products from the reactions of codeine in urine adulterated with pyridinium chlorochromate.

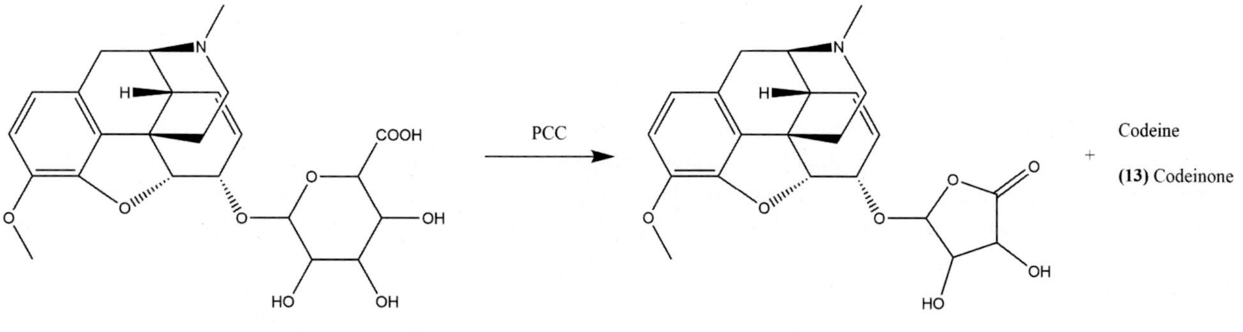

C6G (17) Lactone of C6G

FIGURE 26.7 Oxidation products from the reactions of codeine-6-glucuronide in urine adulterated with pyridinium chlorochromate.

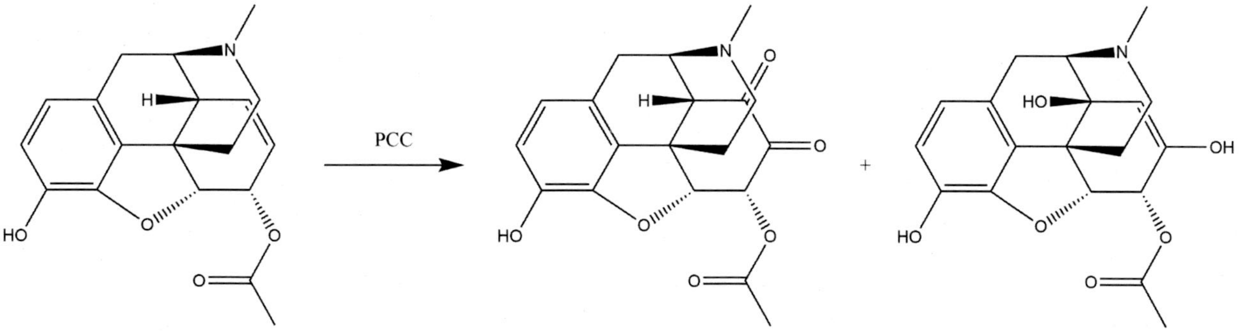

6-MAM (18) 7,8-diketo-6-MAM (19) 7,14-dihydroxy-6-MAM

FIGURE 26.8 Oxidation products from the reactions of 6-monoacetylmorphine in urine adulterated with pyridinium chlorochromate.

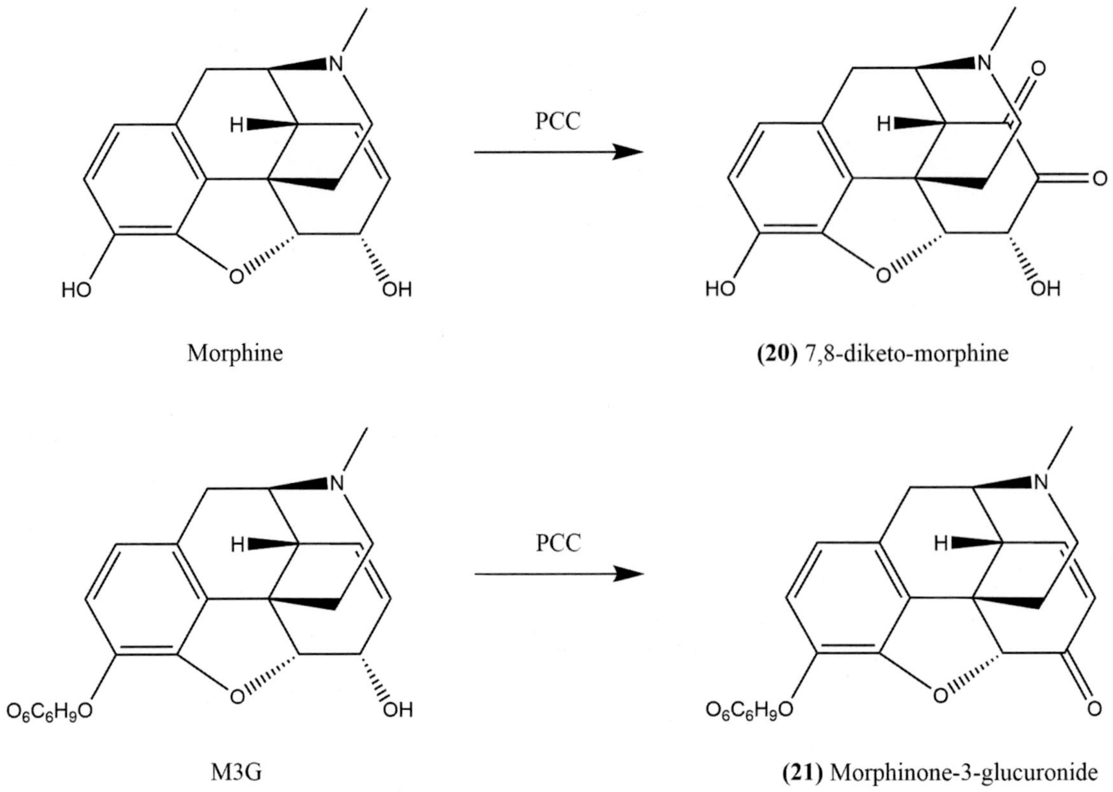

FIGURE 26.9 Oxidation products from the reactions of morphine and morphine-3-glucuronide in urine adulterated with pyridinium chlorochromate.

metabolized and therefore no longer able to be detected. Obtaining follow-up samples is also costly and time consuming for the laboratory. When it comes to clinical toxicology and patient care, being able to identify the actual drug present within a sample is a lot more important than simply identifying that drugs are present or adulteration has occurred.

Immunoassays are the most common drug-screening tool implemented due to the minimal sample preparation required, the ease of automating the process and high sample throughput. On the other hand, they are only available for a limited number of drug classes as they require specific antibodies, which can target the drug analytes of interest, and they are also vulnerable to interference from a number of oxidizing and nonoxidizing chemicals. When the adulteration procedure involves the use of nonoxidizing chemicals such as detergents [126], table salts [100], and zinc [101], the targeted analyte structures remain unchanged, and the use of mass spectrometric techniques, such as LC−MS, will improve sensitivity by reducing the number of false negative screening results. A recent proof-of-concept study using a metabolomics-based untargeted LC−MS approach has shown promise for the detection of adulteration with nitrite based on concentration changes of endogenous urinary metabolites [118]. This metabolomics-based strategy warrants further investigation and may allow for a broad identification range of markers of urinary adulteration to be integrated into routine MS screening procedures.

The use of oxidizing chemicals when adulterating urine samples destroys the drug analytes themselves and can therefore adversely affect the detection of these compounds in both screening and confirmatory tests. An alternative method to traditional adulterant tests for these oxidizing chemicals is the detection of unique oxidation products that form from their reaction with the drug analytes. This is a challenging technique that requires the collection of a lot of background information, however previous studies have shown promise. The main advantage of this type of technique is the ability to identify both the drug taken and the adulterant used, which would make all samples useful, regardless of whether adulteration with oxidizing chemicals was performed or not. Such a technique could be used to augment current screening and confirmatory methods used for the detection of adulterants in many toxicological laboratories.

The main challenge with utilizing oxidation markers is the lack of available reference standards for these compounds. These standards are required to validate and optimize the analytical techniques used for quantification of the relevant markers. The commercial production of these standards will only become available once the structures have been identified. In addition, ongoing research will be required to keep up with the continued emergence of new compounds onto the drug market in order to identify the oxidation products from the reaction with each common oxidizing adulterant. While this would put additional strain on the drug testing industry, the value of these testing methods cannot be overstated and the success of this innovative approach can be invaluable in the era of overzealous pain management and sports doping control.

While GC−MS techniques are the traditional workhorse for urine drug testing, LC−MS techniques may soon overtake them as significant advancements involving MS/MS with accurate mass capabilities have emerged in the last decade [59,127−129]. LC−MS is a more sensitive alternative to GC−MS and offers a number of benefits including reduced sample preparation as derivatization is not required and there are no high temperatures used which can help avoid the degradation of analytes. LC−MS also has the advantage of being able to screen for new emerging compounds, without being limited by the availability of immunoassay reagents. The MRM transitions for many of the oxidization products have already been documented (Table 26.13), meaning that they can be incorporated into existing LC−MS procedures even without the availability of reference standards.

REFERENCES

[1] Bush DM. The US Mandatory Guidelines for Federal Workplace Drug Testing Programs: current status and future considerations. Forensic Sci Int 2008;174(2−3):111−19.
[2] Nolan S. Drug-free workplace programmes: New Zealand perspective. Forensic Sci Int 2008;174(2−3):125−32.
[3] Santoro PE, De Nardis I, Fronterrè P, Felli M, et al. A snapshot of workplace drug testing in Italy. Drug Test Anal 2012;4(2):66−70.
[4] Notarianni LJ, Belk D, Collins AJ. Compulsory random drug testing of prisoners in England and Wales: design flaws in the system. J Clin Forensic Med 1995;2(4):185−8.
[5] Phan HM, Yoshizuka K, Murry DJ, Perry PJ. Drug testing in the workplace. Pharmacotherapy 2012;32(7):649−56.
[6] Shearer DS, Baciewicz CJ, Kwong TC. Drugs of abuse testing in a psychiatric outpatient service. Clin Lab Med 1998;18(4):713−26.
[7] Reisfield GM, Goldberger BA, Bertholf RL. 'False-positive' and 'false-negative' test results in clinical urine drug testing. Bioanalysis 2009;1(5):937−52.
[8] West R, Pesce AJ, Crews B, Mikel C, Rosenthal M, et al. Determination of illicit drug cutoff values in a pain patient population. Clin Chim Acta 2011;412(17−18):1589−93.
[9] Alford DP. Chronic back pain with possible prescription opioid misuse. JAMA 2013;309(9):919−25.
[10] Thevis M, Kohler M, Schanzer W. New drugs and methods of doping and manipulation. Drug Discov Today 2008;13(1−2):59−66.
[11] Bowers LD. The international antidoping system and why it works. Clin Chem 2009;55(8):1456−61.
[12] Kraemer T, Paul LD. Bioanalytical procedures for determination of drugs of abuse in blood. Anal Bioanal Chem 2007;388(7):1415−35.
[13] Gallardo E, Queiroz JA. The role of alternative specimens in toxicological analysis. Biomed Chromatogr 2008;22(8):795−821.
[14] Tsanaclis LM, Wicks JFC, Chasin AAM. Workplace drug testing, different matrices different objectives. Drug Test Anal 2012;4(2):83−8.
[15] Molnar A, Lewis J, Doble P, Hansen G, et al. A rapid and sensitive method for the identification of delta-9-tetrahydrocannabinol in oral fluid by liquid chromatography−tandem mass spectrometry. Forensic Sci Int 2012;215(1−3):92−6.
[16] Molnar A, Lewis J, Fu S. Recovery of spiked Δ9-tetrahydrocannabinol in oral fluid from polypropylene containers. Forensic Sci Int 2013;227(1−3):69−73.
[17] Verstraete AG. Detection times of drugs of abuse in blood, urine, and oral fluid. Ther Drug Monit 2004;26(2):200−5.
[18] Fu S. Adulterants in urine drug testing. In: Makowski GS, editor. Advances in clinical chemistry, vol. 76. Burlington, MA: Academic Press; 2016. p. 123−63.
[19] Fu S, Luong S, Pham A, Charlton N, et al. Bioanalysis of urine samples after manipulation by oxidizing chemicals: technical considerations. Bioanalysis 2014;6(11):1543−61.
[20] Schwartz RH. Urine testing in the detection of drugs of abuse. Arch Intern Med 1988;148(11):2407−12.
[21] Huestis MA, Cone EJ. Differentiating new marijuana use from residual drug excretion in occasional marijuana users. J Anal Toxicol 1998;22(6):445−54.
[22] Cook JD, Caplan YH, Lodico C, Bush DM. The characterization of human urine for specimen validity determination in workplace drug testing: a review. J Anal Toxicol 2000;24(7):579−88.
[23] Fu S, Lewis J. Novel automated extraction method for quantitative analysis of urinary 11-nor-Δ9-tetrahydrocannabinol-9-carboxylic acid (THC-COOH). J Anal Toxicol 2008;32:292−7.
[24] Moeller KE, Lee KC, Kissack JC. Urine drug screening: practical guide for clinicians. Mayo Clin Proc 2008;83(1):66−76.
[25] Allen KR. Screening for drugs of abuse: which matrix, oral fluid or urine? Ann Clin Biochem 2011;48(Pt 6):531−41.
[26] Department of Defense. Instruction 1010.01: Military Personnel Drug Abuse Testing Program (MPDATP). Washington, DC, USA: Department of Defense; 2012.
[27] AS/NZS 4308:2008 Procedures for specimen collection and the detection and quantitation of drugs of abuse in urine. New Zealand: Standards Australia/Standards; 2008.

[28] Department of Health and Human Services Substance Abuse and Mental Health Services Administration, Mandatory Guidelines for Federal Workplace Drug Testing Programs, in Federal Register, Department of Health and Human Services Substance Abuse and Mental Health Services Administration Editor; 2004. p. 19644−73.
[29] Goggin MM, Tann CM, Miller A, Nguyen A, et al. Catching fakes: new markers of urine sample validity and invalidity. J Anal Toxicol 2017;41(2):121−6.
[30] Center for Substance Abuse Prevention, Medical Review Officer Manual for Federal Agency Workplace Drug Testing Programs, Department of Health and Human Services Substance Abuse and Mental Health Services Administration Center for Substance Abuse Prevention, Editor; 2010.
[31] Jaffee WB, Trucci ET, Levy S, Weiss RD. Is this urine really negative? A systematic review of tampering methods in urine drug screening and testing. J Subst Abuse Treat 2007;33(1):33−42.
[32] Wu A. Urine adulteration and substitution prior to drugs of abuse testing. J Clinic Ligand Assay 2003;26(1):11−18.
[33] Dasgupta A. The effects of adulterants and selected ingested compounds on drugs-of-abuse testing in urine. Am J Clin Pathol 2007;128(3):491−503.
[34] Dasgupta A. Beating drug tests and defending positive results: a toxicologist's perspective. New York: Humana Press; 2010.
[35] Dasgupta A. Adulteration of drugs-of-abuse specimens. In: Wong R, Tse H, editors. Forensic science and medicine: drugs of abuse: body fluid testing. Totowa, NJ: Humana Press Inc; 2005. p. 217−32.
[36] Thevis M, et al. Sports drug testing: analytical aspects of selected cases of suspected, purported, and proven urine manipulation. J Pharm Biomed Anal 2012;57:26−32.
[37] Deventer K, et al. Screening for 18 diuretics and probenecid in doping analysis by liquid chromatography−tandem mass spectrometry. Biomed Chromatogr 2002;16(8):529−35.
[38] Holden B, Guice EA. An investigation of normal urine with a creatinine concentration under the cutoff of 20 mg/dL for specimen validity testing in a toxicology laboratory. J Forensic Sci 2014;59(3):806−10.
[39] Wu A, Schmalz J, Bennett W. Identification of UrinAid-adulterated urine specimens by fluorometric analysis. Clin Chem 1994;40(5):845−6.
[40] Ferslew KE, Hagardorn AN, Robert TA. Capillary ion electrophoresis of endogenous anions and anionic adulterants in human urine. J Forensic Sci 2001;46(3):615−26.
[41] Valtier S, Cody JT. A procedure for the detection of Stealth adulterant in urine samples. Clin Lab Sci 2002;15(2):111−15.
[42] Dasgupta A, Wahed A, Wells A. Rapid spot tests for detecting the presence of adulterants in urine specimens submitted for drug testing. Am J Clin Pathol 2002;117(2):325−9.
[43] Paul BD. Six spectroscopic methods for detection of oxidants in urine: implication in differentiation of normal and adulterated urine. J Anal Toxicol 2004;28(7):599−608.
[44] Tsai LS, ElSohly MA, Tsai SF, Murphy TP, et al. Investigation of nitrite adulteration on the immunoassay and GC−MS analysis of cannabinoids in urine specimens. J Anal Toxicol 2000;24(8):708−14.
[45] Paul BD, Martin KK, Maguilo Jr J, Smith ML. Effects of pyridinium chlorochromate adulterant (Urine Luck) on testing for drugs of abuse and a method for quantitative detection of chromium (VI) in urine. J Anal Toxicol 2000;24(4):233−7.
[46] Paul BD, Jacobs A. Effects of oxidizing adulterants on detection of 11-nor-delta9-THC-9-carboxylic acid in urine. J Anal Toxicol 2002;26(7):460−3.
[47] Liu RH, Wu CH, Chen YJ, Chang CD, et al. Intensity of the internal standard response as the basis for reporting a test specimen as negative or inconclusive. J Anal Toxicol 2007;31(9):549−54.
[48] Urry FM, Komaromy-Hiller G, Staley B, Crockett DK, et al. Nitrite adulteration of workplace urine drug-testing specimens. I. Sources and associated concentrations of nitrite in urine and distinction between natural sources and adulteration. J Anal Toxicol 1998;22(2):89−95.
[49] Luong S, Shimmon R, Hook J, Fu S. 2-Nitro-6-monoacetylmorphine: potential marker for monitoring the presence of 6-monoacetylmorphine in urine adulterated with potassium nitrite. Anal Bioanal Chem 2012;403(7):2057−63.
[50] Luong S, Fu S. Detection and identification of 2-nitro-morphine and 2-nitro-morphine-6-glucuronide in nitrite adulterated urine specimens containing morphine and its glucuronides. Drug Test Anal 2013;6(3):277−87.
[51] Pham AQN, Kelly T, Fu S. Urine adulteration: can bleach be used to mask MDMA use? Anal Methods 2013;5(16):3948−55.
[52] WorkCover NSW. Alcohol and other drugs in the workplace. Gosford: WorkCover NSW Publications; 2006.
[53] Substance Abuse and Mental Health Services Administration. About the Division of Workplace Programs (DWP). 2015. Available from: <https://www.samhsa.gov/workplace/about> [cited 12.01.18.].
[54] Tsai JS, Lin GL. Drug-testing technologies and applications. In: Wong RC, Tse HY, editors. Drugs of abuse: body fluid testing. Totowa, NJ: Humana Press; 2005. p. 29−69.
[55] Armbruster DA, Hubster EC, Kaufman MS, Ramon MK. Cloned enzyme donor immunoassay (CEDIA) for drugs-of-abuse testing. Clin Chem 1995;41(1):92−8.
[56] Lequin RM. Enzyme immunoassay (EIA)/enzyme-linked immunosorbent assay (ELISA). Clin Chem 2005;51(12):2415−18.
[57] Warner A. Interference of common household chemicals in immunoassay methods for drugs of abuse. Clin Chem 1989;35(4):648−51.
[58] Cody JT, Valtier S. Effects of Stealth™ adulterant on immunoassay testing for drugs of abuse. J Anal Toxicol 2001;25(6):466−70.
[59] Wu AH, Gerona R, Armenian P, French D, et al. Role of liquid chromatography−high-resolution mass spectrometry (LC-HR/MS) in clinical toxicology. Clin Toxicol 2012;50(8):733−42.

[60] Australian Health Practitioner Regulation Agency. Drug and alcohol screening protocol: registrant information. Melbourne: Australian Health Practitioner Regulation Agency; 2017.

[61] Ambre J, Fischman M, Ruo TI. Urinary excretion of ecgonine methyl ester, a major metabolite of cocaine in humans. J Anal Toxicol 1984;8(1):23−5.

[62] Maurer HH, Sauer C, Theobald DS. Toxicokinetics of drugs of abuse: current knowledge of the isoenzymes involved in the human metabolism of tetrahydrocannabinol, cocaine, heroin, morphine, and codeine. Ther Drug Monit 2006;28(3):447−53.

[63] Boerner U, Abbott S, Roe RL. The metabolism of morphine and heroin in man. Drug Metab Rev 1975;4(1):39−73.

[64] Kraemer T, Maurer HH. Toxicokinetics of amphetamines: metabolism and toxicokinetic data of designer drugs, amphetamine, methamphetamine, and their N-alkyl derivatives. Ther Drug Monit 2002;24(2):277−89.

[65] Fitzgerald RL, Griffin TL, Yun YM, Godfrey RA, et al. Dilute and shoot: analysis of drugs of abuse using selected reaction monitoring for quantification and full scan product ion spectra for identification. J Anal Toxicol 2012;36(2):106−11.

[66] Cone EJ, Lange R, Darwin WD. In vivo adulteration: excess fluid ingestion causes false-negative marijuana and cocaine urine test results. J Anal Toxicol 1998;22(6):460−73.

[67] Finkel K. Water intoxication presenting as a suspected contaminated urine sample for drug testing. South Med J 2004;97(6):611−13.

[68] Thevis M, Maurer J, Kohler M, Schanez W. Proteases in doping control analysis. Int J Sports Med 2007;28(7):545−9.

[69] Wu A, Forte E, Casella G, Sun K, et al. CEDIA for screening drugs of abuse in urine and the effect of adulterants. J Forensic Sci 1995;40(4):614−18.

[70] Pearson SD, Ash KO, Urry FM. Mechanism of false-negative urine cannabinoid immunoassay screens by Visine eyedrops. Clin Chem 1989;35(4):p636−8.

[71] Goldberger BA, Caplan YH. Effect of glutaraldehyde (UrinAid) on detection of abused drugs in urine by immunoassay [letter]. Clin Chem 1994;40:1605−6.

[72] Larson SJ, Holler JM, Magluilo Jr J, Dunkley CS, et al. Papain adulteration in 11-nor-9-tetrahydrocannabinol-9-carboxylic acid-positive urine samples. J Anal Toxicol 2008;32:438−43.

[73] Lin CN, Strathmann FG. Elevated urine zinc concentration reduces the detection of methamphetamine, cocaine, THC and opiates in urine by EMIT. J Anal Toxicol 2013;37(9):665−9.

[74] Uebel RA, Wium CA. Toxicological screening for drugs of abuse in samples adulterated with household chemicals. S Afr Med J 2002;92(7):547−9.

[75] Charlton N, Fu S. Effect of selected oxidising agents on the detection of 11-nor-9-carboxy-Δ-9-tetrahydrocannabinol (THC-COOH) in spiked urine. TIAFT Bulletin 2012;42(2):33−6.

[76] ElSohly MA, Feng S, Kopycki WJ, Murphy TP, et al. A procedure to overcome interferences caused by the adulterant "Klear" in the GC-MS analysis of 11-nor-delta9-THC-9-COOH. J Anal Toxicol 1997;21(3):240−2.

[77] Peace MR, Tarnai LD. Performance evaluation of three on-site adulterant detection devices for urine specimens. J Anal Toxicol 2002;26(7):464−70.

[78] Wu A, Bristol B, Sexton K, Cassella-McLane G, et al. Adulteration of urine by "Urine Luck". Clin Chem 1999;45(7):1051−7.

[79] Dasgupta A, Chughtai O, Hannah C, Davis B, et al. Comparison of spot tests with AdultaCheck 6 and Intect 7 urine test strips for detecting the presence of adulterants in urine specimens. Clin Chim Acta 2004;348(1−2):19−25.

[80] Kuzhiumparambil U, Fu S. Effect of oxidizing adulterants on human urinary steroid profiles. Steroids 2013;78(2):288−96.

[81] Heard K, Mendoza CD. Consequences of attempts to mask urine drug screens. Ann Emerg Med 2007;50(5):591−2.

[82] Ferslew KE, Nicolaides AN, Robert TA. Determination of chromate adulteration of human urine by automated colorimetric and capillary ion electrophoretic analyses. J Anal Toxicol 2003;27(1):36−9.

[83] Cassells NP, Craston DH. The effects of commonly used adulterants on the detection of spiked LSD by an enzyme immunoassay. Sci Justice 1998;38(2):109−17.

[84] Mikkelsen SL, Ash KO. Adulterants causing false negatives in illicit drug testing. Clin Chem 1988;34(11):2333−6.

[85] Tsai SC, ElSohly MA, Dubrovsky T, Twarowska B, Towt J, et al. Determination of five abused drugs in nitrite-adulterated urine by immunoassays and gas chromatography−mass spectrometry. J Anal Toxicol 1998;22(6):474−80.

[86] Luong S, Ung AT, Kalman J, Fu S. Transformation of codeine and codeine-6-glucuronide to opioid analogues by urine adulteration with pyridinium chlorochromate: potential issue for urine drug testing. Rapid Commun Mass Spectrom 2014;28(14):1609−20.

[87] Luong S, Kuzhiumparambil U, Fu S. Elucidation of markers for monitoring morphine and its analogs in urine adulterated with pyridinium chlorochromate. Bioanalysis 2015;7(18):2283−95.

[88] Baiker C, Serrano L, Lindner B. Hypochlorite adulteration of urine causing decreased concentration of delta 9-THC-COOH by GC/MS. J Anal Toxicol 1994;18(2):101−3.

[89] Bronner W, Nyman P, von Minden D. Detectability of phencyclidine and 11-nor-delta 9-tetrahydrocannabinol-9-carboxylic acid in adulterated urine by radioimmunoassay and fluorescence polarization immunoassay. J Anal Toxicol 1990;14(6):368−71.

[90] Chou SL, Giang YS. Influences of seven Taiwan-produced adulterants on the fluorescence polarization immunoassay (FPIA) of amphetamines in urine. Forensic Sci J 2008;7(1):1−12.

[91] Chou SL, Ling YC, Yang MH, Giang YS. Influences of seven Taiwan-produced adulterants on gas chromatographic−mass spectrometric (GC−MS) urinalysis of amphetamines. J Chin Chem Soc 2008;55(3):682−93.

[92] Chou SL, Giang YS. Elucidation of the US urine specimen validity testing (SVT) policies and performance evaluation of five clinical parameters for pre screening adulterants in Taiwan's opiates urinalysis. Forensic Sci J 2007;6(2):45–58.
[93] Smith FP, Reuschel SA, Jenkins KC. ONLINE™ opiate immunoassay evaluation: precision, accuracy and adulterants. Sci Justice 1995;35(1):65–71.
[94] Schwarzhoff R, Cody JT. The effects of adulterating agents on FPIA analysis of urine for drugs of abuse. J Anal Toxicol 1993;17(1):14–17.
[95] Cody JT, Schwarzhoff RH. Impact of adulterants on RIA analysis of urine for drugs of abuse. J Anal Toxicol 1989;13(5):277–84.
[96] Lewis SA, Lewis LA, Tuinman A. Potassium nitrite reaction with 11-nor-delta 9-tetrahydrocannabinol-9-carboxylic acid in urine in relation to the drug screening analysis. J Forensic Sci 1999;44(5):951–5.
[97] Wong R. The effect of adulterants on urine screen for drugs of abuse: detection by an on-site dipstick device. Am Clin Lab 2002;21(1):37–9.
[98] Cody JT, Valtier S, Kuhlman J. Analysis of morphine and codeine in samples adulterated with Stealth. J Anal Toxicol 2001;25(7):572–5.
[99] Burrows DL, Nicolaides A, Rice PJ, Dufforc M, et al. Papain: a novel urine adulterant. J Anal Toxicol 2005;29(5):275–395.
[100] Kim HJ, Cerceo E. Interferences by NaCl with the EMIT method of analysis for drugs of abuse [letter]. Clin Chem 1976;22:1935–6.
[101] Venkatratnam A, Lents NH. Zinc reduces the detection of cocaine, methamphetamine, and THC by ELISA urine testing. J Anal Toxicol 2011;35(6):333–40.
[102] George S, Braithwaite RA. The effect of glutaraldehyde adulteration of urine specimens on syva EMIT II drugs-of-abuse assays. J Anal Toxicol 1996;20(3):195–6.
[103] Edwards C, Fyfe MJ, Liu RH, Walia AS. Evaluation of common urine specimen adulteration indicators. J Anal Toxicol 1993;17(4):251–2.
[104] Murray RL. Creatinine. In: Kaplan LA, Pesce AJ, editors. Clinical chemistry: theory, analysis and correlation. Mosby: University of Michigan; 1989. p. 1015–20.
[105] Schumann GB, Schweitzer SC. Examination of urine, in clinical chemistry: theory. Analysis and Correlation 1989;820–49.
[106] Cody GT. Specimen adulteration in drug urinalysis. Forensic Sci Rev 1990;2:63–75.
[107] National Institute on Drug Abuse, Mandatory Guidelines for Federal Workplace Drug Testing Programs, National Institute on Drug Abuse, editor; 1988. p. 11970–89.
[108] Chaturvedi AK, Sershon JL, Craft KJ, Cardona PS, et al. Effects of fluid load on human urine characteristics related to workplace drug testing. J Anal Toxicol 2013;37(1):5–10.
[109] Sittiwong J, Unob F. Paper-based platform for urinary creatinine detection. Anal Sci 2016;32(6):639–43.
[110] Wyness SP, Hunsaker JJH, Snow TM, Genzen JR. Evaluation and analytical validation of a handheld digital refractometer for urine specific gravity measurement. Pract Lab Med 2016;5:65–74.
[111] Edgell K, Caplan YH, Glass LR, Cook JD. The defined HHS/DOT substituted urine criteria validated through a controlled hydration study. J Anal Toxicol 2002;26(7):419–23.
[112] Dasgupta A. A health educator's guide to understanding drugs of abuse testing. Sudbury, MA: Jones & Bartlett Learning; 2011.
[113] Caitlin D, Cowan D, Donike M, Fraisse D, et al. Testing urine for drugs. Clin Chim Acta 1992;2007:S13–26.
[114] King EJ. Performance of AdultaCheck 4 test strips for the detection of adulteration at the point of collection of urine specimens used for drugs-of-abuse testing. J Anal Toxicol 1999;23(1):72.
[115] Microgenics Corporation. DRI® General Oxidant-Detect® Test. Available from: <https://fscimage.fishersci.com/images/D13677~.pdf>; [cited 17.01.18.].
[116] Alnajjar A, McCord B. Determination of heroin metabolites in human urine using capillary zone electrophoresis with β-cyclodextrin and UV detection. J Pharm Biomed Anal 2003;33:463–73.
[117] Minakata K, Yamagishi I, Kanno S, Nozawa H, et al. Determination of Urine Luck in urine using electrospray ionization tandem mass spectrometry. Forensic Toxicol 2008;26:71–5.
[118] Steuer AE, et al. A new metabolomics-based strategy for identification of endogenous markers of urine adulteration attempts exemplified for potassium nitrite. Anal Bioanal Chem 2017;409(26):6235–44.
[119] Schneider HJ, Arnold K, Schneider TD, Poetzsch M, et al. Efficacy of a polyethylene glycol marker system in urine drug screening in an opiate substitution program. Eur Addict Res 2008;14(4):186–9.
[120] Jones JD, Atchison JJ, Madera G, Metz VE, et al. Need and utility of a polyethylene glycol marker to ensure against urine falsification among heroin users. Drug Alcohol Depend 2015;153:201–6.
[121] Charlton N, Fu S. Analysis of stable oxidation products of THC-COOH following urine adulteration: pyridinium chlorochromate and betadine. In: Forensic and Clinical Toxicology Association (FACTA) Biannual Meeting. Sydney, NSW, Australia; 2013.
[122] Charlton N. Personal communication. Searching for stable oxidation products of carboxy THC after urinary adulteration. Honours Thesis for Bachelor of Science (Honours) in Applied Chemistry—Forensic Science. Sydney, NSW: University of Sydney Technology; 2008.
[123] Charlton N, et al., Formation of oxidation products of (-)-trans-Δ^9-tetrahydrocannabinol-9-carboxylic acid (THCCOOH) after exposure to oxidizing adulterants. In: The International Association of Forensic Toxicologists (TIAFT) Annual Meeting. Geneva, Switzerland (Abstract); 2009.
[124] González-Mariño I, Rodriguez I, Quintana JB, Celar R. Investigation of the transformation of 11-nor-9-carboxy-Δ9-tetrahydrocannabinol during water chlorination by liquid chromatography–quadrupole-time-of-flight-mass spectrometry. J Hazard Mater 2013;261:628–36.
[125] Luong S, et al., Transformation of codeine and codeine-6-glucuronide to opioid analogues by urine adulteration with PCC: an update, in Forensic and Clinical Toxicology Association (FACTA) Biennial Meeting. Sydney, NSW; 2013.

[126] Jones JT, Esposito FM. An assay evaluation of the methylene blue method for the detection of anionic surfactants in urine. J Anal Toxicol 2000;24(5):323–7.

[127] Concheiro M, Simões SM, Quintela O, de Castro A, et al. Fast LC–MS/MS method for the determination of amphetamine, methamphetamine, MDA, MDMA, MDEA, MBDB and PMA in urine. Forensic Sci Int 2007;171(1):44–51.

[128] Kolmonen M, Leinonen A, Kuuranne T, Pelander A, et al. Hydrophilic interaction liquid chromatography and accurate mass measurement for quantification and confirmation of morphine, codeine and their glucuronide conjugates in human urine. J Chromatogr B 2010;878(29):2959–66.

[129] Wissenbach DK, Meyer MR, Weber AA, Remane D, et al. Towards a universal LC–MS screening procedure—can an LIT LC–MS(n) screening approach and reference library be used on a quadrupole-LIT hybrid instrument?. J Mass Spectrom 2012;47(1):66–71.

Chapter 27

When Hospital Toxicology Report Is Negative in a Suspected Overdosed Patient: Strategy of Comprehensive Drug Screen Using Liquid Chromatography Combined With Mass Spectrometry

Ernest D. Lykissa
Expertox Laboratory, Houston, TX, United States

INTRODUCTION

Drug screen performed by most hospital laboratories have many limitations. In general urine specimens are screened for 7–10 commonly abused drugs and if initial testing is negative no further analysis is performed unless a physician questions the validity of the test result. In this protocol, the presence of drugs in urine samples is performed by using cutoffs of detection that attempt to exclude concentrations of drug residues suggestive of accidental exposure, examples such as passively inhaled smoke for marijuana, or employing high cutoff of opiates to ensure that there will be no false positive reporting in the case of poppy seed ingestion by the subject tested.

There are many limitations of current urine toxicology screen performed in most hospital laboratories which include testing for limited drugs and SAMHSA (Substance Abuse and Mental Health Services Administration) mandated cutoffs. Most hospital laboratories do not test for new psychoactive drugs including bath salts (synthetic cathinone) and synthetic marijuana (spices). Palamar [1] based on a survey of 8604 high school seniors reported that 1.1% students abused bath salts in the last year. Another study based on 7805 high school seniors indicated that 2.9% students reported current use of synthetic cannabinoids and 1.4% students reported abusing of synthetic cannabinoids on three or more occasions in the past month [2]. Moreover, drugs frequently involved in date rape situations such as gamma-hydroxybutyric acid (GHB) and ketamine are not tested in most hospital laboratories. Therefore, any overdosed patients exposed to such drugs will be tested negative in hospital toxicology screen.

Case Report

A 34-year-old married woman with no known drug abuse history died in her home. Although she was admitted before to the hospital with cardiac problems, this time she died with cardiovascular failure. Her initial toxicology report was negative. However, suspecting this as a case of homicide, the court ordered further investigation which revealed that her husband chronically poisoned her over a period of 1 year using ketamine by adding ketamine to tea she consumed in the afternoon. The postmortem toxicology analysis performed using gas chromatography/mass spectrometry (GC/MS) revealed ketamine concentration of 2.1 μg/mL, 3 μg/mL, and 1.2 μg/mL in gastric content, blood, and urine, respectively. The most striking forensic findings were cardiac muscle fibrosis, and hyaline degeneration of small arteries in victim's heart, the pathological features of ketamine poisoning already reported in animal studies [3].

PROBLEM WITH LIMITED DRUG TESTING PROTOCOL AND CUTOFF LEVELS

SAMHSA workplace drug testing protocol recommends testing for limited drugs at a predetermined cutoff level. However, private employer may test additional drugs in their workplace drug testing protocol. The initial testing is performed using immunoassays and then specimens tested positive in the initial screen are further tested using either GC/MS or liquid chromatography combined with tandem mass spectrometry (LC−MS/MS) for confirmation. Recommended screening cutoff concentrations of SAMHSA and non-SAMHSA drugs are listed in Table 27.1.

In general drug screening protocol offered by most hospital laboratories only offer testing for 9−10 drugs which include all SAMHSA drugs and some non-SAMHSA drugs. These laboratories may not have capabilities of drug confirmation using GC/MS and LC−MS/MS. As a result negative toxicology report in a suspected drug overdosed patient is observed more often in these hospitals compared to large medical centers and academic medical centers where comprehensive urine drug screen protocols using LC−MS/MS may be available.

Using a predetermined cutoff concentration as recommended also has many limitations. For example, sometimes clinicians order screening of benzodiazepine in urine to ensure patient compliance but cutoff of 200 ng/mL is too high for certain benzodiazepines, particularly if prescribed in low dosages. The benzodiazepine cutoff of 200 ng/mL is intended to identify individuals abusing benzodiazepine. Another common mistake is ordering opiate immunoassay for checking compliance with oxycodone prescription use. Although oxycodone has structural similarity with morphine, it has poor cross-reactivities with antibodies that are used in opiate immunoassays which target morphine. Therefore, opiate immunoassay screen is likely to be negative in a patient taking prescription oxycodone.

Designer drugs were initially synthesized by clandestine laboratories in order to bypass legal consequences of manufacturing and selling illicit drugs. Then in 1986, the US Controlled Substances Act was amended in order to make manufacturing and selling of designer drugs illegal. Other countries also adopted similar laws to ban designer drugs. The common types of designer drugs include amphetamine analogues, opiate analogues (including fentanyl derivatives), piperazine analogues, tryptamine-based hallucinogens, phencyclidine analogues, and GHB analogues. Most commonly encountered designer drugs are structurally related to phenethylamine, opioids, and tryptamines. However, newly introduced designer drugs such as bath salts and spices could not be detected by routine toxicology drug testing due to inability of amphetamine immunoassays to detect bath salts and inability of marijuana assay to detect spices (synthetic marijuana). These drugs are not usually tested in most hospital laboratories.

TABLE 27.1 Cutoff Concentrations of Initial Immunoassay Screening for SAMHSA and Commonly Tested Non-SAMHSA Drugs in Urine

SAMHSA Drugs	Cutoff Concentration	Non-SAMHSA Drugs	Cutoff Concentration
Amphetamine/methamphetamine	500 ng/mL	Barbiturates	200/300 ng/mL
MDMA	500 ng/mL	Benzodiazepines	200 ng/mL
Cocaine (as benzoylecgonine)	250 ng/mL	Methadone	300 ng/mL
Opiates	2000 ng/mL	Propoxyphene	300 ng/mL
Hydrocodone/hydromorphone	300 ng/mL	Buprenorphine[a]	5 ng/mL
Oxycodone/oxymorphone	100 ng/mL	Methaqualone	300 ng/mL
6-Monoacetyl morphine (heroin metabolite)	10 ng/mL	LSD[a]	0.5 ng/mL
Marijuana as THC-COOH	50 ng/mL		
Phencyclidine	25 ng/mL		

LSD, Lysergic acid diethylamide.
[a]Less often tested in 9−10 toxicology drug panel offered by hospital laboratories.

The objective of this chapter is familiarizing readers with limitations of route toxicology analysis because often such toxicology report is negative in a suspected overdosed patient. Therefore, when such report is negative in a suspected overdosed patient, further testing of urine specimens using LC−MS/MS is useful to identify the drug or drugs causing such overdose. If a particular hospital does not have such capacity then urine and/or blood specimen should be sent to a reference laboratory or a larger medical center with capabilities of conducting broad urine drug screening and confirmation as well as quantitation if needed.

Case Report

A year and six months old girl was admitted to the emergency room with seizures and showed severe symptoms of drug-induced toxicity. The attending physician ordered a urine drugs test which showed the presence of methamphetamine above the cutoff levels. The urine subsequently was shipped to our reference laboratory with an order to conduct a confirmation test only for methamphetamine. The confirmation analysis yielded highly elevated methamphetamine and amphetamine (methamphetamine metabolite) concentrations. Upon additional communication with the attending pediatrician, further testing was performed on a hair sample of the child. The results were positive for methamphetamine, amphetamine, heroin, morphine, cocaine, delta-9-tetrahydrocannabinol (Marijuana; THC), carisoprodol, meprobamate, dextromethorphan, and doxylamine. However, when the original urine sample was retested by enzyme immunoassay (EIA) opiates, THC and carisoprodol (Soma) testings were negative because concentrations of such drugs were below the cutoff concentrations of these assays. The child was exposed to multiple drugs but initial hospital toxicology screen was able to detect only amphetamine.

Case Report

A newborn exhibiting symptom of drug intoxication had urine and meconium samples collected and the urine was tested by the hospital laboratory. However, the urine toxicology screen was negative. Suspecting drug overdose, the attending physician sent the specimens to our reference laboratory for further testing. Urine and meconium tests of the neonate child were deemed positive for the presence of phencyclidine, though the screening test by EIA demonstrated immunologic activity below the preemployment drug testing cutoffs. Again this case report shows the limitation of drug testing using immunoassays.

Case Report

A newborn child was born with pronounced drug intoxication symptoms. Meconium was tested with EIA but results were negative. Further testing by LC−MS/MS confirmed the presence of cocaine and THC in the meconium. Again the routine hospital toxicology analysis failed to confirm the presence of these drugs.

These case histories illustrate the conflict that exists in the testing of pediatric samples for drug screening by EIA and employing standardized preemployment cutoff levels by hospital laboratories and/or doctor offices. The issues that arise from these discrepancies are multifold. The hospital laboratory is bound by regulatory restrictions, to report as positive only samples that yield results above the cutoff levels. The reference laboratories may circumvent this restriction with the readily available confirmation re-testing of samples that have yielded positive readings above the absolute negative controls. This practice necessitates the training of laboratorians involved in screening drug testing to scrutinize results associated with neonatal and pediatric specimens for the presence of immunologic activity and submitting them for further testing to a reference Laboratory that is well equipped in technology and analysts that may decipher the true status of a questionable specimen.

In addition, this type of testing often results in subsequent involvement of the legal system. Positive drug findings in children of a very young age result in criminal child endangerment charges bring by the State to the child's caregiver, of. The discrepancies listed earlier provide evidence that such examples may be very difficult to defend in a court hearing, with parents battling each other for child custody and attorneys who are not very adept at scientific explanations of such complexity. Furthermore the laboratory tests may be used by defendants in intoxication cases to disprove testing results that corroborate them abusing various prescribed or illicit drugs during the commission of a crime. As it has become evident recently, the lack of proper documentation for these types of tests, i.e., chain of custody, confirmation tests results, etc. may invalidate hospital acquired analytical results, or rendered them almost impossible to defend upon cross-examination by the defense attorneys. It is in these cases that employing a reference laboratory capable of performing both forensic and clinical testings with appropriate certification may help alleviate some of these problems.

COMMUNICATING WITH PHYSICIAN WHEN TOXICOLOGY REPORT IS NEGATIVE

When a clinician calls the clinical laboratory questioning validity of a negative toxicology report when all clinical features indicate drug overdose, then it is essential for laboratory scientists or pathologists to properly communicate with the attending physician for further testings. In general the following two scenarios are most common.

Scenario A

The physician ordered a 9–10 drug panel drug screen but many drugs escape detection; most commonly certain benzodiazepines and opioids. If the clinician is suspecting a positive test result for certain benzodiazepine, then the best approach should be ordering confirmation of the prescription benzodiazepine using GC/MS or preferably LC–MS/MS to confirm patient compliance with the medication. In our experience lorazepam, alprazolam, and temazepam often escape detection by benzodiazepine screening assay.

If the clinician is expecting positive test result for oxycodone then laboratory scientists/pathologists need to explain to the clinician that oxycodone cannot be detected by opiate immunoassay. If the clinician needs a positive oxycodone screening test result then rerunning the urine sample using oxycodone immunoassay should resolve the issue.

If the clinician is expecting positive test result for methadone, propoxyphene or any other opioids then it is necessary for the laboratory scientist to explain that opiate immunoassays do not detect these opioids. Many clinicians think that if they prescribe opioids then compliance can be verified by ordering opiate screening assay but this is not the case. In this case it is best to order opioid confirmation test for the specific drug.

Scenario B

The clinician is convinced that the patient is overdosed but toxicology screen is negative. The patient is not taking any particular drug such as opioids or benzodiazepines but clinical picture is consistent with drug overdose. In such case best approach should be to send urine specimen to a reference laboratory capable of detecting 100 or more drugs in a comprehensive drug screen using LC–MS/MS. If certain drugs are detected then these drugs can be confirmed and quantified by another chromatographic method if necessary. However, if the patient is young (age 18–29) and male there is a possibility that he may be taking bath salt or spices. Based on a survey of 1600 overdosed patients seeking treatment in nine states, Warrick et al. [4] observed that people who abuse bath salts are predominately male (67.9%) and young (mean age 29.2 years; while 47.9% of overdosed patients were between 18 and 29 years old). In this case the attending physician may order testings for bath salts and spices and the urine specimen must be sent to an appropriate reference laboratory.

Sometimes when marijuana metabolite is detected by immunoassay, the clinician may question the validity of test result because clinical presentation indicates sever overdose not consistent with marijuana overdose alone. In this case most likely agent is spice because almost 50% marijuana abusers also abuse spices [5]. The laboratory scientist/pathologist may suggest to the clinician to send urine sample to a reference laboratory for further testing.

If the patient is a young female and physical examination indicates recent sexual activity but no evidence of struggle then it may indicate drug assisted rape. Specimen must be sent to a forensic laboratory under chain of custody for testing of GHB, ketamine, and Rohypnol.

If the patient is a Native American it may indicate use of peyote cactus during religious ceremony. Mescaline is the active substance. The urine drug screen should be negative. Unfortunately most hospital laboratories do not test for the presence of mescaline in urine.

CHALLENGES OF DETECTING BENZODIAZEPINES

Pain management patients are often prescribed benzodiazepines as well as opiates as a part of their regimen of pain management medications [6]. Patients are tested to ensure compliance with these medications. Usually urine is used as the specimen of choice to test for the presence of prescribed medications and failure to observe the presence of these drugs in urine may have dire consequences including dismissal from the pain management program. In general benzodiazepine immunoassays target a cutoff concentration of 200 ng/mL. Unfortunately, detecting certain benzodiazepines such as clonazepam, estazolam, lorazepam, and alprazolam could be difficult in urine specimens of patients taking these prescription medications due to low concentrations of these drugs in urine as well as poor cross-reactivity with the antibody used benzodiazepine immunoassays which may target a drug different from the prescribed benzodiazepine. Fraser et al. commented that the cutoff values of 200 ng/mL for benzodiazepines were established many years ago when most benzodiazepines were prescribed in 5–20 mg/day dosage. Therefore, this cutoff is too high to establish patient compliance with currently used benzodiazepines [7].

Clonazepam is a Schedule IV medication in the United States which is frequently prescribed for the treatment of anxiety and also epilepsy. Potential problems associated with improper use or abuse of this drug include dependence, suicidal thoughts, depression, sleep disorder, and aggression. As a result patient compliance with this drug is tested using urine specimen. Less than 0.5% clonazepam is excreted unchanged in urine but the major metabolite 7-aminoclonazepam is present in urine in much higher concentration than clonazepam. The cross-reactivity of 7-aminoclonazepam varies significantly with antibodies used in immunoassays for benzodiazepine screening in urine with some immunoassays showing less than 2% cross-reactivity. As a result a urine specimen may be negative by a benzodiazepine immunoassay despite patient compliance with clonazepam. West et al. tested 180 urine specimens collected from patients taking prescription clonazepam and analyzed such specimens using Microgenics DRI benzodiazepine immunoassay at 200 ng/mL cutoff concentration and observed that only 38 specimens (21%) tested positive. In contrast, 126 specimens (70%) showed the presence of 7-aminoclonazepam when tested by LC–MS/MS at 200 ng/mL cutoff. However, the positivity rate was increased to 87% when LC–MS/MS cutoff concentration was lowered to 40 ng/mL. In addition, authors also observed that 7% urine specimens had 7-aminoclonazepam concentrations between 40 and 100 ng/mL. Therefore, DRI benzodiazepine immunoassay is not suitable for determining compliance of patients with clonazepam. The authors also suggested that 200 ng/mL cutoff concentration is too high and recommended a lower cutoff of 40 ng/mL for LC–MS/MS analysis of 7-aminoclonazepam to determine compliance with clonazepam therapy [8].

Flunitrazepam often encountered in date rape is not approved for clinical use in the United States. However, some individuals may have access to the drug, particularly those living in Southern Border States. Although flunitrazepam is a benzodiazepine, some commercial immunoassays for benzodiazepines may have relatively low cross-reactivity with flunitrazepam. In addition, urine concentration of flunitrazepam may not be adequate and it is a challenge for routinely used benzodiazepine assays to detect the presence of flunitrazepam in urine. Forsman et al. using CEDIA (cloned enzyme donor immunoassay) benzodiazepine assay at a cutoff of 300 ng/mL failed to obtain a positive result in urine specimens collected from volunteers after they received a single dose of 0.5 mg flunitrazepam. In addition, only 22 out of 102 urine specimens collected from volunteers after receiving the highest dose of flunitrazepam (2 mg) showed positive screening test result using the CEDIA benzodiazepine assay [9].

Another reason for poor detectability of certain benzodiazepines in urine by benzodiazepine immunoassays is that glucuronide metabolites of lorazepam, temazepam, and even oxazepam have poor cross-reactivity with the antibody. A high sensitivity cloned enzyme donor immunoassay (HS-CEDIA) is commercially available in which beta-glucuronidase is added in the reagent which is capable of hydrolyzing glucuronide metabolites. Darragh et al. evaluated the performance of regular CEDIA, HS-CEDIA, and KIMS (kinetic interaction of microparticle in solution) benzodiazepine immunoassays by comparing their performance with an LC–MS/MS-based method. Of the 299 urine specimens tested 141 specimens (47%) confirmed positive for one or more of the benzodiazepines/metabolites by LC–MS/MS, but HS-CEDIA showed positive screen with only 110 specimens out of 141 confirmed positive (78%). However, the original CEDIA assay only showed positive response with 78 specimens (55% positive) and the KIMS assay showed positive response with 66 specimens (47%). The authors concluded that while the HS-CEDIA provides higher sensitivity than the KIMS and CEDIA assay mostly due to increased capacity of detecting lorazepam, HS-CEDIA assay still missed high percentage of benzodiazepine positive specimens from patients treated for chronic pain. Therefore, LC–MS/MS quantification with enzymatic specimen pretreatment offers superior sensitivity and specificity for monitoring benzodiazepines in patients treated for chronic pain [10].

Dixon et al. attempted to circumvent problem of poor cross-reactivities of glucuronide metabolites with antibody used in the EMIT (enzyme multiplied immunoassay technique) benzodiazepine assay for application of the Vista 1500 analyzer (Siemens Healthcare) by incubating urine specimen with beta-glucuronidase to liberate free benzodiazepine prior to immunoassay. However, despite this attempt, 11 out of 31 urine specimens collected from patients taking benzodiazepines showed negative result. As expected all 31 specimens had benzodiazepine concentrations above 200 ng/mL as determined by LC–MS/MS. The authors concluded that patient compliance with benzodiazepine therapy must be monitored using a chromatographic method, preferably LC–MS/MS [11].

Sometimes point of care (POCT) benzodiazepine assays are used to determine the compliance of pain management patients with benzodiazepine therapy. However, POCT assays for benzodiazepine are also inadequate for determination of patient compliance with benzodiazepine therapy. In one study, the authors observe that POCT assays yielded negative results for patients taking prescription benzodiazepines nearly 20% of the time (98 out of 498 patients). Moreover, POCT assays failed to produce positive results for persons who were shown by LC−MS/MS to be positive for lorazepam or clonazepam. The authors concluded that POCT assays could fail to provide accurate information regarding patient-specific medication use. The authors suggested use of more specific MS for such testing to assure accuracy and also to improve patient safety [12].

CHALLENGES OF DETECTING OPIOIDS

Many opiate immunoassays are not capable of detecting the presence of semisynthetic opiates, such as oxycodone, hydrocodone, hydromorphone, and oxymorphone in urine due to the low cross-reactivity of opiate immunoassays for these opiates. All opiate immunoassays utilize antibody that targets morphine because morphine is the common metabolite of both codeine and heroin. Bertholf et al. analyzed 112 urine specimens collected from patients taking hydrocodone or hydromorphone but were presumptive negative by the Roche Online Opiate II immunoassay calibrated at 300 ng/mL cutoff. However, using GC/MS confirmatory test with a detection limit of 50 ng/mL for both hydrocodone and hydromorphone, one or both of the opiates were detected in 81 out of 112 urine specimens (72.3%) which were initially tested negative by the immunoassay. The authors commented that opiate immunoassays are developed to detect morphine and are most sensitive to morphine and codeine. Although many opiate immunoassays also detect hydrocodone or hydromorphone, sensitivities for these analytes are often much lower, thus increasing the possibility of negative test results even when the drug is present in the urine [13].

Although opiate immunoassays have some cross-reactivity with hydrocodone and hydromorphone, opiate immunoassays in general have very poor cross-reactivities with keto-opioids such as oxycodone and oxymorphone. For example, the Abbott Abuscreen ONLINE opiate assay has only 0.3% cross-reactivity with oxycodone while CEDIA opiate assay has only 3.1% cross-reactivity with oxycodone based on the information provided in package inserts. For detection of oxycodone present in low concentrations, immunoassays that are specifically designed for detecting oxycodone are commercially available and their specificity for oxycodone has been validated. In addition, opiate immunoassays cannot detect other opioids such as methadone, fentanyl, propoxyphene, tramadol, and related drugs. See Chapter 25, Drug Testing in Pain Management, for more details.

Case Report
A 40-year-old man receiving 20 mg oxycodone twice a day for headache, routinely called the clinic stating that he finished his medication faster and needed a refill. His urine drug test was negative using opiate immunoassay and suspecting he was selling oxycodone, he was dismissed from the clinic by the physician. A family member contacted a toxicologist who informed that oxycodone may cause false negative urine opiate drug screen. An aliquot of the original urine specimen was retested using GC/MS and oxycodone level of 1124 ng/mL was confirmed.

DESIGNER DRUGS ARE NOT DETECTED IN ROUTINE URINE TOXICOLOGY

In general hospital toxicology testings cannot detect designer drugs except 3,4-methylenedioxymethamphetamine (MDMA, ecstasy), 3,4-methylenedioxyamphetamine (MDA), and heroin (as 6-monocateylmorphine and/or morphine metabolite). However, most of the newer designer drugs cannot be detected by urine toxicology analysis. Amphetamine like designer drugs can be classified under six broad categories:

1. Phenylethylamine type designer drugs
2. Substituted amphetamine type designer drugs
3. Piperazine type designer drugs
4. 2,5-dimethoxy amphetamine type designer drugs
5. Beta-keto amphetamine type designer drugs (bath salts).

Although these drugs are amphetamine like designer drugs, they have very poor cross-reactivities with amphetamine immunoassays. See Chapter 19, Overview of Common Designer Drugs, for an overview of designer drugs. In recent years a new class of designer drugs has emerged in many countries which are beta-keto (bk) derivatives of amphetamine and related compounds. These drugs include mephedrone (bk-4-methylmethamphetamine), butylone, and methylone (bk-3,4-methylenedioxyamphetamine) [14]. Bath salts have been placed in Schedule I drug recently due to high abuse potential. See Chapter 21, Review of Bath Salts on Illicit Drug Market, for more details.

Since 2008, synthetic marijuana compounds sold as spice, K2, or herbal high are gaining popularity among drug abusers. Although less toxic than bath salts, these compounds are very addictive and may cause severe overdose including death. The first synthetic compound in this category JWH-018 was synthesized by Dr. John W. Huffman at Clemson University to study the effect of this compound on cannabinoid receptors. It has been speculated that someone noticed the paper and copied his method to produce JWH-018 illegally for abuse. Synthetic cannabinoids are called herbal high or legal high and usually plant materials are sprayed with these compounds so that these active compounds are present at the surface of the product for maximum effect. However, pure compounds which are white powders are also available for purchase form black market. Currently more than 100 compounds are available which belong to this class of abuse drugs but most common are JWH-018, JWH-073, JWH-250, JWH-015, JWH-081, HU-210, HU-211 (synthesized in Hebrew University), CP-47,497 (synthesized at Pfizer). Although these compounds are called synthetic cannabinoids, they do not have any structural similarity with marijuana and cannot be detected in urine using marijuana immunoassays. See Chapter 22, Review of Synthetic Cannabinoids on Illicit Drug Market, for more details.

GHB is an endogenous constituent of mammalian brain which is present in nanomolar concentration and acts as a neurotransmitter. Although GHB was found in health food stores before as a safe body building and fat burning compound, 16 cases of adverse effects due to GHB containing health products were reported to the San Francisco Bay Area Regional Poison Control Center from June 1990 to October 1990. Use of GHB caused coma in four patients and tonic−clonic seizure in two patients for dosage ranging from one-fourth of a teaspoon to 4 tablespoon [15]. Following that report, the US Food and Drug Administration (FDA) banned the over-the-counter sale of GHB in November 1990 and now GHB is a Schedule I controlled substance in the United States. The management of patient addicted to GHB is complicated due to high dosage of medications that they require to control withdrawal symptoms [16]. Even 7 days use of GHB may produce severe withdrawal [17]. Unfortunately GHB is used in date rape situation and this drug is not targeted in routine toxicological screen of urine in most hospital laboratories. In addition, analogues of GHB are sold in the underground markets and worldwide. GHB analogues which are found as street drugs include GBL (gamma-butyrolactone), 1,4-BD (1,4-butanediol), GHV (gamma-hydroxy valeric acid), and GVL (gamma-valerolactone). GBL and 1,4-BD are converted endogenously after abuse into GHB and exert similar effects of GHB abuse. Therefore, clinical manifestation and management of 1,4-BD intoxication is similar to GHB intoxication [18]. However, self-reported GHB ingestion was much more common than GBL ingestion but GBL was more commonly found in seized samples. In one report, the authors stated that of the 418 seized samples, 225 were in liquid form out of which 140 contained GBL while only 85 contained GHB [19]. As expected none of these designer drugs are tested in hospital laboratories. For in-depth discussion on date rape drugs, see Chapter 18, Drugs Assisted Sexual Assaults: Toxicology, Fatality, and Analytical Challenge.

TOXICOLOGY SCREEN USING LC−MS/MS

Over the last decade LC−MS or preferably LC−MS/MS has been established as the most reliable methodology in toxicology laboratories for analysis of drugs of abuse as well as therapeutic drugs. In contrast to the "Gold Standard" of the past 30 + years, namely GC−MS, LC−MS/MS has enhanced the confirmation processes by a factor of 5. Moreover, labor-intensive sample preparation and tedious derivatizations which are essential for GC/MS analysis of most drugs are not necessary for LC−MS/MS analysis. In most cases during LC−MS/MS analysis, the compound of interest can be identified by observing the protonated molecular ion (M + 1) in positive ionization mode or as the molecular ion in the negative ionization mode. In contrast to GC−MS methodology which is capable of analyzing only volatile nonpolar compounds, LC−MS/MS method can analyze both polar and thermal labile compounds.

The most common detection technique in LC−MS/MS method is electrospray ionization (ESI) performed in atmospheric pressure. Recently there is a trend of increased use of electrospray interfaces with pneumatically assisted nebulization allowing for increased flow rates and larger solvent volumes. The ESI interface de-solvates the sample in free ions. The ions present in the solution in turn are drawn into a capillary resistive glass tube via an electrostatic potential difference between the capillary and a mesh electrode inside the instrument. From this the first mass spectrometer's quadrupole acts as a molecular prefilter, i.e., in the case of cocaine it selects out of the stream of ions entering the quadrupole only the m/z 304 (cocaine molecular ion is 303 plus a proton) in the positive ionization mode. The second quadrupole then selects out of all the m/z 304 ions (precursor or mother ion) it admits in a relay race fashion, only the 304 ions that will transition to m/z 182 ion (product or daughter ion). Those transition elements are selected and reported as unequivocal evidence of the presence of cocaine in the sample in addition to other transition elements also specific for cocaine. This then comprises the molecular fingerprint of cocaine. In Table 27.2, precursor ions, product ions, and

TABLE 27.2 Precursor Ion, Product Ions, and Computer Tuning Parameters Used in LC–MS/MS Analysis of Commonly Encountered Drugs in Toxicology Analysis Along With Parameters of Several Internal Standards

Drug	Precursor Ion (m/z)	Product Ion (m/Z)	DP EP CE CXP
Amphetamine	136	119, 91	40 10 12 10
Amphetamine-d5 (IS)	141.1	93	40 10 12 10
Methamphetamine	150	119, 91	60 10 15 8
Methamphetamine-d5 (IS)	155.	92	60 10 15 8
MDMA	194	163, 135	100 10 20 10
MDA	180.1	133.1, 105.1	40 10 30 8
MDEA	208.1	163.1, 105.1	100 10 20 12
Ephedrine	166.1	148, 117	40 10 12 10
Pseudoephedrine	166.1	148.1, 133.1	40 10 12 10
Methedrone	194.1	166.1, 146.1	50 10 26 6
Methylone	208.1	160.1, 132.1	60 10 25 12
Methylone-d3 (IS)	211.1	163.1	60 10 25 12
Methylenedioxypyrovalerone (MDPV)	276.1	205.1, 175.1	70 10 30 8
MDPV-d8 (IS)	284.1	134.1	70 10 30 8
Ketamine	238	220, 125	80 10 35 6
Phentermine	150.1	133, 91	
Benzoylecgonine	290.1	168, 105	80 10 37 12
Benzoylecgonine-d3 (IS)	293.1	171.2	80 10 37 12
Cocaethylene	318.2	196.2/82.2	80 10 37 12
Alprazolam	309.1	281.1, 205.1	80 10 35 10
Alpha-OH-alprazolam	325.1	297.1	60 10 40 15
7-Aminoclonazepam	286.1	250.1, 121.1	80 10 37 13
Diazepam	285.1	193, 154	80 10 40 10
Nor-diazepam	271.1	165.1, 140.1	71 10 37 10
Nor-diazepam-d5 (IS)	276.1	140	71 10 37 10
Lorazepam	321.1	275.1, 229.1	80 10 40 10
Midazolam	326.1	291.1, 249.1	101 10 42 22
Alpha-OH midazolam	342.1	203.1, 168.1	60 10 53 11
Oxazepam	287.1	269.2, 241.1	76 10 31 18
Temazepam	301.1	255.1, 177.1	70 10 50 8
Butabarbital	211.1	168, 42	−60 −10 −40 −7
Secobarbital	237.1	194.1, 42.1	−90 −10 −40 −7
Secobarbital-d5 (IS)	242.1	42	−90 −10 −40 −7
Phenobarbital	231.1	188, 42.1	−70 −10 −40 −7
Buprenorphine	468.3	414.2, 55	120 10 52 18

(Continued)

TABLE 27.2 (Continued)

Drug	Precursor Ion (m/z)	Product Ion (m/Z)	DP EP CE CXP
Nor-buprenorphine	414.3	83, 55	120 10 52 18
Codeine	300.1	215.1, 165	100 10 35 12
Codeine-d6 (IS)	306.2	152.2	100 10 35 12
Dihydrocodeine	302.1	201.1, 199.1	100 10 35 12
Morphine	286	165, 152	90 10 75 12
6-Monoacetyl morphine	328.1	211.1, 165	
6-Monoacetyl morphine-d3 (IS)	331.1	165, 93	100 10 49 8
Morphine-d6 (IS)	292.1	152	90 10 75 12
Hydrocodone	300.1	199, 128	100 10 39 14
Hydromorphone	286.1	185, 157	100 10 39 12
Oxycodone	316.1	298.1, 241.1	75 10 40 8
Oxycodone-d6 (IS)	322.1	178.1	75 10 40 8
Oxymorphone	302	227.1, 198.1	90 10 40 8
Methadone	310.1	265, 105.1	60 10 35 14
Methadone-d3 (IS)	313.2	105.1	60 10 35 14
2-Ethylidene-1,5-dimethyl-3,3-diphenylpyrrolidine (EDDP)	278.1	186, 234.1	80 10 60 8
Tramadol	264.1	58.1, 42.2	66 10 103 8
O-desmethyltramadol	250.1	58.1, 42	66 10 103 10
Propoxyphene	340.1	266.1, 91	46 10 60 8
Meperidine	248.1	220.1, 174	86 10 28 10
Meperidine-d4 (IS)	252.2	224.1	86 10 28 10
Acetyl fentanyl	323.1	188.1, 105,1	90 10 31 10
Fentanyl	337.1	188.1, 105.1	90 10 50 8
Fentanyl-d5 (IS)	342.3	105.1	90 10 50 8
Sufentanil	387.1	238, 111.1	46 10 49 10
Propoxyphene	340.1	266.1, 91	
THC-COOH	343.1	299.1, 245.1	−110 −10 −28 −11
THC-COOH-d3 (IS)	346.1	302.1	−110 −10 −28 −11
JWH-018 4-OH pentyl	358.1	155.1, 127.1	100 10 39 12
JWH-018 4-OH pentyl-d5 (IS)	363.1	155.1	100 10 39 12
JWH-073 OH hexyl	372.2	155.1, 127.1	90 10 27 12
JWH-250-4-OH pentyl	386.1	155.1	100 10 39 12
Phencyclidine (PCP)	244.1	159.1, 91.1	56 10 40 8
PCP-d5 (IS)	249.3	96.1	56 10 40 8

CE, Collision energy; *CXP*, cell exit potential; *DP*, de-clustering potential; *EP*, entrance potential.

important compound tuning parameters used forensic drug testing by LC—MS/MS for some commonly encountered drugs in our reference laboratory (Expertox) are listed but this is not a complete list. In our reference laboratory we are capable of identifying over 150 drugs using LC—MS/MS.

This analytical breakthrough has given rise to a new breed of laboratory analysis, as it relates to forensic drugs testings, clinical drugs testing in overdosed patient when initial toxicology testing is negative, as well as pain management patients. A number of laboratories are now offering pain management panels of up to 300 different drug analytes, by a single injection into an LC—MS/MS instrument with a sample preparation of "dilute and shoot." This method enables diluting the sample with mobile phase solvent mixture, adding deuterated internal standards followed by direct injection to the LC—MS/MS analyzer. Theoretically this is a very easy methodology to employ by anyone, at the expense of instrument wear and tear. However, a more conservative method is to employ some mode of extraction of drug present in the urine sample either by liquid—liquid or solid phase extraction. In addition with the incorporation of calibrators and proper quality control, concentrate the extract by evaporation and then reconstitute with mobile phase and inject. It is more time consuming but definitely more scientifically sound approach and the resulting acquired data may yield quantitative results.

In our reference laboratory after adding appropriate internal standards (100 μL) to the urine specimen (1 mL), 1 mL glucuronidase with 1 mL 1 M acetate buffer pH 4.8 is also added followed by incubation for 30 min 60°C. Then 200 μL of 10% ammonium hydroxide is added to the specimen and drugs along with internal standards are extracted using liquid—liquid extraction using 3 mL of extraction solvent (toluene:ethyl acetate:hexane:isopropanol:50:30:15:5 by vol.). After mixing, for 10 min, specimen is centrifuged and 2 mL of top organic layer is removed. Then the supernatant removed from the specimen is dried, reconstituted with organic solvent, and then injected into the LC—MS/MS instrument. We also analyze calibrators (low, medium, and high), positive as well as negative controls along with analysis of donor's urine specimens. Sometimes solid phase extraction is also performed after treatment of urine specimen with glucuronidase. Then solid phase column is washed with water followed by water/methanol and dugs along with internal standards are eluted from solid phase column using 2 mL of dichloromethane/isopropyl alcohol/ammonium hydroxide (75:20:5 by vol.). In Fig. 27.1 chromatogram and mass spectra of oxycodone and the internal standard oxycodone-d6 are presented. The patient was taking oxycodone and the urine oxycodone level was 131 ng/mL as confirmed in our reference laboratory using LC—MS/MS. In Fig. 27.2 chromatogram of and mass spectra of alpha-hydroxy alprazolam and the internal standard are given. The alpha-hydroxy alprazolam level was confirmed at 645 ng/mL.

It is important to note that LC—MS/MS methodology can be applied not only for urine drug analysis but also for drug analysis in blood, oral fluid, meconium, hair, and postmortem tissue specimens. However, dilute and shoot approach is applicable most commonly for analysis of urine and oral fluid by LC—MS/MS.

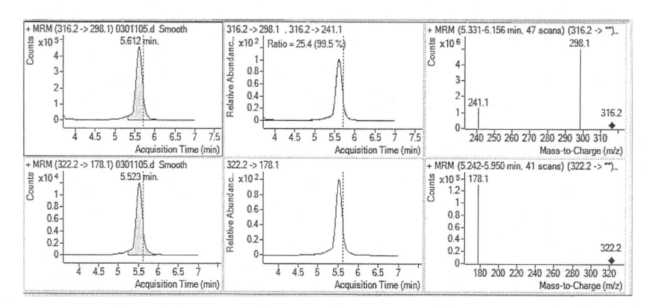

FIGURE 27.1 Chromatograms and mass spectra of oxycodone and the internal standard oxycodone-d6 are presented. The patient was taking oxycodone and the urine oxycodone level of 131 ng/mL was confirmed in our reference laboratory using LC—MS/MS.

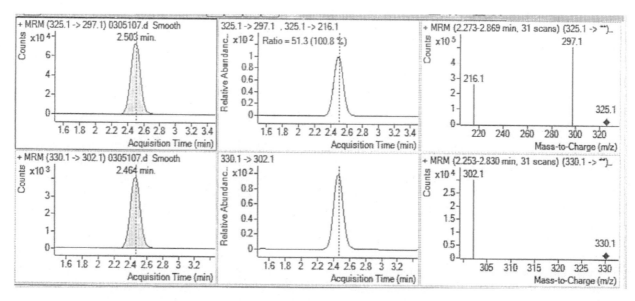

FIGURE 27.2 Chromatogram of and mass spectra of alpha-hydroxy and the internal standards are given. The alpha-hydroxy alprazolam level was confirmed at 645 ng/mL in our reference laboratory using LC–MS/MS.

Dilute and Shoot Versus Sample Preparation

In the medical literature both dilute and shoot as well sample preparation prior to LC–MS/MS analysis has been reported. Dong et al. analyzed 71 stimulants and 7 metabolites in human urine using dilute and shoot approach. The limits of detection varied from 0.1 to 25 ng/mL [20]. Cao et al. described a simple "dilute-and-shoot" protocol for analysis of 78 drugs and metabolites in urine using LC–MS/MS. The assay improved on currently available methods by including glucuronide conjugates, thus allowing direct detection of metabolites that might otherwise be missed by current methods [21]. Kong et al. used a simple sample cleanup procedure using "dilute-and-shoot" approach for analysis of 113 abused drugs and their metabolites using LC–MS/MS. Drugs were separated in a Capcell PakMG-III C-18 reverse phase column using a gradient elution of 1 mM ammonium formate with 0.1% formic acid in water and acetonitrile. The total analysis time was 32 min. The multiple reaction mode using two transitions (e.g., quantifier and qualifier) was optimized for both identification and confirmation [22].

Alcántara-Durán reported a sensitive nanoflow LC combined with high-resolution MS screening method for analysis of 81 drugs. The method was based on the use of a nanoLC column (75 μm × 150 mm, 3 μm particle size, and 100 Å pore) with the nanospray emitter tip integrated so that dead volumes are significantly minimized. Data acquisition method included both full-scan and all ion fragmentation experiments using an Orbitrap analyzer operated in the positive ionization mode. The authors adopted a simple dilute-and-shoot methodology for their analysis and did not observe any matrix effect when a dilution 1:50 was implemented. Despite this high-dilution factor, limits of quantification were still satisfactory, with values below 5 ng/mL for most drugs analyzed [23].

Oral fluid analysis has traditionally required extensive cleanup protocols and sample concentration, which can be avoided by the use of a fast, "dilute-and-shoot" method that performs no considerable sample manipulation. In one study, the authors described a quantitative method for the determination of eight common opioids and associated metabolites (codeine, morphine, hydrocodone, hydromorphone, nor-hydrocodone, oxycodone, noroxycodone, and oxymorphone) in oral fluid after 10-fold dilution of specimen in methanol/water and then direct analysis using LC–MS/MS. In contrast to most published methods of this particular type, this method uses no sample cleanup or concentration and has a considerably faster turnaround time which makes this method ideal for very high-throughput laboratories. Moreover, specimen requirement is only 100 μL. The validated calibration range for this method was 2.5–1000 ng/mL. The authors concluded that the "dilute-and-shoot" method removes the unnecessary, costly, and time-consuming extraction steps found in traditional methods and still surpasses all analytical requirements [24]. However, other authors used solid phase extraction of drugs from oral fluid prior to chromatographic separation and MS analysis. Clauwaert et al. [25] developed a method for determination of cocaine, benzoylecgonine, and cocaethylene in oral fluid by LC-quadrupole-time-of-flight MS after solid phase extraction of drugs from oral fluid.

Other investigators reported extraction of drugs from urine or other biological matrix prior to analysis using LC−MS/MS. Mueller et al. developed a method for analysis of 301 forensically important drugs, e.g., benzodiazepines, hypnotics, opioids, cocaine, amphetamine, cannabinoids, antidepressants, neuro-epileptic, and cardiac drugs in one simple LC−MS/MS analysis. Both blood and urine could be analyzed using this protocol. Samples were extracted either with liquid−liquid extraction or solid phase extraction. Finally drug identification was carried out by library search with a newly developed mass spectra library based on ESI spectra at three different collision energies in positive mode [26]. Teng et al. developed a method for analysis of 151 abused drugs and toxic compounds in human whole blood after solid phase extraction. The authors used LC combined with time-of-flight MS for analysis of these compounds [27]. High-resolution MS is also gaining popularity in toxicology laboratories for drug analysis. For in-depth discussion on this topic, see Chapter 14, High-Resolution Mass Spectrometry: An Emerging Analytical Method for Drug Testing, and Chapter 23, Application of Liquid Chromatography Combined With High-Resolution Mass Spectrometry for Rapid Screening and Confirmation of a Large Number of Drugs in Urine.

CONCLUSIONS

Urine drug testing offered by many hospital laboratories utilize commercially immunoassays and target 9−10 commonly used drugs. However, such testing has many limitations even in detecting a drug in a drug class being tested. For example, benzodiazepine immunoassays at 200 ng/mL may fail to detect therapy with clonazepam, lorazepam, alprazolam, etc. in a complaint patient. Moreover, opiate immunoassay cannot detect the presence of oxycodone and oxymorphone despite structural similarity with morphine, the antibody target of opiate immunoassays. However, major limitations of urine drug testing protocol of hospital laboratories are that toxicology report of drugs overdosed patient may be negative if the patient was exposed to a designer drug which are not routinely tested. However, with development of LC−MS/MS methods for analysis of many abused and toxic chemicals in urine or biological fluids, identification of an illicit drug responsible for overdose is possible. Unfortunately, LC−MS/MS technology is not available in most hospital laboratories. Therefore, when a clinician complains about a negative toxicology report in a suspected overdose patient, specimen must be sent to a reference laboratory for identifying the drug or toxin responsible for such overdose.

REFERENCES

[1] Palamar JJ. Bath salts use among a nationally representative sample of high school seniors in the United States. Am J Addict 2015;24:488−91.
[2] Palamar JJ, Barratt M, Coney L, Martins SS. Synthetic cannabinoids use among high school seniors. Pediatrics 2017;140:e20171330.
[3] Tao Y, Chen XP, Qin ZH. A fatal chronic ketamine poisoning. J Forensic Sci 2005;50:173−6.
[4] Warrick BJ, Hill M, Hekman K, Christensen R, et al. A 9 stage analysis of designer stimulant "bath salts" hospital visits reported to poison control centers. Ann Emerg Med 2013;62:244−51.
[5] Lisi DM. Designer drugs: patients may be using synthetic cannabinoids more than you think. JEMS 2014;39:56−9.
[6] Manchikanti L, Damron KS, McManus C, Barnhill R, et al. Patterns of illicit drug use and opioid abuse in patients with chronic pain at initial evaluation: a prospective observational study. Pain Phys 2004;7:431−7.
[7] Fraser AD, Meatherall R. Comparative evaluation of five immunoassays for the analysis of alprazolam and triazolam metabolites in urine: effect of lowering the screening and GC−MS cut-off values. J Anal Toxicol 1996;20:217−23.
[8] West R, Pesce A, West C, Crews B, et al. Comparison of clonazepam compliance by measuring of urinary concentration by immunoassay and LC−MS/MS in pain management population. Pain Phys 2010;13:71−8.
[9] Forsman M, Nystrom I, Roman M, Berglund L, et al. Urinary detection times and excretion patterns of flunitrazepam and its metabolites after a single oral dose. J Anal Toxicol 2009;33:491−501.
[10] Darragh A, Snyder ML, Ptolemy AS, Melanson S. KIMS, CEDIA and HS_CEDIA immunoassays are inadequately sensitive for detection of benzodiazepines in urine from patients treated for chronic pain. Pain Phys 2014;117:359−66.
[11] Dixon RB, Floyd D, Dasgupta A. Limitation of EMIT benzodiazepine immunoassay for monitoring compliance of patients with benzodiazepine therapy even after hydrolyzing glucuronide metabolite in urine to increase cross-reactivity: comparison of immunoassay results with LC/MS/MS values. Ther Drug Monit 2015;37:137−9.
[12] Mikel C, Pesce AJ, Rosenthal M, West C. Therapeutic monitoring of benzodiazepines in the management of pain: current limitations of point of care immunoassays suggest testing by mass spectrometry to assure accuracy and patient safety. Clin Chim Acta 2012;413:1199−202.
[13] Bertholk RL, Joahannsen LM, Reisfield GM. Sensitivity of an opiate immunoassay for detecting hydrocodone and hydromorphone in urine from a clinical population: analysis of subthreshold results. J Anal Toxicol 2015;39:24−8.
[14] Meyer MR, Wilhelm J, Peters FT, Maurer HH. Beta-keto amphetamines: studies on the metabolism of the designer drug mephedrone and toxicological detection of mephedrone, butylone and methylone in urine using gas chromatography-mass spectrometry. Anal Bioanla Chem 2010;397:1225−33.

[15] Dyer JE. Gamma-hydroxybutyrate: a health food product producing coma and seizure like activity. Am J Emerg Med 1991;9:321−4.
[16] Gonzalez A, Nutt DJ. Gamma hydroxy butyrate abuse and dependency. J Psychopharmacol 2005;19:195−204.
[17] Perez F, Chu J, Bania T. Seven days of gamma-hydroxybutyrate (GHB) use produces severe withdrawal. Ann Emerg Med 2006;48:219−20.
[18] Palmer RB. Gamma-hydroxybutyrolactone and 1,4-butanediol: abused analogs of gamma-hydroxybutyrate. Toxicol Rev 2004;23.21−31.
[19] Wood DM, Warren-Gash C, Ashraf T, Greene SL, et al. Medical and legal confusion surrounding gamma-hydroxybutyrate (GHB) and its precursors gamma-butyrolactone (GBL) and 1,4-butanediol (1,4-BD). QKM 2008;101:23−9.
[20] Dong Y, Yan K, Ma Y, Wang S, et al. A sensitive dilute-and-shoot approach for the simultaneous screening of 71 stimulants and 7 metabolites in human urine by LC−MS−MS with dynamic MRM. J Chromatogr Sci 2015;53:1528−36.
[21] Cao Z, Kaleta E, Wang P. Simultaneous quantitation of 78 drugs and metabolites in urine with a dilute-and-shoot LC−MS−MS assay. J Anal Toxicol 2015;39:335−46.
[22] Kong TY, Kim JH, Kim JY, In MK, et al. Rapid analysis of drugs of abuse and their metabolites in human urine using dilute and shoot liquid chromatography-tandem mass spectrometry. Arch Pharm Res 2017;40:180−96.
[23] Alcántara-Durán J, Moreno-González D, Beneito-Cambra M, García-Reyes JF. Dilute-and-shoot coupled to nanoflow liquid chromatography high resolution mass spectrometry for the determination of drugs of abuse and sport drugs in human urine. Talanta 2018;182:218−24.
[24] Enders JR, McIntire GL. A dilute-and-shoot LC−MS method for quantitating opioids in oral fluid. J Anal Toxicol 2015;39:662−7.
[25] Clauwaert K, Decaestecker T, Mortier K, Lambert W, et al. The determination of cocaine, benzoylecgonine, and cocaethylene in small volume oral fluid samples by liquid chromatography-quadrupole-time-of flight mass spectrometry. J. Anal Toxicol 2004;28:655−9.
[26] Mueller CA, Weinmann W, Dresen S, Schreiber A, Gergov M. Development of a multi-target screening analysis for 301 drugs using a QTrap liquid chromatography/tandem mass spectrometry system and automated library searching. Rapid Commun Mass Spectrom 2005;19:1332−8.
[27] Teng X, Liang C, Wang R, Sun T, et al. Screening of drugs of abuse and toxic compounds in human whole blood using online solid-phase extraction and high-performance liquid chromatography with time-of-flight mass spectrometry. J Sep Sci 2015;38:50−9.

FURTHER READING

Von Seggern RL, Fitzgerald CP, Adelman LC, Adelman JU. Laboratory monitoring of OxyContin (oxycodone): clinical pitfalls. Headache 2004;44:44−7.

Chapter 28

Testing of Drugs of Abuse in Oral Fluid, Sweat, Hair, and Nail: Analytical, Interpretative, and Specimen Adulteration Issues

Uttam Garg[1] and Carl Cooley[2]

[1]Department of Pathology and Laboratory Medicine, Children's Mercy Hospital, and University of Missouri School of Medicine, Kansas City, MO, United States, [2]PRA Health Sciences Lexana, KS, United States

INTRODUCTION

Substance abuse remains a serious problem in the United States and around the world. It is estimated that over 25 million Americans abuse illicit or prescription drugs, resulting in tens of thousands of deaths each year. It is well established that drug abuse negatively impacts public safety and productivity. In the United States alone, substance misuse costs more than $400 billion annually in healthcare costs and lost productivity. Deterrence against substance abuse is generally achieved through patient counseling and drug testing. Since patient counseling and patient questioning is very subjective, drug testing is frequently used to detect drug abuse.

Various specimens that are used for drug testing include urine, blood, oral fluid, meconium, hair, nail, placenta, and cord tissue. Urine is the most commonly used specimen for drug testing as drug concentrations are higher in urine and it is an easy specimen to work with. Blood is also a commonly used specimen, particularly when urine is not available. Alternative specimens particularly oral fluid and hair are increasingly being utilized in drug testing. These specimens offer many advantages over blood and urine such as easy sample collection without a need for a special collection facility or personnel, less intrusion in privacy during sample collection, and reduced potential for sample adulteration. Despite these advantages, alternative specimens present unique analytical and interpretive challenges. Advantages and disadvantages of various specimens are shown in Table 28.1 [1–3], and relative detection windows for various specimen types are shown in Fig. 28.1 [2].

Substance Abuse and Mental Health Services Administration (SAMHSA) has been exploring the possibility of using alternate samples for Federal Workplace Drug Testing Programs. In 2004, SAMHSA proposed the inclusion of oral fluid, hair, and sweat specimens in workplace drug testing programs [4]. However, due to the public and federal agencies' comments and concerns, in 2008, it was concluded that there was insufficient information to include these alternate samples in Federal Workplace Drug Testing Programs [5]. Despite this, SAMHSA has remained committed to monitor developments in alternative specimen testing and recently implemented guidelines to allow oral fluid as an alternate specimen if the donor is unable to provide a urine specimen [6]. SAMHSA's proposed cutoffs are provided in Tables 28.2–28.4. European Workplace Drug Testing Society (EWDTS) recommended cutoffs are provided in Tables 28.5 and 28.6 [7,8].

DRUG TESTING IN ORAL FLUID

Although the terms "oral fluid" and "saliva" are used interchangeably, oral fluid is the preferred term. Oral fluid is constituted of saliva (secretion of salivary glands), mucosal transudate, and crevicular fluid. It contains mostly water (95%) with some solutes such as electrolytes, amylase, glucose, urea, and proteins. The volume of oral fluid secreted varies

TABLE 28.1 Advantages and Disadvantages of Various Alternative Samples in the Detection of Drug Abuse

Specimen	Advantages	Disadvantages
Urine	• Most studied and well-accepted sample type • Laboratory methods well standardized • Less expensive • Point of collection devices available	• Special collection facility required • High potential for sample adulteration and tempering • Supervised collection invasive • Shorter window of detection: 1–3 days • No drug concentration and dose relationship • Drug concentration dependent on hydration status
Blood	• Positive test indicates recent use • Drug concentration correlates better with impairment • Sample is collected under direct supervision • Adulteration unlikely	• Sample collection needs trained personnel • Risks associated with phlebotomy • Point of collection devices for blood not available • Shorter window of drug detection
Oral fluid	• Detects recent drug use • Results may relate to behavior/performance • Simple, noninvasive sample collection • Very low potential for sample adulteration • Point of collection testing available	• Shorter window of detection • Contamination from passive exposure • Expensive and less established analytical methods • In dry mouth syndrome sample may be unavailable • High sample collection variability
Hair	• Long detection window of weeks to months • Sample collection easy and noninvasive • Very low potential for sample adulteration • Sample stable for years • Useful sample can be obtained from mummified or exhumed bodies • Sample recollection from the same source possible	• Does not detect recent drug use • Sample may not be available • Potential hair color bias • Expensive and less established analytical methods • Possible external contamination • Adulteration possible by sample treatment before collection
Sweat	• Provides cumulative drug exposure for few days to few weeks • Collection devices relatively tamper resistant • Easy and noninvasive sample collection	• Large intersubject variation in sweat production • Expensive and less established analytical methods • Point of collection tests not available • Limited collection devices

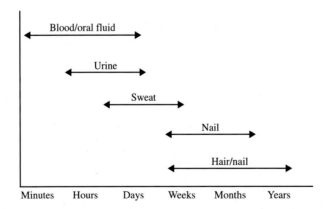

FIGURE 28.1 Relative retention time of drugs in different matrices.

from 0.6 to 1.5 L/day. Oral fluid is the most commonly used alternative specimen for drug testing and offers many advantages over urine and blood (Table 28.1). A major advantage of oral fluid is an easy sample collection without the need for a special specimen collection facility or personnel [9–12]. In forensic situations, the sample can be collected on-site under direct supervision to avoid adulteration and substitution. Also, point-of-collection testing devices are readily available for oral fluid for many drugs. Furthermore, concentrations of drugs in oral fluid correlate better with blood concentrations as compared to urine. Most drugs of abuse can be detected in oral fluid at the low ng/mL range for up to 48 h after dosing.

TABLE 28.2 SAMHSA-Proposed Cutoff for Oral Fluid Testing

Analyte	Concentration (ng/mL)	
	Initial Screening	Confirmatory Cutoffs
Marijuana metabolite(s)		
THC and THC-COOH	4	—
THC	—	2
Cocaine		
Cocaine metabolites	20	
Cocaine or cocaine metabolite (benzoylecgonine)		8
Opiates/opioids		
Opiate metabolites	40	—
Morphine	—	40
Codeine	—	40
6-Monoacetylmorphine	4	4
Phencyclidine (PCP)	10	10
Amphetamines		
Amphetamines (target analyte: methamphetamine)	50	—
Amphetamine	—	50
Methamphetamine (specimen must also contain amphetamine)	—	50
3,4-Methylenedioxymethamphetamine (MDMA)	50	50
3,4-Methylenedioxyamphetamine (MDA)	—	50
3,4-Methylenedioxyethylamphetamine (MDEA)	—	50

Transfer of Drugs from Plasma to Oral Fluid

Most drugs and their metabolites move from capillaries to salivary ducts by passive diffusion through lipid membranes. Drugs can also be incorporated into oral fluid by ultrafiltration and secretion, during oral intake, and smoking and intranasal administration [13–15]. Only free-unionized form of drugs moves from plasma to oral fluid. Historically, it was thought that oral fluid drug concentrations paralleled the free drug concentrations in plasma; however, the relationship among oral fluid and plasma drug concentration is more complex. Many factors including plasma pH, oral fluid pH, drug ionization, protein binding, and the volume of oral fluid affect drug concentrations in oral fluid. Unstimulated oral fluid has a slightly acidic pH [6,7]. When stimulated for collection, oral fluid pH is slightly basic (7.5–8.0) due to an increase in bicarbonate levels. Changes in pH can significantly change the concentration of drugs in oral fluid through ionization and binding effects [11].

Sample Collection and Transportation

A number of commercial devices are available for the collection of oral fluid. Most of these devices have absorbent foam or a pad affixed to a nylon stick. Prior to the sample collection, the donor refrains from drinking, eating, or putting anything in his/her mouth for 10–20 min. The donor is asked to put the device under their tongue or between their gums and cheek for 2–5 min. Due to certain pathological or medical reasons, some donors may not be able to provide an oral fluid sample. For example, persons on anticholinergic drugs may not be able to provide a sufficient quantity of oral fluid. After the sample collection, the absorbent foam or pad may be directly used for on-site testing or placed in

TABLE 28.3 SAMHSA-Proposed Cutoff for Sweat Patch Testing

Analyte	Concentration (ng/patch)	
	Initial Screening	Confirmatory Cutoffs
Cannabinoids		
THC metabolites	4	–
THC	–	1
Cocaine		
Cocaine metabolites	25	–
Cocaine or benzoylecgonine	–	25
Opiates/opioids		
Opiate metabolites[a]	25	–
Morphine	–	25
Codeine	–	25
6-Monoacetylmorphine	–	25
Phencyclidine (PCP)	20	20
Amphetamines		
Amphetamines (target analyte:methamphetamine)	25	–
Amphetamine	–	25
Methamphetamine (specimen must also contain amphetamine)	–	25
3,4-Methylenedioxymethamphetamine (MDMA)	25	25
3,4-Methylenedioxyamphetamine (MDA)	–	25
3,4-Methylenedioxyethylamphetamine (MDEA)	–	25

[a]Laboratories are permitted to initial test all specimens for 6-acetylmorphine (6−AM) at 2.5 ng/patch.

diluents/preservatives for shipping to a testing laboratory. Various commercial devices including Salicule, OralLab, RapiScan, Drugwipe, Cozart, Statsure, and SalivaScreen are available for oral fluid collection.

When sending samples to a laboratory, precautions should be taken for sample handling and transportation. For legal specimens, chain of custody is very important. Legal issues related to chain of custody are the same as those for blood and urine. Samples may be transported at 4°C or room temperature. Most drugs are stable for several days at 4°C or room temperature [16,17].

Specimen Analysis

Once the specimen is collected, it may be tested on-site or sent to the laboratory. Point-of-collection testing devices vary significantly for sensitivity and specificity, and should be evaluated for a specific need. Recently, a review and meta-analysis on 31 studies involving point-of-collection testing devices have been published [18]. Of the 31 studies, 17 (54.8%) were conducted in traffic enforcement settings, 9 (29.0%) exclusively with drug users, 3 (9.7%) exclusively in laboratories, and 2 (6.4%) with drug users and in laboratories. Regarding drugs tested, 18 studies analyzed cannabinoids, 15 opioids, 14 cocaine, 14 amphetamines, and 6 benzodiazepines. There was significant variation among different devices. For example, cannabinoids sensitivity, specificity, and accuracy varied from 7.7% to 100%, 9% to 100%, and 41% to 100%, respectively. For amphetamines, sensitivity, specificity, and accuracy varied from 16.7% to 100%, 33% to 100%, and 57.1% to 100%, respectively. Similarly, there were large performance variations for cocaine,

TABLE 28.4 SAMHSA-Proposed Cutoff for Hair Testing

Analyte	Concentration (pg/mg)	
	Initial Screening	Confirmatory Cutoffs
Cannabinoids		
THC metabolites	1	–
THC-COOH	–	0.05
Cocaine		
Cocaine metabolites	500	–
Cocaine[a]	–	500
Cocaine metabolites[a]	–	50
Opiates/opioids		
Opiate metabolites[b]	200	–
Morphine	–	200
Codeine	–	200
6-Monoacetylmorphine	200	–
6-Monoacetylmorphine (specimen must also contain morphine at a concentration ≥200 pg/mg)	–	200
Phencyclidine (PCP)	300	300
Amphetamines		
Amphetamines (target analyte: methamphetamine)	500	–
Amphetamine	–	300
Methamphetamine (specimen must also contain amphetamine at ≥50 pg/mg)	–	300
3,4-Methylenedioxymethamphetamine (MDMA)	500	300
3,4-Methylenedioxyamphetamine (MDA)	–	300
3,4-Methylenedioxyethylamphetamine (MDEA)	–	300

[a]Cocaine concentration is greater than or equal to the confirmatory cutoff and benzoylecgonine/cocaine ratio is greater than or equal to 0.05, cocaethylene is greater than or equal to 50 pg/mg, or norcocaine greater than or equal to 50 pg/mg.
[b]Laboratories are permitted to initial test all specimens for 6-AM using a 200 pg/mg cutoff.

opioids, and benzodiazepines. Another study that compared several on-site devices found that most devices perform well for methamphetamine and opiates but not for cannabinoids. The performance for cocaine and amphetamine was device dependent [19]. In another study, the point-of-collection testing devices performed well for opiates and amphetamines, with variable performance for cocaine and THC [20].

The laboratory testing generally involves drug screening by immunoassay, followed by confirmation of positive results with definitive methods (such as gas chromatography-mass spectrometry or liquid chromatography-tandem mass spectrometry). Gas chromatography–mass spectrometry (GC-MS) has been widely used for confirmation of drugs in oral fluid [21–26]. In recent years, liquid chromatography–mass spectrometry (LC-MS) methods have been described in the literature [27–32]. Confirmation methods must be evaluated for analytical performance for sensitivity, specificity, precision, linearity, and carry-over monitoring. It is recommended that the measurement of human IgG or albumin should be carried out to rule out possible sample adulteration. Industry cutoffs of 0.5 mg/L for IgG and 0.6 mg/dL for albumin are used as decision points for sample adulteration [4].

TABLE 28.5 EWDTS Proposed Cutoff for Oral Fluid Testing

Analyte	Concentration (ng/mL)	
	Initial Screening	Confirmatory Cutoffs
Cannabinoids		
Cannabis (THC)	10	2
Opiates/Opioids		
Morphine	40	15
6-Acetylmorphine (AM)	4	2
Codeine	–	15
Norcodeine	–	2
Dihydrocodeine	–	15
Methadone (L)	50	–
Methadone or metabolites (D + L)	–	20
Propoxyphene or metabolites	40	5
Cocaine		
Cocaine + metabolites	30	–
Cocaine metabolite (benzoylecgonine)	–	8
Cocaine	–	8
Amphetamines		
Amphetamine group	40	
Amphetamine (D + L)	–	15
Methamphetamine	–	15
3,4-Methylenedioxyamphetamine (MDA)	–	15
3,4-Methylenedioxymethamphetamine (MDMA)	–	15
Benzodiazepines group	10	–
7-Amino-flunitrazepam, 7-amino-clonazepam, 7-amino-nitrazepam, alprazolam, bromazepam, clonazepam, desmethyldiazepam, diazepam, flunitrazepam, flurazepam, lorazepam, lormetazepam, midazolam, nitrazepam, nordiazepam, oxazepam, and phenazepam	–	3
Temazepam	–	10

Comparison of Positivity of Oral Fluid with Blood and Urine

Urine is the most commonly used matrix for drug screening; however, urine drug concentrations do not correlate well with degree of impairment. Blood is the best sample to correlate drug concentrations and impairment, but it is difficult to collect. Due to many advantages, oral fluid is increasingly being used in drug screening. It is, therefore, desirable to compare positivity rates of oral fluid with that of blood and urine. Several studies have been carried out to address this question. When comparing these matrices, it is important to keep in mind that, in general, urine has higher concentrations of drug metabolites whereas oral fluid and blood have higher concentrations of parent drugs; therefore, parent drugs are generally assayed in oral fluid. However, when parent drugs are measured in oral fluid, the questions of passive exposure may be raised. The measurement of drug metabolites in oral fluid is preferred to minimize this argument about passive exposure.

TABLE 28.6 EWDTS Recommended Initial Screening and Confirmation Cutoffs for Hair Testing.

Analyte	Concentration (ng/mg)	
	Initial Screening	Confirmatory Cutoffs
Alcohol		
Alcohol (segments 0–3)	–	EtG: 0.03 FAEE: 0.5
Cannabinoids		
THC	0.1	0.05
THC-COOH	–	0.0002
Cocaine	0.5	0.5
Benzoylecgonine	–	0.05
Cocaethylene	–	0.05
Nor-cocaine	–	0.05
Opiates/opioids	0.2	–
Morphine	–	0.2
Codeine	–	0.2
6-MAM	–	0.2
Amphetamines	0.2	–
Amphetamine	–	0.2
Methamphetamine	–	0.2
3,4-Methylenedioxymethamphetamine (MDMA)	–	0.2
3,4-Methylenedioxyethylamphetamine (MDEA)	–	0.2
3,4-Methylenedioxyamphetamine (MDA)	–	0.2
Methadone	0.2	0.2
EDDP	–	0.05
Buprenorphine	0.01	0.01
Norbuprenorphine	–	0.01
Ketamine	0.5	0.5
Norketamine	–	0.1
Benzodiazepines/z-drugs	0.05	0.05

The correlation of drug concentrations between oral fluid and blood is influenced by several factors such oral fluid pH, volume, drug type, and protein binding. Oral fluid pH is generally acidic. Under acidic conditions, acidic drugs are neutralized and diffused into blood. Therefore, in general, the concentration of acidic drugs is lower in oral fluid as compared to basic drugs. Also, in the absorptive phase, the drug concentrations are higher in oral fluid as compared to blood.

Different studies comparing the drug positivity rate of oral fluid with blood and urine have shown significant differences. In a large study, 77,218 oral fluid samples were tested for amphetamines, marijuana (THC; tetrahydrocannabinol), cocaine, opiates, and phencyclidine (PCP) [33]. Overall oral fluid positivity rate of 5.1% was comparable to the general workforce Drug Testing Index positivity rate for urine of 4.46% ($N > 5,200,000$). In this study, oral fluid samples were screened by enzyme immunoassay for THC (parent drug and metabolite), cocaine metabolites, opiate metabolites, PCP, and amphetamines at cutoffs of 3, 15, 30, 3, and 120 ng/mL, respectively. Immunoassay screen positive samples were confirmed using GC/MS/MS at cutoff values of 1.5, 6, 30, 30, 3, 1.5, 120, and 120 ng/mL for THC (parent drug), benzoylecgonine, morphine, codeine, 6-acetylmorphine (6-AM), PCP, amphetamine, and methamphetamine, respectively. The order of rate of positivity was THC > cocaine > amphetamines > opiates > PCP. THC and cocaine were accounted for 86% of the positive results. The positivity rate for amphetamine and cocaine was 60% higher than urine suggesting that these drugs accumulate more efficiently in oral fluid compared to urine. Another remarkable finding in this study was the presence of 6-AM in 67% morphine-positive samples, indicating the wide use of heroin in the tested population.

However, in several studies, drug positive rates among oral fluid and urine differed significantly. In one study using 114 adult arrestees, the sensitivity for THC detection in oral fluid was only 5% compared to urine. Cocaine and heroin had a sensitivity of 100% and 88%, and specificity of 99% and 100%, respectively [34]. Another study compared 1500 paired oral fluid and urine samples for opiates, cannabis metabolite, cocaine metabolite, amphetamine, and benzodiazepines. Urine had a positivity rate of 3.7% as compared to an oral fluid positivity rate of only 0.5% [35]. Use of different analytical screening methods and cutoffs may be the reason for these differences.

Laloup et al. conducted a study to determine the positivity rate of amphetamine and its metabolites between plasma and oral fluid. Their results showed oral fluid and plasma correlated with amphetamine and 3,4-methylenedioxymethamphetamine (MDMA) with 98% sensitivity and 100% specificity [36]. In another study, paired oral fluid and whole blood samples were analyzed for amphetamines, opioids, cocaine and metabolites, THC, and benzodiazepines [37]. All substances had a correlation between oral fluid and whole blood (except for lorazepam and THC); however, concentrations between the two specimens were not equivalent. This indicates that oral fluid can be used as a qualitative measure of drug use but no estimation should be made on blood concentrations.

Detection of Amphetamines and Other Sympathomimetic Amines in Oral Fluid

Amphetamines are central nervous system (CNS) stimulants which have a high potential of abuse and addiction, though they have limited use in the treatment of certain diseases, such as narcolepsy, obesity, and attention deficit hyperactivity disorder (ADHD). Amphetamine, methamphetamine, 3,4-methylenedioxyamphetamine (MDA), 3,4-methylenedioxymethamphetamine (MDMA) and 3,4-methylenedioxyethylamphetamine (MDEA) have been detected in oral fluid for up to 24 h after a single dose [38]. The concentration of sympathomimetic amines in oral fluid is generally much higher as compared to plasma. In a study on human volunteers, oral fluid methamphetamine concentrations and average area under the curve for a 24-h time period were higher in oral fluid than plasma. The volunteers received four doses of 10 and/or 20 mg S-(+)-methamphetamine. Disposition of methamphetamine in oral fluid was dose-related, though there was high intra and interindividual variability [38].

SAMHSA- and EWDTS-proposed cutoff guidelines for amphetamine, methamphetamine, MDMA, MDA, and MDEA in oral fluid are provided in Tables 28.2 and 28.5. A study involving 4080 drivers to determine the accuracy of the Statsure Saliva Sampler oral collection device with blood amphetamine and methamphetamine levels found that the most accurate cutoff concentrations were 130 and 280 ng/mL, respectively, well above the SAMHSA recommended 50 ng/mL [36]. Various methods including immunoassays [39,40], GC-MS [41], LC-fluorescence [42] LC-MS, and LC-MS/MS [43,44] have been described for the analysis of amphetamines in oral fluid.

Detection of Cannabinoids in Oral Fluid

Cannabinoids are a group of C21 compounds that are found in the plant species *Cannabis sativa* and have psychoactive properties. The major psychoactive cannabinoid is delta-9-tetrahydrocannabinol (THC). Though used through ingestion, it is generally consumed by smoking marijuana. The effect occurs within minutes of smoking and persists for several hours. Psychoactive effects include euphoria and relaxation. Long-term use is associated with loss of short-term memory and intellectual impairment. Although by federal laws, consumption of marijuana is considered illegal, it has some legitimate medical use. Synthetic THC, Dronabinol (Marinol), is used to treat anorexia and nausea in cancer and AIDS patients. In recent years, several states in the United States have legalized the use of marijuana for medical and recreational use.

The major metabolite of THC is 11-nor-delta-9-tetrahydrocannabinol-9-carboxylic acid (THC-COOH). In urine, THC-COOH is the target compound to determine the THC use; however, in oral fluid, the parent compound THC is in a higher concentration. Despite the higher concentration of THC in oral fluid, detection of THC-COOH is still preferred as it minimizes the argument for passive exposure to marijuana in drug-testing cases. It should be noted, though, that passive exposure was not shown to be the cause of positive test results [45]. The presence of THC in oral fluid is mostly from smoking or ingestion as very little transport occurs from blood to oral fluid.

Since THC concentrations are very low in oral fluid, sensitive methods are required for its detection. Various methods including immunoassays, HPLC-MS/MS, and GC-MS have been described for the detection of cannabinoids in oral fluid. After a single dose, THC can be detected on average up to 31 h by enzyme immunoassay (EIA) and 34 h by GC-MS [46]. Maseda et al. [47] described a method for the determination of THC with an electron capture detector (GC-ECD) using pentafluoropropyl derivatives. The method was linear over the range of 5–200 ng/mL with a detection limit of 1 ng/mL. It was demonstrated that for at least 4 h after smoking, the level of THC was sufficient for detection. An LC-MS/MS method was developed to analyze THC, THC-COOH, tetrahydrocannabivarin, cannabidiol, and cannabigerol using 1 mL of oral fluid with quantitation limits of 15 pg/mL for THC-COOH and 0.2 ng/mL for all other analytes [32]. Advantages of LC-MS/MS over GC-MS include simple sample preparation and greater method sensitivity. Besides laboratory analysis, several point-of-collection testing devices for the detection of THC in oral fluid are available. These devices vary significantly in their performance [20,48–50].

Detection of Cocaine in Oral Fluid

Cocaine is an alkaloid present in the leaves of the coca plant. It has potent CNS stimulatory properties with high potential for abuse. Stimulatory effects of cocaine are through inhibition of epinephrine, norepinephrine, and dopamine reuptake at nerve synapses. It is consumed by snorting or through intravenous injection. The freebase form of cocaine, commonly referred to as "crack," is more volatile and easier to smoke. Cocaine and its metabolites including anhydroecgonine, benzoylecgonine, and ecgonine methyl ester are detectable in oral fluid after smoking and intravenous or intranasal administration.

Various studies have looked at the relationship between oral fluid and blood cocaine concentrations. A recent study on drivers suspected of driving under the influence showed a correlation of several drug concentrations between oral fluid and whole blood [37]. The concentration of cocaine was higher in oral fluid as compared to blood. The oral fluid-to-blood concentration ratio for cocaine ranged from 1.2 to 63 (median 17), while the ratio for benzoylecgonine was much lower (median 1.7). This is most likely a result of the acidic metabolite not being able to diffuse into the oral fluid. Methods of analyses for cocaine and its metabolites in oral fluid include immunoassays [51], GC-MS [37,52–54], and LC-MS [55].

Detection of Opiates/Opioids in Oral Fluid

A number of studies have shown the presence of opiates in oral fluid [56–59]. The major opiates studied in oral fluid include codeine, morphine, hydrocodone, hydromorphone, 6-AM (heroin metabolite), fentanyl, and oxycodone. Heiskanen et al. [60] tested oral fluid as an alternative to plasma to monitor opioid therapy in cancer patients dosed with morphine, oxycodone, or fentanyl. The authors observed that oral fluid for morphine and fentanyl correlated well with plasma while oxycodone was approximately 15 times higher. Methadone that is extensively used in the maintenance therapy of heroin addicts is also easily detectable in oral fluid. Heroin is quickly metabolized to 6-AM and then to morphine (half-lives being 2–7 and 6–25 min, respectively) making it difficult to distinguish between heroin or morphine use [60,61]. In a study of methadone-treated patients, Vindenes et al. [61] showed 6-AM was preferentially detected in oral fluid over urine providing a better means to verify heroin relapse versus morphine use. Oral fluid to plasma concentration is higher for most opioids including codeine, methadone, morphine, and tramadol [37].

Methods for oral fluid opioids include immunoassays, GC-MS, and LC-MS/MS. LC-MS/MS is preferred over GC-MS methods due to simpler sample preparation. When performing opiate assay by GC-MS, keto-opiates (hydrocodone, hydromorphone, and oxycodone) pose special problems as they form multiple derivatives that interfere in the analysis of other opiates and are difficult to quantify. Oximization with hydroxylamine or methyoxyamine is used to eliminate this problem [62]. Singh et al. [63] used sodium bisulfite followed by solid-phase extraction to eliminate interference by keto-opiates.

Detection of Phencyclidinein Oral Fluid

PCP is a potent veterinary analgesics and anesthetic, and is a common drug of abuse. The drug is structurally related to ketamine and was developed as an intravenous anesthetic for human use. Overdose of PCP results in lethargy, disorientation, and hallucinations. The half-life of PCP in plasma is 7–50 h with a mean of 18 h. The drug has been detected in oral fluid and there is a good correlation between oral fluid and plasma concentrations. In a study, paired oral fluid and plasma samples were collected from 100 patients who were suspected of using PCP. There was a good correlation of positivity for PCP with 74 oral fluids and 75 serum samples testing positive; however, there was no correlation between PCP concentrations and the severity of intoxication [64]. The frequently used methods for analysis of PCP in oral fluid include immunoassays, GC-MS, and LC-MS [64,65].

Detection of Barbiturates and Benzodiazepines in Oral Fluid

Barbiturates are a class of CNS depressants with sedative and hypnotic properties and have a high potential for abuse. They are prescribed for many disorders, such as migraine headaches, and seizures. Due to their narrow therapeutic window, they have largely been replaced by benzodiazepines as sedative, hypnotic, and anxiolytic compounds. Both barbiturates and benzodiazepines have been detected in oral fluid [66]. In one study on drug facility patients, eight different benzodiazepines were found [67], with concentrations ranging from 0.11 to 121 ng/mL. Alprazolam was the most commonly found benzodiazepine. In another study of database review from a pool of 635,000 oral fluid specimens, 892 were benzodiazepine positive. Nor-diazepam was the most common benzodiazepine; however, alprazolam was not included in the study. Oral fluid to blood concentration ratios vary significantly (1%—30%) among different benzodiazepines [37].

Detection of Other Drugs in Oral Fluid

Besides common drugs of abuse such as amphetamines, barbiturates, benzodiazepines, cocaine, cannabinoids, opiates, and PCP, many other substances including prescription and nonprescription medications have been reported in oral fluid. These substances include antipyrine, caffeine, carbamazepine, cisplatin, cyclosporine, digoxin, ethosuximide, gamma hydroxybutyrate, irinotecan, lithium, methadone, metoprolol, nicotine, oxprenolol, paracetamol, phenytoin, primidone, procainamide, propoxyphene, quinine, sulfanilamide, theophylline, and tolbutamide [68]. Oral fluid is used to detect cotinine, a major metabolite of nicotine. Oral samples may be screened for the presence of cotinine by ELISA and confirmed by mass spectrometry techniques. In one study, the ratio of hydroxycotinine to cotinine was used to distinguish light from heavy pregnant smokers. Average oral fluids to blood concentrations of some drugs are shown in Table 28.7 [13,37].

Adulteration Issues in Oral Fluid Specimens

Although it is difficult to adulterate oral fluid samples due to direct observation of sample collection, questions about sample adulteration have been raised. A number of adulterants are available on the Internet and claim to beat the drug test. The effectiveness of these adulterants is questionable and several studies have investigated the effect of these

TABLE 28.7 Average Oral Fluid to Blood Concentration of Some Drugs [13]

Drug (Type)	Average Oral Fluid to Blood Concentration Ratio
Barbiturates (acidic)	0.3
Buprenorphine	1
Cocaine (basic)	3
Codeine (basic)	4
Delta-9-tetrahydrocannabinol (neutral)	1.2
Diazepam (acidic)	0.01—0.02
Ethanol	1.07
MDMA (basic)	7
Methadone (basic)	1.6
Methamphetamine (basic)	2
Morphine (basic)	0.8

adulterants on oral fluid drug testing. Wong et al. [69] investigated the effects of various adulterants and foodstuffs on an Oratect device and concluded that these products do not interfere in the testing. In addition, two commercial adulterant products "Clear Choice Fizzy Flush" and "Kleen Mouthwash" were also investigated. These products did not destroy the drugs and had no effect on the pH of oral fluid. The effect of these adulterants was no different from a common mouthwash. In another study, mouthwash alteration has been shown to have no effect on oral fluid THC concentration [70]. This study investigated the effect of various solutions (water, milk, or Kleaner) on THC after smoking. The results showed that while there was a decrease in THC concentration immediately after smoking (15 min), the concentrations of THC were still greater than 75 ng/mL, well exceeding the 4 ng/mL cutoff.

In another study, the effects of a number of adulterants were evaluated on the performance of the oral fluid Cozart microplate EIA for opiates. These adulterants had no effect [71]. Although there is no scientific evidence that these adulterants work, SAMHSA recommends the use of human albumin or IgG testing. The absence of human albumin or IgG in the sample suggests sample adulteration. In oral fluid, SAMHSA recommended cutoffs for albumin and IgG are 0.6 mg/dL and 1 mg/L, respectively. Nevertheless, to avoid sample adulteration, the donor should not be allowed to put anything in his/her mouth for 10−20 min prior to sample collection.

DRUG TESTING IN SWEAT

Eccrine glands make up the majority of sweat glands on the skin's surface while apocrine glands are found in specific regions (e.g., axilla and pubic regions). Sweat is predominantly aqueous though some glands on the face and scalp produce sebaceous (oily) secretions. Sweat is produced through two types of mechanisms: insensible (consisting of diffusion) or sensible (caused by activity). Passive diffusion is thought to be the major mechanism of drug transport in sweat. Sweat is slightly acidic with an average pH of 5.8. Exercise can increase the pH to 6.4, and can affect the diffusion and, thus, the concentration of drugs in sweat. Many other factors such as temperature, hydration, other drugs, and diet can affect sweat pH. With the developments of reliable sweat collection technology and sensitive analytical methods, interest in sweat drug testing has been growing. Sweat testing provides a convenient and relatively less invasive method for continuous monitoring of drug abuse for several days to a few weeks. Sweat testing is currently being utilized to monitor drug abstinence in criminal justice and drug rehabilitation programs. Since sweat is collected only once every 1−2 weeks, sweat testing may be more economical than urine testing which requires multiple samples per week. Also, sweat provides cumulative drug exposure over several days to a few weeks [1,72].

Sample Collection and Analysis

Sweat may be collected as liquid perspiration on sweat wipes or with a sweat patch. A sweat patch is affixed to the upper arm or torso like a band-aid, and consists of an absorbent pad sandwiched between a porous inner layer and a semipermeable outer layer which allows oxygen, carbon dioxide, and water vapor to pass while larger nonvolatile molecules adhere to the pad. The protective membrane prevents external contamination of the absorbent pad. They are waterproof, tamper resistant, and comfortable to wear, and are generally acceptable by the sample donors. Several devices are available for sweat collection. One such device called PharmChek is FDA cleared and has been used in a number of studies. Before application of a sweat patch, the skin should be thoroughly cleaned with an organic solvent (such as isopropyl alcohol) to remove any dirt, skin oils, or skin care products (such as lotions and creams). This also helps in removing any external drug contamination. Once the patch is in place, it is difficult to contaminate as the outer polyurethane layer is impermeable to molecules larger than water and is specifically designed to be removed only once. Contaminants applied to the outer layer do not reach the collection pad [73].

Once the sweat sample has been collected, the patch is removed, observed for any possible tampering and then sent to a laboratory for drug testing. In the laboratory, the adsorbent pad is removed and drugs are eluted into a buffer. The eluent is tested for drugs by immunoassays or chromatographic techniques such as GC-MS or LC-MS/MS [74,75]. Various drugs including amphetamines, cannabinoids, cocaine, opiates, and PCP have been detected in sweat. There is no point-of-collection testing device available for sweat testing. SAMHSA-proposed screening and confirmation cutoffs in sweat for various drugs of abuse are shown in Table 28.3. The cutoffs are significantly lower in sweat as compared to urine.

Detection of Amphetamines and Other Sympathomimetic Amines in Sweat

The presence of amphetamine, methamphetamine, and other sympathomimetic drugs in sweat has been reported. Fay et al. [76] dosed volunteers with 10, 20, and 25 mg methamphetamine and collected sweat using a PharmChek sweat patch. The drugs were eluted from the collection patch and tested using EIA and GC-MS. Sweat primarily contained the parent drug methamphetamine, although amphetamine was also detected. In another study on 180 drivers who failed the field sobriety tests, sweat samples collected from their foreheads with a fleece moistened with isopropanol showed a positive predictive value of 90% for amphetamines by GC-MS [77]. Samyn et al. [78] collected sweat for 5 h from recreational MDMA-users who were administered 75 mg MDMA. The average concentration of MDMA was 25 ng/wipe.

Both immunoassays and chromatographic methods are used for the analysis of amphetamines in sweat. An HPLC-MS/MS method was developed by Koster et al. [75] to monitor amphetamine, methamphetamine, MDMA, MDA, and MDEA in PharmChek sweat patches. The method provided a linear range of 3–300 ng/patch and established stability at ambient temperatures for 7 days. To avoid the analytical issues of methamphetamine formation from pseudoephedrine and/or phenylpropanolamine during sample analysis, the presence of amphetamine in methamphetamine positive samples is needed.

Detection of Cannabinoids in Sweat

The presence of cannabinoids has been reported in sweat [79]. In one study, the positivity rate of urine, oral fluid, and sweat was compared in injured drivers. Out of the 198 drivers tested, 22 were positive for urine THC-COOH, 14 were positive for THC in oral fluid (1–103 ng/Salivette), and 16 were positive for sweat collected by forehead wipe (4–152 ng/pad). THC metabolites, 11-hydroxy-THC and THC-COOH were not detected in oral fluid or sweat [80]. Sweat testing has overall good positive predictive values. In a study, the positive predictive value of sweat wipe analysis with GC-MS was over 90% for cocaine and amphetamines and 80% for cannabinoids [77].

Detection of Cocaine in Sweat

Cocaine and its metabolites have also been detected in sweat. In contrast to urine, sweat primarily contains the parent drug, cocaine, and ecgonine methyl ester in much higher concentrations as compared to benzoylecgonine [81,82]. Kacinko et al. [82] performed a thorough study on subjects with a previous history of cocaine use. In this study, 15 volunteers were studied for 12 weeks in a closed research unit. The first 3 weeks were used as a washout period to excrete any previously self-administered drug. On the 4th week, the participants received three subcutaneous injections of cocaine hydrochloride (75 mg/70 kg). Three weeks later, the participants received three additional doses of cocaine hydrochloride (150 mg/70 kg). These doses were comparable with the usual dose administered by cocaine users. Using PharmChek sweat patches, 1390 samples were collected before, during, and after the controlled cocaine dosing, followed by an analysis on GC-MS. At a cutoff of 2.5 ng/patch, 24% of patches were positive, whereas at a cutoff of 25 ng/patch (SAMHSA-proposed cutoff) only 7% of patches were positive for cocaine. The study also analyzed other cocaine metabolites such as cocaethylene, norcocaine, norcocaethylene, m- and p-hydroxycocaine, and m- and p-hydroxybenzoylecgonine. Pyrolysis products of crack cocaine anhydroecgonine methylester, and ecgonidine have also been monitored in sweat patches. In a study conducted with 180 subjects, 92% were positive for cocaine use and, of those, 58% had crack metabolites, anhydroecgonine methylester being the most predominate analyte [83].

Detection of Opiates/Opioids in Sweat

Various opioids including morphine, codeine, and hydrocodone have been reported in sweat. Fogerson et al. [73] used PharmChek sweat collection patches with EIA and GC-MS for opiates analysis on human subjects injected with known doses of codeine and heroin. The authors observed that sweat primarily contains parent opiates heroin and codeine, and lipophilic metabolite 6-monoacetylmorphine (6-MAM); however, heroin is not stable at room temperature, thus making it less ideal for long-term monitoring. The EIA in which morphine was used as a calibrating analyte showed cross-reactivity with codeine of 588%, hydrocodone of 143%, diacetylmorphine of 28%, and 6-MAM of 30%. Using 215 patches from 95 subjects and receiver operator characteristics curves, the optimal cutoff was 10 ng/mL morphine equivalents. At this cutoff concentration, the assay had a diagnostic sensitivity of 86.9% and a diagnostic specificity of 92.8% as compared to GC-MS. The positive predictive value at a prevalence of 50% was 86%.

Detection of Other Drugs in Sweat

In addition to the above-mentioned drugs, many other drugs and their metabolites including atomoxetine, barbiturates, benzodiazepines, buprenorphine, bupropion, carbamazepine, clozapine, ethanol, methadone, methadone metabolite

2-ethylidene-1,5-dimethyl-3,3-diphenylpyrrolidine (EDDP), 6-MAM, nicotine, PCP, phenytoin, and phenobarbital have been detected in sweat [74,75,84].

Issues of Special Interest in Sweat Drug Testing

Rarely, some subjects can develop an allergic reaction to the sweat patch. As described above, changes in sweat pH can affect drug concentrations. Questions about sweat patch adulteration have been raised. Once the patch has been properly applied, it is difficult to contaminate the patch as the outer polyurethane layer is impermeable to molecules larger than water. Various studies have shown that the contaminants applied to the outer layer of the sweat patch do not reach the collection pad. Many products available on the Internet claim to beat a sweat test but there is no evidence that these products are effective. Fogerson et al. investigated the effect of 18 adulterants on opiates detection [73]. In this study, the sweat patches were spiked with adulterants and air dried. The patches were then spiked with heroin, codeine, or morphine and incubated at 37 °C for 7 days. Most of the adulterants tested did not have any significant effect. Also, many of these adulterants caused visible irritation, making their use evident; however, injecting an adulterant using a hypodermic needle is a possibility. Therefore, it is recommended that before analysis, the outside layer of the patch is inspected for possible tampering.

HAIR DRUG TESTING

Beginning in the 1960s, hair has been used for the evaluation of chronic exposure to heavy metals such as arsenic, cadmium, and mercury. Since that time, it has been known that drugs accumulate in hair, but due to lack of sensitive methods, the use of hair for detection of drug abuse did not become available until the late 1970s and early 1980s. Using radioimmunoassay in 1979, Baumgartner et al. [85] reported the detection of morphine in hair. Since then, detection of various drugs of abuse including amphetamines, cocaine, cannabinoids, and PCP has been reported in hair. Baumgartner et al. [85] also reported that the presence of morphine along the hair shaft correlates with the time of drug use.

Hair analysis is useful when an assessment of drug abuse over a longer time window ranging from days to years is desired. Hair analysis is useful to detect drug use for pre-employment, random screening, return-to-duty or follow-up testing. However, hair is not a suitable specimen for reasonable suspicion or cause as it takes 4–10 days for drugs and drug metabolites to appear in hair. Hair consists of hair follicle and hair shaft. In the keratin matrix of hair, drugs are incorporated permanently in contrast to blood and urine where they are eliminated within a few hours to days. Circulating drugs in blood capillaries first incorporate in hair follicles and then get entrapped in the core of the shaft as the hair grows out of the follicle. In hair, drugs can also incorporate from sweat and sebum, and from the external environment. The physiochemical properties of a drug can greatly influence how readily it is incorporated into the hair. The more neutral/basic lipophilic the substance is, the easier it incorporates into the hair, while acidic hydrophilic drugs do not diffuse as readily [86]. Once the hair shaft is protected from light, heat, and moisture, the drugs are stable for hundreds of years. For example, cocaine metabolites have been detected in hair from more than 4000-year-old mummies [87,88]. Since hair grows at a relatively constant rate, sectional analysis of hair can be used to roughly calculate the time of drug use and to verify drug abstinence. Studies on patients from rehabilitation or treatment centers have shown higher drug concentration in terminal parts and low drug concentration in proximal parts of hair [89–91].

Fetal hair has been used to detect maternal drug abuse during pregnancy. Fetal hair grows in the last trimester and positive neonatal hair indicates maternal use after pregnancy [92,93]. Usefulness of hair drug analysis in determination of compliance, alcohol abuse, doping practices, and the reissuance of a driver's license have been reported [94]. Today, hair analysis is routinely used for the detection of drugs of abuse, pharmaceuticals, and environmental contaminants in a variety of applications such as forensic science, traffic violation, correctional facilities, occupational medicine, prenatal exposure, and clinical toxicology [94].

Specimen Collection and Analysis

Hair sample collection is not well standardized and varies significantly among collection facilities. The best hair sample is likely from the area at the back of the head called the vertex posterior. Hair in this area has less variability in growth and number of hairs in the growth phase. Also, hair in this area is less influenced by age and sex [8,95]. Besides the site of sample collection, sample size also varies significantly among laboratories, from a few hairs to several hundred milligrams. Since hair grows at a rate of approximately 0.5 in./month, SAMHSA proposed the use of hairs ~1.5 in.

long which represents a time period of ~90 days [4]. Hair should be cut close to the root as the distance from the hair root reflects the time elapsed since drug use. Also, the root end must be clearly indicated.

Various methods have been described in the literature for the analysis of drugs in hair. The common steps to these procedures are extensive washing of hair to remove external drug contamination, cutting or grinding hair into small pieces, extracting drugs in organic solvents, and analysis using immunoassay and chromatographic methods [86,96]. The decontamination step to remove external contamination is an important step in hair drug analysis. Although there is no consensus on a specific procedure, various aqueous and organic solvents have been studied for decontaminating hair. These washing solvents include methanol, dichloromethane, dichloromethane/water, 2-propanol, 2-propanol/phosphate buffer, and ethanol/phosphate buffer. Decontamination with organic solvents followed by aqueous washes is recommended. Despite the most comprehensive and complex decontamination procedures, measurable concentrations of some drugs have been reported [8,97]. Once the hair is washed, drugs are extracted using various extraction solvents and buffers. A wide range of extraction procedures involving methanol, aqueous acids, urea and thioglycolate solution, supercritical fluid extraction, ionic liquids, enzymatic digestion, and digestion with aqueous sodium hydroxide have been described [96,97].

Common methods of drug hair analyses include highly sensitive immunoassays such as radioimmunoassays, enzyme immunoassays, and fluorescent immunoassays, as well as chromatographic methods, GC-MS, and LC-MS. Tandem mass spectrometry (MS/MS) using gas or liquid chromatography has emerged in recent years as it has lower limits of detection and greater specificity compared to single mass analyzers. SAMHSA- and EWDTS-proposed screening and confirmatory cutoffs for hair drug testing are shown in Tables 28.4 and 28.6 [4,8]. Like urine, the guidelines for hair testing require the presence of amphetamine in methamphetamine positive samples, and the presence of morphine in 6-MAM positive samples [4].

Analysis of Amphetamines and Other Sympathomimetic Amines in Hair

Amphetamine, methamphetamine, and other sympathomimetic amines have been reported in hair using various methods including immunoassays, GC, GC-MS, and LC-MS/MS [98–107]. Rothe et al. [107] collected hair samples from subjects with a self-reported history of amphetamine or MDMA use. Using GC-MS selected ion monitoring, the samples were analyzed for amphetamine, methamphetamine, MDA, MDMA, and MDEA. Amphetamine was detected in 17 out of 20 samples, but methamphetamine was not detected in any sample. Despite significant interindividual variations, there was a correlation between increased ecstasy abuse and the concentration of MDA, MDMA, and MDEA in the proximate 3 cm segments. Cooper et al. [106] analyzed amphetamine, methamphetamine, MDA, MDMA, and MDEA in hair samples collected from 100 subjects who were recruited from the "dance scene" in and around Glasgow, Scotland. The hair samples were analyzed in two 6 cm segments or in full, ranging from 1.5 to 12 cm depending on the length of the hair. Of the 139 segments analyzed, 77 (52.5%) were positive for amphetamines. Despite more than 50% concordance between the drugs consumed and the drugs detected in hair, there was no correlation between the drug levels and reported number of "ecstasy" tablets consumed. In one study, seven volunteers with a history of stimulant use received four doses of 10 mg of sustained release methamphetamine within 1 week. Weekly head hair samples were collected by shaving. After 3 weeks, four volunteers received four doses of 20 mg doses, followed by weekly sample collections. Maximum concentration of methamphetamine occurred from 1–2 weeks after both doses. The amphetamine/methamphetamine ratio ranged from 0.07 to 0.37 [108].

Analysis of Cannabinoids in Hair

THC and its major metabolite THC-COOH have been analyzed in hair. In contrast to urine, THC is present in much higher concentrations in hair than THC-COOH. Despite lower concentrations, it is preferred that THC-COOH is measured along with the parent drug. Detection of THC-COOH in hair provides a better proof of drug use as it is not present in the smoke or dust and is an in vivo metabolite of THC.

Immunoassays and chromatographic methods have been described for the assay of THC and THC-COOH in hair. Various extraction methods are utilized for extracting THC and its metabolite. Alkaline extraction with sodium hydroxide has been quite successful. Han et al. [109] used gas chromatography-negative chemical ionization tandem mass spectrometry (GC-MS/MS-NCI) to analyze THC and THC-COOH in hair. About 20 mg samples were incubated in 1 mL of 1 M sodium hydroxide for 30 min at 85°C, followed by extraction with hexane:ethyl acetate and a derivatization with pentafluoropropionic anhydride/pentafluoropropanol (PFPA-PFP-OH). THC and THC-COOH were measured in known cannabis users at ranges of 7.52–60.4 ng/mg and 0.1–11.7 pg/mg, respectively. Kintz et al. [110] used GC-MS/NCI for the detection of THC-COOH in hair. In their method, hair samples were decontaminated with methylene chloride and pulverized in 1 mL of 1 N sodium hydroxide for 30 min at 95°C. The extract was acidified and THC-COOH was extracted using n-hexane/ethyl acetate. Samples were derivatized with PFPA-PFP-OH. Several other methods for cannabinoids including THC, cannabidiol (CBD), cannabinol (CBN), and THC-COOH have been described [1,2].

Analysis of Cocaine in Hair

Cocaine and its metabolites have been detected in hair using various techniques such as immunoassay, GC-MS, and LC-MS/MS. Due to higher sensitivity, speed, and low cost, immunoassays are frequently employed for screening cocaine in hair. Screen positive samples are confirmed by GC-MS or LC-MS/MS. Baumgartner et al. [111] used RIA for the analysis of cocaine metabolite, benzoylecgonine, in the hair collected from patients undergoing drug rehabilitation.

Despite large variation, correlation between the concentration of cocaine in hair and the amount of cocaine consumed has been reported. Graham et al. [112] reported benzoylecgonine concentration of 0.64—29.1 ng/mg in heavy users and 0.032—1.21 ng/mg in occasional users. Also, the correlation between time elapsed after use and cocaine concentration has been reported. Seven subjects who used cocaine for 2—12 months were tested before and after 2 months of abstinence. Benzoylecgonine concentrations estimated by RIA were 0.6—6.4 and 0.3—0.5 ng/mg hair, respectively. Hair samples from infants whose mothers used cocaine during pregnancy had benzoylecgonine concentration of 0.2—2.75 ng/mg [112]. The drug was not detected after 10 weeks which corresponded to the loss of their fetal hair. A study on maternal urine and neonatal hair testing for cocaine showed hair analysis for gestational cocaine exposure which had a sensitivity of 88%, specificity of 100%, and positive and negative predictive values of 100% and 85%, respectively [113]. In another study, median cocaine concentration was tenfold higher in hair of the mothers compared with the neonates (3.56 ng/mg vs 0.31 ng/mg of hair). Segmental analysis has been used and correlated with the timing of drug use. A study on a cocaine abuser showed the disappearance of cocaine in the hair sections closest to the root after 3 months of abstinence [89].

The measurement of cocaine metabolites is preferred to rule out external contamination. Various methods for the detection of cocaine and its metabolites have been described [114,115]. A highly sensitive GC-MS/MS method for the simultaneous determination of cocaine, anhydroecgonine methylester, ecgonine methylester, and cocaethylene metabolites has been reported. The limit of detection for these compounds was 0.005, 0.050, 0.025, 0.005 ng/mg hair, respectively [115].

Analysis of Opiates/Opioids in Hair

Various opioids including codeine, morphine, hydrocodone, hydromorphone, oxycodone 6-AM, heroin, and buprenorphine have been detected in hair. In earlier studies, Baumgartner et al. [85] reported the presence of heroin and morphine in hair samples collected from heroin addicts. Strikingly, all hair samples were positive as compared to only a 30% positivity rate in the urine samples. This is likely due to chronic versus acute use. The authors also noted a good correlation between drug concentration and duration of drug use. However, in a study on 20 subjects taking part in a heroin maintenance program and receiving drug doses of 30—800 mg/day, no significant correlation was found between heroin dose and the concentration [116].

Various methods including radioimmunoassay, enzyme immunoassays, GC-MS, and LC-MS/MS have been reported in the literature for analysis of opiates in hair. Baumgartner et al. [85] used radioimmunoassays for the assay of morphine and heroin. Cooper et al. [105] evaluated a Cozart microplate EIA method for opioid detection in hair. They collected 164 hair specimens from volunteers and from drug-related deaths. The drugs were extracted using methanol at 60°C. The methanolic extract was dried and derivatized with N,O-bis(trimethylsilyl)trifluoroacetamide (BSTFA) and analyzed with GC-MS. Using a cutoff of 200 pg/mg hair, Cozart EIA had a sensitivity and specificity of 98% and 93%, respectively. Lachenmeier et al. [117] used oral fluid microplate enzyme immunoassays for hair opiate testing and were found to have good correlation with GC-MS. Jones et al. [118] described a solid-phase extraction GC-MS method for simultaneous determination of several opioids including codeine, morphine, hydrocodone, hydromorphone, 6-AM, and oxycodone in hair and oral fluid. Hair was powdered using stainless steel balls and drugs were extracted using a buffered, aqueous matrix using a solid-phase cartridge. The extracts were derivatized with methoxime/BSTFA to eliminate interference from the keto-opiates. Fernandez Mdel et al. [119] described an LC-MS/MS method for a simultaneous analysis of 33 basic drugs including amphetamines, cocaine, opiates, opioids, and metabolites collected from forensic and clinical cases.

Analysis of Phencyclidine in Hair

PCP and its metabolites have been reported in hair. Baumgartner et al. [120] used radioimmunoassay for the detection of PCP in hair from subjects admitting PCP use. All seven subjects tested positive for PCP by hair analysis. Only one out of seven subjects tested positive with urine analysis, indicating chronic use and a wider window of detection through hair analysis. PCP concentrations ranged from 0.3 to 5.2 and 0.3 to 2.8 ng/mg in unwashed and washed hair samples, respectively. Also, there was a correlation between duration of use and the drug concentration. In another study on psychiatric patients, hair analysis detected PCP in 11 out of 47 patients while blood and urine were negative for all patients [121]. These studies demonstrate the longer stability and wider window of detection for PCP in hair samples.

Analysis of Other Drugs in Hair

The presence of drugs, other than those described above, has been reported in the literature. These drugs include barbiturates, benzodiazepines, buprenorphine, cotinine ethanol metabolites, fentanyl, γ-hydroxybutyrate, ketamine, methadone, methylphenidate, nicotine, PCP, quetiapine, tramadol, and Z-drugs [94,95,97,122].

Environmental or External Contamination of Hair

Questions about external drug contamination are frequently raised, particularly by the subjects who had a positive drug screen. Various hair washing combinations of aqueous and/or organic solvents are used to decontaminate hair of any drug from passive exposure. Despite extensive washing, studies have shown that external contamination for certain drugs cannot be completely removed. In two studies, hairs were incubated with cocaine followed by extensive washing. These washings, however, did not completely remove the cocaine contamination [123,124]. It has been proposed that detection of cocaine metabolites may differentiate the active use of cocaine from external contamination or passive exposure. Koren et al. [125] showed that pyrolysis of crack results in the accumulation of cocaine in hair, but not its metabolite benzoylecgonine. On the other hand, in cocaine users both cocaine and benzoylecgonine are detectable in hair. Many studies, however, have shown the presence of drug and/or drug metabolites after external contamination [97,126,127]. Also, it is important to note that intense washing procedures can lead to the removal of drugs from hair, thus increasing the chance of false negative results [126,128,129]. Studies on different wash solutions for the removal of cocaine and its metabolites, benzoylecgonine, norcocaine, ecgonine methylester, and cocaethylene, showed that different wash solutions lead to significant differences in the measured concentrations of analytes [124,128].

Certain shampoos contain a significant amount (1%–3%) of cannabinoids. In a study, analysis of a Cannabio shampoo revealed the presence of THC, cannabidiol (CBD), and cannabinol (CBN) at concentrations of 412, 4079, and 380 ng/mL, respectively [130]. To explore the possibility of false positive results due to the use of Cannabio shampoo, in the same study, three subjects washed their hair with Cannabio shampoo once daily for 2 weeks. The hairs from these subjects were analyzed for THC, CBD, and CBN and none tested positive. Also in this study, the hairs were incubated in water/Cannabio shampoo (20:1, v/v) for 30 min, 2 h, and 5 h. The samples that incubated for 30 min were negative for THC, CBD, and CBN, but the samples that were incubated for 2 h and 5 h tested positive for CBD and CBN but not for THC [130]. The other strategy to rule out external contamination is to analyze samples for marijuana metabolite. Uhl and Sachs [131] reported a case in which hair samples from a couple living together tested positive for THC and CBN. The male subject used marijuana, whereas the female subject did not. Only the hair sample from the male subject tested positive for THC-COOH, whereas the sample from the female subject tested negative (LOQ 0.1 pg/mg). SAMHSA guidelines proposed the hair analysis for THC-COOH in THC positive samples [4].

As described above, for cocaine and cannabinoids, similar questions about external contamination have been raised about other drugs including amphetamine, opioids, and PCP. In a study in New Zealand, hair samples were collected from children removed from clandestine laboratories. The samples were analyzed for methamphetamine and amphetamine with 38 out of 52 (73%) hair samples testing positive for methamphetamine (>0.1 ng/mg), and of these positive samples, 34 were also positive for amphetamine. Amphetamine was not detected in any case without the presence of methamphetamine [132]. The presence of PCP has also been reported in external contamination. The detection of PCP metabolites may be helpful to rule out passive exposure. Nakahara et al. [133] reported the simultaneous detection of PCP and its two major metabolites, 1-(1-phenylcyclohexyl)-4-hydroxypiperidine and trans-1-(1-phenyl-4-hydroxycyclohexyl)-4′-hydroxypiperidine (t-PCPdiol), by GC-MS. The authors recommend the use of t-PCPdiol as an indicator of active PCP use. SAMHSA-proposed guidelines recommend analysis of PCP without any mention of the PCP metabolite. In a study, ratios of hydromorphone to morphine and hydrocodone to codeine were used to distinguish heroin use from external contamination. Hair samples after external contamination with street heroin were negative for hydromorphone and hydrocodone. The authors concluded that hydromorphone and hydrocodone are solely formed during body passage and metabolite ratios can be used to distinguish morphine/heroin use from external contamination [134].

Although questions regarding external hair contamination are valid, it seems that the use of effective wash procedures, appropriate cutoffs, metabolite ratios, and metabolite analyses can distinguish external drug contamination from active drug use.

Hair Color

Many studies have shown that different colored hair incorporates drugs to a different degree. This raises the question of bias based on hair color. Melanin that determines the hair color is a polyanionic polymer of eumelanin and pheomelanin. Eumelanin concentration is highest in black color, whereas pheomelanin is highest in red color. White hairs have little to no melanin. It is postulated that these pigments have different affinities for different drugs, resulting in different drug concentrations; this effect has been studied in both animals and humans. Long-Evans rats provide an ideal model for the study of drug incorporation as it produces both black and white hair. Borges et al. [135] studied amphetamine incorporation in the hair of Long-Evans rats. Black hair had a significantly higher concentration of amphetamine compared to white hair. Amphetamine concentrations in black and white hair were 6.44 and 2.04 ng/mg, respectively. The Long-Evans rat model has also been used to study the incorporation of PCP into different colors of hair. PCP levels were significantly higher in pigmented hair (14.33 ng/mg) as compared to nonpigmented hair (0.47 ng/mg) [136]. In a recent study, Long-Evans rats were administered 25B-, 25C- or 25I-NBOMe, new psychoactive substances derived from phenethylamines. The drug doses ranged from 30 to 300 µg/kg over a period of 10 days. Rats were shaved prior to their first dose and reshaved after the 10-day period. Black and white hairs were separated and analyzed using an LC-MS/MS method. A dose-dependent concentration increase was observed in the black hair while, for white hair, only 25B- and 25C-NBOMe were quantifiable at higher doses [137].

Several human studies have also shown the influence of different colored hair on the incorporation of different drugs. Kronstrand et al. [138] measured methamphetamine and amphetamine levels in patients receiving selegiline, the drug which metabolizes to methamphetamine and amphetamine. The drugs were analyzed in pigmented and nonpigmented hairs from gray-haired patients. Methamphetamine and amphetamine had higher concentration in pigmented hairs than white hair. Goldberger et al. [91] found higher concentrations of cocaine in the Africoid group as compared to the Caucasoid group. There was significantly more cocaine in black hair than in brown or blonde hair per mg of cocaine dosed [139]. Using human volunteers, codeine incorporation in different colors of hair has also been studied [140]. The volunteers were given 30 mg of codeine three times a day for 5 days. Codeine concentrations in black, brown, blonde, and red hair were 1429 + 249, 208 + 17, 99 + 10, and 69 + 11 pg/mg (mean + SE), respectively. These differences were attributed to differences in melanin concentrations, and when the results were normalized to melanin concentrations, these differences attenuated. The effect of hair color on cocaine incorporation has also been studied in vitro. Incubation of hair with cocaine and benzoylecgonine resulted in higher concentration of benzoylecgonine in colored hair as compared to blonde hair. The concentration of benzoylecgonine was black > brown > blond hair [141].

Contrary to the above-mentioned studies, a number of studies did not find a significant difference between different hair colors, race, or ethnicity. Schaffer et al. [142] studied cocaine incorporation in blonde, auburn, brown, and black hair. They did not find any difference among different hair colors; however, when hair was permed, there was an increase in cocaine uptake. The authors postulated that porosity, not hair color, may be responsible for the observed differences. A study on 1852 applicants for employment that classified themselves as "black" or "white" showed no significant difference between hair and urine drug test results for these two classes [143]. In another study, hairs from pre-employment applicants were used for the detection of amphetamines, cocaine, and cannabinoids. The drug positive rates for hair and urine samples were compared in various ethnic and racial groups. These authors concluded that the differences were largely due to drug preferences among the various groups, and not due to hair color [144].

Hair Adulteration Issues

Due to direct supervision, the chances of adulteration during hair collection are minimal; however, concern remains regarding hair treatment before collection. There are numerous sites on the Internet that sell products which claim to beat or pass the hair drug test. Although a number of these products claim to work 100%, there is little scientific evidence to back up their claim of effectiveness. Ultra Clean shampoo is one of these products. The Internet site for Ultra Clean shampoo claims, "Hair purifying shampoo is designed for extremely swift yet elegant removal of all toxin residues. Within 10 min, you will be in the 'Clean Zone' for up to 4 h. It is safe for the scalp and all hair types." Rohrich et al. [107] investigated the effect of Ultra Clean shampoo on removing THC, cocaine, amphetamine, MDA, MDMA, MDE, heroin, 6-MAM, morphine, codeine, dihydrocodeine, and methadone from hair samples. In this study, postmortem hair samples from subjects with a known history of drug abuse were used. The samples were analyzed with and without the Ultra Clean shampoo treatment. Although samples tested positive after the Ultra Clean shampoo treatment, there were slight decreases in drug concentrations: cocaine—5% ($n = 10$), 6-MAM—9% ($n = 12$), morphine—26% ($n = 12$), and THC—36% ($n = 4$). The authors concluded that a single use of Ultra Clean shampoo will not cause a negative result. Pritchett and Phinney [145] found that chemical straightening products significantly decreased cocaine, benzoylecgonine, and cocaethylene from positive hair samples by 70%—95% after a single use. These findings were similar for fortified hair standards reference material. The authors speculate that the high pH of the straightener could cause hydrolysis or decomposition of the drugs.

DRUG TESTING IN NAIL

Both hair and nails are keratinized matrices which accumulate xenobiotics. When hair is not available, the nail has been proposed as an alternative matrix. Nails grow at a rate of 1.9—4.4 mm/month (mean 3.47 mm/month for fingernails) and 0.9—2.1 mm/month (mean 1.62 mm/month for toenails). Like hair, nails can provide a drug detection window of several months. As compared to hair, nail is less prone to drug destruction from exposure to sunlight, heat, and water. For drug analysis, nails are first washed to remove any external contamination followed by drug extraction with organic solvents. Krumbiegel et al. [146] collected hair and nail samples from 70 postmortem cases and analyzed them for 89 drugs using liquid-chromatography-quadrupole time-of-flight mass spectrometry. There was 90% agreement between hair and nail samples.

In contrast to hair, nail formation does not exclusively occur from the root. The growth of nail takes place in two different areas; approximately 80% of the nail matrix originates from the nail root and 20% is produced from the nail bed during growth. The incorporation of drug in nail takes place from sweat as well as from handling of powders, liquids, and tablets. Therefore, nail analysis may offer advantage over hair testing especially where hair has been treated with destructive hair cosmetic products and procedures such as hair bleaching and dying. In one study, the hair (20 mg) and nail samples (20 mg) were ground using a ball mill and were extracted for 18 h. Then the authors analyzed extracts using ultra-high-performance liquid chromatography tandem mass spectrometry and included 76 substances in the analysis. Whole nail and hair samples from seven autopsy cases were also included in the study. The authors observed several drugs in hair and nail specimens, but actual drugs concentrations in hair and nail were not comparable. Previous heroin intake could be assumed from the detection of 6-MAM and morphine in only nail segment in one case. The authors concluded that nail specimens may serve as an alternative matrix for the detection of long-term consumption of a wide range of drugs but external contamination may affect the analysis of nail clippings [147].

Shu et al. collected nail specimens by clipping 2—3 mm nail from all fingers (100 mg specimen) and analyzed 10,349 nail specimens (collected from high-risk cases over a period of 3 years) using both immunoassays (initial testing) and LC-MS/MS. Amphetamine and methamphetamine were present in 14% specimens (concentration range: 40—572, 865 pg/mg), but only 0.3% specimens were positive for MDMA. Cocaine and related compounds were present in 5% specimens (concentration range: 20—265,063 pg/mg), while overall opioid concentrations ranged from 40 to 118,229 pg/mg of nail. The most prevalent opioid detected was oxycodone (15.1% specimens) followed by hydrocodone (11.2%) compared with 1.0%—3.6% for the others including morphine, codeine, hydromorphone, methadone, EDDP, and oxymorphone. The positivity rate of tetrahydrocannabinol metabolite was, however, highest (18.1%) with a concentration range of 0.04—262 pg/mg of nail. Other drugs found were barbiturates, benzodiazepines, ketamine, meperidine, tramadol, zolpidem, propoxyphene, naltrexone, and buprenorphine [148].

CONCLUSIONS

In summary, alternate samples such as oral fluid, sweat, hair, and nail provide unique opportunities in the detection of drug abuse. Due to a number of advantages offered by these matrices, drug testing interest in these matrices is growing. Each matrix provides a unique perspective and opportunity in drug abuse detection. Nail testing is getting more popular because nail is a useful specimen in men who may have short hair or for postmortem analysis. Besides many advantages, these matrices present unique analytical and interpretive challenges.

REFERENCES

[1] Garg U. Hair, oral fluid, sweat and meconium testing for drugs of abuse: advantages and disadvantages. In: Dasgupta A, editor. Handbook of drug monitoring methods: therapeutics and drugs of abuse. Totowa, NJ: Humana Press; 2008. p. 337–64.

[2] Garg U. Testing of alternative specimens for drugs of abuse: Analytical, interpretive and sample adulteration issues. In: Dasgupta A, editor. Critical issues in alcohol and drug testing. Washington, DC: AACC Press; 2009. p. 229–44.

[3] Garg U, Ferguson AM. Alternate specimens for drugs-of-abuse testing: preanalytical and interpretative considerations. In: Barbarajean M, Bissell MG, Kwong TC, Wu A, editors. Clinical toxicology testing: a guide for laboratory professionals. Chicago, IL: CAP Press; 2012. p. 71–80.

[4] Department of Health and Human Services Administration, Substance abuse and mental health services administration mandatory guidelines for federal workplace drug testing programs, Federal Register/vol. 69, No. 71. 2004:19673–732.

[5] Department of Health and Human Services Administration, Substance abuse and mental health services administration mandatory guidelines for federal workplace drug testing programs, Federal Register/vol. 80, No. 94. 2015:28054–101.

[6] Prego-Meleiro P, Lendoiro E, Concheiro M, Cruz A, Lopez-Rivadulla M, de Castro A. Development and validation of a liquid chromatography tandem mass spectrometry method for the determination of cannabinoids and phase I and II metabolites in meconium. J Chromatogr A 2017;1497:118–26.

[7] Brcak M, Beck O, Bosch T, Carmichael D, Fucci N, George C, et al. European guidelines for workplace drug testing in oral fluid. Drug Test Anal 2017.

[8] Salomone A, Tsanaclis L, Agius R, Kintz P, Baumgartner MR. European guidelines for workplace drug and alcohol testing in hair. Drug Test Anal 2016;8:996–1004.

[9] Cone EJ. Oral fluid testing: new technology enables drug testing without embarrassment. J Calif Dent Assoc 2006;34:311–15.

[10] Dyer KR, Wilkinson C. The detection of illicit drugs in oral fluid: another potential strategy to reduce illicit drug-related harm. Drug Alcohol Rev 2008;27:99–107.

[11] Garg U, Presley L. Role of saliva in detection of substance abuse. In: Wong D, editor. Saliva diagnostics. Ames, IA: Wiley-Blackwell Press; 2008. p. 169–79.

[12] Groschl M. Saliva: a reliable sample matrix in bioanalytics. Bioanalysis 2017;9:655–68.

[13] Drummer OH. Drug testing in oral fluid. Clin Biochem Rev 2006;27:147–59.

[14] Cone EJ, Huestis MA. Interpretation of oral fluid tests for drugs of abuse. Ann N Y Acad Sci 2007;1098:51–103.

[15] Aps JK, Martens LC. Review: the physiology of saliva and transfer of drugs into saliva. Forensic Sci Int 2005;150:119–31.

[16] Cone EJ, Menchen SL. Stability of cocaine in saliva. Clin Chem 1988;34:1508.

[17] Ventura M, Pichini S, Ventura R, Zuccaro P, Pacifici R, de la Torre R. Stability studies of principal illicit drugs in oral fluid: preparation of reference materials for external quality assessment schemes. Ther Drug Monit 2007;29:662–5.

[18] Scherer JN, Fiorentin TR, Borille BT, Pasa G, Sousa TRV, von Diemen L, et al. Reliability of point-of-collection testing devices for drugs of abuse in oral fluid: a systematic review and meta-analysis. J Pharm Biomed Anal 2017;143:77–85.

[19] Walsh JM, Flegel R, Crouch DJ, Cangianelli L, Baudys J. An evaluation of rapid point-of-collection oral fluid drug-testing devices. J Anal Toxicol 2003;27:429–39.

[20] Walsh JM, Crouch DJ, Danaceau JP, Cangianelli L, Liddicoat L, Adkins R. Evaluation of ten oral fluid point-of-collection drug-testing devices. J Anal Toxicol 2007;31:44–54.

[21] Moore C, Coulter C, Rana S, Vincent M, Soares J. Analytical procedure for the determination of the marijuana metabolite 11-nor-Delta9-tetrahydrocannabinol-9-carboxylic acid in oral fluid specimens. J Anal Toxicol 2006;30:409–12.

[22] Moore C, Rana S, Coulter C. Simultaneous identification of 2-carboxy-tetrahydrocannabinol, tetrahydrocannabinol, cannabinol and cannabidiol in oral fluid. J Chromatogr B Analyt Technol Biomed Life Sci 2007;852:459–64.

[23] Huestis MA, Cone EJ. Methamphetamine disposition in oral fluid, plasma, and urine. Ann N Y Acad Sci 2007;1098:104–21.

[24] Bahmanabadi L, Akhgari M, Jokar F, Sadeghi HB. Quantitative determination of methamphetamine in oral fluid by liquid–liquid extraction and gas chromatography/mass spectrometry. Hum Exp Toxicol 2017;36:195–202.

[25] Langel K, Gunnar T, Ariniemi K, Rajamaki O, Lillsunde P. A validated method for the detection and quantitation of 50 drugs of abuse and medicinal drugs in oral fluid by gas chromatography–mass spectrometry. J Chromatogr B Analyt Technol Biomed Life Sci 2011;879:859–70.

[26] Strano-Rossi S, Leone D, de la Torre X, Botre F. Analysis of stimulants in oral fluid and urine by gas chromatography–mass spectrometry II: pseudophedrine. J Anal Toxicol 2010;34:210–15.

[27] Oiestad EL, Johansen U, Christophersen AS. Drug screening of preserved oral fluid by liquid chromatography-tandem mass spectrometry. Clin Chem 2007;53:300–9.

[28] Teixeira H, Proenca P, Castanheira A, Santos S, Lopez-Rivadulla M, Corte-Real F, et al. Cannabis and driving: the use of LC-MS to detect delta9-tetrahydrocannabinol (delta9-THC) in oral fluid samples. Forensic Sci Int 2004;146(Suppl):S61–3.

[29] Allen KR, Azad R, Field HP, Blake DK. Replacement of immunoassay by LC tandem mass spectrometry for the routine measurement of drugs of abuse in oral fluid. Ann Clin Biochem 2005;42:277–84.

[30] Samyn N, Laloup M, De Boeck G. Bioanalytical procedures for determination of drugs of abuse in oral fluid. Anal Bioanal Chem 2007;388:1437–53.

[31] Fiorentin TR, D'Avila FB, Comiran E, Zamboni A, Scherer JN, Pechansky F, et al. Simultaneous determination of cocaine/crack and its metabolites in oral fluid, urine and plasma by liquid chromatography–mass spectrometry and its application in drug users. J Pharmacol Toxicol Methods 2017;86:60–6.

[32] Desrosiers NA, Scheidweiler KB, Huestis MA. Quantification of six cannabinoids and metabolites in oral fluid by liquid chromatography–tandem mass spectrometry. Drug Test Anal 2015;7:684–94.

[33] Cone EJ, Presley L, Lehrer M, Seiter W, Smith M, Kardos KW, et al. Oral fluid testing for drugs of abuse: positive prevalence rates by Intercept immunoassay screening and GC-MS-MS confirmation and suggested cutoff concentrations. J Anal Toxicol 2002;26:541–6.

[34] Yacoubian Jr GS, Wish ED, Perez DM. A comparison of saliva testing to urinalysis in an arrestee population. J Psychoactive Drugs 2001;33:289–94.

[35] Casolin A. Comparison of urine and oral fluid for workplace drug testing. J Anal Toxicol 2016;40:479–85.

[36] Gjerde H, Langel K, Favretto D, Verstraete AG. Detection of illicit drugs in oral fluid from drivers as biomarker for drugs in blood. Forensic Sci Int 2015;256:42–5.

[37] Langel K, Gjerde H, Favretto D, Lillsunde P, Oiestad EL, Ferrara SD, et al. Comparison of drug concentrations between whole blood and oral fluid. Drug Test Anal 2014;6:461–71.

[38] Schepers RJ, Oyler JM, Joseph Jr RE, Cone EJ, Moolchan ET, et al. Methamphetamine and amphetamine pharmacokinetics in oral fluid and plasma after controlled oral methamphetamine administration to human volunteers. Clin Chem 2003;49:121–32.

[39] Laloup M, Tilman G, Maes V, De Boeck G, Wallemacq P, Ramaekers J, et al. Validation of an ELISA-based screening assay for the detection of amphetamine, MDMA and MDA in blood and oral fluid. Forensic Sci Int 2005;153:29–37.

[40] Nieddu M, Burrai L, Baralla E, Pasciu V, Varoni MV, Briguglio I, et al. ELISA detection of 30 new amphetamine designer drugs in whole blood, urine and Oral fluid using Neogen(R) "amphetamine" and "methamphetamine/MDMA" kits. J Anal Toxicol 2016;40:492–7.

[41] Kankaanpaa A, Gunnar T, Ariniemi K, Lillsunde P, Mykkanen S, Seppala T. Single-step procedure for gas chromatography–mass spectrometry screening and quantitative determination of amphetamine-type stimulants and related drugs in blood, serum, oral fluid and urine samples. J Chromatogr B Analyt Technol Biomed Life Sci 2004;810:57–68.

[42] Concheiro M, de Castro A, Quintela O, Lopez-Rivadulla M, Cruz A. Determination of MDMA, MDA, MDEA and MBDB in oral fluid using high performance liquid chromatography with native fluorescence detection. Forensic Sci Int 2005;150:221–6.

[43] Mortier KA, Maudens KE, Lambert WE, Clauwaert KM, Van Bocxlaer JF, Deforce DL, et al. Simultaneous, quantitative determination of opiates, amphetamines, cocaine and benzoylecgonine in oral fluid by liquid chromatography quadrupole-time-of-flight mass spectrometry. J Chromatogr B Analyt Technol Biomed Life Sci 2002;779:321–30.

[44] Wood M, De Boeck G, Samyn N, Morris M, Cooper DP, Maes RA, et al. Development of a rapid and sensitive method for the quantitation of amphetamines in human plasma and oral fluid by LC-MS-MS. J Anal Toxicol 2003;27:78–87.

[45] Niedbala RS, Kardos KW, Fritch DF, Kunsman KP, Blum KA, Newland GA, et al. Passive cannabis smoke exposure and oral fluid testing. II. Two studies of extreme cannabis smoke exposure in a motor vehicle. J Anal Toxicol 2005;29:607–15.

[46] Niedbala RS, Kardos KW, Fritch DF, Kardos S, Fries T, Waga J, et al. Detection of marijuana use by oral fluid and urine analysis following single-dose administration of smoked and oral marijuana. J Anal Toxicol 2001;25:289–303.

[47] Maseda C, Hama K, Fukui Y, Matsubara K, Takahashi S, Akane A. Detection of delta 9-THC in saliva by capillary GC/ECD after marihuana smoking. Forensic Sci Int 1986;32:259–66.

[48] Crouch DJ, Walsh JM, Flegel R, Cangianelli L, Baudys J, Atkins R. An evaluation of selected oral fluid point-of-collection drug-testing devices. J Anal Toxicol 2005;29:244–8.

[49] Cirimele V, Villain M, Mura P, Bernard M, Kintz P. Oral fluid testing for cannabis: on-site OraLine IV s.a.t. device versus GC/MS. Forensic Sci Int 2006;161:180–4.

[50] Speedy T, Baldwin D, Jowett G, Gallina M, Jehanli A. Development and validation of the Cozart DDS oral fluid collection device. Forensic Sci Int 2007;170:117–20.

[51] Niedbala RS, Kardos K, Fries T, Cannon A, Davis A. Immunoassay for detection of cocaine/metabolites in oral fluids. J Anal Toxicol 2001;25:62–8.

[52] Campora P, Bermejo AM, Tabernero MJ, Fernandez P. Quantitation of cocaine and its major metabolites in human saliva using gas chromatography-positive chemical ionization–mass spectrometry (GC-PCI-MS). J Anal Toxicol 2003;27:270–4.

[53] Yonamine M, Tawil N, Moreau RL, Silva OA. Solid-phase micro-extraction-gas chromatography-mass spectrometry and headspace-gas chromatography of tetrahydrocannabinol, amphetamine, methamphetamine, cocaine and ethanol in saliva samples. J Chromatogr B Analyt Technol Biomed Life Sci 2003;789:73–8.

[54] Cognard E, Bouchonnet S, Staub C. Validation of a gas chromatography–ion trap tandem mass spectrometry for simultaneous analyse of cocaine and its metabolites in saliva. J Pharm Biomed Anal 2006;41:925–34.

[55] Clauwaert K, Decaestecker T, Mortier K, Lambert W, Deforce D, Van Peteghem C, et al. The determination of cocaine, benzoylecgonine, and cocaethylene in small-volume oral fluid samples by liquid chromatography-quadrupole-time-of-flight mass spectrometry. J Anal Toxicol 2004;28:655–9.

[56] Piekoszewski W, Janowska E, Stanaszek R, Pach J, Winnik L, Karakiewicz B, et al. Determination of opiates in serum, saliva and hair addicted persons. Przegl Lek 2001;58:287–9.

[57] Speckl IM, Hallbach J, Guder WG, Meyer LV, Zilker T. Opiate detection in saliva and urine—a prospective comparison by gas chromatography–mass spectrometry. J Toxicol Clin Toxicol 1999;37:441–5.

[58] Kidwell DA, Holland JC, Athanaselis S. Testing for drugs of abuse in saliva and sweat. J Chromatogr B Biomed Sci Appl 1998;713:111–35.

[59] Lo Muzio L, Falaschini S, Rappelli G, Bambini F, Baldoni A, Procaccini M, et al. Saliva as a diagnostic matrix for drug abuse. Int J Immunopathol Pharmacol 2005;18:567–73.

[60] Heiskanen T, Langel K, Gunnar T, Lillsunde P, Kalso EA. Opioid concentrations in oral fluid and plasma in cancer patients with pain. J Pain Symptom Manage 2015;50:524–32.

[61] Vindenes V, Yttredal B, Oiestad EL, Waal H, Bernard JP, Morland JG, et al. Oral fluid is a viable alternative for monitoring drug abuse: detection of drugs in oral fluid by liquid chromatography–tandem mass spectrometry and comparison to the results from urine samples from patients treated with methadone or buprenorphine. J Anal Toxicol 2011;35:32–9.

[62] Broussard LA, Presley LC, Tanous M, Queen C. Improved gas chromatography-mass spectrometry method for simultaneous identification and quantification of opiates in urine as propionyl and oxime derivatives. Clin Chem 2001;47:127–9.

[63] Singh J, Burke RE, Mertens LE. Elimination of the interferences by keto-opiates in the GC-MS analysis of 6-monoacetylmorphine. J Anal Toxicol 2000;24:27–31.

[64] McCarron MM, Walberg CB, Soares JR, Gross SJ, Baselt RC. Detection of phencyclidine usage by radioimmunoassay of saliva. J Anal Toxicol 1984;8:197–201.

[65] Fritch D, Blum K, Nonnemacher S, Haggerty BJ, Sullivan MP, Cone EJ. Identification and quantitation of amphetamines, cocaine, opiates, and phencyclidine in oral fluid by liquid chromatography–tandem mass spectrometry. J Anal Toxicol 2009;33:569–77.

[66] Fritch D, Kardos K. Personal communication, barbiturate intercept micro-plate EIA kit information. Bethlehem, PA: OraSure Technologies; 2007.

[67] Ngwa G, Fritch D, Blum K, Newland G. Simultaneous analysis of 14 benzodiazepines in oral fluid by solid-phase extraction and LC-MS-MS. J Anal Toxicol 2007;31:369–76.

[68] Kaufman E, Lamster IB. The diagnostic applications of saliva—a review. Crit Rev Oral Biol Med 2002;13:197–212.

[69] Wong RC, Tran M, Tung JK. Oral fluid drug tests: effects of adulterants and foodstuffs. Forensic Sci Int 2005;150:175–80.

[70] de Castro A, Lendoiro E, Fernandez-Vega H, Lopez-Rivadulla M, Steinmeyer S, Cruz A. Assessment of different mouthwashes on cannabis oral fluid concentrations. Drug Test Anal 2014;6:1011–19.

[71] Cooper G, Wilson L, Reid C, Baldwin D, Hand C, Spiehler V. Validation of the cozart microplate EIA for analysis of opiates in oral fluid. Forensic Sci Int 2005;154:240–6.

[72] Gambelunghe C, Rossi R, Aroni K, Bacci M, Lazzarini A, De Giovanni N, et al. Sweat testing to monitor drug exposure. Ann Clin Lab Sci 2013;43:22–30.

[73] Fogerson R, Schoendorfer D, Fay J, Spiehler V. Qualitative detection of opiates in sweat by EIA and GC-MS. J Anal Toxicol 1997;21:451–8.

[74] Gentili S, Mortali C, Mastrobattista L, Berretta P, Zaami S. Determination of different recreational drugs in sweat by headspace solid-phase microextraction gas chromatography mass spectrometry (HS-SPME GC/MS): application to drugged drivers. J Pharm Biomed Anal 2016;129:282–7.

[75] Koster RA, Alffenaar JW, Greijdanus B, VanDerNagel JE, Uges DR. Application of sweat patch screening for 16 drugs and metabolites using a fast and highly selective LC-MS/MS method. Ther Drug Monit 2014;36:35–45.

[76] Fay J, Fogerson R, Schoendorfer D, Niedbala RS, Spiehler V. Detection of methamphetamine in sweat by EIA and GC-MS. J Anal Toxicol 1996;20:398–403.

[77] Samyn N, De Boeck G, Verstraete AG. The use of oral fluid and sweat wipes for the detection of drugs of abuse in drivers. J Forensic Sci 2002;47:1380–7.

[78] Samyn N, De Boeck G, Wood M, Lamers CT, De Waard D, Brookhuis KA, et al. Plasma, oral fluid and sweat wipe ecstasy concentrations in controlled and real life conditions. Forensic Sci Int 2002;128:90–7.

[79] Saito T, Wtsadik A, Scheidweiler KB, Fortner N, Takeichi S, Huestis MA. Validated gas chromatographic-negative ion chemical ionization mass spectrometric method for delta(9)-tetrahydrocannabinol in sweat patches. Clin Chem 2004;50:2083–90.

[80] Kintz P, Cirimele V, Ludes B. Detection of cannabis in oral fluid (saliva) and forehead wipes (sweat) from impaired drivers. J Anal Toxicol 2000;24:557–61.

[81] Liberty HJ, Johnson BD, Fortner N. Detecting cocaine use through sweat testing: multilevel modeling of sweat patch length-of-wear data. J Anal Toxicol 2004;28:667–73.

[82] Kacinko SL, Barnes AJ, Schwilke EW, Cone EJ, Moolchan ET, Huestis MA. Disposition of cocaine and its metabolites in human sweat after controlled cocaine administration. Clin Chem 2005;51:2085–94.

[83] Liberty HJ, Johnson BD, Fortner N, Randolph D. Detecting crack and other cocaine use with fastpatches. Addict Biol 2003;8:191–200.

[84] De Giovanni N, Fucci N. The current status of sweat testing for drugs of abuse: a review. Curr Med Chem 2013;20:545–61.

[85] Baumgartner AM, Jones PF, Baumgartner WA, Black CT. Radioimmunoassay of hair for determining opiate-abuse histories. J Nucl Med 1979;20:748–52.

[86] Vogliardi S, Tucci M, Stocchero G, Ferrara SD, Favretto D. Sample preparation methods for determination of drugs of abuse in hair samples: a review. Anal Chim Acta 2015;857:1–27.

[87] Rivera MA, Aufderheide AC, Cartmell LW, Torres CM, Langsjoen O. Antiquity of coca-leaf chewing in the south central Andes: a 3,000 year archaeological record of coca-leaf chewing from northern Chile. J Psychoactive Drugs 2005;37:455–8.

[88] Cartmell LW, Aufderhide A, Weems C. Cocaine metabolites in pre-Columbian mummy hair. J Okla State Med Assoc 1991;84:11–12.

[89] Felli M, Martello S, Marsili R, Chiarotti M. Disappearance of cocaine from human hair after abstinence. Forensic Sci Int 2005;154:96–8.

[90] Musshoff F, Lachenmeier K, Wollersen H, Lichtermann D, Madea B. Opiate concentrations in hair from subjects in a controlled heroin-maintenance program and from opiate-associated fatalities. J Anal Toxicol 2005;29:345–52.

[91] Goldberger BA, Darraj AG, Caplan YH, Cone EJ. Detection of methadone, methadone metabolites, and other illicit drugs of abuse in hair of methadone-treatment subjects. J Anal Toxicol 1998;22:526–30.

[92] Garcia-Bournissen F, Rokach B, Karaskov T, Koren G. Methamphetamine detection in maternal and neonatal hair: implications for fetal safety. Arch Dis Child Fetal Neonatal Ed 2007;92:F351−5.

[93] Garcia-Bournissen F, Rokach B, Karaskov T, Koren G. Cocaine detection in maternal and neonatal hair: implications to fetal toxicology. Ther Drug Monit 2007;29:71−6.

[94] Villain M, Cirimele V, Kintz P. Hair analysis in toxicology. Clin Chem Lab Med 2004;42:1265−72.

[95] Kintz P. Value of hair analysis in postmortem toxicology. Forensic Sci Int 2004;142:127−34.

[96] Restolho J, Barroso M, Saramago B, Dias M, Afonso CA. Development, optimization, and validation of a novel extraction procedure for the removal of opiates from human hair's surface. Drug Test Anal 2015;7:385−92.

[97] Cuypers E, Flanagan RJ. The interpretation of hair analysis for drugs and drug metabolites. Clin Toxicol (Phila) 2017;1−11.

[98] Nakahara Y. Detection and diagnostic interpretation of amphetamines in hair. Forensic Sci Int 1995;70:135−53.

[99] Vinner E, Vignau J, Thibault D, Codaccioni X, Brassart C, Humbert L, et al. Hair analysis of opiates in mothers and newborns for evaluating opiate exposure during pregnancy. Forensic Sci Int 2003;133:57−62.

[100] Takayama N, Iio R, Tanaka S, Chinaka S, Hayakawa K. Analysis of methamphetamine and its metabolites in hair. Biomed Chromatogr 2003;17:74−82.

[101] Sweeney SA, Kelly RC, Bourland JA, Johnson T, Brown WC, Lee H, et al. Amphetamines in hair by enzyme-linked immunosorbent assay. J Anal Toxicol 1998;22:418−24.

[102] Quintela O, Bermejo AM, Tabernero MJ, Strano-Rossi S, Chiarotti M, Lucas AC. Evaluation of cocaine, amphetamines and cannabis use in university students through hair analysis: preliminary results. Forensic Sci Int 2000;107:273−9.

[103] Miki A, Katagi M, Tsuchihashi H. Application of EMIT d.a.u. for the semiquantitative screening of methamphetamine incorporated in hair. J Anal Toxicol 2002;26:274−9.

[104] Lendoiro E, Jimenez-Morigosa C, Cruz A, Paramo M, Lopez-Rivadulla M, de Castro A. An LC-MS/MS methodological approach to the analysis of hair for amphetamine-type-stimulant (ATS) drugs, including selected synthetic cathinones and piperazines. Drug Test Anal 2017;9:96−105.

[105] Cooper G, Wilson L, Reid C, Baldwin D, Hand C, Spiehler V. Validation of the Cozart microplate ELISA for detection of opiates in hair. J Anal Toxicol 2003;27:581−6.

[106] Cooper GA, Allen DL, Scott KS, Oliver JS, Ditton J, Smith ID. Hair analysis: self-reported use of "speed" and "ecstasy" compared with laboratory findings. J Forensic Sci 2000;45:400−6.

[107] Rohrich J, Zorntlein S, Potsch L, Skopp G, Becker J. Effect of the shampoo ultra clean on drug concentrations in human hair. Int J Legal Med 2000;113:102−6.

[108] Polettini A, Cone EJ, Gorelick DA, Huestis MA. Incorporation of methamphetamine and amphetamine in human hair following controlled oral methamphetamine administration. Anal Chim Acta 2012;726:35−43.

[109] Han E, Park Y, Kim E, In S, Yang W, Lee S, et al. Simultaneous analysis of Delta(9)-tetrahydrocannabinol and 11-nor-9-carboxy-tetrahydrocannabinol in hair without different sample preparation and derivatization by gas chromatography-tandem mass spectrometry. J Pharm Biomed Anal 2011;55:1096−103.

[110] Kintz P, Cirimele V, Mangin P. Testing human hair for cannabis. II. Identification of THC-COOH by GC-MS-NCI as a unique proof. J Forensic Sci 1995;40:619−22.

[111] Baumgartner WA, Black CT, Jones PF, Blahd WH. Radioimmunoassay of cocaine in hair: concise communication. J Nucl Med 1982;23:790−2.

[112] Graham K, Koren G, Klein J, Schneiderman J, Greenwald M. Determination of gestational cocaine exposure by hair analysis. JAMA 1989;262:3328−30.

[113] Katikaneni LD, Salle FR, Hulsey TC. Neonatal hair analysis for benzoylecgonine: a sensitive and semiquantitative biological marker for chronic gestational cocaine exposure. Biol Neonate 2002;81:29−37.

[114] Bermejo AM, Lopez P, Alvarez I, Tabernero MJ, Fernandez P. Solid-phase microextraction for the determination of cocaine and cocaethylene in human hair by gas chromatography−mass spectrometry. Forensic Sci Int 2006;156:2−8.

[115] Cognard E, Rudaz S, Bouchonnet S, Staub C. Analysis of cocaine and three of its metabolites in hair by gas chromatography−mass spectrometry using ion-trap detection for CI/MS/MS. J Chromatogr B Analyt Technol Biomed Life Sci 2005;826:17−25.

[116] Kintz P, Bundeli P, Brenneisen R, Ludes B. Dose-concentration relationships in hair from subjects in a controlled heroin-maintenance program. J Anal Toxicol 1998;22:231−6.

[117] Lachenmeier K, Musshoff F, Madea B. Determination of opiates and cocaine in hair using automated enzyme immunoassay screening methodologies followed by gas chromatographic−mass spectrometric (GC-MS) confirmation. Forensic Sci Int 2006;159:189−99.

[118] Jones J, Tomlinson K, Moore C. The simultaneous determination of codeine, morphine, hydrocodone, hydromorphone, 6-acetylmorphine, and oxycodone in hair and oral fluid. J Anal Toxicol 2002;26:171−5.

[119] Fernandez Mdel M, Di Fazio V, Wille SM, Kummer N, Samyn N. A quantitative, selective and fast ultra-high performance liquid chromatography tandem mass spectrometry method for the simultaneous analysis of 33 basic drugs in hair (amphetamines, cocaine, opiates, opioids and metabolites). J Chromatogr B Analyt Technol Biomed Life Sci 2014;965:7−18.

[120] Baumgartner AM, Jones PF, Black CT. Detection of phencyclidine in hair. J Forensic Sci 1981;26:576−81.

[121] Sramek JJ, Baumgartner WA, Tallos JA, Ahrens TN, Heiser JF, Blahd WH. Hair analysis for detection of phencyclidine in newly admitted psychiatric patients. Am J Psychiatry 1985;142:950−3.

[122] White RM. Drugs in hair. Part I. Metabolisms of major drug classes. Forensic Sci Rev 2017;29:23—55.
[123] Blank DL, Kidwell DA. External contamination of hair by cocaine: an issue in forensic interpretation. Forensic Sci Int 1993;63:145—56, discussion 157—60.
[124] Stout PR, Ropero-Miller JD, Baylor MR, Mitchell JM. External contamination of hair with cocaine. evaluation of external cocaine contamination and development of performance-testing materials. J Anal Toxicol 2006;30:490—500.
[125] Koren G, Klein J, Forman R, Graham K. Hair analysis of cocaine: differentiation between systemic exposure and external contamination. J Clin Pharmacol 1992;32:671—5.
[126] Romano G, Barbera N, Lombardo I. Hair testing for drugs of abuse: evaluation of external cocaine contamination and risk of false positives. Forensic Sci Int 2001;123:119—29.
[127] Wang X, Drummer OH. Review: interpretation of drug presence in the hair of children. Forensic Sci Int 2015;257:458—72.
[128] Paulsen RB, Wilkins DG, Slawson MH, Shaw K, Rollins DE. Effect of four laboratory decontamination procedures on the quantitative determination of cocaine and metabolites in hair by HPLC-MS. J Anal Toxicol 2001;25:490—6.
[129] Hill V, Cairns T, Schaffer M. Hair analysis for cocaine: factors in laboratory contamination studies and their relevance to proficiency sample preparation and hair testing practices. Forensic Sci Int 2008;176:23—33.
[130] Cirimele V, Kintz P, Jamey C, Ludes B. Are cannabinoids detected in hair after washing with Cannabio shampoo? J Anal Toxicol 1999;23:349—51.
[131] Uhl M, Sachs H. Cannabinoids in hair: strategy to prove marijuana/hashish consumption. Forensic Sci Int 2004;145:143—7.
[132] Bassindale T. Quantitative analysis of methamphetamine in hair of children removed from clandestine laboratories—evidence of passive exposure? Forensic Sci Int 2012;219:179—82.
[133] Nakahara Y, Takahashi K, Sakamoto T, Tanaka A, Hill VA, Baumgartner WA. Hair analysis for drugs of abuse. XVII. Simultaneous detection of PCP, PCHP, and PCPdiol in human hair for confirmation of PCP use. J Anal Toxicol 1997;21:356—62.
[134] Madry MM, Bosshard MM, Kraemer T, Baumgartner MR. Hair analysis for opiates: hydromorphone and hydrocodone as indicators of heroin use. Bioanalysis 2016;8:953—64.
[135] Borges CR, Wilkins DG, Rollins DE. Amphetamine and N-acetylamphetamine incorporation into hair: an investigation of the potential role of drug basicity in hair color bias. J Anal Toxicol 2001;25:221—7.
[136] Slawson MH, Wilkins DG, Foltz RL, Rollins DE. Quantitative determination of phencyclidine in pigmented and nonpigmented hair by ion-trap mass spectrometry. J Anal Toxicol 1996;20:350—4.
[137] Nisbet LA, Venson R, Wylie FM, Scott KS. Application of a urine and hair validated LC-MS-MS method to determine the effect of hair color on the incorporation of 25B-NBOMe, 25C-NBOMe and 25I-NBOMe into hair in the rat. J Anal Toxicol 2017;41:559—65.
[138] Kronstrand R, Ahlner J, Dizdar N, Larson G. Quantitative analysis of desmethylselegiline, methamphetamine, and amphetamine in hair and plasma from Parkinson patients on long-term selegiline medication. J Anal Toxicol 2003;27:135—41.
[139] Ursitti F, Klein J, Sellers E, Koren G. Use of hair analysis for confirmation of self-reported cocaine use in users with negative urine tests. J Toxicol Clin Toxicol 2001;39:361—6.
[140] Rollins DE, Wilkins DG, Krueger GG, Augsburger MP, Mizuno A, O'Neal C, et al. The effect of hair color on the incorporation of codeine into human hair. J Anal Toxicol 2003;27:545—51.
[141] Reid RW, O'Connor FL, Crayton JW. The in vitro differential binding of benzoylecgonine to pigmented human hair samples. J Toxicol Clin Toxicol 1994;32:405—10.
[142] Schaffer M, Hill V, Cairns T. Hair analysis for cocaine: the requirement for effective wash procedures and effects of drug concentration and hair porosity in contamination and decontamination. J Anal Toxicol 2005;29:319—26.
[143] Hoffman BH. Analysis of race effects on drug-test results. J Occup Environ Med 1999;41:612—14.
[144] Kelly RC, Mieczkowski T, Sweeney SA, Bourland JA. Hair analysis for drugs of abuse. Hair color and race differentials or systematic differences in drug preferences? Forensic Sci Int 2000;107:63—86.
[145] Pritchett JS, Phinney KW. Influence of chemical straightening on the stability of drugs of abuse in hair. J Anal Toxicol 2015;39:13—16.
[146] Krumbiegel F, Hastedt M, Tsokos M. Nails are a potential alternative matrix to hair for drug analysis in general unknown screenings by liquid-chromatography quadrupole time-of-flight mass spectrometry. Forensic Sci Med Pathol 2014;10:496—503.
[147] Krumbiegel F, Hastedt M, Westendroft L, Niebel A, Methling M, Parr MK, et al. The use of nails as ana alternative matrix for the long eterm detection of previous drug intake: validation of sensitive UHPLC-MS/MS methods for the quantification of 76 substances and comparison of analytical results for drugs in nails and hair samples. Forensic Sci Med Pathol 2016;12:416—34.
[148] Shu I, Jones J, Jones M, Lewis D, Negrusz A. Detection of drugs in nails: three yera experience. J Anal Toxicol 2015;39:624—8.

Chapter 29

Advances in Meconium Analysis for Assessment of Neonatal Drug Exposure

Steven W. Cotten
Department of Pathology and Laboratory Medicine, University of North Carolina at Chapel Hill, Chapel Hill, NC, United States

INTRODUCTION

Conventional toxicology testing focuses on urine drug screening in adult populations. Applications of such testing often include emergency medicine, preemployment drug screening, and pain management testing. For these applications, laboratory testing plays a central role in employment or medical management of patients. Another application of toxicology testing is assessment of in utero drug exposure in newborns. This population comes with unique preanalytical and analytical challenges for laboratories providing these testing services. The specimen types available for analysis and the metabolites present in the specimens are two contributing factors that make neonatal toxicology testing different than the adult population.

Testing specimens from the neonate seeks to evaluate prenatal drug exposure during pregnancy. Traditional drug screening in adults can use either verbal interrogation for drug use, self-reporting behavior, or biological specimen analysis to provide the necessary information for establishment of compliance or abuse. For neonates, clinicians and healthcare providers rely on maternal self-reported information, maternal specimens, medications lists, and analysis of specimens from the neonate. To complicate the picture, interpretation of drug testing results may be left to those outside the immediate care team such as social services workers.

In addition to the analytical challenges that are outlined in the following chapter, neonatal specimens have unique preanalytical issue as well as postanalytical complexities. Some major preanalytical issues related to specimen collection, stability, and processing can influence test accuracy. Neonatal specimens can be accessed by family members of the infant due to extensive unsupervised time after delivery but before discharge. Collection of urine samples from diapers can be challenging and sample volume is often low, limiting the ability to confirm presumptive positive results. Complete collection of meconium is also challenging and requires combining samples from multiple diapers over possible several days. This can be difficult to manage multiple teams of health-care providers across multiple shifts.

MECONIUM AS SPECIMEN

Several types of specimens are available for analysis including, urine, serum, hair, umbilical cord, and meconium. Neonate urine has a short window of detection for drug exposure going back only the last 3–5 days prior to delivery. Neonate hair begins to form during the last trimester and may capture exposure during the last 3–4 months. Umbilical cord tissue (UCT) has also gained popularity as a specimen type going back several months however challenges remain regarding adequate detection of certain drug classes. Additionally, the detection of epidural medications administered during delivery complicate interpretation. For these reasons, meconium remains the specimen of choice for assessment of neonatal drug exposure due to its high sensitivity and long window of detection over the last 20 weeks.

Meconium refers to the fecal matter passed usually within the first 48 h after birth and is characterized as being odorless, dark green/black in color, and having a thick tar-like appearance. The formation of meconium begins at the 12th week of gestation and continues to accumulate until delivery. Deposition of meconium occurs via swallowing of amniotic fluid and shedding of cells in the developing gastrointestinal (GI) tract. Drugs may be deposited in the GI tract after crossing the placenta into the fetus blood stream. Primary or secondary pass metabolism may occur in the mother and additional metabolism of drugs may occur in the placenta. The continuous process of accumulation contributes to the long window of detection for drug exposure during pregnancy.

The collection of meconium is considered a passive, noninvasive process. Specimen collection can usually occur within 48 h after birth with 99% of full-term infants passing their entire meconium during this timeframe. However, drug exposed neonates are slower to pass meconium with the median time to first stool being day 3 and 90% by day 12 post birth. Collection of the entire meconium is considered best practice for maximum sensitivity for drug detection but limitations related to logistical recovery of meconium from diapers often limit complete collection.

Over the last 10 years, a variety of new methods related to the extraction and analysis of drugs and alcohol from meconium have been published. The following chapter will review the analytical methods published over the last decade that primarily involved the development of novel liquid chromatography combined with tandem mass spectrometry (LC/Ms/MS) analysis of meconium for clinical laboratories. These studies reflect the current focus on illicit opiates, opiates for medication maintenance, alcohol abuse, nicotine abuse, antiretroviral (ARV) medication, and cannabinoid exposure which may be coupled to increase opioid abuse and cannabinoid legalization in the United States. To a lesser extent, antidepressants, antipsychotics, cocaine, and benzodiazepines have been investigated and methods are published for their analysis in meconium. The heterogeneity of the analysis is illustrated by the various methods employed and the limit of quantitation's reported for similar drugs from each protocol. This diverse method performance may be due to extraction, chromatographic, and ionization efficiencies. Instrumentation may also vary in analytical sensitivity which further contributes to protocol variation.

ANALYSIS OF ALCOHOL AND FATTY ACID ETHYL ESTERS

Alcohol use in women of childbearing age has risen slightly over the last decade. Data from 2015 demonstrated that alcohol consumption by women varied by state and ranged from 34% (West Virginia) to 73% (District of Columbia) [1]. Among pregnant women, 1 in 10 reported any alcohol use in the last 30 days and 1 in 33 reported binge drinking in the last 30 days. Exposure to alcohol in utero can results in fetal alcohol spectrum disorder which can manifest growth impairment, abnormal facial features, increased risk for mental disorders, and social impairments [2]. Analysis of fatty acid ethyl esters (FAEEs), ethyl glucuronide (EtG), and ethyl sulfate (EtS) in meconium have been used to evaluate the degree of prenatal alcohol exposure. Various cutoffs have been proposed that utilize individual FAEEs or a cumulative concentration. Initially a total cutoff for positive exposure of 50 ng/g in meconium was first proposed by several studies but this cutoff was not evidence based. An evidence-based cutoff of 2 nmol/g has been proposed and recommended by several international consensus groups [3]. This cutoff uses total amount from seven FAEE (palmitic, palmitoleic, stearic, oleic, linoleic, linolenic, and arachidonic acid ethyl esters) but excludes lauric and myristic acid ethyl esters.

Morini et al. [4] have reported analysis of EtG and EtS from two separate cohorts. Six and seven point calibration curves from 1 to 500 ng/g were used to establish the analytical measuring range. Recovery ranged from 72.1% to 95.6% for EtS and 78.7%–96.8% for EtG. Separation was performed via LC and detection was performed using multiple reaction monitoring in negative mode over 9 min. Comparison of quantitative EtG and EtS to total FAEE showed no correlation.

Kwak et al. [5] have reported an improved liquid chromatography/mass spectrometry (LC/Ms) method of analysis of nine FAEE. The method has notable changes from a previous reference method. Instead of 1 g, 0.5 g of meconium is used. Mobile phase of 0.01% formic acid and 90% acetonitrile resulted in better peak shape. Analytical runtime was reduced from 21 to 15 min. Instead of 50 μL, 20 μL injections were used, thus improving the limit of quantitation. When applied to a cohort of 81 women who reported alcohol consumption during pregnancy, the total FAEEs from the new method ranged from 0 to 37.8 nmol/g.

Hastedt et al. [6] published a gas chromatography/mass spectrometry (GC/Ms) method using headspace solid phase microextraction for seven FAEE. Oven temperatures ramped from 70 to 300°C at 10°C per minute for a total runtime of 30 min. Positive samples were identified in 20% of cases which is three times more prevalent than previous studies within the same population.

Himes et al. [7] have published another LC/Ms/MS method that for analysis of nine FAEEs, EtS, and ethyl EtG. Supported liquid extraction (SLE) was used to extract FAEE while solid phase extraction (SPE) analysis using the Evolute-AX chemistry was utilized for EtG and EtS. Analysis of FAEE and EtG and EtS was performed separately. Instability of FAEEs in 50%–100% aqueous mobile phase makes it challenging for simultaneous normal reverse phase conditions required for EtG and EtS, due to their high polarity and rapid elution in low organic conditions. Separate chromatographic conditions were required for optimal peak height and retention on the column. EtG and EtS were analyzed in negative mode while FAEE were analyzed in positive mode. Methods for analysis of FAEE, EtG, and EtS in meconium are summarized in Table 29.1.

TABLE 29.1 Analysis of Ethyl Glucuronide Ethyl Sulfate and Fatty Acid Ethyl Esters in Meconium

Analytes	N	Sample Weight (g)	Extraction	Method of Analysis	LOQ	Reference
Ethyl glucuronide, ethyl sulfate	64	0.2	Amino-propyl silica SPE	LC/ESI MS/MS	1 ng/g Ethyl Sulfate 5 ng/g EtG	[4]
Ethyl laurate Ethyl myristate Ethyl palmitate Ethyl palmitoleate Ethyl oleate Ethyl linoleate Ethyl linolenate Ethyl stearate Ethyl arachidonate	81	0.5	Liquid/Liquid	LC/ESI MS/MS	0.02–0.27 nmol/g	[5]
Ethyl myristate Ethyl palmitate Ethyl oleate Ethyl linoleate Ethyl stearate Ethyl linolenate	122	0.05	Liquid/Liquid Extraction + HS-SPME	GC/MS	21 ng/g 17 ng/g 19 ng/g 36 ng/g 10 ng/g 44 ng/g	[6]
Ethyl laurate Ethyl myristate Ethyl linolenate Ethyl palmitoleate Ethyl arachidonate Ethyl linoleate Ethyl palmitate Ethyl oleate Ethyl stearate Ethyl sulfate Ethyl glucuronide	13	0.1	SLE for (FAEEs) SPE Evolute-AX for (EtG/S)	LC/ESI/ MS/MS	50 ng/g 25 ng/g 25 ng/g 25 ng/g 25 ng/g 25 ng/g 50 ng/g 25 ng/g 50 ng/g 2.5 ng/g 5 ng/g	[7]

LQQ, Lower limit of quantitation; *LC/ESI/MS/MS*, liquid chromatography combined with tandem mass spectrometry with electrospray ionization; *GC/Ms*, Gas chromatography/mass spectrometry; *SPE*, solid phase extraction; *HS-SPME*, headspace solid-phase microextraction; *SLE*, supported liquid extraction.

ANALYSIS OF AMPHETAMINE

Although amphetamine (AMP) abuse is currently not as prevalent when compared to opiate abuse in the United States, emerging data from 2018 suggests AMP abuse is increasing. There is currently limited data investigating the effects of in utero exposure to AMP, methamphetamine (MAMP), and 3,4-methylenedioxymethamphetamine (MDMA) alone or in conjunction with other drugs. Kelly et al. [8] from the National Institutes of Health (NIH) have developed and validated an LC/Ms/MS method for analysis of 10 AMPs and metabolites in meconium. The challenge for analysis of both parent and metabolite compounds is due to the extreme variation in concentration. AMP and MAMP have been reported to be present at concentrations exceeding 1000 ng/g in meconium samples. Minor metabolites may only be present at much lower concentrations. This makes creating a single calibration curve spanning several factors of 10 challenging. To accommodate this, two calibration curves utilizing two different injection volumes were validated. The first curve ranged from 1.25 to 2500 ng/g and used a 10 μL injection. The second curve ranged from 125 to 10,000 ng/g and used a 1 μL injection. Using liquid–liquid extraction for 30 min followed by SPE cleanup, the calculated extraction efficiencies ranged from 68.8% to 86.8%. Liquid chromatographic separation occurred over 15 min followed by an 18 min equilibration. Total runtime was approximately 33 min. The lower limit of quantitation (LOQ) ranged from 1.25 to 40 ng/g. The elevated limit of quantitation for 4-hydroxy-3-methoxyamphetamine (HMA) was attributed to background interference present in blank meconium with the HMA qualifier transition ion.

One year later, work from the same group was published that applied this method to 43 meconium specimens to investigate the deposition of various metabolites including para-hydroxyamphetmaine (pOHAMP), para-hydroxy methamphetamine (pOHMAMP), HMA, 4-hydroxy-3-methoxymethamphetamin (HMMA), norephedrine in meconium [9]. Analysis revealed that after MAMP and AMP, pOHMAMP was the most abundant metabolite present with 86% of samples being positive for this compound. Norephedrine and HMMA were present in 25% and 18% of samples, respectively. There was a correlation between MAMP concentrations with AMP and pOHMAMP. There was no correlation of MAMP concentration with HMMA or norephedrine.

This group later expanded upon their earlier work by applying the method for additional AMP metabolites to 132 meconium samples irrespective of initial immunoassay screening results for AMPs [10]. Meconium collected from the infants of three sets of women was evaluated. The first set was negative self-reported MAMP use with positive immunoassay results. The second was positive self-reported MAMP use with negative immunoassay results. The third group was positive self-reported MAMP use with positive immunoassay results. Addition of pOHMAMP increased positive detection by 14%. For six samples it was the only AMP metabolite present in the meconium specimen. The detection of pOHMAMP was loosely correlated with continued MAMP use in the third trimester and occurred at higher concentrations. Detection of pOHMAMP was also associated with lower birth weight and younger maternal age although not in a linear fashion. Despite increases in detection of MAMP exposure, the addition of pOHMAMP did not increase the confirmation rate of positive immunoassay results. The biotransformation of MAMP to AMP and pOHMAMP by CYP2D6 may play a role in variability of detection. The authors acknowledge that CYP2D6 has greater than 70 known allelic variants and its activity is increased in pregnancy. Additionally, fetal CYP2D6 activity is known to be present but its activity toward MAMP and its related metabolites is unknown. Methods for analysis of AMP and related compounds in meconium are summarized in Table 29.2.

TABLE 29.2 Analysis of Amphetamine and Related Compounds in Meconium

Analytes	N	Sample Weight (g)	Extraction	Method of Analysis	LOQ (ng/g)	Reference
AMP MAMP, MDMA MDA MDEA HMA HMMA pOHAMP pOHMAMP Norephedrine	1	1	Liquid/Liquid + SPE Strata-XC	LC/APCI MS/MS	1.25−40	[8]
AMP MAMP, MDMA MDA MDEA HMA HMMA pOHAMP pOHMAMP norephedrine	43	1	Liquid/Liquid + SPE Strata-XC	LC/APCI MS/MS	8 pOHMAMP 12.5 pOHAMP 12.5 Norephedrine	[9]
Amphetamine Methamphetamine, MDMA MDA MDEA HMA HMMA pOHAMP pOHMAMP norephedrine,	132	1	Liquid/Liquid + SPE Strata-XC	LC/ESI MS/MS	1−5	[10]

ANALYSIS OF ANTIRETROVIRALS

ARV medications have become increasingly prescribed to prevent the transmission of HIV from mother to baby during delivery. Globally, sub-Saharan Africa accounts for 91% of all pregnant women living with HIV [11]. In the United States approximately 9000 HIV-infected women give birth each year. Administration of ARV medication to HIV-infected pregnant women reduces transmission of HIV to the neonate to less than 2%. Due to the large number of compounds analyzed, a sophisticated extraction process was developed. Meconium samples underwent two 10 min liquid−liquid extractions with methanol that were pooled and evaporated under nitrogen to near dryness. The samples were then reconstituted in 3 mL of water pH 2.9 and subject to SPE. The SPE (Strata-X) is targeted at basic drugs of abuse and contains are proprietary polymer. When samples were applied to the column, the resulting eluate was collected and retained for further extraction. SPE columns were washed with 5% methanol then drugs eluted with 0.025% formic acid in acetonitrile. After elution, the SPE columns were reequilibrated to pH 1.1 and the initial elution was applied. The second SPE step was necessary because tenofovir was not retained on the Stata-X column at pH 2.9. The eluate was acidified to pH 1.1 and applied to the column. After washing with water, target compounds were released with the elution solvent and combined with the eluate from the first SPE. Two injections, one in positive mode and one in negative mode were required for optimal sensitivity. Most analytes were optimized and acquired in positive mode except azidothymidine (AZT, also known as zidovudine), AZT-glucuronide, efavirenz (EFV), and d4T. Compounds in positive mode were separated over 34 min. The second injection separated compounds over 16 min. Both conditions used a C18 column. Extraction efficiencies ranged from 55.6% to119.5% for all compounds except tenofovir which had reduced recovery despite a separate SPE cleanup of 32.8%−44.0% across the linear range.

The method was later applied to 598 meconium samples collected from babies born to HIV positive women enrolled in the Pediatric HIV/AIDS Cohort Study [12]. Results of the study evaluated detection of ARV medication prescribed in the second and third trimesters separately. Only 6 out of 107 samples with second trimester ARV prescriptions were positive in the assay. This may support findings that suggest meconium is best for detection of third trimester drug exposure and has limited detection of exposure in the second trimester. AZT, which could come from maternal prescription during pregnancy or prophylactic administration to the infant at birth, was investigated separately. Neonatal prophylaxis was administered in 98.7% cases. Meconium was positive for AZT in 94.5% of infants exposed to AZT by any route. The addition of AZT-glucuronide did not increase detection rates for exposure as all specimens positive for AZT-glucuronide were also positive for AZT. There was no detection of ARV medications in 107 meconium samples collected from babies born to mothers receiving third trimester ARV. The authors suggest that this high number of negative results may be a result of poor medication compliance, earlier delivery, lower gestational age, maternal−fetal metabolism, or shorter exposure duration. Methods for analysis of ARVs in meconium are summarized in Table 29.3.

TABLE 29.3 Analysis of Antiretrovirals in Meconium

Analytes	N	Sample Weight(g)	Extraction	Method of Analysis	LOQ (ng/g)	Reference
Abacavir Atazanavir Darunavir Efavirenz Emtricitabine Lamivudine Lopinavir Nelfinavir Nevirapine Raltegravir Ritonavir Saquinavir Stavudine Tenofovir Zidovudine ABC-carboxylate ABC-glucuronide NFV hydroxyl-tert-butylamide AZT-glucuronide	32 (2013) 598 (2015)	0.25	Liquid/Liquid + SPE Strata-X	LC−ESI MS/MS	10−500	[11]* [12]*

Source: Refs. [11] and [12] are published by the Dr. Huestis's research group, one in 2013 (Ref. [11]) and follow-up paper in 2015 (Ref. [12]).

ANALYSIS OF CANNABINOIDS

The use of cannabis and tetrahydrocannabinol (THC) extracts has increased in the United States as medical marijuana and legalization become more prevalent at the state level despite federal prohibition. Marijuana continues to be the most widely used illegal drug with approximately 22.2 million users per month [13]. Data on use during pregnancy is estimated at 1 in every 25 in the United States. Despite such high use there is limited information regarding the short and long term effects of cannabis exposure during fetus development. However, some associations have been found between marijuana use and developmental hyperactivity, low birth weight, and premature birth [14]. Peat et al. [15] published a GC method for analysis of 11-carboxy-delta-9-tetrahydrocannabinol (THC-COOH) in meconium using a liquid−liquid extraction. One gram of meconium is used in the extraction and the sample must be aliquoted then frozen at −20°C overnight to increase extraction efficiency. Two separate acetonitrile extractions are performed which are combined then dried down. The sample is alkalinized with potassium hydroxide (KOH) and sodium hydroxide (NaOH) followed by a 15 min extraction with hexane: ethyl acetate. The organic layer is discarded and the aqueous layer is acidified then reextracted again for 15 min with hexane: ethyl acetate. The organic phase is retained and evaporated to dryness followed by derivatization with N,O-bis(trimethylsilyl)-trifluoroacetamide and trimethylchlorosilane. This complex extraction partitions the THC-COOH into the aqueous layer under alkaline conditions then into the organic layer under acidic conditions. Glucuronide metabolites are also extracted and hydrolyzed under acidic conditions. The concentration of the organic extract prior to derivatization and injection takes approximately 30 min. Separation via GC takes 12 min with elution of THC-COOH occurring between 8 and 8.5 min. The method is linear from 1 to 500 ng/g THC-COOH. No data was given on the application of the method with a sample set of meconium specimens.

Prego-Meleiro et al. [16] have described an LC/Ms method for analysis of eight compounds in meconium. One major difference between this method and previous methods is the direct inclusion of glucuronide conjugates in the analytical method. Previous published methods consistently included a hydrolysis step to liberate glucuronide conjugates as a means to increase sensitivity. This method seeks to achieve similar detection of cannabinoid exposure in meconium without a hydrolysis step. A methanol extraction was performed for 30 min on 0.25 g of meconium followed evaporation to dryness and resuspending in 200 μL of methanol first to solubilize the cannabinoids followed by addition of water and formic acid. The extract was subject to cation exchange SPE then dried down. The dried eluent was resuspended in water: Acetonitrile (60:40). Separation of target compounds was achieved using a C18 column over 7.5 min. All data were acquired in positive mode. During development, several SPE columns, elution buffers, and different pH values were evaluated. The wide range of characteristics of the target compounds such as polar di OH-THC and glucuronides compared to nonpolar compounds such as cannabidiol (CBD) and THC make a single extraction challenging. Mixed mode columns provided a balance between sensitivity and recovery but certain columns clogged due to the meconium matrix. The larger particle size of the Oasis MCX was found to perform best for this application. When the method was applied to 19 authentic specimens, cannabinoid metabolites were detected in four specimens. THC, diOH-THC, CBD, cannabinol (CBN), and THC-COOH-glucuronide were found in all specimens. THC-COOH was found in three of the four positive specimens. OH-THC was found in one specimen of four specimens. THC-glucuronide was not detected in any specimen. More studies utilizing this panel of compounds are needed to determine which metabolites are required for maximum sensitivity but preliminary results suggests THC-glucuronide and OH-THC do not improved detection rates. Methods for analysis of THC and related compounds in meconium are summarized in Table 29.4.

ANALYSIS OF COCAINE

Cocaine use by the general population has changed drastically over the last 100 years. Beginning as a completely unregulated compound in the early 1900s to a schedule I drug has seen an ebb and flow of popularity for abuse. The present opiate addiction crisis was preceded by cocaine abuse in the 1980s. Much of the original focus on assessment of in utero drug exposure arose from evaluating cocaine exposure during this time period. Several methods were published recently detailing updated methods of evaluation of cocaine and its relevant metabolites in meconium. Gray et al. [17] described a method for analysis of 20 drugs, 4 of which were cocaine and relevant metabolites. Cocaine, benzoylecgonine, cocaethylene, and meta-hydroxy-benzoylecgonine (mOHBE) were quantitated in a single injection via LC/Ms along with 14 other drugs of abuse in meconium. Target compounds were extracted from 0.25 g of meconium for 30 min using 1 mL of methanol with 0.1% formic acid. After centrifugation, the supernatant was dried down under nitrogen and reconstituted in phosphate buffer pH 6.0. The sample was applied to a conditioned CleanScreen DAU SPE column and after cleanup the resulting eluate was dried down after adding one drop of 1% HCl to prevent evaporation of volatile compounds. Samples were reconstituted in 200 μL of 1 mM ammonium formate pH 3.4. Extracted samples

TABLE 29.4 Analysis of THC and Related Compounds in Meconium

Analytes	N	Sample Weight(g)	Extraction	Method of Analysis	LOQ (ng/g)	Reference
11-Carboxy-delta-9-tetrahydrocannabinol	Not given	1	Liquid/Liquid	GC/MS	10	[15]
Tetrahydrocannabinol THC-glucuronide Hydroxy-tetrahydrocannabinol Di-hydroxy-tetrahydrocannabinol Carboxy-tetrahydrocannabinol THC-COOH glucuronide Cannabinol, CBD	19	0.25	Liquid/Liquid + SPE Oasis MCX	LC/MS/MS	10 (THC-COOH glucuronide) 4 all others	[16]

TABLE 29.5 Analysis of Cocaine and Related Compounds in Meconium

Analytes	N	Sample Weight(g)	Extraction	Method of Analysis	LOQ (ng/g)	Reference
Cocaine Benzoylecgonine Cocaethylene mOHBE	5	0.25	Liquid/Liquid SPE CleanScreen DAU	LC/ESI MS/MS	1 1 1 5	[17]
Cocaine Benzoylecgonine Cocaethylene Anhydroecgonine methyl ester	50	0.3	Liquid/Liquid Disposable Pipet Extraction	GC/MS	10 15 10 10	[18]

were separated over 18 min with cocaine metabolites eluting between 8.4 and 13.4 min. Cocaine, benzoylecgonine, and cocaethylene were linear from 1 to 500 ng/g while mOHBE was linear from 5 to 500 ng/g. All drugs were acquired in positive mode. Degradation of up to 30% was observed for target compounds at room temperature for 24 h. Five authentic specimens were used to confirm the applicability of the method. Cocaine metabolites were found in two of the five meconium specimens. Benzoylecgonine was present in the highest concentration followed by mOHBE, cocaine, and cocaethylene.

Mozaner Bordin et al. [18] have described another analysis technique that specifically addressed issues related to low sample volume and high solvent requirements for extraction and analysis of meconium. To achieve success with a low sample volume, a disposable pipet extraction (DPX) was employed that reduces requirements for large amounts of sample and solvents. Notably, this had been applied to other analytes and specimen types but never meconium. A methanol extraction was performed on 0.3 g of meconium for 20 min. The supernatant was collected a subject to DPX and eluted three times with dichloromethane/2-propanol/ammonium hydroxide (78: 20: 2, v/v/v). Eluate was dried then reconstituted in acetonitrile and derivatized with N-methyl-N-(trimethylsilyl) trifluoroacetamide (MSTFA). Four cocaine metabolites (cocaine, benzoylecgonine, cocaethylene, and anhydroecgonine methyl ester) were separated over 15.5 min via GC. The method was applied to 50 authentic meconium specimens. Thirteen specimens were positive for at least one cocaine metabolite. Two specimens were only positive for benzoylecgonine. No specimens were positive for cocaethylene. The inclusion of cocaethylene or anhydroecgonine methyl ester metabolites did not improved detection rates for cocaine exposure. In contrast to the work done by Gray et al. cocaine was the most prevalent metabolite and found at the highest concentration in the meconium samples. This discrepancy does highlight the impact that extraction techniques contribute toward detection and recovery of target metabolites. Methods for analysis of cocaine and related compounds in meconium are summarized in Table 29.5.

ANALYSIS OF NICOTINE

Nicotine use during pregnancy remains one of the most common detrimental drugs abused. Prevalence varies by state but can be greater than 25% of pregnant women in some areas (West Virginia). Despite its legal status, tobacco use remains very dangerous and is associated with negative fetal outcomes. The use of nicotine during pregnancy has recently been estimated to double the risk of stillbirth [19]. Much of the focus nationally has been on the increased prevalence of opioid abuse over the last 10 years but more methods have been published for nicotine analysis in meconium than any other drug class reviewed. Gray et al. [20] evaluated 125 paired hydrolyzed and nonhydrolyzed specimens for nicotine and metabolites. No significant difference was observed in detection rates with the addition of hydrolysis. Nicotine was the most prevalent compound in meconium followed by 3OH-cotinine, cotinine, nornicotine, and norcotinine. The inclusion of nornicotine and norcotinine in the panel did not improve detection rates as they were never the only compounds detected in any specimen. The inclusion of 3OH-cotinine and cotinine did improve detection rates due to 3OH-cotinine only ($n = 3$) and cotinine only ($n = 1$) specimens found in the meconium study set. Interestingly there was a high abundance (18.6%) of nonhydrolyzed nicotine only positive samples in the study. This is significant because most immunoassays are targeted toward cotinine and are unreactive to nicotine at physiologically relevant levels. Therefore, the authors suggest that application of cotinine immunoassays for meconium analysis may fail to detect one in four tobacco exposed neonates. Grey et al. at the NIH have published two method of analysis of nicotine and its metabolites. The meconium extractions are identical but the analysis is performed on different instrumentation. One method utilized atmospheric pressure chemical ionization and the other used electron spray ionization. Lower limits of detection were similar between both methods.

Separately, Marin et al. [21] published a method for analysis of nicotine, cotinine, 3OH-cotinine, norcotinine and anabasine. A methanol extraction was followed by SPE with a reverse phase and ion exchange component similar to Gray et al. Comparable to previous publications, nicotine and 3OH-cotinine were the most prevalent compounds and were present in similar concentrations. Anabasine was not detected in any sample.

A very comprehensive extraction method was published by Xia et al. [22] that uses four separate sample preparation, extraction, and cleanup steps. The first step involved digestion of the meconium sample with 5 N KOH to effectively dissolve and homogenize the sample. The authors suggest this step is crucial to prevent clumping and instead form a well-dispersed mixture prior to extraction with acidified methanol. Additionally, the KOH can hydrolyze any glucuronide metabolites present in the sample. The digestion was followed by extraction with methanol and SLE with a Chem-Elute SLE column. The resulting eluent was subject to additional liquid−liquid extraction followed by SLE with a CleanScreen SPE column. The sensitivity of the method was improved for cotinine and 3OH-cotinine by approximately a factor of 10 compared to previously published methods with LOQs in the 0.2 ng/g range. The authors detected nicotine in approximately 76% of meconium samples despite the fact that only 22% of women self-reported smoking or exposure to secondhand smoke.

Two methods utilizing GC have been published that utilize novel up front extraction techniques for meconium. The first method by Sant'Anna et al. [23] applied accelerated solvent extraction (ASE) to the meconium sample as a means to isolate target compounds under controlled evaporative temperature and pressure. The extract was then subject to SPE using a BondElute SPE column. After dry down, the reconstituted sample was analyzed via GC coupled to a nitrogen phosphorous detector (GC-NPD). When the method was applied to 16 authentic meconium samples with known smoking exposure, 14 out of 16 were positive for nicotine and 16 out of 16 were positive for cotinine. Nicotine concentrations were approximately 5 × higher than cotinine concentrations. The LOQ for nicotine in the method was 40 ng/g which may explain the negative nicotine results for two study samples. The other GC method from Mozaner et al. [18] used DPX as previously described in the cocaine analysis section. Application of the method to 35 meconium samples identified 25 positive for cotinine and 11 positive for nicotine. This is in disagreement with LC/Ms methods that report a higher prevalence of nicotine followed by 3OH-cotinine accumulation in meconium. The various detection rates and measured concentrations highlight the method dependence of metabolite-specific recovery for tobacco exposure. Methods for analysis of nicotine and related compounds in meconium are summarized in Table 29.6.

TABLE 29.6 Analysis of Nicotine and Related Compounds in Meconium

Analytes	N	Sample Weight (g)	Extraction	Method of Analysis	LOQ (ng/g)	Reference
Nicotine Cotinine 3OH-cotinine Norcotinine Nornicotine	215	0.5	Liquid/Liquid SPE CleanScreen DAU	LC/APCI MS/MS	5 1.25 1.25 1.25 5	[20]
Nicotine Cotinine 3OH-cotinine	5	0.25	Liquid/Liquid SPE CleanScreen DAU	LC/ESI MS/MS	1–5	[17]
Nicotine Cotinine 3OH-cotinine, norcotinine, anabasine	19	0.25	Liquid/Liquid SPE Trace-B	LC/ESI MS/MS	4 2 4 2 2	[21]
Nicotine, cotinine 3OH-cotinine	374	0.5	Base Digestion Liquid/Liquid SLE Chem-Elute SPE CleanScreen DAU	LC/ESI MS/MS	3.15 0.23 0.30	[22]
Nicotine Cotinine	16	0.5	ASE SPE Bond-Elut Certify	GC-NPD	40 5	[23]
Nicotine Cotinine	50	0.3	Liquid/Liquid Disposable Pipet Extraction	GC/MS	20 10 10	[18]

ANALYSIS OF OPIATES

Opiate abuse has risen dramatically over the last 20 years as opioid prescribing practices increased [24]. Rising prescribing has been coupled with increased overdoses and abuse by various populations including pregnant women. A variety of methods for assessment of in utero opioid exposure in meconium have been previously published using a variety of methods such as LC/Ms, GC/Ms, and high-performance liquid chromatography with diode array detection [25–28]. More recent methods have exclusively relied on LC coupled with tandem mass spectrometry. Kacinko et al. [29] published the first method for quantitation of buprenorphine and metabolites in meconium. The method utilized two 0.25 g of aliquots of meconium, one that underwent enzymatic hydrolysis for 3 h at 60°C to cleave glucuronide metabolites to estimate total drug and one nonhydrolyzed specimen to estimate free drug. The resulting supernatant was subject to SPE using a CleanScreen DAU followed by LCMS analysis. Twenty microliters of reconstituted extract was separated over 13.5 min on a Synergi Polar-RP 80A column. Spectral data were acquired by atmospheric ionization in positive ion mode. Fragmentation patterns and deuterization locations for buprenorphine, norbuprenorphine, and their respective internal standards impacted the choice of quantifier and qualifier ions selected for the method. Deuterization of buprenorphine occurs on the cyclic side chain and is lost during fragmentation preventing comparison to the similar transition ion from the undeuterated buprenorphine molecule. As a result, the quantifier ion for norbuprenorphine is 17 mass units away from buprenorphine instead of the 4 deuterated units. The extraction efficiency ranged from 85% to 88% for buprenorphine and 77% to 91.5% for norbuprenorphine. When the method was applied to a study cohort of 10 meconium samples, total buprenorphine ranged from 24.3 to 296.8 ng/g and total norbuprenorphine ranged from 323.9

TABLE 29.7 Analysis of Opioids in Meconium

Analytes	N	Sample Weight (g)	Extraction	Method of Analysis	LOQ (ng/g)	Reference
Buprenorphine Norbuprenorphine Glucuronides	10	0.25	Sonication Glusulase SPE CleanScreen DAU	LC/APCI MS/MS	20	[29]
6MAM Morphine Codeine Hydromorphone Hydrocodone Oxycodone Methadone EDDP Buprenorphine Norbuprenorphine	5	0.25	Liquid/Liquid SPE CleanScreen DAU	LC/ESI MS/MS	1–5	[17]
Buprenorphine Norbuprenorphine	Not given	0.25	Bead Homogenization SPE Strata XL-C	LC/MS/MS	20	[30]

6MAM, 6-Monoacetylmorphine, a metabolite of heroin.

to 1880.2 ng/g. Significant differences were seen between total and free buprenorphine concentrations but not between total and free norbuprenoprhine concentrations. Infants with neonatal abstinence syndrome (NAS) had higher buprenorphine concentrations but no relationship was seen with norbuprenorphine.

An alternative multidrug class method containing opiates published by Grey et al. [17] has been previously described in other sections of this chapter. The method was developed for AMPs, cocaine, nicotine, and 10 opiates. 6-Monoacetylmorphine, morphine, codeine, hydromorphone, hydrocodone, oxycodone, methadone, and 2-ethylidene-1,5-dimethyl-3,3-diphenylpyrrolidine (EDDP) which could all be separated in a single chromatographic run. However, adequate sensitivity below 100 ng/g could not be achieved for buprenorphine and norbuprenorphine. Increasing the sample injection volume from 5 to 10 μL improved sensitivity but led to saturation of the mass detector from other drugs. A separate method was developed specifically for buprenorphine and norbuprenorphine that was shorter and used a short isocratic separation of the compounds over 5 min using 75% mobile phase B.

Another method for buprenorphine and norbuprenorphine has been published recently by Marin et al. [30]. This method uses homogenization with stainless steel beads followed by enzymatic hydrolysis at 70°C for 1 h. After centrifugation, the supernatant is applied to a Strata XL-C SPE column for sample cleanup. Compounds were separated over 5 min using a gradient of 0.1% formic acid in acetonitrile. No application of the method to authentic specimens was described. Methods for analysis of opioids in meconium are summarized in Table 29.7.

ANALYTICAL PERFORMANCE AND OUTCOME STUDIES

Several studies have evaluated the sensitivity of meconium versus other neonatal specimens to detect various drug classes as well as correlation with NAS diagnosis. Although meconium remains the gold standard for neonatal drug testing, UCT has emerged as a possible alternative specimen to meconium. Several studies have compared drugs detected in paired meconium and UCT samples for both immunoassay and MS analysis. Montgomery et al. [31] evaluated 118 paired meconium and UCT specimens and evaluated drugs detected via enzyme-linked immunosorbent assay for AMPs, cannabinoids, cocaine, and opiates for both specimen types. Although the overall agreement for both positive and negative results was reported as greater than 90% for all drug classes, agreement for detection of positive specimens specifically was much lower. UCT showed agreement with positive results 87% for AMPs, 100% for cocaine, 61% for cannabinoids, and 63% for opiates. This suggests that although UCT may have an improved window of detection compared to urine, for maximum sensitivity meconium is still the specimen of choice. Similar results have also

TABLE 29.8 Positive Drug Class Agreement Between Meconium and Umbilical Cord Tissue

	Montgomery et al. [31]	Colby JM [32]	Labardee et al. [33]
Amphetamines (%)	87	40	NR
Benzodiazepines (%)	NR	35	100
Cannabinoids (%)	61	61	NR
Cocaine (%)	100	100	90
Opiates (%)	63	83	60
Oxycodone (%)	NR	NR	40
Methadone (%)	NR	NR	100

NR, Not reported.

been published from several other comparison studies between meconium and UCT [32,33]. An analysis of several published comparison studies illustrates that unpredictable agreement when the two specimens types are compared. With the exception of cocaine, there appears no discernable trend in agreement for drug detection. This speaks to the challenging nature of the specimen types, the various extraction techniques used and analytical methods employed which impart method variation. Positive drug class agreement between meconium and UCT is summarized in Table 29.8.

CONCLUSION

Assessment of in utero drug exposure continues to be a challenging area for laboratory medicine. Meconium as a specimen type remains the gold standard due to its long window of detection and superior sensitivity from the accumulation of most drugs. Conventional laboratory analysis for drugs of abuse has relied upon liquid−liquid extraction followed by GC coupled with MS. Over the last decade much of the published methods have shifted from single drug quantitation using GCMS to bundled panels encompassing many drugs of abuse that are analyzed using LCMS utilized either one or two injections. Several drugs, specifically buprenorphine and norbuprenorphine, remain difficult to incorporate into comprehensive drug panels without losing the sensitivity required for detection in meconium. However, significant improvements have been made regarding sample extraction and cleanup which facilitate increased throughput and improved recovery. These newer methods published over the last decade have improved the utility of meconium testing and have made its analysis more readily achievable for labs that have implemented GCMS or LCMS instrumentation.

REFERENCES

[1] Data & Statistics|FASD|NCBDDD|CDC (n.d.), <https://www.cdc.gov/ncbddd/fasd/data.html> [accessed 02.06.18.].
[2] Burd L, Hofer R. Biomarkers for detection of prenatal alcohol exposure: a critical review of fatty acid ethyl esters in meconium. Birt Defects Res A Clin Mol Teratol 2008;82(7):487−93.
[3] Chan D, Bar-Oz B, Pellerin B, Paciorek C, et al. Population baseline of meconium fatty acid ethyl esters among infants of nondrinking women in Jerusalem and Toronto. Ther Drug Monit 2003;25(3):271−8.
[4] Morini L, Marchei E, Pellegrini M, Groppi A, et al. Liquid chromatography with tandem mass spectrometric detection for the measurement of ethyl glucuronide and ethyl sulfate in meconium: new biomarkers of gestational ethanol exposure? Ther Drug Monit 2008;30(6):725−32.
[5] Kwak HS, Kang YS, Han KO, Moon JT, et al. Quantitation of fatty acid ethyl esters in human meconium by an improved liquid chromatography/tandem mass spectrometry. J Chromatogr B Analyt Technol Biomed Life Sci 2010;878(21):1871−4.
[6] Hastedt M, Krumbiegel F, Gapert R, Tsokos M, et al. Fatty acid ethyl esters (FAEEs) as markers for alcohol in meconium: method validation and implementation of a screening program for prenatal drug exposure. Forensic Sci Med Pathol 2013;9(3):287−95.
[7] Himes SK, Concheiro M, Scheidweiler KB, Huestis HA. Validation of a novel method to identify in utero ethanol exposure: simultaneous meconium extraction of fatty acid ethyl esters, ethyl glucuronide, and ethyl sulfate followed by LC−MS/MS quantification. Anal Bioanal Chem 2014;406(7):1945−55.

[8] Kelly T, Gray TR, Huestis MA. Development and validation of a liquid chromatography-atmospheric pressure chemical ionization–tandem mass spectrometry method for simultaneous analysis of 10 amphetamine-, methamphetamine- and 3,4-methylenedioxymethamphetamine-related (MDMA) analytes in human meconium. J Chromatogr B Analyt Technol Biomed Life Sci 2008;867(2):194–204.

[9] Gray TR, Kelly L, LaGasse LL, Smith LM, et al. Novel biomarkers of prenatal methamphetamine exposure in human meconium. Ther Drug Monit 2009;31(1):70–5.

[10] Gray TR, LaGasse LL, Smith LM, Derauf C, et al. Identification of prenatal amphetamines exposure by maternal interview and meconium toxicology in the infant development, environment and lifestyle (IDEAL) study. Ther Drug Monit 2009;31(6):769–75.

[11] Himes SK, Scheidweiler KB, Tassiopoulos K, Kacanek D, et al. Pediatric HIV/AIDS cohort study, development and validation of the first liquid chromatography–tandem mass spectrometry assay for simultaneous quantification of multiple antiretrovirals in meconium. Anal Chem 2013;85(3):1896–904.

[12] Himes SK, Tassiopoulos T, Yogev R, Huestis MA. Pediatric HIV/AIDS Cohort Study (PHACS), antiretroviral drugs in meconium: detection for different gestational periods of exposure. J Pediatr 2015;167(2):305–11.

[13] Results from the 2015 National survey on drug use and health: detailed tables, SAMHSA, CBHSQ (n.d.), <https://www.samhsa.gov/data/sites/default/files/NSDUH-DetTabs-2015/NSDUH-DetTabs-2015/NSDUH-DetTabs-2015.htm> [accessed 02.12.18.].

[14] NIDA. Abuse, Substance Use While Pregnant and Breastfeeding (n.d.), <https://www.drugabuse.gov/publications/research-reports/substance-use-in-women/substance-use-while-pregnant-breastfeeding> [accessed 02.12.18.].

[15] Peat J, Davis B, Frazee C, Garg U. Quantification of 11-carboxy-delta-9-tetrahydrocannabinol (THC-COOH) in meconium using gas chromatography/mass spectrometry (GC/MS). Methods Mol Biol Clifton NJ 2016;1383:97–103.

[16] Prego-Meleiro P, Lendoiro E, Concheiro M, Cruz A, et al. Development and validation of a liquid chromatography tandem mass spectrometry method for the determination of cannabinoids and phase I and II metabolites in meconium. J Chromatogr A 2017;1497:118–26.

[17] Gray TR, Shakleya DM, Huestis MA. A liquid chromatography tandem mass spectrometry method for the simultaneous quantification of 20 drugs of abuse and metabolites in human meconium. Anal Bioanal Chem 2009;393(8):1977–90.

[18] Mozaner Bordin DC, Alves MNR, Cabrices OG, de Campos EG, et al. A rapid assay for the simultaneous determination of nicotine, cocaine and metabolites in meconium using disposable pipette extraction and gas chromatography–mass spectrometry (GC–MS). J Anal Toxicol 2014;38(1):31–8.

[19] Varner MW, Silver RM, Rowland Hogue CJ, Willinger M, et al. Eunice Kennedy Shriver National Institute of Child Health and Human Development Stillbirth Collaborative Research Network, association between stillbirth and illicit drug use and smoking during pregnancy. Obstet Gynecol 2014;123(19):113–25.

[20] Gray TR, Magri R, Shakleya DM, Huestis MA. Meconium nicotine and metabolites by liquid chromatography–tandem mass spectrometry: differentiation of passive and nonexposure and correlation with neonatal outcome measures. Clin Chem 2008;54(12):2018–27.

[21] Marin SJ, Christensen RD, Baer VL, Clark CJ, et al. Nicotine and metabolites in paired umbilical cord tissue and meconium specimens. Ther Drug Monit 2011;33(1):80–5.

[22] Xia Y, Xu M, Alexander RR, Bernert JT. Measurement of nicotine, cotinine and trans-3′-hydroxycotinine in meconium by liquid chromatography–tandem mass spectrometry. J. Chromatogr B Analyt Technol Biomed Life Sci 2011;879(22):2142–8.

[23] Sant'Anna SG, Oliveira CDR, de EM, Diniz A, et al. Accelerated solvent extraction for gas chromatographic analysis of nicotine and cotinine in meconium samples. J. Anal. Toxicol 2012;36(1):19–24.

[24] 2017-cdc-drug-surveillance-report.pdf (n.d.), <https://www.cdc.gov/drugoverdose/pdf/pubs/2017-cdc-drug-surveillance-report.pdf> [accessed 03.13.18.].

[25] Stolk LM, Coenradie SM, Smit BJ, van As HL. Analysis of methadone and its primary metabolite in meconium. J Anal Toxicol 1997;21(2):154–9.

[26] Gareri J, Klein J, Koren G. Drugs of abuse testing in meconium. Clin Chim Acta Int J Clin Chem 2006;366(1–2):101–11.

[27] ElSohly MA, Stanford DF, Murphy TP, Lester BM, et al. Immunoassay and GC–MS procedures for the analysis of drugs of abuse in meconium. J Anal Toxicol 1999;23(6):436–45.

[28] Gray TR, Choo RE, Concheiro M, Williams E, et al. Prenatal methadone exposure, meconium biomarker concentrations and neonatal abstinence syndrome, Addict. Abingdon Engl 2010;105(12):2151–9.

[29] Kacinko SL, Shakleya DM, Huestis MA. Validation and application of a method for the determination of buprenorphine, norbuprenorphine, and their glucuronide conjugates in human meconium. Anal Chem 2008;80(1):246–52.

[30] Marin SJ, McMillin GA. Quantitation of total buprenorphine and norbuprenorphine in meconium by LC–MS/MS. Methods Mol Biol Clifton NJ 2016;1383:59–68.

[31] Montgomery D, Plate C, Alder SC, Jones M, et al. Testing for fetal exposure to illicit drugs using umbilical cord tissue vs meconium. J Perinatol 2006;26(1):11–14.

[32] Colby JM. Comparison of umbilical cord tissue and meconium for the confirmation of in utero drug exposure. Clin Biochem 2017;50(13–14):784–90. Available from: https://doi.org/10.1016/j.clinbiochem.2017.03.006.

[33] Labardee RM, Swartzwelder JR, Gebhardt KR, Pardi JA, et al. Method performance and clinical workflow outcomes associated with meconium and umbilical cord toxicology testing. Clin Biochem 2017;50(18):1093–7. Available from: https://doi.org/10.1016/j.clinbiochem.2017.09.016.

Chapter 30

Analytical True Positive Drug Tests Due to Use of Prescription and Nonprescription Medications

Matthew D. Krasowski[1] and Tai C. Kwong[1,2]

[1]Department of Pathology, University of Iowa Carver College of Medicine, Iowa City, IA, United States, [2]Department of Pathology and Laboratory Medicine, University of Rochester Medical Center, Rochester, NY, United States

INTRODUCTION

Widespread substance abuse has led to increasing use of urine drug testing in many clinical services [1,2]. The last two decades have seen a significant rise in prescription drug misuse (especially opioids and benzodiazepines), leading to dramatic increases in drug overdoses and deaths [3,4]. For some patients, misuse of prescription medications evolves into the abuse of illicit street drugs such as heroin. For the purposes of discussion in this chapter, the term "nonmedical drug use" will be used to encompass the misuse of either prescription medications or illicit street drugs, the majority of which are scheduled compounds (controlled substances) in the United States.

Urine drug testing is widely used in clinical medicine [1]. In emergency medicine, urine drug testing is often used to rapidly test for the presence of drugs that may account for symptoms such as altered mental status or unconsciousness. Other clinical areas may use urine drug testing to screen for nonmedical drug use as a risk factor for substance abuse, child mistreatment, and drug diversion. Examples include urine drug testing within pediatrics and obstetrics. Substance abuse treatment programs often use urine drug testing to verify abstinence. Lastly, healthcare providers prescribing controlled substances may employ urine drug testing to verify adherence with prescribed medications (e.g., ensure the patient is not diverting medication to someone else) and detect unauthorized use of other controlled substances [2,5]. Urine drug testing provides an objective means to document a patient's drug use, and a positive result is interpreted as an evidence of drug exposure.

Urine drug testing typically consists of an initial test, usually an immunoassay, which if positive, can be confirmed by a second test [2,6,7]. A confirmed positive result is an analytical true positive result and constitutes laboratory evidence that the individual has been exposed to the drug. Immunoassays may utilize polyclonal or monoclonal antibodies, with a trend toward monoclonal antibody-based designs [6]. Many hospital-based clinical laboratories perform immunoassay urine drug-testing panels targeted toward commonly abused drugs or drug classes such as amphetamines, benzodiazepines, cannabinoids, cocaine, opiates, synthetic opioids (e.g., methadone), and phencyclidine. Confirmation assays typically use chromatography and mass spectrometry–based methods such as gas chromatography/mass spectrometry (GC/MS), liquid chromatography-mass spectrometry (LC/MS), or liquid chromatography–tandem mass spectrometry (LC/MS/MS) [2,7]. Confirmatory methods may also be required for the detection of drugs for which screening immunoassays do not currently exist (e.g., zolpidem or many "designer" drugs). While an increasing number of clinical laboratories are using mass spectrometry–based methods, relatively few hospital-based clinical laboratories perform this testing with rapid turnaround time. Clinical services may also elect not to perform confirmatory testing due to factors such as cost or slow turnaround time. There are also some clinical laboratories (especially reference laboratories) that skip immunoassay screens for urine drug testing and instead directly analyze by a mass spectrometry–based method such as liquid chromatography/time-of-flight–mass spectrometry (LC/TOF-MS) [8,9].

A positive result on urine drug testing, even if confirmed by confirmation testing, does not provide direct information on the person's history and pattern of drug use [1]. In fact, a positive drug test result could be the consequence of many causes other than nonmedical drug use. Interpretation of a positive result should take into consideration that there are causes other than nonmedical drug use that can produce analytical true positive results. For example, the individual is not a drug abuser, but has used a medication which contains the drug to be tested or a medication which is metabolized to the drug. Other causes include passive exposure to the drug and consumption of food items containing the drug as in the case of positive opiate test result due to the consumption of poppy seed-containing food products. Thus in the context of using urine drug testing to detect nonmedical drug use, this analytical true positive result is a clinical false positive because it wrongly implies nonmedical drug use. Having knowledge of those medications that contain drugs to be tested for, an understanding of the pathways of drug metabolism, as well as a full medical review, will ensure correct interpretation of a clinical false positive result and not wrongly implicate nonmedical drug use.

ANALYTICAL TRUE POSITIVE VERSUS CLINICAL FALSE POSITIVE RESULTS

The assumption underlying interpretation of a drug test result is the accuracy and reliability in the analysis and identification of the drug or drug metabolite present in the urine. The standard urine drug-testing protocol involves an initial test using a panel of immunoassays [1,10]. The accuracy of an immunoassay is determined by the immuno-specificity of the assay antibody, and immunoassays in general do not have strict specificity for the target drug or drug metabolites [1,10,11]. Immunoassays directed at classes of drugs (e.g., benzodiazepines or opiates) intentionally have broader specificity to capture multiple drugs and metabolites within the class. This can make interpretation challenging as multiple drugs could be responsible for a positive screening result [1,10–12]. Many immunoassays have measurable reactivities with structurally similar compounds. For example, amphetamines immunoassays may detect sympathomimetic amines including ephedrine, phentermine, and pseudoephedrine or other compounds such as a metabolite of labetalol (medication used for hypertension) structurally similar to amphetamine [13–15]. Other immunoassays may detect substances that are structurally unrelated to the target analyte (e.g., phencyclidine assays often cross-react with high concentrations of dextromethorphan) [1,10–12]. Thus an initial positive result by immunoassay could be a false positive result. The awareness of imperfect immunoassay specificity has led to two important principles of urine drug-testing interpretation:

1. The positive result of an immunoassay is only a presumptive positive result.
2. A definitive identification of the drug or the metabolite must be based on a second test, the confirmation test.

Therefore laboratory scientists rely on the confirmation to further verify the initial positive screening results. Accordingly, clinicians making decisions based on a drug test result do so because of the confidence they have that a positive result confirmed by the laboratory is a true positive. But a confirmed positive result is only an analytical true positive, while it can also be a clinical false positive, thus confounding interpretation of the positive test result. This is because a confirmed analytical true positive result only documents the individual has been exposed to the drug but offers no inference regarding the nature of the exposure or the reason for the positive test. The individual may have a valid explanation for having the positive drug test; for example, the positive codeine and morphine results can be attributed to the use of a prescription medication containing codeine. The converse of this is when the individual cannot offer a credible explanation for the positive result, in which case the test result is not only an analytical true positive, but it is also a clinical true positive in the sense that the drug test has identified nonmedical drug use. Thus in the interpretation of a positive result, particularly a confirmed positive result, the possibility that the result being a clinical false positive should be taken into consideration. Recognizing a positive result to be a clinical false positive eliminates the wrongful implication that the individual has engaged in nonmedical drug use, which can have serious downstream consequences for the patient. This can be accomplished, in some cases, by resorting to additional laboratory testing (e.g., 6-acetylmorphine testing for morphine positive results to definitively rule in heroin use), having a thorough knowledge of the causes of clinical false positives, and including a thorough medical review conducted by a qualified healthcare provider in the program.

TRUE POSITIVE RESULTS DUE TO THE USE OF MEDICATIONS

Many reports of clinical false positive results can be found in the literature. Causes of these clinical false positives can be grouped into four general categories: (1) use of a medication that contains the target drug, (2) use of a medication that is metabolized to the target drug, (3) environmental (passive or secondary) exposure, and (4) consumption of a food item that contains or is contaminated with the target drug. This chapter will focus on clinical false positives caused by medications (categories 1 and 2 above), with examples listed in Tables 30.1 and 30.2. Clinical false positives of each of the major drug classes of amphetamines, cocaine, marijuana, and opiates will be discussed in details in the following sections.

TABLE 30.1 Reported Examples of Clinical False Positive Results[a]

Causes of Clinical False Positive	Drug Confirmed by Analytical Method
1. Use of prescription medications containing target drug(s): • Acetaminophen with codeine • Robitussin-DAC • Adderall, Vyvanse • Dronabinol (Marinol)	• Codeine ± morphine • Codeine ± morphine • Amphetamine • THC-COOH
2. Use of prescription medication which is metabolized to target drug(s): • Selegiline • Clobenzorex	• L-Methamphetamine, L-amphetamine • D-Amphetamine
3. Environmental: • Passive inhalation/secondary smoke	• THC-COOH
4. Consumption of food products which contain or are contaminated with target drug(s): • Poppy seeds • Hemp products	• Morphine (and codeine) • THC-COOH

[a]Medical Review Officer Manual for Federal Agency Workplace Drug Testing Programs, effective October 1, 2017 (https://www.samhsa.gov/sites/default/files/workplace/mro-guidance-manual-oct2017_2.pdf, accessed 1/14/2018). The drugs confirmed represent the most common parent drug and/or metabolites detected.
Source: Inclusion of SAMHSA content does not constitute or imply endorsement or recommendation by the Substance Abuse and Mental Health Services Administration, the U.S. Department of Health and Human Services, or the U.S. Government.

TABLE 30.2 Examples of Amphetamines-Containing Products[a]

	Amphetamines-Containing Products
Substances known to contain D-amphetamine or D,L-amphetamine	Adderall Dexedrine DextroStat
Substances known to contain D-methamphetamine	Desoxyn
Substances known to contain L-methamphetamine	Vicks VapoInhaler

[a]Medical Review Officer Manual for Federal Agency Workplace Drug Testing Programs, effective October 1, 2017 (https://www.samhsa.gov/sites/default/files/workplace/mro-guidance-manual-oct2017_2.pdf, accessed 1/14/2018).
Source: Inclusion of SAMHSA content does not constitute or imply endorsement or recommendation by the Substance Abuse and Mental Health Services Administration, the U.S. Department of Health and Human Services, or the U.S. Government.

AMPHETAMINES

Amphetamine and methamphetamine exist in two optical isomeric forms (enantiomers) designated as D- (dextro) or L- (levo) and alternatively as S(+)- and R(−)-, respectively. The D- and L-isomers have very different pharmacological actions: the D-isomer has a strong central nervous system (CNS) stimulant effect and, therefore a high abuse potential; the L-isomer has a peripheral vasoconstrictive property, and is not a drug of abuse. For example, L-methamphetamine, which has a much lower potency as a CNS stimulant than D-methamphetamine is available as a nonprescription nasal inhalant (Vicks VapoInhaler) [16].

Amphetamines clinical false positive results can be due to the use of medications containing amphetamine or methamphetamine, or medications containing substances which are metabolized to amphetamine or methamphetamine.

Nonprescription Medication Containing Methamphetamine

The only nonprescription medication that contains L-methamphetamine is Vicks VapoInhaler. Although most amphetamines immunoassays target D-methamphetamine detection, and despite these assays being relatively stereospecific for D-methamphetamine, some of these assays do have sufficient reactivity toward L-methamphetamine and can produce positive results if L-methamphetamine is present in high concentration [16,17]. Moreover, standard GC/MS and LC/MS/MS confirmation methodologies do not distinguish between D- and L-isomers and will confirm the presence of methamphetamine regardless of which isomer is present [16]. Therefore an individual who has used Vicks VapoInhaler might test positive for methamphetamine (and possibly also amphetamine as a metabolite of methamphetamine): an analytical true positive confirmed by GC/MS, but a clinical false positive in the context of suspected abuse of methamphetamine. The only laboratory test that can determine the source of a suspected clinical false positive methamphetamine is chiral analysis, which can verify the enantiomeric identity of the confirmed methamphetamine, although chiral analysis is not routinely available by most clinical laboratories that offer amphetamine confirmation analysis [16,18]. In chiral analysis, optically active derivatizing reagents convert D- and L-enantiomers into derivatives which can be chromatographically separated prior to mass spectrometric analysis [19]. If L-methamphetamine is present at greater than 80% (not 100% because Vicks VapoInhaler may be contaminated by a trace amount of D-methamphetamine), the result is confirmed as a clinical false positive methamphetamine since it is consistent with Vicks VapoInhaler use, not methamphetamine abuse [20].

Prescription Medications Containing Amphetamine or Methamphetamine

Amphetamine and methamphetamine are listed by the Drug Enforcement Agency (DEA) as Schedule II controlled substances. Therapeutic uses of amphetamine include treatment of attention deficit disorder with hyperactivity, narcolepsy, and obesity. Methamphetamine is much less commonly used as a prescription drug, being mainly used for treatment of obesity. Pharmaceutical preparations of these drugs have a specific enantiomeric identity: methamphetamine is D-methamphetamine (e.g., Desoxyn); amphetamine, however, is available as D-amphetamine (e.g., Dexedrine, Destrostat), as well as a mixture of D- and L-isomers (e.g., Adderall). Lisdexamfetamine (Vyvanse) is a prodrug that is metabolized to amphetamine in the body after administration [21]. Examples of medications containing D-amphetamine, racemic amphetamine, and D-methamphetamine are listed in Table 30.2.

Illicit methamphetamine and amphetamine products consist mostly of the D-enantiomer but, depending on the starting materials used by clandestine laboratories, significant amounts of the L-enantiomer may be present. All pharmaceutical and illicit products can produce positive amphetamines results by immunoassay as well as analytical true positive results by the standard GC/MS and LC-MS/MS confirmation tests. Chiral analysis is needed to distinguish between the two enantiomers and determine their relative percentages to aid in determining the source of the amphetamine/methamphetamine positives [19]. Chiral analysis is not routinely performed due to the extra complexity and cost.

Substances Known to Metabolize to Methamphetamine and Amphetamine

There are medications that metabolize to methamphetamine or amphetamine (Table 30.3). If a patient is on one of these medications, the positive amphetamine/methamphetamine result is a clinical false positive. For example, selegiline (Eldepryl, Zelapar), a drug used in the treatment of Parkinson's disease, is metabolized to L-methamphetamine (and L-amphetamine) without any racemization during metabolism [22]. It is unlikely, however, that patients on selegiline would test positive by amphetamines immunoassay because: (1) immuno-specificity of most amphetamines immunoassay is mostly directed toward D-methamphetamine, not L-methamphetamine and (2) the concentrations of L-methamphetamine and L-amphetamine derived from selegiline metabolism are relatively low. Should there be a positive amphetamines immunoassay test, the enantiomeric analysis should show the presence of only the L-enantiomer if the patient has taken only selegiline [19]. Clobenzorex is a prescription drug that is metabolized to D-amphetamine. Therefore a patient who has taken clobenzorex should have only D-amphetamine in the urine as determined by enantiomeric analysis. The presence of L-amphetamine in the urine disproves the patient's claim that he or she has taken only clobenzorex [23].

COCAINE

The only prescription medications containing cocaine are topical local anesthetic products mainly used in otolaryngology and ophthalmologic surgeries. Clinical false positive results for cocaine metabolite have been reported due to legitimate medical use of cocaine-containing preparations during surgeries and for skin suturing. Patients who had these procedures tested positive for benzoylecgonine (one of the main metabolites of cocaine) up to 2−3 days following the procedure [24−26]. The topical use of cocaine-containing products has declined significantly over the last 20 years, in large part due to the risks of diversion and regulatory burden with maintaining cocaine products (currently regulated as

TABLE 30.3 Examples of Substances Which are Metabolized to Amphetamine or Methamphetamine[a]

	Products
Substances known to metabolize to methamphetamine (and amphetamine)	Benzphetamine (Didrex) Dimethylamphetamine Selegiline (Eldepryl)
Substances known to metabolize to amphetamine	Amphetaminil Clobenzorex Ethylamphetamine Fenethylline Mefenorex

[a]Medical Review Officer Manual for Federal Agency Workplace Drug Testing Programs, effective October 1, 2017 (https://www.samhsa.gov/sites/default/files/workplace/mro-guidance-manual-oct2017_2.pdf, accessed 1/14/2018).
Source: Inclusion of SAMHSA content does not constitute or imply endorsement or recommendation by the Substance Abuse and Mental Health Services Administration, the U.S. Department of Health and Human Services, or the U.S. Government.

Schedule II controlled substances). There is no prescription medication containing cocaine that would be prescribed in the clinic setting. Overall, clinical false positives for cocaine should be very uncommon. However, analytical true positive may occur after drinking coca tea which contains cocaine (see Chapter 32: Miscellaneous Issues: Paper Money Contaminated with Cocaine and Other Drugs, Cocaine-Containing Herbal Teas, Passive Exposure to Marijuana, Ingestion of Hemp Oil and Occupational Exposure to Controlled Substances for details).

MARIJUANA

Marijuana (cannabis) is on the DEA list of controlled substance as a Schedule I drug, although multiple states within the United States have legalized marijuana for medical use and, in some cases, for recreational use as well [27,28]. A synthetic Δ^9-tetrahydrocannabinoid (THC), dronabinol, is available as a prescription drug under the trade name Marinol (a Schedule III controlled substance). Dronabinol is prescribed for the treatment of nausea and vomiting associated with cancer chemotherapy, appetite stimulation in AIDS patients, and the management of glaucoma. Since dronabinol is THC; it will undergo the same metabolic pathway as the THC in marijuana to Δ^9-tetrahydrocannabinoid carboxylic acid (THC-COOH), the major metabolite that is the target analyte for all cannabinoid initial and confirmation tests [29]. Therefore all standard testing for marijuana use, based on the detection of THC-COOH, cannot distinguish between marijuana or dronabinol use. As a consequence, a patient on dronabinol, if given a urine drug test, will have an analytical true positive result for THC-COOH that is also a clinical false positive result for marijuana use. In order to resolve the issue of marijuana versus dronabinol use as the cause of a positive drug test, it has been proposed that Δ^9-tetrahydrocannabivarin (THCV) may be useful as a marker of marijuana use. THCV is a naturally occurring component of cannabis, whereas dronabinol, being a synthetic product, does not contain THCV. THCV is a C3 homolog of THC, and both compounds are metabolized to their respective carboxylic acid metabolite: THCV-COOH and THC-COOH [30]. Therefore the presence of THCV (and its metabolite THCV-COOH) in a urine specimen is considered as a proof that the patient has used marijuana, with or without Marinol. However, analysis for THCV is not routinely available clinically.

OPIATES

Consumption of poppy seeds causing positive opiates drug test result is a well-recognized issue in drug test interpretation [7,31]. In November 1998, the Federal Government (DHHS: Department of Health and Human Services) increased the screening cutoff of opiate immunoassays from 300 to 2000 ng/mL in an attempt to eliminate false positive opiate results due to ingestion of poppy seed products. Poppy seeds often contain morphine and codeine in sufficient amounts to result in positive urine opiate immunoassay screens, especially at the 300 ng/mL cutoff that is commonly used in clinical testing. The issue of positive urine opiate testing resulting from poppy seed consumption has especially attracted attention for perinatal and neonatal drug testing, with nonmedical use wrongly attributed to pregnant women who consumed poppy seed−containing foods [7,31]. Some of these cases have led to inappropriate child protection actions such as temporary loss of custody for the mother.

TABLE 30.4 Examples of Opiates-Containing Products[a] and Other Opioid Containing Medications

Drug	Prescription Products	Nonprescription Products[b]
Codeine	Ambenyl with Codeine	Kaodene with Codeine OTC
	Codimal DM Syrup	
	Fioricet and Codeine	
	Fiorinal with Codeine	
	Guiatuss AC	
	Phenaphen with Codeine	
	Robitussin-DAC	
	Triacin-C	
	Tylenol with Codeine	
Morphine	Avinza	Donnagel-PG[c]
	Astramorph PF	Infantol Pink[c]
	Depodur	Kaodene with Paregoric[c]
	Duramorph	Quiagel PG[c]
	Kadian	
	MS Contin Tablets	
	Oramorph SR	
	Roxanol	
	Paregoric[c]	
Oxycodone	OxyCotonin	
	Oxydose	
	Roxicodone	
	Percodan	
Hydrocodone	Vicodin	
	Vicoprofen	
	Zydone	

[a]Medical Review Officer Manual for Federal Agency Workplace Drug Testing Programs, effective October 1, 2017 (https://www.samhsa.gov/sites/default/files/workplace/mro-guidance-manual-oct2017_2.pdf, accessed 1/14/2018).
[b]Nonprescription products are antidiarrheal medications. Nonprescription sale is prohibited in some states.
[c]Contain opium.
Source: Inclusion of SAMHSA content does not constitute or imply endorsement or recommendation by the Substance Abuse and Mental Health Services Administration, the U.S. Department of Health and Human Services, or the U.S. Government.

Medications Containing Morphine or Codeine

A frequently encountered cause of clinical false positive morphine or codeine is the use of a medication that contains morphine or codeine. Many drug products contain morphine or codeine, some of which are nonprescription medications. Some products containing opiates are listed in Table 30.4.

Metabolites of Opiates

A clinical false positive interpretation may also occur if the presence of an unexpected opiate is not recognized as a metabolite of the prescription opiate the patient is on, and these metabolites are themselves also potential drugs of

TABLE 30.5 Opiates and Metabolites

Opiate	Opiate Metabolite(s)
Morphine	Hydromorphone
Codeine	Morphine
	Hydrocodone
Hydrocodone	Hydromorphone
	Dihydrocodeine
Oxycodone	Noroxycodone
	Oxymorphone

abuse. Thus it is important to understand the metabolic pathways of the opiates: morphine is a metabolite of codeine; hydromorphone and dihydrocodeine are the metabolites of hydrocodone; oxymorphone is a metabolite of oxycodone; and hydrocodone and hydromorphone are the metabolites found in the presence of very high codeine and morphine concentrations[32,33]. Common opiates and their metabolites, including minor ones, are shown in Table 30.5. Therefore without knowing that the presence of the unexpected opiates (e.g., hydromorphone) is due to a metabolic pathway of the prescribed medication (morphine), the positive result (hydromorphone) may be mistakenly interpreted as an evidence of illicit use of another opiate (hydromorphone, Dilaudid)—a clinical false positive.

CONCLUSIONS

A patient who is on a medication containing the drug to be tested for, or a substance that is metabolized to the drug, will give a positive drug screen. This is an analytical true positive result, but clinically, it is a clinical false positive because the patient has not engaged in nonmedical drug use as implied by the positive test result. Eliminating wrong interpretations requires knowing those medications which can cause clinical false positive results, having additional testing (e.g., enantiomeric analysis of amphetamines) by the laboratory, and a thorough medical review by a qualified healthcare provider.

REFERENCES

[1] Moeller KE, Lee KC, Kissack JC. Urine drug screening: practical guide for clinicians. Mayo Clinic Proc 2008;83:66−76.
[2] Kwong TC, Magnani B, Moore C. Urine and oral fluid drug testing in support of pain management. Crit Rev Clin Lab Sci 2017;54:433−45.
[3] Pergolizzi Jr. JV, LeQuang JA, Taylor Jr. R, Raffa RB, Group NR. Going beyond prescription pain relievers to understand the opioid epidemic: the role of illicit fentanyl, new psychoactive substances, and street heroin. Postgrad Med 2018;130:1−8.
[4] Wilkerson RG, Kim HK, Windsor TA, Mareiniss DP. The opioid epidemic in the United States. Emerg Med Clin North Am 2016;34:e1−e23.
[5] Schwarz DA, George MP, Bluth MH. Toxicology in pain management. Clin Lab Med 2016;36:673−84.
[6] Melanson SE. The utility of immunoassays for urine drug testing. Clin Lab Med 2012;32:429−47.
[7] Tenore PL. Advanced urine toxicology testing. J Addict Dis 2010;29:436−48.
[8] McMillin GA, Marin SJ, Johnson-Davis KL, Lawlor BG, Strathmann FG. A hybrid approach to urine drug testing using high-resolution mass spectrometry and select immunoassays. Am J Clin Pathol 2015;143:234−40.
[9] Wu AH, Gerona R, Armenian P, French D, Petrie M, Lynch KL. Role of liquid chromatography-high-resolution mass spectrometry (LC-HR/MS) in clinical toxicology. Clin Toxicol (Phila) 2012;50:733−42.
[10] Melanson SE, Baskin L, Magnani B, Kwong TC, Dizon A, Wu AH. Interpretation and utility of drug of abuse immunoassays: lessons from laboratory drug testing surveys. Arch Pathol Lab Med 2010;134:735−9.
[11] Krasowski MD, Siam MG, Iyer M, Pizon AF, Giannoutsos S, Ekins S. Chemoinformatic methods for predicting interference in drug of abuse/toxicology immunoassays. Clin Chem 2009;55:1203−13.
[12] Krasowski MD, Pizon AF, Siam MG, Giannoutsos S, Iyer M, Ekins S. Using molecular similarity to highlight the challenges of routine immunoassay-based drug of abuse/toxicology screening in emergency medicine. BMC Emerg Med 2009;9:5.
[13] Duenas-Garcia OF. False-positive amphetamine toxicology screen results in three pregnant women using labetalol. Obstet Gynecol 2011;118:360−1.

[14] Gilbert RB, Peng PI, Wong D. A labetalol metabolite with analytical characteristics resembling amphetamines. J Anal Toxicol 1995;19:84–6.
[15] Yee LM, Wu D. False-positive amphetamine toxicology screen results in three pregnant women using labetalol. Obstet Gynecol 2011;117:503–6.
[16] Smith ML, Nichols DC, Underwood P, Fuller Z, Moser MA, Flegel R, et al. Methamphetamine and amphetamine isomer concentrations in human urine following controlled Vicks VapoInhaler administration. J Anal Toxicol 2014;38:524–7.
[17] Magnani B. Concentrations of compounds that produce positive results. In: Shaw L, Kwong T, editors. The clinical toxicology laboratory: contemporary practice of poisoning evaluation. Washington, DC: AACC Press; 2001. p. 482–97.
[18] Cody JT. Determination of methamphetamine enantiomer ratios in urine by gas chromatography–mass spectrometry. J Chromatogr 1992;580:77–95.
[19] Cody JT. Important issues in testing of methamphetamine enantiomer ratios in urine by gas chromatography-mass spectrometry. In: Liu RH, Goldberger B, editors. Handbook of workbook drug testing. Washington, DC: AACC Press; 1995. p. 239–88.
[20] Services DoHaH, Medical Review Officer Guidance Manual for Federal Workplace Drug Testing Programs, 2017. https://www.samhsa.gov/sites/default/files/workplace/mro-guidance-manual-oct2017_2.pdf.
[21] Boellner SW, Stark JG, Krishnan S, Zhang Y. Pharmacokinetics of lisdexamfetamine dimesylate and its active metabolite, D-amphetamine, with increasing oral doses of lisdexamfetamine dimesylate in children with attention-deficit/hyperactivity disorder: a single-dose, randomized, open-label, crossover study. Clin Ther 2010;32:252–64.
[22] Mahmood I. Clinical pharmacokinetics and pharmacodynamics of selegiline. An update. Clin Pharmacokinet 1997;33:91–102.
[23] Cody JT, Valtier S. Amphetamine, clobenzorex, and 4-hydroxyclobenzorex levels following multidose administration of clobenzorex. J Anal Toxicol 2001;25:158–65.
[24] Bralliar BB, Skarf B, Owens JB. Ophthalmic use of cocaine and the urine test for benzoylecgonine. N Engl J Med 1989;320:1757–8.
[25] Patrinely JR, Cruz OA, Reyna GS, King JW. The use of cocaine as an anesthetic in lacrimal surgery. J Anal Toxicol 1994;18:54–6.
[26] Altieri M, Bogema S, Schwartz RH. TAC topical anesthesia produces positive urine tests for cocaine. Ann Emerg Med 1990;19:577–9.
[27] Brooks E, Gundersen DC, Flynn E, Brooks-Russell A, Bull S. The clinical implications of legalizing marijuana: are physician and nonphysician providers prepared?. Addict Behav 2017;72:1–7.
[28] Hall W, Lynskey M. Evaluating the public health impacts of legalizing recreational cannabis use in the United States. Addiction 2016;111:1764–73.
[29] McGilveray IJ. Pharmacokinetics of cannabinoids. Pain Res Manag 2005;10(Suppl A):15A–22A.
[30] ElSohly MA, deWit H, Wachtel SR, Feng S, Murphy TP. Delta9-tetrahydrocannabivarin as a marker for the ingestion of marijuana versus marinol: results of a clinical study. J Anal Toxicol 2001;25:565–71.
[31] Lachenmeier DW, Sproll C, Musshoff F. Poppy seed foods and opiate drug testing—where are we today? Ther Drug Monit 2010;32:11–18.
[32] Cone EJ, Heit HA, Caplan YH, Gourlay D. Evidence of morphine metabolism to hydromorphone in pain patients chronically treated with morphine. J Anal Toxicol 2006;30:1–5.
[33] Oyler JM, Cone EJ, Joseph Jr. RE, Huestis MA. Identification of hydrocodone in human urine following controlled codeine administration. J Anal Toxicol 2000;24:530–5.

Chapter 31

Analytical True Positive: Poppy Seed Products and Opiate Analysis

Amitava Dasgupta
Department of Pathology and Laboratory Medicine, University of Texas McGovern Medical School at Houston, Houston, TX, United States

INTRODUCTION

Poppy plants (*Papaver somniferum*, *Papaver paeoniflorum*, *Papaver giganteum*) are herbaceous annual plants that can grow almost anywhere. Of all the different species, *P. somniferum* is the most popular due to its beautiful flowers. Moreover, seeds are used for making muffins, breads, and other food. The stems and pods are often used for flower arrangements. Opium is found in the latex (a milky fluid) of unripe pods of poppy plants 1−3 weeks after flowering. The milky fluid is collected by incision of green seed pods. More than 30 alkaloids have been isolated from *P. somniferum* out of which three alkaloids morphine, codeine, and noscapine (antitussive) are used directly in therapy. Thebaine is a biosynthetic intermediate of the morphine pathway which is used by the pharmaceutical industry for synthesis of oxycodone, oxymorphone, buprenorphine, and naloxone, an opiate antagonist [1]. Major alkaloids found in *P. somniferum* and their uses are listed in Table 31.1. Chemical structures of morphine, codeine, and thebaine are given in Fig. 31.1.

HISTORY OF POPPY PLANT AND OPIUM

Opium use was known to ancient cultures. For example, the Sumerian clay tablet (approximately 2100 BC) is considered as one of the oldest recorded medical prescription document in the ancient world. It appeared that opium was referred in such document. The Ebers Papyrus from 1552 BC described a blend of substances including opium which was used to sedate children. Hippocrates probably prescribed poppy juice as a purgative, a narcotic, and possibly a cure for leucorrhea. It is assumed that Galen popularized opium in Rome but was also aware of potential for abuse. During 8th century traders from the Near East introduced opium to India and China, while in Europe opium was introduced

TABLE 31.1 Major Alkaloids Found in *Papaver somniferum* and Their Use

Poppy Alkaloid	Pharmaceutical Use	Comment
Morphine	Narcotic analgesic	Major alkaloid of opium which is also a controlled substance.
Codeine	Analgesic and antitussive	Codeine after oral administration is converted into morphine by liver enzymes most noticeably CYP2D6. Therefore codeine is a prodrug of morphine. Codeine is also a controlled substance.
Thebaine	Precursor of opioids	Thebaine is a biosynthetic intermediate of the morphine which is used by the pharmaceutical industry for synthesis of oxycodone, oxymorphone, buprenorphine, and naloxone. Thebaine is also a controlled substance.
Noscapine (narcotine)	Antitussive	This compound may have anticancer properties. Noscapine is not a controlled substance.
Papaverine	Antispasmodic	Papaverine is not a controlled substance.

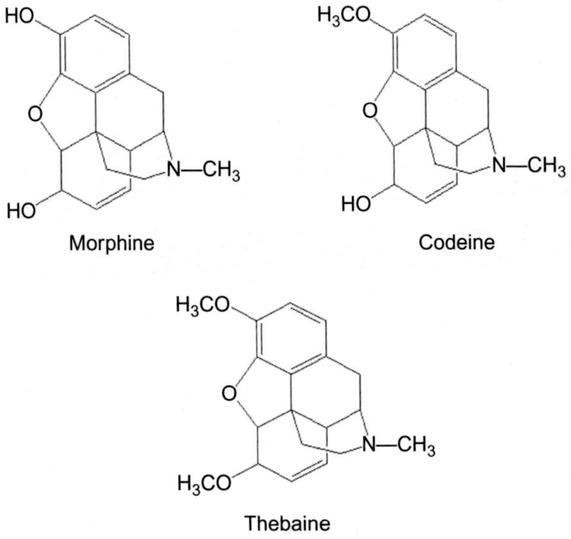

FIGURE 31.1 Chemical structures of narcotic alkaloids from poppy: morphine, thebaine, and codeine.

between 10th and 13th century. An alcoholic tincture of opium known as "laudanum" was developed in the 17th century by Thomas Sydenham, an English physician. Although laudanum was commonly consumed with whiskey, in early 20th century, it was used for preparing patients for surgery in North America [2].

Along with opium use, opium addictions also become a significant health hazard. Opium addiction was wide spread in China and following a 1799 ban on opium in China smuggling of opium became a common industry that led to the opium war between the British and China in 1839. During the 19th century, opium was grown in the United States and was also imported. In addition to indiscriminate medical use, opiates were available in the United States in myriad tonics and patent medicines, and smoking in opium dens was unhindered, resulting in an epidemic of opiate addiction by the late 1800s. The generous use of morphine in treating wounded soldiers during the Civil War in the United States also produced many addicts. Heroin was first synthesized in 1874. Heroin was considered as a highly effective medicine for treating cough, chest pains, and discomfort of tuberculosis. Moreover, heroin was also considered as a nonaddictive drug.

OPIUM AND POPPY PLANTS: LEGAL ISSUES

In 1875 the city of San Francisco adopted an ordinance in order to prohibit smoking of opium. In 1905 the United States Congress banned opium and in 1906 the Pure Food and Drug Act required accurate labeling of all patent medicines containing opium. The Harrison Narcotics Act of 1914 taxed and regulated the sale of narcotics and prohibited giving maintenance doses to addicts who made no effort to recover, leading to the arrest of some physicians, and the closing of maintenance treatment clinics. Since then, numerous laws attempting to regulate import, availability, use, and treatment have been passed. Opium and its constituent chemicals are listed as Schedule II drugs, while heroin is a Schedule I drug in the United States. Opium poppy for legal pharmaceutical purposes is grown in various countries under government license but very little is produced in the United States. The countries which can legally import crude opium to the United States include India, Turkey, Spain, France, Poland, Hungary, and Australia. India is the largest importer of opium to legal pharmaceutical markets of the world. Unfortunately, large quantities of opium are still grown worldwide mostly for use in the illegal manufacture of heroin. The sale of poppy seed from *P. Somniferum* is banned in Singapore and Saudi Arabia.

VARIATION OF MORPHINE AND OTHER ALKALOIDS CONTENT IN POPPY SEEDS

In general, morphine is the most abundant alkaloid found in raw opium (4%−21%), followed by noscapine (4%−8%), codeine (0.8%−2.5%), papaverine (0.5%−2.5%), and thebaine (0.5%−2%). For example, per gram of Indian opium contains approximately 125.0 mg of morphine and 36.6 mg of codeine. However, Turkish opium contains 171.7 mg of morphine per gram of opium and 17.7 mg of codeine per gram of opium [3]. Because poppy seeds itself does not contain latex, older botanical literature reported that poppy seeds contain no alkaloid. However, this old assumption is not true because various amounts of morphine and codeine have been detected in poppy seeds.

The concentrations of morphine and codeine can vary widely among different brands of poppy seeds which are related to variety of poppy plant, geographical location, and time of harvest as well as processing. However, morphine and codeine content of poppy seeds are significantly lower compared to their content in opium. Traditionally ripe seeds are manually shaken so that seeds fall out of the holes below the many layered stigma. However, when seeds are machine processed, they may get contaminated with unripe capsules which are removed later. However, washing the poppy seeds can significantly reduce the morphine content indicating that morphine is located on the surface of the seed [4]. Trafkowski et al. reported that the morphine content varied from less than 0.1 to 450 mg/kg of seed while codeine content varied from less than 0.1 to 57.1 mg/kg of seed. The thebaine content varied from 0.3 to 41 mg/kg of seed, while noscapine and papaverine content varied from 0.84 to 230 mg/kg and 0 to 67 mg/kg, respectively [5].

Report from the American Spice Trade Association, New York, indicated that Australian, Dutch, and Turkish poppy seeds variety represented approximately 94% of the US market of poppy seeds. The morphine content of the Australian poppy seeds ranges from 90 to 200 μg of morphine per gram of poppy seed, while Dutch and Turkish poppy seeds contain only 4–5 μg of morphine per gram of poppy seed [6]. Pelders et al. studied morphine and codeine content of poppy seeds originated from Australia, Hungary, Czech Republic, Spain, Turkey, and the Netherlands. The authors found highest amount of morphine and codeine in poppy seeds imported from Spain [7]. In general, there is a wide variation of morphine and codeine content in commercially available poppy seeds [8]. Another report indicated that typical morphine and codeine content of poppy seeds available in the US market were 18.6 and 2.3 μg/g in poppy seeds, respectively [9].

Although morphine content of the Netherland poppy seeds is low, another report indicated that morphine content was 100 mg/kg of poppy seeds [10]. Zentai et al. analyzed 737 poppy seed samples in Hungary and detected morphine in 726 out of 737 specimens. Codeine and thebaine were detected in 61.3% and 63.0% of samples analyzed while noscapine was present in only 6.2% of specimens analyzed. Although average morphine and codeine content were 18.7 and 3.6 mg/kg, respectively, highest morphine content was 533.0 mg/kg, while highest codeine content was 60.0 mg/kg indicating wide variation of morphine and codeine content in various specimens of poppy seeds [11]. Morphine and codeine content of poppy seeds originating from different countries are summarized in Table 31.2.

TABLE 31.2 Examples of Morphine and Codeine Content of Various Poppy Seeds Originated From Different Countries

Country of Origin	Morphine Content (mg/kg)	Codeine Content (mg/kg)	References
Australia	90	6.5	[7]
Australia	200	Not reported	[6]
Australia	325	Not reported	[7]
Spain	251	57.1	[7]
Hungary	46	3.7	[7]
Hungary	6.9	Not reported	[9]
Hungary	533	60	[11]
Denmark	8.4	Not reported	[9]
Czech Republic	2	0.5	[7]
Turkey	27	15.5	[7]
The Netherlands	4	0.4	[7]
The Netherlands	100	Not reported	[10]
India	167	Not reported	[10]

FOODS CONTAINING POPPY SEEDS

Many ethnic foods contain poppy seeds in various amounts but in Western countries common foods containing poppy seeds are breads, muffins, pastries, and cakes. Poppy seeds are often combined with blueberries or lemon in some muffin and cake for better taste. Fillings in pastries sometimes contain finely ground poppy seeds mixed with sugar, butter, or milk. Poppy paste is used in preparing some ethic foods. Poppy seeds are also used in preparing certain salad dressings. Although infrequently consumed, poppy tea is a narcotic analgesic tea which is usually brewed from the dried *P. somniferum* plants. Tea can also be prepared from seed pods or seeds. The tea is usually bitter in taste and the flavor is sometimes improved by mixing it with coffee, honey, or lemon juice.

Reduced Morphine Content in Poppy Seeds During Food Processing

Morphine content of poppy seed is significantly reduced during food processing. Washing can significantly remove morphine from poppy seeds. Studies have shown that washing poppy seeds with hot water is more effective than washing with cold water because washing with hot water (60°C) may reduce morphine content by approximately 70% but cold water washing reduces morphine content approximately 30%. However, boiling water is not more effective than hot water for reducing morphine content of poppy seeds. Longer washing time can further reduce morphine content of poppy seeds. However, when poppy seeds are used for decoration, such washing step is avoided because strewing of buns demands poppy seeds with certain adhesive properties which may be lost during washing. In addition to washing, drying as well as grinding of poppy seeds also reduces morphine content. Baking poppy seed—containing cakes and buns also further reduces morphine content [12].

When a person eats poppy bun, poppy muffin, or poppy cakes, the person consumes only about 3 g of poppy seeds which is not a health risk because morphine content is significantly reduced during food processing. For example, in one study, the authors reported that median morphine content of poppy seeds was 6.8 mg/kg which was reduced to 3.9 mg/kg during grinding and heating steps. Finally in the finished food sold in bakeries, the morphine content was less than 1 mg/kg [13]. Food processing including baking is capable of reducing morphine content of poppy seeds from 0.002% to 0.0002% in topping on a poppy seed bun, indicating food processing (washing, baking, etc.) may reduce morphine concentration of poppy seeds by 90% [4].

Simple grinding can reduce morphine content by 34%. In addition, the authors also observed that for poppy cakes, only 16%—50% of morphine and 10%—50% of codeine present in original poppy seeds can be recovered, but for poppy buns baked at high temperature (220°C), only 3% of morphine and 7% of codeine could be recovered. The authors concluded that baking temperature has a significant influence on reducing morphine content of poppy seeds because observed reduction of morphine at baking temperature of 135°C was relatively low (around 30%) but when baking temperature was 220°C, higher reduction in morphine content could be observed. The authors advised consumers and bakers to wash poppy seeds with water before direct use or before grinding seeds prior to baking. Moreover, it is advisable to use high temperature for baking in order to reduce morphine and codeine content in finished products. It is possible to remove 100% of morphine by superior food-processing technique including baking at high temperature [14]. Morphine and codeine are found in poppy seeds both as free and bond forms. In one study, the authors observed that washing removed 45.6% free morphine and 48.4% of free codeine [15]. Loss of morphine in various stages of food processing involving poppy seeds is summarized in Table 31.3.

Most morphine and codeine are lost from poppy seeds during food processing. Therefore cases involving toxicity due to consumption of poppy seed—containing food are rarely reported in the literature. However, drinking poppy tea, eating poppy seeds, or food containing high amount of poppy seeds may cause severe toxicity. Although not commonly consumed, drinking poppy tea may cause opiate addiction because poppy tea is often prepared from the pods of *P. somniferum* plant, which contains much higher amounts of morphine and codeine compared to poppy seeds. Moreover, preparing such tea in hot water effectively extracts morphine and codeine along with other alkaloids present in poppy pods.

DANGERS OF CONSUMING POPPY TEA

Drinking poppy tea may cause opiate addiction [16]. In one study, the authors reported that some patients suffered from opiate dependency due to consumption of poppy tea. Patients reported onset of action 15 min after consuming the tea and the effect may last up to 24 h [17]. It may be necessary to treat such dependance with methadone or buprenorphine. Nanjayya et al. described a case of an 82-year-old woman who was dependent on poppy tea for approximately 55 years. As access to poppy tea became more problematic in India due to legal restriction, she was brought by her family

TABLE 31.3 Reduction of Morphine Content of Poppy Seeds During Food Processing

Food Processing Step	Reduction in Morphine Content	Mechanism	References
Washing with cold water	Approximately 30%	Washing removes free morphine and codeine from poppy seeds.	[14,15]
Washing with hot water (60°C)	Approximately 70%	Higher temperature may be more effective in removing free morphine and codeine from poppy seeds.	[12,15]
Grinding of poppy seeds in poppy mills	Average 34%	This may be related to oxidation of morphine in the by phenol oxidase in the presence of caffeic acid.	[4,12,14]
Baking temperature of 135°C	Approximately 30%	Thermal degradation of morphine.	[4,14]
Baking temperature of 220°C	80%–90%	Higher rate of morphine degradation at elevated temperature.	[4,14]

member to an addiction recovery center for treatment for her severe withdrawal symptoms. She was successfully treated on buprenorphine maintenance therapy [18]. Seyani et al. described a case of a 64-year-old woman who was admitted to the hospital due to respiratory arrest after drinking poppy tea. A naloxone infusion was able to reverse her opiate toxicity and she made good recovery. It was subsequently discovered that she had brewed tea from poppy buds which she picked from a nearby commercial poppy farm. She learnt that practice of preparing tea from poopy buds when she was in Afghanistan [19].

Significant amounts of morphine may be found in blood and urine of subjects consuming poppy tea. A baker consumed poppy tea prepared from seeds and experienced tonic-clonic seizure and delirium. His business partner informed that he was purchasing 25 kg of poppy seeds per week whereas only 3 kg was required for bakery. The concentration of morphine in his blood was almost 3.0 mg/L. The patient admitted drinking about 2 L of poppy tea made from 4 kg of seeds. When a typical tea was prepared the morphine concentration was 0.14 mg/mL indicating that he was consuming approximately 280 mg morphine a day [20]. Van Thuyne et al. reported that the morphine content of poppy tea prepared from two specimens of a different species of poppy (*Papaveris fructus*) contained 10.4 µg/mL (tea A) and 31.5 µg/mL (tea B) morphine. After administration of tea A in five healthy volunteers, the maximum morphine level in urine was 4.34 (4300 ng/mL) in one volunteer but after consuming tea B the maximum urinary concentration of morphine was 7.4 µg/mL (7400 ng/mL) [21].

Fatality has also been reported due to consumption of poppy seed tea. A death has been reported after repeated consumption of poppy tea in a person. The amount of morphine was 0.259 µg/mL in the typical tea specimen analyzed [22]. Steentoff et al. reported seven deaths in Denmark during the period of 1982–85 due to consumption of opium tea [23]. Bailey et al. reported a case of a 42-year-old Caucasian male who died as a result of drinking poppy tea along with ingesting phenazepam. The deceased had a history of ordering medication over the Internet. The postmortem blood showed the presence of morphine (116 ng/mL), codeine (85 ng/mL), and thebaine (72 ng/mL), all due to consumption of poppy seed tea prior to death. The blood level of phenazepam was 386 ng/mL [24].

Dangers of Opium-Containing Foods

Rachacha, also called "rach" or "low-grade opium," is a homemade preparation using decoction of poppy heads (*P. somniferum*) which is then transformed into a black paste by allowing water to evaporate. This product can be consumed directly or may also be infused in herbal tea. This product is similar to opium tea. In one study, the authors reported death of a 30-year-old man and a 28-year-old woman after ingesting rachacha balls along with alcohol in one evening with a friend. The black paste found in the scene was identified as rachacha and analysis of the black paste showed the presence of morphine (5.2%) and codeine (0.51%) along with saccharose and glucose but other opium alkaloids were not detected. Toxicological analysis showed the presence of morphine and codeine in the blood, urine, and bile with an absence of 6-monoacetyl morphine (6-MAM; also known as 6-acetyl morphine) indicating opiate abuse but not heroin abuse. High levels of morphine and codeine found in the gastric content of both man (morphine: 10, 130 ng/mL; codeine: 4015 ng/mL) and woman (morphine: 74, 1000 ng/mL; codeine: 10, 650 ng/mL) indicated that cause of death was opiate overdose. Concurrent consumption of alcohol certainly played a role in death of both man and woman [25].

Martinez et al. reported death of a 32-year-old Caucasian male who worked in a legal poppy field in Spain. He was found unresponsive in a poppy field. His friends reported that the victim suffered from epilepsy. Some tools (wood stick, razor blade, fabric handle, and paper) were found beside his body but no drug or medicine was discovered. Autopsy performed 15 h after death showed the presence of morphine (0.13 mg/L in blood, 4.50 mg/L in urine, 0.13 mg/L in vitreous, and 6.60 mg/L in gastric content) as well as codeine (0.48 mg/L in blood, 0.88 mg/L in urine, 0.137 mg/L in vitreous, and 1.50 mg/L in gastric content) and thebaine in various biological matrixes. Moreover, morphine, codeine, and thebaine were also detected in tools found beside his body. Metabolites of cocaine and cannabis were also detected in the deceased. Apparently, the victim stole poppy capsules and ingested and unknown quantity of latex in order to experience euphoria [26].

TOXICITY FROM CONSUMING POPPY SEEDS OR POPPY SEED−CONTAINING FOOD

Although it is extremely unlikely that eating poppy seed−containing food may cause adverse effect, toxicity may result from consuming high amount of poppy seeds or poppy seed−containing food. A mother had given her 6-month-old infant 75 mL of strained milk of baking poppy seeds with an intention to put the child to sleep. The mother prepared the milk following a recipe from an old cookbook and the milk was prepared from a mixture of 200 g of poppy seeds in 50 mL of milk. After drinking the strained milk, the infant suffered from respiratory depression requiring hospitalization. Suspecting opiate toxicity due to presence of poppy seeds in the milk, the physician treated the infant with naloxone, an antidote for opiate toxicity. Urine toxicological analysis showed high level of morphine (18 mg/L) confirming opiate toxicity. It was estimated that the infant consumed approximately 30 mg of morphine [27].

In another case, a person reported "dim feeling in the head," vomiting and hangover like feeling on the next day after eating a pasta dish strewn with a mixture of poppy seeds and sugar. The person consumed approximately 16 mg of morphine and 3 mg of codeine [14]. Death due to complications of bowel obstruction following raw poppy seed ingestion has also been reported. A 54-year-old woman with intractable vomiting was found unresponsive at home and later pronounced dead. At autopsy, cast like large bowel obstruction composed of poppy seeds was identified. The cause of death was determined to be complications of bowel obstruction due to eating a large amount of poppy seeds. The authors further commented that concomitant use of other central nervous system depressant medications such as opiates, benzodiazepines, or barbiturates, or illicit drugs along with poppy seeds may cause severe toxicity [28]. Even consuming large amounts of poppy seed may cause opiate addiction. Kaplan described a case of a woman who became addicted to poppy seeds consuming 2 kg of poppy seeds per week. She would fill her mouth with poppy seeds and sucked seeds to experience tingling sensation in her body followed by feeling of euphoria [29].

EFFECT OF POPPY SEED−CONTAINING FOOD ON OPIATE DRUG TESTING

Eating poppy seed−containing foods may cause positive, opiate test result during immunoassay screening and then positive morphine and codeine are confirmed using gas chromatography−mass spectrometry (GC-MS) or a suitable confirmation method. In general, morphine concentrations are higher than codeine concentrations after eating poppy seed−containing foods. Wide range of morphine concentrations (up to 18,000 ng/mL) have been reported by authors in volunteers after consuming various amounts of poppy seeds or poppy seed products. However, most of these studies were not controlled for food processing−related loss of morphine and codeine in poppy seeds. In some studies, volunteers consumed equivalent amount of unprocessed poppy seeds expected to be present on poppy seed−containing foods. As a result urine morphine levels were much higher than expected in real life situation after eating poppy seed cakes, bagels, or muffins [4].

The original cutoff for opiate screening in workplace urine drug testing was 300 ng/mL but in November 1998, the cutoff was increased to 2000 ng/mL to circumvent false positive opiate workplace drug testing result after consuming poppy seed−containing products. Although current SAMHSA (Substance Abuse and Mental Health Services Administration) guidelines for Federal (US Government) workplace drug testing mandates 2000 ng/mL as the screening cutoff of opiates in urine drug testing, some private employers may still use 300 ng/mL as the screening cutoff for opiates in workplace drug testing. Fraser and Worth reported that following the new guideline codeine and morphine positive results were reduced from 7.1% positive in 1994−96 to 2.1% in 1998 [30]. Although positive opiate test result at of 300 ng/mL may result from eating poppy seed−containing food within 24 h after ingestion, it is very unlikely to test positive at 2000 ng/mL after eating 1–2 pieces of poppy seed cake, muffin, or bagel. Therefore when the opiate level in the screening test is more than 2000 ng/mL, the consumption of poppy seed–containing food can be excluded [31].

High Morphine Levels After Eating Poppy Seeds

Because untreated poppy seeds contain significantly higher amounts of morphine and codeine, than processed poppy seed–containing foods, in some studies the authors reported very high amounts of morphine and codeine in urine after volunteers consumed unprocessed poppy seeds. An early report of excretion of morphine in urine after consumption of poppy seed product was published in 1985. The morphine content of the poppy seeds used in the study varied from 4 to 200 μg/g and maximum morphine concentration in urine was 18 μg/mL (18,000 ng/mL) in volunteers who consumed poppy seeds [32].

Hill et al. investigated the effect of ingesting a high amount of poppy seed on the urinary concentrations of morphine in volunteers as well as concentrations in hair. The poppy seed study was performed using Australian poppy seed because it contains the highest amount of morphine of any poppy seed available in the US market. Ten subjects (six males, four females) ingested 150 g of poppy seeds over a period of 3 weeks. Urine specimens were collected on days when volunteers consumed poppy seeds while hair specimen was collected in the fifth week of the study. The range among the 10 subjects of the highest urine value for each subject was 2929–13,857 ng/mL for morphine and 208–1174 ng/mL for codeine (determined by GC-MS). Moreover, urinary morphine levels remained above the 2000 ng/mL for as long as 10 h. Hair morphine levels were 0.05–0.48 ng/10 mg of hair [6].

Thevis et al. reported highest urine morphine level of 10,040 ng/mL 6 h postingestion in one volunteer out of nine volunteers (seven males, two females) who ingested poppy seed–containing cakes. However, the authors used poppy seeds containing very high amount of morphine (151.6 mg/kg of seeds) and volunteers consumed much higher amounts of poppy seeds (more than 12 g) than expected in a typical poppy seed cake [9]. In another study, six volunteers consumed 4 g of poppy seeds equivalent to the amount of poppy seeds present on two bagels. Urine specimens were collected at various times after ingestion and were analyzed for morphine and codeine content. The authors observed that all opiates were excreted fast after ingestion of bagels by all subjects and peak concentrations were reached within 6–8 h. However, no specimen was tested positive 24 h after ingestion of poppy seed bagels. Significant intraindividual differences were observed in excretion of opiate [7].

Hayes et al. also studied the effect of ingesting poppy seeds on morphine and codeine levels in urine. The poppy seeds used for this study contained 17–294 μg/g of morphine and 3–14 μg/g of codeine. The subjects ingested 40 g of poppy seeds that contained approximately 2.5 mg of morphine and 0.16 mg of codeine. The immunoassay screening of specimens by EMIT (enzyme multiplied immunoassay technique) showed typical positive screen (>300 ng/mL cutoff) up to 24 h of ingestion of poppy seed and peak concentration of total morphine ranged from 700 to 2635 ng/mL as measured by the GC-MS. The highest morphine level of 2635 ng/mL in one volunteer was observed 3 h postingestion [32]. In another study, using poppy seeds containing 17.4–18.6 μg/g of morphine and 2.3–2.5 μg/g of codeine, the authors reported highest morphine concentration of 4.5 mg/L (4500 ng/mL) in one subject 5 h after ingestion of poppy seeds [8].

Smith et al. using 22 volunteers who consumed two 45 g poppy seed doses 8 h apart (each dose containing 15.7 mg morphine and 3 mg codeine) observed that 26.6% urine specimens (out of 391 specimens), were positive for morphine by GC-MS at 2000 ng/mL cutoff but 83.4% specimens were positive at 300 ng/mL cutoff. In specimens morphine concentrations varied from <300 to 7522 ng/L. No specimen was positive for codeine at 2000 ng/mL but 20.2% specimens showed codeine concentration above 300 ng/mL [33].

Lower Morphine Level in Subjects After Eating Processed Poppy Seed–Containing Foods But High Level After Eating Poppy Seeds

Meadway et al. studied opiate concentrations following ingestion of poppy seed products. The maximum morphine and codeine concentrations in the seeds used for the study were 33.2 and 13.7 μg/g, respectively. Following consumption of bread rolls (0.76 g seed covering per roll) by four subjects, all urine specimens analyzed by EMIT (Dade Behring EMIT II opiate screen) were negative except for one subject who consumed two bread rolls. The subject showed positive opiate screen in urine up to 6 h with maximum morphine and codeine concentrations of 832.0 and 47.9 ng/mL, respectively. However, all four subjects showed positive urine opiate screen up to 24 h after ingestion of poppy seed cake containing an average 4.69 g of seed per slice [34]. This study clearly demonstrated that the urinary concentration of codeine and morphine after consumption of poppy seed products should not cause positive opiate screening test result at 2000 ng/mL cutoff. In another study, the concentrations of morphine and codeine were 2797 and 214 ng/mL, respectively, in one healthy volunteer who ingested three poppy seed bagels. Opiate was present in the urine 22 h postingestion but combined morphine and codeine was significantly below 2000 ng/mL [35]. Although in this study, the concentration of morphine and codeine exceeded 2000 ng/mL 3 h postingestion, the volunteer consumed three bagels. Usually people eat one or two bagels.

In another study, four volunteers ate three poppy seed bagels each. Neither morphine nor codeine was detected in oral fluids (limit of detection 3 ng/mL). However, trace amounts of codeine were detected (less than 4 0 ng/mL) in urine specimens and highest morphine concentration of 603 ng/mL was observed 1–4 h postingestion. Lowest morphine level was 314 ng/mL. This experiment clearly shows that after eating poppy seed bagels, urine opiate levels were significantly below 2000 ng/mL cutoff. However, in this study, authors also clearly showed that eating unprocessed poppy seeds along with poppy seed bagel could produce very high morphine level in urine. When three volunteers ate one poppy seed bagel (820 mg of poppy seed) and then ingested an unlimited amount of poppy seeds in 1 h (volunteer 1 ingested 14.82 g seeds, volunteer 2 ingested 9.82 g seeds, and volunteer 3 ingested 20.82 g seeds), the urine morphine level in volunteer 3, exceeded 10,000 ng/mL approximately 2.4 h after ingestion. Urine specimens remained positive up to 8 h at 2000 ng/mL. In addition, the oral fluid tested positive up to 1 h after ingestion of poppy seed bagels and poppy seeds at a 40 ng/mL cutoff (highest morphine: 205 ng/mL). This study indicates that eating poppy seed bagels may cause positive opiate screening at 300 ng/mL cutoff but not at 2000 ng/mL cutoff. However, eating poppy seeds along with bagel may cause very high urine morphine levels [36].

ElSohly designed an experiment to determine urinary morphine and codeine level after realistic consumption of poppy seed–containing food. Two male and two female volunteers participated in four protocols of eating poppy seed–containing rolls and each protocol was separated at least by 1 week. Subjects ingested one, two, or three poppy seed rolls each containing 2 g of Australian poppy seeds (108 mg morphine/kg of seed). In the fourth protocol subjects ingested two rolls per day for four consecutive days. The highest concentration of morphine was observed 3–8 h after ingestion or in the first void specimen. Out of 264 urine specimens collected from these volunteers, only 16 specimens exceeded 300 ng/mL limit using immunoassays. The highest observed morphine concentration was 954 ng/mL but all positive specimens were negative within 24 h. When one volunteer consumed a poppy seed cake containing 15 g of poppy seed, highest morphine level of 2010 ng/mL was observed 9 h postingestion [37]. This study clearly shows that eating realistic amount of poppy seed–containing baked food such as roll, muffin, cake should not cause positive opiate test result at 2000 ng/mL cutoff except for eating food containing very high amounts of poppy seeds.

The wide variations of reported morphine and codeine values after ingestion of poppy seed are due to various amounts of poppy seed ingested by subjects as well as due to significant variation in morphine and codeine content of poppy seeds originating from different countries. In one study, volunteers ingested one or two curry meals prepared from washed poppy seeds with estimated consumption of 200.4–1002 μg of morphine and 95.9–479.5 μg codeine. The authors reported that urinary morphine levels varied between 120 and 1270 ng/mL and codeine level varied between 40 and 730 ng/mL [15].

In Germany, a blood level of free morphine should be over 10 ng/mL in drivers may be considered as driving under influence. Moeller et al. studied blood and urine morphine levels after consumption of poppy seed products. Five volunteers ate different kinds of poppy seed rolls or poppy seed cakes but all blood specimens analyzed postingestion was negative for free morphine by both immunoassays or GC-MS (values below 10 ng/mL). However, GC-MS analysis of urine showed both morphine (range: 147–1300 ng/mL) and codeine (range: 11–36 ng/mL). The authors conclude that eating poppy seed–containing food should not result in positive blood test for free morphine [10].

In one study, the authors clearly demonstrated that urine as well as oral fluid morphine and codeine levels should be significantly lower after eating poppy seed–containing food compared to ingesting raw poppy seeds. In first part of the experiment, 12 volunteers consumed poppy seed roll and in the second experiment the same individuals consumed equivalent amount of raw poppy seeds containing approximately 3.3 mg of morphine and 0.6 mg of codeine in each dose. As expected urinary morphine levels varied from 155 to 1408 ng/mL (as determined by GC-MS) after eating poppy seed rolls but much higher urine morphine levels were observed (294–4213 ng/mL) when same subjects in a later experiment consumed equivalent amount of raw poppy seeds. As expected, urinary codeine concentrations were significantly lower (range: 140–194 ng/mL) after eating poppy seed rolls compared to after eating raw poppy seeds (range: 121–664 ng/mL). Similar trend was also observed with oral fluids with much lower morphine and codeine levels after consuming poppy seed rolls (morphine: 7–143 ng/mL up to 1.5 h, codeine detected only in 5.5% specimens) compared to eating raw poppy seeds (morphine: 7–600 ng/mL, codeine: 8–112 ng/mL; 0.25–3 h after consumption) [38]. Highest morphine concentrations after ingestion of raw poppy seeds or poppy seed–containing products as reported by various authors are summarized in Table 31.4.

TABLE 31.4 Highest Levels of Morphine in Urine After Consuming Poppy Seed–Containing Food or Poppy Tea as Reported in Some Published Studies

Ingested Product	Morphine (ng/mL)	Comments	References
Poppy seeds	18,000	High morphine in urine is due to ingestion of unprocessed poppy seed.	[32]
Poppy seeds	13,857	High morphine in urine is due to ingestion of unprocessed poppy seed. Highest codeine level was 1174 ng/mL.	[6]
Poppy seeds	1180	Although subjects ingested poppy seeds, relatively lower morphine level may be due to lower amount of morphine in seeds.	[7]
Poppy seed cake	10,040	Although subjects ate cake, high morphine level was due to use of poppy seeds with very high morphine content in baking such cakes. Moreover, each subject consumed more than 12 g of poppy seeds while typical poppy seed cake contains 3–4 g of poppy seeds.	[9]
Poppy seeds	2635	Although subjects ate poppy seeds, morphine highest morphine level just exceeded 2000 ng/mL.	[39]
Poppy seeds	4500	Higher urine morphine related to ingestion of unprocessed poppy seeds.	[8]
Poppy seeds	7522	High morphine level in subjects is related to eating unprocessed poppy seeds.	[33]
Herbal tea containing *Papaveris fructus*	7400	High morphine level as expected from consuming poppy tea.	[21]
Poppy seed–containing bread rolls	832	Urine morphine level was well below 2000 ng/mL (morphine: 832 ng/mL, codeine: 48 ng/mL) indicating that eating poppy seed–containing bread should not cause positive opiate test at 2000 ng/mL cutoff.	[34]
Poppy seed bagel	2797	Although highest morphine level (2797 ng/mL) and codeine level (214 ng/mL) were observed 3 h postingestion, after 22 h morphine level was reduced to 676 ng/mL and codeine level was reduced to 16 ng/mL indicating that urine would have tested negative at 2000 ng/mL cutoff.	[35]
Poppy seed bagel	603	Volunteers ate three poppy seed bagels but opiate screen was negative at 2000 ng/mL.	[36]
Poppy seed cake containing 15 g of seed	2010	Total opiate level exceeded 300 ng/mL cutoff for approximately 24 h with one specimen showed 1010 ng/mL of morphine and 78 ng/mL of codeine 9 h postingestion. However, when subjects ate up to three poppy seed rolls each containing 2 g of Australian poppy seeds, the highest urinary morphine level was 954 ng/mL.	[37]
Poppy seed roll or cake	1300	Maximum codeine level was 36 ng/mL indicating that eating poppy seed–containing food should not cause false positive test result with opiate at 2000 ng/mL cutoff.	[10]
Poppy seed roll versus eating poppy seeds	1408 (poppy seed roll) 4213 (raw poppy seeds)	Highest codeine after eating poppy seed roll was 194 ng/mL but after eating raw poppy it was 664 ng/mL. This study clearly shows that morphine and codeine levels should be significantly lower after eating poppy seed–containing food compared to ingesting raw poppy seeds.	[38]

In another study, 17 healthy volunteers consumed 45 g of raw poppy seed doses each containing 15.7 mg morphine and 3.1 mg codeine. Oral fluid specimens ($n = 459$) were collected before and up to 32 h after the dose. All oral fluid specimens screened positive for opiates for 0.5–13 h at Draeger 20 ng/mL cutoff for morphine. Maximum oral fluid morphine and codeine concentrations were 177 and 32.6 ng/mL, respectively [40].

Since the amount of morphine is very low in cooked poppy seed–containing foods, impairment is unlikely after eating such foods. Meneely reported that seven volunteers who ingested 25 g of poppy seed each baked into cakes showed opiate positive urine at 300 ng/mL shortly after consuming poppy seed products but none of the volunteer exhibited symptoms of opiate impairment based on series of standardized drug recognition evaluation test [41].

BROWN MIXTURE AND OPIATE LEVELS IN URINE

Brown mixture is used as a cold remedy and is a legal prescription drug in Taiwan. Each tablet contains 2.5 mg opium powder (10%–15% morphine, approximately 30 μg morphine per tablet). Following consumption of brown mixture significant amounts of morphine and codeine can be observed in urine. With consumption of one tablet each by volunteers, maximum urinary morphine concentration of 379 ng/mL was observed while consumption of six tablets by volunteers, maximum urinary concentration of morphine was 2525 ng/mL. The authors concluded that after consumption of brown sugar it is unlikely that urinary morphine concentration would be greater than 4000 ng/mL [42].

POPPY SEED DEFENSE

Because poppy seed defense can be used by illicit drug users, it is important to differentiate heroin abusers from subjects consuming poppy seed–containing food. Heroin has a very short plasma half-life (approximately 5 min) and is quickly metabolized to 6-MAM by esterases in the liver, plasma, and erythrocytes. As a result heroin may not be present in urine. Heroin may be detected in blood for a short time but has a longer window of detection in oral fluid. Jenkins et al. showed that heroin can be detected in saliva up to 60 min after intravenous administration and even up to 24 h after smoking heroin base [43]. Therefore it is unlikely to determine heroin abuse by detecting the presence of heroin in urine. However, 6-MAM, which is eventually metabolized to morphine and is conjugated with glucuronic acid (morphine-3-glucuronide, major inactive metabolite, morphine-6-glucuronide, minor active metabolite), can be detected in urine for 24 h or less after last heroin abuse and is considered as a biomarker of heroin abuse because morphine-3-glucuronide in urine may originate from heroin abuse or morphine present in the poppy seeds (codeine present in poppy seed may also metabolized into morphine), but the presence of the heroin-specific metabolite 6-MAM can only be explained due to abuse of heroin [44]. Borriello et al. commented that 6-MAM is a biomarker of heroin abuse in both urine and blood with 100% sensitivity for urine and 95% sensitivity for blood [45].

Acetyl Codeine

A major limitation of 6-MAM is its short half-life and may often be absent in urine collected from heroin abusers due its short window of detection. Therefore acetyl codeine has been suggested as a marker for heroin abuse. Codeine is present naturally in opium which is converted into acetyl codeine, a synthetic byproduct during acetylation of morphine to produce illicit heroin. However, acetyl codeine is not a better marker than 6-MAM for heroin abuse. In one study, the authors analyzed 100 criminal justice urine specimens containing >5000 ng/mL of morphine and detected acetyl codeine only in 37 specimens with concentrations ranging from 2 to 290 ng/mL (median: 11 ng/mL). However, 6-MAM was also present in these specimens with much higher concentrations than acetyl codeine (range: 49–12 ng/mL, 600 ng/mL; median: 740 ng/mL). Of 63 specimens negative for acetyl codeine, 36 specimens showed the presence of 6-MAM at concentrations ranging from 12 to 4600 ng/mL (median: 124 ng/mL). The authors commented that when detected, acetyl codeine concentration was much lower than 6-MAM (average 2.2% of 6-MAM concentration). Therefore due to very low concentration acetyl morphine in urine of heroin abusers, this marker is less reliable than 6-MAM to confirm heroin abuse in workplace drug testing or criminal justice system. However, if present in urine, detection of acetyl codeine further validates heroin abuse [46]. In one study, the authors reported that average half-life of acetyl codeine in urine is 237 min and as a result peak acetyl codeine concentration is observed 2 h after administration and the detection window is only 8 h [47].

Acetyl codeine is also present in oral fluid. Phillips and Allen analyzed 513 oral fluid specimens using liquid chromatography linked to atmospheric pressure ionization tandem mass spectrometry (LC-MS-MS) and observed detectable amounts of one or more opiates in 297 specimens. Out of these specimens, morphine, codeine, 6-MAM, acetyl codeine, and heroin were detected in 97%, 82%, 77%, 55%, and 45%, respectively. However, nine specimens showed detectable amount of heroin but no acetyl codeine. The authors concluded that acetyl codeine detection of acetyl codeine in oral fluid indicates heroin abuse but has limited use [48].

Morphine/Codeine Ratio

When no 6-MAM or acetyl codeine is present in urine or blood, morphine/codeine ratio may be used to establish or rule out heroin abuse. Moriya et al. proposed following criteria for heroin abuse when no 6-MAM is present in urine [49]:

- Detectable level of free morphine (unconjugated morphine) exists in urine and concentration of total morphine is greater than 10,000 ng/mL.
- Detectable amount of codeine is present in urine.
- The morphine-to-codeine ratio is greater than 2 for both free and total amounts of morphine and codeine.

Ceder and Jones analyzed blood and urine specimens from 339 driving under the influence of drug suspects and additional 882 blood specimens where urine specimens were not available. The authors identified 6-MAM in only 16 out of 675 blood specimens analyzed (2.3% positive for 6-MAM). In contrast, the authors detected 6-MAM in 212 out of 339 urine specimens analyzed (62% positive for 6-MAM). The authors concluded that only 2.3% heroin abusers were identified based on 6-MAM in blood but 62% heroin abusers can be identified using urine drug testing. However, the authors identified 85% of heroin abuser by using a morphine/codeine ratio above 1 in blood. In addition, the authors further commented that morphine/codeine ratio of <1 may also indicate prescription use of codeine but not heroin abuse [50].

The distribution of morphine and codeine is similar in blood and urine. Therefore morphine-to-codeine ratio over in urine should also indicate heroin abuse. In another study, based on a large population of forensically examined autopsy cases ($n = 2438$), the authors concluded that morphine-to-codeine ratio great than one in blood and or urine is indicative of heroin abuse [51]. Bu et al. analyzed urine specimens collected from heroin abusers and compared results with 21 volunteers who ingested codeine phosphate oral solution. The authors observed that when free morphine (unconjugated) concentration exceeded 64.2 ng/mL in urine, it indicated possibility of heroin abuse at a 95% confidence interval. Moreover, when free urinary morphine concentration exceeded 64.2 ng/mL and free morphine-to-codeine ratio exceeded 1.16, it indicated heroin abuse [52].

If the total codeine to total morphine ratio in urine is <0.5 and total morphine concentration in urine is >200 ng/mL, codeine may be excluded as the source of morphine in urine. The likely source of morphine in such urine may be due to heroin abuse. Conversely, a total morphine concentration in urine < 200 ng/mL and a total codeine to total morphine ratio > 0.5 indicates codeine ingestion [53]. Codeine is not classified as prohibited drug by the World Anti-Doping Agency (WADA) but morphine, a metabolite of codeine, is a prohibited drug during competition. Metabolism of codeine to morphine is also affected by polymorphism of CYP2D6 gene encoding CYP2D6 enzyme. Urine sample of a soccer player tested positive for morphine but the player denied taking any narcotics including morphine but admitted taking several acetaminophen-codeine tablet (300 mg acetaminophen, 10 mg codeine) day before drug testing due to tooth pain. The observed morphine-to-codeine ratio was 1.03 in urine. The disciplinary committee concluded morphine in his urine was due to codeine intake and was cleared of any wrong doing. The disciplinary committee's decision was based on the fact that the athlete could be an ultrarapid CYP2D6 metabolizer where morphine/codeine ratio may slightly be above 1 (in this case 1.03) [54]. Although after codeine intake morphine/codeine ratio in blood or urine should be below 1, He et al. showed that in ultrarapid metabolizer, morphine-to-codeine ratio in blood may slightly exceeds the value of 1 (highest value 1.060) when codeine is consumed approximately 24 h ago but the ratio should be below 1 when codeine is taken 12 h before the testing [55].

Opium Alkaloids as Biomarker of Poppy Seed Ingestion

Reticuline is a precursor of major opium alkaloids thebaine, morphine, and papaverine. Reticuline is a minor component of opium but is not present in heroin or poppy seed. Therefore reticuline has been suggested as a marker to differentiate between opium and poppy seed consumption as well to differentiate opium abuse from prescription use of codeine. In one study, reticuline was detected in all 291 specimens collected from opium users and the percent concentration ratios of reticuline/morphine were higher (2%–12%) than such ratio in opium (0.01%–3%) [56]. Noscapine and papaverine are poppy alkaloids but these compounds cannot differentiate poppy seed consumption heroin abuse because these alkaloids are also present in illicit heroin preparations. Nevertheless, noscapine or papaverine has not yet been determined in urine or blood samples after poppy seed consumption [53].

Thebaine is a natural constituent of poppy seed. Therefore positive thebaine findings can be attributed to an ingestion of opium preparations or poppy seed products. Moreover, thebaine itself is not found in illicit heroin because it is converted primarily to thebaol and acetylthebaol during preparation of illicit heroin under the action of acetic anhydride and thus cannot be detected in urine after street heroin administration. In addition, thebaine is not available as pharmaceutical products and it is also not a metabolite of heroin, morphine, codeine, or ethylmorphine. As a result the presence of thebaine confirms consumption of poppy seed products or opium but absence of thebaine in urine does not rule out poppy seed consumption [53]. In one study, thebaine was detected in concentrations ranging from 2 to 91 ng/mL in volunteers after consumption of 11 g of poppy seed [57].

ATM4G: A Biomarker of Heroin Abuse

Recently, ATM4G (acetylated-thebaine-4-metabolite glucuronide), a new marker for the discrimination between street heroin consumption and poppy seed ingestion has been described by Chen et al. This new biomarker of heroin abuse is due to the presence of thebaine in opium which is acetylated during illicit manufacture of illicit heroin. Then this acetylated product undergoes molecular rearrangement. After heroin abuse, this product is metabolized into ATM4G. When 22 specimens collected from heroin abusers which were tested negative for 6-MAM by GC-MS (reporting threshold: 10 ng/mL) were analyzed by LC-MS/MS, peak corresponding to ATM4G was identified in 16 out of 22 specimens. In contrast, 6-MAM was detected by LC-MS/MS only in three specimens with concentration >1 ng/mL. The authors concluded that ATM4G is a novel urine biomarker to identify heroin abusers [58].

In another study, when authors analyzed urine specimens after eating poppy seed products, ATM4G, papaverine, noscapine, and 6-MAM could not be detected in urine but only morphine and codeine were present. In contrast, when urine specimens collected from suspected heroin abusers were analyzed, ATM4G could be detected in 9 out of 43 cases but 6-MAM and acetyl codeine were present in 7 urine specimens. The authors concluded that ATM4G could be detected in urine samples in which neither 6-MAM nor 6-AC could be detected. The authors concluded that ATM4G is a specific biomarker of street heroin abuse [59]. Biomarkers useful in differentiating poppy seed ingestion from heroin abuse are listed in Table 31.5.

TABLE 31.5 Biomarkers Useful in Differentiating Poppy Seed Ingestion From Heroin Abuse

Biomarker	Recommended Value	Comments
6-Monoactetyl morphine (6-MAM)	Any level at or above cutoff concentration of 10 ng/mL in urine	Usually window of detection is 8 h in urine and may not be present in urine of heroin abuser. However, 6-MAM is more commonly detected in urine compared to blood.
Acetyl codeine	Any concentration that can be confirmed	Because of low concentration in urine and short window of detection (8 h), it may not be present in urine of heroin abusers. However, its presence along with 6-MAM confirms heroin abuse.
Morphine	>2000 ng/mL	Excludes the consumption of poppy seeds (poppy seed defense).
Morphine/codeine ratio	>1	The value >1 in blood or urine rules out poppy seed defense or prescription codeine use.
Reticuline	Reticuline to morphine concentration ratio in urine of opium users varied from 2% to 12%	Reticuline is not present in heroin or poppy seed but can only be found in opium. Therefore the presence of reticuline in urine confirms opium abuse.
Thebaine	Typical concentration in urine 2–91 ng/mL after eating poppy seed–containing food	Specific marker for ingestion of poppy seed but may not be present in urine specimens of all individual consuming poppy seeds. Therefore absence does not rule out poppy seed ingestion.
ATM4G (acetylated-thebaine-4-metabolite glucuronide)	Presence in urine confirmed by LC-MS/MS	A marker of heroin abuse but not present in poppy seeds or pharmaceutical opiates.

CONCLUSIONS

Eating poppy seed–containing food may lead to positive opiate screening test results at 300 ng/mL cutoff level but very unlikely to cause positive opiate urine drug screen at 2000 ng/mL cutoff. The presence of 6-MAM or acetyl codeine in urine indicates heroin abuse and is not consistent with the consumption of poppy seed–containing food. However, in the absence of 6-MAM, morphine-to-codeine concentration ratio in blood or urine >1 is also not consistent with ingestion of poppy seed–containing food but is indicative of possible heroin abuse. Although the presence of thebaine confirms consumption of poppy seed–containing food, thebaine is not always detected in urine after eating poppy seed–containing food. Therefore absence of thebaine in urine does not rule out poppy seed defense. More recently, ATM4G, a novel metabolite detected in urine specimens of heroin abusers, has been proposed as a specific biomarker of heroin abuse because this metabolite is not found in urine after eating poppy seed–containing food or taking prescription opiates.

REFERENCES

[1] Schmidt J, Boettcher C, Kuhnt C, Zenk MH. Poppy alkaloid profiling by electrospray tandem mass spectrometry and electrospray FT-ICR mass spectrometry after [ring-13C]-tyramine feeding. Phytochemistry 2007;68:189–202.

[2] Stefano GB, Pilonis N, Ptacek R, Kream RM. Reciprocal evaluation of opiate science from Medical and cultural perspective. Med Sci Monit 2017;23:2890–6.

[3] Reid RG, Durham DG, Boyle SP, Low AS, Wangboonskul J. Differentiation of opium and poppy straw using capillary electrophoresis and pattern recognition techniques. Anal Chim Acta 2007;605:20–7.

[4] Lachenmeier DW, Sproll C, Musshoff F. Poppy seed foods and opiate drug testing-where are we today?. Ther Drug Monit 2010;32:11–18.

[5] Trafkowski J, Madea B, Musshoff F. The significance of putative urinary markers of illicit heroin use after consumption of poppy seed products. Ther Drug Monit 2006;28:552–8.

[6] Hill V, Cairns T, Cheng CC, Schaffewr M. Multiple aspects of hair analysis for opiates: methodology, clinical and workplace population, codeine and poppy seed ingestion. J Anal Toxicol 2005;29:696–703.

[7] Pelders M, Ross JJ. Poppy seeds: difference in morphine and codeine content and variation in inter-and intra individual excretion. J Forensic Sci 1996;41:209–12.

[8] Pettitt B, Dyszel SM, Hood L. Opiates in poppy seeds: effects on urinalysis after consumption of poppy seed cake filling. Clin Chem 1987;33:1251–2.

[9] Thevis M, Opfermann G, Schanzer W. Urinary concentrations of morphine and codeine after consumption of poppy seeds. J Anal Toxicol 2003;27:53–6.

[10] Moeller MR, Hammer K, Engel O. Poppy seed consumption and toxicological analysis of blood and urine samples. Forensic Sci Int 2004;143:183–6.

[11] Zentai A, Szeitzne-Szabo SM, Szabo IJ, Ambrus A. Exposure of consumers to morphine from poppy seeds in Hungary. Food Addit Contam A 2012;29:403–14.

[12] Sproll C, Perz RC, Buschmann R. Guidelines for reduction of morphine in poppy seeds intended for food purposes. Eur Food Res Technol 2007;226:307–10.

[13] Kniel B. Morphin in backwaren: Fakten aus der praxis contra theorie der risikobewertung. BMI Aktuell 2006;1/2016:2–4.

[14] Sproll C, Perz RC, Lachenmeier DW. Optimized LC/MS/MS analysis of morphine and codeine in poppy seed and evaluation of their fate during food processing as a basis of risk analysis. J Agric Food Chem 2006;54:5292–8.

[15] Lo DS, Chua TH. Poppy seeds: implications of consumption. Med Sci Law 1992;32:296–302.

[16] Unnithan S, Strang J. Poppy tea dependance. Br J Psychiatry 1993;163:813–14.

[17] Braye K, Harwood T, Inder R, Beasley R, Robinson G. Poppy seed tea and opiate abuse in New Zealand. Drug Alcohol Rev 2007;26:215–19.

[18] Nanjayya SB, Murthy P, Chand PK, Kandaswamy A, et al. A case of poppy tea dependance in an octogenarian lady. Drug Alcohol Rev 2010;29:216–18.

[19] Seyani C, Green P, Daniel L, Pegden A. An interesting case of opium tea toxicity. BMJ Case Rep 2017;28 2017.

[20] King M, McDonough MA, Drummer OH, Berkovic SF. Poppy tea and the baker's first seizure. Lancet 1997;350:716.

[21] Van Thuyne W, Van Eenoo P, Delbeke FT. Urinary concentrations of morphine after the administration of herbal tea containing *Papaveris fructus* in relation to doping analysis. J Chromatogr B Analyt Technol Biomed Life Sci 2003;785:245–51.

[22] Sigillata T. Poppy seed tea can kill you. http://scienceblogs.com/terrasig/2009/03/31/poppy-seed-tea-can-kill [accessed 03.07.17]2017.

[23] Steentoff A, Kaa E, Worm K. Fatal intoxication in Denmark following intake of morphine from opium poppies. Z Rechtsmed 1988;101:197–204.

[24] Bailey K, Richards-Waugh L, Clay D, Gebhardt M, et al. Fatality involving the ingestion of phenazepam and poppy seed tea. J Anal Toxicol 2010;34:527–32.

[25] Monteil-Ganiere C, Gaulier JM, Chopineaux D, Barrios L, et al. Fatal anoxia to rachacha consumption: two case reports. Forensic Sci Int 2014;245:e1–5.

[26] Martinez MA, Ballesteros S, Almarza E, Garijo J. Death in a legal poppy field in Spain. Forensic Sci Int 2016;265:34–40.

[27] Hahn A, Michalak H, Begemann K, Meyer H, et al. Severe health impairment of a 6-week old infant related to ingestion of boiled poppy seeds. Clin Toxicol 2008;46:607.
[28] Schuppener LM, Corliss RF. Death due to complications of bowel obstruction following raw poppy seeds ingestion. J Forensic Sci 2018;63:614—18.
[29] Kaplan R. Poppy seed dependence [Letter to the editor]. Med J Aust 1994;161:176.
[30] Fraser AD, Worth D. Experience with a urine opiate screening and confirmation cutoff of 2000 ng/mL. J Anal Toxicol 1999;23:549—51.
[31] Stefanidou M, Athanaselis S, Spillopoulou C, Dona A, et al. Biomarkers of opiate use. Int J Clin Pract 2010;64:1712—18.
[32] Hayes LW, Krassolt WG, Mueggier PA. Concentrations of morphine and codeine in serum and urine after ingestion of poppy seeds. Clin Chem 1987;33:806—8.
[33] Smith ML, Nichols DC, Underwood P, Fuller Z, et al. Morphine and codeine concentrations in human urine following controlled seeds administration of known opiate content. Forensic Sci Int 2014;241:87—90.
[34] Meadway C, George S, Braithwaite R. Opiate concentrations following the ingestion of poppy seeds products-evidence for poppy seed defense. J Forensic Sci 1998;96:29—38.
[35] Struempler RE. Excretion of codeine and morphine following ingestion of poppy seeds. J Anal Toxicol 1987;11:97—9.
[36] Rohrig TP, Moore C. The determination of morphine in urine and oral fluid following ingestion of poppy seeds. J Anal Toxicol 2003;27:449—52.
[37] ElSohly HN, ElSohly MA, Stanford DF. Poppy seed ingestion and opiates urinalysis: a closer look. J Anal Toxicol 1990;14:308—10.
[38] Samano KL, Clouette RE, Rowland BJ, Sample RH. Concentrations of morphine and codeine in paired oral fluid and urine specimens following ingestion of a poppy seed roll and raw poppy seeds. J Anal Toxicol 2015;39:655—61.
[39] Fritschi G, Prescott WR. Morphine levels in urine subsequent to poppy seed consumption. Forensic Sci Int 1985;27:111—1117.
[40] Concheiro M, Newmeyer MN, de Costa JL, Flegel R, et al. Morphine and codeine in oral fluid after controlled poppy seed administration. Drug Test Anal 2015;7:586—91.
[41] Meneely KD. Poppy seed ingestion: the Oregon perspective. J Forensic Sci 1992;37:1158—62.
[42] Liu HC, Ho HO, Liu RH, Yeh GC, Lin DL. Urinary excretion of morphine and codeine following the administration of single and multiple doses of opium preparations prescribed in Taiwan as "brown mixture". J Anal Toxicol 2006;30:225—31.
[43] Jenkins AJ, Oyler JM, Cone EJ. Comparison of heroin and cocaine concentrations in saliva with concentrations in blood and plasma. J Anal Toxicol. 1995;19:359—74.
[44] von Euler M, Villen T, Svensson JO, Stahle L. Interpretation of the presence of 6-monoacetylmorphine in the absence of morphine-3-glucuronide in urine samples: evidence of heroin abuse. Ther Drug Monit 2003;25:645—8.
[45] Borriello R, Carfora A, Cassandro P, Petrella R. Clinical and forensic diagnosis of very recent heroin intake by 6-acetylmorphine immunoassay test and Lc-MS/MA analysis in urine and blood. Ann Clin Lab Sci 2015;45:414—18.
[46] O'Neal CL, Poklis A. The detection of acetyl codeine and 6-acetylmorphine in opiate positive urine. Forensic Sci Int 1998;95:1—10.
[47] Brenneisen R, Hasler F, Wursch D. Acetyl codeine as a urinary marker to differentiate the use of stress heroin and pharmaceutical heroin. J Anal Toxicol 2002;26:561—6.
[48] Phillips SG, Allen KR. Acetyl codeine as a marker of illicit heroin abuse in oral fluid samples. J Anal Toxicol 2006;30:370—4.
[49] Moriya F, Chan KM, Hashimoto Y. Concentrations of morphine and codeine in urine of heroin abusers. Legal Med 1999;1:140—4.
[50] Ceder G, Jones AW. Concentration ratios of morphine to codeine in blood of impaired drivers as evidence of heroin use and not medication with codeine. Clin Chem 2001;47:1980—4.
[51] Konstantinova SV, Normann PT, Arnestad M, Karinen R, et al. Morphine to codeine concentration ratio in blood and urine as a marker of illicit heroin use in forensic autopsy samples. Forensic Sci Int 2012;217:216—21.
[52] Bu J, Zhan C, Huang Y, Shen B, et al. Distinguishing heroin abuse from codeine administration in the urine of Chinese people by UPLC-MS-MS. J Anal Toxicol 2013;37:166—74.
[53] Mass A, Madea B, Hess C. Confirmation of recent heroin abuse: accepting the challenge. Drug Test Anal 2017; July 6 [e-pub ahead of print].
[54] Seif-Barghi T, Moghadam N, Kobarfard F. Morphine/codeine ratio, a key in investigating a case of doping. Asian J Sports Med 2015;6:e28798.
[55] He YJ, Brockmoller J, Schmidt H, Roots I, et al. CYP2D6 ultrarapid metabolism and morphine/codeine ratio in blood: was it codeine or heroin? J Anal Toxicol 2008;32:178—82.
[56] Al-Amri AM, Smith RM, El-Haj BM, Juma'a MH. The GC-MS detection and characterization of reticuline as a marker for opium use. Forensic Sci Int 2004;142:61—9.
[57] Cassella G, Wu AH, Shaw BR, Hill DW. The analysis of thebaine in urine for the detection of poppy seed consumption. J Anal Toxicol 1997;21:376—83.
[58] Chan P, Braithwaite RA, George C, Hylands PJ, et al. The poppy seed defense: a novel solution. Drug Test Anal 2014;6:194—201.
[59] Mass A, Kramer M, Sydow K, Chen PS, et al. Urinary excretion study following consumption of various poppy seed products and investigation of the new potential street heroin marker ATM4G. Drug Test Anal 2017;9:470—8.

Chapter 32

Miscellaneous Issues: Paper Money Contaminated With Cocaine and Other Drugs, Cocaine Containing Herbal Teas, Passive Exposure to Marijuana, Ingestion of Hemp Oil, and Occupational Exposure to Controlled Substances

Amitava Dasgupta
Department of Pathology and Laboratory Medicine, University of Texas McGovern Medical School at Houston, Houston, TX, United States

INTRODUCTION

Analytical true positive and clinical false positive result in workplace drug testing or legal drug testing may have devastating consequence on the subject. One of such scenarios is confirmation of benzoylecgonine in the urine specimen by gas chromatography/mass spectrometry (GC/MS) or liquid chromatography combined with tandem mass spectrometry (LC−MS/MS) due to drinking some imported herbal tea such as Health Inca tea or Coca-de-Mate which contains cocaine. In contrast, people may defend positive workplace drug testing results by providing several excuses such as handling paper money contaminated with drug or being present in a concert where many people were smoking marijuana and positive test result was due to passive inhalation of marijuana. Ingesting hemp oil which may contain trace amount of Δ^9-tetrahydrocannabinol (THC) (marijuana) may also be cited as an explanation of positive marijuana test result. In this chapter these topics are discussed.

US PAPER MONEY CONTAMINATED WITH COCAINE AND OTHER DRUGS

Cocaine and other drugs have been isolated from US paper money as well as paper money from Europe and other parts of the world. The most commonly present contamination in banknotes is cocaine while other abused drugs reported with much lower frequencies. Oyler et al. examined 1 dollar bills from several big cities in the United States for the presence of cocaine. The authors extracted dollar bills using methanol. Presence of cocaine was confirmed by GC/MS. The authors found cocaine in 74% of the bills in amounts above 0.1 μg. Moreover, 54% of currency showed cocaine concentration above 1.0 μg. The highest amount of cocaine found was 1327 μg in a 1 dollar bill [1]. Negrusz et al. analyzed ten $20 bills collected from Rockford, IL, and four $1 bills collected from Chicago, IL. The concentration of cocaine varied from 10.02 to 0.14 μg in $20 bills and 2.99 μg to none detected in 1 dollar bills. Overall 92.8% of all bills analyzed were contaminated with cocaine [2].

Jenkins reported the analysis of 10 randomly collected US $1 bills from each of five cities (Baltimore, Chicago, Denver, Honolulu and San Juan, Porto Rico) for cocaine, 6-acetylmorphine, morphine, codeine, methamphetamine, amphetamine, and phencyclidine. Bills were immersed in acetonitrile for 2 h in order to extract these drugs followed by confirmation of these drugs using GC/MS. The authors analyzed a total of 50 one dollar bills in the study and observed the presence of cocaine in 46 one dollar bills (92% 1 dollar bill positive for cocaine with concentration of cocaine varied from 0.01 to 922.7 μg per bill). The dollar bill containing highest amount of cocaine was collected from Honolulu. Heroin was detected in seven bills (14%) with highest amount of 168.5 μg detected in one bill collected from Baltimore. In addition, 6-monoacetyl-morphine (a metabolite of heroin and a marker of heroin abuse) and morphine were detected in three bills (6%), one obtained from Chicago and another from Baltimore (not the same bill containing high amount of heroin). In addition, amphetamine was detected in one bill (2%) and methamphetamine in 3 bills (6%). The author further reported that phencyclidine in two bills (4%) both collected from Denver. This report demonstrates that although cocaine is the major contamination of US paper currency, other abused drugs may also be present [3].

Zuo et al. described a nondestructive gas chromatographic method using a fast ultrasonic extraction for determination of cocaine in the US paper currencies. The authors reported the presence of cocaine in 67% of circulated banknotes collected in Southeastern Massachusetts. The amount of cocaine varied widely from 2 ng to 49.4 μg per bill (highest cocaine observed in a $5 bill). Interestingly $5, 10, 20, and 50 denominations contain higher amount of cocaine than $1 and 100 denominations of all banknotes [4].

In a large study, Jourdan et al. analyzed 4174 currency notes collected from 66 cities in 43 states including District of Columbia. The authors initially used GC/MS with an ion-trap detector for analysis of possible presence of cocaine in these bills. Deuterated cocaine (d_3) was used as the internal standard and the limit of detection was 0.1 ng/banknote. Later the authors also used liquid chromatography combined with mass spectrometry with electrospray ionization (LC−MS) as well as LC−MS/MS. Electrospray ionization operating in positive ionization mode was also used for LC−MS/MS analysis. The authors observed no cocaine in uncirculated bills collected directly from the US Bureau of Engraving and Printing. In contrast, 97% bills in circulation showed the presence of cocaine with an average of 2.34 ng of cocaine per bill across all denominations ($1, $5, $10, $20, $50, and $100). However, amount of cocaine found in bills in circulation was relatively low as probability of a currency note containing 300 ng/banknote of cocaine was 0.09% but probability of encountering a bill containing <10 ng of cocaine was 79.6% [5].

European paper money is also contaminated with cocaine. In one study the authors used nondestructive and environmentally friendly GC−MS−MS for analysis of cocaine in Euro banknotes. The authors observed the presence of cocaine in all 16 banknotes they analyzed (100% positive for cocaine) with the highest amount of 889 μg in one 10 euro bill [6].

In a recent study the authors analyzed 65 euro banknotes (denomination 5, 10, 20, and 50 euro) collected from different districts of Berlin, Germany, and observed the presence of cocaine in all banknotes (100% contaminated). The authors reported that amount of cocaine was below 1 μg/banknote in 25 banknotes but amount of cocaine exceeding 11 μg/banknote in 40 banknote. Highest amount of cocaine was 1.12 mg in one 5 euro currency note. Two 10 euro notes also showed high amount of cocaine (55 μg/banknote). The authors also observed that after release new 20 euro denominations were contaminated within a few weeks [7].

Lavins et al. analyzed 165 randomly collected paper currency notes from 12 US cities and 4 foreign countries (Columbia, Qatar, New Zealand, and India). The authors collected 125 one dollar bills from various US cities, 40 foreign currency bills (200 pesos notes from Columbia, 1 riyal notes from Qatar, 5 dollar currency notes from New Zealand, and 50 rupee notes from India) and analyzed these currency notes for the presence of THC, cannabinol, and cannabidiol. Uncirculated dollar bills were used as the control where no drug was detected. The authors observed that for US$1 bill ($n = 125$), THC was present in only two bills (1.6%) with concentrations 0.085 and 0.146 μg/banknote. The authors detected the presence of cannabinol in 13 dollar bills (10.31%) with a range of values between 0.014 and 0.774 μg/banknote. Cannabidiol was present in two bills with concentrations of 0.032 and 0.086 μg. For foreign currencies ($n = 40$), no THC, cannabinol, or cannabidiol was detected in any currency collected from Qatar and India. Moreover, no cannabidiol was detected in any currency collected from New Zealand. However, THC and cannabinol were detected in 9 currency notes collected from New Zealand with THC amount ranging from 0.026 to 0.065 μg/banknote and cannabinol level ranging from 0.061 to 0.197 μg/banknote [8].

Although cocaine is the major contaminant of paper currencies, the presence of benzoylecgonine, a metabolite of cocaine, has also been detected in paper currencies. In one study using LC−MS/MS, the authors detected the presence of cocaine in all euro banknotes analyzed (uncirculated banknotes containing no drug were used as negative controls). The median amount of cocaine in euro banknotes analyzed was 106 ng/banknote. Moreover, benzoylecgonine was also detected in these banknotes (median concentration: 43 ng). Benzoylecgonine found in banknotes may be due to spontaneous hydrolysis of cocaine into benzoylecgonine. The highest amount of cocaine (12.4 μg) was present in one 10 euro note. In addition to cocaine, the authors also detected the presence of heroin (median value: 41 ng/banknote), 6-monoacetyl-morphine (median value: 15.5 ng/banknote), and morphine (16.5 ng/banknote) in over 90% banknotes. The presence of 6-monoacetyl-morphine or morphine in these banknotes may be due to hydrolysis of some heroin present in these notes. Heroin was predominantly detected in 5 euro notes but the highest amount of heroin (507 ng) was detected in on a 10 euro note. That banknote also showed the presence of 6-monoacetyl morphine (301 ng). Traces of methamphetamine (median: 7 ng/banknote) was detected in 53 out of 64 banknote analyzed. Only 3 notes showed methamphetamine contamination over 100 ng/banknote. The authors reported that 17 specimens showed the presence of 3,4-methylenedioxymethamphetamine, MDMA (median value: 9 ng/banknote). However, only 4 banknotes tested positive for THC. In general, small denominations (5, 10, and 20 euro bills) showed much higher amounts of illicit drugs than larger denominations such as 50, 100, 200, and 500 euro notes. This may be due to higher circulation rates of smaller denominations and/or a more frequent use in illicit drug dealings. In addition, analyzed euro notes from smaller denominations showed much rawer surfaces than bigger denominations which may explain why illicit drugs may stick to these bills more effectively than bills with higher denominations. In bills will rougher surface, fibers may spread apart producing interstices that may trap small molecules more effectively [9].

Luzardo et al. analyzed 120 euro notes in circulation (30 each of 5, 10, 20, and 50 euro notes) collected from tourist areas of Canary Island using LC–MS/MS for the presence of 21 drugs and metabolites and observed that 92.5% euro bills analyzed showed the presence of one or more illicit drugs and/or metabolites. As expected cocaine was the most frequently detected drug present in 87% of banknotes analyzed with median level of 190 ng/note (highest level: 15.0 μg/banknote). Interestingly cocaethylene, an active metabolite, only detected in people who abuse alcohol and cocaine together was present in 37 out of 120 banknotes (highest level: 8.1 ng/banknote) analyzed. In addition, the authors detected methadone in 12 banknotes (10%), heroin in 9 banknotes (7.5%), 6-monoacetyl-morphine in 13 banknotes where no heroin residue was detected (12%), various opiates in 42 banknotes (35%), amphetamine in 39 banknotes (32.5%), methamphetamine in 54 banknotes (45%), MDMA in 15 banknotes (12.5%), MDA (3,4-methylene-dioxyamphetamine) also in 15 banknotes (12.5%), MDEA (3,4-methylenedioxyethylamphrtamine) in 9 banknotes (7.5%), and benzodiazepines in 18 banknotes (15%) [10]. Highest amounts of cocaine and other drugs reported in various studies in the United States and foreign banknotes are listed in Table 32.1.

Currency Contaminated With Drugs: Clean Money Versus Drug Money

Cocaine is the major contaminant of paper currencies, various other drugs have been detected in lower frequencies as contaminants of banknotes. Therefore, it may be speculated that money involved in illegal drug transactions (drug money) should contain higher amount of cocaine or other illicit drugs than money in general circulation (clean money). On the basis of the United Nations Convention against Illicit Traffic in Narcotic Drugs and Psychotropic Substances, 1988 (the Vienna Convention), money involved in drug dealing can be seized. Using technique such as thermal desorption atmospheric pressure mass spectrometry, a large number of banknotes can be scanned rapidly to establish background trace amount of illicit drug level in banknotes of general circulation. Then comparing levels of illicit drugs in suspected drug money with background contamination in clean money may provide clue if money was used in drug dealing. Although paper currencies associated with criminal cases usually contain 50–1000 times more drug than currencies in the general circulation, there is no clear-cut cutoff value to establish if paper currency is clean money or drug money. In 1994, the US 9th Circuit Court of Appeals determined evidence of banknotes contaminated with illicit drugs is not admissible as evidence unless it can be determined that traces of drugs are markedly higher than from those expected from banknotes from a particular geographical area. Therefore, fingerprint detected in drug money may provide a better evidence of involvement of a person in crime [11].

Various levels of cocaine from nanogram to microgram quantity per bill have been reported by various authors. Some authors have suggested a 100 ng/banknote, as the background level of cutoff point for banknotes in general circulation, while other authors have proposed a cutoff value of 188.5 ng/banknote for cocaine. Therefore, a 5 Euro banknote containing 15.0 μg/banknote must be involved in snorting cocaine or in drug deal. Interestingly, the authors also indicated that in their study, Euro banknotes more frequently involved in drug deal or snorting cocaine were higher denomination banknotes (10, 20, and 50 Euro banknotes) [10]. In one study, the authors observed that banknotes found in a plastic bag containing 5 kg of cocaine which was seized along with drug traffickers from Rio de Janeiro, Brazil, contained on average 3425 μg/banknote, clearly showing that such money was drug money. Another banknote seized straight from a drugs dealer contained 4725 μg/banknote indicating that money involved in criminal drug dealings contain significantly higher drug level than money in general circulation [12].

In general, defendants commonly claim large quantities of banknotes in their possession represent a lifetime's savings, a gift, or an inheritance but the contamination patterns in large quantities of banknotes may determine the difference between drug money and lifetime's saving. In one study, the authors concluded based on a probabilistic model that there is only 0.3% chance that a bundle of 100 notes from the general banking population contains more

TABLE 32.1 Highest Amounts of Cocaine and Other Drugs Found in Various Denominations of Currency Notes as Reported in Various Studies

Country of Origin	Denomination	Drug Detected	Highest Amount Detected	Reference
United States	$1	Cocaine	1327 μg/banknote	[1]
United States	$1	Cocaine	2.99 μg/banknote	[2]
United States	$20	Cocaine	10.02 μg/banknote	[2]
United States	$1	Cocaine	922.7 μg/banknote	[3]
United States	$1	Heroin	168.5 μg/banknote	[3]
United States	$1	Morphine	5.51 μg/banknote	[3]
United States	$1	6-Monoacetyl-morphine	9.22 μg/banknote	[3]
United States	$1	Phencyclidine	1.87 μg/banknote	[3]
United States	$1	Amphetamine	0.85 μg/banknote	[3]
United States	$1	Methamphetamine	0.60 μg/banknote	[3]
United States	$5	Cocaine	49.4 μg/banknote	[4]
United States	$10	Cocaine	46.2 μg/banknote	[4]
United States	$20	Cocaine	9.2 μg/banknote	[4]
United States	$50	Cocaine	15.9 μg/banknote	[4]
United States	$100	Cocaine	8.87 μg/banknote	[4]
Spain	1 Euro	Cocaine	234 μg/banknote	[6]
Spain	10 Euro	Cocaine	889 μg/banknote	[6]
Spain	20 Euro	Cocaine	76.7 μg/banknote	[6]
Spain	50 Euro	Cocaine	67.8 μg/banknote	[6]
Germany	5 Euro	Cocaine	1.12 mg/banknote	[7]
Germany	10 Euro	Cocaine	55.0 μg/banknote	[7]
United States	$1	Tetrahydrocannabinol	0.146 μg/banknote	[8]
United States	$1	Cannabinol	0.774 μg/banknote	[8]
United States	$1	Cannabidiol	0.086 μg/banknote	[8]
New Zealand	5 dollar	Tetrahydrocannabinol	0.065 μg/banknote	[8]
New Zealand	5 dollar	Cannabinol	0.197 μg/banknote	[8]
Europe[a]	10 Euro	Cocaine	12.4 μg/banknote (BE also present: 4.2 μg/banknote)	[9]
Europe[a]	10 Euro	Heroin	507 ng/mL (6-MAM also present: 301 ng/banknote)	[9]
Europe[a]	10 Euro	MDMA	115 ng/banknote	[9]
Europe[a]	Euro (denomination not stated)	Tetrahydrocannabinol	>3 μg/banknote	[9]
Canary Island	5 Euro	Cocaine	15.0 μg/banknote	[10]
Canary Island	20 Euro	Methamphetamine	5.5 ng/banknote	[10]
Canary Island	50 Euro	Amphetamine	13.9 ng/banknote	[10]
Canary Island	20 Euro	Alprazolam	4.8 ng/banknote	[10]
Canary Island	5 Euro	Heroin	150.6 ng/banknote	[10]

6-MAM, 6-Monoacetyl-morphine; BE, benzoylecgonine; MDMA, 3,4-methylebedioxymethamphetamine.
Canary Island is a Spanish archipelago.
[a]Banknotes were collected from National Bank of Luxembourg but the authors stated that these banknotes were originated from eight different euro emitting countries.

than six heroin contaminated specimens. This is due to poor transfer of heroin from a contaminated banknote to a noncontaminated banknote during mechanical counting or other handlings of money in bank. As a result only approximately 2% of banknotes in circulation are contaminated with heroin [13]. However, this model may not be valid for differentiation of drug money from clean money in case of contamination with cocaine because approximately 87%—100% of banknotes (depending on studies) may be contaminated with cocaine. In contrast prevalence of heroin contamination in paper money vary from 2% to 14% (based on multiple studies) although in one study authors observed the presence of heroin or its degradation product 6-monoacteyl morphine or morphine(heroin in the presence of moisture breaks down to these products) in over 90% banknotes analyzed [9].

Mechanism of Contamination

Various mechanisms have been proposed to explain contamination of cocaine and other drugs in paper currencies. One possibility is that the green ink on the US Federal Reserve note never solidifies. This can be demonstrated by rubbing an old Federal Reserve note on a white piece of paper and noting the resulting trace of green pigment to the white surface. Therefore, this sticky surface may allow adherence of various oils (from human sebaceous gland), and various environmental dirt and grime including residue amounts of abused drugs during circulation of dollar bills. Therefore, direct contact between currency and drug or drug contaminated surface explains why paper currency may be contaminated with drugs. This might occur through the use of a rolled up dollar bill to snort cocaine, through the handling of drugs and money during an illicit transaction or through innocent transfer via mechanical currency counters in bank. However, it is unlikely that illicit drug trade have actually handled the number of bills that are currently circulating with the cocaine contamination. Therefore, one means of contamination for significant portion of paper currency is related to the use of mechanical currency counters. Swipe from the interior of the mechanical currency counter in a bank showed the presence of cocaine explaining how uncontaminated fresh dollar bills may be contaminated with cocaine during counting of such bills using a mechanical currency counter [5].

The prevalence of cocaine as a contaminant of paper money may be partly due to widespread abuse of cocaine through snorting where a rolled up paper money may be used and another factor may be crystal size of cocaine molecule that may stick easily to the fabric of paper money. If larger amount of cocaine is present in a contaminated banknote, it is capable of contaminating a very large amount of banknotes through mechanical transfer (machine counting or hand counting) but other illicit drugs are not so widely spread through mechanical transfer. This may be due to crystal size of cocaine molecule (1.4 μm) that can be retained by fabric of paper money [14].

Although amphetamine is also abused in the powder form, the crystal size of amphetamine is larger than cocaine and as a result retention of amphetamine by banknotes appears to be less effective. Although lysergic acid diethylamide (LSD) in its pure form is a crystalline solid, it is normally abused in a solution and soaked into paper "tabs" for use. Therefore, LSD contamination has not been reported. The simple physical transfer of cannabis in its various forms (e.g., marijuana, sinsemilla, buds, hashish, hashish oil, etc.) from the fingers of individuals handling the material to currency notes may be simplest explanation of cannabinoid contamination of paper currency. However, THC is usually abused in compressed (tablet or resin) forms which are less likely to form particulate matters which may be retained by banknotes explaining low prevalence of THC contaminated money. Moreover, several contaminated illicit drugs such as THC, heroin, and cocaine are known to readily degrade in the presence of moisture in air. If these parent compounds are detected in banknotes, then recent contact of banknotes with these drugs may be suspected. When heroin degrades, it may be possible to establish the presence of crude heroin by identifying more stable opiates such as narcotine and papaverine which often are found in heroin because these products often persist following acetylation of opium. Degradation product such as benzoylecgonine may also been sought for cocaine. However, it is important to know that traces of illicit drugs on banknotes could arise from contact with drug itself, a contaminated hand or another contaminated item such as mechanical money counting device. Thus it is possible that an individual who has no personal involvement with drugs has contaminated banknotes in his or her possession. The mere presence of a drug is insufficient evidence to link that individual with drug abuse or drug trafficking [11].

Can Handling Contaminated Paper Money May Cause Positive Drug Screen?

ElSohly investigated whether individuals who handled cocaine contaminated paper money would test positive by urinalysis. Two dollar bills were immersed in dry powdered cocaine and then shaken free of loose cocaine. One individual then handled the money several times during the course of the day. Analysis of urine samples collected over a period of approximately 24 h after handling the contaminated money revealed that the maximum concentration of benzoylecgonine observed was

72 ng/mL. This value was significantly below the cutoff level of benzoylecgonine screening assay (300 ng/mL). The value was even lower than the US military recommended cutoff of 100 ng/mL. Therefore, handling cocaine contaminated paper money should not cause positive drug testing results. Similarly paper money contaminated with marijuana should not cause false positive urine drug screen [15]. It is also unlikely that handling paper money contaminated with other illicit drugs may result in false positive urine drug test.

HERBAL TEAS CONTAMINATED WITH COCAINE

The drinking of coca tea is common among people in South America. The tea is usually packed in tea bags containing approximately 1 g of plant material. Drinking such tea is equivalent to consumption of cocaine along with other alkaloids present in the tea. In one study authors found an average of 5.11 mg of cocaine per tea bag of coca tea originated from Peru and average 4.86 mg of cocaine per tea bag in Bolivian coca tea. When tea was prepared, one cup of Peruvian coca tea had an average of 4.14 mg of cocaine while one cup of Bolivian tea had an average of 4.29 mg of cocaine. When one volunteer drank one cup of Peruvian tea, a peak benzoylecgonine concentration of 3940 ng/mL was observed 10 h post consumption. Similarly, consumption of one cup of Bolivian tea by a volunteer resulted in a peak benzoylecgonine concentration of 4979 ng/mL 3.5 h after consumption of tea. These results clearly indicate that significant amounts of cocaine can be found in coca tea from Bolivian and Peru and drinking such tea may cause a significant amount of benzoylecgonine in urine. Therefore drinking such tea will cause positive results in drug testing [16].

Although US custom regulations require that no cocaine should be present in any herbal tea, literature references indicate that some health Inca tea sold in the United States contains cocaine. Jackson et al. reported urinary concentration of benzoylecgonine after ingestion of one cup of health Inca tea by volunteers. Benzoylecgonine was detected up to 26 h post ingestion. Maximum urinary benzoylecgonine concentration ranged from 1400 to 2800 ng/mL after ingestion of health Inca tea. The total excretion of benzoylecgonine in 36 h ranged from 1.05 to 1.45 mg which correlated with 59%–90% of the ingested cocaine dose [17].

Coca tea or Mate de Coca is commercially available tea made from coca leaves (*Erythroxylum coca*). This tea is available in South America and also may be found in the United States. Mazor et al. studied the effect of drinking coca tea on excretion of cocaine metabolite in urine. Five healthy volunteers consumed coca tea and underwent serial urine testing for cocaine metabolites using the fluorescence polarization immunoassay. Each participant showed positive urine sample (over 300 ng/mL cutoff for cocaine metabolite) by 2 h after drinking coca tea, and urine specimens from three out of five volunteers showed positive test results up to 36 h. Mean benzoylecgonine concentration was 1777 ng/mL (range: 1065–2495). The authors concluded that coca tea ingestion resulted in positive urine test for the cocaine metabolite [18]. Turner et al. reported positive tests for cocaine metabolite in subjects after drinking Mate de Coca tea. Tea was prepared by allowing one Mate de Coca tea bag to be immersed in 250 mL boiling water for 25 min. The bag was removed and squeezed in tea to drain additional water. A 5 mL sample was taken for analysis and volunteers drank the rest. Urinary samples were collected at 2, 5, 8, 15, 21, 24, 43, and 68 h after drinking tea. All urine samples tested positive for benzoylecgonine, the metabolite of cocaine by immunoassay. The amount of cocaine in tea was estimated to be 2.5 mg [19]. ElSohly et al. also reported the presence of cocaine in coca tea [20]. Amount of cocaine found in various coca teas are listed in Table 32.2.

TABLE 32.2 Cocaine Content in Various Coca Teas		
Tea	Average Cocaine Content	Reference
Peruvian coca tea (per tea bag)	5.11 mg/tea bag	[16]
Bolivian coca tea (per tea bag)	4.86 mg/tea bag	[16]
Peruvian tea (one cup)	4.14 mg	[16]
Bolivian tea (one cup)	4.29 mg	[16]
Mate de Coca (one cup)	2.5 mg	[19]
Coca tea	2.12 mg (tea sample)	[20]

Bieri et al. studied distribution of cocaine in wild *Erythroxylum* species and reported that 23 out of 51 species investigated contained cocaine. The highest amount of cocaine was found in *Erythroxylum laetevirens*. The authors reported that cocaine content of *E. laetevirens* was similar to cultivated species. Mate de Coca, also known as health Inca tea, is mainly composed of pure coca leaves and contain cocaine [21].

There is an isolated report of blue cohosh being contaminated with cocaine. Blue cohosh is an herbal preparation traditionally used to induce labor in pregnant women or in nonpregnant women to cause a menstrual period. A female infant was born to a 24-year-old woman who consumed blue cohosh at the recommendation of her obstetrician for inducing labor. The infant just after birth developed multiple problems, including perinatal stroke and toxicological analyses of urine and meconium showed the presence of benzoylecgonine (confirmed by GC/MS). Analysis of blue cohosh bottle collected from the mother also confirmed the presence of cocaine. The author concluded that blue cohosh ingested by the mother was contaminated with cocaine [22].

Distinguishing Coca Leaves Chewing From Cocaine Abuse

Coca leaves have traditionally been used for chewing or brewing tea for centuries among the indigenous population of the Andean region. There is traditional belief that coca chewing causes no harm but helps in overcoming altitude sickness, suppressing fatigue and helping in digestion. Although cocaine is a controlled substance, in some South American countries possession of unprocessed coca leaves may remain generally legal (but not possession of cocaine), or tolerated. For example, in Argentina, law allows chewing coca leaves or drinking coca tea and it is not restricted to native people. Moreover, coca leaf chewing or drinking coca tea is still legal in Bolivia. Therefore, it is important to distinguish between cocaine abuses which are illegal in these countries from legal chewing of coca leaves.

Rubio et al. analyzed 38 urine specimens collected from cocaine abusers and 24 urine specimens collected from people who chewed coca leaves (once a week to several times a day; subjects living in Argentina), and observed the presence of hygrine and cuscohygrine (cuscohygrine is unstable and degrades to hygrine) in urine specimens of all coca leaf chewers but such compounds were absent in urine of cocaine abusers. Sometimes coca leaves were chewed along with alkali in order to improve extraction of cocaine from coca leaves but these two compounds were present in urine regardless of use of alkali during coca chewing. In addition to these biomarkers in urine specimens of individuals chewing coca leaves, cocaine, its metabolite methyl ecgonine (benzoylecgonine was not targeted), and cinnamoylcocaine were also detected. Although only one specimen out of 38 collected from cocaine abusers showed the presence of cinnamoylcocaine, the authors commented that it is not a good biomarker of chewing coca leaves because cinnamoylcocaine can be found in seized cocaine and also in urine specimens of some cocaine abusers. The authors concluded that hygrine and cuscohygrine may be used as biomarkers to distinguish between chewing coca leaves and abusing cocaine. These two alkaloids are present in coca leaves and are absorbed during coca leaf chewing but are lost during processing of cocaine from coca leaves [23].

Hair analysis for the presence of cocaine, metabolites of cocaine, and other alkaloids present in coca leaves is also useful to discriminate between chewing coca leaves and abusing cocaine. The authors analyzed hair from 26 Argentinean coca chewers and 22 German cocaine abusers for the presence of cocaine, nor-cocaine, benzoylecgonine, ecgonine methyl ester, tropacocaine, cocaethylene, cinnamoylcocaine, hygrine, and cuscohygrine by using hydrophilic interaction liquid chromatography in combination with triple-quad mass spectrometry and hybrid quadrupole time-of-flight mass spectrometry. However, due to lack of reference standard, only qualitative data were obtained for hygrine and cuscohygrine metabolites.

In all coca chewers hygrine (only qualitative data), cuscohygrine (range: 0.028–24.7 ng/mg of hair), and M2 metabolite of cuscohygrine (only qualitative data) were detected along with cocaine (0.085–75.5 ng/mg of hair), benzoylecgonine (0.046–36.5 ng/mg of hair), ecgonine methyl ester (0.018–5.6 ng/mg of hair) along with nor-cocaine, cinnamoylcocaine, and tropacocaine. Cocaethylene was also present in some specimens indicating that some people who chewed coca leaves also consumed alcohol. In cocaine abusers, hair cocaine levels varied from 0.39 to 38.7 ng/mg of hair. In addition, cocaine metabolites benzoylecgonine, ecgonine methyl ester, and nor-cocaine were detected in hair of all subjects. Cocaethylene was observed in some species. However, no metabolite of cuscohygrine was detected in any specimen. However, traces of hygrine and cuscohygrine were detected in some specimens. The authors concluded that ratio of cuscohygrine to cocaine, cinnamoylcocaine to cocaine, and ecgonine methyl ester to cocaine can differentiate between subjects who chewed coca leaves versus cocaine abuser. Moreover, anhydroecgonine methyl ester, a pyrolysis product of cocaine, may be a marker for crack cocaine abuse [24]. Anhydroecgonine, formed during pyrolysis of cocaine during smoking, can also be considered as potential marker for cocaine abuse. Biomarkers used for separating coca leaf chewers from cocaine abusers are listed in Table 32.3.

TABLE 32.3 Biomarkers to Differentiate Between Chewing Coca Leaves and Abusing Cocaine

Biomarker	Comments
Hygrine	Always present in coca chewers but usually absent in cocaine abusers or found in trace levels.
Cuscohygrine	Always present in coca chewers but usually absent in cocaine abusers or found in trace levels.
Cuscohygrine M2 metabolite	Always present in coca chewers but never found in cocaine abusers.
Cuscohygrine/cocaine ratio	Value >0.01 indicate coca chewing. In one study the mean ratio was 0.098 in coca chewers compared to mean ratio of 0.004 in cocaine abusers.
Cinnamoylcocaine/cocaine ratio	Value >0.02 indicate coca chewing because the mean ratio was 0.087 in coca chewers compared to mean ratio of 0.009 in cocaine abusers.
Ecgonine methyl ester/cocaine ratio	Value >0.015 indicate coca chewing because the mean ratio was 0.057 in coca chewers compared to mean ratio of 0.0088 in cocaine abusers.
Anhydroecgonine and anhydroecgonine methyl ester	These compounds may be potential biomarkers of smoking cocaine (crack cocaine).

Herbal Tea and Contamination/Adulteration

There is an isolated report of positive benzodiazepine urine drug screen in a 20-year-old female athlete who denied taking any drug. Later the presence of diazepam was confirmed in her urine. On further questioning she admitted taking a Chinese herbal product obtained from her chiropractor in Ohio. Further analysis using GC/MS confirmed the presence of diazepam in the Chinese herb sold as Miracle Herb. The Ohio chiropractor obtained that herb from supplier at Huchinson, Kansas [25].

Contamination and adulteration of herbal supplements especially traditional Indian and Chinese remedies is a serious problem. There are several reports of contamination of Indian Ayurvedic medicines with heavy metals such as lead, arsenic, and mercury. Some traditional Chinese medicines are also adulterated with various prescription medications. In addition herbal supplements manufactured in Asian countries may also be contaminated with parasites, microbes, fungi, mold, and other substances. Taking such adulterated herbal supplements may cause adverse effects including life-threatening conditions [26]. A detailed discussion on this topic is beyond the scope of this book.

PASSIVE EXPOSURE TO MARIJUANA

Passive inhalation of marijuana smoke is a popular defense for positive workplace marijuana test result. It is usually assumed that during smoking marijuana cigarette, 23%–30% of THC is degraded due to pyrolysis, while the active smoker ingests about 20%–37% of THC. The remaining 40%–50% of THC is released to the environment as sidestream smoke. Nevertheless, it is very unlikely that passive inhalation may cause a positive marijuana test result because values are below the cutoff levels of drug testing (50 ng/mL screening cutoff, 15 ng/mL confirmation cutoff). The US Department of Transportation indicated that Medical Review Officers (MRO) should not recognize passive drug exposure as a legitimate medical explanation for a positive test [27,28].

Although substantial amount of THC is released as sidestream smoke, it is highly diluted by the room air as it is incorporated in aerosol particles. Niedlbala et al. studied the effect of passive inhalation of marijuana on urine and oral fluid testing using high marijuana containing cigarettes. In Study 1 four smokers smoked THC mixed with tobacco (39.5 mg THC) in an unventilated eight passenger van and four volunteers were passive smoker. In Study 2 four volunteers smoked marijuana only (83.2 mg THC). Oral fluid was collected using Intercept Oral Specimen Collection Device (OraSure Technology, Bethlehem, PA). Participants were allowed to go outside the van 60 min after exposure. Oral fluid collected at baseline (30 min before exposure), 0min, 15min, 45 inside the van and 1, 1.25, 1.5, 1.75, 2, 2.5, 3.5, 4, 6, and 8 h outside the van. Oral fluid collection continued till 72 h after exposure. Urine specimens were also collected. Oral fluids were tested for THC metabolite (11-nor-9-carboxy-Δ^9-tetrahydrocannabinol-9-carboxylic acid: THC-COOH) using Cannabinoids Intercept Micro-Plate enzyme immunoassay with a cutoff of 3 ng/mL (confirmation: 2.0 ng/mL). For urine specimens 50 ng/mL cutoff was used. All urine specimens tested negative (50 ng/mL cutoff) for all passive smokers (GC/MS showed THC metabolite concentration in the range of 5.8–14.7 ng/mL 6–8 h after exposure). In Study 1 where oral fluid was collected in the van, some subjects showed positive response due to contamination of the device with THC smoke but in the Study 2 when all oral fluid specimens were collected outside the van, no positive specimens were observed. On the other hand, all smokers showed significant THC in both oral fluid and urine as expected. The mean urinary concentration was 75 ng/mL 4 h after smoking [29].

In general, it is assumed that THC can be detected in urine after 1 h of exposure and peak THC-COOH concentration is observed 6–8 h after exposure. Nevertheless the urine concentration of THC-COOH should be below 15 ng/mL. Any value over 15 ng/mL indicates very high exposure to passive marijuana smoke and it is virtually impossible that the person was not aware that smokers in the immediate vicinity were smoking marijuana. It has been suggested that values below 15 ng/mg of creatinine can be used as a cutoff value to distinguish passive smoking of marijuana from active smoking. Oral fluid is not an appropriate matrix to evaluate passive exposure to marijuana smoke [30].

Hair analysis of drugs of abuse is sometimes applied to workplace drugs testing, child protection cases, and other legal drug testing scenarios. Usually THC is the analyte of choice during analysis of hair specimens because concentration of THC-COOH is usually low. Nevertheless THC-COOH is not present in marijuana smoke but is incorporated from bloodstream to hair after smoking marijuana because THC-COOH is a major metabolite of THC. Therefore, THC may be positive in hair due to passive exposure to marijuana smoke but the presence of THC-COOH in hair confirms marijuana abuse. In one study the authors described a case where a couple living together in an apartment were prosecuted for concealment among other crimes. The male admitted smoking marijuana several times a day but his female companion denied any consumption. Although hair specimens of both of them confirmed the presence of THC and cannabinol, only hair of the male subject showed the presence of THC-COOH at a level >6.6 pg/mg of hair. Negative result of THC-COOH in hair of the female subject confirmed that the presence of THC in her hair was due to passive exposure to marijuana smoke [31].

Usually concentration of THC-COOH in hair is very low thus detecting such low level is analytically challenging. Thieme et al. commented that although gas chromatography-negative chemical ionization-tandem mass spectrometry or GC-MS-MS coupling may be used for analysis of THC-COOH in hair, combination of liquid chromatography with tandem mass spectrometry is a superior analytical approach for measuring THC-COOH levels in hair. The authors also developed an LC–MS/MS operated in negative electrospray ionization mode for quantitation of THC-COOH in hair after selective methylation of 9-carboxylic acid group in the THC-COOH molecule. The authors used methyl iodide as the methylating agent in the presence of solid sodium carbonate and were able to quantitate THC-COOH as THC-9-carboxymethy ester at a level below 100 fg/mg of hair. The authors used fragmentation reactions m/z 357–325 and m/z 357–297 for their analysis [32].

THC is first hydrolyzed to 11-hydroxy-THC (THC-OH) which is an active metabolite. Then this compound is further oxidized to THC-COOH, the major inactive urinary metabolite. THC and its metabolites subsequently undergo Phase II biotransformation to glucuronide conjugates. Pichini et al. proposed that THC-COOH glucuronide is a hair biomarker of cannabis consumption and described a sensitive ultra-performance LC–MS/MS for analysis of THC-COOH glucuronide in hair at a limit of quantitation of 0.25 pg/mg of hair. The authors used amiodarone as the internal standard [33].

Alternative hair biomarker for differentiating passive exposure to marijuana from active smoking has also been proposed. One possibility is to use Δ^9-tetrahydrocannabinolic acid A (THCA-A), the nonpsychoactive biogenic precursor of THC as a biomarker of passive exposure to marijuana. THCA-A is not incorporated in relevant amounts into hair via bloodstream but can be incorporated through sidestream exposure or direct contact of hair with a contaminated hand. In one study, the authors showed that when participants rolled a marijuana joint for 5 consecutive days, the concentration of THCA-A in hair ranged from 15 to 1800 pg/mg of hair and concentrations of THC ranged from <10 to 93 pg/mg of hair. Four weeks after exposure THCA-A was still detected (range: 4–57 pg/mg of hair) in hair of 9 out of 10 volunteers while THC was still detected (range: <10–17 pg/mg of hair) 5 out of 10 volunteers. The authors concluded that part of the THC as well as major part of THCA-A found in routine hair analysis are derived from external contamination. Therefore, the presence of THCA-A in THC positive hair specimens in children or partners of cannabis users indicate passive exposure to cannabis due to close body contact [34].

Washing Hair With Cannabio Shampoo

Cannabis plants are used in preparing shampoo (cannabio shampoo) but it usually contains <1% THC. In one study, the authors observed that one commercially available cannabio shampoo showed the presence of 412 ng/mL of THC, 4078 ng/mL of cannabidiol, and 380 ng/mL of cannabinol based on GC/MS analysis. In order to investigate whether washing hair with cannabio shampoo may cause false positive hair testing for marijuana, three subjects washed their hair daily using cannabio shampoo for 2 weeks. After that hair specimens were collected from all subjects but no THC, cannabidiol and cannabinol were detected in any hair specimen indicating that use of cannbio shampoo for normal hygiene practice should not cause false positive drug testing result. Even when drug-free hair specimens were incubated with 10 mL water/cannabio shampoo (20:1 v/v) for 5 h, no THC detected in the hair specimen (although small amounts of cannabidiol was detected) again indicating that washing hair with cannabio shampoo should not cause false positive THC hair testing result [35].

INGESTION OF HEMP OIL

Industrial hemp is part of a number of varieties of *Cannabis sativa* L. which is cultivated for industrial and agricultural purpose mainly for seeds and fiber. Cannabis is the only plant genus that contains cannabinoids. Many cannabinoids have been identified from the hemp plant including THC and cannabidiol. THC is psychoactive while cannabidiol is anti-psychoactive and counteracts the effect of THC. Industrial hemp is characterized by low THC and high cannabidiol content. Cultivation of hemp in the United States is illegal unless a special permission is obtained from US Food and Drug Administration (FDA). Marijuana is produced from a different variety of *C. sativa* which is high on THC but low on cannabidiol. The THC content is usually <1% and the current legal level of cultivation in Europe and Canada is 0.3%. In contrast, *C. sativa* plants cultivated for purpose of producing marijuana contain 2%–5% THC (per dry weight) in female flowers although THC content may be as high as 20%.

The THC level is so low in hemp products that one cannot get high from ingesting these products. In addition, cannabidiol present in the hemp product counteracts the effect of marijuana. Moreover washed industrial quality hemp seeds contain no or very little THC coming from the resin sticking to the seed. Hemp oil contains high amounts of essential fatty acids, linoleic acid, and alpha-linoleic, acid and thus may have health benefits [36]. Currently use of hemp oil is legal but use of cannabis flowers essential oil is illegal (except for states where recreational use of marijuana is legal) in the United States because the cannabis flower essential oil is produced from flowering buds of cannabis plant and is considered as marijuana products in the United States. However in some European countries such as Switzerland, England, France, and the Netherlands, production of cannabis flower oil is legal.

It is possible that seeds and oils of hemp may contain a measurable amount of THC. In 1997 a survey found that the THC content of hemp oil was 11−117 μg/g (parts per million) [37]. These seeds were imported from China. THC concentrations as high as 3568 μg/g were found in Swiss hemp oil products because growing higher THC containing hemp plant (THC content over 0.3%) is legal in Switzerland. Moreover proper cleaning of the seeds may not have been undertaken [38]. Lehman et al. (from Switzerland) reported 3−1500 μg/g of THC in 25 different cannabis seed oil [39]. Since 1998, more thorough seed drying and cleaning process were undertaken that significantly reduced the level of THC in hemp products. The majority of THC is present on the outside of seed hulls. Now the typical levels of THC content in hemp oil and hulled seeds are 5 and 2 μg/g, respectively. In 2000 oil containing 10−20 μg/g of THC and hull seeds containing 2−3 μg/g were reported in the United States in 2000 [40].

Studies conducted in 1998 or earlier reported significant amounts of THC metabolites after consumption of hemp oil because THC content in most hemp oil exceeded 50 μg/g, but now the THC content in most commercial hemp oil as low as 5 μg/g. Therefore later studies concluded that ingestion of reasonable amount of hemp oil should not cause positive THC results in workplace drug testing. Costantino et al. in 1997 observed that in volunteers consuming 15 mL of hemp oil 2 urine specimens (out of 18 specimens) collected from volunteers screened >100 ng/mL cutoff for THC metabolite while 7 screened positive using the cutoff of 50 ng/mL and 14 screened positive using 20 ng/mL cutoff [41]. Struempler et al. also reported in 1997 positive cannabinoid workplace drug testing following ingestion of commercially available hemp oil preparation. The first specimen testing negative was 53 h after ingestion [42]. Alt and Reinhardt also reported in 1998 the presence of THC metabolites 80 h post ingestion of 40−90 mL hemp seed oil by volunteers [43]. Authors also reported the presence of THC in hemp food products like hemp bar, hemp flour, and hemp liquor. However a report from Switzerland indicated the presence of very high amount of THC after ingestion of 11 or 22 g of cannabis seed oil due to very high THC content (1500 μg/g) of the oil. Urine samples were positive up to 6 days after ingestion of oil. The THC metabolite concentrations (THC-COOH) ranged from 5 to 431 ng/mL (analyzed by GC/MS) [39]. Another report from Switzerland confirmed THC poisoning in four patients after eating salad preparation containing hemp oil because the concentrations of THC in hemp oil far exceeded the recommended tolerance dose [44]. Swiss Federal Office of Public Health issued a warning concerning consumption of hemp oil after publication of this report.

Hemp Ale, an alcoholic beverage formulated, brewed, and bottled, by Frederick Brewing Company (Frederick, MD) did not contain any THC because the hemp seeds were subjected to two wash cycles before brewing so that vegetative material that potentially may contain THC was removed. Gibson et al. reported the absence of THC in Hemp Ale drink and concluded that ingestion of moderate amount of drink is not sufficient to produce a cannabis positive drug screen [45]. To the knowledge of the author this drink is no longer available on the consumer market.

More recent reports from the United States indicate that drinking moderate amounts of hemp oil should not cause THC positive results in workplace drug testing. Leson et al. reported that consumption of 125 mL of hemp oil (equivalent to ingestion of 0.6 mg of THC; a very high consumption) produced highest THC metabolite concentration of 5.2 ng/mL (determined by GC/MS). This value was below the 15 ng/mL cutoff of GC/MS confirmation. However several urinary specimens collected from volunteers after consumption of hemp oil tested positive by the immunoassay screen at 20 ng/mL cutoff and one specimen tested positive at 50 ng/mL cutoff. Therefore programs rely on immunoassay screening tests only and using a lower screening cutoff may still encounter occasional positives in subjects ingesting hemp food because significant amount of THC present in hemp food samples may exist as THC acid A or B. These species are metabolized analogous to free THC but are not quantitated in the GC/MS confirmation of THC-COOH which is derived from free THC [46]. Another study published in 2006 concluded that consumption of hemp oil currently produced by US manufacturers according to the guidelines of the manufacturers should not lead to positive tests for THC in plasma or whole blood [47].

Edible Marijuana Products

Legal use of marijuana for medical and recreational purpose is expanding worldwide. Currently, some use of marijuana is legal in 26 states in the United States, District of Columbia, and in multiple countries (Canada, Australia, Israel, and the Netherland). In addition, recreational use of marijuana is legal in four US states (Colorado, Washington, Oregon, and Alaska) and recently bill to legalize marijuana has been passed in four more states (California, Nevada, Massachusetts, and Maine). Moreover, marijuana is the most commonly abused drug in the United States.

Expansion of legal marijuana market has substantially increased variety of cannabinoid products available in the market including edible cannabis products, most commonly baked foods (brownies, cookies, etc.), candies beverages, and cooking ingredients such as butter, oil, etc. In general lower levels of THC and its metabolites are detected in blood and urine after ingestion of marijuana containing products compared to when marijuana is smoked. This is due to lower bioavailability of THC after eating marijuana containing food compared to smoking of marijuana joint. Nevertheless, workplace drugs testing result may be positive after eating marijuana containing products. In one study, volunteers ate two brownies which contained 1.6 g of marijuana plant material or brownie containing no marijuana (placebo). The authors confirmed the presence of substantial amounts of THC-COOH in urine (108–436 ng/mL) of subjects who ate marijuana containing brownies over a period of 3–14 days. No positive result was observed as a result of placebo brownies [48].

Vandrev et al. studied blood and oral fluid disposition of THC and its metabolites in human after eating cannabis brownies containing 10, 25, or 50 g THC. The low dose (10 mg THC) matched the current unit dose adopted by the state of Colorado for edible marijuana that can be sold legally to adults. Interestingly, after consumption of 10, 25, or 50 g marijuana containing brownies, mean Cmax for THC in whole blood were 1, 3.5, and 3.3 ng/mL, respectively, and no blood THC level exceeded 5 ng/mL. In two participants, no THC was detected in blood following the ingestion of brownie containing 10 mg THC. However, levels of active metabolite THC-OH and inactive metabolite THC-COOH were higher than THC levels in blood and subjects experienced effects of marijuana after eating brownies. Significant impairment was also observed in subjects who ingested brownies containing 25 and 50 mg THC containing brownies. The authors concluded that window of THC detection ranged from 0 to 22 h for whole blood and 1.9–22 h for oral fluid. Subjective drug and cognitive performances effects were generally dose dependent, peaked 1.5–3 h post exposure and lasting for 6–8 h. Compared with inhalation, levels of cannabinoids in blood and oral fluid following oral cannabis administration are low and variable among subjects [49].

Although edible marijuana products produce lower blood THC levels than smoking similar amounts of cannabis, sometimes people may consume substantial amount of edible marijuana products that may cause overdose. Moreover, in states where marijuana is legal, for employees covered by federal regulations, marijuana use both on and off job is prohibited because marijuana is still a Schedule I drug. Moreover, most workers compensation statue allow reduced benefits when a worker is under influence of alcohol or illicit drug during accident. Two samples must be obtained because second specimen may be needed for confirmation. For marijuana, urine THC-COOH level does not prove acute impairment but whole blood level of 5 ng/mL could be used as an indication for impairment. Studies have shown that THC level in serum may exceed 5 ng/mL after consuming eating edible marijuana.

Favrat described two cases of health subjects who were occasional user of marijuana without psychiatric history who developed transient psychiatric symptoms after oral administration of marijuana. In one subject after drinking decoction containing 16.5 mg of THC showed impairment up to 4 h and THC blood level of 6.2 ng/mL 1 h after drinking THC containing decoction [50]. In one case report where a 19-year-old man jumped of fourth floor balcony and died after eating marijuana containing cookies, had blood THC level of 7.2 ng/mL, and THC-COOH level of 49 ng/mL. The autopsy was performed 29 h after death. The authors commented that legal whole blood limit of THC for driving in Colorado is 5.0 ng/mL and the male was intoxicated with marijuana when he jumped from the balcony and died. Therefore, the death was linked to marijuana consumption since the state of Colorado legalized recreational marijuana use in 2012 [51].

Unwanted Exposure of Marijuana Containing Products to Children

Unwanted exposure to marijuana containing products to children may cause serious toxicity including admission to hospital. This may be particularly a problem in states where marijuana is legal for recreational purpose. In one study, the authors observed that after legalization of recreational marijuana in Colorado, the mean rate of marijuana-related visits to children's hospitals increased from 1.2 per 100,000 population 2 years prior to legalization of marijuana to 2.3 per 100,000 population 2 years after legalization of marijuana. In addition, 48% exposures were related to edible marijuana products [52].

OCCUPATIONAL EXPOSURE TO CONTROLLED SUBSTANCES

Personnel working in law enforcement, crime laboratories, or manufacturing plants for opioid medications are exposed to controlled substances as part of their occupation. Therefore, it is important to know whether such occupation exposure may cause false positive workplace drug testing results. In one study, the authors observed no urinary benzoylecgonine level in laboratory management personnel not working with drug samples but a urinary benzoylecgonine of 227 ng/mL was observed in one narcotics criminalist who was working on a case involving 2 kg of cocaine hydrochloride in the narcotics laboratory. A maximum urinary benzoylecgonine of 1570 ng/mL was observed in the urine specimen of one narcotic criminalist who was sampling a case containing 50 kg of cocaine hydrochloride over a period of 3 h. The authors commented that proper ventilation of the crime laboratory and use of gloves, face masks, and goggles are essential to minimize cocaine exposure in crime laboratory personnel [53]. In another study, the authors determine urinary benzoylecgonine levels in personnel preparing cocaine training aids for a military working dog program. Out of 233 urine specimens analyzed, benzoylecgonine was detected in 88 specimens. Two specimens showed benzoylecgonine concentrations above 100 ng/mL (138 and 460 ng/mL) as confirmed by GC/MS analysis [54].

Effect of occupational exposure to methamphetamines in laboratory personnel during preparation of training materials for drug detection dogs has also been studied. Urine samples were collected from individuals on the day before, the day of, and the day after handling up to 500 g of methamphetamine. A total of 101 urine specimens were analyzed for methamphetamine and its metabolite amphetamine using GC/MS. The mean methamphetamine concentration was 48 ng/mL (range: 1.6–262 ng/mL) in urine samples collected during and after handling methamphetamine. None of the urine specimen showed any detectable amount of amphetamine. The authors concluded that occupational exposure to methamphetamine may cause small but detectable amounts of methamphetamine in urine [55]. Because amphetamine must be confirmed in a methamphetamine positive sample to confirm methamphetamine use in workplace drug testing, occupational exposure to methamphetamine should not cause positive workplace drug testing result.

A 46-year-old male police officer in charge of the destruction of the seized drug and a 37-year-old female clerk (a couple) were arrested and subjected to investigation on the charges of drug trafficking. The couple was exploiting their administrative position to make money by reselling seized drug. A 30-year-old police informer was selling drugs for them but later denounced the trafficking. The laboratory was requested to analyze hair specimens collected from the couple, the informer, and also 11 other police officers of the same unit in order to compare the results as external contamination was proposed to account for the positive test results. The results reveal an occasional abuse of heroin by the arrested police officer (6-monoacetyl-morphine: 0.5 ng/mg per hair, morphine: 0.2 ng/mg of hair) and the clerk (6-monoacetyl-morphine: 0.8 ng/mg per hair, morphine: 0.4 ng/mg of hair) but no drug was detected in hair of 11 police officers working in the same unit as well as the police informer. The authors concluded that occupational exposure of drugs to officers of the drug enforcement administration should not cause positive drug testing result using hair specimen [56].

The protection of workers from potential harmful effects of active pharmaceutical ingredients poses a significant challenge for drug manufacturing industry. Occupational exposure to fentanyl due to both inhalation and dermal exposure has been reported in workers of fentanyl manufacturing facility. Body locations showing the highest level of fentanyl contamination in these workers were identified as hands, neck, and lower arms. The uptake of fentanyl by the body was evidenced by detection of small amounts of fentanyl in urine specimens of these workers. The authors concluded that in most workers, dermal pathway is the primary route of fentanyl exposure [57]. Occupational allergic contact dermatitis caused by oxycodone has also been reported [58].

CONCLUSIONS

Handling paper money contaminated with illicit drugs should not cause any false positive test results during workplace drug testing. Similarly passive inhalation of marijuana and ingestion of hemp products most likely should not lead to any false positive results but eating edible marijuana products in states where use of recreational use of marijuana is legal may cause positive workplace drug testing result. In addition, drinking coca tea is most likely to cause positive rest result in workplace drug testing. Toxicologists, MRO, and prosecutors should be aware of these popular excuses in individuals tested positive for drugs of abuse.

REFERENCES

[1] Oyler J, Darwin WD, Cone EJ. Cocaine contamination of United States paper currency. J Anal Toxicol 1996;20:213–16.
[2] Negrusz A, Perry JL, Moore C. Detection of cocaine on various denominations of United States currency. J Forensic Sci 1998;43:626–9.
[3] Jenkins AJ. Drug contamination in US paper currency. Forensic Sci Int 2001;121:189–93.
[4] Zuo Y, Zhang K, Wu J, Rego C, et al. An accurate and nondestructive GC method for determination of cocaine on US paper currency. J Sep Sci 2008;31:2444–50.
[5] Jourdan TH, Veitenheimer AM, Murray CK, Wagner JR. The quantitation of cocaine on US currency: survey and significance of the levels of contamination. J Forensic Sci 2013;58:616–24.
[6] Esteve-Turrillas FA, Armenta S, Moros J, Garrigues S, et al. Validated non-destructive and environmental friendly determination of cocaine in Euro bank notes. J Chromatogr A 2005;1065:321–5.
[7] Abdelshafi NA, Panne U, Schneider RJ. Screening for cocaine on Euro banknotes by a highly sensitive enzyme immunoassay. Talanta 2017;165:619–24.
[8] Lavins ES, Lavins BD, Jenkins AJ. Cannabis (marijuana) contamination of United States and foreign paper currency. J Anal Toxicol 2004;28:439–42.
[9] Wimmer K, Schneider S. Screening for illicit drugs on Euro banknotes by LC–MS/MS. Forensic Sci Int 2011;206:172–7.
[10] Luzardo OP, Almeida M, Zumbado M, Boada LD. Occurrence of contamination by controlled substances in Euro banknotes from Spanish archipelago of Canary Island. J Forensic Sci 2011;56:1588–93.
[11] Sleeman R, Burton F, Carter J, Roberts D, et al. Drugs on money. Anal Chem 2000;72:397A–403A.
[12] Almeida VG, Gassella RJ, Pacheco WF. Determination of cocaine in real banknotes circulating at the state of Rio de Janeiro, Brazil. Forensic Sci Int 2015;251:50–5.
[13] Ebejer KA, Winn J, Carter JF, Sleeman R, et al. The difference between drug money and "lifetime's savings". Forensic Sci Int 2007;167:94–101.
[14] Carter JF, Sleeman R, Parry J. The distribution of controlled drugs on banknotes via counting machine. Forensic Sci Int 2003;132:106–12.
[15] ElSohly MA. Urinalysis and casual handling of marijuana and cocaine. J Anal Toxicol 1991;14:46.
[16] Jenkins AJ, Llosa T, Montoya I, Cone EJ. Identification and quantitation of alkaloids in coca tea. Forensic Sci Int 1996;77:179–89.
[17] Jackson GF, Saady JJ, Poklis A. Urinary excretion of benzoylecgonine following ingestion of health Inca tea. Forensic Sci Int 1991;49:57–64.
[18] Mazor SS, Mycyk MB, Wills BK, Brace LD, Gussow L, Erickson T. Coca tea consumption causes positive urine cocaine assay. Eur J Emerg Med 2006;13:340–1.
[19] Turner M, McCrory P. Time for tea anyone? Johnston A. Br J Sports Med 2005;39:e37.
[20] ElSohly MA, Stanford DF, ElSohly HN. Coca tea and analysis of cocaine metabolite. J Anal Toxicol 1986;10:256.
[21] Bieri S, Brachet A, Veuthey J, Christen P. Cocaine distribution in wild *Erythroxylum* species. J Ethnophramacol 2006;103:439–47.
[22] Finkel RS, Zarlengo KM. Blue cohosh and perinatal stroke. N Eng J Med 2004;351:302–3.
[23] Rubio NC, Strano-Rossi S, tabernero MJ, Gonzalez JL, et al. Application of hygrine and cuscohygrine as possible biomarkers to distinguish coca chewing from cocaine abuse on WDT and forensic cases. Forensic Sci Int 2014;243:30–4.
[24] Rubio NC, Hastedt M, Gonzalez J, Pragst F. Possibilities for discrimination between chewing of coca leaves and abuse of cocaine by hair analysis including hygrine, cuscohygrine, cinnamoylcocaine and cocaine metabolite/cocaine ration. Int J Legal Med 2015;129:69–84.
[25] Eachus PL. Positive drug screen for benzodiazepine due to a Chinese herbal product. J Athl Tran 1996;31:165–6.
[26] Posadzki P, Watson L, Ernst E. Contamination and adulteration of herbal medicinal products (HMPs): an overview of systematic review. Eur J Clin Pharmacol 2013;69:295–307.
[27] Mule SJ, Lomax P, Gross SJ. Active and realistic passive marijuana exposure tested by three immunoassays and GC/MS. J Anal Toxicol 1988;12:113.

[28] Department of Transportation. Part 40-procedures for transportation workplace drugs and alcohol testing. Federal Registrar 2000;65:79510–641.
[29] Niedbala RS, Kardos KW, Fritch D, Kunsman K, Blum KA, Newland GA, et al. Passive cannabis smoke exposure and oral fluid testing II. Two studies of extreme cannabis smoke exposure in a motor vehicle. J Anal Toxicol 2005;29:607–15.
[30] Berthet A, De Cesare M, Favrat B, Sporkert F, et al. A systematic review of passive exposure to cannabis. Forensic Sci Int 2016;269:97–112.
[31] Uhl M, Sachs H. Cannabinoids in hair: strategy to prove marijuana/hashish consumption. Forensic Sci Int 2004;145:143–7.
[32] Thieme D, Sachs H, Uhl M. Proof of cannabis administration by sensitive detection of 11-nor-Delta (9)-tetrahydrocannabinaol-9-carbozylic acid in hair using selective methylation and application of liquid chromatography-tandem multistage mass spectrometry. Drug test Anal 2014;6:112–18.
[33] Pichini S, Marchei E, Martello S, Gottardi M, et al. Identification and quantification of 11-nor-9-carboxy-Δ^9 tetrahydrocannabinol-9-carboxylic acid glucuronide (THC-COOH-glu) in hair by ultra-performance liquid chromatography tandem mass spectrometry as a potential hair biomarker of cannabis use. Forensic Sci Int 2015;249:47–51.
[34] Moosmann B, Roth N, Auwarter V. Hair analysis of Δ^9 tetrahydrocannabinolic acid A (THCA-A) and Δ^9 tetrahydrocannabinol (THC) after handling cannabis plant material. Drug Test Anal 2016;8:128–32.
[35] Cirimele V, Kintz P, Jamey C, Ludes B. Are cannabinoids detected in hair after washing with Cannbio shampoo? J Anal Toxicol 1999;23:349–51.
[36] Schwab US, Callaway JC, Erkkilä AT, Gynther J, Uusitupa MI, Järvinen T. Effects of hempseed and flaxseed oils on the profile of serum lipids, serum total and lipoprotein lipid concentrations and hemostatic factors. Eur J Nutr 2006;45:470–7.
[37] Bosy TZ, Cole KA. Consumption and quantitation of Δ^9 tetrahydrocannabinol in commercially available hemp seed oils and products. J Anal Toxicol 2000;24:562–6.
[38] Mediavilla V, Derungs R, Kanzig A, Magert A. Qualitat von hanfsamenol aus der Schweiz. Agraforschung 1997;4:449–51.
[39] Lehman T, Sager F, Brenneisen R. Excretion of cannabinoids in urine after ingestion of cannabis seed oil. J Anal Toxicol 1997;5:373–5.
[40] Cole K. Division of Forensic Toxicology, Armed Forces Institute of Pathology, Rockville, MD, Personal communication; 2000.
[41] Costantino A, Schwartz RH, Kaplan P. Hemp oil ingestion causes positive urine tests for delta-9-tetrahydrocannabinol carboxylic acid. J Anal Toxicol 1997;21:482–5.
[42] Struempler RE, Nelson G, Urry TM. A positive cannabinoid workplace drug test following the ingestion of hemp seed oil. J Anal Toxicol 1997;21:283–5.
[43] Alt A, Reinhardt G. Positive cannabis results in urine and blood samples after consumption of hemp food products. J Anal Toxicol 1998;22:80–1.
[44] Meier H, Vonesch HJ. Cannabis poisoning after eating salad. Schweiz Med Wochenschr 1997;127:214–18.
[45] Gibson CR, Williams R, Browder R. Analysis of Hemp Ale for cannabinoids. J Anal Toxicol 1998;22:179.
[46] Leson G, Pless P, Grotenhermen F, Kalant H, ElSohly MA. Evaluating the impact of hemp food consumption on workplace drug tests. J Anal Toxicol 2001;25:691–8.
[47] Goodwin RS, Gustafson RA, Barnes A, Nebro W, Moolchan ET, Huestis MA. Delta(9)-tetrahydrocannabinol, 11-hydroxy-delta(9)-tetrahydrocannabinol and 11-nor-9-carboxy-delta(9)-tetrahydrocannabinol in human plasma after controlled oral administration of cannabinoids. Ther Drug Monit 2006;28:545–51.
[48] Cone EJ, Johnson RE, Paul BD, Mell LD, et al. Marijuana-laced brownies: behavioral effects, physiological effects, and urinalysis in humans following ingestion. J Anal Toxicol 1988;12:169–75.
[49] Vandrey R, Hermann ES, Mitchell JM, Bigelow GE, et al. Pharmacokinetic profile of oral cannabis in humans: blood and oral fluid disposition and relation to pharmacodynamic outcomes. J Anal Toxicol 2017;41:83–99.
[50] Favrat B, Menetrey A, Augsburger M, Rothuizen LE, et al. Two cases of cannabis acute psychosis following the administration of oral cannabis. BMC Psychiatry 2005;5:17.
[51] Hancock-Allen JB, Barker L, VanDyke M, Holmes DB. Notes from the field: death following ingestion of an edible marijuana product—Colorado, March 24. Morb Mortal Wkly Rep 2015;64:771–2.
[52] Wang GS, Le Lait MC, Deakyne SJ, Bronstein AC, et al. Unintentional pediatric exposures to marijuana in Colorado, 2009–2015. JAMA Pediatr 2016;170:e160971.
[53] Le SD, Taylor RW, Vidal D, Lovas JJ, et al. Occupational exposure to cocaine hydrochloride involving crime lab personnel. J Forensic Sci 1992;37:959–68.
[54] Gehlhausen JM, Klette KL. Urine analysis of laboratory personnel preparing cocaine training aids for a military working dog program. J Anal Toxicol 2001;25:637–40.
[55] Stout PR, Horn CK, Klette KL, Given J. Occupational exposure to methamphetamine in workers preparing training AIDS for drug detection dogs. J Anal Toxicol 2006;30:551–3.
[56] Villain M, Muller JF, Kintz P. Heroin markers in hair of a narcotic police officer: active or passive exposure? Forensic Sci Int 2010;196:128–9.
[57] Van Nimmen NE, Poels KL, Veulemans HA. Identification of exposure pathways for opioid narcotic analgesics in pharmaceutical production workers. Ann Occup Hyg 2006;50:665–77.
[58] Wootton C, English JS. Occupational allergic contact dermatitis caused by oxycodone. Contact Dermatitis 2012;67:383–4.

Chapter 33

Abuse of Magic Mushroom, Peyote Cactus, LSD, Khat, and Volatiles

Amitava Dasgupta
Department of Pathology and Laboratory Medicine, University of Texas McGovern Medical School at Houston, Houston, TX, United States

INTRODUCTION

Magic mushroom is a common name given to various species of psychedelic fungi, which are used for recreational purpose. Hallucinogenic mushroom, cactus such as peyote cactus, and plants were used to induce altered state of consciousness in healing ritual and religious ceremonies during ancient pre-Columbian Mesoamerican culture. Maya and Aztec people used magic mushroom peyote cactus and other psychedelic plants/fungus in group ceremonies to achieve intoxication. Archeological evidences indicated use of magic mushroom even in 3000 BC and use of peyote cactus in 5000 BC. Today local shamans and healers still use these psychoactive compounds during healing and or religious ceremonies [1]. Under the influence of magic mushroom and peyote cactus, people see images and feel sensations that are unreal. Emotional mood swings and feeling of spiritual energy may also occur under the influence of hallucinogens. These agents produce their biochemical effect by disrupting the interaction of nerve cells and the neurotransmitter serotonin. Despite abuse, hallucinogens are not routinely tested in workplace drug testing. Khat abuse is common among certain ethnic population. Lysergic acid diethylamide (LSD) was a popular drug of abuse in 1960s and is still abused today. Solvent and glue sniffing are also of significant concerns especially among younger people. Death may result from such abuse. Street names of magic mushroom, peyote cactus, LSD, and khat are listed in Table 33.1.

MAGIC MUSHROOM

Magic mushrooms (psychoactive fungi) that grow in the United States, Mexico, South America, and many other parts of the world contain the hallucinogenic compounds psilocybin and psilocin. Psilocybin and psilocin are tryptamines because they have an indole ring structure, a fused double ring consisting of a pyrrole ring and a benzene ring joined to an amino group by two carbon side chains [2]. Active ingredients of magic mushroom are classified under Schedule I controlled substances with no known medical use but these agents have a high abuse potential. Unlawful possession of a Schedule I controlled substance is a felony by law in the United States. In 2005 the UK government changed the law and included dried or cooked psychoactive mushrooms under Class A drugs which are similar to Schedule I drugs in the United States.

In general most species of magic mushrooms appear as brown or dark tan in color. Magic mushrooms are eaten raw, cooked with food or as dried. These mushrooms can be mistaken for other nonhallucinogenic mushrooms or even poisonous mushrooms such as Amanita class. Death from ingestion of Amanita mushroom due to mistaken identity as magic mushroom has been reported [3]. It is dangerous practice to eat wild mushrooms because there are other reports of severe toxicity due to ingestion of toxic wild mushroom as a result of misidentification [4].

Prevalence of Magic Mushroom Abuse

Magic mushrooms were widely abused in the United States during 1960s when LSD was very popular. However, magic mushroom is very popular in rave parties and college parties even today. Hallock et al. based on a survey of 409 college

TABLE 33.1 Street Names of Magic Mushrooms, Peyote Cactus, LSD, and Khat

Abused Agent	Common Street Names
Magic mushroom	Magic mushroom, sacred mushroom, musk, Mexican mushroom, fly-agaric, God's flesh
Peyote cactus	Buttons, mesc, peyote, cactus
LSD (lysergic acid diethylamide)	Acid, blotter, doses, trip
Khat	Chat, khat kat, kafta, qat, qaad, quaadka ghat, mirra, Abyssinian tea, Somali tea, Arabian tea

students reported that 29.5% of responded experimented with psilocybin-containing hallucinogenic magic mushrooms. The number one cause of using magic mushroom was to achieve mystical experience. Interestingly, users did not believe that abuse magic mushroom may negatively impact their academic performance, physical well-being, as well as mental health because they did not believe that magic mushroom use may cause addiction. Magic mushroom users were also prone to abuse other drugs including cocaine, ecstasy, opiates, prescription drugs, and LSD [5]. Another study indicated that an estimated 21 million people in the United States used magic mushrooms in the past while an estimated 11 million people experimented with mescaline. The authors commented that psychedelics continue to be widely used in the United States. Common reasons for such uses include curiosity, mystical experiences, and introspection. Rates of use of psychalics are higher in men than women. Interestingly, psilocybin (magic mushroom) use is more common in younger adults while use of LSD and mescaline are more common among older adults [6].

Riley and Blackman surveyed 174 participants from United Kingdom in 2004 when the sale of magic mushroom was not prohibited. Participants reported infrequent but heavy consumption of magic mushroom (47% used magic mushroom 4–12 times in a year) and the average consumption in one setting was about 12 g. Although hallucinations were experienced by majority of participants, feeling of altering perceptive (41%–74%) and experiencing closer to nature (49%) were also common. However, a significant number of participants also felt negative experience including paranoia (35%) and anxiety (32%) [7].

Active Ingredients of Magic Mushroom

Psilocybin and psilocin are found in over 150 species of mushrooms. The structures of psilocybin and psilocin are given in Fig. 33.1. The group of psychoactive mushrooms includes species of the genera of *Conocybe*, *Gymnopilus*, *Panaeolus*, *Pluteus*, *Psilocybe*, and *Stropharia* [8]. *Panaeolus cyanescecens* usually contains the high amounts of psilocybin and psilocin, while the most common type of magic mushroom in Germany is *Psilocybe semilanceata*, which contains highly variable amounts of psilocybin (<0.003%–1.15%) and psilocin (0.01%–0.90%) [9]. The psilocybin contents in *Psilocybe cubeneis* was 0.44%–1.35% in the cap of the mushroom and 0.05%–1.27% in the stem while the corresponding psilocin concentrations were 0.17%–0.78% and 0.09%–0.30%, respectively, according to a published report. In general hallucinogenic alkaloids are found more in the cap of the mushroom than the stem [10]. Active ingredients and duration of action of magic mushroom are given in Table 33.2.

Pharmacology and Toxicology of Psilocybin

After oral administration, psilocybin, the major active ingredient of magic mushroom (*O*-phosphoryl-4-hydroxy-*N,N*-dimethyl-tryptamine), is rapidly dephosphorylated into psilocin (*N,N*-dimethyl-tryptamine), an active metabolite under acidic environment of the stomach and in the intestinal mucosa by alkaline phosphatase (and other nonspecific esterases). Conversion of psilocybin into psilocin also occurs during hepatic first-pass metabolism. Psilocybin could therefore be considered as a "prodrug" where most pharmacological activities are due to active metabolite psilocin. Both psilocybin and psilocin are structurally related to neurotransmitter serotonin. Psilocybin and psilocin in their pure form are white crystalline powders. While psilocybin is water soluble, and psilocin is more soluble in lipids. As a result psilocin readily crosses the blood–brain barrier where it exerts psychedelic effect which is similar to LSD [11].

Psychedelic effect of magic mushroom requires ingestion of over 15 mg of psilocybin which produces plasma psilocin levels of 4–6 ng/mL. The onset of psychedelic effect after ingestion of magic mushroom is 20–40 min and maximum effect is observed within 60–90 min. The duration of effect is 4–6 h after oral ingestion while most effects

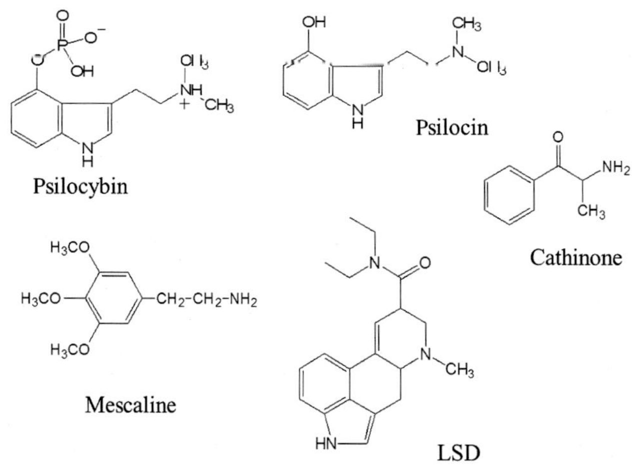

FIGURE 33.1 Chemical structures of psilocybin, psilocin, mescaline, cathinone, and LSD.

TABLE 33.2 Active Ingredients and Duration of Action of Magic Mushrooms, Peyote Cactus, LSD, and Khat

Abused Agent	Active Ingredient	Duration of Action
Magic mushroom	Psilocybin and psilocin, but psilocin is also the active metabolite of psilocybin and is responsible for most psychedelic effects	The duration of effect is 4–6 h after oral ingestion while most effects disappear within 6–8 h.
Peyote cactus	Mescaline	The highest psychedelic effect may be achieved within 2 h of ingestion but the effect may last as long as 8 h.
LSD (lysergic acid diethylamide)	LSD	The duration of effects: 9–10 h.
Khat	Cathinone (major active ingredient)	The stimulatory effects appear approximately 30 min after initiation of chewing khat plant and the effects may last up to 3 h.

disappear within 6–8 h. In general no effect is observed after 24 h [12]. In another study when healthy subjects ingested 0.224 mg/kg body weight of psilocybin (10–20 mg), peak plasma level of psilocin was observed at 105 min (the mean value), and average plasma concentration was 8.2 ng/mL. The average bioavailability of psilocin was 52.7% after oral administration of psilocybin according to this study [13].

Passie et al. detected psilocin in human plasma 30 min after oral administration of psilocybin. Maximum plasma levels were observed between 80 and 105 min while detection window was approximately 6 h. The full psychedelic effects occurred with oral dosage of psilocybin between 8 and 25 mg. The average half-life of psilocin in plasma after oral administration of psilocybin was 2.5 h [14]. In another report the concentration of psilocin in serum was 52 ng/mL and psilocin concentration in urine was 1760 ng/mL in one person who consumed magic mushroom [15].

Psilocin is conjugated in the liver by the action of UDP-glucuronosyltransferase into psilocin-*O*-glucuronide and in this form 80% of psilocin is excreted from the body in urine. In addition, psilocin is also subjected to metabolism through methylation and deamination of into 4-hydroxyindole-3-yl-acetaldehyde. Further oxidation of this metabolite produces 4-hydroxyindol-3-acetic acid and 4-hydroxytryptofol. These minor metabolites account for approximately 4% metabolism of psilocin. Both psilocin (90%–97%) and psilocybin (3%–10%) are detected in human urine mostly conjugated with glucuronic acid. Only 3%–10% psilocin and psilocybin are detected in unmodified form in urine. The majority of pharmacologically active compounds are excreted from the body (mostly urine but also bile and feces) within 3 h after oral administration of magic mushroom. Although these compounds are almost completely eliminated from the body within 24 h, trace amount may be detected in urine even after a week [16].

Poisoning as well as organ damage from use of magic mushroom is common but fatality is less common. Satora et al. reported that in the autumn of 2004 several people were admitted in the hospital after ingesting magic mushroom to experience hallucinatory sensations. Three people had visual hallucination while the fourth person had both visual and auditory hallucinations followed by psychosis [17]. Borowiak et al. described a case where an 18-year-old man suffered from Wolff–Parkinson–White syndrome, arrhythmia, and myocardial infarction as a result of magic mushroom poisoning (*P. semilanceata*) [18]. Raff et al. reported renal failure in a 20-year-old woman who consumed magic mushroom [19]. Bickel et al. reported severe rhabdomyolysis, acute renal failure, and posterior encephalopathy after abuse of magic mushroom (*Psilocybe cubensis*) [20].

There are several reports of fatality due to ingestion of magic mushroom. One person died 6–8 h after ingestion of unknown quantity of magic mushroom. Postmortem toxicology analysis showed very high plasma concentration of psilocin (4000 ng/mL). Another person who was a heart transplant recipient died after consumption of magic mushroom. This individual showed a much lower psilocin concentration of 30 ng/mL in postmortem plasma. However, tetrahydrocannabinol level of 4 ng/mL was also reported in plasma although no other drug or alcohol was present [21]. A fatal case of magic mushroom poisoning in a 27-year-old male has also been reported from Japan [22].

Some of the species of magic mushroom contain phenylethylamine which may cause cardiac toxicity. Beck et al. reported that the amount of phenylethylamine may vary much more than psilocybin in magic mushrooms. Three young men were hospitalized due to adverse reaction from using magic mushroom. The highest amount of phenylethylamine found in the mushroom (*P. semilanceata*) was 146 μg/g wet weights [23].

Mechanism of Action of Active Ingredients of Magic Mushroom

Psilocin, the major metabolite of psilocybin, and psilocybin have predominant agonist activity on serotonin (5-hydroxytryptamine: 5-HT) receptors located within and beyond the central nervous system (CNS) as well as several transporters (serotonin and norepinephrine). Psilocybin and psilocin bind with high affinity to 5-HT$_{2A}$ receptor and such activation occurs in several brain network hubs where cells become more sensitive through depolarization. However, these compounds also bind with 5-HT$_{2C}$, 5-HT$_{1A}$, and 5-HT$_{1D}$ receptors with lower affinity. Although these compounds exhibit no apparent affinity for dopamine D$_2$ receptor, administration of haloperidol, a D$_2$ receptor antagonist, reduces psilocybin-induced psychotomimesis, raising the possibility of a dopaminergic neuronal transmission involvement [24].

In vitro studies indicate that ketanserin, a 5-HT$_{2A}$ agonist, prevents most psychotic symptoms associated with magic mushroom ingestion indicating that serotonergic system is responsible for psychedelic effect of magic mushroom mediated through activation of 5-HT$_{2A}$ receptor [25]. Study with rat model indicated that effect of psilocin was more intense in male rats than female rats indicating that there may be sex difference in serotonergic response after ingestion of magic mushroom [26].

Therapeutic Potential of Psilocybin

Dysregulations in the serotonin system are associated with alteration in stress hormones, such as cortisol, and mood disorders. Psilocybin has structural similarity with serotonin, and after administration of psilocybin in a clinically controlled situation, executive control network of the brain is activated with subsequent increased control over emotional process and relief of persistent negative emotion. Preliminary data indicates potential therapeutic use of psilocybin in treating drug and alcohol addictions. Psilocybin also has a low risk of dependence [27]. Studies have shown that single oral doses of psilocybin reduced anxious depressed mood, existential distress, and improved quality of life in test subjects [28]. Preliminary results also show that psilocybin may be effective in treating drug-resistant depression [29].

Analysis of Active Ingredients of Magic Mushroom

Psilocin is not routinely tested in urine or blood and as a result toxicology report may be negative in a person who is exposed to magic mushroom. A 28-year-old man with a history of drug and alcohol abuse presented multiple times to the hospital over a period of 2 months (three emergency department visits, two medical flood admission and one intensive care admission), but his toxicology results in urine conducted using gas chromatography/mass spectrometry (GC/MS) were always negative. His symptoms included altered mental status, vomiting, diaphoresis, and mydriasis. The patient later admitted to the nurse that he used magic mushroom. The authors concluded that in the absence of confirmatory tests, but supported by exclusionary and anecdotal data, most likely symptoms of the patient was consistent with magic mushroom toxicity [30].

However, psilocin has a low cross-reactivity with the fluorescence polarization immunoassay (FPIA) of amphetamine for application the AxSYM analyzer (Abbott Laboratories, Abbott Park, IL). Tiscione and Miller described a case of a 20-year-old male driver who drove of the road and struck a light pole and was taken into the hospital. The subject indicated to the emergency personnel that he recently consumed mushroom. Suspecting a case of driving under influence, urine specimen was collected for toxicological analysis. The initial screening with immunoassay showed positive result with amphetamine/methamphetamine FPIA assay. However, analysis of urine specimen using GC/MS did not show the presence of amphetamine, methamphetamine, or related drugs, such as MDMA (3,4-methylenedioxy-methamphetamine) or MDA (3,4-methylenedioxy amphetamine). Further investigation identified the presence of psilocin. Using psilocin standard, the authors determined that at a concentration of 50 μg/mL, the cross-reactivity of psilocin with the amphetamine immunoassay was 1.3%. The urine concentration of psilocin decreased rapidly despite the fact that the specimen was stored at 4°C indicating that psilocin is unstable in urine specimen [31].

In general, high-performance liquid chromatography or GC/MS is used for identification and quantification of psilocybin and psilocin in biological specimens. Because psilocin is found mostly in the conjugated form, hydrolysis of the glucuronide derivative prior to extraction is recommended. In one report the authors performed enzymatic hydrolysis using beta-glucuronidase to liberate free psilocin from its conjugated form in urine and then used MSTFA (*N*-methyl-*N*-trimethylsilyltrifluoroacetamide) for derivatization of psilocin to its trimethylsilyl derivative in order to determine the concentration of psilocin by GC/MS [32]. Psilocin can also be analyzed by GC/MS after derivatization with BSTFA reagent (*N,O*-bis-trimethylsilyl-acetamide containing 1% trimethylchlorosilane) to covert psilocin to its trimethylsilyl ester (psilocin-1,4-di-trimethylsilyl) (Paul DB unpublished data). The mass spectrum of derivatized psilocin is given in Fig. 33.2. In the mass spectrum, a distinct molecular ion peak at *m/z* 348 was observed. The base peak at *m/z* 290 (M^+ −58) was due to loss of $N(CH_3)_2CH_2$ group from the side chain of the derivatized psilocin.

Psilocin and psilocybin can also be determined using liquid chromatography combined with tandem mass spectrometry (LC-MS/MS). For LC-MS/MS method derivatization prior to analysis is not required. Kamata et al. used LC-MS/MS for analysis of psilocin and psilocybin in magic mushroom samples. The authors used octadecysilyl (ODS) column for liquid chromatographic analysis. The authors observed wide variation in psilocybin (0.18−3.8 mg/g of dry weight) and psilocin (0.60−1.4 mg/g of dry weight) content in four samples of magic mushrooms analyzed [33]. Anastos et al. also used high-performance liquid chromatography for analysis of psilocin and psilocybin in magic mushroom samples. The authors used 4-hydroxyindole as the internal standard [34].

Martin et al. developed a valid method for quantitation of psilocin in plasma using a liquid chromatography−electrospray ionization/tandem mass spectrometry method. The authors used solid-phase extraction using ascorbic acid and nitrogen for drying to protect unstable analyte. A reverse phase C-18 which was heated to 40°C during analysis and 10 μL of the specimen was injected in the column. Separation was achieved with a 10% mobile phase A (methanol with 0.1% formic acid) 90% mobile phase B (2 mM ammonium acetate buffer with 0.1% formic acid, pH 3) at a flow rate of 0.2 mL/min for 7 min, switching to 90% mobile phase A for 4 min to remove lipophilic impurities. The

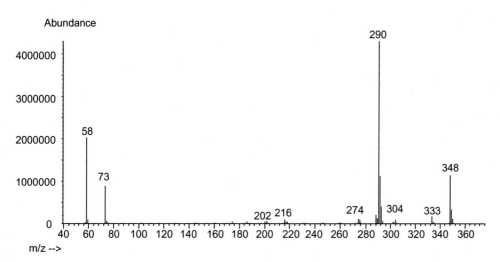

FIGURE 33.2 Mass spectrum of psilocin-1,4-di-trimethylsilyl (trimethylsilyl derivative of psilocin). *Figure courtesy of Dr. Buddha D. Paul, OAFME-Forensic toxicology division of Armed Forces Institute of Pathology, Rockville, MD (now retired).*

TABLE 33.3 Analytical Methods for Analysis of Active Ingredients of Magic Mushrooms, Peyote Cactus, LSD, and Khat

Abused Agent	Analytical Methods
Magic mushroom	Although psilocin at high concentration cross-reacts with antibody of FPIA amphetamine immunoassay, this assay is not sensitive enough for screening of psilocin in urine. In general, both GC/MS- and LC-MS/MS-based assays are used for detection and confirmation of psilocin and psilocybin in biological matrix.
Peyote cactus	Mescaline, the active ingredient of peyote cactus can be detected in urine using a biochip array immunoassays. However, both GC/MS- and LC-MS/MS-based assays are used for detection and confirmation of psilocin and psilocybin in biological matrix.
LSD	Various immunoassays are commercially available (CEDIA, EMIT, KIMS, etc.) for screening in urine. However, both GC/MS- and LC-MS/MS-based assays are used for detection and confirmation of psilocin and psilocybin in biological matrix.
Khat	Cathinone cross-reacts with one antibody used in biochip array immunoassays and also can be analyzed by both GC/MS and LC-MS/MS. However, abuse of bath salts (synthetic cathinones) is significantly more widespread than khat abuse. In some published protocol for detection and confirmation of bath salts in urine, cathinone can also be measured.

column was equilibrated for 10 min and as a result total run time was 21 min. The retention time of psilocin was 6 min. The mass spectrometer was operated in multiple reaction monitoring mode. The authors used deuterated psilocin (psilocin-d_{10}) as the internal standard. The ions monitored were m/z 205 > 58 (quantifier) and m/z 205 > 160 (qualifier) for psilocin. The ions monitored for the internal standard were m/z 215 > 66 (quantifier) and m/z 215 > 164 (qualifier). The limit of detection was 0.1 ng/mL and the limit of quantitation was 0.34 ng/mL. Processed samples were stable in the auto-sampler (cooled at 12°C) for at least 26 h. However, specimens stored at room temperature showed a continuous decline in psilocin concentration leading to loss of 90% in 1 week. Storage in a refrigerator at 4°C improved specimen stability almost for 7 days if fluoride was added as a preservative but freezing of blood samples lead to a not reproducible loss of psilocin. The authors suggested that enzymes involved in psilocin degradation are released from hemolysis that may occur during freezing [35].

Martin et al. also reported a valid method for quantitation of psilocin, bufotenine (a compounds secreted from glands present on skin of toad; genus *Bufo* which has hallucinogenic activity), LSD and its metabolites in serum and urine using a liquid chromatography—electrospray ionization/tandem mass spectrometry method after solid-phase extraction [36]. Analysis of psilocin, bufotenine, LSD, and its metabolites in hair has also been reported [37]. Analytical methods for analysis of active ingredients of magic mushroom are listed in Table 33.3.

USE OF PEYOTE CACTUS

Peyote cactus (*Lophophora williamsii*) is a small spineless cactus that grows in the Southwestern part of the United States and Mexico. Peyote cactus is small and round without sharp spines and the top of the cactus which is above ground is called "crown." The crown consists of disc-shaped buttons which contain the psychoactive compound mescaline (3,4,5-trimethoxyphenethylamine). Chemical structure of mescaline is given in Fig. 33.1. Mescaline can be extracted from the peyote cactus or buttons can be chewed or soaked in water to produce psychoactive liquid. Native North American Indian recognized the psychotropic properties of peyote as long as 5700 years ago. El-Seedi et al. analyzed two archeological specimens of peyote buttons from the collection of the White Museum in San Antonio using carbon dating and reported that the buttons dated back to 3780−3660 BC. The authors also extracted alkaloids from the buttons and demonstrated the presence of mescaline the active ingredient of peyote [38]. Mescaline is classified as a Schedule I controlled substance but approximately 300,000 members of Native Americans Church can ingest peyote cactus legally as a religious sacrament during all night prayer in the Native American Church based on 1994 the American Indian Religious Freedom Act. The members can attend the prayer ceremonies as regularly as 2−3 times a week or infrequently as once a year. Based on a study of three groups of Navajo Native American Indians, age 18−45 (group 1: 61 Native American Indian church members who regularly ingested peyote, group 2: 36 individuals with pattern of alcohol dependence, and group 3: 79 individuals reporting minimal use of peyote, alcohol, or other substances), the authors found no evidence of psychological or cognitive deficits among Native American Indians using peyote regularly in a religious setting [39].

Native Americans not only used peyote for centuries and for its hallucinogenic effects but also for its medicinal properties. It can be used for treating snake bites, burns, wounds, and rheumatism, as well as for toothache, fever, and scorpion stings [40].

Prevalence of Peyote Cactus Abuse

In general use of peyote cactus is low in the United States. According to a survey completed by 886,088 individuals (surveyed for most years 1995–2010), peyote use among American Indian and general US population remained stable between 1% and 2% [41]. Exposure to peyote is rarely reported to poison control centers. In 2007 a total of 116 peyote/mescaline exposures were reported to US poison control centers out of more than 2.8 million total reported exposures. Therefore exposure to peyote cactus represented only 0.004% of all exposures reported. Out of these 116 exposures, only 32 patients (28%) were treated in health care facilities but no major toxicity or death was reported [42]. Interestingly among youths aged 12–17 years, lifetime use of peyote has been reported to be significantly less (0.4%) compared to other hallucinogens such as ecstasy (3.2%), LSD (3.1%), or psilocybin-containing mushrooms (2.1%) [40].

Based on the review of California Poison Control System database for years 1997–2008, Carstairs and Cantrell identified 31 cases of single substance abuse involving peyote cactus. The majority of these individuals were male (84%) and ages ranged from 14 to 59 years. Thirty (97%) exposures were through oral route: 23 ingested plant material, 6 boiled peyote buttons and drank the resulting tea, and 1 ingested a tablet containing mescaline. The commonly reported symptoms included hallucination, tachycardia, agitation, and mydriasis. One subject reported vomiting. Although five subjects were managed at home, others received medical attention in health care facilities indicating dangers of peyote cactus abuse [40].

Pharmacology and Toxicology of Mescaline

Mescaline [2-(3,4,5-trimethoxyphenyl) ethanamine], a natural alkaloid present in peyote cactus, is a hallucinogen with psychoactive properties. It can be ingested orally for hallucinogenic effects similar to LSD which include deeply mystical feelings, but LSD is approximately 4000 times more potent than mescaline in producing altered state of consciousness [43]. Although mescaline has the lowest potency among naturally occurring hallucinogens (30 times less potent than psilocybin), a full dose (200–400 mg) has a long duration of action. Mescaline is rapidly absorbed from the gastrointestinal tract and onset of effect may be observed within 30 min of ingestion. The highest psychedelic effect may be achieved within 2 h of ingestion but the effect may last as long as 8 h. The plasma half-life of mescaline is approximately 6 h [44]. Higher dosage of mescaline is needed to produce the psychedelic effect not only due to weak potency of mescaline as a hallucinogen but also due to polar nature of the mescaline molecule. Because of poor lipid solubility, mescaline does not easily pass the blood–brain barrier [45]. Active ingredient and duration of action after abuse of peyote cactus are given in Table 33.2.

Although mescaline is metabolized by the liver enzymes, approximately 87% of ingested dosage is excreted in urine within 24 h while 92% is excreted within 48 h [46]. A small amount of mescaline is oxidatively deaminated into 3,4,5-trimethoxyphenyacetic acid [47]. Another minor metabolite of mescaline in human is 3,4-dimethoxy-5-hydroxyphenyethylamine.

After ingestion, mescaline can be detected in blood, urine, and hair. A mother found her boy of 17 years in his bedroom along with a plastic glass containing dark green liquid (tea) with strong bitter odor. The analysis of the liquid by GC/MS showed mescaline concentration of 1.3 mg/L. However, no other drug was detected indicating that the tea was prepared from peyote cactus. No mescaline was detected in the urine specimen but the analysis of hair showed the presence of mescaline at a concentration of 1 ng/mg of hair in the portion of the hair nearest to the scalp. Based on these results, the authors confirmed abuse of peyote cactus in this boy [48].

Mescaline, the active ingredient of peyote, may cause fetal abnormalities [49]. Although rare, severe toxicity, and even death may occur from mescaline overdose. One person who died under the influence of mescaline showed 9.7 µg/mL of mescaline concentration in blood and 1163 µg/mL in urine [50]. Henry et al. described distribution of mescaline in postmortem tissue in an individual who died due to multiple gunshot wounds. The concentrations of mescaline were 2.95 µg/mL in blood, 2.36 µg/mL in vitreous, 8.2 mg/kg in the liver, and 2.2 mg/kg in the brain [51].

Botulism causes skeletal muscle weakness due to bacterial exotoxin that irreversibly blocks the release of acetylcholine from presynaptic motor neurons. Thirteen church members ingested peyote from a communal jar during a ceremony and 2–4 days afterward, three men developed severe clinical symptoms due to botulism. The peyote recovered from the jar containing water was analyzed by the Center of Disease Control and Prevention, Botulism Laboratory,

Atlanta, Georgia, and type B botulinum toxin was identified in the peyote cactus. Fortunately no patient died [52]. Food-borne botulism is due to toxin produced by harmful bacteria *Clostridium botulism* and such bacteria grow in anaerobic condition. Storing peyote cactus with water in a closed jar may provide such condition for bacteria to grow.

Nolte and Zumwalt reported a case where a 32-year-old Native American man with a history of alcohol abuse ingested peyote tea. Later he developed respiratory distress and suddenly collapsed. His antemortem blood specimen showed the presence of mescaline (0.48 μg/mL) and the concentration of mescaline in urine was 61 μg/mL. Other than a trace amount of chlordiazepoxide, no other drug or ethanol was detected in postmortem blood [53].

Mechanism of Action of Mescaline

Although weaker than LSD and MDMA in potency, mescaline is also a serotonin receptor agonist with affinity for $5-HT_{1A}$, $5-HT_{2A}$, $5-HT_{2B}$, and $5-HT_{2C}$ receptors [54,55]. However, mescaline is a nonselective serotonin receptor agonist. Interestingly, biochemical and neural patterns are common to both mescaline and morphine. The study has indicated an association between mescaline-induced aggression and abnormal electrical activities in amygdale [56].

Laboratory Determination of Mescaline

The sole active ingredient of peyote cactus is mescaline. Currently there is one commercially available biochip array immunoassay (drugs of abuse array V assay by Randox Corporation) capable of detecting the presence of mescaline in urine specimens. This assay has 11 different antibodies capable of detecting various bath salts (two antibodies), benzylpiperazines, mescaline, various phenylpiperazine (two antibodies), salvinorin, and various synthetic cannabinoids (four antibodies targeted to different compounds). These antibodies are polyclonal antibodies which are immobilized on a chemically modified biochip. This is a competitive immunoassay where drugs if present in the urine specimen compete with horseradish peroxidase-labeled analyte (conjugate) for binding sites on the immobilized polyclonal antibody specific for the drug. After incubation, a single reagent containing 1:1 mixture of luminol/enhancer and peroxide solution is added and chemiluminescent signal is detected (horseradish peroxidase-labeled analyte bound to antibody provides the signal).

Battle et al. evaluated performance of this assay to detect the presence of mescaline in urine by analyzing 526 presumptive positive and 198 randomly selected negative specimens (source of specimens department of defense). The authors also used GC/MS method for confirmation of mescaline if present in the urine specimen. The cutoff level of the immunoassay as suggested by the manufacturer is 6 ng/mL. For GC/M analysis, mescaline was extracted from 1 mL of urine after addition of the internal standard (deuterated mescaline: mescaline-d_9) using solid-phase extraction. The drug along with the internal standard was eluted using 3 mL methylene chloride/2-propanol/ammonium hydroxide (78:20:2 by vol). Then mescaline along with the internal standard was derivatized using pentafluoropropionic anhydride. The mass spectrometer was operated using selected ion monitoring mode scanning m/z 181.1, 194.1, and 357.1 for mescaline and m/z 190.1, 203.1, and 366.1 for the internal standard. For quantification, m/z 181.1 was used for mescaline and m/z 190.1 for the internal standard. The limit of quantification was 1 ng/mL. Using GC/MS the authors were unable to confirm the presence of mescaline in any urine specimen presumptive positive by the immunoassays. The authors concluded mescaline use among US military personnels is rare. Therefore routine monitoring of mescaline during drug of abuse testing is not warranted at this time [57].

There are other reports of GC/MS analysis of mescaline in biological matrix without derivatization. Henry et al. used a GC with nitrogen phosphorus detector as well as GC/MS for analysis of mescaline in postmortem blood and tissue. The authors extracted mescaline from blood or tissue specimens using butyl chloride. For identification and qualitative analysis of mescaline, the authors operated mass spectrometer in the selected ion monitoring mode. The primary ion used for positive identification was m/z 182 with secondary ions at m/z 167 and 181. For quantitative analysis, the authors used GC combined with nitrogen phosphorus detector [51]. The electron impact mass spectrum of mescaline is shown in Fig. 33.3. A distinct molecular ion peak was observed at m/z 211.

Mescaline can also be analyzed by LC-MS/MS. In one study the authors developed a protocol for analysis of mescaline in urine using LC-MS/MS. The authors used reverse phase C-18 column (5-μm particle size) and methanol gradient in ammonium acetate buffer for the mobile phase. The authors also used deuterated mescaline as the internal standard. Among 462 urine samples collected from young people with alcohol or drug problem, 32% specimens were positive for illicit drugs but none for mescaline [58]. Analytical methods available for analysis of active ingredient of peyote cactus are given in Table 33.3.

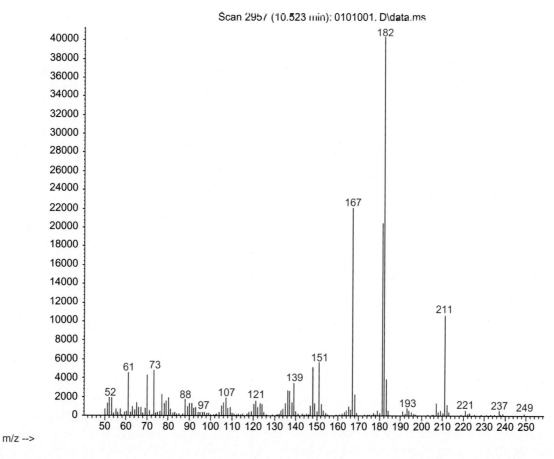

FIGURE 33.3 Mass spectrum of mescaline. *Figure courtesy of Dr. Buddha D. Paul, OAFME-Forensic toxicology division of Armed Forces Institute of Pathology, Rockville, MD (now retired).*

ABUSE OF LYSERGIC ACID DIETHYLAMIDE

LSD is a semisynthetic substance derived from lysergic acid, a natural product found in parasitic rye fungus (*Claviceps purpurea*). The molecule consists of an indole system with a tetracycline ring. LSD was first synthesized by Albert Hofmann, a scientist at the Sandoz AG Pharmaceutical Company (Basel, Switzerland) in 1938, while searching for pharmacologically active derivative of lysergic acid. He accidentally discovered dramatic psychological effect of LSD in 1943. Toward the end of 1960s, people used LSD for recreational and spiritual purposes leading to a "psychedelic movement." Although use of LSD has declined after that, it is still abused today. LSD is a very potent hallucinogen (D-isomer) where an oral dose of 75–150 μg can produce altered state of consciousness. The optimum dosage was suggested being 100–200 μg [59]. The chemical structure of LSD is given in Fig. 33.1.

After oral administration, LSD is absorbed and effects are observed 30–45 after ingestion. In one study, the authors showed lower blood level of LSD if ingested (160 μg) after meal. The peak plasma concentration was reached after 60 min (mean value: 2.75 ng/mL) in subjects who ingested LSD after full meal. The peak concentration was also reached after 60 min in subjects who ingested LSD without meal but the peak LSD level was significantly higher (mean value: 4.9 ng/mL). The duration of effects was 9–10 h [60]. Dolder et al. studied pharmacokinetics of LSD after oral administration using 16 healthy subjects (8 males and 8 females). The oral bioavailability of LSD was determined to be 71%. After ingestion of LSD, maximum plasma concentrations of LSD (mean: 4.5 ng/mL) were reached 1.5 h (median value) after administration (range: 0.5–4 h). Then LSD concentration declined following the first-order kinetics with a mean half-life of 3.6 h for first 12 h. Only 1% LSD excreted in urine was unchanged while the major metabolite detected in urine was 2-oxo-3-hydroxy-lysergic acid diethylamide. The authors also observed a close relationship between plasma LSD level and its psychotropic effects [61].

In another study involving 24 subjects who ingested 100 μg LSD and 16 subjects who ingested 200 μg LSD, the authors observed maximum plasma concentration values (range:1.2−1.9 ng/mL) 1.4 h after administration of 100 μg of LSD. Interestingly peak plasma concentrations (range: 2.6−4.0 ng/mL) was also reached 1.5 h after ingestion of 200 μg LSD by volunteers. The mean plasma half-life was 2.6 h. The subjective psychotropic effects lasted for 8.2 h (mean value) after ingestion of 100 μg LSD but such effects lasted for 11.6 h (mean value) after ingestion of 200 μg of LSD. Subjective peak effects were reached 2.8 and 2.5 h after administration of 100 and 200 μg of LSD, respectively [62].

Similar to active ingredients of magic mushroom and peyote cactus, LSD also binds to serotonin receptor. Studies have shown that serotonin 2A receptor ($5-HT_{2A}$) is critically involved in the formation of visual hallucinations and cognitive impairment in LSD-induced states. Moreover activation of $5-HT_{2A}$ receptor by LSD leads to a hippocampal-prefrontal cortex-mediated breakdown of inhibitory processing, which might subsequently promote the formation of LSD-induced visual imageries [63]. Duration of action of LSD is given in Table 33.2.

Analysis of Lysergic Acid Diethylamide and Its Metabolite

Immunoassays are commercially available as screening tests for the presence of LSD in urine. For example, cloned enzyme donor immunoassay (CEDIA) for LSD assay utilizes a monoclonal antibody capable of detecting LSD and the assay has a cutoff concentration of 0.5 ng/mL. CEDIA for LSD can be used either as a positive/negative screening mode or for obtaining semiquantitative values. Other immunoassays such as enzyme multiplied immunoassay technique (EMIT), kinetic interaction of micro-particles in solution (KIMS) are also commercially available. These assays can be easily adopted on automated analyzers commonly used in clinical laboratories. However, initial screening results must be confirmed by another analytical method such as LC-MS/MS or GC/MS because LSD immunoassays are subjected to interferences. Ritter et al. observed a high rate (4.2%) of positive results for LSD using EMIT assay in 1898 urine specimens that were submitted primarily from psychiatric patients for drugs of abuse testing. Specimens that tested positive by EMIT subsequently tested negative by two radioimmunoassays designed for detecting LSD in urine. In addition, randomly selected EMIT positive LSD urine specimens also tested negative when tested by GC/MS [64].

Several drugs such as ambroxol, fentanyl, and sertraline are known to cross-react with CEDIA assay for screening of LSD in urine [65]. In one report, the authors observed significant discrepancy between KIMS and CEDIA assay (cutoff concentration: 5 ng/mL LSD for both assays) for screening of the presence of LSD in urine specimens. However, the CEDIA assay showed overall a satisfactory total concordance of 92.3% with LC-MS assay. Two specimens which were positive for LSD in the CEDIA assay but negative using LC-MS assay contained fentanyl [66].

It is important to confirm initial positive test result for LSD by a chromatography-based confirmation method. Moreover chromatographic methods are also capable of detecting various LSD metabolites in urine. Iso-LSD, a diastereomer of LSD, is formed as an impurity during synthesis of LSD under basic condition. This compound can be used as a biomarker of LSD abuse. The main metabolite of LSD is 2-oxo-3-hydroxy-lysergic acid diethylamide (O-H-LSD) which is usually measured in urine for documentation of LSD abuse. Another metabolite N-desmethyl-LSD (nor-LSD) is also measured in urine. Minor metabolites present in urine after LSD abuse include nor-iso-LSD, lysergic acid ethylamide (LAE), lysergic acid ethyl-2-hydroxyethylamide (LEO), trioxylated-LSD, di-hydroxy-LSD, and glucuronides of 13- and 14-hydroxy-LSD [67].

Francom et al. described a protocol for analysis of LSD in urine at a concentration as low as 0.5 ng/mL using GC/MS. After addition of deuterated LSD as the internal standard to the urine, LSD if present along with the internal standard were extracted using n-butyl chloride at pH 8 followed by derivatization using N,O-bis(trimethylsilyl) trifluoroacetamide. The trimethyl silyl derivative of LSD as well as the internal standard was analyzed using a fused silica capillary column and electron impact ionization mass spectrometry operated in selected ion monitoring mode. The ions monitored for LSD were m/z 395.2 (the molecular ion of N-trimethylsilyl derivative of LSD), 293.2, and 268.2. The molecular ion of tri-deuterated LSD trimethylsilyl derivative (m/z 398.3) was also monitored. After administration of 7.5 μg of LSD in two human volunteers, in the urine of subject A, LSD was detected up to 8 h and in subject B up to 10 h after which that the LSD level was below the detection limit (0.5 ng/mL) of the assay. The maximum concentration of LSD detected in subject A was 0.9 ng/mL but in subject B maximum value was 1.6 ng/mL [68]. Paul et al. also described a protocol for analysis of LSD after converting LSD into trimethylsilyl derivative. The authors used LSD analog lysergic acid N-methyl-N-propylamide (LAMPA) as the internal standard. The procedure allowed detection of LSD concentrations as low as 20 pg/mL. Quantitation of LSD was linear over the concentration range 50−2000 pg/mL [69]. Sklerov et al. reported analysis of underivatized LSD in urine by gas chromatography−ion trap tandem mass spectrometry following a single-step solid-phase extraction of 5 mL urine [70].

LSD and its metabolites can be analyzed in both blood and urine. Cheung et al. developed a protocol for analysis of LSD and its metabolites (iso-LSD, nor-LSD, and O-H-LSD) in blood and urine using ultra-performance LC/MS/MS. The limit of detection in blood was 5 pg/mL (LSD and iso-LSD) or 10 ng/mL (nor-LSD and O-H-LSD). In urine, detection limit was 10 pg/mL for all analytes [71]. An LC-MS/MS method has also been developed for the simultaneous determination of LSD, iso-LSD, nor-LSD, and O-H-LSD in blood, urine, and vitreous humor samples. The limit of quantification was 20 pg/mL for both LSD and nor-LSD in blood, urine, and vitreous humor. No significant interfering substance or ion suppression was observed for LSD, iso-LSD, and nor-LSD analysis. The method was successfully applied to the forensic determination of postmortem LSD levels in the biological fluids of a multidrug abuser where LSD could be detected in vitreous humor [72].

Jang et al. described an LC-MS/MS method for analysis of LSD and its major metabolite O-H-LSD in both urine and hair using deuterated LSD (LSD-d_3) as the internal standard. The target analyte was extracted from hair using methanol at 38°C for 15 h while for urine (100 μL urine supplemented with 50 μL of 1 ng/mL LSD-d_3 and 300 μL of distilled water and 300 μL of 0.1 M sodium carbonate buffer; pH: 9.5) liquid–liquid extraction was performed using dichloromethane/isopropyl alcohol (1:1 by vol). The chromatographic separation was achieved by using a C-18 reverse phase column (2.1 × 100 mm, particle size 1.8 μm). The mass spectrometer was operated in electrospray ionization positive ion mode. The precursor ion monitored for LSD was m/z 324.1 and the product ions were m/z 229.9 and 226.0. For the O-H-LSD, the precursor ion monitored was m/z 356.2 while product ions were m/z 236.8 and 222.1. The precursor ion monitored for LSD-d_3 was m/z 327.3 while product ion was m/z 226. The limit of detection in hair was 0.25 pg/mg for LSD and 0.5 pg/mg for O-H-LSD. The detection limits in urine were 0.01 and 0.025 ng/mL for LSD and O-H-LSD, respectively. The authors applied this method to detect LSD in two legal cases of LSD ingestion. In these two cases, LSD concentrations in hair were 1.27 pg/mg and 0.95 pg/mg, respectively. No LSD metabolite in hair was detected in subject 2, but the concentration was below the limit of quantitation in subject 1. In urine sample collected 8 h after LSD ingestion in subject 1, the concentrations of LSD and O-H-LSD were 0.48 and 4.19 ng/Ml, respectively. For subject 2, urine specimen collected 3 h after ingestion showed LSD and O-H-LSD concentrations of 2.70 and 25.2 ng/mL. As expected, concentrations of O-H-LSD were significantly higher than LSD in urine [73]. There are other chromatography-based published methods for analysis of LSD and its metabolites in blood and urine. Analytical methods for analysis of LSD in biological matrix are given in Table 33.3.

ABUSE OF KHAT

Khat (*Catha edulis*) is a flowering evergreen shrub cultivated as a bush or small tree which is native to Ethiopia, East Africa, and the Southern Arabian Peninsula. The leaves of khat tree have an aromatic odor and slight sweet taste. Its young bud and tender leaves contain amphetamine-like psychoactive substance which upon chewing produces stimulation and euphoria. Historical evidences indicate that khat use existed in ancient Ethiopia (Abyssinia) as early as 13th century and khat leaves were introduced in Yemen in early 15th century. Therefore cultivation of khat plant started earlier than cultivation of coffee plants. Khat leaves can be chewed or leaves may be dried and powdered to make tea known as Abyssinia, African, or Arabian tea. The dried powder can also be consumed as a paste with honey [74]. Moreover, alcoholic extracts of khat are sold as "herbal high."

Khat use is a part of Yemen culture where approximately 90% males and 50% females chew khat leaves on a regular basis. Even 15%–20% children below the age of 12 chew khat plants. Global distribution of khat has increased significantly in last three decades where khat is available in both European and US market. Although in the past most consumers of khat in the Western countries were immigrants from East Africa or Middle-East, in last 10 years khat has emerged in illicit drug markets of the United States, Canada, New Zealand, Norway, and Hong Kong [75]. The World Health Organization considers khat as a drug of abuse. Khat is banned in European countries, United States, and Canada while it is lawful in Yemen, Somalia, and Ethiopia [76].

The fresh khat leaves contain many compounds including alkaloids, tannins, flavonoids, terpenoids, sterols, glycosides, amino acids, vitamins, and minerals. Of these chemicals, the most important group of compounds responsible for stimulating activity is phenyl-alkylamines which constituent cathinone (S(-)-alpha-aminopropiophenone), nor-pseudoephedrine, and norephedrine. Cathinone is a sympathomimetic amine and has structural similarity with amphetamine. Besides phenyl-alkylamines, the other major alkaloids found in khat leaves include cathedulins (62 compounds known in this calls isolated from fresh khat leaves) [77]. Chemical structure of cathinone is given in Fig. 33.1. Khat is known as "Herbal Ecstasy" because of its CNS stimulant effects similar to amphetamine.

Khat: Pharmacology and Toxicology

The active ingredients of khat are released during chewing. However, cathinone due to higher lipid solubility is more psychoactive than other components after khat abuse. Cathinone is degraded with time to cathine (inactive metabolite) and norephedrine by enzymatic actions. The degradation of cathinone is accelerated due to exposure to sunlight and heat. As a result after cultivation, khat plants are wrapped with banana leaves to reduce degradation. Usually 100–200 g of fresh leaves wrapped in banana leaves are chewed for getting stimulatory effect of cathinone present in khat leaves [78].

Cathinone is present in varying amounts (78–343 mg) in 100 g of khat leaves. The stimulatory effects of khat appear approximately 30 min after initiation of chewing khat plant and the effect may last up to 3 h. Approximately 90% of active compound present in leaves are released during chewing. The main route of absorption of cathinone is through oral mucosa (approximately 60% of absorption) and the rest of cathinone is absorbed through the gastrointestinal tract [79]. A typical session of khat chewing is similar to the effect of 5 mg amphetamine [80].

Halket et al. studied plasma concentration of cathinone after consuming khat (0.8 mg/kg). The mean peak cathinone level of 83 ng/mL was observed 1.5–3.5 h after chewing khat leaves [81]. Cathinone is extensively metabolized into cathine and norephedrine while only 7% of dose is excreted unchanged in urine [82]. Cardiovascular complications from cathinone use are similar to the complications observed with amphetamine abuse [83]. Moreover, increased blood pressure and increased heart rates are also observed in individuals abusing khat [84]. Ali et al., based on a study involving 934 patients who were khat chewers, concluded that khat chewing is associated with increased risk of stroke and death [85]. In general effects of khat on peripheral nervous system could include constipation, urine retention, and acute cardiovascular effects while effects on CNS could include increased alertness, dependence, and psychiatric symptoms [86].

Cathinone, the active ingredient of khat, is also sold in form of capsules and are also abused. Hagigat capsules (containing 200 mg cathinone produced illicitly) have been available in streets in Israel as a natural stimulant and aphrodisiac. Bentur et al. studied data of 34 consecutive patients (age 16–54) who consumed half to six Hagigat capsules. The major complications of use of cathinone were myocardial ischemia, pulmonary edema, and intracerebral hemorrhage. The authors concluded that exposure to illicitly synthesized cathinone may cause serious cardiovascular and neurological toxicity even in young subjects [87]. Active ingredient and duration of action of khat are listed in Table 33.2.

Mechanism of Action

Cathinone has similar CNS stimulatory and sympathomimetic effects characterized by increased blood pressure, heart rate, mydriasis, and hyperthermia similar to amphetamine [88]. All cathinones including natural cathinone and its synthetic derivatives (bath salts) are inhibitors of monoamine transporters, but their selectivity for serotonin receptors, norepinephrine transporter, and dopamine transporters vary substantially. The mechanism of cathinone on neurotransmission is to trigger presynaptic dopamine release and reduce the reuptake of dopamine which is similar to mechanism by which amphetamine exerts its pharmacological effects. Cathinone although binds to dopamine and serotonin receptors, it has the highest affinity for norepinephrine receptors. Moreover, cathinone has also been shown to induce serotonin release and inhibits its reuptake [89].

Laboratory Analysis of Cathinone

Abuse of bath salts (synthetic cathinone) is more widespread than khat abuse. Although cathinone cross-reacts with one antibody used in biochip array immunoassays (Randox Corporation), some of the chormatographic methods for analysis of bath salts also include cathinone as an analyte. Chromatographic-based assays offer more sensitivity and specificity than biochip array immunoassays for detection of bath salts in urine. In general three synthetic cathinone, methylone, mephedrone, and 3,4-methylenedioxy pyrovalerone (MDPV) currently make up approximately 98% of all bath salts abuse. In one published report, the authors developed a GC/MS protocol for analysis of seven amphetamine-type stimulants and 22 cathinones including three metabolites in urine. Using this protocol, cathinone if present in the urine specimen can also be analyzed [90]. Abuse of synthetic cathinones is a serious public health issue because fatalities have been reported in the literature due to abuse of bath salts. The death was attributed to hyperthermia, hypertension, cardiac arrest, and more in general to the classical serotonin syndrome [91]. For in-depth coverage of this topic please see Chapter 21: Review of bath salts on illicit drug market. Analytical methods for analysis of active ingredients of khat in biological fluids are listed in Table 33.3.

ABUSE OF VOLATILES AND GLUE

The general populations are exposed to many volatile compounds which are used in common household products including cleaning agents, adhesives, paints, and cosmetic products. The occupational exposures to various solvents are also troublesome but in most developed countries, such exposures are limited due to stringent regulations. However, there are certain volatile compounds which are listed under drug class of abused inhalants. These compounds usually have low vapor pressure and high volatility at room temperature that enables use of these solvents as euphorigenic inhaled agents. These volatiles classified as abused inhalants include aromatic hydrocarbons, aliphatic hydrocarbons, halocarbons, halogenated ethers, nitrous oxide, and alkyl nitrites [92].

Prevalence of Solvent Abuse

Various readily available household and office products are abused including glue, adhesives, nail polish, nail polish remover, cigarette lighter fluid, butane gas, air fresheners, deodorant, hair spray, pain relieving spray, typewriter correction fluid, paint thinners, paint removers, and a variety of other agents. These household and office products contain toxic solvent such as toluene (paint, spray paint, adhesives, paint thinner, shoe polish), acetone (nail polish remover, typewriter correction fluid, and markers), hexane (glue, rubber cement), chlorinated hydrocarbon (spot and grease removers), xylene (permanent markers), propane gas (gas to light the grill, spray paints), butane gas (lighter fluid, spray paint), and fluorocarbons (hair spray, analgesic spray, refrigerator coolant such as Freon). However, toluene is found in many household products such as glues and thinners which are widely abused due to psychoactive property of toluene. Chemical compositions of commonly abused inhalants are summarized in Table 33.4.

Although solvent (inhalant) abuse is common among adolescents not only in the United States but also worldwide, this problem is often overlooked. In the United States, approximately 20% of adolescents have tried inhalants at least once by the time they had reached eighth grade. Analysis of data from the US Poison Control Centers from 1993 to

TABLE 33.4 List of Commonly Abused Inhalants and Their Composition

Product	Solvent Found
Spray paint	Propane, butane, toluene, hydrocarbon, fluorocarbons
Hair spray	Propane, butane, fluorocarbons
Analgesic spray	Fluorocarbons, isopropyl alcohol
Paint thinner	Toluene, methyl chloride, methanol
Shoe polish	Toluene
Cleaner, disinfect	Xylene
Nail polish remover,	Acetone, toluene
Lighter fluid	Butane
Domestic fuel	Propane, butane, isooctane
Film cleaner, correction fluid	1,1,1-Trichloroethane, acetone
Adhesive glue	Toluene, xylene, ethylbenzene, hexane
Air freshener	Fluorocarbons
Rubber cement, marker	Toluene, hexane, methyl chloride, acetone
Spot remover, degreasers	Chlorinated hydrocarbons
Gasoline	Combination of aliphatic and aromatic hydrocarbons and other volatile organic compounds
Refrigerator fluid	Trichlorofluoromethane (Freon)
Metal cleaner	n-Propyl bromide

2008 showed 35,453 cases of toxicity due to abuse of inhalants. Prevalence was highest among children 12−17 years and peaked in 14 years old. Inhalant abuse was more common in boys (73.5%) than girls indicating that boys may pursue riskier usage behavior. More than 3400 different products were involved in inhalant abuses. Propellants, gasoline, and paint were the most frequent product categories. Butane, propane, and air fresheners had the highest fatality rates [93]. Although solvent abuse is common among adolescents, solvent abuse by pregnant woman has also been reported. Such abuse in pregnancy is associated with severe maternal and neonatal sequela [94].

Although drug abuse has been known for over 2000 years, abuse of solvents is a relatively new phenomenon. The first documented case was in 1946 when a boy being treated for psychotic symptoms at a hospital admitted to the attending physician that he chronically and uncontrollably inhaled gasoline for its intoxicating effect. A decade later the press from major American cities started reporting on intentional glue sniffing among youths. At that point medical professionals started to become aware that solvents may have euphorigenic properties and could possibly produce dependence. Solvent abuse may induce psychotic symptoms including hallucination and delusion [92].

Pharmacology and Toxicology of Abused Solvents

Inhalants are abused either by breathing directly from a container or by soaking a rag with the solvent and then placing it over the nose and mouth. Moreover abusers also pour the solvent in a plastic bag and then breathe fumes. Abuse of inhalant can produce euphoria like other abused drugs. When an abuser becomes hypoxic by rebreathing from a bag, the euphoric effect may even intensify. In general one of the commonly abused volatile toluene can be detected by humans at a very low concentration of 11 ppm (parts per million) and low-detectable concentration is also likely for other abused solvents. However, solvent abusers who intentionally inhale volatiles for intoxication usually expose themselves for approximately 15 min to extremely high vapor concentrations up to 15,000 ppm [92].

Toluene has the well-documented pharmacological and toxicological profile of all solvents studied. Inhalation of toluene vapor causes euphoria followed by CNS depression. Toluene is metabolized by the liver cytochrome P-450 mixed function oxidase into benzoic acid and hippuric acid which are excreted in urine. The most common clinical presentation of toluene intoxication is muscular weakness due to hypokalemia which may cause respiratory depression and may be life-threatening. Renal tubular acidosis may also occur due to toluene poisoning. Liver injury and rhabdomyolysis is also common feature of toluene poisoning [95].

Toluene is an aromatic hydrocarbon found in gasoline, acrylic paints, varnishes, paint thinners, adhesive, and shoe polish. Fatalities due to abuse of products containing toluene have been reported. A middle-aged man was found dead in his house after varnish application by spray. The blood test revealed the presence of toluene and the cause of death was established as cardiopulmonary failure caused by toluene [96]. The toluene concentration over 2000 ppm is considered dangerous to life and health. Argo et al. reported death of a 15-year-old boy due to ingestion of varnishes diluting solvent after failing an examination. Volatile organic compounds such as toluene and xylene were detected in gastric content, blood, liver, and kidney [97]. In another report, the authors described two cases of fatal intoxication with toluene due to glue sniffing. Using gas chromatography with flame ionization detector and GC/MS with headspace method, toluene was detected in biological specimens. Highest concentrations were observed in liver and brain in both cases (13.82−20.97 μg/g). Based on this data, the cause of death in both cases was determined to be due to toluene poisoning [98].

In one study the authors investigated 318, 939 exposures and concluded that exposure to hydrocarbons which are systematically absorbed and have low viscosity such as benzene, toluene, xylene, halogenated hydrocarbon, kerosene, and lamp oil caused the highest hazard values. Moreover, the risk associated with hydrocarbon exposure is greater among children and adolescents [99]. Fatalities may occur from abusing any solvent. Seetohul reported a case of death from abusing diethyl ether [100]. Sugie et al. reported three cases of sudden death due to inhalation of portable cooking stove fuel (case 1), cigarette lighter (case 2), and liquefied petroleum gas (case 3). Butane was found in tissues of cases 1 and 2 while propane was the major compound present in case 3 [101]. Alunni et al. reported death from butane abuse in two teenagers. An 11-year-old boy died after inhaling deodorant aerosol. The toxicological analysis detected the presence of butane at concentrations of 22 mg/L in the peripheral blood, 97 mg/L in the gastric fluid, and 8 mg/kg in the lung. The authors also detected limonene (gastric fluid) and decamethylcyclopentasiloxane (gastric fluid and fatty tissue). Identification of these chemicals may help in identifying particular brand of deodorant because these compounds are not present in all brands. A 16-year-old girl with unregulated type 1 diabetes also died after deodorant abuse. Butane was identified in postmortem peripheral blood at a concentration of 18.1 and 27.3 mg/L in the gastric fluid. One limonene was identified in gastric fluid. It emerged that the victim has suffered the seizure after inhaling from a tissue soaked in deodorant [102].

Solvent abusers often present themselves with nonspecific symptoms, but long-term abusers may come to the hospital with a wide range of neuropsychiatric symptoms. The exact mechanism of action for the volatiles and inhalants has

TABLE 33.5 Adverse Effects of Solvent and Inhalant Abuse

Toxicity	Agents
Cardiotoxicity	Inhalants such as fluorocarbons, gasoline, benzene, toluene, butane, propane (acute effect of inhalants on the heart may be fatal) are cardiotoxic.
Pulmonary	Some hydrocarbons may cause chemical pneumonitis.
Renal	Toluene, hydrocarbons.
Hepatic	Halogenated hydrocarbons such as carbon tetrachloride, trichloroethane, toluene are liver toxins.
Teratogenic	Inhalants due to lipophilic nature cross the placenta.
Neurologic	Various inhalants such as toluene, hexane, methyl isobutyl ketone.

not been completely understood yet. It is possible that volatiles in general produce a slowing of axonal ion channel transport by increasing fluidity of membranes. Another possibility is that volatiles potentiate hyperpolarization of gamma-aminobutyric acid receptors (GABA receptors). At the primary site of action of inhalants due to their lipophilic nature can quickly produce CNS depression which results in slurred speech, diplopia, ataxia, visual hallucination, and disorientation. Further CNS depression can cause respiratory arrest, seizures, and coma. The most serious consequence of solvent abuse is death which may occur secondary to aspiration or asphyxia. Nearly 50% of deaths from solvent abuse are due to sudden sniffing death syndrome. When an acutely intoxicated abuser is startled, a burst of catecholamines may trigger ventricular fibrillation [103]. Adverse effects of solvent abuse are listed in Table 33.5.

Laboratory Detection of Solvent Abuse

Laboratory tests are helpful for diagnosis of solvent abuse in a suspected patient. As a result, diagnosis is based on clinical presentation and high index of suspicion. In general complete blood count, electrolyte, acid−base assessment, and hepatic and renal chemistry profiles are ordered. Blood alcohol should be ordered as many solvent abusers also abuse alcohol. Urine drug screen should also be ordered as illicit drugs are often detected in solvent abusers. Treatment is mostly supportive. For forensic investigation, headspace gas chromatography is useful to identify individual solvent. Detection can be carried out using flame ionization detector or mass spectrometry [104]. Antonucci et al. described a protocol for analysis of 10 volatile compounds (benzene, toluene, ethylbenzene, ortho-xylene, meta-xylene, para-xylene, methyl tert-butyl ether, ethyl tert-butyl ether, 2-methyl-2-butyl methyl ether, and diisopropyl ether) in urine of children using headspace gas chromatography/mass spectrometry [105].

CONCLUSIONS

Abuse of magic mushroom, peyote cactus, khat, LSD, and solvent are problematic. Psilocybin and psilocin are active components of magic mushroom which can be identified by various chromatographic methods including GC/MS and LC-MS/MS. The active component of peyote cactus is mescaline which cross-reacts with some amphetamine/methamphetamine immunoassay. A specialized immunoassay marketed by the Randox Corporation (drugs of abuse array V biochip immunoassay) is also useful to detect the presence of mescaline in urine because a specific antibody is immobilized on the biochip capable of detecting mescaline. Chromatographic methods are also available for detection of mescaline in various biological matrixes. The active component of khat plant is cathinone which cross-reacts with various amphetamine/methamphetamine immunoassays. Synthetic derivatives of cathinone are sold as bath salts and some of these compounds can be detected by the drugs of abuse array V biochip immunoassay. Chromatographic methods are also available. Although immunoassays are available for screening of the presence of LSD in urine, false-positive tests may occur. Therefore chromatographic methods must be used to confirm the presence of LSD in urine specimens tested positive for LSD using an immunoassay. Moreover, chromatographic methods are also applicable for analysis of LSD in blood and other body fluids. For detecting solvent abuse, headspace gas chromatography coupled with flame ionization detector or mass spectrometry is the preferred analytical method.

REFERENCES

[1] Carod-Artal FJ. Hallucinogenic drugs in pre-Colombian Mesoamerican culture. Neurologia 2015;30:42–9.
[2] Tittarelli R, Mannocchi G, Panato F, Romolo FS. Recreational use, analysis and toxicity of tryptamines. Curr Neuropharmacol 2015;13:26–46.
[3] Madhok M. Amanita bisporgera: Ingestion and death from mistaken identity. Minn Med 2007;90:48–50.
[4] Madhok M, Scalzo AJ, Blume CM, Neuschwander-Tetri BA, et al. Amanita biosporigera ingestion: mistaken identity, dose related toxicity, and improvement despite severe hepatotoxicity. Pediatr Emerg Care 2006;22:177–80.
[5] Hallock RM, Dean A, Knecht ZA, Spencer J, Taverna E. A survey of hallucinogenic mushroom use, factors, related to usage and perceptions of use among college students. Drug Alcohol Depend 2013;130:245–8.
[6] Krebs TS, Johansen PO. Over 30 million psychedelic users in the United States. F1000 Research 2013;2:98.
[7] Riley SC, Blackman G. Between prohibitions: patterns and meaning of magic mushroom use in the UK. Substance Use & Misuse 2008;43:55–71.
[8] Reingardiene D, Vilcinskaite J, Lazauskas R. Hallucinogen mushrooms. Medicine (Kaunas) 2005;41:1067–70 [article in Lithuanian].
[9] Musshoff F, Madea B, Beike J. Hallucinogenic mushrooms on the German market-simple instructions for examination and identification. Forensic Sci Int 2000;113:389–95.
[10] Tsujikawa K, Kannamori Y, Iwata Y, Ohmae Y, et al. Morphological and chemical analysis of magic mushrooms in Japan. Forensic Sci Int 2003;138:85–90.
[11] Anastos N, Barnett NW, Lewis SW, Gathergood N, et al. Determination of psilocin and psilocybin using flow injection and tris (2,2'-bipyridyl) ruthenium (II) chemiluminescence detection respectively. Talanta 2005;67:354–9.
[12] Hasler F, Grimberg U, Benz MA, Huber T, et al. Acute psychological and psychological effects of psilocybin in healthy humans: a double blind, placebo controlled dose-effect study. Psychopharmacology (Berlin) 2004;172:145–56.
[13] Hasler F, Boiurquin D, Brenneisen R, Bar T, et al. Determination of psilocin and 4-hydroxyindole -3-acetic acid in plasma by HPLC-ECD and pharmacokinetic profile of oral and intravenous psilocybin in man. Pharm Acta Helv. 1997;72:175–84.
[14] Passie T, Seifert J, Schneider U, Emrich EM. The pharmacology of psilocybin. Addict Biol 2002;7:357–64.
[15] Sticht G, Kaferstein H. Detection of psilocin in body fluids. Forensic Sci Int 2000;113:403–7.
[16] Tyls F, Palenicek T, Horacek J. Psilocybin-summary of knowledge and new perspectives. Eur Neuropsychopharmacol 2014;24:342–56.
[17] Satora L, Goszcz H, Ciszowski K. Poisonings resulting from the ingestion of magic mushroom in Krakow. Przegl Lek 2005;62:394–6.
[18] Borowiak KS, Ciechanowski K, Waloszczyk P. Psilocybin mushroom (*Psilocybe semilanceata*) intoxication with myocardial infarction. J Toxicol Clin Toxicol 1998;36:47–9.
[19] Raff E, Halloran PF, Kjellstrand CM. Renal failure after eating "magic" mushrooms. Can Med Asso J 1992;147:1339–441.
[20] Bickel M, Ditting T, Watz H, Roesler A, et al. Severe rhabdomyolysis, acute renal failure and posterior encephalopathy after magic mushroom abuse. Eur J Emerg Med 2005;12:306–8.
[21] Lim TH, Wasywich CA, Ruygrok PN. A fatal case of magic mushroom ingestion in a heart transplant recipient. Intern Med J 2012;42:1268–9.
[22] Gonomori K, Yoshioka N. The examination of mushroom poisonings at Akita University. Leg Med (Tokyo) 2003;5(Supply1):S83–6.
[23] Beck O, Helander A, Karlson-Stiber C, Stephansson N. Presence of phenylethylamine in hallucinogenic Psilocybe mushroom: possible role in adverse reactions. J Anal Toxicol 1998;22:45–9.
[24] Dinis-Oliveira RJ. Metabolism of psilocybin and psilocin: clinical and forensic toxicological relevance. Drug Metab Rev 2017;49:84–91.
[25] Stebelska K. Fungal hallucinogens psilocin, ibotenic acid and muscimol: analytical methods and biological activities. Ther Drug Monit 2013;35:420–42.
[26] Tyls F, Palenicek T, Kaderabek L, Lipski M, Kubesova A, Horacek J. Sex differences and serotonergic mechanisms in the behavioral effects of psilocin. Behav Pharmacol 2016;27:309–20.
[27] de Veen BT, Schellekens AF, Verheij MM, Homberg JR. Psilocybin for treating substance use disorders? Expert Rev Neurother 2017;17:2013–212.
[28] Sellers EM. Psilocybin: good trip or bad trip. Clin Pharmacol Ther 2017;102:580–4.
[29] Patra S. Return of psychedelics: psilocybin for treatment of resistant depression. Asian J Psychiatr 2016;24:51–2.
[30] McClintock RL, watts DJ, Melanson S. Unrecognized magic mushroom abuse in a 28 year-old man. Am J Emerg Med 2008;26(972):e3–4.
[31] Tiscione NB, Miller MI. Psilocin identified in a DUID investigation. J Anal Toxicol 2006;30:342–5.
[32] Grieshaber AF, Moore KA, Levine B. The detection of psilocin in human urine. J Forensic Sci 2001;46:627–30.
[33] Kamata T, Nishikawa M, Katagi M, Tsuchihashi H. Liquid chromatography-mass spectrometric and liquid chromatographic-tandem mass spectrometric determination of hallucinogenic indoles psilocin and psilocybin in magic mushroom samples. J Forensic Sci 2005;50:336–40.
[34] Anastos N, Lewis SW, Barnett NW, Sims DN. The determination of psilocin and psilocybin in hallucinogenic mushrooms by HPLC utilizing a dual reagent acidic potassium permanganate and tris (2,2;-bipyridyl) ruthenium (II) chemiluminescent detection system. J Forensic Sci 2006;51:45–51.
[35] Martin R, Schurenkamp J, Pfeiffer H, Kohler H. A validated method for quantitation of psilocin in plasma by LC-MS/MS and study of stability. Int J Legal Med 2012;126:845–9.
[36] Martin R, Schurenkamp J, Gasse A, Pfeiffer H, et al. Determination of psilocin, bufotenine, LSD and its metabolites in serum and urine by SPE-LC/MS/MS. Int J Legal Med 2013;127:593–601.
[37] Martin R, Schurenkamp J, Gasse A, Pfeiffer H, et al. Analysis of psilocin, bufotenine and LSD in hair. J Anal Toxicol 2015;39:126–9.
[38] El-Seedi HR, DeSmet PA, Beck O, Possnert G, Bruhn JG. Prehistoric peyote use: alkaloids and radiocarbon dating of archaeological specimens of Lophophora from Texas. J Ethnopharmacol 2005;101:238–42.

[39] Halpern JH, Sherwood AR, Hudson JI, Yugerlum-Todd D, Pope Jr. HG. Psychological and cognitive effects of long term peyote use among Native Americans. Biol Psychiatry 2005;58:624–31.
[40] Carstairs SD, Cantrell FL. Peyote and mescaline exposures: a 12 year review of a statewide poison center database. Clin Toxicol (Phila) 2010;48:350–3.
[41] Prue B. Prevalence of reported peyote use 1985–2020 effects of the American Indian Religious Freedom Act of 1994. Am J Addict 2014;23:156–61.
[42] Bronstein AC, Spyker DA, Cantilena LR, Green JL, et al. Annual report of the American Association of Poison Control Center's National Poison Data System (NPDS): 25th annual report. Clin Toxicol (Phila) 2008;46:927–1057.
[43] Halpern JH. Hallucinogens and dissociative agents naturally growing in the United States. Pharmacol Ther 2004;102:131–8.
[44] Nichols DE. Hallucinogens. Pharmacol Ther 2004;101:131–81.
[45] Laing RR, Beyerstein BL. Hallucinogens: A forensic drug handbook. MA: Academic Press,; 2003.
[46] Charalampous KD, walker KE, Kinross-Weight J. Metabolite fate of mescaline in man. Psychopharmacologia 1966;9:48–63.
[47] Musacchio JM, Goldstein M. The metabolism of mescaline 14-C in rats. Biochem Pharmacol 1967;16:963–70.
[48] Gambelunghe C, Marsili R, Aroni K, Bacci M, Rossi R. GC-MS and GC-MS/MS in PCI mode determination of mescaline in peyote cactus tea and biological matrices. J Forensic Sci 2013;58:270–8.
[49] Gilmore HT. Peyote use during pregnancy. S D J Med 2001;54:27–9.
[50] Reynolds PC, Jindrich EJ. A mescaline associated fatality. J Anal Toxicol 1985;9:183–4.
[51] Henry JL, Epley J, Rohrig TP. The analysis and distribution of mescaline in postmortem tissues. J Anal Toxicol 2003;27:381–2.
[52] Hashimoto H, Clyde VJ, Parko KL. Botulism from peyote [letter]. N Eng J Med 1998;339:203–4.
[53] Nolte KB, Zumwalt RE. Fetal peyote ingestion associated with Mallory-Weiss lacerations. West J Med 1999;170:328.
[54] Monte AP, Waldman SR, Marona-Lewicka D, Wainscott DB, Nelson DL, Sander-Bush E, et al. Dihydrobenzofuran analogues for hallucination-4. Mescaline derivatives. J Med Chem 1997;40:2997–3008.
[55] Karaki S, Becamel C, Murat S, Mannoury la Cour C. Quantitative phospho proteomics unravels biased phosphorylation of serotonin 2 A receptor at Ser280 by hallucinogenic versus nonhallucinogenic agonists. Mol Cell Proteomics 2014;13:1273–85.
[56] Kovacic P, Somanathan R. Novel, unifying mechanism for mescaline in the central nervous system: electrochemistry, catechol redox metabolite, receptor, cell signalling and structure activity relationship. Oxid Med Cell Longev 2009;2:181–90.
[57] Battle D, Barnes AJ, Castaneto MS, Martin TM, Klette KL, Huestis MA. Urine mescaline screening with a biochip array immunoassay and quantification by gas chromatography-mass spectrometry. Ther Drug Monit 2015;37:805–11.
[58] Bjornstad K, Helander A, Beck O. Development and clinical applications of and LC-MS-MS method for mescaline in urine. J Anal Toxicol 2008;32:227–31.
[59] Hoffer A. LSD: A review of its present status. Clin Pharamcol Ther 1965;6:183–225.
[60] Upshall DG, Wailling DG. The determination of LSD in human plasma following oral administration. Clin Chim Acta 1972;36:67–73.
[61] Dolder PC, Schmid Y, Haschke M, Rentsch KM, et al. Pharmacokinetics and concentration effect relationship of oral LSD in humans. Int J Neuropsychopharmacol 2015;19(1):1–7.
[62] Dodler PC, Schmid Y, Steuer AE, Kraemer T, et al. Pharmacokinetics and pharmacodynamics of lysergic acid diethylamide in healthy subjects. Clin Pharmacokinet 2017;56:1219–30.
[63] Schmidt A, Muller F, Lenz C, Dodler C, et al. Acute LSD effects on response inhibition neural networks. Psychol Med 2017; Oct 2 [e-pub ahead of print].
[64] Ritter D, Cortese CM, Edwards LC, Barr JL, et al. Interference with testing for lysergic acid diethylamide. Clin Chem 1997;43:635–7.
[65] Citterio-Quentin A, Seidel E, Ramuz L, Parant F, et al. LSD screening in urine performed by CEDIA® LSD assay: positive interference with sertraline. J Anal Toxicol 2012;36:289–90.
[66] Gawinecka J, Muller DM, von Eckardstein A, Saleh L. Pitfalls of LSD screening assays: comparison of KIMS and CEDIA immunoassays with LC-MS. Clin Chem Lab Med 2017;55:e10–12.
[67] Canezin J, Cailleux A, Turcant A, Le Bouil A, et al. Determination of LSD and its metabolites in human biological fluids by high-performance liquid chromatography with electrospray tandem mass spectrometry. J Chromatogr B Biomed Sci Appl. 2001;765:15–27.
[68] Francom P, Andrenyak D, Lim HK, Bridges RR, et al. Determination of LSD in urine by capillary column gas chromatography and electron impact mass spectrometry. J Anal Toxicol 1988;12:1–8.
[69] Paul BD, Mitchell JM, Burbage R, Moy M, et al. Gas chromatography-electron impact mass fragmentometric determination of lysergic acid diethylamide in urine. J Chromatogr 1990;529:103–12.
[70] Sklerov JH, Kalasinsky KS, Ehorn CA. Detection of lysergic acid diethylamide in urine by gas chromatography-ion-trap tandem mass spectrometry. J Anal Toxicol 1999;23:474–8.
[71] Chung A, Hudson J, McKay G. Validated ultra-performance liquid chromatography-tandem mass spectrometry method for analyzing LSD, iso-LSD, nor-LSD and O-H-LSD in blood and urine. J Anal Toxicol 2009;33:253–9.
[72] Favretto D, Frison G, Maietti S, Ferrara SD. LC-ESI-MS/MS on an ion trap for the determination of LSD, iso-LSD, nor-LSD and 2-oxo-3-hydroxy-LSD in blood, urine and vitreous humor. Int J Legal Med 2007;124:259–65.
[73] Jang M, Kim J, Han I, Yang W. Simultaneous determination of LSD and 2-oxo-3-hydroxy LSD in hair and urine by LC-MS/MS and its application to forensic cases. J Pharm Biomed Anal 2015;115:138–43.
[74] Alles GA, Fairchild MD, Jensen M. Chemical pharmacology of Catha edulis. J Med Pharm Chem 1961;3:323–52.

[75] Anderson D, Beckerleg S, Hailu D. Regulating Khat dilemmas and opportunities for the international drug control system. Int J Drug Policy 2009;20:509—13.
[76] El-Menyar A, Mekkodathil A, Al-Thani H, Al-Motarreb A. Khat use: history and heart failure. Oman Med J 2015;30:77—82.
[77] Sikiru L. Flowers of paradise (Khat: catha edulis): psychosocial, health and sports perspective. Afr Health Sci 2012;24:69—83.
[78] Yousef G, Huq Z, Lambert T. Khat chewing a cause of psychosis. Br J Hosp Med 1995;54:322—6.
[79] Feyissa AM, Kelly JP. A review of the neuropharmacological properties of khat. Prog Neuropsychopharmacol Biol Psychiatry 2008;32:1147—66.
[80] Ishraq D, Jiri S. Khat habit and its health effect. A natural amphetamine. Biomed Papers 2004;148:11—15.
[81] Halket JM, Karasu Z, Murray-Lyon IM. Plasma cathinone levels after chewing khat leaves (Catha edulis Forsk). J Ethnopharmacol 1995;49:111—1113.
[82] Tonnes SW, Kauert GF. Excretion and detection of cathinone, cathine and phenylpropanolamine after khat chewing. Clin Chem 2002;48:1715—19.
[83] Kuczkowski KM. Herbal ecstasy: cardiovascular complications of Khat chewing in pregnancy. Acta Anaesthesiol Belg 2005;56:19—21.
[84] Al-Motarreb A, Baker K, Broadley KJ. Khat: pharmacological and medical aspects and its social use in Yemen. Phytother Res 2002;16:403—13.
[85] Ali WM, Zubaid M, Al-Motarreb A, Singh R, Al-shereigi SZ, Shehab A, et al. Association of khat chewing with increased risk of stroke and death in patients presenting with acute coronary syndrome. Mayo Clin Proc 2010;85:974—80.
[86] Nasi Tajure W. Chemistry, pharmacology and toxicology of Khat (catha edulis forsk). Addict Health 2011;3:137—9.
[87] Bentur Y, Bloom-Krasik A, Raikhlin-Eisenkraft B. Illicit cathinone (Hagigat) poisoning. Clin Toxicol (Phila) 2008;46:206—10.
[88] Valente MJ, Guedes de Pinho P, de Lourdes Bastro M, Carvalho F, Cavalho M. Khat and synthetic cathinones: a review. Arch Toxicol 2014;88:15—45.
[89] Feng LY, Battulga A, Han E, Chung H. New psychoactive substances of natural origin: a brief review. J Food Drug Anal 2017;25:461—71.
[90] Alsenedi KA, Morrison C. Determination and long-term stability of twenty nine cathinones and amphetamine-type stimulants (ATS) in urine using gas chromatography-mass spectrometry. J Chromatogr B Analyt Technol Biomed Life Sci 2018;1076:91—102.
[91] Zaami S, Giorgetti R, Pichini S, Pantano F, et al. Synthetic cathinones related fatalities: an update. Eur Rev Med Pharmacol Sci 2018;22:268—74.
[92] Beckley JT, Woodward JJ. Volatile solvents as drugs of abuse: focus on the cortico-mesolimbic system. Neuropsychopharmacology 2013;38:2555—67.
[93] Marsolek MR, White NC, Litovitz TL. Inhalant abuse monitoring trends by using poison control data 1993—2008. Pediatrics 2010;125:906—13.
[94] Scheeres JJ, Chudley AE. Solvent abuse in pregnancy: a perinatal prospective. J Obstet Gynaecol Can 2002;24:22—6.
[95] Camara-Lemarroy CR, Rodriguez-Gutierrez R, Monreal-Robles R, Gonzalez-Gonzalez JG. Acute toluene intoxication —clinical presentation, management and prognosis: a prospective observational study. BMC Emerg Med 2015;15:19.
[96] Prayulsatein W. J Med Assoc Thai 2013;96:1242—4.
[97] Argo A, Bongiorno D, Bonifacio A, Pernice V, et al. A fatal case of a paint thinner ingestion: comparison between toxicological and histological findings. Am J Forensic Med Pathol 2010;31:186—91.
[98] Diurendic-Brenesel M, Stojilkovic G, Pilia V. Fatal intoxication with toluene due to inhalation of glue. J Forensic Sci 2016;61:875—8.
[99] Cobaugh DJ, Seger DL, Krenzelok EP. Hydrocarbon toxicity: an analysis of AAPCC TESS data. Przegl Lek 2007;64:194—6.
[100] Seetohul LN, De Paoli G, Maskell DO. Volatile substance abuse: fatal overdose with dimethyl ether. J Anal Toxicol 2015;39:415.
[101] Sugie H, Sasaki C, Hashimoto C, Takeshita H, et al. Three cases of sudden death due to butane or propane gas inhalation: analysis of tissues for gas components. Forensic Sci Int 2004;143:211—14.
[102] Alunni V, Gaillard Y, Castier F, Piercecchi-Marti MD, et al. Death from butane inhalation abuse in teenagers: two new case studies and review of literature. J Forensi Sci 2018;63:330—5.
[103] Anderson CE, Loomis GA. Recognition and prevention of inhalant abuse. Am Fam Physicians 2003;68:869—74.
[104] Da Broi U, Colatutto A, Sala P, Desinan L. Medical legal investigation into sudden deaths linked with trichloroethylene. J Forensic Leg Med 2015;34:81—7.
[105] Antonucci A, Vitali M, Avino P, Manigrasso M, et al. Sensitive multiresidue method by HS-SPME/GC-MS for 10 volatile organic compounds in urine matrix: a new tool for biomonitoring studies on children. Anal Bioanal Chem 2016;408:5789—800.

Chapter 34

Performance Enhancing Drugs in Sports

Brian D. Ahrens and Anthony W. Butch
UCLA Olympic Analytical Laboratory and Department of Pathology & Laboratory Medicine, David Geffen School of Medicine at UCLA, Los Angeles, CA, United States

INTRODUCTION

Doping is an integral part of competitive sports and dates back to ancient Greek Olympic events [1]. Over the years, the word doping has come to mean the use of drugs, ergogenic substances, or artificial methods for the purpose of improving athletic performance. Around 800 BC, Greek Olympic athletes and Roman gladiators consumed mushrooms, plants, and herbs in an attempt to gain a competitive edge and mask pain [2]. The practice of doping continued to evolve and by the 1800s athletes participating in endurance events such as cycling used a mixture of strychnine, caffeine, cocaine, and alcohol to gain a competitive edge [3].

Modern Olympic Games made their debut in 1896 and additional drugs such as ephedrine and heroin were used to increase speed and endurance, and allow athletes to compete while injured [1]. The practice of doping quickly spread throughout sporting events, with each athlete and coach having their own unique magic mixture. Cocaine and heroin use diminished in the 1920s when they became available only by prescription. In the late 1930s, amphetamines replaced strychnine as the favored stimulant. Anabolic steroids were first used by Soviet Olympic athletes in the mid-1950s (substances similar to the hormone testosterone) to promote muscle mass and power in weightlifting and bodybuilding events. The use of anabolic steroids rapidly spread to other sporting events and was widely used by athletes from many countries. After the collapse of the German Democratic Republic in 1990, documents were recovered describing an extensive program of steroid administration to young athletes to enhance performance in a wide variety of sporting events [4]. Today, it is generally believed that anabolic steroids along with other performance enhancing substances are still widely used by Olympic athletes.

The commercialization of Olympic Games has nurtured the "win at all costs" attitude and has promoted the use of doping agents by Olympic athletes. The generous financial rewards and celebrity status attached to winners have placed medical and ethical issues secondary to winning. No longer is competing considered the most important aspect of Olympic Games. It is regrettable that this attitude has now spread to professional athletes, college athletes, and even high-school athletes, along with noncompeting males and females seeking to obtain a "ripped" body appearance.

THE ANTI-DOPING MOVEMENT

The International Amateur Athletic Federation was the first Sport Federation to prohibit the use of stimulants in 1928 [5]. The ban on doping quickly spread to other International Federations. However, this ban did not deter athletes from doping because drug testing was not available at that time. The invention of synthetic hormones and the availability of anabolic steroids in the 1950s made the problem worse, and made steroid use more pervasive within the international sporting community. The death of a Danish cyclist during the 1960 Olympic Games in Rome due to amphetamine use heightened public awareness regarding these agents and their potential harmful side effects [5]. Two years later, the Council of Europe published a list of banned substances that included narcotics, stimulants, alkaloids, and some hormones. France, Belgium, and Greece soon adopted national anti-doping legislation. In 1966, doping tests were introduced by the International Cycling Union and Fédération Internationale de Football Association at their World Championships [1]. The widely publicized death of Tommy Simpson in the Tour de France in 1967 due to the illegal use of amphetamines prompted the International Olympic Committee (IOC) to establish a Medical Commission to deal with the problem of sports doping [6]. The IOC created the first prohibited list of substances for Olympic Games that included stimulants, sympathomimetic amines, and narcotic analgesics [1]. Drug testing was subsequently introduced at the Winter Olympic Games in Grenoble and the Summer Olympic Games in Mexico City in 1968.

Despite the adoption of anti-doping legislation and routine drug testing by International Sports Federations, doping continued to be problematic. This was due to inadequate drug testing methods and the fact that athletes quickly learned how to beat early drug tests that could not detect low urinary drug concentrations. Another problem was that specific drugs were abused before testing methods were developed. For example, anabolic steroids were widely used in the early 1970s at a time when reliable testing methods were not available. Techniques to detect anabolic steroids were developed a few years later and this class of compounds was added to the prohibited list in 1975. Erythropoietin (EPO) and human growth hormone (hGH) are other examples of substances that were abused prior to the introduction of routine testing methods.

Public attention to doping cases and scandals continued to mount in the 1980s and 1990s. Ben Johnson tested positive for an anabolic steroid at the Seoul Olympic Games in 1988 and was stripped of his gold medal. In 1998, a police raid during the Tour de France discovered a huge supply of prohibited performance enhancing drugs, drawing further attention to the worldwide doping problem. This scandal prompted the IOC to convene a World Conference on Doping in Sports. The outcome of this conference was the creation of the World Anti-Doping Agency (WADA) in 1999. The mission of WADA is to combat doping in sports and provide unified standards for doping control. At the time, Sports Federations and governments often had divergent anti-doping policies resulting in a wide variety of athlete sanctions that did not always withstand legal challenge. The WADA is designed to work independently of the IOC, governments, and Sports Federations, but is structured on equal representation by stakeholders. The WADA has developed a single set of rules, the *Code*, which deals with all aspects of sports doping [7]. Each national anti-doping organization is responsible for developing a drug testing program in accordance with the *Code*, and must have testing performed by WADA-accredited laboratories. There are 32 WADA-accredited laboratories worldwide, with two located in the United States [8]. Each WADA laboratory must be able to detect and identify substances on the prohibited list as mandated by the World Anti-Doping Code International Standard for Laboratories (ISL).

WADA PROHIBITED LIST

The violation of any of the anti-doping rules established by the WADA is considered a doping offense. Not only does this include the presence of a prohibited substance or its metabolite(s) in an athlete's body, but also includes the attempted use of a prohibited substance or a prohibited method (blood doping, gene doping, etc.). Refusing to submit a specimen, not being available for testing, tampering or attempting to tamper with the doping process, possessing or trafficking of prohibited substances/methods, and administration or attempted administration of a prohibited substance or prohibited method are also considered doping violations.

The WADA prohibited list is organized by general categories of substances (stimulants, diuretics, etc.) and methods (Table 34.1). The prohibited list is updated at least annually and new classes and individual substances are being continually added to the list. For instance, LGD-4033 and RAD140 were added as additional examples of selective androgen receptor modulators (SARMS) within the prohibited class anabolic agents while ethanol was removed from the 2018 WADA prohibited list [9]. Not all classes of compounds are banned throughout the year. For example, stimulants, narcotics, cannabinoids, and glucocorticosteroid agents are only prohibited during sport competitions (Table 34.1). Other substances such as beta-blockers are only prohibited in particular sports such as archery, golf, and shooting, and in some cases only during competition.

In 2017, 322,050 urine specimens (Olympic and non-Olympic sports) were tested by WADA-accredited laboratories, with 4596 (1.43%) samples testing positive for a prohibited substance (adverse analytical finding) [10]. Anabolic agents were the most commonly detected group of substances, followed by diuretics, stimulants, and hormone and metabolic modulators (Table 34.2). The remaining prohibited substances and methods accounted for <19% of all reported adverse analytical findings (Table 34.2).

OVERVIEW OF THE TESTING PROCEDURE

The process begins with supervised collection of urine. A witnessed collection helps ensure that the specimen is not adulterated or substituted with fake or drug-free urine. The urine is divided into an "A" and "B" bottle that is securely sealed (tamper-evident tape is used to detect opening of the bottles), packaged, certified by both parties (collector and athlete), and is then shipped to a WADA-accredited testing laboratory. Specific steps in the testing sequence are illustrated in Fig. 34.1. Once delivered to the laboratory, the urine specimens are unpacked and inspected for leakage and

TABLE 34.1 WADA 2018 List of Prohibited Substances and Methods

Prohibited Substances

1. Anabolic agents
   a. Exogenous anabolic steroids
   b. Endogenous anabolic steroids
   c. Other anabolic agents

2. Hormones and related substances
   a. Erythropoiesis stimulating agents
   b. Peptide hormones and modulators
   c. Growth factors and modulators

3. Beta-2 agonists

4. Hormone and metabolic modulators
   a. Aromatase inhibitors
   b. Selective estrogen receptor modulators
   c. Other anti-estrogenic substances
   d. Agents modifying myostatin function
   e. insulins and metabolic modulators

5. Diuretics and other masking agents

6. Stimulants (in competition only)

7. Narcotics (in competition only)

8. Cannabinoids (in competition only)

9. Glucocorticosteroids (in competition only)

10. Beta-blockers (in competition primarily, some sports)

Prohibited Methods

1. Manipulation of blood and components

2. Chemical and physical manipulation

3. Gene doping

evidence of tampering. Chain of custody documentation is initiated, which records who has contact with the specimens, the procedures performed on aliquots that are made, and the secure location of the original specimens after aliquoting takes place. The pH and specific gravity of the urine are recorded (as additional evidence supporting that the urine is genuine and has not been adulterated), and aliquots from the "A" bottle are screened for a wide variety of banned substances. If a prohibited substance is identified during the screening procedure, confirmation testing is performed on an additional aliquot from the same "A" bottle. If the confirmation test is positive for the same banned substance(s) or metabolite(s), then it is considered positive and is called an adverse analytical finding. A documentation package detailing all aspects of the testing, including chain of custody, is prepared and submitted to the responsible authority (testing authority).

The testing authority determines the need for confirmation testing of the "B" bottle based upon the substance detected in the "A" bottle. For example, an athlete may test positive for a substance that they are allowed to use for a preexisting medical condition. In these cases, if a therapeutic use exemption has been approved by the sports organization then a doping violation has not occurred and testing of the "B" bottle would be unnecessary. On the other hand, the presence of a substance such as an exogenous anabolic agent almost always requires "B" bottle testing unless the athlete confesses to the doping violation. When "B" bottle testing is performed, the athlete or designated representative is permitted to witness the opening of the bottle to verify that it was collected from him/her and is also allowed to watch the entire testing process. If the "B" bottle testing confirms the presence of the prohibited substance identified and confirmed by "A" bottle testing, then a "B" bottle documentation package is prepared and submitted to the testing authority for sanctioning of the athlete. The supervised collection of urine specimens into tamper-evident containers, the chain of custody documentation,

TABLE 34.2 Adverse Analytical Findings Reported in 2017

Substance	% of Total[a]
Anabolic agents	44.0
Diuretics and masking agents	15.0
Stimulants	14.0
Hormone and metabolic modulators	8.0
Glucocorticosteroids	5.0
Beta-2 agonists	4.0
Cannabinoids	4.0
Hormones and related substances	3.0
Narcotics	2.0
Beta-blockers	0.3
Chemical and physical manipulation	0.02
Alcohol	0.0
Enhancement of oxygen transfer	0.0

[a]The total may include findings from athletes that have therapeutic use exemption approval.

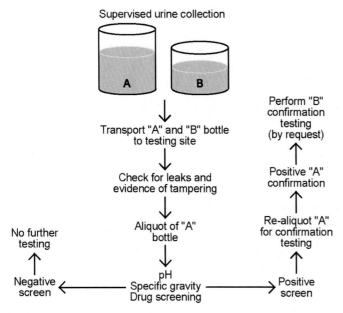

FIGURE 34.1 A simplified version of the testing procedure used by WADA-accredited laboratories. Chain of custody documentation is maintained throughout the entire process.

"A" and "B" bottle confirmatory testing, and the opportunity to observe "B" bottle testing are some of the aspects that distinguish WADA-accredited anti-doping laboratories from routine clinical laboratories.

GENERAL TESTING METHODS

The preferred specimen for identifying WADA prohibited substances in most cases is urine. A large volume of urine is easy to collect and is viewed as less invasive by athletes. Additionally, most target compounds remain in the bloodstream for short periods of time and are rapidly metabolized and excreted into the urine. By analyzing urine, it is possible to detect both the parent drug and metabolites for several hours after the drug has been removed from circulation.

In order to screen for all classes of substances on the WADA prohibited list and provide confirmation testing, a minimum of 60 mL of urine is needed in the bottle "A," which in most cases is easily obtained from an athlete. A minimum of 30 mL of urine is needed for bottle "B" confirmation testing.

Depending on the screening method and expected target analyte concentration, anywhere from 1−20 mL of urine is needed for testing. Each analytical procedure is designed to maximize recovery of specific compounds in the class of substances and remove other compounds that might interfere with detection of the target compounds. The amount of sample preparation varies by compound class from none for immunoassays to detect human chorionic gonadotropin (hCG) [11], to simple liquid/liquid extraction for the analysis of diuretics by liquid chromatography tandem mass spectrometry (LC−MS/MS) [12]. Sample preparation for isotope ratio analysis of steroids often requires a multistep procedure incorporating solid-phase extraction, an LC cleanup step, and derivatization of functional groups for gas chromatography isotope ratio mass spectrometry (GC/IRMS) analysis. The various testing methods produce turnaround times that vary from 1 h to several days. A summary of testing methods and cleanup procedures for specific substances and classes of compounds is given in Table 34.3.

In addition to the ISL, the WADA has produced several Technical Documents (TDs) which provide guidelines for specific aspects of the testing procedure and accompanying documents [13]. TDs provide requirements that must be followed by accredited laboratories along with specific recommendations in order to harmonize testing procedures among laboratories. For example, the WADA Technical Document TD2018MRPL defines minimum detection limits (also called minimum required performance limits) that each laboratory must validate and routinely obtain for compounds in each of the classes of prohibited substances (Table 34.3). Although individual laboratories are free to set their own criteria for various aspects of testing procedures, the criteria used must be at least as stringent as those set forth by the WADA in specific TDs and the ISL. In addition to TD2018MRPL, other TDs have been produced that deal with the detection of endogenous steroids, 19-norandrosterone and EPO, and information that must be provided in laboratory documentation packages and chain of custody documents. There is also a TD that specifies tolerance windows for retention times and relative ion intensities for chromatographic separation methods and mass spectrometry detection [14].

Chromatography, using either a gas or a liquid as the mobile phase, separates target compounds from one another and from interfering compounds as a function of time. The volatility, polarity, and three-dimensional structure of target molecules are important properties that influence partitioning of compounds between the mobile and stationary phases and determine the amount of time they are retained on the column. As illustrated in Table 34.3, some substances such as beta-2 agonists can be routinely detected using both gas chromatography combined with tandem mass spectrometry

TABLE 34.3 Testing Procedures and Performance Limits for Prohibited Substances

Prohibited Substances	Isolation Technique[a]	Instrumentation	MRPL[b] (per mL)
Anabolic agents	SPE, LL	GC−MS/MS, LC−MS/MS	0.2−5 ng
Hormones	SPE, LL, DS	LC−MS/MS	1−2 ng
Beta-2 agonists	LL	GC−MS/MS, LC−MS/MS	20 ng
Hormone modulators	SPE, LL, DS	GC−MS/MS, LC−MS/MS	0.05−200 ng
Diuretics	SPE, DS	LC−MS/MS	2−200 ng
Stimulants	SPE, LL, DS	GC/MS, LC−MS/MS	100 ng
Narcotics	SPE, LL	GC−MS/MS, LC−MS/MS	2−50 ng
Cannabinoids	SPE, LL	GC−MS/MS, LC−MS/MS	1 ng
Glucocorticosteroids	LL	LC−MS/MS	30 ng
Beta-blockers	LL	LC−MS/MS	100 ng
Erythropoietic proteins	UC, IEF, PAGE	Chemiluminescent	−
hCG	−	Immunoassay	5 mIU
Testosterone	SPE, LC	IRMS	−

[a]DS, Direct or dilute and shoot; IEF, isoelectric focusing; LL, liquid−liquid extraction; PAGE, polyacrylamide gel electrophoresis; SPE, solid-phase extraction; UC, ultra-centrifugation.
[b]MRPL is the minimum required performance limit and can differ among substances within the class of compounds.

TABLE 34.4 Identification Criteria for Prohibited Substances by GC–MS/MS and LC–MS/MS

Retention Time		Relative Ion Intensity By Percent of Base Peak	
Absolute	Relative		
1% or 0.1 min	0.5%–1%	>50%	± 10% (absolute)
		25%–50%	± 20% (relative)
		<25%	± 5% (absolute)

(GC–MS/MS) and LC–MS/MS, whereas some substances are typically analyzed by one or the other. Both techniques are used to identify anabolic steroids since the soft ionization employed in electrospray LC–MS/MS can lead to low sensitivity for some steroids while the chromatographic behavior of other steroids is poor when analyzed by GC–MS/MS. Some compounds can only be detected by a specific separation method or can be detected at lower concentrations using one of the separation methods. For example, although GC- and LC-based methods can be used to detect stanozolol, GC–MS/MS can detect lower concentrations of the compound compared with LC–MS/MS. In contrast, tetrahydrogestrinone can only be detected by LC-based methods since it is unstable and lost during the most frequently employed chemical derivatization required for GC-based analysis.

Single-stage mass spectrometry measures the relative abundance of individual ion fragments. In contrast, MS/MS measures the relative abundance of pairs of precursor ion fragments and subsequent product ions derived from the precursor ions. The detection of target compounds is achieved by comparing the retention time and relative intensities of ion fragments using single-stage mass spectrometry or the relative intensities of precursor/product ion pairs from a sample to those obtained from analysis of a reference material using MS/MS. Defined criteria that must be met for retention time and relative ion intensities are provided in Table 34.4 [14].

The likelihood of a target compound and an interfering substance having the same precursor/product ion pair is considerably lower than the likelihood that the two spectra will contain the same fragment ion. This makes MS/MS data considerably easier to interpret and thus is amenable to software automation. Some automation for GC–MS is possible for analysis of stimulants and narcotics because full spectra instead of selected ion fragments are acquired, making it easier for the software to distinguish between target compounds and interfering substances. In addition, the detection of stimulants, narcotics, cannabinoids, and beta-blockers is achieved almost exclusively by identifying the parent compounds. In contrast, anabolic agents are detected by identifying parent compounds and a wide array of metabolites, with some metabolites being common to multiple parent drugs. The detection of anabolic agents is also complicated by the fact that many of these steroids are normally synthesized by the body and typically appear in all urine samples. Unique features associated with GC–MS detection of anabolic agents are described in the next section.

ANABOLIC AGENTS

Anabolic agents are the most commonly detected class of substances reported to the WADA in 2017 (Table 34.2). Anabolic agents are divided into steroids with anabolic androgenic properties and nonsteroidal agents with anabolic properties (clenbuterol and the selective androgen receptor modulators). Anabolic androgenic steroids can be endogenously produced by the body and appear in all urine specimens or are exogenous and only appear in the urine after administration of a banned substance or its metabolic precursor. Minor modifications of the endogenous steroid testosterone can result in numerous exogenous steroid preparations with potent androgenic and anabolic properties. As shown in Fig. 34.2, the basic steroid carbon skeleton of testosterone can be modified by the addition/removal of a methyl group and/or double bonds, or the addition of ring structures. The sheer number of combinations that can be produced makes steroid analysis very challenging for anti-doping laboratories. Among all anabolic agents, clenbuterol was the most commonly detected anabolic agent followed by stanozolol, nandrolone, methandienone, and drostanolone. These exogenous anabolic agents accounted for more than 60% of the anabolic agents reported in 2017 (Table 34.5). Administration of testosterone or another closely related endogenous steroid as determined by GC/IRMS analysis accounted for an additional 8.7% of reported anabolic agents. The ability to detect testosterone administration is enhanced by longitudinal studies involving multiple testing of individual athletes over a period of months (see longitudinal steroid profiling section).

Analytical protocols for detecting exogenous and endogenous steroids in urine are identical until the point of data analysis. In brief, urine aliquots undergo enzymatic hydrolysis to free glucuroconjugates, followed by solid-phase

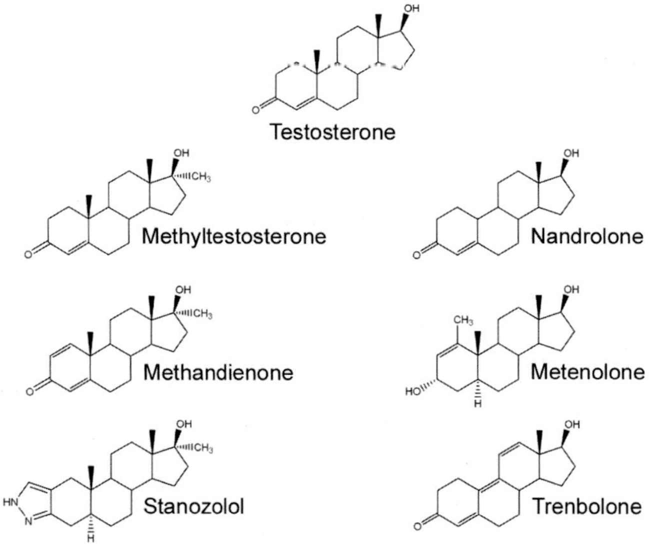

FIGURE 34.2 Examples of some chemically modified analogues of testosterone. Methylation or de-methylation, addition and removal of double bonds, and the addition of ring structures results in testosterone analogues with altered chemical and anabolic properties.

TABLE 34.5 Endogenous and Exogenous Anabolic Agents Reported as an Adverse Analytical Finding in 2017

Anabolic Agent	% of Total[a]
Clenbuterol	16.2
Stanozolol	15.7
Nandrolone	11.3
Testosterone	8.7
Methandienone	7.3
Drostanolone	6.6
Oxandrolone	4.6
Boldenone	4.6
Dehydrochloromethyltestosterone	4.6
Trenbolone	3.6
Metenolone	3.5
Ostarine	2.6
Mesterolone	2.0
Clostebol	1.5
Methasterone	0.7
Fluoxymesterone	0.7
1-Androstenedione	0.6
Desoxymethyltestosterone	0.6
Methyltestosterone	0.6
LGD-4033	0.6

[a]The total number of adverse analytical findings is 1813. Only substances that represent >0.5% of the total are displayed. Some adverse analytical findings correspond to multiple measurements on the same athlete.

extraction to recover the relatively nonpolar steroids. A derivatization step to the corresponding per-trimethylsilyl steroid is performed for analysis by GC–MS/MS in order to improve chromatographic and mass spectral properties. LC–MS/MS analysis does not require steroid derivatization. The characteristic mass spectrum of a compound is important for identification, and derivatization of a steroid generally produces a spectrum containing fragments of larger size. Larger fragments make it easier to distinguish steroids from other compounds with similar retention times. Typically 20–30 min are required to acquire data for a GC–MS/MS-based steroid analysis. The collected data are reduced to dedicated windows comprised of selected time slices and mass-to-charge (m/z) ion transitions corresponding to expected elution times and mass spectral precursor and product ion pairs for the target steroids.

Ion transition chromatograms for the bis-trimethylsilyl derivatives of 17α-methyl-5α-androstane-3α,17β-diol and 17α-methyl-5β-androstane-3α,17β-diol, metabolites which are present individually or together after administration of the steroids mestanolone, methyltestosterone, and oxymetholone are shown in Fig. 34.3. Ion transition chromatograms for m/z transition 435.3–255.1 and m/z transition 255.1–199.1 are shown in the left-hand panels for a negative urine (A), a positive urine (B), and a reference material (C). The analyst would identify the co-eluting peaks for m/z transition 435.3–255.1 and m/z transition 255.1–199.1 in both the sample and reference standard at a retention time of 11.69

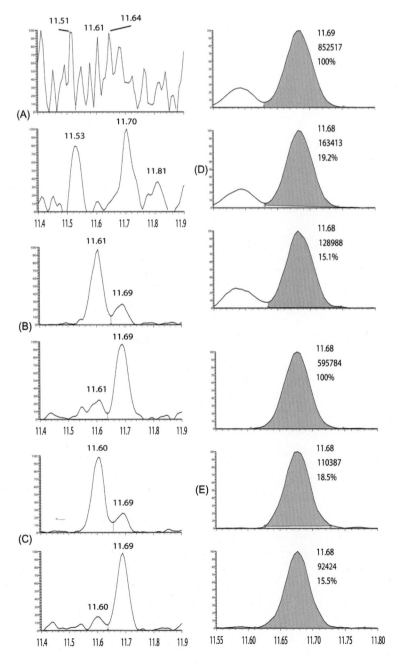

FIGURE 34.3 GC–MS/MS screening data for the bis-trimethylsilyl derivatives of the exogenous steroid metabolites 17α-methyl-5α-androstane-3α,17β-diol and 17α-methyl-5β-androstane-3α,17β-diol in (A) a negative urine, (B) a positive urine, and (C) reference material. Left panels show screening chromatograms of m/z transitions 435.3–255.1 upper and 255.1–199.1 m/z lower. The x-axis is retention time (min) and the y-axis is ion transition intensity. The right panels show confirmation chromatograms of m/z transitions 435.3–255.1 upper, 435.3–199.1 middle, and 435.3–213.1 lower for (D) a reference material and (E) a positive urine.

minutes. The negative urine does not contain peaks at this retention time, but has several background peaks that routinely must be distinguished from true co-eluting peaks by the analyst. If a comparison of retention times and relative ion transition ratios between sample and reference standard is not conclusive, the sample would be re-injected using a different GC–MS/MS method that monitors more ion transitions, or utilizes a GC temperature program optimized for the compound of interest. Upon review of this data, the sample of interest would be submitted for confirmation testing. Confirmatory testing would utilize the same type of data but with an additional m/z transition acquired from a new aliquot of the athlete's sample and the reference material as shown for 17α-methyl-5β-androstane-3α,17β-diol in the right-hand panels of Fig. 34.3. When compared against the reference material (D), the retention times and relative intensities of each of the three m/z transitions in the sample (E) are required to be within the WADA-stipulated tolerance ranges in order to consider the sample positive (Table 34.4). The presence of exogenous steroid metabolites such as 17α-methyl-5α-androstane-3α,17β-diol and 17α-methyl-5β-androstane-3α,17β-diol is prohibited by WADA at any concentration [9].

Testosterone to Epitestosterone Ratio

Detecting the use of testosterone and other endogenous steroids provides a unique challenge because endogenous steroids are always present in urine samples. Therefore, it is important to detect changes in the normal steroid excretion profile in order to identify endogenous steroid use. Although the parameters monitored between laboratories may vary, they usually involve the measurement of urinary steroid concentrations and the ratio of specific steroids. For example, an epitestosterone concentration of 100 ng/mL in a urine sample from a male would be unremarkable whereas a concentration >200 ng/mL would trigger additional analysis to assess its origin [15].

Since the early 1980s, the urinary ratio of testosterone glucuronide to its inactive 17-epimer, epitestosterone glucuronide (testosterone to epitestosterone (T/E) ratio), has been the primary parameter used to identify testosterone use. Interestingly, this parameter is defined as the calibrator or calibration curve-corrected peak height or area ratio of testosterone and epitestosterone and not a concentration-based ratio [15]. At the outset, a population-based T/E ratio of 6:1 was established as the cutoff, with values >6.0 indicating testosterone use. The cutoff was lowered to 4.0 by WADA in 2005. It is well known that some athletes have T/E ratios >4.0 while others have very low T/E ratios in the absence of testosterone use. The range and distribution of T/E ratios for 17,813 athletes is shown in Fig. 34.4. The majority of athletes have T/E ratios between 0.5 and 1.5. Natural-log transformation of the data reveals a bimodal distribution, with one group having a median T/E ratio of 0.1 and the other with a median T/E ratio of 1.3 (Fig. 34.4, inset). There is considerable evidence indicating that a deletion polymorphism in the uridine diphospho-glucuronosyl transferase (UGT) 2B17 gene is responsible for athletes in the group with the lower T/E ratio [16,17]. The UGT2B17 gene codes for an enzyme that conjugates glucuronide to testosterone. In the absence of the enzyme less testosterone glucuronide is formed resulting in less conjugated testosterone appearing in the urine [16]. Additional studies support the concept that a T/E ratio cutoff of 4.0 will not detect testosterone use in individuals homozygous (del/del) for the UGT2B17 deletion [17]. The del/del genotype occurs in 66.7% of Korean men and 9.3% of Swedish men [18]. For athletes with the del/del

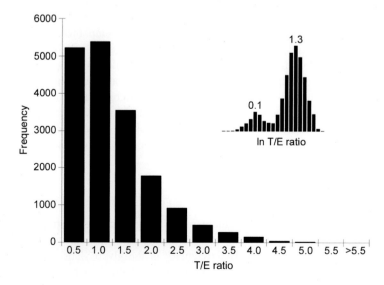

FIGURE 34.4 Distribution of urinary T/E ratios for 17,813 male and female athletes screened at the UCLA Olympic Analytical Laboratory. Only 69 samples had a T/E ratio >4.0 (0.39%). One sample had a T/E ratio between 5.0 and 5.5, and six samples had a T/E ratio >5.5. Samples with a confirmed T/E ratio >4.0 were tested by IRMS and only those samples found to be negative for synthetic testosterone are included in the graph. T/E ratios on the *x*-axis represent upper cutoff values for each bar. Natural-log transformation of data from 9780 male urine samples reveals two distinct populations with median T/E ratios of 0.1 and 1.3 (inset).

genotype and those with naturally elevated T/E ratios, other parameters are often used to detect testosterone use, such as longitudinal steroid profiling and GC/IRMS.

Longitudinal Steroid Profiling

Another approach to detect testosterone use that has gained widespread acceptance is the longitudinal study of urinary steroid concentrations. The concept is based on the observation that the T/E ratio for a single individual male will vary by <30% whereas variability between athletes can be considerably larger. Longitudinal profiling consists of a minimum of three urine collections from the same athlete at different times in order to establish baseline concentrations for various androgens and the T/E ratio. Suspicious results differing significantly from baseline are proof of synthetic testosterone use. A Bayesian statistical test using both population and individual athlete data appears to be the best method for detecting T/E ratio outliers [18]. It has been further shown that assessment of longitudinal data is enhanced by incorporating UGT2B17 genotype-based cutoffs [17]. In 2014 WADA introduced a steroidal module to the Athlete Biological Passport which considers several biomarkers of steroid doping in order to more effectively target samples for GC/IRMS analysis.

Isotope Ratio Mass Spectrometry

Urine samples with T/E ratios >4.0 are suspicious for testosterone use and are typically submitted for additional testing by IRMS. IRMS can measure very small differences in the $^{13}C/^{12}C$ ratio of testosterone and various steroid metabolites. The ^{13}C content of endogenously produced testosterone is influenced by plant and animal sources consumed in the diet. In contrast, synthetic testosterone is produced from plant precursors such as stigmasterol that contain less ^{13}C. This produces a smaller $^{13}C/^{12}C$ ratio for synthetic testosterone compared to natural testosterone, which can be measured by IRMS. In IRMS analysis, a combustion oven is placed at the end of the GC column and the separated steroids are combusted to CO_2 and water [19]. The water is removed and the CO_2 enters the mass spectrometer where the relative amounts of the different isotopic species of CO_2 (m/z 44, 45, and 46 ions) are measured. Because all compounds undergo combustion and are measured as CO_2, any interfering compound that has the same retention time as the target steroid will be measured. For this reason, all compounds that are analyzed must be completely separated from each other, and any interfering compounds must be removed during the extraction phase or during GC separation. An extensive cleanup procedure is often used in order to remove interfering substances and provide good chromatographic separation of steroids.

The relative amounts of m/z 44, 45, and 46 ions are used to determine the $^{13}C/^{12}C$ ratio of each target compound. Although m/z 44 ion comes only from $^{12}C^{16}O_2$, m/z 45 ion is derived from a mixture of $^{13}C^{16}O_2$ and $^{12}C^{17}O^{16}O$ and it is necessary to correct the abundance of m/z 45 ion for the contribution of ^{17}O-bearing CO_2 [20]. The instrument software uses the ratio of m/z 46 ion (from $^{12}C^{18}O^{16}O$, $^{13}C^{17}O^{16}O$, and $^{12}C^{17}O^{17}O$) to m/z 44 ion to correct for m/z 45 ion coming from $^{12}C^{17}O^{16}O$. Because the observed differences in $^{13}C/^{12}C$ ratios are quite small, results are expressed using δ-notation where $\delta^{13}C$ in units per thousand (‰) is calculated as $[(R_{Sample}/R_{Standard}) - 1] \times 10^3$ and expressed as δ‰. The R corresponds to the $^{13}C/^{12}C$ ratio of the compound in the sample and in an international standard, Pee Dee Belemnite $CaCO_3$. The δ‰ for target compounds including urinary testosterone or its metabolites such as androsterone and etiocholanolone, or 5α-androstane-3α,17β-diol and 5β-androstane-3α,17β-diol is compared to the $\delta^{13}C$ value for a compound in the steroid pathway not altered by testosterone use [21–23]. Endogenous reference compounds that are not altered by testosterone use include pregnanediol, pregnanetriol, 11β-hydroxyandrosterone, 11-ketoetiocholanolone, and 5α-androst-16-en-3α-ol [16]. Each laboratory is required to establish their own reference population comprised of negative urines from at minimum 20 males and 20 females. One required measure of method performance which each laboratory must achieve is a standard deviation for all endogenous reference compounds—target compounds combinations in their reference population <1.2‰. Uniform criteria for evaluation of IRMS data and determination of endogenous steroid use were first implemented by WADA in 2014 and further revised in 2016 [24]. Because IRMS is extremely labor-intensive, the test is not used as a screening method and is typically used to identify testosterone use in athletes with an elevated T/E ratio.

ERYTHROPOIESIS STIMULATING AGENTS

Erythropoiesis stimulating agents (ESAs) are abused by athletes because they increase the oxygen-carrying capacity of the blood by stimulating red blood cell production [25]. EPO is a protein hormone (165 amino acids) primarily synthesized by the kidneys that contains 3 N-glycan and 1 O-glycan side chain. Heterogeneity in the glycosylation pattern results in multiple molecules with different charges and molecular weight. The glycosylation pattern of recombinant

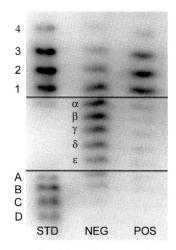

FIGURE 34.5 Image of an IEF gel containing a urine sample positive for recombinant EPO. The standard is a mixture of recombinant human EPO (Epogen) and darbepoetin (Aranesp), both from Amgen. The recombinant human EPO standard migrates to the basic region (labeled 1–4) whereas darbepoetin has extra N-glycosylation sites making it appear in the acidic region (labeled A–D). The middle region (labeled α–ε) contains endogenously produced EPO. Note that for the negative urine sample the bands in the endogenous region are more intense than bands in other regions. In contrast, the positive sample contains bands in the basic region that visually appear to be more intense than any band in the endogenous region.

forms of EPO (rEPO) is determined by the cell type used to express the protein and can differ significantly among formulations. Several modified forms of rEPO and analogues have been developed for clinical use.

There are currently two electrophoretic methods to detect administration of ESAs: isoelectric focusing (IEF) [26] and polyacrylamide gel electrophoresis (PAGE) incorporating either sodium dodecyl sulfate (SDS) or Sarcosyl (SAR) in the sample and running buffers [27,28]. IEF separates molecules based on charge differences that result from various glycosylation patterns and can separate several forms of rEPO from natural human EPO. SAR-PAGE and SDS-PAGE separate the various forms of EPO based on molecular weight differences related to the extent of glycosylation.

Urinary concentrations of EPO are very low, so samples are concentrated 100-fold by ultrafiltration prior to IEF or PAGE. After focusing under denaturing conditions, the proteins are transferred to a polyvinylidene difluoride (PVDF) membrane and incubated with a mouse monoclonal anti-human EPO antibody. The EPO–antibody complex is then disrupted and the monoclonal antibody from the first blot is transferred to a second PVDF membrane. The locations of the monoclonal antibody are identified using a biotinylated anti-mouse secondary antibody followed by streptavidin horseradish peroxidase and a chemiluminescent detection system.

In order to standardize the methods, WADA created a TD in 2004 and revised versions in 2007 and 2014 that deal with critical reagents and criteria to identify rEPO by IEF and PAGE in urine and blood [29]. Identification criteria for IEF can differ depending upon both the matrix and technique used for detection. For example, the identification of rEPO in urine requires the presence of at least three consecutive bands in the basic region of the gel and two of these bands must be more intense than any bands in the endogenous region. When the IEF technique is applied to a blood sample, the two most intense bands in the basic region must be approximately twice or more intense than any bands in the endogenous region. An image of an IEF gel is shown in Fig. 34.5. It is important to note that the positive urine contains at least three consecutive bands in the basic region (bands 1, 2, and 3) that appear to be more intense than any of the bands in the endogenous region (middle region labeled α–ε) from visual inspection. The intensity of the bands in the positive urine would be quantitated by densitometry to confirm that they meet the WADA criteria for an adverse analytical finding. In contrast, the interpretation of results from SDS-PAGE or SAR-PAGE regardless of the sample matrix utilized involves a more straightforward comparison of the location of a single or mixed band between an athlete's sample and a reference material.

Doping with rEPO can also be detected indirectly by measuring selected blood parameters such as hemoglobin concentration, red blood cell count, hematocrit, mean cell volume, mean cell hemoglobin, mean cell hemoglobin concentration, percent and absolute reticulocyte count, EPO, and soluble transferrin receptors [30]. Several sport federations currently utilize population-based cutoffs for these parameters in order to determine if an athlete is allowed to compete. Because many of these parameters exhibit large between-individual variation, the effectiveness of using population-based cutoffs is somewhat limited. To improve the sensitivity of this approach for detecting rEPO use, multiple blood samples are collected and used to establish an individual athlete's baseline values for the various parameters. There is considerably less variability in these marker concentrations over time for an individual compared to the variability between athletes. This approach, under review by WADA and sport federations since 2002, was put into regular use by the International Cycling Union [31] in 2008. Together with WADA this panel of biomarkers of blood doping was formalized as the first module of the Athlete Biological Passport. This module has been used since then to target athlete urine samples for ESA analysis.

GROWTH HORMONE

hGH is a peptide hormone (22 kDa) normally excreted by the pituitary gland. Its potential to increase muscle mass and decrease fat mass makes it attractive as a performance enhancing agent. Its use among athletes is believed to be widespread and it is often used in combination with testosterone. The hGH and testosterone promote muscle growth through different mechanisms. The development of a method to detect hGH use has been hindered by the fact that recombinant hGH is identical to the 22 kDa natural form. hGH also has a short half-life of 24–36 h, limiting the window of detection [32].

Because hGH is found in very small concentrations in urine, blood is the specimen of choice for the analysis. There are currently two strategies for detecting hGH in blood samples. The indirect approach involves measuring multiple hGH-dependent markers such as insulin-like growth factor-1 (IGF-1) and IGF-binding proteins [33,34], IGF-1 and type 3 pro-collagen (P-III-NP) [33–36], or IGF-binding proteins and P-III-NP [34]. The advantage of this approach is that these markers have a longer half-life and the potential for less variability compared to hGH. WADA published guidelines for the detection of hGH use utilizing an age-adjusted, sex-specific discriminant function of the serum concentrations of P-III-NP and IGF-1. This function is used to calculate the GH-2000 score which is then compared to assay and sex-specific decision limits. Confirmation of adverse findings for hGH using this approach requires the use of two different pairs of assays for the determination of P-III-NP and IGF-1 concentrations [37].

The other approach directly measures hGH and takes advantage of the fact that in addition to the major monomeric 22 kDa form, naturally produced hGH can also be detected in the blood in several other molecular forms (dimers, oligomers, 20 kDa monomer, and various fragments). Administration of the 22 kDa recombinant form of hGH results in an increase in the concentration of this molecular species relative to the other circulating forms [38]. Two separate immunoassays are performed to detect exogenous use of hGH [39]. One immunoassay measures all forms of hGH (total hGH assay) and the other immunoassay only measures the 22 kDa form. The ratio of 22 kDa to total hGH is then used to determine a doping violation. Reformulated test kits with improved detection limits have recently been produced and validated by WADA and were made available to WADA-accredited laboratories early in 2008.

SUPPLEMENTS AND PROHORMONES

Dietary supplements, particularly products containing what would be classified as prohormones (substance with the potential to be metabolized to a naturally produced hormone), continue to offer a challenge to the anti-doping community. If there are no drug claims, the US Food and Drug Administration regulates dietary supplements as foods [40]. This means that supplement manufacturers have less stringent production requirements than drug manufacturers. This often results in dilution of the supplement and/or contamination with another substance(s) that was previously manufactured using the same manufacturing vessels and equipment. In addition, some supplements are inadvertently or intentionally mislabeled, or are labeled in a confusing manner as to their true ingredients. For instance, the supplement 6-OXO from Ergopharm lists 4-etio-allocholen-3,6,17-trione on the label. Although this compound does not appear on the WADA prohibited list it is the same as 6-OXO-androstenedione, which is banned because it is a prohormone of testosterone and has structural similarity to androstenedione. Supplements containing prohormones often result in an elevated T/E ratio or produce an adverse analytical finding for a urinary metabolite on the WADA prohibited list. Another example is Animal Pak from Universal Nutrition that lists bovine colostrum as one of the more than 50 ingredients. The majority of athletes do not realize that colostrum is a rich source of IGF-1, a WADA prohibited substance.

How big a problem is dietary supplement use in sports? At the 2000 Olympics in Sydney, more than 2000 athletes reported taking either a supplement or medication, and more than 500 athletes reported taking more than five supplements per day. Even at the college level most athletes report taking at least one or more supplements a day. Surprisingly, a study of 634 nonhormonal supplements purchased from 13 countries found that 94 (14.8%) of the supplements contained anabolic androgenic steroids not listed on the label [41]. The majority were prohormones, with dehydroepiandrosterone being the most common. In addition, 10 of the supplements contained testosterone. Table 34.6 lists some supplements that have been analyzed and found to be contaminated with a prohibited substance.

TABLE 34.6 Primary Ingredients in Some Dietary Supplements Shown to Be Contaminated With Banned Substances

Amino acids
Vitamins
Proteins
Creatine
Pyruvate
Fatty acids
Enzymes
Carnitine
Ribose
Caffeine
Herbal extracts

To make matters worse, there is also considerable batch-to-batch variability in the level of supplement contamination. The concentration of the prohibited substance in the dietary supplement, the amount of supplement ingested, and the time between ingestion and collection of urine are all important variables that influence the results of a urine drug screen. Advancements made by laboratories in recent years in the effort to extend detection windows for prohibited substance abuse including significant increases in method sensitivity and identification of longer lasting metabolites have also increased the likelihood that the presence of trace contaminants in supplements taken by an athlete will result in a positive test. Unfortunately, the only way for an athlete to avoid a positive urine drug test from a dietary supplement is to not take supplements at all. According to the WADA *Code*, each athlete has the responsibility to ensure that no prohibited substance enter his/her system.

FUTURE CHALLENGES

There are several future challenges in sports drugs testings which are addressed in this section.

Designer Steroids

Designer steroids are synthetic anabolic steroids that are produced by chemical modification of another steroid. They are not naturally produced by the body and many combinations of chemically modified analogues of testosterone can be produced. The rings of testosterone can be modified at numerous carbon sites to produce steroids with different chemical properties (see Fig. 34.2 for examples). The intent is to produce steroid analogues with altered absorption or catabolism properties, increased potency, or chemical properties that prevent detection by standard testing methods. Combinatorial chemistry can also be used to bypass using a four-ring steroid structure as the starting compound and has the potential to generate thousands of different molecules with potential anabolic properties. Although certain modifications will lead to a molecule without biological activity, there is still only a modest understanding of the relationship between structure and biological action for anabolic steroids. Since selected ion monitoring is typically used to identify known anabolic steroids, a different approach is needed to detect unknown designer steroids. Several strategies have been tested, such as the ability of a urinary extract to bind androgen receptors in a bioassay [42]. Androgen receptors can also be used in a solid-phase or liquid format to capture unknown molecules with anabolic activity for subsequent analysis [43]. Another approach would be to move away from selected ion monitoring of urinary extracts to acquisition of full-scan information. This information would allow analysis of fragmentation patterns to identify unknown designer steroids. IRMS analysis could also be used to identify compounds with nonphysiological $^{13}C/^{12}C$ ratios, which would then prompt further analysis of suspicious urine samples. All of the above approaches continue to be being investigated by WADA-accredited laboratories as tools to rapidly detect unknown anabolic agents.

Alternative Matrices

While the advantages associated with the use of urine are likely to ensure it will be the matrix of choice for the majority of drug testing, other matrices including serum, plasma, and dried blood spots are now being explored. For some time, it has been recognized that the retrospectivity of serum exceeds that of urine for the detection of the ESA methoxypolyethylene glycol-epoetin beta [44]. In addition tests for insulin, growth hormone, hemoglobin-based oxygen carriers, and the Athlete Biological Passport rely upon blood as a matrix. As blood sampling becomes more frequent the invasiveness of the collection process may be more easily justified by adding utility to the sample itself. The collection of dried blood spots offers yet another approach. While this matrix offers advantages over blood itself, being less invasive and easier to collect and transport, the volume of the sample may limit its utility. Plasma and dried blood spots have already been applied to testing for testosterone esters and SARMS, respectively [45,46]. One possible application for blood and dried blood spots in the future is complementary determination of drugs and metabolites which are prohibited only during competition.

Results Management

It has always been the case that the role played by laboratories engaged in sport drug testing extends beyond the testing itself to include interaction with clients in support of test results. With the advent of enhanced detection capabilities and a rapidly evolving list of prohibited substances, drug testing laboratories are routinely asked to provide guidance to testing authorities both prior to and during litigation proceedings. For example, although anabolic agents are prohibited when detected at any concentration, a question often posed to the laboratory is whether the amount and combination of metabolites detected could be consistent with use of a contaminated supplement, recent administration, or long-term elimination from an administration which took place months ago. Although laboratories are not required by WADA to quantify the amount of parent drug or metabolite present, the ability to effectively support results often requires that this knowledge is available. In some instances the presence or absence of a particular metabolite in an athlete's urine can determine whether a doping violation has occurred. WADA has recently initiated efforts to address this particular issue by specifying which target drugs and metabolites must be monitored in the initial testing procedures of all laboratories.

CONCLUSIONS

The problem of doping in sporting events continues to receive much public attention, despite the fact that the use of performance enhancing agents has been with us for many centuries. Central to the anti-doping movement is the WADA, which provides a set of rules governing all aspects of sports doping. WADA-accredited laboratories must continually update and develop new testing methods and procedures in order to detect new substances with the potential to enhance athletic performance. It is a formidable task to develop and validate testing procedures as quickly as prohibited substances are being manufactured and introduced into the sports community. Current challenges involve developing simple and harmonized methods to detect a plethora of new designer drugs including steroids, cannabimimetic, and hormone modulators in urine and a variety of alternative matrices. The next challenge facing anti-doping labs will be to develop analytical methods to detect targeted gene and cell doping.

REFERENCES

[1] De Rose EH. Doping in athletes—an update. Clin Sports Med 2008;27:107–30.
[2] Landry GL, Kokotaio PK. Drug screening in athletic settings. Curr Probl Pediatr 1994;24:344–59.
[3] Baron DA, Martin DM, Magd SA. Doping in sports and its spread to at-risk populations: an international review. World Psychiatry 2007;6:118–23.
[4] Franke WW, Berendonk B. Hormonal doping and androgenization of athletes: a secret program of the German Democratic Republic government. Clin Chem 1997;43:1262–79.
[5] Ljungqvist A. Acute topics in anti-doping: brief history of anti-doping. Med Sport Sci 2017;62:1–10.
[6] Bowers LD. Abuse of performance-enhancing drugs in sports. Ther Drug Monit 2002;24:178–81.
[7] World Anti Doping Agency. The code, <https://www.wada-ama.org/sites/default/files/resources/files/wada-2015-world-anti-doping-code.pdf> [accessed January 2018].
[8] World Anti Doping Agency. Accredited laboratories, <https://www.wada-ama.org/en/resources/laboratories/list-of-wada-accredited-laboratories> [accessed January 2018].

[9] World Anti Doping Agency. The 2018 prohibited list, <https://www.wada-ama.org/sites/default/files/prohibited_list_2018_en.pdf> [accessed January 2018].
[10] World Anti Doping Agency. 2017 Adverse analytical findings reported by accredited laboratories, <https://www.wada-ama.org/sites/default/files/resources/files/2017_anti-doping_testing_figures_en_0.pdf> [accessed September 2018].
[11] Stenman U, Unkila-Kallio L, Korhonen J, Alfthan H. Immunoprocedures for detecting human chorionic gonadotropin: clinical aspects and doping control. Clin Chem 1997;43:1293–8.
[12] Deventer K, Van Eenoo P, Delbeke FT. Simultaneous determination of beta-blocking agents and diuretics in doping analysis by liquid chromatography/mass spectrometry with scan-to-scan polarity switching. Rapid Commun Mass Spectrom 2005;19:90–8.
[13] World Anti Doping Agency. International Standard for Laboratories, <https://www.wada-ama.org/sites/default/files/resources/files/isl_june_2016.pdf> [accessed January 2018].
[14] World Anti Doping Agency. Identification criteria for qualitative assays incorporating chromatography and mass spectrometry, <https://www.wada-ama.org/sites/default/files/resources/files/td2015idcr_-_eng.pdf> [accessed January 2018].
[15] World Anti Doping Agency. Endogenous anabolic androgenic steroids measurement and reporting, <https://www.wada-ama.org/sites/default/files/resources/files/wada-td2016eaas-eaas-measurement-and-reporting-en.pdf> [accessed January 2018].
[16] Jakobsson J, Ekström L, Inotsume N, Garle M, Lorentzon M, Ohlsson C, et al. Large differences in testosterone excretion in Korean and Swedish men are strongly associated with a UDP-glucuronosyl transferase 2B17 polymorphism. J Clin Endocrinol Metab 2006;91:687–93.
[17] Schulze JJ, Lundmark J, Garle M, Skilving I, Ekström L, Rane A. Doping test results dependent on genotype of uridine diphospho-glucuronosyl transferase 2B17, the major enzyme for testosterone glucuronidation. J Clin Endocrinol Metab 2008;93:2500–6.
[18] Sottas PE, Saudan C, Schweizer C, Baume N, Mangin P, Saugy M. From population to subject-based limits of T/E ratio to detect testosterone abuse in elite sports. Forensic Sci Int 2008;174:166–72.
[19] Brand WA. High precision isotope ratio monitoring techniques in mass spectrometry. J Mass Spectrom 1996;31:225–35.
[20] Santrock J, Studley SA, Hayes JM. Isotopic analyses based on the mass spectrum of carbon dioxide. Anal Chem 1985;57:1444–8.
[21] Becchi M, Aguilera R, Farizon Y, Flament MM, Casabianca H, James P. Gas chromatography/combustion/isotope-ratio mass spectrometry analysis of urinary steroids to detect misuse of testosterone in sport. Rapid Commun Mass Spectrom 1994;4:304–8.
[22] Shackleton CH, Phillips A, Chang T, Li Y. Confirming testosterone administration by isotope ratio mass spectrometric analysis of urinary androstanediols. Steroids 1997;62:379–87.
[23] Piper T, Mareck U, Geyer H, Flenker U, Thevis M, Platen P, et al. Determination of 13C/12C ratios of endogenous urinary steroids: method validation, reference population and application to doping control purposes. Rapid Commun Mass Spectrom 2008;22:2161–75.
[24] World Anti Doping Agency. Detection of synthetic forms of endogenous anabolic androgenic steroids by GC/C/IRMS, <https://www.wada-ama.org/sites/default/files/resources/files/wada-td2016irms-detection_synthetic_forms_eaas_by_irms-en.pdf> [accessed January 2018].
[25] Badia R, De La Torre R, Segura J. Erythropoietin: potential abuse in sport and possible methods for its detection. Biol Clin Hematol 1992;14:177–84.
[26] Lasne F, de Ceaurriz J. Recombinant erythropoietin in urine. Nature 2000;405:635.
[27] Kohler M, Ayotte C, Desharnais P, Flenker U, Lüdke S, Thevis M, et al. Discrimination of recombinant and endogenous urinary erythropoietin by calculating relative mobility values from SDS gels. Int J Sports Med 2008;29:1–6.
[28] Reichel C, Abzieher F, Geisendorfer T. SARCOSYL-PAGE: a new method for the detection of MIRCERA- and EPO-doping in blood. Drug Test Anal 2009;1:494–504.
[29] World Anti Doping Agency. Harmonization of analysis and reporting of erythropoiesis stimulating agents (ESAs) by electrophoretic techniques, <https://www.wada-ama.org/sites/default/files/resources/files/WADA-TD2014EPO-v1-Harmonization-of-Analysis-and-Reporting-of-ESAs-by-Electrophoretic-Techniques-EN.pdf> [accessed January 2018].
[30] Sharpe K, Ashenden MJ, Schumacher YO. A third generation approach to detect erythropoietin abuse in athletes. Haematologica. 2006;91:356–63.
[31] Union Cycliste Internationale. The Athlete Biological Passport—ABP, <http://www.uci.ch/clean-sport/the-athlete-biological-passport-abp/> [accessed January 2018].
[32] Wu Z, Bidlingmaier M, Dall R, Strasburger CJ. Detection of doping with human growth hormone [Letter]. Lancet 1999;353:895.
[33] Kicman AT, Miell JP, Teale JD, Powrie J, Wood PJ, Laidler P, et al. Serum IGF-1 and IGF binding proteins 2 and 3 as potential markers of doping with human GH. Clin Endocrinol 1997;47:43–50.
[34] Kniess A, Ziegler E, Kratzsch J, Thieme D, Muller RK. Potential parameters for the detection of hGH doping. Anal Bioanal Chem 2003;376:696–700.
[35] Dall R, Longobardi S, Ehrnborg C, Keay N, Rosen T, Jorgensen JOL, et al. The effect of four weeks of supraphysiological growth hormone administration on the insulin-like growth factor axis in women and men. J Clin Endocrinol Metab 2000;85:4193–200.
[36] Erotokritou-Mulligan I, Bassett EE, Kniess A, Sonksen PH, Holt RIG. Validation of the growth hormone (GH)-dependent marker method of detecting GH abuse in sport through the use of independent data sets. Growth Hormone IGF Res 2007;17:416–23.
[37] World Anti Doping Agency. Human growth hormone (hGH) biomarkers test for doping control analyses, <https://www.wada-ama.org/sites/default/files/resources/files/wada-guidelines-for-hgh-biomarkers-test-v2.0-2016-en.pdf> [accessed January 2018].
[38] McHugh CM, Park RT, Sonksen PH, Holt RIG. Challenges in detecting the abuse of growth hormone in sport. Clin Chem 2005;51:1587–93.
[39] Saugy M, Robinson N, Saudan C, Baume N, Avois L, Mangin P. Human growth hormone doping in sport. Br J Sports Med 2006;40(Suppl. 1):i35–39.

[40] Gregory AJM, Fitch RW. Sports medicine: performance enhancing drugs. Pediatr Clin N Am 2007;54:797−806.
[41] Geyer H, Parr MK, Mareck U, Reinhart U, Schrader Y, Schanzer W. Analysis of non-hormonal nutritional supplements for anabolic-androgenic steroids—results of an international study. Int J Sports Med 2004;25:124−9.
[42] Death AK, McGrath KCY, Kazlauskas R, Handelsman DJ. Tetrahydrogestrinone is a potent androgen and progestin. J Clin Endocrinol Metab 2004;89:2489−500.
[43] Nielen MWF, Bovee TFH, van Engelen MC, Rutgers P, Hamers ARM, van Rhijn JHA, et al. Urine testing for designer steroids by liquid chromatography with androgen bioassay detection and electrospray quadrupole time-of-flight mass spectrometry identification. Anal Chem 2006;78:424−31.
[44] Dehnes Y, Hemmersbach P. Effect of single doses of methoxypolyethylene glycol-epoetin beta (CERA, Mircera) and epoetin delta (Dynepo) on isoelectric erythropoietin profiles and haematological parameters. Drug Test Anal 2011;3:291−9.
[45] Forsdahl G, Vatne HK, Geisendorfer T, Gmeiner G. Screening of testosterone esters in human plasma. Drug Test Anal 2013;5:826−33.
[46] Thomas A, Geyer H, Schänzer W, Crone C, Kellman M, Moehring T, et al. Sensitive determination of prohibited drugs in dried blood spots (DBS) for doping controls by means of a benchtop quadrupole/Orbitrap mass spectrometer. Anal Bioanal Chem 2012;403:1279−89.

Index

Note: Page numbers followed by "*f*" and "*t*" refer to figures and tables, respectively.

A

A118G polymorphism, 107
A-796,260, 280*f*
AB-001, 279*f*
AB-CHMINACA, 282*f*
AB-FUBINACA, 282*f*, 308
AB-PINACA, 282*f*, 307
Acamprosate, 50, 51*t*
AccuTOF-DART, 290, 297–298
Acetaldehyde, 4, 8, 17, 25–26, 31
Acetaminophen, 337*t*
Acetic acid, 25–26
Acetone, 17, 29–30
Acetyl codeine, 458–459, 460*t*
Acetylcholine (ACh), 49
6-Acetylmorphine (6-AM), 122*t*, 189, 190*t*, 192*t*, 325*f*, 328*t*, 354, 453–454
ACN9, 49
4-AcODiPT (4-Acetoxy-*N*,*N*-diisopropyltryptamine), 250
4-AcO-DMT (4-acetoxy-*N*,*N*-dimethyltryptamine), 250
Adamantyl-indazole-3-carboxamide, 282*f*, 308–309
Adamantyl-indole-3-carboxamide, structures and analysis of, 300–301
ADB-BICA (1-benzyl-*N*-(1-carbamoyl-2,2-dimethylpropan-1-yl)-1H-indole-3-carboxamide), 274–277, 281*f*, 304
ADB-FUBINACA, 282*f*
ADB-PINACA, 282*f*
Adderall, 210
ADH genes, 47
Adultacheck devices, 377–378
Age-related dementia, alcohol consumption and, 9
AKB-48, 273, 283*f*, 309
Alanine (ALT) aminotransferase, 33
Alanine aminotransferase (ALT), 25–26
Albumin, 33
Alcohol analysis, 17, 430
 in body fluids, 20–21
 in expired breath, 21–22
 factors influencing the effects of alcohol, 17–19
 storage of samples for, 20
 in whole blood versus serum or plasma, 19
Alcohol biomarkers, 25, 30–36
 clinical applications of, 37*t*
 indirect, 33

overview of, 25–26
Alcohol dehydrogenase (ADH), 18–19, 27, 45–47, 124, 226
Alcohol dependence, genetic markers of, 46–48
 acetylcholine (ACh), 49
 ACN9, 49
 alcohol and aldehyde dehydrogenases, 46–47
 dopamine (DA), 47–48
 GABA, 48–50
 glutamate, 49
 neuropeptide Y (NPY), 49
 neurotransmitter systems, 47
 opioids, 50
 serotonin, 49
Alcohol dependence treatment, 51
 alcohol detoxification, 50
 pharmacogenetics of, 50–51
Alcohol metabolism, 27–28, 45, 45*f*
Alcohol toxicity, 13
Alcohol Use Disorder Identification Test (AUDIT), 38
Alcoholic beverages, alcohol content of, 2–3, 2*t*
Alcoholism, 43, 46, 52
 factors contributing to, 44*f*
Aldehyde dehydrogenase (ALDH), 19, 27, 45–47, 226
Aldehyde dehydrogenase gene, polymorphism of, 5
Alfentanil, 252–253
Alkaloids, in *Papaver somniferum*, 449*t*
α-hydroxyalprazolam, 198–199
Alpha-pyrrolidinophenone-derived designer drugs, 241–242
Alprazolam, 341, 394–395, 414
Alzheimer's disease, 9
AM-679, 274*f*
AM-694, 274*f*, 277–279, 287–288
AM-695, 287
AM-1220, 277*f*
AM-1220 Azepan isomer, 277*f*
AM-1241, 274*f*, 277–279
AM-1248, 279*f*, 301
AM-2201, 277*f*, 289, 302
AM-2201 *N*-4-OH, 277*f*
AM-2201 *N*-6-OH, 277*f*
AM-2232, 277*f*
AM-2233, 274*f*, 277–279, 287–288

American Academy of Pain Medicine, 343
American Board of Forensic Toxicology (ABFT), 84
American Geriatric Society, 343
American Indian Religious Freedom Act, 482
American Pain Foundation (APF), 343–344
American Pain Society, 343
7-Amino analogs, 197–198
7-Aminoclonazepam, 198–199, 395
7-Aminoflunitrazepam, 231
Amino-methyl-oxobutan-indazole-3-carboxamide, 282*f*, 306–308
Amitriptyline, 175–178, 177*f*, 200
Ammonium cerium nitrate, 371, 371*t*
Ammonium hydroxide, 400
Amoxapine, 200
Amphetamine (AMP), 108–109, 153, 189, 190*t*, 250, 305, 321, 325*f*, 328*t*, 329–330, 366–367, 421, 443–444, 463–464
 analogues, 392
 analysis of, 431–432
 in hair, 418
 detection of
  in oral fluid, 412
  in sweat, 415–416
 evaluation of interferences for, 329*t*
 nonprescription medication containing methamphetamine, 443–444
 prescription medications containing amphetamine/methamphetamine, 444
 substances known to metabolize to methamphetamine and amphetamine, 444
 therapeutic use, 211*t*
D,L-Amphetamine, 193
Amphetamine immunoassay, interferences in, 129–135
 dietary weight loss products and positive amphetamine, 133–134
 drugs that may cause positive amphetamine screening, 134–135
 interferences of sympathomimetic amines with amphetamine immunoassay, 130–131, 132*t*
 Vicks inhaler and false positive in amphetamine immunoassays, 131–133
Amphetamine immunoassays, 74–75, 124–125, 207–208
Amphetamine/methamphetamine assay, 122*t*, 192*t*

Amphetamine-derived designer drugs, 237–240, 238f
Amphetamine-type stimulants (ATS), 74–75, 108–109, 207, 360–361, 362t, 367t
  additional considerations, 212
  false-positive immunoassay results, 209
  GC/MS and LC-MS/MS confirmation procedures, 210
  immunoassay cutoff concentrations and cross-reactivity, 208–209
  interference and false-positive methamphetamine, 210–211
  isomer resolution, 211–212
  issues of cross-reactivity, 207–208
  minimizing false-positive immunoassay results, 209–210
  true-positive results, 210
Amphetaminil, 211t
AMPPPCA, structures and analysis of, 284f, 310–311
AMT (2-(1H-indol-3-yl)-1-methyl-ethylamine), 250
Anabasine, 436
Anabolic agents, 500–504
  isotope ratio mass spectrometry (IRMS), 504
  longitudinal steroid profiling, 504
  testosterone to epitestosterone ratio, 503–504
Anabolic steroids, 495–496, 507
Analyte–adulterant complexes, 365
Analytical methods, in drugs of abuse analysis, 157, 346–356
  gas chromatography (GC), 160
  ionization, 161–163
    atmospheric pressure chemical, 163
    chemical, 162
    electron, 162
    electrospray, 162–163
  liquid chromatography (LC), 160–161
  mass spectrometry based methods versus immunoassays, 168–169
  quadrupole mass spectrometry, 163–165
    triple, 164–165
  quantitative measurement, 165–168
    sensitivity, 165–167
    specificity, 167–168
  specimen introduction, 159
  specimen preparation, 157–159
Analytical true positive drug tests, 441
  amphetamines, 443–444
    nonprescription medication containing methamphetamine, 443–444
    prescription medications containing amphetamine or methamphetamine, 444
    substances known to metabolize to methamphetamine and amphetamine, 444
  cocaine, 444–445
  due to the use of medications, 442–443
  marijuana, 445
  opiates, 445–447
    medications containing morphine or codeine, 446

    minor metabolites of opiates, 446–447
    versus clinical false positive results, 442
Anesthetic agents, 362t
Anhydroecgonine methyl ester (AEME), 215–216, 220–221, 435–436
Anion gap, 63
ANKK1 gene, 47–48
4-ANPP, 185f, 186t
Antibody specificity, 124
Antibody–antigen interactions, 365
Antidepressants, 189, 200–201
Anti-doping movement, 495–496
Antihistamines, 225
Antiretrovirals, analysis of, 433
5-APB (1-benzofuran-5-ylpropan-2-amin), 250
6-APB (6-(2-aminopropyl)benzofuran), 250
APICA, 279f, 300–301
APINACA, 283f
Apolipoprotein J (Apo J)
  sialic acid and sialic index of, 37
Aqueous acid–base purification, 190–191
Area under the curve (AUC), 165
Arsenic, 337t
ARV medications, 433
Ascend multianalyte immunoassay technique, 348f
Aspartate aminotransferase (AST), 25–26, 33
ATM4G, 460, 460t
Atmospheric pressure chemical ionization (APCI), 163, 193
Atomoxetine, 134–135
Attention deficit hyperactivity disorder (ADHD) medication, 134–135
Azidothymidine (AZT), 433

B

Barbiturates, 122t, 199
  detection of, in oral fluid, 414
Base extraction, 190–191
Bath salts, 207, 274, 391–392, 394, 396–397
  cathinone analogs (CA). See Cathinone analogs (CA)
BB-22, 279f
BB-22 3-carboxyindole, 299–300
BDB (R, S-benzo-dioxolylbutanamine), 237
Beckman amphetamine assay, 124–125
Beckman Coulter Synchron AMP assay, 134
Beckman Synchron immunoassay reagents, 134
Benziisothiazolinone, 378
Benzodiazepine (BZD), 50, 76, 122t, 189, 197–199, 225, 247, 354, 362t, 367t, 414
  challenges of detecting, 394–396
  detection of, in oral fluid, 414
  POCT assays for, 395–396
  in urine, 392
1,4-Benzodiazepines (diazepam), 197–198
Benzodifurans, 250
Benzoyl indole cannabimimetics, 277
Benzoylecgonine, 74, 96, 176f, 190t, 192t, 215–216, 219, 325f, 328t, 419, 434–436, 444–445, 464–465, 467–469, 474
3-Benzoylindole, 288–289

Benzoylindoles, structures and analysis of, 277–289
Benzphetamine, 211t
Benzylpiperazines, 240–241
Beta-cyclodextrins (β-CD), 267
Betadine, 371, 371t, 379–381
Beta-hexosaminidase (β-Hex), 25–26
Beta-ketone amphetamine analog, 260
Beta-keto-type designer drugs, 238f, 242–243
BIM-018, 283f
BIM-2201, 283f
Binge drinking, 5, 6t, 12
Biomarkers of alcohol abuse, 13–14
BioSite Incorporated (Triage TOX Drug Screen), 135–136
Blood concentration of ethanol (BAC) or drug, 85–86
Blood ethanol concentrations, 17–18, 18t
Blood hematocrit (HCT), 268
Blood specimen, 405, 406t, 410
Blood/breath ratio, 22
Blood-based matrices, 158
Blue cohosh, 469
Body stuffing, 334–336, 335f
Botulism, 483–484
Brady Notice, 87–88
Brady v. Maryland, US Supreme Court 373 U. S. 83 (1963), 87–89
Brain damage, alcohol abuse and, 10–12
8-Bromo-2,3,6,7-benzo-dihydro-difuran-ethylamine, 250
4-Bromo-2,5-dimethoxy-β-phenethylamine (2C-B), 243–245, 245f
Bromocriptine, 50–51, 51t
Bromo-DragonFly, 250
(1-(4-Bromofuro[2,3-f] [1]benzofuran-8-yl) propan-2-amine, 250
Brown mixture and opiate levels in urine, 458
Buflomedil, 184f
Bufotenin, 482
Buprenorphine, 122t, 195–196, 329–330, 437–438
Bupropion, 200, 260
1,4-Butanediol, 225–226, 226f
2,3-Butanediol, 64–66
Butt Wedge, 360
Butylone, 242–243, 244f, 260
Butyrylcholinesterase (BchE), 105–106, 106t
BzODZ-Epyr, structures and analysis of, 274–277, 281f, 305

C

C957T polymorphism, 109
Cancer, alcohol consumption and, 8
Cannabicyclohexanol, 285f, 312
Cannabicyclohexanol + C2 variant, 312
Cannabidiol (CBD), 434, 464
Cannabimimetics, 273–274. See also Synthetic cannabinoids

Cannabinoids, 153, 470–471
    analysis of, 434
    in hair, 418–419
    detection of, in oral fluid, 412–413
    detection of, in sweat, 416
    natural, 73
    receptors (CB1 and CB2), 248, 273
    synthetic, 73–74
Cannabio shampoo, 471–472
Cannabis, 321, 360–361, 362t, 367t, 472
Capillary electrophoresis, 378
Carbazoles, 311
Carbohydrate-deficient transferrin (CDT), 13, 25–26, 36–37
    and GGT/gamma-CDT, combination of, 36–37
Carbon monoxide, 337t
11-Carboxy-delta-9-tetrahydrocannabinol, 434
4-Carboxyhexafluorobutyryl chloride (4-CB), 47
Cardiac arrhythmia, 340
Cardiovascular disease, alcohol abuse and, 12
Carfentanil, 252–253
Carfentanyl, 253f
Catalase, 3–4
Catechol-*O*-methyltransferase (COMT), 48, 112, 239–240
Catha edulis, 260
Cathinone, 260, 479f, 487–488
    laboratory analysis of, 488
Cathinone analogs (CA), 247–250, 259
    bioanalytical approaches, 260–268, 265t
        in hair, 267
        in miniaturized dried samples, 268
        in oral fluid, 267–268
        in urine, 266–267
        using hematic matrices, 264–266
    chemical structures of, 264f
    pharmacology and toxicology, 259–260
CB-13, 286f, 313–314
2C-B-Fly, 250
2C-C-NBOMe (2-(4-Iodo-2,5-dimethoxyphenyl)-*N*-[(2-methoxyphenyl)methyl]ethanamine), 250
2C-E (2,5-dimethoxyphenethylamine). *See* 4-Ethyl-2,5-dimethoxy-β-phenethylamine (2C-E)
Central nervous system (CNS)–depressant drugs, 340
Cerebral edema, 334, 334f
Cerebrovascular atherosclerosis, 217
Certification and accreditation standards, 84
Certified reference material (CRM), 173, 361
Chemical ionization (CI), 131t, 162
Chemical manipulation, 360
Chemical warmers, 352
Chemiluminescent immunoassays (CLIA), 123–124
"China white", 252–253
Chlordiazepoxide, 197–199
Chlorodifluoroacetic anhydride (CIF2AA), 210
Chromatographic peaks, 159, 166f

Chromatography system, 325–326
Chronic injection drug use, 336f
2C-I (2-(4-iodo-2,5-dimethoxyphenyl)ethan-1-amine), 250
Citalopram, 200
5Cl-AKB-48, 283f, 309
Clean money versus drug money, 465–467
Clinical and Laboratory Standard Institute (CLSI), 146
Clinical and medico-legal urine drug testing programs, 362t
Clinical false positive results
    versus analytical true positive results, 442
Clinical Laboratory Improvement Amendments (CLIA), 84
Clobenzorex, 211t, 444
Clonazepam, 197–198, 231, 394–395
Cloned enzyme donor immunoassay (CEDIA), 123, 349–350, 350f, 361, 366–367, 395, 486
Club drugs, 74–75
Cobas Integra analyzer, 135
Coca leaves, chewing, 469
Coca tea, 468
Coca-de-Mate, 463
Cocaethylene (CE), 215–216, 219, 434–435, 469
    role in toxicity, 219–220
Cocaine, 74, 106, 108–109, 189, 260, 305, 321, 360–361, 362t, 367t, 444–445, 495
    abuse, 469
    administration, routes of, 215
    analysis of, 434–436
        in hair, 419
    analytical methods, 220–221
    as benzoylecgonine, 122t
    combined toxicity of cocaine and alcohol, 218–219
    detection of
        in oral fluid, 413
        in sweat, 416
    metabolism, 215–216
    smoking and toxicity, 217–218
    toxicity, 216–217
    US paper money contaminated with, 463–468
Cocaine immunoassay, interferences with, 136
Codeine, 168, 189, 190t, 328t, 329–330, 341, 383f, 438, 446, 446t, 449t, 451–452, 459
Codeine/morphine, 192t
Codeine-6-glucuronide (C6G), 361
    urine adulterated with pyridinium chlorochromate, 383f
Codeinone, 381
Collaborative study on genetics of alcoholism (COGA), 49
College of American Pathologists (CAP), 84
Collision-induced dissociation (CID), 164
Common designer drugs, 237
Common postmortem toxicology samples, 339
    blood, 339
    tissue, 339
    urine, 339

vitreous fluid, 339
Competition immunoassays, 122, 123t
Compliance testing, 168–169
Confirmatory methods, 351–352, 365, 441
Conjugated glucuronides, 197–198
"Controlled substances", 247
    occupational exposure to, 474–475
Coroners, 333, 342
Correctional Service of Canada (CSC)
    program, 150
    and diluted urine specimens, 152–153
    specimen processing in, 150–151
Cotinine, 435–436
3OH-cotinine, 436
CP-47,497, 273, 285f, 312
CP-47,497 analog V, 285f
CP-47,497 analog VII, 285f
CP-47,497 analog VIII, 285f
CP-47,497-C7, 312
CP-47,497-C8, 273, 285f, 312
CP-55,490, 312
CP-55,940, 285f
CPE (cannabipiperidiethanone), 278f, 290, 298
Crack cocaine, 219
"Crack", 215
Crawford v. Washington (2004), 92–94
Creatinine as a biomarker of diluted urine, 149–150
Creatinine clearance, 149
Cross-reactants, 124
CSC dilution protocol, 150
CSC urine drug testing protocol, analysis of specimens in, 151f
Cushing's syndrome, 12
Cyanide, 337t
Cyclohexylmethyl metabolite, 308
Cyclohexylphenols, structures and analysis of, 312–313
CYP enzymes, 353
CYP2B6, 230
CYP2C9, 230
CYP2D6, 104–105, 240, 245–246, 353–354, 432, 459
CYP2E1, 18–19
CYP3A4, 230, 353
Cytochrome P450 (CYP), 103–104
    2D6, 27
    2E1, 27–28
    isozymes, 3–4, 50

D

DART-ORBITRAP method, 290
DAT1 SNPs, 110
Data-dependent acquisition techniques, 183
Data-independent acquisition techniques, 183
"Date rape" drug, 231
DAU assay components and protocols, 122t, 125–126
Daubert standard for expert testimony/opinion, 79–80
Daubert v. Merrill Dow Pharmaceuticals, 79–80

Death investigation, forensic toxicology in, 333
  common postmortem toxicology samples, 339
  computed tomography in, 334–336
  determining cause of death, 340–341
  determining the manner of death, 341
  medicolegal death investigation, 333–334
  ordering toxicology, 339–340
  role of autopsy in, 336–337
  toxicology reports, 340
  toxicology specimen collection at autopsy, 338
Defense Office of Hearings and Appeals (DOHA), 98–99
Dehydronorketamine, 230
Delta opioid receptor gene (OPRD1), 108
Delta-9-tetrahydrocannabinol (THC), 412, 418
  chemical structures of, 72f
Department of Defense (DoD) laboratory certifications, 84
Department of Health and Human Services (DHHS), 71, 363
Deprenyl, 211t
Designer drugs, 75, 392, 507
Designer stimulants, 75
Desipramine, 200
Desmethyl-flunitrazepam, 231
Desoxyephedrine, 352
Detergent/soap, 372t
Dexmedetomidine, 50, 51t
Dextromethorphan, 136
Diagnostic and Statistical Manual of Mental Disorders (DSM), 43
Diaxolobenzodiazepines (midazolam), 197–198
Diazepam (Valium), 50, 51t, 230, 328t, 470
Dietary weight loss products and positive amphetamine, 133–134
Diethylene glycol (DEG), 25–26, 30, 59, 60t, 61
  metabolism of, 61f
1,1-Difluoroethane, 340
Dihydrocodeine, 446–447
7,14-Dihydroxy-6-MAM, 381–382
Dihydroxylated metabolite of STS-135 (M21), 280f
3,4-Dihydroxymethamphetamine (HHMA), 228, 240
Di-iodo-THC-COOH, 380t
7,8-Diketo-6-MAM, 381–382
7,8-Diketo-morphine, 381–382
"Dilute-and-shoot" methods, 361–363, 401
Diluted urine, criteria for, 150
Dilution protocol and positive test results, 153–154
2,5-Dimethoxy-4-methyl-β-phenethylamine (2C-D), 243–245, 245f
2,5-Dimethoxy-4-propylthio-β-phenethylamine (2C-T-7), 243–245, 245f
2,5-Dimethoxy phenethylamine-type designer drugs (2CS), 238f, 243–246
Dimethyl amphetamine, 211t
Dimethylamylamine (DMAA), 133–134

N-(1,2-Dimethylpropyl)-cinnolinamine, 306–307
N,N-Dimethyltryptamine (DMT), 250
Diode-array detection (DAD), 287
Dipropylene glycol, 60t
Direct analysis in real time (DART), 178–179, 182
Direct analysis in real time coupled with a hybrid ion trap (DART-LTQ ORBITRAP) method, 288
Direct biomarkers, 25–26, 26t
Direct current (DC) balance, 163–164
Direct ethanol measurements, 31
Direct ionization techniques, 178–179
Discovery requests, 86, 88–91
Disease burden, alcohol consumption and, 25
Disialotransferrin, 36
Disposable pipet extraction (DPX), 435–436
Distilled liquors, 2
District Court of King County for the State of Washington, 97–98
Disulfiram, 50–51, 51t
Dopamine (DA), 47–48, 260
Dopamine D$_4$ receptor (DRD4), 48
Dopamine receptor genes, 109–110
Dopamine receptors, 47–48, 109
Dopamine transporter (SLC6A3), 48
Dopamine transporter gene (DAT1/SLC6A3), 110–111
Dopamine β-hydroxylase (DβH), 48, 112
Doping, 495–496
  sports doping, 127
Doping control method, 327
Doxepin, 200
Drano (sodium hydroxide), 372–375, 372t
DRD1 polymorphisms, 110
DRD2 gene, 47–48
DRD2 polymorphisms, 109
DRI benzodiazepine immunoassay, 395
DRI oxycodone assay, 135
Dried blood spots (DBS), 254, 268
Dried plasma spot (DPS) samples, 268
Driving under influence of drug (DUID), 294
Dronabinol, 445
Drug Abuse Warning Network (DAWN), 215, 221
Drug addiction, 103
Drug of abuse (DOA) testing, 71
  amphetamine-type stimulants, 74–75
  benzodiazepines, 76
  cocaine, 74
  designer stimulants, 75
  natural cannabinoids, 73
  opiates and opioids, 75–76
  synthetic cannabinoids, 73–74
  in the United States, 71–73
Drug-assisted sexual assaults, 225
  flunitrazepam (Rohypnol), 230
    analysis, 231
    association with drug-facilitated sexual assault, 231
    pharmacology, 231
    sources, 230
    toxicity, 231

γ-hydroxybutyric acid (GHB) and analogs, 225–229
  abuse, 226–227
  analysis, 227
  association with drug-facilitated sexual assault, 227–228
  pharmacology, 226
  sources, 225–226
  toxicity, 227
ketamine, 229–232
  abuse, 229–230
  analysis, 230
  association with drug-facilitated sexual assault, 230
  pharmacology, 230
  sources, 229
  toxicity, 230
3-4-methylenedioxymethamphetamine, 228
  abuse, 228
  analysis, 228–229
  association with drug-facilitated sexual assault, 229
  pharmacology, 228
  sources, 228
  toxicity, 229
nonbenzodiazepine sedative-hypnotics, 231
  analysis, 232
  association with drug-facilitated sexual assault, 232
  pharmacology, 231–232
  sources, 231
  toxicity, 232
Drug-facilitated crime, 225
Drug-facilitated sexual assault (DFSA), 225
  flunitrazepam association with, 231
  ketamine association with, 230
  3-4-methylenedioxymethamphetamine (MDMA) association with, 229
  nonbenzodiazepine sedative-hypnotics association with, 232
  γ-hydroxybutyric acid (GHB) association with, 227–228
Drunken Monkey Hypothesis, 1
DSM-IV, 38
DSM-V, 38
Dynorphins, 108

E

Eccrine glands, 415
Ecgonidine, 215–216
Ecgonine methyl ester (EME), 105–106, 215–216, 361
Ecstasy, 74–75, 91, 129, 212–213, 250
Efavirenz (EFV), 124, 135–136
Electron impact. *See* Electron ionization (EI)
Electron ionization (EI), 162
Electronic capture and retention of documents, 87
Electronic circular dichroism (ECD), 311
Electrospray ionization (ESI), 162–163, 193, 397–400
Endogenous cannabinoids, structures and analysis of, 312

Endogenous opioid ligands, 108
Endogenous opioid receptors, 107–108
  delta opioid receptor gene (OPRD1), 108
  endogenous opioid ligands, 108
  kappa opioid receptor gene (OPRK1), 107–108
  mu opioid receptor gene (OPRM1), 107
Endorphins, 108
Energy drinks, 5
Enkephalins, 108
Entactogens, 250
Entecavir, 175–178, 177f
Enzo[d]imidazole, structures and analysis of, 309
Enzyme immunoassay (EIA), 123–124, 361
Enzyme-linked immunosorbent assay (ELISA), 123–124, 361
Enzyme-multiplied immunoassay technique (EMIT), 123, 125, 136–137, 349, 349f, 361, 375, 395, 486
Ephedrine, 131
Ephedrine propyl carbamate, 131
Erythropoiesis stimulating agents (ESAs), 504–506
Erythropoietin (EPO), 496, 505
  recombinant forms of, 504–505, 505f
Erythroxylon coca, 74
Erythroxylum coca, 468
Erythroxylum laetevirens, 469
Escherichia coli, 341
Estazolam, 394–395
2-Ethyl-5-methyl-3,3-diphenylpyrrolidine (EMDP), 196
4-Ethyl-2,5-dimethoxy-β-phenethylamine (2C-E), 243–245, 245f, 250
4-Ethylthio-2,5-dimethoxy-β-phenethylamine (2C-T-2), 243–245, 245f
Ethyl glucuronide (EtG), 25–26, 32, 430, 431t
Ethyl sulfate (EtS), 25–26, 32, 430
Ethylamphetamine, 211t
Ethylene glycol (EG), 3, 25–26, 28–30, 59, 60t
  autopsy findings, 337t
  challenges and pitfalls in the analysis of, 64–67
  laboratory analysis of, 63–64
  mechanism of toxicity, 62
  metabolism of, 61f
  patient management, 67
  pharmacodynamics and pharmacokinetics, 59–64, 61t
  poisoning, clinical and laboratory assessment of, 62–63
  toxic and critical values, 64
Ethylene glycol monobutyl ether (EGBE), 59, 60t
Ethylene glycol monoethyl ether, 60t
Ethylene glycol monomethyl ether (EGME), 59, 60t
2-Ethylidene-1,5-dimethyl-3,3-diphenylpyrrolidine (EDDP), 76, 105, 196, 438
Ethylone, 242–243, 244f, 260

Eumelanin, 421
Euro banknotes, 465
European Monitoring Centre for Drugs and Drug Addiction (EMCDDA), 247–249, 259, 277
European paper money, 464
European Workplace Drug Testing Society (EWDTS), 405
Exculpatory evidence, 88
Expert witness, 82–84
  curriculum vitae (CV), 82
  preparation for testimony, 83–84
  versus expert consultant, 83

F

4-FA (4-fluoroamphetamine), 250
Famprofazone, 211t
Fatty acid ethyl esters (FAEEs), 32–33, 430
  analysis of, 430
  FAEE synthase, 32
FDU-PB-22, 279f
Federal Bureau of Investigation (FBI) laboratory, 92
Federal Rules of Evidence, 80–82
Federal workplace drug testing, 208–209
Federation of State Medical Boards (FSMB), 343–344
Fencamine, 211t
Fenethylline, 211t
Fenproporex, 211t
Fentanyl, 195–196, 253f
Fentanyl analogs, 184–185, 186t
Fentanyl exposure, 474–475
Fentanyl-containing prescription drugs, 252–253
Fetal alcohol spectrum disorders (FASD), 25
Fetal alcohol syndrome (FAS), 12, 25
Fetal hair, 417
Finasteride, 50, 51t
5F-AB-PINACA, 282f
5F-ADB, 283f, 310
5F-AKB-48, 283f, 309
5F-AMB, 283f, 310
5F-AMPPPCA, 284f
5F-NPB-22, 284f
5F-PB-22, 279f
5F-PB-22 3-carboxyindole, 299–300
5F-SDB-005 (CBL(N)-2201), 283f, 310
5-HTTLPR, 111
Flow injection analysis, 182
1-(5-Fluoropentyl)-3-benzoylindole, 288–289
1-(5-Fluoropentyl)-indole, 1-(5-fluoropentyl)-indole-3-carboxaldehyde, 288–289
Flunitrazepam (Rohypnol), 197–198, 225, 230, 395
  analysis, 231
  association with drug-facilitated sexual assault, 231
  chemical structure of, 226f
  pharmacology, 231
  sources, 230
  toxicity, 231

Fluorescence polarization immunoassay (FPIA), 122–123, 131–133, 136–137, 361, 481
Fluoro isobutyryl fentanyl (FiBF), 185f
4-Fluoroamphetamine, 207
4-Fluoromethamphetamine (4-FMA), 207, 250
Fluoroquinolone antibiotics, 135
Fluoxetine, 200
Fomepizole, 3, 67
Forensic Autopsy Performance Standards, 334
Forensic toxicology, case law pertaining to, 88–98
Fourier transform ion cyclotron resonance (FT-ICR), 178
Fragrance Powder, 287, 308
French paradox, 8
Frye standard for scientific evidence/examination, 79
FUB-AMB, 281f, 303–304
FUBIMINA, 309
FUB-NPB-22, 284f
FUB-PB-22, 279f, 300
Furanyl fentanyl (FF), 185f, 186t
Furfenorex, 211t

G

G protein–coupled receptors, 47–48
$GABA_A$ receptors, 231
Gabapentin, 51
Gamma-aminobutyric acid (GABA), 47–50, 225, 338
γ-butyrolactone (GBL), 225–227, 226f, 397
Gamma-CDT equation, 36–37
Gamma-glutamyltransferase (GGT), 13, 25–26, 35–36
Gammahydroxybutyric acid (GHB), 225–229, 252, 253f, 338, 391–392, 397
  abuse, 226–227
  analysis, 227
  association with drug-facilitated sexual assault, 227–228
  pharmacology, 226
  sources, 225–226
  toxicity, 227
Gas chromatography (GC), 13, 21, 66, 74–75, 160, 173, 189, 346, 351
Gas chromatography with chemical ionization mass spectrometry (GC-CI-MS) method, 288
Gas chromatography–mass spectrometry (GC-MS), 21, 63–64, 121, 131–133, 143, 150, 168, 190–193, 209, 220, 264, 277, 321, 359, 392, 409, 430, 443–444
Gastric alcohol dehydrogenase, 18
GC/tandem mass spectrometry (GC-MS/MS), 95
GC-MS confirmation cutoff concentrations, 190t
German Society of Toxicology and Forensic Chemistry (GTFCh), 312
Giglio v. United States, US Supreme Court 450 U.S. 150 (1972), 87, 89
Glucose-6-phosphate dehydrogenase (G-6-PD), 349

Glucuronidation via UDP
  glucuronyltransferases,
  239–240
Glutamate, 49
Glycerol dehydrogenase, 64
Glycoaldehyde, 29–30
Glycolic acid, 29–30, 59, 62
Glycols, 59
  challenges and pitfalls in the analysis of,
    64–67
  laboratory analysis of, 63–64
  mechanism of toxicity, 62
  patient management, 67
  pharmacodynamics and pharmacokinetics,
    59–64
  poisoning, clinical and laboratory assessment
    of, 62–63
  toxic and critical values, 64
Golden-seal tea, 372t
Growth hormone, 506
Guidelines for alcohol use, 5

H

Hair analysis, 469, 471
Hair drug testing, 406t, 417–422
  amphetamines and other sympathomimetic
    amines, 418
  cannabinoids, 418–419
  cocaine, 419
  environmental/external contamination of
    hair, 420–421
  hair adulteration issues, 421–422
  hair color, 421
  opiates/opioids, 419
  phencyclidine, 420
  specimen collection and analysis,
    417–418
Hallucinogens, 362t
Harrison Narcotic Act, 343
Headspace gas chromatography coupled with
  mass spectrometry and using solid-phase
  microextraction (HS-SPME-GC-MS)
  method, 279, 289, 296–297, 312
Health Inca tea, 463, 469
Heavy drinking, 5, 6t
  definition, 6t
  health hazards of, 10–13
    adverse effects, 11t, 12–13
    brain damage, 10–12
    fetal alcohol syndrome, 12
    increased risk of cardiovascular disease
      and stroke, 12
    liver damage, 10
Hemodialysis, 67
Hemp Ale, 472–473
Hemp oil, ingestion of, 472–474
  edible marijuana products,
    473–474
  unwanted exposure of marijuana containing
    products to children, 474
Heptafluorobutyric anhydride (HFBA), 210
Herbal Ecstasy, 487

Herbal teas
  and contamination/adulteration, 470
  contamination with cocaine, 468–470
    distinguishing coca leaves chewing from
      cocaine abuse, 469
Heroin, 334–336, 352, 396, 450, 460,
  464–465, 467, 495
Heterogeneous immunoassays, 346
High performance LC, 190
High sensitivity cloned enzyme donor
  immunoassay (HS-CEDIA), 395
High-density lipoprotein cholesterol (HDL
  cholesterol), 8, 35
High-performance liquid chromatography
  (HPLC), 160
High-performance liquid chromatography
  coupled with a diode-array detector (HPLC-
  DAD), 299
High-performance supercritical fluid
  chromatography (HPSFC), 266
High-resolution mass spectrometry (HRMS),
  134–135, 173, 248, 259, 321, 323t
  acquisition modes and types of data,
    182–186
    data-dependent acquisition techniques,
      183
    data-independent acquisition techniques,
      183
    targeted, semitargeted, and nontargeted
      analyses, 183–186
  for drug testing, 326–331
  features of immunoassays, 326
  fundamentals of, 174–175
  history and fundamentals of, 321–326
    chromatography system, 325–326
    direct injection of urine, 324
    Orbitrap analyser, 324
    TOF analyzer, 323
  key benefits of, 175–178
  limitations of, 178
  mass accuracy, 174–175
  mass resolution and mass resolving power,
    177–178
  resolving power (RP) of, 322f
  technological aspects, 178–182
    Orbitrap analyzer, 181–182
    time-of-flight (TOF) mass spectrometry,
      179–181
High-risk drinking, 6t
High-throughput screening, 327
Hillbilly Heroin, 195
Home-brew testing, 33
Homogeneous immunoassays, 346, 347f,
  349–351
  cloned enzyme donor immunoassay
    (CEDIA), 349–350, 350f
  enzyme-multiplied immunoassay technique
    (EMIT), 349, 349f
  kinetic interaction of microparticles in
    solution (KIMS), 350–351
Household chemicals, 360
HPLC with charged aerosol detection (HPLC-
  CAD) method, 296, 300

HS-SPME-GC-MS method, 289
HTR2A A-1438G variant, 111
HU-210, 286f, 313
HU-211, 286f
HU-308, 286f, 313
HU-311, 286f
Human carboxylesterase 1 (hCE1 or CES1),
  105
Human chorionic gonadotropin (hCG), 499
Human growth hormone (hGH), 496
Human liver microsome (HLM) incubations,
  309
Human urine integrity test parameters, 376t
Human urine metabolome database, 328t
Hydrocodone, 168, 195, 340, 353–354, 438,
  446t
Hydrocodone/hydromorphone, 122t
Hydrogen sulfide, 337t
Hydrolysis, defined, 159
Hydromorphone (Dilaudid), 168, 195, 340,
  396, 438, 446–447
Hydrophilic interaction chromatography,
  160–161
4-Hydroxy-2-nonenal (HNE), 27
4-Hydroxy-3-methoxyamphetamine (HMA),
  228, 431
4-Hydroxy-3-methoxymethamphetamine
  (HMMA), 240, 431–432
8-Hydroxy-7,8-dihydrocodeinone, 381
m-Hydroxy benzoylecgonine, 194–195
14-Hydroxycodeinone, 381
4″-Hydroxycyclohexyl metabolites, 308
2-Hydroxyethoxyacetic acid (HEAA), 30, 61
Hydroxyethyl radicals (HER), 27–28
3-Hydroxyflunitrazepam, 231
3-Hydroxyindazole, 306–307
5-Hydroxyindole-3-acetic acid (5-HIAA),
  31–32
4-Hydroxymethamphetamine, 105
5-Hydroxytryptamine (5-HT) transporters, 45
5-Hydroxytryptophol (5-HTOL), 25–26,
  31–32
Hydroxynorephedrine, 212
11-Hydroxy-THC (THC-OH), 471
Hypochlorite-based bleach, 366–368, 367t

I

IgA, 33
IgG, 33, 121, 409, 415
IgM, 33
Imipramine, 200
Immunalysis Corporation (Cannabinoids
  (THC/CTHC) Direct ELISA Kit), 135–136
Immunoassays, 76, 85, 121, 129, 273–274,
  346, 442
  DAU assay components and protocols,
    125–126
  in drugs of abuse testing, 121–122
  false-positive test results in other drugs-of-
    abuse immunoassays, 136–137
  formats of immunoassay design,
    122–124

polyclonal versus monoclonal antibody, 124
interferences in amphetamine immunoassay, 129–135
  dietary weight loss products and positive amphetamine, 133–134
  interferences of sympathomimetic amines with amphetamine immunoassay, 130–131, 132t
  other drugs that may cause positive amphetamine screening, 134–135
  Vicks inhaler and false positive in amphetamine immunoassays, 131–133
interferences with cocaine immunoassay, 136
interferences with opiate immunoassay, 135
interferences with tetrahydrocannabinol immunoassay, 135–136
limitation of, 124–125
in point of care drug testing, 347–349
qualitative versus quantitative reporting, 126
specimen type, 126–127
sports doping, 127
Immuno-chromatographic assay, 347–348
Immunoturbidimetric assay, 350–351
In vitro adulteration, 359–360, 364, 375, 378–379
In vivo adulteration, 360, 363
and urine dilution, 363–364
Inca Tea, 195
Indazoles, 273–274, 306–311
  adamantyl-indazole-3-carboxamide, 308–309
  amino-methyl-oxobutan-indazole-3-carboxamide, 306–308
  AMPPPCA, 310–311
  enzo[d]imidazole, 309
  5F-ADB and 5F-AMB, 310
  naphthalenyl-indazole, 310
  NPB-22, 311
  SDB-005, 310
Indirect biomarkers, 13–14, 25–26, 26t, 28f, 31, 33
Indirect markers, 31, 33
Individualized quality control plan (IQCP), 146
Indolalkylamines, 250
Indoles, 273–306
  adamantyl-indole-3-carboxamide, 300–301
  ADB-BICA (1-benzyl-*N*-(1-carbamoyl-2,2-dimethylpropan-1-yl)-1H-indole-3-carboxamide), 304
  benzoylindoles, 277–289
  BzODZ-Epyr, 305
  JWH-210, 305–306
  MDMB-CHMICA, MDMB-FUBINACA, and FUB-AMB, 303–304
  naphthalenyl-indole-3-carboxylates, 299
  naphthoylindoles, 289–296
  NNEI, 304
  phenylacetylindoles, 296–299
  piperazinylindoles, 299
  quinolin-8-yl-indole-3-carboxylates, 299–300

tetramethylcyclopropylmethanone-pentylindoles, 301–303
Inhalants, abuse of, 489t, 490, 491t
Insulin-like growth factor-1 (IGF-1), 506
Int8 VNTR, 110
4-Iodo-2,5-dimethoxy-β-phenethylamine (2C-I), 243–245, 245f
Ion suppression, 166–167, 212, 324–326, 487
Ion transition chromatograms, 502–503
Ionization, 161–163
Isoelectric focusing (IEF), 505
Isopropanol, 13, 20–21, 25, 28, 30, 415–416
  measurement of, 17
Isotope ratio mass spectrometry (IRMS), 504
Isotopes, 174

J

Joint Commission on Accreditation of Healthcare Organizations, 344
Judicial proceeding, 94
JWH-007, 277f, 292t
JWH-015, 277f, 289–291, 292t, 397
JWH-016, 277f
JWH-018, 273, 277f, 288–289, 302, 397
JWH-018 6-OH-indole, 277f
JWH-018 *N*-4-OH, 277f, 292–293
JWH-018 *N*-5-OH, 277f, 292–294
JWH-018 *N*-COOH, 277f, 292–294
JWH-018-*N*-(5-OH-pentyl), 291
JWH-018-*N*-pentanoic acid, 291
JWH-019, 277f, 289–291, 293, 296
JWH-019-*N*-5-OH-indole, 277f, 291
JWH-022, 277f, 289, 291–292
JWH-030, 284f, 311
JWH-073, 273, 277f, 289–290, 294
JWH-073 *N*-3-OH-butyl, 277f
JWH-073 *N*-butanoic acid, 292–294
JWH-073 N-COOH, 277f, 292–293
JWH-073-6-OH-indole, 277f
JWH-073-*N*-(4-OH-butyl), 291
JWH-081, 277f, 289–290, 293, 295–296, 397
JWH-122, 277f, 289–290, 292–296, 292t, 302
JWH-122 *N*-4-pentenyl, 277f
JWH-122 *N*-5-OH, 277f, 293
JWH-122-*N*-5-OH-urine, 291
JWH-200, 273, 274f, 277f, 289–292, 292t, 295–296, 314
JWH-200-5-OH-indole, 291
JWH-201, 278f, 289
JWH-203, 249f, 278f, 296–297, 299
JWH-210, 274–277, 289–290, 292, 302
  chemical structure of, 282f
  structure and analysis of, 305–306
JWH-250, 278f, 290–291, 296–298, 306, 397
JWH-251, 278f, 296–298
JWH-253, 278f
JWH-302, 278f
JWH-307, 284f, 311
JWH-398, 277f, 289, 293
JWH-398 *N*-5-OH, 277f
JWH-398-*N*-5-OH pentyl, 291

K

Kappa opioid receptor gene (*OPRK1*), 107–108
Kappa receptor gene (*OPRK1*), 50
Ketamine, 122t, 225, 229–232, 252, 253f, 391
  abuse, 229–230
  analysis, 230
  association with drug-facilitated sexual assault, 230
  chemical structure of, 226f
  pharmacology, 230
  sources, 229
  toxicity, 230
Khat (*Catha edulis*), 479t
  abuse of, 487–488
  analytical methods for analysis of, 482t
  laboratory analysis of cathinone, 488
  mechanism of action of, 488
  pharmacology and toxicology, 488
  street names of, 478t
K-hole, 229–230
Kinetic interaction of microparticles in solution (KIMS), 123, 350–351, 361, 486
Korsakoff's syndrome, 10–12
Kumho Tire Co. v. Carmichael, 79–80

L

Labetalol, 130, 134, 442
Laboratory developed tests (LDTs), 169
Laboratory information systems (LIS), 121
Laboratory Medicine Practice Guidelines (LMPG), 130, 207–208
Laboratory records and the court, 82
Laboratory records package, 82
Laboratory testing, role of, 45
Laboratory testing for alcohol, 13
Lactate dehydrogenase, 20–21, 25–26
Lateral-flow immunoassay, 347–348, 347f, 348f
Laudanum, 449–450
LC combined with tandem mass spectrometry (LC-Ms/MS) method, 264, 267–268
LC/MS–MS methods, 352
LC-QTOF-MS method, 266, 293, 296, 302
LC-quadrupole time of flight (LC-QTOF), 266
Legal aspects of drug testing, in US and military courts and civil courts, 79
  case law pertaining to forensic toxicology, 88–98
    adherence to laboratory procedures and guidance documents, 97–98
    challenges to the admissibility of evidence and inference of guilt, 94–97
    evidentiary disclosure and discovery requests, 88–91
    role of laboratory personnel and laboratory reports, 91–94
  elements of laboratory operations, 84–88
    Brady notices, 87–88
    compliance and documentation, 87
    discovery requests, 86
    laboratory certification and accreditation, 84

Legal aspects of drug testing, in US and
military courts and civil courts (*Continued*)
personnel certifications and licensures,
84–85
presence of alcohol or drug and
impairment, 85–86
specimen security, chain of custody, and
testing procedures, 85
expert witness in the courtroom, 82–84
expert witness preparation for testimony,
83–84
expert witness versus expert consultant, 83
laboratory records and the court, 82
marijuana decriminalization and medical
use, 98–99
scientific evidence, 79–82
Daubert standard for expert testimony/
opinion, 79–80
Federal Rules of Evidence, 80–82
Frye standard for scientific evidence/
examination, 79
Levetiracetam, 51
Lipid markers, 35
Liquid chromatography (LC), 160–161, 173,
189, 260
LC-HRMS, 173
for urine drug testing, 326–327, 329–331
Liquid chromatography combined with mass
spectrometry (LC-MS), 150, 208, 220, 277,
321, 346, 359, 409, 430
Liquid chromatography combined with mass
spectrometry/tandem mass spectrometry,
193–195
Liquid chromatography combined with tandem
mass spectrometry (LC–MS/MS), 85, 121,
208, 273–274, 292t, 302, 359, 392, 409,
430, 499
of mescaline, 484
toxicology screen using,
397–402
Liquid chromatography linked to atmospheric
pressure ionization tandem mass
spectrometry (LC-MS-MS), 458–459
Liquid chromatography with electrospray
ionization mass spectrometry (LC-ESI-MS/
MS), 291–292
Liquid chromatography-chemiluminescence
nitrogen detection (LC-CLND) method, 287,
289, 297
Liquid–liquid extraction (LLE), 158, 190, 254
Lisdexamfetamine, 211t, 444
Liver damage, 33
alcohol abuse and, 10
Lofentanyl, 253f
Longitudinal steroid profiling, 504
Lorazepam, 225, 231, 341, 394–395
Lysergic acid diethylamide (LSD), 95, 122t,
467, 477, 479f, 479t, 482
abuse of, 485–487
analytical methods for analysis of, 482t
street names of, 478t
Lysergic acid *N*-methyl-*N*-propylamide
(LAMPA), 486

M
M-3-glucuronide, 328t
MAB-CHMINACA, 282f, 308
Macrocytosis, 35
Magic mushroom, 477–482
abuse, prevalence of, 477–478
active ingredients of, 478, 479t
mechanism of action of, 480
analysis of active ingredients of, 480–482
analytical methods for analysis of, 482t
pharmacology and toxicology of psilocybin,
478–480
street names of, 478t
therapeutic potential of psilocybin, 480
Magnetic sector, 178
Malondialdehyde (MDA), 27
MAM-2201, 277f, 292–294
MAM-2201 *N*-4-OH, 277f
MAM-2201 *N*-COOH, 277f
Maprotiline, 200
Marijuana, 73, 122t, 189, 394, 397, 434, 445,
472
passive exposure to, 470–472
Marijuana products, edible, 473–474
Mass accuracy, 174–175
Mass filters, 351
Mass spectrometric techniques, 191, 375, 378
Mass spectrometry (MS), 189, 346, 351
Mass transition, 164
Mate de Coca, 468–469
Matrix effect, 212
Matrix-assisted laser desorption ionization,
178–179
Matrix-assisted laser desorption ionization-
time-of-flight mass spectrometry (MALDI-
TOF-MS) method, 287
MBDB (*R*, *S*-*N*-methyl-benzo-
dioxolylbutanamine), 237
MDA/MDMA/MDEA, 192t
MDMB-AMB, 274–277
MDMB-CHMCZCA, 284f
MDMB-CHMICA, 274–277, 281f, 303–304
MDMB-FUBINACA, 274–277, 281f,
303–304
Mean corpuscular volume (MCV), 13, 25–26
Meconium analysis, for neonatal drug
exposure, 429
alcohol and fatty acid ethyl esters, 430
amphetamine (AMP), 431–432
analytical performance and outcome studies,
438–439
antiretrovirals, 433
cannabinoids, 434
cocaine, 434–436
nicotine, 436
opiates, 437–438
Meconium and umbilical cord tissue
positive drug class agreement between,
438–439, 439t
Meconium as specimen, 429–430
Medical examiner (ME) system, 333
Medically important alcohol, 26, 28–30
Medicolegal death investigation, 333–334

Medium- to low-resolution tandem mass
spectrometry (Ms/MS) systems, 248
Mefenorex, 211t
MEKC-DAD method, 305
MEKC-MS/MS method, 288, 301
Melanin, 421
Melatonin, 6
Mellanby effect, 17–18
Membrane-bound COMT (MB-COMT), 48
Memorandum for record (MFR), 87
5-MeODMT (5-methoxy-*N*, *N*-
dimethyltryptamine), 250
5-MeO-DPT ("Foxy-Methoxy", 5-methoxy-*N*,
N-dipropyltryptamine), 250
3-MeO-PCE, 253f
3-MeO-phencyclidine, 253f
4-MeO-phencyclidine, 253f
Meperidine (Demerol), 195–196
Mephedrone, 242–243, 248–249, 260
MEPIRAPIM, 278f, 299
Mescaline, 394, 479f, 482, 484
laboratory determination of, 484
mechanism of action of, 484
pharmacology and toxicology of, 483–484
Metabolic acidosis, 29–30, 60–62, 67
Metabolic enzymes, 103–106, 104t
Metabolism of alcohol, 4–5
Metabolizing enzymes, 48
Meta-hydroxy-benzoylecgonine (mOHBE),
434–435
Methadone, 75–76, 122t, 195–196, 305, 325f,
328t, 438
Methadone metabolite, 153–154
Methamphetamine (MAMP), 105, 189, 190t,
209–210, 211t, 250, 325f, 328t, 329–330,
334–336, 352, 418, 421, 431, 444, 463–464
nonprescription medication containing,
443–444
occupational exposure to, 474
prescription medications containing, 444
Methamphetamine propyl carbamate, 131
Methanandamide, 285f, 312
Methanol, 25, 28–29
measurement of, 17
Methanol toxicity, 29
Methaqualone, 122t
Methedrone, 288, 327, 398t
Methiopropamine, 207, 212
Methoxetamine, 253f
1-(4-Methoxyphenyl)piperazine (MeOPP), 240,
242f
6-Methyl-2-{-[(4-methylphenyl)amino]-1-
benzoxazin-4-one} (URB-754), 311
4-Methylamphetamine, 207
6-*O*-Methylcodeinone, 381
1-Methylcyclohexanyl-3-cinnolinamine,
306–307
1-Methylcyclohexanyl-*N*-(valinamidyl)-
cinnolinamine, 306–307
1-Methylcyclohexanylindazole, 306–307
Methylecgonidine (MED), 215–218
3,4-Methylenedioxyamphetamine (MDA), 129,
130t, 189, 207, 212, 237, 396, 412, 481

1-(3,4-Methylenedioxybenzyl)piperazine (MDBP, MDBZP), 240–241
3,4-Methylenedioxyethylamphetamine (MDEA), 189, 190t, 207, 237, 412
3,4-Methylenedioxymethamphetamine (MDMA), 74–75, 105, 122t, 129, 130t, 189, 190t, 207, 212, 225, 228, 237–239, 250, 366–367, 379f, 396, 412, 431, 464–465, 481
  abuse, 228
  analysis, 228–229
  association with drug-facilitated sexual assault, 229
  pharmacology, 228
  sources, 228
  toxicity, 229
Methylenedioxypyrovalerone, 248–249, 398t
3,4-Methylenedioxypyrovalerone (MDPV), 241–242, 260, 288, 488
Methylenedioxy-substituted amphetamines, 237
4-Methylethcathinone (4-MEC), 260
3-Methylfentanyl (TMF), 252–253
3-Methylhistidine, 378
N-Methyl-N-tert-butyl-dimethylsilyl-trifluoroacetamide (MTBSTFA), 199
Methylone, 242–243, 244f, 260, 288
Methylphenidate, 110, 130, 136, 330, 420
Micellar electrokinetic chromatography (MEKC), 287
Micellar electrokinetic chromatography method coupled to tandem mass spectrometry (MEKC–Ms/MS), 266–267
Michaelis–Menton kinetics, 3
Microextraction by packed sorbent (MEPS), 254
Microgenics, 349–350
Microgenics Corporation (Cedia Dau MultiLevel THC), 135–136
Microgenics DRI benzodiazepine immunoassay, 395
Microgenics DRI General Oxidant-Detect Test, 378
Microparticle-enhanced immunoassay (MEIA), 123–124
Microsomal alcohol oxidizing system (MEOS), 3–4
Military Personnel Drug Abuse Testing Program (MPDATP), 359
Miniature ion trap mass spectrometry system, 301–302
Miniaturized dried sampling, 268
Moderate drinking, 6–9, 6t
  mental health benefits of, 9
  physical health benefits of, 8–9
  and reduced risk of cardiovascular diseases, 6–8
Modern TOF mass spectrometers, 181
6-Monoacetylmorphine (6-MAM), 168, 338, 340, 361, 382f, 383f, 416, 438, 453–454, 458, 460, 460t, 464–465
Monoamide oxidase, 48

Monoamine neurotransmitter system, 108–112
  catechol-O-methyltransferase, 112
  dopamine receptor genes, 109–110
  dopamine transporter gene (DAT1/SLC6A3), 110–111
  dopamine β-hydroxylase (DβH), 112
  monoaminergic-related genes, 111
  norepinephrine transporter gene (SLC6A2), 112
  serotonin receptor genes (HTR, multiple genes), 111
  serotonin transporter gene (5-HTT/SLC6A4), 111
Monoamine oxidase (MAO), 245–246
Monoaminergic-related genes, 111
Monoclonal antibody, 121, 124
Monohydroxylated of STS-135 (M25), 280f
Monosialotransferrin, 36
Morphine, 75, 107, 189, 190t, 202t, 325f, 328t, 329–330, 340, 382f, 438, 446–447, 446t, 449t, 460t
  evaluation of interferences for, 329t
  high morphine levels after eating poppy seeds, 455
  lower morphine level in subjects after eating processed poppy seed–containing foods, 455–458
  medications containing, 446
  in poppy seeds, 450–451
  in urine, 458
Morphine/codeine ratio, 459, 460t
Morphine-3-glucuronide (M3G), 190, 361, 384f
  urine adulterated with pyridinium chlorochromate, 384f
Morphine-6-glucuronide (M6G), 361, 382f
Morphinone-3-glucuronide, 381–382
Mu opioid receptor gene (OPRM1), 107
Multiple reaction monitoring (MRM), 164, 165f, 248, 254–256, 287
MultiStix (Bayer), 377
Muscarine ACh receptor subtype M2 (CHRM2), 49
Myocardial infarctions (MI), 216–217

N

Nabilone, 286f
N-acetyl-β-hexosaminidase, 34–35
Nail, drug testing in, 406t, 422
Naloxone, 135, 453
Naloxone glucuronide, 135
Naltrexone, 50–51, 51t, 422
NAME' Forensic Autopsy Performance Standards, 340
Naphthalene-1-yl(1-pentyl-1H-indol-3-yl) methanone, 277f
Naphthalene-1-yl(1-pentyl-1H-indol-4-yl) methanone, 277f
Naphthalene-1-yl(1-pentyl-1H-indol-5-yl) methanone, 277f
Naphthalene-1-yl(1-pentyl-1H-indol-6-yl) methanone, 277f

Naphthalene-1-yl(1-pentyl-1H-indol-7-yl) methanone, 277f
Naphthalenyl indazole, structures and analysis of, 310
Naphthalenyl-indole-3-carboxylates, structures and analysis of, 299
2-(1-Naphthoyl)-1-pentylindole, 293
3-(1-Naphthoyl)-1-pentylindole, 293
4-(1-Naphthoyl)-1-pentylindole, 293
5-(1-Naphthoyl)-1-pentylindole, 293
6-(1-Naphthoyl)-1-pentylindole, 293
7-(1-Naphthoyl)-1-pentylindole, 293
Naphthoylindoles, 248, 289–296
Naphthoylpyrroles, 248, 311
Naphthylmethylindoles, 248
Napthylmethylindenes, 248
Narcotic analgesics, 189, 195–196
National Association of Medical Examiners (NAME), 334
National Institute for Drug Abuse (NIDA) Guidelines, 71
National Institute of Alcohol Abuse and Alcoholism (NIAAA), 5
National Laboratory Certification Program (NLCP), 150
Natural cannabinoids, 73
Navy-Marine Corps Court of Criminal Appeals, 94
N-benzylpiperazine (BZP), 240, 250
NBOMes, 245–246
N-chloroMDMA, 379, 380t
N-desmethylvenlafanine, 200
Near-infrared spectroscopy (NIR) method, 295
Near-patient-testing (NPT), 121, 141
Negative toxicology report, 391
  challenges of detecting benzodiazepines, 394–396
  challenges of detecting opioids, 396
  communicating with physician, 394
  designer drugs are not detected in routine urine toxicology, 396–397
  problem with limited drug testing protocol and cutoff levels, 391–394
  toxicology screen using LC–MS/MS, 397–402
    dilute and shoot versus sample preparation, 401–402
Neonatal abstinence syndrome (NAS), 105, 437–438
Neonatal drug exposure, meconium analysis for assessment of. See Meconium analysis, for neonatal drug exposure
Neuropeptide Y (NPY), 47, 49
Neurotransmitter systems, 47
New benzoylindole-type synthetic cannabinoid (NE-CHMIMO), 274f, 288
New psychoactive substances (NPS), 237, 247, 252–253, 287, 321
  bioanalytical strategies for, 254–256, 255t
  cathinone analogs (CA), 248–250
  phenethylamines (PA), 250
  synthetic cannabinoid receptor agonists (SCRA), 248

New psychoactive substances (NPS) (*Continued*)
  tryptamines (TA), 250−251
Nicotine, analysis of, 436
7-Nitro compounds, 197−198
2-Nitro-6-monoacetlymorphine (2-nitro-MAM), 381
2-Nitro-M6G, 380t
2-Nitro-MAM, 380t, 381
2-Nitromorphine, 380t, 381
NM-2201, 279f
N-methyl-D-aspartate (NMDA), 230
  NMDA receptor, 49
NNEI, 281f, 304
NNL-1, structure and analysis of, 286f, 313
NNL-2, 278f
Nonbenzodiazepine sedative-hypnotics, 225, 231
  analysis, 232
  association with drug-facilitated sexual assault, 232
  pharmacology, 231−232
  sources, 231
  toxicity, 232
Nonmedical drug use, 441−442
Nonoxidative metabolism, 27
Non-SAMHSA drugs, confirmation testing for, 195−202, 202f
  antidepressants, 200−201
  barbiturates, 199
  benzodiazepines (BZD), 197−199
  narcotic analgesics, 195−196
  Z-drugs, 201−202
Nonsteroidal antiinflammatory drugs, 136−137
11-Nor-9-carboxy-tetrahydrocannabinol, 361
Noradrenaline, 260
Norbuprenorphine, 196, 437−438
Norcotinine, 436
11-Nor-delta-9-tetrahydrocannabinol (THC)-9-carboxylic acid, 189, 192t
Nordiazepam, 50, 197−198, 414
Norephedrine, 212, 431−432
Norepinephrine transporter gene (SLC6A2), 112
Norfentanyl, 196, 253f
Norfluoxetine, 201, 201f
Norketamine, 230
Normal phase chromatography, 160−161
Normetanephrine, 378
Nornicotine, 436
Norpseudoephedrine, 210−212, 487
Norsurfentanyl, 253f
Nortriptyline, 200
Noscapine (narcotine), 449t
Novel psychoactive substances (NPS), 173, 259
NPB-22, structures and analysis of, 284f, 311
Nuclear magnetic resonance (NMR), 273−274, 301, 311
Nutritional status and liver damage, general markers of, 33

O

O-desmethylvenlafaxine (ODMV), 200
Olanzapine, 341
Oleamide, 285f, 312
Ondansetron, 50, 51t
On-site testing, 141
Opiate drug testing
  effect of poppy seed−containing food on, 454−458
Opiate immunoassay, interferences with, 135
Opiates, 122t, 153, 305, 337t, 360−361, 362t, 367t, 445−447
  abuse, 107−108, 431, 437−438, 453−454
  analogues, 392
  analysis of, 437−438
  medications containing morphine or codeine, 446, 446t
  minor metabolites of, 446−447
  opioids and, 75−76
  detection of
    in oral fluid, 413
    in sweat, 416
Opioid analgesics, metabolism of, 353−354, 353f
Opioids, 50, 189, 247
  detection of
    in oral fluid, 413
    in sweat, 416
  challenges of detecting, 396
  in pain management, 344−346
Opium, 449
  legal issues, 450
Opium alkaloids as biomarker of poppy seed ingestion, 459−460
Opium-containing foods, dangers of, 453−454
Optimal amphetamine immunoassay, 130
Oral fluid, drug testing in, 405−415, 406t
  adulteration issues, 414−415
  amphetamines and other sympathomimetic amines, 412
  barbiturates and benzodiazepines, 414
  cannabinoids, 412−413
  cocaine, 413
  comparison of positivity of oral fluid with blood and urine, 410−412
  opiates/opioids, 413
  phencyclidine, 413−414
  sample collection and transportation, 407−408
  specimen analysis, 408−409
  transfer of drugs from plasma to oral fluid, 407
Oral fluid analysis, 267−268
Oral fluid testing using POCT devices, 145−146
Orbitrap analyzer, 174, 178, 181−182, 324, 324f
  cutaway view of, 182f
  performance of, 181t
Osmolal gap, 62−63, 66
Over-the-counter (OTC) drugs, 209
Oxalic acid, 29−30, 59, 62−64

Oxazepam, 50, 197−198, 325f, 328t
  evaluation of interferences for, 329t
Oxidative metabolism, 18−19, 27, 29, 215−216, 352−353
Oxidizing agents, detection of, 379−382
  amphetamine-type substances, 379
  cannabinoids, 379−381
  opiates/opioids, 381−382
Oxycodone, 75, 122t, 195, 392, 396, 400, 438, 446t
  with homogeneous enzyme immunoassay (HEIA) oxycodone, 135
OxyContin, 195, 343
Oxymorphone, 75, 195, 396
Oxytourism, 344

P

Pain management, drug testing in, 343
  analytical methods for, 346−356
    challenges in drug testing, 355−356
    confirmatory methods, 351−352
    general considerations on immunoassays, 346−347
    homogeneous immunoassays, 349−351
    immunoassays used in point of care drug testing, 347−349
    metabolism of opioid analgesics, 353−354
    specimen tampering, 352
  opioids, 344−346
Pain management programs, 75
Palmitamide, 285f, 312
Papain/meat tenderizer, 372t
Papaver somniferum, 449
  alkaloids found in, 449t
Papaverine, 449t, 450−451, 459−460
Paperspray, 182
Para-hydroxyamphetmaine (pOHAMP), 431−432
Para-hydroxymethamphetamine (pOHMAMP), 431−432
Paraphernalia, 91, 341
Paroxetine, 200
Party pills. *See* Phenethylamines (PA)
PB-22, 279f, 299
PB-22 *N*-pentanoic acid, 299−300
Peak symmetry, 166
Pentafluoropropionic anhydride (PFPA), 484
Pentasialotransferrin, 36
Penten-4-yl-AB-PINACA, 306−307
N-Penten-4-yl-JWH-210, 306
1-Pentylindazole-3-carboxaldehyde, 306−307
1-Pentylquinoline, 306
Periodate, 210−211
Peroxide/peroxidase system, 370
  effects of, 370, 370t
Peroxisomal catalase, 4
Peyote cactus, 394, 477, 479t, 482−484
  abuse, prevalence of, 483
  analytical methods for analysis of, 482t
  laboratory determination of mescaline, 484
  mechanism of action of mescaline, 484

pharmacology and toxicology of mescaline, 483–484
street names of, 478t
Pharmacodynamics, 30–31, 50, 59–64, 248
Pharmacogenetics, 104, 112
Pharmacogenomics of drugs of abuse, 103
  endogenous opioid receptors, 107–108
    delta opioid receptor gene (OPRD1), 108
    endogenous opioid ligands, 108
    kappa opioid receptor gene (OPRK1), 107–108
    mu opioid receptor gene (OPRM1), 107
  metabolic enzymes, 103–106, 104t
  monoamine neurotransmitter system, 108–112
    catechol-O-methyltransferase, 112
    dopamine receptor genes, 109–110
    dopamine transporter gene (DAT1/SLC6A3), 110–111
    dopamine β-hydroxylase (DβH), 112
    monoaminergic-related genes, 111
    norepinephrine transporter gene (SLC6A2), 112
    serotonin receptor genes (HTR, multiple genes), 111
    serotonin transporter gene (5-HTT/SLC6A4), 111
Pharmacokinetics, 50, 59–64
  of alcohol, 2–5, 4t
PharmChek, 415–416
Phencyclidine (PCP), 93, 122t, 136, 189, 190t, 192t, 229, 253f, 362t, 367t, 368
  analogues, 392
  analysis of, in hair, 420
  detection of, in oral fluid, 413–414
Phenethylamines (PA), 250, 251f
Phenylacetylindoles, 248
  structures and analysis of, 296–299
Phenylephrine, 212
Phenylpiperazines 1-(3-chlorophenyl) piperazine (mCPP), 240–241, 242f
Phosphatidyl alcohol, 4–5
Phosphatidylethanol (PEth), 25–27, 35
p-hydroxy benzoylecgonine, 194–195
Piperazine-derived designer drugs, 238f, 240–241
Piperazines, 240, 254, 267–268, 392
Piperazinylindoles, structures and analysis of, 299
Plasma, 19
p-methoxy methamphetamine (PMMA), 240, 250
Point of care (POCT) benzodiazepine assays, 395–396
Point of care assays, 346
Point of care drug testing, immunoassays used in, 347–349
Point of care testing (POCT) devices, 121, 126, 141
  design of, 141
  guidelines for using, 146
  oral fluid testing using, 145–146
  reliability and effectiveness of, 141–146

Point-of-collection testing devices, 405–406, 408–409, 413
Polyacrylamide gel electrophoresis (PAGE), 505
Polyclonal antibody, 121, 124
Polyclonal versus monoclonal antibody, 124
Polyethylene glycols, 59, 60t
Polypropylene glycol, 59
Polyvinylidene difluoride (PVDF), 505
Poppy plant and opium
  history of, 449–450
  legal issues, 450
Poppy seeds
  defense, 458–460
    acetyl codeine, 458–459
    ATM4G, 460
    morphine/codeine ratio, 459
    opium alkaloids as biomarker of poppy seed ingestion, 459–460
  effect of, on opiate drug testing, 454–458
  reduced morphine content in, during food processing, 452
  toxicity from, 454
  variation of morphine and other alkaloids content in, 450–451
Poppy tea, dangers of, 452–454
Postmortem computed tomography (PMCT), 334–336
Postmortem urine, 339
Potassium chlorate, 371, 371t
Potassium chromate, 371t
Potassium dichromate, 371, 371t
Potassium nitrite, 382f
Potassium perchlorate, 371, 371t
Potassium permanganate, 371, 371t, 376
PPA(N)-2201, 282f
Prealbumin, 33
Precursor ions, 193–194
Pregabalin, 330
Prenylamine, 211t
Pressurized liquid extraction (PLE), 267
Prodynorphin (PDYN), 50, 108
Prohormones, 506–507
1,2-Propanediol, 66–67
1,3-Propanediol, 63–64, 66–67
Propofol, 229
Propoxyphene, 122t, 325f, 328t
Propylene glycol (PG), 25, 30, 59, 60t, 61–62
  metabolism of, 61f
Propylhexedrine, 212
Prosecution consultants, 83
Protein precipitation (PP), 158, 254
Pseudocholinesterase, 105–106
Pseudoephedrine, 131
Psilocin (4-hydroxy-N,N-dimethyltryptamine), 250, 477–478, 479f, 480–481
  quantitation of, in plasma, 481–482
Psilocybe cubeneis, 250, 478
Psilocybin (4-phosphoryloxy-N,N-dimethyltryptamine), 250, 477–478, 479f, 481
  pharmacology and toxicology of, 478–480
  therapeutic potential of, 480

Pulmonary edema, 334, 334f, 336–337, 488
Purdue Frederick Company, 343
Putrescine, 338
Pyridinium chlorochromate (PCC), 369, 369t, 383f, 384f
Pyrroles, 273–274, 311
Pyrrolidinophenone-derived designer drugs, 238f, 241–242

Q

Quadrupole ion-trap-TOF, 181
Quadrupole mass analyser, design of, 163f
Quadrupole mass spectrometry (MS)-based drugs, 157
Quadrupole MS-based testing, 157
Quadrupole time-of-flight mass spectrometry (QTOF-Ms), 254
Quadrupole TOFs (QTOFs), 180f, 181
  key characteristics of, 181t
Quality assurance (QA) programs, 20, 84
Quinolin-8-yl-indole-3-carboxylates, structures and analysis of, 299–300
Quinolones, 135

R

R (-)-α-methoxy-α-trifluoromethylphenylacetyl chloride (R-MTPAC), 210, 212
R,S-3′,4′-methylenedioxy-α-pyrrolidinopropiophenone (MDPPP), 241–242
R,S-40-methyl-α-pyrrolidinohexanophenone (MPHP), 241–242, 264f
R,S-4′-methoxy-α-pyrrolidinopropiophenone (MOPPP), 241–242, 243f
R,S-4-methylthioamphetamine (4-MTA), 240
R,S-4′-methyl-α-pyrrolidinopropiophenone (MPPP), 241–242
R,S-parametoxy methamphetamine (PMMA), 240, 241f
R,S-para-methoxyamphetamine (PMA), 240, 241f
R,S-α-pyrrolidinopropiophenone (PPP), 241–242
Radio frequency (RF) voltage, 163–164
Radioimmunoassay (RIA), 123–124, 346, 366–367
Ramifentanyl, 253f
Ranitidine, 124–125, 130, 134
RapidFire tandem mass spectrometry (RF-MS/MS) methods, 295
RCS-4, 274f, 277–279, 287–288
RCS-4-N-5-OH, 274f
RCS-8, 278f, 296–297
Reactive oxygen species (ROS), 27–28
Receiver Operating Characteristic (ROC) curve analysis, 209
Recombinant DNA technology, 349–350
Remifentanil, 252–253
Resolving power (RP), 174, 322, 322f
Restriction fragment length polymorphism (RFLP), 109

Resveratrol, 9
Reticuline, 459, 460t
Reversed phase chromatography, 160–161, 324–326
Rheumatoid arthritis, alcohol consumption and, 8–9
Rifampicin, 135
Roche Abuscreen ONLINE assays, 361
Rubbing alcohol, 28

S

Salicylates, 337t
Salsolinol, 4
Salted-out assisted enzymatic hydrolysis, 291
Salted-out assisted liquid–liquid extraction (SALLE), 291
Sandwich immunoassays, 122, 346
Sarcosyl (SAR), 505
Scientific Working Group for Forensic Toxicology (SWGFT) guidelines, 303
Scientific Working Group for the Analysis of Seized Drugs (SWGDRUG), 247
Screened (presumptive) positive, 189
Screening reagents, 189
SDB-005, structures and analysis of, 283f, 310
Selected ion monitoring (SIM), 164, 173, 191, 248, 321
Selective serotonin reuptake inhibitors (SSRIs), 200
Selegiline, 211t, 421
Semitargeted/suspect screening, 183–184
Serotonin, 49, 108–109, 200
Serotonin 2A receptor (5-HT2A), 480, 486
Serotonin receptor genes (HTR, multiple genes), 111
Serotonin receptors, 250, 488
Serotonin transporter (5-HTT), 49, 108–109, 260
Serotonin transporter gene (5-HTT/SLC6A4), 111
Sertraline, 200, 486
Serum cholinesterase, 105–106
Siemens Syva EMIT II Plus Amphetamine assay, 134
Signal-to-noise (S/N) ratio, 165–166
Single nucleotide polymorphisms (SNPs), 46, 49, 104, 107
  in the human OPRK1 gene, 107–108
Single-stage mass spectrometry, 500
Skin popping, 336
SLC6A4 polymorphism, 111
Smith Amendment, 99
Smoking, 6–7, 215
  and coronary heart disease, 6–7
Sodium dodecyl sulfate (SDS), 505
Sodium hypochlorite (NaOCl), 379f
Sodium iodate, 371, 371t
Sodium metaperiodate, 371, 371t
Sodium periodate, 209
Solid phase extraction (SPE), 158–159, 190–191, 254, 430
Solid-phase microextraction (SPME), 266
Soluble COMT (S-COMT), 48

Solvent abuse
  adverse effects of, 491t
  laboratory detection of, 491
  prevalence of, 489–490
Special Access Programs (SAP), 98–99
Specimen processing in the correctional service of Canada program, 150–151
Spectrophotometric analysis, 378
Spice products, 277, 287, 292, 297, 305
Sports, performance enhancing drugs in, 495
  anabolic agents, 500–504
    isotope ratio mass spectrometry (IRMS), 504
    longitudinal steroid profiling, 504
    testosterone to epitestosterone ratio, 503–504
  anti-doping movement, 495–496
  erythropoiesis stimulating agents (ESAs), 504–506
  future challenges, 507–508
    alternative matrices, 508
    designer steroids, 507
    results management, 508
  general testing methods, 498–500
  growth hormone, 506
  supplements and prohormones, 506–507
  testing procedure, 496–498
  WADA prohibited list, 496
Sports doping, 127, 495–496, 508
Stearamide, 285f, 312
Stimulant amines, 130, 207–208
Streptococcus pneumoniae, 341
Stroke, 9, 12
STS-135, 279f, 280f, 301
Styrene-divinylbenzene sulfonic acid (SDVBS) polymer, 190–191
Substance Abuse and Mental Health Services Administration (SAMHSA), 5, 71, 84, 145, 189, 359, 361, 391, 405
  confirmation testing, 190–193
    gas chromatography/mass spectrometry, 190–193
    non-SAMHSA drugs, 195–202
  general confirmation requirements, 194–195
  liquid chromatography combined with mass spectrometry/tandem mass spectrometry, 193–195
Succinic acid, 226, 338
Sufentanil, 252–253
Suicide, 341
Sulfentanyl, 253f
Supercritical fluid chromatography coupled with an electron-spray ionization tandem mass spectrometer (SFCESI-MS/MS) method, 293
Supplements and prohormones, 506–507
Supported liquid extraction (SLE), 158, 430
Surface enhanced Raman spectroscopy (SERS) method, 295–296
Sweat, drug testing in, 406t, 415–417
  amphetamines and other sympathomimetic amines, 415–416
  cannabinoids, 416
  cocaine, 416

  issues of special interest, 417
  opiates/opioids, 416
  sample collection and analysis, 415
Sympathomimetic amines, 130–131
  analysis of, in hair, 418
  detection of, in sweat, 415–416
Synthetic cannabinoid receptor agonists (SCRA), 247–248, 249f
Synthetic cannabinoids, 73–74, 273, 391, 397
  carbazoles, 311
  CB-13, 286f, 313–314
  classical synthetic cannabinoids, 313
  cyclohexylphenols, 312–313
  endogenous cannabinoids, 312
  indazoles, 306–311
    adamantyl-indazole-3-carboxamide, 308–309
    amino-methyl-oxobutan-indazole-3-carboxamide, 306–308
    AMPPPCA, 310–311
    benzo[d]imidazole, 309
    5F-ADB and 5F-AMB, 310
    naphthalenyl-indazole, 310
    NPB-22, 311
    SDB-005, 310
  indoles, 274–306
    adamantyl-indole-3-carboxamide, 300–301
    ADB-BICA (1-benzyl-N-(1-carbamoyl-2,2-dimethylpropan-1-yl)-1H-indole-3-carboxamide), 304
    benzoylindoles, 277–289
    BzODZ-Epyr, 305
    JWH-210, 305–306
    MDMB-CHMICA, MDMB-FUBINACA, and FUB-AMB, 303–304
    naphthalenyl-indole-3-carboxylates, 299
    naphthoylindoles, 289–296
    NNEI, 304
    phenylacetylindoles, 296–299
    piperazinylindoles, 299
    quinolin-8-yl-indole-3-carboxylates, 299–300
    tetramethylcyclopropylmethanone-pentylindoles, 301–303
  NNL-1, 286f, 313
  pyrroles, 311
    naphthoylpyrroles, 311
  URB-class, 311
  WIN-55,212-2, 314
Synthetic/semisynthetic opioids, 362t
Syva EMIT, 143

T

T1246C, 110–111
T265C, 110–111
Table salt, 360, 364t, 372–375, 372t
Tapentadol, 195–196
TaqIA, 109
Technical Documents (TDs), 499
Teenage drinking, 10
Temazepam, 50, 197–198, 354
Tert-butyl metabolite, 308

Testing Designated Positions (TDP), 98
Testosterone to epitestosterone ratio, 503–504
Tetraethylene glycol, 378
Δ⁹-Tetrahydrocannabinoid, 445
Tetrahydrocannabinol, 166, 305, 361, 434, 480
Δ⁹-Tetrahydrocannabinol, 463–464, 467, 470–473
Tetrahydrocannabinol immunoassay, interferences with, 135–136
Δ⁹-Tetrahydrocannabinolic acid A (THCA-A), 471
Δ⁹-Tetrahydrocannabivarin (THCV), 445
Tetramethylcyclopropylmethanone-pentylindoles, structures and analysis of, 301–303
Tetrasialotransferrin, 36
THC-COOH, 73, 135–136, 153, 190–191, 294, 296, 298, 328t, 366, 368–369, 379, 471
THC-COOH-glucuronide, 361
Thebaine, 449, 449t, 450f, 451, 453–454, 460, 460t
Thiamine deficiency, 10–12
Thin-layer chromatography (TLC), 287
Thiopental, 229
THJ-018, 283f, 310
THJ-2201, 283f, 310
Threshold accurate calibration (TAC), 212, 287
Time-of-flight (TOF) analyzer, 323, 323f
Time-of-flight (TOF) mass spectrometry, 174, 178–181
  key characteristics of, 181t
  performance of, 181t
Tolmetin, 124–125, 136–137
Toluene, 489–490
Topiramate, 48–51, 51t
Total protein, 33
Toxic alcohols, 26, 28–30
Toxicology specimen collection at autopsy, 338
Tramadol, 105, 125, 195–196, 344, 413, 420, 422
Transferrin glycosylation, 31
Trazodone, 130, 134, 200, 240–241
Triazolobenzodiazepines (alprazolam), 197–198
2,2,2-Trichloroethyl chloroformate, 210
Tricyclic antidepressants (TCAs), 35–36, 200
Triethylene glycol, 60t, 378
Trifluoroacetic anhydride (TFAA), 210
N-Trifluoroacetyl-1-prolyl chloride (TPC), 210
Trifluoroacetylated fluoxetine, 201, 201f
1-(3-Trifluoromethylphenyl)piperazine (TFMPP), 240, 242f
Triglycerides, 32, 64
Triple quadrupole mass spectrometry, 164–165
Triple-stage quadrupole mass spectrometer, 351, 351f
Trisialotransferrin, 36
True positive results due to the use of medications, 442–443
Tryptamine-based hallucinogens, 392
Tryptamines (TA), 250–251, 252f
Type 2 diabetes, alcohol consumption and, 8

U

UHPLC-MS/MS analysis, 288
Ultra Clean shampoo, 421–422
Ultra-high-performance liquid chromatography (UHPLC), 266, 325–326
Ultra-high-performance liquid chromatography–tandem mass spectrometry (UHPLC–Ms/MS), 254, 266
Ultra-performance liquid chromatography coupled with time of flight mass spectrometry (UPLC-TOF-MS), 277–279, 289, 292, 302, 311
Ultra-performance liquid chromatography-electrospray ionization tandem mass spectrometry (UPLC-MS/MS), 273–274, 313
Umbilical cord tissue (UCT), 429, 438–439
Underage drinking, 6t
United States of America v. Baker et al., US Eighth Circuit Court of Appeals, 855 F2d 1353 (8th c.1988), 92
United States of America v. McKinney, US Eighth Circuit Court of Appeals, 631 F2d 569 (8th c.1980), 92
United States of America v. Washington, US Fourth Circuit Court of Appeals, 498 F3d 225 (4th c.2007), 93–94
United States v. Agurs, US Supreme Court 427 U.S. 97 (1976), 89
United States v. Campbell, (Campbell II), US Court of Appeals for the Armed Forces, 52 MJ 386 (CAAF 2000), 95–96
United States v. Green, US Court of Appeals of the Armed Forces, 55 MJ 76 (CAAF 2001), 96–97
United States v. Harcrow, US Court of Appeals for the Armed Forces, 66 MJ 154 (CAAF 2008), 94
United States v. Israel, US Court of Appeals for the Armed Forces, 60 MJ 485 (CAAF 2005), 90
United States v. Jackson, US Court of Appeals for the Armed Forces, 59 MJ 330 (CAAF 2004), 90
United States v. Magyari, US Court of Appeals for the Armed Forces, 63 MJ 123 (CAAF 2006), 92–93
UR-144, 273, 280f, 301–302
  and pyrolysis product, 280f
UR-144 N-pentatonic acid, 302
UR-144-4-OH, 280f
UR-144-5-OH, 280f
UR-144-pentanoic acid, 280f, 301
URB-597, 285f, 311
URB-754, 285f, 311
URB-class, structures and analysis of, 311
Uric acid, 378
UrinAid (glutaraldehyde), 372–375, 374t
Urinary bladder distension, 334, 334f
Urinator, 360
Urine adulterants, 360, 364t, 366, 366t
Urine drug immunoassay screening, 190t
Urine drugs test (UDT), 321, 345–346, 352, 355–356, 361–364, 402, 441
  counteracting urine adulteration, 375–379
    capillary electrophoresis and mass spectrometric techniques, 378
    color tests, 376
    dipstick devices, 376–378
    immunoassays, 378
    polyethylene glycol urine marker system, 378–379
    spectrophotometric methods, 378
    urine integrity tests, 375–376
  effects of chemical adulterants on, 365–375
    hypochlorite-based bleach, 366–368
    nitrite, 368–369
    nonoxidizing adulterants, 372–375
    other oxidants, 371–372
    peroxides, 370–371
    pyridinium chlorochromate (PCC), 369–370
  HRMS for, 326–331
  in vitro chemically adulterated urine, 364
  in vivo adulteration and urine dilution, 363–364
  mass spectrometry in, 321
  urine substitution, 363
Urine Luck, 360, 369
Urine specimen, in the detection of drug abuse, 406t, 410
Urine substitution, 359–360, 363
Urobilin, 378
US Centers for Disease Control and Prevention (CDC), 25
US Controlled Substances Act, 392
US Court of Appeals for the Armed Forces (CAAF), 88, 90
US military courts, 79
US paper money
  contamination with cocaine and other drugs, 463–468
    clean money versus drug money, 465–467
    mechanism of contamination, 467
    positive drug screen, 467–468
Utero drug exposure, 429, 434–435, 439
3'-UTR VNTR polymorphism, 110

V

Variable number of tandem repeats (VNTR), 48
Venlafaxine, 175–178, 177f, 200
Vibrational circular dichroism (VCD), 311
Vicks inhaler, 131–133, 443–444
Vinegar, 364t, 372–375, 372t
Visine eye drops, 364t, 372–375, 372t
Volatiles and glue, abuse of, 489–491
  laboratory detection of solvent abuse, 491
  pharmacology and toxicology of abused solvents, 490–491
  prevalence of solvent abuse, 489–490
Volumetric absorptive microsampling (VAMS), 254, 268

W

Washington State Toxicology Laboratory (WSTL), 97–98
Wernicke's encephalopathy, 10–12
Wernicke–Korsakoff syndrome, 10–12
White heroin, 252–253
Whizzinator, 360, 363
Widmark method, 20
WIN-48,098, 274*f*
WIN-55, 212-2, 314, 314*f*
Workplace drug testing, 73, 129–131, 361
Workplace drug testing protocol, 392
World Anti-Doping Agency (WADA), 359, 459, 496, 505
  prohibited list, 496, 497*t*

X

Xenobiotics, 103, 267, 422
XLR-11, 273, 280*f*, 301–302
XLR-11 6-OH-indole, 301
XLR-11 pyrolysis product, 280*f*
XLR-11-4-OH, 280*f*, 303
XLR-12, 280*f*, 301

Z

Zaleplon (Sonata), 201, 225, 231
Z-drugs, 189, 201–202, 231–232, 420
Zero-order kinetics, 3, 18–19, 27, 31
Zidovudine. *See* Azidothymidine (AZT)
Zinc, 372*t*, 375, 384
Zolpidem (Ambien), 201, 225, 231
  chemical structure of, 226*f*
Zopiclone (Lunesta), 201, 225, 231, 330
  chemical structure of, 226*f*

Printed in the United States
By Bookmasters